Sound Systems: Design and Optimization

Third Edition

Sound Systems: Design and Optimization provides an accessible and unique perspective on the behavior of sound systems in the practical world. The third edition reflects current trends in the audio field thereby providing readers with the newest methodologies and techniques.

In this greatly expanded new edition, you'll find clearer explanations, a more streamlined organization, increased coverage of current technologies and comprehensive case studies of the author's award-winning work in the field.

As the only book devoted exclusively to modern tools and techniques in this emerging field, *Sound Systems: Design and Optimization* provides the specialized guidance needed to perfect your design skills.

This book helps you:

- Improve your design and optimization decisions by understanding how audiences perceive reinforced sound.
- Use modern analyzers and prediction programs to select speaker placement, equalization, delay and level settings based on how loudspeakers interact in the space.
- Define speaker array configurations and design strategies that maximize the potential for spatial uniformity.
- Gain a comprehensive understanding of the tools and techniques required to generate a design that will create a successful transmission/reception model.

Bob McCarthy is the Director of System Optimization at Meyer Sound and President of Alignment & Design, Inc. As a developer of FFT analysis systems, he pioneered the methods for tuning modern speaker systems that have since become standard practice in the industry. He is the foremost educator in the field of sound system optimization and has conducted training courses worldwide for over thirty years. Bob received the USITT Distinguished Achiever in Sound Design in 2014. His clients have included esteemed companies such as Cirque du Soleil and Walt Disney Entertainment, as well as many of the world's best sound designers, including Jonathan Deans, Tony Meola, Andrew Bruce and Tom Clark.

Sound Systems: Design and Optimization

Modern Techniques and Tools for Sound System Design and Alignment

Third Edition

Bob McCarthy

LONDON AND NEW YORK

Third edition published 2016

Published 2019 by Routledge
2 Park Square, Milton Park, Abingdon, Oxon, OX14 4RN
52 Vanderbilt Avenue, New York, NY 10017

Routledge is an imprint of the Taylor & Francis Group, an informa business

First published 2006 by Focal Press
Second edition published 2009 by Focal Press

Library of Congress Cataloging-in-Publication Data
McCarthy, Bob.
Sound systems : design and optimization : modern techniques and tools for sound system design and alignment / by Bob McCarthy. — Third edition.
pages cm
Includes bibliographical references and index.
ISBN 978-0-415-73099-0 (hardback) — ISBN 978-0-415-73101-0 (pbk.) — ISBN 978-1-315-84984-3 (ebk) 1. Sound—Recording and reproducing. I. Title.
TK7881.4.M42 2016
621.389'3—dc23
2015016842

ISBN: 978-0-415-73099-0 (hbk)
ISBN: 978-0-415-73101-0 (pbk)

Typeset in Giovanni
By Apex CoVantage, LLC

Printed and bound in Great Britain by
TJ Books Limited, Padstow, Cornwall

Dedication

To the woman who knew me back when this journey began, and is still there through three editions, the love of my life,
Merridith.

In memoriam

During the course of writing the first edition our field lost one of its most well-loved and respected pioneers, Don Pearson a.k.a. Dr Don. I had the good fortune to work with Don and receive his wisdom. He was there when it all began and is still missed. More recently we lost Tom Young, and Mike Shannon who contributed so much to the industry and to me personally. And finally, I lost my brother Chris who was an inspiration to all who knew him.

CONTENTS

PART II • Design

PART III • Optimization

PREFACE

This book is about a journey. On the one hand, the subject is the journey of sound as it travels through a sound system, then through the air, and inevitably to a listener. It is also a personal journey, my own quest to understand the complicated nature of this sound transmission. The body of this text will detail the strictly technical side of things. First, however, I offer you some of the personal side.

I was supposed to build buildings. Unbeknownst to me at the time, this calling was derailed on February 9, 1964 by the appearance of the Beatles on *The Ed Sullivan Show*. Like so many of my generation, this landmark event brought popular music and an electric guitar into my life. I became a great enthusiast of live concerts, which I regularly attended throughout my youth at any chance presented. For years, it remained my expectation that I would enter the family construction business. This vision ended on a racetrack in Des Moines, Iowa on June 16, 1974. The experience of hearing the massive sound system at this Grateful Dead concert set my life in a new direction. On that day I made the decision that I was going to work in live concert sound. I wanted to help create this type of experience for others. I would be a mix engineer and my dream was to one day operate the mix console for big shows. I set my sights on preparing for such a career while at Indiana University. This was no simple matter because there was no such thing as a degree in audio. I soon discovered the Independent Learning Program. Under the auspices of that department, I assembled a mix of relevant courses from different disciplines and graduated with a college-level degree in my self-created program of audio engineering.

FIGURE 0.1
Ticket stub from the June 16, 1974 Grateful Dead concert in Des Moines, Iowa that led to my life of crime

By 1980, I had a few years of touring experience under my belt and had moved to San Francisco. There I forged relationships with John Meyer, Alexander Yuill-Thornton II (Thorny) and Don Pearson. These would become the key relationships in my professional development. Each of us was destined to stake our reputations on the same piece of equipment: the dual-channel FFT analyzer.

I would like to say that I have been involved in live concert measurement with the dual-channel FFT analyzer from day one, but this is not the case. It was day two. John Meyer began the process on a Saturday night in May of 1984. John took the analyzer, an analog delay line and some gator clips to a Rush concert in Phoenix, Arizona, where he performed the first measurements of a concert sound system using music as the source with audience in place. I was not destined to become involved in the project until the following Monday morning.

From that day forward, I have never been involved in a concert or a sound system installation without the use of a dual-channel FFT analyzer. I haven't mixed a show since that day, resetting my vision to the task of helping mix engineers to practice their art. For Don, John, Thorny and many others, the idea of setting up a system without the presence of the FFT analyzer was unthinkable. Seeing a sound system response in

FIGURE 0.2
The author with the prototype SIM analyzer with the Grateful Dead in July 1984 at the Greek Theater in Berkeley, California (Clayton Call photo)

high resolution, complete with phase, coherence and impulse response, is a bell that cannot not be un-rung. We saw its importance and its practical implications from the very beginning and knew the day would come when this would be standard practice. Our excitement was palpable, with each concert resulting in an exponential growth in knowledge. We introduced it to everyone who had an open mind to listen. The first product to come from the FFT analysis process was a parametric equalizer. A fortuitous coincidence of timing resulted in my having etched the circuit boards for the equalizer on my back porch over the weekend that John was in Phoenix with Rush. This side project (a bass guitar preamp) for my friend Rob Wenig was already six months late, and was destined to be even later. The EQ was immediately pressed into service when John nearly fell over as he saw that it could create the complementary response (in both amplitude and phase) to what he had measured in Phoenix. The CP-10 was born into more controversy than one might imagine. Equalization has always been an emotional "hot button" but the proposition that the equalizer was capable of counteracting the summation properties of the speaker/room interaction was radical enough that we obtained the support of Stanford's Dr Julius Smith to make sure that the theory would hold up.

Don Pearson was the first outside of our company to apply the concepts of in-concert analysis in the field. Don was owner of Ultrasound and was touring as the system engineer for the Grateful Dead. Don and the band immediately saw the benefit and, lacking patience to wait for the development of what would become the Meyer Sound SIM System, obtained their own FFT analyzer and never looked back. Soon thereafter, under the guidance of San Francisco Opera sound designer Roger Gans, we became involved with arena-scale performances for Luciano Pavarotti. We figured it was a matter of months before these techniques would become standard operating procedure throughout the industry. We had no idea it would take closer to twenty years! The journey, like that of sound transmission, was far more complex than we ever expected. There were powerful forces lined up against us in various forms: the massive general resistance of the audio community to sound analyzers and the powerful political forces advocating for alternate measurement platforms, to name a few.

In general, the live sound community was massively opposed to what they conceptualized as an analyzer dictating policy to the creative forces involved in the music side of the experience. Most live concert systems of the day lacked complexity beyond piles of speakers with left and right channels. This meant that the process of alignment consisted of little more than equalization. Because all of the system calibration was being carried out at a single location, the mix position, the scientific and artistic positions were weighing in on the exact same question at the same point in space. Endless adversarial debate about what was the "correct" equalization ensued because the tonal balancing of a sound system is, and always has been, an artistic endeavor. It was an absurd construct. Which is better—by ear or by analyzer?

FIGURE 0.3
November 1984 photo of Luciano Pavarotti, Roger Gans, the author (back row), Drew Serb, Alexander Yuill-Thornton II and James Locke (front row) (Drew Serb photo)

This gave way to a more challenging and interesting direction for us: the quest beyond the mix position. Moving the mic out into the space left us with a terrible dilemma: The new positions revealed conclusively that the one-size-fits-all version of system equalization was utter fantasy. The precision tuning of parametric filters carried out with great care for the mix position had no justification at other locations. The interaction of the miscellaneous parts of the speaker system created a highly variable response throughout the room. Our focus shifted from finding a perfect EQ to the quest for uniformity over the space.

This would require the subdivision of the sound system into defined and separately adjustable subsystems, each with individual level, equalization and delay capability. The subsystems were then combined into a unified whole. The rock and roll community was resistant to the idea, primarily because it involved turning some of the speakers down in level. The SPL Preservation Society staunchly opposed anything that might detract from the maximum power capability. Uniformity by subdivision was not worth pursuing if it cost power (pretty much nothing else was either). Without subdivision, the analysis was pretty much stuck at the mix position. If we are not going to change anything, why bother to look further?

There were other genres that were open to the idea. The process required the movement of a microphone around the room and a systematic approach to deconstructing and reconstructing the sound system. We began developing this methodology with the Pavarotti tours. Pavarotti was using approximately ten subsystems, which were individually measured and equalized and then merged together as a whole. Our process had to be honed to take on even more complexity when we moved into the musical theater world with Andrew Bruce, Abe Jacob, Tony Meola, Tom Clark and other such sound designers. Our emphasis changed from providing a scientifically derived tonal response to maximizing consistency of sound throughout the listening space, leaving the tonal character in the hands of the mix engineer. Our tenure as the "EQ police" was over as our emphasis changed from tonal quality to tonal equality. The process was thus transformed into *optimization*, emphasizing spatial uniformity while encompassing equalization, level setting, delay setting, speaker positioning and a host of verifications on the system. A clear line was drawn between the artistic and the scientific sectors.

In the early days, people assessed the success of a system tuning by punching out the filters of the equalizer. Now, with our more sophisticated process, we could no longer re-enact before-and-after scenarios. To hear the "before" sound might require repositioning the speakers, finding the polarity reversals, setting new splay angles, resetting level and time delays, and finally a series of equalizations for the different subsystems. Lastly, the role of the optimization engineer became clear: to ensure that the audience area receives the same sound as the mix position.

In 1987, we introduced the Source Independent Measurement system (SIM). This was the first multichannel FFT analysis system designed specifically for sound system optimization (up to sixty-four channels). It consisted of an analyzer, multiple mics and switchers to access banks of equalizers and delays. All of this was under computer control, which also kept a library of data that could be recalled for comparison of up to sixteen different positions or scenarios. It thereby became possible to monitor the sound system from multiple locations and clearly see the interactions between subsystems. It was also possible to make multiple microphone measurements during a performance and to see the effects of the audience presence throughout the space.

This is not to say we were on Easy Street at this point. It was a dizzying task to manage the assembly of traces that characterized a frequency response, which had to be measured in seven separate linear frequency sections. A single data set to fully characterize one location at a point in time was an assembly of sixty-three traces, of which only two could be seen at any one time on the tiny 4-inch screen. Comparison of one mic position to another had to be done on a trace-by-trace basis (up to sixty-three operations). It was like trying to draw a landscape while looking through a periscope.

The multichannel measurement system opened the door to system subdivision. This approach broke the pop music sound barrier with Japanese sensation Yuming Matsutoya under the guidance of Akio Kawada, Akira Masu and Hiro Tomioka. In the arenas across Japan we proved that the techniques we had developed for musical theater and Pavarotti (level tapering, zoned equalization and a carefully combined subsystem) were equally applicable to high-power rock music in a touring application.

The introduction of the measurement system as a product was followed by the first training seminar in 1987. A seminal moment came from an unexpected direction during this first seminar as I explained the process of system subdivision and mic placement for system optimization. Dave Robb, a very experienced engineer, challenged my mic placement as "arbitrary." In my mind, the selection was anything but arbitrary. However, I could not, at that moment, bring forth any objective criteria with which to refute that assertion. Since that humiliating moment, my quest has been to find a defensible methodology for every decision made in the process of sound system optimization. It is simply not enough to know *that* something works; we must know *why* it works. Those optimization methodologies and an accompanying set of methods for sound system design are the foundation of this book.

I knew nothing of sound system design when this quest began in 1984. Almost everything I have learned about the design of sound systems comes from the process of their optimization. The process of deconstructing and reconstructing other people's designs gave me the unique ability/perspective to see the aspects that were universally good, bad or ugly. I am very fortunate to have been exposed to all different types of designs, utilizing many different makes and models of speakers, with all types of program materials and scales. My approach has been to search for the common solutions to these seemingly different situations and to distill them into a repeatable strategy to bring forward to the next application.

Beginning with that very first class, with little interruption, I have been optimizing sound systems and teaching anybody who wanted to attend my seminars everything I was learning. Thorny, meanwhile, had moved on and founded a company whose principal focus was sound system optimization services using the dual-channel FFT systems. Optimization as a distinct specialty had begun to emerge.

The introduction of SIA-SMAART in 1995 resulted from the collaboration of Thorny and Sam Berkow with important contributions by Jamie Anderson and others in later years. This low-cost alternative brought the dual-channel FFT analyzer into the mainstream and made it available to audio professionals at every level. Even so, it took years before our 1984 vision of the FFT analyzer, as standard front-of-house equipment, would become reality. Unquestionably, that time has arrived. The paradigm has reversed to the point where tuning a system without scientific instrumentation would be looked at with as much surprise as was the reverse in the old days.

Since those early days we have steadily marched forward with better tools—better sound systems, better sound design tools and better analyzers. The challenge, however, has never changed. It is unlikely that it will change, because the real challenge falls mostly in the spatial distribution properties of acoustical physics. The speakers we use to fill the room are vastly improved and signal-processing capability is beyond anything we dreamed of in those early days. Prediction software is now readily available to illustrate the interaction of speakers, and we have affordable and fast analyzers to provide the on-site data.

And yet we are fighting the very same battle that we have always fought: the creation of a uniform sonic experience for audience members seated everywhere in the venue. It is an utterly insurmountable challenge. It cannot be achieved. There is no perfect system configuration or tuning. The best we can hope for is to approach uniformity. I believe it is far better to be coldly realistic about our prospects. We will have to make decisions that we know will degrade some areas in order to benefit others. We want them to be informed decisions, not arbitrary ones.

This book follows the full transmission path from the console to the listener. That path has gone through remarkable changes along its entire electronic voyage. But once the waveform is transformed into its acoustic form it enters the very same world that Jean-Baptiste Joseph Fourier found in the eighteenth century and Harry Olson found in the 1940s. Digital, schmigital. Once it leaves the speaker, the waveform is pure analog and at the mercy of the laws of acoustical physics. These unchanging aspects of sound transmission are the focus of 90 per cent of this book.

Let's take a moment to preview the challenges we face. The primary player is the interaction of speakers with other speakers, and with the room. These interactions are extremely complex on the one hand, and yet can be distilled down to two dominant relationships: relative level and relative phase. The combination of two related sound sources will create a unique spatial distribution of additions and subtractions over the space. The challenge is the fact that each frequency combines differently, creating a unique layout. Typical sound systems have a frequency range of 30 to 18,000 Hz, which spans a 600:1 ratio of wavelengths. A single room, from the perspective of spatial distribution over frequency, is like a 600-story skyscraper with a different floor plan at every level. Our job is to find the combination of speakers and room geometry that creates the highest degree of uniformity for those 600 floor plans. Every speaker element and surface will factor into the spatial distribution. Each element plays a part in proportion to the energy it brings to the equation at a point in the space. The combined level will depend upon the relationship between the individual phase responses at each location at each frequency. How do we see these floor plans? With an acoustic prediction program we can view the layout of each floor, and compare them and see the differences. This is the viewpoint of a single frequency range analyzed over the entire space. With an acoustic analyzer we get a different view. We see a single spot on each floor from the foundation to the rooftop through a piece of pipe as big around as our finger. This is the viewpoint of a single point in space analyzed over the entire frequency range.

This is a daunting task. But it is comprehensible. This book will provide you with the information required to obtain the X-ray vision it takes to see through the 600-story building from top to bottom, and it can be done without calculus, integral math or differential equations. We let the analyzer and the prediction program do the heavy lifting. Our focus is on how to read X-rays, not on how to build an X-ray machine.

The key to understanding the subject, and a persistent theme of this book, is sound source identity. Every speaker element, no matter how big or small, plays an individual role, and that solitary identity is never lost. Solutions are enacted locally on an element-by-element basis. We must learn to recognize the individual parties to every combination, because therein lie the solutions to their complex interaction.

This is not a mystery novel, so there is no need to hide the conclusion until the last pages. The key to spatial uniformity is control of the overlap of the multiple elements. Where two elements combine they must be in phase to obtain spatial uniformity. If the elements cannot maintain an in-phase relationship, then they must decrease the overlap and subdivide the space so that one element takes a dominant role in a given area. There are two principal mechanisms to create isolation: angular separation and displacement. These can be used separately or in combination and can be further aided by independent control of level to subdivide the room. This is analogous to raising children: If they don't play well together, separate them. The interaction of speakers to the room is similar to the interaction of speakers with other speakers. Those surfaces that return energy back toward our speakers will be the greatest concern. The strength of the inward reflections will be inversely proportional to our spatial uniformity.

There is no single design for a single space. There are alternate approaches and each involves tradeoffs. There are, however, certain design directions that keep open the possibility of spatial uniformity and others that render such hopes statistically impossible. A major thrust of the text will be devoted to defining the speaker configurations and design strategies that maximize the potential for spatial uniformity.

Once designed and installed, the system must be optimized. If the design has kept the door open for spatial uniformity, our task will be to navigate the system through that door. The key to optimization is the knowledge of the decisive locations in the battle for spatial uniformity. The interactions of speakers and rooms follow a consistent set of spatial progressions. The layering of these effects over each other provides the ultimate challenge, but there is nothing random about this family of interactions. It is logical and learnable. Our measurement mics are the information portals to decipher the variations between the hundreds of floor plans and make informed decisions. Our time and resources are limited. We can only discern the meaning of the measured data if we know where we are in the context of the interaction progressions.

We have often seen the work of archeologists where a complete rendering of a dinosaur is created from a small sampling of bone fragments. Their conclusions are based entirely on contextual clues gathered from the knowledge of the standard progressions of animal anatomy. If such progressions were random, there would be nothing short of a 100 per cent fossil record that could provide answers. From a statistical point of view, even with hundreds of mic positions, we will never be able to view more than a few tiny fragments of our speaker system's anatomy in the room. We must make every measurement location count toward the collection of the data we need to see the big picture. This requires advance knowledge of the progression milestones so that we can view a response in the context of what is expected at the given location. As we shall see, there is almost nothing that can be concluded from a single location. The verification of spatial uniformity rests on the comparison of multiple locations.

This book is about defined speakers in defined array configurations, with defined optimization strategies, measured at defined locations. This book is not intended to be a duplication of the general audio resource texts. Such books are available in abundance and it is not my intention to encompass the width and breadth of the complete audio picture. My hope is to provide a unique perspective that has not been told before, in a manner that is accessible to the audio professionals interested in a deeper understanding of the behavior of sound systems in the practical world.

There are a few points that I wish to address before we begin. The most notable is the fact that the physical realities of loudspeaker construction, manufacture and installation are largely absent. Loudspeakers are described primarily in terms of acoustic performance properties, rather than the physical nature of what horn shape or transducers were used to achieve it. This is also true of electronic devices. Everything is weightless, colorless and odorless here. The common transmission characteristics are the focus, not the unique features of one model or another.

The second item concerns the approach to particular types of program material such as popular music, musical theater or religious services, and their respective venues such as arenas, concert halls, showrooms or houses of worship. The focus here is the shape of the sound coverage, the scale of which can be adjusted to fit the size of the venue at the appropriate sound level for the given program material. It is the venue and the program material taken together that create an application. The laws of physics are no different for any of these applications, and the program material and venues are so interchangeable that attempts to characterize them in this way would require endless iterations. After all, the modern-day house of worship is just as likely to feature popular music in an arena setting as it is to have speech and chant in a reverberant cathedral of stone.

The third notable aspect is that there are a substantial number of unique terminologies found here and, in some cases, modification of standard terminologies that have been in general use. In most cases the conceptual framework is unique and no current standard expressions were found. The very young field of sound system optimization has yet to develop consistent methods or a lexicon of expressions for the processes shown here. In the case of some of these terms, most notably the word "crossover," there are compelling reasons to modify the existing usage, which will be revealed in the body of the text.

The book is divided into three parts. The first part, "Sound systems," explores the behavior of sound transmission systems, human hearing reception and speaker interaction. The goal of this part is a comprehensive understanding of the path the signal will take, the hazards it will encounter along the way and how the end product will be perceived upon arrival at its destination. The second part, "Design," applies the properties of the first part to the creation of a sound system design. The goals are comprehensive understanding of the tools and techniques required to generate a design that will create a successful transmission/reception model. The final part, "Optimization," concerns the measurement of the designed and installed system, its verification and calibration in the space.

ABOUT THE THIRD EDITION

From the viewpoint of my publisher, Focal Press, this is indeed the third edition of *Sound Systems: Design and Optimization*. From my perspective it feels more like the thirtieth edition, because I have been writing about these same subjects for thirty+ years. You might think I would have figured this subject out by now but I can assure you I am still learning. This field of work continues to evolve as we get better tools and techniques, which is exactly what I find most interesting about it. Study in this field is a moving target as new technology opens doors and removes obstacles and excuses. The more I learn about this, the more I realize how much I have to learn.

Adding new areas is the easy part of creating a new edition. It's what to do with the previous material that presents a challenge. There are two ways to approach the old material: innocent until proven guilty, or the opposite. The former approach leaves things in unless they are conclusively out of date or irrelevant. The latter approach throws the old material out unless it can prove it is still current practice and up to date.

I studied the later editions of several other authors and noticed a troubling trend. Although new information was added in later editions, a lot of old information remained in place. Seeing vacuum tube circuits from the 1960s in a current-day pro audio text was a tipping point for me. The decision was made to trim out the old to make room for the new. If it's the way we do things now, it's in. If we've moved on, it's out.

The surprise for me was how much we have moved forward in this time, which meant entire chapters were bulldozed and rebuilt. One of the hardest decisions was to let go of the perspective sidebars that colored the previous editions with the wisdom and insight of so many of my friends and colleagues. The bottom line is that there is simply too much new information to be added. I take comfort in knowing that there are many other places where those voices can be heard and that optimization is now firmly ensconced as part of the audio landscape.

There have been no updated laws of physics and our audio analyzers still compute things the same way as they did in 1991. But today's analyzers are faster, easier and able to multitask, which means we can get much more done in a short time. We can tune methodically, and methods are what this book is about. We have far better loudspeakers, processors and steadily better rooms to work in. All this leads to the primary goal of this third edition: current methodologies and techniques for sound system design and optimization.

ACKNOWLEDGEMENTS

The development of this book spans more than thirty years in the field of sound system optimization. Were it not for the discoveries and support of John and Helen Meyer, I would have never become involved in this field. They have committed substantial resources to this effort, which have directly helped the ongoing research and development leading up to this writing. In addition, I would like to acknowledge the contribution of every client who gave me the opportunity to perform my experiments on their sound systems. Each of these experiences yielded an education that could not be duplicated elsewhere. In particular I would like to thank David Andrews, Peter Ballenger, Nick Baybak, Mark Belkie, Mike Brown, Andrew Bruce, John Cardenale, Tom Clark, Mike Cooper, Jonathan Deans, François Desjardin, Steve Devine, Martin Van Dijk, Steve Dubuc, Duncan Edwards, Aurellia Faustina, T. C. Furlong, Roger Gans, Scott Gledhill, Michael Hamilton, Andrew Hope, Abe Jacob, Akio Kawada, Andrew Keister, Tony Meola, Ben Moore, Philip Murphy, Kevin Owens, Frank Pimiskern, Bill Platt, Marvic Ramos, Harley Richardson, Paul Schmitz, David Sarabiman, Pete Savel, Rod Sintow, Bob Snelgrove, David Starck, Benny Suherman, Leo Tanzil and Geoff Zink, all of whom have given me multiple opportunities through the years to refine the methods described here. Special thanks to Mr O at the National Theatre of Korea, Mr Song at Dongseo University, Mr Lee and others at the LG Art Center.

I have also learned much from other engineers who share my passion for this field of work and constantly challenge me with new ideas and techniques. This list would be endless and includes but is not limited to Brian Bolly, Michael Creason, Ales Dravinec, Josh Evans, Peter Grubb, Glenn Hatch, Luke Jenks, Miguel Lourtie, Karoly Molnar, John Monitto, Matt Salerno, John Scandrett and Robert Scovill.

I would also like to thank Meyer Sound for sponsoring my seminars and Gavin Canaan and others for organizing them. I am grateful to everyone who has attended my seminars, as the real-time feedback in that context provides a constant intellectual challenge and stimulation for me. My fellow instructors in this field have contributed much collaborative effort through discussion and the sharing of ideas. Notable among these are Jamie Anderson, Oscar Barrientos, Timo Beckman, Sam Berkow, Harry Brill, Richard Bugg, Steve Bush, Jim Cousins, Pepe Ferrer, Michael Hack, Mauricio Ramirez, Arthur Skudrow, Hiro Tomioka, Merlijn Van Veen and Jim Woods.

Thanks to Daniel Lundberg for his creation of the uncoupled array calculator based on data from my previous edition. This tool is a mainstay of my design process.

A huge majority of the knowledge, data (and graphics) in this book comes from two sources: Meyer Sound's SIM3 Audio Analyzer and their MAPP Online™ platform. I would have nothing to write about without these tools. My gratitude goes to everyone who contributed to their creation, including (but not limited to) John and Helen Meyer, Perrin Meyer, Dr Roger Schwenke, Fred Weed, Todd Meier, Mark Schmeider, Paul Kohut and the late Jim Isom.

The following figures contain data from my earlier publications at Meyer Sound and their permission is gratefully acknowledged: Figures 3.9, 4.24, 4.27, 13.12 to 13.16. The data presented in Figures 1.1, 1.4, 1.11 and 12.11 were created using the calculations by Mauricio Ramirez. The 3-D wraparound graphics (Figure 12.8) were adapted from the animations created by Greg Linhares. Merlijn Van Veen contributed a mix of core calculations, data and graphics that went into the construction of Figures 1.12, 2.2, 2.5 to 2.12, 3.30, 3.37, 3.42, 4.8, 8.11, 8.14 and 14.10. John Huntington contributed some of the photographs used in the section break pages, specifically Section 1 (panels 1, 5 and 7) and Section 3 (panel 4).

Thanks go to all of the people who aided in the process of bringing this edition to physical reality such as my editor Megan Ball and Mary LaMacchia at Focal Press. Additional thanks go to Margo Crouppen for her support throughout the entire publishing process.

I received proofing and technical help for my previous editions from Jamie Anderson, Sam Berkow, David Clark, John Huntington, Mauricio Ramirez and Alexander (Thorny) Yuill-Thornton. Harry Brill Jr., Richard Bugg, Philip Duncan, John Huntington, Jeff Koftinoff and Mauricio Ramirez contributed to the third edition.

Merlijn Van Veen deserves special recognition for his enormous contributions to this edition. He hung with me at every step of the way, pushing me to clarify language, methodology and backing up my calculations. He also contributed greatly to the graphics, providing both material support and advice on how to best convey the information. I learned much from Merlijn in the course of this writing and his mark has clearly been left in these pages.

Finally, there is no one who comes close to contributing as much support as my wife Merridith. She has read every page and seen every drawing of every edition and been absolutely tireless in her efforts. Each edition has been an endurance test lasting over a year and she has stuck with me through each, swayed by my promises that this would be the last one (fooled her again). This book would not have been possible without her support as my agent, manager, copy editor, proofreader and cheerleader.

PART I

Sound systems

CHAPTER 1

Foundation

We begin with the establishment of a firm foundation upon which to build the structure for the study of sound system design and optimization. Here we standardize definitions and terminology for usage throughout this book. If this is not your first day in audio you will already understand many of the concepts in this chapter, because much of this is the universal foundation material found in books, the Internet and the tribal knowledge passed down from elders on the road. I have, however, selectively edited the list of fundamentals to those concepts pertinent to modern-day design and optimization. We won't cover the Doppler effect, underwater acoustics, industrial noise suppression and any other areas that we can't put to immediate practical use.

foundation n. solid ground or base on which a building rests; groundwork, underlying principle; body or ground upon which other parts are overlaid.
Concise Oxford Dictionary

The next section will read somewhat like a glossary, which is traditionally placed at the rear of the book, and the last place you would normally read. These are, however, the first concepts we need to establish and keep in mind throughout. The foundation we lay here will ease the building process as we progress upward in complexity.

We begin with the foundations of this book.

Sound

Sound is a vibration or mechanical wave that is an oscillation of pressure (a vibration back and forth) transmitted through some medium (such as air), composed of frequencies within the range of hearing.

System

A system is a set of interacting or interdependent components forming an integrated whole. A sound system consists of a connected collection of components whose purpose is to receive, process and transmit audio signals.

The basic components consist of microphones, signal processing, amplifiers, speakers, interconnection cabling and digital networking.

Design

Design is the creative process of planning the construction of an object or system. We design sound systems in rooms by selecting the components, their function, placement and signal path.

Optimization

Optimization is a scientific process whose goal is to achieve the best result when given a variety of options. In our case, the goal is the maximization of sound system performance in conformance with the design intent. And do we ever have a variety of options! The primary metric for optimization is uniformity of response over the space.

1.1 UNIVERSAL AUDIO PROPERTIES

Let's define the universal properties within our limited field of study: the acoustical and analog electrical behavior of sound and its mathematical renderings in digital form.

1.1.1 Audio

Audio is a stream of data beginning and/or ending as sound. The audible version connects directly to our ears through the air. Audio can also exist in an encrypted form that cannot be heard until decoded. The monumental breakthrough of Edison's phonograph was encoding audio into a groove on a lacquer cylinder for playback through a mechanical decoder (a diaphragm attached to a moving needle). Encoded audio exists in many forms: magnetic flux (in tape, transformers, microphones or loudspeakers), electronic signal in a wire and even as a digital numerical sequence.

Audio stream oscillations can be rendered as a sequential set of amplitude values over time. Analog audio renderings are a continuous function (i.e. the amplitude and time values are infinitely divisible). Digital audio renderings are finitely divisible (i.e. a single amplitude value is returned for each block of time).

1.1.2 Frequency (f) and time (T)

Frequency (f or Hz) is the number of oscillations completed in one second (the reciprocal of the time period). Period (T) is the time interval to complete one cycle. Frequency is cycles/second and time is seconds/cycle ($f = 1/T$ and $T = 1/f$). Either term describes an oscillation, the choice being for convenience only. Fluency in the translation of time and frequency is essential for design and optimization (Fig. 1.1). Period formulas for T are computed in seconds, but in practice we almost always use milliseconds (1 ms = 0.001 of a second).

1.1.3 Cycle

A cycle is a completed oscillation, a round trip that returns to the starting state of equilibrium. The distinction between cycle and period is simply units. A period is measured in time (usually ms) and a cycle is measured in completed trips. A cycle at 250 Hz has a period of 4 ms. One cycle @125 Hz (or two cycles @250 Hz) have 8 ms periods. We often subdivide the cycle by fractions or degrees of phase, with 360° representing a complete cycle. It is common to use the term "cycle" when dealing with the phase response, e.g. 1 ms @250 Hz, which is ¼ cycle (90°).

1.1.4 Oscillation

Oscillation is the back and forth process of energy transfer through a medium. This may be mechanical (e.g. a shaking floor), acoustical (e.g. sound in the air) or electromagnetic (e.g. an electronic audio signal). The oscillating matter's movement is limited by the medium and returns to equilibrium upon completion. Energy transfer occurs *through* the medium.

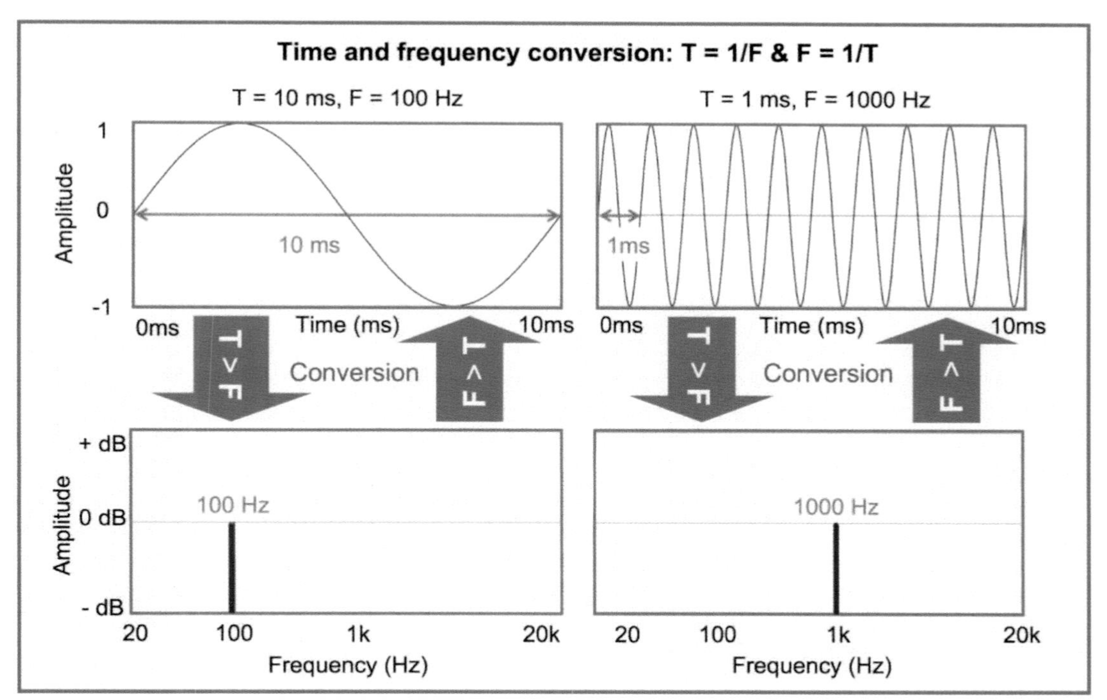

FIGURE 1.1
Relationship of time and frequency

1.1.5 Amplitude (magnitude)

Amplitude is the quantitative measure of oscillating energy, the extent of mechanical displacement (m, cm, etc.), acoustical pressure change (SPL), electrical voltage change (V), magnetic flux (B) and others. Amplitude values can be expressed linearly (e.g. volts) or logarithmically as a ratio (the dB scale). Amplitude is one of the more straightforward aspects of audio: bigger is bigger. Amplitude (black T-shirt) and magnitude (lab coat) are interchangeable terms. We will introduce audio amplitude in various forms next and then cover the scaling details (such as the dB scale) in section 1.2. It may be necessary to bounce between those sections if you are completely unfamiliar with these scales.

1.1.5.1 DC POLARITY (ABSOLUTE POLARITY)

DC (direct current) polarity is the signal's directional component (positive or negative) relative to equilibrium. Electrical: +/- voltage, acoustical +/- pressure (pressurization/rarefaction), etc. This is applicable in strict terms to DC signals only, because AC signals have both positive and negative values. A 9 V battery illustrates the electrical version. Connecting the battery to a speaker illustrates the acoustical version, because it only moves in one direction.

1.1.5.2 ABSOLUTE AMPLITUDE

Absolute amplitude is the energy level relative to equilibrium (audio silence). Electrical audio silence is 0 VAC, whether DC is present or not. Only AC can make audio. DC moves a speaker but unfortunately the only sound it can make is the speaker burning. Acoustic systems are referenced to changes above or below the ambient air pressure (air's equivalent for DC). Absolute amplitude values cannot be less than zero, because we can't have less movement than equilibrium. A "-" sign in front of an amplitude value indicates negative polarity. Relative amplitude values are more common in audio than absolute ones.

1.1.5.3 RELATIVE TO A FIXED REFERENCE

Audio levels change on a moment-to-moment basis. Therefore most amplitude measurements are relative to a reference (either fixed or movable). Examples of fixed references include 1 V (electrical) or the threshold of human hearing (acoustical) (Fig. 1.2). The reference level can be expressed in various units and scales, as long as we agree on the value. An amplitude value of 2 volts can be expressed as 1 volt above the 1 V volt reference (linear difference) or twice the reference level (linear multiple). Many reference standards for audio are specified in decibel values (dB), which show amplitude changes in a relative log scale (like our hearing).

One volt, 0 dBV and +2.21 dBu are the same amount of voltage, expressed in different units or scales. A musical passage with varying level over time can be tracked against the fixed reference, e.g. a certain song reaches a maximum level of 8 volts (+18 dBV, +20.21 dBu) and an acoustical level of 114 dB SPL (114 dB above the threshold of hearing).

1.1.5.4 RELATIVE TO A VARIABLE REFERENCE (AMPLITUDE TRANSFER FUNCTION)

We can monitor the amplitude of constantly changing signals in second-cousin form, i.e. relative to a relative (Fig. 1.3). We compare signal entering and exiting a device, such as music going through a processor. The relative/relative measurement (the 2-channel output/input comparison) is termed "transfer function measurement," the primary form of analysis used in system optimization. Frequency response amplitude traces in this book are relative amplitude (transfer function) unless specified otherwise.

Let's return to the above example. The music level is changing, but the output and input waveforms track consistently as long as the processor gain remains stable. If output and inputs are level matched (a 1:1 ratio), the device has a transfer function voltage gain of unity (0 dB). If the voltage consistently doubles, its transfer function gain is 2× (+6 dB). We can span the electronic and acoustic domains by comparing the processor output with the sound level in the room. This reveals a voltage/SPL tracking relationship, such as +0 dBV (1 V) creates 96 dB SPL (and +6 dBV (2 V) creates 102 dB SPL etc.).

The beauty of transfer function measurement is its ability to characterize a device (or series of devices) with random input material across multiple media, so long as the waveforms at both ends are correlated. This will be covered extensively in Chapter 12.

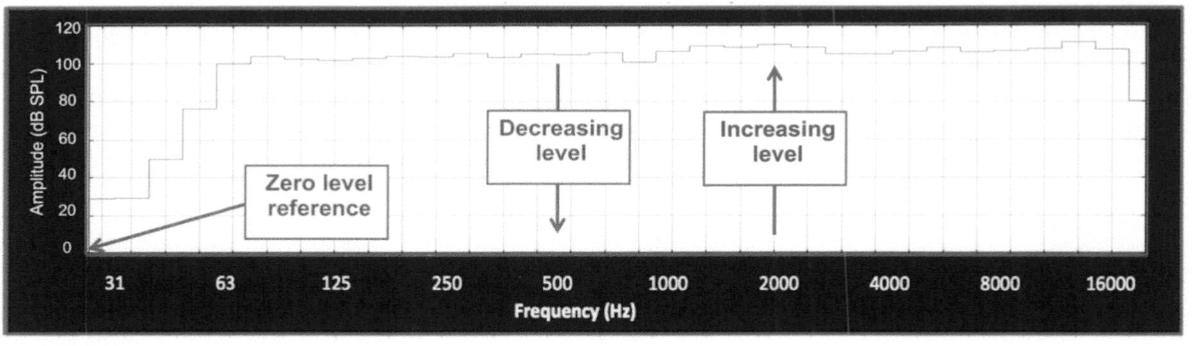

FIGURE 1.2
Absolute amplitude vs. frequency. Amplitude is referenced to 0 dB SPL at the bottom of the vertical scale.

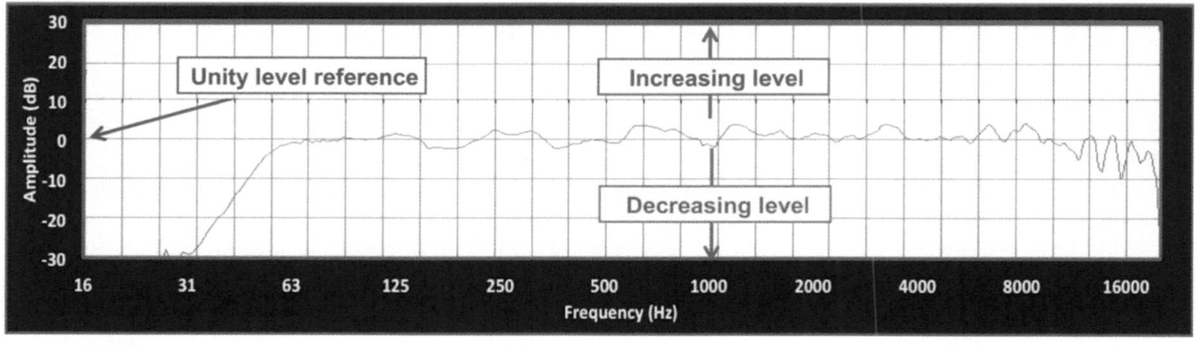

FIGURE 1.3
Transfer function amplitude vs. frequency. Amplitude is referenced to unity gain at the center of the vertical scale.

1.1.5.5 PEAK (PK) AND PEAK-TO-PEAK (PK–PK)

The peak (pk) amplitude value is the signal's maximum extent above *or* below equilibrium whereas peak-to-peak (pk–pk) is the span between above *and* below values. Any device in the transmission path must be capable of tracking the full extent of the pk–pk amplitude. Failure results in a form of harmonic distortion known as "clipping" (because the tops of the peaks are flattened). The waveform seen on an oscilloscope or digital audio workstation is a representation of the pk–pk values.

1.1.5.6 RMS (ROOT MEAN SQUARED)

The rms value (root-mean-squared) is the waveform's "average-ish" amplitude. The rms calculation makes AC (+ and -) equivalent to DC (+ or -). For example a 9 V_{DC} battery and 9 V_{RMS} generator supply the same power. We use rms instead of simple averaging because audio signals move both above and below equilibrium. A sine wave (such as the AC line voltage) averages to zero because it's equally positive and negative. Sticking your fingers in a wall socket provides a shocking illustration of the difference between "average" and rms. Kids, don't try this at home.

The rms value is calculated in three steps: (s) squaring the waveform, which enlarges and "absolutes," making all values positive, (m) finding the squared waveform's mean value and (r) taking the square root to rescale it back to normal size. Its proper full name is "the root of the mean of the square." Note that rms values are strictly a mathematical rendering. We never hear the rms signal (it would be less recognizable as audio than an MP-3 file). The waveforms we transmit, transduce, render digitally and hear are peak–peak.

1.1.5.7 CREST FACTOR

Crest factor is the ratio between the actual amplitude traced by the waveform (the peak or crest) and the heat-load-simulating rms value (Fig. 1.4). For a DC signal (such as a battery) the difference is nothing: crest factor of 1 (0 dB). The simplest audio signal, the sine wave, has a crest factor of 1.414 (3 dB). Complex signals can have vastly higher peak/average ratios (and higher crest factors). Pink noise (see section 1.1.8.3) is approximately 4:1 (12 dB), whereas transient signals such as drums can have 20 to 40 dB.

1.1.5.8 HEADROOM

Headroom is the remaining peak–peak amplitude capability before overload at a given moment, the reserved dynamic range, and our insurance policy against overload. Every electronic or electromagnetic device has its

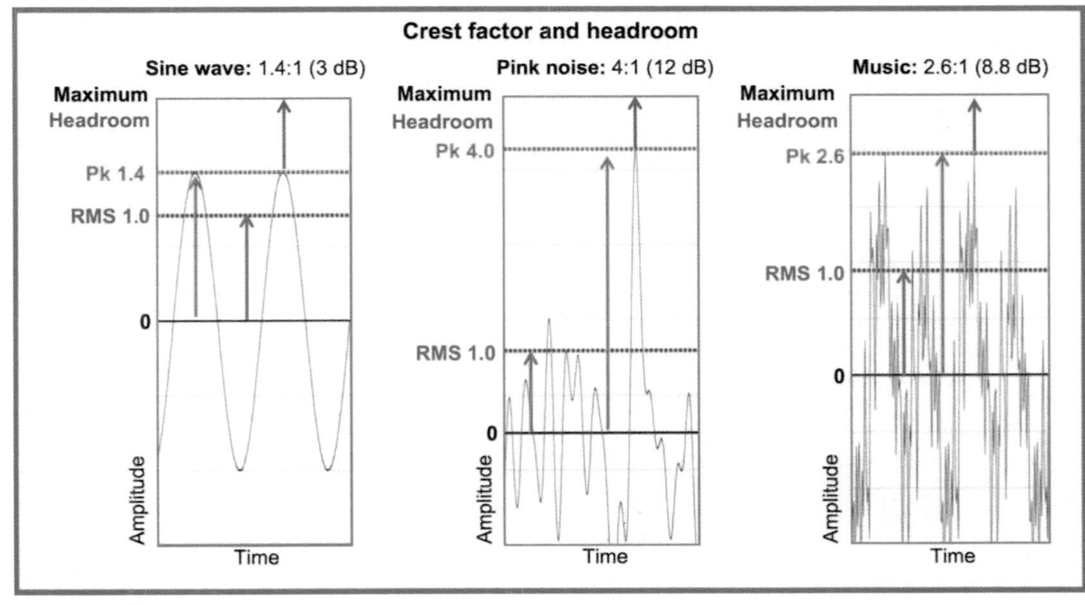

FIGURE 1.4
Crest factor examples with different signals

amplitude upper limit. Linear audio transmission requires the entire extent of the peak–peak signal to pass through without reaching the device's upper limit (no clipping, limiting or compression). Headroom is the remainder between the device's limits and the signal's positive or negative peak. This has historically had a mysterious quality in part because of slow metering ballistics that fall short of tracking the peak–peak transient values. This leaves engineers concerned (rightfully) about potential clipping even when meters indicate remaining dynamic range. An oscilloscope demystifies headroom/clipping because it displays the peak–peak waveform. Digital headroom is the remaining upper bits in the rendering of the pk–pk waveform.

1.1.6 Phase

Phase is the radial clock that charts our progress through a cycle. A completed cycle is 360°, a half-cycle is 180° and so on. The phase value is calculated in reference to a specific frequency. There is not a limit to the phase value, i.e. we can go beyond 360°. The phasor radial positions of 0°, 360° and 720° are equivalent but the phase delay is not, revealing that things have fallen one and two cycles behind respectively. Two race cars with matched radial positions will cross the finish line together, but there is a million dollar difference between 0° and 360°. This makes a difference to sound systems as well, as we will see.

1.1.6.1 ABSOLUTE PHASE

Absolute phase is the value at a given moment relative to a stationary time reference, typically the internal clock of an analyzer. Yes, you read that correctly. Absolute phase is *relative* to a reference, such as the start of the measurement period, which becomes the 0 ms time and 0° phase reference. We don't need to see the absolute phase numbers even though our analyzers compute them. It's like having a wristwatch with only a second hand, which won't help us get to the gig on time. Our analyzers show *relative* phase (a comparison between two channels of internal absolute phase calculations). Note that the term "absolute phase" is often misapplied for the concept of absolute polarity (section 1.1.7.1).

1.1.6.2 RELATIVE (PHASE TRANSFER FUNCTION)

As stated above, relative phase is the difference between two absolute phase values (Fig. 1.5). Relative phase (in degrees) is the only version of phase response shown in this book, so we can proceed to shorten the relative phase phrase to simply "phase." A series of phase values over frequency taken together create a phase slope that can be translated to phase delay (section 1.1.6.6). Phase is a radial function that can set our heads spinning when it comes to reading the 2-D charts. The response often contains "wraparound," which is how we display phase response that exceed the limits of the 360° vertical scale (see section 1.3.4).

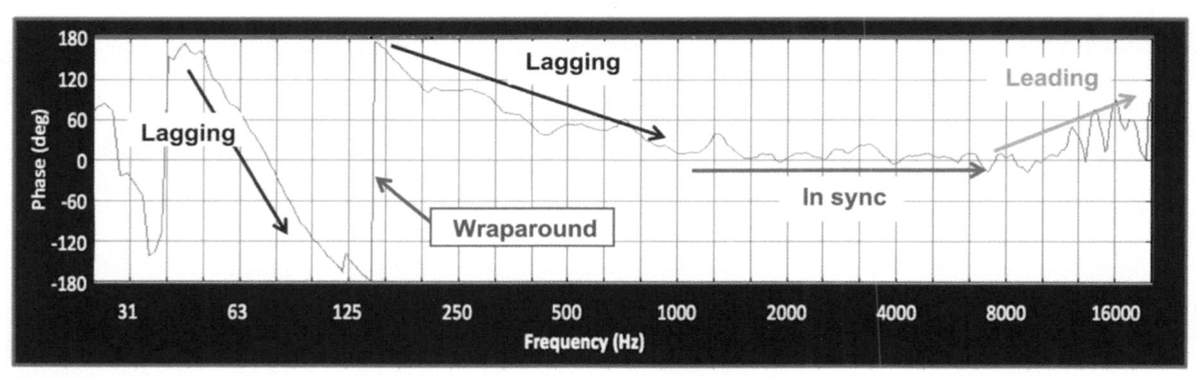

FIGURE 1.5
Transfer function phase vs. frequency. Phase is referenced to unity time (0°) at the center of the vertical scale.

1.1.6.3 PHASE SHIFT

The "shift" in question here is phase change over frequency (Fig. 1.6). This can be stable and constant (such as phase shift caused by a filter) or unstable and variable (such as wind). Our practical concern is frequency-dependent delay (i.e. different frequencies shifted by different amounts). A system with such phase shift has a temporally stretched transient response (often termed "time smearing"). A translation example: The rise and fall of a drum hit would be rounded and expanded because parts of the transient are behind others. The secondary concern regarding phase shift is compatibility with other devices that share common signals and sum together (either in the air or inside our gear). An example is the combination of two different speaker models, with unmatched phase shift characteristics. When summed together we get a phase shift between the phase shifts, which we term the phase offset (the next topic).

1.1.6.4 PHASE OFFSET

Phase offset is the favored term here for phase differences between two measured systems (Fig. 1.7) and a famous audio sorority (Δ-Phi). A known phase offset (in degrees over a frequency span) can be converted to time offset (or vice versa). Phase offset is put to practical use when correlated sources are summed. With known phase and level offsets we can precisely predict the summed response at a given frequency. Phase offset requires a frequency specification and is therefore preferred for frequency-dependent time differences, such as the crossover between an HF and LF driver.

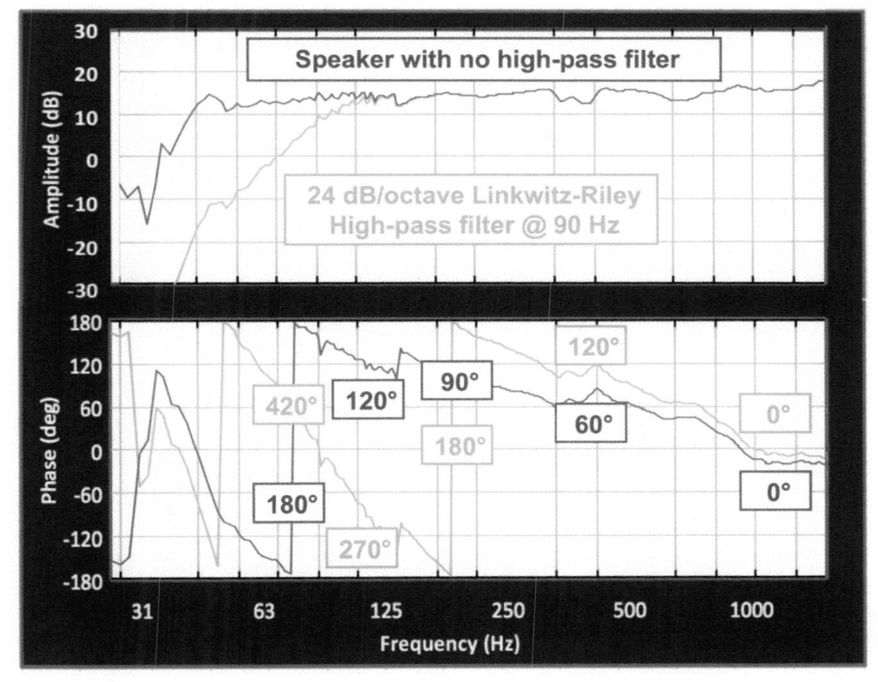

FIGURE 1.6
Example of phase shift added by a 24 dB/octave Linkwitz-Riley filter @90 Hz. The filter only affects the amplitude response below 100 Hz but the phase is shifted over a much wider range.

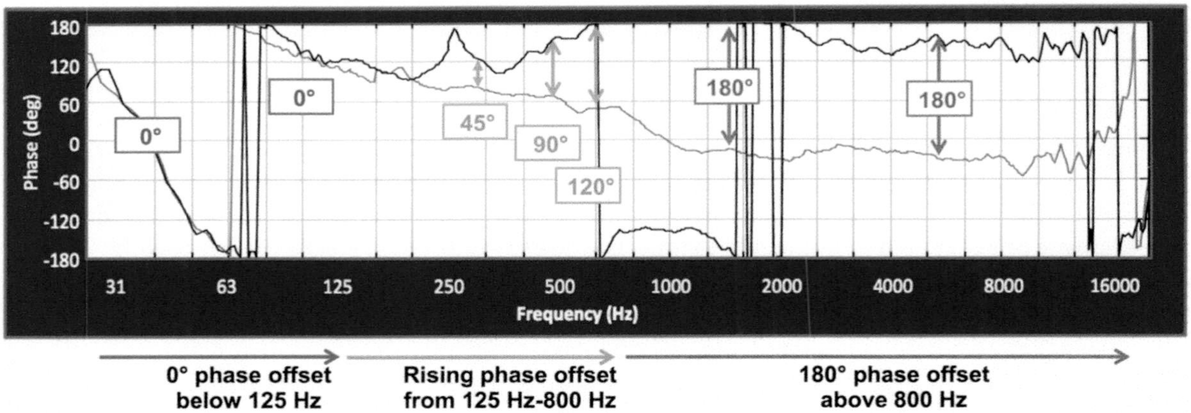

0° phase offset below 125 Hz
Rising phase offset from 125 Hz-800 Hz
180° phase offset above 800 Hz

FIGURE 1.7
Phase offset example showing two different speaker models. The LF ranges match but the HF ranges are offset by 180°.

We don't use the terms "phase offset" and "phase shift" interchangeably here. The distinction is drawn as follows: Phase shift occurs inside a single device or system. Phase offset is *between* devices or systems. A filter creates phase shift. Moving two speakers apart creates phase offset.

1.1.6.5 TIME OFFSET

Time offset (in ms) is a frequency-independent measure for propagation paths (Fig. 1.8). Latency in the signal path or different arrival times between a main and delay speaker are practical examples of time offset. Time offset is our preferred term to describe frequency-independent time differences. Frequency-dependent time offsets are better described by phase delay (below). Time offset (in ms) can be translated to phase offset (in degrees for a given frequency), and vice versa. A fixed-time offset at all frequencies creates an increasing amount of phase offset as frequency rises.

We are very concerned with time offsets within the signal path, particularly in the analog electronic and digital paths where even small amounts are very audible. Strategies for managing time offset between speakers play a big part in system optimization.

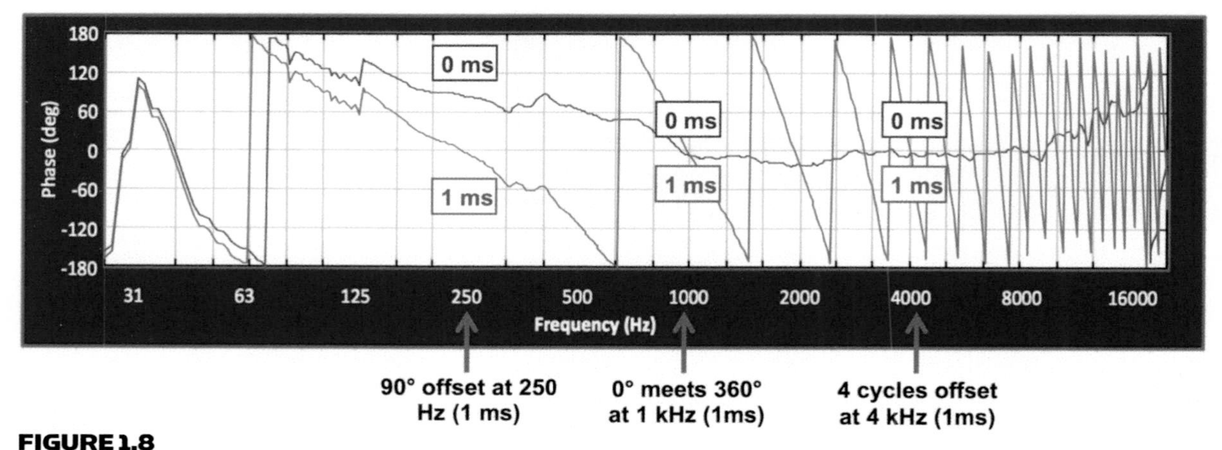

FIGURE 1.8
Time offset example showing 1.0 ms between two otherwise matched speakers. The phase offsets are 90° @250 Hz, 360° @1 kHz and 1440° @4 kHz.

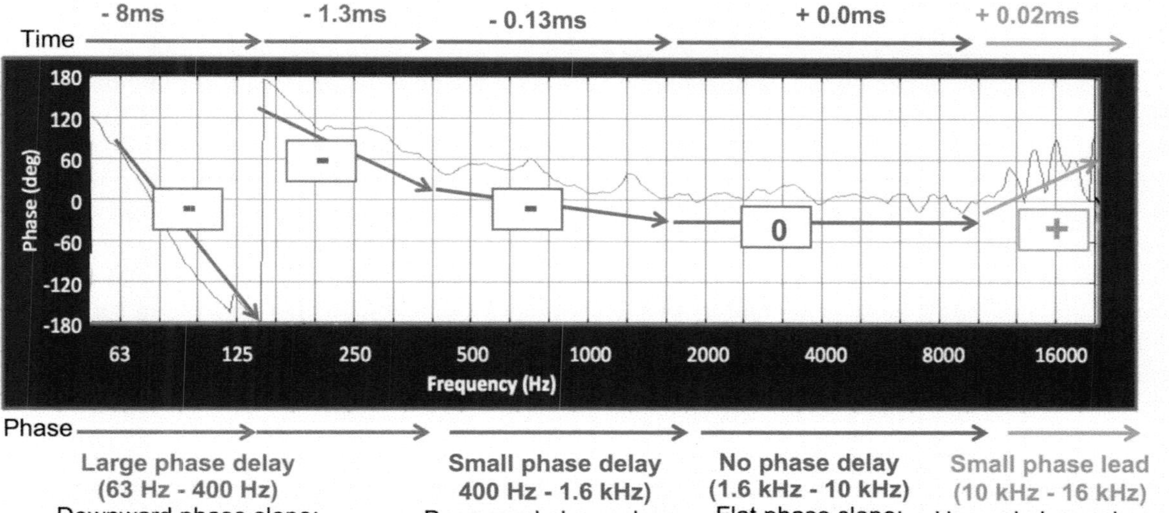

FIGURE 1.9
Example of frequency-dependent phase delay. The LF range has a downward phase slope, indicating phase delay behind the HF region.

1.1.6.6 PHASE DELAY

Phase delay, the time-translated version of phase shift, is a metric used to describe systems with frequency-dependent delay (Fig. 1.9). Phase delay is computed by finding the phase shift over a range of frequencies (the phase slope). The phase value at two frequencies must be known to make the calculation. Phase delay is mostly interchangeable with the term "group delay," an unimportant distinction for optimization decisions.

Any frequency band-limited device will exhibit some phase delay, i.e. some frequencies are transmitted later than others. This includes every loudspeaker not known to marketing departments. Unless extraordinary measures are performed, real-world loudspeakers exhibit increasing phase delay in the LF range. Speaker models have substantially different amounts of phase delay. This can cause compatibility issues when combined, because a single time offset value cannot synchronize the two speakers over the full spectrum. Phase delay is often used to characterize and remedy phase offsets between speakers during optimization.

1.1.7 Polarity

Polarity is a binary term representing the positive vs. negative amplitude orientation of the waveform. Systems with "normal" polarity (+) proceed first in a positive direction followed by negative and back to equilibrium (whereas reverse polarity systems do the opposite). Loudspeakers with a positive voltage applied show normal polarity as a forward movement (pressurization) and negative polarity as a rearward movement (rarefaction).

Polarity has a storied history in professional audio. The foremost concerns JBL founder James B. Lansing, who in 1946 chose to reverse the polarity of his speakers to ensure incompatibility with those of his former employer Altec Lansing. The polarity war lasted over forty years. We all lost.

Our standard line level connector also lacked a polarity standard (the XLR connector was pin 3 hot in the USA and pin 2 hot for Europe and Japan). We had actually entered the digital audio age before an analog standard of pin 2 hot was established worldwide.

The term polarity is often mistakenly substituted for "phase" because a polarity reversal creates a unique case of phase shift, where all frequencies are changed by 180°. Polarity, however, has no time offset or phase delay and is frequency independent. Phase shift (described above) is frequency dependent and is caused by delay in the signal.

It is interesting to note that we have more terms for polarity than options: +/-, normal or reversed, non-inverting or inverting, right or wrong.

1.1.7.1 ABSOLUTE POLARITY

The "absolute" here refers to a system's net polarity orientation to the original source, often referred to by audiophiles as "absolute phase." A positive absolute polarity value is claimed if the waveform at the end of the chain matches the polarity of the original. Let's use an example to follow absolute polarity from start to finish. A kick drum moves forward (+). The positive pressure wave moves a microphone diaphragm inward (+). The console and signal processing maintain normal polarity to the power amplifier inputs (+). The amplifier output swings positive voltage on the hot terminal (+) and the speaker moves forward creating a positive pressure wave (+). Success! But some big assumptions have been made. Perfect mics, electronics and speakers that have flat phase responses. Pssst! They don't. There is frequency-dependent delay in the speakers, enough to make parts of them 180° different than others. Polarity is suddenly ambiguous. High-pass filters, low-pass filters, reflex box tuning and LF driver radiation properties all add up to frequency-dependent phase delay. Who's right? The phase response of the waveform at the end of the chain does not match the original, which means the polarity can no longer be called absolute.

I will not enter the fray as to whether humans can detect absolute polarity. Yet one thing is clear: A perfectly flat amplitude and phase transmission system moves us closer to reproducing the original drum sound, possibly to the point where polarity *is* absolute enough for us to make some conclusions. Nonetheless it is gospel among audiophiles that this is *hugely* important. Here is the best part about that. Audiophiles listen to recordings. Recordings can be assembled from hundreds of tracks recorded at different times, from different studios, transposed across different media, manipulated from here to kingdom come in digital audio workstations, sent to a mastering lab across the country, transcribed to a lacquer mother, a metal stamper and then pressed to vinyl, played by a turntable cartridge and finally through speakers that have drivers on the sides and rear to give that awesome surroundoscopic envelopment and create stereo everywhere. Can you please tell me what exactly is the absolute reference for my polarity?

1.1.7.2 RELATIVE POLARITY

Relative polarity is the one that counts: the polarity relationship between audio devices (Fig. 1.10). There is no disputing the fact that summing devices with different relative polarities causes cancellations and other unexpected outcomes. Relative polarity errors are best prevented by ensuring all devices are matched (and normal).

1.1.7.3 POLARITY AND PHASE DELAY COMBINED

Systems that exhibit frequency-dependent phase delay (e.g. speakers and most filters) can have their phase response further modified by a polarity change. The resulting phase values over frequency are a combination

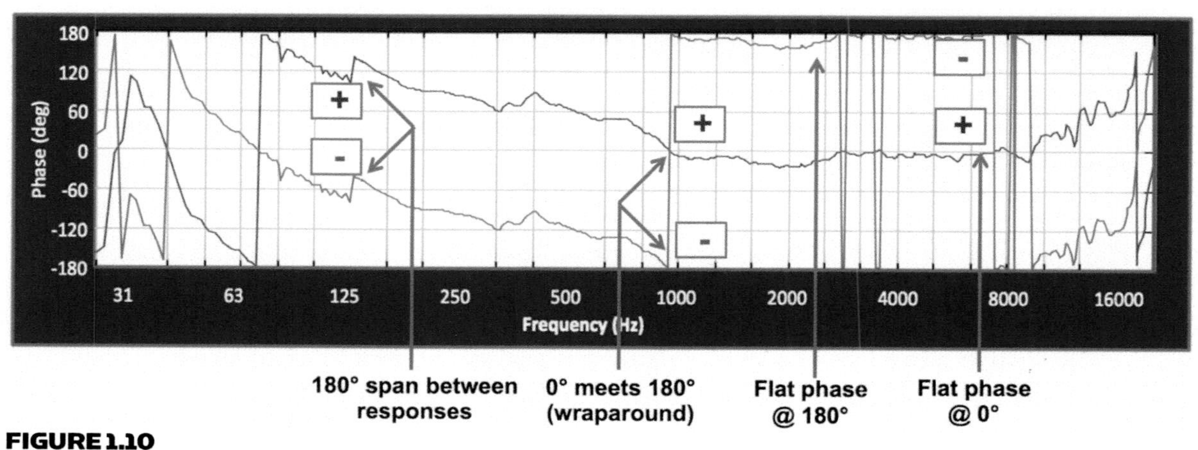

FIGURE 1.10
Example of polarity and reverse polarity over frequency

of these two modifiers (phase delay is frequency dependent whereas polarity reversal changes all frequencies by 180°). This finds practical use in acoustical crossover filter and delay settings, and steering cardioid subwoofer arrays.

1.1.8 Waveform

The waveform is the cargo transmitted through our delivery network: the sound system. It's the shape of amplitude over time, the fingerprint of an audio signal in all its simplicity or complexity. A sine wave is waveform simplicity. "In-a-Gadda-da-Vida" is a 17-minute complex waveform with many frequencies and three chords.

The waveform audio stream is a continuous function, a chronological series of amplitude values. Complex waveforms result from combinations of sine waves at various frequencies with unique amplitude and phase values. We can mathematically construct the waveform once we know the amplitude and phase values for each frequency. Conversely, we can deconstruct a known waveform into its component frequencies and their respective amplitude and phase values.

We are all familiar with the "synthesizer," a musical device that creates complex waveforms by combining individual oscillators at selectable relative levels and phase. We had an early MOOG synthesizer at Indiana University (SN# 0005) that required patch cables to mix oscillators together to build a sound. This was raw waveform synthesis. Its inverse (the deconstruction of a complex signal into its sine components) is the principle of the Fourier Transform, the heart of the audio analyzer used in optimization.

1.1.8.1 SINE WAVE

A sine wave, the simplest waveform, is characterized as a single frequency with amplitude and phase values. A "pure" sine wave is a theoretical construct with no frequency content other than the fundamental. Generated sine waves have some unwanted harmonic content (harmonics are linear multiples of the fundamental frequency) but can be made pure enough for use as a test signal to detect harmonic distortion in our devices.

The sine wave is the fundamental building block of complex audio signals, combinations of sine waves of different frequencies with individual amplitude and phase values.

The steady state of the sine wave makes it suitable for level calibration in electronic systems (analog or digital) whose flat frequency response allows for a single frequency to speak for its full operating range. Speakers cannot be characterized as a whole by a single sine wave because their responses are highly frequency dependent and readings can be strongly influenced by reflections.

1.1.8.2 COMBINING UNMATCHED FREQUENCIES

Multiple frequencies coexist in the waveform when mixed together (Fig. 1.11). The amplitude and phase characteristics of the individual frequencies are superpositioned over each other in the waveform. The amplitude rises in portions of the waveform where signals are momentarily phase matched, thereby increasing the crest factor beyond the 3 dB of the individual sine wave.

It is possible to separate out the individual frequencies with filters or by computation (e.g. the Fourier Transform). The relative phase of mixed frequencies affects the combined waveform amplitude but does not affect that of the individual frequencies. In other words, 10 kHz cannot add to, or cancel, 1 kHz regardless of relative phase. The combined waveform, however, differs with relative phase.

1.1.8.3 WHITE AND PINK NOISE

There are a few special waveforms of recurring interest to us. White noise is arguably the most natural sound in the world: all frequencies, all the time, with random phase and statistically even level. White noise is the product

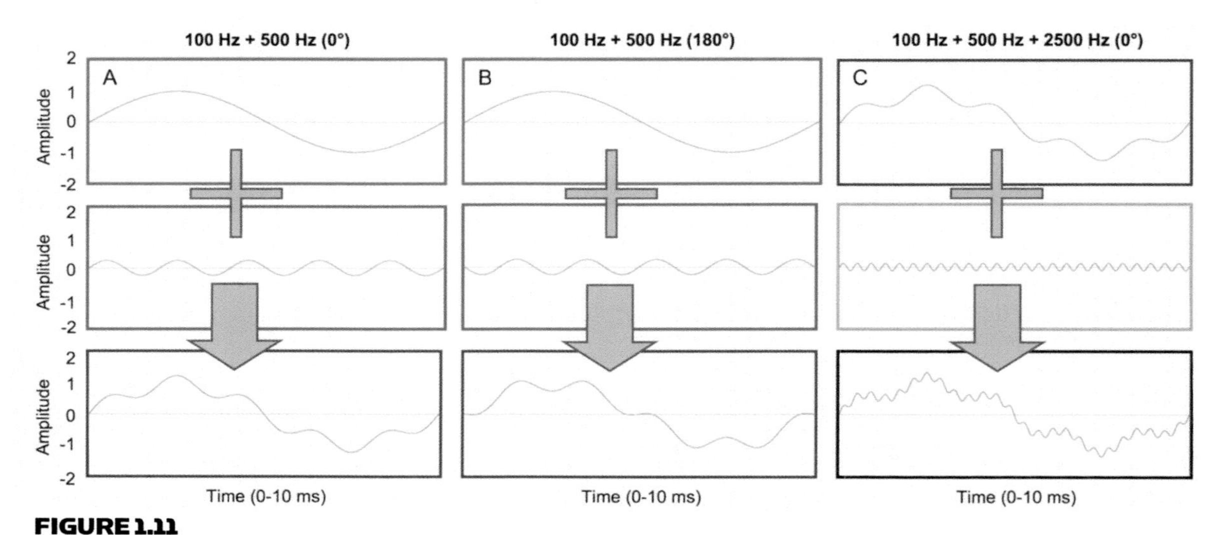

FIGURE 1.11
Waveform mixing: the combination of signals with unmatched frequencies

of random molecular movements in our electronics, well known as the noise floor and the sound of radio mic disaster. The energy is spread evenly over the linear frequency range, so half of the humanly audible energy is below 10 kHz and half above. White noise is perceived as spectrally tilted toward the HF (and called "hiss" because our ears respond on a log basis over frequency).

Pink noise, the most common audio spectrum test signal, is doctored white noise, filtered to sound even to our ears. High frequencies are attenuated at 3 dB/octave. This "logs" the linear noise and reallocates the energy to equal parts/octave, ⅓ octave, etc.

1.1.8.4 IMPULSE

We just met random noise: all frequencies, equal level, continuous with random phase. The impulse is another special waveform: all frequencies, equal level, one cycle, in phase. Think of an impulse as all frequencies at the starting line of the racetrack. The starter pistol goes off and everybody runs one lap and stops. In fact, a starter pistol is an acoustic impulse generator used by acousticians for analysis. Listening to an impulse in a room reveals the timing and location of reflection paths. The impulse generator that won't get you in trouble with airport security is two hands clapping.

1.1.8.5 COMBINING MATCHED FREQUENCIES

Signals with matched frequencies merge to create a new waveform differing only in amplitude from the original signals that comprise it (Fig. 1.12). The resulting waveform depends upon its relative amplitude and phase characteristics, and may be greater, lesser or equal to the individual contributors. The combination of equal level-matched frequency signals will generally be additive when the phase offset is < ±120° and subtractive when between ±120° and ±180°. It is not possible, *post facto*, to separate out the original signals with filters or computation. The combined waveform of two amplitude- and phase-matched signals is indistinguishable from that of a single signal at twice the level. Conversely, the combined waveform of two amplitude-matched signals with 180° phase offset is indistinguishable from unplugged.

1.1.8.6 COMBINING UNCORRELATED WAVEFORMS (MIXING)

Unmatched (uncorrelated) complex waveforms merge together to create a new waveform with a random association to each of the original parts. The combination of unmatched complex waveforms (multiple frequencies) must be evaluated on a moment-to-moment basis. The obvious example of uncorrelated signals

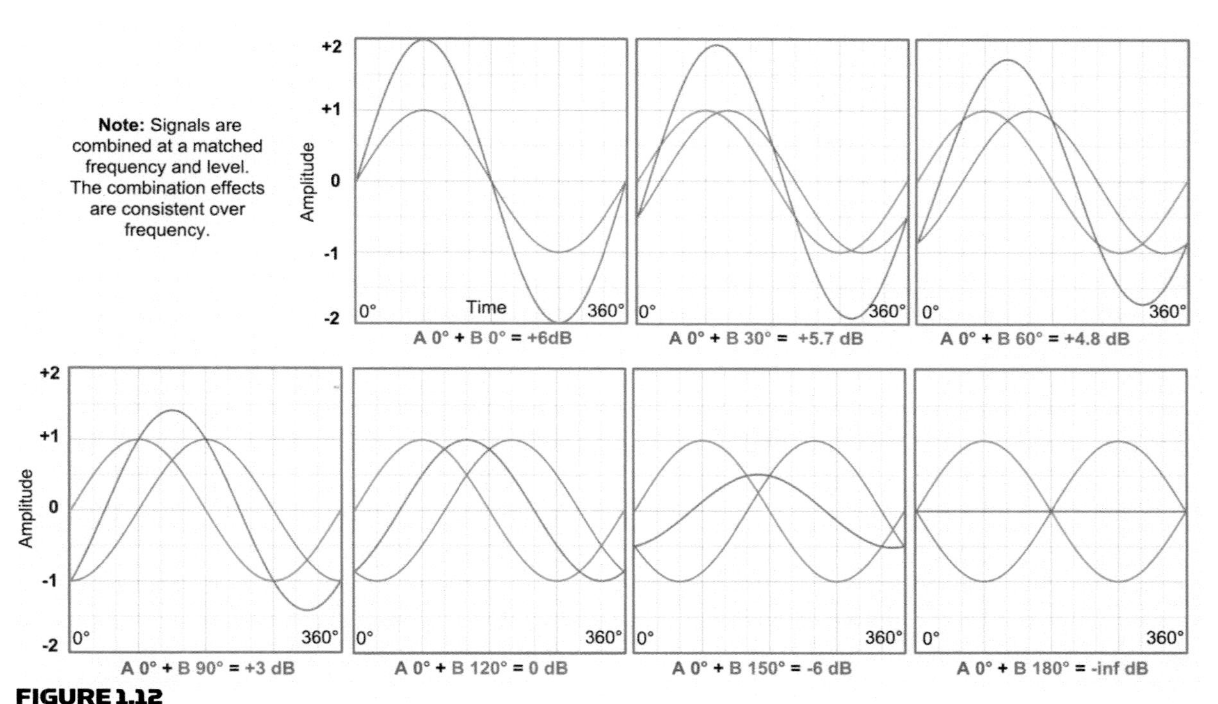

FIGURE 1.12
Combining signals with matched frequencies (correlated summation): The effects of relative phase on the combined responses are shown

is two different music streams: frequencies are in phase one moment and out of phase the next. These could be different songs, different instruments playing the same song or even violins playing the same parts in a symphony. In all cases the relationship between the signals is unstable and therefore incapable of consistent addition or subtraction at a given frequency. Combining unmatched waveforms is the essence of "mixing," and thus is separate from the combinations of matched waveforms (correlated summation), a primary concern of system optimization.

1.1.8.7 COMBINING CORRELATED WAVEFORMS (SUMMATION)

The combination of matched (correlated) complex waveforms also creates a new waveform, but with an orderly and predictable relationship to the original parts. Phase-matched signals create a combined waveform similar in shape to the individuals but with higher amplitude. If there is time offset between the signals, the new waveform will be modified in a stable and predictable way, with alternating additions and subtractions to the amplitude response over frequency (a.k.a. "comb filtering").

An example is copies of the same music stream combined in a signal processor. With no time offset, the combination will be full-range addition. Comb filtering is created by time offset between the signals. The interaction of speakers carrying the same signal in a room is more complex because the interaction must be evaluated on a frequency-by frequency and location-by-location basis (covered in depth in Chapter 4).

1.1.8.8 ANALOG FORM

Oscillation is a continuous function. We cannot get from Point A to Point B without passing through all the points in between. That is the essence of analog audio: the motion of a tuning fork, string, phonograph needle, speaker cone and more. A song, in analog form, is a series of movements in the waveform over time, always going one way or the other (+/-, in/out, etc.). If the movement stops, so does the song. If we can trace this continuous movement on one device and transfer it to another audio device, we will recognize it as the same song, even if one version came from magnetic flux (a cassette tape) and another came from the mechanical motion of a needle in a groove. Analog audio is like a continuous drawing exercise, a transcription that never lifts the pencil from the paper.

Transferring a waveform between electrical, magnetic and acoustic transmission mediums is like redrawing that pencil sketch with a different medium such as paint, stone or whatever. Notice that the term "medium" has much the same meaning in both fields.

1.1.8.9 DIGITAL FORM

Digital audio is a non-continuous function. We can only get from Point 0 to Point 1 without evaluating any points in between. Listening to 0s and 1s sounds pretty boring, even for people at raves. Digital waveforms are copies of analog waveforms, but the operation differs from the transduction process between analog mediums discussed above. Analog-to-digital converters slice the continuous signal into tiny bits (ba-da-boom), each of which represent the best fit for the momentary amplitude value. The faithfulness of the digital rendering depends on how finely we slice both amplitude and phase (time). Amplitude resolution is defined by the number of bits (24-bit is the current standard), and temporal resolution (a.k.a. sample rate) is typically 48 kHz (approximately .02 ms).

It's as if we have a photograph of our original pencil drawing. If you look close enough at the photograph you can see that there are no continuous lines, just lots of little dots. That is the essence of digital. Audio pixels. The beauty of it is that once we have the digital copy we can send it around the world without changing it. As long as we are very careful, that is. This is not even hypothetically true of an analog signal because all audio mediums have some form of degradation (distortion, frequency response variation, phase shift, etc.). Bear in mind that we can never hear digital audio. We only get that pleasure after the waveform is converted back to analog, because at the end of the day we have to use the analog air medium to get to our analog ears. This time the conversion requires our scribe to pick up the pencil again and draw one continuous connecting line between every dot in our digital photograph of the original line drawing.

1.1.9 Medium

Analog audio waveforms propagate through a medium. Within this book air molecules will be the acoustical medium and electronic charge and magnetic flux serve as the electromagnetic medium. Each medium has unique properties such as transmission speed, propagation characteristics, loss rate, frequency response, dynamic range and more. Digital audio (between analog conversions) is transmitted over a medium (not through it).

1.1.9.1 PROPAGATION SPEED

Propagation through a medium is a chain reaction. Each energy transfer takes time (an incremental latency) so the more media we go through, the longer it takes. Propagation speed is constant over frequency but variable by medium. Electromagnetic propagation is so fast we are mostly able to consider it to be instantaneous. Acoustic propagation speed is related to the medium's molecular density (higher density yields higher speeds). Sound propagates through a metal bar (very dense) faster than water (medium density), which is faster than air (low density). Sound propagation is the same speed, however, for heavy metal bars and air shows (ba-da-boom).

1.1.9.2 WAVELENGTH (λ)

An audio frequency has physical size once it exists within a transmission medium. The wavelength (λ) is the transmission speed/frequency, or transmission speed × period (T). Wavelength is inversely proportional to frequency (becomes smaller as frequency rises).

Audible wavelengths in air range in size from the largest intermodal-shipping container to the width of a child's finger, a 1000:1 range. Why should we care about wavelength? After all, no acoustical analyzers show this, and no knobs on our console can adjust it. In practice, we can be blissfully ignorant of wavelength, as long as we use only a single loudspeaker in a reflection-free environment. Good luck getting that gig. Wavelength is a decisive parameter in the acoustic summation of speaker arrays and rooms. Once we can visualize wavelength, we can move a speaker and know what will happen.

Transmission mediums				
Medium	Transmission	Speed (m/s)	Wavelength @ 1kHz	Transducers
Air	Pressure change	342	0.342	Microphone, loudspeaker
Water	Pressure change	1484	1.484	Hydrophone, loudspeaker
Iron	Mechanical vibration	5102	5.102	Accelerometer, vibrator
Electrical	Electrical charge	200k+	200+	Transformer, mic, speaker
Magnetic	Magnetic flux	200k+	200+	Transformer, mic, speaker

FIGURE 1.13
Basic properties of audio transmission mediums

1.1.9.3 TRANSDUCTION, TRANSDUCERS AND SENSITIVITY

Transduction is the process of waveform conversion between media (Fig. 1.13). Transducers are media converters. Examples include acoustic to electromagnetic (microphones), and vice versa (speakers). The amplitude values and wavelength in one media (e.g. pressure for acoustical) are scaled and converted to another media (e.g. voltage for electromagnetic). The scaling key for transduction is termed "sensitivity," which inherently carries units from both sides of the conversion. Microphone sensitivity links output voltage to input pressure (SPL), with the standard units being mv/pascal. Speaker sensitivity relates acoustic output pressure to the input power drive. The standard form is dB SPL@1 meter with 1-watt input.

1.2 AUDIO SCALES

This book is full of charts and graphs, all of which have scales. The sooner we define them the easier it will be to put them to use.

1.2.1 Linear amplitude

Amplitude is all about size. There are a million ways to scale it (or should I say there are 120 dB ways). Linear level units are seldom used in audio even though they correspond directly to electrical and acoustic pressure changes in the physical world (our level perception is logarithmic). Many engineers go their entire careers without thinking of linear sound level. Can we do a rock concert with a sound system that can only reach 20 pascals (120 dB SPL)? My mix console clips at 10 V_{RMS}. Is this normal? (Yes, that's +20 dBV.) I know a guy who sent +42 dBV @60 Hz as a test tone to some people he didn't like. You might recognize the linear version of that: 120 V_{RMS} @60 Hz. Voltage is found in many areas outside the audio path, so it helps to have bilingual fluency between linear and log.

Let's count in linear. Incremental changes from 1 volt to 2, 3 and 4 volts are sequential linear changes of +1 volt. If we started at 101 V and continued this linear trend we would see 101 V, 102, 103 and 104 V.

Now let's count the same voltage sequence in log (approximately): 1 V to 2 V (+6 dB), 2 V to 3 V (+4 dB), 3 V to 4 V (+2 dB). The total run from 1 V to 4 V is 12 dB. By contrast, the entire 4-volt sequence starting at 101 V would not even total 0.5 dB. Linear amplitude scales include volts (electrical), microbars or pascals (acoustical), mechanical movement (excursion), magnetic flux and more.

1.2.2 Log amplitude (20 $\log_{10}$ decibel scale)

The log scale characterizes amplitude as a ratio (dB) relative to a reference (fixed or variable) (Fig. 1.14). Examples include dBV (electrical) and dB SPL (acoustical). Successive doublings (+6 dB) of sound pressure are perceived as equal increments of change (a log scale perception). Therefore acoustic levels (and the electronics that drive them) are best characterized as log. Successive linear doublings of 1 microbar (to 2, 4 and 8 microbars) would be successive changes of approximately 6 dB (94, 100, 106 and 112 dB SPL).

Decibel scale ratio reference							
$20*\log_{10}$ (Output / Input)				$10*\log_{10}$ (Output / Input)			
Pressure (SPL), voltage (V), current (I)				Power (P)			
Gain +		Loss -		Gain +		Loss -	
Log (dB)	Ratio (Out/In)	Log (dB)	Ratio (Out/In)	Log (dB)	Ratio (Out/In)	Log (dB)	Ratio (Out/In)
0.0	1.00	0.0	1.00	0.0	1.00	0.0	1.00
1.0	1.12	-1.0	0.89	0.5	1.12	-0.5	0.89
2.0	1.26	-2.0	0.79	1.0	1.26	-1.0	0.79
3.0	1.41	-3.0	0.71	1.5	1.41	-1.5	0.71
4.0	1.59	-4.0	0.63	2.0	1.59	-2.0	0.63
5.0	1.78	-5.0	0.56	2.5	1.78	-2.5	0.56
6.0	2.00	-6.0	0.50	3.0	2.00	-3.0	0.50
7.0	2.24	-7.0	0.45	3.5	2.24	-3.5	0.45
8.0	2.51	-8.0	0.40	4.0	2.51	-4.0	0.40
9.0	2.82	-9.0	0.35	4.5	2.82	-4.5	0.35
10	3.16	-10	0.32	5.0	3.16	-5.0	0.32
12	4.00	-12	0.25	6.0	4.00	-6.0	0.25
14	5.00	-14	0.20	7.0	5.00	-7.0	0.20
15	5.63	-15	0.18	7.5	5.66	-7.5	0.18
18	8.00	-18	0.13	9.0	8.00	-9.0	0.13
20	10	-20	0.10	10	10	-10	0.10
26	20	-26	0.05	13	20	-13	0.05
32	40	-32	0.025	16	40	-16	0.025
38	80	-38	0.013	19	80	-19	0.013
40	100	-40	0.010	20	100	-20	0.010
60	1,000	-60	0.001	30	1,000	-30	0.001
80	10,000	-80	0.0001	40	10,000	-40	0.0001
100	100,000	-100	0.00001	50	100,000	-50	0.00001

FIGURE 1.14
Decibel scale reference table showing ratio conversions for the 20 log and 10 log scales

Audio system voltage level reference							
Audio signal level			Mic level	Line level	Speaker level		
Voltage (Volts)	dBV	dBu	250 Ω (Watts)	1k Ω (Watts)	16 Ω (Watts)	8 Ω (Watts)	4 Ω (Watts)
89	39	41.2			512	1024	2048
63	36	38.2			256	512	1024
45	33	35.2			128	256	512
32	30	32.2			64	128	256
22	27	29.2			32	64	128
16	24	26.2		250 m	16	32	64
11	21	23.2		125 m	8	16	32
8.0	18	20.2		63 m	4	8	16
5.6	15	17.2		32 m	2	4	8
4.0	12	14.2		16 m	1	2	4
2.8	9	11.2		8 m	500 m	1	2
2.0	6	8.2		4 m	250 m	500 m	1
1.4	3	5.2		2 m	125 m	250 m	500 m
1.0	0	2.2		1 m	63 m	125 m	250 m
707 m	-3	-0.8			31 m	63 m	125 m
500 m	-6	-3.8			16 m	31 m	63 m
356 m	-9	-6.8			8 m	16 m	31 m
250 m	-12	-9.8			4 m	8 m	16 m
178 m	-15	-12.8			2 m	4 m	8 m
125 m	-18	-15.8			1 m	2 m	4 m
89 m	-21	-18.8				1 m	2 m
63 m	-24	-21.8					1 m
45 m	-27	-24.8					
32 m	-30	-27.8	250 m				
22 m	-33	-30.8	125 m				
16 m	-36	-33.8	63 m				
11 m	-39	-36.8	32 m				
8.0 m	-42	-39.8	16 m				
5.6 m	-45	-42.8	8 m				
4.0 m	-48	-45.8	4 m				
2.8 m	-51	-48.8	2 m				
2.0 m	-54	-51.8	1 m				
1.4 m	-57	-54.8					
1.0 m	-60	-57.8					
707 μ	-63	-60.8					
500 μ	-66	-63.8					
356 μ	-69	-66.8					
250 μ	-72	-69.8					
178 μ	-75	-72.8					
125 μ	-78	-75.8					
89 μ	-81	-78.8					
63 μ	-84	-81.8					
45 μ	-87	-84.8					
32 μ	-90	-87.8					
22 μ	-93	-90.8					
16 μ	-96	-93.8					
11 μ	-99	-96.8					
8.0 μ	-102	-99.8					

FIGURE 1.15
Analog electronic operational levels: mic, line and speaker levels

1.2.2.1 ELECTRONIC DECIBEL (DBV AND DBU)

One of the most heavily enforced standards in professional audio is to never have just one standard (Fig. 1.15). The dB scale for voltage has at least twenty. Only two are used enough any more to be worth memorizing: dBV (1 volt standard) and dBu (0.775 volt standard). The difference between them is a constant 2.21 dB (0 dBV = +2.21 dBu and 0 dBu = -2.21 dBV). There is an extensive history regarding dB voltage standards going back to the telephone, which you can read somewhere else when you have trouble sleeping.

The dB scale is favored because our audio signals are in a constant state of change. We can't control the music or the musicians. We are constantly riding current levels relative to each other, to a moment ago or to the legal level allowed before the police shut us down. The dB scale is complicated but is easier than linear when trying to monitor relative levels that have a 1,000,000:1 ratio.

We should at least know the maximum voltage level for our equipment, which is usually around 10 volts (i.e. +20 dBV, +22.21 dBu). The noise floor should be in the -100 dBV range. In between we find the "nominal" value of 0 dBV (or dBu), the placement target for the audio mainstream. This leaves 20 dB of headroom above and the noise 100 dB below.

$$Level(dBV)=20\times \log10 Level\ 1/1\ V$$

$$Level(dBu)=20\times \log10 Level\ 1/0.775\ V$$

1.2.2.2 ACOUSTIC DECIBEL (dB SPL)

The common term for acoustic level is **dB SPL** (sound pressure level), the measure of pressure variation above and below the ambient air pressure.

$$Level(dBSPL)=20\times \log10\ P/0.0002$$

where P is the RMS pressure in microbars (dynes/square centimeter).

The reference standard is 0 dB SPL, the threshold of the average person's hearing (Fig. 1.16). The limit of audibility approaches the noise level of the air medium, i.e. the level where the molecular motion creates its own random noise. It is comforting to know we aren't missing out on anything. The threshold of pain is around 3 million times louder at 130 dB SPL. The threshold of audio insanity has reached 170 dB SPL by car stereo fanatics.

The following values are equivalent expressions for the threshold of hearing: 0 dB SPL, 0.0002 dynes/cm^2, 0.0002 microbars and 20 micropascals (μPa). dB SPL is log and all the others are linear. For most optimization work we need only deal with the log form.

dB SPL subunits

dB SPL has average and peak levels in a manner similar to the voltage units. The SPL values differ, however, in that there may be a time constant involved in the calculation.

- **dB SPL peak**: The highest level reached over a measured period is the peak (dB SPL_{pk}).
- **dB SPL continuous (fast)**: the average SPL over a time integration of 250 ms. The fast integration time mimics our hearing system's perception of SPL/loudness to relatively short bursts. It takes about 100 ms for the ear to fully integrate the sound level.
- **dB SPL continuous (slow)**: An extension of the integration time to 1 second that is more representative of heat load and long-term exposure.
- **dB SPL LE (long term)**: This is the average SPL over a very long period of time, typically minutes. This setting is used to monitor levels for outdoor concert venues that have neighbors complaining about the noise. An excessive LE reading can cost the band a lot of money.

Acoustical operating levels						
Listening level (dB SPL) Cont.	Peak	Ear level range	Comments	High level music	Medium level music	Low level music
136	148	X-X				
130	142		Pain threshold			
124	136		EDM			
118	130					
112	124		High level music			
106	118					
100	112		Medium level music	X		
94	106					
88	100			N	X	
82	94		Low level music	O		
76	88			M	N	X
70	82			I	O	
64	76		Speech @ 0.5 meter	N	M	N
58	70		Speech @ 1 meter	A	I	O
52	64		Noisy HVAC	L	N	M
46	58				A	I
40	52		Typical HVAC		L	N
34	46		Quiet room	N		A
28	40		VERY quiet room	O		L
22	34		Recording studio	I	N	
16	28			S	O	
10	22		Dream on!	E	I	N
4	16				S	O
0	12	N	Brownian motion		E	I
-6	6	Z	Noise floor of air			S
-12	0		and ear			E

FIGURE 1.16
Acoustical operational levels: quiet, loud and insane

dB SPL weighting

There are filtered versions of the SPL scale (Fig. 1.17) that seek to compensate for the level-dependent frequency response variations in human hearing perception (the "equal loudness curves" described in section 5.1).

- **dB SPL Z ("Z" weighting)**: This is a recent trend to designate the unweighted response. Easier to spell.
- **dB SPL A ("A" weighting)**: Corresponds to the ear's response at low levels. LF range is virtually nonexistent. Applicable for noise floor measurements. Often used as a maximum SPL specification for voice transmission. Subwoofers on or off goes undetected with A-weighted readings.
- **dB SPL B ("B" weighting)**: Corresponds to the ear's response at intermediate levels. LF range is rolled off but not as much as A weighting (-10 dB @60 Hz). Applicable for measurements using music program material.
- **dB SPL C ("C" weighting)**: Corresponds to the response of the ear at high levels. Close to a flat response. Applicable for maximum-level measurements. Used as a specification for full-range music system transmission levels. Subwoofers on or off will have a noticeable effect when C weighting is used.

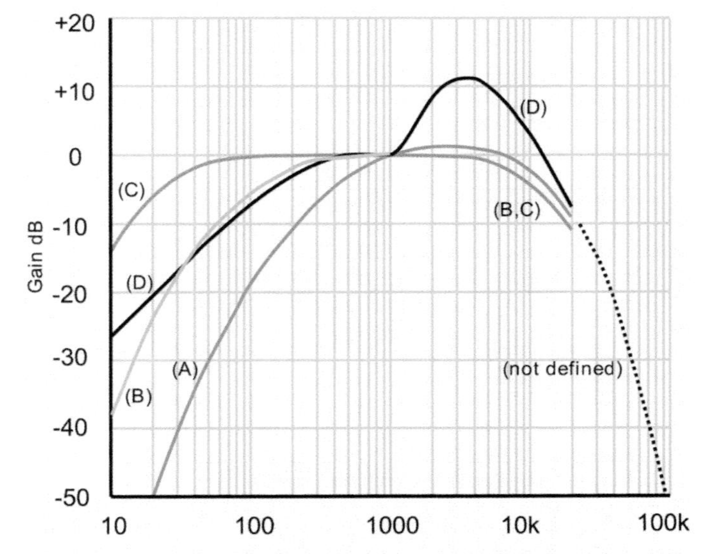

FIGURE 1.17
Frequency response curves for ABC and D weighting. Note: D weighting is shown here but not typically used for our applications (graphic by Lindosland @ en.wikipedia, public domain, thank you).

1.2.3 Power (10 $\log_{10}$ decibel scale)

Power is derived from a combination of parameters (e.g. voltage and current or pressure and surface area). The 10 $\log_{10}$ formula is the log conversion of power ratios (also Fig. 1.14). It's rarely used in system optimization because the analyzers monitor singular parameters such as voltage and SPL rather than power (in watts). It's better to prioritize our limited memory space for the 20 $\log_{10}$ formula and leave the 10 $\log_{10}$ for Google.

1.2.4 Phase

The standard scale for phase spans from 0° to 360°. The most common display places 0° at center and ±180° on the upper and lower extremes, but other options are available. Phase scaling is circular and therefore very different from amplitude. When we get more amplitude we simply expand the scale. We don't go up to 10 volts and then start over at 0 if we go higher. Phase is different because it tops out at 360°. When phase shift exceeds 360° it wraps around as an alias value within the 360° limits (e.g. 370° reappears as 10°). This is similar to an automobile race where cars on the lead lap are indistinguishable from those a lap behind. We only know who will win the race from other evidence, such as watching the entire race, not just the last lap. For audio phase our evidence will be similarly provided, but in this case it's about watching the phase values over the whole frequency response rather than a single frequency.

Phase serves us poorly as a unit of radial measure. Radians are the more "calculation-friendly" unit for radial angle found inside formulas that require accurate phasor positioning. Radians are rarely used by audio engineers for system optimization (or in a sentence for that matter). A complete cycle (360°) has a value of 2 π radians so beware of radian cancellation if two sources fall π radians out of sync (180°).

We learn to memorize the 360° phase scale even though it seems an arbitrary division for a circular function. It is a vestige of a merciful rounding error by ancient Egyptian mathematicians. Let's be grateful it's not 365° with a leap-degree every four cycles.

1.2.5 Linear frequency axis

Linear frequency scaling shows equal spacing by bandwidth (unequal spacing in octaves). The linear scale is an annoying, reality-based construction that corresponds to how frequency and phase interact in the physical world. The spacing between 1 kHz, 2 kHz, 3 kHz and 4 kHz (consecutive bandwidths of 1 kHz) is shown as equal spacing. Phase, the harmonic series and comb filter spacing all follow the linear frequency axis. The frequency resolution of the Fourier Transform (the math engine of our analyzer) is linear. For example, a set of 100 Hz wide filters with 100 Hz spacing is termed "100 Hz resolution."

1.2.6 Log frequency axis

A log frequency scale shows equal spacing in *percentage* bandwidth (octaves), and unequal spacing in bandwidth. This corresponds closely to our perception of frequency spacing. The equal linear spacing between 1 kHz, 2 kHz, 3 kHz and 4 kHz are percentage bandwidths of 1 octave, ½ and ⅓ octave respectively. A true log frequency response is made of log spacing of log filters. For example, a set of ⅓ octave filters at ⅓ octave spacing is termed "⅓ octave resolution."

1.2.7 Quasi-log frequency axis

The quasi-log frequency scale is a log/linear hybrid (log spacing of linear bandwidths). Stretching a single linear response over the full audio range is not practical for optimization because there is not enough data in the lows and/or too much in the highs. Instead a series of (typically) eight octave-wide linear sections are spliced together to make a quasi-log display. This is implemented in almost every modern analyzer used for optimization. (Full details are in Chapter 12.) Each octave of the quasi-log frequency response is derived from log spacing of the linear resolution (e.g. 1/48 octave spacing of 48 data points). This is termed "48 points/octave resolution."

1.2.8 Time

Tick, tick, tick. It seems strange to have to write that time is linear (evenly spaced increments) and not log (proportionally spaced increments). This is only mentioned because the frequency response effects of time offsets are entirely linear but are perceived by our log brains. So let's put this to rest: There is no log time.

1.3 CHARTS AND GRAPHS

Let's put the scales together to make the charts and graphs we use for design and optimization. Fluency in reading these graphs is a mandatory skill in this field. The graphs are 2-D, with an *x*-axis and *y*-axis. We generally find frequency and time on the *x*-axis and amplitude, phase and coherence on the *y*-axis.

1.3.1 Amplitude (*y*) vs. time (*x*)

Amplitude vs. time is a peak–peak waveform tracing commonly seen on oscilloscopes or digital audio editors (Fig. 1.18). The amplitude scale can be linear or log but time is only linear.

1.3.2 Amplitude (*y*) vs. frequency (*x*)

Absolute level over frequency is used to check the noise floor, the incoming spectrum, harmonic distortion, maximum output and more. The *y*-axis shows level against a fixed standard. We can observe the individual channels used to make transfer function computations (next item).

1.3.3 Relative amplitude (*y*) vs. frequency (*x*)

Relative level over a quasi-log frequency scale is the most common graph in system optimization (Fig. 1.19). This transfer function response is used for level setting, crossover alignment, equalization, speaker positioning and more. The *y*-axis scales unity level to the center and shows gain above and loss below (in dB). The quasi-log frequency scale is preferred because of its constant high resolution (24 to 48 points/octave) and close match to human hearing perception. An alternative option, the linear frequency scale can help identify time-related mechanisms (such as phase, comb filtering, reflections, etc.).

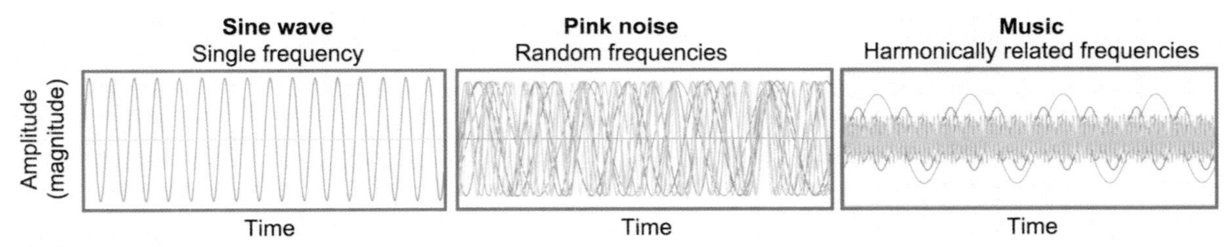

FIGURE 1.18
Amplitude vs. time plots for various waveforms

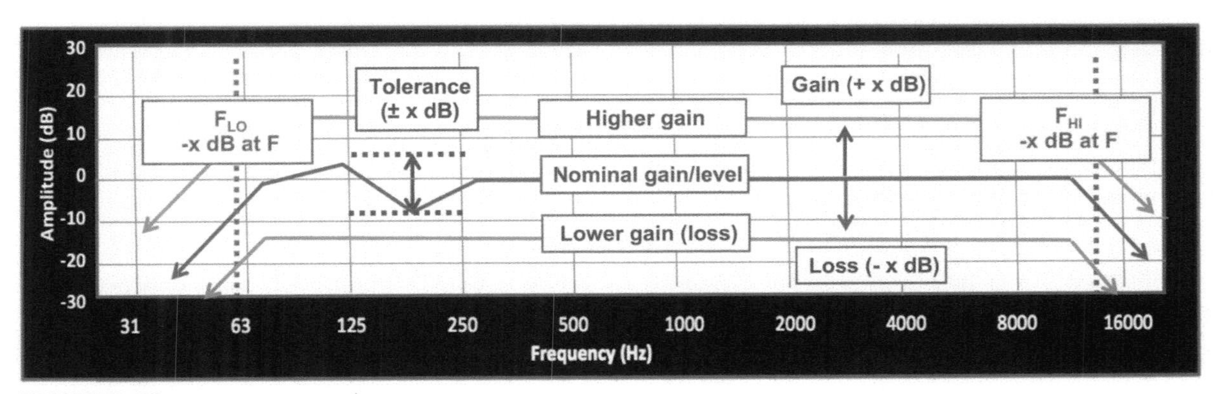

FIGURE 1.19
Introduction to the relative amplitude vs. frequency (quasi-log) display

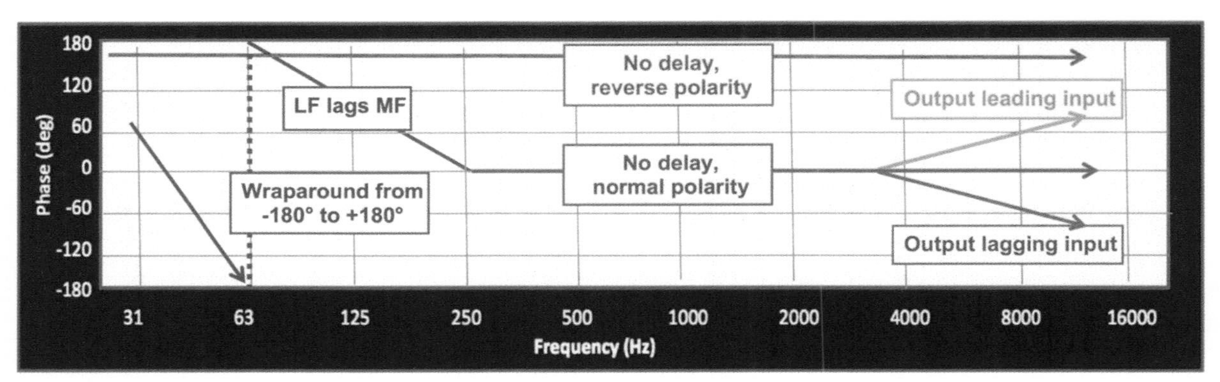

FIGURE 1.20
Introduction to the relative phase vs. frequency (quasi-log) display

1.3.4 Relative phase (y) vs. frequency (x)

This is our standard phase display (Fig. 1.20). The y-axis shows a 360° span. The vertical center is typically 0° but can be normalized around any phase value. A flat phase response (horizontal line) indicates zero phase shift and zero time offset over the frequency range shown. Variations from flat phase response indicate some time offset, either full band (i.e. latency) or frequency–dependent (phase delay). A downward slope (left to right) indicates positive delay whereas an upward slope indicates negative delay.

A constant phase rotation at linear frequency intervals indicates latency (frequency-independent delay). A quasi-log display shows the slope steepening with frequency (a linear function in a log display). Comb filtering appears as increasingly narrowing spacing of peaks and dips as frequency rises (again, linear function, log display). Filters create frequency-dependent delay. The phase response of a filter with a given Q will maintain the same shape (slope) as frequency rises.

Note: Don't freak out if you don't understand phase yet. This section is intended to provide just enough info to help read phase traces as we go along. If phase were easy, this book would be five pages long.

Relative phase on a linear frequency scale is less popular, but more intuitive than log. The phase slope over a linear frequency scale clearly reveals the relationship of phase and time (a linear mechanism on a linear scale). A flat phase response indicates no time difference at any frequency, just as the log display. Latency creates a constant phase slope as frequency rises. Comb filtering appears as consistently spaced peaks and dips as frequency rises. The term "comb filtering" comes from its linear frequency scale appearance.

1.3.5 Impulse response: relative amplitude (y) vs. relative time (x)

The impulse response is a favorite of the modern analyzer (Fig. 1.21). We can find delay offsets between speakers with extreme accuracy in seconds. Follow the dancing peak, read the number in ms. Done! Magic! And it seems like magic to most of us, even more so when we stop to think about what this computation is.

The FFT analyzer impulse response is a mathematical construction of the picture we would see on a hypothetical oscilloscope (amplitude vs. time) with a hypothetical single pulse. In practice we get relative amplitude vs. relative time. The full story on what's under the hood will have to wait until section 12.12. For now we will focus on how to read it.

Relative level (y-axis) is not like our amplitude over frequency. The vertical center is silence, not unity gain. Unity gain (normal polarity) is shown as an upward vertical peak at a value of 1 (0 dB) and positive gain is a bigger peak (and loss is smaller). A downward peak indicates polarity inversion.

Time (x-axis) is relative, a comparison of output–input arrival times. A centered impulse indicates synchronicity. The peak moves rightward when the output is late and vice versa. Yes, it is possible to have the output before the input in our measurements, because we can delay signals inside the analyzer.

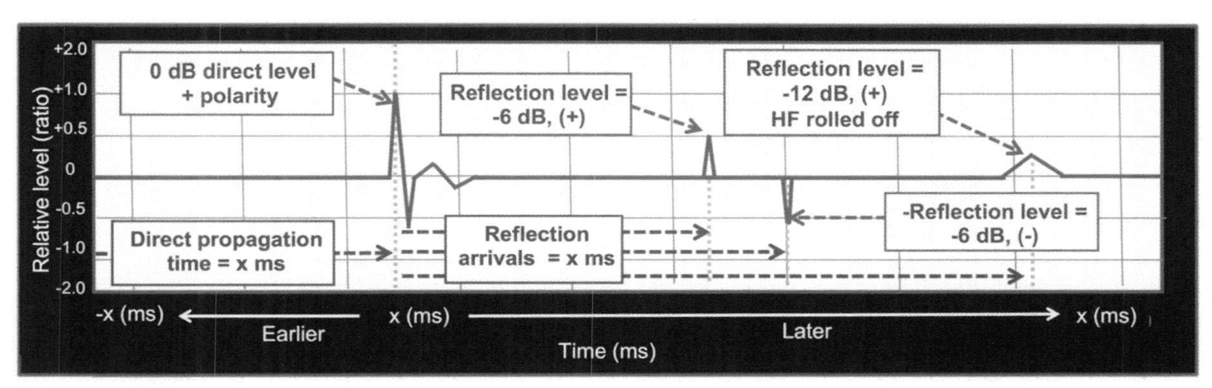

FIGURE 1.21
Introduction to the impulse response display

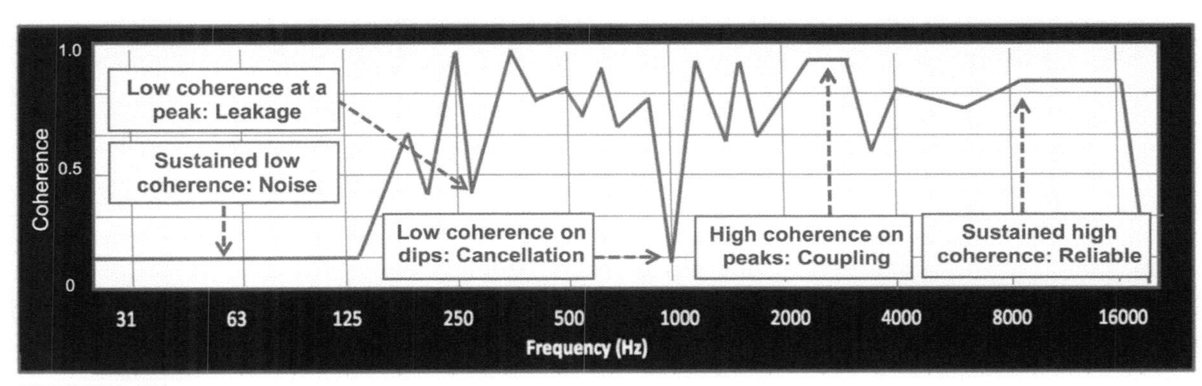

FIGURE 1.22
Introduction to the coherence vs. frequency display

A perfectly flat frequency response (amplitude and phase) makes a featureless impulse shape (straight single line, up and down). The pulse will have ringing, overhang and various other distortions to its shape if the measured device's response is *not* flat. Reflections appear as secondary impulses on the display. Like phase, this is enough information to enable us to read the displays going forward.

1.3.6 Coherence vs. frequency

The coherence function is a data quality index that indicates how closely the output signal relates to the input (Fig. 1.22). Amplitude and phase data are deemed reliable when coherence is high and unreliable when low. Coherence alerts one to take the wooden shipping cover off the speaker front instead of boosting the HF EQ. Yes, a true story.

Coherence is derived from averaging dual-channel frequency responses and is indicative of data stability. A value from 0 to 1 is awarded based on how closely the individual samples match the averaged value in amplitude and phase. Details are in section 12.11.

1.4 ANALOG ELECTRONIC AUDIO FOUNDATION

1.4.1 Voltage (*E* or *V*)

Voltage, electrical pressure, can be characterized linearly (in volts) or logarithmically in dB (dBV, dBu, etc.). The electronic waveform is a tracing of voltage vs. time. Voltage is analogous to acoustical pressure.

1.4.2 Current (*I*)

Current is the density of signal flow through an electronic circuit. As current increases, the quantity of electron flow rises. Analog audio transmission between electronic devices (except amplifiers to speakers) requires minimal current flow.

1.4.3 Resistance (*R*)

Resistance restricts current flow in an electronic circuit. As resistance rises, the current flow (for a given voltage) falls. Resistance is frequency independent (impedance is not).

1.4.4 Impedance (*Z*)

Impedance is the frequency-dependent resistive property of a circuit, a combination of resistance, capacitance and inductance. Impedance ratings are incomplete without a specified frequency. Output/input impedance ratios play a critical role in the interconnection of analog electronic devices, amplifiers and speakers, determining the maximum quantity of devices, cable loss and upper and lower frequency range limits.

An 8 Ω speaker (nominal impedance) illustrates the difference between impedance and resistance. The DC resistance is 6 Ω. The lowest impedance in its operating range is around 8 Ω. Impedance rises (and output level falls) at frequencies above its operating range.

1.4.4.1 CAPACITANCE (CAPACITIVE REACTANCE)

Capacitors pass signal across conductive, but unconnected, parallel plates. DC cannot flow across the gap in the plate. An AC signal, however, can flow across the plate (resistance falls as frequency rises). Capacitors approximate an open circuit (maximum impedance) to DC and short circuit (minimum impedance) to AC. Capacitance in series rolls off the LF response (resistance is inversely proportional to frequency). Parallel capacitance (such as between wires of an audio cable) rolls off the HF via a shunt path to ground.

1.4.4.2 INDUCTANCE (INDUCTIVE REACTANCE)

Inductor coils resist voltage changes in the signal. An unchanging signal (DC) passes freely but the inductor becomes increasingly resistant as frequency rises (the rate of electrical change increases). The inductor's response is the opposite of a capacitor: maximum impedance to AC signals and the minimum impedance to DC. Series inductance increasingly rolls off the HF response. Parallel inductance shunts the LF to ground.

1.4.5 Power (P)

Electrical power (in watts) is the combined product of voltage, current and impedance. Ohm's law expresses the relationship of these three parameters (Fig. 1.23). Most electronic transmission involves negligible power (low voltage, low current and high impedance). Speaker-level transmission requires substantial power (high voltage, high current and very low impedance). Acoustical power, also in watts, is produced by pressure, surface area and acoustic impedance (inertance). Our ears (and microphones) are pressure sensors, not power sensors, which characterize sound level by pressure only (SPL).

1.4.6 Operating levels

Signal levels are divided into three categories by voltage and impedance. A lucky waveform can experience all three (return to Fig. 1.15) if beginning at mic level, going through a "pre" amp to line level and rising to speaker level in a power amplifier.

1.4.6.1 MIC LEVEL

Mic-level signals are typically generated by small passive devices such as microphone coils, guitar pickups, phonograph cartridges, etc. Mic level is a matter of necessity, not choice. There are few advantages and many disadvantages to operating in the microvolt range. Signals are vulnerable to induced noise and other complications

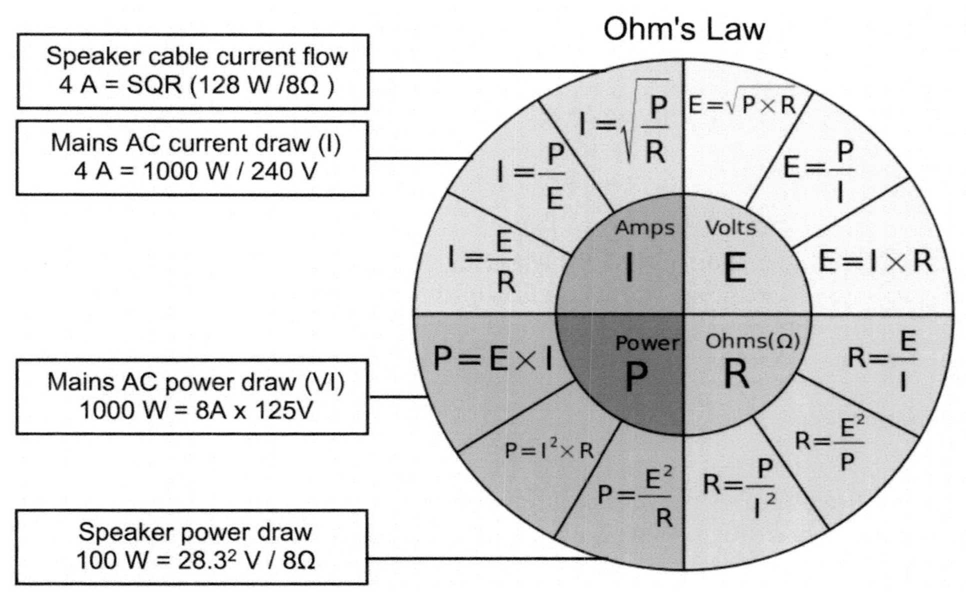

FIGURE 1.23
The Ohm's law pie chart with some examples applicable to audio systems (chart authored by Matt Rider, http://commons.wikimedia.org/wiki/File:Ohm's_Law_Pie_chart.svg)

relating to extremely low voltage and current flow (e.g. jumping connector and relay gaps). The winning strategy is to preamplify mic-level signals to line level as soon as possible. The worst-case scenario, unbalanced, high-impedance mic level (e.g. a guitar pickup), requires the shortest possible preamp path. Mic-level sources generate signals in the μV to 100 mV range with nominal source impedances of a few hundred ohms (microphones) and a few k Ω (pickups).

1.4.6.2 LINE LEVEL

Active devices, such as consoles, processor, instrument direct outs, playback equipment and more, usually generate line-level signals. This is the standard operating range, with nominal levels in the 1 V range, and maximum levels over 10 V. Balanced line-level low-impedance outputs (150 Ω typical) driving high-impedance inputs (10 kΩ typical) should have good noise immunity and minimal loss.

1.4.6.3 SPEAKER LEVEL

Power amplifiers are the exclusive generators of speaker-level signals. There is no "nominal" speaker level. Instead, speaker level ratings refer to the maximum power capability (watts). Power ratings require two known parameters: voltage and impedance. For example, a 100-watt amplifier with an 8 Ω speaker load can generate 28.3 volts, whereas a 400-watt amp will reach 56.6 volts.

1.4.7 Analog electronic audio metrics

This is an overview of some relevant standard specifications for professional-grade analog audio devices (Fig. 1.24).

1.4.7.1 FREQUENCY RESPONSE/RANGE (±dB/Hz)

There are two main categories for frequency response: range and deviation. Range is the spectral area between the limits, typically the half-power points (-3 dB points) at the LF and HF extremes. Response deviations are given as ±×dB for the area within the optimal range. Electronic device deviations are normally <±1 dB.

1.4.7.2 MAXIMUM LEVEL (dBV, dBu)

Any electronic device has a maximum voltage limit: the clipping point, typically specified at 1 kHz. This parameter helps ensure compatibility between interconnected devices to ensure full dynamic swing through the signal chain. It is preferable to have all devices clip at around the same level, lest the voltage swing become limited by the weakest link.

1.4.7.3 NOISE FLOOR (dB)

Let's measure a device unplugged (or with its input shorted). What's left is the device's residual self-generated noise floor. This is typically given as the worst case across the spectrum and falls into two categories: hum (harmonic multiples of the line frequency) and noise (white noise). Noise floor measurements are usually shown as "A"-weighted values.

1.4.7.4 DYNAMIC RANGE (dB)

Dynamic range is the span between the maximum-level capability and noise floor. Devices with a single gain stage have straightforward dynamic range specifications. Devices with multiple gain stages can vary their dynamic range by offsetting the internal settings and thereby decrease the maximum capability and/or increase the noise floor. In other words, cranking up the input while cranking down the output is likely to change the dynamic range through the device.

1.4.7.5 POLARITY (+/-)

Every device has one of two polarities: right (normal) or wrong (inverted). This should be a non-issue for balanced-input/balanced-output line-level devices. Polarity must be specified for any device with unbalanced inputs (e.g. DJ mixers) or outputs (e.g. power amplifiers).

1.4.7.6 VOLTAGE GAIN (SENSITIVITY)

A device passing audio has voltage gain, even if it's loss. Sound strange? The term "voltage gain" connotes the output/input voltage ratio. It's positive gain when the output is larger than the input and negative gain when less. Unity (0 dB) is the expected default gain in line-level devices. Power amplifiers require positive voltage gain, specified directly in dB or as sensitivity, i.e. the input voltage level required to reach full output power.

1.4.7.7 MAXIMUM POWER RATING (WATTS)

This is the maximum amount of power a device output can drive to another device input (the load). This specification is only used for power amplifiers and the speakers they drive.

1.4.7.8 TOTAL HARMONIC DISTORTION (THD)

Harmonic distortion is the addition of uninvited frequencies to the original waveform. Harmonics are linear multiples of the original transmitted frequency. Every transmission system adds distortion; the question is how much, and the answer is given in percentage of total harmonic distortion (%THD). Analog electronic devices normally exhibit a fairly consistent percentage THD over level and frequency. Therefore the specification is normally given at 1 V_{RMS} (line level) or rated output power (amplifiers) at 1 kHz.

1.4.7.9 INTERMODULATION DISTORTION (IMD)

Audio signals are generally (and hopefully) more complex than a simple sine wave. Therefore our system must remain linear while reproducing complex waveforms as well as simple sinusoids. Intermodulation distortion has the potential to arise when two or more sine tones are simultaneously reproduced. Where harmonic distortion generates spurious frequencies by multiplication, intermodulation does so by addition and subtraction. IMD products are difference tones related to the linear spacing between the signals. For example, a mix of 60 Hz and 1 kHz would potentially show IMD products at 940 Hz and 1060 Hz. In this case 60 Hz is, in essence, modulating 1 kHz, hence the name. Loudspeaker motion must track complex waveforms so IMD testing is one way to separate the men from the boys (or the under-seat FX generators from real speakers).

Analog electronic metrics				
Category	Measurement	Typical	Results	Applications
Range/tolerance	Amplitude vs. frequency	±3 dB points relative to nominal, ±1 dB	F_{LOW} to F_{HI} ± x dB	Sets the operating range for design
Maximum input/output	Sine wave input at maximum output	Clip point of either input or output	x dBV, dBu or volts for the input and output	Provides design info regarding interconnection and cable runs
Noise	Residual noise with no input (or shorted input)	White noise and hum	x dB (A weighted)	Provides performance criteria for design
Phase	Phase shift or phase delay over frequency	< ± 60° within operating frequency range	± x degrees (or ms) over a specified frequency range	Provides performance criteria for design
Distortion	THD, IMD, TIM	< 0.05 % at nominal level	x % @ nominal level (e.g. 1v) at 1 kHz or other freq	Provides performance criteria for design
Impedance	Input and output	Input > 5kΩ, Output < 250Ω	Input = x Ω, Output = x Ω	Provides design info regarding interconnection, cable runs, and parallel options

FIGURE 1.24
Standard metrics for the evaluation of analog electronic devices

The Society of Motion Picture and Television Engineers (SMPTE) standard for IMD testing is 60 Hz and 7 kHz mixed at a 4:1 ratio (-12 dB).

1.5 DIGITAL AUDIO FOUNDATION

Digital audio is the numerical rendering of an analog waveform constructed from a series of evenly spaced, end-to-end time records. The data are transmitted in non-continuous packets and reassembled for further processing, re-transmission or conversion to analog.

1.5.1 Numerical value (analogous to amplitude)

The waveform is traced as a series of numerical values, i.e. bytes (the more modern term is "octets" because bytes most commonly are 8-bit words), spaced evenly in time (the sample rate). The number can be referred to a fixed-point standard or floating-point. Fixed-point values can be linear or log (in dB), relative to the full-scale reference.

1.5.1.1 FIXED-POINT

"Fixed-point" describes the process of carrying the amplitude value as an integer, limited by the possible bit combinations. Increments are $2n^{th}$ power, with n being the number of bits. Sixteen-bit audio uses a sign bit (+/-) and 15-signal level bits for a total of 65,536 possible values (from 0000000000000000 to 1111111111111111). Twenty-four-bit has 16,777,216 iterations. Fixed-point topology has an absolute upper limit, known as "full-scale digital," and a lower limit, set by the "bit depth."

1.5.1.2 FLOATING-POINT

Floating-point math renders the amplitude value as an integer of fixed length and an exponent and direction (+/-) vector. The multiplication capability allows numbers to exceed the fixed-point topology's simple binary bit limits. Floating-point systems can carry "out of scale" numbers for internal processing functions, but must revert to "in scale" at the point of D/A conversion or transmission to a fixed-point audio device. Floating-point numbers have virtually unlimited size, but not unlimited precision. The precision is limited by the resolution of the mantissa (the numbers that precede the exponent, e.g. 1.3×10^6 vs. 1.33×10^6 vs. $1.33333333333333333333 \times 10^6$).

Hey kids, what's the highest number in the universe? The fixed-point child answers "10 gazillion." The floating-point kid answers "$10^{GAZILLION}$."

1.5.2 Current, resistance, impedance and power

Digital audio can be transmitted in electrical form but we won't be using an ohm meter to check continuity. There is voltage and current flow between digital devices, but they relate to the transmission of bits, not the waveform. In essence, an electronic digital audio system has two electrical states (e.g. 0 or +5 V_{DC}), in contrast with the continuous 120 dB range of analog. Digital audio connections are typically singular, point-to-point and made through a cable with a characteristic impedance of 75 Ω or 100 Ω. The key is to connect compatible devices with an impedance-matched cable under its specified maximum length.

1.5.3 Digital audio metrics

Here is an overview of the established parameters for digital audio devices (Fig. 1.25).

1.5.3.1 SAMPLE RATE

Sample rate sets the frequency range upper limit. The highest usable frequency must not exceed half the sample rate (the Nyquist frequency) to prevent calculation errors. Therefore steep filters are employed around the Nyquist frequency to prevent aliasing errors. A 48 kHz sample rate yields a frequency range up to 24 kHz and a 96 kHz sample rate yields 48 kHz of audio bandwidth.

1.5.3.2 FULL-SCALE DIGITAL

The maximum numerical value for the device at the conversion point or transmission to another device is termed "full-scale digital." The maximum value for a 24-bit signal is 16,777,215. Full-scale digital is equivalent to the analog clip point.

1.5.3.3 LEAST SIGNIFICANT BIT

Semiconductors and other analog electronic components constantly emit some level of molecular level noise. Not so with the 1s and 0s of digital. The digital floor is the sound of the lowest bit's inability to make a firm decision, a rounding error known as quantization noise. The least significant bit (LSB) has the difficult task of determining whether to round the tiniest signals up or down. Quantization noise is the sound of flip-flopping, which can be more disturbing to listeners than old-fashioned analog hiss. Adding low-level white noise to the signal, a process known as "dithering," can mask audible quantization noise. Today's 20-bit and 24-bit systems have moved the quantization noise under the analog noise floor, which reduces the need for dithering.

1.5.3.4 BIT DEPTH

Bit depth is the analog for dynamic range. In fixed-point digital this is the level difference between the maximum capability (full scale) and the noise floor (the LSB). We approximate dynamic range as bit number × 6 dB. Each time we add a bit to the resolution we are able to slice the finest increment in half, which is equivalent to 6 dB in terms of amplitude. For example, 16-bit audio can reach a dynamic range of 96 dB (16 × 6 dB), which is less than a well-designed analog circuit. By contrast, a 24-bit system has a potential of 144 dB, a very high bar for an analog circuit to reach.

1.5.3.5 dB FULL SCALE (dBFS)

Recall that sensitivity is the conversion factor for acoustic/electric transduction. dBFS plays this role between analog and digital. The process has three stages: voltage to number to voltage (dBV to full scale to dBV). dBFS is the answer to the question: "How many volts equals all bits set to 1?" Let's set an example A/D converter to full scale at 14.14V^{PK} (10 V_{RMS} for a sine wave). This is +20 dBV = 0 dBFS, a sensible choice that matches the digital maximum to the typical analog maximum. It's easy to get confused about this because our analog side thinks of 0 dB as "nominal" (with 20 dB of headroom remaining for peaks) whereas the digital world sees 0 dBFS as maximum with 20 dB of legroom below this to reach the "nominal" value (see Fig. 3.29 for reference). Confusion on this concept can cost a pile of dynamic range. We can keep it simple by following the industry standard for dBFS, which is . . . just kidding. There are too many to count.

Digital audio metrics				
Category	Measurement	Typical	Results	Applications
Sample frequency	Number of samples taken for A/D conversion	48 kHz, 96 kHz	HF limits of 22 kHz and 44 kHz respectively	Higher resolution more closely approximates the analog signal but also requires more data transfer speed
Bit depth	Number of bits sets dynamic range	20-24 bit	Seconds, octave bands 125 Hz to 8 kHz	Bit depth sets the digital dynamic range limit. Ideally this will equal or exceed the analog limits
dB FS	Sensitivity conversion factor for voltage to full scale digital	No standard	x dBV, dBu or Volts = Full scale digital	Standardized dB FS values enables maximization of dynamic range of both analog and digital systems
Format	The standard packet for audio file transmission	AES3 or multichannel compatible versions	Standard AES3 64-bit audio packet	Industry wide

FIGURE 1.25
Standard metrics for the evaluation of digital audio devices

RMS and peak

Recall that the digital numerical sequence traces the waveform's peak–peak structure. There is no such thing as an rms audio waveform, either analog or digital: ONLY peak (negative and positive). Rms values are mathematical calculations for heat dissipation or integrations of perceived loudness. Are we worried about cooking the numbers? That's the accountant's job. Loudness doesn't matter in the numbers game. It is simply a matter of fitting the waveform within the limits of the counter (the digital pipe). Our concern is headroom, the remainder between the positive or negative peak and the full-scale number. Now how do we relate them?

dBFS meets the sound system

The key is to make sure we match the reference points within our terminology. We have to link nominal with nominal and/or maximum with maximum. 0 dBV is nominal line level and +20 dBV is maximum analog. 0 dBFS is maximum digital and -20 dBFS is "nominal." The mistake would be linking this 0 dB analog to this 0 dB digital because they are nominal and maximum respectively. We can link maximum to maximum with settings of +20 dBV = 0 dBFS, more typically expressed as 1 V_{RMS} (0 dBV) = -20 dBFS.

We can reasonably expect to find systems in the range of -16 to -20 dBFS.

1.6 ACOUSTICAL FOUNDATION

We now enter the world of acoustics. Because we are only covering overwater acoustics we can boil this down to two things: air and surfaces. Air, the medium, is the easy one, with relatively few parameters in play. The surface part is the challenge: floors, ceilings, walls, I-beams, movie screens, balcony fronts and even our paying patrons. Anything that the sound hits, bends around, reflects off or pushes its way through before it has decayed below our hearing threshold goes in the category of acoustic surfaces.

1.6.1 Sound propagation

Sound moves through the air medium as pressure variations propagating spherically from the source at a speed of approximately 1 meter every 3 ms (1.1 ft/ms). The surface area increases and sound pressure level falls as the wave propagates outward. There is no distance limit to the propagation but the frictional and pressure losses eventually decrease the pressure variations to the point of inaudibility and noise floor of the air itself (Fig. 1.26).

1.6.1.1 ACOUSTIC AMPLITUDE (SPL)

Sound pressure level (SPL) is analogous to voltage in the electronic waveform. The standard free-field loss rate is 6 dB/doubling of distance from the source.

1.6.1.2 ACOUSTIC CURRENT (SURFACE AREA)

The waveform stretches over an increasingly large area as sound propagates outward. Acoustic power remains constant (neglecting frictional heat loss) so pressure falls as surface area expands.

1.6.1.3 ACOUSTIC IMPEDANCE (INERTANCE)

Acoustic impedance is the static pressure resisting the transmission of our waveform through the elastic medium of air. Speakers must generate actual power to overcome this resistance (unlike microphones, which can just sense the pressure). Air impedance varies slightly over temperature and altitude, but not enough to make us design mountaintop sound systems different from valley sound systems. The acoustic impedance in outer space is infinitely low (a vacuum), where it's easy for speakers to move but hard to make noise.

1.6.1.4 ACOUSTIC POWER (WATTS)

Acoustic power is analogous to electric power derived from the acoustic properties of pressure, surface area and inertance (voltage, current and impedance). The acoustic power generated by a loudspeaker dissipates gradually by friction heat loss but otherwise remains nearly constant as it propagates spherically outward. It's easy to miss this because our ears sense pressure (which falls rapidly over distance), not power. A stun grenade has almost as much far-field acoustic power as near-field, but proximity makes a big difference in how we experience it (175 dB SPL peak @1 m).

The difference in acoustic power between senders and receivers is found in the surface area (acoustic current), not the SPL. A speaker generating 94 dB SPL at 10 m fills a surface area of 1256 sq/m. By contrast the mic senses the same SPL over the surface area of its diaphragm.

1.6.2 Direct sound

The term "direct sound" refers to the first arrival, free and clear of any surface reflection paths. It's a straight path unless bent by wind, refraction or diffraction around a relatively small object. The direct sound path is the most important trajectory in sound system design due to its strong effect on perceived uniformity. Analyzers used for system optimization are highly focused on the direct sound response and early reflections, with varying degrees of inclusion of the late reflections.

Acoustical propagation		
Parameter	**Properties**	**Notes**
Spherical propagation	Propagates at equal speed and loss rate in all directions	Directional speakers have an initial pressure imbalance over surface area, which holds over distance
Speed	Set by elasticity of air, approximately 344 m/sec	Changes slightly with temperature and ambient pressure
Loss rate	Pressure falls as surface area expands. 6 dB loss (pressure) for each doubling of distance	Changes in the HF range with humidity. Affected by boundaries and reflections
Power	Power required to overcome the inertance of the medium (acoustic resistance)	Remains constant over distance except for frictional losses. Pressure loss is traded for expanded surface area
Dynamic limits	Reduced linearity at extreme high pressures. Brownian noise (air molecule noise floor). Cannot transmit in a vacuum.	The negative side of the waveform at 194 dB SPL is -1 atmosphere. Therefore essentially a vacuum for the negtive side of the waveform, causing acoustical clipping
Reflection	Sound bounces off large surfaces (relative to wavelength)	Angle of incidence = angle of reflection (like a mirror)
Refraction	Propagation path bends when passing through medium at different speeds	Air temperature affects speed so sound bends when passing through cold and warm air gradients
Diffraction	Large wavelengths bend around small objects or through openings	Otherwise we could hear our neighbor's television.

FIGURE 1.26
Parameters of acoustic propagation

1.6.3 Reflected sound

Reflections are the direct sound's children. They are undeniably related, always late and interrupt speech. Although there is only a single direct sound path, the number of reflected sound paths can be almost infinite. Acousticians and sound engineers categorize reflected sound into "early" and "late" differently. The acoustician's focus is on the long-term statistical reflection model, the room's unitary decay characteristic. They love reflections because they would not have a job without them. The sound engineer sees the direct sound distribution as primary and hopes the reflections don't do too much damage.

1.6.3.1 EARLY REFLECTIONS

Early reflections arrive close behind the direct sound. This small but critical minority of the overall reflections has strong effects on tonal perception and acoustic gain. Their paths are analyzed individually more than collectively (especially by sound engineers).

1.6.3.2 LATE REFLECTIONS

Late reflections are the tail end of the sound. There's lots of gray area around the division between early and late reflections (see section 5.3). For now let's simply say that early reflections fuse closely enough with the direct signal to be perceived as adding loudness. Late reflections only stretch time, which is characterized statistically. Ideally the late reflection decay character matches the venue's program material.

1.6.4 Room acoustical metrics

The parameters of room acoustics are well established. These metrics describe the room "unplugged," i.e. without a sound system. The numbers are useful to room designers, especially for this endangered species: the room without a sound system. Nonetheless, this is the bottom line for sound system design and optimization: These numbers have very little bearing on our decision-making process. Knowing "strength" or "bass ratio" or whether the decay is double sloped changes nothing for us. We still point the speakers at the audience, not the walls. More on this topic as we go (Fig. 1.27).

1.6.4.1 REVERB TIME (RT 60)

Reverb time is the interval for a signal to decay 60 dB after cessation. This is the oldest and most common singular room characterization metric. The measurement was conducted historically by exciting the room with a loud impulse such as a balloon or acoustic (starter pistol) and charting the decay response. This can now be done computationally by a variety of methods and analyzers.

1.6.4.2 REVERB TIME (T30, T20, T15)

We can measure a shorter sample of time and then extrapolate the results to the RT60 time frame. T30 cuts the time in half, T20 to ⅓ and so on. The assumption is that the decay trend of the first 15, 20 or 30 dB would continue with a longer measurement. The shortened decay measurements can increase confidence that we are above the room's noise floor, which can be a challenge for full 60 dB decays.

1.6.4.3 EARLY DECAY TIME (EDT)

Early decay time (EDT) measurements use a different starting point (the first reflection) than the RT series (direct sound arrival). As the name implies, this is a statistical analysis of the early reflections. The published value is derived from 15 dB of decay and extrapolated to the 60 dB equivalent. EDT highlights the different perspectives of acousticians and sound engineers. An EDT measurement characterizes a room without direct sound and a free-field speaker measurement characterizes the direct sound without a room.

1.6.4.4 NOISE FLOOR (NOISE CRITERIA)

The noise floor for rooms is generally specified with an "NC" rating. This stands for noise criteria, which relates to its perceived loudness. NC ratings are one of the few acoustical levels you'll see specified without dB in the answer,

Acoustical metrics				
Category	Measurement	Inclusive	Results	Applications
RT 60	60 dB of measured decay	Direct sound to 60 dB decay	Seconds, octave bands, 125 Hz to 8 kHz	Optimal times are application dependent
T 10	Initial 10 dB decay extrapolated to the 60 dB decay rate	-5 dB to -15 dB decay		
T 20	Initial 20 dB decay (to 60 dB rate)	-5 dB to -25 dB decay		
T 30	Initial 30 dB decay (to 60 dB rate)	-5 dB to -35 dB decay		
Early decay time (EDT)	10 dB decay from first reflection (60 dB rate)	0 dB to -10 dB decay		Favors early arrivals, so this is one of the more relevant acoustical numbers when sound systems are used.
Clarity (C50)	Ratio of early to late sound energy	Before and after 50 ms	0 dB is equal. + x dB indicates early is greater than late	Used for vocal intelligibility analysis. Positive clarity (in dB) aids intelligibility
Clarity (C80)	Ratio of early to late sound energy	Before and after 80 ms		Used for musical applications. -2 and below are favored for musical support

FIGURE 1.27
Standard metrics for the evaluation of room acoustics

e.g. a quiet room has an NC 20 rating. The NC ratings are derived from octave bands from 63 to 8 kHz and can be integrated to a single value that can be described as "A overweighted" at lower levels, meaning it is even less sensitive to the LF range. The weighting is level dependent (like the equal loudness contours). An NC rating of 25 corresponds to approximately 35 dB A. The primary applications are noise control (heating, ventilation, and air conditioning (HVAC), isolation, etc.).

1.6.5 Speaker acoustical metrics

There are also well-established parameters to describe the free-field acoustical behavior of loudspeakers (exclusive of the room) (Fig. 1.28). Only the most basic are discussed here as this will be covered in detail in section 2.7.

1.6.5.1 FREQUENCY RESPONSE/RANGE

There are two main categories for frequency response: operating range and tolerance. Range limits are less standardized for speakers than simple electronics, with some manufacturers using 3 dB, 4 dB, 6 dB and even 10 dB down. The most logical limit choice is -6 dB because it's the lowest value that can still combine to a unity crossover.

Tolerance is given as ± × dB within the operational range. Speaker tolerances are often < ±4 dB, but this spec is not standardized and may have unspecified frequency resolution (⅓ octave, octave, etc.).

1.6.5.2 MAXIMUM SPL

Maximum SPL specifications separate speakers by operating level. A single frequency was sufficient for evaluating the maximum level on electronic equipment. But what can 1 kHz tell us about our subwoofer, or any speaker for that matter? A speaker's maximum level needs to be evaluated on a case-by-case (frequency-by-frequency) basis. It is obvious that a single maximum SPL number cannot accurately characterize a speaker, but that doesn't mean that you won't see it on spec sheets.

SPL ratings are standardized around a 1 m distance (even if actually measured at 2 m or 4 m). The standard input signal is maximum-level continuous pink noise. From here it gets sketchy. How much distortion is allowable? Compression? Limiting? No standard. Do we know the speaker will survive another 3 seconds? In the end we get some numbers: maximum continuous, peak, "music," "program" and more. Various spectral weighting curves "A" and "C" or "Z" tailor the SPL specification over frequency.

It is important to understand that a meter reading of 120 dB SPL does not mean a speaker is generating 120 dB SPL at all frequencies. The 120 dB value is the integration of all frequencies (unless otherwise specified) and no conclusion can be made for a particular frequency range. Maximum SPL for a particular frequency range can be measured by limiting the input drive to the desired spectrum, a practice known as banded SPL measurements. The same data cannot be attained by analysis with post-process band limiting. Energy is spread over the full band when a full-range signal is applied to a device. Band-limited measurements show lower maximum levels for a given band if the device is reproducing frequencies outside of the measured band.

1.6.5.3 COVERAGE ANGLE

Coverage angle is the radial spread between the -6 dB points (referenced at an equal distance to the on axis point), normally given for the vertical and horizontal planes. No known speaker can maintain a constant coverage angle over frequency, and yet the determination of "nominal" coverage angle is not standardized. The 1 kHz coverage angle is not applicable as the nominal value for a speaker. Instead we use the average coverage angle over the upper several octaves (devices with a fairly constant coverage angle in the HF range) or the narrowest range (wavelength-proportional coverage angle). This spec helps match the speaker to the coverage target shape and for array building.

1.6.5.4 BEAMWIDTH

Beamwidth is coverage angle over frequency. Coverage angle, as a single number to describe a speaker, can be insufficient for many applications. Beamwidth plots allow us to compare the frequency dependent coverage shape of different speakers (which may have the same "nominal" coverage angle). This spec greatly aids the process of matching speakers to the desired coverage shape and array building.

1.6.5.5 SENSITIVITY (1 WATT/1M)

Sensitivity is the electrical and acoustical conversion factor, given as the speaker's SPL at 1 m with 1 watt drive level. Sensitivity ratings help users pair generic speakers and amplifiers. Industry trends show increasing preference for manufacturer-selected pairings or self-powered speakers, leaving sensitivity concerns to the manufacturers.

Loudspeaker metrics				
Category	**Measurement**	**Inclusive**	**Results**	**Applications**
Range/tolerance	Amplitude vs. frequency	Typically -6 dB points relative to nominal	F_{LOW} to $F_{HI} \pm x$ dB	Application dependent. e.g. Fills can have reduced LF range, higher tolerance
Maximum SPL	Full range level at maximum output at 1m	Max level with < 3 dB of limiting (Not standardized)	Seconds, octave bands, 125 Hz to 8 kHz	Provides power scale for designing with the speaker
Coverage angle	Amplitude vs. angle	0 dB to -6 dB in the HF range	x degrees over a nominal range	Provides coverage data for speaker selection (solo and array applications)
Beamwidth	Coverage angle vs. frequency	0 dB to -6 dB in octave or 1/3rd octave bands	Graph of coverage angle vs. frequency	Provides coverage data particularly relevant for array applications
Phase	Phase shift or phase delay over frequency	No industry standard	± x degrees (or ms) over a specified frequency range	Provides performance and compatibility data for speakers
Sensitivity	Speaker efficiency	Calculated on a per driver basis	x dB (1 watt, 1 meter)	Provides info for pairing speakers with amplifiers
Rated power	Power capability of the speaker	Maximum level before fire for each driver	x watts (peak, continuous, burst, long term)	Provides info for pairing speakers with amplifiers
THD	Distortion over level	Calculated on a per driver basis	x % THD at F_X at x dB SPL	Provides info for whether a speaker sounds like s#&%.

FIGURE 1.28
Standard metrics for the evaluation of loudspeakers

1.6.5.6 MAXIMUM POWER RATING (WATTS)

This is the maximum power the speaker can dissipate before being sent to the smoking section, with various iterations such as continuous, music, FTC, maximum and marketing watts. One could assume that an 800 watt speaker could be run at full level with an 800 watt amp, but your mileage may vary (a lot). This spec is intended to help match generic speakers and amps.

1.6.5.7 TOTAL HARMONIC DISTORTION (THD)

Harmonic distortion level in speakers is highly variable over level and frequency, which makes this a much more difficult specification to monitor. Many manufacturers provide little or no data on this, and if you measure their speakers you will know why.

1.6.6 Combined loudspeaker/room acoustical metrics

And finally we have metrics that describe the performance of the speakers in the room. These include frequency response, direct/reverberant ratio, intelligibility and others. These are intended as final measures of installed system/room performance (Fig. 1.29).

1.6.6.1 FREQUENCY RESPONSE

This is the combined frequency response of the speaker (direct sound) and the room (reflected sound). Analyzers differ in their approach to reflected response inclusion. Single-spectrum linear analyzers use a single time record. Reflection inclusion is the same for all frequencies within the specified time period and includes more wavelengths of reflected sound as frequency rises.

Quasi-log FFT analyzers use a series of different length time records that vary with frequency (LF are longest). Reflection inclusion is proportional to wavelength, i.e. the response includes the direct sound and a given number a wavelengths beyond. This is consistent with our hearing system's perceived tonal fusion zone.

1.6.6.2 DIRECT/REVERBERANT RATIO

Direct to reverberant ratio is exactly what the name connotes. Higher ratios correlate to greater clarity and increased signal/noise. Results vary greatly over frequency and therefore are poorly suited to being represented as a single number value. The HF will normally have a much higher ratio than the LF. Published D/R specifications standardize around the 1 kHz band. There are various versions C50, C7 and C35 but all incorporate only this single range. An additional limitation to D/R ratio is that it incorporates very few sound system performance parameters. Distortion and gross mismanagement of the optimization settings are just a few things that could slip under this metric's radar.

1.6.6.3 CRITICAL DISTANCE

Critical distance links a location in the room to direct-to-reverberant ratio. It is the distance from the loudspeaker where the D/R reaches unity. Closer locations are presumed to have more direct (and equivalent reverb) and more distant locations the opposite. Some engineers find value in staking out this position but I can't offer any guidance on how to find it or what to do with it. The reason is that D/R ratio is a number with 1000 faces (20 kHz/20 Hz) and therefore critical distance is in 1000 places. The distance that puts 1 kHz in critical condition leaves 10 kHz with a scratch and 100 Hz with a toe tag. One-size-fits-all numbers can't speak for our full-range systems and one-place-fits-all locations are just a version of the same thing. I have never been able to place a mic in a room and say, "As you can see, we are now at the critical distance."

1.6.6.4 COHERENCE

Coherence is the data quality metric used for system optimization. It is a standard feature of all the dual-channel FFT analysis systems. It is covered in detail (section 12.10) but we will touch on it briefly here. Coherence is able to discern signal/noise ratio vs. frequency (typically 48 point/octave). This allows us to see the degrading effects of reflections, air loss and wind on S/N ratio in high resolution. It also detects distortion, compression and combing between elements on the speaker side. A coherence factor of 50% indicates 0 dB S/N ratio at that location at that frequency.

1.6.6.5 INTELLIGIBILITY

Intelligibility is a complex metric that seeks to correlate system performance to its intelligible speech transmission capability. This is a book of its own (or books since there is so much debate about the various methods). There is widespread agreement on one item: More work needs to be done to find a metric that reliably correlates to intelligibility while incorporating both the room and the sound system. It's a tremendously complex task, but the fairest grade we can give to these systems at this time is "incomplete."

%Alcons (percentage articulation loss of consonants)

D/R ratio and EDT data are combined to create a quality index based on the percentage of lost consonants. Acceptable loss rates are expected to be <10% for general PA applications. Only a ⅓ octave band centered at 2 kHz is included in the data so its results are not valid for James Earl Jones.

Speech Transmission Index (STI)

STI uses a dedicated test signal that mimics speech. Scores are on a scale of 0 to 1. Human speech results from fundamental frequencies that are modulated by our vocal cords and mouth movements. The STI test signal is modulated to mimic this response. The transmission index scores the system by how much of the modulation depth is lost in transit.

Rapid Speech Transmission Index (RASTI)

RASTI is based on the STI principles but packaged as a commercially available test system. RASTI is simpler to operate and interpret than STI but at the cost of reduced spectral representation (two octave bands centered at 500 Hz and 2 kHz). RASTI is economical and has enjoyed widespread usage. Nonetheless it has serious shortcomings for those of us who design systems that span beyond the telephone range. Distortion, equalization, timing errors, compression and limiting can all affect (or not affect) the readings in unexpected ways. For example, limiting reduces the score, but clipping does not. Let's book Spinal Tap and turn it up to 11!

Loudspeaker/room combination metrics				
Category	**Measurement**	**Typical**	**Results**	**Applications**
Maximum SPL	Maximum level for the whole system	Too loud	x dB SPL at a given position, typically FOH	Match power scale to program material. Evaluate need for delays, fills etc.
Frequency response	Spectral response	Many specs say ± 2 dB though unrealistic	Seconds, octave bands, 125 Hz to 8 kHz	Overall curve is artistically determined. Used to evaluate acoustic treatment, speaker aiming, needs for fills etc.
Level uniformity	Relative level over the seating area	Many specs say ± 2 dB though unrealistic	x dB over and under the target	Evaluation of level distribution, need for fills and speaker aiming
Direct/Reverb ratio	Various methods. Most common in the modern era is coherence	Venue dependent	> 50%-80% in favorable situations.	Used to evaluate acoustic treatment, speaker aiming, needs for fills etc.
Intelligibility	STI		Rating system of 1 to 5	Evaluation of speech quality for the combination of room and speakers
% ALCons	Modulation transfer function (loss of modulation depth)	Loss of 10 % or less is considered acceptable for PA applications	x % ALCons	Evaluation of speech quality for the combination of room and speakers

FIGURE 1.29
Standard metrics for the evaluation of loudspeaker/room combination

SII

SII, as the name connotes, is 2nd-generation STI. This is currently under standards development and has the promise of superior correlation to our experience than the previous generations. It's not perfect but has addressed many of the shortcomings of the previous methods. SII has an expanded bandwidth (150 Hz to 8.5 kHz) and higher frequency resolution than previous methods. The effects of reverberation, noise and distortion are also included. Cancel Spinal Tap.

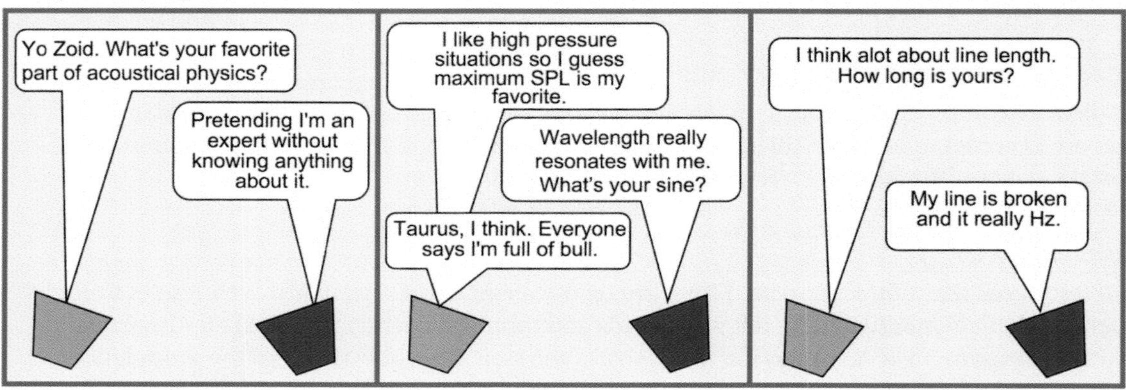

FURTHER READING

http://en.wikipedia.org/wiki/A-weighting.
Jones, R., Speech Intelligibility Papers: www.meyersound.com/support/papers/speech.
http://en.wikipedia.org/wiki/Intelligibility_(communication).
http://en.wikipedia.org/wiki/Speed_of_sound.

CHAPTER 2

Classification

What are sound systems made of? How do we classify the "system" into component parts for discussion? The simplest distillation might be signal sources, processors, transmitters and the connections between them. Now let's slice it a little finer. Sources include signal emitters such as the voice, musical instruments, recorded playback, etc. The human voice will tell you straight from the source that they don't want to be categorized as part of the sound system. They are the source for our sources: microphones and direct interfaces that give us custody of the signal. These in turn become sources for the next stage, the mix console, which in turn becomes the source for the system's signal processing. The signal processing includes the filters, delays, level distribution and more that prepare the signal for transmission. The transmission system consists of the amplifiers and loudspeakers to put the signal into the air toward the listeners. We can slice it finer yet, which is precisely the focus of this chapter, an inventory of the system components.

> *class n. 1. division according to quality. 2. group of persons or things having some characteristic in common. classify v.t. arrange in classes; assign to a class.*
> **Concise Oxford Dictionary**

2.1 MICROPHONES

Microphones are acoustic to electronic waveform media converters, i.e. transducers. The various microphone types differ in the intermediate stages of the conversion process. There are two main branches of the microphone family tree: dynamic (moving coil in a static magnetic field) and condenser (moving charged capacitive plate in an electric field). In both cases the diaphragm tracks the waveform's acoustic response but the transduction path to electricity differs. From there the mics differentiate into directional types, diaphragm sizes, materials and physical arrangements.

Most books focus on the primary application of microphones: recording and reinforcing sound sources. They are at the beginning of the signal path, serving as our original waveform sources. This book, by contrast, focuses primarily on the measurement microphones at the end of the signal path serving as surrogate ears for acoustical analysis. Nonetheless, we will briefly cover the microphones in general.

2.1.1 Microphone types

2.1.1.1 DYNAMIC

Dynamic microphones are the mechanical inverse of a standard loudspeaker (Fig. 2.1). The diaphragm has an attached coil, which floats inside a static magnetic field. The acoustic waveform creates a tracking movement in the diaphragm, which moves the coil in the gap between the opposing magnetic fields. The current induced in the coil is a transcription of the acoustic waveform. Voltage fluctuations in the coil are the electronic waveform we carry forward through the signal path. Dynamic mics, like their namesake the dynamo, are electrical power generators (albeit microscopic amounts) and therefore do not require external power.

A variety of factors converge on the dynamic mic to make life difficult in the VHF range: the mechanics of moving a coil on a diaphragm, reactive losses in the wire, the extremely low levels generated by the mic and more. This renders the dynamic mic unsuitable for applications requiring VHF extension such as cymbals and acoustic measurement.

Except for people saying "Check, 1,2" into an SM58 at front of house (FOH), I have never seen anyone use a dynamic mic as the test reference mic for system optimization (and hope I never will). Therefore this will be the end of our discussion regarding dynamic mics that are classified as velocity microphones.

2.1.1.2 CONDENSER

Condenser microphones have a variety of subsets and names including capacitance, electret and more (also Fig. 2.1). The basic scheme is the same: The waveform is transcribed by variable capacitance. The diaphragm must be a conductive metal (or coated with one) because it will act as one of the charged parallel plates of a capacitance circuit. The second plate is fixed in position and charged with a DC voltage. When no acoustic signal is present the DC voltage becomes the ambient-level reference point. When the diaphragm moves, the spacing between the plates changes, resulting in a variable capacitance between them. The variations are a transcription of the waveform and create an audio signal (AC) that can later be separated from the DC polarizing voltage. Condenser mics require an external DC power supply, commonly known as "phantom power" (48 volts DC) to charge the plates.

Condenser mics have higher cost and lower durability than dynamic mics, but the extended HF range and minimal coloration weigh in their favor. Condensers have the mechanical advantage of moving only a lightweight diaphragm, with no coil attached. Second, the phantom power allows for on-board circuitry to minimize reactive losses and maximize signal/noise ratio. For this reason the condenser mic is the preferred choice for applications that require VHF extension such as cymbals and acoustic measurement.

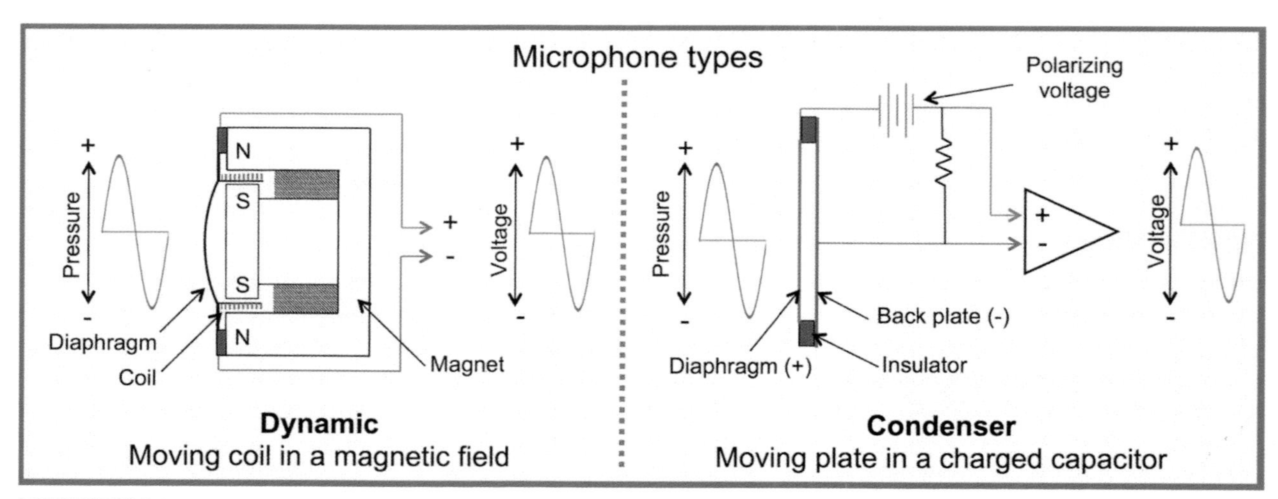

FIGURE 2.1
Microphone types

2.1.2 Microphone coverage patterns

Coverage pattern is an independent variable in microphone construction. In other words, dynamic or condenser mics can be configured in any of the basic patterns described below. The term "coverage pattern" refers to the mic's directional response, i.e. the differences between the on-axis response and those at other angles of incidence. In gross terms, there are mics that are largely indifferent to direction (omnidirectional) and those that have strong orientation in particular directions (Fig. 2.2).

2.1.2.1 OMNIDIRECTIONAL

An omnidirectional microphone is one whose frequency response is indifferent (or nearly so) to its angular orientation to a sound source. The mics are sealed so no sound reaches the rear of the diaphragm. Therefore there is no cancelling counterforce to the diaphragm movement. A pressure change from any direction can move the diaphragm, which is why these are known as "pressure-operated" microphones.

You might think an ideal omni mic would have no difference in its frequency response over angle, but such a thing is not possible or even desirable. A mic can only stay fully omnidirectional if its diaphragm is infinitely small. As diaphragm size rises, the frequency range of omnidirectionality falls. There is another side effect of diaphragm size: noise. As diaphragm size falls, the mic's self-noise rises. It is wise to consider the real-world tradeoff between best directionality and noise floor for a given application.

Omni mics have limited use in sound reinforcement because they have high leakage and low gain before feedback. They are the standard measurement mic for system optimization. All acoustic frequency response traces in this book come from omnidirectional condenser mics. There are more details in Chapter 12.

2.1.2.2 UNIDIRECTIONAL

A unidirectional microphone has substantial response variations over angle. Directional mics have a path for sound to reach both the front and rear of the diaphragm. The output signal is the pressure difference between the front and rear of the diaphragm. These are termed "pressure-gradient" microphones.

Bi-directional microphone (figure 8)

A mic whose diaphragm is equally open on both front and back is termed "bi-directional." Sound sources traveling toward the front or back of the mic move the diaphragm and create an electrical waveform. The frequency response may be similar in front and back but polarity is reversed for signals from the rear. Sources at the side of the mic are rejected because there is no pressure difference between the front and back of the diaphragm. The result is a "figure 8" pattern. This is the most common configuration for ribbon mics.

Cardioid, hypercardioid and supercardioid microphone

Consider the approach to the back side of the omni and bi-directional mics just discussed. The omni is sealed behind the diaphragm and the bi-directional is fully open to both front and back. Both mics cover the area in front and behind. The bi-directional rejects the sides. What would happen if we made a hybrid of those two: a mic that was open in the front and partially open in the rear? The response would be a hybrid that strongly rejects the rear and to a lesser extent the sides. This is the cardioid family of microphones.

Cardioid mics enable a portion of the sound to reach the back of the diaphragm through side and rear ports. We'll examine this in three parts. We first follow a sound wave's pressurized (+) portion arriving from behind. Sound first enters the port and later moves around to the microphone front. Two paths converge: Positive pressure from behind pushes the diaphragm outward, while positive pressure from the front pushes it inward. It's a perfect polarity reversal cancellation, right? Not quite, because perfect cancellation requires unity gain and exactly 180° phase offset. We have neither. The rear path to the diaphragm is shorter than the front. The clever remedy is adding a tiny labyrinth in the mic housing between the port and the diaphragm rear. This delays the rear arrival until it times out perfectly to maximize cancellation. However, we lose a few dB by taking the port path compared with the front.

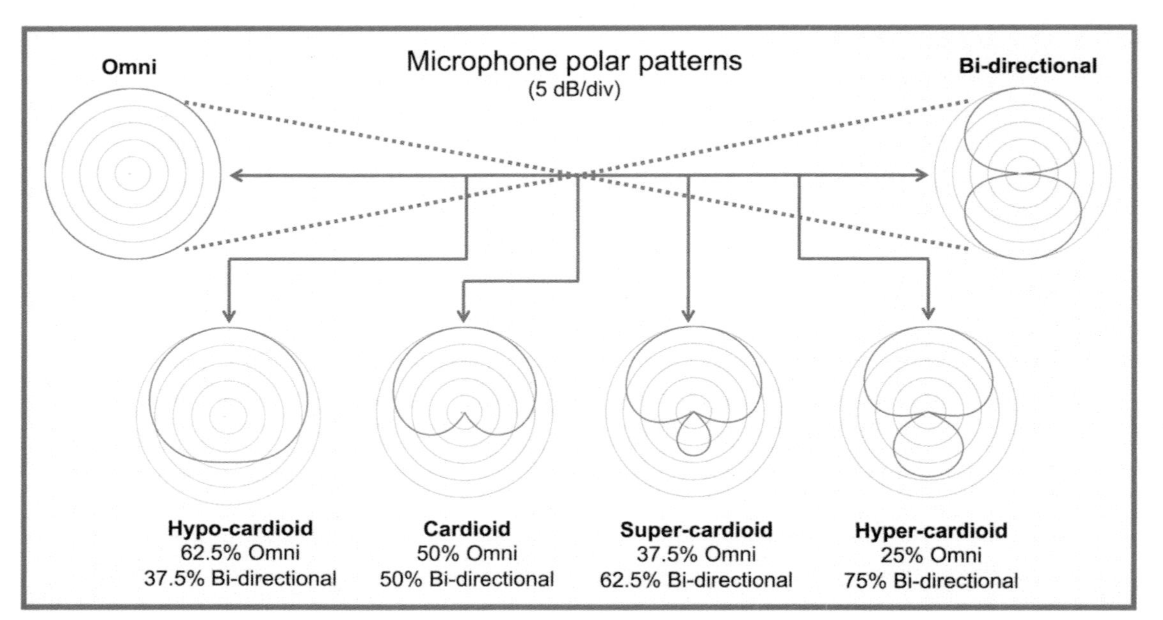

FIGURE 2.2
Microphone coverage patterns

That means we don't see as deep a cancellation as theoretically possible but can often achieve 20 dB or more in practice.

By contrast, sound originating in front arrives first (and loudest) at the diaphragm front, takes longer to get to the port and then falls further behind (in time and level) going through the labyrinth. The signal at the back of the diaphragm is in no condition to counter the force at the front. It's too little, too late.

The difference between cardioid, hypercardioid and supercardioid is in the port configuration, which varies the proportions of the two elements: the omni action and the bi-directional action. Cardioid has a lesser proportion of the bi-directional (hence the suppressed back lobe) whereas the hyper- and supercardioid mics have more (note the resemblance to the bi-directional figure 8 pattern).

2.2 INPUTS AND OUTPUTS (I/O)

Most electronic devices in the microphone-to-speaker pathway share common input and output stage features (Fig. 2.3). Between the I/O sections is the unique circuitry separating mix consoles from equalizers etc. We're only classifying configurations now. We'll be connecting them together and transmitting signal in the next chapter.

2.2.1 Analog input

An analog input receives the output waveform (voltage vs. time) of a preceding analog device. The stage must be appropriately gain-ranged to track the input signal's full dynamic range without clipping or adding excessive noise. Console inputs are typically scaled (or variably scalable) for microphone level (<0.1 V) and/or line-level input signals (<20 V). Analog inputs are normally configured as high-impedance balanced-line differential drive.

2.2.2 Digital input

A digital input is a numerical encoder receiving the output stream of a preceding digital audio device. The input stage tracks the sequence of incoming signal values. It must be configured to read the encoded waveform, i.e. sender and receiver must agree on the data transmission protocol.

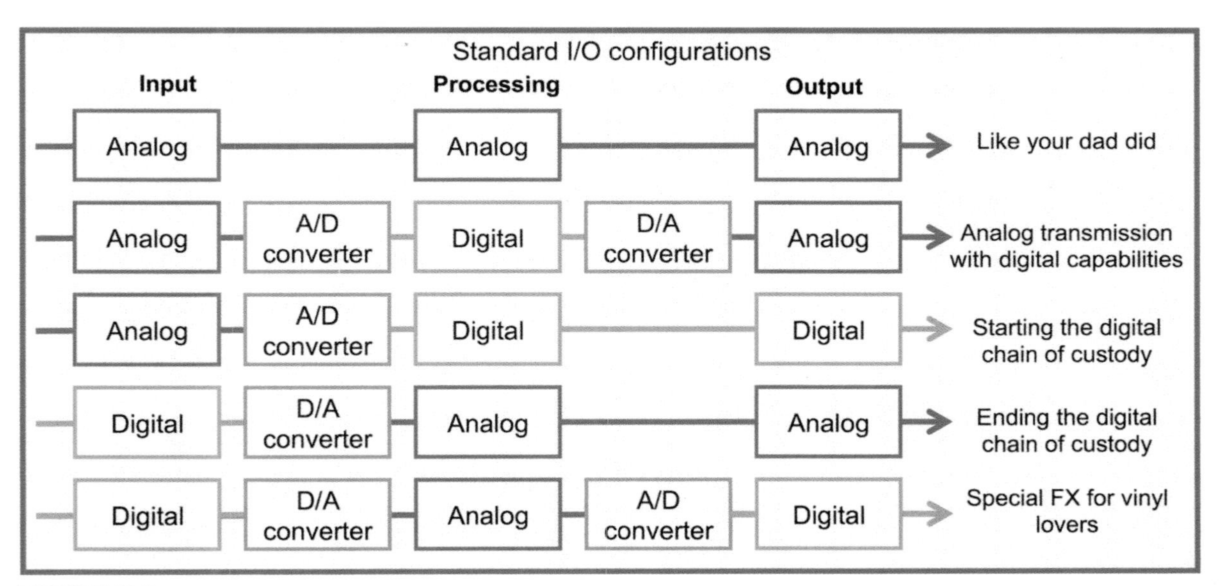

FIGURE 2.3
Input and output configurations

2.2.3 Analog processing

Analog processing circuitry modifies the signal's level, frequency response, dynamic response, phase response and more. This ranges from the simple resistor–capacitor filter to multi-stage equalization, compressors, limiters, routers, mix buses, matrices, analog delays and so forth. Multi-stage processing schemes must carefully monitor internal voltage gain to preserve dynamic range (preventing inter-stage clipping or excess noise).

2.2.4 Digital signal processing (DSP)

Digital signal processing (DSP) modifies the signal's level, frequency response, dynamic response, phase response and more. This ranges from digital equivalents of standard analog circuits as well as other exotics beyond analog's practical scope. Because internal digital processing signals exist only as numerical values, the gain range restrictions are not analogous to internal analog voltage transmission (which is limited by DC rail voltages). Put simply, analog inter-stage gain rules don't apply to internal DSP stages, which can take advantage of expanded bit depth and/or floating-point calculations. Values must return to the standard gain range when the output stage is reached for transmission to the next device or analog conversion.

2.2.5 Analog output

An analog output transmits the waveform to a following analog input stage. The output stage must have low-enough source impedance and high-enough voltage swing to drive the next stage without clipping or adding excessive noise. Analog outputs are normally balanced, push–pull drive. Single-ended (unbalanced) outputs should be avoided except for very short runs.

2.2.6 Digital output

A digital output is a numerical encoder that sends an output stream to a following digital audio device. The output stage tracks the sequence of values from the preceding internal processing stage. The output stage must be appropriately configured to provide a readable encoded waveform to the following device, i.e. the sender and receiver must agree on the data transmission protocol.

2.2.7 Analog to digital (A/D) converter

The A/D converter is the interim stage between an analog input and digital processing stage. A/D converters numerically transcribe the waveform into a chronological string of amplitude over time values. The analog side of the equation must be appropriately scaled to the digital side to preserve dynamic range and minimize noise.

2.2.8 Digital to analog (D/A) converter

The A/D converter's counterpart is the D/A converter, which mathematically constructs an analog waveform (voltage/time) from a set of numerical values in evenly spaced increments of time. The same scaling considerations apply to both converter types.

2.2.9 Input/output (I/O) combinations

When CDs first came out they had designations such as AAD or ADD, which indicated the original recording format (analog or digital), original mastering format (analog or digital) and final mastering (digital only, because it was a CD). We can do this with audio processing devices, except the designators are for input, processing and output.

The standard I/O configurations are:

- AAA: analog in, processing and output. Simple, old school. No latency, or boot up and you can bang on it when it doesn't pass signal.
- ADA: analog in, digital processing and analog out. A digital island in the analog world. DSP capabilities with analog hookup on both ends. A common format for stand-alone delays, electronic crossovers and early DSPs.
- ADD: analog in, digital processing and digital out. The entry port into the digital custody chain. Mix consoles and system DSPs will have at least some inputs for analog.
- DDD: digital all the way. An intermediary path with analog provided by others. Modern mixing consoles and system DSPs will have at least some inputs for digital.
- DDA: a digital processor with analog out (e.g. a power amplifier with digital input and processing). Others would be analog line-level output options on a DSP or mixer.
- DAD: an analog processor nested in a digital I/O. There are few (if any) applications for this configuration in the field of system optimization. On the artistic side there may be special effects that only analog can do. This functions as equivalent to an analog insert patch within the digital signal path.

2.3 MIX CONSOLE (DESK)

The mixer is where individual input signals from mics, line-level devices and digital devices are combined to form output signals: a.k.a. mixes. Decisions such as instrumental balance are purely artistic/subjective and therefore not within our scope. Delivery of the console's output waveform is our primary concern, so we will study the mixer as a signal source. The console outputs are our sound system inputs. We must know our source to assure that the sound system can fulfill its delivery task (Fig. 2.4).

2.3.1 Input stage

The mixer has a variety of inputs optimized for different levels and signal types. Microphone inputs with high-gain preamp stages, line-level inputs, and digital inputs. Each input channel contains a suite of signal processing functions such as level control, equalization, etc.

2.3.2 Mix stage(s)

The singular input channel can be routed to a mix bus to join other channels. This mix channel can have its own signal processing, which applies globally to all the mixed inputs contained there. The output waveform of this mix bus (a submix) moves on to become an input to the next mix bus level where it joins other submixes. This can go on indefinitely until eventually the output stage is reached.

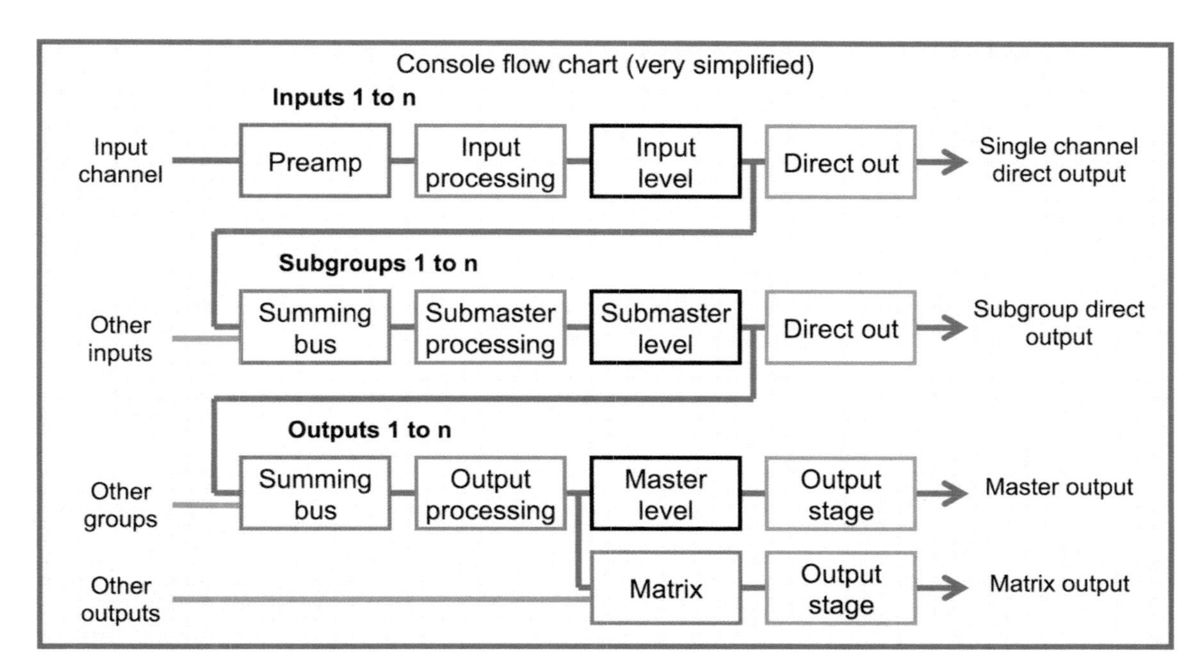

FIGURE 2.4
Mix console flow chart

2.3.3 Matrix

The matrix is a grid-like mixing bus with *x*-axis inputs and *y*-axis outputs (or vice versa), and level controls at the intersections. The mix engineer can send whatever, wherever. This is relevant to design and optimization strategy because the matrix often performs the channel division that feeds the sound system, e.g. a theatrical system with separate stereo music and mono vocal speaker systems. The matrix links music inputs to their appropriate outputs, and voice inputs to voice outputs. The matrix can also bridge parts of the system that share signals, such as delay speakers fed with summed left, right and voice.

2.3.4 Output stage

The mixer output stage is the entry point to the sound system, a key location in the design and optimization process. This is the finished product for transmission, the reference point for the sound system: our input waveform. The waveform leaves the mixer output as a work of art and enters the sound system as a scientific reference standard. This milestone will be referred to here as the "Art/Science Line."

2.4 SIGNAL PROCESSORS

The mixer feeds the speaker system's signal processors. The processor's role is to pre-compensate the waveform for the effects that will occur as soon it enters the speakers and room. The tasks are: EQ, level setting, delay, frequency division (crossovers) and more. In the past this was done with a series and parallel configuration of separate devices, complete with patch-bays and hordes of wires and connectors. In the future this will be a small number of devices connected with a few network cables. Update: The future is now, but we still need to know enough about the past to be able to navigate existing systems.

The processor's inner workings and practical application will be covered later. However, its overall configuration follows the same input–processing–output topology as the mix console described above. Modern systems are almost exclusively "all-in-one" format digital signal processors operating DDD, DDA, ADD or ADA.

A sound system may carry out its processing with multiple devices, thereby creating various series and parallel combinations, of AAA, ADA and other formats. Inter-stage and inter-device gain structure must be carefully monitored in such cases.

2.4.1 Filters

Filters (analog or digital) come in many sizes and shapes, and can be combined to create equalizers, frequency dividers and more. They can be basically classified by filter type and parameter variability.

2.4.1.1 SHELVING

Shelving filters have a transition point known as the cutoff frequency (or "corner" frequency) (Fig. 2.5). The range above (or below) the cutoff frequency can be adjusted upward or downward, in bulk (hence the term shelf), from the remainder range. Three parameters are in play: cutoff frequency, slope and level. A high shelf raises the entire range above the cutoff frequency by the same amount. The greater the slope, the sharper the transition between the ranges above and below the cutoff frequency. For example, a high shelf with a cutoff frequency of 4 kHz set to -6 dB will reach -6 dB (the shelf level) at some frequency above 4kHz and remain so as frequency rises. The choice of slope will affect how close to 4 kHz the shelf level is reached.

2.4.1.2 HIGH PASS/LOW PASS

High-pass (HPF) and/or low-pass filters (LPF) also have a "corner" frequency (Fig. 2.6). The range above (or below) the cutoff frequency is increasingly attenuated. The loss rate is set by the slope (6 dB/octave, 12 dB/octave, etc.). The precise amplitude and phase response around the cutoff frequency is a product of the filter slope (in 6 dB/octave increments) and its exact topology (Bessel, Butterworth, elliptical, Linkwitz-Riley and so on).

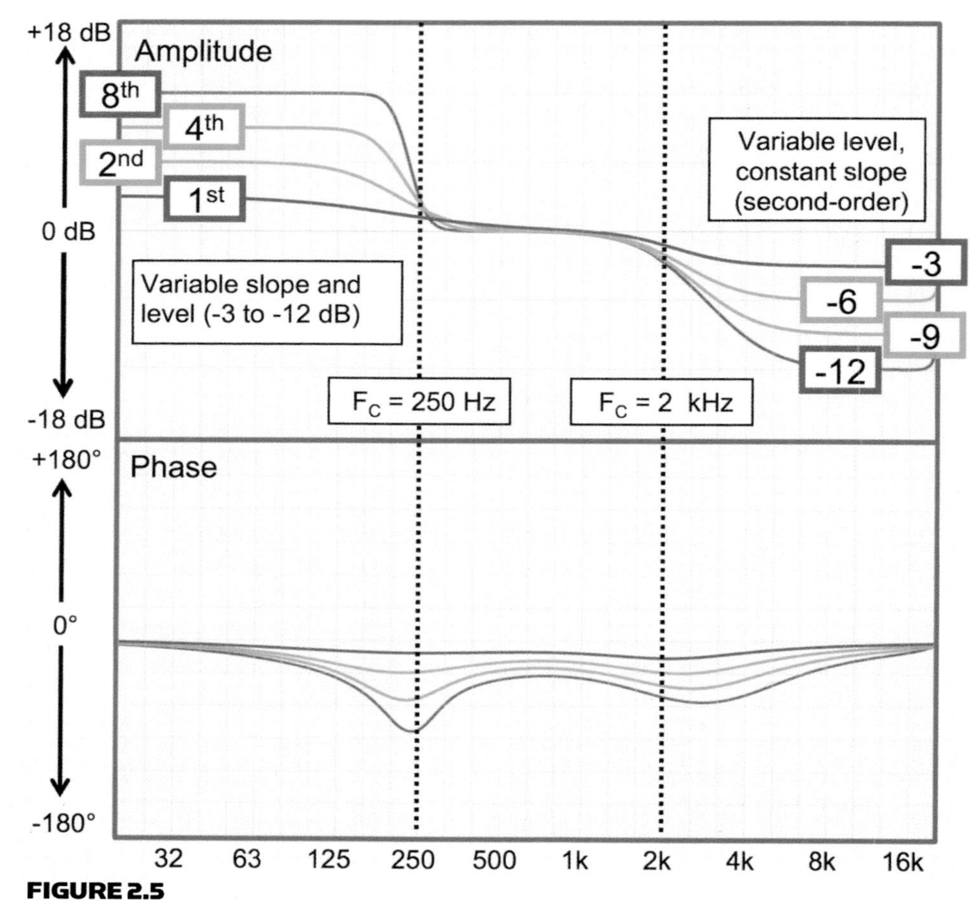

FIGURE 2.5
Shelving filter responses of various orders

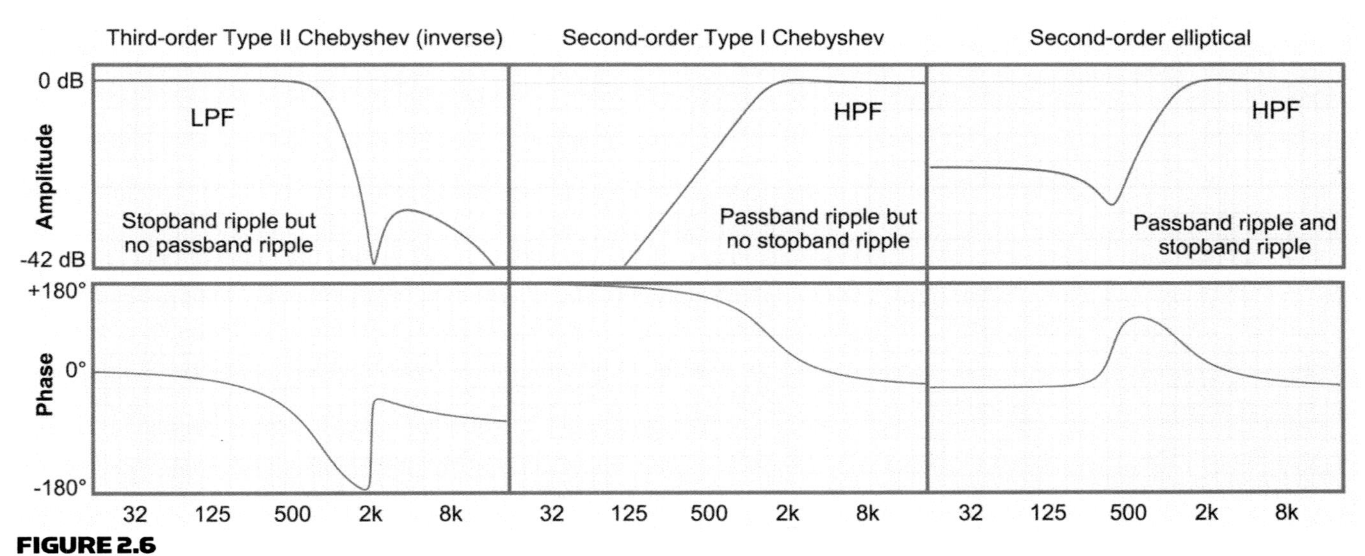

FIGURE 2.6
Three examples of HPF and LPF filters commonly used in spectral crossovers

2.4.1.3 BAND-PASS FILTER AND STATE-VARIABLE FILTERS (SVF)

Band-pass and state-variable filters are rarely used as standalone filters. These filters, however, are the standard building blocks inside graphic and parametric equalizers (Fig. 2.7). Unlike the filters discussed above, these have a center frequency with two corner frequencies. The band-pass filter leaves the center frequency above all others. Frequencies inside the corners (-3 dB points) are considered inside the bandwidth. Those outside the corners fall at

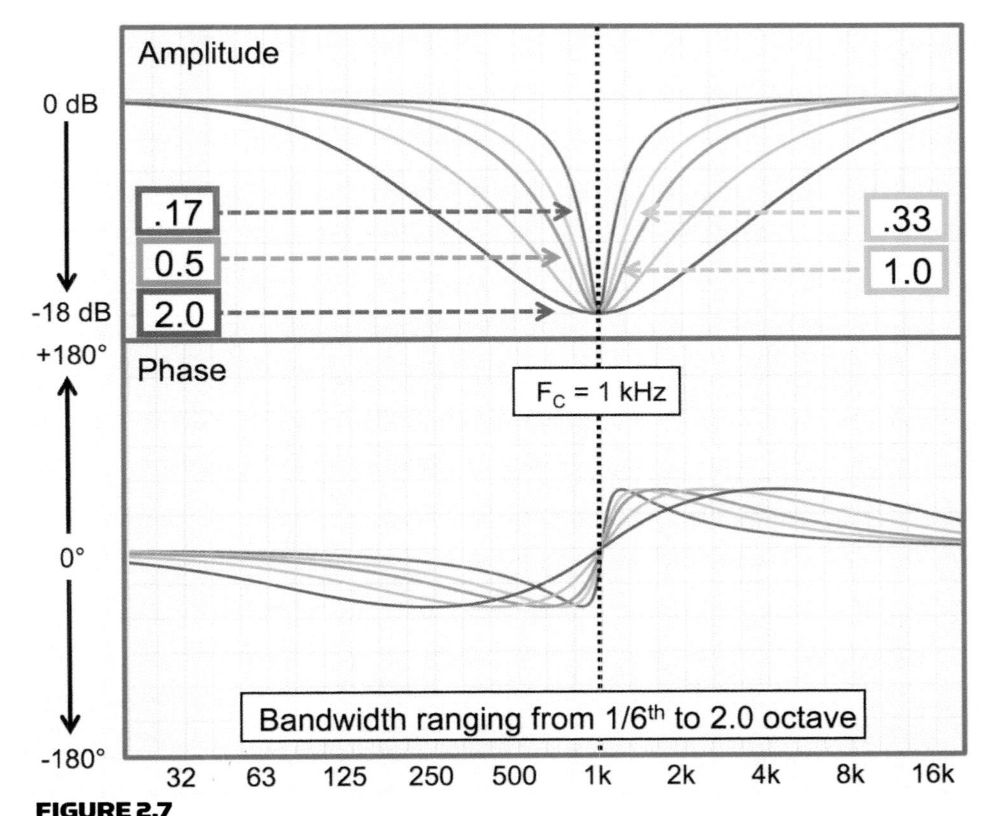

FIGURE 2.7
State-variable filter (adjustable frequency, bandwidth and level) showing different bandwidth settings. Affected frequency range is greatest with wide filters, whereas phase shift is greatest with narrow filters. These are the building blocks of the parametric equalizer.

Bandwidth vs. Q	
BW (Octaves)	Q (rounded)
2	0.7
1.4	1
1	1.4
0.7	2
0.5	3
0.35 (1/3rd)	4
0.25 (1/4th)	6
.167 (1/6th)	9
0.125 (1/8th)	12
0.08 (1/12th)	18

FIGURE 2.8
Bandwidth vs. Q conversion reference (after Rane Corp., www.rane.com/library.html)

the rate consistent with the filter slope, which can be first order (6 dB/octave), second order (12 dB/octave) and so on. The state-variable filter is a band-pass filter with adjustable center frequency and bandwidth.

Bandwidth and Q

There are two commonly used terms for filter width: percentage bandwidth and "quality factor" (Q). Both refer to the frequency range between the -3 dB points compared with the center frequency level. Neither provides a truly accurate representation of the filter as implemented. Why? What if a filter has no -3 dB point, such as a 2.5 dB boost? This requires a brief peek under the hood. The signal in an equalizer follows two paths: a direct line from input to output and a filter section in a feedback loop between them. The level control for each band determines how much of the filtered signal we are feeding back (positively or negatively) into the direct signal. This adds a positive (boost) or a negative (cut) summation to join the full-range direct signal. Bandwidth or Q specifications are derived from the internal filter shape (a band-pass filter) before it has summed with the unfiltered signals. The bandwidth reading on the front panel typically reflects the filter before summation, because its actual bandwidth in practice will change as level is modified. Some manufacturers use the measured bandwidth at maximum boost (or cut) as the front panel marking. This is the setting that most closely resembles the internal filter slope.

The main difference between the two standards is one of intuitive understanding. Manufacturers seem to favor Q, which plugs directly into filter design equations. Most audio system operators have a much easier time visualizing 1/6 of an octave than they do a Q of 9 (Fig. 2.8).

2.4.1.4 ALL-PASS FILTERS

Phase alignment circuitry is an additional parameter in some frequency dividers. This can be standard signal delay or as a specialized form of phase filter known as an **all-pass** filter (Fig. 2.9). Standard delay can compensate for the mechanical offset between high and low drivers. The all-pass filter is a tunable delay that can be set to a particular range of frequencies. The bandwidth and center frequency are user-selectable. All-pass filters are often used in controlled applications such as dedicated speaker controllers and active speakers. The all-pass has gained popularity of late as an additional tool for system optimization. There are some promising applications such as the modification of one speaker's phase response in order to make it compatible with another. There is also exciting potential for low-frequency beam steering in arrays by selectively applying delay to some elements. Such tools require far greater skill for practical application than traditional filters. Such exotic solutions should not take precedence over the primary goals of uniformity over the space. The belief that an all-pass filter tuned for the mix position will benefit the paying customers is as unfounded as any single-point strategy. It will be a happy day in the future when we have speaker systems that are so well optimized that the only thing left to do is to fine tune all-pass filters.

2.4.1.5 FIR AND IIR FILTERS

Finite impulse response filters (FIR) and infinite impulse response filters (IIR) are exotic options for specialty applications. The math is far too complex for me to offer a simplified distillation of what goes on under the hood. Some of the potential applications are a bit clearer, most notably beam steering for speaker arrays. I don't have enough experience measuring or optimizing systems driven by these engines to offer any further insight at the time of this writing.

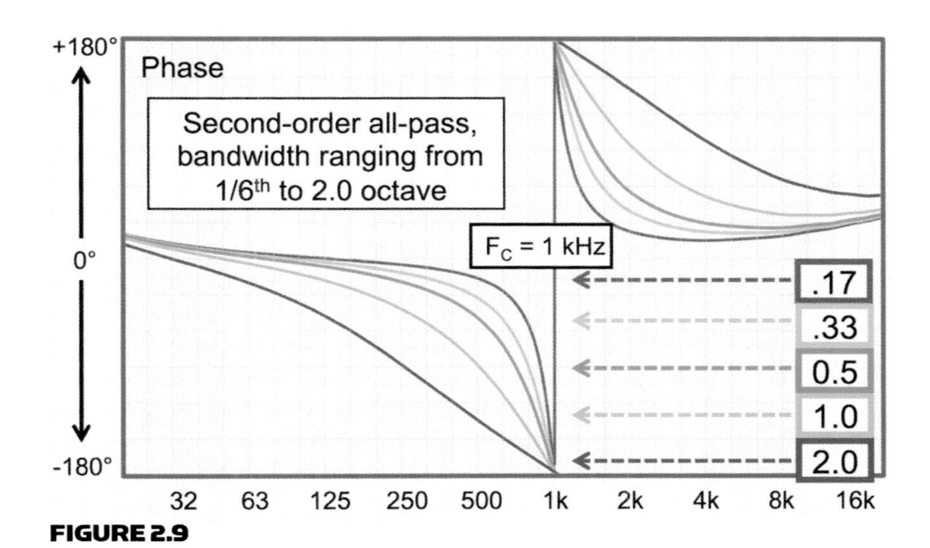

FIGURE 2.9
All-pass filter family

2.4.2 Equalizer

Equalizers are built from collections of the filters discussed above in series and parallel configurations. Parametric and graphic equalizers use a parallel arrangement of band-pass filters in the feedback loop of an internal unity gain amplifier stage. This configuration allows for the filter to be neutral, boost or cut within the passband while leaving outside areas unaffected. HPF, LPF and shelving filters are typically configured serially. Equalizers may contain any or all of the above because they are used for artistic tone shaping as well as system optimization.

2.4.2.1 PARAMETRIC EQUALIZER

Parametric equalizers use a parallel configuration of state-variable filters (SVF). Users can select center frequency, level (±) and bandwidth. Multiple SVF stages can be employed, allowing us to have more filters than should ever be used. The parametric equalizer is a mainstay of system optimization.

2.4.2.2 GRAPHIC EQUALIZER

Graphic equalizers use a parallel arrangement of log-spaced band-pass filters (typically ⅓ octave). Users are restricted to selecting level (±) of the fixed set of center frequencies and fixed bandwidth. The front panel layout creates an impression that the amplitude settings of the individual filters correspond to the combined filter shape, hence the name "graphic equalizer." In practice, the interaction between neighboring filters can substantially alter the combined response, a reality not reflected in the front panel graphic. Graphic equalizers are inferior to the parametric for system optimization. They are best suited for artistic shaping, where their inaccuracy and lack of flexibility is outweighed by ease of operation

2.4.2.3 MULTI-BAND EQUALIZER

There is a relatively new filter type that has become increasingly popular for system optimization, mastering and other applications (Fig. 2.10). This topology is a combination of the shelf filters detailed above. The spectrum can

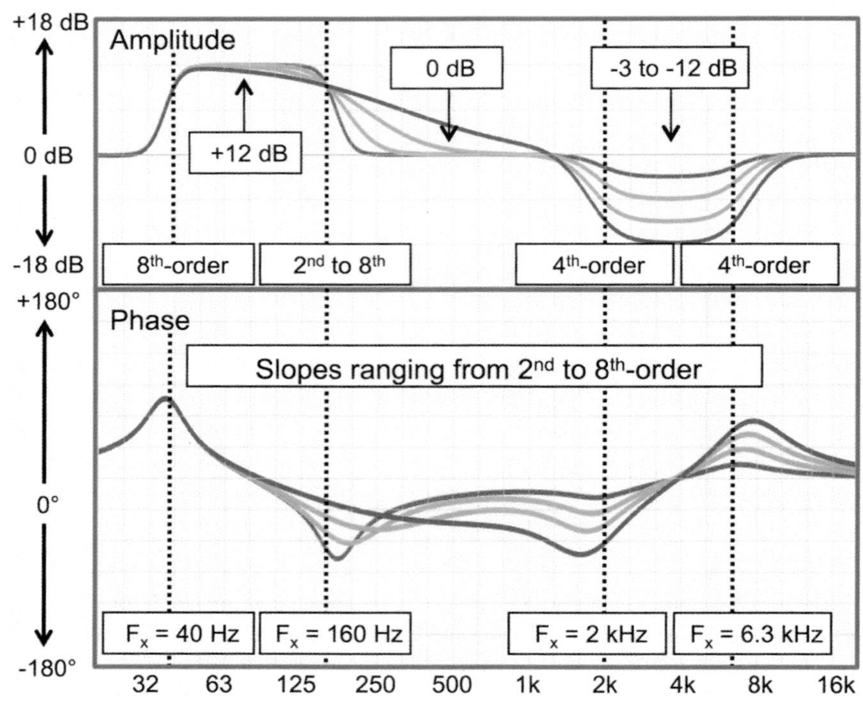

FIGURE 2.10
Multi-band equalizer family of curves

be separated into multiple bands with user-variable corner frequencies and slopes. The result is that regions of the spectrum can be raised or lowered in flat sections rather than having a distinct center frequency. The multi-band equalizer simulates the capabilities of adjusting level and slope in the sections of a multi-way speaker system. Users select the crossover frequency and slope between sections and then adjust the levels as desired.

2.4.3 Delay line

The term delay "line" comes from the original analog version, where the signal moved through a series of capacitive circuits, called a bucket brigade (like the ancient fire department). Analog units are still in use for guitar pedals, but in our digital world the "line" is storage in a series of memory buffers. The digital delay line has three defining parameters: minimum delay, maximum delay and incremental resolution.

2.4.3.1 MINIMUM DELAY (LATENCY)

This is a digital processor, so we know it will have some fixed latency. If the user interface says "0 ms" then we have to assume this means latency+0 ms or as I term it "0 msR" (milliseconds relative). For example, a device with 1.5 ms latency has this value as the 0 msR reference. The relative aspect may be unimportant, as long as all paths in the system carry the same base latency. This cannot be assumed, however. There can be time offsets (a) if you have analog devices in your system whose latency is microscopic compared with the digital units, (b) if there is more than one model of digital device in your system and (c) if a device has variable latency. Any of these factors can cause trouble if and when signals from the different devices combine.

2.4.3.2 MAXIMUM DELAY

The limits to this are strictly practical and economic. Each frame of delayed audio costs memory. A stand-alone delay line will have fixed memory allocation and therefore a fixed time limit. There are two basic schemes to manage this in the delay line sections of DSPs: (a) allocate a fixed limit for all channels (used for fixed-topology DSPs) or (b) allow user selection of limit ranges on a case-by-case basis (used in variable-topology DSPs).

2.4.3.3 DELAY RESOLUTION (INCREMENTS)

The digital nature of the delay line sets the limit for the incremental resolution at 1/clock frequency. The finest time slices for a 96 kHz digital audio device are .01041667 ms, which we will mercifully round to 0.01 ms (10 μsec). In practice most user interfaces will allow the maximum incremental flexibility because giving users this option does not require additional memory. However, some manufacturers coarsen the resolution as the settings rise (e.g. small increments for settings between 0 and 10 ms and then coarse increments above that). Maximum resolution at all settings is strongly preferred for system optimization.

2.4.4 Frequency divider

The signal processor receives a full-range waveform. The large cone drivers and small compression drivers that follow can't individually reproduce the entire spectrum. The job of dissecting the waveform into the optimal ranges for each transducer falls on the frequency divider, commonly known as an electronic crossover. The goal is to parse the waveform electronically, deliver it to the speakers and then recombine it in the air. The frequency range where recombination occurs is known as the acoustical crossover.

A frequency divider has two (or more) sections that contain high-pass and/or low-pass filters (Fig. 2.11). We break down a particular filter into two parts: the passband (the unaffected area) and the stopband (the attenuated range). The cutoff frequency is the transition point between bands. The passband frequency range is below the stopband in a LPF and above the stopband in a HPF.

Frequency dividers are classified by their passband and stopband behavior, with the gross attenuation shape (the slope) of the stopband being the most prominent feature. The standard slope options are increment multiples of 6 dB/octave (12 dB, 18 dB, 24 dB and so on). The secondary classification level is the filter topology, many of which carry the names of mathematicians that originated the circuit design. The topologies differ in the amplitude and phase details in both the passband and stopband.

There is no "best" slope or topology choice in the abstract. It is possible to create symmetric divisions, which, if recombined electrically, would sum back perfectly to flat amplitude and phase. It is also possible to create asymmetric hybridized combinations that would create cancellation and phase shift in the crossover region if summed electrically. These facts do not conclusively make one combination better than another because of

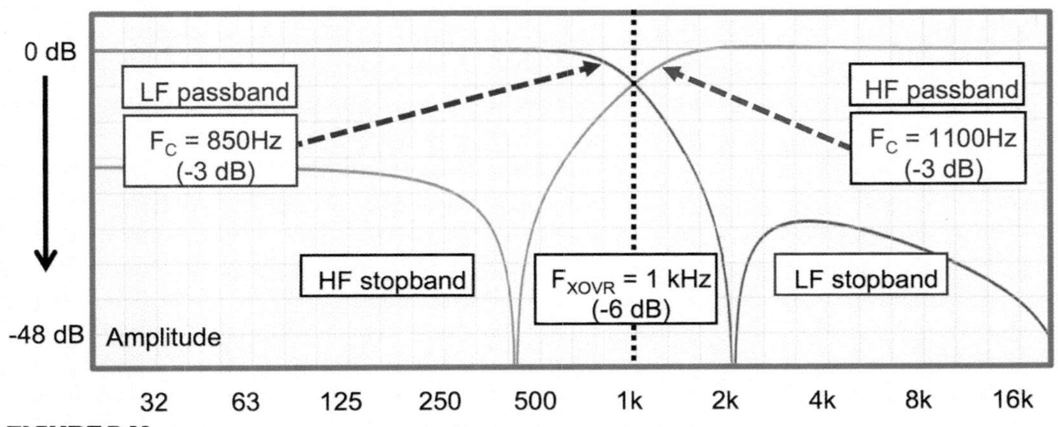

FIGURE 2.11
Frequency divider

one critical fact: We will be combining two distinct acoustical devices, not re-summing the electrical signal. The result that matters is the *combined* amplitude and phase responses of the frequency divider *and* the individual loudspeakers. Every loudspeaker has its own amplitude and phase behavior, and we can be quite sure that the upper region of a low driver is not a symmetrical match to the lower region of a high driver. Therefore we can only consider the purely electronic behavior of a frequency divider up to a point. In the end we will need to mate it with actual loudspeaker devices (covered later in section 4.3.4).

2.4.4.1 CUTOFF FREQUENCY (CORNER FREQUENCY)

The cutoff frequency is the break point between the passband and the stopband, where the response has been attenuated -3 dB (most topologies) or -6 dB (Linkwitz-Riley).

2.4.4.2 ELECTRONIC CROSSOVER FREQUENCY

The frequency range where the low-pass and high-pass bands are at equal amplitude is the electronic crossover point. The crossover point is determined solely by amplitude and does not necessarily have a matched phase response. This level equality is only electrical and does not factor in the sensitivity of the different loudspeakers involved. In practice the electronic crossover may or may not match the acoustical crossover frequency (the range where the LF and HF transducers are matched in level). The electronic crossover is the first step toward an optimized acoustical crossover.

2.4.4.3 SLOPE (FILTER ORDER)

The slope is the attenuation rate for the filter, given in dB/octave. The slope does not reach its full loss rate immediately at the cutoff frequency and does not necessarily maintain this rate to infinity. Therefore we can term this the "nominal" rate. Slope choices run in 6 dB increments (orders): 6 dB (first order), 12 dB (second order) and so on. The gap between a first-order and second-order response begins slowly and then widens further away from the cutoff. This trend continues for the higher orders.

An acoustical crossover made from low-order slopes will have the most gradual transition between the elements being separated and combined. Conversely a high-order slope can complete the transition between two notes on the tempered music scale. There are advantages and disadvantages to low- and high-order slopes in practice. The process of selection for a particular application must include the loudspeaker devices being combined.

2.4.4.4 TOPOLOGY

The high- and low-pass filters that create these slopes were originally developed with resistors, capacitors and inductors. The most basic forms were passive resistor/capacitor/inductor circuits simply referred to as an "x order" high pass or low pass. We now have more advanced analog and digital options courtesy of the esteemed law firm, Bessel, Chebyshev, Butterworth and Linkwitz-Riley (Fig. 2.12).

Why do we need alternative topologies? In short, there are variations in the phase response as well as some artifacts that affect the passband range nearby the cutoff frequency and finally the stopband response several octaves away. In general the artifacts become more severe as filter order rises. The price for a fast getaway is increased phase shift, amplitude peaks near the cutoff frequency (passband ripple) and comebacks in the stopband response (stopband ripple). Here is a driving analogy: We tend to push a bit to the left just before taking a sharp right turn. Our filter is creating a downward turn and the sharper it gets, the greater the over-steering errors.

High pass/low pass

These are the simplest of the circuits. The first-order version is a single RC circuit, second order is two of these in series and onward. The standard high- and low-pass filters have a gradual transition between the passband and the stopband. The overlap area between combined elements is likely to be highest with this filter type. In practice we rarely see standard high- and low-pass filters used beyond second order.

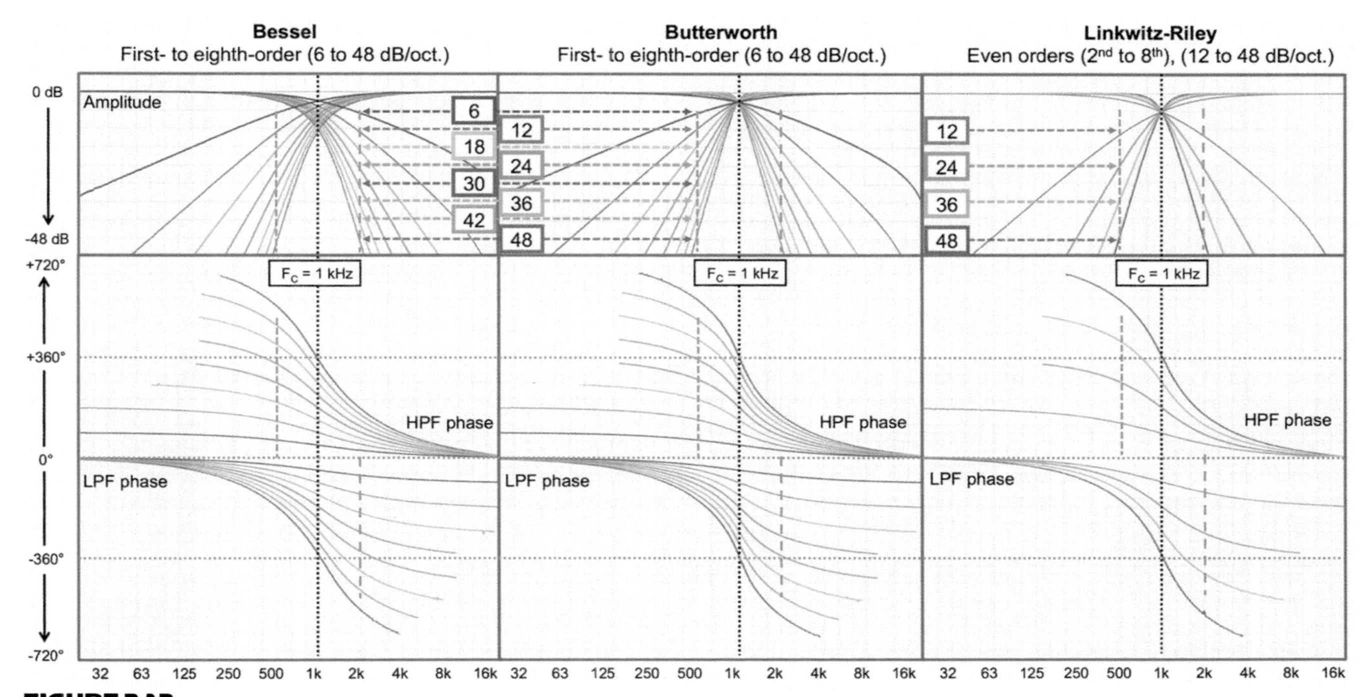

FIGURE 2.12
Various filter topologies used in spectral crossovers

Butterworth

Originated by Stephen Butterworth in 1930, this filter achieved increased sharpness in the transition range with minimal artifacts in the passband or stopband. The Butterworth filter can be employed for both even and odd orders, and is used fairly commonly. His wife is famous for pancakes of the first order.

Bessel

The Bessel filter family has a more gradual-level transition and minimal phase delay. Applications where overlap can be tolerated can successfully employ Bessel filters.

Chebyshev I and II

The Chebyshev filter family has an extremely steep transition between bands. There is substantial ripple (peaks and dips near the cutoff frequency) in both types. Chebyshev Type I has ripple in the passband (a particularly undesirable feature for our application) and the Type II creates ripple in the stopband. After a certain drop in level the response rises back slightly and then largely levels off.

Elliptical

Elliptical filters are even steeper than the Chebyshev's and have substantial ripple (peaks and dips near the cutoff frequency) in both the passband and the stopband. The possible advantages to this extremely sharp transition must be weighed against the certain disadvantages of response ripple.

Linkwitz-Riley

The Linkwitz-Riley (L-R) filter topology, arguably the most popular for frequency dividers, is created by cascaded pairs of Butterworth filters. A second-order L-R filter is made from a cascaded pair of first-order Butterworth filters. This gives some insight into why the L-R filters use the -6 dB point (2*-3 dB) as their slope reference. The L-R filters continue the low ripple features of the Butterworth but allow for steeper slopes. The phase responses of the high- and low-pass sections are complementary, allowing for a matched set of HP and LP filters to be recombined as a flat electrical summation. Note: The second-order version requires a polarity reversal between the sections to combine in phase. The fourth-order (24 dB/octave) L-R filter is a very common slope/topology combination in pro audio. It's a cascaded pair of second-order Butterworth filters. This is commonly used in applications where minimal overlap is desired. Recall that the end product is a phase-aligned acoustical crossover between two unmatched acoustical devices. The fact that the L-R filters can combine to a perfect electrical summation does not assure perfect acoustical summation.

The L-R family extends even to eighth order (48 dB/octave) and sixteenth order (96 dB/octave), which are derived from pairs of fourth- and eighth-order Butterworth filters respectively. These come in handy for applications when having only 95 dB of rejection is just not enough.

2.4.5 Output-level control

This is a gain control circuit that adjusts the output level. Digital versions can be counted on to be non-continuous but usually in sufficiently small enough steps (typically 0.1 dB) and reliably indicated (in dB) on the user interface. Analog versions can range from precise to highly variable in accuracy and continuity (model to model, channel to channel). The user interface notations for analog level controls range from accurate (in dB) to "what were they thinking?" (0–10, tick marks of unknown parentage and more). Analog-level controls are often log taper (wide range, low accuracy but fairly evenly spread) or in special cases linear (high accuracy over a 10 dB range for 120° of rotation and then highly compressed scaling for the remainder of the rotation).

2.4.6 Limiters and compressors

Limiters and compressors keep the amplitude within the operational limits of the system. They monitor the waveform and actively attenuate the signal as needed, thereby reducing the upper limits of dynamic range. It's an automated version of a hand on the fader, ready to attenuate the level a bit if things get too loud. Buzz-kill in a box.

Compression differs from limiting primarily in the time constants involved in the simulation of the moving hand and the leniency of the decision process. As a general rule, limiters are faster and stiffer in carrying out their attenuation than compressors, which are slow and spongy. Either of these can be employed as a full-range dynamic control or in selective frequency bands. Either can be employed for artistic dynamic shaping but limiters do the lion's share of loudspeaker protection. Our focus will be on protection and therefore limiters take center stage.

2.4.6.1 LIMITER PARAMETERS

Limiters are voltage regulation devices. The protection they offer to speakers may be power (voltage × current) reduction but the actions conducted are strictly in the voltage domain. The power side of the equation is predicted, based on known voltage and expected current (or impedance). A digital limiter operates on the amplitude values, the equivalent of converted voltage.

A limiter circuit in its resting state should be a unity voltage gain device. When engaged, the output level is capped and the net gain is less than unity. Simply put, the input signal rises but the output remains the same: a net loss of voltage gain, typically shown in the user interface as "gain reduction."

Threshold

When the waveform exceeds the threshold voltage, the limiter readies for action. The action may be quick, perhaps even as fast as the circuits can react: an action known as *peak limiting*. The primary mission of peak limiting is the prevention of mechanical failure due to over-excursion or other extreme stresses. Alternately there may be a timing component added to the determination, waiting to see if the signal stays too long above the threshold before throttling down the level. This type of limiting seeks to monitor the average power dissipation rather than the instantaneous. Known as *rms limiting*, the primary mission is to prevent the overheating of driver coils.

Attack and release

Attack and release are terms to describe the limiter's start and stop timing. Once the threshold has been breached the first timing sequence begins: Limiting starts if the signal stays above for too long. The next decision is when to stop limiting. If the limiting stops at the first moment the waveform falls under the threshold it may be forced back into limiting a moment later when the waveform rises again. Guitar players would recognize this amplitude-modulated sound as a "tremolo" effect, which we only want added to our main system if we are doing surf music. Therefore the release time for a limiter is normally much longer than the attack time. We want to make sure that the "all clear" has sounded before we release the limiter.

Expected peak limiter onset time is in the <20 ms range and the release in >100 ms range. An rms limiter will engage in the >100 ms range and have a slower release time.

2.4.6.2 LIMITER FUNCTIONS

We are focusing on limiters tasked with speaker system protection rather than outboard special effects in the artistic sphere. In this regard limiters have two prime directives: prevent mechanical failure (peak limiters) and prevent overheating (rms limiters).

A speaker can suffer mechanical failure in a single excursion. If pushed too far outward the former jumps out of the gap. Too far inward bottoms it out on the magnet structure. Either of these can bend the former and begin the funeral arrangements. We can break glue bonds, tear paper, shatter titanium, over-stress rubber or whatever materials make up the moving structure of the speaker. How can we protect a loudspeaker from these dangers? Appropriately scaled amplifiers and peak limiters.

A speaker coil is a tiny, fragile, thin wire. It can be overheated, melted and burned through. It's wound in an extremely tight spiral, which can be shorted, turn by turn, if the insulation is melted. Heat is the enemy. We quantify heat dissipation in AC circuits as the rms value. The limiting circuits designed to prevent overheating set their threshold based on the calculated thermal limits of the speaker. These are rms limiters.

In the olden days speaker protection was handled by the power amplifiers. How was that? This was accomplished by grossly under-powering the speakers, so the signal did not have enough peak power to stress the mechanics or enough long-term power to burn the coil. The drivers themselves were very high distortion so the sound of driving the amplifiers into clipping was not nearly as apparent then as would be nowadays. The result was distorted audio pretty much all the time, but the show did go on and we made it to the end.

In the modern era we have seen extreme inflation in amplifier output power capability and a lesser amount in the world of loudspeaker coils and mechanics. We also have higher expectations of linearity (i.e. low distortion). If left to their own devices, or rather, without any additional devices, the amplifiers will win the game. The device that allows us to get away with massively overpowering the speakers is the limiter (Fig. 2.13).

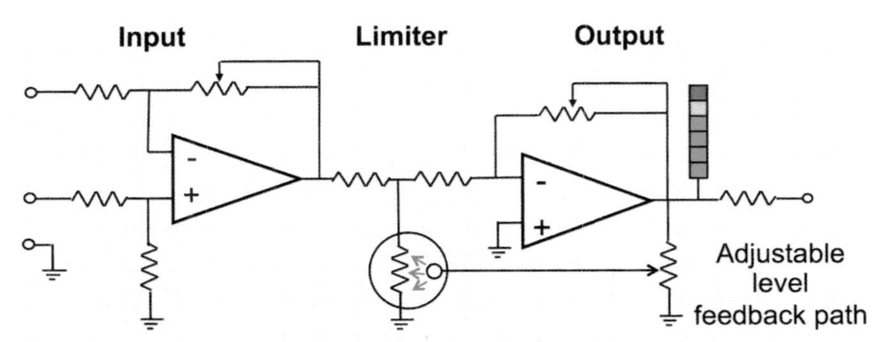

FIGURE 2.13
Simplified schematic for a limiter circuit

2.4.6.3 LIMITER PLACEMENT

There are two options: before the frequency divider or after (pre- or post-crossover).

Pre-crossover

This placement drives the limiter with a full-range input signal (Fig. 2.14). This compresses the entire spectrum even if only one frequency range exceeds the threshold. This would be appropriate for loudspeaker protection only if the speaker was full range (or a passive crossover). Multi-band limiters exist that separate the signal into bands with individual thresholds and then recombine the signals. These are commonly used in broadcasting to maximize transmission density but are not suitable for sound reinforcement systems.

Placing a limiter before an active frequency divider can significantly reduce performance by gross reduction of dynamic range. In practice this can actually endanger, rather than protect, the loudspeaker system. How can this happen? As the system is pushed and limiting increases, the crest factor (the peak to rms ratio) falls. The signal density increases (square waves are more dense than sine waves), the heat load rises and yet the system doesn't feel much louder. The system then gets pushed harder: a vicious cycle.

Full-range limiting doesn't match our primary goal (individual speaker protection) because it penalizes all for the violations of one. Unless the device separates the spectrum into bands for individual limiting we can expect that the entire response will be limited regardless of whether it is the high-hat or the kick drum that is pushing the edge.

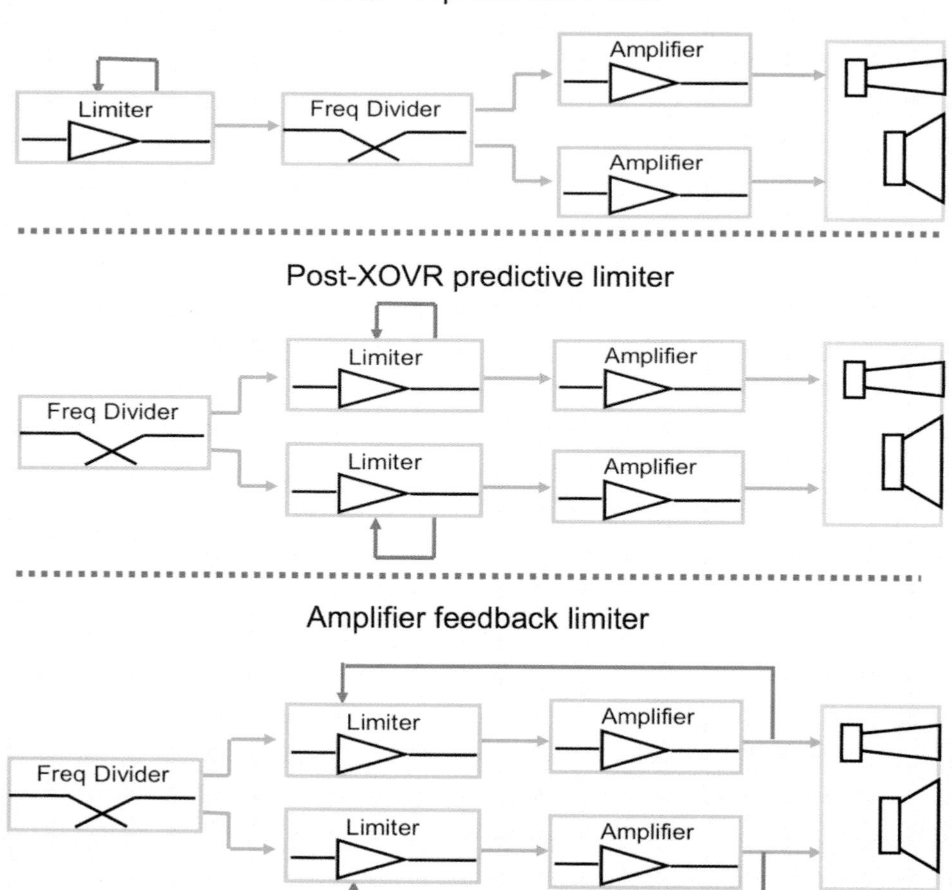

FIGURE 2.14
Limiter flow chart. Pre- and post-crossover locations and amplifier feedback type are shown.

Post-crossover

Post-crossover placement allows the limiter to be precisely tailored to the amplifier and speaker downstream (also Fig. 2.14). We can maximize dynamic range by limiting only the speakers in the system currently facing overload conditions. If the high drivers are being pushed we want to maximize their protection, not the subwoofers.

2.4.6.4 LIMITER FEEDBACK LOOP

Let's visualize the signal path as A–B–C (Input–Limiter–Output). All are unity gain. The output has a feedback loop path back to the limiter. If the level at C is too high the gain at B is reduced. Once the level at C falls below the threshold set at B, the gain at B goes back to unity.

There are several schemes for the placement of the C–B feedback loop, some of which are completed within the limiter circuit and others, which include entire devices downstream such as the power amplifiers. It is the final stage, the power amplifier output voltage, that needs to be tamed. Therefore we must either measure that directly or have a method of accurately estimating it.

To visualize this let's return to the A–B–C model and say we have set a threshold +10 dBV as the highest allowable level. Note that, because C has unity gain, it would not matter if we closed the feedback loop earlier, say at the connection point between B and C. The result would be the same because the input and output levels at C are the same. The limiter threshold is set at +10 dBV and the maximum level at the output of B or C is +10 dBV.

Now let's put 20 dB (10×) of gain in the C section. Now it matters whether we reference the input or output of C. If we use the output of C, the same final output level (+10 dBV) is maintained. The maximum level internally at B is -10 dBV but the final output at C is still +10 dBV. There is gain in the signal chain, but it is taken into account by this scheme.

If we use the internal reference point (B) there will be 20 dB more level at the final output than is found in the feedback loop. The level at B would be clamped at +10 dBV whereas C would be putting out +30 dBV!

There are several possibilities from here. Fire is one. Alternatively we can keep the B reference point, but factor in the gain of the C section as a known and throttle down the limiter threshold by 20 dB. Result is -10 dBV at B and +10 dBV at C. We can take off the safety goggles now.

If we don't know about the gain, or it is subject to change without notice, then our limiter scheme becomes guesstimation, a scary prospect for our speaker system. The simple rule for accurate limiting: Variable gain stages must be inside the loop, but known, fixed gain stages can be after the loop.

2.4.6.5 AMPLIFIER NEGATIVE FEEDBACK (NON-PREDICTIVE-KNOWN)

This is a fully inclusive feedback approach that returns the amplifier outputs to a line-level processor post-crossover limiter stage. The power amplifier returns are padded down to line level and threshold is scaled accordingly. Amplifier voltage gain is therefore a known parameter even when user adjusted. Amplifier clipping is factored into the feedback level to the limiter. This approach is built around a known processor and speaker, and an unknown (and variable) amplifier. This is used less in the current era due to the prevalence of self-powered systems and systems with amplifiers dedicated for use with a particular processor and speaker.

2.4.6.6 DEDICATED AMPLIFIER (PREDICTIVE-KNOWN)

Dedicated amplifiers use a semi-inclusive feedback approach that returns the inputs of a known amplifier to a post-crossover limiter stage. This is standard practice in self-powered speakers and speaker systems with dedicated amplifier/processing included. The actual feedback loop may be earlier in the signal path than the amplifier inputs, provided the gain downstream is fixed and known. If the amplifier level controls are user adjustable, the feedback loop must be post-gain. Amplifier clipping (if allowed) might not be sensed if the loop point is before the final amplifier stage. Modern systems favor this approach for its maximization of dynamic range, ease of use and reliability.

2.4.6.7 NON-DEDICATED AMPLIFIER (PREDICTIVE-UNKNOWN)

A non-inclusive closed-loop feedback approach can be employed with generic amplifiers. Users must provide the threshold settings for protection. The limiting circuit may be located in the amplifier or upstream in the processing. The threshold settings must consider the open variables listed here in signal path order: crossover settings (affects excursion/heat load), processor output gain (affects threshold settings), amplifier voltage gain (threshold settings), amplifier power capability (affects excursion/heat load), passive speaker-level components and finally the loudspeaker itself (rms and peak capabilities). If users modify these parameters the limiter setting calibration will be offset accordingly (up or down). Clipping (if allowed) occurs outside the loop and is not factored into the feedback level to the limiter.

Needless to say, this scheme has the most open variables of all the options. For some this makes it all the more attractive, whereas others (like myself) see a can of worms.

2.4.7 Line driver (distribution amplifier)

Line drivers and distribution amplifiers are pass-through signal boosters, used primarily for long analog runs. They have a low-impedance output drive to minimize line loss and maximize noise immunity. Distribution amplifiers are active splitters with adjustable output levels. These were widely used back in the days when consoles and crossovers could not drive long lines or lacked sufficient output matrix flexibility. The digital era has made these totally obsolete.

2.4.8 Dedicated speaker controller

Some manufacturers offer a line-level device (analog or digital) with specialized response and limiter circuits to work with one or more particular speaker models (Fig. 2.15). A given speaker paired with a dedicated controller and a standardized amplifier becomes a "system." The controller contains the familiar functions: level setting, frequency divider, equalization, phase alignment and limiters, all preconfigured and under one roof. Many units utilize negative feedback returns from the amplifier outputs for limiting.

Recent trends have lessened their popularity. Third-party generic DSPs provide most of the control functionality, often with recommended factory settings. Advantages include relatively low cost and high flexibility. Disadvantages include a lack of standardized response due to intended, or unintended (i.e. screw-up), customization of the settings. Even this approach is waning in favor of putting the speaker control circuitry in the same chassis as the amplifier, which may be in a rack or in the speaker enclosure. User tip: Never assume that the correct configuration has been loaded in.

2.4.9 Dedicated amplifiers

Dedicated amplifiers are specialized to work with one or more particular speaker models, which join together to make a "system" (again Fig. 2.15). This is the dedicated speaker controller just described above, joined to an amplifier with known parameters and an appropriate power scale. Limiting schemes can be assumed to know the amplifier loop gain. A manufacturer may make multiple speaker models using the same dedicated amplifier platform with selectable configurations in a firmware library. User tip (again): Never assume that the correct configuration has been loaded in.

2.4.10 Active (self-powered) speakers

The ultimate version of the dedicated speaker controller is a frequency divider, limiter, delay line, level control, equalizer and power amplifier in a single unit directly coupled to the speaker itself (yet again Fig. 2.15). The self-powered or "active" speaker has a closed set of variables: known drivers, enclosure, known physical displacement, maximum excursion and dissipation. They can be internally optimized for linear amplitude and phase response, standardized polarity and maximum protection over their full dynamic range.

Active speakers maximize flexibility for system subdivision, i.e. the number of active speakers equals the number of channels and subdivision options. Externally powered speakers, by contrast, often share up to four drivers per power amplifier channel and thereby reduce subdivision flexibility.

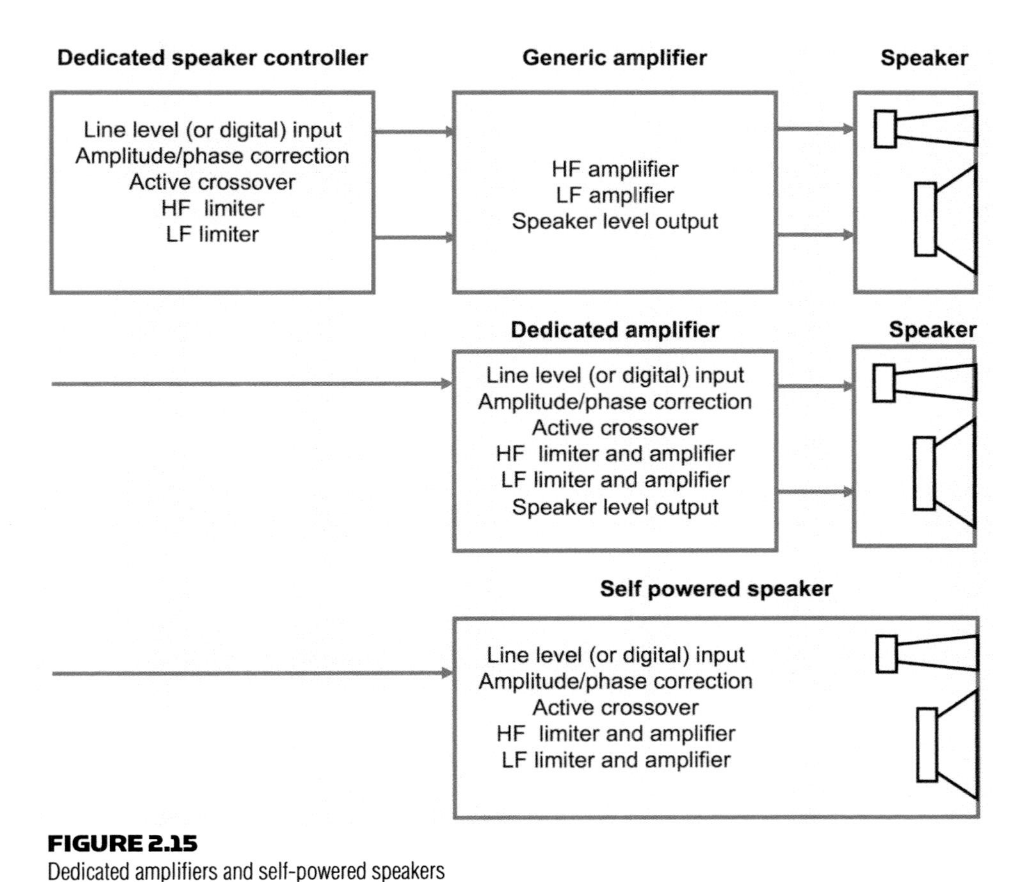

FIGURE 2.15
Dedicated amplifiers and self-powered speakers

We will leave advocacy for active or externally powered (passive) loudspeakers to the manufacturers. The principal differences will be noted when they affect our optimization and design strategies. We consider the "speaker system" to be complete with frequency divider; limiters power amplifiers and drivers inclusive regardless of whether they are all in the same enclosure.

2.5 DIGITAL NETWORKS

Digital audio networks are a transportation and distribution system. The audio is moved in packaged form and is not available for operations such as equalization, mixing, etc. We never much liked the equalization and mixed-in noise that wire added to our analog audio, so we won't complain. The network moves audio packets from port to port where they can be unpacked and become available for audio operations.

It takes a keen grasp of the obvious to see audio systems becoming fully networked from microphone to speaker. Let's classify the major parts in broad terms.

2.5.1 Network structure

The network is the local family of connected talker and listeners. We have devices that connect between different networks (routers), devices that move data between multiple devices in the same network (switches) and those that connect one talker/listener to another (interface or port). The most basic classification between network types is between the local area network (LAN) and the wide area network (WAN). The LAN is a private, walled-in structure where traffic can be controlled to suit our needs. The WAN is the wild world outside the LAN such as shared fiber optic lines between buildings or connections to the Internet. Professional digital audio networks have urgent timing considerations and therefore require the use of LAN topology to manage transmission. Any show-critical connections will be inside the LAN, whereas broadcast streaming of the show can go out to the WAN.

2.5.2 Network speed/bandwidth

Networks are classified by how much data they can move in a second. A network operating at 1 Mb/second (also termed a 1 MHz bandwidth) can transmit one file with a size of 1 Mbit or 10 files sized at 100 kb/second.

- Ethernet (10BASE-T): a 10 Mb/sec network widely used for digital audio from the late 1990s.
- Ethernet (100BASE-X): a 100 Mb/sec network in use in older systems.
- Gigabit (1000BASE-X): a 1000 Mb/sec network most commonly implemented currently.
- 10 Gigabit: a 10,000 Mb/sec network that will be in use (or superseded) by the time you read this.

2.5.3 Network devices

The devices that connect networks are roughly classified by their transportation and distribution roles (Fig. 2.16).

- Interface (port): Connection points between devices inside the network, which function as combined input and output.
- Router: the connection point between networks. A router must sometimes convert data between different network configurations.
- Hub: A non-discriminating device capable of connecting multiple devices within a single network. The signal sent to one interface (port) is transmitted to all interfaces. Switches have made hubs largely obsolete in current systems.
- Switch: A discriminating device capable of connecting multiple devices within a single network. The signal sent to one interface (port) is passed through to approved interfaces (ports) only.
- Repeater: an interface (port) replicator. Signal is received at one port and retransmitted verbatim from another.
- Network adaptor (interface controller): A data packaging and translation device that enables a computer to communicate over a specific data standard such as Ethernet.

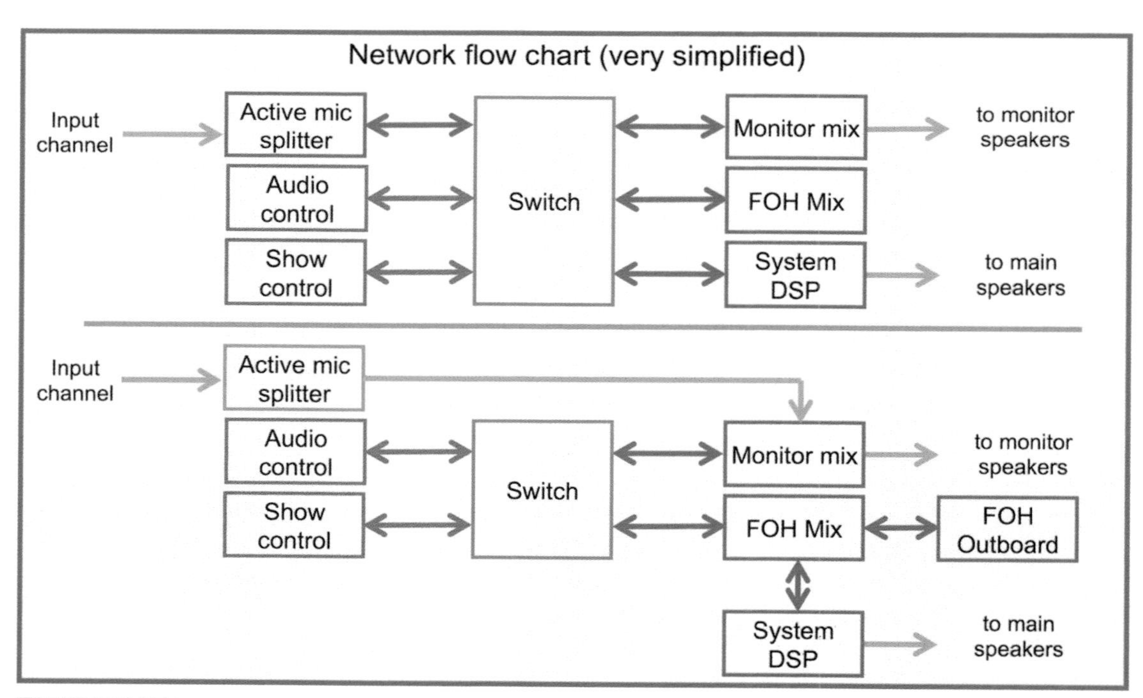

FIGURE 2.16
Digital audio network flow diagram showing two different network topologies to accomplish the same signal flow

2.6 POWER AMPLIFIERS

A power amplifier is a voltage and current booster with an extremely low impedance output. The power amplifier receives the input waveform and steps up the voltage (hence the term "amplifier") and the current (hence the term "power"). The power amplifier can be simplified down to a three-stage device, input, processing and output, just as we did previously for the line-level devices (section 2.2). The input stage is essentially the same as the previously described devices (line-level analog or digital). The processing module can vary in function from nothing to a full suite of signal processing functions (equalization, delay, frequency dividers, limiters and more). The final stage is the power module where voltage and current are stepped up to speaker level.

Amplifiers are typically configured as AAA, ADA or DDA (input, processing and output stage).

2.6.1 Amplifier classes

Power amplifiers are separated by their output-stage topology. This internal structure has minimal implications on wiring or speaker/amp pairing. Some people prefer the sound quality of particular amplifier classes. Others like the fact that Class D amplifiers weigh a fraction of the others (Fig. 2.17).

AMPLIFIER CLASSES

- Class A: Output transistors conduct for the full waveform (+ and -). This is popular for audiophiles due to low distortion but too inefficient for pro audio.
- Class B: Output transistors conduct for 50% of the waveform (separate sets for + and -). This highly efficient design suffers from "crossover distortion" at the transition between + to - devices. The distortion percentage is greatest with low-level signals.
- Class AB: a hybrid of Class A and Class B. At low levels the amplifier operates Class A, which keeps the crossover distortion low. At high levels the Class B devices are added, which provides efficient amplification of high-level signals. This is a common amplifier configuration for sound reinforcement.
- Class D analog: The amplifier controls the output via pulse modulation of the input signal. These are termed "switching" amplifiers because the pulses are fully on or off. We are familiar with pulse modulation because it is also used to digitize audio signals, but a Class D amplifier is not inherently a digital audio amplifier. Although the pulses control the voltage and current flow at the output, the audio signal is continuous (analog) with a high-frequency modulating carrier wave, in effect "digitally controlled analog." The ultrasonic carrier signal is filtered away at the output. Class D has high efficiency and minimum weight, a very attractive combination for the mobile sector of pro audio.

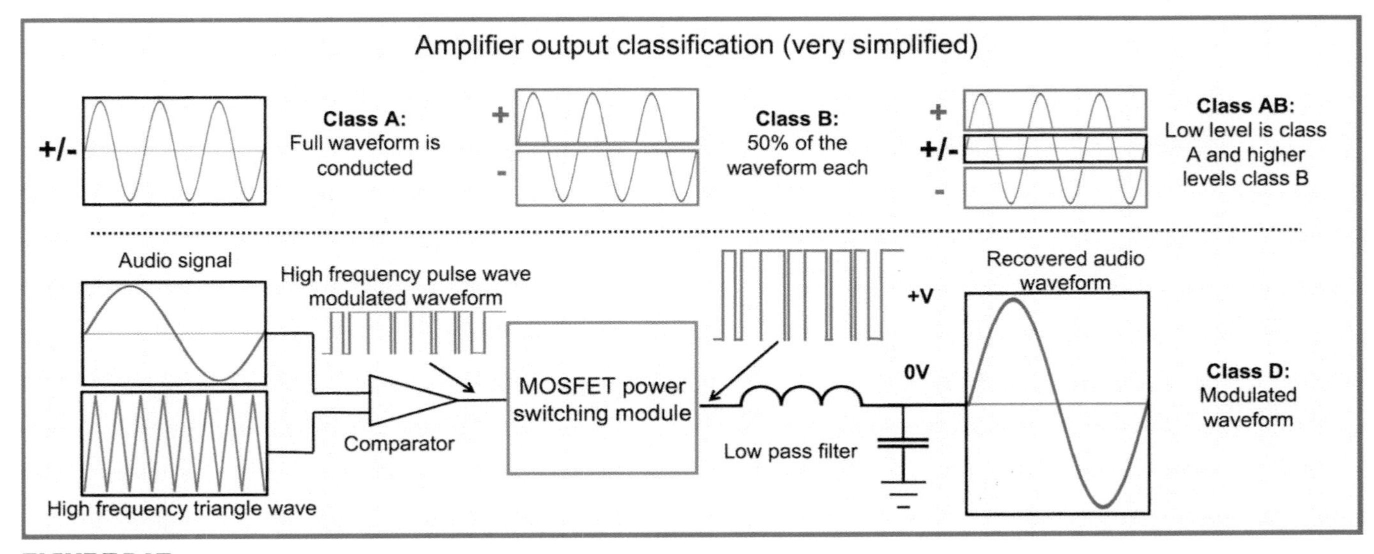

FIGURE 2.17
Amplifier output classes (graphic derived from www.learnabout-electronics.org)

- Class D digital: The input stage is a finite resolution digital audio driver stage with the D/A conversion conducted in the output stage. The conversion requires some shaping functions to reduce the quantization noise and distortion that would occur if the digital signal drove the output directly.
- Bridged mode: This wiring configuration that can be employed on any of the above output classes to double the voltage across the load. It is basically the same principle as the analog balanced line applied to speaker level. A two-channel amp is set to mono with outputs configured as a differential push–pull drive. One output is polarity reversed, usually with an internal switch. The speaker is loaded across the two hots (+ and -) rather than one hot and ground. This doubles the voltage and halves the effective load impedance, and greatly increases the power capability. Bridged mode is an expensive way to increase power (uses two channels) and has the potential for more cable loss (due to lowered impedance). This is by no means standard practice but it's out there as an option.

2.6.2 Amplifier configurations

We are basically down to three types of amplifiers in the trade: generic, externally paired and internal (Fig. 2.18).

- Generic: stand-alone units that will take on any speaker that walks in the door. Basic versions have a flat response and level control, and perhaps a few minor user features. This was the previous industry standard but very few of the top professional systems use generics any more. Leaving the pairing of amplifiers and speakers as an open variable can lead to unexpected outcomes. Few of us have the time, patience or spirit of adventure to hear your custom speakers with exotic amplifiers. Industry trends are moving away from this approach in favor of arranged marriages.
- Paired external: stand-alone amps with dedicated presets implanted in the processing stage. These may include crossovers, filters, delays, limiters and more. They are manufactured for (or by) speaker companies to pair with one or more speaker models. The external speaker cables open many possibilities: short circuits, variable load impedance, etc. Therefore the output stage is not different from the generic.
- Internal: a 'mate for life' pairing found in self-powered (active) speakers. Amplifier and speaker are in a closed loop. A known load allows for a reduction in the power amplifier circuitry (e.g. less need for short circuit protection). The processing options are similar to the paired external.

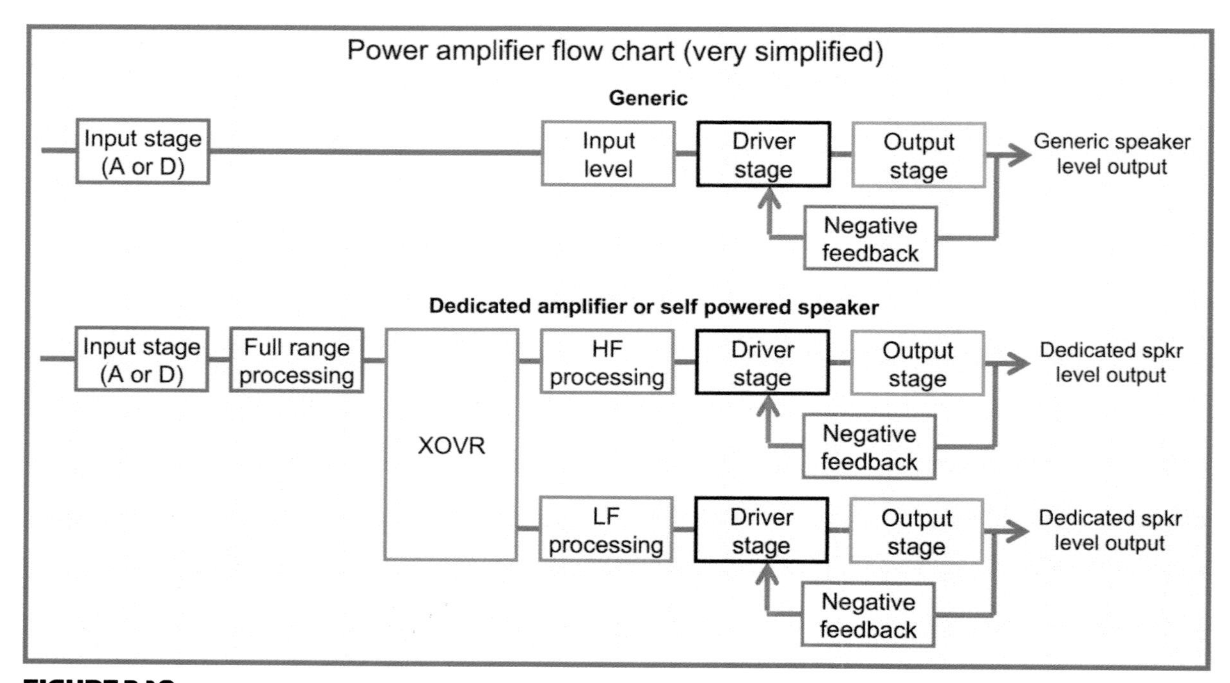

FIGURE 2.18
Amplifier stages and flow chart

2.6.3 Standard amplifier specifications

Many power amplifier specifications are similar to line-level devices (THD, frequency range, noise, dynamic range, input impedance, etc.). Let's move on to the unique specifications for amplifiers.

SPECIALTY SPECIFICATIONS FOR POWER AMPLIFIERS

- Power rating: the output wattage for a given load impedance. This is usually given for 1 kHz @1% THD. It can be given with burst and continuous values. At high impedances the continuous and burst values will be close. Amplifiers are available in a continuous range of power levels, so there are no clear classifications to divide them in this way.
- Minimum load impedance: Amplifier economics favor low minimum impedance. If an amp can drive 2 Ω then we need fewer amplifiers to drive our speakers. This sounds like a good idea except that it usually doesn't sound good as a practice. Cable losses, distortion, loss of load control (damping) and smoke inhalation are potential downsides to driving very low-impedance loads.
- Damping factor (DF): a quality metric of amplifier load control (the damping of motion errors). The DF specification is based on the ratio of the lowest allowable load impedance (typically 4 Ω or 2 Ω) over the source impedance. Values over 20 are considered acceptable for sound reinforcement systems and 50 is considered high fidelity. We will expand on this in section 3.9.4.3.
- Sensitivity/gain: the rms input voltage required to reach full power. Gain specifications describe the ratio of output to input voltage in linear terms (20×, 30×, etc.) or log (26 dB, 30 dB, etc.).

2.7 LOUDSPEAKERS

Loudspeakers are electronic to acoustic media converters, i.e. transducers. The various speaker types differ in the intermediate conversion stages. There are two main branches of the loudspeaker family tree: dynamic (moving coil in a static magnetic field) and other (electrostatic, ribbon, planar, piezoelectric, flat panel and many more). In professional audio's olden days, when tube amplifiers were king and mix engineers wore lab coats, speakers were mostly the dynamic type: paper cone drivers and various lightweight materials for compression drivers. More exotic speakers were reserved for the hi-fi crowd. In the modern world of digital audio and polo shirts, the speakers are still mostly the dynamic type. Better paper cones, metal diaphragms and composite substitutes, better magnets, coils and suspensions, but basically the same game. I will leave it to others to explain the exotic stereophile breeds.

2.7.1 Loudspeaker components

All loudspeakers covered here have a magnet, coil, former, flexible surround, movable surface and a frame. The coil is wound around the former, which is attached to the diaphragm (the surface that converts the waveform into acoustics). The surround holds the assembly in place freeing the coil to move back and forth in the magnetic gap (Fig. 2.19).

2.7.1.1 CONE DRIVERS

Cone drivers are named for their diaphragm shape, which widens with distance from the magnet. Standard cone diaphragms are paper or high-tech composites with two surround attachments. The inner surround, the "spider," attaches near the cone bottom to precisely control coil movement in the gap. The outer surround attaches the cone top to the frame. Drivers are classified by diaphragm diameter (typically 4″ to 18″).

2.7.1.2 DOME DRIVERS

These are named for the diaphragm's outwardly visible dome shape, which is the side from which the primary sound propagates. The standard dome driver has only the outer surround attachment as described above (no spider). They are classified by diaphragm diameter (typically 0.75″ to 4″). Common materials include lightweight metals (titanium, beryllium, aluminum and more), plastics, fabrics and various combinations.

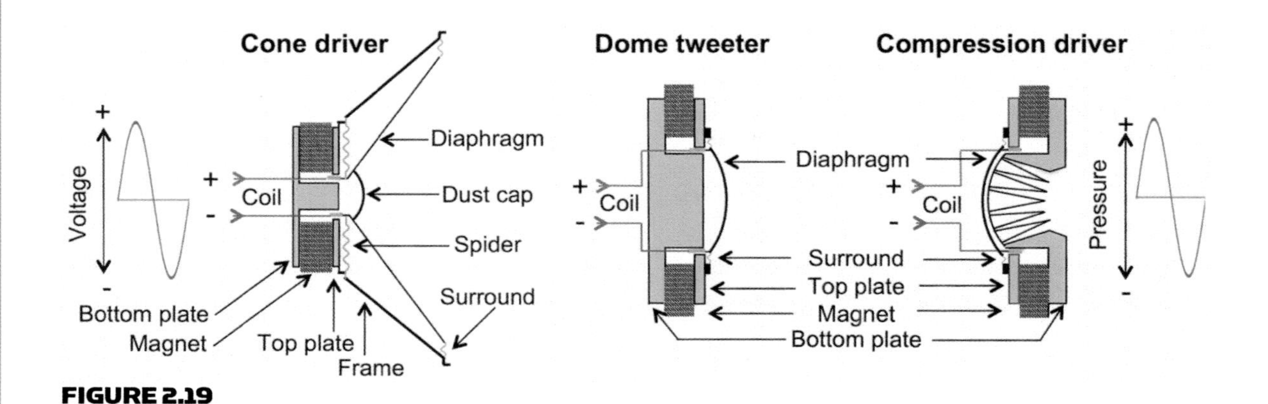

FIGURE 2.19
Loudspeaker/driver types

2.7.1.3 COMPRESSION DRIVERS

The name comes from the diaphragm compressing sound into a small chamber before exiting into an outer chamber (such as a horn or ribbon emulator). The compression driver uses an inverted domed diaphragm, i.e. the direction of propagation is on the concave side. Sound is funneled into a chamber smaller than the dome surface area, compressing the air. Drivers are classified by diaphragm diameter (sometimes improperly by throat diameter). Common sizes range from 1″ to 4″, while materials include lightweight metals (titanium, beryllium, aluminum and more).

2.7.1.4 RIBBON DRIVERS

The diaphragm is a flat, thin, metallized surface suspended in a magnetic field. The diaphragm is itself electrically conductive, or has metal foil inside a non-conductive material such as Mylar©. The ribbon is moved forward and rearward in the magnetic field.

2.7.2 Horns

The horn is the oldest sound system component. The original version was a funnel coupled to a human mouth, the source of the term "loud speaker." It's still in use by cheerleaders and as a South African torture instrument called the vuvuzela.

2.7.2.1 HORN BASICS

A horn is an acoustic pressure adaptor and wavefront-shaping device (Fig. 2.20). Horn terminology harks back to its human-driven origin: the throat (the entrance) and mouth (the exit). The signal at the throat is high pressure in a small surface area. As sound moves forward from throat to mouth the surface area gradually expands and the pressure decreases. To begin to characterize any horn requires (at a minimum) the surface areas of the throat and mouth, the length between them and the mathematical equation that describes the slope along the length (in both planes). The geometry that describes the shape-shifting between the throat and mouth is termed the horn flare or flare rate. This can range from a simple linear expansion (the cone) to equations using half the Greek alphabet. The simplest one that we still might find in the field is the exponential taper (or exponential-ish variations thereof).

There are two main reasons to attach a horn to a driver: It gets louder and puts more of the sound where we point it. This is admittedly a very unscientific way of stating this, so let's step it up a level. As per Bjorn Kolbrek, the two reasons are improved loading of the driver (i.e. increased efficiency) and directivity control. Both the loading and the directional focusing create a louder on axis response than the raw driver.

Size matters

Well-engineered horns are sound reflectors that guide the sound to create a wavefront with a directional pattern stamped on it that will continue after it has left the mouth. The horn flare (in both planes) is the driving force in

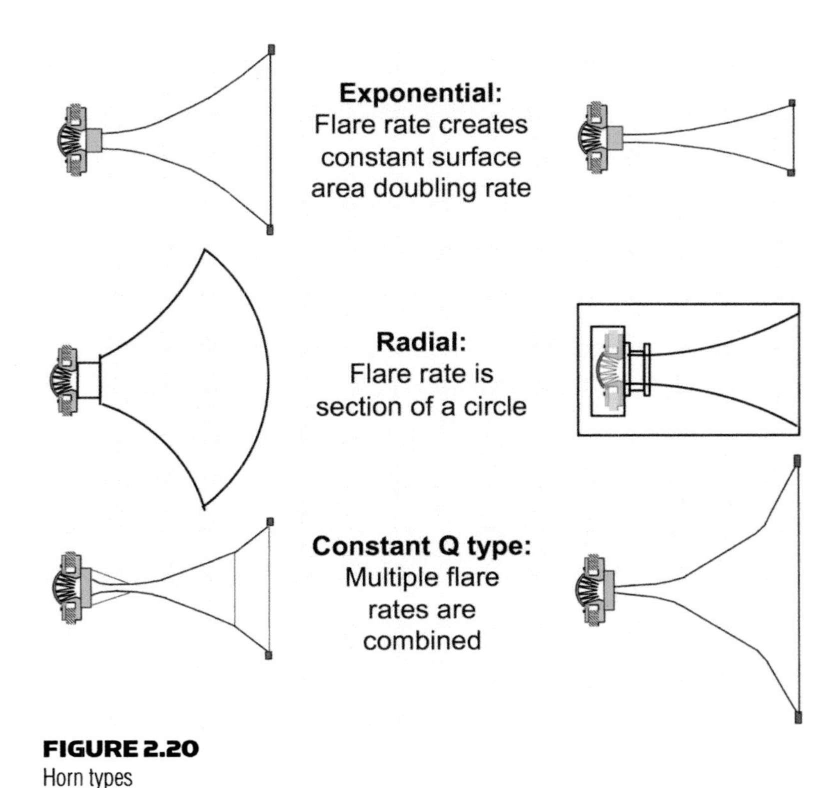

FIGURE 2.20
Horn types

directional control. This is scalable. In other words, a big horn and a small horn made from the same equation will have the same pattern. The size of sound however does not scale. One kilohertz has fixed size (neglecting temperature, atmospheric pressure changes, etc.) and therefore it fits very differently into a small horn than a big one. This is a Goldilocks aspect for any real physical horn: It is too big and too small and just right, depending on the real physical size of a frequency, i.e. wavelength. Too big means there are reflection paths that create cancellations, thereby reducing efficiency and disturbing the efforts to consistently guide the sound into a given shape. Too small means the guidance tool (the horn flare) is too small to effectively steer the wavelengths and does not "couple" with the direct sound, resulting in reduced efficiency. Just right means we realize the efficiency gains and have consistent steering without cancellation lobes. Therefore any given horn has a limited working range that must be respected.

2.7.2.2 CUTOFF FREQUENCY

The cutoff frequency is the lowest range where the horn increases efficiency and maintains directivity. The driving forces are a combination of horn length, mouth size and flare rate. It is generally agreed that the horn length should be a minimum of ½ λ and the mouth circumference should be at least 1 λ. If you want big deep low end, get a big deep horn.

2.7.2.3 MACRO SHAPE (STRAIGHT AND FOLDED HORNS)

Typical MF and HF horns proceed from throat to mouth with minimal bending. This is not so easily done with LF horns because they won't fit in the truck. Most low-frequency horns are folded type, i.e. the taper rate is the result of creative carpentry and engineering.

The folded horn is the solution to a maze. The quest is to minimize the overall length of the speaker enclosure while maintaining the length and taper of a long horn (thereby giving us high efficiency at a low frequency). Consider the tuba, which is actually a horn with a length of several meters, but coiled up to make it humanly manageable. If we made speaker enclosures out of brass we might do the same. Instead we make wooden or molded shapes with hard bends (folds) that create the horn taper (mostly quasi-exponential). In practice the folded horn performs well for only a limited frequency span, because the folding geometry is subject to stray paths when a wide range of wavelengths are transmitted.

2.7.2.4 FLARE TYPES

This is not a history book so we will not be going through all the horns nobody uses any more. The horn flare rate affects this calculation and the math complexity rises exponentially (or hyperbolically).

- **Exponential and quasi-exponentials**: Surface area expands as an exponential growth curve. The beamwidth tends to narrow with frequency. There are flare variations closely related to the exponential curve, such as the tractrix and others that can produce some improvement in beamwidth consistency.
- **Radial family**: The flanks of the flare are sections of a circle (hence the term "radial"). The radial horn, or variations thereof, is often used successfully to create wide patterns.
- **Constant directivity, Constant Q and related**: Flares are compound curves, i.e. the equation changes over the length, often segmented from a series of flat sections rather than a continuous curve. The goal is a constant (flat) beamwidth but the extent to which it is achieved is highly variable. Variations of this family are HF horn mainstays for modern sound systems.

2.7.3 Speaker mounting types

The driver has a raw response, but rarely will we see an un-mounted driver in service on a show. These are the major categories in driver mounting (Fig. 2.21).

2.7.3.1 FRONT-LOADED

A front-loaded driver is flush-mounted to the enclosure with a flat front surface. The enclosure provides no directional steering beyond the flat baffle reflection. Internal volume and reflex port tuning affect the LF range lower limit. The raw speaker's directional pattern is only minimally affected by the enclosure. The beamwidth of a front-loaded cone driver is predictable and varies with driver diameter. It's widest in the LF range and narrows with frequency,

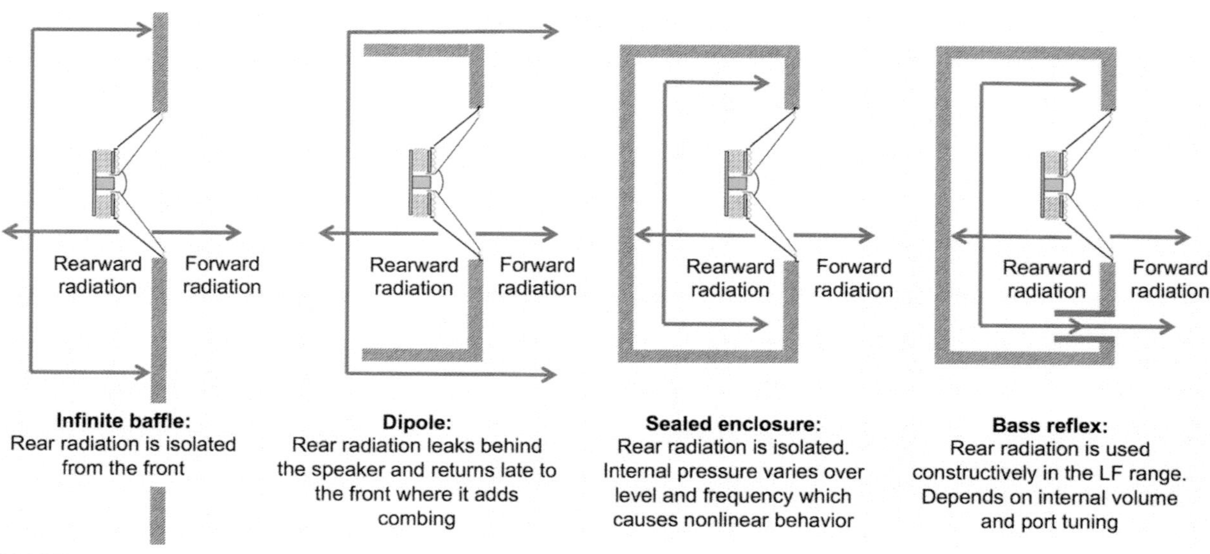

FIGURE 2.21
Typical speaker mounting types

a phenomenon termed "proportional directivity." The pattern will be 90° when speaker diaphragm and transmitted wavelength are equal size. At lower (larger) frequencies the coverage will be wider than 90° and vice versa.

There are three main options in the front-loaded woofer enclosure: the dipole (open back), ported (the bass reflex) and sealed (the infinite baffle).

2.7.3.2 DIPOLE

Ever wonder what to call the speaker mounting method used in a Fender guitar amp? Looking for a term more scholarly than "open-backed" or "really bad idea"? It's the dipole speaker enclosure: a speaker with baffles on the sides (straight out, or bent) and open in the back. A baffle is a wing-type extension beyond the speaker frame designed to reduce or prevent the back radiation from reaching the front. Dipole baffles delay the meeting for their length, but the waves still come together eventually.

The summed radiation pattern is a figure 8, with a huge back lobe. This has extremely few applications in professional sound systems, a rare find beyond the guitar line. You may encounter this in a cinema surround where folks see a very strong, reverse-polarity, delayed reflection as an enhancement. Not kidding, it's in the THX Cinema spec. There are even home speakers that take this concept further by placing a polarity-reversed driver on the rear of a sealed cabinet to replicate the open-back radiation. Combing is usually free but some folks are willing to pay extra for it!

2.7.3.3 INFINITE BAFFLE

A ceiling speaker is an example of a baffle mount, where the surface extends in all directions and seals off the rear of the driver. The baffle length cannot actually be infinite, but is effectively so when the rear propagation path is sealed off. The internal volume affects speaker performance due to asymmetry in pressure between the sealed rear and the open front. Pressure rises in the sealed space when the speaker pushes inward (rarefaction), but not on the rest of the planet. This won't make our ears pop but it's the same mechanism, on a moment-to-moment basis. The greater the excursion, the larger the effects, and the lower the frequency, the larger the excursion. Put those together and we have a speaker with reduced efficiency in the low end. In professional systems we can expect to see the infinite baffle mostly in cases where a cone driver is not expected to reproduce lows, such as a midrange driver.

2.7.3.4 BASS REFLEX (VENTED BOX)

Let's keep the basic structure of the sealed enclosure and add an opening (known as the reflex port or duct). The enclosure is often termed a "vented box" although the port is not much of a cooling mechanism. The pressure relief valve aspect is one feature of the venting, but the main event is increased low-frequency efficiency by additional radiation. The port is a tuned acoustic device: a Helmholtz resonator. Blow across the top of a beer bottle to hear a Helmholtz resonator. The internal volume (one of the Helmholtz parameters) rises as you drink the beer. The resonant frequency goes down, and is inversely proportional to sobriety. The other scaling parameter is the mouth shape (length and cross-sectional area), with larger again bringing the resonance down. The two factors play together to make one resonance. The bass reflex system is a resonance dance, a duet between the raw driver (known as the free air resonance) and the enclosure/port resonance. As we know, the driver's rear radiation from the driver is reverse polarity from the front. Signal exiting the port will be reversed as well then, right? Not necessarily. There are two competing resonant frequencies when the enclosure port resonance is offset to be lower than the driver resonance. The result is 180° of phase delay in the frequency range between the resonances. The signal exiting the port gets 180° + 180° (rear radiation reversal + resonant phase delay) and emerges in phase (+360°) with the front radiation. The efficiency in this range goes up significantly. Bear in mind that phase delay is the price for added SPL. The added signal is a lap behind in the race, which means a transient pulse is stretched in time. Pro audio applications favor the benefits of added SPL and prevail over the added phase delay. It would be very rare to see a non-vented front-loaded low-frequency driver in practice.

At frequencies below the port resonance the tuning breaks down, leaving the driver "unloaded" (as if there was no enclosure). Efficiency drops dramatically leaving the speaker vulnerable to over-excursion. Frequencies below the port resonance must be filtered out to reduce risk of damage. Wherever we see a bass reflex cabinet we should expect to see a high-pass filter to protect it. At frequencies above the resonance range, the port becomes acoustically resistant and reduces its forward radiation. Pick up the beer bottle again, find the resonance and then sing a higher pitch. The bottle no longer sings along. The port becomes less relevant as frequency rises.

What's the right enclosure and port size for an 18″ speaker? Sorry, not so easy. The answer requires a known speaker, known enclosure, desired cutoff frequency, etc., a set of calculations known as the Thiele/Small parameters. Enjoy!

2.7.3.5 MANIFOLD

The term manifold means many. We use it to describe multiple drivers converging to a single path or conversely when a single driver takes multiple paths. We can say there are manifold manifolds here (Fig. 2.22).

Convergence

Multiple drivers can be combined into a common chamber, e.g. two or more compression drivers that feed a common horn throat (with equal path lengths). As long as the combination point precedes the horn flare entrance the coverage pattern should be the same as with a single unit, but with increased SPL capability (and other artifacts such as increased distortion). A second example is a set of multiple low drivers placed as close as possible together in a common chamber. Electro-Voice implemented their Manifold Technology™ by reverse-mounting four low drivers in the open center of single cabinet. In summary, the converging manifold creates increased SPL with minimal changes in coverage pattern. A large-scale subwoofer version of a converging manifold is the TM array, described in section 10.1.3.

Divergence

Let's go the other direction and split the signal from a compression driver into multiple paths. These then exit separately to simulate the response of multiple sources. An example is a single compression driver with multiple horn throat paths. Each exits separately and then acoustically combines to create a horn array response (which alters the directional pattern from that of a single horn). This method is used in the HF sections of modern line array speakers, where the exits can be vertically stacked to create a coupled line source array response from a single driver. This also emulates the response of a ribbon driver.

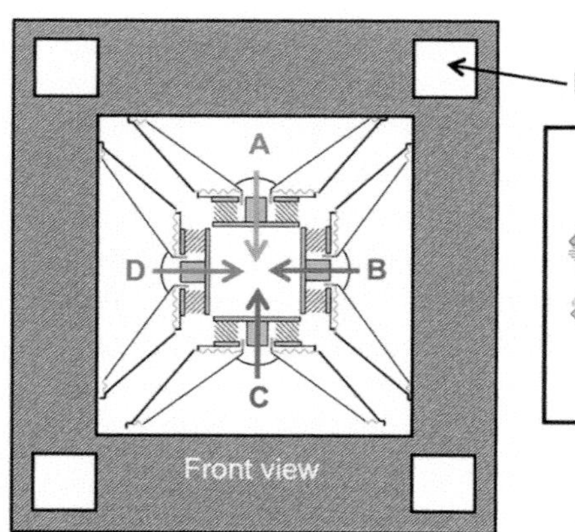

Reflex port

Side view

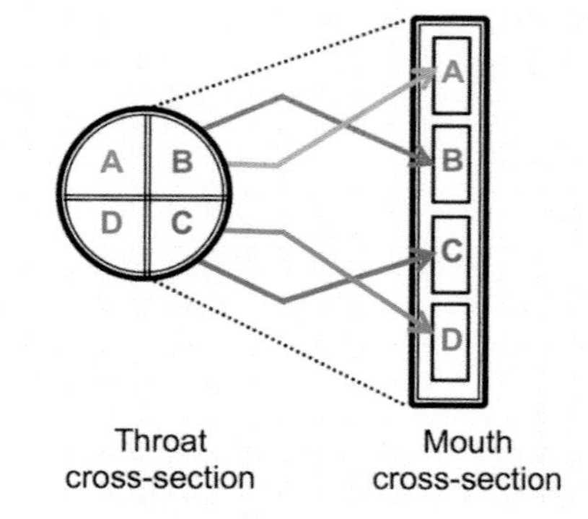

Convergence manifold: Four cone drivers using rear radiation in a single enclosure to quadruple the pressure

Convergence manifold: Two compression drivers coupled on to a single horn throat to double the pressure

Divergence manifold: Compression driver mounted on throat that is divided into four equal length paths to the waveguide mouth, which emulates the behavior of a ribbon driver

FIGURE 2.22
Speaker manifolds, convergence and divergence

2.7.3.6 PARABOLIC DISH

We are accustomed to seeing parabolic reflectors applied to optics and radio transmission. The same principle can be applied to acoustics. A source is placed at the focal point of a parabolic reflector and aimed into the dish. This concentrates a beam in the opposite direction. The same effect can be obtained by placing closely spaced speakers in the parabolic dish surface to create a direct sound field that simulates the focused parabolic reflection.

2.7.4 Frequency range classification

Every loudspeaker has an optimal range of operation. Low-frequency loudspeakers require large, stiff diaphragms to reproduce large wavelengths at high power. The mass limits the acceleration and therefore the HF response. Conversely, high-frequency drivers need small size and low mass to reproduce small wavelengths at high power, which limits their LF response. There are practically no single speakers capable of operating over our full range of frequency (30 Hz to 18 kHz). Therefore the spectrum is divided into ranges of operation for different transducers.

2.7.4.1 FULL-RANGE

The term "full-range" is used for speakers that cover the full vocal range, generally considered 60 Hz to 18 kHz. This speaker is not expected to cover the full range of our hearing, so the term can be confusing. Full-range systems are commonly two-way or three-way and rarely four-way.

2.7.4.2 LOW-FREQUENCY RANGE (SUBWOOFERS)

The range between 30 Hz and 100 Hz is generally considered the LF or subwoofer range. This may overlap the full-range speaker in the 60 Hz to 100 Hz range unless high-pass and low-pass filters are employed.

2.7.4.3 SPECIALTY RANGE

This refers to a variety of unconventional speakers with limited range, such as mid-bass (60–160 Hz), infrasubs (<30 Hz), MF and HF supplemental systems (tweeter banks and stand-alone horns).

2.7.5 Spectral division classifications

Speakers divide up the spectrum. This is how we divide up dividing the spectrum (Fig. 2.23).

2.7.5.1 ONE-WAY

We lack a standard terminology for single-driver loudspeakers covering the full spectrum. We can call it "one-way" and be consistent with the rest of the spectral division types. My general inclination is "no-way" but sometimes we are stuck with them in ceilings or small fills. Typically these systems are low power and contain an 8" or smaller driver.

2.7.5.2 TWO-WAY (PASSIVE)

Passive crossovers are the low-cost option for two-way systems. A single-cone driver is commonly combined with a compression driver or dome tweeter. The internal crossover network is comprised of passive components such as resistors, capacitors and inductors. A perfect mechanical alignment is required to achieve a phase-aligned crossover because delaying one of the drivers is not an option. The choice is often between having the best phase alignment or a speaker that won't fall apart.

Another down side to the passive crossover is limited dynamic capability. The limiting protection is full range (see pre-crossover limiting in section 2.4.6.3). In passive systems, the harmonics generated by clipping are passed through the crossover and sent to the high driver. This contrasts to an active two-way system where the harmonics stay in the low driver, which filters them due to its natural HF roll-off. The only advantage to passive two-way systems is cost savings. The disadvantages scale with size and role in the system design. Passive boxes have their most benign use as fill systems expected to run under clipping. The phase offsets are generally more manageable in smaller passive boxes.

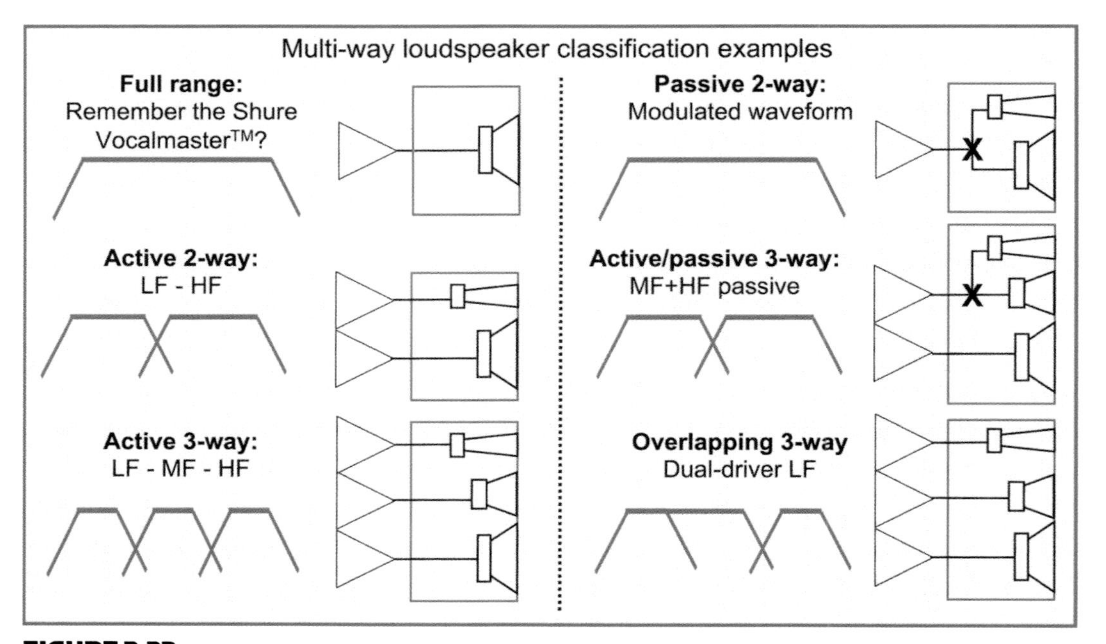

FIGURE 2.23
Active and passive multi-way configurations, two-way and three-way

2.7.5.3 TWO-WAY (ACTIVE)

This very standard configuration commonly uses an LF cone and HF compression driver or dome tweeter. The two-way family generally scales the two drivers together, i.e. smaller LF drivers pair with smaller HF drivers. LF drivers usually range from 5″ to 15″ whereas the accompanying HF drivers would be around 0.75″ to 4″ respectively. Crossover frequency falls as the size scales up, generally ranging between 4 kHz and 800 Hz over the scaling. LF operating range also falls as the size scales up, generally ranging from 100 Hz to 40 Hz.

There are innumerable variations by make and model. The driver loading (horn- or front-loaded) will affect the scale ratios between components and crossover frequency. The front-loaded woofer and medium-sized horn are the standard pairing, e.g. a 12″ and 3″ driver set with a 1200 Hz crossover. We can expect a larger HF driver and lower crossover if the LF driver is horn-loaded. As HF horn size rises, the cutoff frequency falls, allowing for a lower crossover (and a larger HF driver). If the HF driver has a small horn or a dome tweeter, the LF driver will be scaled down and crossover rises. These are basic trends and there are exceptions (both good and bad). My all-time favorite two-way configuration was the old Cerwin-Vega box with an 18″ and a piezoelectric tweeter, passive! Not.

2.7.5.4 SEPARATED THREE-WAY (ACTIVE, PASSIVE OR HYBRID)

Here is another standard configuration, most commonly with LF and MF cones and HF compression driver or dome tweeter. Each driver covers a distinct range of the split spectrum. The three-way family follows the scaling trends described above with the MF driver reducing the spectral load of one or both of the others. LF drivers usually range from 10″ to 15″, MF cone drivers in the 5″ to 10″ range (or a 4″ compression driver) with accompanying HF drivers around 0.75″ to 4″ respectively. Crossover frequencies reflect the scalar and loading relationships between the respective pairings as detailed above.

Again there are innumerable variations by make and model. There are three main versions: fully active three-way, fully passive three-way and hybrid active two-way with one channel split passively. The latter is usually grouped with an active LF–MF crossover and the passive MF–HF. A fully passive three-way box is hard to consider as a professional device.

2.7.5.5 OVERLAPPING THREE-WAY (ACTIVE, PASSIVE OR HYBRID)

Once exotic, this is now a fairly common configuration. The physical appearance is two matched LF drivers and a single HF, but the system operates as a specialized three-way. The two LF drivers operate together at low frequencies where the wavelength is large compared with the driver spacing, yielding cancellation-free response (and some LF directional control). A single speaker covers the remaining range to the HF driver. The LF–MF transition operates as a cross-"over and out." The cross-out frequency is determined by driver spacing. The MF–HF crossover would be the standard two-way. There are only two driver sizes, so their scaling relationship follows the two-way model described above.

In high-power systems these are normally fully active three-way, with budget models using an active/passive hybrid. The fully passive three-way is common for over thirty years in small-format fill speakers.

2.7.5.6 FOUR-WAY (ACTIVE, PASSIVE OR HYBRID)

We can go on forever, four-way, five-way, etc. Four-way systems are fairly rare, with their most common version being essentially a subwoofer and a three-way system in the same enclosure. All of the same logic applies here as outlined above.

2.7.6 SPL classifications

We can roughly classify speakers by their maximum peak SPL capability (Fig. 2.24). How much SPL do we need? The very non-scientific equation would be "program material × distance." Heavy metal at 10 meters might have equivalent needs to "Kum Ba Ya" at 1000 meters. If we double the distances we need twice the SPL (+6 dB) to keep pace. The definition of "loud enough" is beyond the scope of any book. Because one man's main is another man's frontfill, we will lay this out in gross steps of 10 dB.

Maximum SPL over distance by program material

Max dBSPL (Peak)	Distance (meters) 1.0	1.4	2.0	2.8	4.0	5.7	8.0	11	16	23	32	45	64	90	128
154	154	151	148	145	142	139	136	133	130	127	124	121	118	115	112
151	151	148	145	142	139	136	133	130	127	124	121	118	115	112	109
148	148	145	142	139	136	133	130	127	124	121	118	115	112	109	106
145	145	142	139	136	133	130	127	124	121	118	115	112	109	106	103
142	142	139	136	133	130	127	124	121	118	115	112	109	106	103	100
139	139	136	133	130	127	124	121	118	115	112	109	106	103	100	97
136	136	133	130	127	124	121	118	115	112	109	106	103	100	97	94
133	133	130	127	124	121	118	115	112	109	106	103	100	97	94	91
130	130	127	124	121	118	115	112	109	106	103	100	97	94	91	88
127	127	124	121	118	115	112	109	106	103	100	97	94	91	88	85
124	124	121	118	115	112	109	106	103	100	97	94	91	88	85	82
121	121	118	115	112	109	106	103	100	97	94	91	88	85	82	79
118	118	115	112	109	106	103	100	97	94	91	88	85	82	79	76
115	115	112	109	106	103	100	97	94	91	88	85	82	79	76	73
112	112	109	106	103	100	97	94	91	88	85	82	79	76	73	70
Distance (feet)	3.3	4.6	6.5	9.3	13	19	26	37	52	74	105	148	209	296	419

130	Aural Overload	124	High level	112	Medium level	100	Low level

FIGURE 2.24
Maximum SPL over distance by program material

- **Class 1**: speakers with maximum SPL capability in the 110–119 dB range. Low-power systems usable as mains in very limited applications such as speech reinforcement in small spaces or background music (BGM). Their primary role is fill systems for Class 2 and Class 3 mains, e.g. small-format speakers with 4″ or 5″ woofers and dome tweeters. Class 1 coverage patterns are typically wide in both planes.
- **Class 2**: speakers with maximum SPL capability rounded in the 120–129 dB range. Medium power systems usable as mains in limited applications such as reinforcement in small to medium spaces. Also used as fill systems for Class 3 and Class 4 mains, e.g. small-format speakers with 5″ or 8″ woofers and dome tweeters. Class 2 coverage patterns are typically wide in one or both planes.
- **Class 3**: speakers with maximum SPL capability of 130–139 dB range. High-power systems usable as mains for many applications in small to medium spaces. Also used as fill systems for Class 4 mains, e.g. medium-format speakers with 6.5″ to 12″ woofers and 2″ to 3″ compression drivers. Class 3 coverage patterns have no predominant trend.
- **Class 4**: speakers with maximum SPL capability of 140 dB and above. Very high-power systems usable as mains or fills in medium to large spaces, e.g. large-format speakers with 10″ to 15″ woofers and 3″ compression drivers. Class 4 coverage patterns are typically narrow in at least one plane.

2.7.7 Beamwidth classifications

Beamwidth is constructed by a series of coverage angle measurements over frequency (horizontal and vertical) (Fig. 2.25). Two beamwidth shapes yield repeatable success in system optimization: constant (plateau) and proportional.

2.7.7.1 PLATEAU (CONSTANT Q, CONSTANT DIRECTIVITY)

The plateau beamwidth pattern is notable for maintaining a constant angle over a multiple octave range including the upper limits of the speaker's range. Each plane (horizontal and vertical) has two defining values: nominal angle (the average angle of the plateau) and cutoff frequency (where the plateau begins). The HF horn is the foremost plateau-shaping device; therefore it is common to see the plateau begin around the crossover point. As a general trend, larger speakers can have a longer plateau than smaller units. If we took two identical speakers and shrunk one, the plateau would stay the same angle (formed by horn flare rate) but reduce the frequency range (smaller horn size moves the cutoff frequency upward).

A larger horn size allows for a flare rate capable of increased directivity and range (smaller nominal beamwidth and lower cutoff frequency).

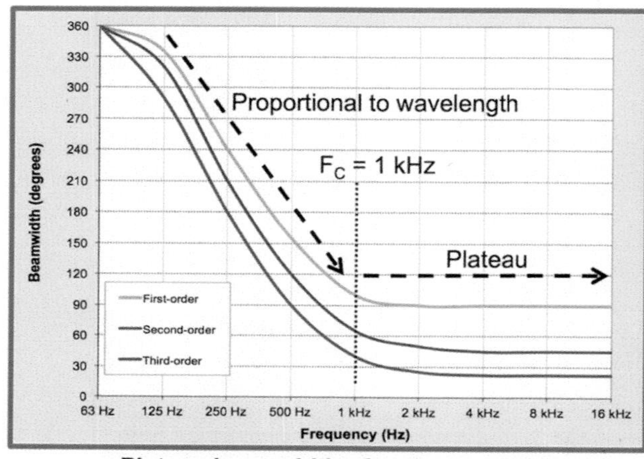

Plateau beamwidth of various orders
Constant coverage angle in the HF range

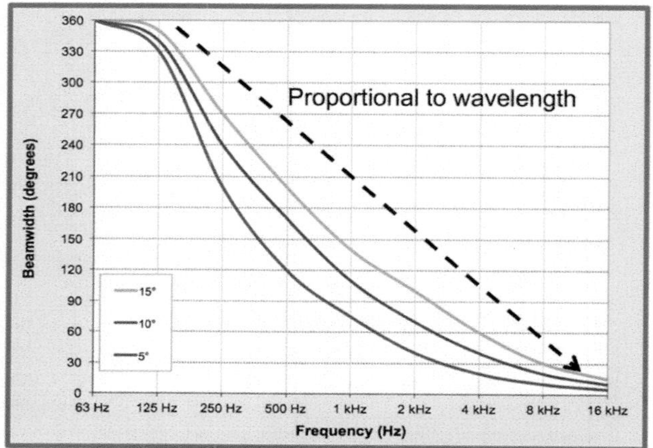

Proportional beamwidth, third-order
Coverage narrows with frequency

FIGURE 2.25
Beamwidth examples for various speakers

2.7.7.2 PROPORTIONAL BEAMWIDTH (PROPORTIONAL Q)

The proportional beamwidth narrows as frequency rises (coverage angle is proportional to wavelength). This is also often termed proportional Q (Q is yet another term for coverage pattern, with higher Q factor denoting narrower coverage). Speakers are wide (LF), medium (MF) and extremely narrow (HF). A single speaker may be 360° @100 Hz, 90° @1 kHz and 7° @10 kHz. There's no nominal angle, so we use the narrowest region (also the highest frequency) as our design and optimization value. As a general trend, larger speakers can have a narrower final value than smaller units. The beamwidth shape results from proportional directivity in cone drivers and ribbon-type behavior in the HF range (an actual ribbon or ribbon-emulating design). Proportional beamwidth is a typical vertical plane response in the modern line array speaker.

2.7.8 Coverage pattern classifications

We can roughly classify speakers by coverage angle as wide, medium or narrow. A nominal value is used for flat beamwidth speakers and minimum angle is used for diagonal (proportional beamwidth) models.

- **First-order plateau (>60°)**: speakers with a wide pattern consistent over the HF and MF range. First-order speakers generally have the longest plateau range because the controlled HF and MF region is not much narrower than the LF range. This is the most common shape for solo speakers or uncoupled arrays.
- **Second-order plateau (20°-60°)**: speakers with a medium pattern consistent over the HF and (maybe) MF range. Large-format speakers can maintain a flat, narrow beamwidth over an extended range. Small-format second-order speakers generally have a small plateau range (high cutoff frequency due to small horn size). This is a common shape for coupled point source arrays of moderate quantity, or as specialty fills.
- **Third-order plateau (<20°)**: speakers with a narrow pattern consistent over the HF and (maybe) MF range. These are specialty devices. Known systems that can create this shape include parabolic dish speakers or have very large horns. There are no small-format versions of this (unless you consider that the current tools are tiny compared with gigantic horns of the audio bronze age). This is a specialty shape for hostile acoustic environments, and/or extremely long distances.
- **Third-order proportional beamwidth (<20°)**: speakers with an ever-narrowing pattern over frequency (described above in section 2.7.7.2). The beamwidth plots resemble a diagonal line. These are the vertical building blocks for the modern "line array" type speaker. This is a specialty shape for speakers operating in coupled point source arrays in potentially large quantities. They are also used as specialty delays and sometimes frontfills (at great risk of missing the target).

FURTHER READING

http://www.learnabout-electronics.org/Amplifiers/amplifiers56.php.
http://en.wikipedia.org/wiki/Horn_loudspeaker.
http://en.wikipedia.org/wiki/Microphone.

CHAPTER 3

Transmission

> ***transmission** n.*
> *transmitting or being transmitted; broadcast program.*
> ***transmit** v.t.* **1.** *pass on, hand on, transfer, communicate.* **2.** *allow to pass through, be a medium for, serve to communicate (heat, light, sound, electricity, emotion, signal, news).*
> Concise Oxford Dictionary

We are now ready to transmit the waveform through the system. We've already met the sources and receivers, so we can proceed to make the connections. Our goal is straightforward: move the waveform from sender to receiver as fast and faithfully as possible. Our challenge is to minimize unsolicited modification of the waveform: added noise, frequency response changes, distortion, dropouts, excess latency and alien invaders.

We start with the transmitted product: the signal or waveform. The journey begins at the source, travels through (or over) the medium and arrives at the receiver. The acoustic waveform is pressure variations over time and our focus will be transmission through air. The senders are acoustic sources and loudspeakers, whereas the receivers are ears and microphones. The analog electronic version is voltage over time, which propagates through the electromagnetic medium. Device outputs and inputs are the sources and receivers. In the digital realm a numerical encoding of the waveform is transmitted *over* a medium. The distinction between transmitting through a medium and over one might seem like splitting hairs but the two transmission forms are almost entirely distinct.

The remaining major pieces are transmission translation devices. In the analog world these are transducers, whose role is to change between mediums. The digital corollary is the converter, which moves us between the analog and digital forms.

Our journey goes from Point A to Point A: acoustic sender to acoustic receiver. There are lots of transmission paths to get there (Fig. 3.1).

- Acoustic to acoustic: sound source to our ears, non-stop flight.
- Acoustic to analog to acoustic: add layovers at a mic, analog electronics and speaker.
- Acoustic to analog to digital to analog to acoustic: insert a digital audio into the chain.
- Acoustic to analog to digital to network digital to analog to acoustic: move the digital audio over the net.

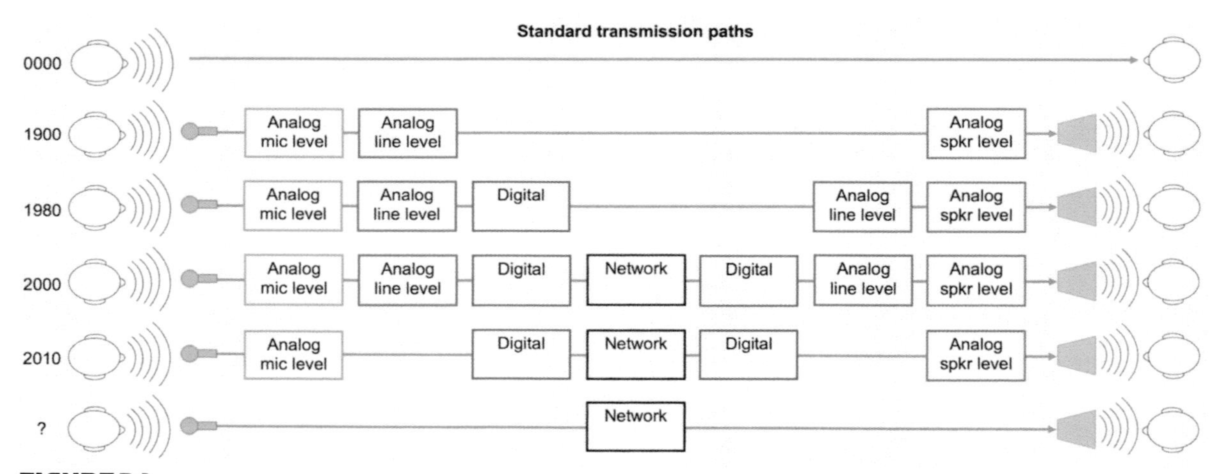

FIGURE 3.1
Transmission paths

3.1 THE ANALOG AUDIO PIPELINE

Prepare yourself for the "mother of all analogies." We are going to use plumbing as the conduit for learning audio transmission in its various forms: analog, digital and networked.

Direct current (DC) electrical circuits are often introduced to beginners in terms analogous to the flow of fluids through a pipe. Voltage is the water pressure at the source. Resistance relates to the pipe width and current is the liquid's flow rate. Let's modify that analogy to our favorite AC signal: audio transmission. In our case the pipeline is the containment vessel for the flowing waveform. The pipe casing represents the transmission system's outer limits, because the waveform can't expand beyond it. Our goal is to move the waveform through the pipe using as much of its open area as possible but never touching the casing. The pipe will vary in material and size as we go along, as will our flowing waveform. We want to move the waveform through a medium and/or transcribe it between mediums without modification. Three principal hazards lie in the analog path: distortion, noise and frequency response modification.

Let's examine a section of audio pipe. It can pass waveforms of limited size (amplitude) and no more. Anything larger hits the edges (clipping). There may be leaks in the casing that spill out (level loss) or allow other signals in (noise). The pumps (active electronics) keep things moving but also add noise. This internal circuit noise scales up or down with the pipe diameter. There are mechanical properties in the pipe casing, the hardness of the edges, the way it bends, the connectors, etc., which affect some portions of our waveform more than others (the frequency-dependent effects).

Three lessons so far: an oversized waveform hits the walls, an undersized one is diluted with noise, and a pipeline made out of poor materials will be inefficient, noisy and exhibit frequency-dependent effects.

3.1.1 Upper and lower limits

The pipeline analogy can be applied to a variety of audio media. The acoustics pipeline limits are pressure levels where transmission becomes increasingly non-linear with level. Distortion rises gradually with level and then increases sharply as the edge is neared. Visualize a slightly expandable hose under high pressure. By contrast, the analog electronic pipeline has a rigid casing set by the DC rail voltage (e.g. ±15 V_{DC}). Hitting it results in hard clipping of the waveform. Transformers and tape saturate the magnetic flux density, whereas speakers and microphones gradually reach the mechanical limits of their moving parts. These are softer forms of clipping. An overactive phonograph needle bursts the pipe and leaks into the next groove, and then it just won't play-just won't play-just won't play.

The air pipeline is almost the same size everywhere (smallest on Mt. Everest). The analog electronic pipe comes in various sizes, roughly classified as mic, line and speaker level. The waveform can never equal or exceed the DC rail voltages (or more accurately about 0.75 volts less than the rail). Rail voltages of 15 V_{DC} leave us with an absolute maximum AC swing of ±14.25 V_{PK}. For a sine wave that's 10 V_{RMS} also known as +20 dBV (+22 dBu). The DC rails (the pipeline size) are the primary factor in setting the maximum output capability of line-level devices. It is difficult in practice to get through an entire device without some pinch points so not all devices will necessarily make it up to the maximum theoretical level, but this is the trend. A thick pipe gives us more room than a thin one (e.g. 18 V_{DC} vs. 12 V_{DC} rails).

If ±14 V is the standard line-level pipe, then what is mic level? That's a tricky one. There are lots of mic-level devices with line-level pipe, e.g. a mic preamp, the output of which needs to swing to full line level. Mic level only exists because microphones, guitar pickups, etc. put out extremely small amounts of level (not because we have a small DC power supply). A mic-level device such as a guitar pedal can get away with a 9 V battery (DC rails of around ±3.5 volts) because the audio waveform won't exceed that pipe size. We will need bigger plumbing later to step it up to line level.

There is a 6 dB difference in maximum swing between line-level pipes with internal diameters of ±11 V and ±23 V varieties (DC rails between 12 V and 24 V). Substantial, but not exactly a game changer. Power amplifiers are another story: Line-level input pipe connects to a much bigger one on the output.

3.1.2 Impedance: low *Z* source to high *Z* receiver

Let's send some signal into the pipe and see what we get at the other end. The source must be able to maintain enough flow (low impedance) to keep the pressure up in the medium. The receiver needs minimal flow (high impedance) to follow the pressure changes (the waveform transcription pattern).

The reversal, a high-impedance source feeding a low-impedance receiver, is far from lossless, dynamically limited and may possibly fail completely. Visualize this: We can drive a garden hose (or ten) with a fire hydrant with no worries about pressure loss (low Z source to high Z receiver). On the other hand, a garden hose could never generate enough flow to fill up a fire hose (high Z source to low Z receiver).

The acoustic pipeline is driven by low-impedance devices (natural or loudspeakers) to our high-impedance ears (or microphones). The analog electronic pipeline has several low Z source to high Z receiver iterations: mics to preamps, line outputs to line inputs and amplifiers to speakers.

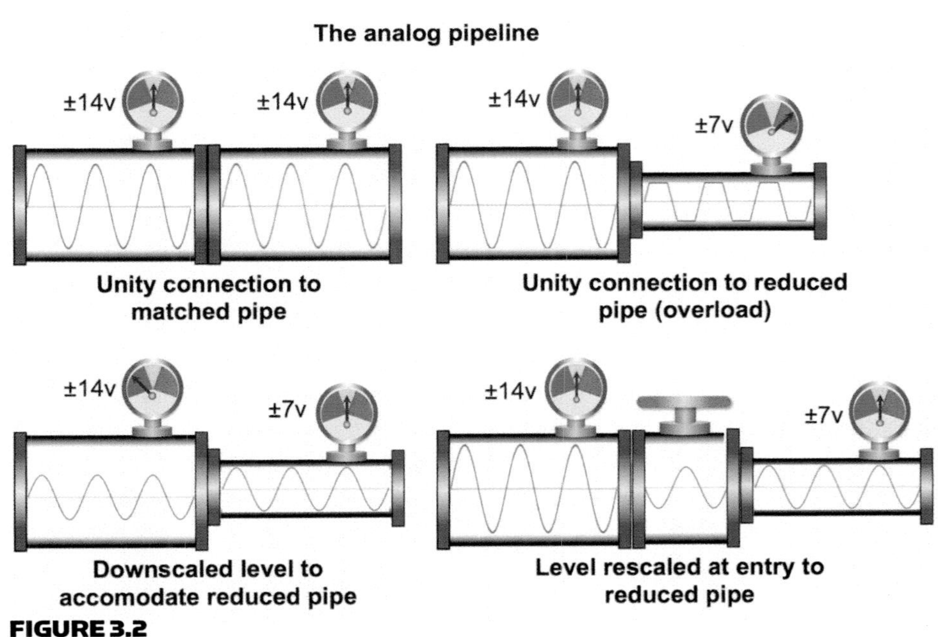

FIGURE 3.2
Introducing the analog audio pipeline

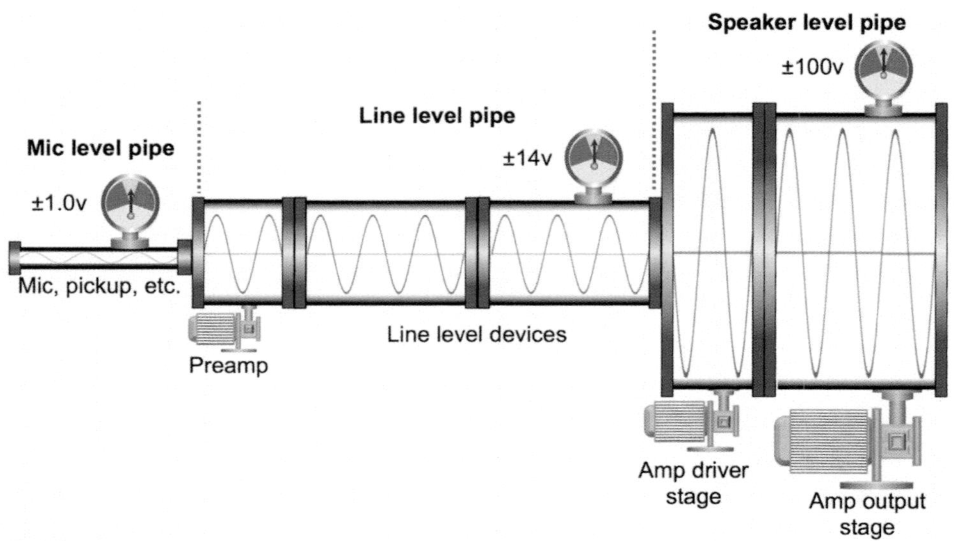

FIGURE 3.3
Analog pipeline: mic, line and speaker levels

3.1.3 Interstage connections

Let's connect two matched line-level pipe sections in series (Fig. 3.2). The outer limit hasn't changed so the waveform fits just the same (a unity gain connection). We just need a stronger pump to keep the pressure up in the longer pipe.

If we connect our 14 V pipe to a 7 V pipe the outer edges of our waveform (anything >7 V) will be clipped at the entry to the new, smaller section. We need to rescale the waveform downward to fit it into the 7 V pipe (a 50% reduction, -6 dB). Either we limit the waveform to 7 V for the entire pipe or run two different scales to properly size the waveform for the current pipe size. We waste 6 dB of headroom if we don't rescale, because the 14 V pipe section will perform as a 7 V section (but with 14 V section noise floor). The best noise immunity comes from rescaling as we go. Mic level, line level and speaker level are the major scaling break points but smaller increments can be used as well (Fig. 3.3).

A mic preamp's role is to rescale from mic to line level. This first circuit on a mix console prevents mic-level signals from being moved through line-level pipes where they would drown in the noise.

3.1.4 Summing

What happens when we join two 14 V pipes (A and B) at a junction that flows into a single 14 V pipe (AB)? Matched waveforms will cause pipe AB to handle 2× the voltage of A (or B) individually (Fig. 3.4). Therefore we will overflow the AB section if we exceed 7 V at A (or B). The solution is easy here: Rescale both A and B down by 50% (6 dB) at the junction and AB will have the same amount of headroom as the individual sections.

We can assume that the A and B signals are different, because there is no good reason to sum identical analog electronic signals. This leads to a wild-card situation where a new combined waveform is created. A moment of amplitude and phase matching can double the individual levels. But it can also be far less, perhaps even perfect cancellation, if we have a moment of equal amplitude and opposite (180°) phase. The average combination over time of two uncorrelated signals at equal level will be about 1.4× (+3 dB) above the amplitude of the individuals. It is certain that the new waveform will be subject to change without notice and the looming possibility of a momentary 6 dB increase. We must assume the worst to ensure our combined signal is safe from overload. This is where the concept of "headroom" comes into play. We build in headroom wherever we have unknown dynamic conditions. Whereas unity gain was the logical choice for connections involving a single channel stream, this is not

Summing the pipeline

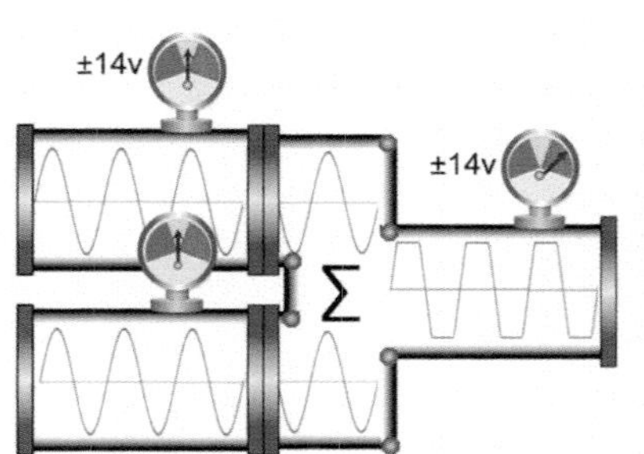

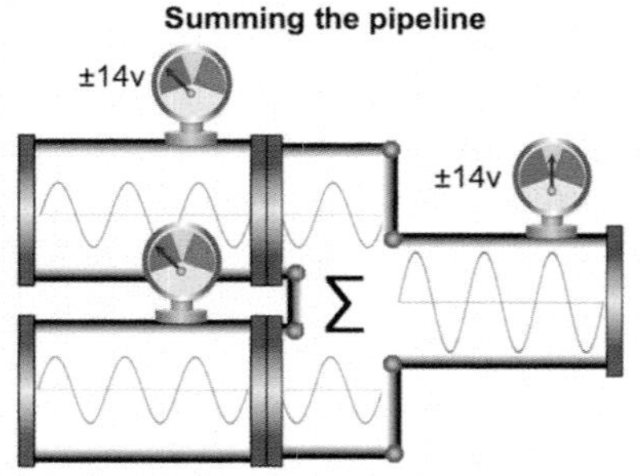

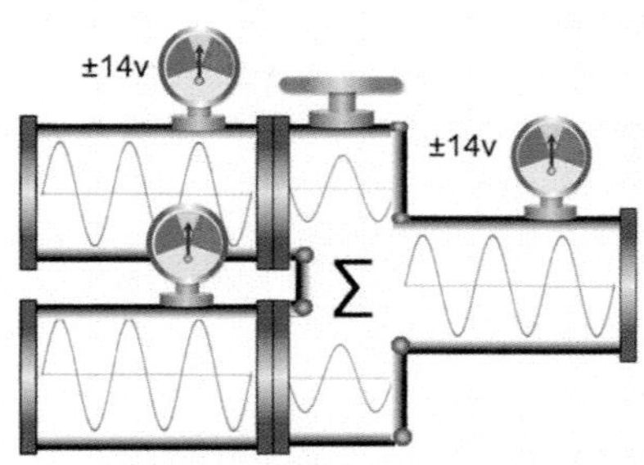

FIGURE 3.4
Analog pipeline summing scenarios

necessarily the case for summing junctions, which can be easily flooded if the multiple streams that feed a junction point are equal in size to the output. Visualize a plumbing system with six 1″ pipes joined together into a 1″ output pipe. This is a recipe for overflow, which is why no plumber would ever allow it. Downstream output pipes are made larger in the world of water. In the line-level audio world we stick with the same pipe size through the entire analog signal chain from the mic preamp to the power amp drive stage. Stepping up from 15 V_{DC} to 18 V_{DC} rails is only 2 dB, less than the gain we get from adding one input. We often connect 100 same-sized line-level pipes through multiple junctions (it's called a mix console). At +3 dB gain (1.4×) per doubling of inputs, we could gain 20 dB (100×). How can we move this variable-sized payload through the fixed-size pipeline system? The key is to scale down the level at the summing junctions. Send optimally loaded pipes toward the junction but throttle down the level at the summing junction to size the combined waveform for the output pipe. This in turn becomes an input pipe to the next stage, which is rescaled down again to keep the output in bounds. We seek consistent dynamic range through the various stages, even as signal density increases and the risk of huge pk–pk transient surges becomes greater. The upper side of dynamic range is headroom, and the lower side is noise floor. We preserve headroom and minimize noise by rescaling levels at the appropriate places.

3.1.5 Headroom

Headroom is the answer to the question: How close are we to overload? For studio mastering of recorded material there's no mystery. Mastering engineers can find the maximum pk–pk levels reached in a track with an oscilloscope. Levels are set to ensure the final mix has the hottest possible level that fits cleanly in the pipe. Headroom is a real-time ongoing guesstimate for live sound, because we don't know what the next moment will bring. Therefore we need plenty of it. Competent operation can prevent a single channel from being driven into the rails but combinations of signals can lead to unexpected levels. Two properly scaled signals can combine to a level out of scale, due to momentary waveform summing (the reason we downscale at the entry to the summer). Returning to our plumbing analogy, we have a valve located at the summing junction to regulate flow into the output pipe. Reducing the gain here also helps preserve the noise floor because the inputs are scaled down along with their noise. Think of the alternative: operating all of our inputs at very low levels (and tons of headroom) so that we still have headroom after summing. This would leave us doing a unity sum of lots of noisy channels. The signal would rise, as would the noise. Instead it may be better to reduce the individual inputs, which enter cleanly into the summer. There are so many places where the pipe can be overloaded: single-stream gain stages or any summing stage (mix buses, aux buses, matrices, etc.).

3.2 ACOUSTIC TO ACOUSTIC TRANSMISSION

Let's begin with an acoustic sender to an acoustic receiver. Commonly known as "natural" sound, this is everybody's favorite except that it leaves us unemployed. We start here because this is the original short-circuit transmission path. All other transmission scenarios are inserted inside this structure.

The senders are musical instruments, voices, anything that makes noise. The receivers are ears and microphones. The medium is air.

3.2.1 Propagation

Propagation requires an elastic medium, in this case air. Air is more complex than our other mediums. It is noticeably frequency dependent, begins in unique shapes, goes in all directions, bends around things and changes with the weather. Traveling through a wire, by contrast, is indoor living at its finest.

Ready to get theoretical? An infinitely small sound source will propagate spherically as an ideal point source. The sound will propagate equally outward in all directions (omnidirectionality) and will exhibit an SPL loss rate of 50% (6 dB) for every doubling of distance. Last week we loaded our infinitely small speakers into the truck but nobody could find them when we got to the gig.

3.2.1.1 SPHERICAL WAVES

Our sound sources, voice, musical instruments and loudspeakers alike, have size. This means they are not absolutely ideal point sources. Nonetheless, we will start with the construct that a single sound source propagates spherically, losing approximately 6 dB/doubling of distance. The directional propagation pattern will be subject to a host of steering options but for now let's suffice to say that once we get far enough away from the source, it will propagate with a constant directional pattern over distance (Fig. 3.5).

3.2.1.2 PLANAR WAVES

Planar waves, a second type of acoustic wavefront, are created when the source is stretched over an area. This is found at the mouth of a horn, where the wavefront has been shaped by the expanding horn taper. The loss rate and directionality diverge greatly from the spherical radiation model inside the area of the horn where planar radiation occurs. Once the sound has propagated some distance out of the horn, however, the behavior reverts to spherical propagation. The horn's directional shaping will be maintained but the loss rate reverts to 6 dB/doubling.

3.2.1.3 INFINITE LINE SOURCE

Lastly we can consider the result of another theoretical construct, a source of infinite length, or an infinitely long array of closely spaced sources: the infinite line source. Not available in stores! Whereas the planar wave stretches its width in two dimensions, the infinite line source stretches in just one. In that dimension we can theoretically

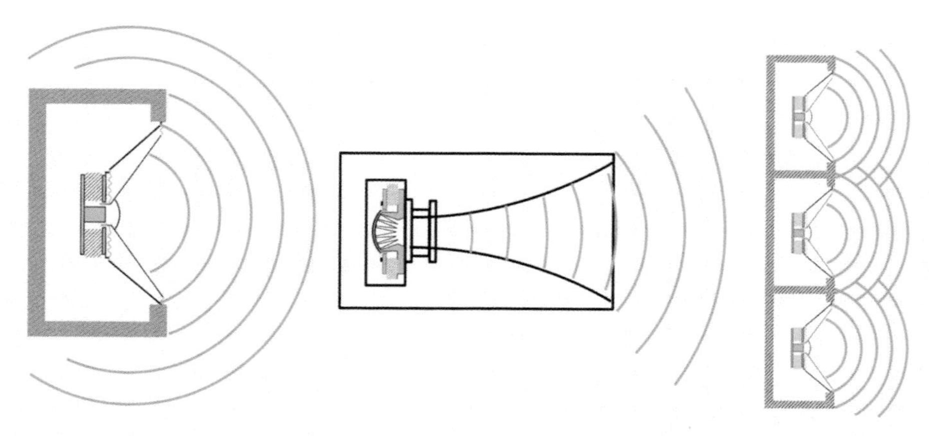

FIGURE 3.5
Wave propagation: spherical, planar and infinite line source

see a reduced loss rate of just 3 dB/doubling of distance. In practice we can never finish unloading an infinite line source from the truck. A finite line source can create reduced loss propagation for a limited distance, and then reverts to the spherical propagation model of 6 dB/doubling. The term "cylindrical wave" has been used by some to describe the early-stage propagation in the finite line source (where a 3 dB loss rate can be observed), but there is no scientific consensus on this. The analyzers we use for system optimization (and our ears) are indifferent to wave type.

Wave propagation mathematics are extremely complex and way beyond the scope of this book (or author). Our analyzer will never say, "Sensors indicate a planar wave, captain." Here is what we must know: As long as our head is not in a horn or the near field of an array we will be measuring spherical propagation.

3.2.2 Speed

The formula for sound speed in air is $c = (331.4 + 0.607 * \text{temperature } (°\text{C}))$ in m/sec. The imperial version is $c = (1052 + 1.1 * \text{temperature } (°\text{F})$ in ft/sec.

For example at 22°C:

$$c = (331.4 + 0.607 * 22) \text{ meters/second}$$

$$c = 344.75 \text{ meters/second}$$

Wrapping our brains around the sound speed formula is difficult because the parameters are out of scale to our practical applications: 334.75 meters is too big to visualize and increments of one second are laughably imprecise. That's a rate of three football fields/second. Does it help to rescale it down to 0.344.75 meters/millisecond? It's easier to invert the rate to ms/meter instead. This reduces to 2.94 ms/meter @22°C, and can be rounded to 3 ms/meter. For those using the English system (feet) there are three options: (a) learn the metric system, (b) ms/foot, which rounds to 0.9 ms/foot and (c) feet/ms, which rounds to 1.1 ft/ms.

We can also use a standard room accessory as a visual aid: seats. Row spacing is typically about 3 ms (English and metric people are the same size). This underscores the advantage of visualizing distances in ms, rather than meters, feet or mega parsecs. We'll use timing when we get to the tuning so let's save the conversion. I cannot emphasize enough the advantage of visualizing distance in milliseconds, because this unit is required for understanding phase and frequency, speaker interaction and reflections.

3.2.3 Wavelength

Wavelength is proportional to the medium's unique transmission speed. Transmission speed through air is among the slowest, which means the wavelengths in this medium are on the small side (less than ¼ their size in water and 1/500,000 their size in an analog wire). When the medium is changed, the transmission speed and all the wavelengths will change with it.

The wavelength formula is

$$\lambda = c/F$$

where λ is the wavelength in meters, c is the medium's transmission speed, and F is frequency (Hz).

A typical sound reinforcement system transmits over at least a 600:1 range of wavelengths: 18,000 Hz down to 30 Hz. A wavelength at 30 Hz is in fact 600× larger than its 18 kHz partner, ranging in size from trucks to fingernails. Wavelength is particularly difficult to visualize because (a) sound is invisible and (b) the mental challenge of simultaneous transmission of something spanning a 600:1 size ratio. It is, however, immensely helpful to develop this skill because the actual physical size of sound is a key factor in its behavior. A handy reference chart is provided in Fig. 3.6.

Wavelength reference chart

Frequency (Hz)	Period (ms)	Wavelength (Room temp) (m)	(ft)	Comparable size
20	50.00	17.24	56.56	
25	40.00	13.79	45.07	Intermodal shipping container
32	31.75	10.94	35.77	
40	25.00	8.62	28.17	Band gear truck length
50	20.00	6.90	22.54	1/2 size intermodal container
63	15.87	5.47	17.89	Gas guzzling SUV length
80	12.50	4.31	14.09	Full Size car length
100	10.00	3.45	11.27	Compact car length
125	8.00	2.76	9.01	Too wide for the truck
160	6.25	2.15	7.04	Shaquille O'Neal
200	5.00	1.72	5.63	Average height
250	4.00	1.38	4.51	Shoulder height
315	3.17	1.09	3.58	
400	2.50	0.86	2.82	
500	2.00	0.69	2.25	Arm's length
630	1.59	0.55	1.79	
800	1.25	0.43	1.41	
1,000	1.00	0.34	1.13	Elbow to fist
1,250	0.80	0.28	0.90	Man's foot
1,600	0.63	0.22	0.70	Woman's foot
2,000	0.50	0.17	0.56	Eight fingers
2,500	0.40	0.14	0.45	
3,150	0.32	0.11	0.36	CD/DVD
4,000	0.25	0.086	0.28	Four fingers
5,000	0.20	0.069	0.23	
6,300	0.16	0.055	0.18	
8,000	0.13	0.043	0.14	Two fingers
10,000	0.10	0.034	0.11	
12,500	0.08	0.028	0.09	
16,000	0.06	0.022	0.07	One finger
20,000	0.05	0.017	0.06	

FIGURE 3.6
Wavelength reference chart

3.2.4 Low *Z* to high *Z*: the pressure follower

An acoustic source (or loudspeaker) propagates sound into a very low-impedance medium: air. The ear (or microphone) senses the propagating waveform without sucking all the air out of the room. In short, lots of power is required to push the waveform through the medium, but not to sense its presence and send a transcription to our brain. Ears (and microphones) present a high acoustic impedance load and can sense the pressure without dragging down the source.

3.2.5 Dynamic range

3.2.5.1 UPPER LIMITS

At extremely high sound pressure levels (>120 dB SPL) the air medium becomes increasingly non-linear but does not hit a hard limit until 194 dB SPL peak. This is equivalent to one atmosphere, which means that on the rarefaction side (low pressure) we have literally run out of air (sooner at the Telluride Bluegrass Festival). The medium has physical limits, not just our ears, so the quest for infinite SPL has an end game at 194 dB SPL peak. The medium gradually saturates, so the highest distortion will occur at the closest locations during the loudest passages. The compression chamber and horn throat have some of the highest levels of air medium distortion in our sound system.

3.2.5.2 LOWER LIMITS

The lower limit is the Brownian noise, which is around the same loudness as our hearing threshold (0 dB SPL). This will be our deepest foray into molecular physics. Nothing in nature actually sits still, and air molecules are no exception. Brownian motion is statistically random and therefore creates white (not brown) noise. Brown was the scientist who quantified it.

3.2.6 Loss rate (inverse square law)

The inverse square law is the common name for the loss rate of 6 dB/doubling of distance associated with spherical wave propagation (Fig. 3.7). This hypothetical standard rate, termed the "free field" response, requires some special conditions that can never quite perfectly be met, but nonetheless is the best starting reference for mapping out level over distance. Deviations above or below the free-field loss rate are noted. Air is not a lossless medium in the HF range, i.e. the uppermost range falls at a higher rate than free field. Reflected energy adds to the direct sound, which (mostly) causes the loss rate to decrease. If a particular reflection adds in phase, the loss rate is slowed. If the reflection is out of phase, the loss rate accelerates. Coupled speaker arrays (sources close together) can decrease the loss rate (as a combination), but only in the near field.

dBSPL loss over distance		
dB loss	(m)	(ft)
-0	1	3.3
-3	1.4	4.6
-6	2.0	6.5
-9	2.8	9.3
-12	4.0	13
-15	5.7	19
-18	8.0	26
-21	11	37
-24	16	52
-27	23	74
-30	32	105
-33	45	148
-36	64	209
-39	90	296
-42	128	418

FIGURE 3.7
Sound propagation loss over distance

Inverse square law does not discriminate. It governs loudspeakers, musical instruments and humans the same. The only exception is babies in airplanes (0 dB loss rate). The free-field loss rate should be committed to memory in at least 3 dB increments (1 dB increments are best, which can be reviewed in Fig. 1.14).

Let's see what happens when we move away from a typical directional source. The HF region maintains free-field behavior for the longest distance before giving in to the additions/subtractions of reflections. The LF drops at a noticeably reduced rate because its wide coverage quickly involves the floor, ceiling and walls. Two factors can extend the free-field behavior in the HF range by reducing the reflected energy: increased source directionality and surface absorption. Air loss, by contrast, causes the HF range to deviate from free field response because the loss rate is accelerated there. A directional speaker in a dry room (acoustically) on a humid day (environmentally) will have an extended range of free-field behavior in the HF range. This is true of loudspeakers and natural sources.

LOSS RATE SUMMARY

- **Free-field standard**: 6 dB loss/doubling of distance.
- **Air loss**: linearly exceeds the standard loss rate.
- **Reflections (near field)**: minimally below standard loss rate.
- **Reflections (far field)**: maximally below standard loss rate.
- **Speaker arrays (near field)**: maximally below standard loss rate.
- **Speaker arrays (far field)**: minimally below standard loss rate.

We just saw the conditions attached to the inverse square law. Is this affected by the sound source coverage pattern? Does an omnidirectional device drop off at a different rate than a directional one? The answer is a slightly qualified no. Once we are far enough away for the source's coverage pattern to stabilize, the dropoff rate will follow the inverse square law. If the front/back level ratio is 10 dB at 1 meter it will be the same at 2 m and 4 m. Spherical wave propagation is an equal-opportunity pressure dropper.

How do we know we've gone far enough to have a stable loss over distance and angle? One way is to keep moving outward until we see the inverse square law loss rate. This is a circular argument. We can go a level deeper by considering the factors at work. A tiny point source like a clicker reaches maturity at a very close distance. A giant Japanese taiko drum requires a long distance by comparison. Spread sources (e.g. a piano) or those with multiple acoustic outputs (e.g. bagpipes) also add complexity. A typical two-way speaker is both spread and multiple outputs. A 5-meter stack of speakers is a very spread source.

3.2.7 Transmission paths

The acoustic transmission path isn't always a straight line. Here are some possibilities (Fig. 3.8).

3.2.7.1 RAY-TRACING MODEL

Sound propagation paths are typically computed by the ray-tracing method. Sound propagates from the source in straight lines, like rays of sunlight. In free field, the sound continues outward and steadily loses level. The relative level values for each ray are adjusted to mimic the coverage pattern shape of the particular speaker.

When the ray strikes a surface it is reflected like light on a mirror (angle of incidence = angle of reflection). Rays and reflections from additional speakers intersect and pass through those of the original source. This is the approximation of sound transmission and summation properties used in most modeling programs.

3.2.7.2 DIFFUSION

The ray-tracing model holds up well for smooth surfaces large enough for the longest wavelengths to reflect like a mirror. By contrast, complex irregular surfaces scatter the sound in different directions over frequency, a process known as **diffusion**, a uniform field of non-uniformity. Surfaces with raised and lowered areas of various sizes and angles present a variable face to sound waves, also of variable size. The statues on the walls of Boston Symphony Hall are a famous example although the original modeling was anatomical rather than acoustic. Venues without a resident sculptor can use a variety of commercially available engineered diffusive surfaces designed to specific dimensional ratios.

3.2.7.3 DIFFRACTION

In some cases sound waves will bend around the surface, rather than bouncing off, a process termed **diffraction**. Tall walls line the sides of the freeway but somehow we still manage to hear the cars: Diffraction bends it over the wall. Thanks to the miracle of diffraction I hear the sound of fire trucks as if they are right outside my open 9th floor window! Sonic diffraction over walls and through openings creates a secondary virtual transmission source. I have to stick my head outside the window (the other side of the diffractor) to localize the fire truck on the ground.

Only diffraction's most basic properties fit within our scope: the ratio of the opening and/or barrier size relative to wavelength. Small barriers (relative to λ) are passed by with minimal effect. Large barriers (relative to λ) reflect

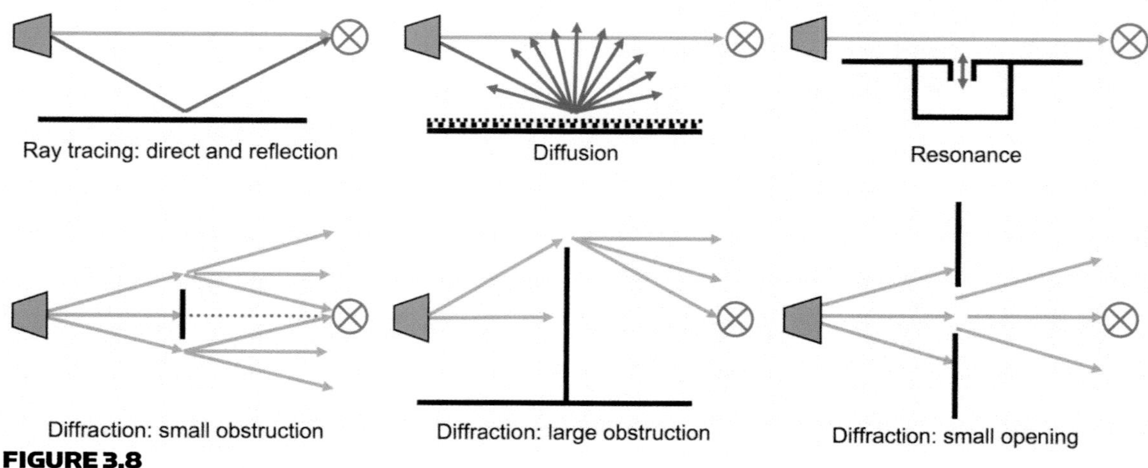

FIGURE 3.8
Acoustic transmission paths: ray tracing, diffusion, resonance and diffraction

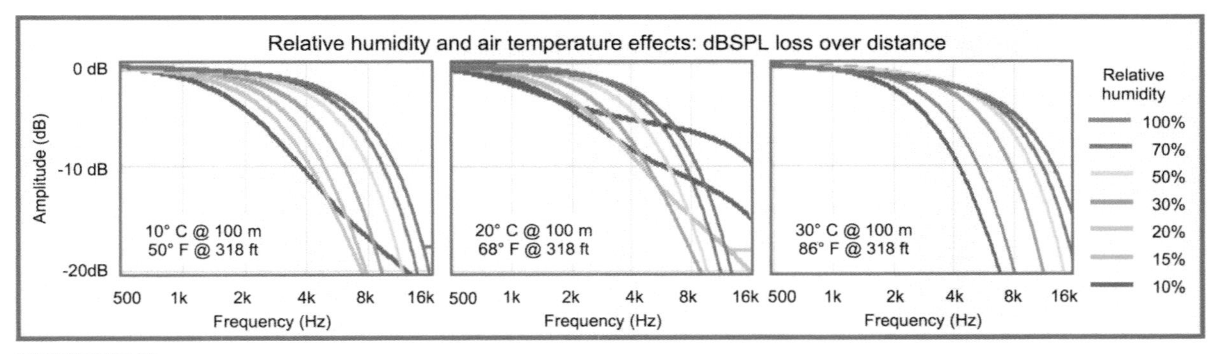

FIGURE 3.9
Humidity effects on acoustic transmission (courtesy of Meyer Sound)

a substantial portion and leave a shadow area of low-level sound behind the barrier (the fire truck in the bedroom). Any actual barrier has frequency-dependent diffractive properties due to fixed opening size and variable λ. Small pillars block only high frequencies while others bend around. A 30 m diameter pillar will block nearly everything.

Diffraction creates huge problems for audio installations. Diffraction is what makes people think it's OK to put an I-beam in front of our speaker. They can still hear it, so what's the problem? It's thanks to diffraction that management thinks we should mix the show through a window opening. Curse you, diffraction!

3.2.7.4 RESONANCE

Room dimensions can create a sympathetic spacing relationship with a particular set of wavelengths, resulting in **resonance**. Room and internal cavity resonances present a decay characteristic problem by prolonging certain frequency ranges. "Helmholtz resonators" are external room-coupled chambers that provide effective absorption of low and mid-range frequencies.

3.2.8 Dynamic effects

Perfectly still air is an imperfect transmission medium. How do changes in the weather affect sound transmission? The answer is blowing in the wind.

3.2.8.1 HUMIDITY EFFECTS

We have previously discussed the air medium's accelerated rate of HF loss but not the primary mechanism that modifies it: humidity (Fig. 3.9). The general trend is increasing HF loss as the air gets drier. This holds until we get some strange variations at extremely dry air at cold temperatures. We really don't need to factor in temperature unless we are doing an outdoor New Year's Eve concert in Oslo.

The frequency response effects of the air are similar to the action of a low-pass filter. The filter slope steepens as humidity falls and the corner frequency falls as transmission length increases.

There is a major takeaway here. Humidity effects on the HF response are easy to hear, and easy to act on. They affect only the very top end so a broad HF boost or cut can return things back to normal. This is one of the easiest parts of system optimization to do without an analyzer.

3.2.8.2 TEMPERATURE EFFECTS

The strongest temperature effect is on sound speed (Fig. 3.10). Hot air is fast air. The effect on the direct sound arrival from a single source is probably undetectable because we don't really know if the sound got to us sooner or later than "normal." Sound always is behind light so even a natural audio track is always out of sync to the visual. The question is simply how much later the sound arrives.

Temperature change moves into audibility when multiple path lengths are involved, because the summation is modified by the speed change. Reflections and delay speakers are prime examples. How much does the sound

Temperature effects on sound speed						
°C	C Sound speed	Δ	ΔC	°F	C Sound speed	Δ
Metric		Temp	Sound speed	English		Temp
Temp (C°)	Speed (m/sec)	change (C°)	change (%)	Temp (F°)	Speed (ft/sec)	change (F°)
37.8	354.3	15.6	2.7%	100	1162	28
35.6	353.0	13.3	2.3%	96	1158	24
33.3	351.6	11.1	1.9%	92	1153	20
31.1	350.3	8.9	1.5%	88	1149	16
28.9	348.9	6.7	1.2%	84	1144	12
26.7	347.6	4.4	0.8%	80	1140	8
24.4	346.2	2.2	0.4%	76	1136	4
22.2	**344.9**	**0.0**	**0.0%**	**72**	**1131**	**0**
20.0	343.5	-2.2	-0.4%	68	1127	-4
17.8	342.2	-4.4	-0.8%	64	1122	-8
15.6	340.8	-6.7	-1.2%	60	1118	-12
13.3	339.5	-8.9	-1.6%	56	1114	-16
11.1	338.1	-11.1	-2.0%	52	1109	-20
8.9	336.8	-13.3	-2.4%	48	1105	-24
6.7	335.4	-15.6	-2.8%	44	1100	-28
4.4	334.1	-17.8	-3.2%	40	1096	-32
2.2	332.7	-20.0	-3.6%	36	1092	-36
0.0	331.4	-22.2	-4.1%	32	1087	-40

FIGURE 3.10
Temperature effects on sound speed

speed change over temperature? Not much. If we did a concert that went from a sauna to ice cold, the sound speed would fall by 10%. It takes a rise of 5.6°C (10°F) to increase the sound speed by a mere 1% (a 100 ms path would now become a 99 ms path).

Temperature change adjusts sound speed by a given percentage over distance. Two paths of different lengths arrive with a time offset. Shifting temperature changes the arrival by the same percentage (a ratio of the lengths) but by different net amounts of time (the difference between the lengths). The time offset changes. That is the key concept. As temperature rises the transit time between mains and delays shrinks, and sound reaches the walls and back sooner than before. The reality is the wavelength of sound is rescaling with temperature. It's hard to visualize, because wavelengths of sound are . . . invisible. Instead we can do a reality polarity reversal and visualize temperature as rescaling our drawings of the sound system and the room, with wavelengths staying the same. In that parallel universe the room shrinks as it warms up. Either way the reality is that a hot room fits fewer wavelengths of a given frequency than a cool one.

The longest paths we hear indoors are low-frequency reflections. They accumulate the most change over temperature. This is where our ears and analyzers will find the most detectible temperature-related changes.

How much will temperature change modify the response? Not enough to fix a bad sound design or break a good one. One immediate concern is delay settings. A quick math exercise will put this in perspective.

An example main + delay system has respective paths of 100 ms and 20 ms. Let's add 80 ms to the delay speaker to synchronize the arrivals (100 ms to 20 ms). It's perfect for one seat and then time offset errors begin to accumulate as we move around. Sad but true. Later the temperature rises by 11.2°C (20°F). This is a BIG temperature change and accelerates the sound speed by 2%. The new paths are 98 ms and 19.6 ms respectively, which would yield an offset of 78.4 ms. The previous delay setting of 80 ms is now in error by 1.6 ms. How bad is that? The location of precise synchronicity will have moved by about one seat.

The temperature effects on reflections are similar, but multiplied tremendously in quantity. Both the direct and reflected sound paths accelerate as a room warms up. We can do a similar exercise as before. Let's use those same numbers as the direct sound and reflection paths (direct is 20 ms and the reflection is 100 ms, so the time offset is 80 ms). This creates a comb filter frequency of 12.5 Hz (peaks at 25 Hz, 37.5 Hz, 50 Hz and so on). If we go through the same 2% temperature shift, the time offset changes again to 78.4 ms. Now the comb filter frequency is 12.75 Hz (peaks at 25.5 Hz, 38.26 Hz, 51 Hz). Not exactly a game changer, eh?

The paths must be extremely long to be worth monitoring for delay speakers and many long reflection paths for us to hear the sound change solely because of temperature.

3.2.8.3 WIND EFFECTS

Sound moves through air. Wind moves air. Sound moving through moving air is a mess. There is no simplistic way to describe the wind's effects at any particular moment, but we can explore the principal mechanisms (Fig. 3.11). Sound travels at 1234 kilometers/hour (767 mph). Gale force winds begin in the 50–60 km/hour range, which could add or subtract 5% to the sound speed, if the wind direction matches the sound propagation direction. Wind and sound moving in the same direction is simple enough to visualize, but a crosswind is less so. All this blows out the window when we remember that sound transmits spherically through the air, which is impossible for wind. If wind moved spherically away from a source it would leave a vacuum. Wind moves along in a direction and air is pulled in behind it to fill the lower pressure. Because sound propagates spherically, we will have every relationship possible to the wind direction, all the time. Our sound waves are always moving with, against and across the wind.

The concert with the gale force winds was cancelled so let's consider a moderate breeze (20–28 kph). Now we have a ±2–3% change in sound speed, which is in the range we find with changing temperature. The effect is not uniform over the space, even if the breeze were rock-steady, because of the propagation issue. If we consider just the path that arrives at our ears then we can run through some possibilities.

Bending

A baseball pitcher throws straight pitches (fastballs) and curve balls. For the fastball he aims at the target and sends it. For the curve ball he aims somewhere else, knowing that the path will have a bend in it. A third pitch is the changeup, an unexpectedly slow pitch. A speaker in still air throws the sound in a straight line: fastballs. It starts throwing the tricky stuff when it gets windy. A crosswind causes our sound path to bend like a curveball. Our sound went to a neighbor and we got someone else's. This wouldn't matter a bit if the wind was constant, but it never is, so our location gets a variety pack of other people's sound. It sounds as if somebody is wiggling our speaker. There will be pitch shifting due to the Doppler effect (the changeup) and frequency response changes from hearing different parts of the speaker's polar response.

Comb filtering wind

A complex speaker array has a fine grain of detailed amplitude and phase interactions between the elements (a.k.a. combing). These are measurable in the best of times, and audibly degrading in the worst. Our brains do

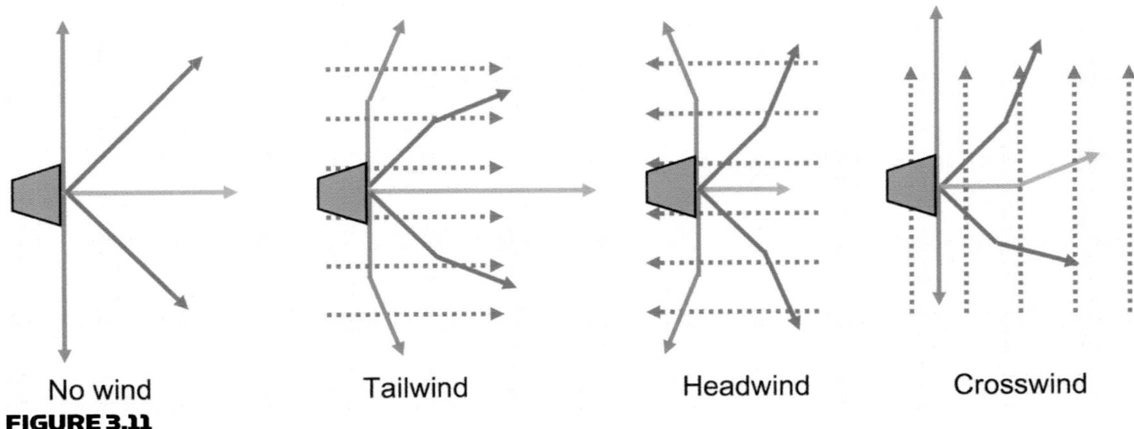

FIGURE 3.11
Wind effects on acoustic transmission

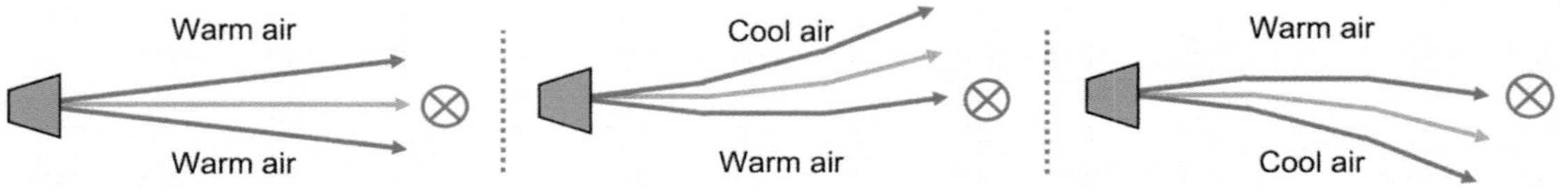

Warm air
Cool air
Warm air
Warm air
Warm air
Cool air

FIGURE 3.12
Refraction effects on acoustic transmission

a remarkable job of adapting to stable combing effects and just as remarkable a job at detecting changing combing effects. Ask anyone who has worked in musical theater. The effect is profound enough to have a name: flanging. Bending the paths from a single source will flange the reflection paths, but this is a minor issue compared with an array. Bending the paths from an array puts fine-grained flanging in motion at our location, and differently at everyone else's.

Indoor wind

We can encounter two types of wind in indoor spaces. The first comes from heating, ventilation and air conditioning (HVAC) blowers, the effects of which can be seen on our analyzers and heard with our ears. The result is indistinguishable from a steady breeze of outdoor wind (modulating phase response, moving impulse response and decreased coherence in the HF). The second is moving speakers, such as a swinging cluster shortly after it is raised into place. This is not really wind, of course, but it walks likes a wind and talks like a wind, so we have to wait for things to settle before we can make a good measurement. Outdoors the wind may actually be moving the speakers, in which case the solution is to quickly secure them (for safety, not phase data).

3.2.8.4 REFRACTION

Acoustic **refraction**, like its counterpart in light, is the bending of a sound transmission passing through layers of media (Fig. 3.12). This only substantially affects large-scale outdoor applications. Sound speed varies over the thermal layers in the atmosphere. The normal atmospheric pattern is cool air over warm air (i.e. slow air over fast air). This causes upward bending in proportion with the speed difference. Warm over cool, known as an inversion layer, bends the sound downward. Consider the distant train horn that is normally faint feeling like it is close by when there is an inversion. It's not that the engineer sometimes blows harder. Large-scale outdoor concerts can sometimes encounter mysterious disappearances of their sound, and surprise landings into unhappy neighborhoods.

3.2.8.5 COMBINED EFFECTS

It's very difficult to ascertain the independent action of these factors in the field. Temperature change is often combined with humidity variations. There is also little need to precisely do so. We can't do much about the wind variations, but the temperature and humidity effects are stable enough to enable corrective action. Because the response change occurs almost entirely in the VHF region, its effects are among the easiest to identify with an analyzer or by ear. The solution is the same regardless of its precise cause.

3.3 ACOUSTIC TO ANALOG ELECTRONIC TRANSMISSION

We interrupt this transmission by inserting a microphone into the path to capture the waveform and convert it to electrical form. We now focus on acoustic to electrical transduction.

3.3.1 Transduction: acoustic to electronic

Transduction is a medium conversion. The acoustic/electric conversion factor is **sensitivity**, given as the open circuit output voltage in mV/Pa. A pascal (Pa) is a unit of pressure, a measure of force over surface area (one newton/square meter). Audio world translation: 94 dB SPL. It's a more practical reference point than 0 dB SPL for the simple fact that we can easily create it and measure it. Standard mic calibrators (piston phones) generate a 1 kHz sine wave at 1.0 Pa or 10 Pa (94 and 114 dB SPL respectively).

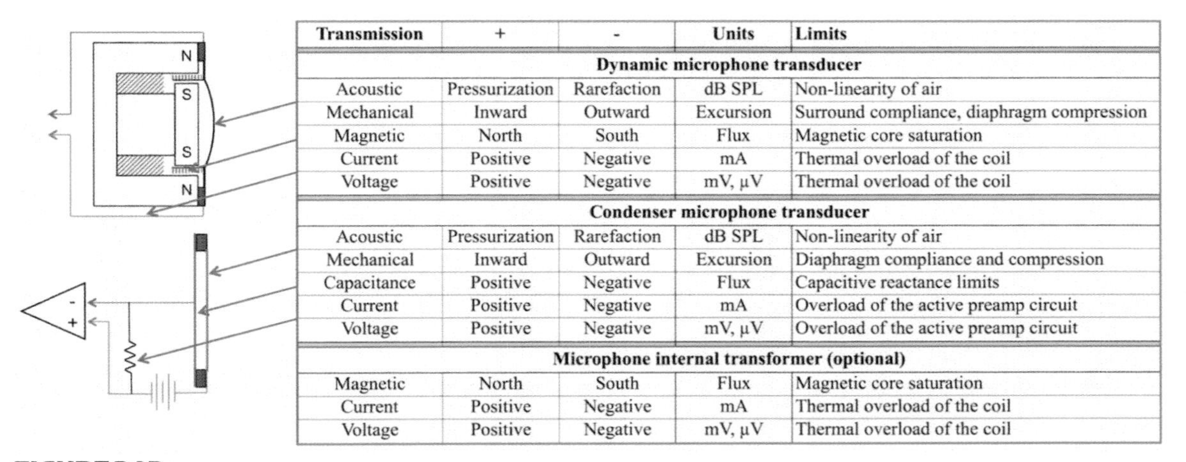

Transmission	+	-	Units	Limits
		Dynamic microphone transducer		
Acoustic	Pressurization	Rarefaction	dB SPL	Non-linearity of air
Mechanical	Inward	Outward	Excursion	Surround compliance, diaphragm compression
Magnetic	North	South	Flux	Magnetic core saturation
Current	Positive	Negative	mA	Thermal overload of the coil
Voltage	Positive	Negative	mV, μV	Thermal overload of the coil
		Condenser microphone transducer		
Acoustic	Pressurization	Rarefaction	dB SPL	Non-linearity of air
Mechanical	Inward	Outward	Excursion	Diaphragm compliance and compression
Capacitance	Positive	Negative	Flux	Capacitive reactance limits
Current	Positive	Negative	mA	Overload of the active preamp circuit
Voltage	Positive	Negative	mV, μV	Overload of the active preamp circuit
		Microphone internal transformer (optional)		
Magnetic	North	South	Flux	Magnetic core saturation
Current	Positive	Negative	mA	Thermal overload of the coil
Voltage	Positive	Negative	mV, μV	Thermal overload of the coil

FIGURE 3.13
Microphone transduction reference

Sensitivity sets the mic's acoustic/electrical operating range (Fig. 3.13). The maximum output voltage can't exceed the line-level limit (10 V_{RMS}) and most mics will flatten their diaphragms and/or saturate their electronic components long before that. A hypothetical mic (maximum level of 1 V, sensitivity of 1 V/Pa and 120 dB dynamic range) could operate from 94 dB SPL to -26 dB SPL. We have little use for such a mic but we can see the trend. Let's step sensitivity down in 20 dB steps to find a practical level: 100 mV/PA (114 dB SPL to -6 dB SPL), 10 mV (134 dB SPL to +14 dB SPL) and 1 mV/Pa (154 dB SPL to +34 dB SPL). Low sensitivities are well suited to close mic'ing, recording rocket launches and gun shots. Raising the sensitivity limits our ability to track extreme SPL sources but gives a better operating voltage for the more reasonable ones (less gain required, better noise immunity). System optimization measurement mics should handle peaks around 130–140 dB SPL and typically fall in the 2.5 mV to 30 mV/Pa range.

Some mics go through an additional transduction before leaving the chassis: an internal transformer. The transformer is a go-between for the electromagnetic capsule or internal preamp and mixer input stage. The transformer steps down the voltage (increasing current), which lowers the mic's source impedance into the preamp. The internal mic transformer's primary advantages are the capability to drive longer lines, lower common mode rejection ratio (CMRR) and capability to drive lower input impedances than a transformerless version. The disadvantages are reduced sensitivity, increased distortion, phase shift and frequency response variations. Industry trends favor reduced length in microphone cable runs to preamps and splitters near the stage. Therefore the primary advantage of transformered mics (long distance) has reduced practical application.

3.3.2 Low *Z* to high *Z*: the pressure converter

We previously discussed the pressure follower: low *Z* acoustic sources transmitting to high *Z* receivers. The same principles apply to microphones and ears. Filling a room with mics doesn't load down the PA.

Let's take a momentary detour to consider why microphones and their acoustic counterparts, speakers, have such a different perspective on power. Consider how we don't need a 2-way or 3-way microphone but this is absolutely required for speakers. The difference is that speakers generate power and microphones simply sense it. They are, quite literally, along for the ride. A water analogy: It requires very little energy for a swimmer to bob up and down as a big wave passes, but it took the engine of a large boat to create it.

Recall that acoustic power (in watts) is SPL over surface area. The surface area of a microphone diaphragm is so small that it's pointless to model a microphone as a power transfer device, although technically it is. We model it more simply as a pressure transfer. The mechanical movement created by the pressure wave is converted to electrical power, which we first see as voltage and current at some given electrical impedance. We don't need to consider the microphone output signal in power terms as long as our preamp input can receive it at a high enough impedance to accurately track the waveform voltage.

The voltage levels for passive dynamic mics can be as low as 1–100 microvolts. Active condenser mics may have an internal preamp that boosts the signal before we ever see it. The level variations we encounter here are so great that we must be ready for anything. This is why the mix console input stages are capable of voltage gains from -20 dB to +70 dB.

3.3.3 Dynamic range

Upper and lower limits are found on either side of the transduction. Diaphragm size sets the lower limits on the acoustic side: Larger ones have lower noise floor (and reduced VHF range extension). The acoustical upper limits are mechanical: excursion for the surround and flattening and/or warping for the diaphragm.

On the electromagnetic side we encounter the noise floor of active and passive electronic components. Electromagnetic interference (EMI) can be induced in the coil or magnet. Magnetic saturation sets an upper limit in the initial transduction or secondary internal transformer. An active internal preamp is also a potential analog electronic limit (especially a high-gain unit). A dying battery reduces the upper limits as the rail voltages collapse.

3.4 ANALOG ELECTRONIC TO ANALOG ELECTRONIC TRANSMISSION

We enter the world of analog-to-analog connections. Mics to preamps, preamps to line level and line level to line level. There are two principal forms: device to device (external) or stage to stage (internal).

3.4.1 Propagation

Audio signal flow requires an elastic transmission medium. Conductive metal is elastic to electronic charge and therefore a suitable medium. Voltage is the waveform's amplitude (magnitude) parameter. Common mediums for electronic transmission include metal wires, traces and connectors made of copper, steel, aluminum, tin, silver, gold, lead and more. The transmission speed through a shielded cable is approximately 2/3 the speed of light (about 200,000 km/sec), 600,000× faster than sound (0.33 km/sec). Should we worry about latency in analog electronic cables? It takes 133 km of cable to match the 1.5 ms latency of a typical A/D converter. Long analog lines are effectively extinct, long ago replaced by digital transmission. Transmission time between analog electronic devices can be considered negligible.

Analog electronic is not a lossless medium. Even a small length of wire has loss as well as some effect on the frequency response. The losses are proportional to cable length and can be minimized with good wiring practices. Transmission paths are also vulnerable to noise intrusion, which can be induced both electrically and acoustically into the signal.

3.4.2 Low *Z* to high *Z*: the voltage follower

We revisit the concept of low *Z* source and high *Z* receiver in the electronic medium (Fig. 3.14). The central circuit topology for analog-to-analog transmission is the voltage follower. The source has enough reserve current capability (this is where the power component comes in) to deliver the waveform voltage to the receiver. An impedance mismatch of 1:10 (the receiver impedance is 10× the source) or better is generally practiced.

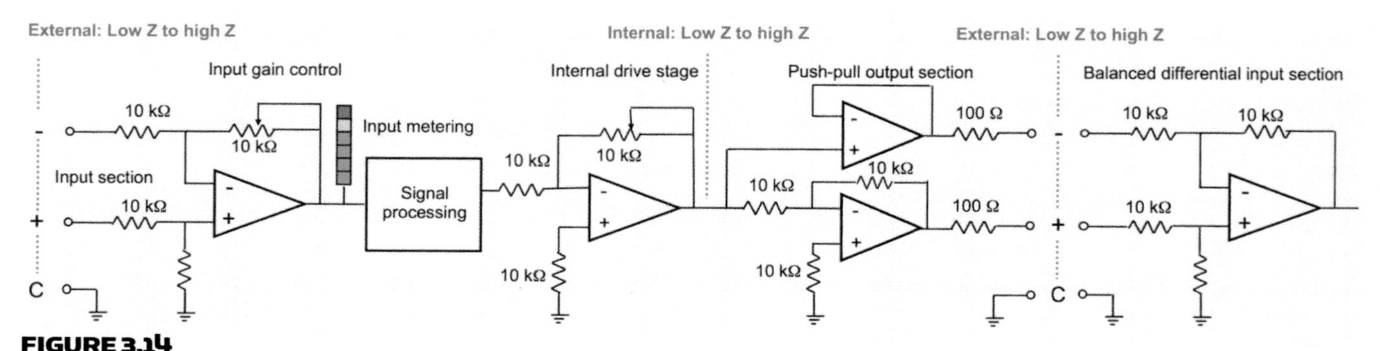

FIGURE 3.14
The voltage follower: low *Z* to high *Z*

3.4.2.1 MICROPHONES TO MIC PREAMPS

The expected source impedance of microphones is 150 Ω to 250 Ω. They drive inputs that nominally fall in the 3 kΩ range, exceeding the 1:10 impedance ratio. Mic preamp specifications sometimes confuse things by stating they have "250 Ω" inputs (which actually refers to their ability to accept mics with a 250 Ω source impedance). If the input impedance were actually 250 Ω there would be an insertion loss of 6 dB.

3.4.2.2 LINE OUTPUT TO LINE INPUT (OP-AMPS)

A typical configuration is 50 Ω to 150 Ω outputs driving 5k Ω to 20 kΩ inputs. The typical driving devices are operational amplifiers (op-amps) in a balanced (push–pull) configuration. Op-amps are the building blocks of modern analog electronic circuit design, generic voltage amplifiers configured to perform functions such as gain adjustment, filtering, delay and more. An op-amp's nearly infinite voltage gain is kept under control by returning signal back to the input in a negative feedback loop. The gain is set by the ratios of resistance within the feedback loop and driving the input. Op-amps have low-impedance outputs and range from high to *extremely* high-impedance inputs. A short-proofing series resistor on the output limits the current and sets the source impedance. Series resistors on the receiving op-amp set the input impedance. An input specification of "10 kΩ, 5 kΩ balanced" means each leg of the balanced input is 5 kΩ.

3.4.2.3 INTERNAL STAGE-TO-STAGE

Analog signals move similarly between internal stages, but under controlled conditions that don't require short-proofing resistors etc. Internal stage–stage has it easy compared with the external version, which must be ready for anything the outside world might throw at it.

3.4.3 Dynamic range

3.4.3.1 UPPER LIMITS

The DC power supply sets the analog audio pipeline's upper limits (section 3.1). Many mic-level, and most line-level devices, have the same rail voltages (±12–18 V_{DC}) and therefore top out at similar voltage. This is a rock-solid limit, resulting in hard clipping at overload (Fig. 3.15).

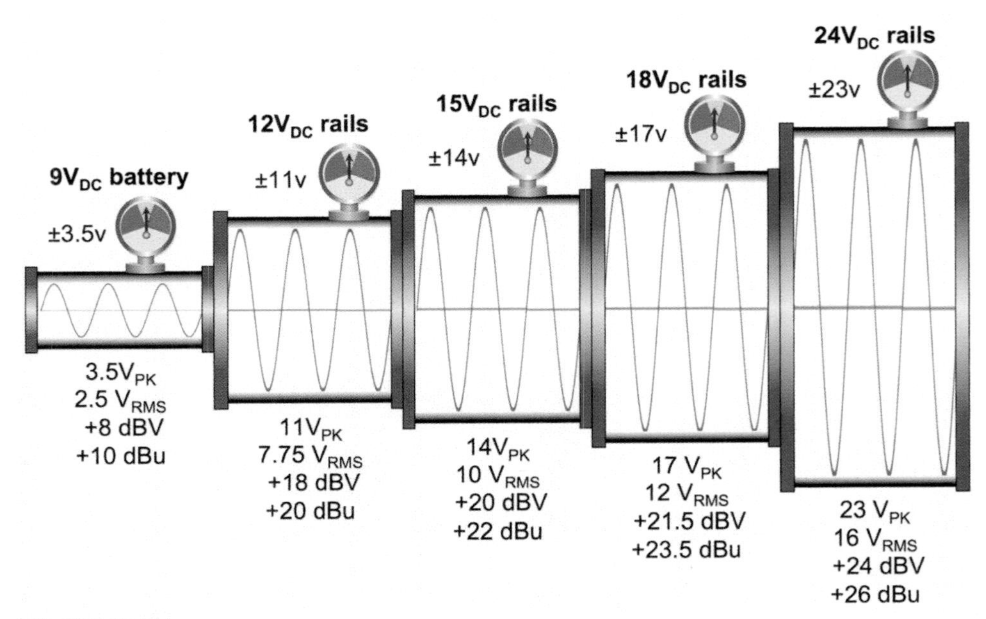

FIGURE 3.15
DC rail voltage and maximum output capability for the analog electronic pipeline

3.4.3.2 LOWER LIMITS

Noise sources include the random noise emitted by any electronic circuit, induced noise from other circuits (e.g. crosstalk, RFI and EMI) and ground loops. The principal frequency response challenges are capacitive and inductive reactance, active and passive filters within the circuit's passband or even outside it.

3.4.3.3 STAGES

- Summing stage: The power supply rail voltage range for line-level devices is rarely any different for summing stages than inputs and outputs. Therefore the same upper limits apply. Summing stages may need gain reduction to prevent overload of the combined signal.
- Active gain stage: Gain (or loss) is usually managed by fixed (or variable) control of an op-amp's feedback loop. Mic preamps can have extremely high gain, while most others don't exceed +10 dB.
- Passive attenuation stage: A passive attenuation stage shunts the signal to ground through a variable resistor (potentiometer). The output of this stage can run from unity (wiper at top of resistor) to full attenuation (wiper at ground).

3.4.4 Loss rate

The loss rate for mic- and line-level transmissions can be negligibly small. In modern practice it would be unusual to lose even 1 dB in a line-level transmission. Three impedance factors are involved in the equation: output (source), input and cable. These three together are the *total* impedance through which the signal flows. The loss is $20 * \text{Log}_{10}$ (total impedance/input impedance).

$$\text{Line loss (dB)} = 20 * \text{Log}_{10} \frac{\text{output} + \text{cable} + \text{input}}{\text{input}}$$

$$\text{Line gain (dB)} = 20 * \text{Log}_{10} \frac{\text{input}}{\text{output} + \text{cable} + \text{input}}$$

The loss formula is the standard form but it becomes clearer if we turn it upside down and calculate the circuit gain (which at most is 0 dB, a ratio of 1:1). The decisive ratio is input impedance over the total (output + cable + input). When output and cable are both zeros we get the lossless ratio: 1:1. This helps to see that anything added by the output or the cable brings us down.

Figure 3.16 is a set of line-loss scenarios, ranging from theoretical, good practice (low Z to high Z), old school (600 Ω) and bad practice (high Z to low Z). The key to minimizing loss is making the input impedance much

Insertion loss examples						
	Sending (Ω)			Receiving (Ω)	Result	
Description	Output	Cable	Input	Input	Gain ratio	Loss (dB)
Lossless line (theoretical)	0.1	0	1000	1000	1.00	0.00
Matched output, cable and input impedance	1000	1000	1000	1000	0.33	-9.54
Low Z output to High Z input (no cable loss)	100	0	10000	10000	0.99	-0.09
Low Z output to Low Z input (no cable loss)	100	0	100	100	0.50	-6.02
High Z output to Low Z input (no cable loss)	10000	0	100	100	0.01	-40.09
Passive 600Ω line, 10m cable	600	0.5	600	600	0.50	-6.02
Microphone to mixer, 10m cable	250	0.5	3000	3000	0.92	-0.70
Active line output to input, 10m cable	150	0.5	10000	10000	0.99	-0.13
Active line output to input, 100m cable	150	5	10000	10000	0.98	-0.13
Active line output to input, 1000m cable	150	50	10000	10000	0.98	-0.17
Active line output to 2 inputs (10kΩ), 10m cable	150	0.5	5000	5000	0.97	-0.26
Active line output to 10 inputs (10kΩ), 10m cable	150	0.5	1000	1000	0.87	-1.22

FIGURE 3.16
Cable and insertion loss reference for balanced analog lines (mic and line level)

higher than the combination of output and cable impedance. Why not raise the input impedance to 1 megΩ and further minimize loss? The biggest reason is that such a minuscule amount of current flow reduces noise immunity and CMRR. The industry is trending in favor of low-output impedance, which minimizes losses even when driving multiple inputs. Each additional input reduces the effective impedance and increases the line loss.

Interconnection loss is easily restored, requiring only later-stage voltage gain. The main side effect is added noise, not decreased maximum capability. Interconnection loss here is far less critical than in speaker cables, where restoration requires power (not just gain) and results in tangible loss of SPL capability.

3.4.5 Analog cable and connectors

3.4.5.1 STANDARD CABLE CONFIGURATIONS

Analog audio cables are typically 2-wire or 3-wire with a few 4-wire and 5-wire exceptions.

- 2-wire: signal (+) and shielded common.
- 3-wire: signal (+), signal (−) and shielded common.
- 4-wire: signal (+), signal (−), common and shielded ground.
- 5-wire: 2× signal (+), 2× signal (−) and shielded common (e.g. StarQuad™).

3.4.5.2 CABLE IMPEDANCE

Cable impedance is combined DC resistance and AC reactance (capacitance and inductance). DC resistance acts as a full-range attenuator while capacitance in series blocks lows and favors highs. Inductance does the opposite. In parallel they both reverse again: Series inductance and parallel capacitance are both HF filters. Inductance is a minor consideration in line-level cable runs.

Capacitance between conductor and shield creates a shunt path (parallel) for high frequencies, effectively lowering the input impedance. Line loss increases with frequency. Cable manufacturers provide capacitance specifications (typically in picofarads/meter). Professional audio cables should be less than 250 pf/m. A 100 m line driven by a low-impedance source (<250 Ω) has <1 dB of HF loss. We have an updated name for cable runs long enough for us to care about capacitance: analog backup.

3.4.5.3 SHIELDING

An audio line's common wire is spread to surround the conductors (rather than running side by side with them). EMI must pass through this outer wire, termed the "shield," before reaching the internal conductors. The EMI is

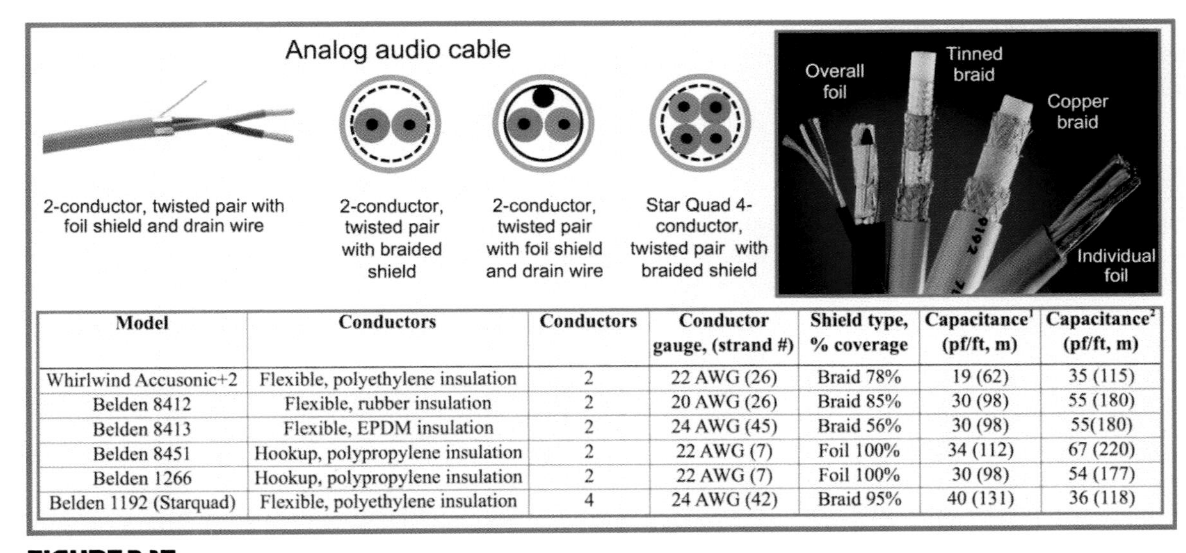

Model	Conductors	Conductors	Conductor gauge, (strand #)	Shield type, % coverage	Capacitance[1] (pf/ft, m)	Capacitance[2] (pf/ft, m)
Whirlwind Accusonic+2	Flexible, polyethylene insulation	2	22 AWG (26)	Braid 78%	19 (62)	35 (115)
Belden 8412	Flexible, rubber insulation	2	20 AWG (26)	Braid 85%	30 (98)	55 (180)
Belden 8413	Flexible, EPDM insulation	2	24 AWG (45)	Braid 56%	30 (98)	55(180)
Belden 8451	Hookup, polypropylene insulation	2	22 AWG (7)	Foil 100%	34 (112)	67 (220)
Belden 1266	Hookup, polypropylene insulation	2	22 AWG (7)	Foil 100%	30 (98)	54 (177)
Belden 1192 (Starquad)	Flexible, polyethylene insulation	4	24 AWG (42)	Braid 95%	40 (131)	36 (118)

FIGURE 3.17
Analog audio cable reference (Belden.com photo)

shunted through the shield to ground, which keeps it from reaching the conductors. Typical forms are fine wire braid or metalized foil (with an additional wire for termination). The braided shield has superior EMI rejection but has generally higher cost.

3.4.5.4 TWISTED PAIR

A twist in the plot of balanced audio cable construction is that the conductors don't simply run in parallel inside the shield. They are twisted around each other, which reduces capacitance and improves EMI rejection. EMI suppression is most effective when the wire is centered within the shield. It's impossible with a pair, but is closely simulated in the twisted quad version. This uses four conductors (two twisted pairs carrying the same signal) that give each signal line a more centered location in the shield (Fig. 3.17).

3.4.5.5 LINE-LEVEL CONNECTOR

- **XLR**: This is the standard 3-pin (and case) connector. Pin 1: common, pin 2: signal +, pin 3: signal. [If you haven't already met the XLR connector, read the Yamaha *Sound Reinforcement Handbook* (Davis and Jones) and come back here later.]
- **RTS phone jack**: Ring-tip-sleeve configured plug is extremely rare now due to its unrivaled unreliability. This is a connector that fails because you moved it, or because you didn't. The most likely place to find this would be in an analog patch-bay. Sleeve: common, tip: signal +, ring: signal -.
- **Phoenix**: a hybrid connector/terminal block with a variable quantity of screw terminals. Popular on modern multichannel audio devices due to its small footprint compared with individual XLRs. Sadly, there is no standard wiring configuration. As with any screw terminal device, the exposed cables are vulnerable to loosening. Best for applications requiring minimal insertions such as permanent installs or inside racks.

3.4.6 Mic- and line-level interconnection schemes

3.4.6.1 OUTPUTS

- **Active balanced (push–pull)**: 3-wire configuration with common (C) and two signals (+ and –). Negative output is polarity inverted to drive the following differential input. The output impedance is kept under 250 Ω to minimize cable loss and allow it to drive multiple devices.
- **Transformer balanced**: 3-wire configuration, wired the same as the active balanced, with advantages in isolation and disadvantages in linearity, noise, distortion, etc.
- **Active unbalanced**: 2-wire configuration with common (C) and one signal (+). The standard configuration for speaker level and sometimes used for very short line-level runs (e.g. mix console insert patches to external gear) and non-professional applications.

3.4.6.2 INPUTS

- **Active balanced (differential)**: 3-wire configuration with a common (C) and two signals (+ and –), a "differential amplifier" whose output is the summation of the two inputs. The audio signal is polarity reversed twice (in the source signal and at the input stage) resulting in phase-matched summation (+6 dB). Noise induced in the lines is polarity reversed only once (at the input stage) resulting in phase cancellation, a process termed "common mode rejection" (Fig. 3.18A).
- **Transformer balanced**: A balanced transformer can be substituted for the active input with advantages in isolation between systems. Transformers are prone to EMI induction, distortion and more but these may be lesser evils when isolation problems arise (Fig. 3.18B).
- **Active unbalanced**: The unbalanced input is a 2-wire configuration with a common (C) and one signal (+). This input configuration yields the difference between the input waveform and common. This semi-professional configuration lacks the common mode rejection noise immunity of the balanced line and should be used only for very short runs such as mix console insert patches to external gear (Fig. 3.18C).

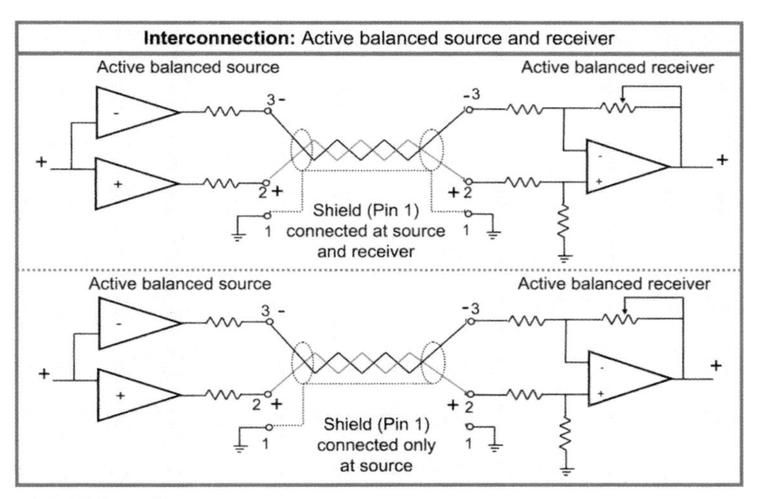

FIGURE 3.18A
Analog connection schemes: active balanced (after Giddings, 1990, pp. 219–220)

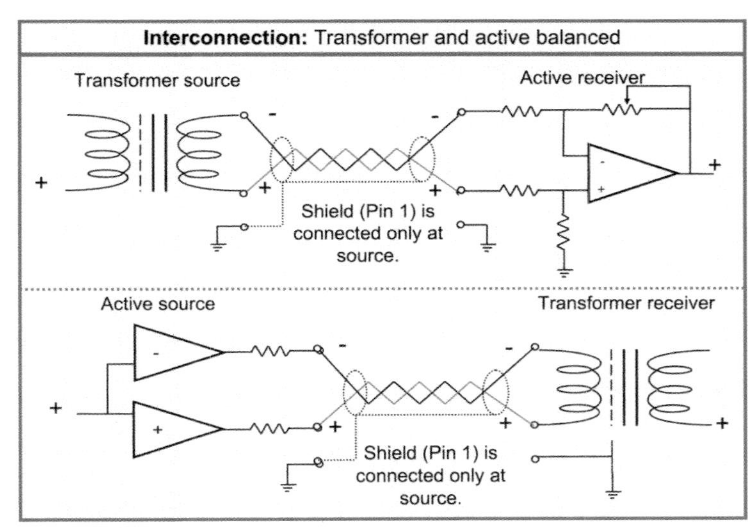

FIGURE 3.18B
Analog connection schemes: active transformer balanced (after Giddings, 1990, pp. 221–223)

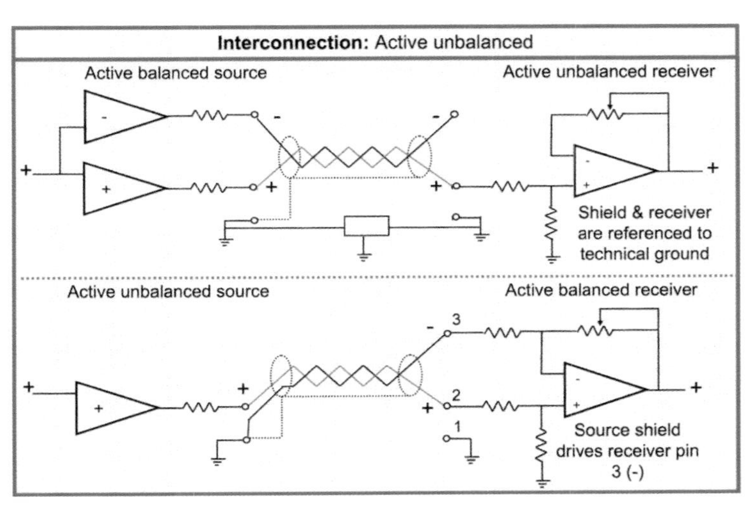

FIGURE 3.18C
Analog connection schemes: active unbalanced (after Giddings, 1990, pp. 226–229)

3.5 ANALOG TO DIGITAL TRANSMISSION

3.5.1 Analog to digital (A/D) conversion

It's time to convert the analog electronic signal to the digital domain. A/D conversion is an encoding process that requires strict standardization or we won't be able to convert it back when we are done. It's like loading a truck: We need a well-organized pack to make sure everything gets there safely and comes out in the right order when we arrive.

The A/D conversion is the first step in the packaging. We convert the continuous analog waveform into a series of still pictures. It's like making a movie, but our frame rate is thousands of times faster.

Linear pulse code modulation (LPCM) is the most common output of A/D converters in professional audio. It is uncompressed audio and its variations include well-known standards such as wav (Compact Disk), Blu-Ray, DVD Audio and our own AES3. Pulse code modulation is a three-step conversion process: sampling, quantization and encoding.

- Sampling: Capture the waveform.
- Quantization: Determine the amplitude value.
- Encoding: Package it up in binary form and ship it.

The sample is an audio snap shot (amplitude vs. time). When the clock says "go" we take the picture. Each picture is a close-up of less than a half-cycle at any frequency. The quantization process categorizes the amplitude value and sends it to the encoder, which strings the amplitude values together.

3.5.2 Sample rate

There are four sample rates in common use. The AES3 standard can read all of them:

- 44.1 kHz: Compact Disk Audio standard (16 bit).
- 48 kHz: older pro audio, 16-bit to 20-bit.
- 96 kHz: common in current pro audio, usually 20-bit or 24-bit.
- 192 kHz: supported by DVD Audio and a few others.

3.5.3 Word-clock

The word-clock sets the start time of each sample. Word-clock frequency and "sample rate" are interchangeable terms. It's a pulse, not a sine wave, so the transmission line needs an extended bandwidth to maintain sharp transitions. The "word" referenced here is the "audio block," the complete data set collected each time the clock strikes. The ideal word-clock has perfect interval uniformity. Interval variations cause the sample to be taken early or late, introducing amplitude value errors. A similar error would occur if we planned to accurately sample the noontime temperature but instead drifted between 9:00 a.m. and 3:00 p.m. The data are compromised, as they would be if we sampled the right time but with a dodgy thermometer. The standard term for word-clock interval instability is jitter.

3.5.4 Time slot (bit clock)

Time slots order the 64 individual bit operations inside a single word-clock interval. The internal bit clock ticking 64× faster than the word-clock manages this (64 time slots @48 kHz = 3072 kb/sec = 6.144 MHz bit clock). Structurally the bit clock is characterized by its transition frequency and coding protocol. AES3 supports bi-phase mark coding (BMC) and 1-bit coding or can use an embedded SMPTE clock.

3.5.5 Latency

The captured analog data is held in a time buffer to process the digital conversion. The device latency results from the time spent in this holding cell and some extra time required to perform other vital operations. The conversion latency varies from model to model but is typically a fixed quantity for a given device. There are digital processing devices with variable latency out there but the variable part most likely results from user-settable processor settings rather than the A/D converters.

3.5.6 Bit-depth (fixed point)

There are three bit-depths in common use (Fig. 3.19). The AES3 standard can read all of them:

- 16-bit: 96 dB dynamic range, Compact Disk Audio standard.
- 20-bit: 120 dB dynamic range.
- 24-bit: 144 dB dynamic range.

3.5.7 dB full scale (dBFS)

Whenever we change mediums we need a conversion factor, the sensitivity. In the move from acoustic to electronic we found the conversion in mv/Pa. Here we get mv/111111111111111111111111. Looks silly but we need to determine how much voltage will top out the digital counter. It's another pipe connection: analog to digital. If the sensitivity is set too low we'll never use the converter's upper bits, the easy way to get 16-bit performance at a 20-bit price. If we go too high we top out the digital side with unusable headroom in the analog source. How do we determine the right number?

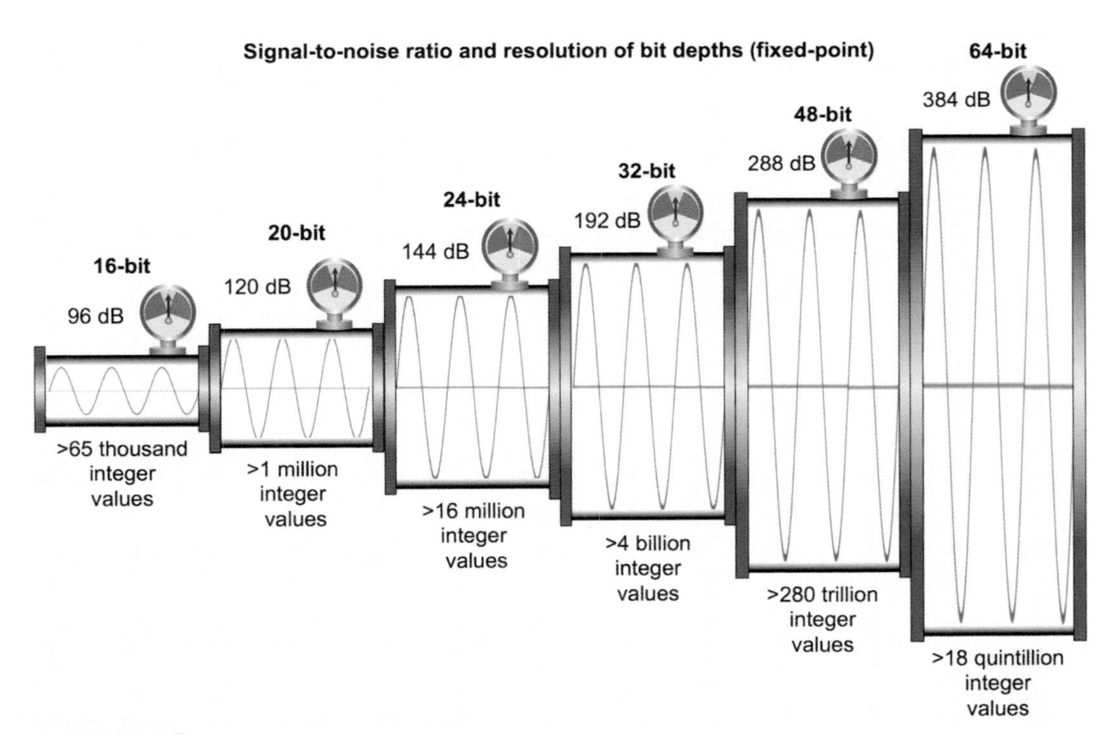

FIGURE 3.19
Bit-depth and dynamic range

Find the analog device's maximum output capability. Freshen up on dBu and dBV, peak and RMS, and prepare to say goodbye to analog (for a while). Determine the maximum peak voltage, because converters ONLY read peak.

We can read specs or simply put in a sine wave and find out for ourselves. Use an analyzer to find the maximum output level (clip point). Don't use VU meters because they read RMS in analog world and peak in digital world. (See verification test in section 13.3 for maximum I/O capability.)

- Look at the ratings on the A/D converter. What are the units? Peak or RMS? (dBu is *only* RMS.) How much flexibility do we have? 10 dB steps, 3 dB?
- Match the dB/FS number to the analog device's maximum capability, making the analog and digital clip points the same. The pipeline external diameters are now matched.
- Anyone concerned about analog device headroom can calibrate dB/FS to a lower voltage. We lose one bit of digital resolution for every 6 dB of added analog headroom.

Let's run through an example. A typical line-level device has a maximum capability of +23 dBV_{PK} (+20 dBV_{RMS}, +22 dBu). If dB/FS was set to "nominal" line level (0 dBV_{RMS}) then the top three bits will never be used. All bits are used if -20 dBFS is set to 0 dBV (or 0 dBFS = +20 dBV).

3.5.8 Quantization

Let's play an age-old game called "20 Questions." Think of a number between 0 and 1,048,575. I can win every time as long as I follow this methodical sequence:

- Q 1: Is it more than 524,288? No.
- Q 2: Is it more than 262,144? Yes.
- Q 3: Is it more than 393,216? Yes.

In just three questions I've narrowed the range down eight-fold to 131,072 possible values (>393,216 and <524,288). Each question splits the remainder in half. By the time I ask the 20th question only two possibilities remain. The three answers so far have been no, yes and yes. Translation: 0–1–1. This is analog to digital conversion in a nutshell: 20-bit audio is 20 questions. The game restarts every time the word-clock ticks to the next sample.

The logic: Start with the entire amplitude range. Positive voltage is at the top, zero at center and negative at the bottom. The present sample value is *somewhere* in there.

- Q 1: Is it in the upper half (1) or lower (0)? This is the sign bit. Positive voltage is up and negative is down. Take the winning half and move to Q 2.
- Q 2: Is it in the upper half (1) or lower (0)? This is the highest level bit. If "1" then the signal is within ½ of the full level (the uppermost 6 dB of dynamic range). Take the winning half and move to Q 3.
- Q 3: Keep asking until the bits run out (the limit of our approximation).

Audio is not as simple as the loudest signal is all 1s (the maximum binary number) and silence is all 0s. We have positive and negative analog voltage values so we use the first bit to determine the sign (+ is 1 and - is 0). So the maximum 20-bit signal could be twenty 1s (+) or a 0 followed by nineteen 1s (–). Silence is either sign followed by nineteen 0s (Fig. 3.20).

STEP-BY-STEP EXAMPLE

96 kHz, 24-bit, 0 dBFS = +20 dBV_{RMS} ($10V_{RMS}$, $14.14V_{PK}$)

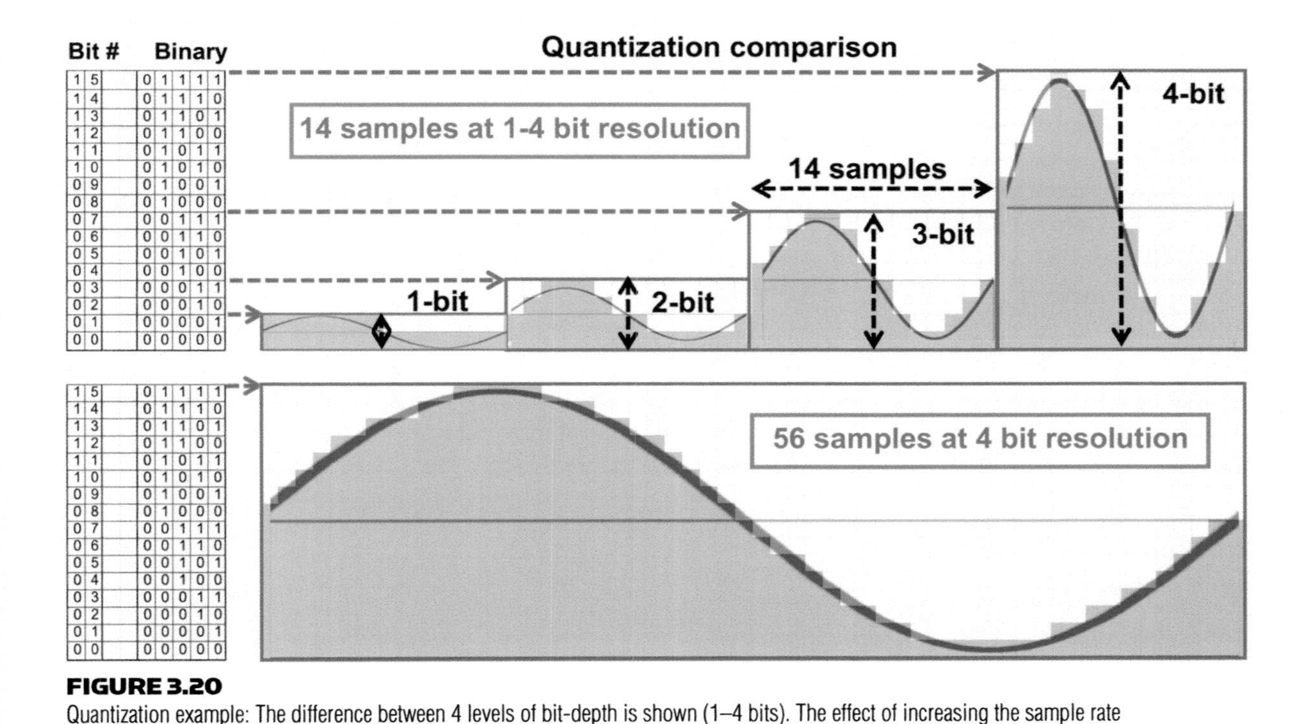

FIGURE 3.20
Quantization example: The difference between 4 levels of bit-depth is shown (1–4 bits). The effect of increasing the sample rate 4× is also shown (56 samples compared with 14).

Disclaimer: This simplified illustration is not necessarily bit-by-bit correct for any commercial converter.

1. Clock strikes 1: Open time window for .01 ms and capture the sample [voltage = +4.5V_{PK}].
2. Sign bit: Is voltage positive? Yes. [Bit_0 = 1.]
3. Most significant bit: Is voltage >0.5 × (½) scale? (7.07 V_{PK}.) No. [Bit_1 = 0.]
4. Next significant bit: Is voltage >0.25 × (¼) scale? (3.53 V_{PK}.) Yes. [Bit_2 = 1.]
5. Next significant bit: Is voltage >0.375 × (⅜) scale? (5.3 V_{PK}.) No. [Bit_3 = 0.]
6. Next significant bit: Is analog value >0.3125 × (5⁄16) scale? (4.42V_{PK}.) Yes. [Bit_4 = 1.]
7. Continue the process of bracketing in the value until the last bit. At the least significant bit approximation errors are the greatest (as a percentage of signal to noise). The conversion process may dither (add white noise) to minimize the objectionable sound of digital indecision: "quantization noise."
8. Send the 20-bit numerical value to the encoder and repeat.

3.5.9 Encoding

Encoding is a packaging process. We wrap the numbers up with a set of instructions for whoever finds them. The standard professional audio encoder is AES3 (and multichannel variations), which has a sufficient mix of flexibility and rigidity to allow it to cross platforms and even play with consumers. Both the sample rate and bit-depth are variable; we just label the data package with the ones we used.

3.5.10 Error and approximations

There are several avenues for error, ranging from show-stopping to just degrading.

- **Dropout**: no explanation needed (except to management). If the clock can't be read, the show stops.
- **Jitter**: time instability in the clock. Pulses arrives early or late and cause ambiguity or delay the sample processing. We can fall in and out of sync with other devices.

- **Reflections**: If the cable is not properly impedance balanced the clock pulse can be reflected back up the line. This has the potential to make a false triggering.
- **Aliasing**: a decoding error that results from ambiguous identification. This is usually caused by frequency content above the Nyquist frequency.
- **Interpolation**: Quantization is finite approximation of an infinitely divisible analog signal. Rounding errors are inherent to the process. We should also face the fact that there are rounding errors in our analog systems as well. A phonograph needle plowing through (and slightly melting) a groove of pliant vinyl is rounding things off quite literally with each usage. There will be additional rounding errors as we move the signals along and then a final set during the reconstruction at the D/A converter.

3.6 DIGITAL TO DIGITAL TRANSMISSION

Our signal will remain digital until its analog reconstruction. There are places to go and many ways to get there. This section describes internal digital audio transmission, and external transmission between devices. In this form we still recognize the signal as encoded audio and perform functions such as mixing, filtering and delay.

3.6.1 Digital signal processing (DSP)

3.6.1.1 INTERNAL TRANSMISSION

We saw previously that internal analog transmission requires less standardization and protection than outside. The same is true for digital. Once the data stream is safely inside the walls we can do what we want up and until it's time to leave. Details vary from model to model so we'll only touch on some common aspects. Audio packets are first unwrapped, then broken down into individual frames. A 96 kHz digital audio packet (192 frames) is 2 ms of data, which would not work well as our minimum audio increment! A frame-by-frame method gives very usable delay increments of 0.01 ms. Transmission moves through a pipeline of computer memory locations. Our speed limit is the internal clocking rate and the outer limits are available memory locations.

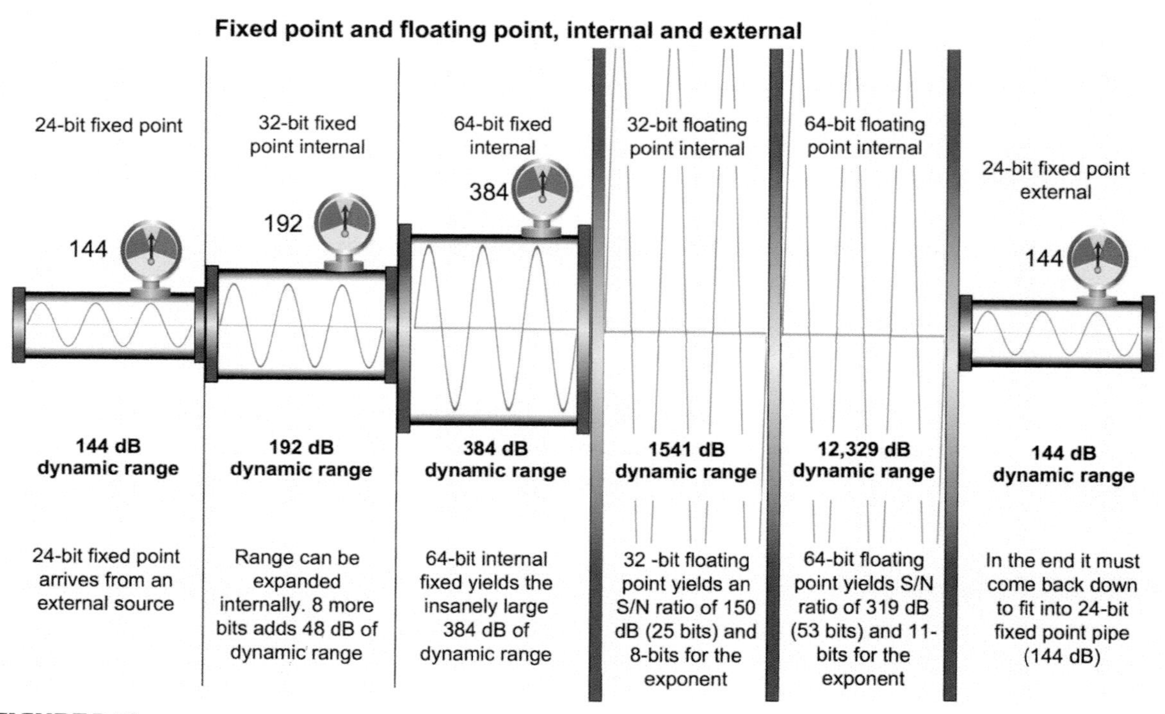

FIGURE 3.21
Fixed point and floating point (internal and external)

3.6.1.2 INTERNAL FIXED AND FLOATING POINT

It's common to see specifications for 32-bit and 64-bit audio processors. This might seem strange for devices with 24-bit A/D converters. The oversized bit-depths are available for internal use only, one of the bonuses of existing exclusively as a pile of numbers. It's a temporarily expandable internal pipeline that runs until we meet the fixed-point bit-depth limits of the D/A converter (Fig. 3.21).

Recall the fixed size of the line-level analog pipeline. The overload risk rises as more signals are summed into the fixed-size pipe. There were real analog mastering mix consoles that addressed this with 120 V_{DC} rails for the internal mix bus: an expandable pipe! Expanded internal bit-depth provides the same function in our DSP without the shock hazard. We must eventually fit the level into our fixed-point D/A converters, but inside the processor we have more freedom in our stage-by-stage dynamics than we ever could with analog. We can safely "overload" summing junctions and trim things back to size later without compromising the signal. By contrast, hitting the side of the internal analog pipeline permanently dents the waveform. Turning it down later won't help.

A 32-bit fixed-point internal bit-depth provides an 8-bit (48 dB) dynamic expansion above the A/D converter's 24-bit maximum. An analog console would need 3000 V_{DC} rails to achieve this. Internal 64-bit adds another 192 dB, so we can really party hard inside the device as long as we don't try to take it outside. Internal fixed point adds bit-depth at the top without adding rounding errors at the bottom (at least until the day of reckoning comes at the D/A converter). Fixed-point expansion requires additional memory resources with associated costs and transmission speed challenges.

Floating-point math renders the amplitude value as an integer of fixed length (25-bit or 53-bit) and an exponent and direction (+/-) vector (8-bit or 11-bit respectively for 32-bit and 64-bit respectively). It is the variable exponent that gives rise to the term "floating" here. The multiplication capability implicit in the integer + exponent format allows numbers to exceed the simple binary bit limits of the fixed-point topology. Therefore a simple list of minimum and maximum values does not apply. Floating-point systems can carry "out of scale" numbers inside their processing engines, needing only to move "in scale" at the point of D/A conversion or transmission to a fixed-point audio device. Floating-point rounding errors are constant over level, because the integer string is the same length regardless of the exponent multiplier. Therefore the rounding errors are the same for loudest signals and lowest level signals.

3.6.1.3 MIX POINTS

Mix points are the digital equivalent of summing buses. Signals are added numerically and follow the same summation progressions as analog (addition/subtraction based on relative amplitude and phase). A new opportunity for rounding errors arises when we mix signals together and create a new one.

3.6.1.4 CONTROL POINTS

Control points manage instructions for operations such as for signal routing, gain (multiplication), delay (sidetracking into a memory bank), filters (multiplication, gain and delay routing) and more. Related operations can be conducted on separate signal streams, like an analog-world VCA (now stands for voltage-controlled arithmetic).

3.6.2 The AES/EBU pipeline

3.6.2.1 OUTPUTS TO INPUTS

We can transmit digital audio between devices without conversion back to analog. We need to follow the instructions and pack it up for transport. AES3 is the standard protocol used for various formats and mediums. The core packaging method is universal within the industry and then applied in various highly compatible forms. Two-channel and multi-channel versions are found in wire and optical formats.

Digital audio transmission is point-to-point communication, with no allowance for splitting output signals to multiple inputs (unlike analog transmission). This is essentially RF communication so some connectors and cabling are the same as used in RF and video.

3.6.2.2 CABLE IMPEDANCE

The standard electrical interconnection between AES output and input is a 75 Ω or 110 Ω cable. These cable impedances sound very high for those of us in the analog frame of mind. Especially for a 1 m cable! How can it be that same impedance for a 10 m cable? The impedance specified for digital transmission is a completely different animal from the length-proportional series resistance, inductance and parallel capacitance that we see adding up in our analog signal cables. The 75 Ω specification is the "characteristic impedance," a function of the cable's cross-sectional anatomy, not its length. The difference between 75 Ω- and 110 Ω-rated cable is the conductor diameter and the spacing and dielectric (insulating) material between it and the shield, not the length. Characteristic impedance is strongest with a single conductor in the center of a circular shield: a coaxial cable. Analog cables also have characteristic impedances (22 AWG is around 70 Ω), but we're not driving 20 kHz bandwidth for a mile or transmitting 6 MHz. That's why we never worry much about characteristic impedance, but the telephone industry certainly did before fiber took over.

A recurring theme in this chapter has been low Z drives high Z. That was audio. This is RF transmission through a cable whose impedance closely matches the source and receiver, the "transmission line." It's a balanced line, but in a totally different way than line-level audio: impedance balanced. The series impedance (resistance + inductance) equals the parallel impedance (resistance + capacitance). This creates a nearly lossless transmission, with minimal reflections in the cable (Fig. 3.22).

In the pro audio analog world we don't think about reflections in a cable, but some of us are old enough to remember them during long-distance phone calls. Reflections become a concern when transmitted wavelength approaches cable length. This is 1156 m @20 kHz for analog electronic circuits, so no problem there. We are working in the 6 MHz range and up here. A reflection in the digital audio transmission line won't cause an audible echo like its analog counterpart. Instead it creates additional data that impersonates the original, a form of aliasing that can cause transmission errors that render data unreadable. Terminating the cable at the characteristic impedance dampens the reflection at the destination.

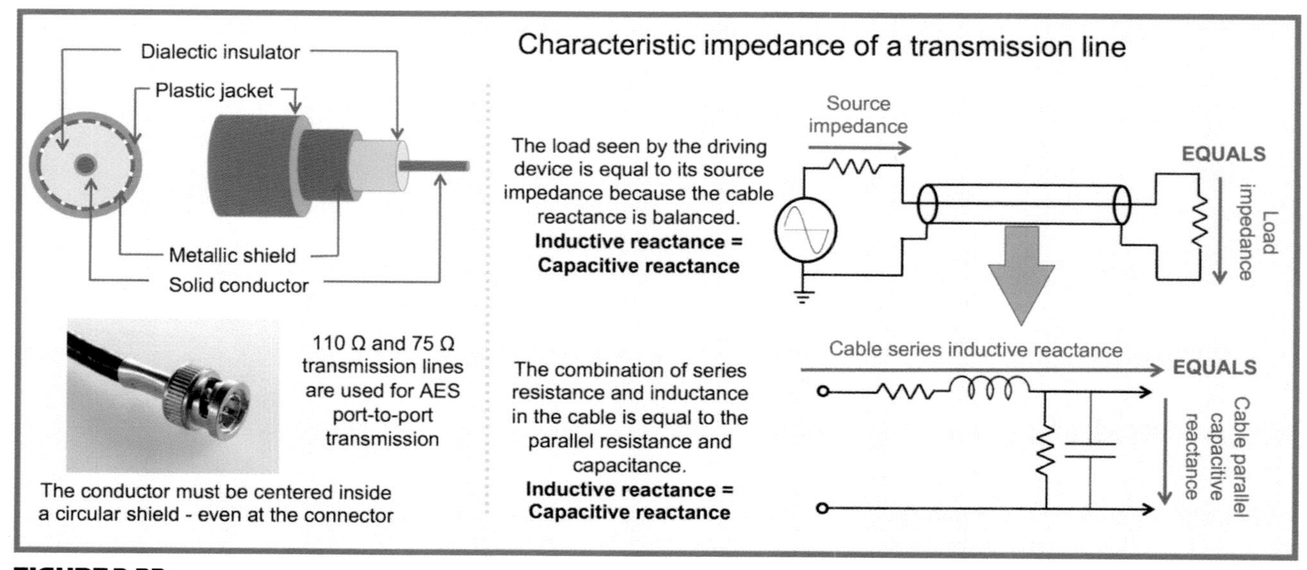

FIGURE 3.22
Characteristic cable impedance (Jonas Bergsten coax photo). Schematic is own work based on drawings by Omegatron.

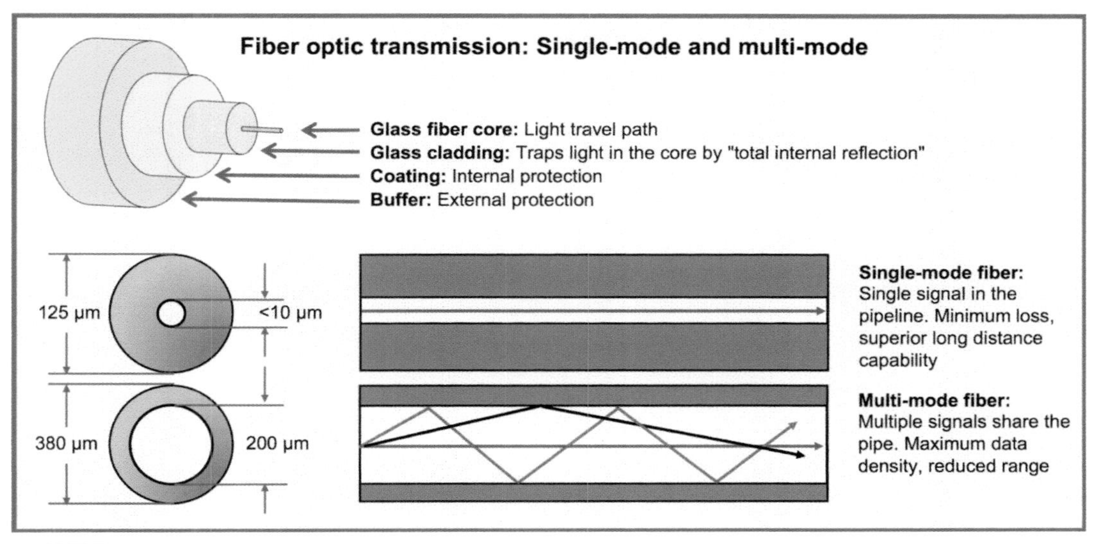

FIGURE 3.23
Optical transmission (self-made graphic based on images by Bob Mellisch and "optical fiber types" by Mrzeon)

3.6.2.3 OPTICAL

Navies have long used wireless communication to send encoded optical digital audio between vessels: A seaman flashes Morse code by opening and closing the shutters on a light. Fiber optic digital audio is much faster, but the concept is the same. When an optical transmission system is on the blink, it's working correctly. Optical digital transmission substitutes light for voltage. The cable has a virtually lossless internally reflective skin. Optical cable can run multiple channels as well by multiplexing (running multiple signals in series and decoding them separately). Optical signals are free of ground looping issues and can be hot-patched without fear of surges. Care must be taken to prevent hard bends in optical cable runs as this breaks the reflective path (Fig. 3.23).

3.6.3 AES-compatible configurations

3.6.3.1 AES/EBU (AES3)

This is a two-channel configuration (nominally stereo but in practice simply two independent channels). Standard AES3 cable is 110 Ω 3-conductor twisted pair with a maximum recommended length of 100 m.

The XLR and RCA connectors themselves underwent analog to digital conversion and were recycled for AES3. I guess this seemed like a good idea at the time but can lead to mis-patching. The connectors fit in the holes, but the cable is not compatible and leads to unexpected results in both directions. Conclusion: Verify your cable.

The standard connector is an XLR (female input, male output) wired the same as its analog ancestor, pin 1: common, pin 2: signal +, pin 3: signal -.

AES3 format structure

Each sample carries a single numerical value in 16 to 24 bits. Each data frame needs identifying markers to keep it in order. It would be very inefficient to individually wrap every frame of every channel. Instead they are grouped in sets of 192 frames, which at 48 kHz amounts to 4 ms of audio (0.02083 ms × 192).

The end product is an "audio block": 2 channels with 192 audio frames and a complete set of packing materials. Let's see how we get there.

The converted audio sample is packed into 32 bits to make an audio frame. The middle 24 bits (the audio bits) are the magnitude values in fixed-point decimal form. Let's call these 24 bits the audio bits and the other 8 the packing bits for ease of discussion. Eight packing bits and 24 audio bits give us a single subframe (one of 2 channels). Join the subframes together to make a complete frame (64 bits of audio data and info). We continue this operation 191 more times and wrap it all up into an audio packet.

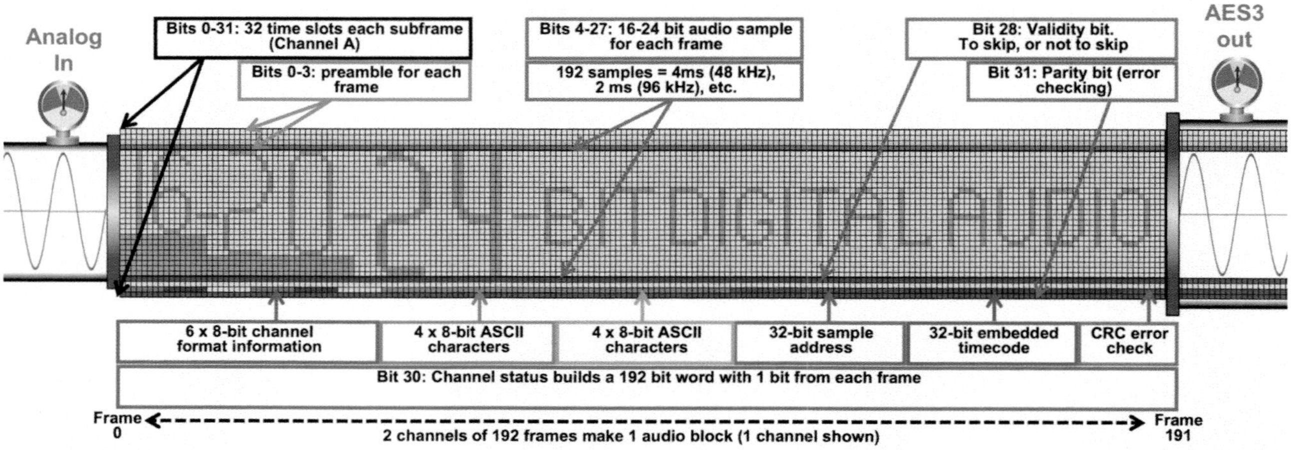

FIGURE 3.24
AES3 frame structure

Each audio block requires 64 serial operations before the next sample (32/subframe). Operations are divided into time slots managed by the bi-phase mark coding clock (BMC), which is much faster than the word-clock. For example for 48 kHz we have 64 bits × 48 kHz = 3072 kb/sec with a bi-phase clock of 6.144 MHz.

How much time is represented by an audio block? This depends on sample rate. 192 frames of 48 kHz audio is 4 ms of audio, a frame rate of 250 Hz. Halve that for 96 kHz (2 ms) and again for 192 kHz (Fig. 3.24).

Let's return to the 32-bit subframe. The audio bits contain 192 unique 24-bit numbers: the audio data stream. We now have a grid of audio values (24 bits amplitude vs. 192 bits time). All formats share the most significant 16 bits and the 20- and 24-bit data fills in the lower bits. The 16- and 20-bit formats send all 0s to the unused lower bits. This grid is the payload for an audio block.

Some of the packing bits are handled differently. They are spread over the 192 frames as 24 words of 8 bytes each (24 × 8 = 192) to tell us the story of this audio block. Audio bits are read vertically, subframe by subframe, and packing instructions are read horizontally over the 192 subframes. This spreads the packing info over the audio block, much higher efficiency than sending complete notations on every frame.

The 8 packing bits are divided in two sections: pre and post. Bits 0–3 are the preamble, preparing to sort the incoming data. Here comes the first frame of channel A. Next is a frame from B etc. The last four bits follow the audio data. Bit 28 checks to see if the data is ready for D/A conversion and puts in a stop-order if not. Bit 31 is the parity bit, the basic transmission error checker.

Channel status bit

The main event for packing is bit 30, the "channel status bit," which has enough detailed information to correctly decode and find this audio needle in a haystack of over 17 million packets. That's over 12 hours of continuous audio at 96 kHz (we can transmit the soundtrack for the *Lord of the Rings* films without repeating the index number). Also included in this bit is embedded SMPTE time code. You might wonder how we get this done in just 8 bits because it takes a 32-bit word to do each time code. It's the payoff for spreading the words over the entire block instead of putting them all on each subframe.

3.6.2.2 AES/EBU-COMPATIBLE FORMATS

- **TOSLINK**: Toshiba Link (TOSLINK) is a short-distance (<10 m) fiber optic version of AES3. Toshiba, Sony Philips Digital In Format (SP/DIF). The connector is the JIS F05.
- **S/PDIF**: The Sony/Philips Digital Interface Format (S/PDIF) is a short-range, unbalanced consumer implementation of AES3 with a lower voltage reference standard. The wired version uses 75 Ω coaxial cable with RCA connectors and the optical version uses the TOSLINK connector.
- **ADAT Lightpipe (ADAT Optical Interface)**: A multichannel optical format initially implemented by Alesis for its ADAT recorders but also used by others. The connector is the same as TOSLINK and S/PDIF but the protocol

is not compatible. The Lightpipe has a fixed capacity but can be resource allocated by number of channels at a given sample rate. For example, 8 channels at 48 kHz, 4 channels at 96kHz.

- **MADI (AES10)**: Multichannel Audio Digital Interface (MADI) can transmit multiple channels of uncompressed AES-compatible audio. The exact channel count will vary by sample rate but 56–64 channels of 48 kHz of 24-bit audio are normal. The channel count is reduced by half at 96 kHz. The connections and cabling for both the electrical and optical formats are the same as the 2-channel version (AES3) so the cable and connector savings are substantial. MADI adheres to the AES10 standard and has widespread manufacturer support. Cable lengths can reach a maximum recommended length of 100 to 3000 meters.

3.7 DIGITAL NETWORK TRANSMISSION

We enter the digital world in a different stream than the audio I/O described above. If you got into audio because you wanted to immerse yourself in the most un-artistic, non-musical, nothing to do with your ears parts of the trade, this is your place. Digital networks are strictly a numbers game. How fast can we reliably move them? There is not even the goal of improving the waveform, just cloning it and moving it around.

Some audio network devices are common to the Information Technology (IT) sector, whereas others are uniquely adapted to us. Troubleshooting skills on a home computer network really help a lot here.

Let's clarify what's in the network and what's out. Devices that send and receive previously digitized audio packets are in. The A/D and D/A converters wrap and unwrap the audio packets from outside the network. The main "in network" devices are network adaptors, switches, clocks and more.

3.7.1 Network overview

It's not an analogy to envision a digital network as a pipeline. It literally is. Networks are just transportation systems. They don't give a %^&# about whether it's going to a tweeter or Twitter. We want our networks to do two things: pack it and ship it (Fig. 3.25). Network entry points require adaptation hardware (the network adaptor) to package our audio stream for transport. The process reverses on the other side.

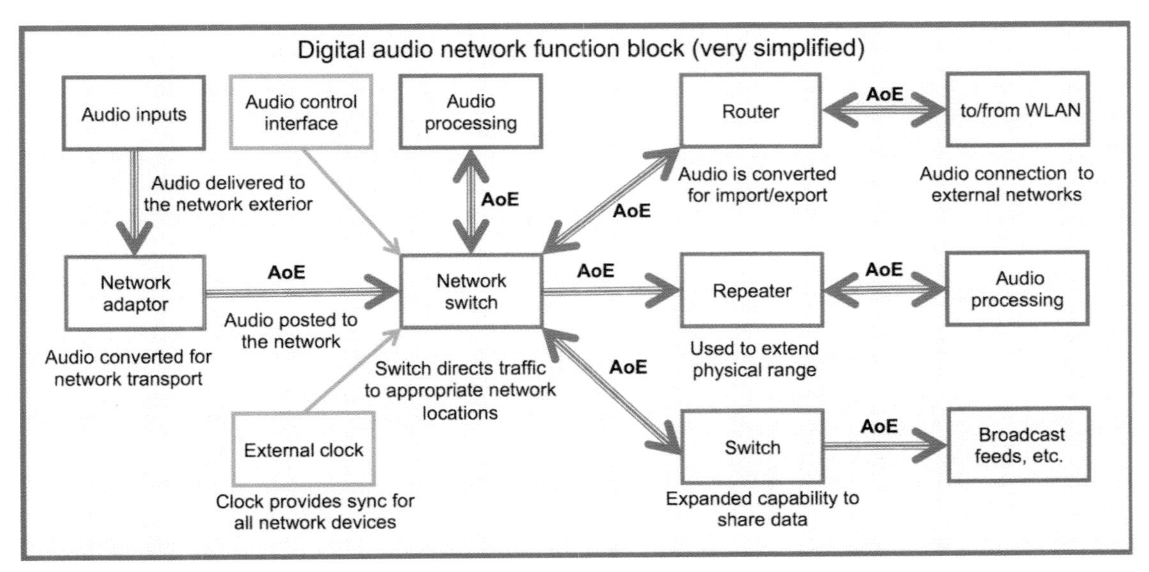

FIGURE 3.25
Digital audio network: simplified function block

Recall that analog audio signal paths are point-to-point(s). An output can drive a single input or be split ("wye'd" apart) to drive additional ones. By contrast, multiple outputs can't be "wye'd" together without some resistive separation or troublesome complications. Analog system designs carefully plan each point-to-point connection in advance and anticipate the waypoints that may require ongoing flexibility. We might find a patch-bay here, which allows optional rerouting and adds a very weak link in reliability terms. AES digital audio is much the same. This is still point to point but even less flexible regarding signal splitting and combining. However, these digital systems can bring large groups of signals together at a common core and reroute signals as needed. There is little need for patch-bays now.

Networks operate differently. Although the physical connections are point to point, the flow of information is not. A signal that appears on the network can be read by all the network devices (if we want it to). We broadcast a signal and then whoever wants it can pick it up.

SOUND NETWORKING CHALLENGES FOR LIVE SOUND APPLICATIONS

- **Low latency**: limited tolerance for signal delay. Excessive latency degrades the relationship of the reinforced audio to the natural sound and falls too far behind the visual.
- **Consistency of clock timing**: All related audio streams must clock together. Our tolerance here is extremely small (by network standards), needing to stay within a single sample (e.g. 0.01 ms for 96 kHz).
- **On-time delivery**: Audio packets must be delivered with enough time to be buffered and unpacked. The stream needs to be consistent enough to prevent dropouts (too late) or buffer overflow (too early).
- **Error-free streaming**: We need a complete set of audio packets with minimal error correction.

Our special transmission needs have always required us to sequester our audio networks into a private setting (our own LAN), with limited broadcast outlets to other LANs or the wide area network.

3.7.2 OSI layer model

The Open Systems Interconnection model (OSI) is a conceptual framework to help visualize the interior design of digital networks. The OSI framework simplifies the geek-speak to aid our comprehension of how data is being moved around. The result is a seven-layer "abstraction model." Raw data from the top layer (such as our audio signal) is passed down layer by layer within our device until it reaches the bottom and is ready for transport.

Digital networks can be visualized as a city of seven-story buildings where people on each floor speak a different language. Seventh-floor residents in our building want to communicate with those of other buildings. Information about the intended destination is added to the message, which is then handed down floor by floor and couriered by ground (the physical layer) to the next building. Each floor adds external packaging to our messaging according to its set of conversion rules (known as protocols). The complete package can be sent out and then unwrapped as it makes its way up to the upper floors of the next building. The message is readable on the 7th floor because they speak the same language (the internal protocol).

The management style in the layers is top-down (Fig. 3.26). Each layer manages the one below it, operates peer to peer with other hosts on their level and takes orders from the one above. Data is in its most raw form at the top level. Each time we move the data down a layer it must be encapsulated to the protocol of the new layer. By the time we reach the bottom layer the data is nested like a seven-level Russian doll. Our user interface is above the highest layer. We will deliver audio packets to the top of the stack and provide a shipping location. The packets are repacked as they move down through the layers and prepare for transmission.

The bottom layer is the physical layer, where we connect electrically or optically between devices. 100-Base-T, USB, Bluetooth, RS-232 and others are physical layer protocols. The physical layer personnel don't read our messages (unlike our government); they just pack them and ship them. We don't care about the shipping details as long as our packets are delivered on time in readable condition.

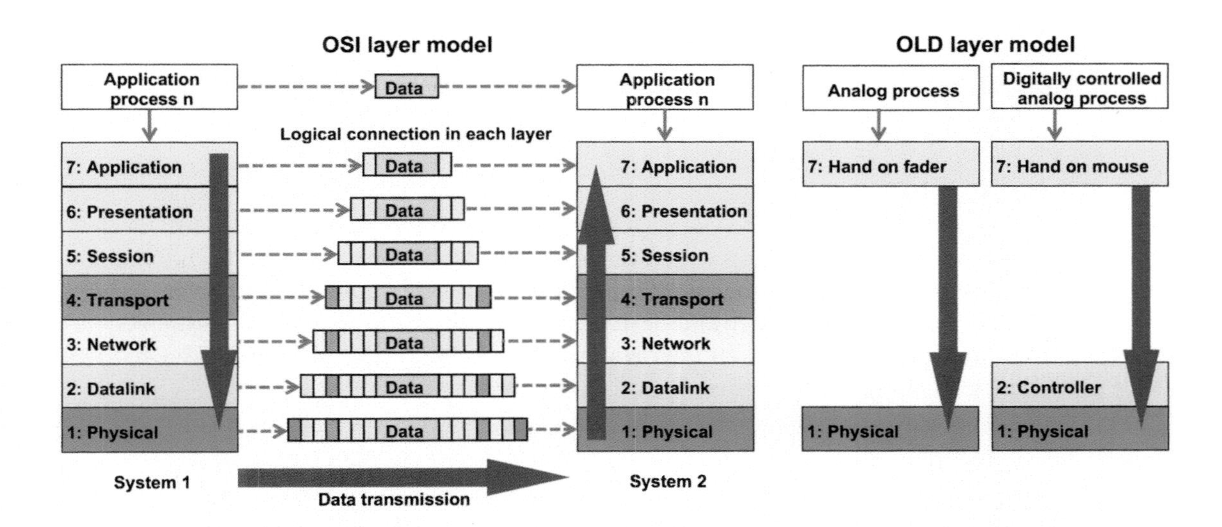

Layer	Units	Application/example	Protocol, hardware	Audio networks
7: Application	Data	Network process to application	SMTP	
6: Presentation		Data representation, encryption and decription	JPEG, ASCII, TIFF, GIF	
5: Session		Interhost communications, managing between applications	Logical ports	
4: Transport	Segments	Host to host, message control (TCP)	TCP, UPD	
3: Network	Packets	Path determination and IP (Logical addressing)	Router	AES67, AVB, Dante (Audinate), RAVENNA, Q-LAN (QSC)
2: Datalink	Frames	MAC and LLC (Physical addressing)	Network switch	AES51, AVB (1722), Cobranet, Ethersound
1: Physical	Bits	Media, signal and binary transmission	Cables, hub, ports	AES50, A-Net (Aviom),RockNet (Riedel), Hydra2 (Calrec)

FIGURE 3.26A AND B
The OSI and OLD layer models (own work based on images posted by Sai Charan Tsalla at www.computersgags.co)

It is important to bear in mind that the OSI layer model is an abstraction, i.e. it is a simplified construction to help us visualize the macro structure and interrelationships of something much more complex. There is some confusion in networking circles between the seven-layer OSI model and the TCP/IP layering model. The terms are not interchangeable. For example, audio transport is on Layer 3 and 4 of the OSI layer model but might be found on Layers 2 and 3 of the TCP/IP model. There are links to both layering models in the reference section of this chapter.

The OSI layer model makes clear how far removed we are from the audio signals we control with our mouse or GUI. Contrast this to the analog model where our hand is in direct contact with a variable resistor controlling the electrical signal flow. This is known as the OLD model (Old Like Dad).

Some commercial audio networking devices operate with Layer 1 protocols, which essentially means they must manage their communication directly using proprietary hardware rather than through general-purpose network devices. We can take more advantage of general-purpose networking devices as we move up the layers but we are also at their mercy. Audio network devices must be very choosy about their partners because time is of the essence, and our margin for error is so small. Layer 2 systems can use audio over Ethernet (AoE) switches and other network hardware. Layer 3 systems are becoming the leaders in audio networking as they able to take advantage of Internet Protocol hardware.

AVB/TSN (discussed in section 3.7.6.5) is based on IEEE Standard 1722–2011, which is a Layer 2 and 3 protocol. However, AVB's Stream Reservation Protocol 'transcends layers' because the frames are only for transporting between ends of a wire and never forwarded or bridged or routed, as if it were Layer 1½!

3.7.3 Network structure

The network is the local family of connected talker and listeners. Local networks connect together to create a global structure.

3.7.3.1 LAN (LOCAL AREA NETWORK)

A private network limits its communication exclusively to designated network devices, i.e. those within its "local area." A local area network (LAN) has a wall of separation (often termed a "fire wall") between it and other LANs or the wide area network (e.g. the Internet). Traffic management is absolutely critical to audio system networks because of the time-sensitive nature of our data stream. Audio devices must be able to talk amongst themselves without having to worry that the show audio will pause for the cash register to sell a souvenir. In short we need a private line, instead of a party line, to ensure on-time delivery.

3.7.3.2 WAN (WIDE AREA NETWORK)

A wide area network (WAN) is a non-private place where audio systems can broadcast to or receive from outside the network. From our LAN (local) perspective, the WAN is anything out there in the wild world beyond our gates. When we send signal out of our network (e.g. a live show streaming to the Internet) we are not directly connected to every interface on the Internet. Instead we travel through an extensive series of LANs (such as our Internet provider) that have opened the appropriate connections to allow our data to stream through to homes in Tasmania. The potential traffic hazards of traveling through the WAN should be immediately apparent, because any single point may lack the speed required to sustain uninterrupted streaming. If you have trouble picturing this, go to youtube.com. If you experience freezes and dropouts and see the word "buffering" on the screen you may be experiencing a WAN traffic jam. You would be nuts to send your show audio out of LAN control until absolutely required, but this does not mean it isn't done. Therefore, in pro audio we must maximize our local control and take the best measures we can to ensure that we receive the priority lane out there on the WAN (Fig. 3.27).

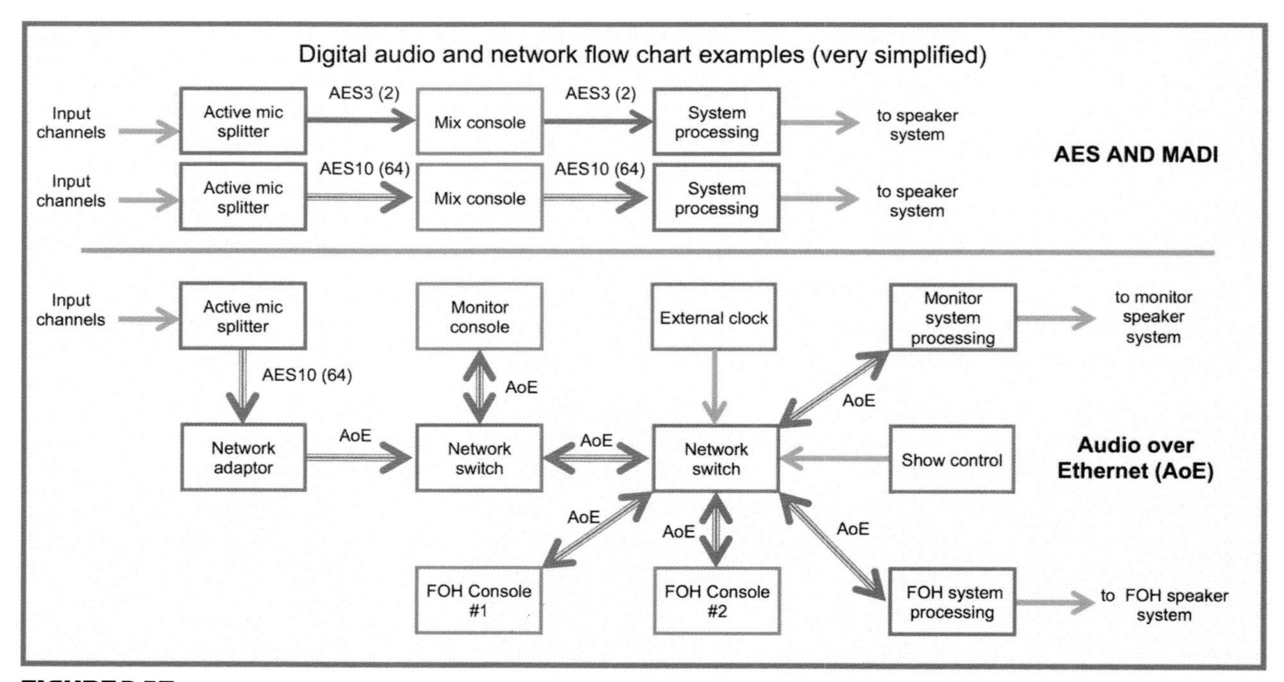

FIGURE 3.27
Digital audio network function block for a concert system (very simplified)

3.7.4 Network speed/bandwidth

How much data can our network (hypothetically) move through the pipe in one second? That is a major way in which networks are classified. A network operating at 1 Mb/second is termed a 1 MHz bandwidth. This network can transmit one file with a size of 1 Mbit or 10 files sized at 100 kb per second.

An example digital audio payload is 24 bits/frame @48 kHz sample rate. This is 1.152 Mb/sec (24*48,000). This is what we are sending down the pipe, along with instructions and packaging.

I remember when my Internet modem upgraded to 14.4 Kb/second! Having dated myself, I'll date this book by stating that we currently see 100 Mb and gigabit network speeds out there. What's a gigabit? That's 1000 Mbits/second. This is 10× faster than 100 Mb (100BASE-X), which is 10× faster than 10 Mb (10BASE-T). If your network is slower than 10BASE-T you might want to move on from your AOL account.

3.7.5 Network devices

Some devices connect between different networks (routers), others move data between multiple devices in the same network (switches) and some connect one talker/listener to another (interface/ports).

- **Interface (port)**: The connection point to other devices on a network. Its function is to send and receive data packets. This is analogous to a combined input AND output in our analog world. The closest thing we have to this is a very old-school intercom, which used the speaker as the mic and vice versa.
- **Router**: Connects a network to a network. My home router connects our local network to the cable company net-doesn't-work. Sometimes routers must convert data between the network configurations, which is analogous to a transducer (converting between mediums). The router manages the data exchange, a controlled point of contact with specific rules to ensure a successful information transfer. Another router analog(y): An audio isolation transformer allows a data (waveform) transfer while leaving the connected systems still safe in their respective orbits.
- **Hub**: Connects multiple devices within a single network. The interface of a given network device connects to an interface in the hub. Data packets sent on the network go into an interface on the hub and are then broadcast to all interfaces. Connected devices may choose to listen or not (like teenagers). A hub is analogous to a mix console summing matrix with every control set to unity: All inputs go to all outputs. This is not desirable in a mixer and there's a similar downside to the network hub: Clogging up the pipeline by sending data where nobody's listening. This is why hubs are now largely obsolete (replaced by switches).
- **Switch**: A discriminating hub. Instead of having relations with everything it meets (like that promiscuous hub) the switch talks/listens only to approved interfaces. It's configured to send packets between specific devices, tracked by their unique MAC addresses. Visualize this as an analog matrix with unity assigns at the desired intersections. Remember that interfaces are both senders and receivers so the matrix analogy doubles up.
- **Repeater**: A port/interface replicator. Signal is received at one interface and retransmitted verbatim from another. Repeaters are used when cable runs are too long for a reliable port-to-port connection. The digital version of the analog line driver.
- **Network adaptor**: a data packaging and translation device that enables a computer to communicate over a specific data standard such as Ethernet. The adaptor packs and unpacks the data to/from the interfaces to the relevant protocol (e.g. TCP/IP). The network adaptor is the interface's translator that speaks the languages of both the local computer and network.

3.7.6 Open and proprietary protocols

The line between open and proprietary is somewhat gray in this realm. After all, a network is a connection of multiple devices, so unless a manufacturer makes everything in the signal chain it pays to publish your protocol and encourage others to make compatible gear. A network that connects to all parts of the system is highly favored, so we don't have to change to another network protocol every time we send to a different manufacturer's gear. Many network manufacturers handle this by establishing partnerships and licensing agreements with others to facilitate network expansion (Fig. 3.28).

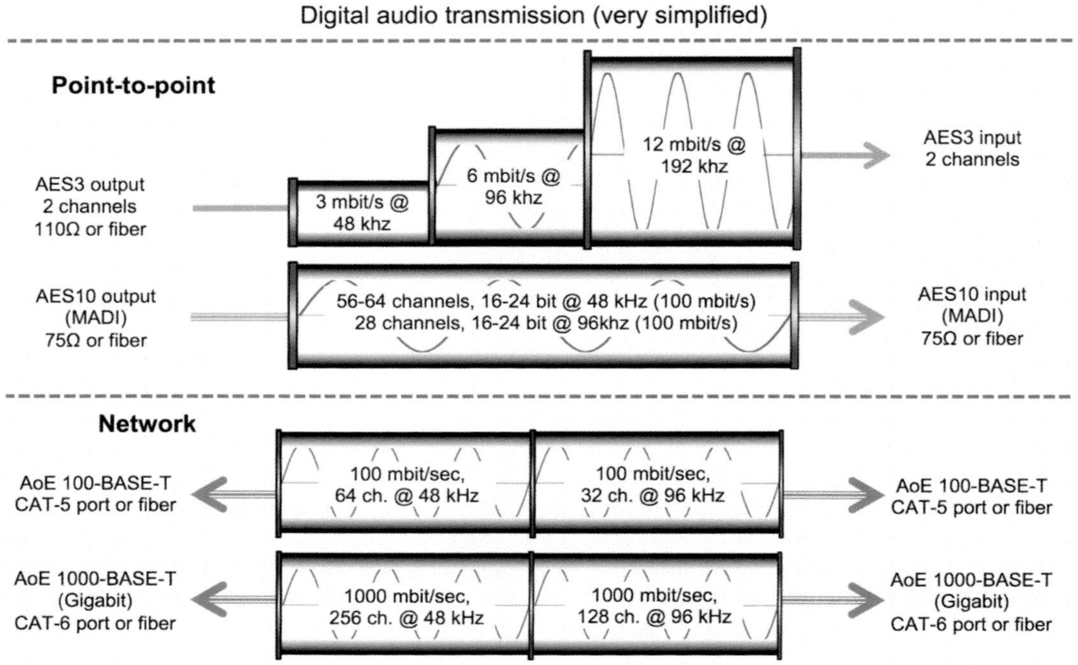

FIGURE 3.28
Digital audio transmission: point to point and network

3.7.6.1 COBRANET

Developed in the 1990s by Peak Audio, CobraNet is a widely used first-generation multichannel Audio over Ethernet (AoE) network protocol. CobraNet uses Ethernet packets instead of TCP/IP (the common type used on the Internet), which prevents it from being able to go through standard routers. Connection is through Cat5 cable and RJ45 connectors. There are Ethernet 10 MB and 100 MB versions. More at www.cobranet.info.

3.7.6.2 DANTE™

Dante is a second-generation Audio over Internet (AoE) system developed by Audinate with support from the Australian government. This is a proprietary protocol with a large number of partner companies. Dante is an IP-based solution (in contrast with CobraNet above) that provides uncompressed multi-channel, gigabit transmission speed with low latency. Because Dante is an IP-based protocol, it can go through Internet hub switches and routers. For more information go to www.audinate.com.

3.7.6.3 AES67

AES67 is a Layer 3 protocol open interoperability standard developed in 2013 to facilitate connections between competing systems such as RAVENNA, Livewire, Q-LAN, Dante and others including AVB. The concept behind AES67 is to establish a standard transport protocol (Real-time Transport Protocol) that the various platforms can offer as an alternative to their proprietary versions. This enables users to interconnect the various commercial networking platforms and prevent format wars from stifling progress in this field. More info at AES.org.

3.7.6.4 Q-LAN (Q-SYSTM)

Q-LAN is QSC's second-generation system, supplanting their RAVE system, which operated under CobraNet. This is gigabit IP-based system. More info at www.qsc.com.

3.7.6.5 AVB (AUDIO VIDEO BRIDGING), A.K.A. TIME-SENSITIVE NETWORKING (TSN)

Wouldn't it be great if a bunch of major players in networking, telecommunications, audio and video all got together to create a common standard for audio and video transmission? Well yes, of course, as long as they get it together to make it happen (which is very much a work in progress at the time of writing). That is the goal of the AVnu Alliance in its effort to standardize audio video bridging (AVB) as the next generation of media networking. Networked professional audio (especially live sound) is much more time and traffic sensitive than most data transfer applications. If the network can't guarantee on-time delivery it puts our show at risk.

I will let Jeff Koftinoff take it from here:

AVB is the name of a collection of Institute of Electrical and Electronics Engineers (IEEE) standards that provide a way of using Ethernet to transport high-fidelity audio and video signals with guaranteed minimum latency, guaranteed bandwidth, and with sample synchronous output at all interfaces on the network. The foundation of AVB includes high-precision generalized PTP (gPTP), stream bandwidth reservation and media-aware Ethernet traffic shaping. This foundation can be used for AVB media transport on any networking layer and can be used for non-media Time-Sensitive Networking (TSN).

AVB is a coordinated effort over multiple industries with far-reaching implications in audio video and system control at both the professional and consumer levels. On the pro audio side we are most interested in its promise of minimum latency and the ability to keep audio samples traveling through our system in sync.

More info at www.avnu.org.

3.8 DIGITAL TO ANALOG TRANSMISSION

We had our fun in the digital world but now it's time to return to planet analog. The converter constructs an analog waveform from the final digital number sequence. We don't need to start from zero on this, because many of the processes are a reversal of the A/D previously discussed.

3.8.1 Getting to "yes"

Recall that our AES audio block has a bit that checks whether the sample is convertible, our version of transportation security. The package is inspected to ensure there's no bomb in there. If it detects a problem that can't be error-corrected then "no sound for you!" As bad as that is, it beats a digital sonic boom.

The signal must be packaged and delivered to the D/A converter in a ready state: A fixed-point numerical series with known format, bit-depth, bit-order, clock frequency, etc. All 192 samples must be in chronological order.

3.8.2 dBFS (dB full-scale)

We need an exit-side sensitivity reference (the inverse of the entry side). It's often a user-settable parameter. At this point full-scale digital *creates* (rather than reads) a certain voltage. It's dBFS on the input side, so maybe we should call it FSdB (probably not, because nobody else does). The dBFS on the A/D and D/A sides don't have to match. Each can be optimized for best conversion with the upstream and downstream devices respectively. In pro audio we can expect to convert into an analog system with standard line-level capacity (+20 dBV) also known as -20 dBFS = 1 V (0 dBV). Maximum analog and maximum digital will match if we target the dBFS to this range. Lower values require added analog gain after conversion to reach full-scale analog (Fig. 3.29).

3.8.3 Interpolation and smoothing

It's time to play "connect the dots." This won't be as simple as the childhood game where we draw a shortest, straightest line between the numbered sequence of dots. Let's begin by understanding what happens if we do nothing, and then move on to how to solve it.

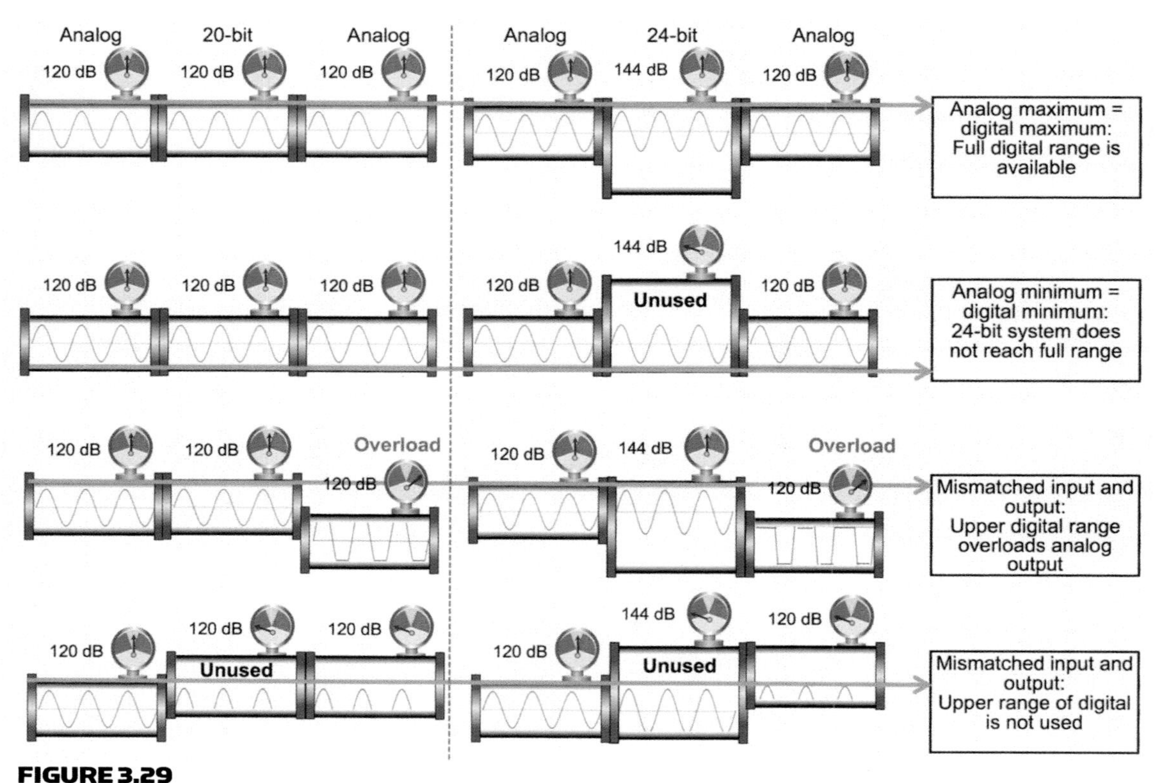

FIGURE 3.29
dBFS scaling for maximum dynamic range

If we follow the digital instructions literally we would hold the present voltage value for the entire sample period and then BOOM, move almost instantly to the next setting (like a square wave). This is called a "step function" because it resembles a stairway. Our analog signal needs a ramp between states, I guess because it's really old. But seriously, we can't ask the analog device to reproduce the sharp transition of a step function. To do so creates harmonics that were not part of the original signal. We are facing aliasing again from the opposite direction because a step function would create the distortion of out-of-band harmonics.

Instead we connect the dots with curves that smooth the transitions into shapes that are organic to harmonic motion in the physical world. We know we are pushing a speaker around so let's round the edges to make it possible to faithfully track. A step function is a recipe for failure for a loudspeaker. It has mass, momentum, compliance, back EMF and more to prevent it from an instantaneous change from rest to full excursion. The bottom line is the speaker won't track the step function so let's help it out as best we can.

The mathematical process is termed the "reconstruction filter," which examines each step transition and applies a smoothing algorithm to best-guess approximate analog signal behavior. Each D/A model approaches this interpolation differently, a "leap of faith" between two known points. This is obviously a critical quality differentiator in the perceived audio quality of converters. How can we tell if the interpolation was a good guess? We would need to compare an analog original with its analog to digital to analog copy (Fig. 3.30). This includes two conversions and everything between, so it's difficult to isolate one single cause, but the differences (known as quantization errors) can be found as long as we are just passing the signal through the digital system (and not mixing, filtering, etc.). Everybody interpolates differently, so from a marketing perspective we can go with the slogan: "My error's better than your error."

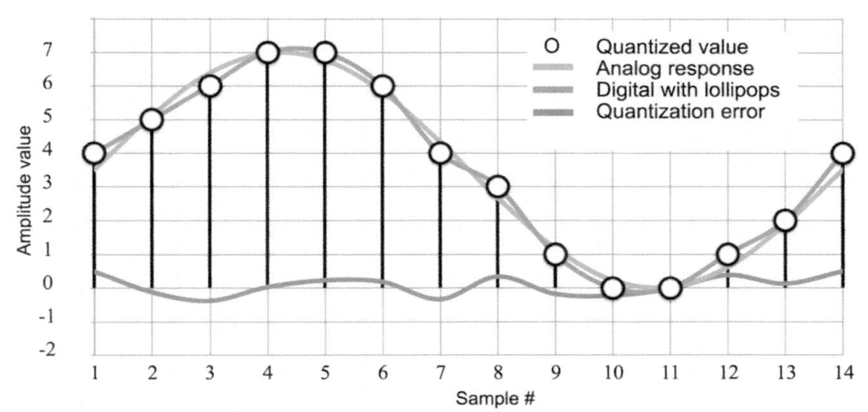

FIGURE 3.30
Comparison of an analog source signal to its digitized and reconstructed copy. The quantization error is seen as the difference.

3.9 ANALOG VOLTAGE TO ANALOG POWER TRANSMISSION

We have gone as far we can go in the one-dimensional world of transmitting voltage and numerical values. The next stage requires power, and power requires current. Lots of it. The power amplifier is properly named. It amplifies power even at unity voltage gain. The volt driving a 10 kΩ line-level input is a different animal than the volt driving a 4 Ω speaker (2500× more current). The power amp rescales the line-level waveform to the much larger speaker-level piping and adds a vastly more powerful pumping system. This is what it takes to push our waveform into a power-hungry acoustical generator (our speaker load). The pipeline connection points are found inside the power amplifier in two stages. The driver stage rescales the voltage to the high-voltage pipe size and the output stage supplies the high-current pump.

3.9.1 Power amplifier stages

3.9.1.1 POWER SUPPLY

The power supply is not an audio stage but rather a ready reserve of current and voltage available for transmission. This is the pump that maintains pressure in the pipeline with a solid block of DC electrical potential. The amplifier's output signal is a modulation of the DC potential transmitted to the speaker load. The modulator is the audio waveform. The power supply must have instantaneous and reserve energy to track the audio waveform's variable demands. The supply converts energy from the AC mains into a stored reserve of DC power. We can overrun its capacity by asking for too much voltage (voltage clipping), too much current (current clipping). Voltage clipping is as simple as reaching the limits of the DC rails (the outer diameter of the audio pipeline). Current clipping occurs when the power supply runs out of storage capacity, which collapses the rails and clips at a lower voltage than previously (sucking out more water than the pump can supply). The power supply may be capable of short-term power bursts but collapses when the waveform demands sustained high levels. This is a sign of high rails but insufficient storage capability. Amplifiers with unbelievably high power specs need to be examined as to how much long-term capability they can sustain (Fig. 3.31).

Exceeding the output device current capability can also cause current clipping and possibly runaway thermal conditions if the devices are not sufficiently protected.

3.9.1.2 INPUT/PROCESSING STAGE

The input stage is standard line-level analog (or digital). User-adjustable parameters such as sensitivity (gain controls) and more are found here. Digital versions may convert after the input (pre-driver stage) or at the output stage. The input has only one important function for us: sensitivity, which we can use to unify speaker system gain structure.

Voltage		DC Rails	Current (Amperes)		Impedance	Power (Watts)	
rms	pk	±VDC*	rms	pk	Ω	rms	pk
14	20	20	1.8	2.5	8	25	50
28	40	40	3.5	5.0	8	100	200
42	60	60	5.3	7.5	8	225	450
57	80	80	7.1	10.0	8	400	800
71	100	100	8.8	12.5	8	625	1250
85	120	120	10.6	15.0	8	900	1800
99	140	140	12.4	17.5	8	1225	2450
113	160	160	14.1	20.0	8	1600	3200
14	20	20	3.5	5.0	4	50	100
28	40	40	7.1	10.0	4	200	400
42	60	60	10.6	15.0	4	450	900
57	80	80	14.1	20.0	4	800	1600
71	100	100	17.7	25.0	4	1250	2500
85	120	120	21.2	30.0	4	1800	3600
99	140	140	24.8	35.0	4	2450	4900
14	20	20	7.1	10.0	2	100	200
28	40	40	14.1	20.0	2	400	800
42	60	60	21.2	30.0	2	900	1800
57	80	80	28.3	40.0	2	1600	3200
71	100	100	35.4	50.0	2	2500	5000
85	120	120	42.4	60.0	2	3600	7200

FIGURE 3.31
DC rail voltage/power rating and amplifier VI reference

3.9.1.3 DRIVER AND OUTPUT STAGE

The driver stage links line and speaker level, bringing together the supply, input and output side of the power stage. This is where the input waveform modulates the supply into the power stage to create the output waveform. The driver stage steps up the voltage and drives the output transistors, which provide massive current gain and ship the signal out to the load. Negative feedback from the output is used to stabilize the power stage and reduce distortion, source impedance and (most importantly) protects the amplifier and speaker.

3.9.2 Polarity

Professional amps with analog inputs are balanced input/unbalanced output. A polarity links output (+) to input (+). The standard is "Pin 2 Hot" (output terminal tracks input pin 2). Any amp with digital inputs or analog models manufactured after 1990 follows this standard so this no longer requires constant vigilance. Keep on the lookout for old Crest, BGW or other American antiques.

3.9.3 Amplifier gain/sensitivity

We can safely assume an amplifier's voltage gain to be greater than unity, but how much greater is anybody's guess. There is no AES standard, not even two. Different manufacturers use different gains and even single manufacturers vary gain from model to model. We can put 1 volt in but we have no idea what we will get out. Seriously.

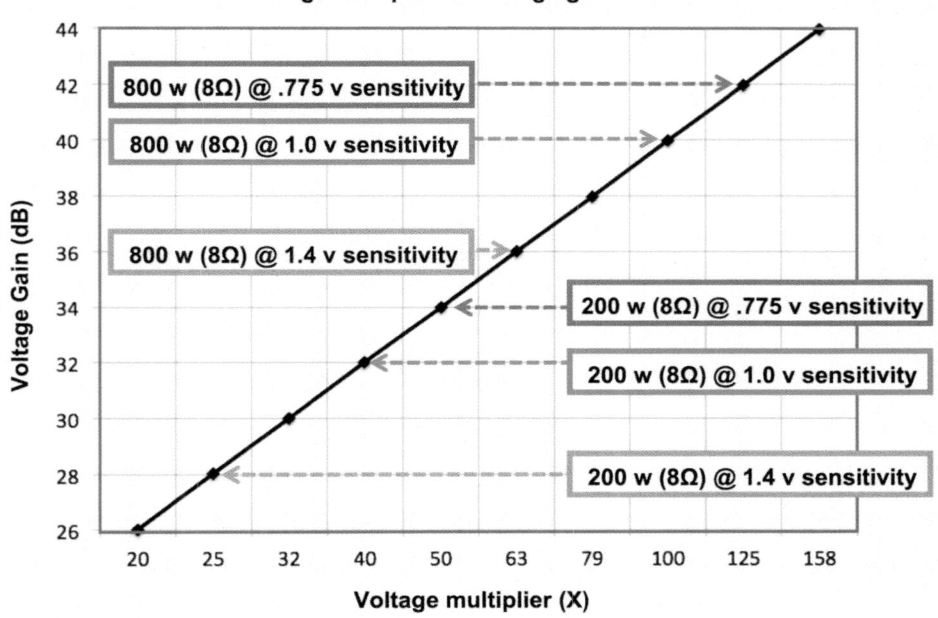

FIGURE 3.32
Power amplifier sensitivity/voltage gain decoder

3.9.3.1 VOLTAGE GAIN

- **Voltage gain (linear)**: expressed as the output/input multiplier; 20 V output with 1 V input is a gain of 20×.
- **Voltage gain (log)**: expressed in log form as the output/input multiplier; +26 dBV output with 0 dBV input is a gain of 26 dB. This is another application of the $20*\text{Log}_{10}$ (output/input) formula.

Note that the linear and log examples shown above match: 20× gain = +26 dB gain.

3.9.3.2 SENSITIVITY

A third way of expressing amplifier gain uses the transducer conversion term: sensitivity. Unlike microphones or speakers, an amplifier does not do a medium conversion; rather it converts electrical voltage to electrical power. In this case the "sensitivity" denotes the input level required to achieve full output level (Fig. 3.32). Let's say we have an amplifier that has a maximum output level of 40 V. It reaches full output when driven with 2 volts.
This would be termed a sensitivity rating of 2 V (+6 dBV) to reach full-rated power. Note that this is once again the same amplifier (20×, +26 dB voltage gain).

3.9.3.3 COMPARING SENSITIVITY AND VOLTAGE GAIN

If the previous sections on voltage gain and sensitivity left you scratching your head, don't feel alone, but rest assured there's a decoder ring (Fig. 3.33). Part one is easy: Linear and log voltage gain are related by the $20*\text{Log}_{10}$ (output/input) equation. Voltage gain is straightforward. It doesn't matter if the amplifier rating is 10 or 1000 W. It is output/input, period. Sensitivity is the outlier (and more common specification).

Sensitivity is the input drive required to reach the rated output power (the clip point). The difference between driving an 8 W and 800 W amp shows up now. If both have the same sensitivity (e.g. 1 V), they will both clip

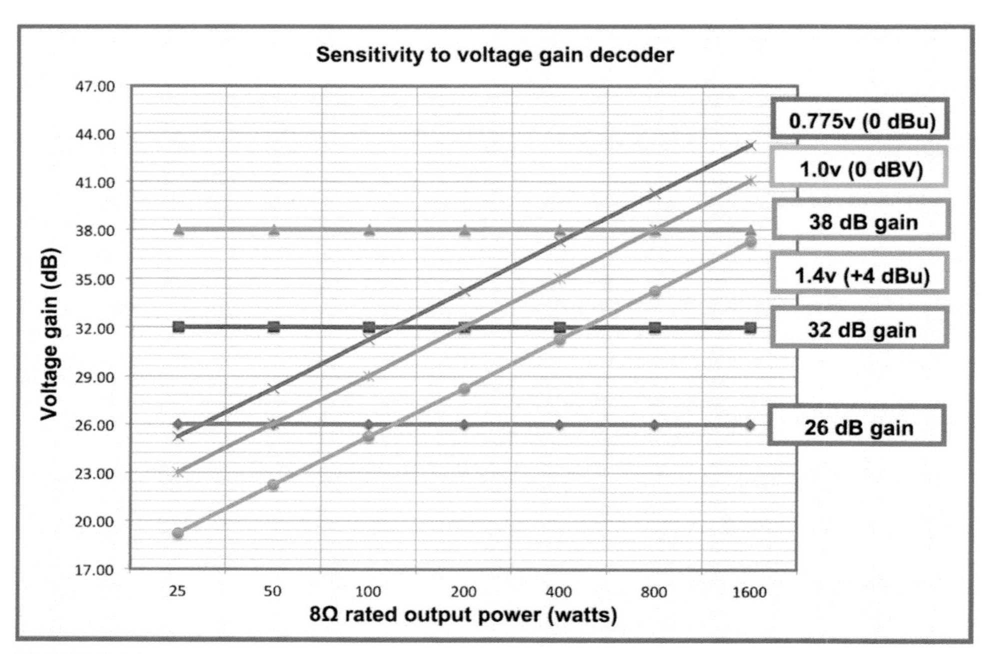

FIGURE 3.33
Amplifier sensitivity decoder

at the same drive. The 800 W amp will need 20 dB more internal voltage gain to get there. By contrast, if we make the voltage gain the same for both models, it will take 20 dB more input drive to get the 800 W amplifier up to the rails.

Part of what makes this confusing is amplifiers are voltage to wattage converters. Next add the brain pretzel of using the $20*\text{Log}_{10}$ scale for voltage and the $10*\text{Log}_{10}$ scale for wattage. Few manufacturers publish a maximum voltage rating, so we have to figure it out ourselves by Ohm's law from the rated wattage and impedance. Use the 8 Ω rating because it should have sufficient current reserves to prevent sagging in the maximum voltage level.

Voltage = √(power/resistance). So 8 watts @8 Ω requires 8 volts. 800 watts @ 8 Ω requires 80 volts. If the sensitivity for both is 1 volt (0 dBV) then the voltage gains are 8× (18 dB) and 80× (38 dB) respectively. If we go the other way and pick a single voltage gain of 20× (26 dB) then the drive levels required to reach full power would be 0.4 volts (-8 dBV) and 4 volts (+12 dBV) respectively.

Which is better, matched sensitivity or matched voltage gain? Matched sensitivities keep inputs matched while outputs diverge. There is advantage in all amplifiers having comparable headroom (proximity to overload). Little amps on little speakers would reach their limits at the same drive level as big amps on big speakers. We shouldn't need to drive the mains 20 dB harder than the frontfills, right?

Recall the fixed-size line-level pipeline. Analog pipe is limited by power supply voltage and digital is typically standardized to a single dBFS for a given installation. The advantage to similar drive levels is it maintains uniform headroom in the line-level pipe to cleanly drive all amps to full power.

Matched voltage gain has merit within the complement of a bi-amplified or tri-amplified speaker system, where signals from HF and LF amps will be recombined acoustically. Acoustical crossover setting is easier when amps have matched voltage gain. For example, if we're using factory-recommended DSP settings or a dedicated speaker controller, the settings assume matched amplifier gains. If unmatched, the relative levels at the acoustical crossover will be offset, causing a crossover frequency shift (see Fig. 4.27).

Why would voltage gains be unmatched? HF and LF drivers driven by amps with different maximum power ratings are unmatched if the inputs are sensitivity based. Continuing from the above example, a 400 W LF amp paired with a 100 W HF amp would have a gain offset of 6 dB at the crossover if the amps were sensitivity based (compared with unity if voltage gain based).

Limiter action is based on voltage gain. Estimates of speaker power dissipation (wattage) are based on the voltage seen in the limiter feedback loop. Accurate power estimation relies on a known voltage gain between the limiter and amplifier output. Unknown voltage gain yields an uncalibrated limiter, which will either decrease the dynamic range (excess limiting) or not provide the needed protection.

3.9.3.4 LEVEL CONTROLS

We're not quite done with voltage gain and sensitivity. Amplifiers have at least one user-settable control: level. Users might be discouraged from using this control if it were labeled "Sensitivity/voltage gain de-calibrator." If we're lucky, the panel settings will tell us how far from the known settings we are (and not just tick marks or clicks).

Level controls attenuate the signal internally and serve as sensitivity/voltage gain adjusters. They don't decrease the output power capability as long as we can reach maximum before clipping the line-level drive signal. Recalibrating the accelerator on your car does not change its maximum speed as long as it's still possible to reach the floor.

There's no standard approach to displaying gain change. Most are in increments of dB attenuation with 0 dB at the top (as if this was unity gain!). A tiny minority is marked in actual voltage gain. It can be done!

Most engineers are reticent to leave a system with different amplifier gains for fear of someone de-calibrating the levels afterward. This is justified due to the dubious capability to restore tampered settings. Fortunately, modern trends allow users to control the levels digitally and store the settings securely. Reasons for level adjustment include noise reduction, gain tapering and crossover alignment.

3.9.4 Low(er) *Z* to low *Z*: power transmission

Relatively speaking, the connection between amp and speaker is another low-impedance output driving a high-impedance load, but we're grading on a curve here. Although 8 Ω would not seem to qualify as a high-impedance load, it does when compared with the extremely low 0.01 Ω source impedance of the amplifier. This familiar impedance scaling allows the amplifier to minimize its heat load and maximize power transfer to the speaker. The low overall impedance allows lots of current to flow, and therefore lots of power.

3.9.4.1 SOURCE IMPEDANCE

This is a serial load that separates (barely) the output transistors from direct contact with the speaker. This is *very* low impedance, typically substantially less than 0.1 Ω. Efficient power transfer relies on the source impedance being very low compared with the load impedance. For example, 0.1 Ω looks very small compared with 8 Ω. The total load the amplifier sees is the serial combination of the source and load impedance (so for our example we have a total load of 8.1 Ω of which around 1.25% is lost in the source impedance). The source impedance is fixed, but the load is variable. If the load is 4 Ω, the total load is 4.1 Ω (a 2.5% loss). With a 2 Ω load (which becomes 2.1 Ω) we are facing a 5% loss (0.5 dB).

We have not yet added speaker cable impedance, which is simply additional serial source impedance. A fixed-cable impedance will create more loss as load impedance falls (see section 3.9.6).

3.9.4.2 LOAD IMPEDANCE

The load impedance is outside the amplifier but greatly affects what happens as a system. Current demands rise as load impedance falls. Insertion losses (source and cable impedance) will also rise as load impedance falls. Maximum voltage swing will also fall once the current demands have exceeded supply capability.

The actual load impedance seen by an amplifier is far from the simplified single numbers we often discuss (such as 8 Ω). Impedance is a frequency-variable parameter, and it is *highly* variable. The specified impedance is higher than the DC resistance (typically 6 Ω for an 8 Ω-rated speaker) and lower than *most* of the speaker's frequency range. The impedance at the resonant peak exhibits a characteristic rise (5× or more than nominal). The impedance will return to the nominal range above resonance and eventually climb steeply off the chart as frequency rises. The insertion losses described here will vary with frequency, with the greatest losses in the range with the lowest impedance. The loss is likely to be increasingly frequency dependent when insertion loss is high (e.g. low load impedance and a long skinny cable). Keeping the source impedance low relative to the load impedance will minimize the frequency-dependent aspect.

3.9.4.3 DAMPING FACTOR (DF)

Damping factor (DF) is a quality metric of amplifier load control (the damping of motion errors). Speaker motion follows the amplifier output signal, but also pushes energy back into the line, making it prone to overshoot (especially massive speakers such as subwoofers). The DF specification is based on the ratio of the lowest allowable load impedance (typically 4 Ω or 2 Ω) over the source impedance. Values over 20 are considered acceptable for sound reinforcement systems and 50 is considered high fidelity. Typical manufacturing specifications (without cabling) show values ranging from 150 to unbridled fantasy.

DF is an overrated rating. Only an amplifier with extremely short cable runs can achieve anything close in practice to the specified numbers. Although the amplifier output impedance is often under 0.05 Ω, the cable resistance is much greater and becomes the dominant player, making the generic DF spec academic. Cable resistance adds to the source impedance and reduces the ratio. Best-case DF is a short, thick cable to an 8 Ω speaker. Worst-case is a long, thin run to a 2 Ω load. This might help visualize damping factor: (a) Write your name on a piece of paper. (b) Tape the pen to a meter-long bamboo stick and trace over your signature. Not so easy, eh? That was a short run at 8 Ω. (c) Now make the pole 4× thicker and 10× longer and try again. That's a long cable with a 2 Ω load. Nothing makes the case for self-powered speakers like DF.

3.9.5 Output power capability

We have reached the output power stage. The pipeline is about to get much wider and able to handle a lot more pressure (voltage) and flow (current) so that we can transmit some serious power.

An amplifier's rated power is the maximum output (in watts) at onset of clipping (<1 %THD). Ratings are given for various output impedances: 8 Ω (standard) and possibly 16 Ω, 4 Ω, 2 Ω and 1 Ω. Here's an example power specification for an actual power amp: 8 Ω, 800 watts, 4 Ω, 1250 watts, 2 Ω, 2000 watts. Notice we get more total power output when we drive a lower impedance. The amp price is the same whether we drive it at 2 Ω or 8 Ω but get 2.5× the power at the low impedance. This factors so heavily into amplifier and speaker resource allocation that I've coined the term "econΩics" for the decision-making process. We'll return to that shortly, but first let's look at the wattage numbers again: 800 watts @8 Ω, 1250 watts @4 Ω, 2 kW @2 Ω. We're halving the resistance (and then again) but we're not seeing the power double (or double again). What is going on here? It's time to put Ohm's law to use again.

We have two of the parameters we need to analyze this with Ohm's law: power and resistance. With known power and resistance we can solve for voltage by the following formula: $E = \sqrt{(P \times R)}$. Let's solve for the 800 watts @ 8 Ω: $E = \sqrt{800 \times 8}$, $E = \sqrt{6400}$, $E = 80$ volts. Current flow can be solved by $I = \sqrt{P/R}$. This yields the following: $I = \sqrt{800/8}$, $I = \sqrt{100}$, $I = 10$ amps. The amplifier puts out 80 volts at 10 amperes into 8 Ω to create 800 watts.

When the impedance drops in half (8 Ω to 4 Ω) the current requirements will double, if (a very critical if) the voltage remains the same. Only then will power double. If the power supply cannot deliver the required current then the voltage cannot be reached to obtain the power doubling.

3.9.6 Speaker cable

Speaker cables are prone to substantial loss because they move real power. The cable impedance can be significant in proportion to the speaker load. Loss rate depends primarily upon three factors: cable length (proportional), conductor diameter (inverse proportional) and load impedance (inverse proportional).

Speaker cable costs money and adds weight so we are mindful of the tradeoffs. Money saved on cable costs power on the amplifier and speaker end. Three decibels of cable loss can be thought of as the speaker running at 70% capacity. Consider money saved on cable against the loss in your speaker investment (Fig. 3.34).

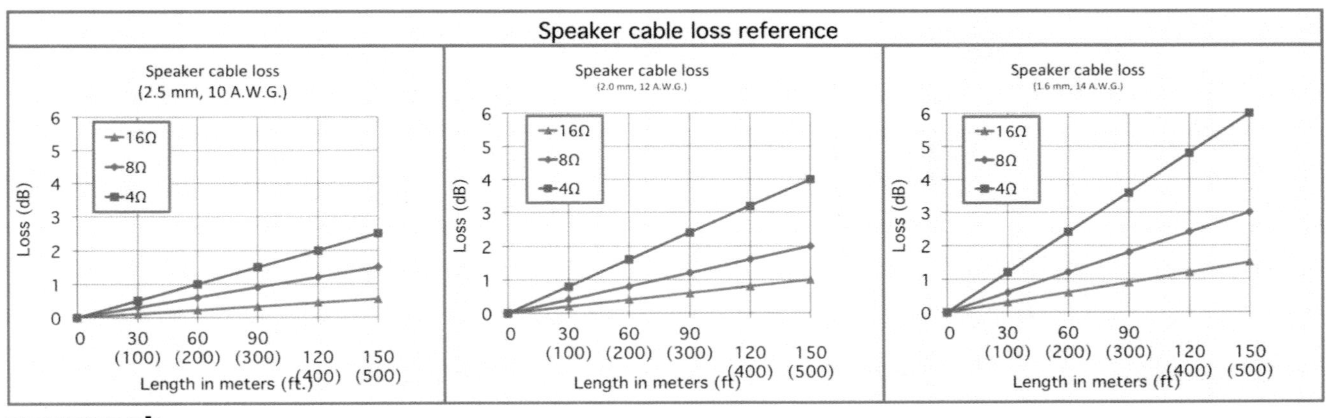

FIGURE 3.34
Speaker cable transmission loss reference (after Giddings, 1990, pp. 333–335)

We often face the choice of driving a speaker pair with a single run (split at the end) or as two home runs. It comes down to the quantity of copper either way. One big cable is the same as two smaller ones when they add to the same cross-section. Adding a second cable of the same size is 1.4× the diameter. In American Wire Gauge (AWG) this is equivalent to three numbers (2 @14 AWG wires are equivalent to 1 @11 AWG).

Speaker cable losses are challenging to measure directly, due to the high voltage levels and because the speaker must be attached to the amplifier to properly monitor the load effects. In practice we often simply measure the results in the speaker's acoustic response. Adjustments for unmatched responses are made at the power amplifier level controls.

3.9.7 Speaker connectors

These connectors move a lot of current, ranging from bare wire to military-grade waterproof. Here are some of the most common.

- **Terminal block**: a simple attachment of stripped wire to a screw-down terminal, cheap and highly vulnerable to vibration, stress and mis-wiring. Exposed wires can potentially short out the amp. These should be avoided in portable systems. We are most likely to encounter this at distribution panels and in "contractor" versions of power amps and speakers.
- **Banana plug**: The fancy-pants name for the banana plug is "5-way binding post." The male connector has two friction-fit banana-shaped terminals that wiggle their way out of the female. The second way is a version of a terminal block, another unreliable friction fit. This connector is becoming ancient history (applause). I never figured out the fifth way, one of life's great mysteries.
- **Speak-on™**: Neutrik's industry-standard connector is a creative mechanical engineering feat with locking connector that leaves no terminals exposed on either end (and saved the lives of countless amplifiers). Amps and speakers share the same chassis mount connector while the cable has two mating inline connectors. It is applicable for both portable and install.

3.10 ANALOG POWER TO ACOUSTIC TRANSMISSION

3.10.1 Transduction process

We're at the end of the electronic chain. The final step is conversion of electrical power to acoustical energy. It's a multi-stage transduction process: electrical to magnetic to mechanical to acoustic. This book is focused on loudspeaker application not their interior workings, so this won't be an extensive treatise.

Electrical current flows through the voice coil suspended in the gap between two opposing magnetic fields. Current in the wire creates a magnetic field of its own, which is attracted to (or repelled from) the fixed magnet, which creates the mechanical force that moves the coil up and down in the gap. The moving coil is attached to the moving diaphragm, which tracks the audio signal and creates the acoustic transmission.

3.10.2 Limits

We return again to our pipeline analogy. The pipe's outer limits are the speaker's maximum excursion, controlled by the compliance of the spider and surround. Distortion rises as mechanical resistance increases near the physical limits. If completely overrun, the former can bottom out against the magnet or move too far forward and jump out of the gap. In either case there is high risk of bending the fragile former and destroying the speaker. The surround and spider can fatigue over time and gradually lose their ability to control the motion. There are plenty of other ways to mechanically destroy a speaker such as breaking glue bonds, over-flexing the wires that connect to the coil, shattering metal diaphragms and endless more. Innovative audio engineers find new ways to destroy speakers every day!

Heat is the other enemy, the by-product of electrical current flowing through the voice coil wire. Excess heat can melt the insulation, shorting the circuit turn by turn. A shorted turn reduces the impedance, beginning a speaker coil death spiral. Each shorted turn further lowers the impedance, increasing the current and raising the risk of shorting more turns and/or endangering the power amp. Heat can also warp the coil on the former, leading to scraping in the gap. The circuit will be opened and the speaker silenced if the wire's melting point is reached.

Speaker coil impedance rises with temperature. A heated coil has lower efficiency than a cool one and therefore our speakers can lose level over the course of a show when driven hard. If you've ever thought a system seems to be losing impact as the show goes on, you may have been right. The show begins on cool, low-impedance drivers, but does not necessarily end that way.

3.10.3 Power to SPL sensitivity (1 W/1 m)

The standard speaker sensitivity rating is the answer to the question, "How much SPL at 1 meter do I get with 1 watt drive?" The two different log equations are found on the opposite side of this transduction. SPL generated at further distances is extrapolated from here by the 6 dB/doubling loss rate ($20*\text{Log}_{10}$). Power consumption is extrapolated at the 3 dB rate ($10*\text{Log}_{10}$). A speaker with a sensitivity of 100 dB (1 W/1 m) creates 103 dB at 2 watts drive and 120 dB with 100 watts. This doesn't mean that it will necessarily generate 130 dB when driven at 1000 watts. It may generate smoke instead.

What does sensitivity tell us? A driver with a sensitivity of 97 dB (1 W/1 m) needs twice the power to match SPL with a 100 dB sensitivity driver. If both can reach the same maximum SPL, then the high-sensitivity driver is a bargain due to reduced amplifier cost. But sensitivity is only a starting reference point. It won't tell us the maximum capability. A Ferrari and Prius can have matched speeds in first gear but their maximums are miles apart.

Matched sensitivity does not guarantee matched response at a given frequency for drivers covering different spectral ranges. Because SPL values are full-range integrations, we can't assume that sensitivity-matched LF and HF drivers will be equal at the intended crossover frequency.

Sensitivity also doesn't contribute much to amplifier/speaker pairing because we need to know the speaker's maximum capability. Current industry practice is for manufacturers to state the maximum power capability and recommended amplifier range. Engineers then specify amps with much higher ratings.

The 1 W/1 m rating fails to factor in amplifier voltage gain, because it's based on output power only. Therefore sensitivity-matched speakers driven with sensitivity-unmatched amplifiers (yes it's the same term but a different conversion) still have matched 1 W/1 m values but unmatched SPL levels. Where does speaker impedance fall in this mix? If an 8 Ω speaker and a 16 Ω speaker have matched sensitivity they will need different amplifier sensitivity (voltage gain) to achieve a matched 1 watt output. Dizzy yet?

Let's add another layer (or twenty). Chances are high that we will be listening to an array of speakers: complex mixtures of active multi-way speakers with different sensitivities, impedances, drive levels, amplifiers and amounts of acoustic addition over frequency. The question we really want answered is, "How many dB SPL can I get out of the console?"

Here's the good news. You probably don't ever need to ponder a sensitivity spec unless you are designing loudspeakers for manufacture. Pair the speaker with the right amp based on maximum capability ratings and adjust amplifier sensitivity (voltage gain) by measuring the frequency response. I've designed and tuned for thirty years without making a single decision by analyzing the 1 W/1 m sensitivity spec (Fig. 3.35).

Loudspeaker sensitivity reference														
Amplifier power rating			94 dB 1w /m SPL @ 1m			100 dB 1w /m SPL @ 1m			106 dB 1w /m SPL @ 1m			112 dB 1w /m SPL @ 1m		
16Ω	8Ω	4Ω	16Ω	8Ω	4Ω	16Ω	8Ω	4Ω	16Ω	8Ω	4Ω	16Ω	8Ω	4Ω
1024 W	2048 W	Fire	124	127	130	130	133	136	136	139	142	142	145	148
256 W	512 W	1024 W	118	121	124	124	127	130	130	133	136	136	139	142
64 W	128 W	256 W	112	115	118	118	121	124	124	127	130	130	133	136
16 W	32 W	64 W	106	109	112	112	115	118	118	121	124	124	127	130
4 W	8 W	16 W	100	103	106	106	109	112	112	115	118	118	121	124
1 W	2 W	4 W	94	97	100	100	103	106	106	109	112	112	115	118
0.5 W	1 W	2 W	91	94	97	97	100	103	103	106	109	109	112	115
0.25 W	0.5 W	1 W	88	91	94	94	97	100	100	103	106	106	109	112

FIGURE 3.35
Loudspeaker sensitivity reference

3.10.4 Voltage to SPL sensitivity (dB SPL/volt)

We've wired up a pile of speakers and are ready to start the show. How do we ensure we get the most out of our system and still have something worth packing in the truck when it's over? Amps and speakers are paired, cables run and crossover levels set to achieve the desired combined frequency response. Limiters are calibrated to the maximum sustainable mechanical and heat stress.

We return for one final round of the pipeline. The key is to get the signal cleanly through the line level, speaker level and acoustic pipelines without a clog point. The console drive is voltage or "faux voltage" (digital signal) and will max out around 10 V (+20 dBV). We must make sure all of the various signals going to different speakers (highs and lows, mains and fills) are still operating linearly until the final stage.

The answer to the "How much SPL at the console question?" can be boiled down to a different form of sensitivity: the SPL/console ratio. This sounds ridiculous but you've been doing it forever. Drive the console to the nominal line level (0 dBV) and measure the SPL at the console. That's your baseline dB SPL/volt value. If everything downstream is running perfectly, you may have up to 20 dB of headroom before you hit the wall.

Now let's relate this to self-powered speakers. The 1 W/1 m sensitivity rating is truly academic in a system with line-level inputs. The dB SPL/volt figure illustrates the speaker's potential at a nominal drive level.

3.10.5 Coverage angle characterizations

Now that sound is moving through the air we need to characterize its spatial response. There are a great variety of ways of describing the response, some of which have become outdated in the modern era of acoustic prediction software. Inevitably we use coverage angle representations to aid our process of speaker selection, splay angle, spacing and level scaling. The data that aids this process most directly is now favored in the industry (Fig. 3.36).

3.10.5.1 POLAR PLOTS AND ISOBARIC CONTOURS

There's a top-secret conspiracy by the bottled-water industry to accelerate global warming so they can tap the melting Antarctic ice floes (a polar plot). The other polar plot is a log rendering of level over angle (Fig. 3.37). Measure a speaker on axis and make this the reference or "normalized" response. Now spin the speaker and chart its loss (or gain) in dB. The result is plotted over a series of log-spaced concentric circles. An omnidirectional speaker (or mic) looks circular, while increasingly directional responses show log-scaled shrinkage in the attenuated coverage angles. There is nothing inherently wrong with traditional polar patterns. But they are easily confused with the spatial response we find in a room. Rooms are linear (not log) so we cannot take a polar plot and lay it over a room drawing and match their shapes. Losses of 6, 12 and 18 dB show as equal-spaced contours on a polar plot, but represent successive reductions to ½, ¼ and ⅛ size in a room.

Isobaric contours are pressure maps. We most often see them during the weather forecast. These are the most common rendering in prediction programs with a color series to denote level. The unity-level coverage shape can be seen directly by following any color transition line. This is the standard view of speaker coverage over space.
I was able to convince the publisher of the first edition of this book to make the first full-color book in pro audio because of the importance of isobaric contour color mapping.

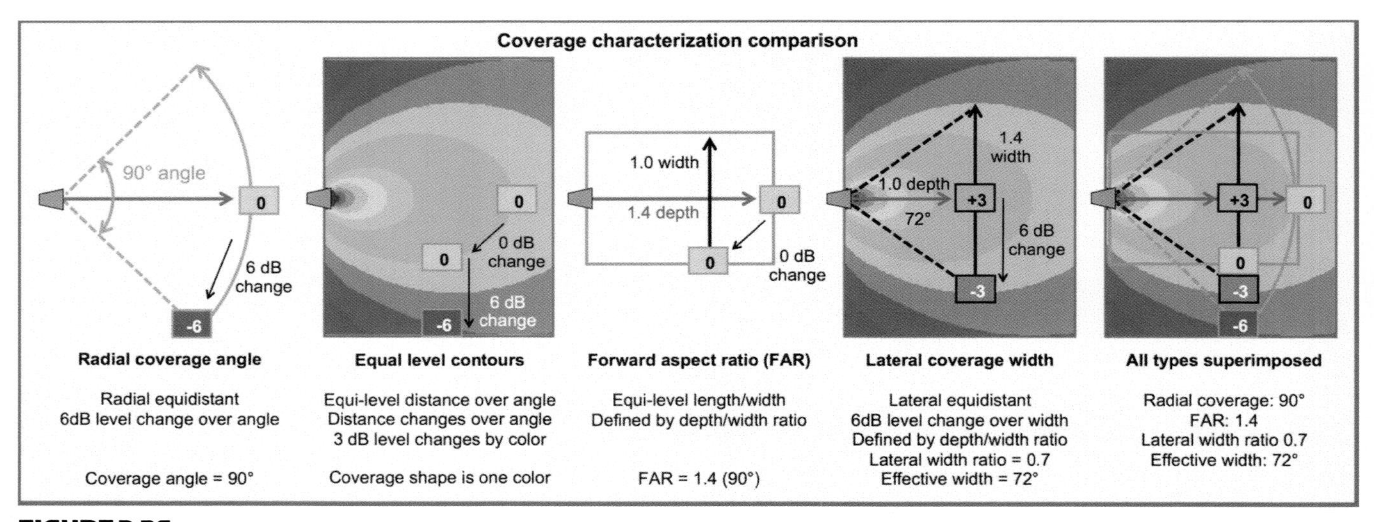

FIGURE 3.36
Coverage angle characterization comparison

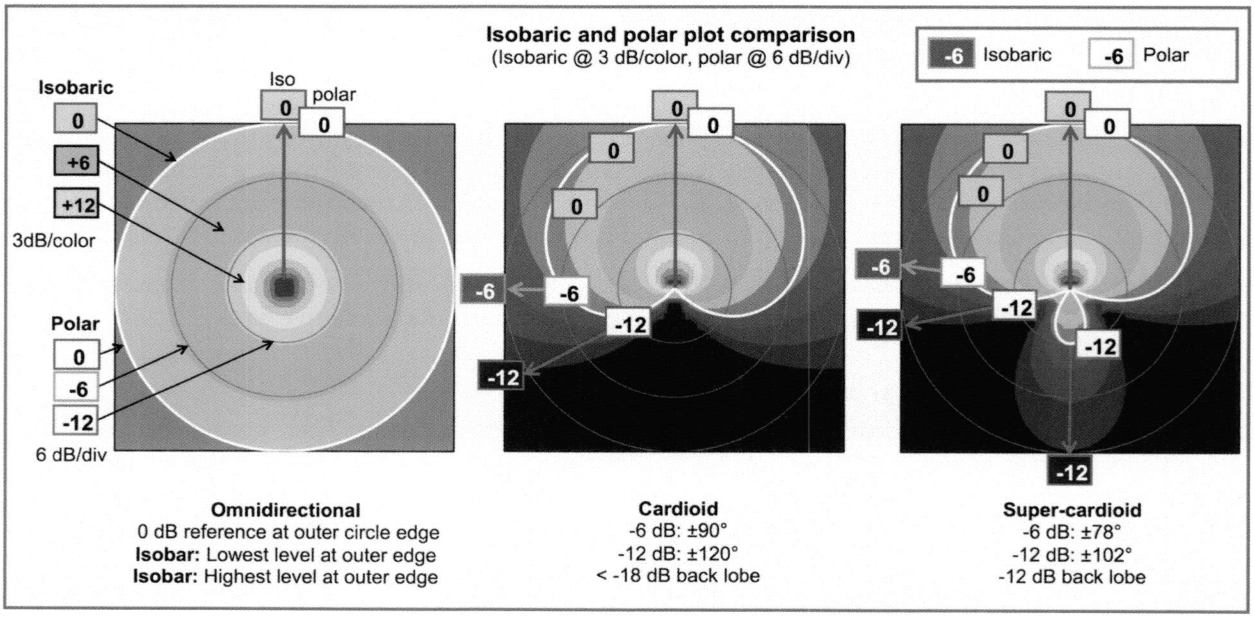

FIGURE 3.37
Comparison of superimposed polar and isobaric renderings of the same data (polars courtesy of Merlijn van Veen)

3.10.5.2 RADIAL COVERAGE ANGLE

Coverage angle is the radial area between off-axis points (6 dB down from on axis). The loss is attributable entirely to angle because all points along the arc are equidistant (Fig. 3.38). Coverage angle is yet another single value for something that varies widely with frequency. What range should we use as the "nominal" value? How about 1 kHz since we use it for electronic devices? Probably a poor choice. The key is to establish the range where listeners would perceive the transition from good to poor coverage. HF roll-off is an unambiguous clue that we are either far away or not in the coverage area of a sound source. Midrange roll-off is harder to wrap our ears around if we are still getting a strong HF response. The landmark to look for is the beamwidth plateau, the range where a steady roll-off is maintained. A well-designed system will reach the plateau in the midrange and carry it through the HF. The "nominal" angle is the one that you would use to achieve unity splay with a second unit (2× 60° speakers at 60° will have 0 dB at the center).

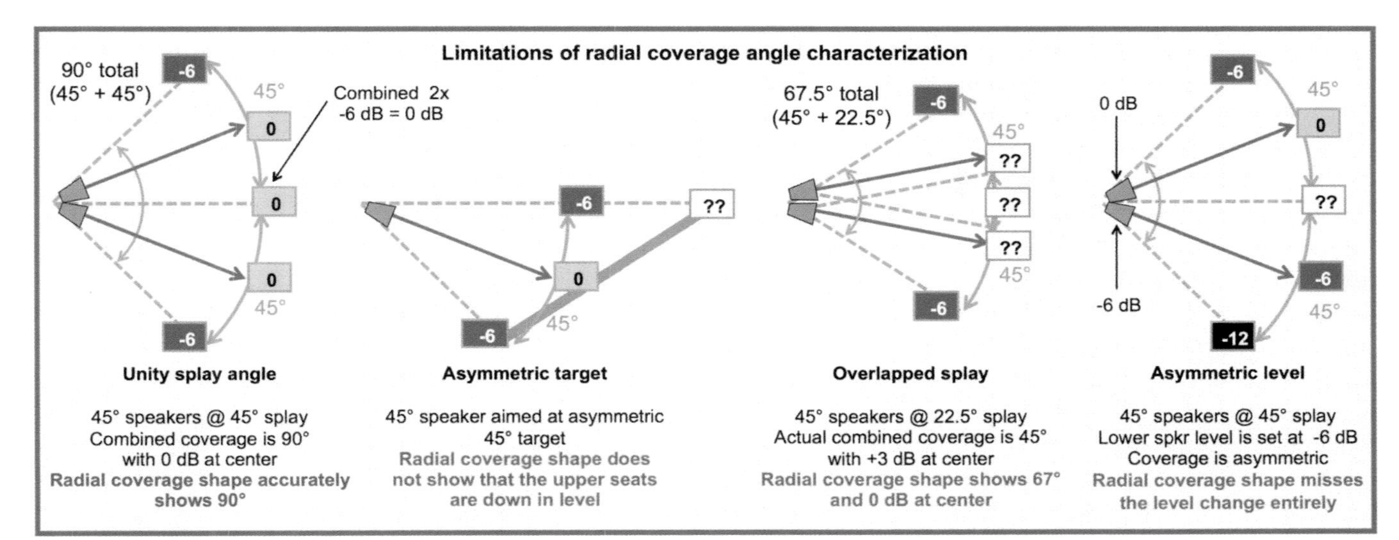

FIGURE 3.38
Limitations of radial coverage angle characterization

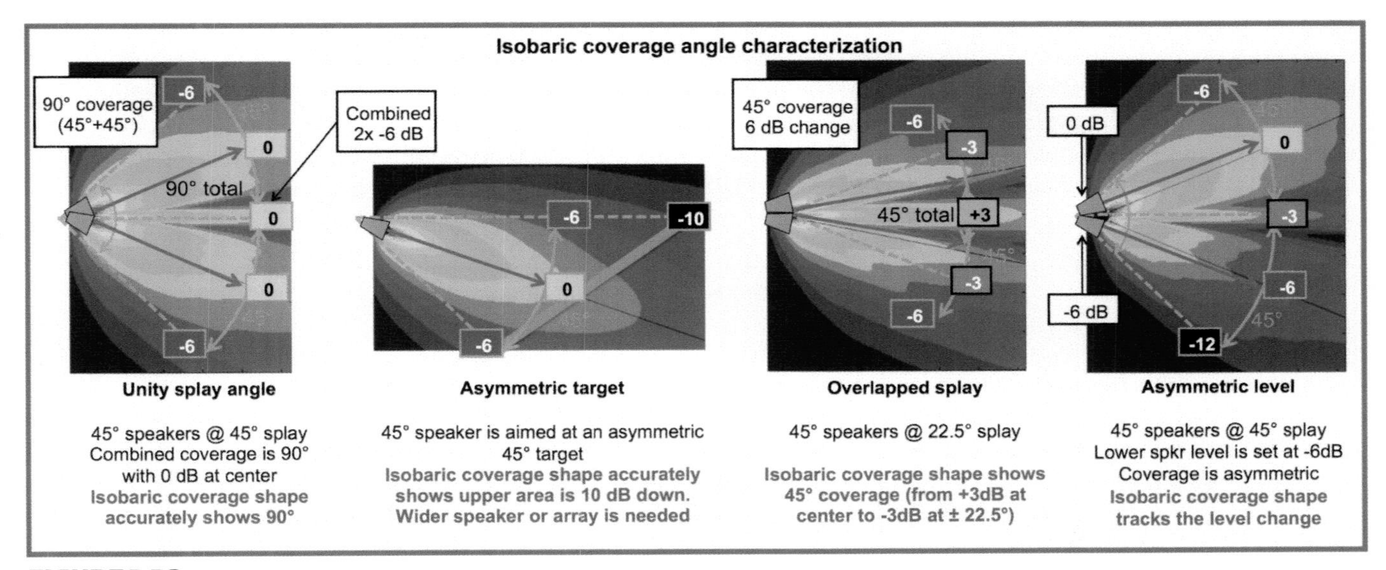

FIGURE 3.39
Speaker coverage shapes: isobaric contours

The radial shape can give some misleading impressions worth addressing. The first is seeing the solid outer edges of the shape as hard edges of sound quality. Sound is not even across the arc and it does not cut off sharply at the edge. Inside the edge is -5 dB and outside of it is -7 dB. Second, bear in mind that if we are different distances from the outer edges then we are at different levels. Third, note that the radial coverage angle has no level scaling. The shape continues to infinity, unchanged by distance or relative level, e.g. the coverage angle lines of the frontfill run all the way to the back of the room even though they are not heard there.

3.10.5.3 SPEAKER COVERAGE SHAPE

The distinction between "coverage pattern" and "coverage shape" is unique to this and my other books. I use "shape" in reference to the unity gain line found in isobaric pressure mapping (the standard representation in prediction programs). The coverage shape is rescaled over distance (it expands) and with relative level (it contracts if a speaker is turned down). The coverage shape links to the coverage angle's milestones (on and off axis).
The off-axis point connects to an on-axis point at 2× the distance. The relationship of "on-axis far to off-axis near" will be a recurring theme here (Fig. 3.39).

3.10.5.4 FORWARD ASPECT RATIO (FAR)

The coverage shape can be conceptualized as a rectangle with proportional length and width. Architects characterize room shape by aspect ratio (length by width) and we can do the same with the speakers we are putting in there. The rectangle includes only the frontal lobe of the response, hence the name "forward aspect ratio (FAR)." The FAR shape connects on-axis far (length) to off-axis near (width). The aspect ratio rises as speaker coverage angle narrows (Fig. 3.40).

The street term for this is "throw." Long-throw speakers reach to the back, whereas short-throw speakers play to the front. How is this quantified? How far does a speaker throw anyway? Infinity actually, unless something stops it or it runs out of air. Aspect ratio gives us a level scalable shape. A speaker with a high FAR will throw . . . far.

3.10.5.5 LATERAL COVERAGE ANGLE

The standard coverage angle shape is radial but we very often find ourselves with flattened coverage targets. Consider an overhead speaker covering a flat floor. The listener directly underneath is more on axis and closer than any other. Using the response here as the reference and moving away, we will lose 6 dB as a result of combined angular and distance loss. A speaker with a radial coverage angle of 90° will achieve less than this across a flat coverage line due to the added distance loss (e.g. if it lost 2 dB by distance it could only lose 4 dB by angle for a combined 6 dB loss). The result is a notable shrinkage of coverage pattern with higher percentage losses at large radial angles.

3.10.5.6 COVERAGE WIDTH (LATERAL ASPECT RATIO)

Coverage width is a variation on the lateral coverage shape used for flat-line coverage targets (Fig. 3.41). The coverage area is characterized in distance (meters or feet) instead of angle. Coverage width is the answer to the question, "How much coverage do I have across a flat line with a 90° speaker at 3 m?" Answer: 4.2 m. A closer target or narrower speaker covers a proportionally smaller target. Coverage width is used to determine spacing between speakers and to help determine the need for fills etc. An example application is shown in Fig. 3.42.

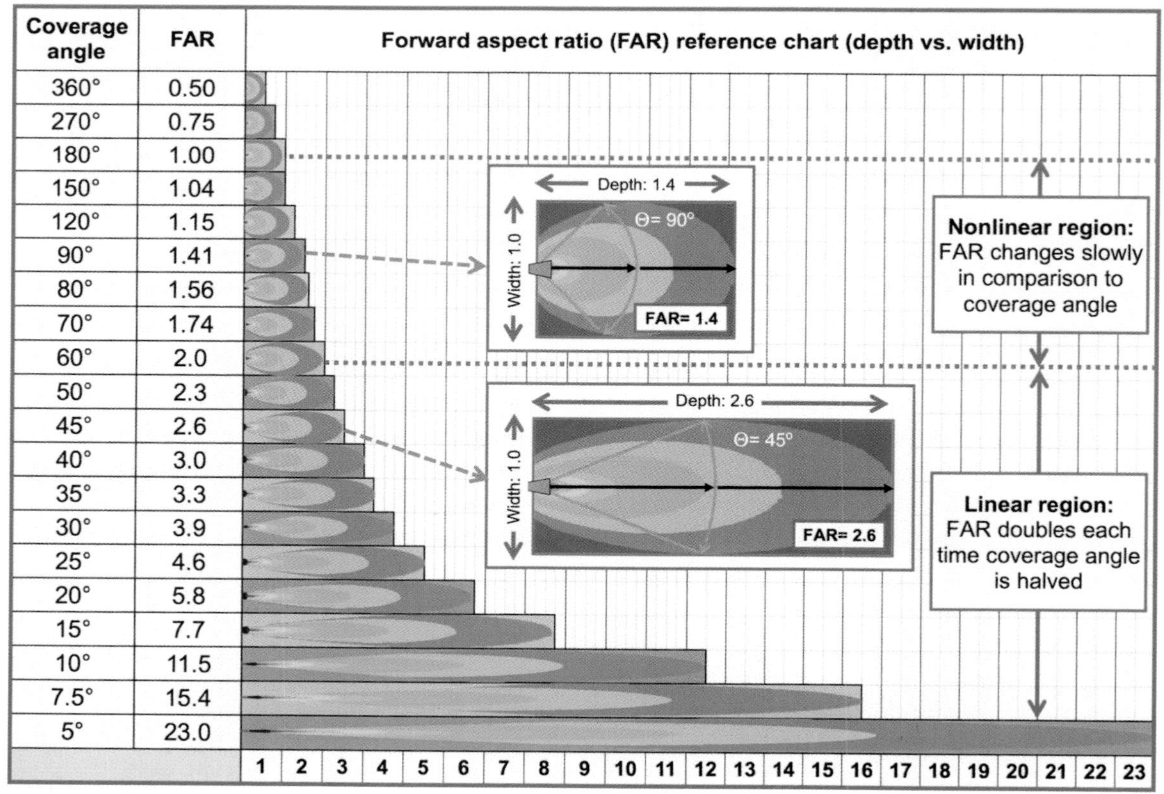

FIGURE 3.40
Speaker coverage shapes: forward aspect ratio (FAR)

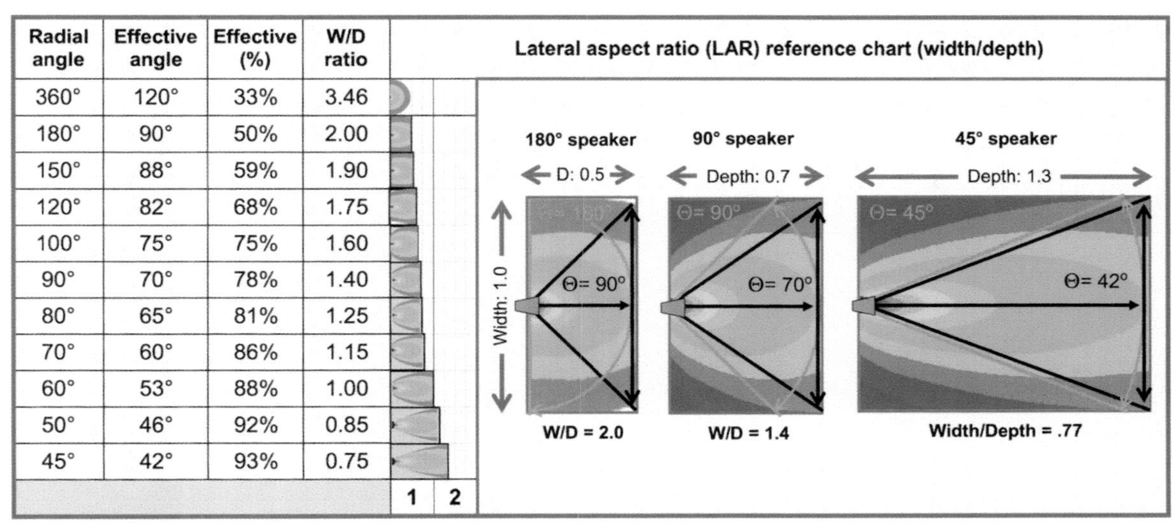

Radial angle	Effective angle	Effective (%)	W/D ratio
360°	120°	33%	3.46
180°	90°	50%	2.00
150°	88°	59%	1.90
120°	82°	68%	1.75
100°	75°	75%	1.60
90°	70°	78%	1.40
80°	65°	81%	1.25
70°	60°	86%	1.15
60°	53°	88%	1.00
50°	46°	92%	0.85
45°	42°	93%	0.75

FIGURE 3.41
Speaker coverage shapes: lateral aspect ratio

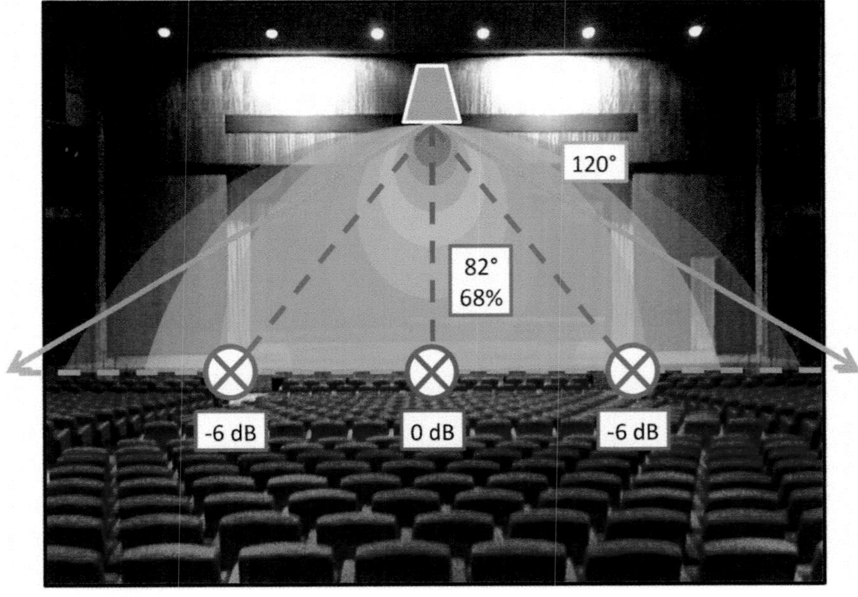

FIGURE 3.42
Lateral aspect ratio example application, center downfill (photo and graphic courtesy of Merlijn van Veen)

3.11 TRANSMISSION COMPLETE

It's been a long journey through the variations of the modern transmission system. The entire path was analog when I began my pro audio career and will be almost all digital by the end. Presently there remain many variations that complicate the process. Many current systems have generations of technology still in the signal chain. Some designers favor gadgetry and complexity over simplicity. We may encounter any and all of the transmission variations, transductions and numerical conversions outlined in this chapter on any given day. Hopefully this information will help maintain a vision of the signal as it moves along from start to finish. Transmission complete.

Trap 'n Zoid by 6o6

Trap 'n Zoid by 6o6

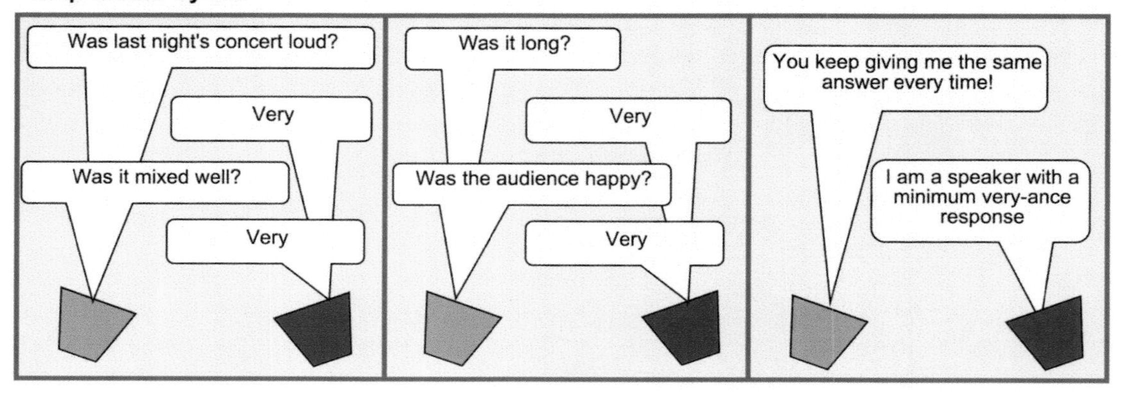

FURTHER READING

Giddings, P. (1990), *Audio System: Design and Installation*, Sams.
Huntington, J. (2012), *Show Networks and Control Systems*, Zircon Designs Press.
http://www.soundonsound.com/sos/apr03/articles/digitalclocking.asp.
http://www.cobranet.info/node/2.
http://en.wikipedia.org/wiki/AES3.
http://en.wikipedia.org/wiki/Quantization_(signal processing).
http://en.wikipedia.org/wiki/Digital_signal_processing.
http://en.wikipedia.org/wiki/ATA_over_Ethernet.
http://en.wikipedia.org/wiki/Digital_audio.
http://en.wikipedia.org/wiki/Optical_fiber_cable.
http://en.wikipedia.org/wiki/TOSLINK.
https://garyherlache.wikispaces.com/Digital+Communications+Fall+2009+Fiber+basics.
http://en.wikipedia.org/wiki/Characteristic_impedance.
http://en.wikipedia.org/wiki/OSI_model.
http://en.wikipedia.org/wiki/Acoustic_shadow.
http://www.avnu.org.
http://www.computersgags.co/2015/01/the-7-layers-of-osi-model.html.
http://www.belden.com.
http://www.sfu.ca/sonic-studio/handbook/Sound_Propagation.html.

CHAPTER 4

Summation

summation n. addition, finding of total or sum; summing up.
sum n. & v. total amount resulting from addition of items.
Concise Oxford Dictionary

An old truism of audio wisdom states, "point the loud side toward the audience." It's good advice for a single speaker. With two or more, a powerful new force arises that can steer the sound to parts unknown: summation. It's hard enough to visualize the shape of sound from a single speaker, but we must dig much deeper to understand combinations. The interactions differ at every frequency and every location. Although complex, they are predictable and we have the decoder keys: level and phase offset.

Understanding summation is a big challenge but an even bigger payoff. The understanding of level and phase offset gives us a working crystal ball. I have measured sound systems for thirty years and never seen a single exception to the rules of summation that follow. We can demystify what happens wherever low drivers meet high drivers, speakers meet speakers or speakers meet walls.

This chapter draws a continuous line between three of our sound system's most interactive and volatile aspects: the spectral divider (crossover), the speaker array and the room. Summation behavior of these seemingly distinct entities so closely relate that they can be lumped into a single category: acoustical crossovers. The crossover is where correlated waveforms of equal level come to meet. When they do, whether they come from different driver types, different speaker cabinets or off the walls, they follow the rules of summation.

4.1 AUDIO ADDITION AND SUBTRACTION

Summation occurs whenever and wherever two or more matched frequency audio signals are combined. There are only three possible outcomes: the summed signal level will be greater, lesser or the same as the individual signals.

4.1.1 Correlated summation criteria

We can evaluate summation at this moment and get one answer. We'll get a completely different answer moments later if we're combining unrelated signals. Only correlated signals sum in a stable fashion, creating a consistent and predictable frequency response and spatial distribution. Optimization strategies are based around the stable results of correlated summation. Correlated signals maintain a linear level and phase relationship, i.e. they sing the same song.

4.1.1.1 SOURCE CORRELATION

Correlated waveforms are descendants of a common original source (Fig. 4.1). In genetic terms, they are children of the same parental waveform. If they are identical twins, the summation will behave like simple mathematical addition: 1 + 1 = 2. If there are differences between them, we evaluate the interaction in complex form: amplitude and phase. The response is stable, predictable, measurable and possibly treatable with delay, equalization and/or other alignment procedures. By contrast, the uncorrelated summation of Beethoven's 9th Symphony with Black Sabbath's "Iron Man" would be unstable, because there would not be sustained periods of source matching. A hybrid case between these extremes is stereo, a semi-correlated summation. Only the center-panned parts of the mix will create a stable summation. A mostly mono mix will have higher stability than one with high proportions of separation.

4.1.1.2 DURATION

Summing two copies of a signal doesn't guarantee things get louder. The signals must be close enough in time that they share the space (or the wire etc.). Playing a song through two speakers in parallel makes it twice as loud, but

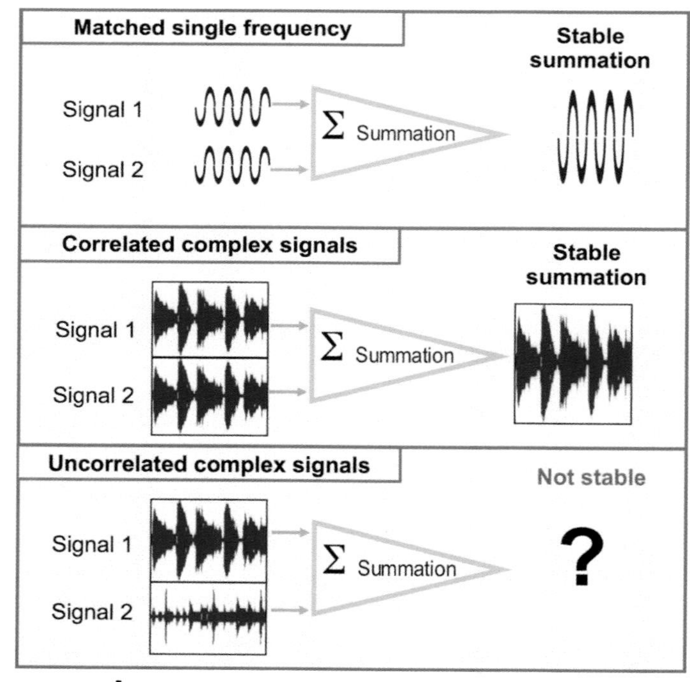

FIGURE 4.1
Source correlation effects on summation: Correlated sources exhibit stable summation over frequency. Uncorrelated sources create a constantly changing summation.

not longer. Playing it twice in series makes it longer, but not louder. When time offset sources are summed, the result is a mixture of serial and parallel effects: somewhat louder, and somewhat longer (Fig. 4.2).

Summing time offset correlated sources yields three possible outcomes: highly overlapped, minimally overlapped or separated. These are not clear distinctions in mathematical terms, but our hearing perception differs along this continuum. Highly overlapped summation is perceived as increasing loudness and modifying the tonal structure. At the other extreme is the non-overlapping summation, which is perceived as a separated, separated, echo, echo. Partial overlap is the stretchy area in between that forces us to consider the source material and the frequencies involved. We'll never perceive separation if the source is continuous noise. We can find separation in a heartbeat by using a transient pulse. The perception threshold changes with music because it mixes continuous and transient signals. The separation threshold gets shorter as frequency rises, e.g. very short reflections can make a high-hat sound like two sources whereas it takes very long reflections to perceive two basses. Our analyzer's quasi-log frequency response is tailored to account for this perception shift by using longer time captures at lower frequencies. Signals that arrive later than our analyzer's capture time (e.g. around 5 ms for the highest frequencies) are treated as separated (deemed to be uncorrelated noise by the analyzer).

4.1.2 Electrical vs. acoustical summation

Although most summation properties apply equally to electrical and acoustical systems there are some obvious differences. Electrical summation has a negligible geometric dimension, whereas acoustical summation can't be considered without it. Electrical additions and subtractions are imprinted on the signal wherever our wire takes it. Complete cancellation is possible. Any single point in the acoustical space will behave like electronic summation, but don't expect to see the same trace somewhere else. Acoustical summation's additions and

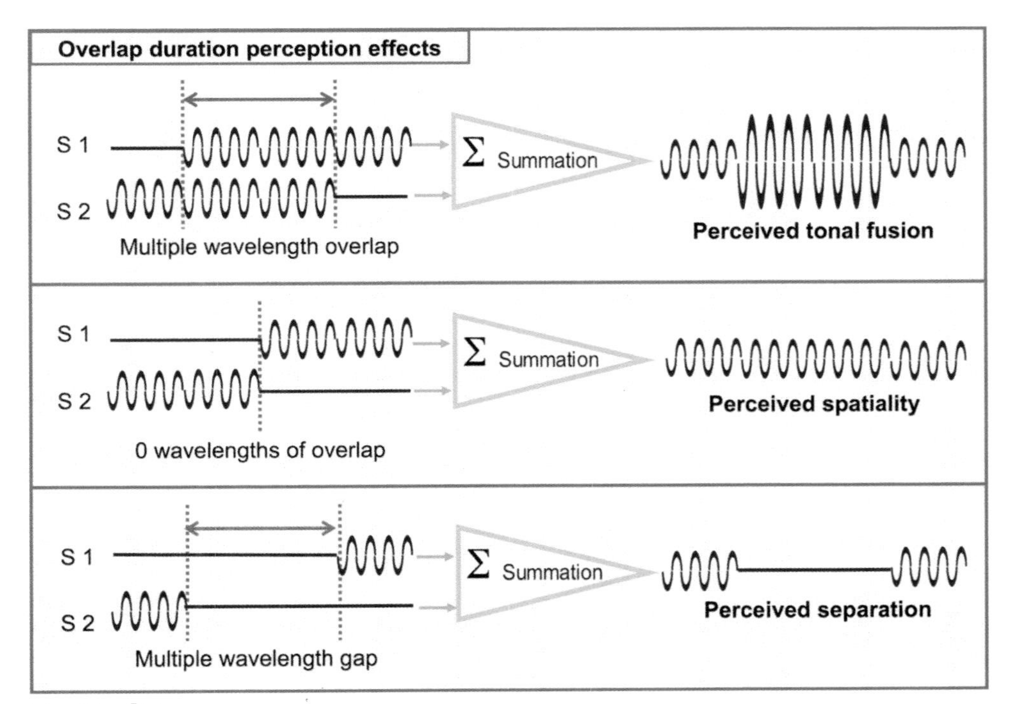

FIGURE 4.2
Summed signal overlap: The perception of summation effects changes with the amount of signal overlap

subtractions play out differently at every location in the room. It's impossible to cancel at one location without adding somewhere else.

4.1.3 Acoustical source direction

Acoustical summations contain signals arriving from waves propagating in different directions (Fig. 4.3). Relative propagation direction doesn't affect sound pressure level at a given location. The particle movement direction (the sound intensity) differs for sources summed from different locations but isn't a factor in the summed tonal response (our concern in system optimization).

Relative source direction has a huge effect upon the *uniformity of distribution* of the summation over the whole of the space. The directional relationship between sound sources will be a prime factor in the *rate of change* of the summation over the space (see section 9.1).

4.1.4 Simple summation math

Let's start with audio addition and subtraction. The simplest form is 1 + 1 = 1 (±1)* with the asterisk reminding us that it depends upon relative phase. The combination of two equal-level signals can double the money, break even or lose it all. These three different outcomes are driven by relative phase values of 0°, 120° and 180° respectively. We'll return to phase shortly. For now let's assume matched phase and isolate the amplitude addition mechanism.

The summation of two synchronized correlated audio sources is:

$$\text{Summation} = 20 \times \text{Log}_{10} \frac{(A + B)}{A}$$

where A is the stronger signal and B is an equal or weaker signal.

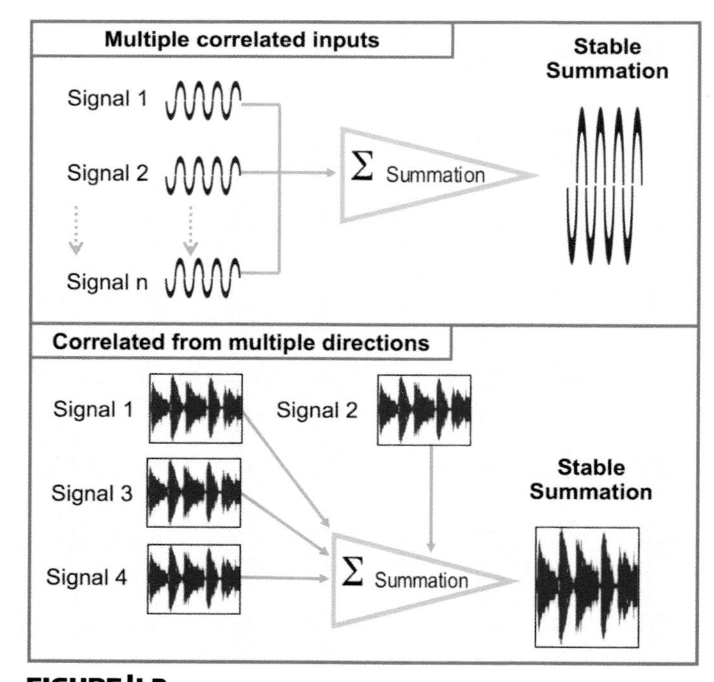

FIGURE 4.3
Multiple input summation: Number of correlated inputs is unlimited. Source direction does not affect summation stability.

This formula requires a log function calculator. Alternatively, we can memorize a reference table but that doesn't help understand it. The problem is not you. It's log. It's great for tracking audio-level changes, but stinks at math. Audio math becomes intuitive after translation back to linear. We don't usually see audio levels expressed in their linear ratio form. If we did, we'd add two equal-level signals as 1 + 1 = 2. Instead we use rote memorization for the same equation in log form: 0 dB + 0 dB = +6 dB. We've memorized (-6 dB) + (-6 dB) = 0 dB, the translation of ½ + ½ = 1. I'll bet you haven't memorized the answer to (-12 dB) + (-12 dB) but you'd have no trouble with ¼ + ¼ = ½ (-6 dB). There's a pattern here: The sum of two level (and phase) matched signals is a doubling (2×, +6 dB).

Things get challenging when the levels don't match. Let's solve 0 dB + (-6 dB). Do you find yourself thinking -3 dB? The linear decoding is 1 + 0.5 = 1.5 (+3.5 dB). Mystery solved! The linear to log-level conversions for a single source are found in Figure 4.4.

The formula A + B = AB finds the gain from summing two phase-matched sources by linear addition using the single source values. All we need is each relative level (in dB) and then linearly convert them. Our example is 100 dB SPL (A) + 98 dB SPL (B) = AB. Step 1 is to note the difference of 2 dB. Source (A) is given a relative value of 0 dB (1) and (B) is -2 dB (0.8). We add them as 1 + 0.8 = 1.8. The decoder table shows that a linear ratio of 1.8 corresponds to a gain of +5 dB. Our summed level (AB) is 100 + 5 (105 dB SPL).

Loss	Gain
Log = Linear	Log = Linear
0 dB = 1.00	0 dB = 1.00
-1 dB = 0.89	+1 dB = 1.12
-2 dB = 0.79	+2 dB = 1.26
-3 dB = 0.71	+3 dB = 1.41
-4 dB = 0.63	+4 dB = 1.59
-5 dB = 0.56	+5 dB = 1.78
-6 dB = 0.50	+6 dB = 2.00
-7 dB = 0.45	+7 dB = 2.24
-8 dB = 0.40	+8 dB = 2.51
-9 dB = 0.35	+9 dB = 2.82
-10 dB = 0.32	+10 dB = 3.16
-12 dB = 0.25	+12 dB = 4.00
-15 dB = 0.18	+15 dB = 5.60
-18 dB = 0.13	+18 dB = 8.00
-20 dB = 0.11	+20 dB = 10.00

FIGURE 4.4
Log/linear ratio decoder (20 Log_{10} scale).

Figure 4.5 is a linear/log conversion (1 dB iterations) of A + B = AB. The beauty of the linear ratio model is we don't have to stop at two. Three boxes at equal level is a linear ratio of 3:1, and four is 4:1. Yes, it is that simple (in linear). The log translations would be +10 dB and +12 dB respectively. We can also add multiple sources at different levels and find the result with ease. Let's try -3 dB + (-6 dB) + (-9 dB) + (-12 dB). The linear ratios are found as 0.7 + 0.5 + 0.36 + 0.25 = 1.81 (+5 dB again).

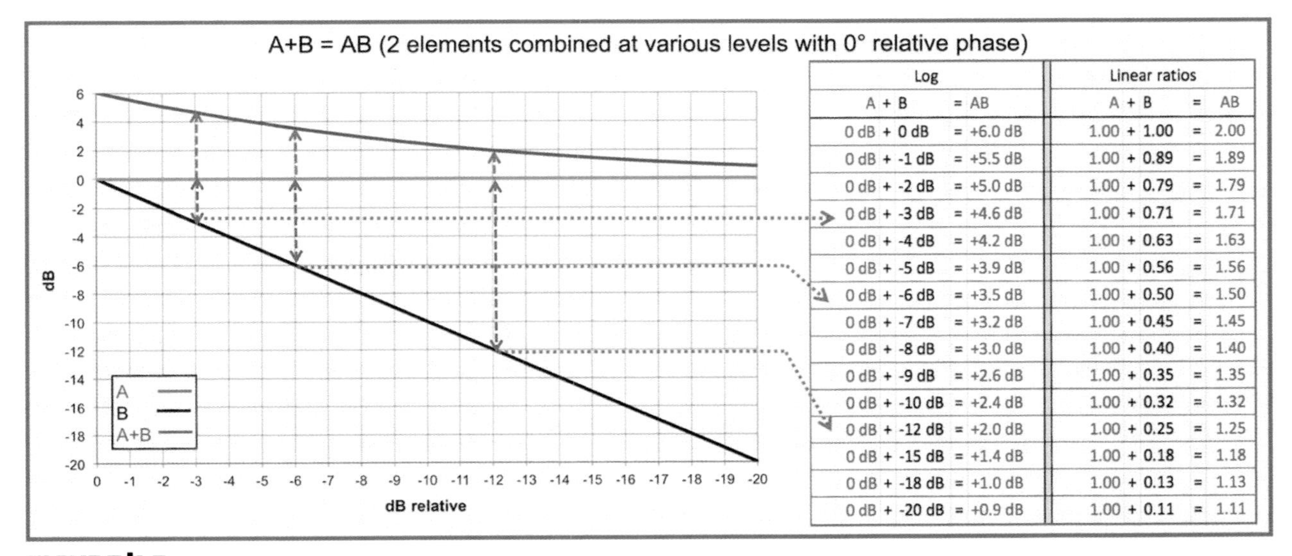

Log		Linear ratios	
A + B	= AB	A + B	= AB
0 dB + 0 dB	= +6.0 dB	1.00 + 1.00	= 2.00
0 dB + -1 dB	= +5.5 dB	1.00 + 0.89	= 1.89
0 dB + -2 dB	= +5.0 dB	1.00 + 0.79	= 1.79
0 dB + -3 dB	= +4.6 dB	1.00 + 0.71	= 1.71
0 dB + -4 dB	= +4.2 dB	1.00 + 0.63	= 1.63
0 dB + -5 dB	= +3.9 dB	1.00 + 0.56	= 1.56
0 dB + -6 dB	= +3.5 dB	1.00 + 0.50	= 1.50
0 dB + -7 dB	= +3.2 dB	1.00 + 0.45	= 1.45
0 dB + -8 dB	= +3.0 dB	1.00 + 0.40	= 1.40
0 dB + -9 dB	= +2.6 dB	1.00 + 0.35	= 1.35
0 dB + -10 dB	= +2.4 dB	1.00 + 0.32	= 1.32
0 dB + -12 dB	= +2.0 dB	1.00 + 0.25	= 1.25
0 dB + -15 dB	= +1.4 dB	1.00 + 0.18	= 1.18
0 dB + -18 dB	= +1.0 dB	1.00 + 0.13	= 1.13
0 dB + -20 dB	= +0.9 dB	1.00 + 0.11	= 1.11

FIGURE 4.5
Level effects on two input summation in log and linear form

Multiple input summation reference		
Linear	Log gain*	Example
1	+0 dB	
2	+6 dB	
3	+10 dB	
4	+12 dB	
5	+14 dB	
6	+17 dB	
8	+18 dB	
10	+20 dB	
12	+22 dB	
16	+24 dB	
20	+26 dB	
24	+28 dB	
32	+30 dB	
* All inputs matched in level and phase		

FIGURE 4.6
Multiple input summation gain

8 speakers @ various levels	A	B	C	D	E	F	G	H	SUM	Log gain	Taper loss
8 @ 0 dB	1.00	1.00	1.00	1.00	1.00	1.00	1.00	1.00	8	+18.1 dB	-0.0 dB
7 @ 0 dB, -1dB	1.00	1.00	1.00	1.00	1.00	1.00	1.00	0.90	7.9	+18.0 dB	-0.1 dB
6 @ 0 dB, -1, 2 dB	1.00	1.00	1.00	1.00	1.00	1.00	0.90	0.80	7.7	+17.7 dB	-0.4 dB
5 @ 0 dB, -1,2,3 dB	1.00	1.00	1.00	1.00	1.00	0.90	0.80	0.70	7.4	+17.4 dB	-0.7 dB
4 @ 0 dB, -1,2,3,4 dB	1.00	1.00	1.00	1.00	0.90	0.80	0.70	0.63	7.03	+16.9 dB	-1.2 dB
3 @ 0 dB, -1,2,3,4,5 dB	1.00	1.00	1.00	0.90	0.80	0.70	0.63	0.55	6.58	+16.4 dB	-1.7 dB
2 @ 0 dB, -1,2,3,4,5,6 dB	1.00	1.00	0.90	0.80	0.70	0.63	0.55	0.50	6.08	+15.7 dB	-2.4 dB

FIGURE 4.7
Log/linear decoder example for combined levels of eight speakers (neglecting phase). Tapering in successive 1 dB increments is shown. The loss from the maximum possible summation is shown in the last column.

Figure 4.6 gives the linear and log gains for quantities up to thirty-two units. A common question about modern speaker arrays concerns the effects of level tapering some of the cabinets. The answer can be easily approximated by linear addition. Un-tapered cabinets count as 1 whereas tapered units are derated to their linear equivalent (e.g. -3 dB = 0.7). An example is shown in Figure 4.7, which can be applied to any array.

4.1.5 Complex summation math

Let's move beyond a one-dimensional perspective to a complex summation model including amplitude and phase. Phase is very important to the outcome of summation but let's not overrate it. It's clearly secondary to amplitude. How so? If my speaker is 100 dB louder than yours, then the phase doesn't mean much, does it? By contrast, when speakers are level matched, the phase response runs the show. It's the tie-breaker. If we have a clear winner in the amplitude category, the phase effects are just details. When levels are close, phase makes the decisions about the signal's frequency response and spatial distribution.

4.1.5.1 PHASE MULTIPLIER

Adding phase to the summation equation requires math beyond our scope here: complex addition of real and imaginary numbers. As long as the levels are matched we model the phase effects as a simple gain stage multiplier that reduces gain as phase offset rises. It's when we mix variable phase offset AND variable level offset together that the math complexity tipping point is reached. As you might imagine, the real equation is a mix of real and imaginary numbers, a clear violation of my minimal math promise. A table of summation outcomes over level and phase is provided in Fig. 4.8 so we can get the answers without digging deeper, but we can peel a few layers off and aid our understanding without making the mathematicians stage a protest outside of the hall.

The summation can be modeled simplistically at the phase extremes by adding a secondary multiplication factor to the linear addition we have performed previously. Recall our simplified linear formula: 1 + 1 = 1 (±1)*, which in alphabetic form is A + B = AB. Let's add in the phase multiplier "a" as (A*a) + (B*b) = AB. Because everything here is relative, we automatically set both "A" and "a" values to 1, reducing our equation to 1 + (B*b) = AB. The multipliers are 1 for 0° and -1 for 180° respectively. Let's do two level-matched summations with phase offsets of 0° and 180° respectively:

Maximum addition (0°): (A*a) + (B*b) is 1 + (1*1) is 1 + 1 = 2 (+6 dB)

Maximum subtraction (180°): (A*a) + (B*b) is 1 + (1*(-1)) is 1–1 = 0 (-∞ dB)

This exercise exposes the role of phase offset in maximum addition (coupling) and subtraction (cancellation).

Now let's factor in level offset by turning "B" down 6 dB:

Maximum addition (0°): (A*a) + (B*b) is 1 + (0.5*1) is 1 + 0.5 = 1.5 (+3.5 dB)

Maximum subtraction (180°): (A*a) + (B*b) is 1 + (0.5*(-1)) is 1–0.5 = 0.5 (-6 dB)

We now see that without matched levels, the effects of phase addition and subtraction are reduced.

Let's take it to the next level (literally) by turning "B" down 12 dB:

Maximum addition (0°): (A*a) + (B*b) is 1+ (0.25*1) is 1 + 0.25 = 1.25 (+2.0 dB)

Maximum subtraction (180°): (A*a) + (B*b) is 1 + (0.25*(-1)) is 1–0.25 = 0.75 (-2.5 dB)

At this point we have enough difference in level to reduce the phase offset effects to less than ±3 dB. We have effectively achieved isolation, which means neither substantial addition nor subtraction will occur, an area of increasing indifference to phase offset.

These simplified equations yield the same results as the complex version at the phase extremes of 0° and 180°. The phase multiplier is a number between 1 and -1 for all other phase values but it varies with relative level so a single number cannot be attached to a particular phase value. Nonetheless we can approximate the multiplier to single numbers for phase offsets of 90° or less and stay within ±0.5 dB of the complex results. If that is not close enough then use the full formula, as follows:

$$\mathrm{Re}_n = 10^{\left(\frac{L_n}{20}\right)} \cdot \cos(\Phi_n)$$
$$\mathrm{Im}_n = 10^{\left(\frac{L_n}{20}\right)} \cdot \sin(\Phi_n)$$
$$\Sigma_{\mathrm{Re}_n} = \mathrm{Re}_A + \mathrm{Re}_B + \ldots + \mathrm{Re}_n$$
$$\Sigma_{\mathrm{Im}_n} = \mathrm{Im}_A + \mathrm{Im}_B + \ldots + \mathrm{Im}_n$$
$$L_\Sigma = 20 \cdot \log\left(\sqrt{\Sigma^2_{\mathrm{Re}_n} + \Sigma^2_{\mathrm{Im}_n}}\right)$$

Simplified phase multipliers

- 0° = 1: B is added at its actual level.
- 30° approximates to 0.92 (±0.15 dB): B is added at a derated value of -1 dB below actual.
- 60° approximates to 0.7 (±0.25 dB): B is added at a derated value of -3 dB below actual.

- 90° approximates to 0.35 (±0.5 dB): B is added at a derated value of -9 dB below actual.
- 120° equals 0 when levels are equal (so B adds nothing) and subtracts slightly as levels offset.
- 150°: B is subtracted from A but the rate is too variable over level to simplify.
- 180 = -1: B is subtracted from A at its actual level.

Now let's compare 1 dB of level offset (alone) with 30° of phase offset (alone) and visualize how the phase offset mimics the level offset effects.

B is offset -1 dB @0°: (A*a) + (B*b) is 1 + (0.89*1) is 1 + 0.89 = 1.89 (+5.5 dB)

B is offset 0 dB @30°: (A*a) + (B*b) is 1 + (1*0.92) is 1 + 0.92 = 1.92 (+5.7 dB)*

*The complex equation results differ by .04 dB.

So adding 30° of phase offset is like turning the B speaker down and extra dB. When this pair of -1 dB effects is combined, the result is equivalent to -2 dB of level offset @0°:

B is offset -2 dB @0°: (A*a) + (B*b) is 1 + (0.8*1) is 1 + 0.8 = 1.8 (+5.2 dB)

B is offset -1 dB @30°: (A*a) + (B*b) is 1 + (.89*0.92) is 1 + 0.82 = 1.82 (+5.2 dB)*

*The complex equation results differ by 0.03 dB.

The intent here is to see how phase drives the summation addition and subtraction without tears. The simplified math eases the visualization of summation when the phase offset is a below 90° or near 180°. The range between is highly volatile in any case, which limits our options there anyway. In short, it's a twisted mess back there.

Obviously there is a continuum of phase offsets (and multiplication values that vary with level offset) between the 0° and 180° milestones. There are two other lesser markers we can use to help the visualization. Addition is assured when the phase offset is <90° AND the level offset is <6 dB. When the phase offset is >120° all combinations of level offset yield some amount of subtraction.

We've taken some of the mystery out of both the amplitude and phase mechanisms, and bracketed the relevant level offset range to around 20 dB. This is compiled into a reference table (Fig. 4.8) of outcomes for two-element summation (1 dB amplitude resolution and 30° phase). It's a fairly complete decoder key for combining sources such as speakers. Notice there's only one way to achieve 6 dB addition (0 dB @0°) whereas there are four ways to get +3 dB (0 dB @90°, -4 dB @60°, -7 dB @30° and -8 dB @0°). On the dark side it's equally challenging to lose more than 6 dB (requires >150° of phase offset).

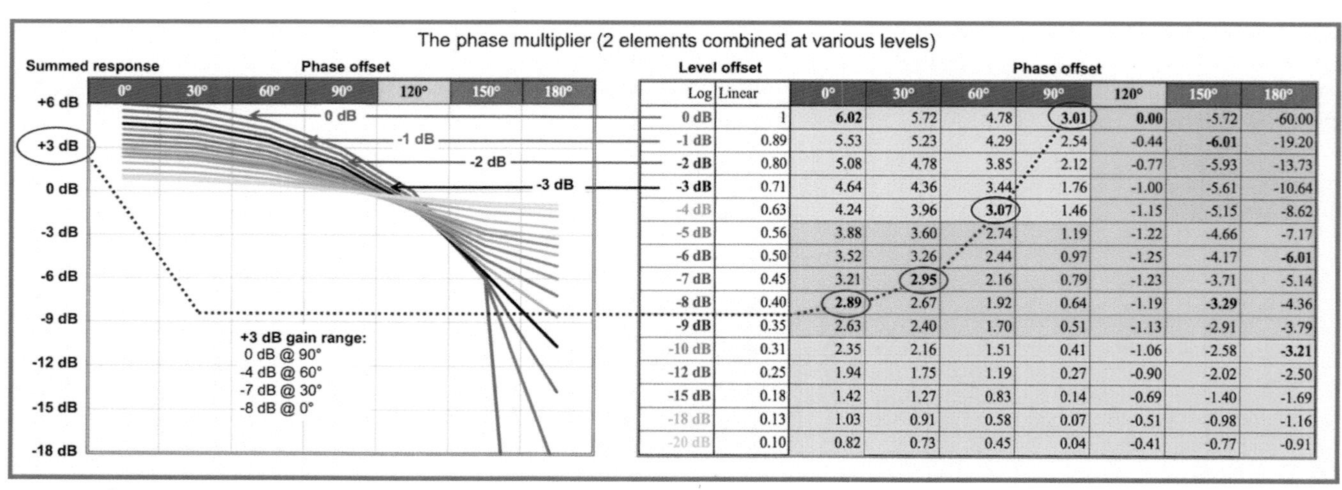

Log	Linear	0°	30°	60°	90°	120°	150°	180°
0 dB	1	6.02	5.72	4.78	3.01	0.00	-5.72	-60.00
-1 dB	0.89	5.53	5.23	4.29	2.54	-0.44	-6.01	-19.20
-2 dB	0.80	5.08	4.78	3.85	2.12	-0.77	-5.93	-13.73
-3 dB	0.71	4.64	4.36	3.44	1.76	-1.00	-5.61	-10.64
-4 dB	0.63	4.24	3.96	3.07	1.46	-1.15	-5.15	-8.62
-5 dB	0.56	3.88	3.60	2.74	1.19	-1.22	-4.66	-7.17
-6 dB	0.50	3.52	3.26	2.44	0.97	-1.25	-4.17	-6.01
-7 dB	0.45	3.21	2.95	2.16	0.79	-1.23	-3.71	-5.14
-8 dB	0.40	2.89	2.67	1.92	0.64	-1.19	-3.29	-4.36
-9 dB	0.35	2.63	2.40	1.70	0.51	-1.13	-2.91	-3.79
-10 dB	0.31	2.35	2.16	1.51	0.41	-1.06	-2.58	-3.21
-12 dB	0.25	1.94	1.75	1.19	0.27	-0.90	-2.02	-2.50
-15 dB	0.18	1.42	1.27	0.83	0.14	-0.69	-1.40	-1.69
-18 dB	0.13	1.03	0.91	0.58	0.07	-0.51	-0.98	-1.16
-20 dB	0.10	0.82	0.73	0.45	0.04	-0.41	-0.77	-0.91

FIGURE 4.8
Summation phase multiplier family of responses for two-element summation. Summation extremes are found at 0 dB, 0° and 180°. Addition and subtraction effects are reduced as level falls. Note that +3 dB summation can be achieved in four ways (0 dB @90°, -3 dB @60°, -7 dB @30° and -8 @0°). (Own work with help from Merlijn van Veen).

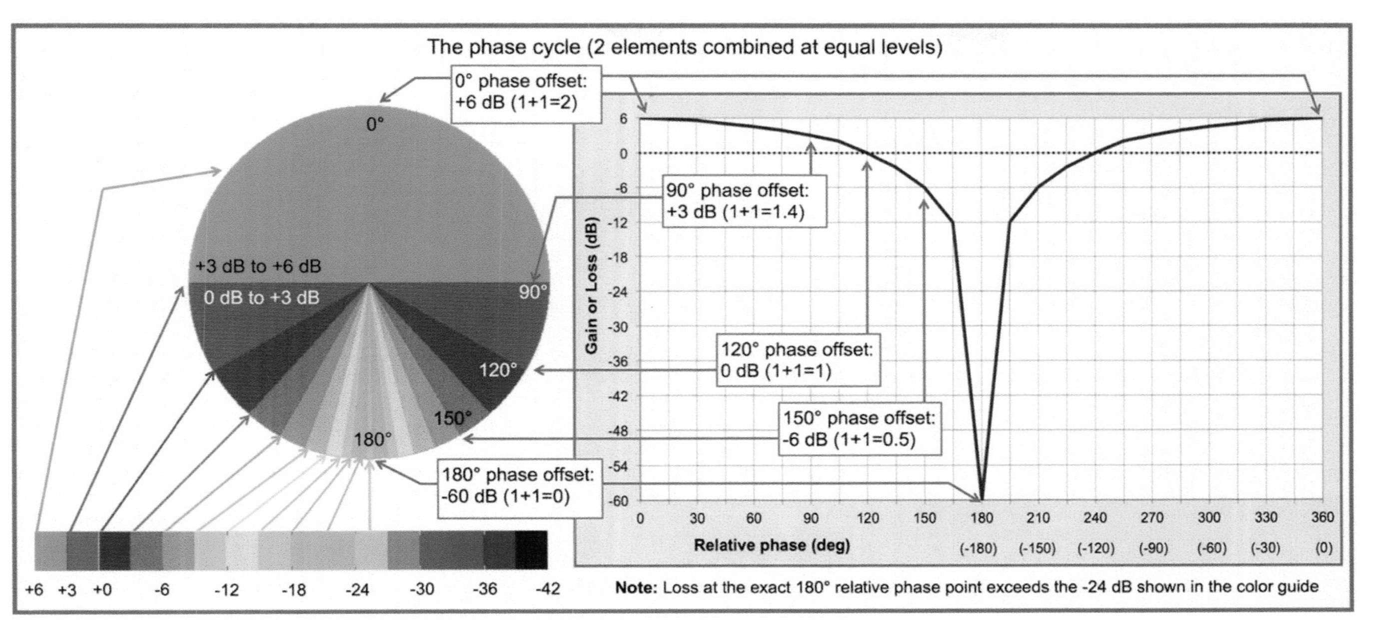

FIGURE 4.9
Phase cycle

We can go beyond two-element summation but must bear in mind that the phase relationships are element to element. An ABC summation contains the phase offsets between AB, BC and AC. Obviously AC is unique and larger. This is beyond a simple table but understanding the two-element summation will get us a long way toward predictable results.

4.1.5.2 PHASE CYCLE EFFECTS ON SUMMATION

Time and phase are related but by no means the same thing. Time marches forward in a straight line whereas phase goes in circles (Fig. 4.9). They start together with time at 0 ms and phase at 0°. Phase passes through 359°, resets to 0° and it's *déjà vu* all over again. We view the phase offset effects on summation as a circular function termed the "phase cycle." The circular nature has important implications.

- The phase cycle is asymmetric, i.e. ⅔ addition and ⅓ subtraction.
- The overall gains and losses are equal but asymmetrically distributed, i.e. additions are small and gradual, whereas subtractions are steep and deep.
- The phase sign (+ or -) does not change the amount of gain or loss, i.e. -90° is the same as +90°. Summation level does not care who came first.
- Summation amplitude is based on radial position in the cycle. For example, sources arriving at 0° and 360° have matching radial positions and add the same as 0° and 0°. There is, however, a limited indifference here, with signal duration putting an end stop on how many cycles we can fall behind and still add level.

4.2 FREQUENCY-DEPENDENT SUMMATION

We're ready to apply summation amplitude and phase effects to the frequency response. As soon as time offset is introduced between sources, there's variable phase offset over frequency. Some frequencies add while others subtract. The resulting sequence of peaks and dips is totally predictable because it follows the exact same script we just went through. We simply have to run the numbers for every frequency instead of just one. It's not as hard as it sounds, once we see the pattern.

Previously our two ingredients were level and phase offset. We'll keep level and trade phase offset for time offset (which leads us to phase offset/frequency). Time offset is linear and creates a linear phase offset over frequency. For example, a 1 ms offset creates a phase cycle spiral turning at a 1 kHz rate. That's 360° @1 kHz, 720° @2 kHz and so on. This also creates 180° @500 Hz and 540° @1500 Hz. Maximum addition is found at every multiple of 1 kHz and maximum subtraction at the linear mid-points between the peaks (500 Hz, 1500 Hz and so on).

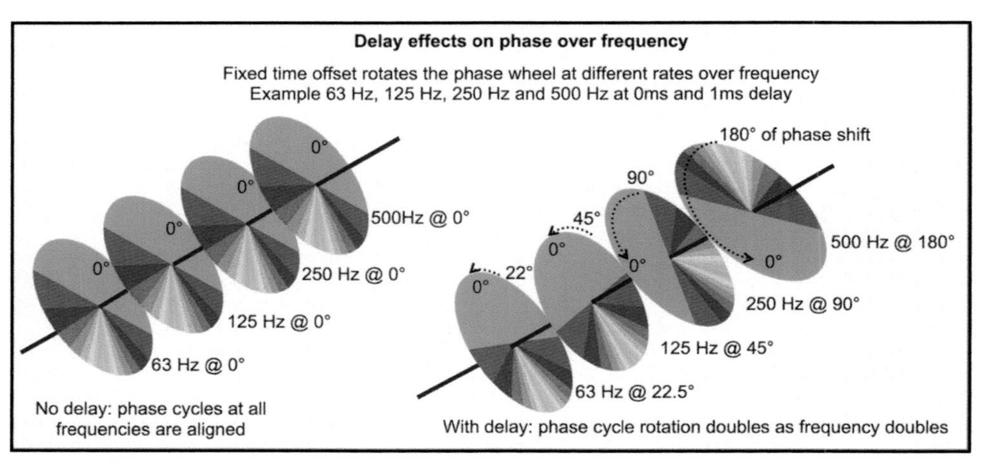

FIGURE 4.10
Time offset effect on the phase cycle over frequency

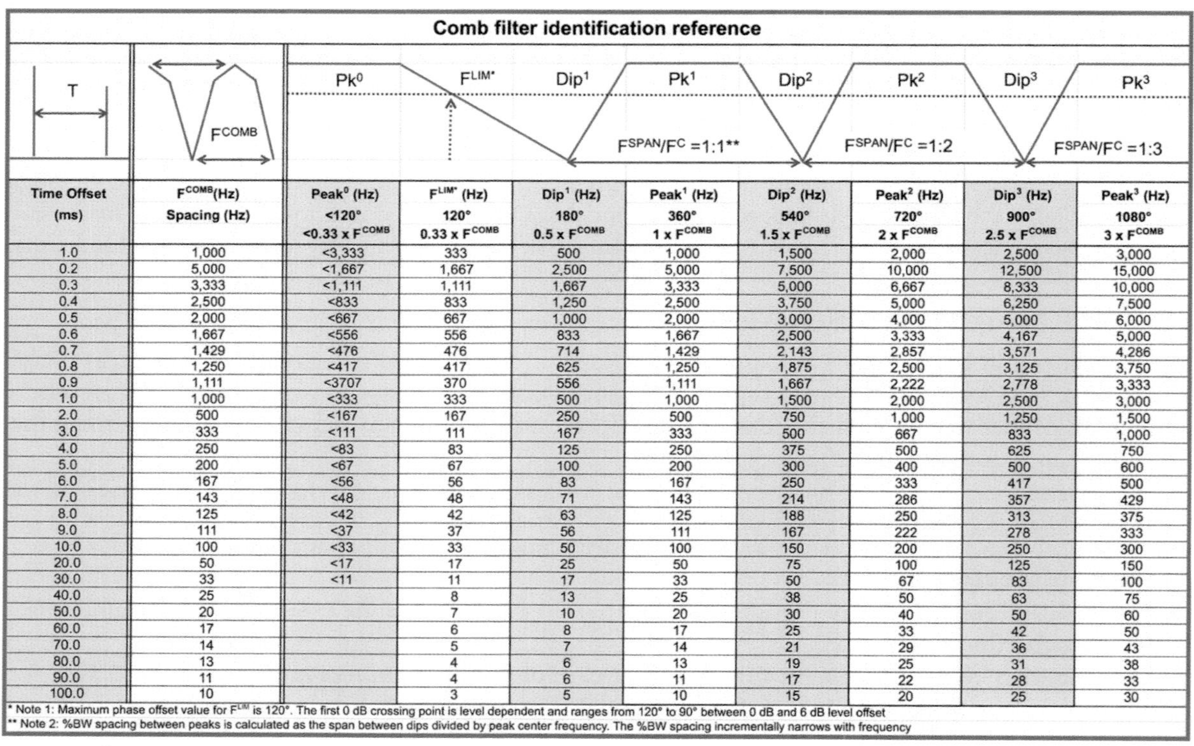

Comb filter identification reference

Time Offset (ms)	F^{COMB}(Hz) Spacing (Hz)	Peak⁰ (Hz) <120° <0.33 x F^{COMB}	F^{LIM*} (Hz) 120° 0.33 x F^{COMB}	Dip¹ (Hz) 180° 0.5 x F^{COMB}	Peak¹ (Hz) 360° 1 x F^{COMB}	Dip² (Hz) 540° 1.5 x F^{COMB}	Peak² (Hz) 720° 2 x F^{COMB}	Dip³ (Hz) 900° 2.5 x F^{COMB}	Peak³ (Hz) 1080° 3 x F^{COMB}
1.0	1,000	<3,333	333	500	1,000	1,500	2,000	2,500	3,000
0.2	5,000	<1,667	1,667	2,500	5,000	7,500	10,000	12,500	15,000
0.3	3,333	<1,111	1,111	1,667	3,333	5,000	6,667	8,333	10,000
0.4	2,500	<833	833	1,250	2,500	3,750	5,000	6,250	7,500
0.5	2,000	<667	667	1,000	2,000	3,000	4,000	5,000	6,000
0.6	1,667	<556	556	833	1,667	2,500	3,333	4,167	5,000
0.7	1,429	<476	476	714	1,429	2,143	2,857	3,571	4,286
0.8	1,250	<417	417	625	1,250	1,875	2,500	3,125	3,750
0.9	1,111	<3707	370	556	1,111	1,667	2,222	2,778	3,333
1.0	1,000	<333	333	500	1,000	1,500	2,000	2,500	3,000
2.0	500	<167	167	250	500	750	1,000	1,250	1,500
3.0	333	<111	111	167	333	500	667	833	1,000
4.0	250	<83	83	125	250	375	500	625	750
5.0	200	<67	67	100	200	300	400	500	600
6.0	167	<56	56	83	167	250	333	417	500
7.0	143	<48	48	71	143	214	286	357	429
8.0	125	<42	42	63	125	188	250	313	375
9.0	111	<37	37	56	111	167	222	278	333
10.0	100	<33	33	50	100	150	200	250	300
20.0	50	<17	17	25	50	75	100	125	150
30.0	33	<11	11	17	33	50	67	83	100
40.0	25		8	13	25	38	50	63	75
50.0	20		7	10	20	30	40	50	60
60.0	17		6	8	17	25	33	42	50
70.0	14		5	7	14	21	29	36	43
80.0	13		4	6	13	19	25	31	38
90.0	11		4	6	11	17	22	28	33
100.0	10		3	5	10	15	20	25	30

* Note 1: Maximum phase offset value for F^{LIM} is 120°. The first 0 dB crossing point is level dependent and ranges from 120° to 90° between 0 dB and 6 dB level offset
** Note 2: %BW spacing between peaks is calculated as the span between dips divided by peak center frequency. The %BW spacing incrementally narrows with frequency

FIGURE 4.11
Comb filter reference chart

Time offset creates variable phase offset and the summation goes by many names, coupling, cancelling and combing being the most relevant for us. Coupling and cancelling can be used selectively to our advantage. Combing is a by-product we do our best to minimize.

Recall that the frequency range for our sound system spans at least a 600:1 ratio of size (18,000 Hz to 30 Hz). A given time offset creates 600× more phase shift at 18 kHz than at 30 Hz. For example 33 ms creates 360° (1 cycle) of phase offset at 30 Hz and 216,000° (600 cycles) at 18 kHz. Think of time offset as a crank-shaft driving 600 different gear ratios. The only time they will all line up together is 0 ms (Fig. 4.10).

The standard time offset story is like a romantic sit-com. They start off together, and then they fight and make up over and over again. The progression moves slowly and we have only a few incidents if the time offset is small.

If large, we see lots of up and downs. Any time offset, big or small, starts the progression. Once the first dip reaches down into our hearing range, the show has begun (Fig. 4.11).

The story begins at the lowest frequency as a happy coupling until phase offset reaches 120°. Next is the first fight (phase offset = 180°), and then we're happy again at 360°. Another low point occurs when 540° is reached and on it goes. If time offset is small, such as 0.1 ms, the initial coupling range will extend all the way to 3 kHz before falling into a hole at 5 kHz and back up again at 10 kHz. By contrast, with 10 ms of time offset, the free coupling zone only lasts up to 30 Hz, with combing slicing up everything above that. Let's name the chapters in the series: coupling ($peak^0$), dip^1, $peak^1$, dip^2 and so on. They are always is the same order. We just have to fill in the frequencies that correspond to each time offset.

Combing is spread evenly through the spectrum, linear spectrum that is. To our log hearing the chapters in the series have different sounds. The coupling zone ($peak^0$) is pure gain. Dip^1 is the biggest hole and $peak^1$ is the most recognizable tonal modifier. The dips and peaks that follow are successively less recognizable as tonal variations. The linear sized/linearly spaced peaks (and dips) are perceived as ever narrowing to our log ears. The linear to log conversion of the spacing between the series of peaks is the reciprocal of the lower peak number ($peak^1$ to $peak^2$ is an octave, $peak^2$ to $peak^3$ is ½ octave and so on). Wide peaks are the easiest to discern tonal character whereas $peak^{100}$ (0.01 octave wide) won't have a perceivable center frequency to our ears.

4.2.1 Summation zones

We've seen how time offset drives the standard summation progression from coupling to the combing milestones dip^1, $peak^1$, etc. Let's add relative level to the mix and create a library of the standard summation responses seen in the wild. Many optimization decisions rely on the proper identification of summation behavior. These summation zones will carry through our discussion of speaker arrays, splay angles, subwoofers and even the room (Fig. 4.12).

4.2.1.1 COUPLING ZONE

The coupling zone should be well known by now: all gain, and no loss because the phase offset is <±120°. Coupling zone range is not limited by level offset but, because addition is minimal with large offsets, we draw the line at 10 dB. The maximum possible gain for two elements is +6 dB (offsets of 0 dB @0°) and minimum is 0 dB (any level offset @120°). The coupling zone is the most effective means of creating acoustic power addition. The effective frequency range is inversely proportional to source displacement (closer spacing extends the range upward). The spatial distribution follows a similar trend: Close sources maintain coupling over a wider area than distant ones. At the minimum, coupling is guaranteed at the exact centerline between two matched speakers (unless one is polarity reversed).

4.2.1.2 CANCELLATION ZONE

The cancellation zone, the dark side of the coupling zone, is defined by its opposing effect: all loss and no gain. The cancellation zone overlaps to create loss by keeping phase offset on the subtractive side. Practical applications include noise suppression and cardioid subwoofers. Strong cancellation (more than -6 dB of reduction) requires near-perfect conditions: very closely matched levels and phase offset within ±30° of 180°. The effect is most commonly achieved with a polarity reversal and delay, which extends the cancellation down to the lowest frequencies.

4.2.1.3 COMBING ZONE

The combing zone merges the coupling and cancellation zones. Close enough levels for a strong effect (within 4 dB) and enough time offset to cycle through the phase offset. The combing zone is evaluated in a "damage control" mode: How deep are the peaks and dips, how much of the frequency range is affected and what happened to my coverage pattern? The combing zone's fatal flaw is that it lacks isolation and yet has full cycles of phase offset. The result is frequency response ripple >12 dB. Combing zone damage is best controlled through strategies that maximize isolation and/or minimize time offsets.

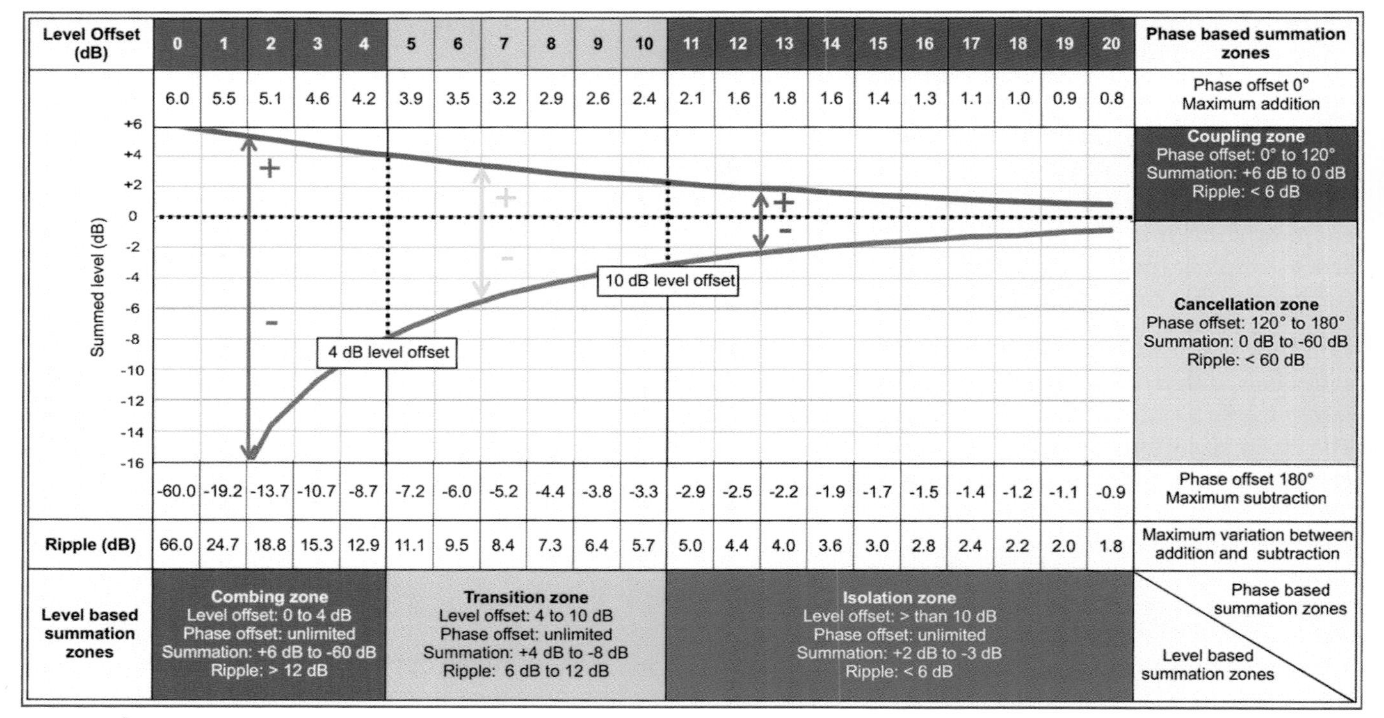

FIGURE 4.12
Summation zones

4.2.1.4 TRANSITION ZONE

As the name suggests, the transition zone bridges the highly overlapped zones (coupling, combing and cancellation) and the refuge of isolation. It's a semi-isolated state bracketed between 4–10 dB of isolation, with full turns of the phase cycle. The result is 6–12 dB of frequency response ripple. The transition zone uses the combing zone's damage control approach, with less severity in the ripple and spatial variations.

4.2.1.5 ISOLATION ZONE

Level offsets beyond 10 dB create enough isolation to minimize phase offset effects. Frequency response ripple is <6 dB. The isolation zone's positive attributes are low ripple and predictable combined coverage patterns. The down side is minimal power addition.

4.2.1.6 SUMMATION ZONES OVER FREQUENCY

Our audio systems may occupy multiple summation zones because they span a 600:1 frequency range. A splayed pair of speakers may have LF coupling and HF isolation due to their variable coverage patterns. Let's evaluate the zones by adding time offset, which drives the range of phase offsets over frequency. We will also add a spectral tilt to the level offset side of the equation: filtering down the HF to observe summation that couples the lows and isolates the highs.

The interaction may progress through four of the five summation categories over the course of a full-range response. Let's use 1 ms of time offset as an example. Coupling zone summation (<=120°) extends to 333 Hz followed by combing zone the rest of the way up. Let's add an HF roll-off filter to one of the sources to create increasing level offset with frequency. The zone progression now goes from coupling to combing to transition (when the filter reaches 4 dB) and finally isolation (10 dB). This zonal progression will be a central theme in this book because it is a standard sequence of events for audio summation interaction.

The series of frequency responses in Figs 4.13 to 4.14 serve as an abridged dictionary of the summation zone family. The actual frequency and level effects are continuously variable. The traces represent major milestones and will be reduced into icons for use in the remainder of this book to designate positions in the hall or around arrays where the conditions that cause them exist. We'll need to become fluent enough with these icons to identify them with our analyzer in the wild. Think of them as "you are here" on the map of our spatial/spectral interactions.

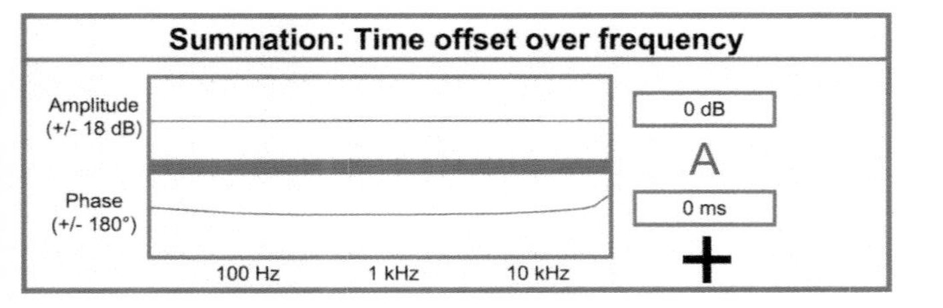

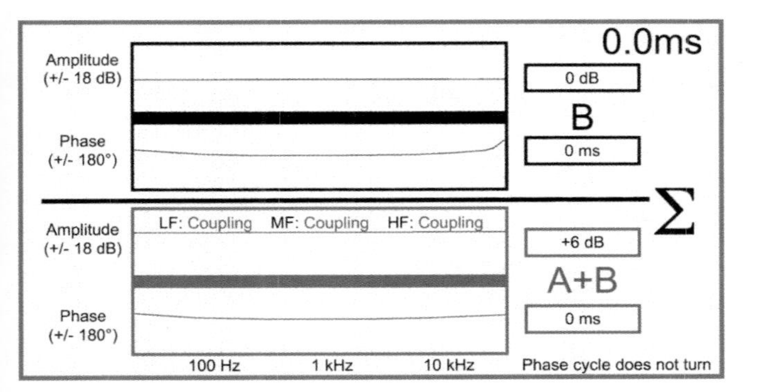

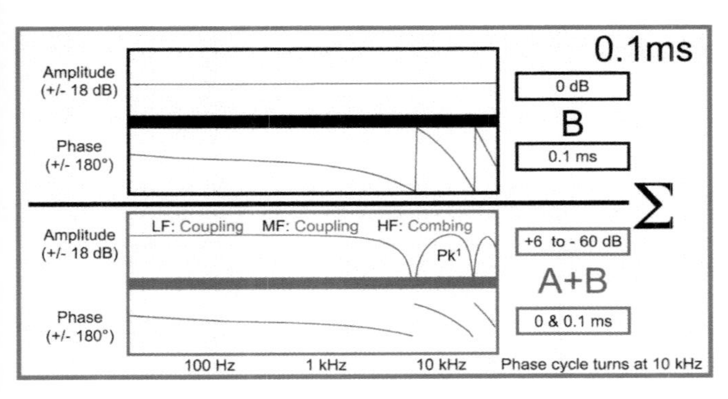

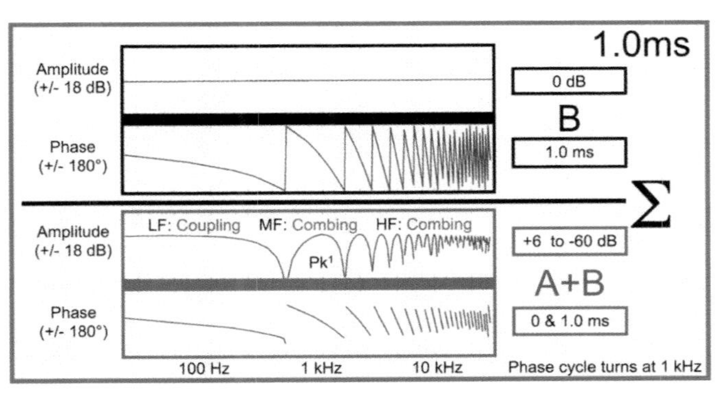

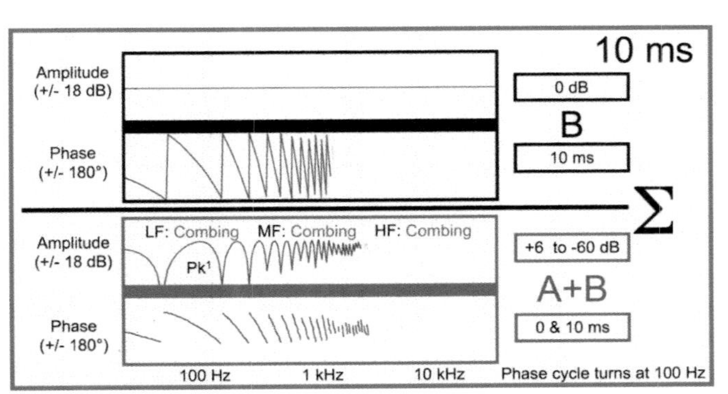

FIGURE 4.13
Summation zones over frequency (A + B) with matched levels and variable time offsets of 0–10 ms

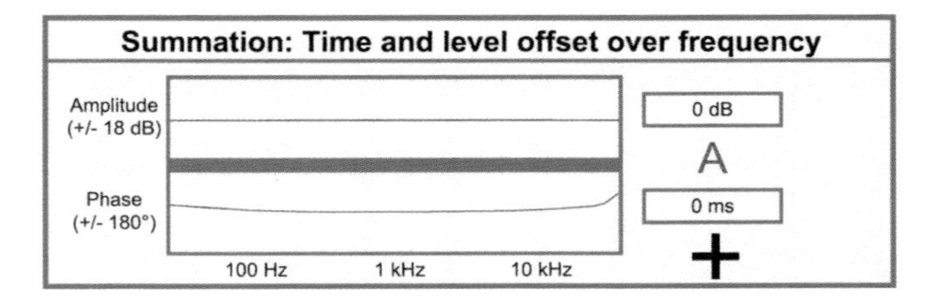

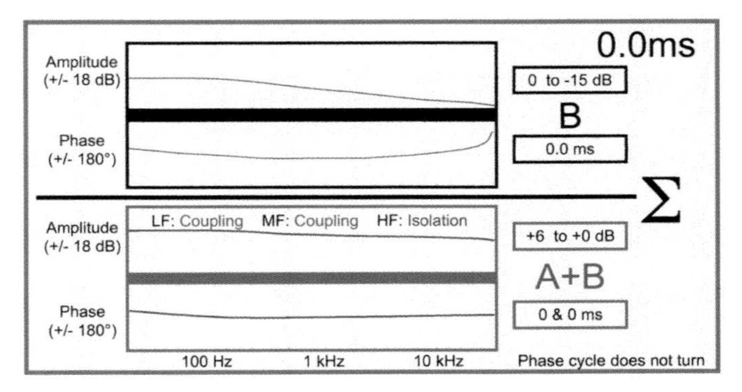

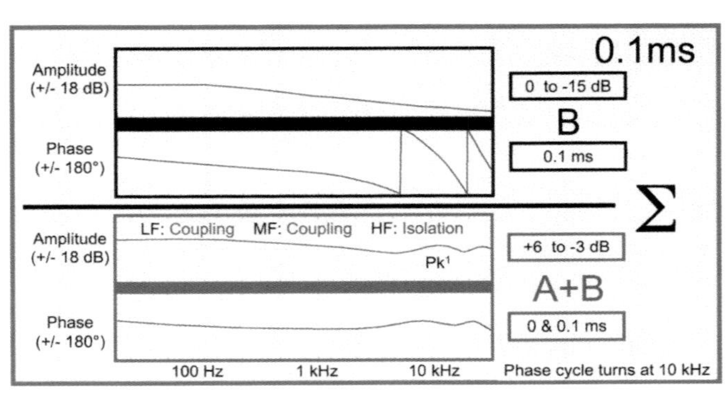

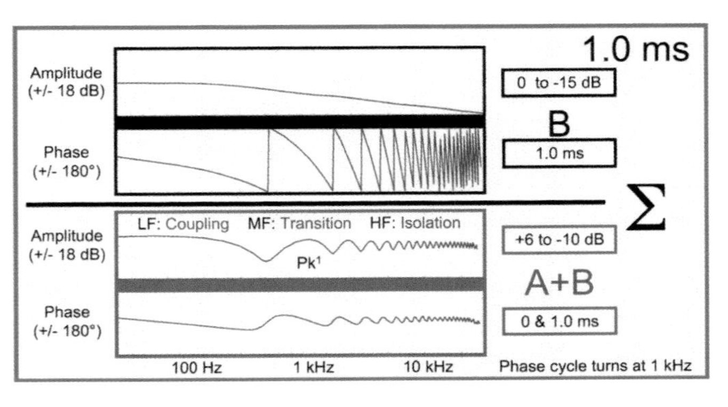

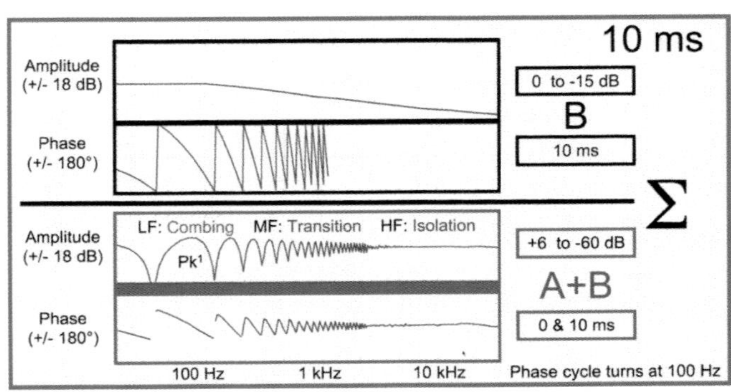

FIGURE 4.14
Summation zones over frequency (A + B) with unmatched levels and variable time offsets of 0–10 ms

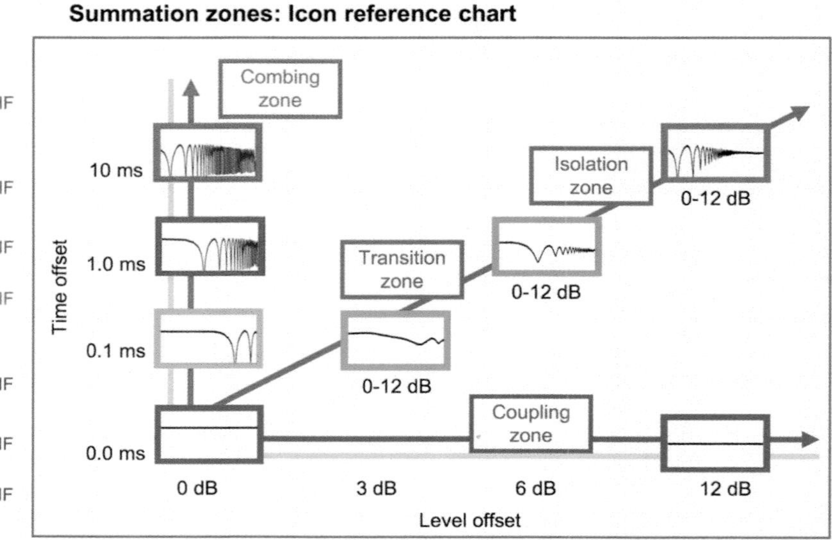

FIGURE 4.15
Summation zone icons displayed as time offset vs. level offset

The icons come together in a single graph positioning the summation zones into the overall perspective of time and level offset (Fig. 4.15). Strategies for design and optimization are drawn from this relationship. Successful strategies transition from the coupling zone (lower left) and move along the path toward level isolation as time offset rises (lower right half). If time offset is allowed to rise without implementing sufficient level offset, we will find ourselves in the volatile combing zone (upper left half).

4.2.1.7 SPECTRAL SUMMATION ZONE SUMMARY

The standard spectral summation zone progression must be fresh in mind.

Summation zone summary

- Comb frequency (F) = 1/time offset.
- Coupling zone lasts until 120° phase offset ($\frac{1}{3}F$ is the end of peak^0).
- Combing zone begins at 120° phase offset if level offset is < = 4 dB. Dip^1 = $\frac{1}{2}F$, peak^1 = F.
- Transition zone begins at 120° phase offset when level offset is between 4 to 10 dB.
- Isolation zone begins at 10 dB level offset regardless of phase offset.

4.2.2 Summation geometry

Acoustic summation is all over the place, literally. It contains spatial aspects not found inside a length of wire, such as individual speaker coverage patterns and displacement between them. Once again it's complicated, but it's not random. We'll see how level offset, time offset and phase offset decide where the sound goes and add geometry to our decoder key.

4.2.2.1 COUPLING AND UNCOUPLING LINES

Here's another icon to help visualize two-element acoustic summation in space (Fig. 4.16). Start with two horizontal dots, enclose them in a solid yellow circle and add a vertical and horizontal line. The dots are the sources, the blue vertical line is the direction of maximum coupling, the red horizontal line is the direction of minimum coupling and the solid yellow areas are the points between these extremes.

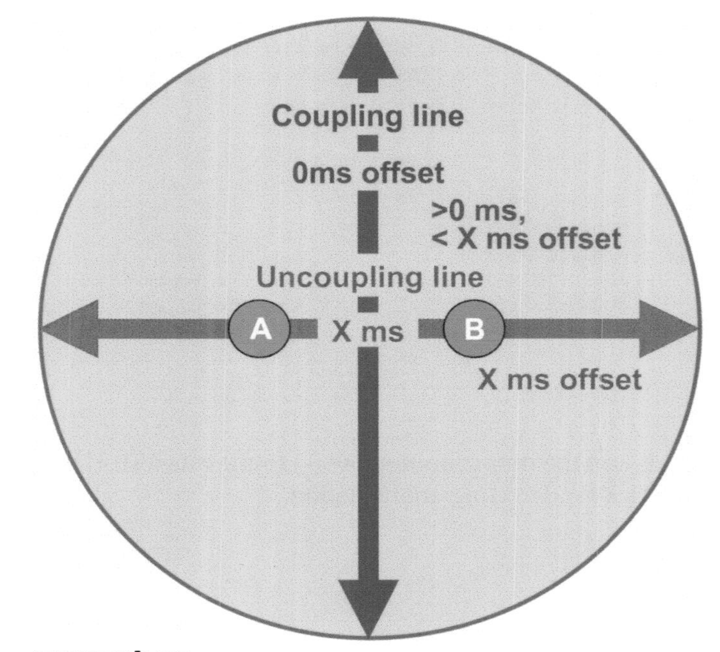

FIGURE 4.16
Summation spatial icon showing coupling and uncoupling lines

This is a guaranteed relationship for two matched sources. Every location on the coupling line (blue vertical) has 0 dB level offset, 0 ms time offset and 0° phase offset, to infinity and beyond. That's +6 dB summation forever. The non-coupling line (red horizontal) can never match that level of performance because the displacement between the sources causes level, time and phase offset. In this direction we know that the addition can be anything less than, but not equal to, +6 dB. We just don't know how much less yet. This line holds the distinction of highest possible level and time offsets, and an infinitely variable phase offset. Offsets in the yellow area between these extremes fall between the minimum and maximum.

Let's put an example number to the distance between the sources: 1 ms. You might not think of 1 ms as a measure of distance yet. The advantage is obvious, because otherwise we have to do a cumbersome conversion from 0.344 meters at 20°C or whatever into the number that tells us what will happen: time offset. By staying with ms we have a distance unit that gives us time offset and level offset without conversion. Case closed.

Let's continue with two sources 1 ms apart. There's nothing new on the coupling line (still no offsets). Along the uncoupling line there's a fixed time offset of (surprise!) 1 ms. This is true no matter how far we travel outward: 2 ms–1 ms, 3 ms–2 ms, 4 ms–3 ms all yield the same answer. The phase offset is also fixed at any given frequency, but different for each. The phase offset is 36° at 100 Hz (coupling), 180° at 500 Hz (dip[1]) and 360° at 1 kHz (peak[1]). The level offset is variable over distance because it's set by the level ratio. If we measure 1 ms away from the near source, the far source is 2 ms away. That's twice the distance (6 dB apart). Further out we find arrivals of 9 ms and 10 ms (1 dB apart). They approach, but never reach 0 dB level offset, but always remain 1 ms apart in time.

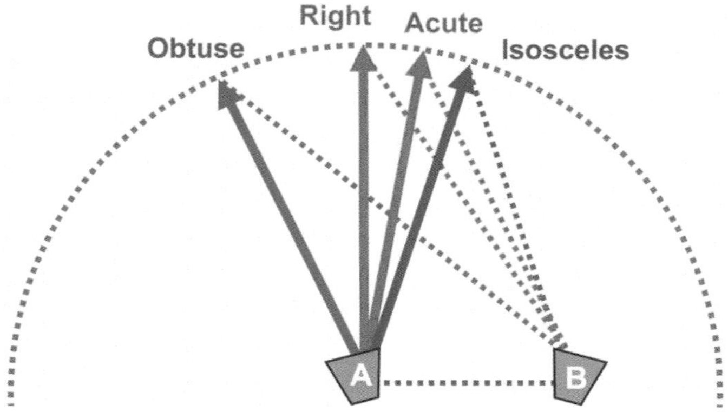

FIGURE 4.17
Summation spatial geometry: triangulation

So there it is: The coupling line (+) points to the loudest points and the non-coupling line (-) to the quietest. The yellow area between the straight-line extremes requires a new trick for decoding: triangulation.

4.2.2.2 TRIANGULATION

The triangle is the elementary shape for two-speaker summation geometry (Fig. 4.17). We standardize the corners as (A) source[1], (B) source[2] and (C) listening location. Transmission paths to the listener are AC and BC. Level and time offset are found by path length comparison.

- Level offset = 20 × Log_{10} (AC/BC).
- Time offset = AC - BC.

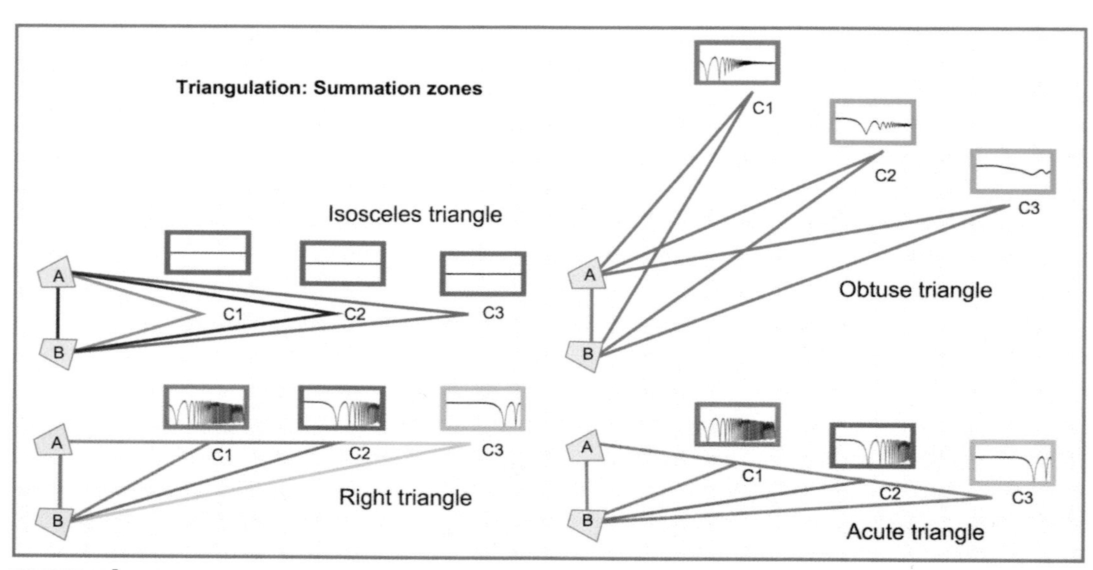

FIGURE 4.18
Summation zone triangulation

Level offset reuses the log/linear level decoder. Find the distance ratio (AC/BC) and read the conversion table. Example: AC = 100 ms and BC = 90ms, a level ratio of 1.1 (1 dB). Time offset is simple subtraction (100–90 = 10 ms). The frequency response at C can be drawn with just this information (and it's not pretty).

We use four triangle types: isosceles, right, acute and obtuse. The isosceles triangle has two sides of equal length. The equal paths are the transmission lines to the listener (AC = BC). The listener's spot is familiar to us already: It's on the coupling line (+).

If we pick a distance from the sources and move in a radius from center (+) outward we will progress through the four triangle types (Fig. 4.18) and then finally reach the horizontal line of minimum coupling (-).

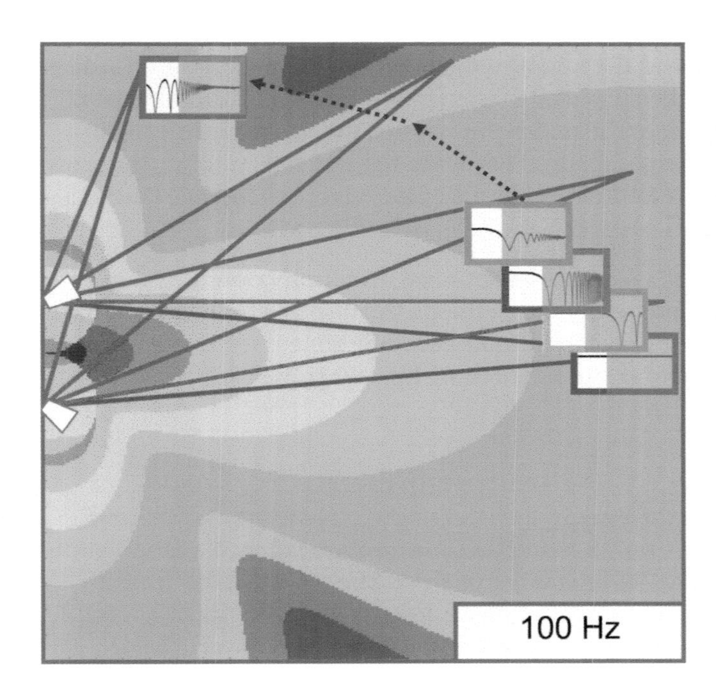

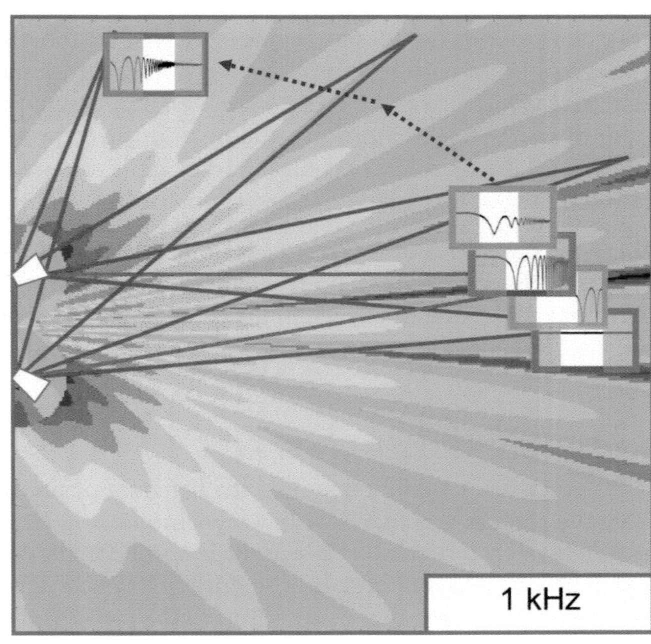

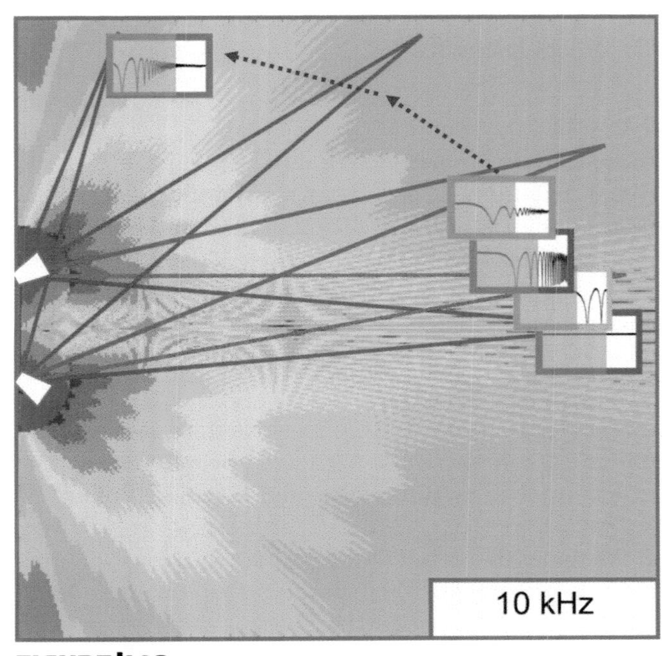

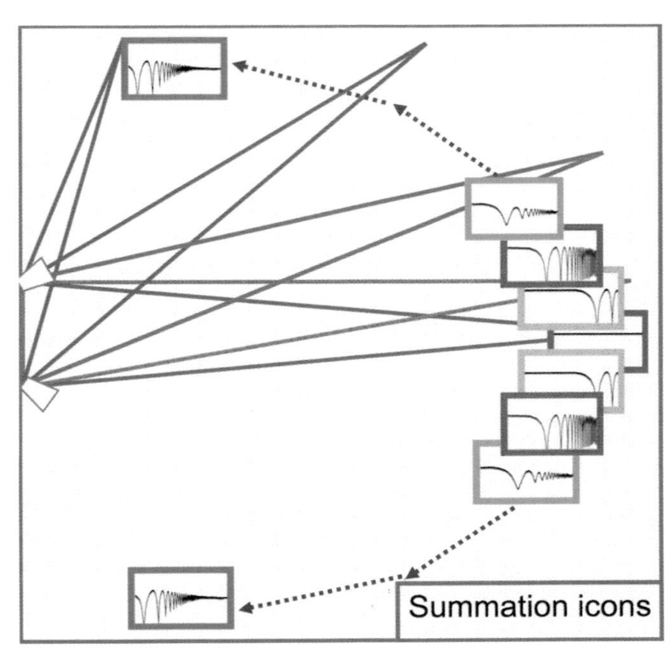

FIGURE 4.19
Summation zone triangulation example

- Isosceles triangle (+): Listener (C) is midway between A and B (AC = BC).
- Acute triangle: Listener (C) is between A and B (AC ≠ BC).
- Right triangle: Listener (C) is directly in front of A (AC ≠ BC).
- Obtuse triangle: Listener (C) is outside A and B (AC ≠ BC).
- Uncoupling line (-): Listener is on the line AB (AC ≠ BC).

Level offset and time offset both increase at each of these steps. The *rate* of change, however, is uneven, falling as we move through the progression. We'll use the right triangle as a marker to separate the volatile areas of rapid change from more stable areas with a slower rate of change. It's not some innate acoustical property of the right triangle but rather a simple, practical one: It's a place we can find.

Let's set up an example (Fig. 4.19): Two sources (A + B) are placed 10 ms apart. The listener (C) will move in a 180° arc around A at a 5 ms distance. We start on the AB line in the middle and end at the AB line on the outside. The center point is 100% coupling. Every other location on the arc puts the listener farther from B, but the rate of change is greatest near the center, and least at the outer extreme. The mid-point is when the listener is directly in front of source A (90° from the AB line). You are now at the right triangle. From the right triangle inward (the acute angle area) the response becomes more volatile and we move toward the combing zone. From the right triangle outward (the obtuse triangle) the response is less volatile and we steer toward the isolation zone.

These are statistics, but we can't fudge them. We can, however, take steps to optimize uniformity over the space by using directional speakers, pointing them the right way and properly spacing them. Any displaced sources will have their maximum rate of change in the acute angle zone, but we only fall into the combing zone if the phase offsets are too high and isolation too low. We use spacing and splay to create isolation by level when we can't beat the phase offset.

4.2.2.3 WAVELENGTH OFFSET

It is a geometric certainty that only the listening positions at the peak of the isosceles triangle will be equidistant from the sources and therefore have synchronized arrivals. It's only a matter of time until we fall into a cancellation as we move off center. More precisely, it is not a matter of *time*—it is a matter of *phase*. The relative delay leads to differences in relative phase over frequency. The losses begin when relative phase exceeds the 120° tipping point. We can predict the spatial distribution of the peaks and dips if displacement and frequency are known. The decisive factor is the ratio of source displacement to wavelength. A displacement of 1 λ draws one picture over the space. A displacement of 2 draws another. The ratio is frequency independent. A similar picture will be drawn at 1 kHz as 100 Hz, provided the displacement is expanded proportionally as frequency falls.

4.2.3 Linking the spectral and spatial summation zones

Recall the standard two-element spectral summation progression: coupling ($peak^0$) to combing (dip^1, $peak^1$, dip^2, etc.). This standard spectral progression links to the spatial distribution. In both cases a given time and level offset create a standard picture. In both cases the pattern scales proportionally (twice the time offset and half the frequency). For example, the spectral pattern for 1 ms time offset at 1 kHz is the same as 2 ms at 500 Hz. The same thing occurs in space: The spatial pattern for 1 ms time offset at 1 kHz is the same as 2 ms at 500 Hz. The beauty of this is we can take the spectral picture, give it a bend and it makes the spatial picture (Fig. 4.20).

Let's do 1 kHz @6 λ displacement:

- Step 1: Capture a summed frequency response of 0 ms + 1 ms (same as Fig. 4.13C). This sets the phase wheel turning at a 1 kHz rate. Crop the frequency response amplitude graph to include 6 kHz and below, and replace the frequency labels with $peak^0$, dip^1, etc. It will have coupling on the left, dip^1 in the middle and half of $peak^1$ on the right.
- Step 2: Mark the lowest frequency (0 Hz) as 0° polar and the upper frequency (1 kHz) as 90° polar.
- Step 3: Bend the frequency response in an arc so that 0 Hz is at the top and 1 kHz points 90° to the right. This is the polar pattern for one quadrant. $Peak^0$ is at 0°, dip^1 at 45° and $peak^1$ is at 90°.
- Step 4: Copy and paste that picture and fill in the other three sections of the circle. This is the complete polar pattern at 1 kHz for a pair of omnidirectional sources spaced 1 ms apart.

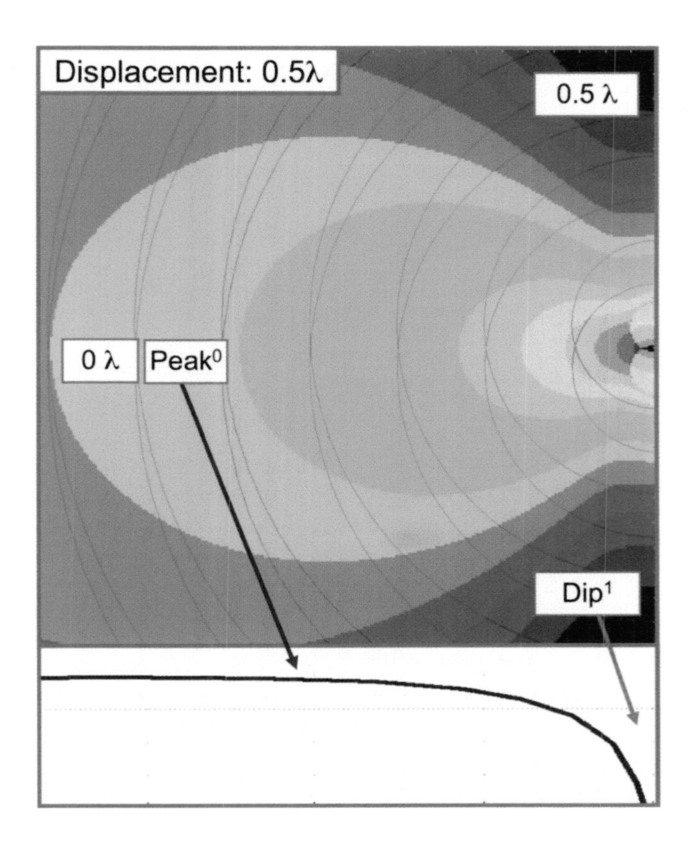

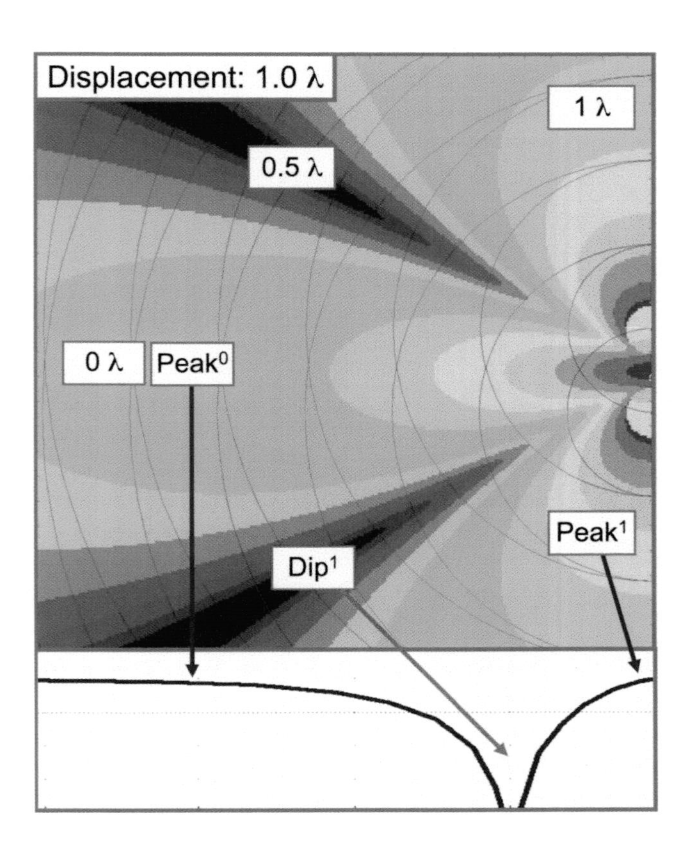

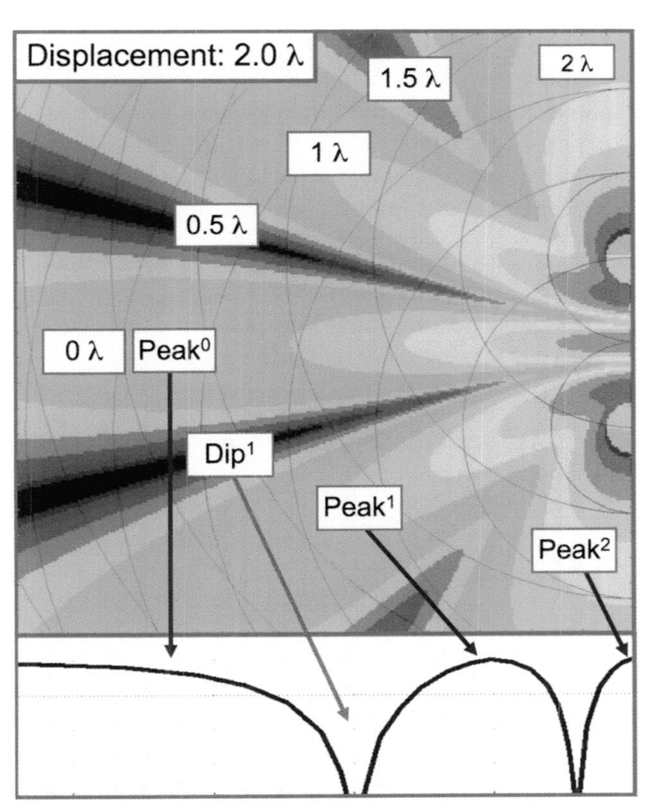

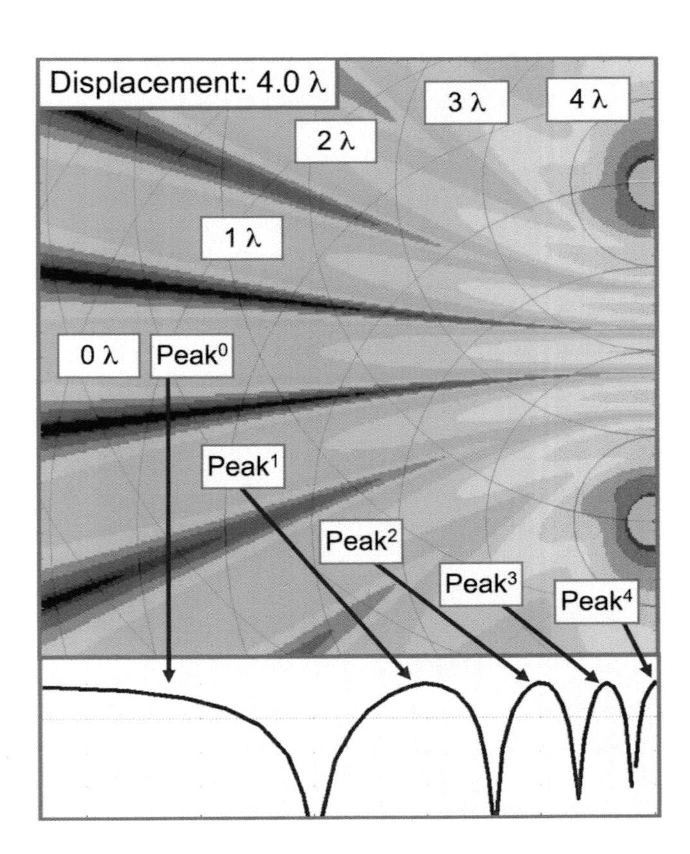

FIGURE 4.20
Spectral and spatial summation are linked by wavelength offset

Angular location of peaks and dips (2 element summation)														
Offset λ		0 λ	0.33 λ	0.5 λ	1 λ	2 λ	3 λ	4 λ	5 λ	6 λ	7 λ	8 λ	9 λ	10 λ
Offset°		0°	120°	180°	360°	720°	1080°	1440°	1800°	2160°	2520°	2880°	3240°	3600°
Peak BW	Peak/dip													
	Peak^0	0°	0°	0°	0°	0°	0°	0°	0°	0°	0°	0°	0°	0°
	F^{120}		90.0	60.0	30.0	15.0	10.0	7.5	6.0	5.0	4.3	3.8	3.3	3.0
	Dip^1			90.0	45.0	22.5	15.0	11.3	9.0	7.5	6.4	5.6	5.0	4.5
Octave	Peak^1				90.0	45.0	30.0	22.5	18.0	15.0	12.9	11.3	10.0	9.0
	Dip^2					67.5	45.0	33.8	27.0	22.5	19.3	16.9	15.0	13.5
1/2 Oct	Peak^2					90.0	60.0	45.0	36.0	30.0	25.7	22.5	20.0	18.0
	Dip^3						75.0	56.3	45.0	37.5	32.1	28.1	25.0	22.5
1/3rd Oct	Peak^3						90.0	67.5	54.0	45.0	38.6	33.8	30.0	27.0
	Dip^4							78.8	63.0	52.5	45.0	39.4	35.0	31.5
1/4th Oct	Peak^4							90.0	72.0	60.0	51.4	45.0	40.0	36.0
	Dip^5								81.0	67.5	57.9	50.6	45.0	40.5
1/5th Oct	Peak^5								90.0	75.0	64.3	56.3	50.0	45.0
	Dip^6									82.5	70.7	61.9	55.0	49.5
1/6th Oct	Peak^6									90.0	77.1	67.5	60.0	54.0
	Dip^7										83.6	73.1	65.0	58.5
1/7th Oct	Peak^7										90.0	78.8	70.0	63.0
	Dip^8											84.4	75.0	67.5
1/8th oct	Peak^8											90.0	80.0	72.0
	Dip^9												85.0	76.5
1/9th Oct	Peak^9												90.0	81.0
	Dip^{10}													85.5
1/10th Oct	Peak^{10}													90.0

FIGURE 4.21
Angular distribution of spatial summation (peaks and dips) for two elements by wavelength offset

The exact same picture is created if we use 500 Hz @2 ms or 2 kHz @0.5 ms. The common thread is the sources are 1 λ apart. When we see dip^1 or peak^1 in our analyzer's frequency response, we know where we are in the room. A different picture is created at 2 λ apart but we can still see the spatial from the spectral (and vice versa). Sources 2 λ apart at 1 kHz (2 ms time offset) create a spectral response with 3 peaks (pk^0 to pk^2) and 2 dips (dip^1 and dip^2). Peak^0 is still at 0° (the center), pk^1 is 45° off center and pk^2 will occupy the 90° polar slot. Dips will be located at 30° and 60° respectively. Everybody got skinnier, as evident in both spectral and spatial responses.

This is an important link. Time offset between sources drives the spectral response and the spatial response. Small time offsets minimize the combing density in both spectral and spatial domains. We can always count on peak^0 at the center, guaranteed coupling. The question is how many degrees (spatial) do we move before finding the octave wide (peak^1). If we've traveled only a few degrees we know our sources are many λ apart and the only hope for combing reduction lies in getting to the isolation zone quickly.

A given pair of sources will have a single time offset value (assuming matched symmetric elements). Time offset over frequency converts to phase offset, which drives the pattern shape. An alternate way of expressing time offset between sources is wavelength displacement, e.g. two speakers are 1 kHz apart (Fig. 4.21).

4.3 ACOUSTIC CROSSOVERS

4.3.1 Acoustic crossover defined

An **acoustic crossover** is defined as the point where correlated signals combine at equal level. In most cases, this is both a particular frequency range and a position in space. An acoustic crossover is the most critical summation junction. It benefits the most from being "in phase" and suffers most from being "out of phase." The crossover location, whether a frequency or a place in the room, is defined by its equality of level sharing. Wherever A meets B at equal level is a crossover. The phase relationship is the wild variable: the one we must tame to make this summation junction function.

Let's augment the terminology by division into two types. The **spectral acoustic crossover** has equal levels at a particular frequency range, the summation of drivers covering different frequency ranges fed by a **frequency (spectral) divider** (section 2.4.4). The **spatial acoustic crossover** has equal levels at a location in the space. It is the summation of sources covering a common frequency range fed by a **spatial divider** (separate speakers and processing channels). A single language is used for the analogous features of both types and the related solutions more easily understood.

The summation zones are common to both types. A primary goal is to place the acoustic crossover in the coupling zone. The secondary goal comes into play when coupling can no longer be maintained: Reach the isolation zone as quickly as possible. Filters accomplish this for spectral crossovers, whereas directional control, splay angle and spacing manage the spatial crossover. In both cases we strive to minimize the combing and transition zones.

4.3.2 Summation zone progressions

Summation zones follow several standard crossover progressions (Fig. 4.22). Let's add A + B and follow them.

- 1-step (coupling (AB)): Speakers with little or no directional control are in extremely close proximity, e.g. subwoofer arrays.
- 2-step (coupling (AB) to cancellation (AB)): coupling on the front side, cancellation in the rear. This is used in cardioid arrays.
- 3-step (isolation (A) to coupling (AB) to isolation (B)): Very achievable in a spectral crossover, but much harder in full-range spatial crossovers. As frequency rises we can expect to see more areas fall into the combing and transition zones, with the realistic goal being minimization, not elimination.
- 5-step (isolation (A) to transition (AB) to coupling (AB) to transition (BA) to isolation (B)): Practically achievable in closely coupled arrays if the coverage angle provides enough isolation to prevent combing.
- 7-step (isolation (A) to transition (AB) to combing (AB) to coupling (AB) to combing (BA) to transition (BA) to isolation (B)): the full progression. This results when isolation is unachievable before combing begins. In the HF this is often the only practical option. Success can be viewed in terms of the share of the progression spent in the combing zone. A progression with more combing zone than coupling and/or isolation zone would rate poorly.

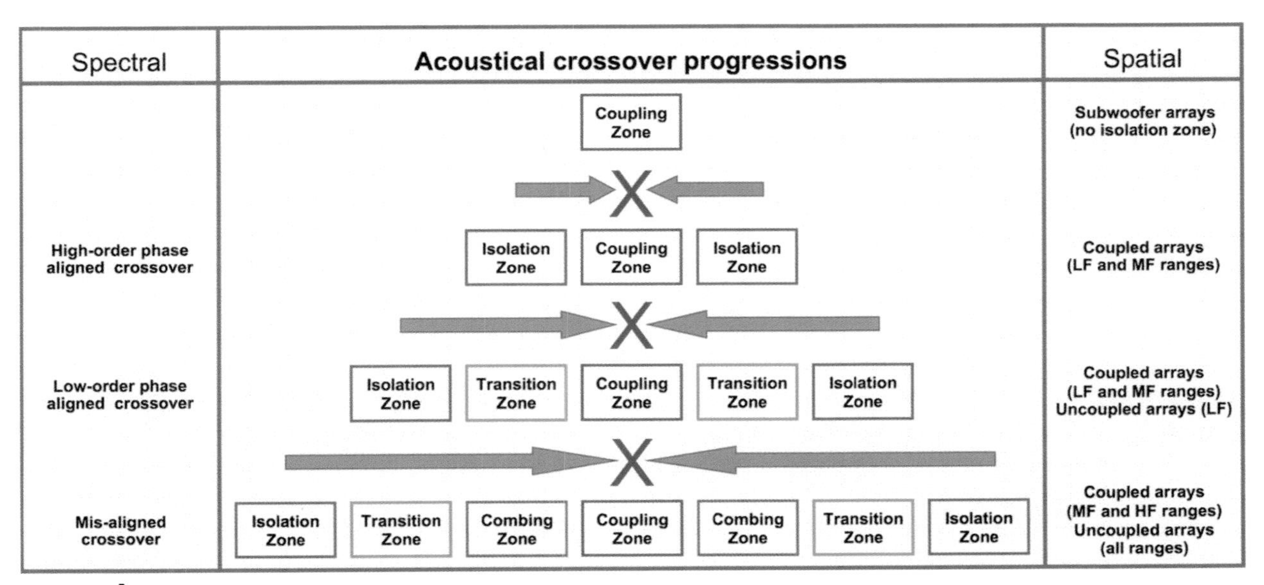

FIGURE 4.22
Summation zone crossover progressions, spectral and spatial

4.3.3 Crossover classifications

4.3.3.1 CLASS

We first classify crossover by the three possible level outcomes to the summation: The crossover range is equal to, greater than or less than the isolated ranges of the individual elements (Fig. 4.23).

Crossover class

- **Unity**: Level through crossover matches the isolated levels (A + B = 0 dB @XAB).
- **Overlapped**: Level through crossover is higher than the isolated levels (A + B > 0 dB @XAB).
- **Gapped**: Level through crossover is lower than the isolated levels (A + B < 0 dB @XAB).

4.3.3.2 SLOPE

Crossover slopes are rated by filter order: first order, second order, etc. The slope steepens as order increases. For spectral crossovers this refers to filters, whereas for spatial crossovers the separation is related to speaker coverage pattern (tighter coverage being higher order).

4.3.3.3 SYMMETRY

We classify crossover symmetry by two possible outcomes: symmetric or asymmetric. Spectral crossovers can have asymmetric filter slopes or filter topologies. Spatial crossovers can be asymmetric by speaker model, splay angles, level and more.

4.3.4 Spectral dividers and spectral crossovers

Let's get practical about the types of spectral crossovers we're likely to find in a modern speaker system. We are either marrying two drivers that already live together in the same box, or arranging one for speakers that are just now meeting.

DRIVER (A AND B) PHYSICAL CONFIGURATION

- Fixed combination: fixed ratio of A and B drivers, known positions, same box.
- Variable combination: variable ratio of A and B drivers, unknown positions, separate boxes.

With fixed combinations we encounter varying levels of manufacturer support.

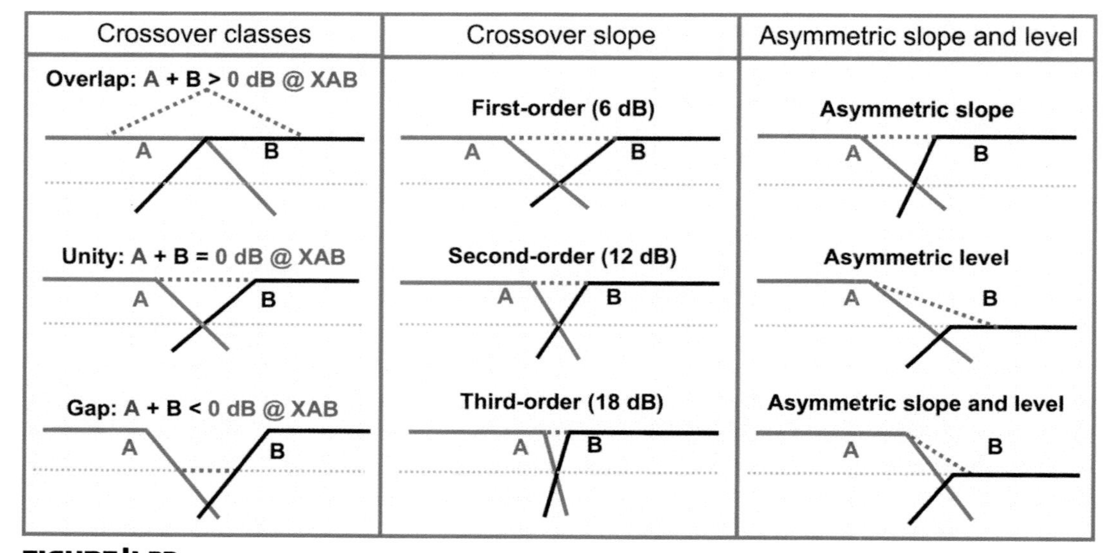

FIGURE 4.23
Spectral crossover classes, slope and asymmetry

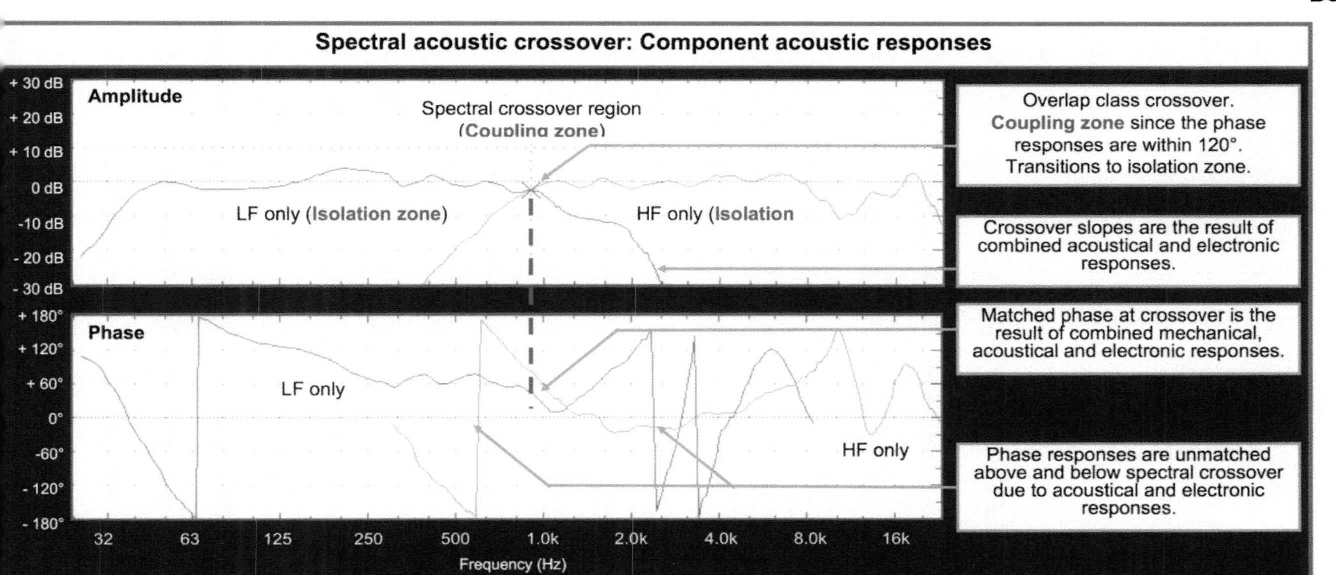

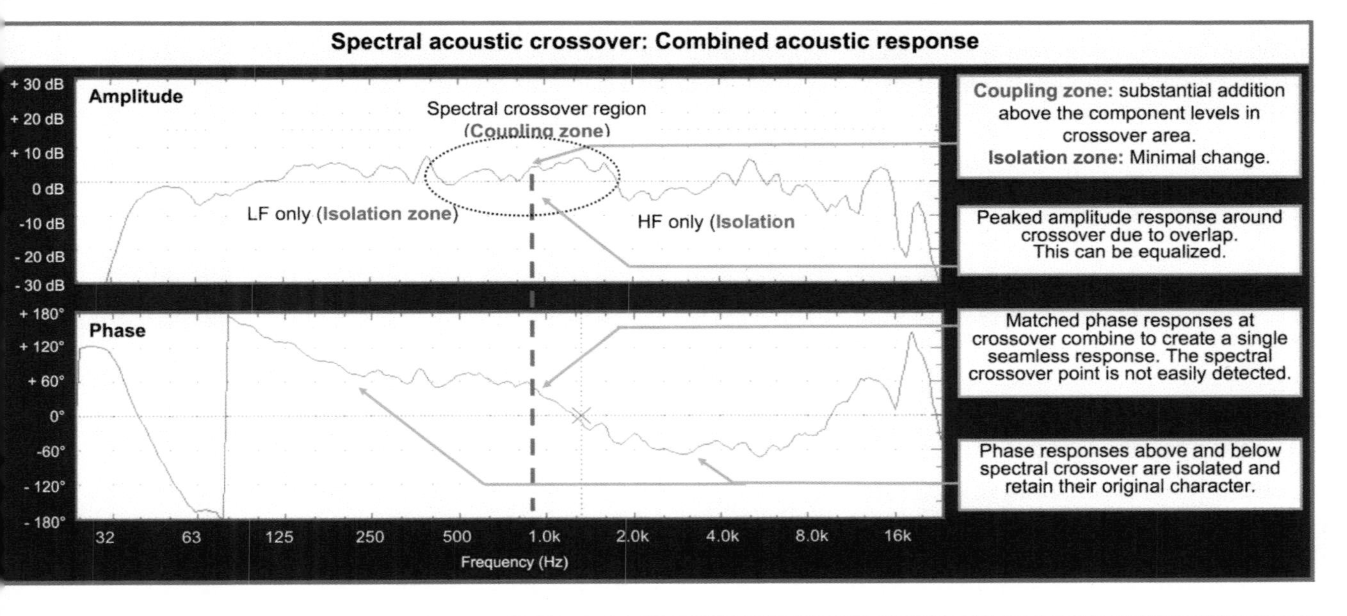

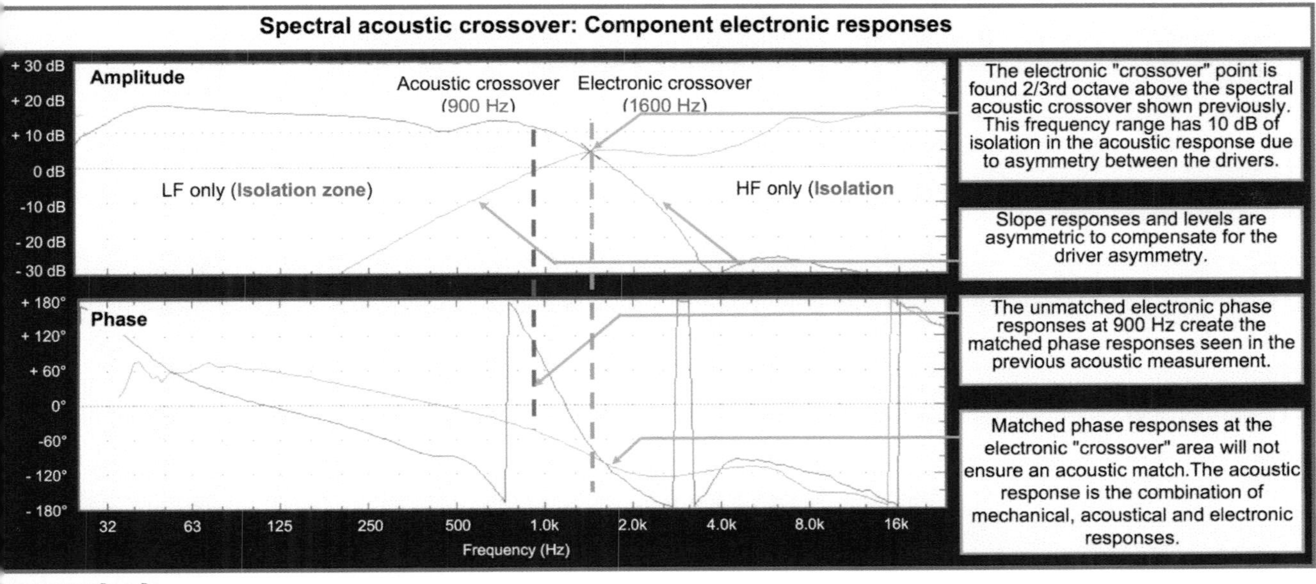

FIGURE 4.24
Spectral crossover example showing individual and combined acoustical and electrical responses

ELECTRONIC CONFIGURATION OF A AND B CHANNELS FOR FIXED COMBINATIONS

- Fixed: Settings are not user settable.
- Programmed: Factory presets are loaded into the crossover (hopefully correctly).
- Suggested: Factory settings are programmed by user into the processor (6 dB more hope required).
- Figure it out yourself: cowboy time!

Spectral crossover analysis of fixed driver combinations can be simply a verification process when factory-set, programmed or suggested settings are used. When no guidance is given we need to evaluate the individual elements and make choices (Fig. 4.24).

We also encounter several levels of manufacturer support with variable combinations, but there are many unknowns, leaving us with more need to customize. We will use a typical example, the combination of subs and full-range boxes with a manufacturer-suggested setting of 100 Hz (Fig. 4.25).

SPECTRAL CROSSOVER CONSIDERATIONS FOR VARIABLE DRIVER COMBINATIONS

- Relative quantity affects crossover frequency (e.g. might be two subs with one top or vice versa). The former raises the crossover and the latter lowers it. Level can be adjusted to match at 100 Hz.
- Driver efficiency affects crossover frequency (e.g. sub might be less (or more) efficient at 100 Hz). The former lowers the crossover and the latter raises it. Level can be adjusted.
- Driver location affects time offset between drivers (e.g. sub might be closer (or farther)). Delay can be set to phase align the crossover.
- Driver response differences affect phase offset at crossover. Phase align the crossover with delay.
- Driver ranges affect the overlap at crossover. (e.g. sub and full range share 60 Hz to 120 Hz). EQ can reduce the summation peak in the overlap area or LPF and HPF filters can reduce the overlap to create unity gain at crossover.

4.3.4.1 UNITY SPECTRAL CROSSOVER (LF+HF)

The standard unity crossover brings LF and HF sections together at -6 dB, 0° phase offset (Fig. 4.26). There is a variety of filter slope and topology combinations that can achieve this. The most straightforward (and popular) topology is the Linkwitz-Riley (L-R). The straightforward feature is that the cutoff frequency specified is the -6 dB point (most other filters use the -3 dB point). If the crossover frequency has already been selected the process takes six steps.

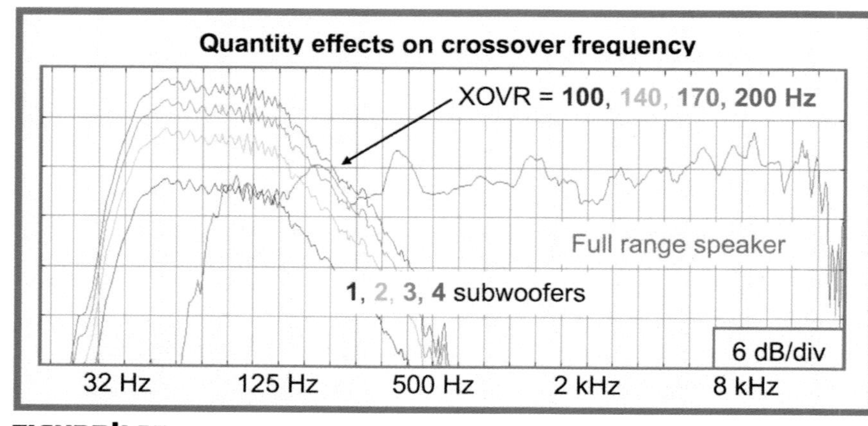

FIGURE 4.25
Quantity effects on crossover frequency (subwoofers vs. mains)

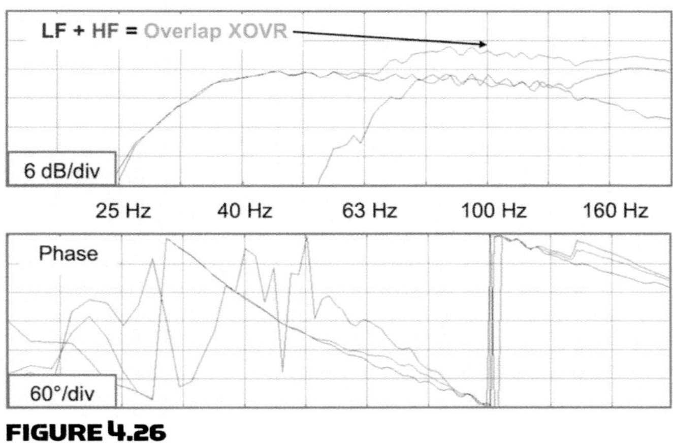

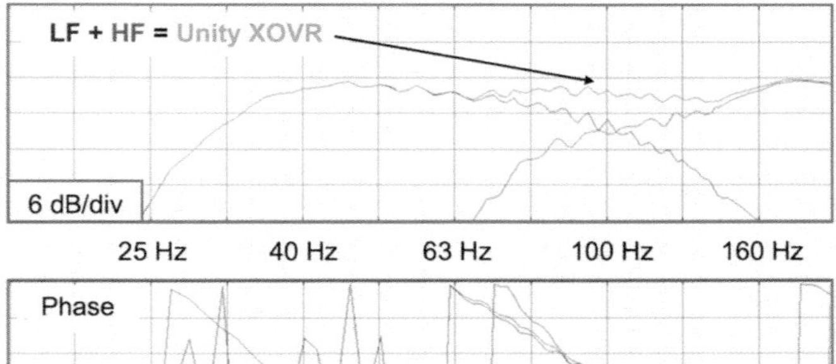

FIGURE 4.26
Overlap and unity spectral crossover examples

Creating a unity gain crossover with L-R filters (LF + HF)

- Determine a reference level: 0 dB.
- Set both the LF and HF channel cutoffs to the target frequency (e.g. 1 kHz).
- Drive each speaker and observe the acoustic response of each individual channel.
- Adjust the relative levels until they match at the crossover target (turn down the louder one).
- Adjust the relative phase until they match at the crossover target (delay the earlier one).
- Drive both channels together. The summed response should match the nominal level and be +6 dB above the individual responses at crossover.

The above holds true regardless of the L-R filter order precisely because the cutoff frequency is the -6 dB point in all cases. The choice of filter order here changes the amplitude response *around* the crossover frequency but not the frequency itself. If the order is changed (or made asymmetric) the phase offset may need adjustment and the behavior around crossover observed to see which gives the smoothest summation.

Other filter topologies (standard LPF, HPF, Bessel, Butterworth, Chebyshev, etc.) require extra work because they use -3 dB as the cutoff specification. An LPF specified at 100 Hz has the same -3 dB point for a first-order slope as an eighth-order one, but a vastly different -6 dB point. To achieve -6 dB at 100 Hz we will need to set the LPF somewhere below 100 Hz, the exact location for which will vary (closer to 100 Hz as filter order rises). The HPF has the same situation in reverse. Electronic settings that would appear to create a gap, such as 80 Hz (LPF) and 125 Hz (HPF), can actually create a unity crossover. As filter order rises, the gap between the electronic settings shrinks but never reaches zero. When asymmetric topologies and/or filter orders are used we must be mindful of where each lands at the -6 dB milestone.

Creating a unity gain crossover with other filter topologies (LF + HF)

- Determine a reference level: 0 dB.
- Set both the LF and HF cutoffs to an octave beyond the target frequency (e.g. 2 kHz for the low and 500 Hz for the HF). This leaves the crossover area free of filter effects for reference.
- Drive each speaker and observe the acoustic response of each individual channel.
- Adjust the relative levels until they match at the crossover target (turn down the louder one).
- Move the LF cutoff frequency down until the response drops 6 dB at the crossover target.
- Move the HF cutoff frequency up until the response drops 6 dB at the crossover target.
- Adjust the relative phase until they match at the crossover target (delay the earlier one).
- Drive both channels together. The summed response should match the nominal level and be +6 dB above the individual responses at crossover.

4.3.4.2 Overlapped spectral crossover (LF + HF)

Overlap-class spectral crossovers are often used to combine subwoofers and full-range enclosures, which may share the 60 Hz to 120 Hz range (again Fig. 4.26). The decision of when and where to use an overlapped crossover instead of unity is not clear-cut.

Overlap and unity crossover considerations

- Overlap increases efficiency and headroom in the affected range (drivers working the same range).
- Overlap narrows the coverage pattern (whether this is desirable is situation dependent).
- Overlap may cause increased combing in some locations (more so if the speakers are far apart).
- Overlap helps the sound image remain spectrally linked (subs not separated apart from mains).
- Operating a speaker into the overlap range may reduce the headroom of the individual drivers. The individual headroom loss may reduce the efficiency gain from the summation in the overlap zone.

The process is close to the unity crossover with an added equalization step.

Creating an overlapped crossover

- Determine a reference level: 0 dB.
- Set both the LF and HF cutoffs to create the target overlap range (e.g. 60–120 Hz). This may be unnecessary if the natural roll-offs give the desired response.
- Drive each speaker and observe the individual channel acoustic response.
- Adjust relative levels until they match in the crossover target range (turn down the louder one).
- Adjust relative phase to match as much of the crossover range as possible (delay the earlier one).
- Driving both channels together should create a +6 dB peak centered in the target range.
- Add a parametric filter (-6 dB at the center of the overlap) to both channels. Adjust bandwidth as required to normalize the response.

4.3.4.3 SELECTING THE CROSSOVER FREQUENCY

It is increasingly rare to encounter systems requiring field selection of the crossover frequency. This should never happen, but it does. Extensive R&D should go into the selection process, an unlikely scenario in the field with minutes ticking down until show time.

Crossover frequency selection considerations

- Optimal frequency range for each driver. Affects power capability, headroom, excursion and more.
- Polar responses of HF and LF components change over frequency. The rate of polar change through crossover may be affected. For example, HF horns tend to become wider as the cutoff frequency is lowered, whereas LF drivers narrow as the cutoff rises.
- Cutoff frequency selection has a significant impact on HF driver excursion. It's beyond the scope of this text, but basically amounts to this: Excursion rises exponentially as frequency drops. Halving the frequency requires quadrupling the excursion to produce the same acoustic power.
- Lowering a crossover frequency may cause the device to operate below the coupling range of the horn, lowering efficiency and requiring still more excursion to achieve the demanded SPL.

It's generally safer to run drivers above their operational range than below. Therefore a good starting point is to observe the upper limits of the raw LF response. The crossover must be below LF limit and see how high it can go. If we start the HF driver with an HPF set an octave below the prospective target we can see if the driver can make the meeting point. Then begins an iterative trial and error in the target range. Different slopes and topologies can be tried until the best fit is found. After aligning the crossover in the on-axis area it is suggested to move off axis to see how well the LF and HF patterns match. There's no way to practically perform a thorough R&D process in the field but this roughs things in when that is all we can do.

4.3.4.4 CROSSOVER AUDIBILITY

We generally seek to suppress clues that alert audiences to the presence of speakers. Violins do not have crossover frequencies. A multi-way speaker can reproduce the full range of the violin, but we must carefully handle the crossover region to ensure a listener does not notice the transition between drivers. One mechanism that can expose the driver transition is displacement. The probability of distinct localization increases as drivers move apart. This is not an issue for integral multi-way enclosures but can be for subwoofers, which are often a large distance apart from the mains. A second factor is overlap, which can expose the crossover transition by having too much or too little overlap. Crossovers with extremely steep filters can transition in a single note of the musical scale. This becomes most audible if the transition moves between elements with large differences in pattern control, e.g. a front-loaded cone driver to a narrow horn. The reverberation character can suddenly change with the transition. On the other hand, with excessive overlap, the filters won't isolate enough to prevent combing around crossover. This leaves the transition exposed by the presence of dips in the response above or below the crossover frequency. A final example

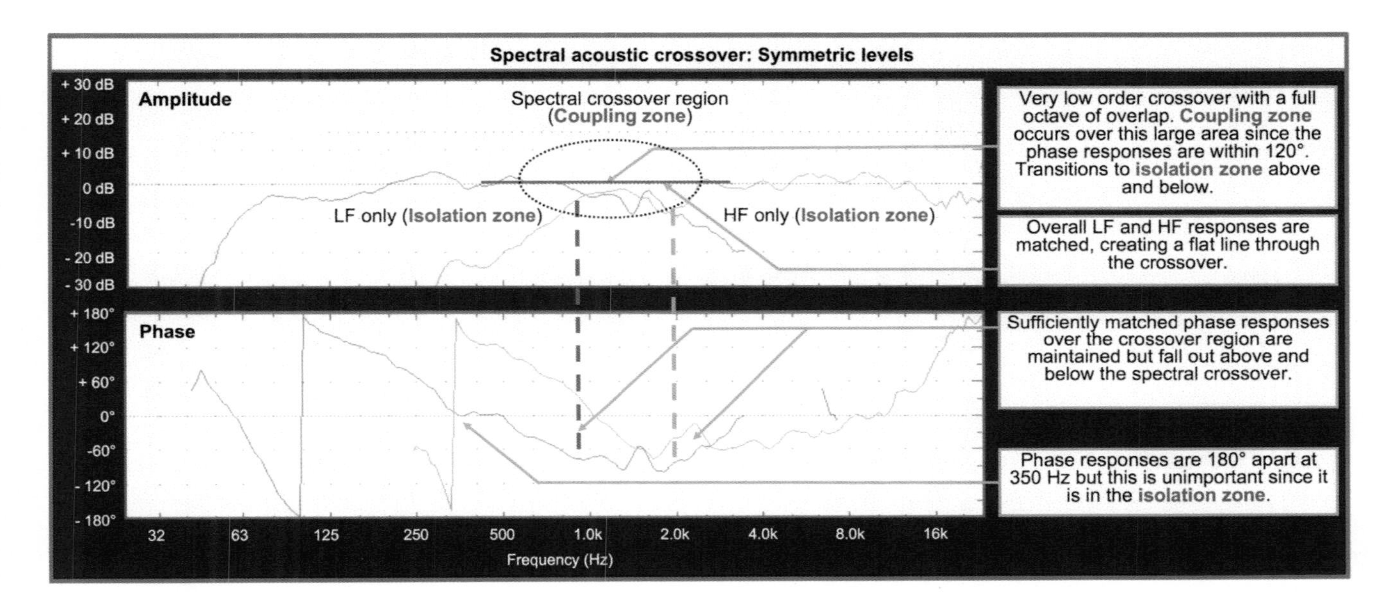

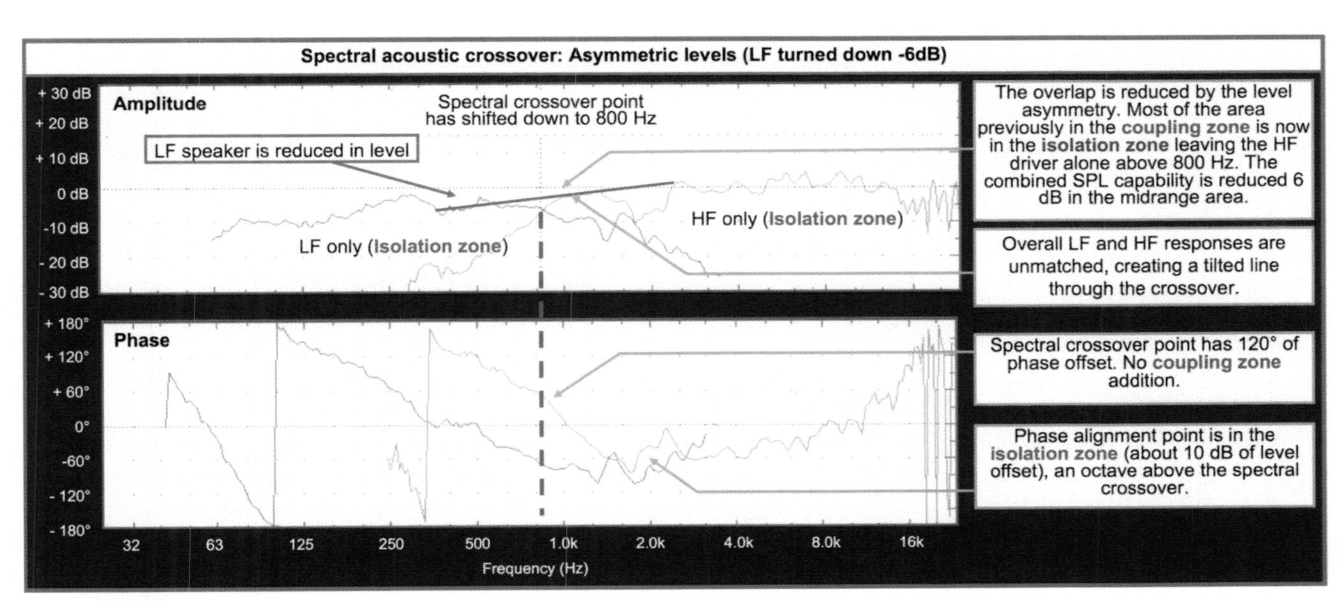

FIGURE 4.27A AND B
Asymmetric-level spectral crossover example

is the combination of both displacement and minimal overlap in the case of grounded subs and flying mains. Both factors emphasize separation, which can be experienced as two distinct sources.

Let's set a goal of minimum "crossover detectability": driver transition without anyone noticing. Minimal displacement, minimal coverage angle transition, minimal combing and more gradual filter slopes have the highest prospects of slipping under our sonar. Our hearing mechanism clues in on abrupt changes in sonic character between notes. Displacement causes an abrupt change in localization. Combing causes an abrupt change in level as one note disappears and the next returns. Coverage angle transition causes a change in the reverberant field, leading the listener to feel as if one frequency is far away in a reverberant space while the next is nearby in a dry space.

4.3.4.5 CROSSOVER ASYMMETRY

Different filter slopes make the transition from coupling to isolation asymmetric. Asymmetric slope rates can be used effectively, but their action must be anticipated. The mixing of even and odd filter orders (e.g. second order and third order) usually requires a polarity reversal and mismatched corner frequencies to achieve a unity crossover result. The most common asymmetric slope choice is a steeper high-pass than low-pass filter. The risks of over-excursion are reduced for the HF and usable power is shared from the LF.

Relative-level settings can also introduce asymmetry in the crossover range, by shifting the crossover frequency and reallocating the division of labor. Dropping the level of the LF driver adds to the burden of the HF and can potentially endanger it. A field example is shown in Fig. 4.27.

4.3.5 Spatial dividers and spatial crossovers

4.3.5.1 SPECTRAL VS. SPATIAL

Let's apply our knowledge of the phase-aligned spectral crossover to coverage over the space: the spatial crossover. The spectral divider split the spectrum and the spectral crossover joined them. The spatial divider splits the coverage and the spatial crossover joins them together.

Let's take a two-way speaker as an illustrative example. The spectral load (e.g. high and low) is split, but the spatial load (the coverage area) is shared equally. Now put two of these in an array. The spatial load is split (e.g. balcony and floor), but the spectral load is shared equally. Our two-way, two-way array contains both species of acoustic crossover, and we'll use the same approach to optimize the response in the most sensitive area: the phase-aligned crossover.

This section illustrates how spatial division, the process of separating the listening area into coverage zones, is so directly analogous to the separation of high, low and midrange drivers. A single set of principles applies to a four-way crossover in a single enclosure (spectral) and a four-element array (spatial). The final piece of the puzzle is the walls of the room. They are the ultimate spatial dividers and their reflections are governed by the same principles. The revision of conventional terminology requires some time to assimilate, but the effort is worthwhile as the mysteries of speaker and room interaction yield their secrets.

Common ground between spectral and spatial acoustic crossovers

- Both interactions are governed by the properties of acoustic summation.
- Optimization strategies are rooted in the same concept: the phase-aligned crossover.

Differences between spectral and spatial acoustic crossovers

- Spatial XOVR is full range. More challenging to achieve isolation without cancellation.
- Phase offset in the spectral divider strongly affects only the crossover frequency range. Phase offset in a spatial divider affects all frequencies.

Analogous functions

- Coupling zone location (speakers at equal level) is analogous to spectral crossover frequency.
- Directivity is analogous to filter slope. Highly directional speakers are like steep filters.

- A change in relative level between speakers shifts the spatial crossover location just as it shifts the spectral crossover frequency.
- Power addition from horizontal or vertical overlap in the spatial crossover is analogous to the addition from overlap in the spectral crossover.

Spatial acoustic crossovers are more complex than their spectral counterparts, yet are variations on the same themes (Fig. 4.28). Place two matched speakers in any orientation and find the mid-point between them. That's the crossover point and the coupling zone for all frequencies. The next part is a little harder. Now we have to find our way out of there to the isolation zone: the position where one speaker is 10 dB more dominant in level. This is easily found for a single frequency by observation of the coverage patterns. The break point into the isolation zone would be as simple as the spectral filter slopes *if* the speaker has the same directional pattern for all frequencies. This is as likely as spotting a unicorn. In the real world the low frequencies will typically overlap much more than the highs, giving us different points in the room for the borders of the coupling, combing, transition and isolation zones over frequency.

4.3.5.2 MULTI-WAY SPATIAL CROSSOVER

Let's expand beyond the two-way (A + B = AB) crossover by adding a third element (C). The summation progresses in two stages as the sound propagates forward: (A + B) + (B + C) = ABC (Fig. 4.29). The adjacent summations occur first and then progress on to the full three-way combination, i.e. element A will close the gap with B before it has met element C and vice versa. At this stage we have two isosceles triangles, side by side. Stack a third one on top and we have combined A + B + C. Notice we've started a pyramid. If the AB and BC gaps closed at 1 m, then the meeting point for the trio ABC will likely be 2 m. The array assumes its permanent far-field shape once all elements have fully summed (at the top of the pyramid).

We can continue the additions for as long as the budget and rigging allows. Each new element expands the base and adds an upward summation layer. Element spacing sets the foundation size of the pyramid base, and the element coverage angle yields the slope of the sides. Narrow elements take longer to meet their next-door neighbors and the next after that, creating steeper triangulation and pushing the far-field response transition into outward. Wide elements close the gap quickly and finish their shaping closer to the sources.

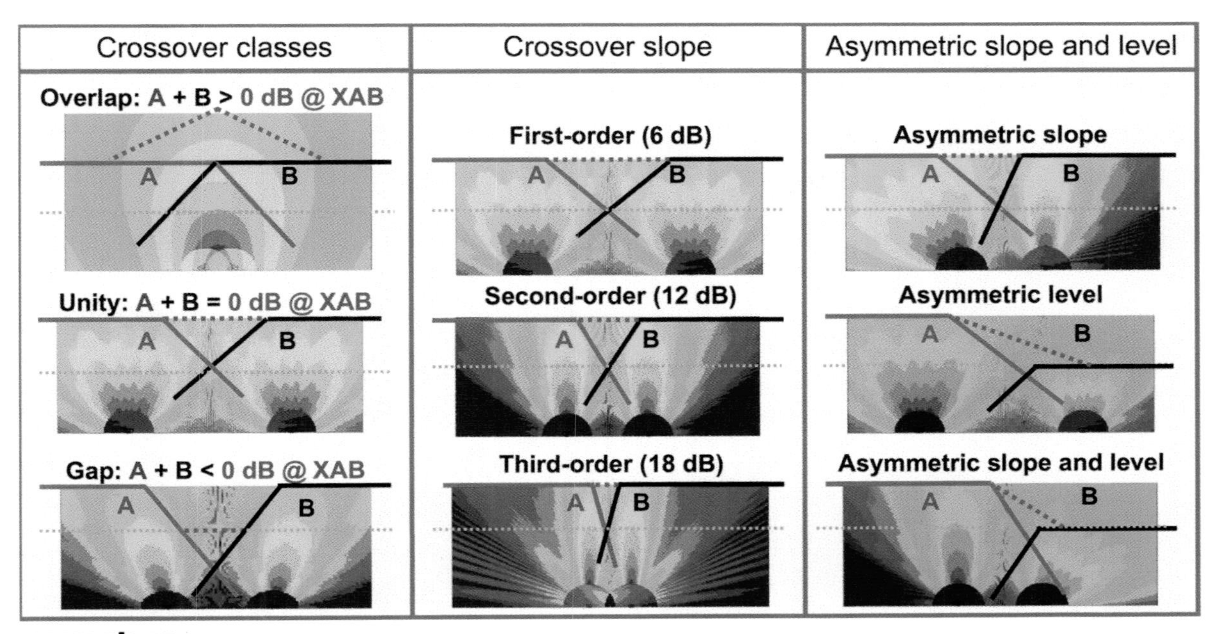

FIGURE 4.28
Spatial crossover classes, slope and asymmetry. Note the analogous relationship to the spectral crossover in Fig. 4.23.

The pyramid is clearly evident when element spacing is close enough to maintain coupling. By contrast, widely spaced elements will comb with their many neighbors rather than couple. The upper summation layers will look more like a fireworks finale than a pyramid. Could we call it comb-bustion? We can't stop the upper layer combing, but we can make it irrelevant by sending in other speakers to take over the coverage. Every seat in the hall can see all of the frontfills. The reason their upper-layer combing doesn't bother us is because we pave over them with the mains.

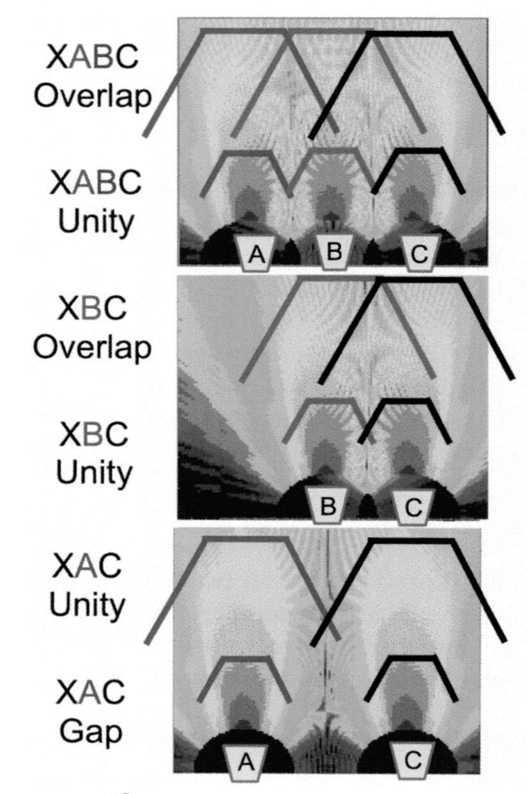

FIGURE 4.29
Multi-way spatial crossover

4.3.5.3 SPEAKER ORDER (CROSSOVER SLOPE)

A big step in transferring our spectral knowledge to the spatial domain is to visualize speaker coverage pattern as a form of spatial filtering (Fig. 4.30). A 360° speaker is as spatially "full range" as one covering 20 Hz to 20 kHz is spectrally. A spectral range of 10 octaves is analogous to a spatial range of 360°. If we straighten out the radial circle we can lay the two shapes side-by-side. A two-way system splits the coverage (2 × 5 octave or 2 × 180°) or we can go three-way (3 × 3.3 octave and 3 × 120°). We can go on slicing like this forever. We know from spectral crossover experience we need steeper filters on a four-way system than a two-way. An eight-way? Even more. Steeper filters isolate the more closely spaced signals. Same with spatial filters. If we're slicing coverage in 20° segments, they need to be steep to prevent excess overlap into neighboring areas. We don't always need to cover 360° any more than we always need to cover 20 Hz–20 kHz,

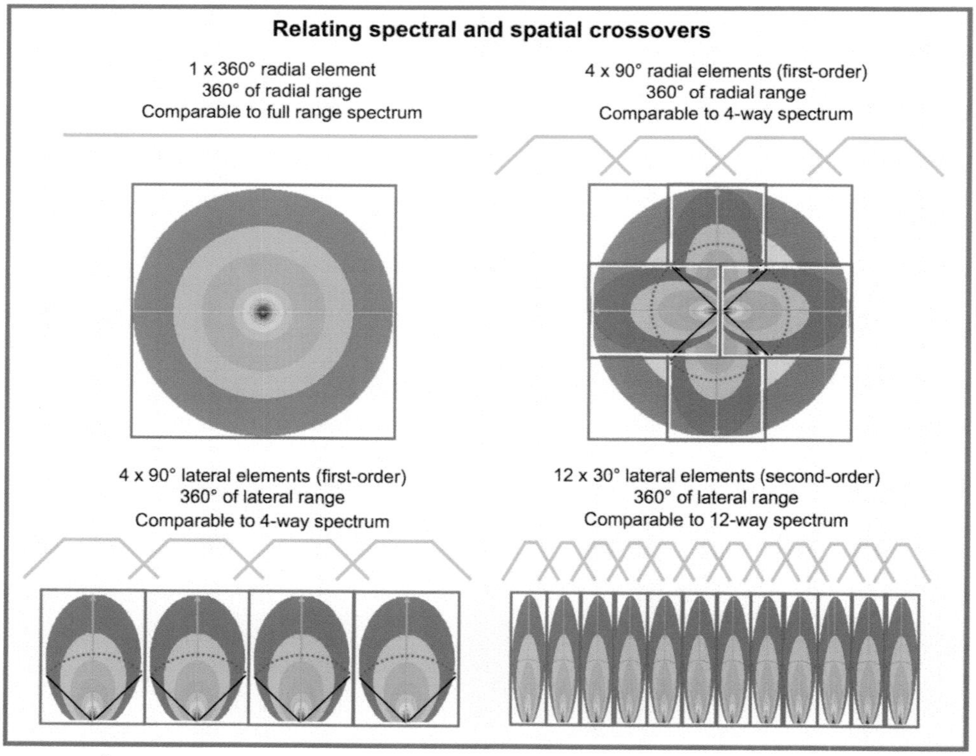

FIGURE 4.30
Linking the spectral and spatial orders: 1 × 360°, 4 × 90°, 12 × 30°

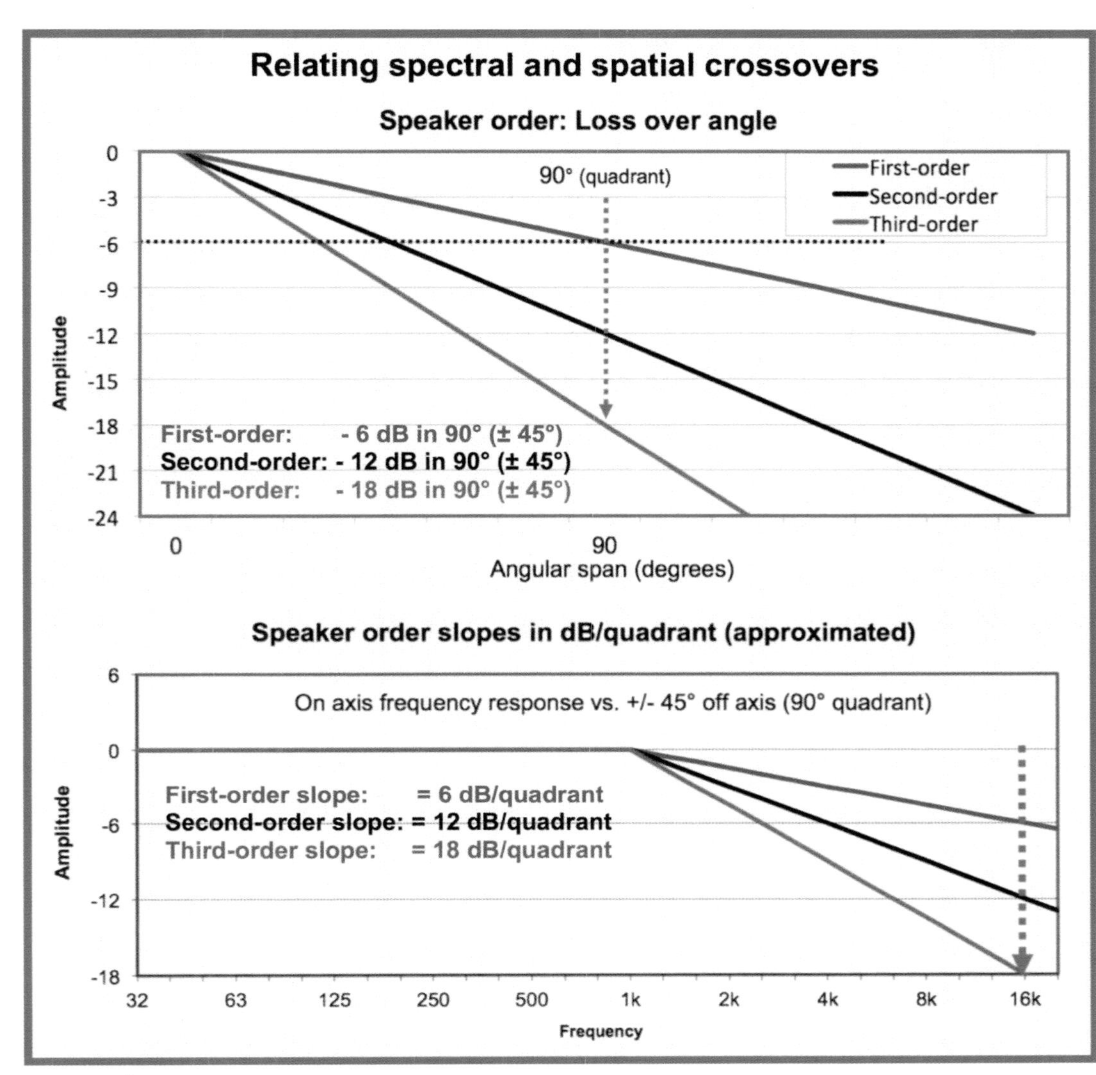

FIGURE 4.31
Linking spectral and spatial crossovers: first order, second order, third order

but if we want to slice things into small bits, then steep filters are the most effective means of getting to the isolation zone and minimizing combing.

We don't use 360° speakers any more than we use a single driver to cover 20 Hz–20 kHz. For purposes of discussion we will divide the space into quadrant slices (90°) and classify coverage from there (Fig. 4.31). A speaker that loses -6 dB in a quadrant is termed a first-order speaker (analogous to -6 dB/octave). One that falls at a rate of 12/dB per quadrant is second order and so on. By classifying the speakers into first, second and third order (as we did filters) we can simplify our discussion of speakers, emphasizing the basic shapes they make rather than whether they are front-loaded, a radial horn, a ribbon, or a retro-encabulator waveguide.

Speaker order classification

- Omnidirectional: 180° to 360° (< 6 dB loss in a 90° quadrant).
- First order: 60° to 180° (around 6 dB loss in 90°).
- Second order: 20° to 60° (around 12 dB loss in 90°).
- Third order: 6° to 20° (around 18 dB loss in 90°).

A given speaker may have different coverage patterns in the vertical and horizontal planes, hence different orders. In addition we know that coverage patterns widen in the LF range, so a single classification will not represent the full range of speaker behavior. We use speaker-order classifications to separate the HF response. In short, we can assume that every speaker will end up omni at the bottom end. The disparity between LF coverage and HF coverage will be greatest in the third-order speakers.

4.3.5.4 UNITY SPATIAL CROSSOVER (A + B)

The unity gain spatial crossover is a mainstay of design and optimization, a principal connection point between speakers in an array and combinations of subsystems (Fig. 4.32). Three main types are found in the room.

Three typical forms of the unity gain spatial crossover (A + B)

- Unity splay angle: -6 dB point is created by angular separation (e.g. a coupled point-source array).
- Unity spacing: -6 dB point is created by lateral or vertical separation (e.g. frontfill spacing).
- Unity distance (forward): -6 dB point is created by doubling distance (e.g. main and delay).

These can be used singly or in combination to create a unity crossover (e.g. frontfills on a curved stage would use splay *and* spacing to create a unity summation).

The standard unity crossover adds A (-6 dB, 0°) + B (-6 dB, 0°) at XAB and sums to 0 dB. The cause of the individual losses may be axial, distance or the combination of both. Summation gain is greatest at XAB and weakens away from there. Individual speakers become stronger as summation gain weakens, creating a consistent level while moving off center. The summed coverage pattern is twice as wide as each individual and is re-evaluated using crossover as the center reference.

If the crossover location has already been selected the process takes five steps.

Creating a unity gain spatial crossover with A and B speakers

- Determine a nominal reference standard: 0 dB.
- Drive A (solo). Observe the acoustic response and find the -6 dB location. This is crossover XAB.
- Drive B (solo). Adjust B until it creates -6 dB at same location (e.g. splay, spacing, level adjust).
- Adjust relative phase until matched at XAB (delay the earlier one).
- Drive A + B. Summed response at crossover should match the nominal individual levels.

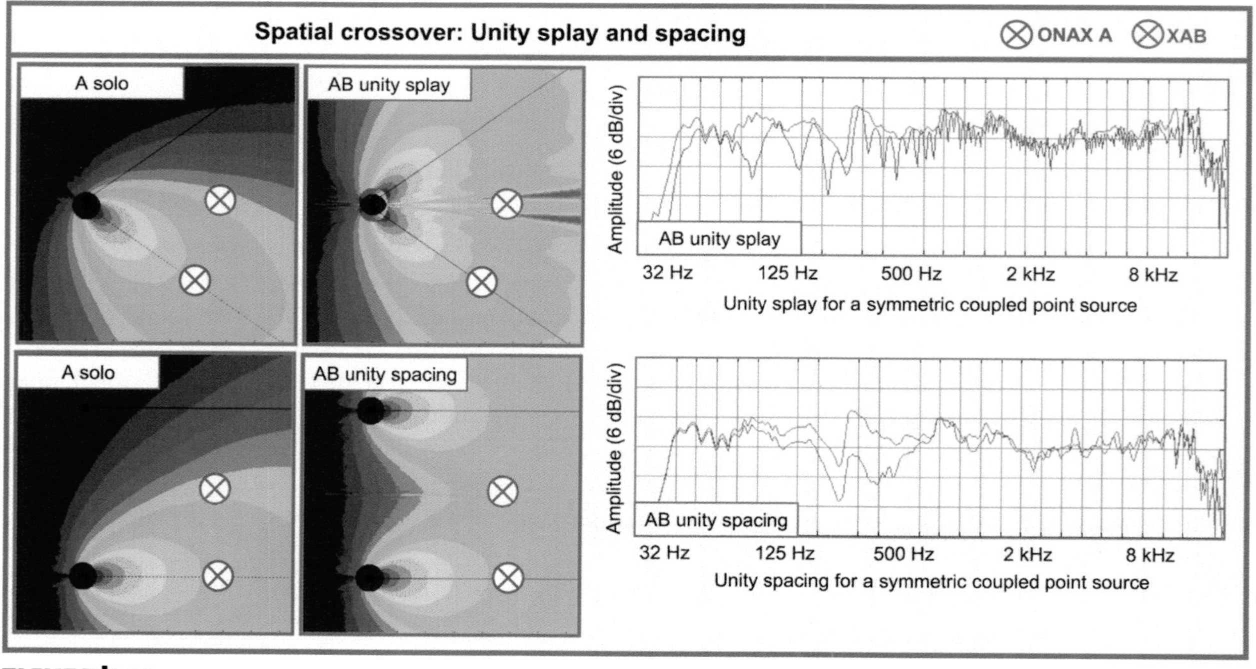

FIGURE 4.32
The unity spatial crossover

4.3.5.5 OVERLAPPED SPATIAL CROSSOVER (A + B)

The overlap crossover adds A (>-6 dB, 0°) + B (>-6 dB, 0°) at XAB and sums to >0 dB (+6 dB max). Summation gain is highest at XAB and lessens off center (Fig. 4.33). The summed coverage pattern is re-evaluated using crossover as the center reference. The combination is wider, narrower or the same as individual elements, depending on overlap percentage (majority overlap narrows, majority isolation widens). Highly overlapped systems might never reach the isolation zone (level offset doesn't reach 10 dB).

Creating an overlap spatial crossover with A and B speakers

- Determine a nominal reference standard: 0 dB.
- Alternate soloing A and B to find the equal-level location (matched at >-6 dB from the reference level). This is crossover location XAB (e.g. if the A and B levels are -2 dB then the summation will be +4 dB).
- Adjust the relative phase until they match at XAB (delay the earlier one).
- Drive both channels together. Summed response should exceed the individual levels by +6 dB.
- If some frequencies are overlapped and others are not, then equalization can be applied to the overlapped ranges (same procedure as overlapped spectral crossovers).

4.3.5.6 GAPPED SPATIAL CROSSOVER (A+B)

The gap crossover adds A (<-6 dB, 0°) + B (<-6 dB, 0°) at XAB and sums to <0 dB (Fig. 4.33 again). Summation gain is greatest at crossover and lessens off center (whereas individual levels get stronger). The gapped zone is defined as the area where levels are below the 0 dB reference. Gaps of 6 dB or more are equivalent to off-axis response. Gap crossovers are used to avoid areas such as balcony fronts.

4.3.5.7 ISOLATION BY ANGLE

We've looked at bringing speakers together. Now we'll evaluate getting them apart. The first way is angular isolation, otherwise known as splay. This method is implemented in the point source array by aiming the speakers apart in front. A symmetric pair will have a centered crossover, which may be overlapped, unity or gapped. The road to isolation (level offset >= 10 dB) begins at crossover and heads toward the outer edge of the coverage. We reach it first in a gap splay, second in a unity splay and last (if at all) in an overlap splay. The coupled point source has a clearly definable isolation zone: A given frequency has a certain angle at which isolation occurs, an angle that holds

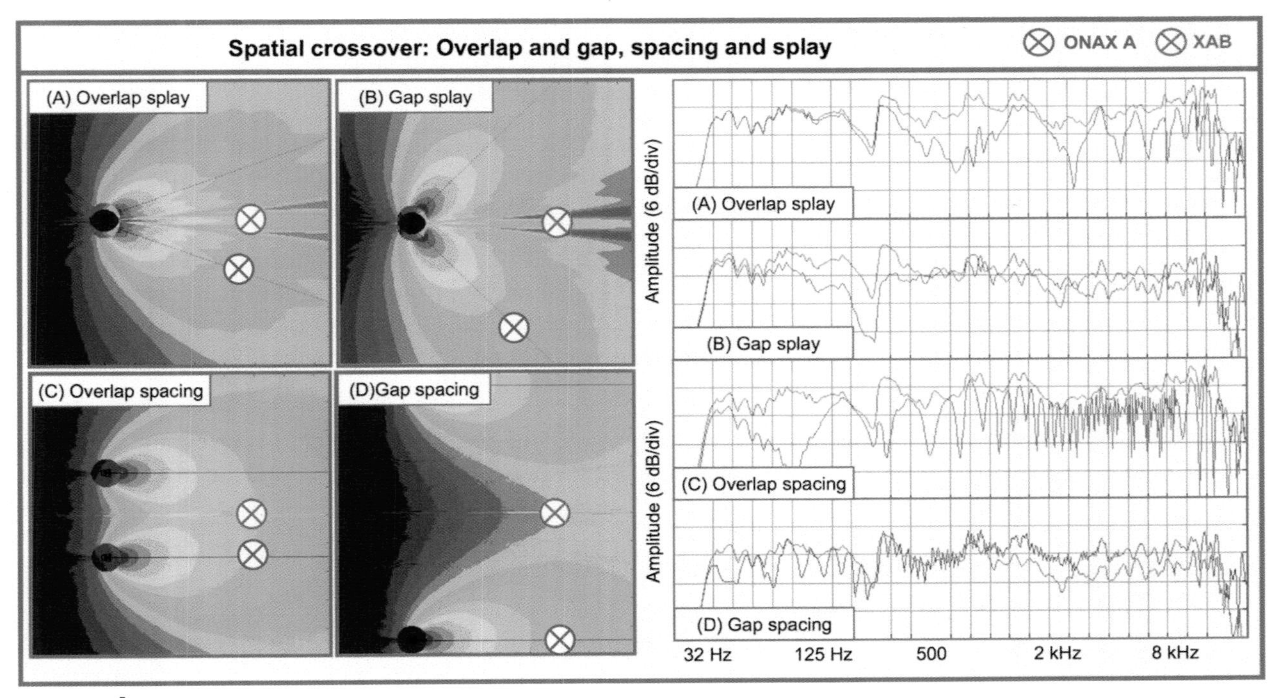

FIGURE 4.33
Overlap and gap spatial crossovers

over distance. For example, if isolation at 8 kHz begins at 20° away from crossover, it will (a) hold the same level of isolation over distance and (b) be more isolated at angles further away from crossover.

4.3.5.8 ISOLATION BY DISPLACEMENT

The next avenue toward isolation is separation/displacement. Move speakers far apart and they must cover a lot of ground before they can connect. This is an uncoupled array configuration, and therefore will morph through different summation zones over distance. A simple example: Two speakers are placed 10 m apart. If we walk a straight line from the front of A (at 1 m) to the front of B (at 1 m) we will move from isolation (A) to a gap (XAB) to isolation (B). The same path at 20 m in front of the speakers will be constant overlap (and combing). At some distance between we can walk a line and go from isolation (A) to unity (XAB) to isolation (B). In front of the unity line is too close (pre-coverage), and past it is too far (post-coverage). Our goal will be to close the gap at the right place, and conduct damage control on the overlap. Wide spacing of narrow elements would maintain isolation the longest, and close spacing of wide elements the least.

4.3.5.9 ISOLATION BY LEVEL

We can shift the summation zone balance by leaving speakers in position and turning one down. The crossover location will move toward the lesser speaker (B). Think of the level reduction as the B speaker yielding territory to A, an asymmetric-level distribution. On the A side we'll see isolation arrive sooner (because B will more quickly fall 10 dB behind), whereas the inverse is true for the B side.

4.3.5.10 ISOLATION BY COMMITTEE

All the isolation mechanisms just listed can be brought together to create the desired shaping. We can move them apart, splay them apart and turn one down. The transition to isolation comes quickly for the dominant speaker. For the smaller one, welcome to life as a fill speaker.

4.3.5.11 CROSSOVER DETECTABILITY

Hiding spatial dividers is challenging because the crossover frequency range extends to the upper limits. It's virtually impossible to transition through crossover without some HF combing. Spatial crossovers can have substantial physical displacement, which makes the combing zone volatile. Transitions between high-order systems with steep angular slopes may be easily detectable but only in a very small portion of the space, whereas low-order systems will be less detectable, yet spread over a wider space. A classic tradeoff.

The most salient contrast to spectral crossovers is that spatial crossovers are detectable only in specific locations. Detectable spectral dividers may be obvious over large areas. Spatial crossovers can be placed on aisles, balcony fronts and other places that render their deficiencies academic. Not so the spectral divider.

The spatial divider gives itself away by shifts in angular position. Our ears can pick up localization clues, which become easier to spot as angular offset rises. Other clues could be mismatched responses in the crossover area. The most obvious is when one speaker has HF range extension well above the other. This must be carefully managed when combining small-format fill speakers with extended VHF with large-format main systems. The mixer won't hear the VHF in the frontfills but the first row surely will.

Another clue comes in the form of the relative distance. Close speakers have superior direct-to-reverberant ratio over distant ones. The level of close delay speakers must be minimized so they don't stand out above the distant main speaker.

4.3.6 The spectral/spatial crossover

Most of the two-way loudspeakers in the world have displaced drivers: HF next to LF or centered between two woofers. The spectral crossover is also a spatial crossover. LF and HF have a crossover frequency and a crossover location (Fig. 4.34). The exception is the coaxial design, where the HF driver is centered inside the LF driver (in which case the displacement is in the depth plane rather than horizontal or vertical).

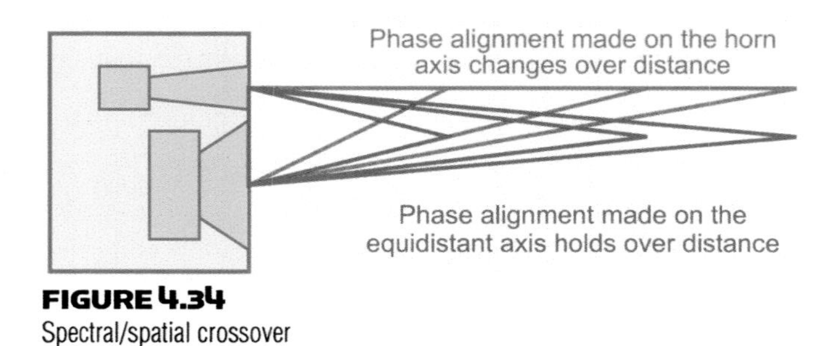

FIGURE 4.34
Spectral/spatial crossover

This adds a second dimension to the quest for combing-free crossover performance. The goal is to stay in the coupling zone until we are out of the angular coverage, i.e. spectral coupling until spatial isolation. We return to the isosceles triangle (our representative of the spatial coupling zone and target for phase alignment). We can view the drivers as a spatial crossover and align the spectral level and phase by the unity spatial crossover procedure outlined previously (4.3.5.4).

Spatial stability is evaluated by measuring off center of the crossover (above and below or side to side). Observe the crossover frequency range and see if the response holds out long enough to reach the spatial coverage edge. This is best done at a distance long enough to represent real-world applications. This allows the time offsets to settle into the range where the speaker will actually be used. Measuring too close can make a perfectly functional spectral/spatial crossover appear troubled.

4.4 SPEAKER ARRAYS

Let's apply our study of summation and the acoustic crossover to the practical construction of speaker arrays. Add two speakers and the sum will depend on their level and phase offsets. Add ten speakers and they will behave exactly as the summation of the summations. The individual element coverage patterns, their displacement, relative angles and levels drive the spatial distribution. If we successfully merge the systems at the spatial crossovers, the rest of the coverage area will become predictable and manageable. All of these factors can be independently controlled in the design and optimization process.

4.4.1 Speaker array types

The pro audio trade news would lead us to believe there are hundreds of different speaker array types, with various trademarked™ names. Others believe that there is only one array type, the line array, and all other configurations have gone the way of the dinosaur.

In practical terms we classify arrays into two families of three types. First we separate coupled from uncoupled and then move on to angular orientation.

Array Type	**Configuration**
Coupled line source	Speakers together in parallel
Coupled point source	Speakers together with outward splay angles
Coupled point destination	Speakers together with inward splay angles
Uncoupled line source	Speakers separated in parallel
Uncoupled point source	Speakers separated with outward splay angles
Uncoupled point destination	Speakers separated with inward splay angles

How do we separate coupled from uncoupled? It's harder than you think. The 600:1 ratio of wavelengths these arrays transmit makes the gray become grayer. The easiest way to clarify is by function.

Coupled array functions	**Uncoupled array functions**
Power gain	Radial coverage expansion
Radial coverage expansion	Lateral coverage expansion
Radial coverage reduction	Forward coverage expansion

Power gain requires coupling zone summation, as does radial coverage narrowing. Radial and lateral coverage expansions require isolation zone summation. The transition zone is the preferred bridge between coupling and isolation whereas the combing zone is to be avoided as much as possible.

Let's return now to the coupled/uncoupled question. As an example we have a straight line of eight subwoofers spaced 1 m apart. On top of each is a small full-range frontfill. We have both a coupled and uncoupled array. The quantity of eight subwoofers adds power gain and narrows coverage. The quantity of eight frontfills creates a lateral coverage expansion. If we need more LF power we add subs. If we need more frontfill coverage we add frontfills. If we need more frontfill power we get a bigger frontfill. Coupled vs. uncoupled.

Any speaker array can create a coupling zone along the coverage centerline (the isosceles triangle). Closely spaced arrays can maintain the coupling for a substantial portion of the spectrum *and* the room. Adding space between the elements lessens coupling in both. We can rescale our arrays, but not the size of 500 Hz. We conclude the obvious: Coupled arrays have superior coupling zone behavior. Duh!

Once we run out of coupling capability we seek isolation, which we can get by angle, spacing or level. Angle is the most effective and long lasting. Isolation by displacement is effective, but as advertisers love to say "for a limited time only!" Level tapering, on its own, is the most limited, and should be considered more as an addendum to isolation than a primary means. We can use them together to great effect.

Which arrays can isolate? The point source is the master of angular isolation. The line source has none. The point destination can provide angular isolation after its beams have passed through the center. This gives it limited applicability. The uncoupled arrays are the winners for displacement isolation. Any array can taper level, but this is not much help without a head start from angle or displacement.

Let's look at the scorecard (Fig. 4.35). One array configuration provides extensive coupling and long-range isolation: the coupled point source. It's no coincidence that this is the main array used in most sound reinforcement systems. The coupled point destination can also provide both but runs into mechanical challenges such as speakers blocking other speakers from crossing through to the opposite side. The coupled line source has no effective isolation mechanism. The result is a concentrated beam focused at infinity.

Array summation properties							
Array type	Isolation method			Summation zones			Range
	Angle	Distance	Level	LF	MF	HF	
Coupled line source	No	No	Limited	Coupling	Coupling	Coupling	Unlimited
Coupled pt. source	Yes	No	Yes	Coupling	Transition	Isolation	Unlimited
Coupled pt. destination	Limited	No	Yes	Coupling	Transition	Isolation	Unlimited
Uncoupled line source	No	Yes	Yes	Combing	Transition	Isolation	Limited
Uncoupled pt. source	Yes	Yes	Yes	Transition	Isolation	Isolation	Limited
Uncoupled pt. destination	No	Yes	Yes	Combing	Combing	Isolation	Short

FIGURE 4.35
Array summation properties

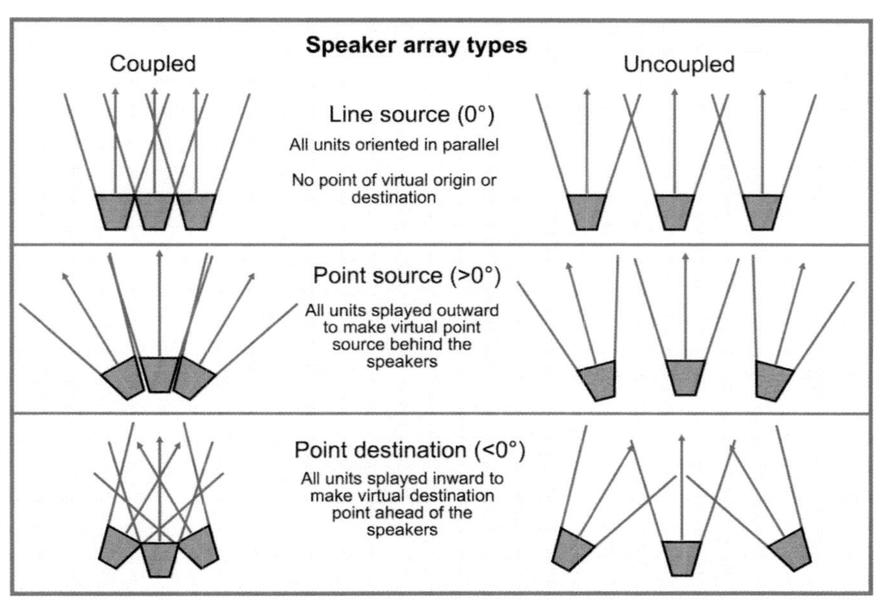

FIGURE 4.36
Speaker array classification

On to the uncoupled arrays where the point source isolates by a combination of displacement *and* angle. This gives it an extensive usable range (both spectral and spatial). The uncoupled point destination is a tug-of-war between isolating displacement and de-isolating inward angle. This is the most limited in range of the uncoupled arrays. The uncoupled line source falls in the middle between the point source and destination.

The next step will be a spatial distribution study of the six array types (Fig. 4.36).

4.4.2 Coupled arrays

The analysis will focus on the far-field response, where the combination of multiple elements is fully matured. Long arrays with many elements may show near-field isolation due to displacement until the gap closes and all elements are summed. The forward distance required to reach maturity will vary with frequency (ranges with wider coverage will close sooner than those with narrow coverage). The response is shown in the 100 Hz, 1 kHz and 10 kHz format, which is used for the rest of this chapter. This series brings together the summation icons presented earlier and gives them context with the spatial crossover locations.

4.4.2.1 COUPLED LINE SOURCE

Coupled line source arrays allow unlimited element quantity at the most limited angle quantity: one. There's no simpler array to describe. No angle details, only element coverage pattern, quantity and displacement. The virtual sound source is elongated, stretched over the array length rather than a single point. The consistent feature of the coupled line source is that overlap class behavior comprises virtually the entirety of the system's response. This is both its most attractive and most ominous feature. The overlap gives it the maximum power addition, but at the cost of minimum uniformity.

Speaker order

The first in this series is Fig. 4.37, where we see the results of a pair of first-order speakers arrayed as a coupled line source. It's a train wreck. The only position to enjoy a ripple-free frequency response is the exact centerline: the

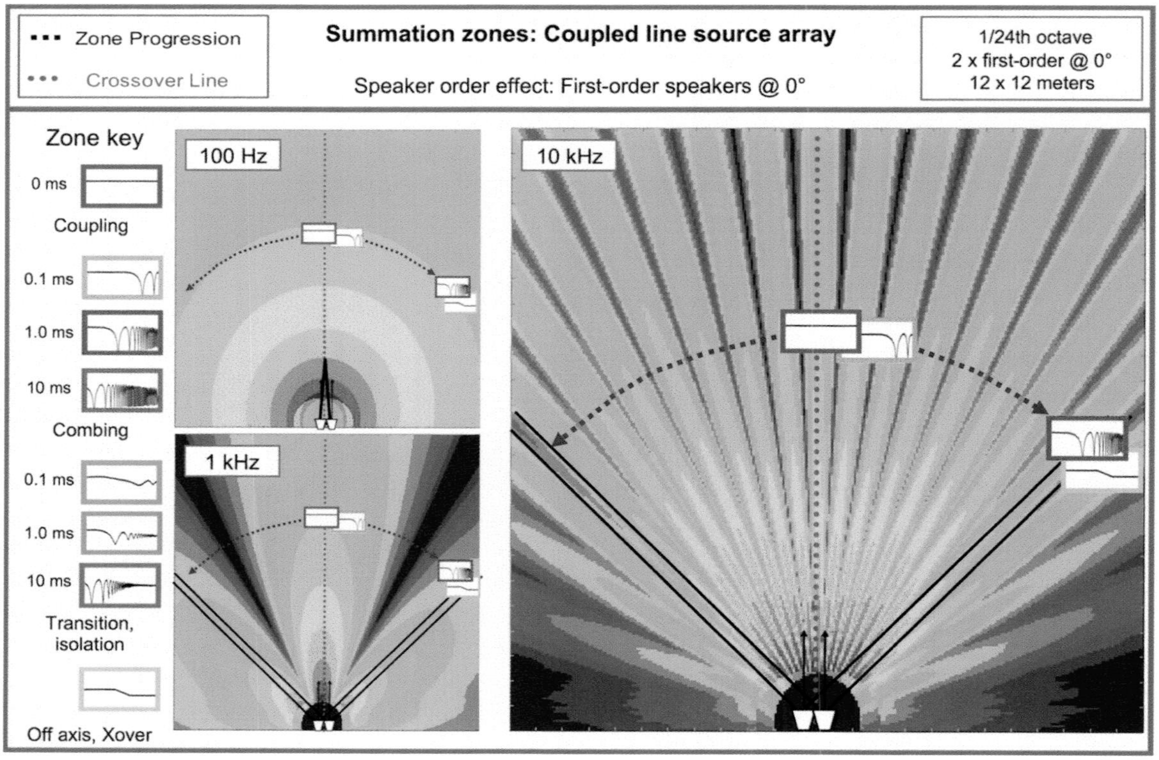

Zone Progression
Crossover Line
Summation zones: Coupled line source array
Speaker order effect: First-order speakers @ 0°
1/24th octave
2 x first-order @ 0°
12 x 12 meters
Zone key
0 ms
Coupling
0.1 ms
1.0 ms
10 ms
Combing
0.1 ms
1.0 ms
10 ms
Transition, isolation
Off axis, Xover
100 Hz
1 kHz
10 kHz

FIGURE 4.37
Summation zone progression factors for the coupled line source array, first-order speakers

Zone Progression
Crossover Line
Summation zones: Coupled line source
Speaker order effect: Third-order speakers @ 0°
1/24th octave
2 x third-order @ 0°
12 x 12 meters
Zone key
0 ms
Coupling
0.1 ms
1.0 ms
10 ms
Combing
0.1 ms
1.0 ms
10 ms
Transition, isolation
Off axis, Xover
100 Hz
1 kHz
10 kHz

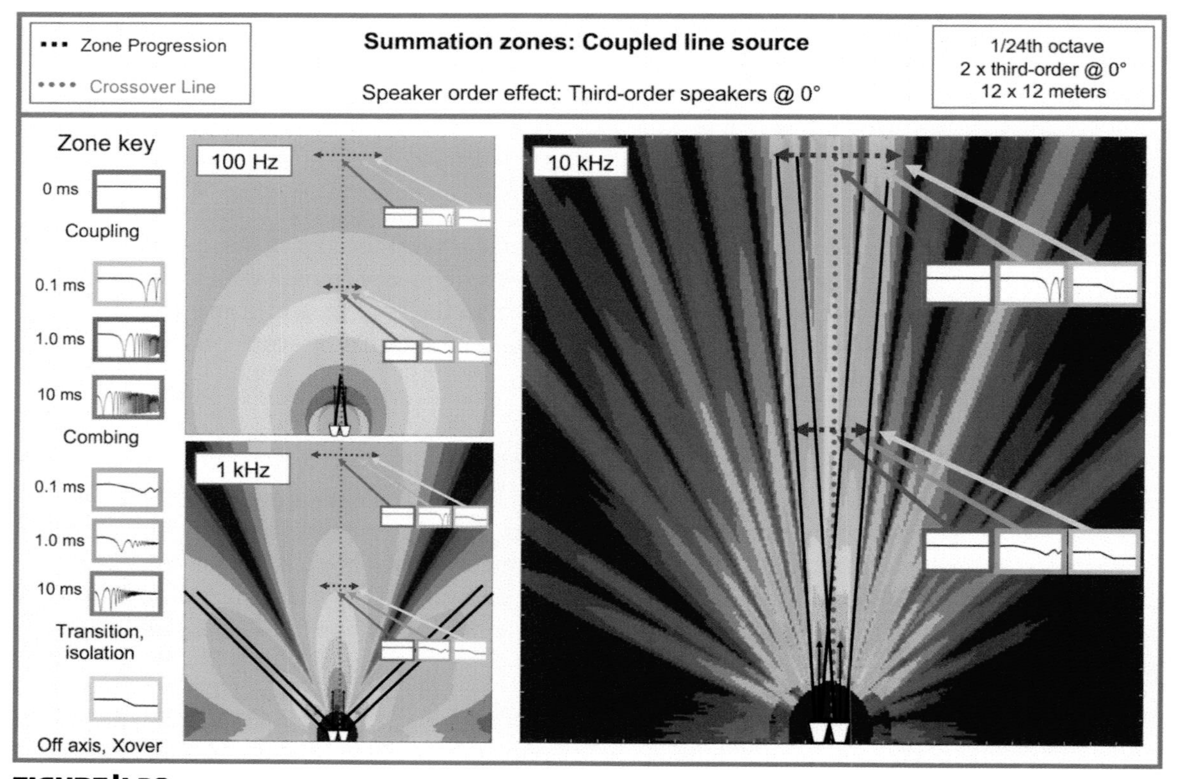

FIGURE 4.38
Summation zone progression factors for the coupled line source array, third-order speakers

spatial crossover, the result of excessive overlap and displacement. The summation zone progressions move from center to left and right. The combing zone dominates in the 10 kHz range, both near and far all the way to the edges. The coupling zone dominates the LF range due to close proximity.

A pair of third-order elements is seen in Fig. 4.38. The individual high-frequency coverage is so narrow that the gap coverage zone can be seen in the near field. This quickly gives way to overlap coverage where three beams can be seen (the main lobe and two side lobes). Notice the lack of uniformity over frequency, with extremely narrow HF and extremely wide LF response (just like the solo element).

Quantity

Maybe we just need more boxes. An additional third-order element is added in Fig. 4.39, extending the gap crossover range into two sections. It does little to address the discrepancy between the HF, MF and LF shapes. The complexity of the triangulation geometry increases with the addition of the third element. There are now multiple triangles stacked together, which increases the overlap with distance, narrowing the coverage angle.

We digress for a moment to address the coupled line source's fundamental property: the pyramid-shaped series of coupling zone summations. The pyramid effect comes from the cascading summations (Fig. 4.40). The phase contours of the three elements converge initially into two zones of addition (and one of cancellation). As we move farther away the three phase responses converge to form a single beam, the pyramid peak. Once the pyramid assembly is complete, the coupled array will assume the characteristics of a single speaker: a definable ONAX, a loss rate of 6 dB per doubling and a consistent coverage angle over distance. These milestones aren't found until the array has fully coupled (the pyramid top), which will make it challenging to place mics for optimization.

An eight-way line source pyramid is shown in Fig. 4.41. The foundation begins with the isolated elements (the gapped crossover) and then continues with seven twoway overlapping crossovers until it eventually converges into a single eight-way overlapped coupling zone summation. The distance to the first pyramid step (the two-way

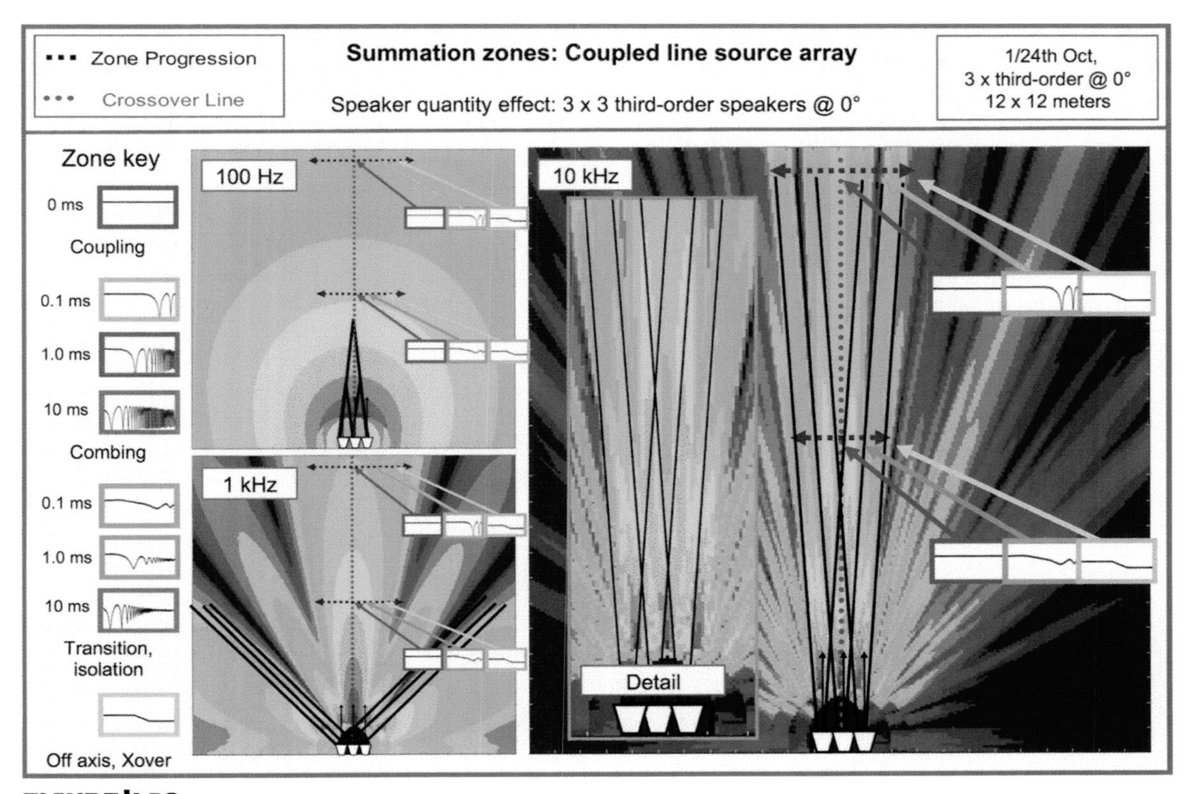

FIGURE 4.39
Summation zone progression factors for the coupled line source array, change of quantity

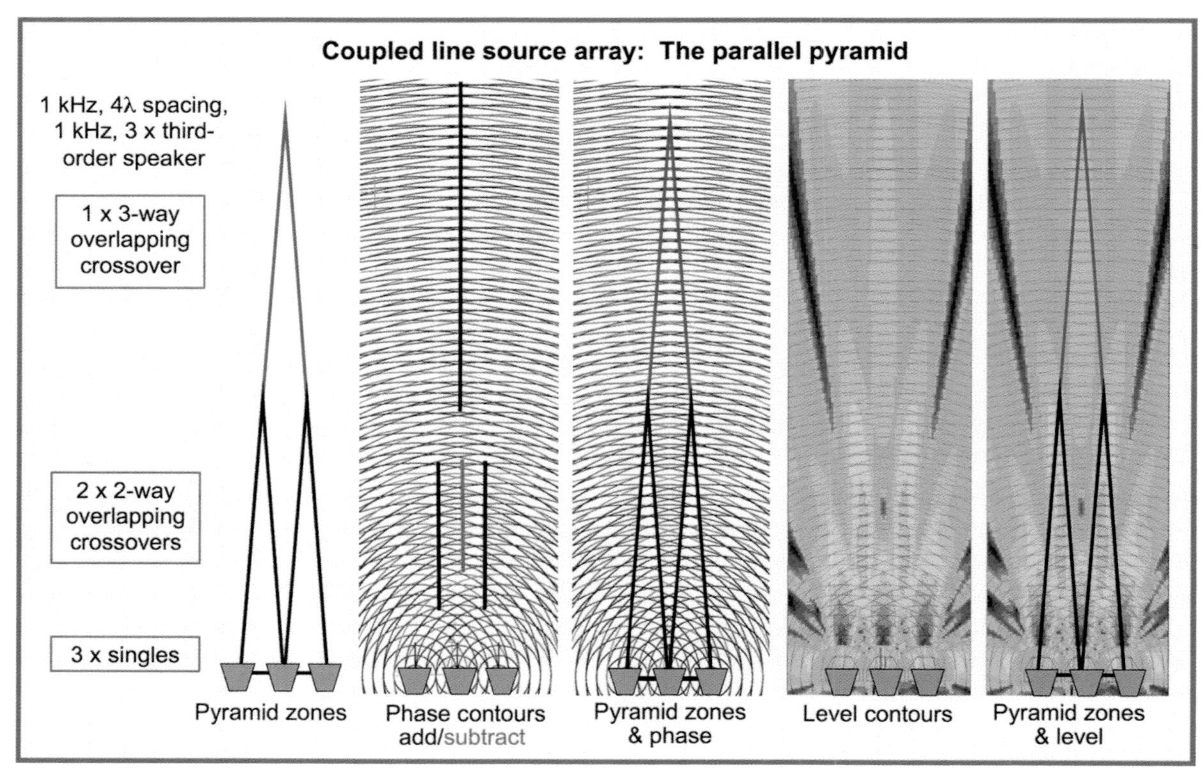

FIGURE 4.40
Parallel pyramid for three elements

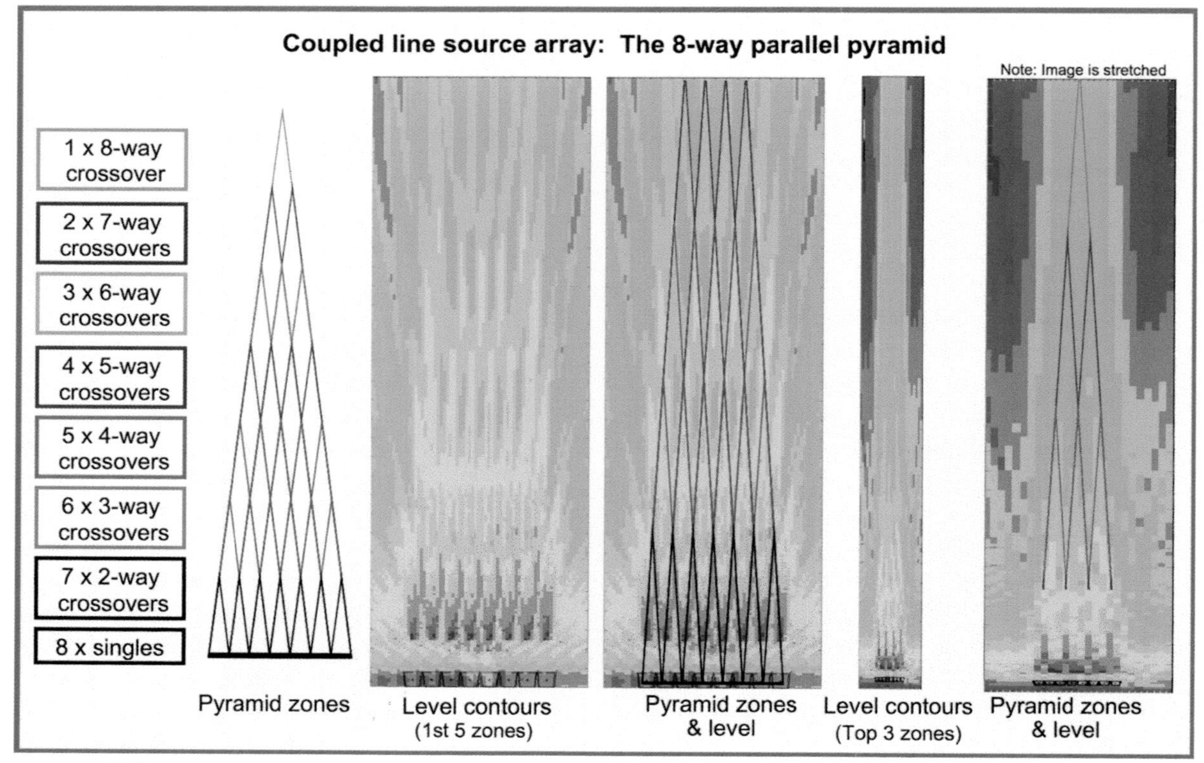

FIGURE 4.41
Parallel pyramid for eight elements

crossover) is controlled by the same factors discussed earlier: element coverage angle and displacement. The distance to each successive step is the same, so the total height is found by multiplying the step height by one less than the element total (e.g. an eight-element pyramid height = distance to first crossover × 7). As elements become more directional or displacement increases, the step height multiplier extends. Because directionality is variable over frequency, the pyramid step height will vary over frequency, which when multiplied can create vast differences in combined shape over frequency.

Combining wide speakers yields less width. Combining narrow speakers yields more narrowing. The more the narrower. If the element starts with wide LF and narrow HF, the array response will end up the same way (relatively to each other). The combination will be proportionally narrower than the single element at every frequency. The coupled line source is an equal opportunity squeezer.

Summation zones

If we hold to the definition of "coupled line source" we have a one-step summation zone progression: coupling. The crossover area (the coupling line) gets the maximum and the sides (non-coupling line) get the minimum. It's all good until somebody gets out of line. I mean out of phase. Real loudspeakers take up space(s). Keeping the phase offset under 120° becomes increasingly challenging as frequency rises. Because there's no road to angular isolation, displacements over 120° push us into the combing zone. It's a race. If we finish the coverage before combing begins we win (i.e. we reach -6 dB off axis before the phase offset is >120°). Let's start with two elements spaced 1 ms apart. Start with F = 33 Hz (T = 30 ms). The phase offset is 1/30 λ (12°). We win. Next: F = 333 Hz (T = 3 ms). The phase offset is 1/3 λ (120°). It's a tie. Things don't look good for 3333 Hz do they? We can't beat the phase offset, which maxes out at 1200°. How about level? Bear in mind that our maximum phase offset (1200°) is along the non-coupling line, which is 90° off axis from crossover. If our speaker has less than 180° of coverage it won't make it to the 90° finish line with enough level to win. The critical location is the radial angle that corresponds to 120° phase offset. The answer for 3333 Hz is ±10° as shown back on Fig. 4.21. A displacement of 3 λ creates 120° phase offset at the radial 10° mark (relative to center between the elements). Raise the frequency an octave and we have only ±5° to work with. The same thing happens if we double the displacement. We've established a sliding scale. For a given displacement we can neutralize the combing zone damage by proportional reduction of the coverage pattern. We maintain our position if we halve the coverage as we double the frequency. It should be obvious that the advantage in combing suppression is balanced by a serious side effect: radically different coverage over frequency. But this is the fate of the coupled line source if it plays with wavelengths too hot to handle. If we keep it on the down low, in the heart of the coupling zone, we can leave coverage out of the equation. But if we keep wide coverage all the way to the top end in a coupled line source we will pay the price.

Coupled line source conclusions

- Coupling frequency limit is set by element displacement ($F^{LIM} = 0.33 \times T$) where T is the displacement in ms. For a 3 ms displacement (about 1 m), $F^{LIM} = (0.33 \times 0.003)$. F^{LIM} = 110 Hz.
- Audible combing effects can be reduced if the coverage pattern narrows as phase offset rises (due to displacement). The maximum comb-free coverage angle = 60°/λ displacement. Example: Two sources 4 λ apart have a maximum coverage angle of 15° (60°/4).
- This array type works for subwoofers but has severe limitations above that range.

We already have a conclusion that eliminates first- and second-order speakers from contention in the coupled line source, but the massive combing we saw in Figure 4.37 should help to alleviate any doubts. First-order elements (≥60° wide) cannot exceed 1 λ displacement. Second-order speakers range down to 20° wide, which corresponds to 3 λ. That's 0.17 ms at 18 kHz (as long as these six words). These elements must be very closely spaced small speakers. This leaves us with proportional beamwidth third-order speakers with their ever-narrowing coverage pattern as frequency rises. There are no rooms that get narrower as frequency rises so this will not provide uniform coverage over frequency. The third-order proportional beamwidth element does have two things going for it: lots of coupling (power gain) and the highest tolerance to overlap. We'll soon put it to use by adding some splay.

4.4.2.2 COUPLED POINT SOURCE

The coupled point source approaches arrays from an entirely different angle (one that's not 0°). Adding splay angle to the equation enables an isolation mechanism that opens up many possibilities for variable array shapes. We can mix speaker orders, splay angles and levels for shaping. The coupled point source has one feature no other array type can duplicate: maintaining a unity class crossover over extended distance. It's not automatic, and is variable over frequency, but no other array can duplicate this for even a single frequency. The coupled point source is the steady long-distance runner. If the coverage angle remains constant over frequency, so will the unity class crossover. That's what we call uniformity of coverage. The gap, unity and overlap areas are all angularly defined and maintain their character over distance.

Let's begin with a first-order pair (90°) at unity splay (Fig. 4.42). The summation zone progression originates at XOVR (coupling zone) and reaches isolation after a brief stay in the combing zone. The progression maintains its angular qualities over distance. Isolation zone behavior is visible in the HF response. The displacement is small enough to limit the MF disturbance to a single -9 dB null before breaking into isolation. The small displacement keeps the LF range entirely in the coupling zone. Note the overall resemblance of the HF, MF and LF shapes, indicative of a similar frequency response across the 180° coverage arc.

Next is a second-order (40°) pair at unity splay (Fig. 4.43). The zone progression is more angularly compressed than the first-order pair. Isolation arrives more quickly due to the steeper spatial filter slope of the second-order speaker. The 40° splay angle results in 0% overlap (unity) @10 kHz, creating a combined shape of 80°. Coverage is highly overlapped at 1 kHz (100° elements at 40° splay), creating more combing and a much wider combined coverage angle of around 140°. The LF response is similar to the first-order scenario.

Next is a third-order proportional beamwidth element with an HF unity splay of 8° (Fig. 4.44). A notable feature of this highly directional element is the visible gap in the near-field HF response. Farther away the crossover has reached unity, which will hold out for an extended range. Recall that the third-order system has the highest LF/MF/HF coverage angle differential, which reduces the unity splay range to a small minority of the spectrum. The overlap percentage exceeds

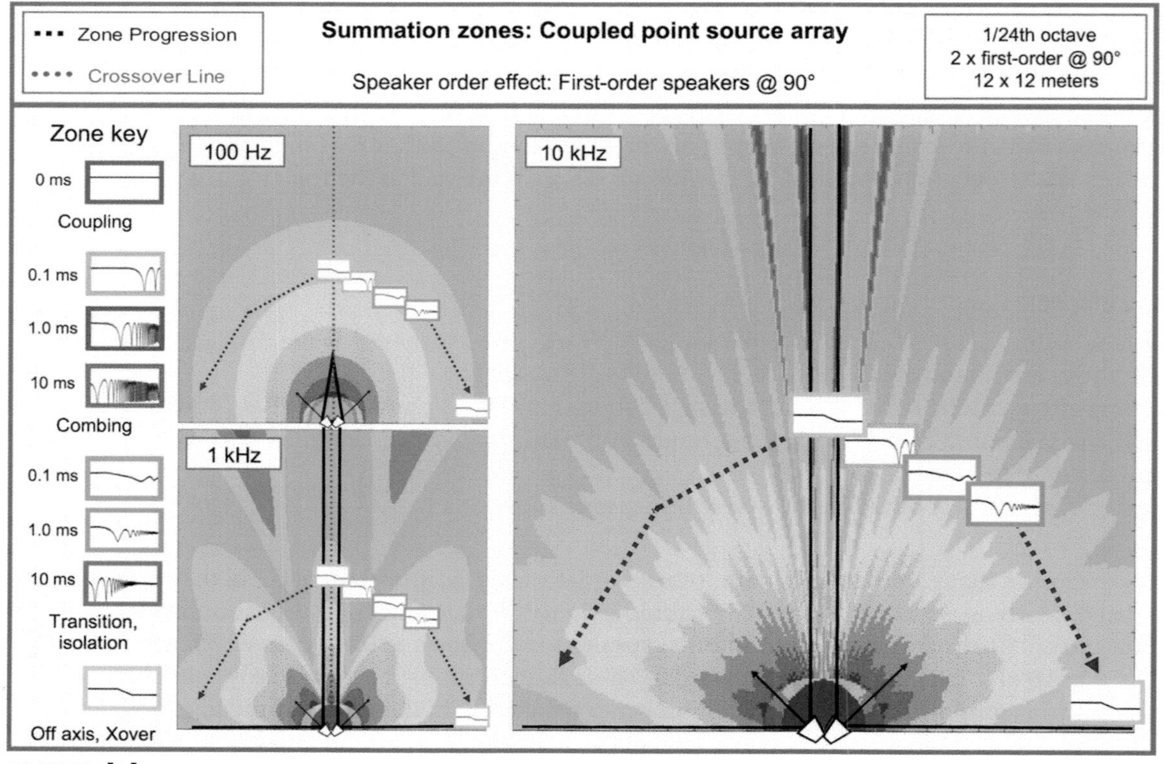

FIGURE 4.42
Summation zone progression factors for the coupled point source array, first-order

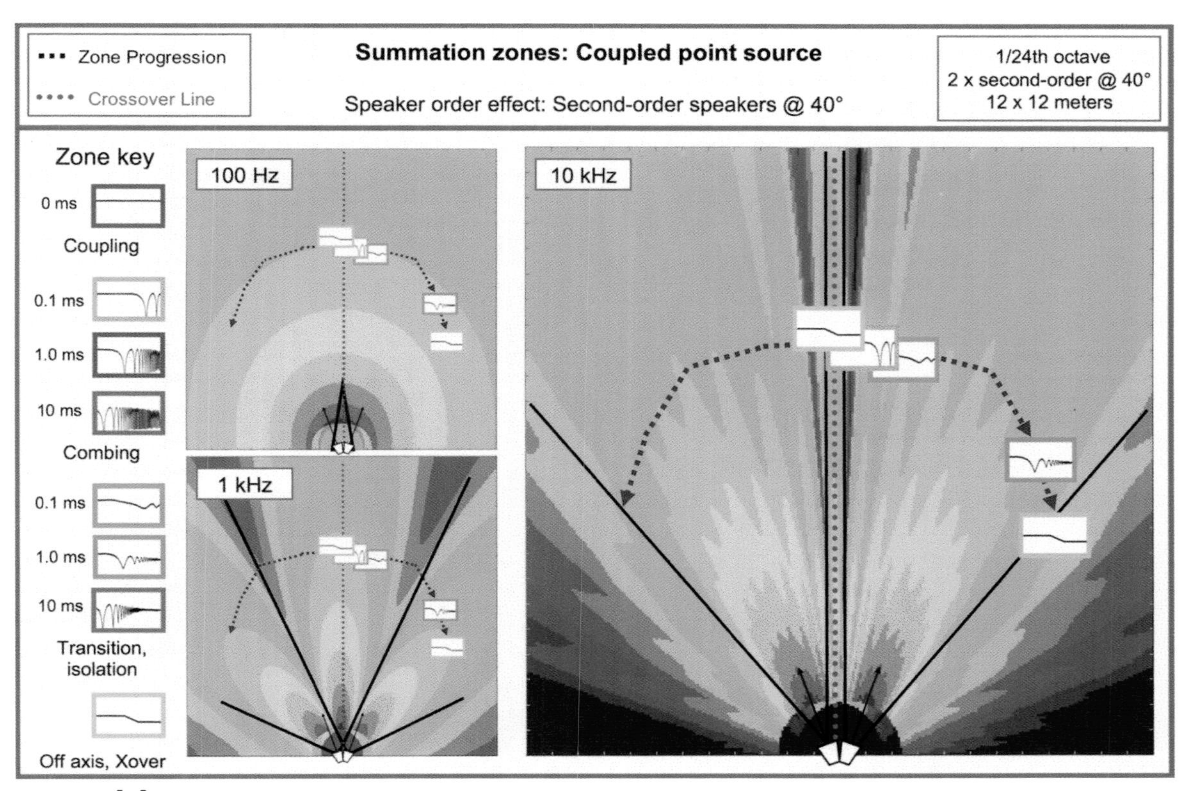

FIGURE 4.43
Summation zone progression factors for the coupled point source array, second-order

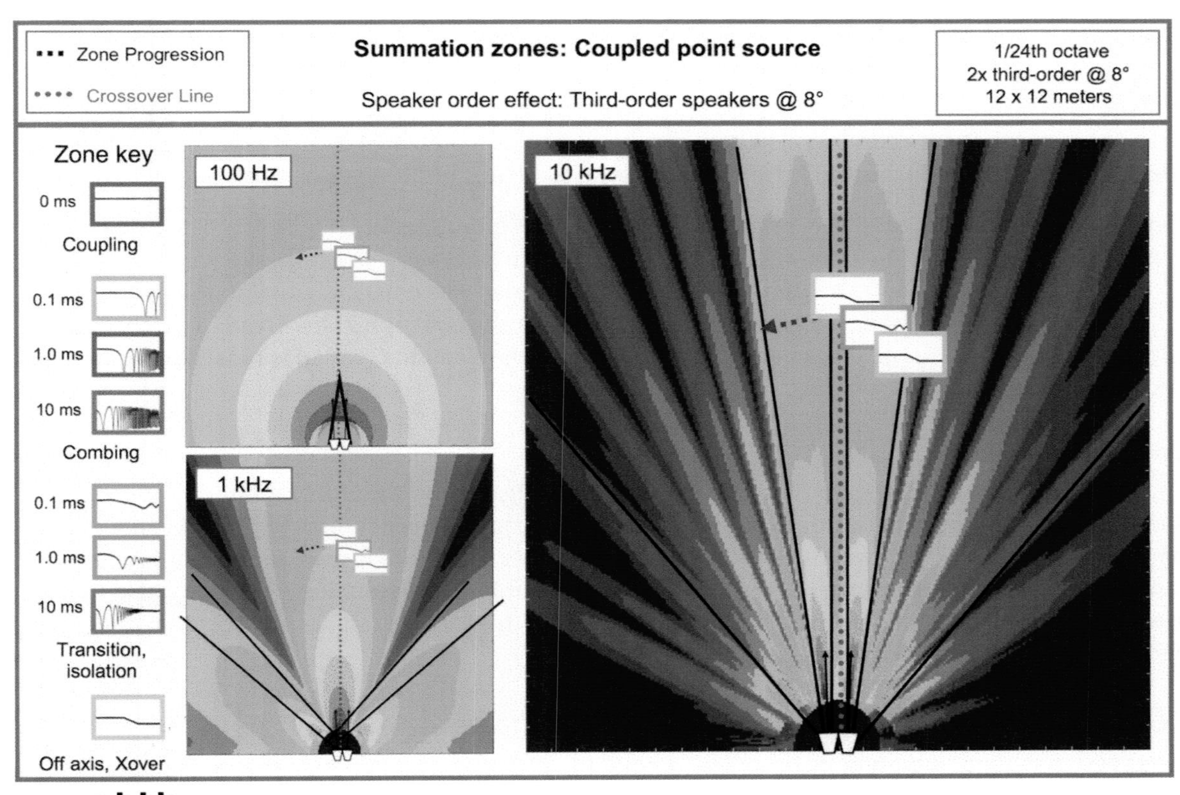

FIGURE 4.44
Summation zone progression factors for the coupled point source array, third-order

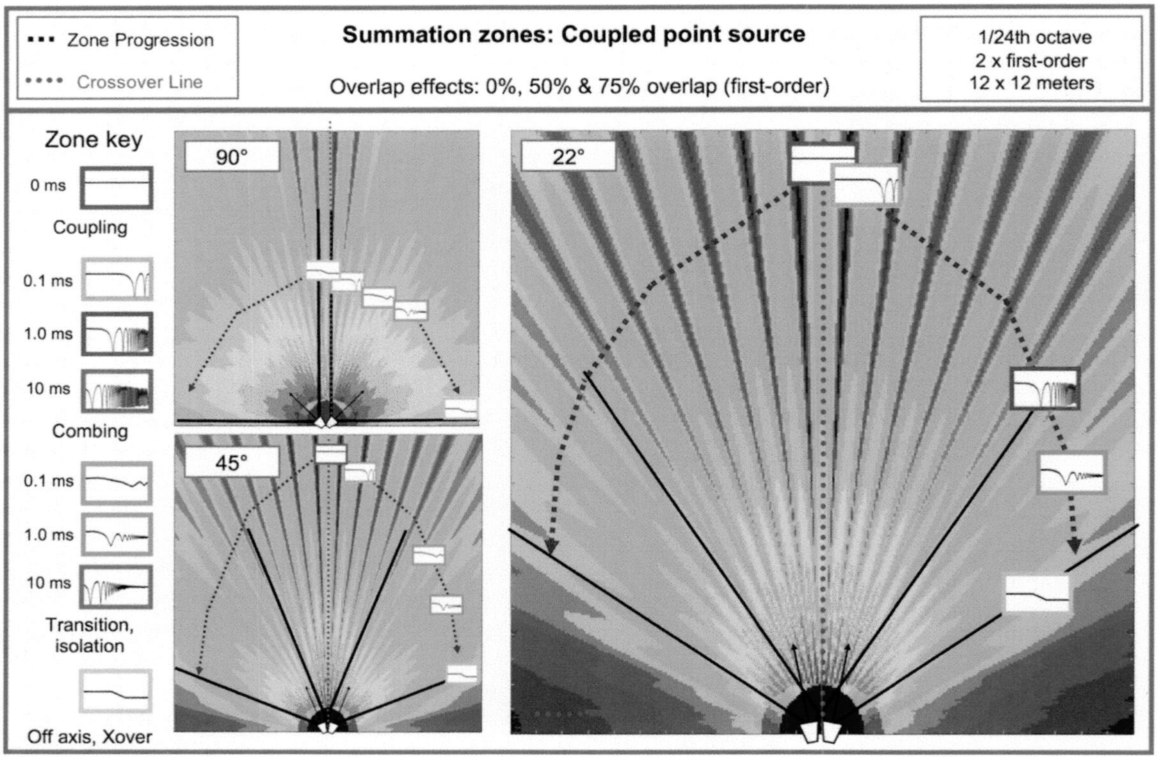

FIGURE 4.45
Summation zone progression factors for the coupled point source array, overlap effects

90% @1 kHz, resulting in a null depth greater than 20 dB. The presence of midrange combing does not eliminate this array from consideration. We will see, later, that an increased quantity of elements will have strong effects in highly overlapped arrays. For now we return to the ratio of LF/MF/HF coverage, where we find that the two-element array has decreased the disparity over frequency. The combined coverage shape widens the HF (dominant zone is isolation) and narrows the LF (dominant zone is coupling). The combined ratio can be represented by this approximation: ½ LF/MF/2 × HF. As we will see in Chapter 9, this will be the guiding principle in third-order speaker applications.

We now focus on the percentage overlap effects in the HF range of a first-order speaker (Fig. 4.45). The isolated (0% overlap) version from Fig. 4.42 is included for reference. The other panels show the response with 50% (45° splay) and 75% overlap (22° splay). The trend is obvious: The overlap percentage is equal to the proportion of the coverage that is in the combing zone, with the remainder being in the isolation zone. Overlap percentage must be carefully controlled whenever displacement is large relative to wavelength. Overlap is most effective when coupling is maintained via small displacements and angular control.

Summation zone progression

It's possible to make the run from coupling to isolation without combing, but we shouldn't admit defeat if we run the full seven-step summation zone progression. We can minimize the combing near crossover but can't always eliminate it. The isolation zone is largest with a unity splay crossover. As speaker order rises, the spatial filtering steepens, increasing the percentage of coverage in the desired zones (coupling and isolation). It's a race again, but angular isolation provides level offset relief when we get in phase offset trouble. We can use any speaker order, choosing to divide the pizza into large or small slices.

4.4.2.3 COUPLED POINT DESTINATION

We don't need to spend much time on the coupled point destination array. Its acoustical behavior is angularly similar to the point source, but with a forward focal point. The location where the speakers cross (the point destination) becomes the substitute point source. It's the same principle as a concave mirror. The chief deficiencies of the coupled point destination are mechanical and practical. There isn't a vortex disturbance in front of the speakers as some might believe. This array is selected only under duress, like an I-beam is blocking the horn of a point source array. We can fire up the welding torch or invert things and make it a point destination that can reach the audience.

For the coupled point destination array

- Functionally equivalent to the coupled point source (splay angle, frequency, etc.).
- Risk of reflecting off neighboring elements rises with quantity and overall splay.
- Impossible to array beyond 90° because elements are aimed through other elements.
- Horn driver placement at the cabinet rear creates unfavorable geometry. Point destination arrays usually have higher displacement than their comparably angled point source counterpart.
- Preferable only when physical logistics win over the point source counterpart.

4.4.3 Uncoupled arrays

We now add the second isolation mechanism: spacing the speakers apart, i.e. uncoupling. We pay a price in power gain but add shaping capabilities not present in the coupled arrays: lateral and forward extension. Uncoupled arrays must be evaluated in progressive stages over distance. The elements begin as clear soloists and then make duos, trios and entire choruses. As we'll see, sweet harmony often ends at the trio and gets worse from there.

RANGE PROGRESSIONS FOR AN EXTENDED SERIES OF UNCOUPLED ELEMENTS

- Pre-coverage: isolation (A) to gap (XAB) to isolation (B) and onward.
- Unity line: isolation (A) to unity (XAB) to isolation (B) and onward.
- Limit line: isolation (A) to overlap (AB) to overlap (ABC) to overlap (BCD) and onward.
- Post-coverage: overlap (AB) to overlap (ABC) to overlap (ABCD) and onward.

The pre-coverage range is used for fill speakers in singular areas. The unity line maintains consistent coverage by the principle of the unity class crossover. The limit line indicates where multiple paths with differing time offsets give the combing zone the majority. The most uniform coverage area lies between the unity and limit lines. In practice we design uncoupled systems to transition coverage to others at the limit line. Beyond the limit line are the tall weeds of the combing zone.

The locations for these milestones are influenced by element coverage (speaker order), spacing and splay. Narrow speakers push the start points deeper, as does wider spacing and outward splay. Wide speakers, closer spacing and inward splay hasten the progression.

The range progressions show a clear favoritism for speaker order. There's advantage to having all frequencies progress through the zones together. We don't want the LF and MF ranges to be in post-coverage combing when the HF range finally hits the unity line. The ideal uncoupled array element would have a flat beamwidth over the entire frequency range. This is unrealistic but we can dream, eh? In practical terms, we seek the longest beamwidth plateau we can get. This is standard operating procedure for first-order speakers. They are, by far, the favored uncoupled element, in large and small formats. Second-order speakers have a greater challenge. A long beamwidth plateau at a narrow angle requires a large horn. We might see these as festival frontfills that have several meters of open security barrier to travel before hitting the audience. Don't expect to see them at *Wicked*. Third-order speakers are a non-starter here.

4.4.3.1 UNCOUPLED LINE SOURCE

The uncoupled line source differs from its coupled cousin in scale only. That's not a small detail, however, because wavelength doesn't scale. The coupling of the parallel pyramid will be restricted to the LF range. Expect highly variable performance above that. An uncoupled array has a continuum of behavior, with tendencies toward coupling as frequency falls and isolation and combing as frequency rises.

We begin with a set of five first-order elements with 3 m spacing (Fig. 4.46). Notice that the response has repeating horizontal themes but is drastically different over depth, a standard feature of all uncoupled arrays. This contrasts with the coupled point source array, which held its angular shape over distance. The gap zone is visible in the

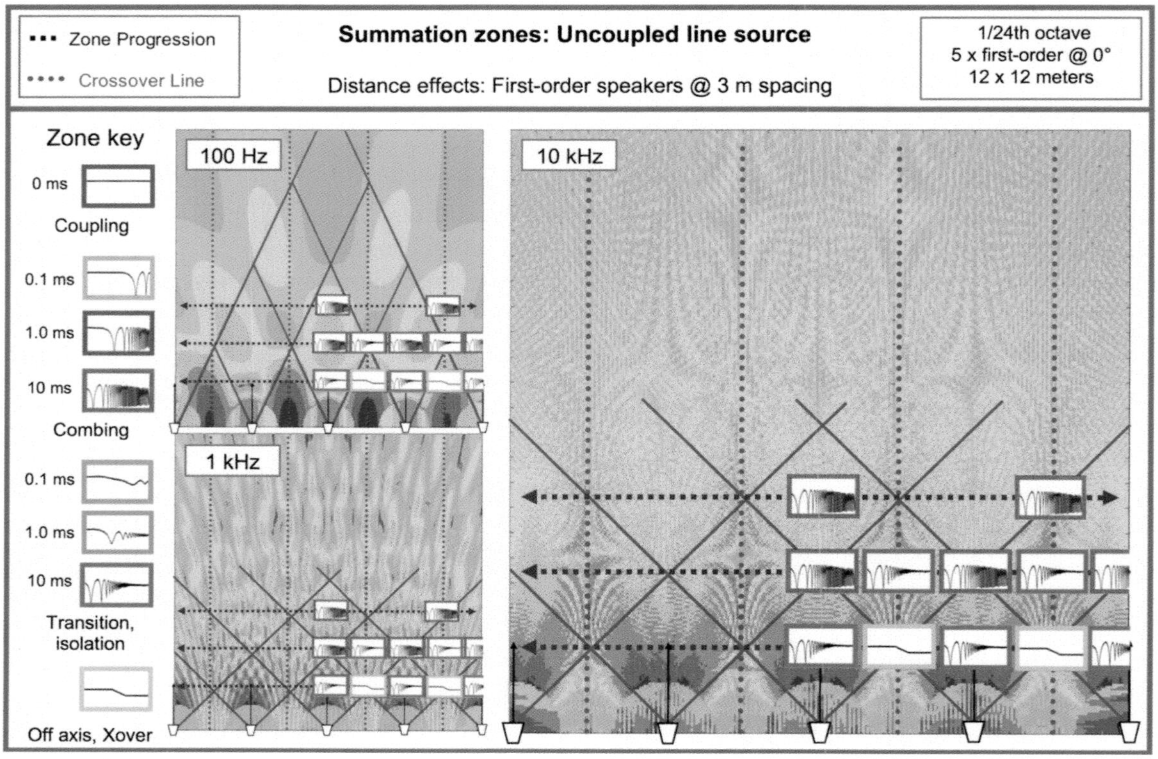

FIGURE 4.46
Summation zone progression factors for the uncoupled line source array, first-order speaker

near-field HF response. The first summation zone progression is the unity class crossover line, which alternates between isolation at ONAX and coupling at XOVR, with transitions into combing between. Positions along this line enjoy the highest uniformity and immunity from ripple. We lose the isolation zone as we move deeper into the shared coverage of two or more (or more again) speakers. The multiple paths created a cascade of combing zone summations. Recall the multiple levels of the parallel pyramid. They are back, but in this case they decimate the response with combing rather than concentrate it with coupling. The weave becomes increasingly dense as we go deeper, with peaks and dips that vary with frequency and location. Maximum uniformity over the width is found in the range between the unity crossover line and the limit line (the depth where three elements converge). Combing becomes increasingly dominant beyond the limit line, which means we need another system to take over the coverage. The relationship between the unity line and limit line is simple for the line source (assuming matched elements and spacing): The limit line is double the distance of the unity line. If we close the gap in 2 m (unity line) then we want to end the party at 4 m (the limit line). Let's turn to the LF response where we can see the parallel pyramid, evidence that our spacing is close enough to maintain coupled line source behavior in this range.

Let's change to a second-order element, which extends the distance to the unity line and enlarges the gap area, but only for the HF range (Fig. 4.47). The MF response (and the MF unity line) is the same as previously, which means our zone transitions are now progressing at different depths. The MF range is already overlapped before the HF has closed the gap. The easiest way to keep the unity line consistent over frequency is to build it with elements that are consistent over frequency (i.e. a flat beamwidth over a wide range). This carries through for all of the uncoupled array configurations.

Round three features a horn-loaded second-order system with increased HF and MF directional control (Fig. 4.48). This extends the crossover progressions in both the HF and MF regions at nearly the same rate. The uniform coverage area (between the unity and limit lines) starts later and has greater depth extension than the first-order system even though the displacement is the same.

First-order elements with an extended beamwidth plateau are well suited for uncoupled line source applications. Second-order systems can also work well provided their directional control extends below the HF range. Third-order elements are unsuitable because of their inconsistent shape.

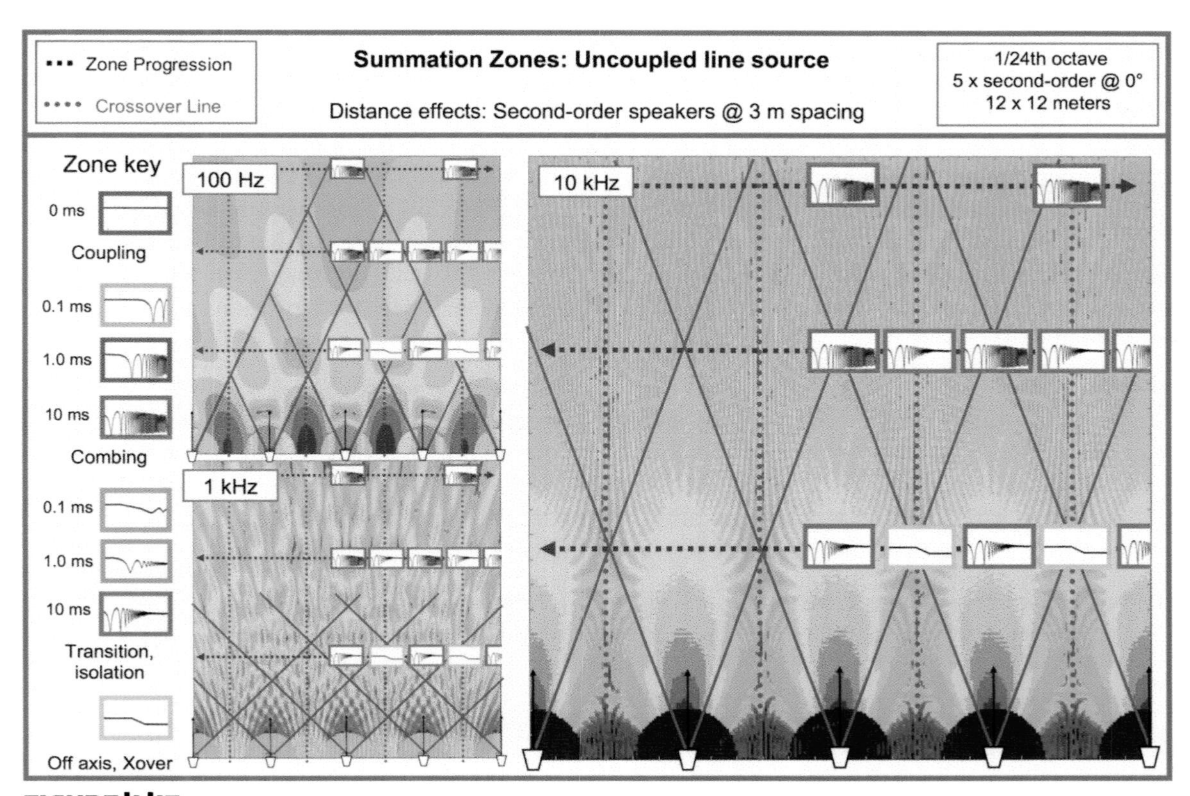

FIGURE 4.47
Summation zone progression factors for the uncoupled line source array, second-order speaker

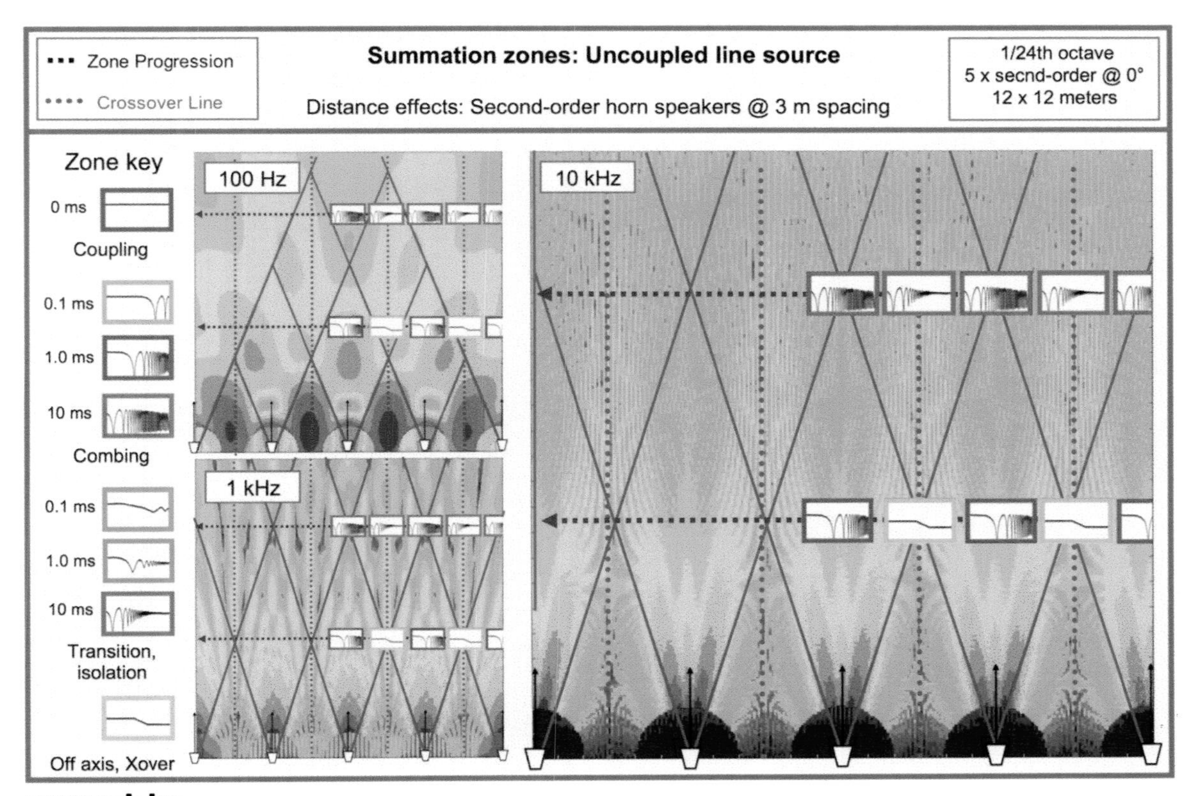

FIGURE 4.48
Summation zone progression factors for the uncoupled line source array, second-order horn-loaded speaker

4.4.3.2 UNCOUPLED POINT SOURCE

The uncoupled point source utilizes both isolation mechanisms: angle and displacement. It doesn't share the foremost feature of its coupled counterpart. It can't maintain a unity class crossover over infinite distance. Think about it. Unity splay in a coupled point source makes the edges touch, forever. Pull them apart and there's a gap, forever. A 3 m displacement and a unity splay angle means they'll never meet (always 3 m apart). It's one way to handle a center aisle! If we want unity somewhere we'll need some angular overlap to compensate for the source displacement. Once the gap is closed, this array will have a greater working depth than a comparably spaced uncoupled line source, because it takes longer for the third element to reach the first (the angles are twice as far apart). The design process includes strategic placement and splay to achieve the desired unity and limit line positions (covered in section 11.5).

We begin again with first-order speakers and resume our 3 m spacing. In the first scenario (Fig. 4.49) we splay the elements with 75% overlap, which extends the gaps and the unity line depth beyond that of the line source. The central area is the most affected because it becomes triple-covered by the middle of the panel. Another notable effect here is the correspondence of the LF response shape to that of the MF and HF shapes over distance. The three ranges have similar overall contours at the unity line (a combination of displacement and angular isolation). The LF response narrows beyond that as it resumes its coupled pyramid behavior.

The second scenario opens the splay angle to 50% overlap (Fig. 4.50). A large isolation zone dominates the HF response panel. The outer elements will never meet because they are angularly isolated and 6 m displaced.

4.4.3.3 UNCOUPLED POINT DESTINATION

The isolation roads are angle and displacement. The uncoupled point destination turns the angle inward, giving us "reverse isolation." Our countermeasure is displacement. The uncoupled point destination is the most spatially variable array type and yet a necessary tool for us. Variability rises as we turn the angles inward, reducing our usable coverage range.

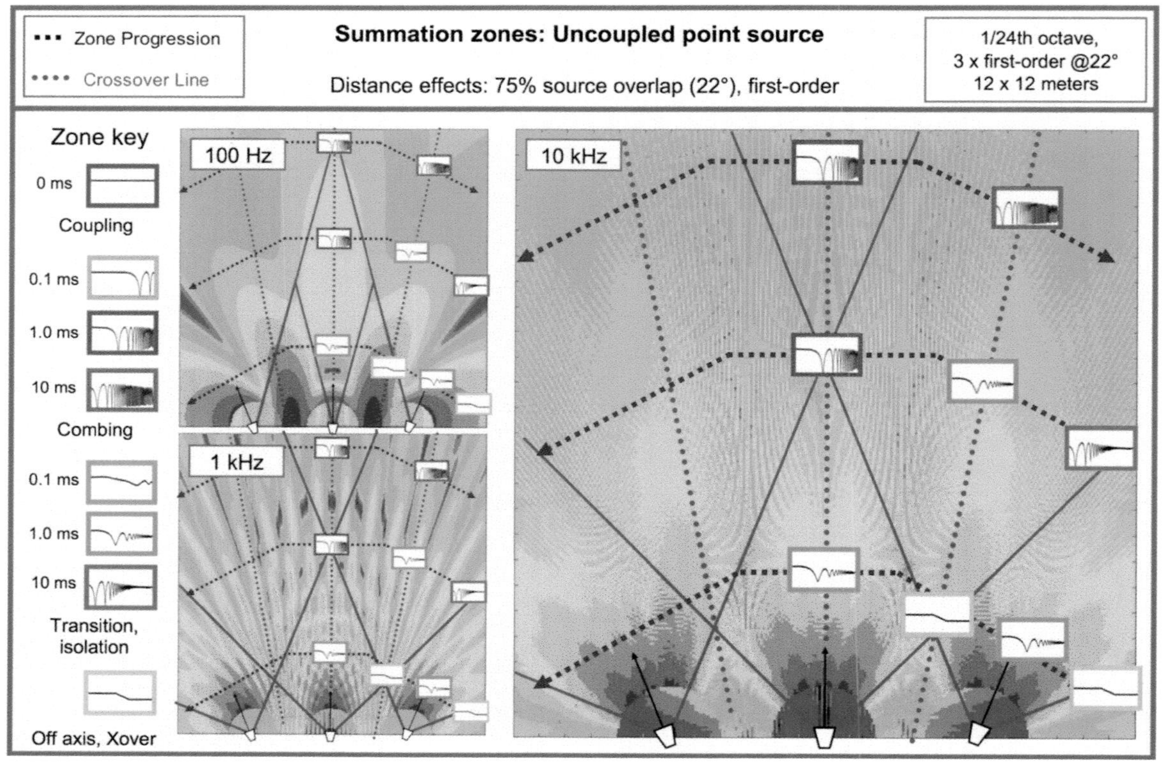

FIGURE 4.49
Summation zone progression factors for the uncoupled point source array, first-order, 75% overlap

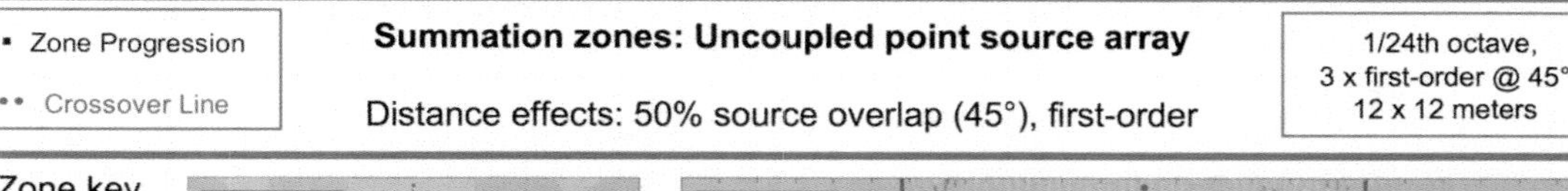

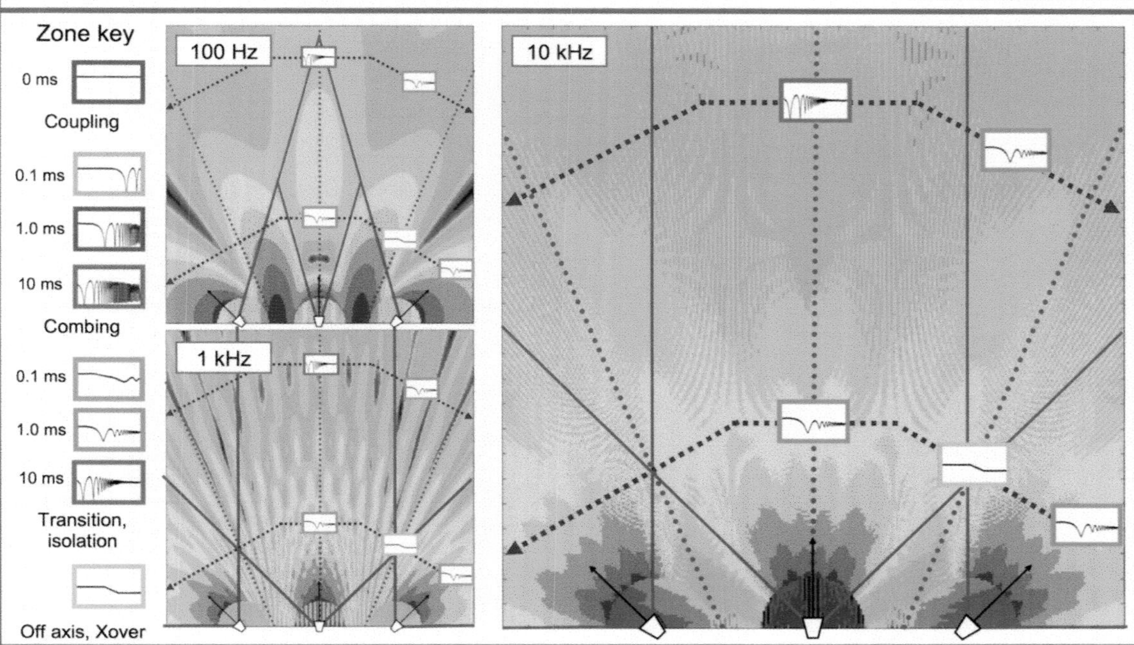

FIGURE 4.50
Summation zone progression factors for the uncoupled point source array, 50% overlap

A symmetric uncoupled point destination has its most uniform response in the isolated pre-coverage area, where speakers are more soloists than array elements. Infill speakers are like this. Each covers its side and we hold our nose at center. Don't you just love that the critics always sit there? Another symmetric version is left and right mains (if turned inward and run mono). The most well-loved symmetric form is the monitor sidefill. Will it make you feel better, or worse, to know this configuration has the statistically highest possible spatial variance? There *is* a worst-case scenario and this is it.

The most common asymmetric version is the main + delay speaker combination. The coverage patterns overlap so the isolation mechanism turns out to be asymmetric level (the delays drop off quickly due to their short doubling distance). They meet at crossover but can continue to be close in level only for a limited range. Other asymmetric versions include various arrays of arrays (e.g. joining the mains to the frontfills in the third row).

We begin with a first-order pair facing 45° inward (Fig. 4.51). The unity crossover (XOVR) is found where the pattern centers intersect (the destination), but this must be a different kind of unity. The other arrays have coupled coupling of OFFAX edges together to build a unity crossover to match the isolated ONAX locations. In this array we have our coupling zone *at* the on-axis center line and the isolation zone seems to have disappeared. What is this breed of unity crossover referenced to? Unity compared to where? The unity location (ONAX) is found half the distance between the speakers and XOVR. Why? We return to the inverse square law: Double the distance and lose 6 dB. The 6 dB we lose between the mid-point (ONAX) and the crossover (XOVR) will be returned to us in coupling zone summation. The mid-point location should be dominated in level by the local speaker, so this will be as good as it gets for isolation zone response. Some familiar behaviors are present: The isolated area (ONAX) is the most stable and the XOVR area the least. Movement between XOVR and ONAX changes the distances between the sources (combing), but does not necessarily give us isolation by angle (we may still be in the coverage of both speakers).

Now that we have two reference points on our map, we can begin to analyze the spatial qualities of this array. The summation zones progress in multiple directions outward from the crossover point. Our area of principal concern centers around ONAX. We can find angularly related off-axis points from here that will define the coverage edges. The standard summation progression holds when traveling between ONAX and XOVR, albeit highly combed. The land beyond XOVR is wild country dominated by combing.

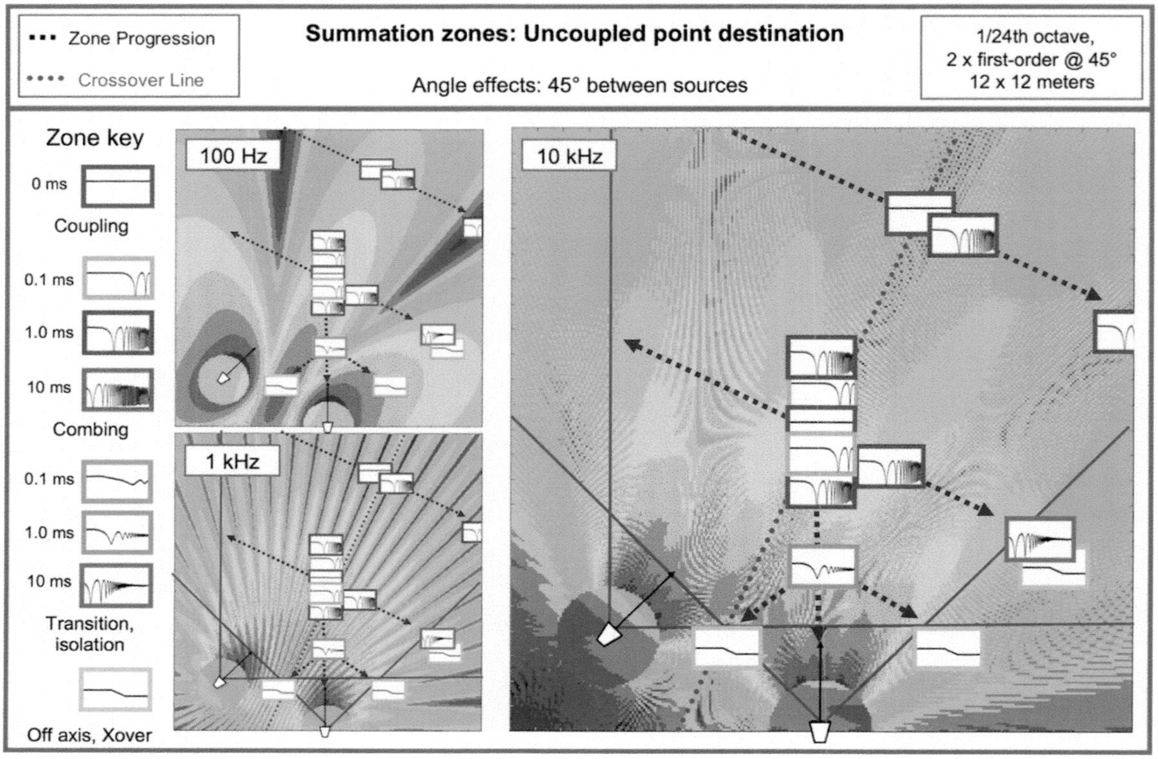

Summation zones: Uncoupled point destination

Angle effects: 45° between sources

FIGURE 4.51
Summation zone progression factors for the uncoupled point destination array, 45° angle effects

We can increase the inward splay angle to 90° (Fig. 4.52). As we approach one speaker we are moving perpendicular to the other. Our reference points are the same: mid-point (isolated) and crossover (coupling). The angular change has a muted effect between our two reference points but has a very pronounced effect on the surrounding areas. As the angle rises, the rate of response change in the peripheral areas increases proportionally. Additionally the proportion of areas with redundant and highly displaced MF and HF coverage rises, creating worst-case scenario combing.

Combing becomes even more widespread with a 135° inward angle (Fig. 4.53) and finally we reach the most inward angle of all: 180°. This array has the dubious distinction of having the most rapid movement into the combing zone and the least prospect of escape. If we are within the coverage of one element, we are within the coverage of the other. It's guaranteed. The rate of change is the highest because a movement toward one element is *de facto* a movement away from the other.

We move on to the asymmetric version of this array. A common application is the delay speaker combined with the mains (Figure 4.54). The speakers are displaced, yet have the same angular orientation, so "on axis" is the same line for both. The delayed speaker is turned down in level, which gives us a range for unity combination. This ONAX reference is forward of the delay, and depends on exactly where we want it to be for the application. We set the level so that the patrons in the delay area (XOVR) have matched combined level to those in the middle of the hall (isolated mains coverage). Time offsets begin to accrue and combing zone interaction takes its toll as we move away from XOVR. Level asymmetry limits the range of combing zone interaction as we depart the crossover area because the main speaker retains level dominance over most areas, due to its longer doubling distance. The combing is inevitable in any case, but, like its symmetric counterpart, rate of change is highly influenced by the angular relationship between the sources. Fig. 4.54 shows the relationship of angle to rate of change for this array. The proportion of combing zone interaction rises as angle increases.

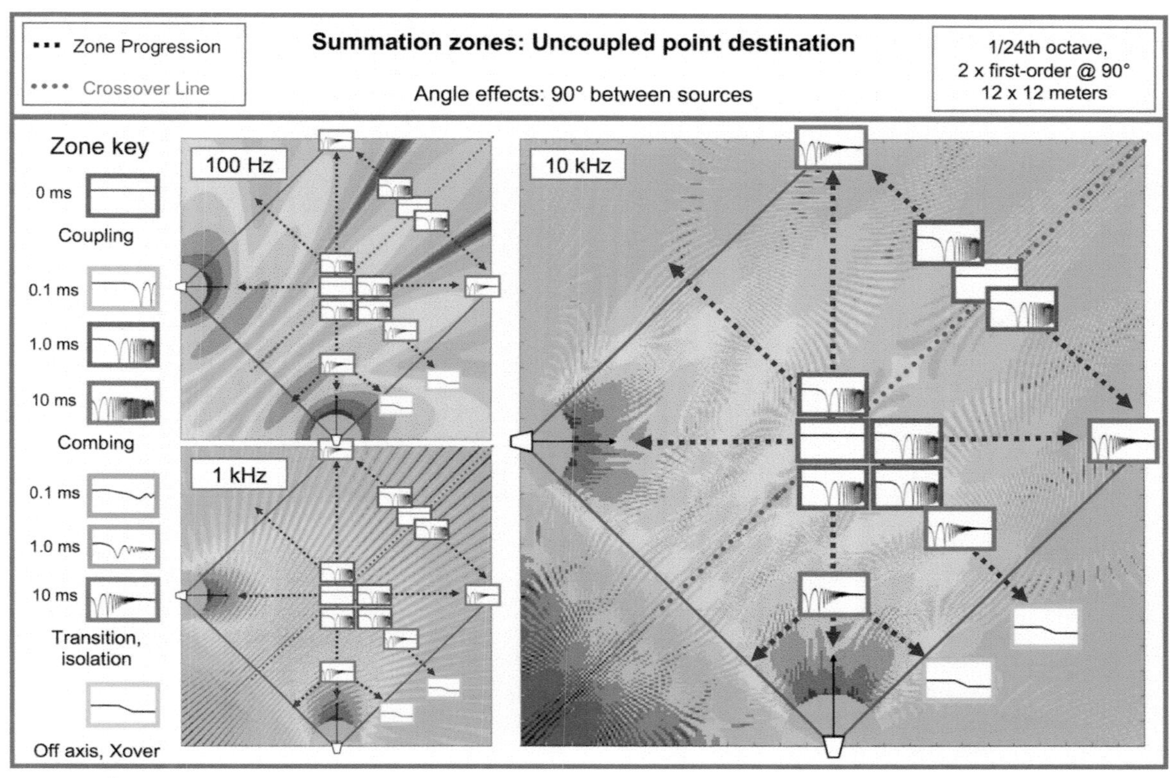

FIGURE 4.52
Summation zone progression factors for the uncoupled point destination array, 90° angle effects

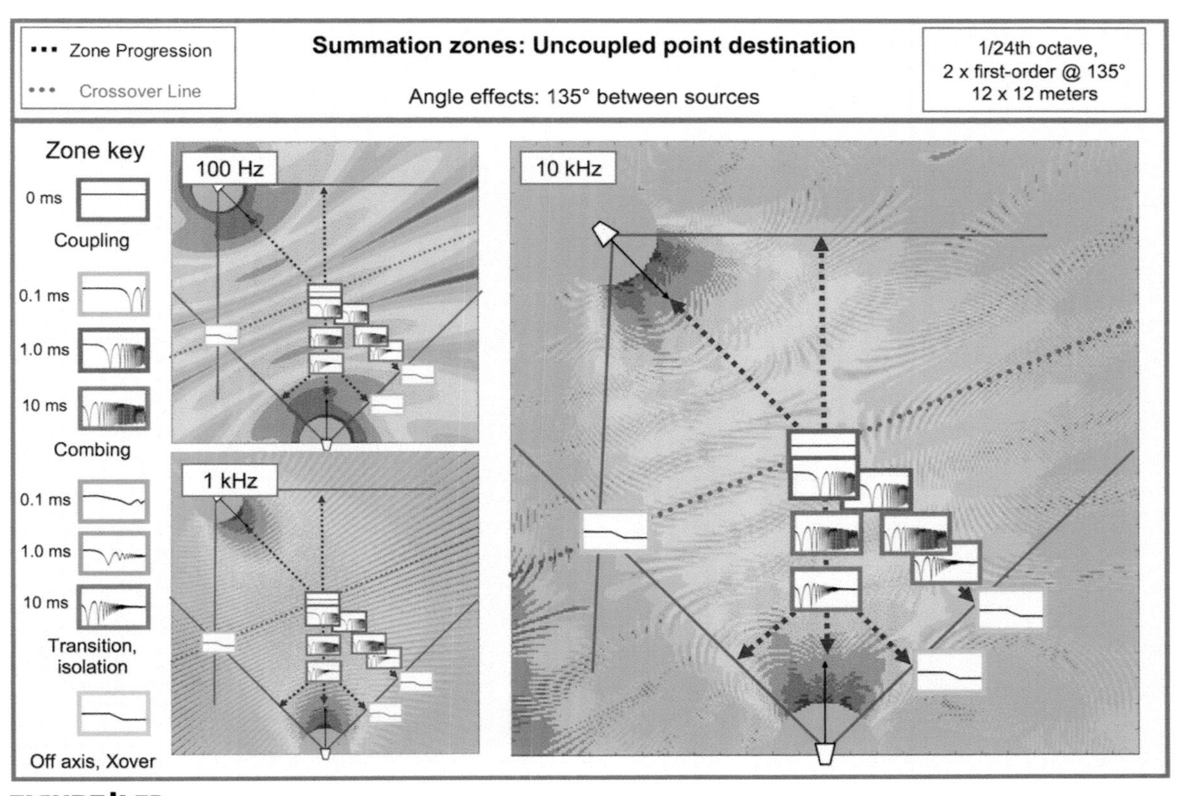

FIGURE 4.53
Summation zone progression factors for the uncoupled point destination array, 135° angle effects

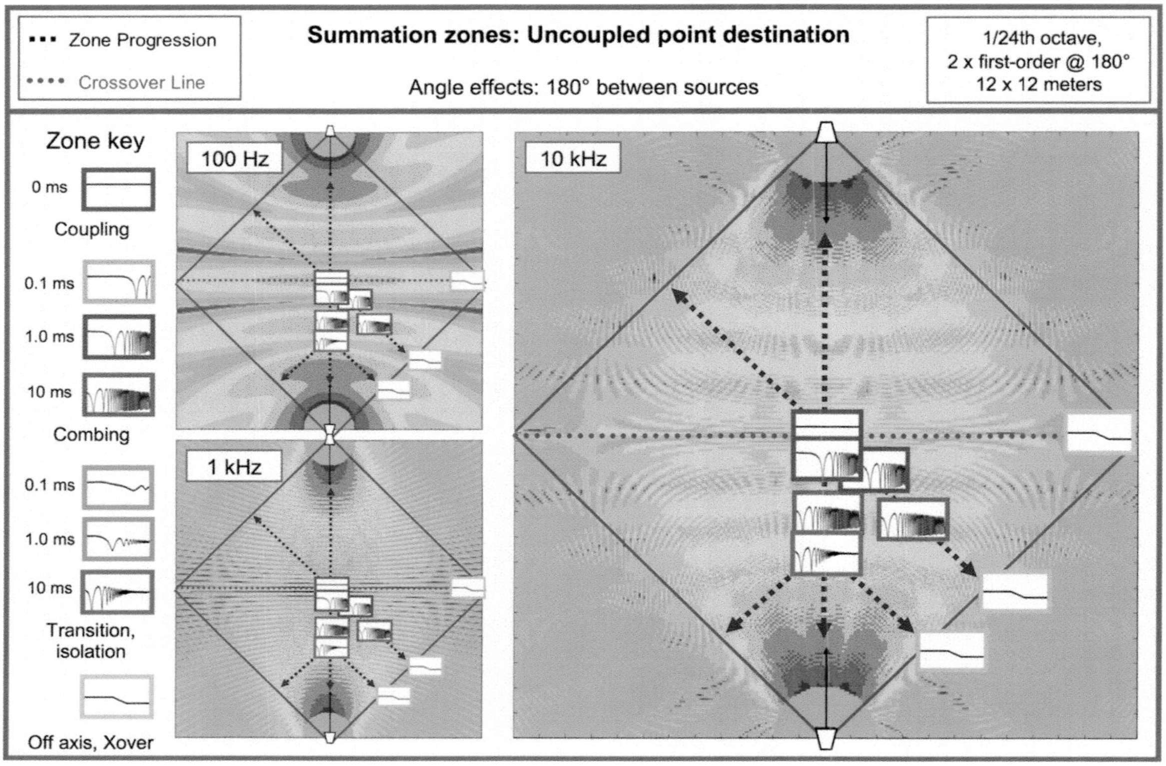

Zone Progression
Crossover Line
Summation zones: Uncoupled point destination
Angle effects: 180° between sources
1/24th octave,
2 x first-order @ 180°
12 x 12 meters
Zone key
0 ms
Coupling
0.1 ms
1.0 ms
10 ms
Combing
0.1 ms
1.0 ms
10 ms
Transition,
isolation
Off axis, Xover
100 Hz
1 kHz
10 kHz

FIGURE 4.54
Summation zone progression factors for the uncoupled point destination array, 180° angle effects

Zone Progression
Crossover Line
Summation zones: Uncoupled point destination
Angle effects: 0°, -9dB with delay
1/24th octave,
2 x first-order @ 180°
12 x 12 meters
Zone key
0 ms
Coupling
0.1 ms
1.0 ms
10 ms
Combing
0.1 ms
1.0 ms
10 ms
Transition,
isolation
Off axis, Xover
100 Hz
-9 dB, 17.45 ms
1 kHz
-9 dB, 17.45 ms
10 kHz
-9 dB, 17.45 ms

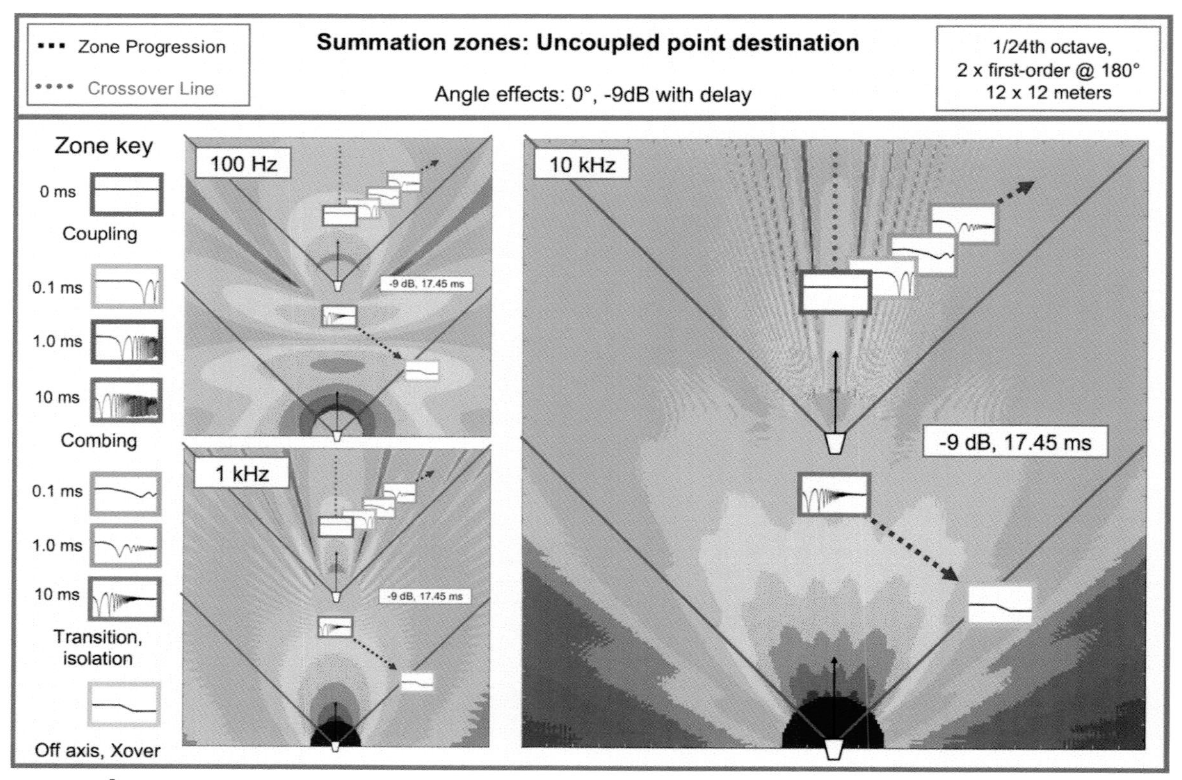

FIGURE 4.55
Summation zone progression factors for the uncoupled point destination array, angle and level effects

Temperature effects on speaker/speaker summation							
Temperature Degrees (C)	Temperature Degrees (F)	Sound speed change (%)	Main Spkr Arrival (ms)	Delay Spkr Arrival (ms)	Synchronised delay time (ms)	Change from room temp setting (ms)	Comb filter frequency (Hz)
10.0	50	2.15%	127.69	25.54	102.15	-2.15	465
11.1	52	1.96%	127.45	25.49	101.96	-1.96	510
12.2	54	1.76%	127.20	25.44	101.76	-1.76	568
13.3	56	1.56%	126.95	25.39	101.56	-1.56	641
14.4	58	1.37%	126.71	25.34	101.37	-1.37	730
15.6	60	1.17%	126.46	25.29	101.17	-1.17	855
16.7	62	0.98%	126.23	25.25	100.98	-0.98	1020
17.8	64	0.78%	125.98	25.20	100.78	-0.78	1282
18.9	66	0.59%	125.74	25.15	100.59	-0.59	1695
20.0	68	0.39%	125.49	25.10	100.39	-0.39	2564
21.1	70	0.20%	125.25	25.05	100.20	-0.20	5000
22.2	72	0.00%	125.00	25.00	100.00	0.00	
23.3	74	-0.20%	124.75	24.95	99.80	0.20	5000
24.4	76	-0.39%	124.51	24.90	99.61	0.39	2564
25.6	78	-0.59%	124.26	24.85	99.41	0.59	1695
26.7	80	-0.78%	124.03	24.81	99.22	0.78	1282
27.8	82	-0.98%	123.78	24.76	99.02	0.98	1020
28.9	84	-1.17%	123.54	24.71	98.83	1.17	855
30.0	86	-1.37%	123.29	24.66	98.63	1.37	730
31.1	88	-1.56%	123.05	24.61	98.44	1.56	641
32.2	90	-1.76%	122.80	24.56	98.24	1.76	568
33.3	92	-1.96%	122.55	24.51	98.04	1.96	510
34.4	94	-2.15%	122.31	24.46	97.85	2.15	465

FIGURE 4.56
Temperature effects on speaker/speaker interaction

4.4.4 Environmental effects

We have previously discussed temperature effects on sound speed (section 3.2.8.2). Components of an acoustical summation will all have their sound speed changed by the same percentage (provided they all undergo the same temperature change). Percentage change is a *ratio*, but we must remember that time offset is a *difference*. Longer propagation paths will accrue more speed change than short ones for a given temperature change. The delay time we set when it was cool, is not cool any more. A reference chart is provided in Fig. 4.56.

4.5 SPEAKER/ROOM SUMMATION

The alternate form of spatial acoustical crossover is the summation of direct and reflected sound. For brevity we shall refer to this as the room (even if outdoors). The summation progressions of direct sound and room reflections are similar to the speaker interactions at the spatial divider discussed previously. This is not surprising, because walls, floors and ceiling are quite literally "spatial dividers." The reflective action of these surfaces will behave as if they were additional speakers adding sound into our listening area. For each speaker/room configuration there are analogous speaker array qualities. Our treatment focuses on these common qualities, and contrasts them as required. The common ground shared between these crossover types is so substantial that our understanding of speaker/speaker summation tells us most of what we need to know about speaker/room summation.

4.5.1 Analogous functions

COMPARE AND CONTRAST SPEAKER AND ROOM SUMMATION

- Both interactions are governed by the properties of acoustic summation.
- Strategies for optimal summation are rooted in the same concepts.
- Room reflections are inherently correlated summations with the direct sound. The wall sings the same song as the sound that hits it.
- Reflections can continue for extended periods as they move from surface to surface around the room. The time offset between direct and reflected sound may exceed the duration of the original signal, thereby moving beyond the overlap duration requirement.

- Few surfaces return all frequencies at matched relative level, phase and reflective angle. Surfaces are so complex that only a rough equivalent of their reflective properties can be asserted here. The complexities create frequency-dependent summation. On a frequency-by-frequency basis, however, the speaker/speaker analogy is generally sufficient.
- Except for the filter effects cited above, the room/speaker crossover is "self-aligning," i.e. it can not be polarity reversed, it is matched in time and level at the crossover (the surface).

ANALOGOUS ORGANS

- Distance between speaker and surface is analogous to mid-point between two speakers.
- Angle between speaker and surface is analogous to half the splay angle between two speakers.
- Coverage angles of the actual speaker and reflected "virtual" speaker match.
- Surface is the spatial crossover (coupling line). Direct and reflected sounds are in phase and can be equal in level (if no absorption).
- Surface absorption is analogous to secondary speaker attenuation.
- Coupling to the surface reflection is analogous to speaker/speaker coupling.

There are practical considerations regarding spatial crossovers located at the building surfaces. Our listeners won't be embedded into floors, walls or, most notably, ceilings, so why should we place them in our discussions? We can't discuss coupling, combing and isolation without the mechanism that drives them: summation from the virtual source behind the walls. The virtual speakers "cross over" into our world at the walls, so we must include these phantom players in the summation game.

Speaker/room summation is modeled by ray tracing the paths of the actual speaker and a "virtual" speaker representing the reflection. A perfect correlation of real and imaginary sources would require a surface with perfect reflection qualities. Evaluation of complex diffuse surfaces is beyond our scope, so this is only an approximation. We initially assume all surfaces to be 100% reflective, such as found in rooms designed by famous architects.

4.5.2 Crossover audibility

The final chapter in crossover audibility is in the room. When we localize the high-hat on the balcony front, we're looking through crossover at the phantom speaker source.

Crossover audibility in the room mirrors the speaker/speaker spatial divider. Risk of detection rises with angular spread. Again we are more sensitive in the horizontal plane due to our binaural localization.

Excess overlap gives away the reflection source position. We can't synchronize at any position except crossover (the surface). The ear perceives two separately localizable sources (a discrete echo), when strong reflections are too long. As with speaker/speaker summation, strong high-frequency content increases the localization risk.

4.5.3 Speaker/room summation types

There are two speaker/room families (coupled and uncoupled) and three relationship types (parallel, outward or inward angled). Sound familiar?

Coupling has the same meaning here. Ground-stacked subwoofers (coupled) are all gain and no loss. Uncoupled surfaces give us combing. The same distance/wavelength factors as speaker/speaker interaction are found here, albeit on a much larger scale (rooms are usually bigger than our speaker arrays).

Note: Unless otherwise specified, this section assumes surfaces to be 100% reflective at all frequencies. The summations are described on a single plane, vertical or horizontal (Fig. 4.57).

4.5.3.1 COUPLED PARALLEL SURFACE

A ground-stacked speaker on a flat floor is such a common array it has its own special name: half-space loading. Half of the spherical radiation is returned in phase by the surface giving us a gain of up to 6 dB. The same surface creates combing above the subwoofer frequency range. This replicates the coupled line source array discussed earlier (Fig. 4.58).

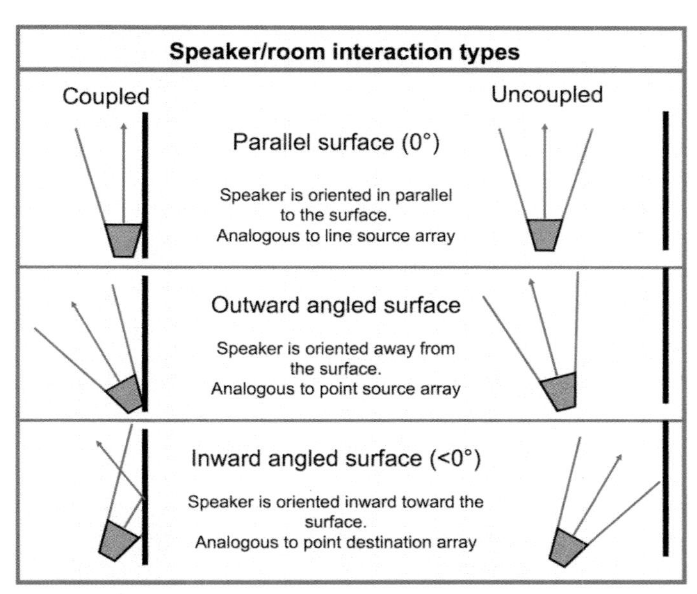

FIGURE 4.57
Speaker/room interaction types

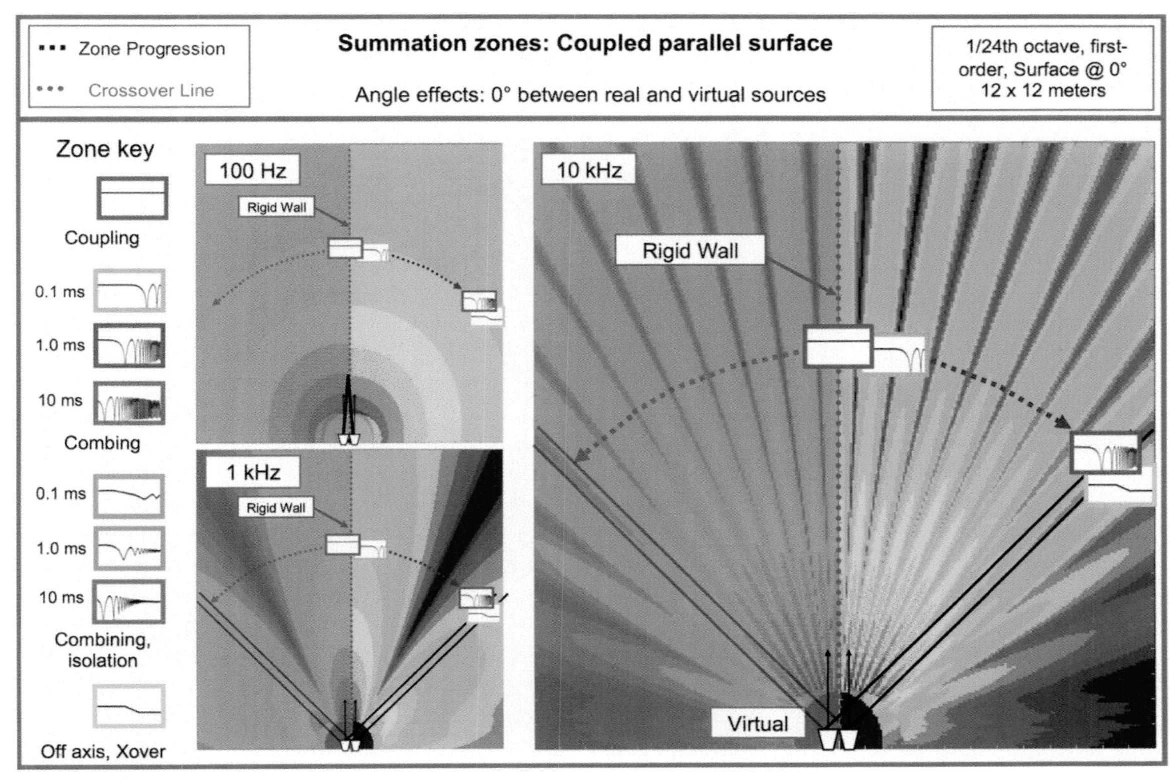

FIGURE 4.58
Summation zone progression factors for the coupled parallel surface

4.5.3.2 COUPLED OUTWARD ANGLED SURFACE

Next we splay a speaker outward from a coupled surface (Fig. 4.59), our analog to the coupled point source (Fig. 4.42). The splay angle from the surface is equal to the splay angle from center to each of the speakers in a coupled point source array. Unity splay would put the wall at the -6 dB point of coverage.

4.5.3.3 UNCOUPLED PARALLEL SURFACE(S)

First in the uncoupled family is the parallel surface as seen in Fig. 4.60. The distance to the surface is analogous to the mid-point between speaker elements in an uncoupled line source array, e.g. 3 m to the wall is equal to 6 m from element to element (3 m each to center). Just as in the array model, the response will degrade steadily over distance. The distance to the walls has the same range-shortening effects on the speaker, as did the displacement between array elements. The example here shows a symmetrical surface pair, with the speaker in the center. Moving the speaker off-center creates an asymmetrical spacing, the near side having a shorter range than far. Note the presence of the parallel pyramid in the LF response, which inevitably appears when the conditions are centered and symmetrical.

4.5.3.4 UNCOUPLED OUTWARD ANGLED SURFACE

A speaker uncoupled from the surface and angled outward (Fig. 4.61) is analogous to the uncoupled point source array (Fig. 4.49). Similar considerations apply but with an additional note of interest. We previously observed that if the splay angle and element angle were equal then the uncoupled point source array would gap for eternity. This can now be used for practical advantage, giving us a means to skim coverage along seating areas without overlapping into a nearby side wall or ceiling. The outward angle surface configuration gives some long-range stability, which was found earlier in the point source arrays.

4.5.3.5 UNCOUPLED INWARD ANGLED SURFACE

Previously we found that the uncoupled point destination array had the highest proportion of combing. It is no surprise that its surface analog does the same. The inward wall (Fig. 4.62) brings on-axis energy back into

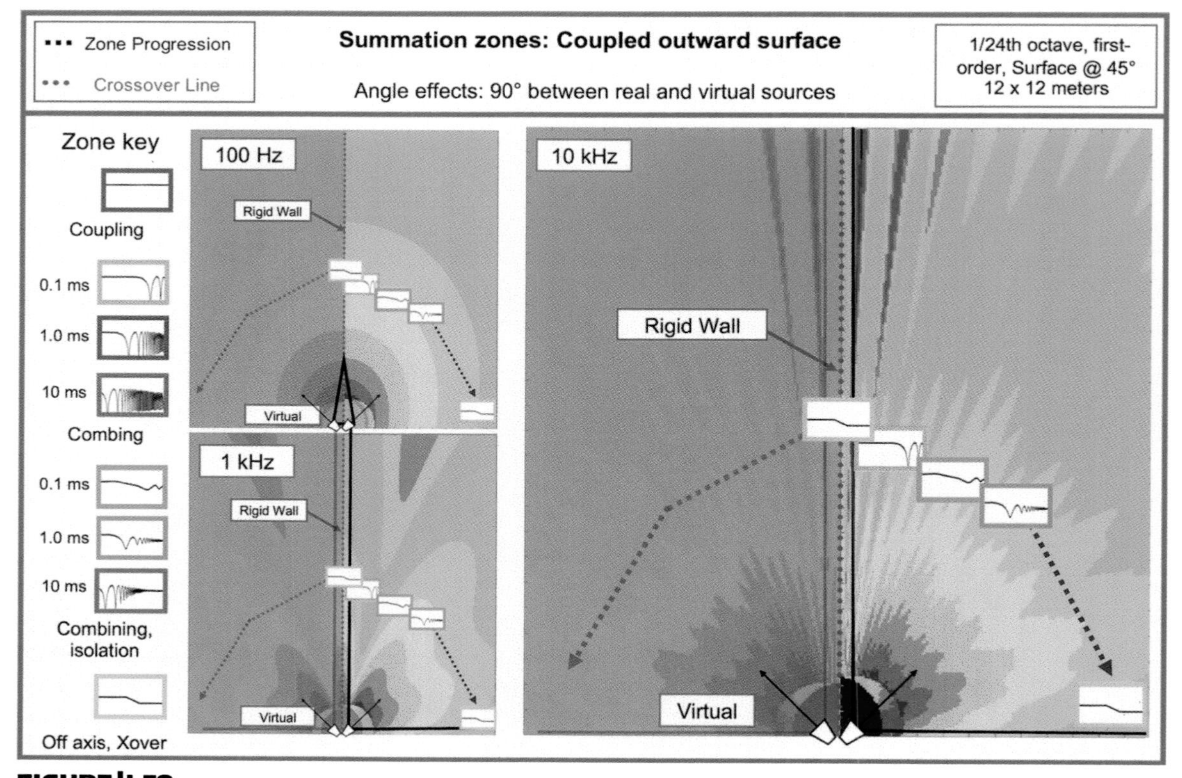

FIGURE 4.59
Summation zone progression factors for the coupled outward surface

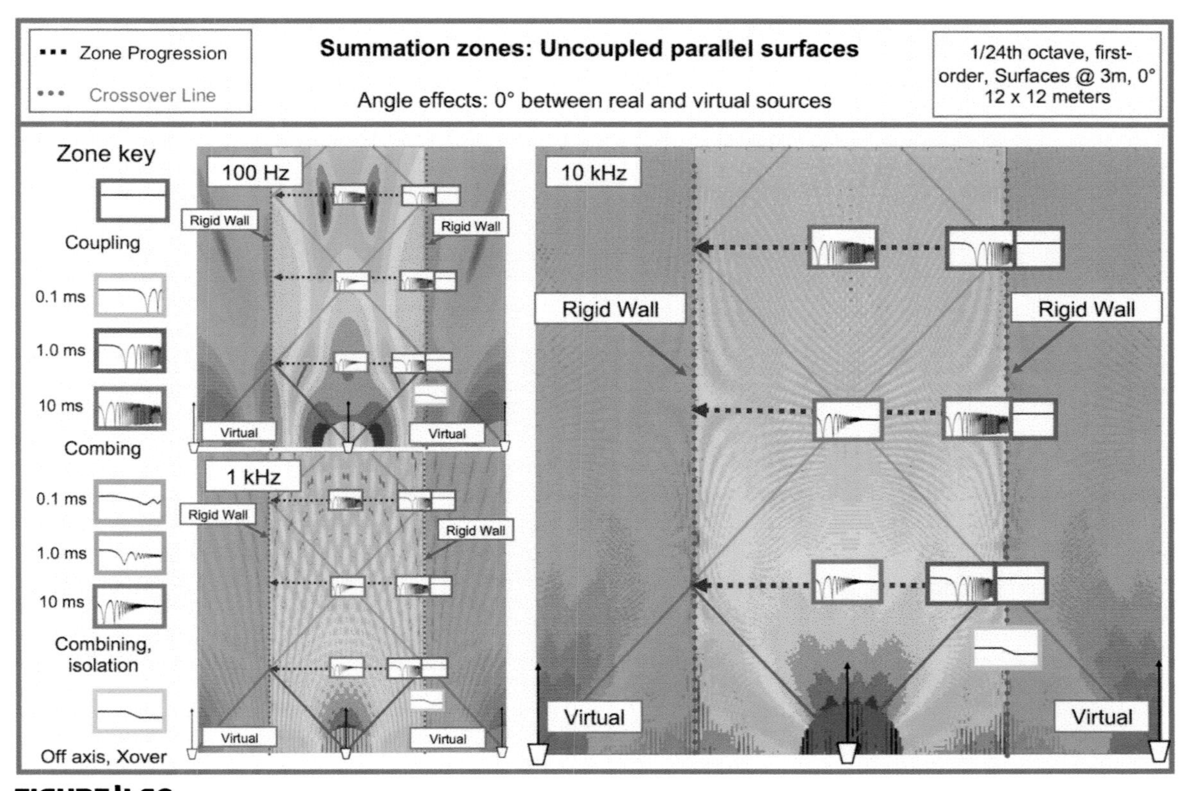

FIGURE 4.60
Summation zone progression factors for the uncoupled parallel surface

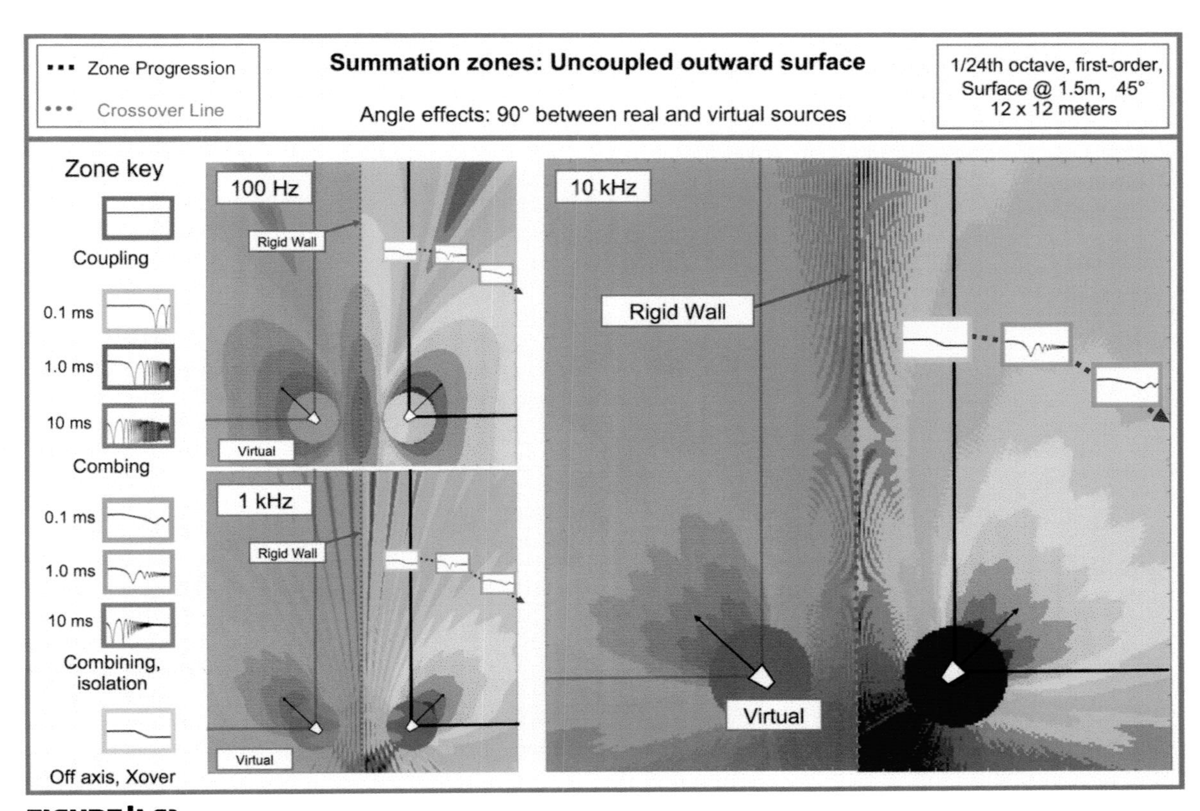

FIGURE 4.61
Summation zone progression factors for the uncoupled outward surface

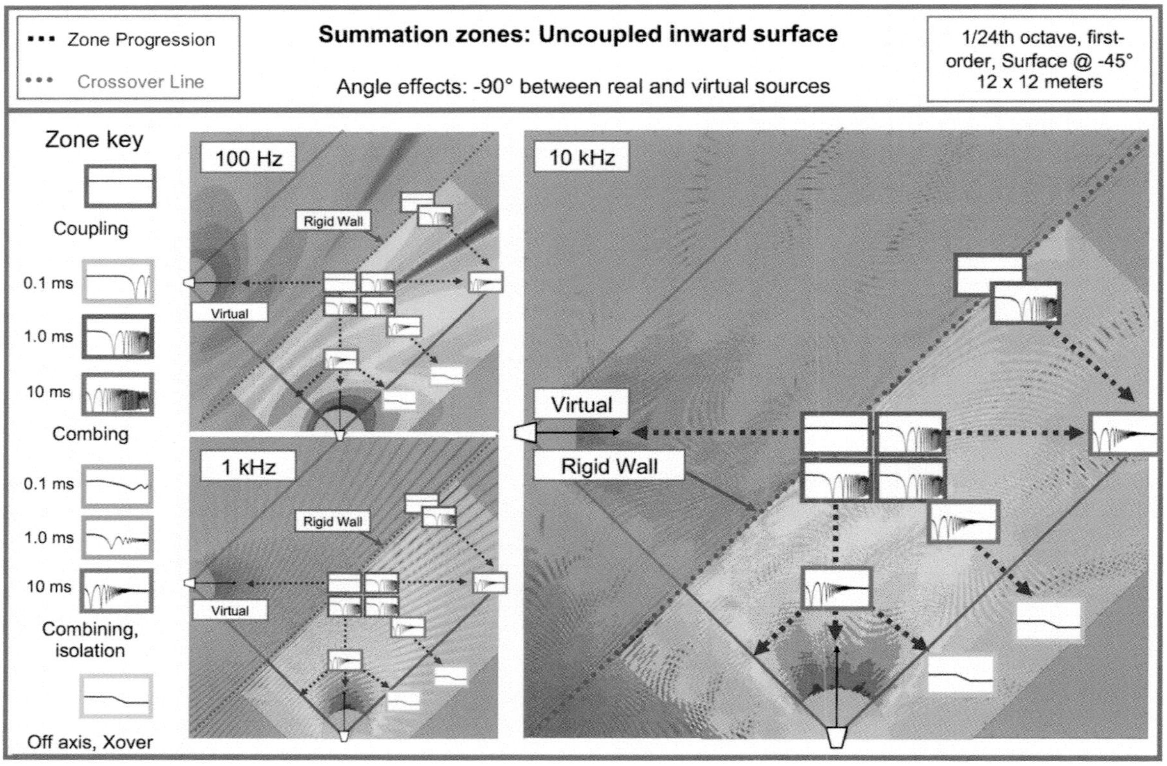

FIGURE 4.62
Summation zone progression factors for the uncoupled inward surface

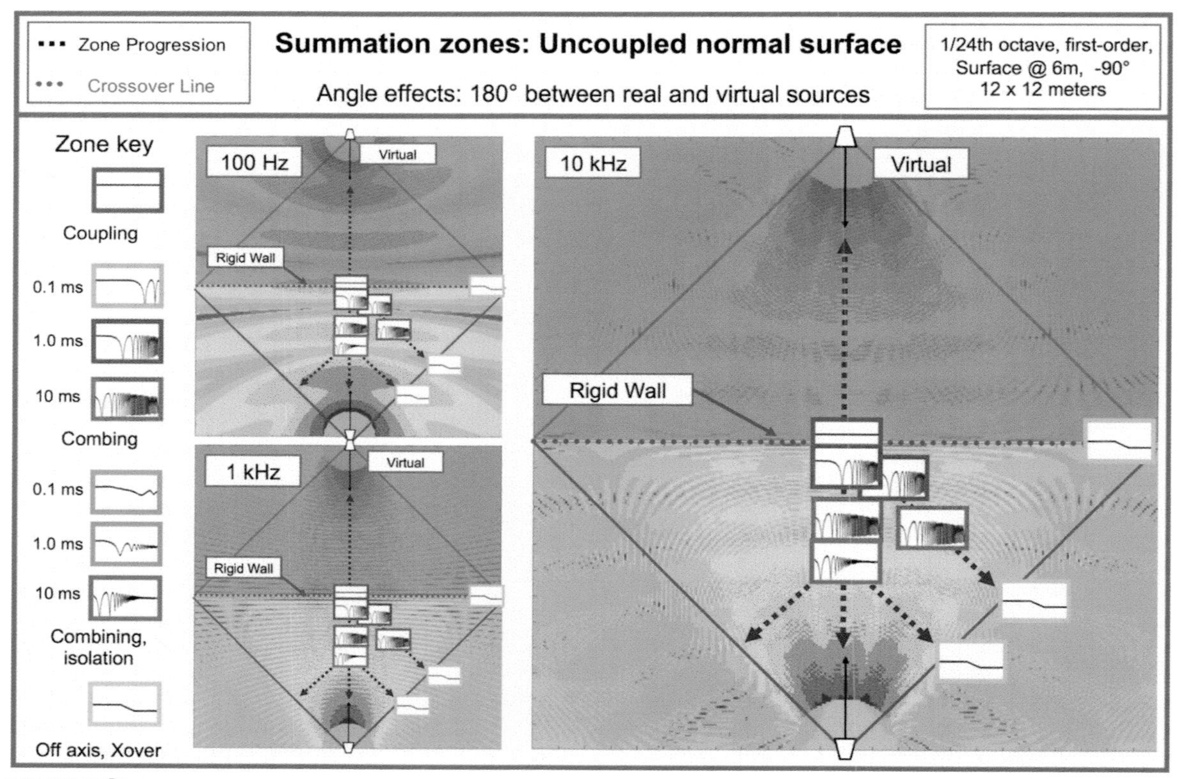

FIGURE 4.63
Summation zone progression factors for the uncoupled normal surface

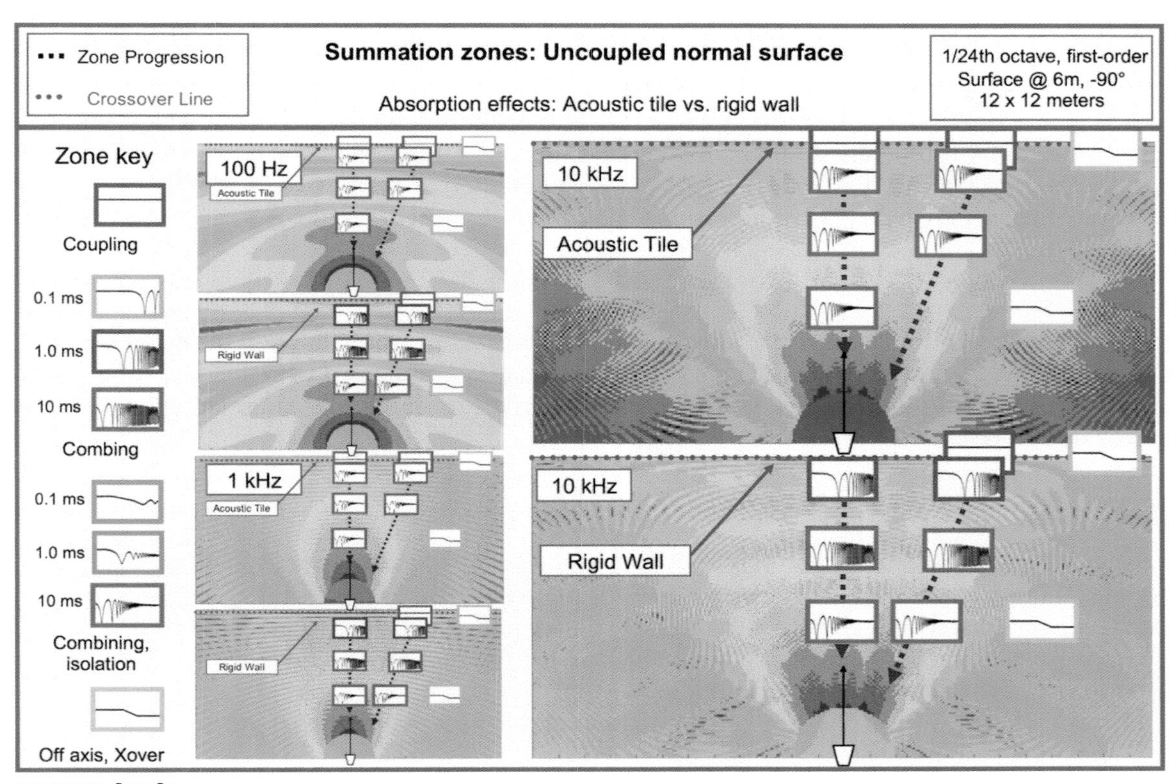

FIGURE 4.64
Summation zone progression factors for the uncoupled normal surface, absorption effects

Temperature effects on speaker/room summation

Temperature (C°)	Temperature (F°)	Temperature change (%)	Sound speed change (%)	Direct sound arrival (ms)	Reflection arrival (ms)	Time offset (ms)	Comb filter frequency (Hz)
10.0	50	-4.13%	2.15%	25.54	127.69	-102.15	9.79
11.1	52	-3.76%	1.96%	25.49	127.45	-101.96	9.81
12.2	54	-3.38%	1.76%	25.44	127.20	-101.76	9.83
13.3	56	-3.00%	1.56%	25.39	126.95	-101.56	9.85
14.4	58	-2.63%	1.37%	25.34	126.71	-101.37	9.86
15.6	60	-2.25%	1.17%	25.29	126.46	-101.17	9.88
16.7	62	-1.87%	0.98%	25.25	126.23	-100.98	9.90
17.8	64	-1.50%	0.78%	25.20	125.98	-100.78	9.92
18.9	66	-1.12%	0.59%	25.15	125.74	-100.59	9.94
20.0	68	-0.75%	0.39%	25.10	125.49	-100.39	9.96
21.1	70	-0.37%	0.20%	25.05	125.25	-100.20	9.98
22.2	72	0.00%	0.00%	25.00	125.00	-100.00	10.00
23.3	74	0.38%	-0.20%	24.95	124.75	-99.80	10.02
24.4	76	0.76%	-0.39%	24.90	124.51	-99.61	10.04
25.6	78	1.14%	-0.59%	24.85	124.26	-99.41	10.06
26.7	80	1.51%	-0.78%	24.81	124.03	-99.22	10.08
27.8	82	1.89%	-0.98%	24.76	123.78	-99.02	10.10
28.9	84	2.27%	-1.17%	24.71	123.54	-98.83	10.12
30.0	86	2.64%	-1.37%	24.66	123.29	-98.63	10.14
31.1	88	3.02%	-1.56%	24.61	123.05	-98.44	10.16
32.2	90	3.40%	-1.76%	24.56	122.80	-98.24	10.18
33.3	92	3.77%	-1.96%	24.51	122.55	-98.04	10.20
34.4	94	4.15%	-2.15%	24.46	122.31	-97.85	10.22

FIGURE 4.65
Temperature effects on speaker/room interaction

the heart of our coverage area. As the angle of the inward surface goes up, so does the need for absorption (or demolition).

The normal surface (a typical back wall) is the most spatially variable configuration for speaker/room summation (Fig. 4.63). It is analogous to the 180° point destination array (monitor sidefill). There is one heartening aspect of our anatomy that helps us to reduce the perceived effects of this summation form: the front/back rejection of our ears. Nonetheless, this type of wall typically needs absorption more than any other.

4.5.4 Absorption effects

As long as we are discussing absorption, we can take a look at the effects that this would have on the normal wall summation. Figure 4.64 shows the results of acoustic tile placed on the surface compared with the rigid surface we have seen up to this point. The results are heartening, with reductions in the combing zone summation in HF and to a lesser extent midrange responses.

4.5.5 Temperature effects

Temperature effects on speaker/room summation (Fig. 4.65) are analogous to those on speaker/speaker summation, but with hundreds of paths rather than just a few. Rising temperature increases the speed for all the sound transmission paths, direct and reflected alike. Relative timing between direct and delayed will change because the reflected paths are longer. Comb filter center frequencies will shift in proportion to the change in time offset. This is easiest to visualize as the entire room shrinking (warmer) or expanding (cooler). The effects are most noticeable in the LF range where (a) we are hearing the summation of the most paths and (b) the summations are more likely to be in the tonal perception zone

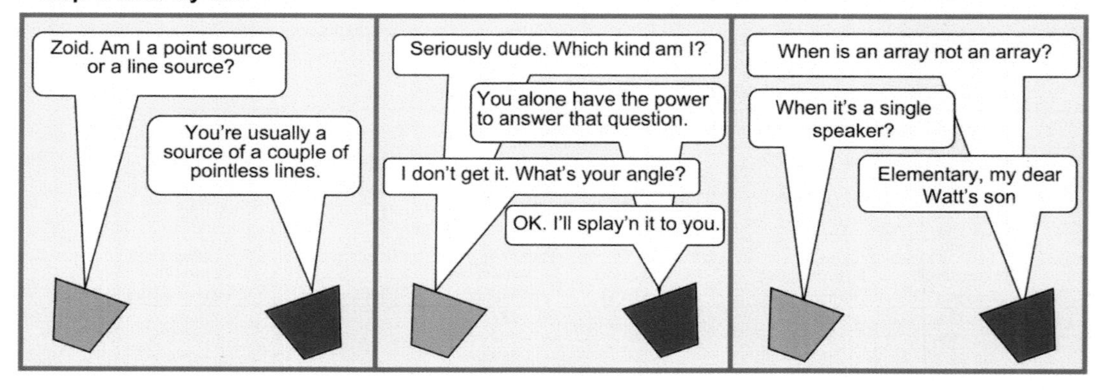

CHAPTER 5

Perception

If a tree falls in the forest and no one hears it, does it make a sound?

We can rephrase this in audio engineering terms:

If an acoustic transmission is not perceived, does the waveform exist?
If there is a concert without an audience, will we get paid?

perception n. act or faculty of perceiving; intuitive recognition; action by which the mind refers its sensations to external object as cause.
perceive *v.t. apprehend with the mind, observe, understand; apprehend through one of the senses; regard mentally in a specified manner.*
Concise Oxford Dictionary

The subjective side of acoustical reception includes the *experience* of human hearing, i.e. **perception**. Unsurprisingly this is a subject of ongoing discovery and debate. We'll limit our scope to areas that affect design and optimization decisions. We're more concerned with perception in *relative* rather than *absolute* terms. For example, the 3 kHz resonance peak in the ear canal is a common anatomical feature for all listeners. Do we design around that? Localization, by contrast, strongly affects our decision process.

These factors directly affect our process: loudness, localization, tonality, echo perception, stereo imaging and how the ear detects a sound system.

5.1 LOUDNESS

There's more to loudness perception than just SPL. The ear integrates level over a period of roughly 100 ms (Everest, 1994, p. 48). Perceived loudness depends upon both level and duration. Short-duration bursts at high levels and longer bursts at lower levels are equally perceived. Two values are commonly given for dB SPL: peak (the maximum pressure regardless of duration) and continuous (the sustained level over a period of more than 100 ms). Crest factor (section 1.1.5.7) is the peak/continuous ratio.

5.1.1 Continuous and peak

A continuous sine wave is minimum peak and maximum duration. The opposite extreme on the dynamic scale is an impulse (maximum peak, minimum duration). Music and speech combine continuous and transient functions. Vowels, closest to sine waves, have low crest factors, whereas consonants are transients. A system that can't reproduce transients loses intelligibility due to articulation loss of the consonants (%ALCONS).

Our sound system must meet the program material's continuous and peak loudness requirements. The limiter's speaker protection role was previously discussed (section 2.4.6). Now we'll investigate how limiters affect perception.

The system will either peak limit or clip when overloaded. Clipping reduces crest factor, saws peaks, adds harmonics and extends the waveform duty cycle. Peak limiting also reduces crest factor, but with less distortion than clipping. Compression reduces crest factor over the long term but may still allow transient bursts to pass. Compression rarely travels alone and is usually accompanied by limiting and/or clipping. Dynamic range ends when peak and continuous capabilities merge to become the same. As crest factor flattens, intelligibility, definition and detail go with it. It's loud though, due to the extended duration that accompanies clipped, limited and compressed signals. In terms of quality, loudness is not all it's cranked up to be.

Our hearing system also has an internal limiter. Our aural muscle, the tensor tympani, contracts at dangerously high levels, which tightens the eardrum and reduces its sensitivity. This reduces the *perceived* loudness. Internal dynamic compression continues until the external level falls. Our ear's onset threshold is more of a slow compressor than peak limiter. High-level transients pass through, but continuous exposure brings it on.

Mix engineers would be wise to consider the action of this muscle. A relentlessly loud mix continuously engages the aural compression. Peak perception is reduced, even if the system has headroom to deliver it. A dynamic mix maximizes perceived loudness by utilizing the full range of our hearing and our sound system.

Sound systems operated in continuous overload may fall into a cycle of compounding compression. Once the system is into clipping and limiting it's much more likely to engage the ear's compression system because continuous levels are high and crest factor low. Dynamic loss can lead to increased drive level in an attempt to restore excitement, but instead increases compression in both the transmission and reception systems. Continual engagement of the tensor tympani strains the muscle, resulting in aural fatigue. A low-power system in continual overload is more fatiguing to the ear system, and more likely to be perceived as louder over the long term, than a dynamic high-power system without compression. Unfortunately a high-power system run into gross overload is the worst of all, and all too common. Such operational matters are strictly under mix engineering control. My musings on this matter will most likely fall on deaf ears.

5.1.2 Equal-loudness contours

Our aural perception system has a non-linear frequency response over level. Our variable sensitivity is characterized by the **equal-loudness contours** (Fig. 5.1). At low levels we tune in to the midrange, which helps to hear what people are saying about us. Things flatten out when it gets loud. Every audio book has this graph so it must be *very* important. Is it relevant to system design and optimization? Very little. The equal-loudness contours are *natural* aspects of perception, not an artificial ingredient introduced by the sound system. Counteracting this with dynamic filters introduces an unnatural element into the equation, a mix engineering "special effect" rather than an optimization choice. It is natural to perceive quiet passages with the normal quiet sound spectrum. Do we want the quiet parts to be perceived as if they were loud? That's an artistic question.

Our optimization goal is equal loudness over the space. The dynamic contours are built into the music, not the sound system. Should we tune differently for louder seats in anticipation of the changing frequency contours? A 20 dB level difference (80–100 dB SPL) would change the perceived response at 30 Hz by a whopping 3 dB. If the front seats are 20 dB too loud then who cares about the equal-loudness contours?

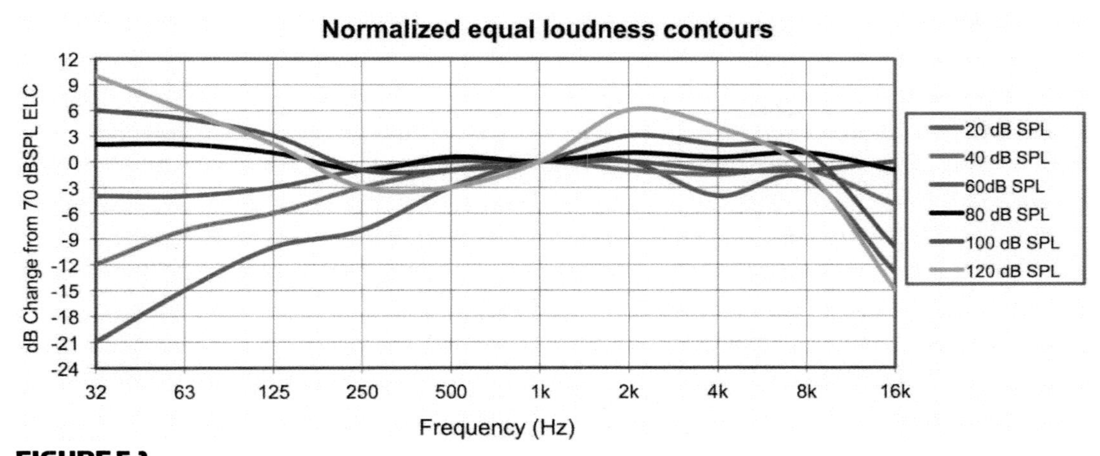

FIGURE 5.1
Equal-loudness contours normalized. The chart shows the perceived frequency response differences over level from the nominal level (after Robinson and Dadson, 1956).

5.2 LOCALIZATION

5.2.1 Sonic image

Sonic image is a sound source's *perceived* location (regardless of its actual one). We often use speakers to create the sonic image where speakers (and sources) aren't. Sonic ventriloquism. A successful magic show depends on speaker placement, relative level and timing. **Sonic image distortion** is a metric for how badly we missed the target, roughly quantified in degrees (the angle between perceived and target positions). The target is often an actual sound source, such as an actor or musician on stage (Figs 5.2 and 5.3). Early and significant source level helps minimize sonic image distortion by decreasing the localization clues to our speakers. If localization were our only concern we would ask the musicians to help by maximizing their stage volume. Thanks, but no. Instead we seek to minimize sonic image distortion with reasonable stage levels.

It is not a given that we always have a natural sonic source to serve as our image anchor. Sometimes there's no appreciable "natural" sound source, e.g. bands with in-ear monitors or sequestered orchestras or theme park character shows where the actors lip-sync to a track. Head-mounted microphones have given theatrical performers the liberty to sing at a whisper level, which decreases the amount they project into the hall.

The first step in sonic image control is to understand how our localization system works. Then we can learn to fool it. The driving forces are relative level and time (surprise!).

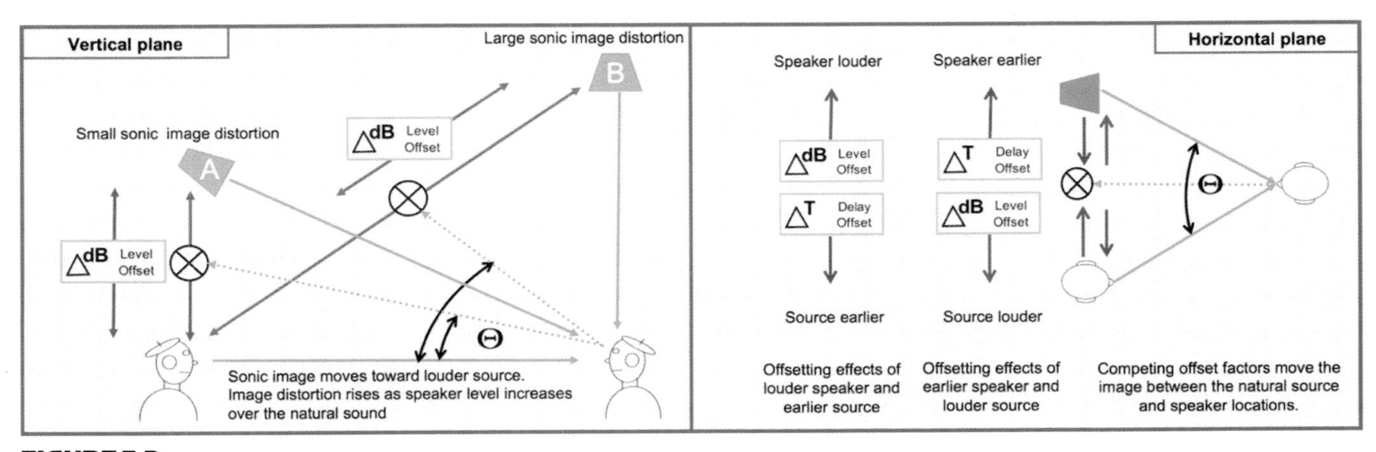

FIGURE 5.2
Vertical (left) and horizontal (right) sonic image distortion and localization factors

5.2.2 Vertical localization

Vincent Van Gogh would have had a difficult time telling whether a sound was coming from above or below him. Why? The famously disturbed painter cut off his outer ear, which removed his primary vertical localization mechanism, the **pinna**. The pinna's contours create a family of reflections as it steers sound into the inner ear, which color the sound on its way in. This tonal signature is encoded in the brain as a map that drives our sense of vertical sonic image (Everest, 1994, p. 54). It would seem that the ear/eyes/brain system learns over time to correlate visual image and sound image. The vertical variations of the pinna's tonal coloration are mapped into memory as an auditory overlay. This response, the head-related transfer function (HRTF), operates independently for each ear. For discussion purposes we will isolate the vertical plane by plugging (not cutting off) one ear.

Localizing the vertical position of an isolated source is easy because a single pinna signature dominates the HRTF. Move it up or down and we can track the changing HRTF cues. Things get more interesting when two matched sources with the same signal are spread in the vertical plane. Two competing and conflicting localization clues have equal priority. The perceived image is a midpoint vertical compromise (Fig. 5.3). It's experienced as spatially spread between the sources rather than as a singular point source (like a solo speaker). This elongation is termed the **apparent source width**. Image elongation is most extreme with equal-level sources and decreases proportionally as level offset rises. Level offset moves the image toward the dominant speaker, following the stronger HRTF localization clues. The ear recognizes this signature as the source direction, albeit with some ambiguity and proportional image elongation.

Vertical localization is a single-channel process, which makes it difficult to discern the arrival order in this plane, especially when close in time (e.g. early reflections and time offset speakers). Level offset dominates vertical localization until the signals are far enough apart in time to be perceived and localized separately.

5.2.3 Front/back localization

The pinna's secondary role is front/back localization (Everest, 1994, p. 53). The pinna provides some high-frequency directionality and adds another HRTF layer to the equation. Front/back localization has a notable link to the measurement mics used for optimization. Omnidirectional microphones have somewhat less front/back rejection than the ear, whereas cardioid microphones have far more. Neither is a perfect modeling of the ear although the omni is far closer.

5.2.4 Horizontal localization

A dual-channel complex audio analyzer (our ears) drives horizontal localization. The spacing between our ears creates a differential spatial sensor. Horizontal position is found by triangulation, i.e. sound is heard in two locations and source location is detected differentially via **binaural localization**.

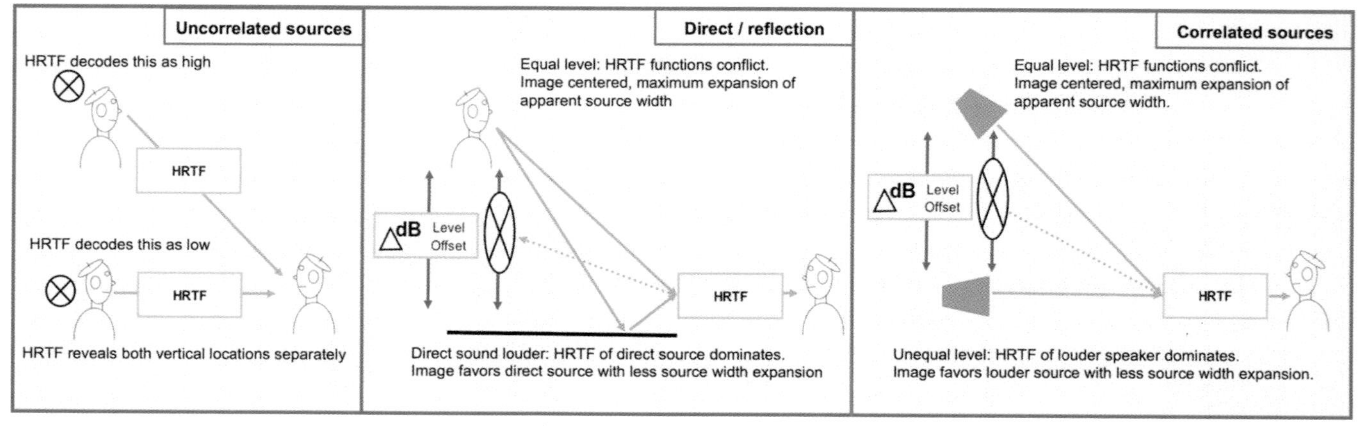

FIGURE 5.3
Vertical localization of two distinct sources (left) in the presence of reflections (center) and in the presence of multiple loudspeakers (right)

5.2.4.1 BINAURAL LOCALIZATION

Our brain analyzes time and level offsets between arrivals. The mechanisms are independent and serve as "second opinions" that may, or may not, concur. The terms **inter-aural time difference** (ITD) and **inter-aural level difference** (ILD) describe our internal analysis mechanisms (Duda, 1998). ITD dominates in low-frequency localization where the wavelengths are large enough to bend around the head and arrive at nearly the same level, yet slightly out of time. The source location is mapped into the brain as polar representations by ITD timing. ILD dominates at high frequencies, because it's hard to get sound through our thick skull. High frequencies refract around the head less, yielding a detectable level offset. ITD and ILD values for a single sound source normally yield the same conclusion, thus confirming the localization.

When a single sound source is placed at the horizontal center, the ILD and ITD values are both zero (Fig. 5.4). The brain decodes its location exactly where we see it: front and center. If the source or receiver move horizontally off center, the closer ear receives an earlier and louder copy of the sound. In this instance, both ITD and ILD values confirm the new location.

Let's add a second sound source. Uncorrelated sources localize separately (e.g. simultaneous conversations around the table). We individually localize participants by their distinct ITD and ILD values, even when simultaneously speaking. If the second source is correlated (e.g. early reflections or a loudspeaker) the localization becomes more complex. Each arrival has independent, potentially conflicting ITD and ILD values. Let's start with a typical early reflection in the horizontal plane. Direct sound is louder and arrives earlier. The sonic image is perceived near the source of the direct signal, albeit with a modified tonal quality and sense of "spaciousness," i.e. the feeling that sound is coming from a general area rather than from a single point (Everest, 1994, pp. 295–301). The direct sound's ITD and ILD are the strongest factors because the direct sound is both earliest and loudest. This continues even as time offset rises until the signals are localized as separate sources (direct sound and discrete echo).

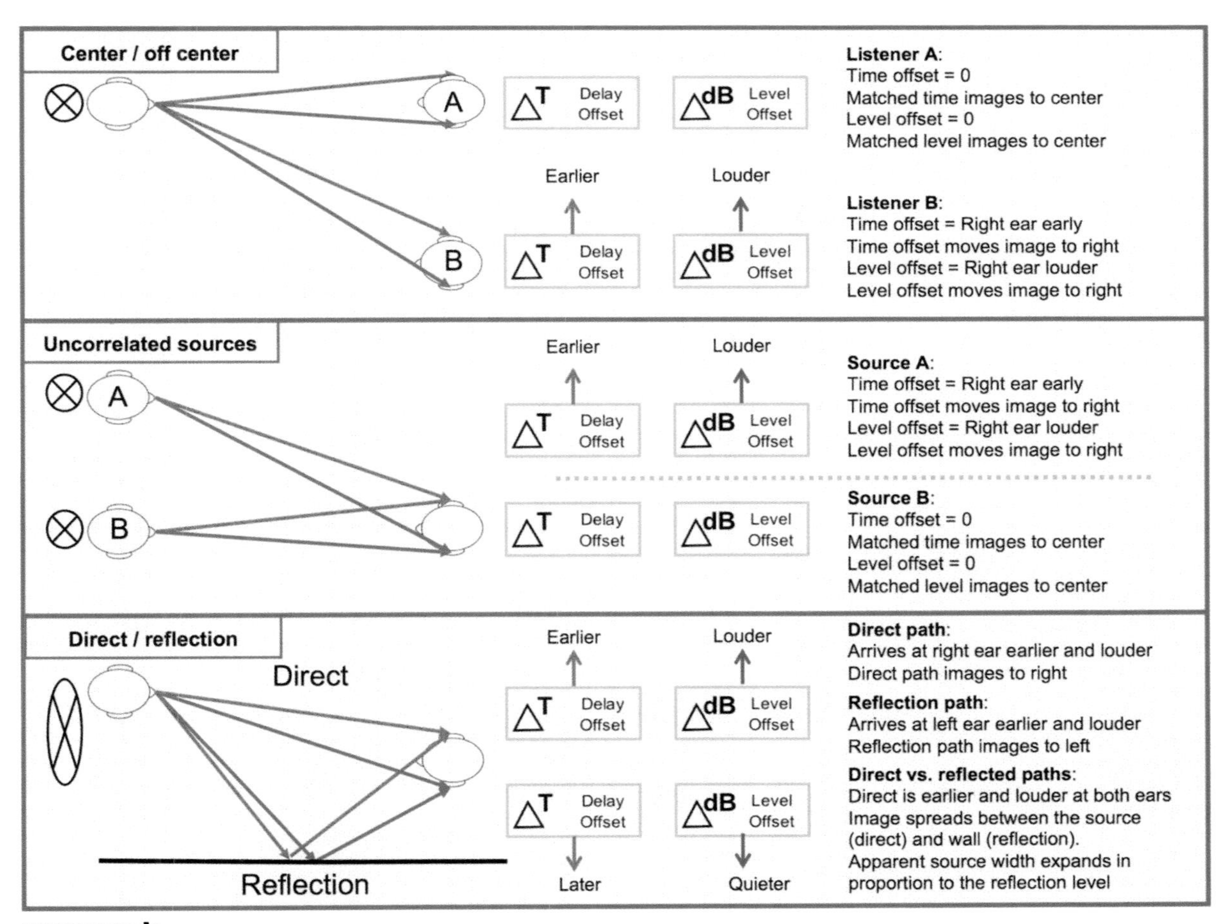

FIGURE 5.4
Horizontal localization of a single source, on and off center (top), uncorrelated sources (middle) and in the presence of a reflection (bottom)

An ambiguous case can arise with an obstructed direct sound path. Direct may be quieter than reflected but might still arrive earlier. ITD votes direct and ILD goes with the reflection. The sonic image compromises somewhere in between. We can use this conflict resolution mechanism to manipulate sonic image. Let's start with a stereo pair.

Take the center seat. Our speakers are matched in level, distance, angle and acoustical environment (Fig. 5.5). When signals are matched, the sonic image appears exactly where there are no speakers: center. Sonic ventriloquism has begun! The ITD and ILD values detect speakers off to the sides, but a compromise center value is reached because they are equal and opposite. If left is louder, the image shifts that way. The ILD value identifies left as "direct" sound and right as "reflection." The localization is ambiguous because the ITD value remains matched, i.e. level says left but time still says center. The result is a compromise location (up to a point). As level offset increases, the image moves leftward and shrinks until it is effectively a single speaker. This is the underlying principle of stereo perception. A mix containing individual signals can route some signals at equal levels, whereas others are louder on one side. Equal-level signals have symmetrically opposite localization values, creating a symmetrically derived center image. Unequal signals have asymmetric localization cues resulting in an off-center compromise image. We'll return to stereo later, but first let's investigate the time factor.

Let's remain at center, reset the system to unity gain between the channels and offset the timing instead. When left is delayed, the ITD values indicate a rightward source, while the ILD values remain centered, a second version of ambiguous localization readings. The ambiguity elongates the apparent source width and prevents the image from moving fully right. As time offset increases, the image moves further rightward until the point of separation, creating a perceptible echo (like a lateral reflection).

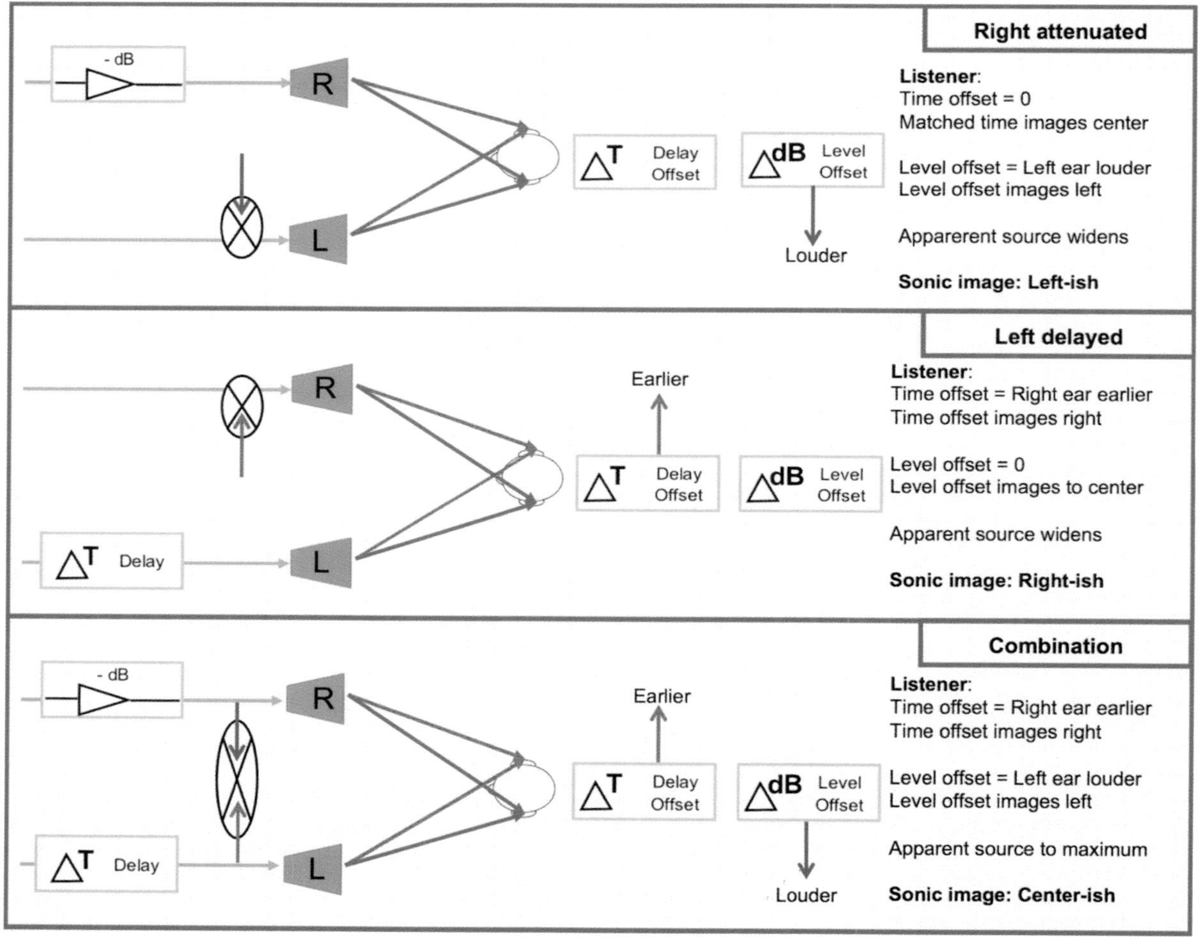

FIGURE 5.5
Horizontal localization between two correlated loudspeakers: level offset effects (top), time offset effects (middle), combined time and level offset effects (bottom)

Horizontal image can be moved by either level or delay offset. The process is termed "panning," short for "**panoramic perception.**" Panning has two forms: level and delay. We can play ILD and ITD effects against each other and create compromise image placements. This is where our sleight of hand takes full force. We can level pan left and time pan right, and keep the image centered or move it anywhere along the horizon. This technique can be used to minimize sonic image distortion.

5.2.4.2 PRECEDENCE EFFECT

The term **precedence effect** (Fig. 5.6) describes how sonic image perception is affected by combined level and delay panning. It's often called the "Haas effect" due to the research credited to Dr Helmut Haas (Everest, 1994, pp. 58–59). Listeners were centered in a stereo field with speakers placed 45° off axis. Time and level panning were used in combination to consistently re-center the image. Naturally equal time and level did the trick. A delayed speaker operated at a higher level also returns the image to center. The later it gets the more help it needs with level. The time window is *very* short. The bulk of the timing activity occurs in the first millisecond and has reached full effect by 10 ms, at which point we will need 10 dB of excess level to bring it back. When the level offset exceeds 10 dB (the isolation zone) the game is over and no amount of delay can center the image.

Our two-channel differential hearing gives us a 10 ms, 10 dB window in which to manipulate sonic image in the horizontal plane. What about the vertical image? We did not get the quad upgrade so we lack time differential in the vertical plane. The HRTF localizes the image in the direction of the level-dominant source. Help comes from an unexpected place, the horizontal system. Recall that each ear localizes the vertical plane independently. The vertical system can accommodate two localization values, as long as they are separate for each ear. Therefore, if sources are displaced both horizontally and vertically we discern their locations independently and a compromise horizontal and vertical value is created (Fig. 5.7). Unless we are at the exact horizontal center between two vertically displaced sources we will receive vertical cues from both sources. The only way to perform pure vertical panning is with level, where signal from one source becomes the dominant HRTF signature. If we're hearing the combination of upper-left and lower-right speakers the precedence effect will compromise the horizontal whereas the two distinct HRTF values will compromise the vertical. The result will be a sound image where no actual speaker exists in either plane. *Voilá*! Now you hear it. Now you don't see it. Note that much of what passes for vertical delay panning in our industry is actually a result of cues taken from the horizontal plane.

An example application is a proscenium theater L/C/R configuration: low sides and high center. Level and delay panning can move the image both vertically and horizontally. The three channels must be well matched in nominal level, frequency response and coverage area for this effect to be perceived beyond a small area. The cost/benefit analysis includes ripple variance vs. image improvement.

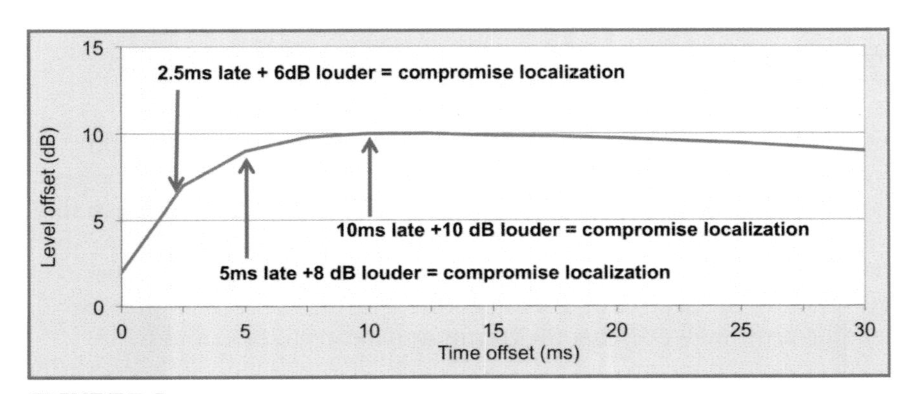

FIGURE 5.6
The precedence effect. The horizontal sonic image is affected by the time and level offsets between sources. The sonic image is centered between the sources with both offsets at 0. The later source must be louder to maintain a central image as time offset rises. The first 10 ms of time offset are the most critical in the balancing act between time and level (after Haas).

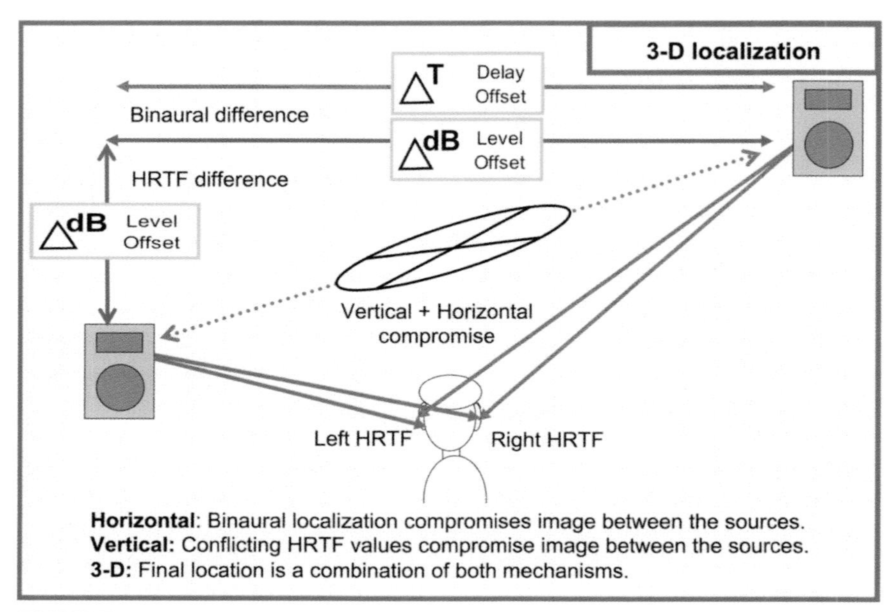

FIGURE 5.7
Combined operations of the vertical and horizontal localization systems. Speakers are horizontally separated with asymmetric vertical orientation. The horizontal localization remains binaural. The conflicting vertical HRTF values reach a compromise that follows the dominant horizontal location.

5.3 TONAL, SPATIAL AND ECHO PERCEPTION

Our ears track the constantly changing tonal shape of music and voice. This changing tone is compared with a history file in our brains. We expect the tonal shape to change *with* the music, not be independent of it. If the reproduction system is transparent, the music passes through without tonal modification. If our sound system has a midrange peak, it is passed on to *all* the music. It's this *relative* tonal shape that concerns us.

A sound source is experienced as a combination of three parts: direct sound, early copies (reflections or correlated sound from other speakers) and late copies (reverberation). All are present unless it's a solo speaker in an anechoic environment. Tonal complexity rises each time a copy is added. The common terms to classify these listening experiences are tonal, spatial and separation (perceptible echo). The lines between these perception zones are mostly gray, gradual and frequency dependent. They can exist singly or in combination (depending on program material, level and time offset). These are variations on the summation themes at the heart of this text. The linkage between summation zones and tonal perception is critical to our optimization and design strategy.

Many audio publications focus heavily on speech perception and vocal intelligibility. We are hard pressed to find statistics for NASTI (nose flute, accordion and snare transmission index) or %ALCONGAS (articulation loss of conga drums) and yet the perception of musical instruments is tremendously important to us. Nothing in this text relates exclusively to the vocal range. We are committed to reducing input inequality. Here we relate the perception issues to specific actions: How can we best match perception with what we can see in the space and on the analyzer? How can we translate this into the actions of optimization? While other texts emphasize the contrast between tonal, spatial and echo perception, we discuss the continuum between them and their common cause.

Echoes are perceived as distinct events, separated from the direct sound by silent space, hence the term "discrete." Physics says otherwise. The separation is frequency proportional, with highs going first and lows remaining linked for longer durations. Direct and reflected sounds never fully separate, but we perceive them that way because the linked remainder is below our hearing range (or that of our speakers). Duration is a key element here. If the duration is infinite, such as a continuous tone, we'll never perceive a discrete echo at any frequency, no matter how much glass surrounds us. Echo perception requires a dynamic change, i.e. some form of transient. An old country song sums it up: "How can I miss you, if you won't go away?"

Let's consider two program material extremes. The monotonous drone of Gregorian chant was created in the most reverberant spaces ever made: stone and glass cathedrals. The chant's nearly infinite duration gives it the maximum tonal perception and immunizes it from disturbance by echo perception. Amen. At the opposite extreme is the pure impulse, rising and falling in a single cycle over a wide frequency band. This signal is the easiest of all to discern echoes, and the hardest to discern tone. The chant is maximum duration and minimum dynamics whereas the impulse is the opposite. Musical signals (hopefully) contain mixtures of transient and steady-state signals.

We regularly transition between tonal, spatial and echo perception without changing our seat because they are highly frequency dependent. The road to echo perception is gradual. It begins at the highest frequencies and peels off successively larger portions of the spectrum. It's widely published that the dividing mechanism is time offset (Fig. 5.8). The first perception zone, the tonal fusion zone, runs for the first 20–30 ms. It follows, then, that problems in this zone would benefit most from a tonal adjustment device: an equalizer. Beyond the 30 ms border we enter the area of spaciousness that continues to around 50 or 60 ms, after which we finally emerge to separation (above 60 ms). This final zone would least benefit from equalization. We should always be suspect of any single number time threshold to describe audio perception. Our hearing spans a 1000:1 time/frequency range (50 ms to .05 ms). Let's consider combined signals with 30 ms of time offset, the limit of the tonal fusion zone. At 33 Hz (30 ms) direct and reflected signals are a single λ apart. The reflected sound would commence at the moment the direct sound finished, extending the perceived duration in a form indistinguishable from two cycles of direct sound. By contrast, the same time offset puts 400 λ between 12 kHz signals. A signal with 200 λ of 12 kHz would have a 200 λ gap between the direct and reflected arrivals. At 30 ms of time offset the HF would be perceived as separated, while 33 Hz would not. The ripple in the summed frequency response will feature an octave-wide peak at 33 Hz. At 12 kHz the peak will be 1⁄400 of an octave wide. Do these look like equal candidates for equalization? The lows are still in the tonal zone, the mids are in the spatial transition zone and the highs have crossed over to separation.

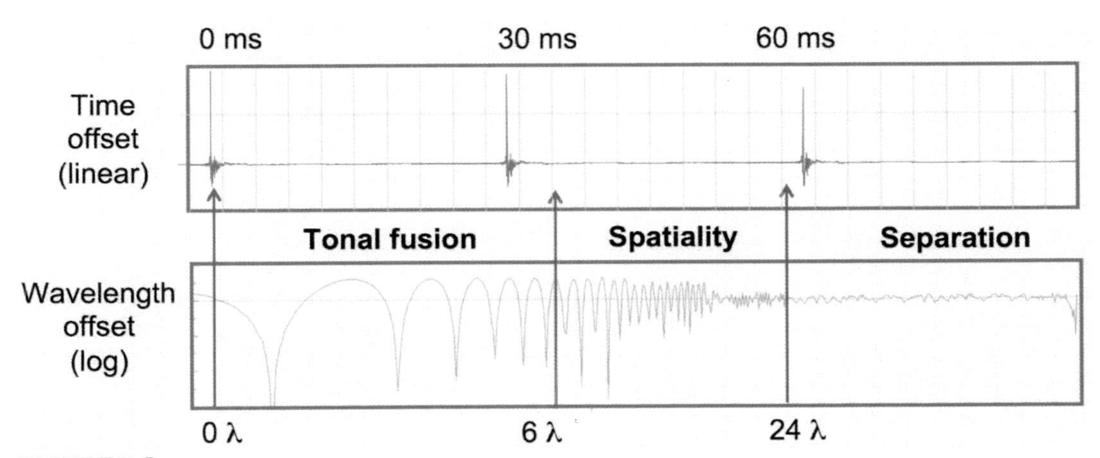

FIGURE 5.8
Comparison of the traditional time-based tonal, spatial and echo perception thresholds to the wavelength proportional model

5.3.1 Tonal perception

The tonal quality is a combination of the direct sound and the early summations (reflections or correlated speakers). The direct-sound tonal character is modified by comb filter ripple in proportion to the strength of the late arrivals. The frequency range where the widest and most tonally dominant filters are found falls as time offset rises. Increasingly narrow filters sweep downward from the upper regions. Eventually the narrowing exceeds our ability to discern distinct peaks and dips. The disturbance does not go away, however, but rather migrates across the tonal perception threshold into spatial perception and eventually echo perception.

The room is dark except for a strand of lights. Our mind focuses on discerning their pattern even though they occupy only a small part of our field of vision. We focus on what we see, not on what we don't see. Our tonal perception is similar. We hear what we hear, not what is missing. Our ears focus on the spectrum's bright spots, the peaks, and relegate the sonic darkness of the dips to the background. The pattern of sonic bright points is the tonal **envelope**, the shape of the spectrum.

The envelope is the spectral shape of the speaker, if alone in a free-field environment. An ideal speaker envelope would be free of tonal character, a flat response with no peaks or dips. Summations add ripple and modify the envelope. Peaked areas stand above the median, and cancellations below. Peaks in the vicinity of the median response will stand out just like we see a hill on a flat horizon. A median response area in a cancellation neighborhood will also stand tall. A peak bordered by deep cancellations is the most extreme case due to the contrast in levels at nearby frequencies.

The ear has limited sensitivity to frequency (pitch) and tonal variation. Very slight pitch variations can be detected (sharp, flat or vibrato) with sensitivity in the 1/100 octave range. Sensitivity to tonal variation connotes identifiable peaks or dips in the response. The threshold is termed the **critical bandwidth** (Everest, 1994, pp. 46–47). Peaks and dips much narrower than critical bandwidth are perceived as adding spatiality or separation rather than identifiable frequency response shapes. Summation ripple has a linear bandwidth that's perceived as progressive narrowing over frequency by our log hearing. Such ripple is typically wider *and* narrower than critical bandwidth, depending on frequency. Small time offsets keep more of the spectrum on the wide tonal perception side and large offsets send more of it into separation.

The frequency response shown on modern high-resolution analyzers is consistent with the perceived tonal envelope (Fig. 5.9). Wide bandwidth peaks are the most identifiable (visually and audibly), even when only a few dB above the median. Very narrow peaks of comparable level are less of a tonal concern. Critical bandwidth is most

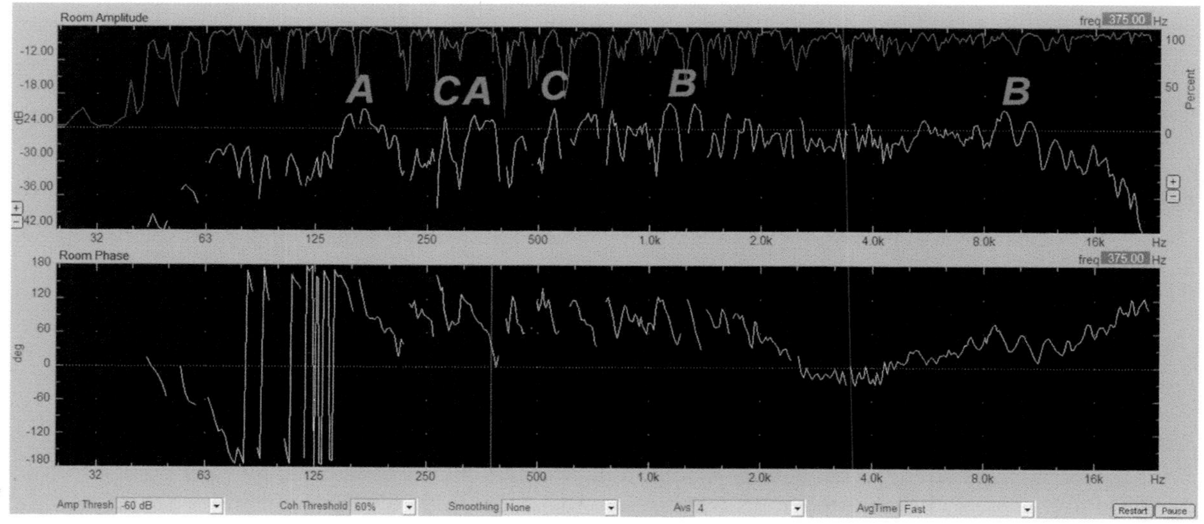

FIGURE 5.9
Tonal envelope example: (A) wide peaks with dips on their sides; (B) wide peaks composed of a series of smaller peaks above the median line; (C) large narrow peaks that are less audible

commonly cited as around 1/6 octave in the mid- and high-frequency ranges, but, like so many aspects of subjective perception, this is not definitive. The ear's low frequency resolution becomes more linear as does the critical bandwidth. This has practical implications to system equalization, which is most effective on the most tonally perceptible anomalies: wide peaks. Narrow peaks and dips may be more effectively treated by solutions other than equalization.

Sensitivity to peaks doesn't connote that dips are inaudible, just less apparent. They are audible by the absence of signal rather than presence. Absent areas are noticeable over time. Portions of the music are missing or out of balance. Some notes stand out; others are lost or unexpectedly colored. We can equalize peaks down to the median level but this does not restore the missing pieces caused by cancellation. These dips still concern us, but are too narrow for effective treatment with tonal solutions, i.e. filters.

5.3.2 Echo perception

There's no dividing line in acoustical physics between tonally perceived early arrivals and discrete echoes. Any secondary arrival appears separated when viewed in the time domain and linked when viewed in the frequency domain. A room reverberation plot shows hundreds of discrete impulses. The log frequency response shows a continuous narrowing of the comb filtering. Nothing on the analyzer indicates a transition into perceived spaciousness or discrete echoes. The dividing line is in our head.

We know tonal modification comes from "early" summation, spatiality from "middle" summation and discrete echoes from "late" summation. How do we define late? It's when we hear it as an echo. We're talking to our own reflection: a circular argument.

It's clear that perception transitions are affected by program material. Gregorian chant (continuous) represents one extreme, while the Drum and Bugle Corps (transient) the other. The key difference is the summation duration. Continuous signals have extended summation duration overlap, and are experienced as tonal coloration. Transient signals have the minimum overlap duration, making them the leaders in perception zone migration. High frequencies cross the line into echo perception first because their transient peaks have the shortest duration.

Time offset effects are linear and our perception is log. Therefore it's fair game to apply a sliding time scale for echo perception. Late for 10 kHz is a very different matter than late for 100 Hz. Signal arriving 5 ms late combs at 1/50 of an octave wide at 10 kHz (50 λ offset). This is far beyond our critical bandwidth (6 λ offset) for tonal variation. What is it then? It has to be something. It's certainly a *potential* echo if the signal is transient. We're certainly not going to equalize it, so, for practical purposes, it's an echo. What's happening at 100 Hz? 100 Hz? Do you read me? We regret to inform you that 100 Hz (half λ offset) has undergone the ultimate tonal variation: cancellation. Its neighbor, 200 Hz, received an octave-wide tonal boost in the process. The sliding scale from tonal to echo is as plain as the logarithmic teeth of our comb filter. Where the teeth are wide we hear tone; when they get narrow we have echo potential.

A time offset of 60 ms is a commonly cited threshold for echo perception. This is another case of the "vocal-centric" perspective giving us a one-size-fits-all value. A time offset of 60 ms would create filtering of approximately 1/24 octave at 400 Hz, right in the heart of the human voice range. The entire voice range is sliced up much finer than critical bandwidth so it makes perfect sense to classify it on the separation side. But the vast majority of a bass guitar's range is below 400 Hz, and the high-hat is miles above this. What if we were to reverse the threshold perspective from time offset to its frequency response effects? Let's consider 1/24 octave (24 λ late) as the threshold instead of the fixed time of 60 ms. This would translate to a sliding time scale with 2 ms at 12 kHz and 240 ms at 100 Hz. This would use a constant bandwidth threshold between time domain and frequency domain solutions. Optimization options for arrivals under the 1/24 octave threshold would include the possibility of equalization, whereas those beyond would be restricted to delay or other solutions. Does it hold up? My own research with hundreds of students supports it. I have listened to summed impulses and continuously adjusted the time offset. The transition to perceptible separation is gradual, beginning with the highest frequencies and steadily adding more of the spectrum with rising time offset. The transient peak showed perceptible separation of the highest frequencies in less than 2 ms. This is consistent with 1/24 octave expectations. As the time is increased to 10 ms the splitting was extremely apparent in the midrange and yet sounded only stretched in the lows. By 25 ms the signals were

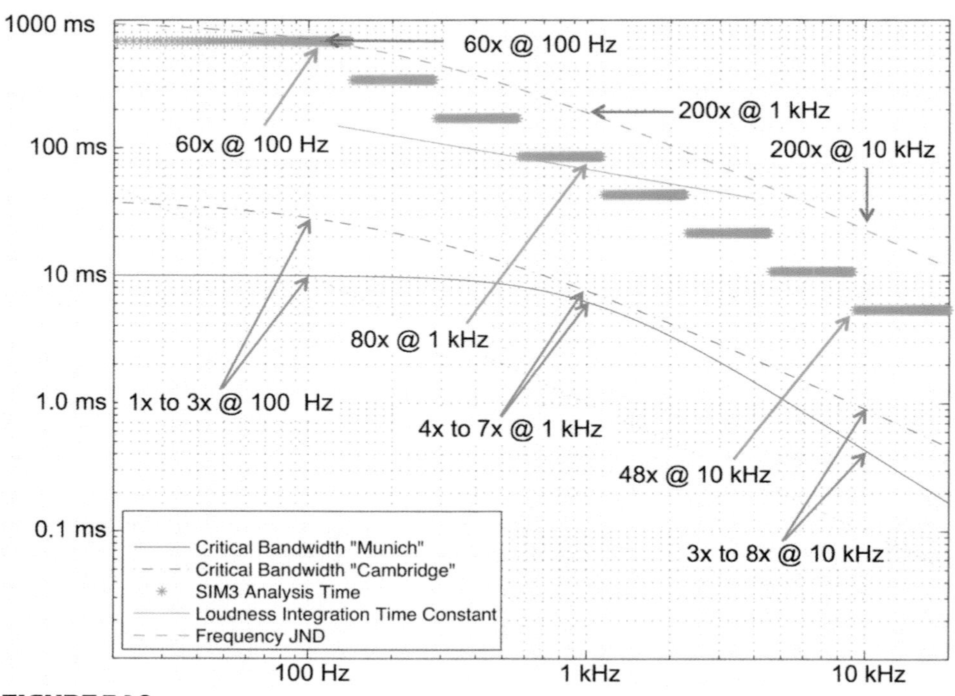

FIGURE 5.10
Time vs. frequency for the critical bandwidth, just noticeable difference and the SIM3 analyzer. The ear's response is quasi-log, mostly log in the mids and highs, and mostly linear in the LF range. The modern FFT analyzer's resolution is sufficient to clearly see any tonal modification trends within the critical bandwidth range (Schwenke and Long).

perceived as strongly separated. But we know they weren't. How could the 30 Hz wavelength separate from itself when it has not completed a single cycle?

This doesn't necessarily conflict with the vocally focused mainstream audiology community findings. It's simply a frequency range extension (above and below). The fixed echo perception threshold of 60 ms corresponds to our sliding 1/24 octave threshold at 400 Hz. The methods are consistent with the midrange vocal perception transition, but differ for the cello and high-hat extremes.

The relationship between critical bandwidth, just noticeable pitch difference and the time record lengths of the modern analyzer is shown in Fig. 5.10. Note that the quasi-log frequency resolution (24th to 48th octave) of the modern analyzer falls in the middle between critical bandwidth and the just noticeable difference.

5.3.3 Perception of spatiality

The gray world between tonal modification and echo separation is termed the zone of perceived spatiality. Instead of perceiving single sources (together or apart) we experience a transitional state of semi-separation. The source image is spatially stretched and expanded between the various arrivals. It's often referred to as "increased spatiality" but could also be termed "I can't figure out where the sound is coming from." The specified time offset range is between 30–60 ms (post-tonal and pre-separation). Because we have already shown the weakness of time as the zone separator, we can call this zone for what it is: the frequency-dependent transition between tone and separation. Some frequencies have migrated over to separation whereas others are still holding out in tonal world. The result is a confused brain with a mixed spectrum of localization clues. The perception of sounding "spacey" and feeling spacey seem pretty well linked here.

We've previously discussed apparent source width. We can only stretch so far between sources before the rubber snaps and we perceive separated sources. The LF range can stretch over larger time offsets (sonic distances) and therefore has greater elasticity than the HF range. Multiple arrivals from different directions, such as room reflections or surround speakers, expand the apparent source width in multiple directions and planes. A complex full-range signal provides a constantly changing mix of tonal fusion, source stretching and separation. Is it any wonder why this zone is termed "spatiality"?

5.3.4 Perception zone progressions

An analyzer with 1/24 octave, or, better, logarithmic frequency display can tell us which perception zone we are in (Fig. 5.8 again). If we follow critical band theory, we can set the tonal threshold to peak spacing of 1/6 octave and under. The echo perception zone (<1/24 octave) is around the upper limits of the analyzer's resolution. If the analyzer is having trouble with the details of the frequency response then so will our ears. A high-resolution analyzer retains important clues in the response that low resolution will smooth over. It's in the fine grain of the frequency response where the perception thresholds are revealed. Wide peaks (tonal zone) appear on the screen as the best candidates for EQ, and they are! Superfine combing (separation zone) appears as the best candidate for solutions other than EQ, and they are! The devil is in the details.

5.4 STEREO PERCEPTION

Everybody loves stereo. It is 100% better than mono unless we are hearing only 50% of the show. We all know the score here. Stereo sound reinforcement systems, like hedge funds, are absolutely great for 1% of the people. No amount of proof about how little stereo is actually experienced in a concert hall or arena will change the fact that it's required on all riders. Nonetheless we will look at how it works (and doesn't) and the side effects of overreaching for it. Stereo Everywhere™ is trademarked by Bose but will require a firmware upgrade to human anatomy to actually be implemented. It's all in your head.

5.4.1 Panoramic field

We all know center is the best seat for stereo with rapidly diminishing quality everywhere else. Let's begin with how stereo is *supposed* to work before moving on to its limitations. The standard stereo signal uses level offset (the ILD portion of our horizontal localization) to place individual signals along the panoramic horizon. Equal level centers the signal, while 10 dB of level offset effectively eliminates the lesser speaker. We can recognize the coupling and isolation summation zones here. Intermediate level offsets place the signal proportionally between center and the dominant side. Let's send five signals into the mix: L, LC, C, RC and R (10 dB, 3 dB and 0 dB level offsets). Our listener also has a secondary localization clue, the ITD, which answers "center" regardless of the level differences. Therefore we can conclude that the mix engineer's intended panoramic placement is derived from the combination of changing ILD and static ITD. It's critical to understand that stereo perception presumes a continuous spread over the horizon. Perceiving left and right speakers as separated sources is a form of multichannel, not stereo.

5.4.1.1 OFF-CENTER EFFECTS

A new set of ILD relationships is found for our five signals when we move one seat to the left. The ILD localization changes the least at the outer extremes (signals L and R) due to the level dominance of the signal at one speaker. L stays at L and R stays at R. All the other channels migrate leftward because our movement brought us closer to the left and farther from the right, which modifies both ILD *and* ITD leftward. The ITD contribution, previously a centering force, is now unambiguously leftward. Visualize a ship full of cannon balls being blown to the port side by a strong wind. The balls had better be fastened down or wind and momentum will combine and tip the stereo

ship over. The "C" channel tacks hard to port because both the ILD and ITD values agree the source is leftward. LC and RC channel information also move leftward, leaving only the R channel to represent the right side.
The tilt increases each time we move further off center. Recall that our sensitive ITD mechanism needs only a few milliseconds to go full tilt.

Let's be absolutely clear here. If the levels for a set of stereo instruments are perfectly balanced for the center location then it is, *de facto*, imperfectly level balanced off center. The guitar solo panned to LC that sits perfectly on top of the mix (at center) is inexplicably under (or over) the optimal level on the side. It's not just level. It's also out of place. There's more. Signals that are losing level at our location are also falling behind in time. They are now perceived as out of balance and out of place. For every seat off the centerline you know one thing for certain: The instruments aren't at the levels or locations you mixed them.

Everybody loves putting precedence effect to use when they think about manipulating the sound image around the stage or under the balcony. But the tables turn on us in the stereo configuration. Instead of precedence helping to hide speakers it does the opposite: It pulls the image on to the closer speaker of the L/R configuration.

Let's go through an example and look at three locations: 0 ms, 2.5 ms and 10 ms off center (Fig. 5.11). These correspond to 0 dB, 6 dB and 10 dB level milestones in the precedence chart. Let's take coverage angle and propagation level loss out of the equation and assume that level differences are only the result of panning at all locations. When we reach 2.5 ms off center (left) it is as if the left speaker is 6 dB louder (more than enough to pull the image there). An instrument panned L, LC or C will appear at the same location: left. It will need to be hard-panned right to have a chance of overcoming the precedence.

Off center effects on stereo panning

Pan	L	R	Listener	Precedence	Perceived level	Image
L	0 dB	-6 dB	Left 10 ms	-10 dB (R)	R is -16 dB	L
			Left 2.5 ms	-6 dB (R)	R is -12 dB	L
			Center (0ms)	0 dB	R is -6 dB	L
			Right 2.5 ms	-6 dB (L)	L= R	C
			Right 10 ms	-10 dB (L)	L is -4 dB	RC
LC	0 dB	-3 dB	Left 10 ms	-10 dB (R)	R is -13 dB	L
			Left 2.5 ms	-6 dB (R)	R is -9 dB	L
			Center (0ms)	0 dB	R is -3 dB	LC
			Right 2.5 ms	-6 dB (L)	L is -3 dB	RC
			Right 10 ms	-10 dB (L)	L is -7 dB	R
C	0 dB	-0 dB	Left 10 ms	-10 dB (R)	R is -10 dB	L
			Left 2.5 ms	-6 dB (R)	R is -6 dB	L
			Center (0ms)	0 dB	L= R	C
			Right 2.5 ms	-6 dB (L)	L is -6 dB	R
			Right 10 ms	-10 dB (L)	L is -10 dB	R
RC	-3 dB	0 dB	Left 10 ms	-10 dB (R)	R is -7 dB	L
			Left 2.5 ms	-6 dB (R)	R is -3dB	LC
			Center (0ms)	0 dB	L is -3dB	RC
			Right 2.5 ms	-6 dB (L)	L is -9dB	R
			Right 10 ms	-10 dB (L)	L is -13 dB	R
R	-6 dB	0 dB	Left 10 ms	-10 dB (R)	R is -4dB	LC
			Left 2.5 ms	-6 dB (R)	L= R	C
			Center (0ms)	0 dB	L is -6dB	R
			Right 2.5 ms	-6 dB (L)	L is -12 dB	R
			Right 10 ms	-10 dB (L)	L is -16 dB	R

FIGURE 5.11
The effects of time offset on stereo perception at off-center seats. Precedence effect skews the panoramic distribution when seats are off center.

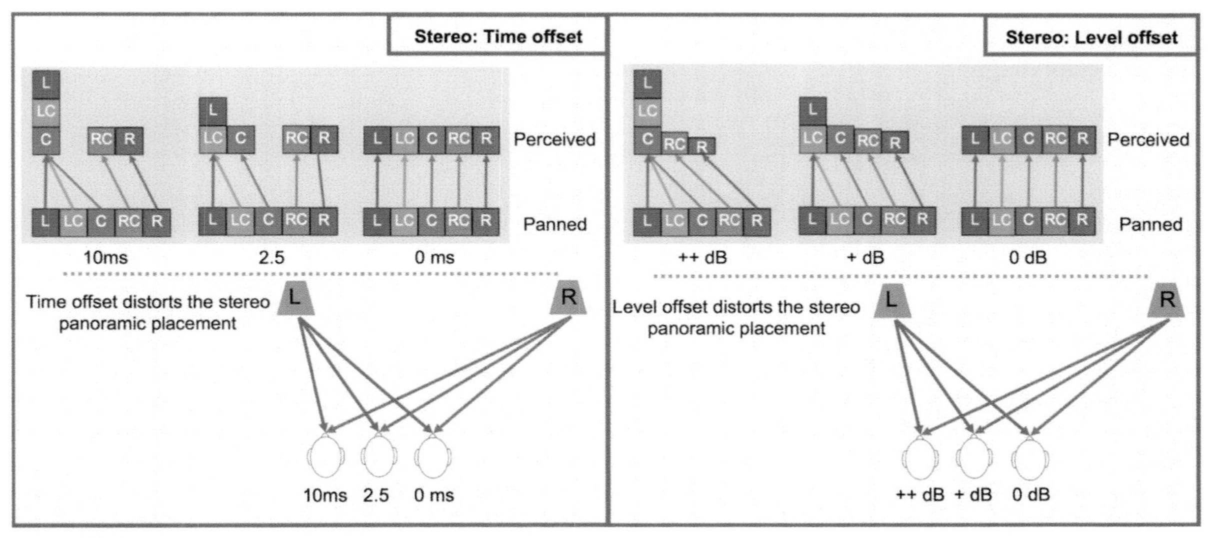

FIGURE 5.12
Time and level effects on stereo perception at off-center seats. Precedence effect and level offset skew the panoramic.

The closer we are to center, the closer we are to the mix. Sound reinforcement systems commonly cover large seating areas that are not even *between* the left or right. They are left of left or right of right. Someone else will have to explain how stereo can even be discussed under these circumstances. How good is the imaging of your car stereo when you are standing outside it? Many theaters and arenas have half their seats outside of the mains. Fan-shaped theaters and houses of worship have even more (and do the craziest things to try to get stereo).

5.4.1.2 ANGLE EFFECTS

The optimal stereo panoramic field angle is generally accepted as 30° from center (an equilateral triangle of left, right and listener). This is easy at home or in a studio where we only care about, the "sweet spot." Overly wide stereo fields suffer from poor image stability as we move away from the center. A narrow stereo field is more stable over location but is "not as fun" due to the reduced angular range. Wide fields create a big effect for a few seats. Narrow fields have a small effect for a lot of seats. Both of these will likely happen in a single room in a concert setting. Close seats may have a wide spread whereas far seats see a narrow one. It's a certainty that the width of the panoramic field will vary over distance and lateral position in the hall. A lack of uniformity of experience is assured.

The angle between L and R compresses and the placement skews as we move off center. There is no perceived left channel when we are placed in front of the left speaker. There is center and right. Once we move leftward outside the L/R pair we actually have the modern political climate: right and far right. Signals panned left, center and soft right appear at the closer right speaker and only extreme hard-panned signals appear in the extreme right channel. All told, we see that a left-panned signal can be perceived on the left, center or right for a given listener, and that the stereo field may be as wide as a canyon or as narrow as a sliver (and skewed to the side).

5.4.1.3 RANGE EFFECTS

The basic layout of a stereo pair is the same for studio and stadium alike. The coverage patterns also look similar, so we can infer that the level offsets will scale up from the home studio. The ILD part of the stereo perception equation is unchanged by scaling. Time offsets, however, increase linearly with scale. Our head size and the wavelength of a given frequency don't rescale when the room gets big. The 10 ms precedence window in the ITD equation doesn't scale up, but the proportion of seats that fall within it shrinks as the room expands. Only a tiny minority of seats falls within 10 ms of both mains in a large space.

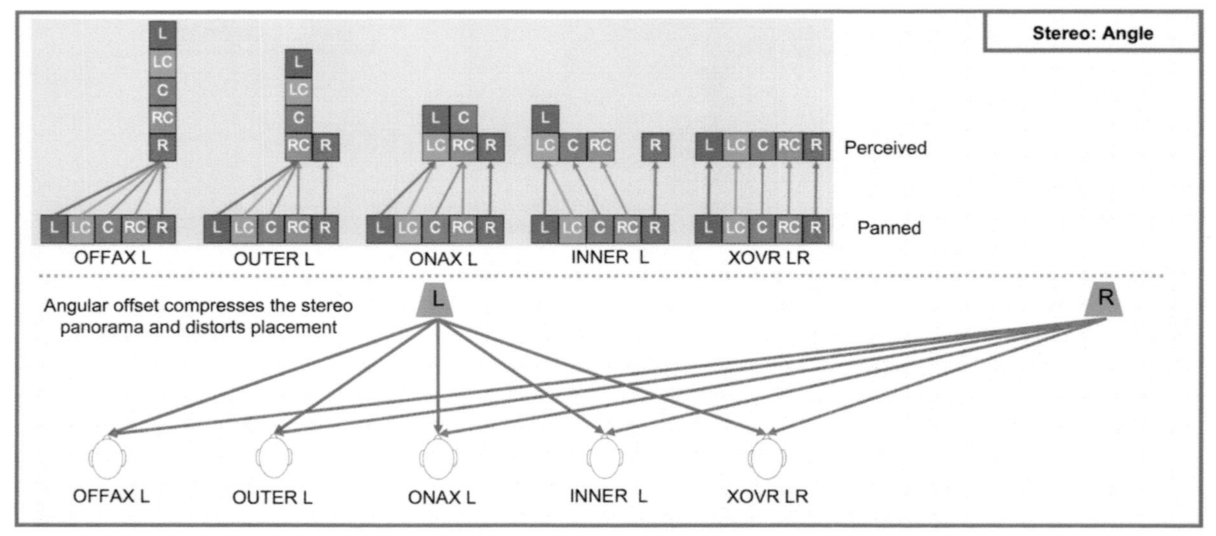

FIGURE 5.13
Stereo perception at off-center seats. Precedence effect level and angle skew the panoramic distribution.

The "stereo field possible" area (Fig. 5.14) is a wedge shape that includes listeners inside the 10 ms ITD window. Both speakers need to cover the wedge, which expands as we move forward in the space. The distance between the speakers controls the wedge size because of the time offset. Wide spacing makes a narrow wedge of perceptible stereo, with a wide angular spread (which accounts for why wide spreads have unstable imaging). Narrow spacing makes a wide wedge of perceptible stereo with a more stable, albeit less dramatic, sound field. Here's how to find the area where stereo perception is possible: predict left + right at 50 Hz. The center lobe encloses the area where <10 ms of offset is found. The nulls on either side are 10 ms.

5.4.2 Stereo side effects

It's depressing but obvious that the panning on your live sound console doesn't translate for much of the room. Off-center listeners hear the show from their side. And yet there are definite up sides to the practice, such as

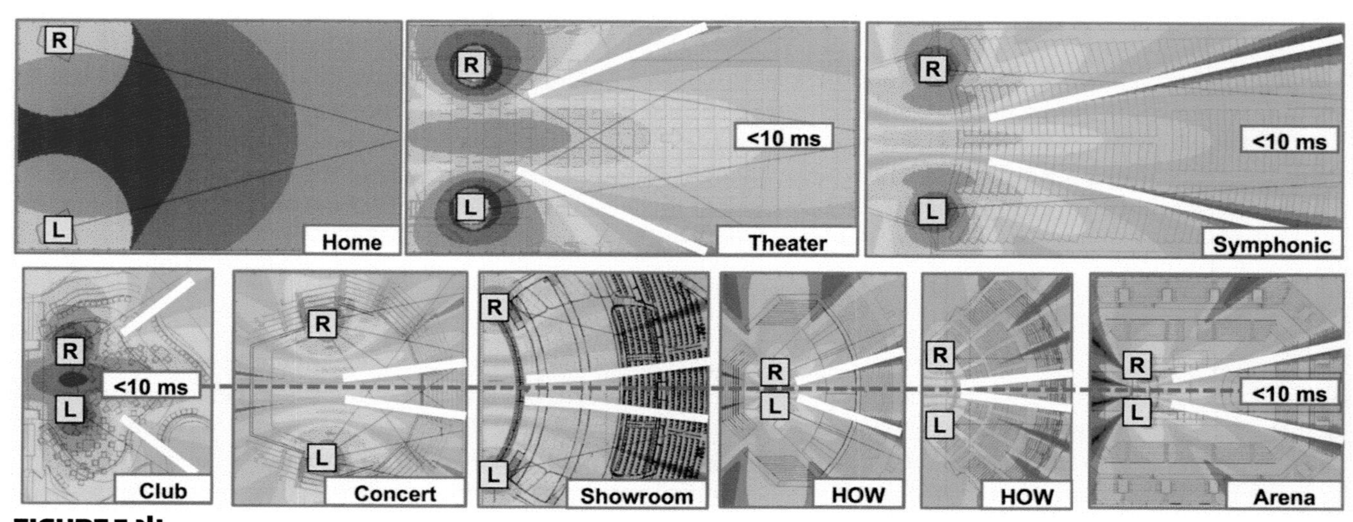

FIGURE 5.14
Examples of stereo panoramic zones in typical halls. The "stereo possible zone" is between the white lines.

dramatic moments where an instrument suddenly appears alone on the opposite side. Like a tom drum. The negative side is decreased intelligibility, increased tonal distortion and echo perception.

We can look at the side effects to see whether stereo is worth the trouble. There's no doubt that it enhances it for some and just as clearly degrades it for others. The side effects are substantial. Any signal assigned to both sides will create a stable comb filter pattern over the space. Listeners closer to center will have tonal variations, whereas those on the sides will hear separated sources (perceptible echo). The extent of the damage depends upon the percentage of coverage overlap.

5.5 AMPLIFIED SOUND DETECTION

Is it live? Or is it Memorex™? This oft-quoted TV commercial featured Ella Fitzgerald singing in a vocal booth while a speaker blared her voice at a wine glass and . . . crash! The wine glass shatters. Because the wine glass could not tell the difference between her natural voice and a speaker then we could be fooled as well (and buy their cassette tape). It would be nice if things were so simple.

We want the audience to focus on the artist, not the loudspeakers, so we work hard to maintain the illusion that listeners are in the performer's direct sound field. We must consider all of the perception clues that might give away the game. It takes only one exposure to break the illusion. One hint might be the hand-held mic. It's possible to make audience members and critics oblivious to the presence of a sound system, in which case we get the best-case review of our show: no mention of the sound system.

5.5.1 Distortion

All forms of distortion give us clues that a speaker is involved. Harmonic distortion is the least obtrusive, because it has the "musical" quality. Intermodulation and digital distortion are far more noticeable. The by-products of this distortion are not harmonically related and are detected even at very low levels.

Operating the system within its limits is, however, a mix engineering issue, and there is no design or optimization strategy immune from this possibility. All we can do is design the system with enough headroom to achieve appropriate levels without overload.

5.5.2 Compression

The absence of limiting and compression may well be more noticeable than its presence. When a singer leans into a mic and screams, the sound system is at extreme risk of detection. Compression and limiting flattens the dynamic response and helps to prevent gross clipping or level surges. Excess limiting can be noticeable as an unnatural dynamic structure. Worst case is an overworked compressor that "breathes," i.e. has audible clamp and release actions. Again, this is an operational issue.

5.5.3 Tonal coloration

If a clarinet does not have the tonal structure expected of a clarinet, then we start to wonder how this could be. Could a strange room reflection cause that? How many drinks did I have? Ah yes! There's a sound system!

A frequency response with substantial peaks and dips attracts our attention. Our memory map contains sonic expectations for a particular instrument or voice. It also remembers what it sounded like two seconds ago. If sound is consistently unnatural, or unnaturally inconsistent, we are prone to suspect a sound system is involved.

5.5.4 False perspective

A violin is a violin. But a violin placed on my shoulder has a very different sound from one that is 30 meters away. We carry sonic perspective as part of our memory map of the violin. A distant violin shouldn't sound like a close one, or vice versa. We become suspicious when the audio perspective doesn't match the visual.

Sound reinforcement systems distort the natural sonic perspective. We can a help a distant source be perceived as near by adding level and extending the high-frequency response. This brings the audience closer to the performance and also allows performers to reach the audience without having to bellow their voices in the manner of the classical theater actors. Performers can act, sing and emote with greater dynamic range. A song can come down to a whisper, but still be heard. In short, with only "natural" sound the actors must speak in an unnatural way to project to the rear. With "unnatural" sound (that's us) the performers can speak in a natural way.

The illusion has limits. We want the audience to suspend disbelief, as they do while immersed in the plot. We can bring audience members and performers acoustically closer than they actually are but not on top of each other. Listeners have acoustic expectations for distant sound in the natural world: low in level, reduced direct-to-reverberant ratio, low-frequency summation from the room and high-frequency air loss. Close sounds are expected to be loud, non-reverberant and flat. Awareness of these trends helps us to remain undetected while moving the perspective.

Level and HF response both rise as we move closer. The LF smooths out and the room falls away. Our sound system can get as close as we want (or closer) without moving. We must prevent the overly intimate "whispering in your ear" experience by never allowing the VHF range to peak above the main body of the response (at least not where anybody can hear it). High-frequency extension intended for distant locations must reach the back without overheating the near seats.

It's also possible to reverse the perspective and make near sources appear distant. This is not desirable in most cases. Nearby off-axis areas have a similar response to distant areas. It feels like we are moving away as we move off axis, even if we're no further from the stage. The lack of HF coverage mimics the air loss and pushes our perspective back. Nobody likes a seat where the performers are visually close and sonically distant.

False perspective can also be detected by overly aggressive low-frequency reduction in the far field. A flat, low end at great distances too closely resembles a near-field response to remain plausible. If the lows are reduced to a level under the midrange, a "telephone" tonal quality arises. Nothing in nature, short of talking through a tube, reduces both highs and lows while leaving the midrange intact. If our system sounds like a telephone, the illusion is broken and the audience will use their telephones to say bad things about us on Tweeter.

5.5.4.1 DUAL PERSPECTIVE

Another way to expose the speakers is dual perspective. We can't be simultaneously near to *and* far from a source, or in a curtained ballroom *and* a stone cathedral. We have a case of dual perspective when the HF is in your face and the LF in the distance.

A common case is underbalcony delays with all the lows cut out. Only the range above 4 kHz actually requires substantial summation gain in a typical mains/delay combination. We only need a dome tweeter to fill in the frequency response. The result is dual perspective: close and intimate HF with distant and reverberant mids and lows. If we add the delay speaker's MF and LF range (even at a reduced level), we add direct sound (with little reverberant) and raise the direct/reverberant ratio over the full range. This brings listeners closer to the stage, over the full range. In practice the mains are typically very LF dominant over the delays anyway. Therefore aggressive LF reduction of the delays has a diminishing return on the combined response.

Another version of dual perspective occurs in a single enclosure, such as a two-way speaker where the LF and HF drivers have very different directivity. A highly directional HF response moves the perspective forward while the wide LF response moves it back. It's natural for this to occur gradually in a room. It's unnatural for it to happen in a single note (as can happen if steep crossover filters are used).

A third variation is displaced speakers covering different frequency ranges (like grounded subwoofers and flown mains). The sonic image can be spectrally split (LF and HF localizing from different locations). This is most audible with large displacement and/or a steep crossover (see sections 4.3.4.4 and 4.3.5.11 on crossover audibility).

5.5.4.2 SONIC IMAGE DISTORTION

Even a perfect copy of the original sound won't fool anybody if it images away from the performer. Displaced sonic image is textbook false perspective. Only strategically placed, timed and level-managed speakers can produce plausible imaging results.

5.5.4.3 SYNCHRONIZATION

Most large-scale events include video projection, which removes any pretense about viewing the actual performer. No need to hide the speakers. Synchronizing sound and video on this scale is an impossible challenge. Video must be delayed but can only sync up at one particular distance. Unsynchronized video/audio decreases speech intelligibility as our subconscious lip reading conflicts with the audio. And looks ridiculous.

FURTHER READING

Duda, R. O. (1998), Sound Localization Research, San Jose State University, http://www-engr.sjsu.edu/~duda/Duda.

Everest, F. A. (1994), *Master Handbook of Acoustics*, TAB Books.

Robinson, D. W. and R. S. Dadson, A re-determination of the equal-loudness relations for pure tones, *British Journal of Applied Psychology*, 7 (1956) 166–181.

Schwenke, R. and Long, B. (2011), *Physiological and Psycho-Acoustic Basis for Multi- Resolution Frequency Response Analysis* (SMPTE presentation).

PART II

Design

CHAPTER 6

Evaluation

Evaluate *v.t. ascertain amount of; find numerical expression for; appraise; assess; to consider or examine something in order to judge its value, quality, importance, extent, or condition.*

Concise Oxford Dictionary

How do we evaluate the performance of our sound system? How do we get beyond grossly subjective opinions such as "good or bad?" There is a wide variety of objective metrics but none provides 100% correlation to our listeners' subjective evaluation of our work. A sound system with stellar specifications does not assure a great show.

The sound system is just one layer of a multi-layer experience, which makes evaluation much more challenging. In the simplest form we have source, sound system and room. How do you listen to a sound system without musical or vocal content and/or without acoustics? As professionals we work hard to separate these layers because it defines our roles: performers, audio engineers and acousticians. Audiences don't care. We know the band was way too loud on stage and that the promoter should never have booked Black Sabbath into Carnegie Hall. We know it wasn't the sound system's fault but the end result is viewed as *our* failure. Who do you think is going to take the fall for a famous band in a hall with a reputation for great acoustics?

Let's further refine the layers in a modern musical performance. We have composition, arrangement, musicians, band gear, stage monitors, monitor engineer, sound system, mix engineer, system engineer and the hall acoustics. All these layers and players must work well together to achieve an optimum result for the artists and audience.

There are objective and subjective means of evaluating the listening experience. The sound engineer can give us SPL, frequency response, distortion, coverage uniformity and other metrics of sound system performance. The acoustician can give us the room data: statistical analysis of a million reflections, a million different ways. Professional and paying customer alike can describe the experience subjectively with any words imaginable, some of which my publisher won't allow. Let's see if we can sum up these multiple perspectives into one sentence:

Sound engineers are from Mars, acousticians are from Venus, audiences are from the Bronx.

The above caption is a twisted reference to a once popular book by John Gray that focuses on the challenges of communicating our needs with other people who don't share our perspective. In Gray's book the parties are men and women. Neither is right nor wrong, but both have very different approaches to life. The key to success, in Gray's view, is to maintain our separation but understand the other's perspective. Then the relationship can meet both parties' needs as effectively as possible. If parties fail to understand each other's perspective a cycle of conflict and blame assessment degrades the potential for a satisfying union. So it can be with sound engineers and acousticians.

A symphony hall acoustician is informed that there is trouble in his recently designed concert hall whenever the sound system is used for pop and jazz concerts. It was known from the start that the hall would host a variety of artists and the owners had been promised perfect acoustics. The hall has been a major success for the symphony orchestra so obviously the problem is not the architectural acoustics. It must be the sound system: the forty-element speaker system, designed for the space by audio engineers. The acoustician proposes a solution: A single omnidirectional point source speaker system, a dodecahedron, is to be hoisted above the stage to replace the existing speakers. This speaker is normally used by acousticians to evaluate the room acoustics. The acoustician was perfectly correct in seeing that the transmission characteristics of the highly directional forty-element speaker system would excite the room very differently from the symphony on stage (for which the acoustics were optimized). The omnidirectional speaker more closely resembles the emission characteristics of natural sound from the stage.

Was this the solution? You know the answer. Feedback and unintelligible sound. The previous sound system alone couldn't be the solution either. Both approaches suffered the same fatal flaw: a mismatch of the emission/transmission system to the reception area. The natural sound transmission model works best when the emitting source is directly coupled to the reception area (the room), which is shaped and conditioned to act as the transmission system. The amplified sound transmission model works best when the sound source is mostly uncoupled from the room and where the room acoustics provide minimal modification of the transmission. Amplified sound transmission into "perfect" symphony acoustics is as mismatched as is a symphony outdoors without so much as a band shell. These are the ultimate impedance mismatches.

6.1 NATURAL SOUND VS. AMPLIFIED SOUND

Which is better, natural sound or amplified sound?

Most people answer "natural sound" without hesitation. The conventional wisdom is that we tolerate "unnatural sound" (speaker systems) because sometimes natural sound is not feasible. In practice the opposite is the case. The principal market for exclusively "natural" sound is limited to extremely small venues or those cases where "unnatural" sound is forbidden by tradition. Natural sound can only thrive when program material, sound source and acoustic space are in perfect harmony, e.g. symphony music in a symphony hall, opera in an opera house or spoken word in a conference room. Take any acoustic signal out of its scale-matched, form-fitted environment and its vulnerability becomes immediately apparent.

If natural sound is so superior, then why has it lost 99% market share? If it's the first choice, then why do so few choose it? Wouldn't we be shocked if twenty people gathered together to hear a speech and there were no loudspeakers? The most famous classical music event of my lifetime was "The Three Tenors" performed in a stadium through a sound system bigger than Woodstock. Hearing such famous tenors unamplified in an opera house would certainly be far preferable acoustically. Stadiums are chosen for their gross revenues, not their gross acoustics. Nothing personal. Just business. An optimized loudspeaker can beat optimized natural sound 99 times out of 100. If you're a sound engineer and don't believe this, it's time to rethink your career choice.

Great news! We just received the sound design contract for Rogers and Hammerstein's *Carousel*. The venue is the 1600-seat Majestic Theater, the same as its 1945 debut, which is highly rated for its excellent acoustics. It's the original orchestrations in the same pit. No stage automation and moving (noise-generating) lights. A revival to rival all revivals. Obviously we don't need a sound system. If we believe natural sound is best, we should resign the project. The original show was composed and staged for natural sound. How can we presume to improve upon this? The following problems arise: The director hates it, the performers hate it, the audience hates it, and the critics hate it (except for the 90-year-old one). The show closes in a week and we'll never work on Broadway again. What's changed? Expectations. Audiences no longer expect natural sound. They expect *magic* sound delivered without the slightest effort required. Sonic "couch potatoes." Get used to it. It's not a fad and it's not just the audience. Performers want magic sound too. They want to act and sing with a wide dynamic range and still be heard at the back. That takes strong magic, but that's our job.

I am not looking for a fight. Acousticians please put down your acoustic pistols. Sound system design work can't move forward if we're working under an inapplicable acoustic construct that pretends we're not in the room. Submitting to the collective group-think of natural sound superiority does nothing to advance our position. Sound system performance is a decisive factor in the audience experience, as long as the room acoustics don't screw it up. A successful outcome is more likely when we express our acoustic requirements realistically, i.e. based on amplified sound transmission.

6.1.1 Contrasting emission, transmission and reception models

Natural sound transmission requires three-part harmony: sound source emission, stage transmission and hall transmission. Direct sound from individual stage performers is accompanied by stage reflections (early) and house reflections (early and late). Early reflections mix the signals together and provide transmission gain, sustain and tonal support whereas late reflections add texture and more sustain. Early reflection transmission gain is absolutely required for performers to achieve adequate loudness and uniform level over the space. Performers and audience share the acoustical space with minimal separation. Reception occurs inside the transmitter: the room. The room has a continuity of sound that goes beyond a particular composition or conductor. The room defines the sound and the sound defines the room.

Amplified sound applications also require three-part harmony, but a very different tune. Each source is captured in isolation by nearby microphones before leaving the stage. Individual signals are transmitted to the mixing console where blending and tonal content are managed. Mixed sources are sent to distinct and separate areas to create matched or customized sound at each location. There are locally distributed versions rather than a single unified source of house sound. Zone separation on stage suits each performer's unique needs, while in the house it fulfills the common needs of audience members listening in unique locations.

Stage source isolation allows us to capture everything and transmit it anywhere. We can send unmatched signals to unmatched areas to provide matched results for the listeners. Recall that, with natural sound, unity of experience is achieved by a *lack* of separation. Amplified sound achieves this by an *abundance* of separation. Amplified sound seeks a sonic continuity over the space for a given performance but this sonic character loads out at 11:00. The local arena rocks on Tuesday and plays basketball on Wednesday. Completely different sound (and sound system). The sound can change day and night between the warm-up band and the headliner. Tonight's sound system blasts into the back wall and the next keeps it on the floor. Any of these will be perceived by our audience as changing "the sound" even though neither changed the room. The room is not defined by the sound and the sound is not defined by the room.

6.1.2 Perspectives: direct sound, early and late reflections

Let's compare and contrast. Direct sound is first (Fig. 6.1). The violins in the orchestra are on house left and violas and cellos are on the right. We localize them as coming from left and right but we don't consider the violins to be the left channel. Their side of the room doesn't get quiet at a whole note rest. All the strings transmit direct sound to the entire room. By contrast, the speaker system divides the direct sound delivery into coverage zones. The left main speaker is the direct sound transmitter for half the house. If it takes a rest we have a big problem. Conclusion: different, but not conflicting. Natural direct sound: Every source is everything to everybody. Amplified sound: A speaker is the spokesperson for all sources for some people.

Early reflections are next (Fig. 6.2). Natural transmission needs gain from strong early reflections but this creates combing. How does it get away with it? Many reflections from many sources in many locations. Each

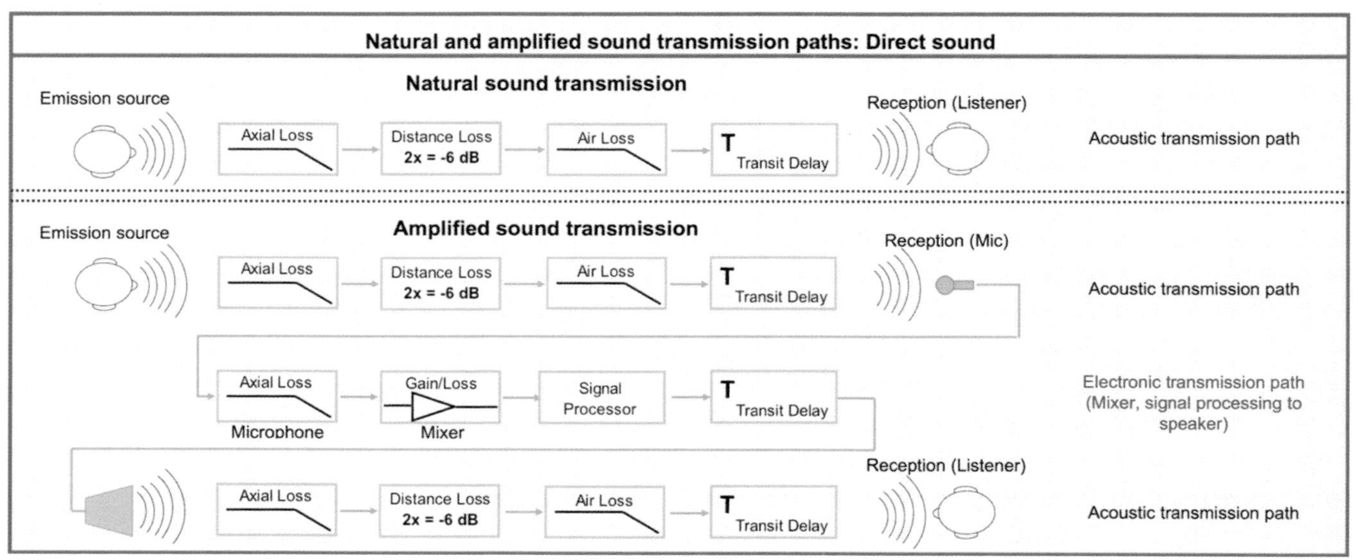

FIGURE 6.1
Direct sound transmission flow block for "natural" sound and "amplified" sound. The natural sound path is potentially filtered by the axial control of the source and HF air loss. The amplified sound has two acoustic paths with similar features. The middle shows the path of the electronic signal, which can compensate for the acoustic transmission effects.

source originates a unique set of early reflections heading toward each seat. Every unique seat gets a unique comb filter response from every unique source. I used "unique" three times in the last sentence because tri-level randomization is the key ingredient to natural sound's uniformity by differentiation. At my seat one violin dips at 250 Hz and the next one dips at 225 Hz. At your seat these might fall at 245 Hz and 220 Hz. Listeners don't hear a precisely repeated comb signature for any two instruments. Every seat gets a slightly different tonal mix of every source.

Our speaker system groups the violins together and transmits them from the same spot (the speaker location). An early reflection of the speaker's signal puts the same comb filter signature on everything the speaker transmits, a decidedly "unnatural" sound. Imagine an orchestra where every musician is placed in an identical private band shell. The band shell's identical combing of each instrument is the dominant tonal feature for the whole orchestra. One poorly placed reflective wall near our speaker will "band shell" every instrument in the sound system. The speaker system also gets transmission gain from the walls (mostly by AC power outlets mounted on them). Strong early reflections give a speaker gain as well, but often at a high price. When used sparingly, such as a subwoofer on the floor, the benefit is well worth the cost. For mids and highs, the combing costs are prohibitive. Conclusion: Different and very conflicting. Natural sound needs strong early reflections and the speakers need to be free of them.

Finally the late reflections arrive. Everybody loves the spatiality and envelopment of a well-behaved reverb tail. The natural model wants more of it, but the amplified model still needs some. The answer is as complex as the reverb tail itself, which takes more than a simple number like RT60 to be characterized. A rich and steadily falling 1.3 second decay in an opera house is very different from the same RT60 in an arena with strong specular reflections spaced widely apart. Why do we see a reverberation unit in the rack at Front of House (FOH) in an arena? It is not because the room is too dry, but because the decay lacks the texture and complexity of a preferred acoustic space. Conclusion: Different and incrementally conflicting. Natural sound needs more late reflections than amplified for a given room size.

6.1.3 Transmission path differences

There are unexpected avenues for early reflections to get into the sound system, even in a dead room: through our open microphones and from our own speakers. Each microphone opens up two paths: a re-entry path from a speaker (house or stage monitor) and a duplicate entry path into other mics. All cause comb filtering with the tonal shapes of an early reflection. The speaker-to-mic re-entry path is also a potential feedback hazard. Once combing is added to a given channel it's difficult to get out and is imprinted on the response sent to the sound system. Each

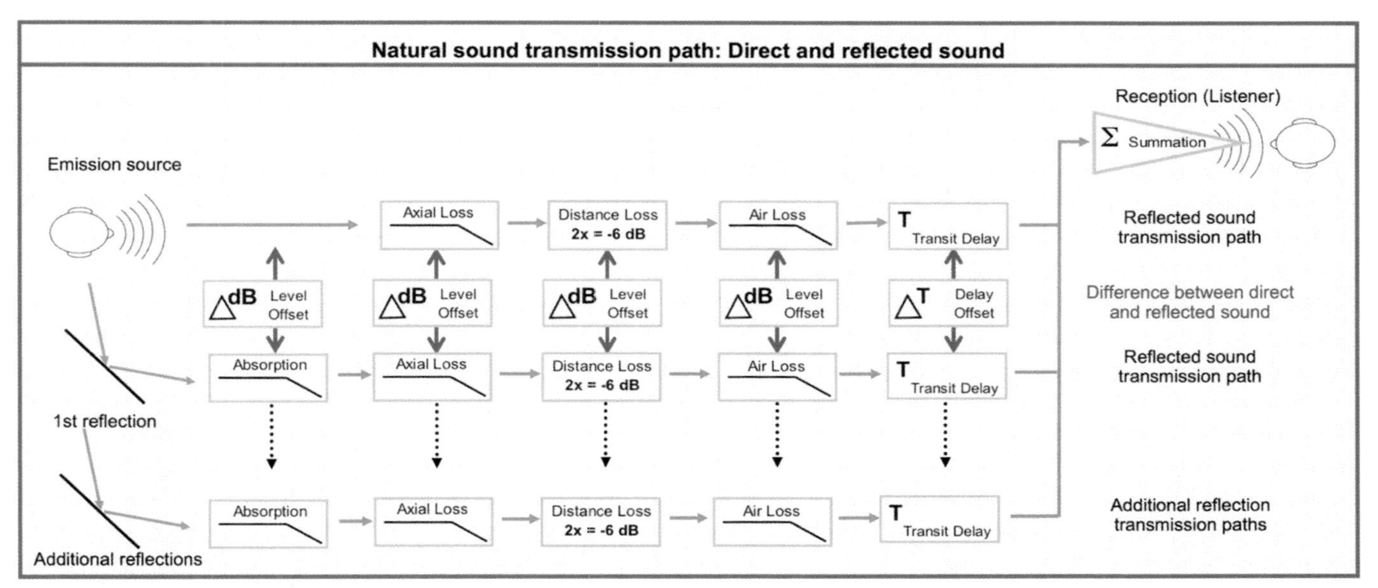

FIGURE 6.2
Natural sound transmission flow block inclusive of room reflections. The listener hears the summation of the direct and reflected paths. The reflections have distinct axial and HF air filter functions, as well as distance loss and transit time. Differences between the paths shown are decisive factors in the listener's experience.

speaker also opens up two paths: re-entry into the stage mics and duplicate entry to the house (when another speaker carries the same signal). Each path is perceived as a reflection, most probably an early reflection (Fig. 6.3).

A hidden multiplication factor is at work: Each re-entry or duplicate summation into a microphone is an embedded reflection in the *source* signal (the direct sound) of the transmission system. When sound from our speaker hits the wall we hear the reflection of a source already containing pre-packaged reflections. This multiplication factor makes us mindful of isolating our mics and minimizing strong reflections anywhere.

The extent to which the sound system enjoys exclusive broadcast rights to the audience depends upon how much stage leakage reaches the house. For the moment we will consider the stage levels negligible in the room.

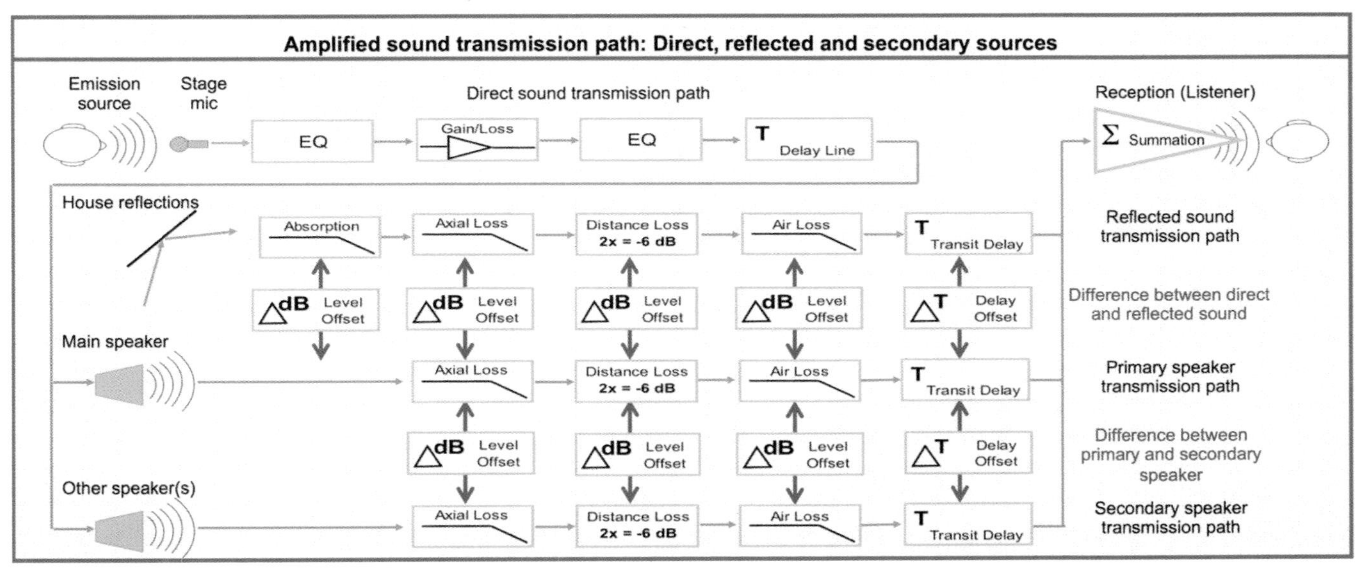

FIGURE 6.3
Amplified sound transmission flow block inclusive of room reflections and secondary loudspeaker sources. The listener hears the summed response of the primary speaker path, its reflections, the secondary speaker's paths and their reflections (not shown). The speaker room/interaction was shown previously in Fig. 6.2. The secondary speaker path is affected by its axial filtering, relative level and time offset. The decisive factors in the listener's experience are the differences between the primary speaker's direct sound, its reflections and those of the secondary speakers.

THE PRINCIPAL SUMMATION RELATIONSHIPS BETWEEN STAGE AND AUDIENCE

- Source/room (stage source reflections with the room/stage surfaces).
- Source/speaker (leakage from stage into the house).
- Speaker/mic (leakage from the speakers into the mics).
- Mic/mic (the summation of open microphone leakage at the console).
- Speaker/speaker (overlap of speaker coverage).
- Speaker/room (room reflections).

Our listening experience is the sum of these summations. They lump together to create the perceived effect, but the techniques for controlling them are distinct. Prospects for success rise if we know which mechanism is responsible for a particular sonic artifact.

Natural sound transmission has only a single avenue of summation: room reflections. The amplified sound system has five more. Therefore the amplified sound model must decrease the role of the room in proportion to the additional summation avenues in order for listeners to perceive an equivalent experience.

Let's follow a single stage-to-house transmission path, a snare drum on stage, through the natural and amplified sound systems (Figs 6.4 and 6.5). The drum is struck and the impulse travels directly to the listener and indirectly (reflected) off the walls. Let's arbitrarily give it this example acoustic response: reverb time of 1 second (-60 dB) with 100 reflection paths spread evenly over that period. Next we acoustically isolate the drum and its transmission into the house. We mic the snare and send it to an omnidirectional speaker that transmits direct sound and 100 reflection paths.

We are perfectly matched so far. Let's add a stage floor reflection. In the natural model this is just one of the 100 paths. In the amplified model it's a multiplier. The reflection enters the mic and is imprinted on the signal sent to the speakers. If the floor reflection is 6 dB down, we can expect to add 90 reflections during the 1 second time period (54 dB of decay would drop it to -60 dB). We are up to 190 audible reflections (100 direct + 90 floor bounce). The drum also enters another stage mic. If this path is 12 dB down, we will gain another 80 or so house reflections (48 dB of audible decay remain, yielding 80% of the reflection paths). At this point, with 270 paths, we have almost tripled the density of *perceived* reverberation without adding plaster. If we remove the acoustic isolation, we add the original natural path and another multiplication path: speaker leakage into the mic.

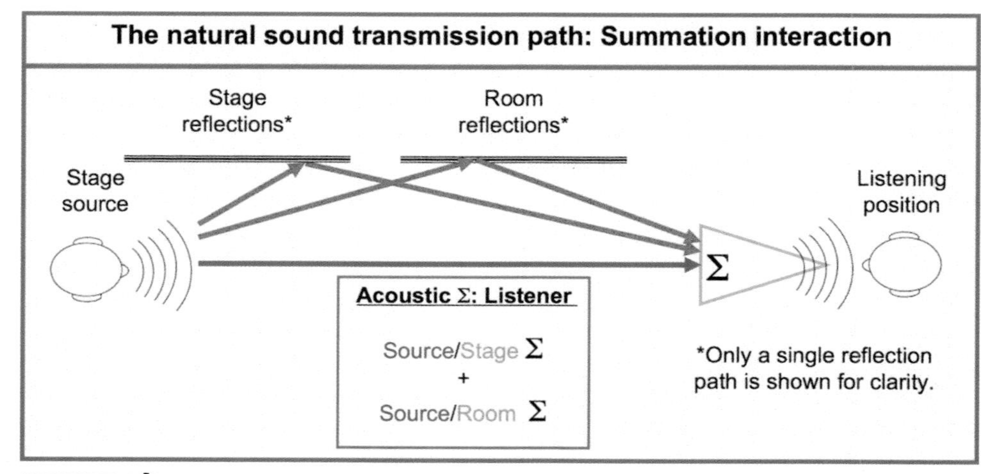

FIGURE 6.4
Natural sound transmission flow block diagram inclusive of all summation path types. Direct sound from the emission source is summed with stage and house reflections. No other paths exist for the single source, although it may be accompanied by other sources that duplicate its signal, e.g. multiple violins or choir members.

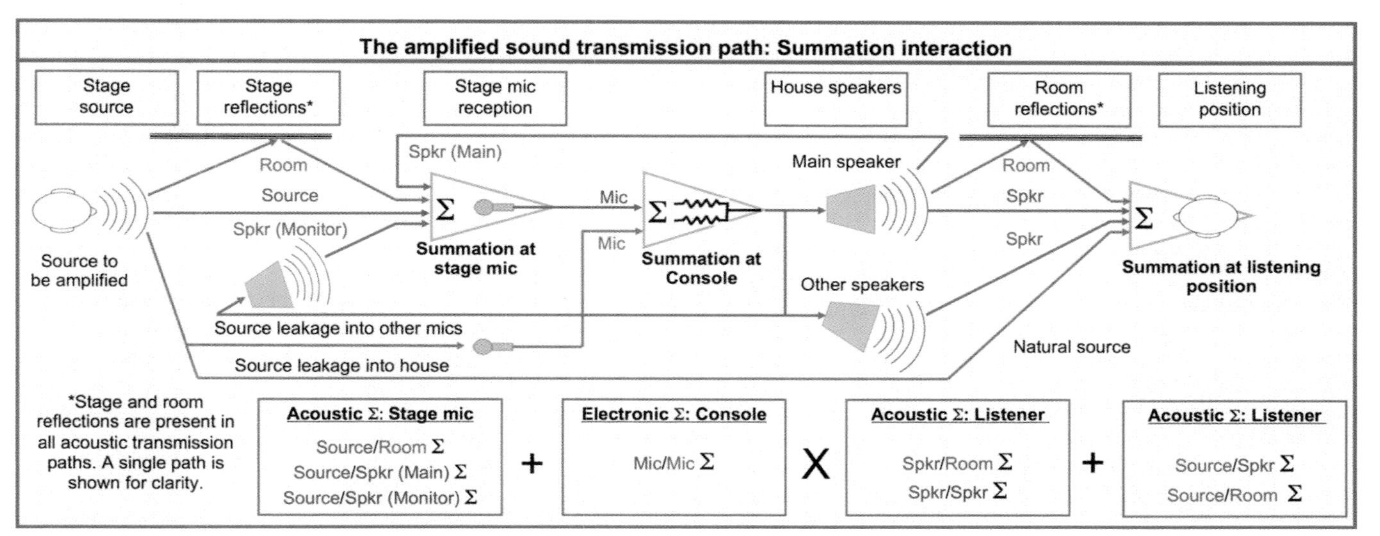

FIGURE 6.5
Amplified sound transmission flow block inclusive of all summation path types. Emission source enters the stage mic and proceeds through the main speaker to the listener. Re-entry summation paths from the main and monitor speakers flow back into the mic. Stage reflections create duplicate entry paths into the mic. Other open mics send duplicates into the mix console. All summations are present in the waveform *entering* the house sound system. Each room reflection of the speaker system signal multiplies all duplicate and re-entry summations contained in the console signal. The final addition is the original source and its own natural reflections.

Hopefully this is a relatively small multiplication. Stage monitors are next and likely to be our biggest multiplier and source of coloration. Let's take an inventory at the mic: direct natural sound, natural floor reflection and leakage from the house and monitor speakers. Leakage from the stage source into other mics (and their floor bounces) is added in the mix console. Getting dizzy yet? Let's send this to the house speaker. We can create 5× the number of reflections without any effort at all. Now imagine more open mics, stage reflections, stage monitors and we are potentially drowning in faux-verberation. Only a fraction of this is actually generated by the main speaker's excitation of the room. What's left to do? Let's add some electronic reverberation!

Openings in the transmission line change the entire perspective about room acoustics. Every re-entry and duplication path into the mics serves as a multiplier of the room's reflection pattern. Reflections accrue only by *addition* with natural acoustics. They accrue by both addition and *multiplication* with amplified sound.

Re-entry or duplicate leakage cannot be removed once it has summed into our transmission line. The waveform traveling to the mic preamp may bear only vague tonal resemblance to the original acoustic source signal due to the combing. The damage is in our source signal. Put in horror movie terms "the killer is in the house," but the problem is we can't get out without bringing him along. These pre-transmission defects are embedded into the speaker signal and distributed equally to all seats.

I had the personal experience of optimizing large-scale sound reinforcement systems for opera megastar Luciano Pavarotti. One might think the biggest challenge would be the horrendous acoustical properties of sports arenas. The far greater challenge was the stage monitor leakage into the singer's mic. When monitor levels rose too high we were left to transmit a vocal channel that was in critical condition before it left the stage.

The implications should be obvious by now. Every effort should be made to reduce the leakage into stage mics. We use directional mics, close placement, noise gates, acoustic baffles and absorbing devices, in-ear monitors and directional main speakers. Even so, there still may be substantial leakage. The transmission signal is already preconditioned with an elaborate reflection pattern, which will be multiplied by the walls. Conclusion: Amplified sound needs fewer reflections *from the room* to create the same perceived reverberation *in the room*.

Let's go through one more thought model to compare and contrast. Every orchestral musician is individually recorded (perfectly) in an anechoic chamber. The bone-dry tracks are mixed together in perfect balance and played back though a single omnidirectional speaker at center stage. All reflections are accrued by addition because there are no re-entry paths. The stage adds its early reflections and the house adds its part. We now have the same number of reflections added to each instrument, as we would with the orchestra live on stage. The difference is that every instrument has the exact same sequence of reflections because they all originate from the exact same location. Let's try again but put a separate speaker on every chair and play back only one instrument in each. We again have the same aggregate number of reflections, but now they are all originating from unique locations and creating unique tonal and decay signatures. It's not amplified vs. natural. It's the funneling down of multiple sources into one pipe that makes the worlds collide.

6.2 ACOUSTICIANS AND AUDIO ENGINEERS

Acousticians and audio engineers are roommates. We don't have to be best friends but we need to get along. We've got pop concerts in symphony halls and opera music in sports arenas. The listening public perceives the *combined* effects of our efforts. Functional relationships are built upon mutual respect and understanding. Our interconnected disciplines must be familiar enough with each other's roles to find creative ways to implement mutually beneficial solutions.

Architectural acoustics is a highly respected profession. Sound system engineering, not so much. Any doubts on this can be clarified at a cocktail party by alternately introducing yourself as an "acoustician" and as a "sound person."

Respect is earned, not freely given. Acousticians were in the room 400 years before us. The first step to gaining respect is to clarify our needs in terms understandable to them. Our roles are separated, but our goals are unified. The means to achieve them are sometimes polar opposites, resulting in a substantial bridge to cross. Acousticians have a long-established language for evaluating the sonic experience. Historically, many of us have deferred to the language of classical acoustics and tried our best to fit our audio planet in the acoustician's universe. Terms that give clear directives for them often leave us blank. Acousticians know a reverb time of 1.4 seconds is better for opera than 1.7 seconds. We see trouble, and more trouble. In either case we design defensively to keep the direct sound off the walls.

Sound engineers maximize control of direct sound transmission, knowing they have minimal control once it hits the wall. The acoustician maximizes control of the walls, knowing they have minimal control of direct sound. We share joint custody of the natural direct sound from stage sources. Otherwise the line is pretty clear. We come first (in chronological order). Our work is done at the time the acoustician clocks in.

We have scattered dialects throughout the audio world. We must translate our needs into the language of the established framework of acousticians. This section is devoted to finding the translation key: a Rosetta Stone for sound systems and architectural acoustics.

6.2.1 Comparing our goals

The best side-by-side evaluation starts with a matched set of specific goals. We will use the esteemed acoustician Leo Beranek's *Music, Acoustics & Architecture* as our reference point. The 1962 book originated a set of evaluative criteria for concert hall performance. Beranek visited halls and measured their acoustic performance, interviewing conductors and critics. The statistical data of fifty-four concert halls was compiled and a categorical performance framework created. Although no singular criterion was found that makes a hall great, certain key trends correlated with success.

Eighteen categories of subjective perception were considered. Some categories were given numerical scoring weights while others were found to be too interdependent to be assigned discrete values. The subjective results were collated with objective measures such as room volume, reverb time, etc. The study correlated subjective and objective data, and the result was a comprehensive assessment of each hall's physical parameters to the listener's experience. From that point forward architectural acoustics moved ahead on a foundation of scientific data.

The eighteen categories remain relevant to our current-day listening experience, whether the transmission path includes speakers, or not. If we achieve the same subjective effect to a blindfolded listener, we have created an equivalent sonic experience. This is the core issue. The physical means for acousticians and audio engineers to achieve matched subjective results are very different. We will have found the translation guide we seek when we can relate the different means of achieving our common goals.

Let's begin by meeting Beranek's subjective parameters (Beranek, 1962, pp. 61–71). I have paraphrased the parameters and generalized them beyond the symphonic experience to be independent of musical genre, venue or sound transmission method.

1. Intimacy: the feeling of proximity to the music.
2. Liveness: fullness of tone in the mids and highs.
3. Warmth: fullness of tone in the lows.
4. Direct sound loudness: Loudness is appropriately scaled to the musical content.
5. Reverberant sound loudness: The reverberation has the appropriate mix of level and duration to provide appropriate additional loudness to the direct signal.
6. Definition, clarity: a clear and distinct sound.
7. Brilliance: bright, clear ringing sound, rich in harmonics.
8. Diffusion: The spatial experience of sound arriving from all directions.
9. Balance: The instruments and voice are in their proper level perspectives. Poor balance favors some instruments over others.
10. Blend: perceived as a harmonious mix of the instruments.
11. Ensemble: how well the musicians can hear themselves. Good ensemble is obtained when the musicians can hear themselves well.
12. Immediacy of response: How well the musicians feel about the responsiveness of the sound. The goal is for the musicians to feel the sound and be able to adapt quickly enough to their changes so as not to disrupt them.
13. Texture: the fine grain of the listening experience. Sound with fine texture has richness and complexity.
14. Freedom from echo: The desired effect is that we do not hear discrete echoes.
15. Freedom from noise: The lowest amount of noise is desired.
16. Dynamic range: the range between the maximum level and the noise.
17. Tonal quality: free from the distortions of peaks and dips in the response over frequency.
18. Uniformity: The extent to which we can create a similar experience for all listeners in the hall.

We have a shared set of goals to keep in mind. Moving forward we will compare and contrast the methods to achieve them. Most importantly we'll investigate areas where methods conflict while seeking common goals.

6.2.2 Characterizing the room

Reverberation results from a mix of volume and absorption. Adding volume increases reverberation. Adding absorption decreases it. Absorption materials are mostly applied to the walls, floor and ceiling. Think of the room as volume, but the absorption as surface area. Reverb falls as percentage of absorptive surface area rises. The late reflection decay that determines the end stop in reverb time is not strongly affected by precisely which particular surfaces are covered. The early reflections, however, are very much affected by the placement of absorption. Stage performers will experience acoustic gain on a lively stage even if the rear wall in the house is absorptive. A dead stage and lively rear wall will have low gain and a potentially troublesome slap back echo. The percentage of reflective/absorptive coverage is critical to late reflection control, while its placement is critical to early reflection control.

Reflective/absorptive surface placement is even more critical to speaker system performance. Relatively small reflective surfaces can have devastating consequences for the sound system if placed just wrong. This is where a potential conflict between the acoustician's plan for early reflection gain and the sound system's need for early reflection immunity comes to the surface. We'll return to this critical issue later.

The elementary acoustic measurement is reverb time (RT_{60}), the time it takes to decay 60 dB from the cessation of sound (section 1.6.4). An omnidirectional source excites the room from various locations and an average RT value obtained. Audio engineers are often surprised that speaker placement has a statistically minor effect, an unthinkable paradigm for sound reinforcement, where moving, panning or tilting a speaker a small amount creates a dramatic change. Why? Consider that an RT of 1.0 sec includes dozens or more reflection paths, including enough round trips to make the starting and ending locations statistically irrelevant. The RT value provides a source-independent generalized room decay characterization. A directional source only changes the numbers slightly. An RT measurement can't tell whether our speaker system is pointed at the audience or the ceiling, because the reflection order doesn't matter. Don't expect the room acoustic measurements to help aim speakers or set your processor.

Other metrics to evaluate early reflections include Early Decay Time (EDT), initial time gap and more. There are fewer leniencies regarding source location for these measurements. We evaluate early reflection arrival *to* any location in the room but not necessarily *from* any location (who cares about early reflections originating from under the balcony). Early reflection data is culled from a reduced statistical basis: smaller time window, fewer paths. Acousticians naturally favor the stage as the primary source location for early reflection data. This has minimal value for our sound design because center stage is the least likely location for the speaker system. Our early reflection concerns are local to each speaker on a very personal one-on-one basis. We do care about early reflections into and from under the balcony. Especially those coming off the ceiling where the underbalcony speaker is hung.

The sound system's perceived tonal response is minimally affected once the reflected signal falls 10 dB (isolation zone summation). Therefore audio engineers have limited application for the last 50 dB of decay shown in an RT plot. Tonal action is in the top left corner of the RT plot, early in time, and close in level. The remainder decay will have negligible influence on equalization, level and delay settings or the fine-tuning of speaker positions. Treatment options for the decay tail (-10 dB to -60 dB) are acoustic solutions (primarily absorption and diffusion), the domain of the acoustician.

There is a time/frequency twist: The relationship of time offset and tonal perception is frequency dependent (longer time at lower frequencies). Tonal perception is strongest at bandwidths of 1/6 octave or less (critical bandwidth) and clearly out of range by 1/24 octave. This leaves a time window of around 5 ms for the 10 kHz frequency range, which lengthens proportionally as frequency falls to around 640 ms at the low end. Equalization is concerned with disturbances within the tonal window. The HF time window is so short that very few early first reflections are included. Secondary paths and late reflection are completely out of the tonal picture. As frequency falls, the time window expands and more of the reflections are included. Even still, the majority of the reflections are outside the window included in optimization decisions.

We've found the principal area where acoustician and audio engineer must work together: strong reflections (0 dB to -10 dB). This small minority wields a lot of potential power. It's highly probable that strong reflections are early and lesser reflections are late but not necessarily assured. Strong late reflections are trouble for acoustician and audio engineer alike. For now we'll assume early to be strong and late to be weak.

6.2.3 Speaker/room relations

Each speaker/room relationship is individually evaluated because speakers are assigned to sections of the room, not the entirety. The natural acoustic model assumes the stage source to be omnipresent, i.e. it reaches every seat in the house. Our speakers divide coverage responsibility into partition zones, with local (early) and global (late) acoustics. Partition shapes are derived by location, coverage pattern and relative levels of the speaker subsystems. Each partition has unique local conditions and its own set of early reflections. Late reflections pass freely through the partitions and remain a shared layer over the whole space. The same rules would apply if the violinists were dispersed into the house to play identical parts in different audience areas.

A frontfill speaker located on center of a high stage lip serves as an example. This speaker most closely approximates the natural stage source (located on stage, wide pattern, line of sound to all listeners). And yet, in practice, the overall room acoustics are less relevant for this speaker than any other. Its transmission will travel no more than four rows before handing over custody to the dominant main speakers. This is the acoustic partition. The mains don't prevent the frontfill speaker from reaching the back wall and exciting the full range of reflection patterns. The mains simply drown them out.

Spatial uniformity is achieved by matching the response within the partitioned areas, and minimizing disturbances in the transition areas. Late reverberation has high spatial uniformity by virtue of its statistical density regardless of the precise aiming of the speakers. Off-axis listening areas feel more reverberant because direct sound is reduced (the reverb level is consistent). This contrasts starkly to the natural acoustic model.

Parameter	Natural	Means to achieve	Sound system
Intimacy	A feeling of proximity to the music, as if listening in a small room. Intimacy is achieved by a pattern of strong early reflections. Early arrivals should be within 5dB and 20 ms of the direct sound.	Strongly opposed	Intimacy is achieved by zonal separation of directionally controlled speakers, which provide matched transmission with high direct/reverb ratio.
Liveness	Considered optimal when the reverberation is long enough to connect the music together without becoming so long that it restricts its ability to change. Symphonic music ranges from 1.5 to 2.2 seconds while opera ranges from 1.5 to 1.7 seconds.	Somewhat opposed	Sound systems connect the music together by mixing it and transmitting it from a small number of locations. Minimal room reverberation provides the needed liveliness for spatial enhancement. Too much reverberation restricts the music's ability to change, especially in the LF range.
Warmth	Achieved by ensuring the LF reverberation time is around 25% longer than the MF and HF. Largely a function of building materials. Plaster and thick wood are recommended	Different but not conflicting	Warmth is achieved primarily by equalization that favors the LF range. System must have sufficient LF power capability to maintain headroom.
Loudness of the direct sound	Sufficient loudness is maintained by arranging the seating as close to the conductor as possible. 60 ft (18m) is seen as the benchmark.	Different but not conflicting	Powerful directional speakers in separate locations provide high level uniform coverage on a scale that the conductor can only dream about.
Loudness of the reverberant sound	The reverberant sound level must scale with venue size. Small rooms should have lower reverb times, while larger rooms need longer decay.	Somewhat opposed	Sound systems for speech and high level pop music must transmit with a high direct/reverberant ratio. Reverb can be added electronically or via variable acoustics to accommodate program material changes. (high ratio for speech and various ratios for music).
Definition, clarity	Clarity is achieved by the optimal mix of intimacy, liveliness and loudness	Somewhat opposed	Clarity is achieved by directional speakers (high direct/reverberant ratios) with defined separation zones.
Brilliance	Brilliance is achieved when the sound is intimate and the reverb time in the high frequency is properly balanced to the midrange.	Different but not conflicting	HF/MF balance is managed with equalization. Spectral uniformity is achieved by zonal separation of directional speakers.
Diffusion	Reflections from diffuse surfaces provide a rich reverberation character. Scattering provides a more gradual and dense reverberation tail.	Strongly Agreed	Same as natural sound
Balance between the instruments	Strong early reflections must have sufficient range and density to balance the intruments at even levels and frequency response. Basses are unbalanced if the room lacks warmth and flutes if too brilliant.	Different but not conflicting	Stage sources are isolated and balanced in the mix electronically. The room does not provide a mixing function unless the sources are transmitted from separate speakers.
Blend of the instruments	A harmonious blend of early reflections is achieved by the positioning and spacing of the instruments.	Different but not conflicting	Instruments are separated on stage and blended in the house via multiple transmission channels. e.g. Stereo and multichannel mixes. The combination occurs in mix console or in the acoustical space.
Ensemble	A sense of ensemble on stage requires the conditions of acoustical intimacy described above. Achieved by strong early stage reflections.	Strongly opposed	Ensemble on stage is achieved by musicians that keep reasonable stage volumes and/or by individually adjustable stage monitor systems.
Response, attack:	The reverberation on stage must be correctly scaled. If too long the room restricts their dynamic changes. If too short the music lacks a continuity between notes.	Somewhat opposed	The initial response and attack comes from individually adjustable stage monitors followed by after effects from the main speakers. Stage must be isolated from the house system leakage (direct and/or reflected) to achieve the desired immediacy of response.
Texture	Fine texture is achieved by carefully spaced and sequenced reflection patterns. Steadily declining level over time.	Different but not conflicting	Texture is controlled globally by acoustic reverberation and selectively on individual signals with electronic reverberation.
Freedom from echo	Freedom from echo is achieved by preventing single reflections to stand out above the decay pattern or focus points to occur from the confluence of multiple reflections.	Strongly Agreed	Speaker systems increase risk of perceptible echoes due to small number of source locations, low reflection density and higher SPL. Concentrated directional beams raise chances of focused reflections.
Freedom from noise	Achieved by isolating the transmission from noise sources such as HVAC, vibration etc.	Strongly Agreed	Same as acoustic model with additions such as electronic noise, feedback, moving lights.
Dynamic range	Dynamic range is maximized by having the gain from strong early reflections and the minimum noise.	Somewhat opposed	Limited at the upper end by two factors: The system's maximum power capability and gain before feedback. Noise limits the bottom.
Tonal distortion	Tonal distortion can be caused by frequency dependent absorption, or via sympathetic vibrations or resonances. The room's tonal content should be frequency neutral (no additions nor subtractions)	Strongly opposed	Tonal distortions unique to electroacoustic systems include comb filtering, harmonic distortion, compression and sonic image distortion.
Uniformity	Uniformity is achieved by coupling all seating areas to the transmission system.	Strongly opposed	Uniformity is achieved by dividing the listening area into isolated transmission zones.

FIGURE 6.6
Comparison of natural and amplified sound models based on Beranek's criteria for subjective evaluation of concert hall sound

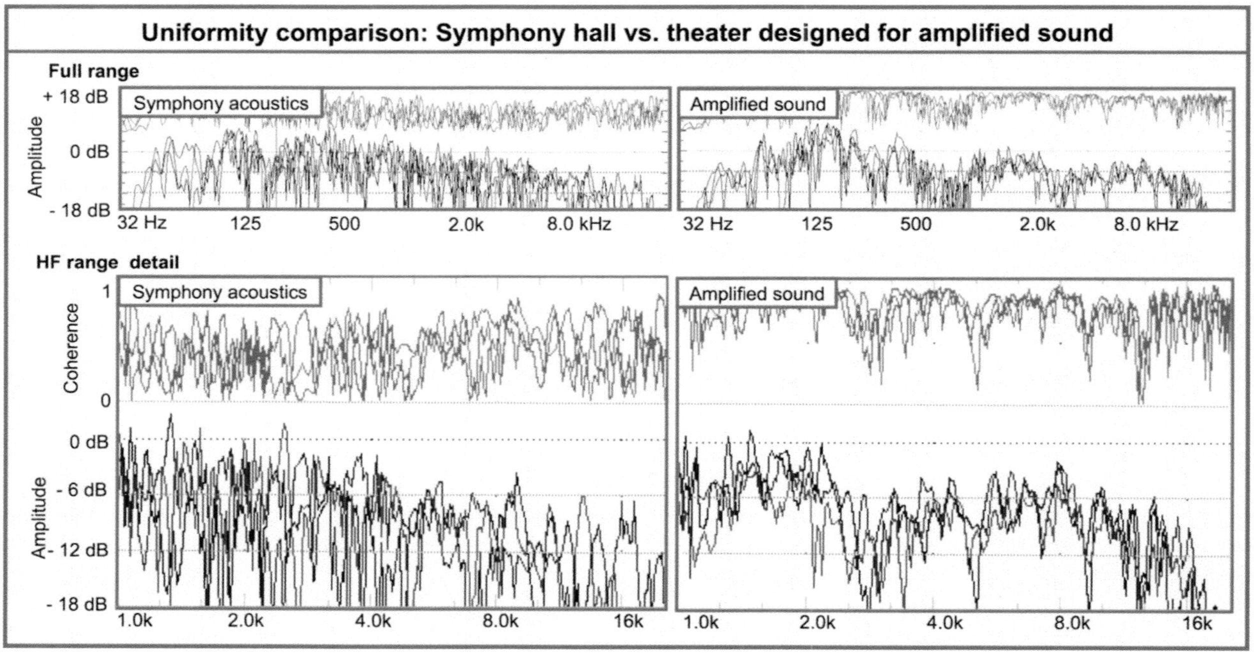

FIGURE 6.7
Comparison of speaker system frequency responses in a highly reverberant symphony hall (left) and venue with acoustical properties more suitable for amplified sound (right)

The means to achieve the goal of uniformity are often diametrically opposed. Coupling, transition and combing zone summation in the natural model accomplish level and spectral uniformity over distance. Front to back level is kept fairly even by different summation zone mixes. Up front the direct sound and early reflections are strongest, but at lower density. At the rear the early arrivals are weaker, but closer together in level and higher in density. Rising reflection density and extended time offsets create combing beyond the ear's frequency resolution, resulting in spectral uniformity and gradual decay of the reverberation tail over the space.

The speaker system model creates uniformity over distance differently, by using coupling (mostly LF) and isolation zone (mostly HF) summation. Directional sources allow partitioning into isolated zones, which can compensate for range differences by selective level settings. Spectral uniformity is also accomplished by isolation zone summation and by minimizing combing. Isolation zone summation is likewise employed to create the gradual decay of the reverberation tail evenly over the space. The speaker system model uses coupling zone summation very sparingly. It's useful only in the frequency ranges (mostly the LF) and locations (the partitions) where it can be employed with minimal combing. The acoustic partitions must be properly joined as phase-aligned spatial crossovers, thereby minimizing the transitional disturbance.

An itemized comparison of approaches to achieve the eighteen perception parameters is shown in Fig. 6.6. There is agreement or non-conflicting differences in nine categories. Commonalities include freedom from perceptible echoes, distortion, noise and desire for warmth, brilliance, texture diffusion and well-behaved reverb tail. Conflicts arise in nine categories with some recurring themes: early reflection strength, source isolation and, to a lesser extent, reverb time.

A field example contrasting halls built for symphonic and amplified sound respectively is shown in Fig. 6.7.

6.3 THE MIDDLE GROUND

Where is the middle ground? Are there mutual solutions? The first strongly conflicting category "ensemble on stage" has long been solved: portable band shell for the orchestra, portable stage monitors for the power trio. That was easy. The following categories require creative solutions: intimacy, tonal variance and uniformity. Each has the same source of conflict: strong early reflections.

The most logical solution is meet in the middle, the "multi-purpose hall" of the 1960s and 1970s. Note that these have all been either re-purposed or de-purposed. Give the orchestra half the reverberation it needs and there will be strong early rejections. These rooms don't work for us either. It's not so much the overall reverb time as the placement of reflective surfaces that create major problems. We must convince the experts to build separate halls for each of us.

> Dear Mayor,
> The citizens of Bikini Bottom want a new symphony hall with perfect acoustics for the orchestra. It will cost tons to build and will lose more money whenever the orchestra plays (correctly priced tickets would be way too expensive). Corporate and private donations will take care of the $hortfall, so no problem there. The orchestra is willing to wear logo-covered suits (like NASCAR drivers). Since the acoustics will be perfect we can book rock concerts (which actually make a nice profit!) and you can do speeches (which makes you a prophet!). By the way, this is cheaper than the football stadium you made us pay for.
> Thanks, SB.

This letter is ridiculous but the economics are real. Symphonies and opera companies are ongoing money losers. There's a good chance of positive cash flow when a rock concert or comedian is booked. Classical music patrons can look down their noses at pop musicians playing their sacred space, but they should appreciate that it helps to subsidize the symphony. Amplified sound can make money. That's why we'll be in every room built from now on.

6.3.1 Variable physical acoustics

The lose/lose path of the middle ground need not be taken. Variable acoustics is the win/win solution. Design the hall with modifiable acoustical features to accommodate the program material. This is now a standard approach for modern symphony hall construction as well as retrofits of existing rooms. The room designs must allow for rapid reconfiguration for acoustic properties appropriate for pipe organ, chamber music, symphony, opera and, finally, sound reinforcement applications. Curtains might drop from winches, wall panels rotate from hardwood (reflective) to soft goods (absorption), and reverberation chambers are opened or sealed. This has the potential to be the *optimized* acoustic design, one that is able to fit the management's needs to fill the hall and also fit the artistic needs of all participants.

A large reverberation range is required to span classical and amplified program material. Priority is typically given to the symphony, even if their share of bookings is small. Addition of variable absorption on the scale required for optimal amplified performance has major economic impact on both construction and operational costs. As a result most such halls are left far more lively than optimal for sound systems. It's extremely unlikely that a hall with optimal symphonic acoustics can convert in a single labor call to one that's too dead for us. We'll take as much absorption as owners are willing to buy us.

6.3.2 Hybrids: combining natural and amplified

The real middle ground is the merger of both natural and amplified sound transmission, the true case of "sound reinforcement." It's a vulnerable situation. Stage levels must be carefully controlled so the speaker system truly plays a supplemental role. If one instrument overpowers the others, the mixer will be obliged to raise the rest in the sound system to keep pace. Things can quickly unravel from there. The sound system needs favorable locations and time alignment to provide plausible sonic imaging to the reinforced instruments. In essence, the sound system needs to be joined to the natural sound as a phase-aligned crossover: meeting in time and in level in the house. The union is lost if either system falls far ahead in either category.

Another hybrid approach is often used in musical theater. A highly directional system is employed exclusively for vocals to maximize intelligibility. A separate stereo system of wide-coverage speakers transmits the music mix, often blending with direct sound from the orchestra pit. Room reverberation is willfully added to the music mix and provides enhanced spatial feeling.

6.3.3 Variable electro-acoustics

We have previously discussed how reflections are modeled as "virtual speakers." The reverse can also happen: Speakers can be modeled as "virtual surfaces" and create the sound of a reverberant space.

Reverberation enhancement systems are distinct from FOH reverb, which only adds reverberation into speakers carrying direct sound. The house reverb adds a realistic tail but is localizable to the speakers (not the surfaces). The singer is in the shower yet we feel perfectly dry. FOH reverb fails to provide the experience of spatial envelopment of a decaying sound field arriving from all directions. By contrast, reverberation enhancement systems increase the actual reverberation time in the room via a complex array of distributed microphones and speakers, which create a multidirectional, diffuse, spatially distributed decay. Microphones over the stage and in the house receive the acoustic signals. This could be direct from a stage source such as a singer, the transmission of the same singer through the sound system or even the off-key singer in the audience. The microphone does not discriminate. The signal re-enters the room via spatially distributed "reverberation source" speakers that return the signal again to the microphones and the process repeats. This is reflection multiplication in the literal sense. Still more multiplication is optionally available in the form of specialized signal processing, further increasing the reflection density. Processing sends unique combinations of multiple microphones to different speakers, further randomizing speaker interactions in the space. This prevents stable summation, creating instead the textured spatial effects we find in the complex reverberation tail of a plaster concert venue. Transmission of this signal by spatially distributed loudspeakers creates the sound of reflective walls in locations where the actual walls may be absorptive (or not there). This approach has stability and credibility limits, assuming we wish the reverberation system undetected. Re-entry summation (speakers into mics) is subject to instability, which in the worst case is runaway feedback. People tend to notice walls that hum, buzz or howl. We must prevent localization to an individual speaker, as that would give away the enhancement system's presence. This is assured by carefully controlled spacing and placement, and visually aided by recessed placement.

The credibility issue links to the concept of suspension of disbelief. Impossible acoustic events arouse suspicion of special effects such as a hidden sound system. Our eyes size up the room and create a scaled expectation of acoustic qualities for the space. Excessive reverberation enhancement pushes beyond the plausible acoustic qualities of the space, leading listeners to suspect trickery. A reverberation enhancement system recirculates all sound sources in the room, including the sounds of audience members. Clapping our hands in a small theater and hearing the reverberation of the Notre Dame Cathedral might force a question: Do you believe you ears, or your lying eyes?

Reverberation enhancement systems move multi-purpose halls out of the fiction section. Halls can be built with reduced physical reverberation and let the electro-acoustic system extend it as needed. Settings are programmable, repeatable and modifiable with a click. Optimal combinations of physical acoustics and enhancement systems can be implemented on a day-by-day, or even cue-by-cue, basis. Both the physical acoustics and the enhancement acoustics must still be expertly controlled. Beginning with good acoustics is essential because the enhancement system only adds layers on top of the original physical acoustics. The optimal approach changes to collaboration between acousticians and our world of microphones, speakers and digital technology. It's increasingly obvious that this is the future of architectural acoustics.

Sound systems can be installed in relatively "dead" rooms and yet have the spatial effects of a "live" space. The presence of reverberation enhancement reduces the room's role in providing diffusion and spatiality for the sound system. The reverberation enhancement system's diffuse spatial field can supplement the speaker system's fully uniform direct-field response. Visualize a wire frame drawing of a theater. As long as we can hang enough reverberation speakers all over the wire frame we've got ourselves "perfect acoustics." If only there was a way to keep the rain out.

The affordable technology of reverberation enhancement redefines the boundary line between architectural acoustics and audio engineering. Spatial enhancement and rich decay character are the most widely sought acoustical properties for amplified sound applications. Those characteristics can now be undersized in the room acoustics and "fixed in the mix." Acousticians' focus on architectural solutions has naturally led them to favor solutions inside their scope of expertise and control; likewise audio engineers. The hall built exclusively for natural acoustics often reduces the audio engineer's options for damage control and emergency mode operation (and truckloads of drapes). The hall with acoustic properties of a pillow lets us mix like we're outdoors on a wind-free day, with all the sound coming at us, rather than around us. Reverberation enhancement moves the line toward erring in favor of excessive absorption, rather than excessive reflections. It's a very substantial scope of work migration toward the audio engineering side, so don't expect it to be welcomed with open arms. Both parties are still very much in the game, however, as the creation of a plausible reverberation character still requires the expertise of those who know what that means: the acousticians. This technology opens up huge possibilities, and its acoustical success will be based upon cooperation between the audio and architectural sides. Reverberation enhancement opens a second avenue for the optimized sound design to meet the optimized acoustic design.

6.4 MOVING FORWARD

We're in the room together, so it's best if we can anticipate each other's needs and perspectives to find the best solutions. Then we can all share the credit rather than pass around the blame. We know what we need, but we can't expect acousticians to read our minds. Let's make it clear by creating a base specification for the architectural acoustic qualities we desire in the hall.

Tips for acousticians designing halls with sound systems

1. Give us a place for left/right mains arrays for any concert/theatrical/performance-type venue. If there is a stage the default thinking should be left/right. Forget the 1962 book about how great center clusters are. Even if a center cluster is in actual fact 100× better, no engineer wants to mix on it and they will go to amazing lengths not to use it. It's fine for a sports arena scoreboard.
2. Limit the horizontal wraparound and the curve on balconies. Wraparound side seating favors a central focal point (e.g. the stage). This is great for visual because the show is in the center. Our speakers are off to the sides, in the face of the wraparound seats. A severely curved balcony gets hot on the outside because the seats get close to the mains. Wraparound can create an occultation on the floor below it. The seats still have visual to the stage but are blocked from the speakers far off to the side. In some cases these seats hear the speakers from the opposite side more than their own.
3. Multi-level wraparounds add another level of challenge because the close seats are spread over an extended vertical range, with lots of obstacles and potential for strong early reflections.
4. Keep the wraparound areas simple. Minimize the staggered, vertical, moving targets, the impossible twisted shapes that require us to break the system into lots of little subsystems.
5. No side boxes. Inset boxes are tough to hit without fills or reflections. Protruding boxes have the same issues and they block other seats and make us need delays.
6. Just say no to hiding the main speakers. Every living person knows speakers exist. Patrons don't care about seeing them, but do care about hearing them. Only the architects have this backwards. As experts in acoustics you know that hiding speakers in soffits behind scrims has no acoustic benefit and 100% probability of degradation. You should be the one leading the "Free the speakers!" campaign.
7. Give us reasonable main speaker positions and make sure people can see them. Every seat that can't see the mains adds cost because we must add fill speakers, channels, conduit, etc. for somebody to sell that seat. Good positions allow us to cover most of the house with the mains and retain a reasonable sound image connection to the show. If pop music is to be performed these must be left/right positions. Resistance is futile.
8. Keep the macro shape simple. Main systems have a limited menu of shapes. Rectangles, trapezoids and fans in the horizontal. Diagonals in the vertical.
9. Think like a speaker. We can't start and stop on a dime, cover two seats, miss two and start again.
10. No battleships. Don't let highly reflective floating structures pop up in the middle of seating areas. We have to cover the seats, and our speakers are not nimble enough to avoid the collateral damage of these reflections.
11. Don't make a fat lively balcony front that focuses back onto the stage or down into the seats. If it must be lively, make it scatter or go upward.
12. Don't put surfaces above the uppermost seats that focus reflections downward. Dead walls, reflected up or scattered, are better than reflected down.

13. Don't make rooms go wide and then get skinny. We must cover the nearby wide area and can't make the coverage shrink for the rear. This is true for both planes. The surface area of our sound waves expands over distance. An expanding room keeps pace with us. A shrinking room creates an unavoidable collision.
14. Suck up as much low end as possible. We'll let you know if we ever encounter a room too tight in the low end.
15. The worst surfaces to make highly reflective (in order): inward angle near the speakers, rear wall, inward side wall.
16. Kill the rear wall (especially down low near the seating). We can't set a range on our speakers and tell them to stop after the last row. A big lively wall right behind where we need to cover leads to two bad outcomes: rear seats uncovered or slap echoes off the rear wall into the house (or both).

A picture is worth a 1000 words and takes up a lot less space. Here are some halls of fame that can illustrate the challenges we face with our sound systems (Fig. 6.8).

By the way, the dodecahedron speaker lasted one listening test and now there are curtains added whenever the speaker system is used. Optimized sound system and optimized acoustics.

FIGURE 6.8 (A)
A large circular room with extensive under- and overbalcony spaces. The first challenge (1) is a high, reflective dome ceiling. There is simply no acoustical upside to a domed ceiling and every seat below gets the downside (pun intended). (2) Extremely deep, low clearance under- and overbalcony areas extend outward beyond the domed part of the room. These spaces are acoustically uncoupled from the central domed part of the room, unable to see the main system, thereby requiring multiple delay rings. (3) The tall, radial balcony front reflects a coherent wavefront back toward the stage. (4) The balcony wraps very far around and is staggered, adding to the challenge of covering all seating areas while avoiding the balcony reflection

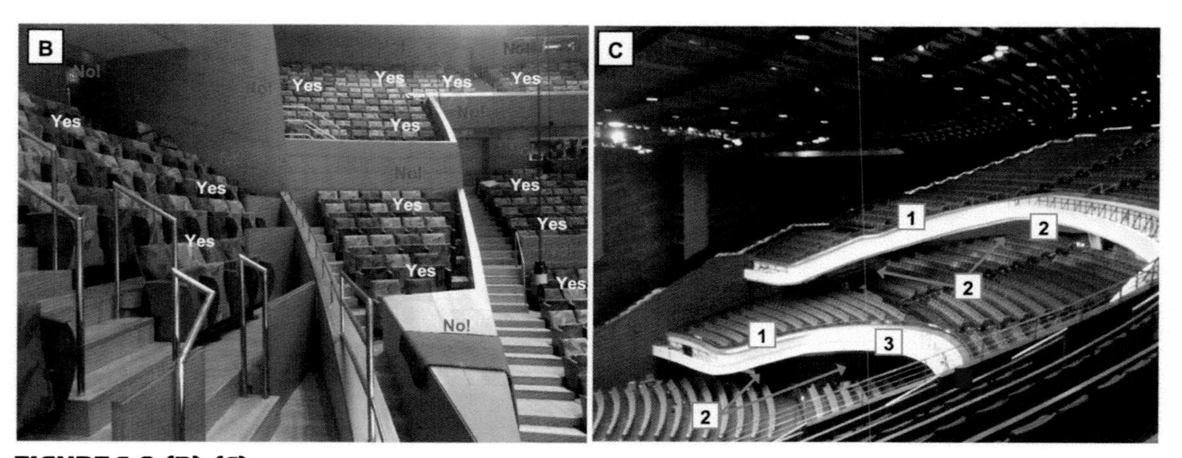

FIGURE 6.8 (B), (C)
(B) A complicated room with a variety of reflective areas mixed in with the target coverage area. Loudspeakers have a limited menu of shapes they can create. They can't stop and start or turn on a dime. Cover here (Yes), then skip this section (No!) and then resume coverage again **(C)** An extreme version of the continuously sloped radial balcony. (1) The extreme curvature, steep downward slope and very low underbalcony clearance create the need for extensive delays (four levels deep in places). (2) The line of sight to L/R positions is blocked in a huge number of seats on the 1F and 2F. (3) The balcony front focuses reflections back on to the early rows and stage

FIGURE 6.8 (D)

Domed walls line the inside of a domed building whose primary function is speech presentations from a podium. (1) All walls are reflective and return multiple strong reflections to every seat from every direction. (2) The seating area narrows at the rear of the room, an impossible feat for sound propagation, ensuring more spill on to the walls. (3) Forcing the sound system to hide in a soffit only makes things worse. But there's more: (4) It's behind the podium.

FIGURE 6.8 (E)

A small minority of seats can have an outsized impact on the sound system. We have to cover EVERY seat so tiny islands of seats are a serious issue. (1) A "president's box" juts out of the side of the balcony (2) blocking seating below and behind. The five seats in the box need to be covered, as do the blocked seats. (3) A side seating area of six seats requires coverage. We must aim the mains there (reflection risk) or add multiple fill speakers. (4) There is a huge flat reflective side wall just above these seats, which means we need specialty coverage with a tight vertical pattern, i.e. we can't hit it with the same speakers that cover the balcony. These twenty-two seats comprise <2% of the seating capacity

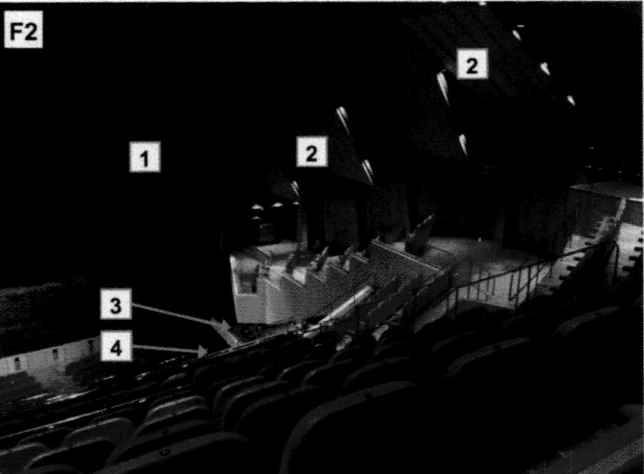

FIGURE 6.8 (F)

Reflecting panels are placed in positions that create trouble for a L/R main system. A center cluster would have a better chance but (please make a note of this) nobody wants center clusters. (1) Huge downwardly angled reflector panels are located very near the L/R main positions. Nothing good can come from this (for the sound system). (2) More downwardly angled reflectors are lurking right over the wraparound balcony. We can't cover those few seats without trouble for a lot of seats. (3) Wraparound is staggered with low clearance and blocks line of sight to 2F seats and (4) also blocks the 1F seating.

FIGURE 6.8 (G), (H)

(G) This room has many large reflective surfaces in locations that will be difficult to avoid. (1) Side balcony boxes are staggered down to make a moving target. (2) Large, flat reflective surfaces are above the staggered boxes. Tiny numbers of seats are nested among giant reflectors. (3) The upper side of the balcony in front of the boxes reflects right into the faces of listeners, giving them a personal combing zone. The extreme wraparound puts the lower boxes right in the face of the left/right mains, which must blow through them to get to the rear of the first floor. (4) The center section of the balcony is extremely thick and reflects down into the underbalcony area. (5) Another giant reflector brings sound down from the ceiling **(H)** Never marry a symphonic orchestra. They will run off with a younger hall and leave you with a lifetime of reflection. Here we have classical leftovers in a room that hosts 100% amplified sound events. The photo approximates the point of view of the house right mains. We can play the yes/no coverage game here as we did in example hall B. (1) There are battleship boxes with a tiny slice of seats protected by reflectors on all sides. (2) There are staggered lower-side boxes whose reflective fronts are on axis to the L/R mains. (3) A very tall and lively back wall is behind the 2F seating. (4) The upper level horizontal width is reduced by a flat reflective wall that we must hit to get to the other seats at that level

FIGURE 6.8 (J), (K)

(J) A complex hall with only the center location available for sound. Seating is 360° around the room and the height of the seating is the ultimate moving target (as are the highly reflective areas). (1) The two balconies are segmented into small sections at different elevations. The 2F side box is the same elevation as the 1F rear center seats (and the same for 3F side and 2F center). (2) Large reflection hazards are found above and below each staggered segmented seating area. (3) Side boxes are two rows deep (a tiny vertical target) with flat walls above and below, reflecting back onto the stage. No simple aiming solutions apply and margin for error is low **(K)** This is similar to the hall (J) above with but with only one balcony and an additional twist: The left side of the room is different from the right. (1) Seating areas are a staggered moving target. (2) There are down-angled reflectors just above the seating. (3) Staggered balcony fronts cannot be easily avoided. Bonus: Sightlines require high speaker trim and many floor seats lack line of sight to the mains

FIGURE 6.8 (L), (M)
(L) (1) A simple, continuous rake in a gentle fan shape, an ideal shape for the modern L/R line array. A single vertical aiming solution works for the whole room. (2) Reflective wall near the L/R main positions would benefit from acoustic treatment. (3) The available position for delays is not deep enough into the room to be useful. Fortunately, they were not needed **(M)** A small arena with uniform seating arrangement. The shape allows for an easy single aiming solution (1) for an uncoupled array spread along the catwalk. (2) Although the room has a domed ceiling, the target shape and speaker locations make it easy to evenly spread direct sound over the whole seating area

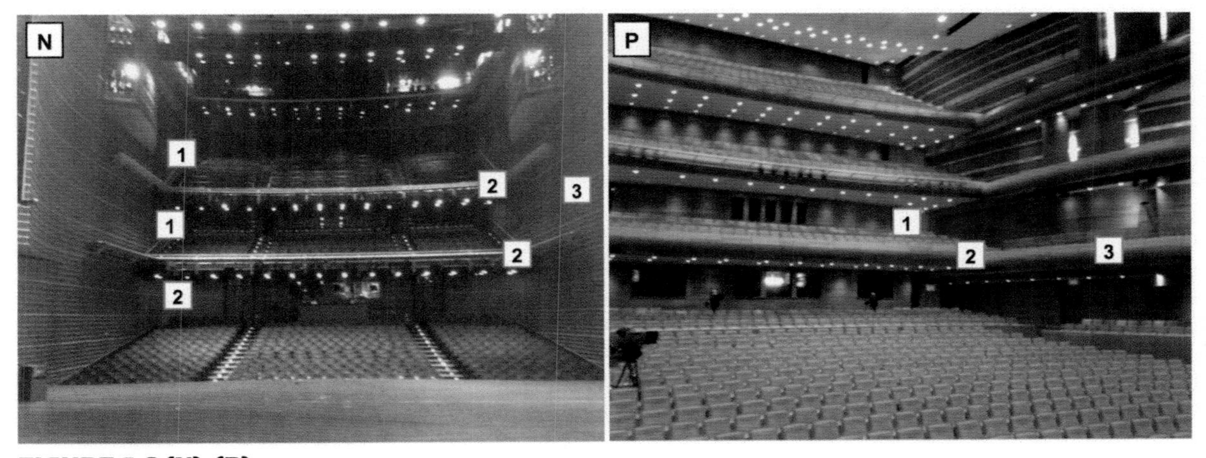

FIGURE 6.8 (N), (P)
(N) A three floor shoebox hall. (1) The side walls are acoustically treated (variable physical acoustics). (2) The balcony clearance ratios leave us with good line of sight, minimal need for delays and minimal impact on how mains are aimed. (3) The balcony front is very thin so there is minimal risk of reflections, which also eases the aiming **(P)** This is a three-balcony version of extended wraparound side seating. (1) Again the balcony clearance ratios are favorable for minimal delay. (2) The balcony fronts are a bit larger but are rounded to scatter the reflections. (3) Wraparound side seating always presents a challenge but this one maintains the same elevation as each balcony (a much easier target for our speakers)

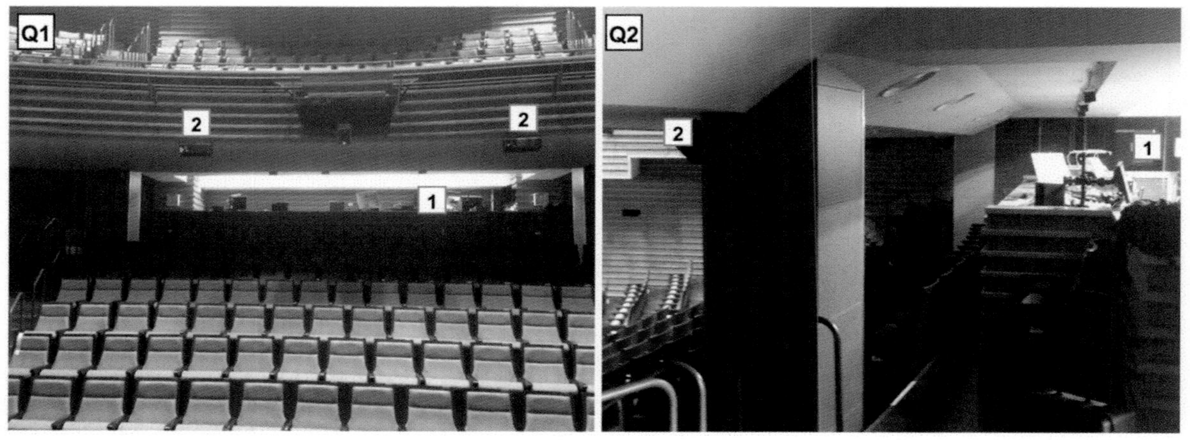

FIGURE 6.8 (Q)
We are happy to know that our sound engineers are safe in their bunker. They won't be able to hear any complaints from the audience (because they can't hear anything at all from there). (1) The booth has no line of sight to the main system and only barely gets coverage from the tiny underbalcony delay speakers. (2) Somebody got paid to design this (and it wasn't a sound engineer). They obviously don't know the first thing about sound system design (because the first thing is NEVER put the sound booth in an underground bunker)

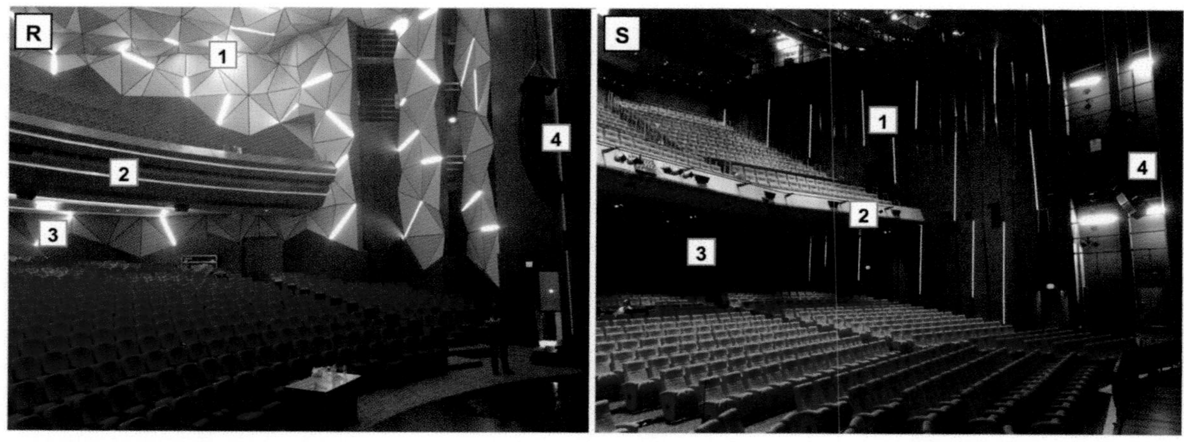

FIGURE 6.8 (R), (S)
(R) A hall built for amplified sound, each with a single balcony. (1) Underbalcony clearance allows for unobstructed line of sound to the L/R mains at all seats and very minimal underbalcony assistance. (2) The balcony front was fairly thick, but diffuse (did not focus a reflection) and has minimal wraparound. Side walls were absorptive in the HF and fairly avoidable due to the lack of shape complexity. (4) The arrays were allowed to be in the open and hung low enough to steer over and under the balcony. **(S)** The hall's acoustic design included a variable acoustic system (electro-acoustic architecture) from the start, along with large-scale sound reinforcement. The RT of the room itself was extremely short but it can be greatly extended without tonal coloration of the sound system. (1) Underbalcony clearance required no delays and had 100% line of sight to the mains. (2) Balcony front was thin and angled to aim the reflection harmlessly upward. (3) Side and rear walls were full-range absorptive. (4) The sound system is openly visible and placed at a favorable height. There were no hard reflective walls near the L/R positions so the sound system had minimal tonal coloration. My favorite room ever (at time of writing).

FURTHER READING

Beranek, L. L. (1962), *Music, Acoustics & Architecture*, John Wiley & Sons.

CHAPTER 7

Prediction

Prediction begins with a vision of design goals and (hopefully) ends with a successfully optimized design. This is true whether it's done with the help of CAD drawings and sophisticated software, a napkin and a protractor or a crystal ball and Tarot reader. Prediction allows us to investigate various scenarios without need of a rigging call. We select speaker models, locations, focus angles and perhaps even processor settings. The predicted design is installed and optimization begins.

predict *v.t. foretell, prophesy.*
Concise Oxford Dictionary

Field measurements may prove or disprove the predictions. Our credibility as designers hinges on whether the aligned system meets the design goals. A successful design needs only fine-tuning during optimization. A failed design requires costly and time-consuming changes as problems are revealed during optimization. Failures may result from inaccurate predictions or mistaken readings of accurate predictions. Either looks the same on our analyzer.

It's a circular relationship: measurement and prediction. After all, prediction data come from laboratory measurement and the system measured in the field comes from prediction. Field measurement tests the design's underlying theory. Successful parts are proved and reinforced. Areas where measurement disproves prediction show we need better predictions or a better understanding of them. The processes elevate each other in an ongoing cycle of discovery.

Prediction and measurement tools must share a common language. Predictions must depict sound propagation as we measure it. Measurements must depict sound as we perceive it. The ear is a very high-resolution analyzer, which sets a high bar for prediction and measurement.

The process is fundamentally the same whether we use a 3-D computer-aided design (CAD) program and acoustic modeling or a pencil and napkin. It's a two-part equation: drawings and predicted acoustic responses. The goal is to determine the best speaker type, placement and aim for the space defined by the drawing. The shape of the space is defined in the drawings, the shape of the speakers by protractors, rulers or prediction programs.

What we need from the drawings

- The macro shape of the building interior.
- Stage location.
- Audience location.
- Sightline exclusions (no speaker zones).
- Where we can place speakers (and/or can't).
- Surface details (reflective areas, absorptive areas).
- Variable configurations (orchestra pit, stage extensions, moving walls, etc.).

What we need from the sound system predictions

- Speaker free-field transmission (coverage and max SPL).
- Speaker/speaker summation.
- Speaker/room summation.
- Transmission medium effects (air).

7.1 DRAWINGS

Drawings range from blueprints to CAD files (.dwg, .dxf, etc.) in 2-D and 3-D formats. The most common 2-D drawing types and their functions are shown in Figure 7.1.

The relationship of the plan and section view is shown in Figure 7.2. If you are working with 3-D CAD drawings you won't need me to explain how to read them.

Parameter	X/Y	Viewpoint	Function
1F Plan	L/W	Looking down on the main floor	Speaker locations, horizontal coverage, aim
2F plan	L/W	Looking down on the upper floor	Speaker locations, horizontal coverage, aim
Reflected Ceiling Plan	L/W	Looking down through the ceiling	Speaker locations, mounting details
Section, Longitudinal	L/H	From horizontal center to side wall	Speaker locations, front/back vertical coverage, aim
Cross section, Transverse	W/H	From front/back center to stage or rear wall	Speaker locations, side/side vertical coverage, aim
Interior Elevation	L/H or W/H	Interior wall surface (side, front or rear)	Speaker locations, mounting details
Exterior Elevation	L/H or W/H	Exterior wall surface (side, front or rear)	Speaker locations, mounting details

FIGURE 7.1
Standard 2-D drawings

7.1.1 Vertical plane

Vertical plane coverage spans top to bottom and front to back. The standard design goal is uniform coverage in this plane. The speaker's vertical pattern and the room's section view are overlaid to analyze coverage. The target is the line connecting the listener head height, shown running front to back at various levels. Coverage gaps in the air above the audience or overlap below the listener's feet are not of concern. Head height wins. The vertical plane is fairly straightforward because the source-to-listener distances shown are generally close to the actual ones. Designing for uniform coverage at the listener level will likely yield the expected result.

7.1.2 Horizontal plane

Horizontal plane coverage spans side/side and front/back. This is much less straightforward than the vertical plane. The overlay of the speaker pattern and the room's plan view show the response as if the speaker were placed at ear height in a flat room. This can be fairly accurate for the frontfills but very unlikely for the mains (because we fly them). The target is the seating area as viewed from above, so there is no indication of head (or speaker) height. Horizontal plane coverage is a continuous function, i.e. it plows through the listeners from front to back. We reach the third row after passing over the second row and so on. We didn't see this in the vertical plane, i.e. we don't reach the first-floor seats after passing through the second floor. Horizontal coverage can appear terrible (and be good) and vice versa. The most common result is that the front seating appears to have severe hot and cold areas when in fact the coverage might be quite uniform. A 2-D horizontal speaker/room overlay will need to factor in the offsetting effects of the vertical plane (much more than vice versa). Horizontal target shapes are areas, whereas vertical targets are lines.

7.1.3 Listener plane

Some prediction programs merge the above planes and display the response on the "listener" plane. This requires 3-D modeling to compute but can be displayed as simple plan and section views. The triangulation paths are factored in and the response is shown side/side and front/back as found at the listener level. This greatly reduces the mental load required for designers to factor in the effects of planar mismatch.

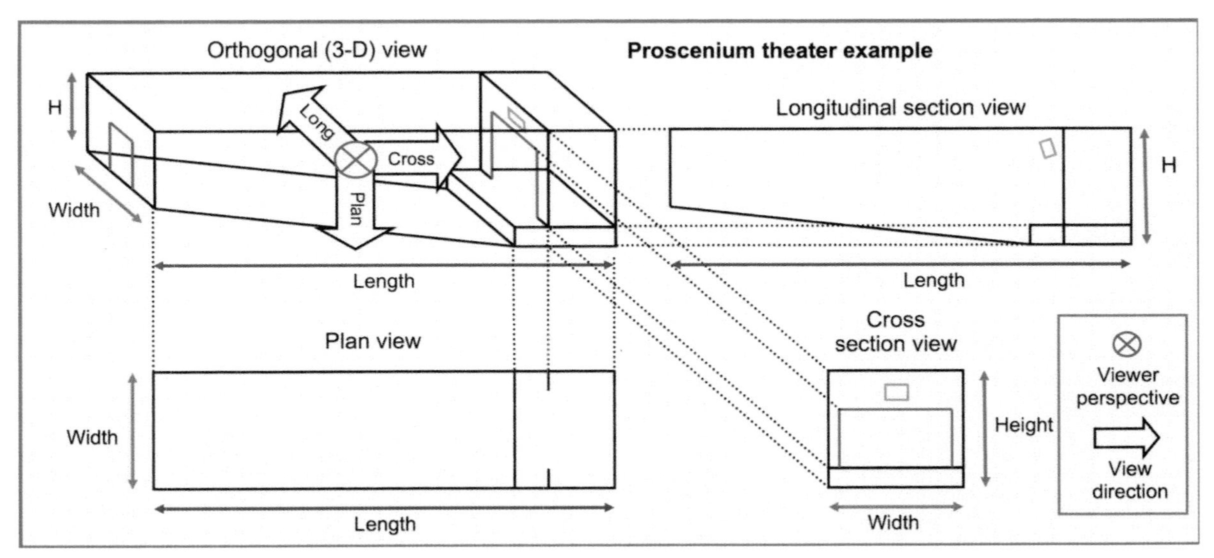

FIGURE 7.2
Relationship of the principal views

7.1.4 Drawings: 2-D in a 3-D world

The 2-D viewpoint is comprised of overlays of vertical and horizontal slices. This has its limitations because sound propagation is spherical and the room is in 3-D. It's not too difficult to visualize coverage accurately when the propagation plan and prediction plane are matched (or nearly so). Let's look at a sample system and see matched, and mismatched, planes in action (Fig. 7.3). In panel A1 we see frontfills and sidefills in plan view. They are both aimed nearly flat to this plane, and the floor in front is also quite flat. Propagation and modeling planes match. The frontfill coverage appears even across the front (and it is). The sidefill appears loud in front (and it is!). In panel A2 we see the underbalcony speakers. They look like frontfills but it seems really loud and uneven in their first row of coverage. It's not. The underbalcony speakers are mounted above the modeling plane and aimed down. This is where we need translation to get an accurate feel. Panel A3 shows a section view of these three subsystems (with the mains off). No translation is needed because the planes are aligned. It looks loud in front and under the balcony because it would be (until we turn the mains on). Let's add the center mains and check the plan view in panel B1. The mains appear to be killing us in the front row. Not likely because they are above the proscenium. Planar mismatch. The sidefills and underbalcony systems also appear loud in front. Sidefill is but underbalcony is not. The final panel (B2) shows an accurate view of the vertical uniformity of the mains, frontfills and underbalcony as they would propagate along the center of the room. The sidefills are closer (and farther), so the relationship to the center mains is not perfectly represented.

The translation process relies on finding the vector transmission path whenever planes are mismatched. The process is shown in Fig. 7.4. Here we have the downfill sections of four clusters in a fan-shaped room. We need to know how these cover the floor so we can select the correct quantity and spacing of the mains. Do we need two, three or four? We know the height because it was determined by video (of course). We also know the frontfills will cover to the third row. We need an arc of coverage at a 5 m depth from the stage (the third row). Let's first look at the downfill in section (A). Everything is clear and simple for both downfill and frontfill. In plan (B) the downfill appears to have coverage gaps at the third row. The downfill main is aimed 40° down, close to the worst-case mismatch of 45°. The vector distance from the mains is 7 m to the third row. We could move speakers back in the plan view by 2 m and then see if they close the gap at the third row. The alternate is to translate the positions to

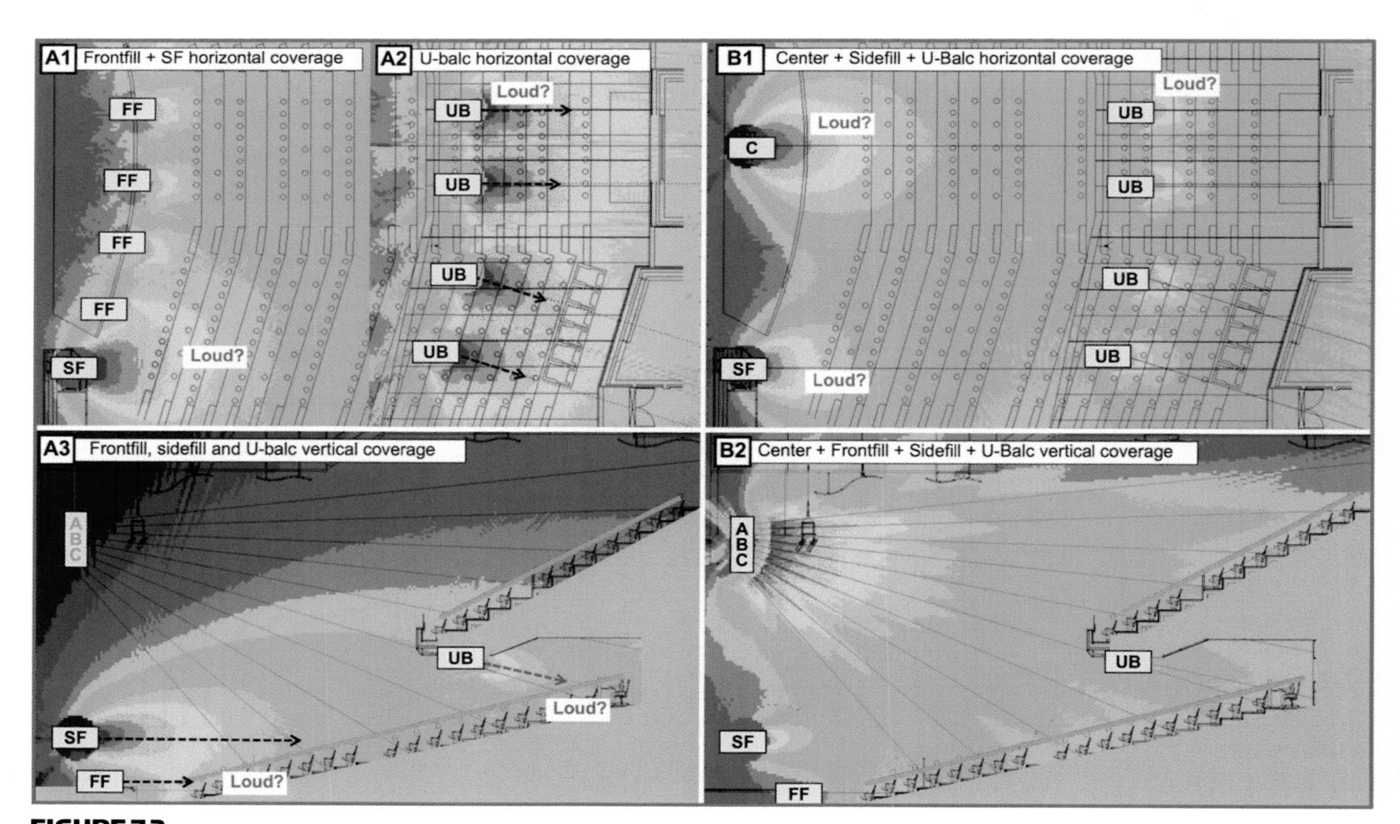

FIGURE 7.3
Example application of speaker system predictions where some are planar matched and others are mismatched

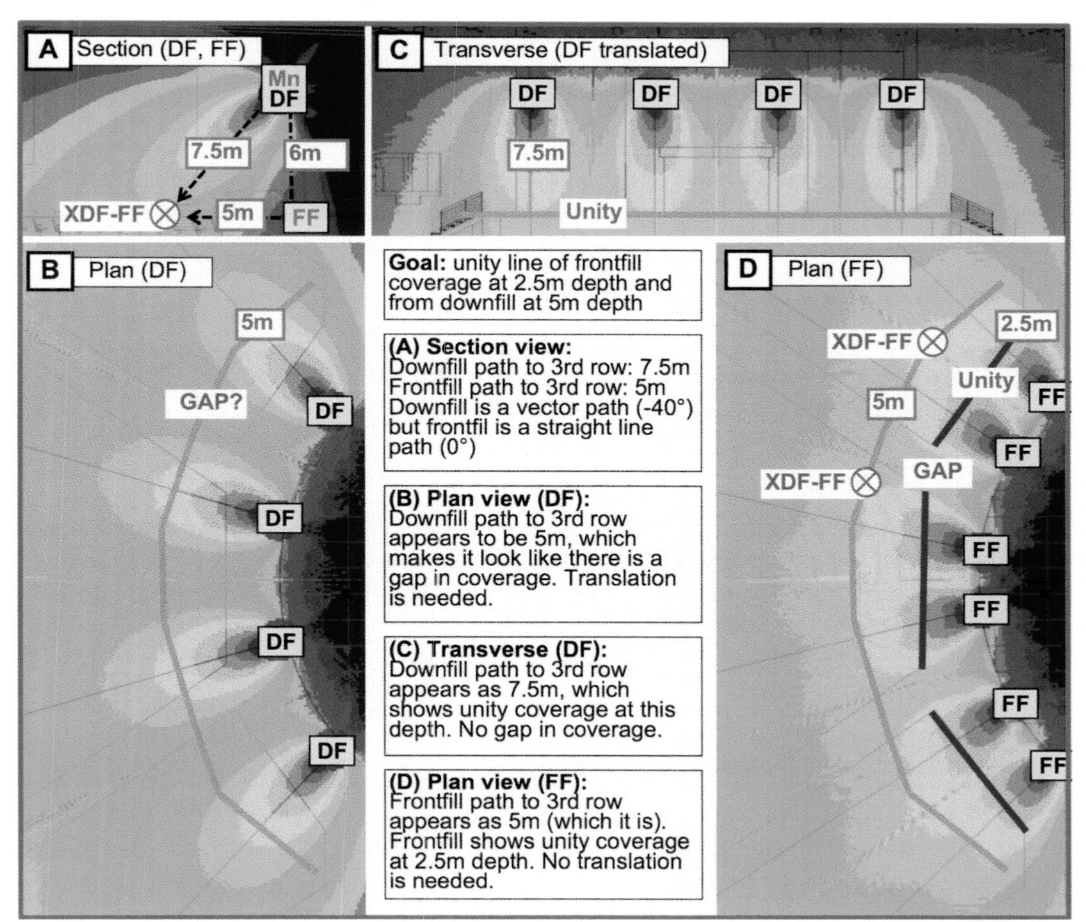

FIGURE 7.4
Translation process for an example downfill system

the transverse view and aim the speakers straight down. The transverse view translation to 7 m (C) shows that the spacing makes an even unity line at the desired range. Case closed; we need four clusters. The final panel (D) shows the frontfill coverage closing the gap at the first row (as designed). We need no translation for that.

7.2 ACOUSTIC MODELING PROGRAMS

Acoustic modeling programs provide advance information about speaker performance in the space. Modeling is based on measured loudspeaker performance, acoustic transmission through air and measured reflective properties of building materials. Those properties are fed into a math engine that simulates the speaker interactions in the space. Prediction accuracy is limited, even with perfect equations and perfect representation of the room dimensions. Modeling is based on a finite number of data points to describe an infinitely divisible entity. A key question is whether the resolution is close enough to our ear's perception capabilities. The acoustic effects of the second coat of paint are inaudible to us, so fuggettaboutit! We only need enough information to select speaker model, quantity, array type, placement, focus angle, signal processing allotment and acoustic treatment. The properties of transmission, summation and reception must be accurately calculated because they play a critical role in these choices.

7.2.1 A very brief history

Modeling programs were initially developed by speaker manufacturers and each maintained closed libraries exclusively for their products. This left consultants (who aspire to be non-partisan to manufacturers) holding protractors and scale rulers while speaker companies dazzled clients with color renderings of speaker performance. To make matters worse (for the consultants) the speaker companies provided predictions to clients for free, creating a paradigm shift in the role of consultants and manufacturers. The lack of standardization created serious (and well-founded) concerns about whether predictions were being generated from acoustical or marketing research. A push for standardization arose in an effort to level the playing field. Standardization would allow designers to access data libraries of competing manufacturers and compare and contrast speaker performance on equal terms.

The foremost of the commercially available acoustic modeling systems is EASE™, originally designed by Dr Wolfgang Ahnert. EASE has an open library of competing speaker manufacturer data and as a result presently enjoys a wide user base. Some companies have agreed on a common data format usable by several new modeling programs from independent companies.

Overall, the programs fall into three basic categories:

- Low resolution 2-D/3-D: 1 octave, 10° angular, optional phase (CLF1).
- Medium resolution 2-D/3-D: 1/3 octave, 5° angular, optional phase (CLF2, EASE™).
- High resolution 2-D: 1/24 octave, 1° angular, phase (MAPP Online™).

7.2.2 Speaker data files

Acoustic modeling begins with the speaker data file obtained in a reflection-free environment, such as an anechoic chamber. The speaker's direct sound transmission is simulated from this data (coverage pattern, frequency response, SPL, etc.). The end product is a table of amplitude and phase values over angle over frequency. File size varies greatly by resolution and rises sharply with 3-D files

Frequency resolution, angular resolution and file size (one plane is assumed symmetric)	
2-D, octave, 10°: 10 × 52 = 520 data points	3-D, octave, 10°: 10 × 614 = 6140 data points
2-D, 1/3 octave, 5°: 30 × 106 = 3180 data points	3-D, 1/3 octave, 5°: 30 × 2522 = 75,660 data points
2-D, 1/24 octave, 1°: 240 × 538 = 129,120 data points	3-D, 1/24 octave, 1°: 240 × 64,080 = 15,379,200 data points

The source data limit the predicted response resolution. Coverage pattern predictions are continuous lines interpolated from individually measured speaker data points. Interpolation "connects the dots" between knowns, a mathematical "leap of faith." Interpolation is a fact of life in acoustic data and our issue is simply a matter of degree (in some cases literally). Accuracy falls as the interpolation span increases. We need enough angular and frequency resolution to make good design decisions.

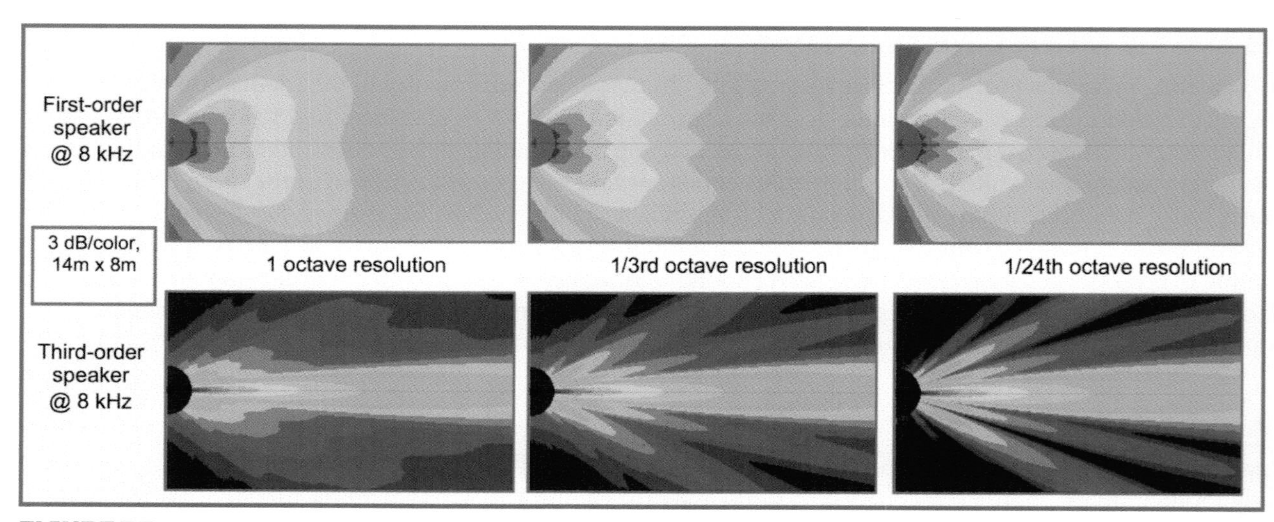

FIGURE 7.5
Frequency resolution issues related to characterization of a single speaker. Top: First-order speaker shows only minor differences between low and high resolution data. Bottom: Third-order speaker shows substantial differences in side lobe details.

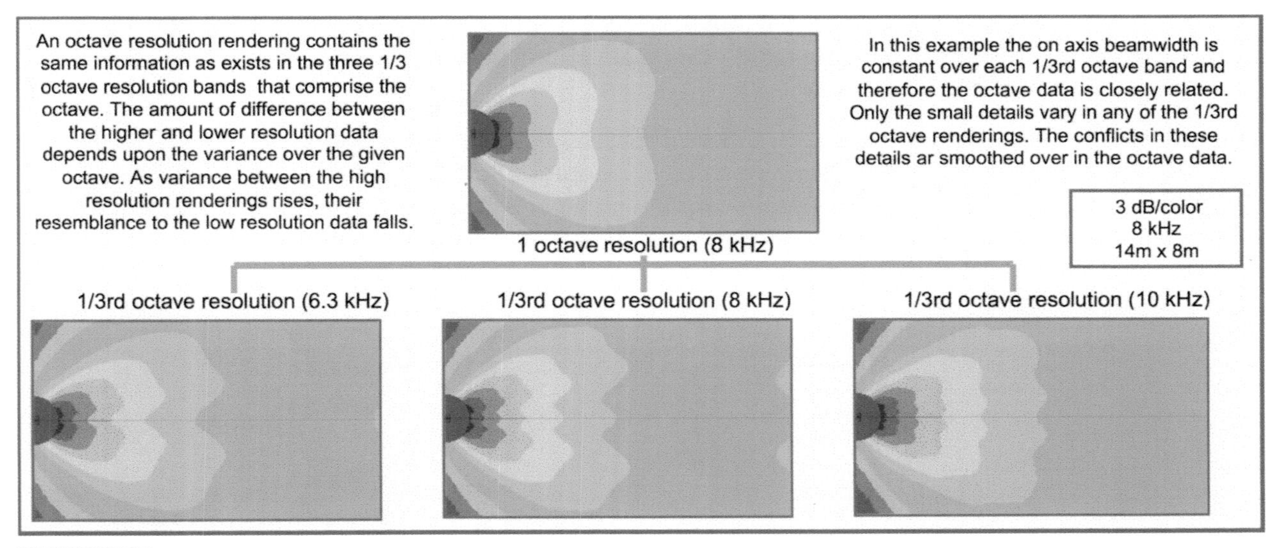

FIGURE 7.6
First-order speaker at octave resolution compared with the ⅓ octave components that comprise the same frequency range. The differences are confined to the details.

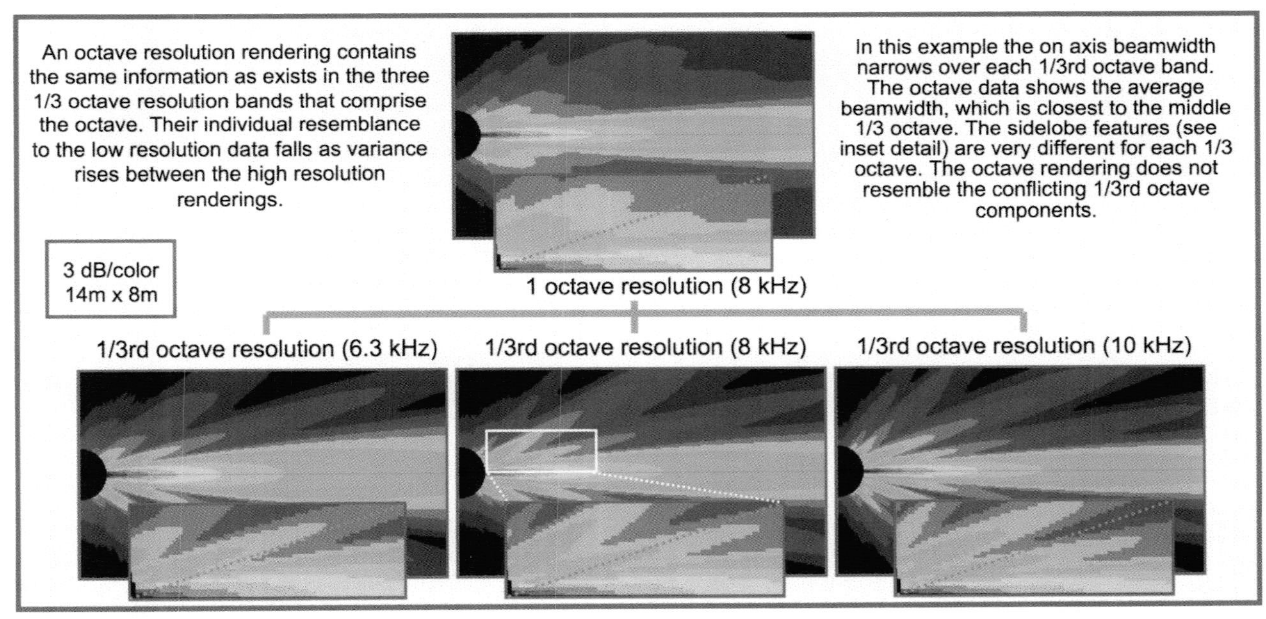

FIGURE 7.7
Third-order speaker at octave resolution compared with the ⅓ octave components that comprise the same frequency range. The differences are found in the side lobe details and in the beamwidth.

7.2.2.1 FREQUENCY RESOLUTION

Frequency resolution sets the layer count for the coverage plot series (1, ⅓ or 1/24 octave). Frequency-dependent variations in coverage are revealed in the layers. The acquisition resolution limits the display resolution. The display resolution can be less than the acquisition by averaging layers together to get a composite display over the desired range (Figs 7.5 and 7.6). Data acquired at ⅓ or 1/24 octave can be shown at any lesser resolution (three slices of ⅓ and twenty-four slices of 1/24 yield octave resolution). By contrast, octave data can't be converted into higher-resolution displays. That's called "making it up" and requires a degree in marketing.

The most forgiving case for low-resolution data is a well-designed low-order speaker with a relatively stable shape over frequency. Increased resolution would show insignificant differences. If the speaker is not well behaved the differences become significant. How can we tell a good speaker from bad? High-resolution data. As speaker directionality increases the shape changes more rapidly and maintaining stability becomes more challenging. The value of hi-res data rises with speaker order (Fig. 7.7).

Low-resolution data smooths over the combing of summed speakers and removes it from view. This creates a false equivalence between design strategies with very different amounts of combing (Figs 7.8 and 7.9). High-res frequency data have enough detail to show audible variations, whereas the low-res predictions remove the evidence.

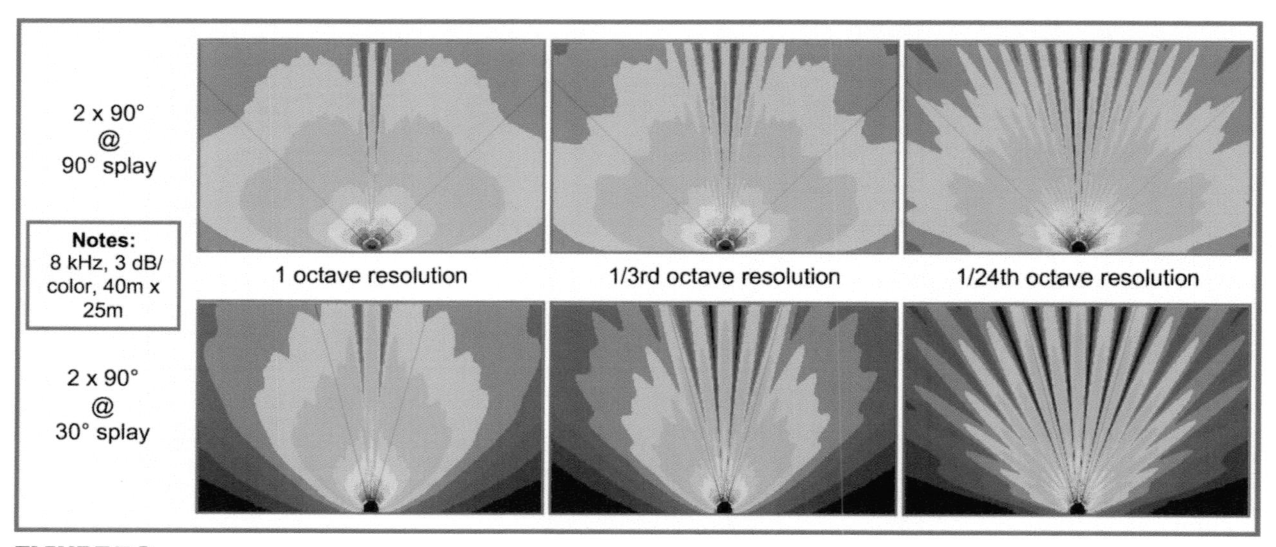

FIGURE 7.8
Frequency resolution issues related to characterization of summation in coupled speaker arrays. Top: Non-overlapping array has minimal comb filter summation. This is best revealed in the high-resolution rendering, where the combing zone and isolation zone interaction can be clearly differentiated. Bottom: The overlapping array appears to have a narrowed shape but there is equivalent combing in the low-resolution prediction. The high-resolution panel shows the combing zone interaction to be dominant over most of the coverage area.

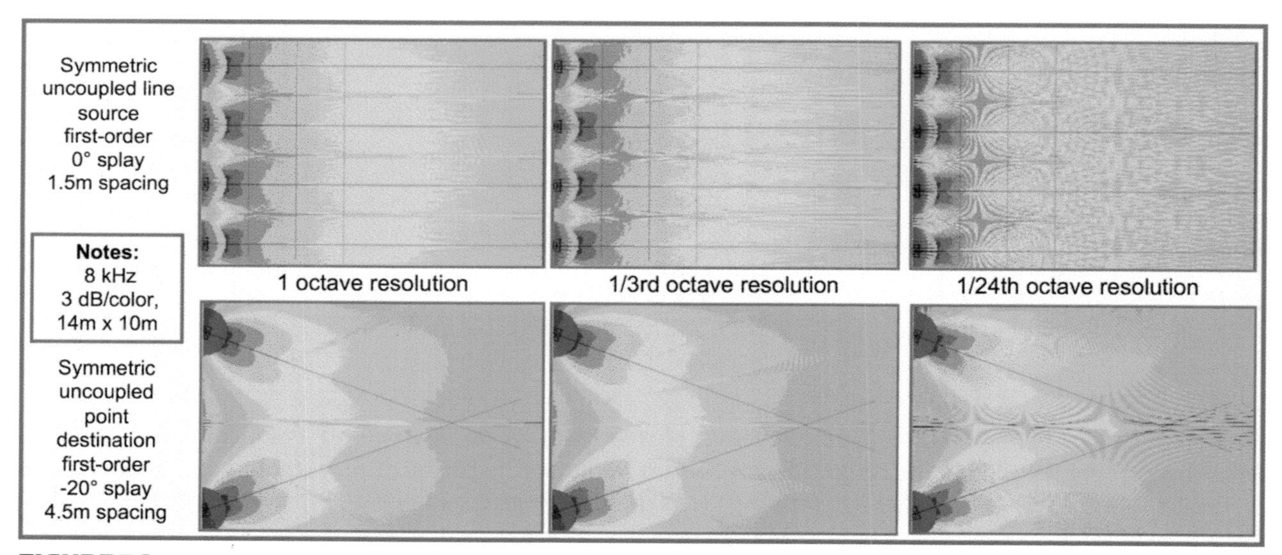

FIGURE 7.9
Frequency resolution issues related to characterization of summation in uncoupled speaker arrays. The lower resolution renderings do not reveal the high level of ripple variance from the speaker interaction. With low-resolution predictions the overlap areas appear to be the most uniform. The high-resolution data reveal the overlap area to be the least uniform.

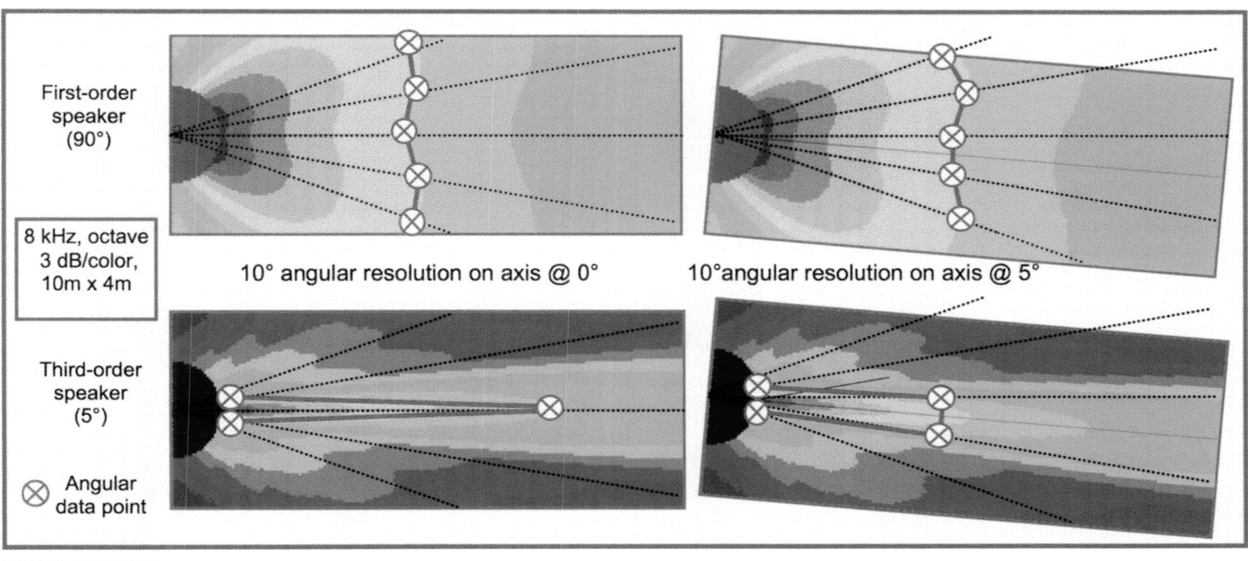

FIGURE 7.10
Angular resolution issues related to characterization of a single speaker. The left panels show the results when the speaker axis is oriented to match the angular acquisition points. The right panels show the speaker axis oriented between the angular acquisition points. Top: First-order speaker shows only minor differences with the angular displacement. Bottom: Third-order speaker shows substantial differences in shape. High-order speakers require high angular resolution for accurate characterization.

7.2.2.2 ANGULAR RESOLUTION

Polar data are a circular or spherical series of evenly spaced points. The spacing is the angular resolution, in degrees. Displayed results include interpolation in order to create smooth continuous curves (Figs 7.10 to 7.11).

Predictions must accurately track the angular changes in the measured speaker. Well-behaved wide-coverage speakers with a low rate of change can be sufficiently characterized with low angular density. A first-order speaker may fit these criteria, but a third-order speaker has too high a rate of change over a small angular area.

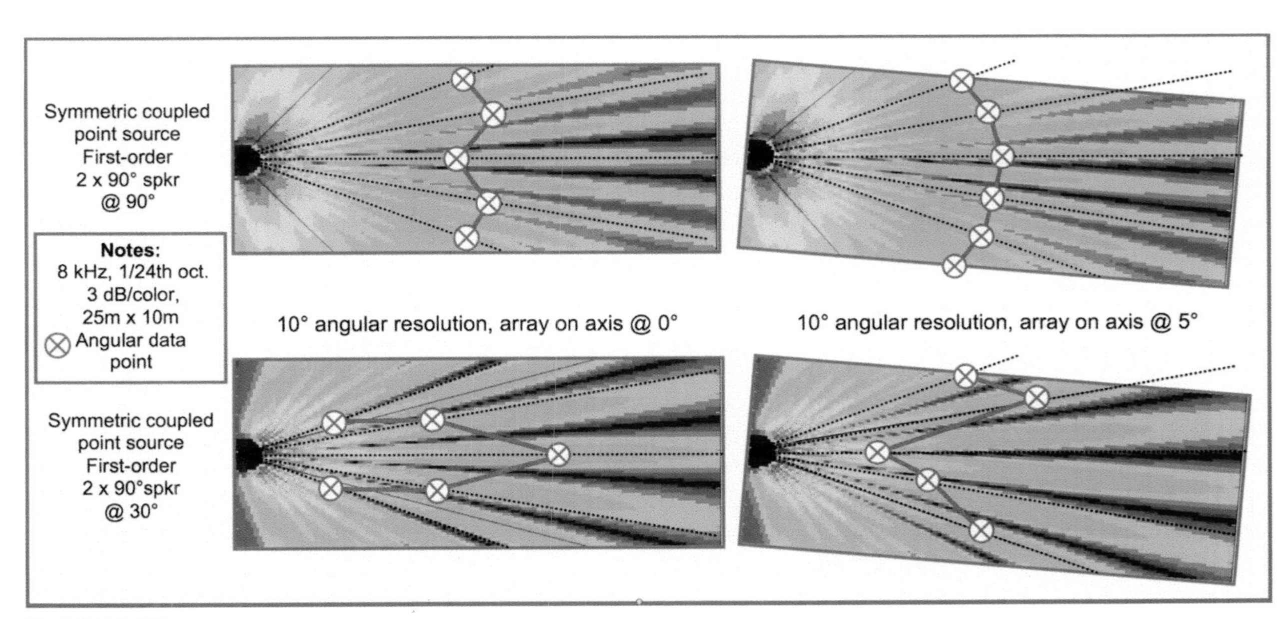

FIGURE 7.11
Angular resolution issues related to characterization of summation in a coupled speaker array. The right panels show the array axis oriented between the angular acquisition points. Top: The non-overlapping array shows only minor differences with the angular displacement. Bottom: The overlapping array shows substantial differences in shape. Overlapping arrays require high angular resolution for accurate characterization.

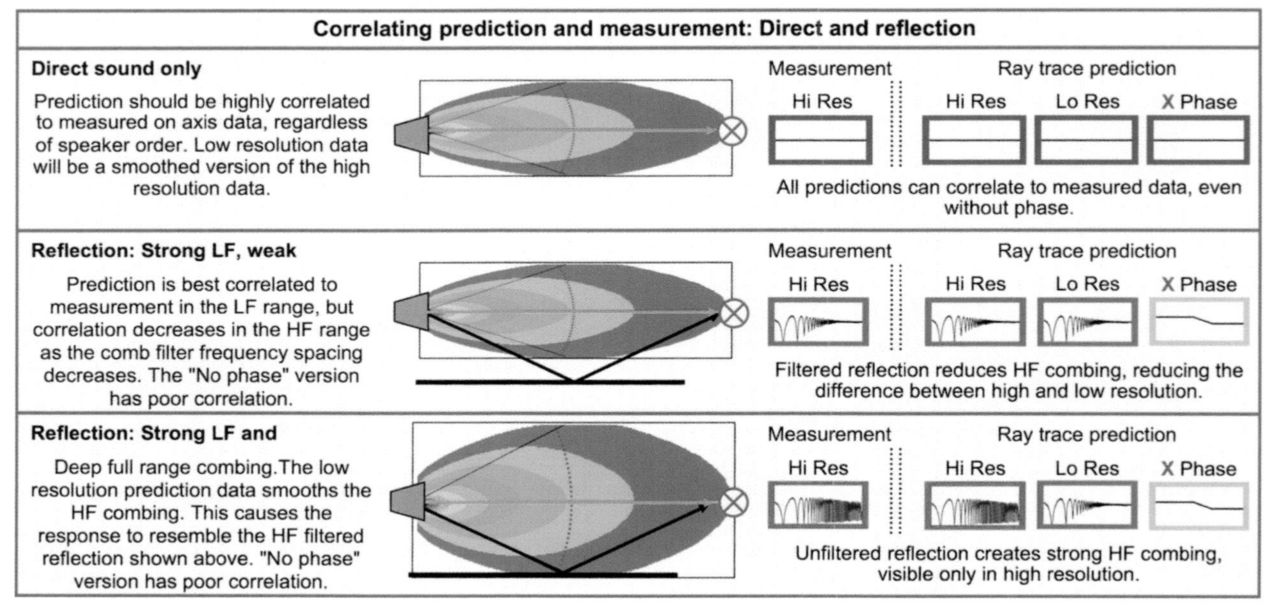

FIGURE 7.12
Comparison of the ray-tracing prediction models with the expected high-resolution measured response. Top: direct sound path on axis response. Middle: filtered reflection. Bottom: unfiltered reflection.

The first-order speaker is not exempt, however. Many first-order speakers have horns with substantial ripple variance within their coverage angle (seen as lobes and fingers). Low angular resolution, especially when coupled with low-resolution frequency acquisition, can create a false equivalence between speakers of vastly different audible quality.

Low-resolution data (angular or frequency) reduces the investigative quality of our predictions, providing smooth optimistic views. High-resolution data can often look like acoustic Armageddon, but we benefit from having the ability to make better informed decisions. Low resolution is a video solution for an audio problem.

7.2.2.3 PHASE RESPONSE

Phase response characterization adds significant complexity to the data interpretation. Is it really necessary? The answer is yes unless we limit our design to the lowest-paying gig in the industry: a single speaker in free field. Phase must be factored in once time offset enters the equation (e.g. multiple speakers, reflections). Summation characterizations without phase data violate the second law of thermodynamics. Energy is created (the positive side of summation) without a compensating loss (the cancellation side). The result is a response that we cannot possibly observe with our ears or analyzer. Therefore we can conclude that phase must be included in the characterization (see section 7.2.5.3).

7.2.2.4 COMMON LOUDSPEAKER FORMAT (CLF)

A format has been created in an effort to standardize acoustic modeling source files. This allows different manufacturers to submit data for use in various third-party prediction programs. In keeping with audio industry standard practice (two standards are better than one) we have two "common loudspeaker formats." CLF1 is low resolution (octave, 10°) and CLF2 is medium resolution (1/3 octave, 5°). Phase response has been added in CLF2 v2. Data files can be created with a set of 2-D or 3-D measurements. Marketing departments can't submit fantasy data. It must be measured in approved facilities and certified under standard conditions.

Both formats can display 3-D spherical models. These may come from actual 3-D measurements or spherical interpolations based on the solo slices of the 2-D files. The margin for error resulting from the combination of low/medium resolution and spherical interpolation must be considered when viewing the results. Notably the program alerts the user when spherical interpolation is taking place.

Acoustic prediction stands to benefit from implementing standards that level the playing field between competitors. The standard must be rigorous enough to differentiate between good- and poor-quality speakers and designs. The low resolution of CLF1 requires us to be cautious about any evaluations made with it. The coarse features of wide-coverage (first-order) speakers are discernible but details of lobes and cancellations will be glossed over. A speaker with many lobes may appear on equal footing with one whose pattern is more uniform. The intention of a standard is to level the playing field without plowing down the goal posts. Consider the modern third-order speakers used in line arrays, which commonly have patterns in the 10° range (±5° from on axis). CLF1 won't take its first data point until the speaker is well past the most critical design milestone, the -6 dB point. How comfortable are you designing around an interpolated -6 dB point?

Octave band frequency resolution also severely limits the application range. We can see single speakers in free field. We can't accurately characterize summed speaker responses with only octave resolution. It plows the combing so we can't see the difference between an optimized design and one with massive combing.

CLF2, comparable (although not compatible) with EASE™, has higher resolution. The change is one of incremental scale. The frequency resolution is 3×and the angular resolution is 2× that of CLF1. It's an obvious improvement, but is it enough?

7.2.2.5 MAPP ONLINE™

MAPP Online is 2-D-only prediction with extremely high resolution and complex data. Speaker data is measured at 1/24 octave, 1° angular resolution in vertical and horizontal slices with the actual measured phase response of each model. Different speaker models can be combined with realistic summations. MAPP has unrivaled resolution and accuracy in its characterization of individual speaker details and their summation. There are, however, three severe limitations: The speaker data library has only Meyer Sound speakers, the room reflection modeling is extremely limited, and all data is 2-D. These limitations obviously preclude its use as a universal tool.

Modern sound designs are reliant on highly directional speakers arrayed in large numbers at overlapping angles. Minute changes of 1° and 2° have very demonstrable effects on array performance. Low-resolution, phase-free data can't provide the critical details needed to make fully informed decisions about such systems. My crystal ball (the old-fashioned form of prediction) says the future holds a standard format with high-resolution complex data. Could that be CLF3?

7.2.3 Acoustic transmission properties

7.2.3.1 DIRECT SOUND TRANSMISSION

Direct sound transmission loss is modeled at the free-field rate (6 dB/doubling). Some programs compute the frequency-dependent HF air absorption loss effects by including user-adjustable temperature and humidity. It's a useful feature but with practical limits. Indoor applications can use a nominal "room temperature" and standard humidity but even these cannot be guaranteed to occur. Outdoor applications have real weather so there's little point in pretending to design the system in anticipation of it.

7.2.3.2 RAY-TRACING MODEL

When the ray strikes a surface it is reflected like light on a mirror (angle of incidence = angle of reflection). Ray tracing is responsible for the bulk of the information driving the predictive design process. The system's response will be measured during optimization, and results will hopefully resemble the prediction in the most important parameters. We can't rely on the prediction model to be perfect, nor do we need it to be. It is be useful to understand where we can expect results to match prediction, and where we need to go it alone and rely exclusively on measured data. Weather-related effects make the case. Even if the prediction program can factor these in it cannot predict the weather. Ongoing monitoring and compensation of weather-related effects will require on-site measurement.

A series of figures illustrates the differences between the ray-tracing prediction model and high-resolution frequency responses we see during optimization. The summation zone frequency response icons introduced in Chapter 4

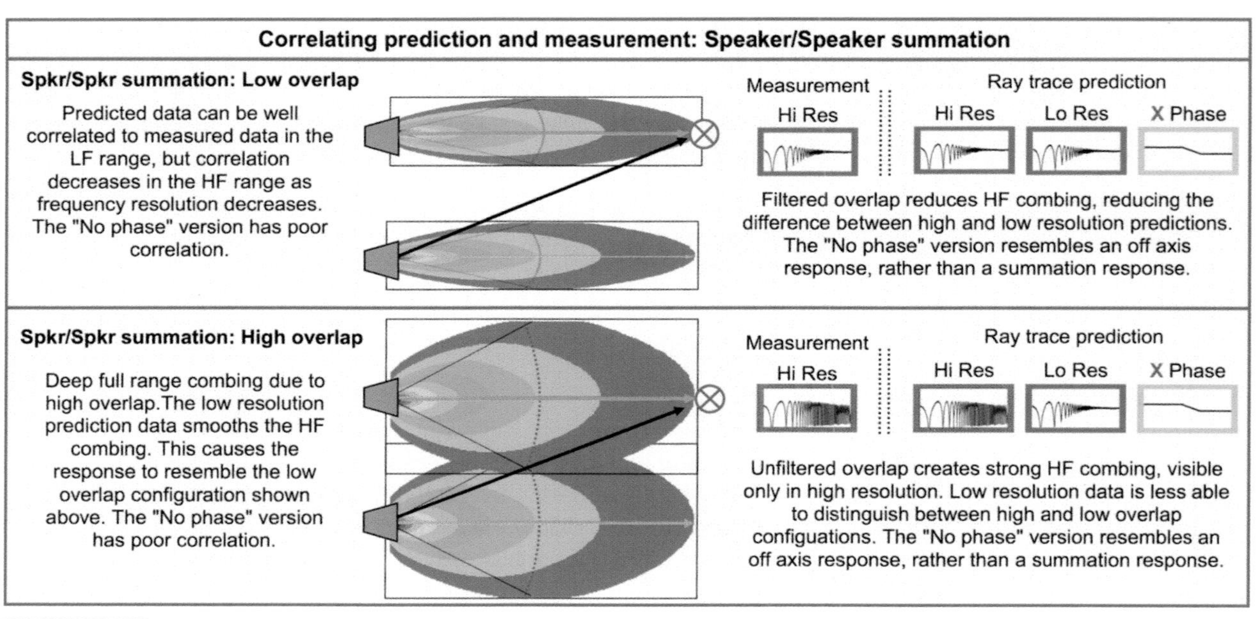

FIGURE 7.13
Comparison of the ray-tracing prediction models with the expected high-resolution measured response. Top: speaker interaction with low percentage of overlap. Bottom: speaker interaction with high percentage of overlap.

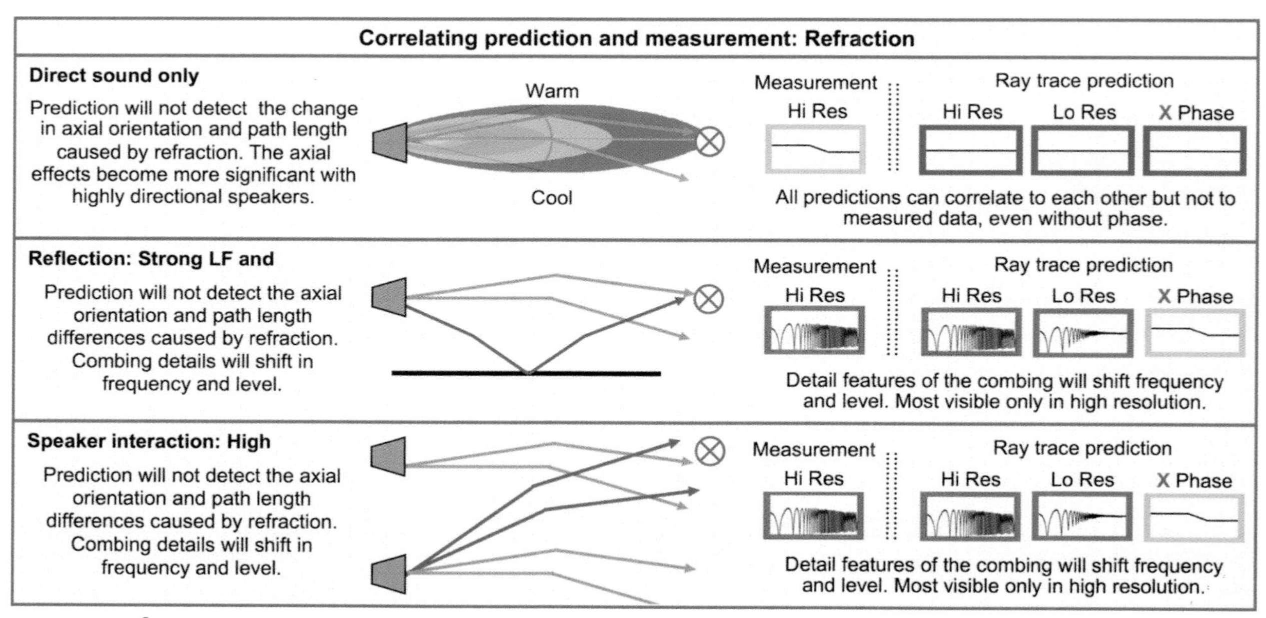

FIGURE 7.14
Comparison of the ray-tracing prediction models with the expected high-resolution measured response. Top: refraction of the direct sound path. Middle: refractive effects on unfiltered reflection. Bottom: refraction of speaker/speaker summation.

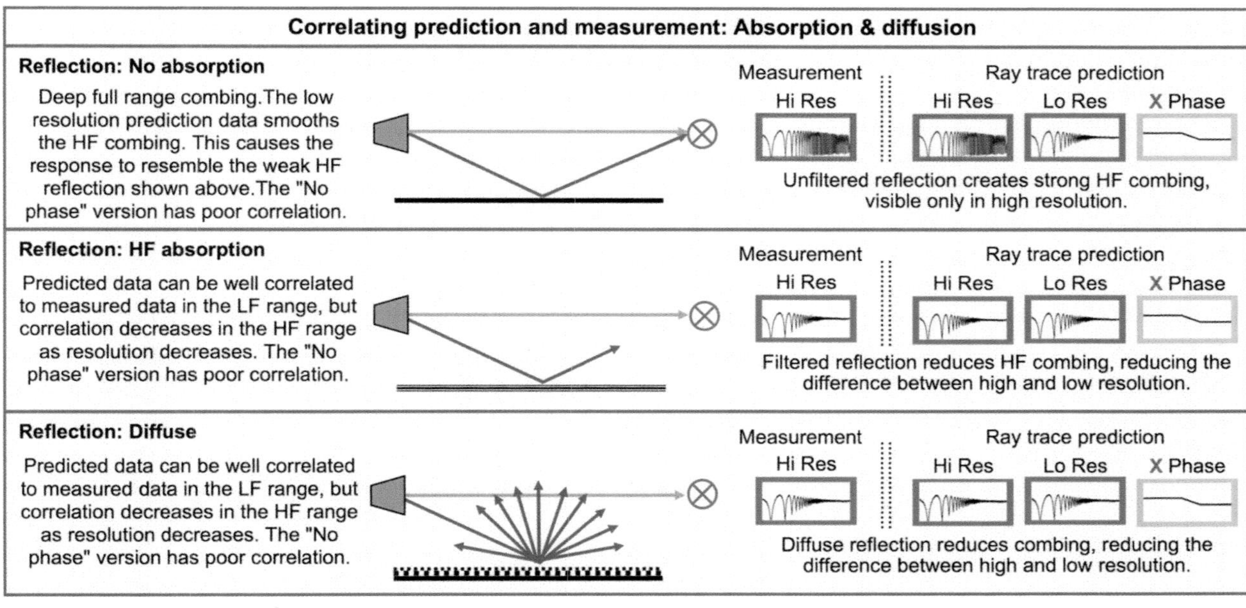

FIGURE 7.15

Comparison of the ray-tracing prediction models with the expected high-resolution measured response. Top: unfiltered reflection from non-absorptive surface. Middle: filtered reflection from absorptive surface. Bottom: unfiltered reflection from diffuse surface.

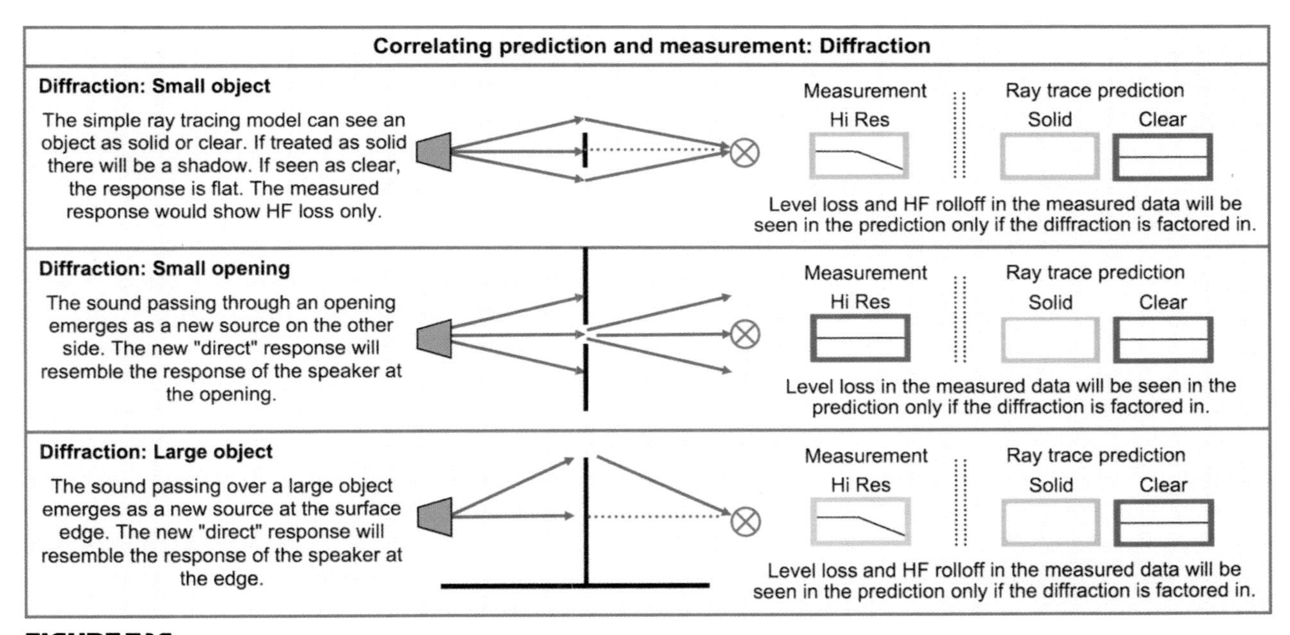

FIGURE 7.16

Comparison of the ray-tracing prediction models with the expected high-resolution measured response. Top: diffraction around a small object. Ray-tracing model may either characterize the object as solid or clear. Middle: diffraction through a small opening. Bottom: diffraction around a large object such as over a wall.

reappear here. The ideal is a perfect match between predicted and measured. Differences are significant, because these summation icons are milestone markers for optimization decisions. If our prediction results stray too far from measured reality we might make poor design decisions that will later be discovered during optimization.

Each figure contains the expected measured result and three versions of the ray-tracing model-predicted results: high resolution, low resolution and phase free. As we will see, the highest correlation will always be the highest resolution, and the lowest is the "no phase" response (Figs 7.12 to 7.16).

Prediction program implications for acoustic transmission modifiers

- **Refraction**: the bending of a sound transmission passing through layers of media. Only relevant to long-distance outdoor applications. Prediction would have to factor in various layers of air temperature and degree of bending this would cause. Extremely doubtful accuracy due to additional outdoor conditions such as wind. There are no sound system design decisions dependent on accurate refraction data.
- **Diffusion**: a complex frequency variable scattering. Extremely complex to model for the very reasons that make them so desirable acoustically. Diffuse reflective surfaces modeled as flat will look worse in the model than the field. It is very doubtful that system design decisions will depend on accurate diffusion data.
- **Diffraction**: sound waves bending around the surface, rather than bouncing off. Any pillars and partial walls will need to have different reflective properties over frequency. Openings must be modeled as secondary sound sources. Advanced predictive modeling incorporating diffraction could characterize I-beams in front of speakers and tell us which seats get what frequencies. Thanks, but no thanks. Preferred solution is a hack saw.
- **Resonance**: sympathetic spacing relationship with a particular wavelength. Are we going to select a different speaker because 275.4 Hz is resonating? The resonant properties will show up when the building is done, our speakers are in place and we are ready to tune.

7.2.4 Material absorption properties

So far we've only considered surfaces with 100% efficiency at reflecting sound waves. Fortunately most surfaces have some frequency-dependent absorptive qualities. Our next consideration is how these properties are represented in prediction and measurement.

The acoustic properties of building materials are compiled into a table of **absorption coefficient** values over frequency (Fig. 7.17). The absorption coefficient indicates the sound energy lost during the transition at the surface. Values range from a maximum of 1.00 (an open window) to 0.00 (100% reflective). We can convert these numbers to dB loss as follows: attenuation is $10*\log_{10}(-\alpha)$ where "α" is the absorption coefficient).

Absorption level loss reference	
Loss (dB) = 10 x log (1-α)	
Absorption	
Value (dB)	Coefficient (α)
-0.10	0.02
-0.25	0.06
-0.5	0.10
-1.0	0.20
-1.5	0.30
-2.0	0.37
-3.0	0.50
-4.0	0.60
-5.0	0.69
-6.0	0.75
-7.0	0.80
-8.0	0.84
-9.0	0.88
-10.0	0.90
Total	1.00

FIGURE 7.17
Absorption coefficient level reference (approximate). The above reference chart converts the absorption coefficient (α) into dB loss for a single reflection. The ray-tracing model computations like this may simulate the loss in the reflected ray as it exits the reflective/absorptive surface.

Published absorption coefficients are usually limited to octave bands ranging from 125 Hz to 4 kHz. They don't factor in the differences between striking a surface at 9.0° and 90°, which can be substantial in some (but not all) materials and variable over frequency. Surface areas that contain mixtures of materials will have varying reflective, absorptive, resonant, diffusive and diffractive qualities over angle of incidence and frequency. The complexity is staggering, even for a simple room. Acoustic modeling with published absorption coefficient data is, at best, a very rough approximation.

7.2.5 Characterizing summation

We cannot overstate the importance of summation management to the optimized design. A modeling program without accurate summation characterization is of extremely limited use. Check that: useless and misleading.

7.2.5.1 FREQUENCY RESOLUTION

First, a quick note about frequency resolution, wavelength and summation. Octave resolution means we see the direct sound and the summation effects of arrivals up to 1 λ behind the direct sound (which creates an octave-wide peak). That's a mere 1 ms at 1 kHz. One-third octave resolution triples that to include signal arriving up to 3 λ behind. Woohoo! This means only a tiny fraction of the summation effects are included in the response. Low resolution gives a free pass to comb filtering (in the prediction) that won't be granted to our listeners. This doesn't mean we need to do everything at 1/24 octave, but we need the option to go there when there's a question. At low resolution it can be very difficult to discern the good, the bad and the ugly.

7.2.5.2 ANGULAR RESOLUTION

Angular resolution also plays a key role in summation decisions. Would you notice somebody changing a splay angle by 5°? Your prediction program might not if its data are derived from 10° angular resolution. It's ridiculous to consider 10° or even 5° angular resolution since modern era speakers have single-digit coverage patterns and rigging that allows for 0.5° changes. Main arrays can contain large quantities of speakers within the span of a few degrees and can create combined patterns that are less than a single degree wide. Small changes in relative angle are critical to the combined directional response, the details of which will be lost without high angular resolution.

By contrast, a first-order speaker can be aimed within 5° or 10° of its optimal angle and few will notice any negative effects on uniformity. Overlapping coupled third-order arrays are aimed with absolute precision and can have the highest levels of uniformity imaginable. An error of a few degrees can throw it all away.

7.2.5.3 PHASE RESPONSE

I wonder how loud it would get if we played a different song through every speaker. That's the answer you get with a speaker modeling prediction that has no phase response. Summation is purely constructive. No cancellation. Even polarity reversals or delay line settings don't matter! There's no application for this and no design strategy that benefits from it. There are two summation families: correlated and non-correlated. We manage the interactions between correlated partners. We don't analyze the interaction of non-correlated speakers because they're random. What design decisions do you make regarding randomly related speakers? I find 2.94 ms of delay sounds best.

7.2.5.4 QUASI-PHASE RESPONSE

An intermediate option is the attribution of a nominal phase response to all speakers (rather than using the actual measured responses of individual models). Flat phase is assumed for all frequencies and the *relative* phase between sources is incorporated into the prediction, based on time offset. This approach shows summation interaction with reasonable accuracy as long as only a single-speaker model is predicted. Combinations of different models (such as mains and a subwoofer, or different full-range models) will not track the measured response found during optimization. Phase compatibility between models is a complex issue that often arises during optimization. Advance notice during the design process may lead to decisions that reduce these challenges during optimization.

Another quasi-phase response limitation is its assumption of phase origination from a single source point. The measured phase response of an actual two-way speaker (non-coaxial) would confirm what we see with our eyes: The HF and LF signals propagate from different physical locations. Therefore even the summation of two identical-model speakers is inaccurate in the acoustical crossover range (in the plane where the HF and LF drivers are displaced). This is the confluence of the spectral and spatial divider as discussed previously in Fig. 4.34.

7.2.6 Putting prediction to work

Acoustic modeling is a design tool. Accurate programs increase the probability of good design and reduce the excuses for a bad one. In the final analysis (pun intended), we will see how they fare during optimization.

It will be a great day when we have readily available, extremely high-resolution 3-D data with measured phase. At the time of writing no available program offers this. In the mean time we have to work with what's available. Each of the programs has its strengths and weaknesses. If we can't have it all, how do we prioritize?

The end goal is successful optimization. Chances of success are greatest if the prediction data reflects what we'll see on the analyzer. This correlation must be prioritized far above presentation quality to boardroom executives. The best choice is the prediction that yields the wisest selection of speaker model, position, array types, signal-processing resources and acoustic treatment. Matters that can be adjusted on site, such as equalization, fine-focus angle and relative level, can be left to the calibration stage.

The prediction task is extremely complex. Layer upon layer of acoustical effects are added to the direct sound propagation. Accuracy inevitably decreases with distance and each additional surface or speaker. The odds are heavily stacked against the speaker/room interaction side of the equation, with its complex acoustical mechanisms and incomplete data available for computation.

THE BEST TO WORST PROSPECTS FOR PREDICTION ACCURACY (IN ORDER)

1. Direct-field response (coverage pattern, frequency response, level, air loss).
2. Speaker/speaker summation effects (coverage pattern, frequency response, level).
3. Speaker/room summation effects (ray-tracing model).
4. Diffraction, diffusion, refraction, resonance, reverberation effects (non-ray-tracing model).

The direct-field response (near and far) of a single speaker has the highest potential for accurate characterization, provided we have enough frequency and angular resolution. This is followed by the direct-field summation response of multiple speakers. Even a huge array is far less complex than speaker/room summation. Last place goes to the non-ray-tracing behavior of speaker/room summation.

We design speaker systems, not symphony halls. The progression falls in the right order for us because the most accurate predictions align with our highest priorities for design and optimization. Our need for precise data declines along with predictive accuracy, leaving the remainder to be measured on site. Speaker model selection, aim angle, array type and signal-processing distribution is primarily based on the direct-field response. If the system can't produce uniform response in free field it's folly to expect room reflections to fill the gaps. It can be helpful to identify surfaces that are potentially troublesome reflection sources. But reflection paths are not revealed until we define the direct sound shape. The source of the trouble is usually their relationship to the direct sound path, not an intricate web of multi-surface paths. We reached the conclusion to avoid aiming at walls and ask for absorption long before modeling programs.

For our application, room acoustics modeling should not be considered on equal footing with direct sound modeling. The best speaker system design will never be found if its own rendering is not accurate, no matter how well the room is predicted. The places where acoustic modification will be most beneficial can be found by looking at where our direct sound spills on surfaces, even if the room is not acoustically modeled. Which comes first, the direct sound or the reflection?

OK. So we can't have a totally accurate room reflection modeling. What's the problem with the inclusion of at least some of the room reflections in our computations of uniformity? Isn't a modeling of questionable accuracy better than none at all?

Consider the following:

- If the speaker data are poor resolution, how can the reflection data be any better?
- Isn't modeling summation without phase like modeling rigging without gravity?

- Best coverage angle is 40° with curtains in. What's best with curtains out?
- We need a 132 dB SPL speaker with curtains in. How about with curtains out?
- A coupled point source array is the best with curtains in. What's the best without?
- There's a 10 m × 20 m glass wall. Is accurate modeling needed to know the solution?
- Wouldn't precise prediction of the direct sound help us steer away from the glass?
- It's hot and humid tomorrow. Should we change the speaker model or focus angle?
- What models and focus angles do I need for a room with an RT of 2.3 seconds?

I can't tell you how to design systems using predictions without enough resolution to discern success from failure or with phase cancellation wished away. I don't know how to do it, so I surely can't teach you. My design perspective always includes the expectation that the system will appear on my analyzer during optimization in high resolution including phase. My analyzer has never found a system that behaved like a low-resolution or phase-free rendering. I have, however, seen systems from many manufacturers that behaved precisely like the high-resolution, phase-inclusive renderings of MAPP Online.

Consider this a confession of author bias or a limitation of my vision. I will struggle with the limitations of 2-D vision over the errors and omissions of low-resolution phase-free data until high-resolution prediction moves into 3-D (and I learn to operate it).

We are ready to apply the prediction process to the optimized design. The next chapters will conduct a thorough search for speaker system designs that provide the maximum uniformity over the listening area. The phase-inclusive, high-resolution prediction model will guide us toward this goal.

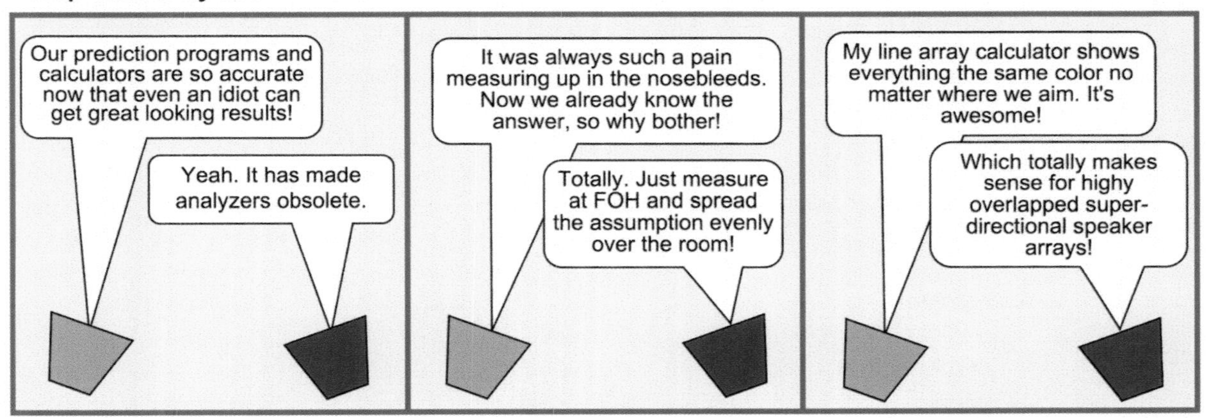

CHAPTER 8

Variation

Speaker coverage has a shape. The audience area has a shape. Sound design is about laying the speaker shapes over the audience. This chapter begins the search for the best speaker shape to fill the room. We start simple, with a single speaker, and set the stage for the next chapter, where we will expand into speaker arrays.

The goal is straightforward: same everywhere. Same level, frequency response, clarity and sonic image location. The applicable principles are the same for single speakers and complex arrays, for a full black-box theatre or a theater full of black boxes. The quest is minimum spatial variance, an objective mission whose success is measured in how low we keep the score. If it sounds good here, it should sound good there. The term variance indicates how closely we approach this desirable, yet ultimately unreachable goal.

> ***variation*** *n. varying, departure from a former or normal condition or action or amount or from a standard or type, extent of this.*
> ***variance*** *n. disagreement, difference of opinion, dispute, lack of harmony.*
> **Concise Oxford Dictionary**

8.1 SINGLE-SPEAKER VARIANCE PROGRESSIONS

We seek to minimize spatial variance in four principal categories:

- Level variance: in dB, e.g. the underbalcony area is -6 dB, compared with the twenty-fourth row.
- Spectral variance: in dB over frequency, e.g. under the balcony the 2–8 kHz range is -6 dB compared with the flat response in the twenty-fourth row.
- Sonic image variance: in offset between perceived image location, e.g. the actors sound as if they are in the roof for seats in the fourth row but image near stage level from the fifteenth row.
- Ripple variance: in dB for summation-related peaks and dips, e.g. the underbalcony area has 12 dB of midrange ripple, compared with 3 dB in the twenty-fourth row.

Some subcategories are subdivisions of major ones or shared between them. Intelligibility variation is affected by ripple (combing and echoes) and spectral (MF or HF loss) and level variation. It's probable that minimum-variance strategies in the principal categories will improve intelligibility. Maximum level capability is a subcategory of level variance, which is preserved only when all speakers maintain a linear response to level (no compression or overload). Speakers must be power scaled for their coverage zone so that all locations reach maximum level together. For example, the underbalcony speakers are scaled to keep up with the mains in their local area. Power scaling is discussed later in Chapter 11. For now we assume appropriate scaling. We also assume all speakers have a common spectral reference point: flat spectral response on axis.

Single-speaker propagation is described by four standard lines of variance in each plane

- Forward-variance line: from 0 dB to -6 dB in a forward direction.
- Radial-variance line: from 0 dB to -6 dB on an equidistant radius.
- Lateral-variance line: from 0 dB to -6 dB on a straight line perpendicular to the speaker.
- Minimum-variance line: from 0 dB to 0 dB in a forward, radial or lateral direction.

Each line has a role to play in the design strategy (Fig. 8.1). For example, the forward-variance line links mains to delays, the radial line links elements of the point source array and the lateral line links uncoupled line sources. The minimum-variance line guides the aiming of single speakers. Forward-, radial- and lateral-variance lines can be linked in combination to create an extended minimum-variance line.

The forward-, radial- and lateral-variance lines are all maximum acceptable variance lines (0 dB to -6 dB). A single speaker's minimum-variance line is the link between the points of equal relative level: the -6 dB ends of the forward-variance and radial- (or lateral-) variance lines. Both ends of these lines are the same, hence minimum variance.

We can now give names to the minimum-variance milestones: $ONAX_{FAR}$ (the -6 dB end of the forward line) and $OFFAX_{NEAR}$ (the -6 dB end of the radial/lateral line). The minimum-variance line connects between them (also Fig. 8.1).

The same principle applies to arrays. The -6 dB ends of the radial- and lateral-variance lines are joined to created a summed connection of 0 dB, which extends the minimum-variance line until it drops 6 dB in a forward, radial or lateral direction (detailed in the next chapter).

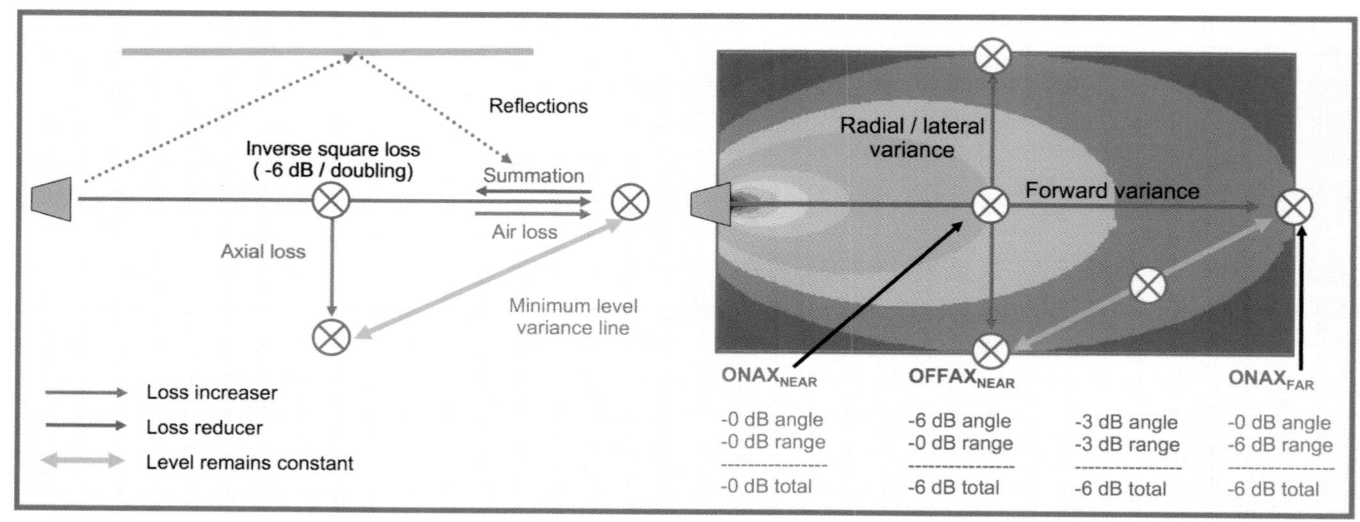

FIGURE 8.1
The level-variance spatial progression. The minimum-variance link between $ONAX_{FAR}$ and $OFFAX_{NEAR}$ is shown.

8.2 LEVEL VARIANCE (SINGLE SPEAKER)

Two mechanisms are in play: coverage pattern and range ratio. Coverage pattern creates a propagation shape. If it's omnidirectional then the only factor is range: closer is louder. A single speaker loses level over distance at the free-field rate in all directions. The natural tendency is for level to fall as we go deeper into the room and/or move outward toward the sides. If the source is directional we get a customized radial shape of uniformity. Tenth-row seats on the side might be as loud as twentieth-row seats in the middle, but then range takes over again. Twelfth-row seats on the side are quieter unless we do something to offset the range effects.

8.2.1 Level-invariant shape (equal-level contour)

We can find a speaker's coverage by circling it and maintaining equal level as we go around. We trace an equidistant circle if the speaker is a perfect omnidirectional radiator. Any other speaker requires us to modify our distance as we go around and create a custom shape that characterizes its unique level-invariant response. This is the equal-level contour map, also known as an isobaric pressure plot. Every speaker has a shape in each plane (and in 3-D for that matter) and it's variable over frequency. We'll concentrate on overall level now and add frequency later. Minimum-level variance is achieved when we place our audience along this shape. That handles the radial distribution but we need to add distance into the equation.

8.2.2 Range ratio

It would be easy to achieve minimum-level variance if our audience was only one row deep: Simply match the speaker's level-invariant shape to the audience shape. When this is not the case we must incorporate the differences in range between our farthest and nearest listener, i.e. the range ratio (Fig. 8.2). This can be expressed as a linear ratio or in log form (dB). A speaker whose farthest seats are twice the range of its nearest has a range ratio of

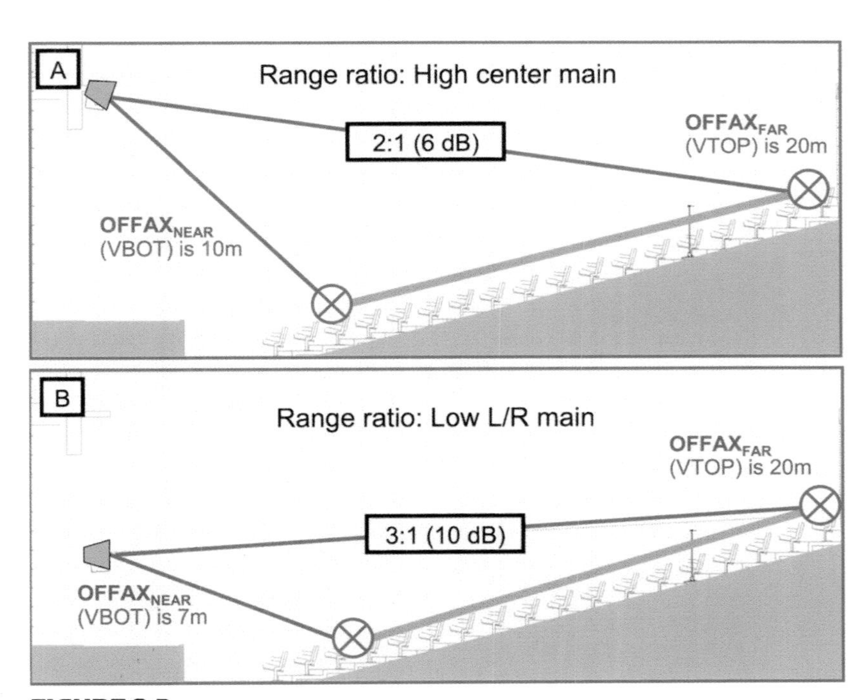

FIGURE 8.2
Range ratio is found from two positions in the same room. The high position (A) yields 4 dB less range ratio than the lower position (B), but with a corresponding tradeoff in sonic image variance.

2:1 (6 dB). This is the listening area's default-level variance. Our mission is to counteract the range difference by aiming/shaping the speaker coverage to create equal level at the far and near locations. One solution is to place the speaker at a location where the front and rear seats are equidistant (reducing the range ratio to 0 dB). This is too often a ridiculously high position in the ceiling. Alternately we can use a speaker with 6 dB of directional shaping, or a combination of both.

Let's use an extreme example for illustration (Fig. 8.3). Place a speaker on the stage lip. It's 4 m to the front row and 32 m to the last. The range ratio is 8:1 (18 dB). We need a speaker with 18 dB of full-range directional shaping to overcome this challenge. This is why we don't stack the PA on the deck any more. Raise the speaker to height typical for L/R mains and we can double our distance to the first row. This reduces the burden to 4:1 (12 dB), which puts us into the realm of practical reality for a speaker array (2:1 is the practical limit for a single speaker). Raising it to a center cluster height reduces the range to 2:1. The lowest range ratio would be achieved with an overhead approach, but that is obviously not viable for reinforcing sound from the stage. As we approach the speaker in the horizontal plane (front to back) we move away from it vertically. The result is offsetting level effects that help us maintain constant level.

Range ratio can also be reduced by partitioning the coverage (lower section of Fig. 8.3). Frontfills extend the starting point for the mains and delays reduce the end, which nets a reduced ratio.

8.3 SPECTRAL VARIANCE

8.3.1 Spectral variance progression

Our level variance findings carry over to spectral variance with one added ingredient: frequency. Sorry, I mean 600 ingredients. We can view this as coverage variance over frequency (spatial/spectrum), or as frequency response variance over coverage area (spectral/spatial). The term spectral variance applies either way, because all the variance forms are "over the space."

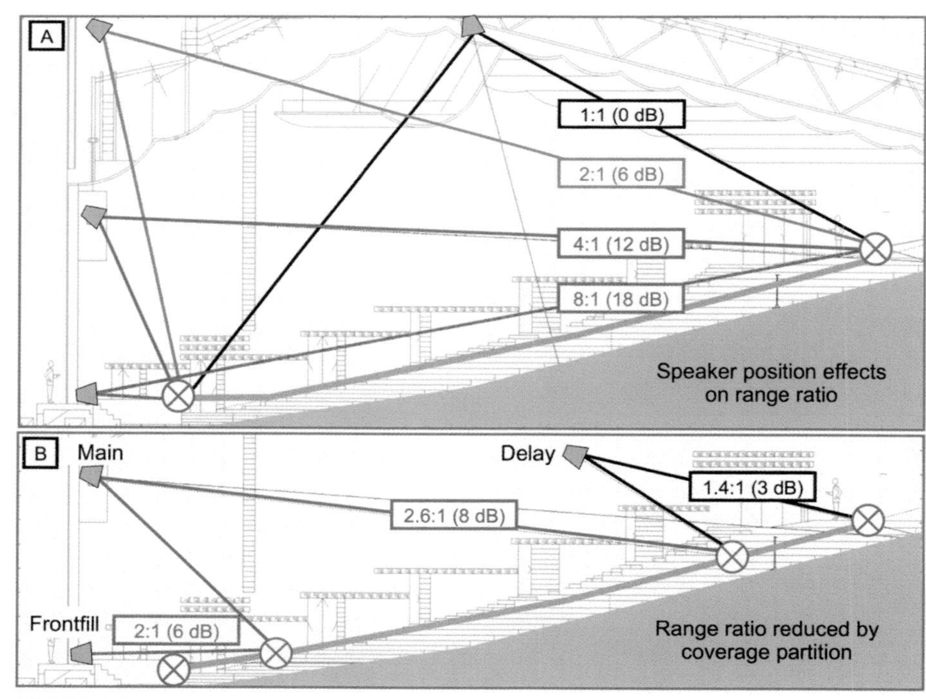

FIGURE 8.3
An extreme example of speaker position effects on range ratio for the same room. (A) Four positions with range ratios of 18 dB to 0 dB. (B) Reducing the main range ratio by coverage partition with frontfills and delays.

Moving forward from $ONAX_{NEAR}$ to $ONAX_{FAR}$ creates a tilted response favoring lows over highs (reflections and air loss). This is an increase in **spectral tilt**. Moving from $ONAX_{NEAR}$ to $OFFAX_{NEAR}$ rolls off the HF more than the LF. Spectral tilt again. The difference in spectral tilt between $ONAX_{NEAR}$ and the other locations is the **spectral variance**. The similarity in tilt between $ONAX_{FAR}$ and $OFFAX_{NEAR}$ is *minimum spectral variance.*

COMPARING SPECTRAL TILT AND SPECTRAL VARIANCE

- Locations with matched flat response: no spectral tilt and no spectral variance.
- Locations with matched HF roll-off or LF boost: spectral tilt but no spectral variance.
- One location has HF roll-off or LF boost: spectral tilt. The other is flat (no tilt): spectral variance.

Our goal is minimum spectral variance, not necessarily minimum spectral tilt. We are not the "flat police." Many engineers favor significant amounts of tilt (understatement). Minimum spectral variance ensures the same spectral tilt is heard everywhere (Fig. 8.4).

8.3.2 Forward tilt

Every sound source on earth exhibits spectral variance as it propagates forward over distance. Even an omnidirectional source in free field will show spectral variance over distance due to air loss attenuation in the HF range. There is simply no way for a single source to maintain its frequency response over distance. Reflections also add spectral variance over distance as they decelerate the loss rate in the LF. The response increasingly tilts with distance relative to the midrange response (LF above and HF below). This behavior is classified as forward spectral variance along the line of $ONAX_{NEAR}$ to $ONAX_{FAR}$ (shown previously in Fig. 8.4 (A)).

8.3.3 Radial/lateral tilt

Directional sources add another layer of spectral variance: radial, i.e. frequency-dependent response over angle. Radial or lateral spectral tilt progresses from $ONAX_{NEAR}$ to $OFFAX_{NEAR}$. Radial spectral variance rises with speaker directionality, while forward spectral variance rises with distance.

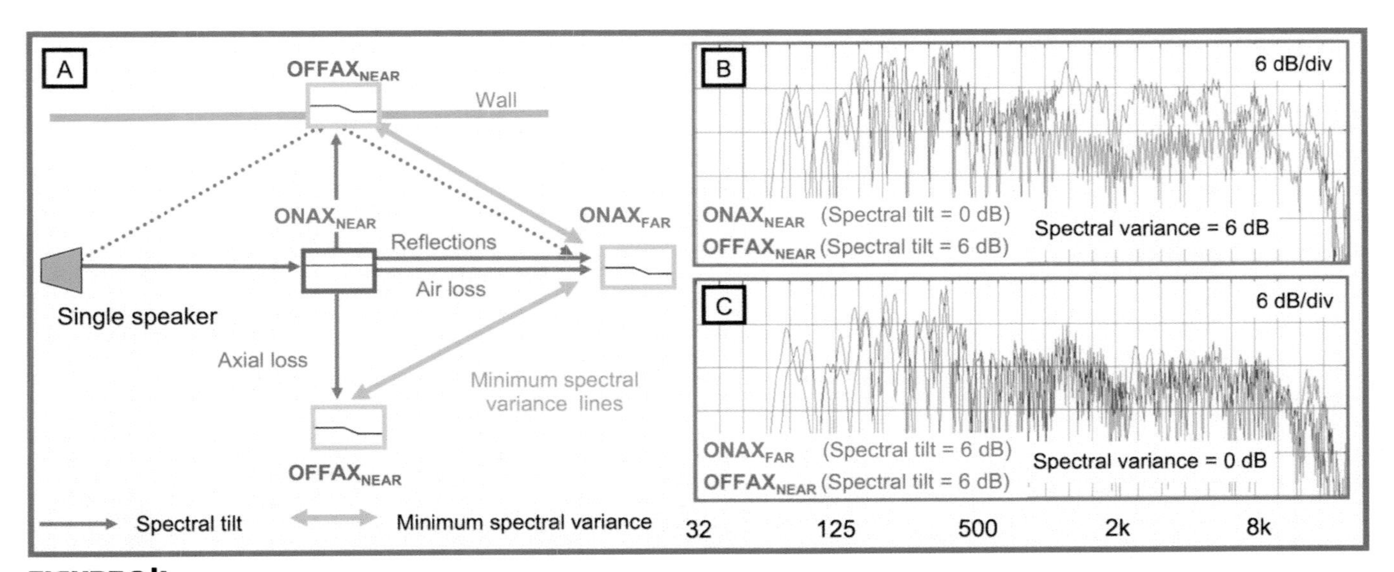

FIGURE 8.4
Spectral tilt and spectral variance. (A) Spatial progressions for spectral tilt and variance. (B) Comparison of $ONAX_{NEAR}$ and $OFFAX_{NEAR}$ shows unmatched spectral tilts and spectral variance of 6 dB. (C) Comparison of $ONAX_{FAR}$ and $OFFAX_{NEAR}$ shows matched spectral tilts and spectral variance of 0 dB.

A typical full-range speaker is wide in the lows and whatever in the highs (Fig. 8.5). A first-order speaker that runs from 360° (LF), 120° (MF), 90° (HF) has low spectral variance compared with a third-order proportional beamwidth speaker that spans 360° (LF), 90° (MF), 5° (HF). This is a fixed parameter for a single speaker, unaffected by equalization, level or delay. Arrays, however, can be constructed with combined coverage shapes that differ from the individual components (discussed in the next chapter).

Our sound design can use the forward and radial/lateral variance progressions in combination to achieve minimum spectral variance within the target listening area. The minimum spectral variance line connects the spectral tilt of $ONAX_{FAR}$ to $OFFAX_{NEAR}$.

8.3.4 Introducing "pink shift"

Astronomers use "red shift" as a range finder for distant sources of light in the universe. Red shift is derived by the Doppler effect and increases with a star's speed away from us. More red shift in the spectrum signals greater distance.

We also have spectral distant ranging psycho-acoustic clues. We expect a near violin to sound brighter than a distant one. We expect the forward spectral tilt to increase with distance and suspect trickery if it doesn't. The response from a flat spectrum source starts off white (un-tilted) and gradually turns pink (tilted) with distance. The reference here is to filtered white noise (equal energy per frequency) compared with "pink noise," a steady 3 dB reduction per octave. This creates equal energy per octave, and balances the noise spectrum for our logarithmic hearing. Spectral tilt is additional "pink shift" added to the response with distance. In natural acoustic transmission this pink shift is directly related to sonic source distance. Our internal sonar system estimates source distance by factoring in our expectations of "pink shift."

Equalization may be applied to reduce pink shift in the response. In effect we are reducing the sonic image range for the audience, bringing them closer. Alternatively one can equalize a system with intentional pink shift, thereby choosing a more distant sonic image range. Flat response is the natural equivalent to close range, while added pink shift is the natural equivalent to distant sound. This is a choice for the engineer to make.

Spectral tilt and pink shift are essentially equivalent terms for the same response shape regardless of whether it resulted from LF summation, HF loss or both.

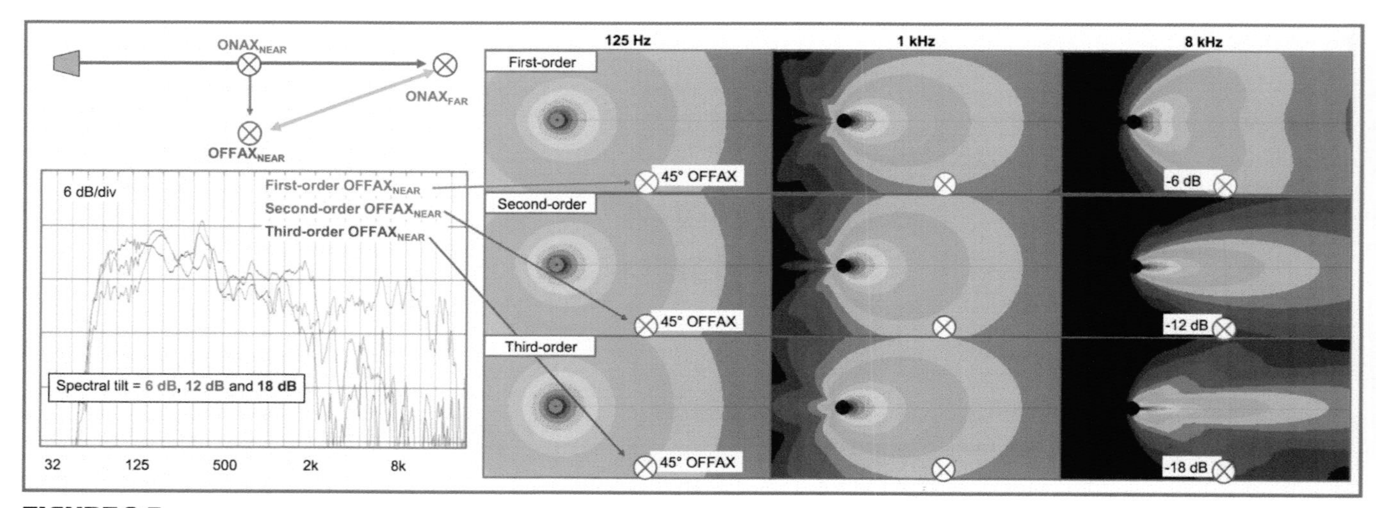

FIGURE 8.5
Radial/lateral spectral tilt. First-, second- and third-order responses are shown. Each frequency response is shown at 45° off the center axis revealing increasing spectral tilt with speaker order. Coverage plots show the increasing difference between the LF, MF and HF shapes as speaker order rises.

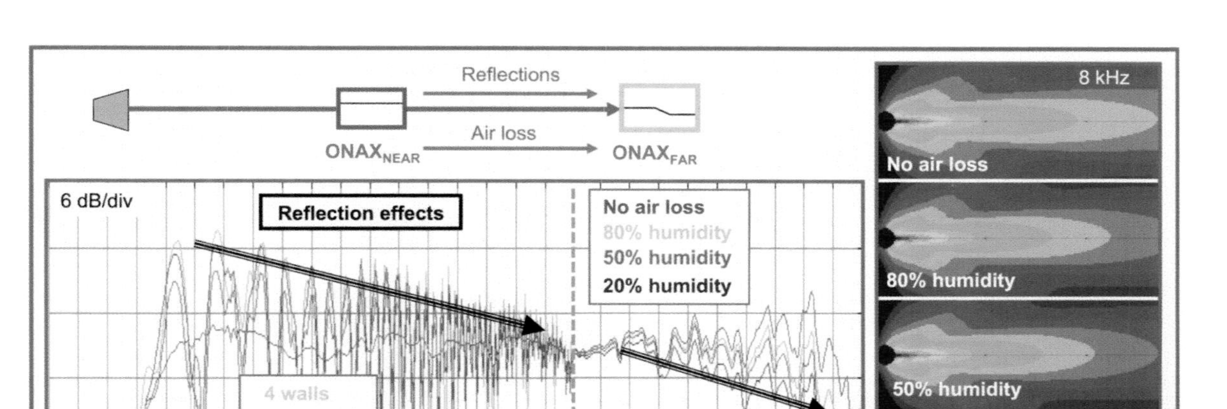

FIGURE 8.6
Pink shift progressions. Room reflections and air loss both show increasing tilt as they are added to the direct signal.

The major trends in the spectral tilt/pink shift distribution over the space are shown in Fig. 8.6. The effects are divided into two types: room reflections that tilt up the lows and air loss that tilts down the highs. Air loss can reduce the coverage shape in the HF range because its loss accrues over distance. The air loss is 2× greater at $ONAX_{FAR}$ than $OFFAX_{NEAR}$ because it has twice as much air travel. This is shown in the coverage plots of Fig. 8.6.

8.3.5 Range ratio effects

Range ratio has implications regarding control of spectral variance. It's easy to control to shape of the HF range, even at high ratios. It's difficult to match that control in the LF (Fig. 8.7). This may result in large-scale spectral variance at close range due to excess pink shift where only the HF is controlled.

Range ratio helps determine when to call for sound reinforcements (pun intended), i.e. downfill, sidefill, infill, etc. We can go it alone with range ratios of 2 or less by offsetting the distance with asymmetric aiming. The axial loss pink shift at $OFFAX_{NEAR}$ may be comparable to the HF air loss and room reflection effects at $ONAX_{FAR}$, but this parity has its limits. Range ratios greater than 2 leave us with heavy pink shift in the nearby off-axis areas. We'll need strong pattern control over a wide spectrum in order to minimize the spectral variance there, a considerable challenge for an array, and virtually impossible for a single speaker. A range ratio of 2 (6 dB) is the tipping point in favor of adding a fill system. Fresh on-axis sound from fill speakers reverses the pink shift.

8.3.6 Aspect ratio, beamwidth and speaker order

The typical speaker is wider in the LF than the HF range. We accept this as normal, but is it ideal? The answer depends on the application. If a speaker lives a life of solitude, never coupling with other speakers, then a flat beamwidth would seem ideal (Fig. 8.8). A different ideal shape emerges for speakers that play well with others. Proportional beamwidth is preferred for large quantities in highly overlapped arrays. The hybrid of these two is a junction of proportional LF beamwidth and constant HF beamwidth. This speaker type can play well enough alone or in limited quantities.

An ideal flat beamwidth speaker would have the same coverage shape at every frequency. Therefore it fits the room just as well at 100 Hz as 10 kHz. Match speaker coverage to the room shape and aim it for minimum variance and we are done. I'm not saying I've seen this unicorn, but we can dream, eh?

Let's focus on the standard tools of the trade rather than an improbable ideal. A typical full-range two-way or three-way speaker is a beamwidth hybrid: a combination of front-loaded LF and horn-loaded HF, with the MF

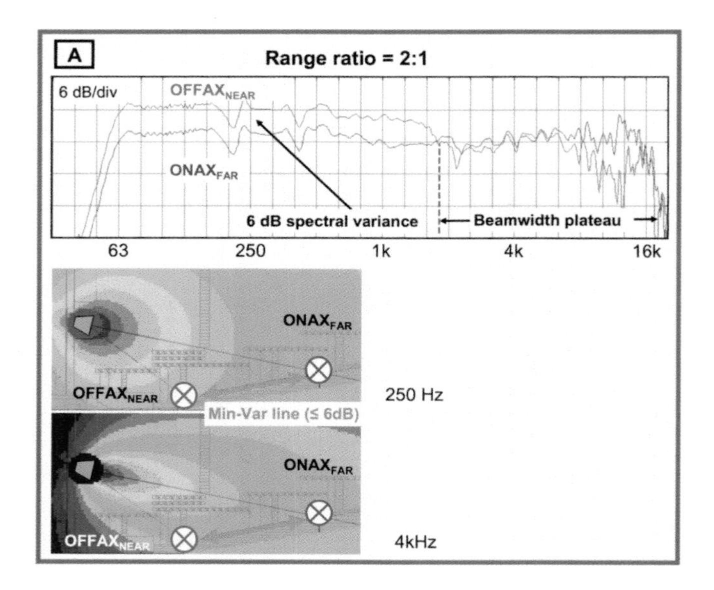

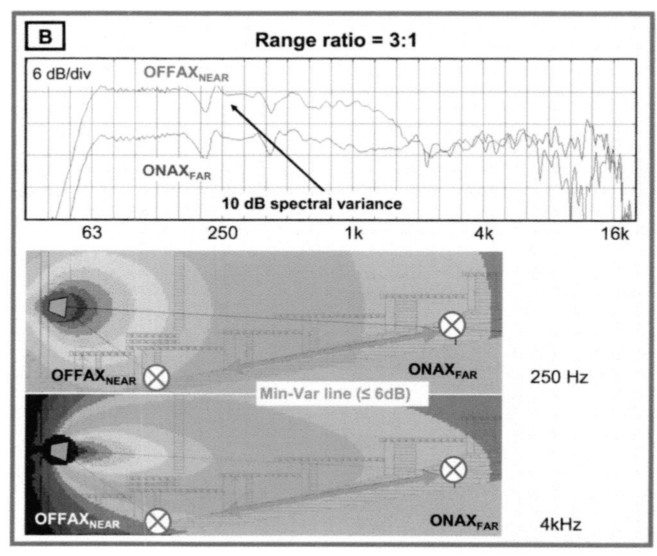

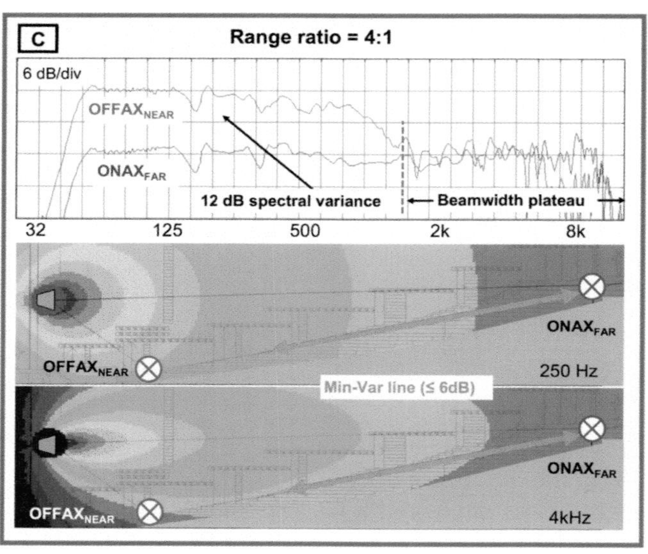

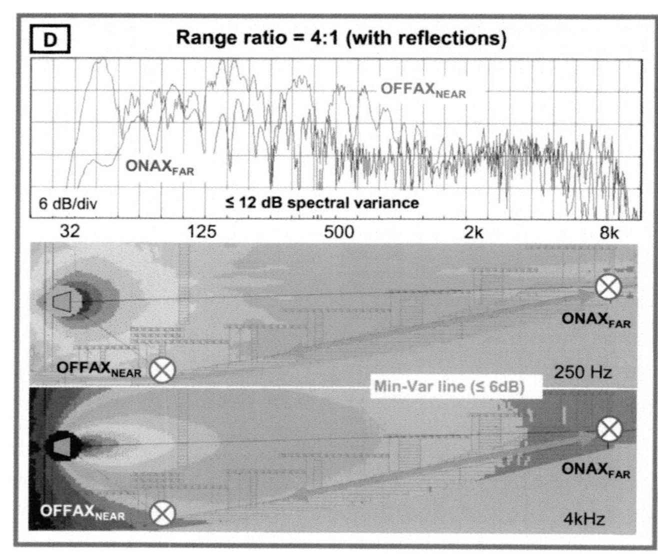

FIGURE 8.7
Example of the range ratio effects on pink shift (spectral tilt). (A) The spectral variance is at the edge of minimum variance (6 dB) for the range ratio of 2:1. (B) The speaker is tilted up to maintain minimum-level variance in the HF range for the 3:1 shape but the spectral variance now reaches 10 dB. (C) The 4:1 range ratio just makes it worse (12 dB). (D) Reflections reduce the ratio (adding more tilt at the rear) but they are not reliable partners for minimum variance.

going either way. Large-format full-range speakers with horn-loaded LF exist, but few are large enough to flatten the beamwidth fully down to their 60 Hz range limits. The standard beamwidth for this family is a diagonal line in the LF (proportional beamwidth) and a flat line in the HF (constant beamwidth). The LF response is standard front-loaded cone driver behavior: a steady rise in directivity until the pattern breaks up into lobes and nulls. Coverage is wide when the transmitted wavelength is large compared with transducer diameter and narrow when the ratio is low. A standard coverage milestone is found when λ and driver size match: 90°. Frequencies below this are wider than 90° and those above are narrower. Bigger drivers hit the 90° beamwidth landmark at a lower frequency than smaller ones.

The other side of the beamwidth hybrid is the horn, which flattens the beamwidth within a given range. Again we're limited at the bottom and top by wavelength vs. transmission device ratios. The horn loses control and beamwidth widens when λ is too large (sooner on highly directional horns than wider ones). The upper end is again limited by lobing. It takes a big horn to go narrow and low but it finishes (lobes) early. A small horn starts and finishes higher.

The hybrid beamwidth is a graft of the diagonal LF and plateau HF. We can define it by the plateau width and corner frequency, e.g. 80° from 1 kHz or 50° from 2 kHz. These two parameters give us enough info to select the right speaker for minimum-variance coverage and aim it.

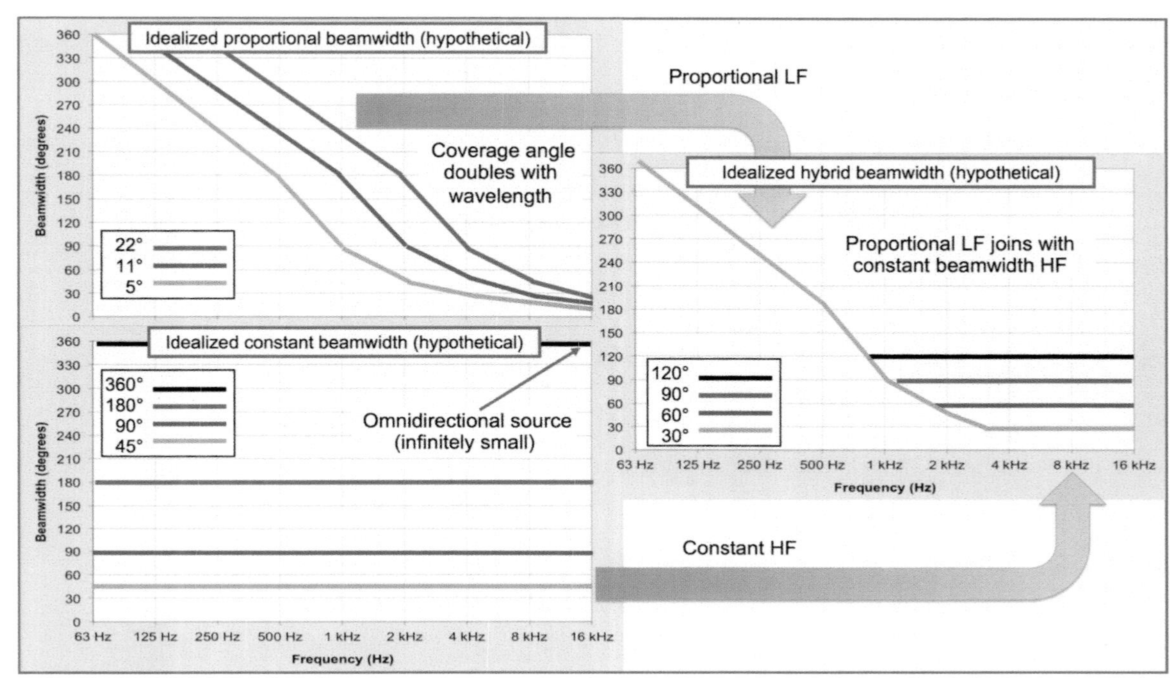

FIGURE 8.8
Hypothetical flat, proportional and hybrid beamwidth trends

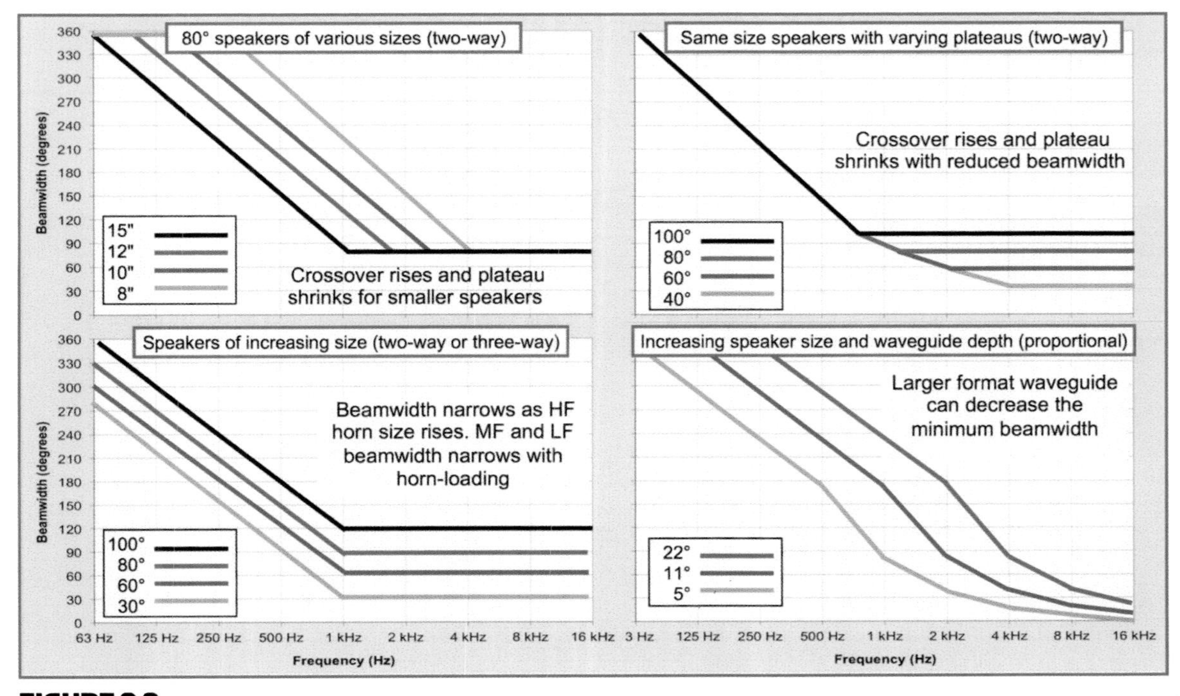

FIGURE 8.9
Simplified trends in the standard beamwidths of single speakers

Let's observe the beamwidth plots for some representative single-speaker elements (Fig. 8.9). As solitary elements their spectral variance is a simple matter of slope.

Five speakers will play a major role in the examples going forward. Their beamwidth (Fig. 8.10) and aspect ratio shapes over frequency (Fig. 8.11) are introduced here.

REPRESENTATIVE EXAMPLE SPEAKER BEAMWIDTH AND FAR

- First-order plateau, 90°, 1 kHz, FAR 1.4.
- Second-order plateau, 40°, 2.5 kHz, FAR 3.0.
- Second-order plateau, 35°, 1.5 kHz, FAR 3.3.
- Third-order proportional, 15°, FAR 8.
- Third-order proportional, 7°, FAR 16.

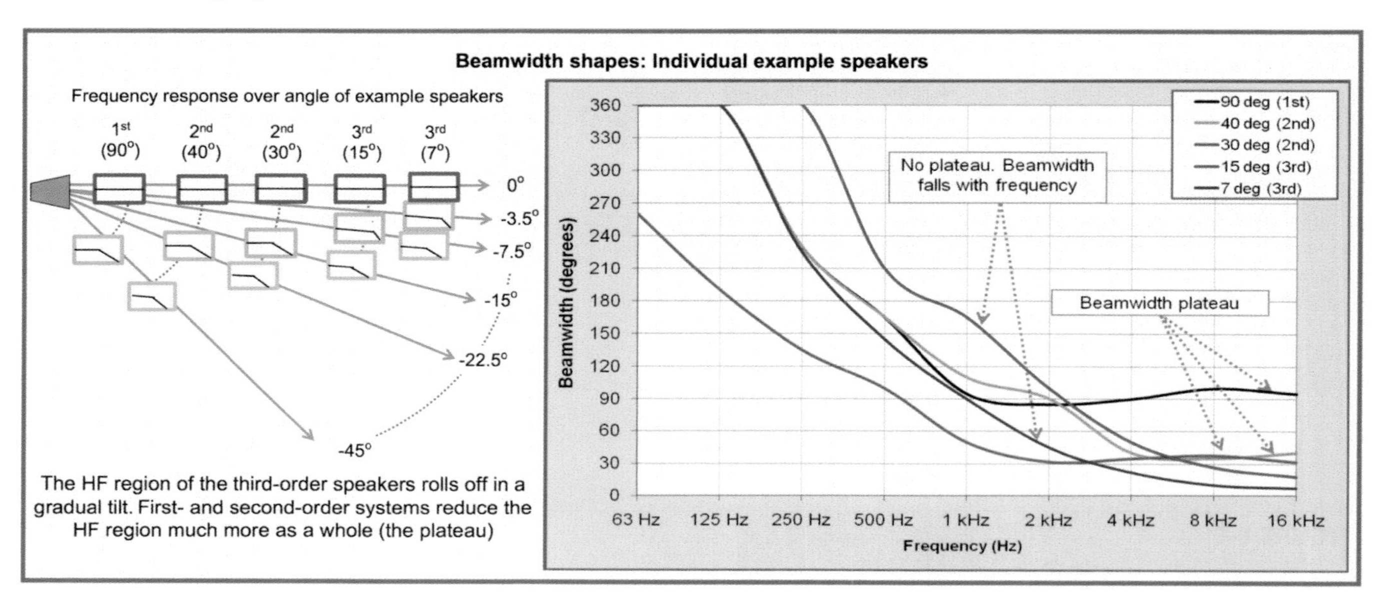

FIGURE 8.10
Beamwidth plots of some representative speakers used in this book. The first- and second-order speakers have beamwidth plateaus whereas the third-order models are wavelength proportional (increasingly narrow with frequency).

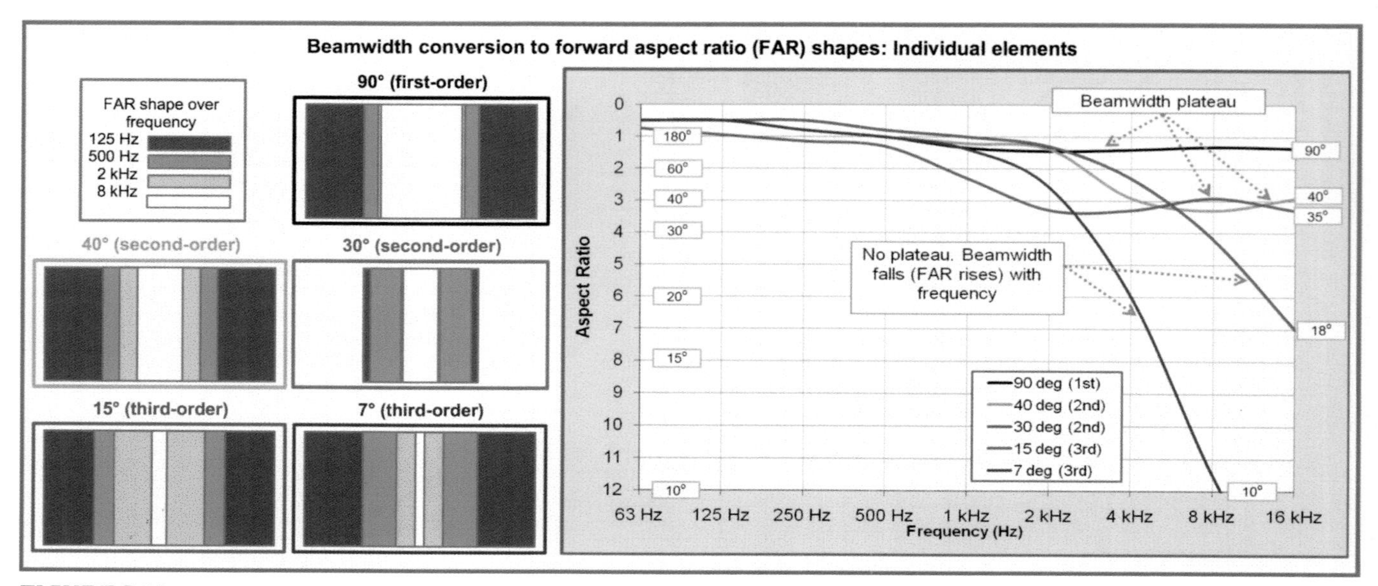

FIGURE 8.11
Aspect ratio over frequency for the same speakers shown in Fig. 8.10. Notice that the continuous narrowing of the third-order speakers is more apparent than in the previous figure. The left side shows each of the five speakers as a set of overlaid FAR shapes over frequency. The first-order shapes are similar over the four ranges whereas the third-order sets reveal huge differences.

8.4 SONIC IMAGE VARIANCE

The surest way to achieve realistic sonic imaging is unplug the sound system and shout. For all other applications we need to consider speaker placement, timing, equalization and level. The perception mechanisms in play here are discussed in Chapter 5. Here we apply it to a stage source and loudspeaker.

8.4.1 Forward range variance

A natural source sounds close in the front rows and distant in the rear (because it is!). This is forward sonic image variance, a natural aspect of sound transmission. One of our sound design goals is to reduce this effect so folks in the rear feel closer than they actually are. Matched spectral tilt (pink shift) creates the perception of equivalent source distance to listeners near and far. Close seats need the least forward movement and rear seats the most. We want to reduce the difference, not eliminate it entirely. Pushing too far will make it sound unnatural (because it is!).

8.4.2 Vertical and horizontal range variance

Moving off axis adds spectral tilt, pushing the perceived image away. Listeners at equivalent-distance ONAX and OFFAX positions do not share equivalent sonic image range. The OFFAX position's excessive pink shift makes a near source sound overly distant, an equal but opposite error to making a distant source sound overly close.

The axial effects on range perception progress equally in the vertical and horizontal planes. Matched spectrum over the vertical and horizontal plane minimizes the differences in perceived range.

8.4.3 Vertical location variance

A single speaker in isolation is guaranteed to image to its actual location (barring smoke and mirrors). If the speaker is reinforcing a stage source, the sound image location is found at or between the two sources (Fig. 8.12). The vertical image is primarily level dependent; therefore the dominant source carries the localization cues. When levels approach unity the image will be centered and stretched between the sources. It's unlikely the relative level between a stage source and single speaker will be constant over the listening area (front to back, top to bottom or side to side). The only way to minimize sonic image variance in such applications is to place the speaker as close as practical to the stage source and operate it at a low relative level. Oh . . . and tell the actors to project!

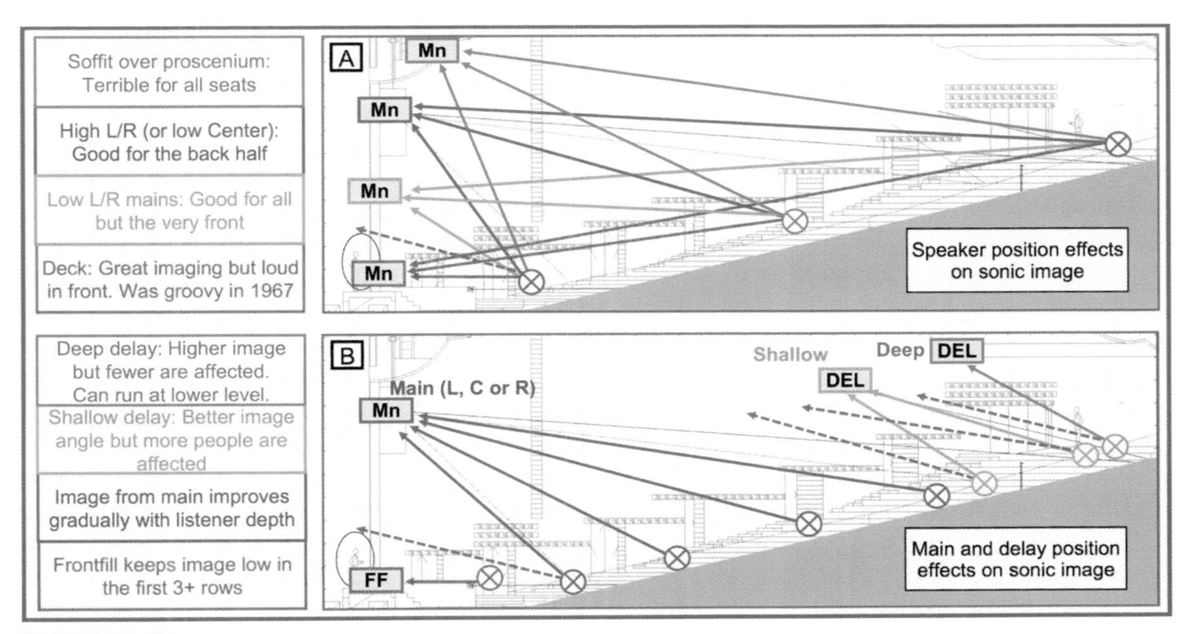

FIGURE 8.12
Sonic image variance in the vertical plane with a sound source at stage level. (A) Sonic image variance rises along with the height of the main system and decreases with depth in the room. (B) Frontfills reduce the sonic image variance of the worst-case-scenario close seating. Delays can reduce level, spectral and ripple variance at a cost of some sonic image variance.

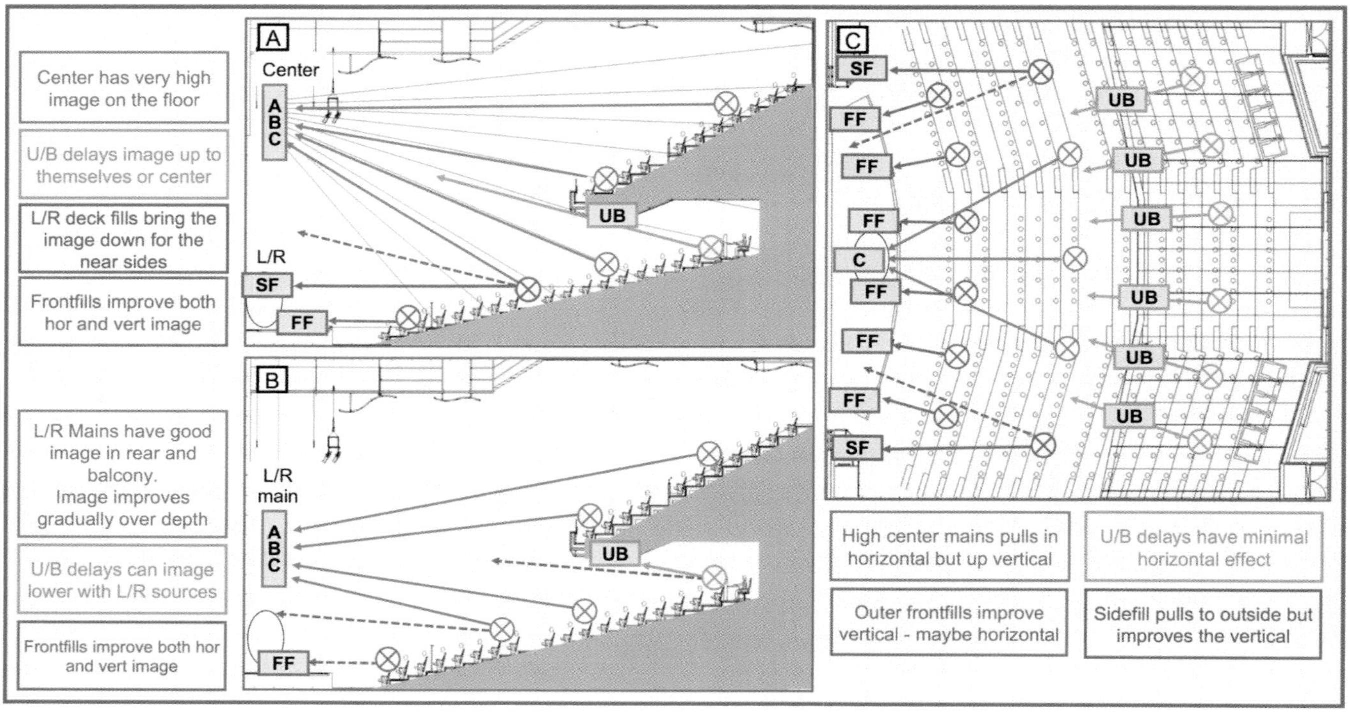

FIGURE 8.13
Sonic image variance example: (A, B) Vertical plane considerations in a hall with center and/or L/R mains. (C) The role of fill speakers in minimizing sonic image variance in both planes.

8.4.4 Horizontal location variance

Horizontal image is both time and level dependent. The earliest source dominates localization when levels are close (precedence effect). Leading by just a few milliseconds has a strong localization effect, which can be reversed by level dominance. It is possible to position the sonic image between the sources inside a 0–7 ms and 0–10 dB window. Level offsets of more than 10 dB (isolation zone) keep the image on the dominant source. Once again, nearby placement and low-level operation of the speaker are critical for minimum sonic image variance in this plane. Our horizontal localization perception mechanism is much more sensitive than the vertical. We detect horizontal location variance easier than vertical, a fact that should be considered when compromises must be made (Fig. 8.13).

8.5 RIPPLE VARIANCE (COMBING)

A single speaker should have just one source of ripple variance: reflections. Its ripple variance progression runs in parallel for multiple reflections, and their added effects are level proportional. The progression finishes at the spatial crossover (the reflecting surface). This is the zero point for ripple variance (coupling), but is adjacent to the area of maximum combing. Ripple depth falls as we move farther away and/or if the surface is absorptive.

8.5.1 Forward ripple variance

Forward ripple variance results from a reflection off a forward surface (such as the rear wall). The path is a straight line from speaker to listener to wall and back again to the listener. In practice our head would block this exact path but the example serves the purpose of analyzing the front to back return paths.

The ripple variance progression (near field of speaker A to rear surface S) is:

ONAX A (isolation) => (transition) => (combing) => XOVR AS (coupling)

A unique cycle is found for each reflection path and the combined effects accumulate at a given location. The zones are proportional to the listening position relative to speaker and surface. The reflection strength at the mid-point location is approximately -10 dB, assuming no absorption. Therefore, from mid-point forward can be considered the isolation zone. The combing zone is reached (-4 dB) at the 77% mark (e.g. 77 ms away from the speaker with the wall at 100 ms). The transition zone falls in the area between 50% and 77%. If the wall is absorptive then isolation and transition zones expand and the combing zone shrinks. An absorption coefficient of 0.37 (2 dB) moves the zone markers to 57% and 88% whereas 0.60 (4 dB) moves the transition zone to 67% and eliminates the combing zone (Fig. 8.14).

The zone progression rate is always longer in the HF range due to air loss, i.e. the isolation zone will be reached deeper in the room due to air loss in the reflection path. Surface absorption can further extend isolation in the HF and possibly other ranges as well (depends on the surface absorption coefficient frequency).

Forward ripple variance is minimized by absorption and pattern control. Absorption is helpful in all cases. Pattern control is effective only when we can avoid the reflecting surfaces and still cover the audience target. Reflective surfaces directly behind the listeners are the most difficult to avoid by pattern control and therefore are the best candidates for heavy absorption. Unfortunately we cannot set a forward range control in our speakers and tell them to cover the last seat and stop before the rear wall.

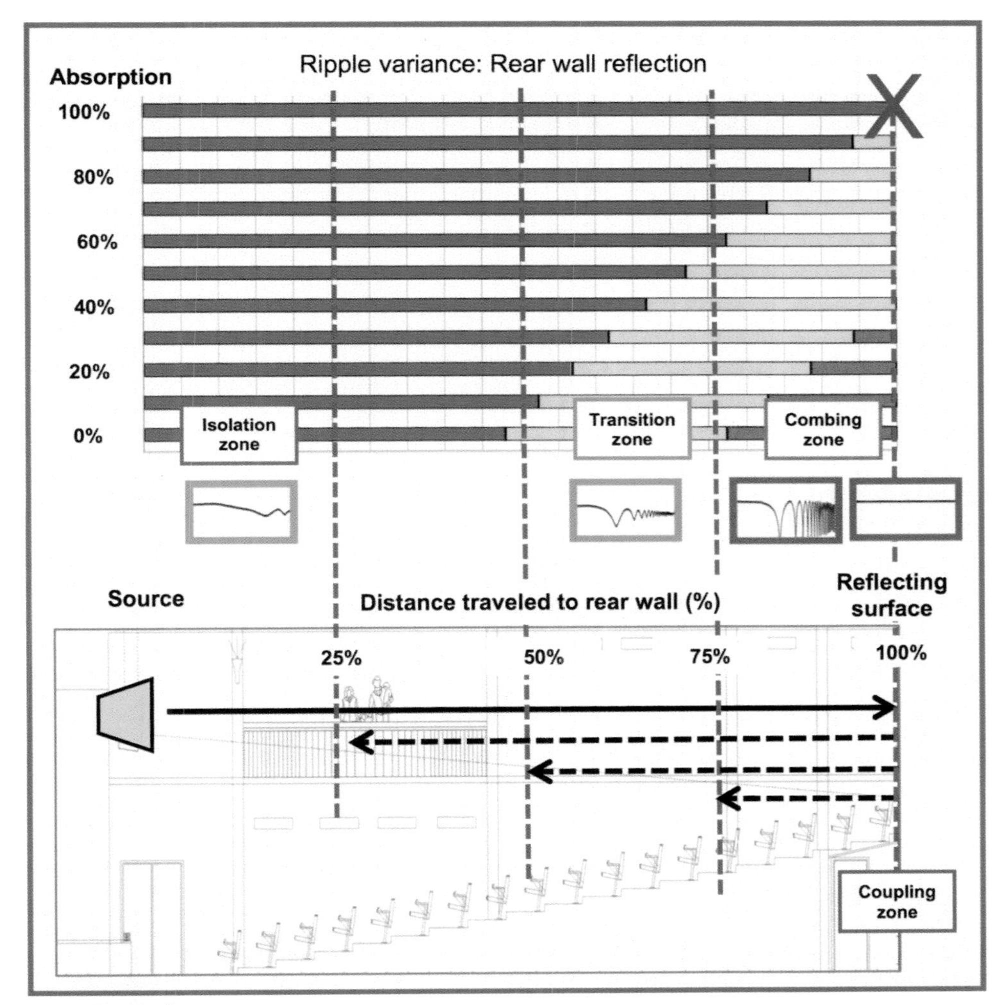

FIGURE 8.14
Forward ripple variance: The standard progression for a speaker facing the back wall is shown.

8.5.2 Lateral ripple variance

Lateral ripple variance results from side wall reflections (horizontal plane). It has several possible progressions depending on how much isolation we have at the ONAX location.

THE LATERAL RIPPLE VARIANCE PROGRESSION (FROM ONAX A TO LATERAL SURFACE S)

- **Isolation @ONAX <4 dB**: ONAX A (combing) => XOVR AS (coupling).
- **Isolation @ONAX <10 dB**: ONAX A (transition) => (combing) => XOVR AS (coupling).
- **Isolation@ONAX >10 dB**: ONAX A (isolation) => (transition) => (combing) => XOVR AS (coupling).

Lateral ripple variance is minimized by absorption and pattern control, just as the forward reflections were. Pattern control has a much higher probability of success here because it's easier to align the speaker coverage pattern to drop off before hitting the surface. The most favorable lateral surface is the outward splay wall (the hall gets wider from front to back) because this follows the speaker's coverage shape. The least is the inward splay wall, which runs directly counter to speaker propagation.

8.5.3 Vertical ripple variance

Vertical ripple variance results from a reflection arriving from above or below. The progressions are the same as the laterals just described but there are some practical differences. Vertical ripple variance is distinctly asymmetric. It's almost certain that we are much farther from the ceiling than the floor and that their acoustical properties are very different. The floor is almost guaranteed to be a hard surface because it must support the audience. Acoustic treatment ranges from zero to carpet, so only the HF range has a chance at absorption. The floor may be covered with highly diffusive seats and people absorbed in the show. The ceiling? Probably not.

The angle of the floor will range from flat to inward splay (known as "raked") to facilitate the visual sense. It's virtually impossible to implement pattern control to reach the audience and avoid the floor. The coverage approach is also asymmetric. Pattern control is the minority player on the underside and we rely primarily on absorption. Top side we have a wide variety of possibilities: close by or far away, inward angle, flat or outward, absorptive material, reflective or diffusive. The worst-case scenario is close, inward and reflective.

8.6 SPEAKER COVERAGE SHAPES

Rooms are highly variable in shape, but are invariable over frequency. Speaker coverage patterns have limited variability in shape but are highly variable over frequency. Our goal of filling the room with even coverage over frequency can be met with a single speaker that matches the room shape and has a constant beamwidth, or via a speaker array that combines to create constant beamwidth. We begin with the single speaker in this chapter and the array in the next. The different ways of characterizing speaker cover were described in section 3.10.5. Here we apply them to minimum variance.

8.6.1 Radial shape

The radial shape is coverage pizza, an equidistant arc from ONAX$_{NEAR}$ (0 dB) to OFFAX$_{NEAR}$ (-6 dB). This is constant distance (not constant level). It's not a minimum-variance radial line but rather the maximum acceptable variance (6 dB variation), the worst we can stand before calling for backup.

There's a lot of open room here. The level along the radial line from ONAX to OFFAX might progress gradually through 0–1–2–3–4–5–6 dB. It could just as easily be 0–0–0–0–0–0–6 dB, 0–5–5–5–5–5–6 dB or even 5–3–1–0–1–3–6 dB. All we actually know is that OFFAX is 6 dB down from the highest level within the arc, which is assumed to be ONAX center.

Is an 80° speaker the best choice for an 80° room? What's an 80° room anyway? Speakers don't fit into rooms like that, even fan-shaped ones (unless we put the speaker back at the apex of the fan behind the stage, ring!!!). The

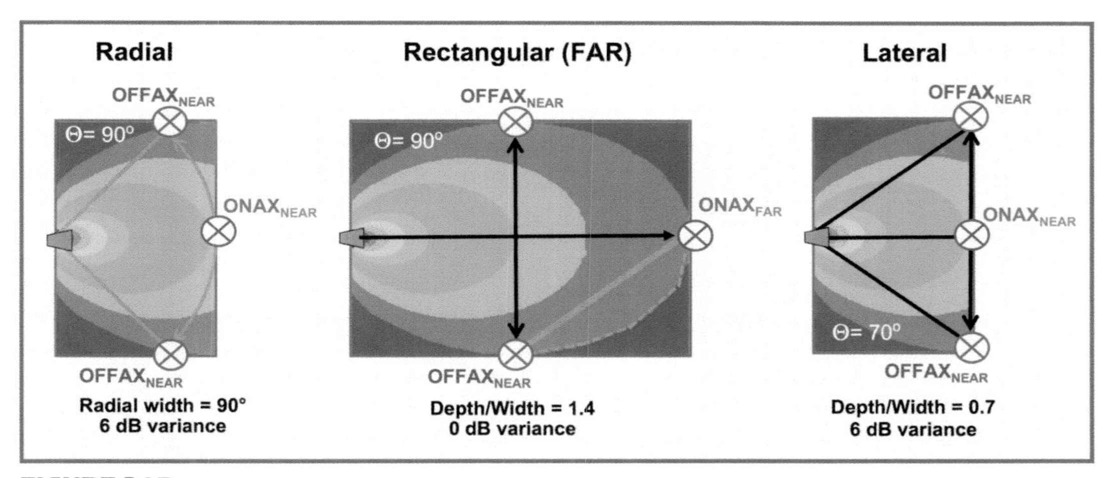

FIGURE 8.15
A comparison of the three speaker coverage shapes: radial, forward aspect ratio (FAR) and lateral aspect ratio (LAR)

radial shape is poorly suited to minimum-variance design for a single speaker because it has no visual link to the minimum-variance shape.

8.6.2 Forward shape (forward aspect ratio)

The forward aspect ratio shape (FAR) outlines the single speaker's minimum-variance shape in the forward direction (exclusive of rearward radiation). It is the rectangulation (depth by width) of $ONAX_{FAR}$ and $OFFAX_{NEAR}$. The minimum-variance line connects between these points and can be simplified as a diagonal but in physical reality is variable and more likely rounded (Fig. 8.15). Speaker shape and room shape are linked via the aspect ratio, i.e. a 2:1 speaker may fit very well in a 2:1 room (a 60° speaker in a 60° room). The FAR helps us put a square peg in a square hole and a rectangular peg in a rectangular hole when we design our systems.

8.6.3 Lateral shape (lateral aspect ratio)

The lateral shape is a straight line of coverage from ONAX to OFFAX. The line is not equidistant (like the radial) or equi-level (like the FAR) because we are moving away from the speaker as we approach the OFFAX position. Therefore the -6 dB point is reached by a combination of radial and forward loss. The lateral shape is relevant to minimum-variance design when the audience shape is a straight lateral line with minimal depth behind the start of coverage (e.g. a frontfill speaker). The speaker coverage width must (at least) match the audience width.

Coverage width is a maximum acceptable variance specification, i.e. 0 dB to -6 dB. If we want to cover 3 meters of audience width we need at least 3 meters of speaker width. How do we get 3 meters of speaker width? They all do it. Somewhere. Just keep moving back till you get it. It's a function of depth/width (see Fig. 3.41).

8.7 ROOM SHAPES

It's tempting to think that the horizontal and vertical planes are simply two versions of the same story. Our approach to coverage belies the fact that these are different animals. The key difference is how we reach the people. In the horizontal plane we plow the coverage through the front rows to the back. The propagation path flows over the shape and the matching of coverage and audience shapes is of great importance. We want wide enough coverage to span the front row, while at the same time not overflowing too much at the rear. This may require a compromise shape that balances underflow and overflow. The vertical plane, by contrast, is only one person deep (lap child excepted). It doesn't matter if our coverage is too narrow in the air above the audience or if we have too much coverage below them in the basement. We need only lay down a line of even coverage at ear level. We evaluate the shapes in fundamentally different ways, and use different versions of the speaker shapes for filling them.

8.7.1 Horizontal

The horizontal shape is evaluated as a solid, a container to fill. The target shape is the audience seating plan (not the walls). The macro shape is depth (length from speaker to last seat), by width. A single speaker perspective distills the shape as follows: distance to the last seat vs. audience width at the mid-point depth. The speaker's coverage shape seeks to approach this dimensional ratio.

A rectangular seating target is the easiest to evaluate because the beginning, middle and ending width are the same. A splayed room (trapezoid) uses the mid-point width as would other variations around the basic rectangle. A narrow, fan-shaped room can be approached this way, but a wide fan can't easily be characterized as having a mid-point width. It can be evaluated as length by radial angle, but this approach is only effective for a single speaker placed at the apex of the fan. That's behind the stage, so forget that. If a fan shape looks like a bad candidate for rectangular approximation then it's probably a bad candidate for a single speaker (a helpful correlation).

8.7.2 Vertical

The vertical shape is evaluated as a coverage line running from the front row to the last seat (Fig. 8.16). The target shape is the audience head height (not the air above or the floor below). The shape is evaluated by coverage angle and range ratio (top seat to bottom seat). If the shape is too complex to be evaluated this way then it's probably a bad candidate for a single speaker (a helpful correlation).

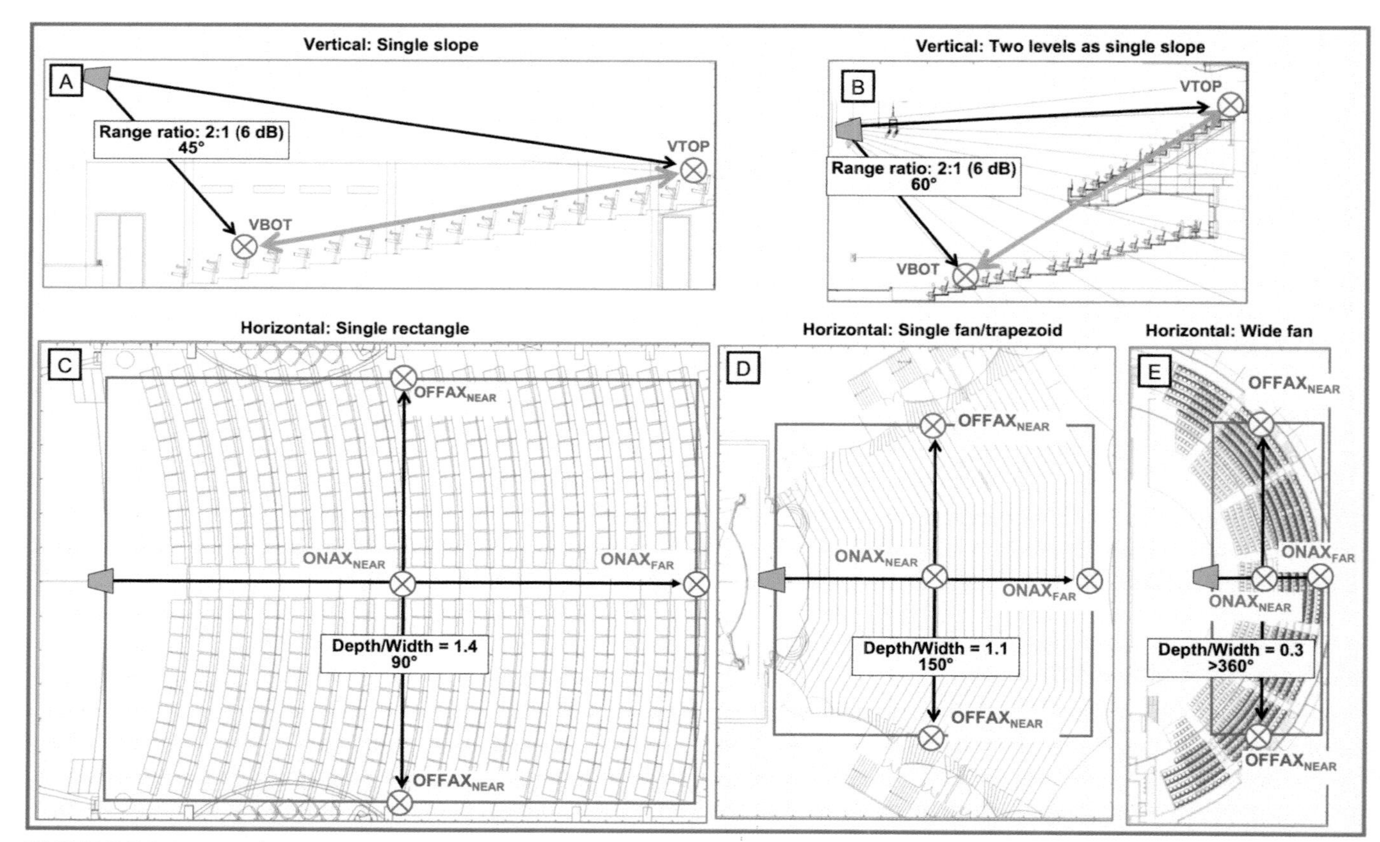

FIGURE 8.16

Examples of horizontal and vertical shape evaluation: (A) Vertical coverage shape connects from the top row to the frontfill transition in a single slope. (B) A single slope approach applied to a room with a relatively short balcony. Deeper balconies require separation. (C) Horizontal coverage is evaluated by the FAR shape for a rectangle and (D) a trapezoid. (E) Overly wide fan shape that cannot be covered by a single speaker.

CHAPTER 9

Combination

9.1 COMBINED SPEAKER VARIANCE PROGRESSIONS

combination n. 1. combining; combined state; combined set of things or persons. 2. united action.
Concise Oxford Dictionary

This chapter explores minimum-variance behavior beyond the single-speaker response. The combined response is the sum of its parts. Therefore, our study of the single-speaker shape was time well spent. Let's begin with a simple question: What is the combined coverage angle of two speakers compared with one? The answer is not so simple. The result can range from double to half of the individual element (Fig. 9.1). The decisive factor is the ratio of two opposing forces: overlap and isolation. These come in two forms: angular and displacement. We reduce overlap by isolating the sources, i.e. pointing them in different directions (angular), moving them apart (displacement), or both. The elements retain much of their individual character when overlap is low, melding into sculpted shapes unique to combined arrays. When overlap is high, the individual elements mostly disappear and merge to resemble a unique (and narrower) single speaker. Combined arrays have characteristics distinct from the elements that comprise them and are highly variable over space and frequency. The shapes are unique, but not random, and therefore understandable, predictable and, most importantly, tunable. Overlap and isolation are the shaping tools. Isolation increases our ability to create complex shapes but has minimal ability to couple for power. There is no power addition without overlap but this carries the risk of combing. This is the Yin and Yang of these powerful forces.

9.1.1 Level variance progressions

We can extend the minimum-level variance line beyond the limits of a single speaker in three directions: forward, radially and laterally (sideways) (Fig. 9.2).

Forward extension is accomplished by adding delayed speakers in the main speaker's on-axis plane. The forward speaker fills in to offset the main's standard progression loss. The combined signals propagate onward from there

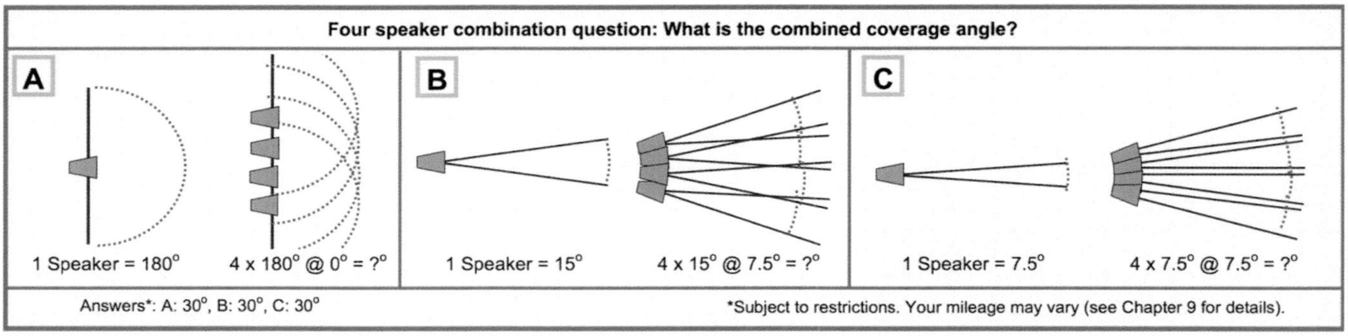

Four speaker combination question: What is the combined coverage angle?

A
1 Speaker = 180° 4 x 180° @ 0° = ?°

B
1 Speaker = 15° 4 x 15° @ 7.5° = ?°

C
1 Speaker = 7.5° 4 x 7.5° @ 7.5° = ?°

Answers*: A: 30°, B: 30°, C: 30°

*Subject to restrictions. Your mileage may vary (see Chapter 9 for details).

FIGURE 9.1
The combination question

and the loss progressions resume (unless another forward extension is added). The minimum-variance (MV) line runs forward (front to back). The forward propagation loss of the individuals (-6 dB) is offset by the summation of the pair (+6 dB). Forward extension has two basic forms: "delay" and "relay." The former is found under and over balconies whereas the latter is found at festivals. The differentiating factor is relative level. We are in the "delay" paradigm when we meet the mains at equal level (or less). The main/delay combination does not exceed 6 dB of addition while also providing increased signal/noise ratio. Forward speakers that dominate the distant mains are power-boosting stations (like cell phone relay towers). Forward extension with delays has a limited practical range whereas extension with relays is limited only by budgets.

Highly overlapped coupled systems create another form of forward extension. The combined pattern is a narrowed version of the individual elements and therefore has more depth for a given width. This gives the system more "throw," a form of forward extension. This is often attributed exclusively to the coupled line source but also applies to the point source and point destination as long as angular overlap is dominant over angular isolation.

Point source arrays create radial extension by fusing speakers at their angular edges. The MV line curves in an arc (from ONAX of one speaker to the next). The individual axial loss (-6 dB) is offset by summation of the pair (+6 dB). The upper limit of radial extension, for obvious reasons, is 360°.

Uncoupled line source arrays are lateral extenders. Sources are spread along a line, as is the coverage. Again the loss of the individuals (-6 dB) is offset by the summation of the pair (+6 dB), but this time the MV line runs straight across (from ONAX of one speaker to the next). Lateral extension accrues incrementally as devices are added and, again, is limited only by our budgets.

These extension mechanisms can also be combined. Underbalcony delays are an example of combined extension: forward (mains + delays) and lateral and/or radial (delays + delays).

9.1.2 Spectral variance progressions

Spectral variation for a single full-range speaker is a fixed parameter. Equalization, level or delay will not change the coverage pattern over frequency. Arrays, however, can be constructed with combined coverage shapes that differ from the individual components (Fig. 9.2). A coupled point source array of high-order speakers can combine to minimum spectral variance by spreading the isolated HF range and narrowing of the overlapping LF range. A very simplified example: A pair of second-order speakers are arrayed at their unity splay angle of 40°, which spreads the HF coverage to 80°. The individual LF coverage angle is far wider than the 40° splay angle. The resulting overlap couples at the center and narrows the LF coverage. The combined shape has lower spectral variance than the individual elements.

SPECTRAL TILT

Speaker arrays share their LF range much more than their HF. The result is spectral tilt in proportion with quantity. Coupled arrays can tilt more evenly over the spectrum than uncoupled arrays, which are limited by ripple variance. We again see spectral tilt in favor of the lows as previously with added reflections (section 8.2.2).

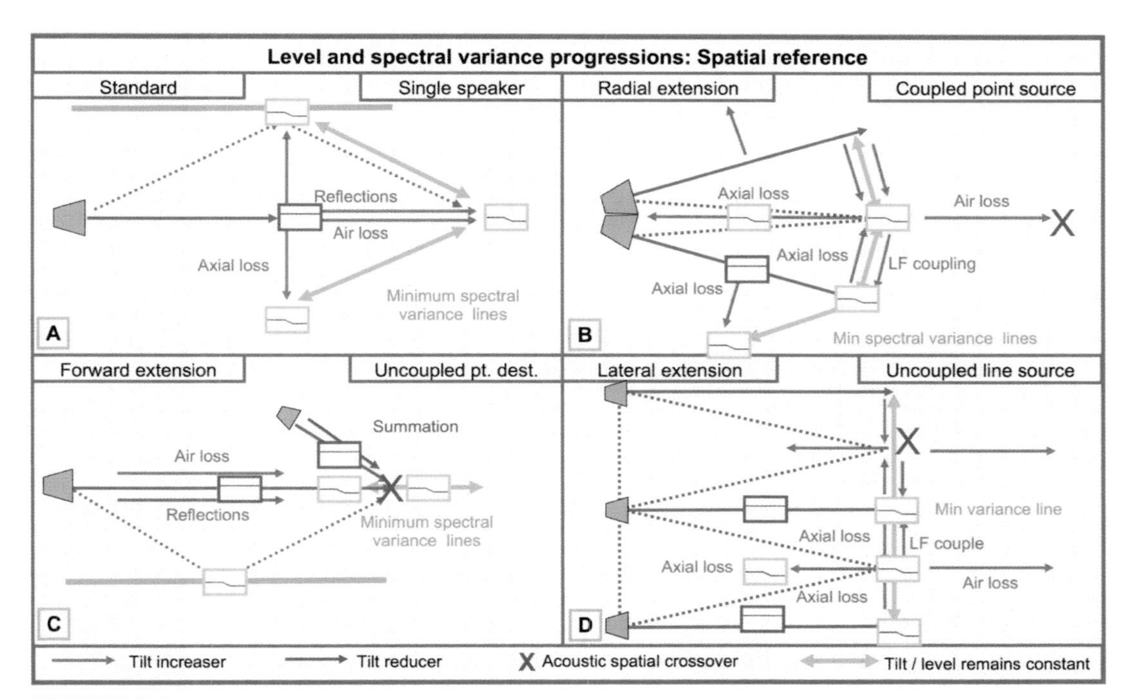

FIGURE 9.2
Summation-related spectral variance progressions

Some representative samples of summation-based mechanisms are shown in Fig. 9.3, where spectral tilting trend lines are compared. The progressions show a consistent trend in favor of LF content over HF. This common trait can be exploited to create a consistent spectral tilt. Because most progressions lead to tilting, the spectral variance is reduced by matching tilts rather than by futile efforts to stop the progression. Tilt can be leveled by equalization to the extent desired. A frequency response need not be flat to be considered a minimum spectral variance. The decisive factor is consistency. Responses tilted in a similar manner are just as matched as flat ones.

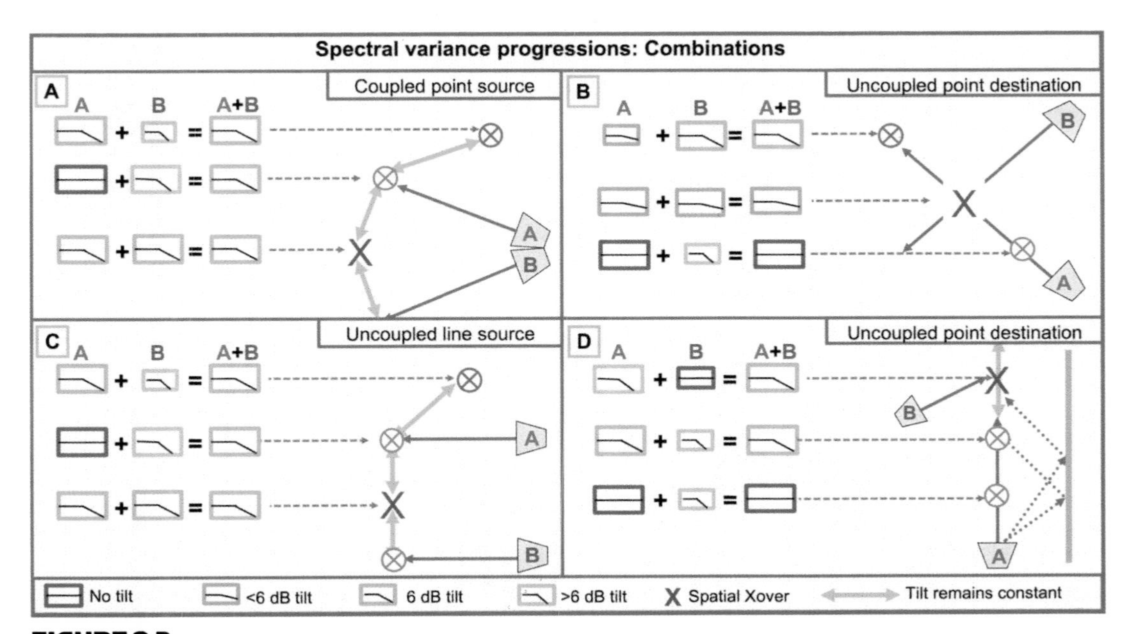

FIGURE 9.3
Spectral variance progressions related to the combinations of speakers. Individual speaker responses are shown with representative tilt and level scaling. Combinations at various locations are shown for comparison.

9.1.3 Ripple variance progressions

The standard ripple variance progression is shown in Fig. 9.4. The cycle is extendible and repeatable for multiple elements but follows a familiar pattern. The center point in the progression is the phase-aligned spatial crossover point (XOVR). This is the point of lowest variance in the progression, and yet it is the center of the area of the highest rate of change in ripple variance. The coupling point is analogous to the point of lowest wind speed in a hurricane: the eye. The area just outside the eye, however, contains the highest variations in wind speed. That is the nature of the XOVR area, and we hope to escape the storm by isolating as quickly as possible.

The ripple variance progression is present in all forms of speaker arrays. Any two speakers in a room will eventually meet somewhere, at least in the LF range, regardless of spacing, relative level or angular orientation. Our focus is on full-range spatial crossover transitions, so we will limit our discussion to those that fall within the coverage edges of the elements. The transitional points will fall in different areas for each array configuration and therein we find the key to managing ripple variance. Some array configurations confine the variance to a small percentage of their coverage. Others fall into the combing zone and never climb out again.

The primary indicators are source displacement and overlap. The least variance occurs when both of these are low. The highest variance occurs when they are both high.

THE RIPPLE VARIANCE PROGRESSION

- **Frequency response ripple**: Ripple increases with overlap and displacement as XOVR is approached.
- **From XOVR**: coupling zone to combing zone to transition zone to isolation zone.
- **From isolation zone**: to coverage edge, or transition zone to combing to coupling at XOVR.

This cycle repeats for each speaker transition until the coverage edge is reached. Each XOVR location will need to be phase aligned. Phase-aligned spatial crossovers cannot eliminate ripple variance over the space. They are simply the best means to contain it.

The progression rate is not equal over frequency. The HF range is the first to enter (and first to leave) each of the zonal progressions. Finalists in both cases are the lows. It's possible, and quite common, to run the HF from XOVR to isolation before the lows have even left the coupling zone. The coupled point source array is the classic example of this. Therefore, our roadmap for the variance progression will need to factor in frequency, since the transitional milestones are not evenly spaced.

RIPPLE VARIANCE GEOMETRY

We have previously discussed how triangulation helps us visualize ripple variance over the space (review Figs 4.16 to 4.18). The four triangle types give strong indication of the spatial behavior of ripple variance. It is given that the ripple pattern is not stable over the space. Our primary concern is identifying areas where combing is strongest (the triangulation method) and most rapidly changing. The rate of change depends upon the spacing and angular

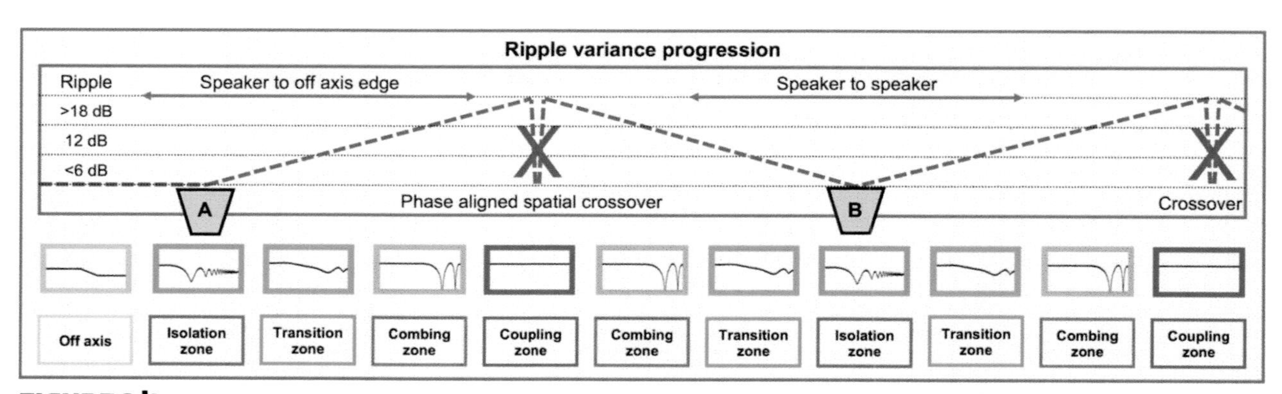

FIGURE 9.4
The standard ripple variance progression from off axis to on axis and through crossover

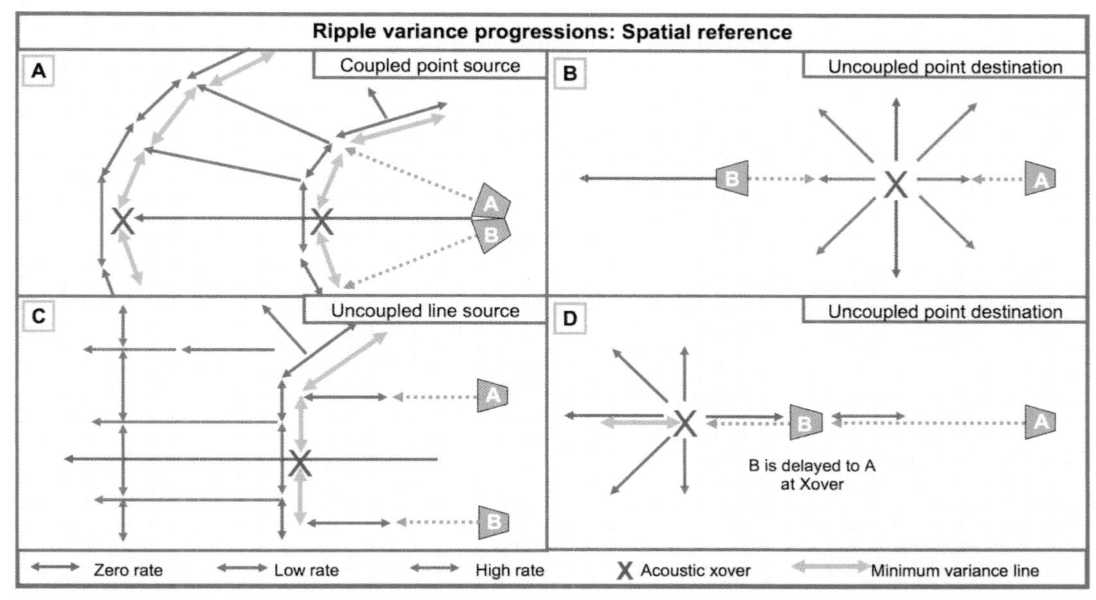

FIGURE 9.5
The spatial reference showing the rate of change for various array configurations

orientation of the sources. The highest rate of change occurs when we are approaching one source while moving away from the other (the line between the sources). The rate decreases when we are moving away (or approaching) from both sources (the uncoupling line). The isosceles triangle, the representative of the coupling zone, has the lowest ripple but will usually have a high rate of change in the right and acute triangle areas nearby. The obtuse triangle will have reduced ripple due to isolation and will enjoy a low rate of change (moving away from both sources). These two scenarios represent the extremes of ripple variance over the space. Example arrays are shown in Fig. 9.5.

9.2 LEVEL VARIANCE OF COUPLED ARRAYS

A series of scenarios follows to illustrate how individual speaker shapes combine into various arrays. Forward aspect ratio (FAR) icons are used in layout form to facilitate visualization of the individual contributors. These are compared with and contrasted to prediction plots of combined arrays in the same configuration. The comparison serves to visually deconstruct the combined prediction response and see the part played by each component. The icons closely match the combined prediction when isolation is dominant and far less so when overlap is dominant (coupling or combing). Coupling modifies the combined shape in a substantial, but consistent manner whereas combing causes frequency-dependent shape-shifting with only vague resemblance to the individuals. We will explore how the shape, spacing, relative level and angle play unique roles in the combination.

9.2.1 Coupled line source

The coupled line source is inherently 100% overlapping (no angular separation and minimal displacement). We start with a minimally displaced third-order system with exclusively coupling zone summation (Fig. 9.6). The aspect ratio icons are overlaid in a vertical line, creating the visual impression of a laterally spread source with a combined coverage angle slightly wider than the elements. The actual combined pattern is narrower than the individual elements because they are most closely phase matched in the center. This shows a limitation in our aspect ratio approach: The simplistic rendering fails to see the phase-driven narrowing of coupling zone summation.

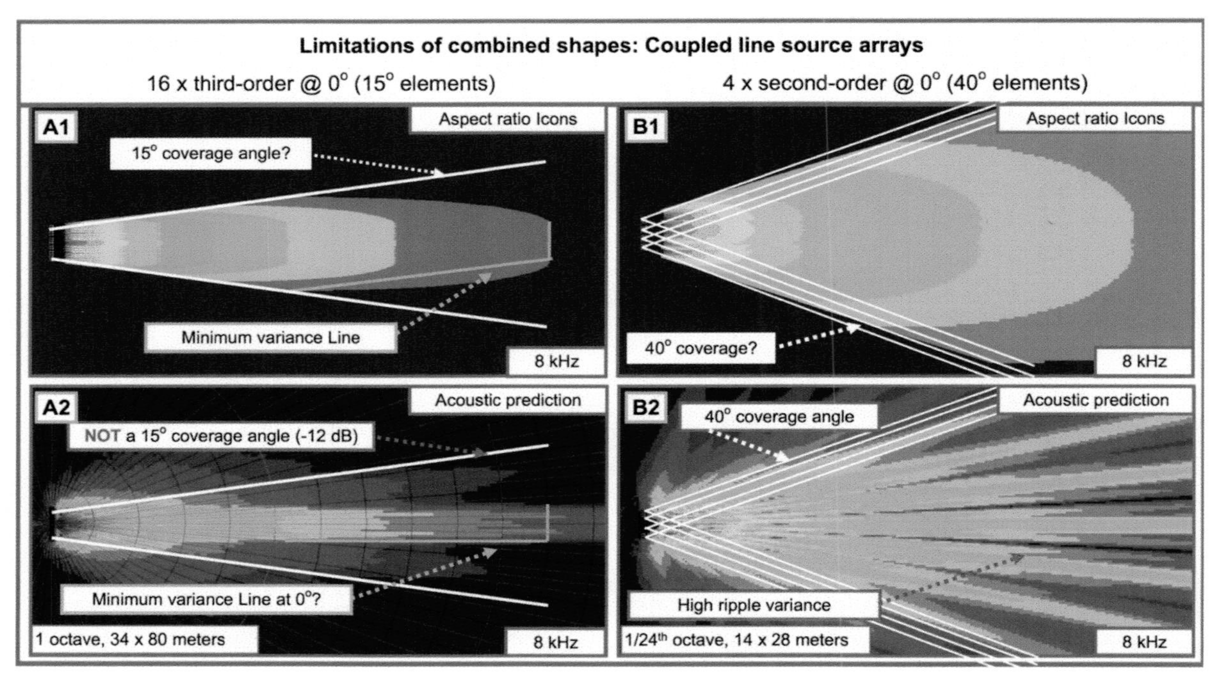

FIGURE 9.6
Minimum-level variance shapes for the symmetric coupled line source array. Left: a sixteen-element third-order speaker array with 0° splay angle and constant level. Right: a four-element second-order speaker array with 0° splay angle and constant level.

Also shown in Fig. 9.6 is a second-order system with enough displacement to create substantial combing. Large displacement and highly overlapped patterns make combing the dominant response over coupling. The resulting pattern has highly variable fingers and lobes over frequency, and therefore lacks resemblance to the individual icon responses. So far the aspect ratio icons have failed to accurately characterize either of the overlap zone behaviors (coupling and combing) because they lack phase information.

Let's look at the coupling zone again and try a different approach with the icons. It's often claimed that arrays combine to resemble a single unified speaker. They can, under the following condition: 100% overlap between closely coupled sources. The composite speaker differs from its elements: it's narrower. Doubling the elements reduces the coverage *shape.* It may be only a slight reduction but could bring it down to half as wide (2× elements = 2× FAR). Additional speakers lengthen the rectangular FAR pattern, without making it wider. Maximum narrowing (without combing) occurs when the sources are ½ λ apart. Smaller displacement results in proportionally less narrowing.

Let's move forward with the ½ λ spacing. Instead of stacking the aspect ratio icons side by side (like the physical speakers) let's try stacking them forward (like the speaker pattern shapes). The forward extension of a fully coupled line source can be visualized by forward stacking of the FAR shapes (Fig. 9.7). A single 180° (FAR = 1) element provides the base shape. Each time a speaker is added to the coupled line source the combined pattern is equivalent to adding another FAR icon forward of the speaker. The FAR doubles with each quantity doubling, and the on-axis power capability rises 6 dB. In each case (two, four and eight boxes) three renderings are seen: the individual elements stacked forward, the combined FAR icon of an equivalent single speaker and a combined acoustic prediction of the given quantity of elements. Note: This effect will not occur until the array is fully coupled, i.e. all of the individual patterns have overlapped. The multi-element assembly process into a combined "single" speaker is shown in Fig. 9.8. Here we see two levels of addition from two pairs of speakers into a trio. Once all elements have overlapped the coverage angle is set and will continue forward in the same shape as a comparable single speaker. The number of assembly steps to reach full coupling (and a constant angular shape) rise with each additional element. Remember that coupled arrays are within partial wavelengths of each other, and uncoupled arrays are multiple λ apart. Displacement must be within ½ λ for FAR multiplication to occur. If not, we get combing instead.

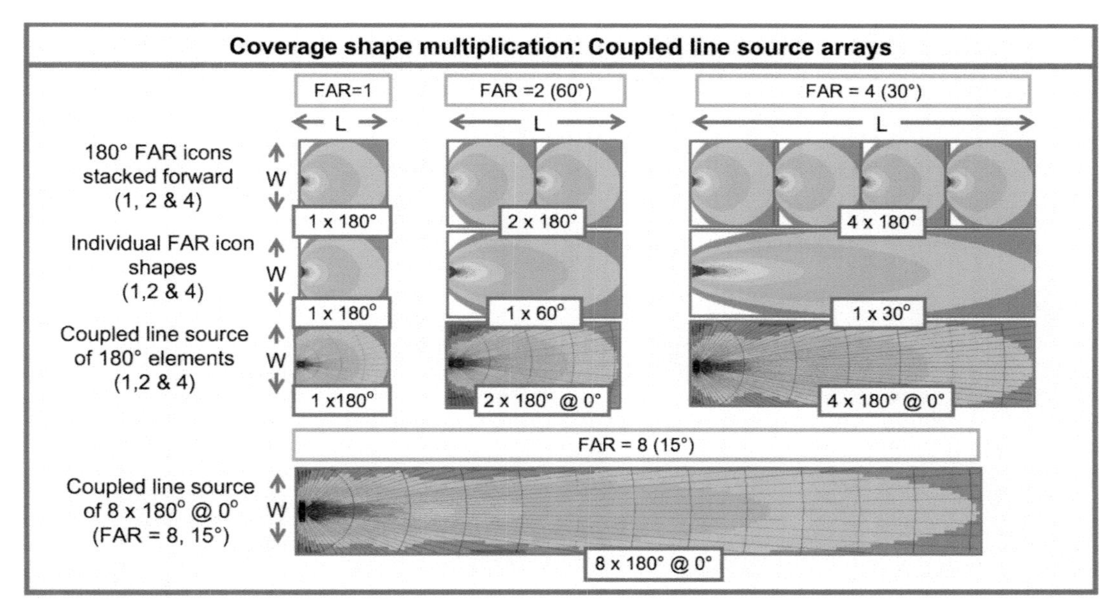

FIGURE 9.7
Quantity effect on the combined aspect ratio for the symmetric coupled line source array. Elements are stacked along the width. Successive quantity doublings cause successive FAR doublings. The combined aspect ratio matches the shape of the individual element aspect ratios placed in a forward line. Coincidence?

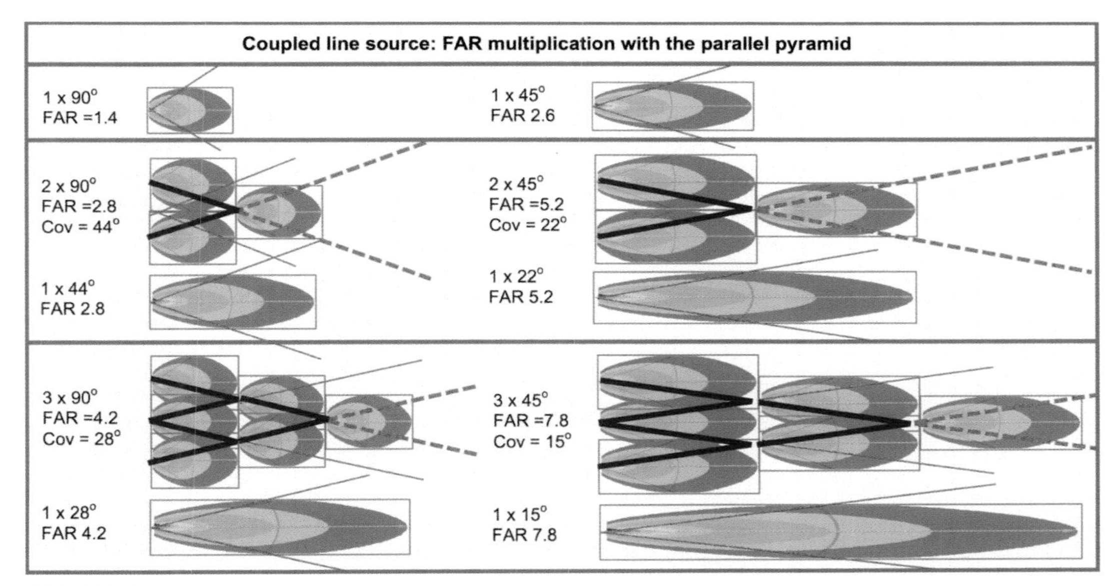

FIGURE 9.8
Combined shape of the coupled line source. The combined coverage angle is found once the array has fully formed at the top of the parallel pyramid.

9.2.2 Coupled point source

Angular separation opens an avenue to isolation, which improves the correlation between the FAR icons and the actual combined shapes. Isolation renders amplitude dominant over phase, allowing the individual icon shapes to hold up. The combined shape is expressed as a coverage angle, in degrees, rather than the rectangular shape of the aspect ratio. The shape of minimum-level variance is now an arc. Our pizza has finally arrived.

The coupled point source spreads energy in radial form (Fig. 9.9). The individual aspect ratio rectangles are spread along the arc like a fan of playing cards. The gaps are filled in by the shared energy and a radial line of

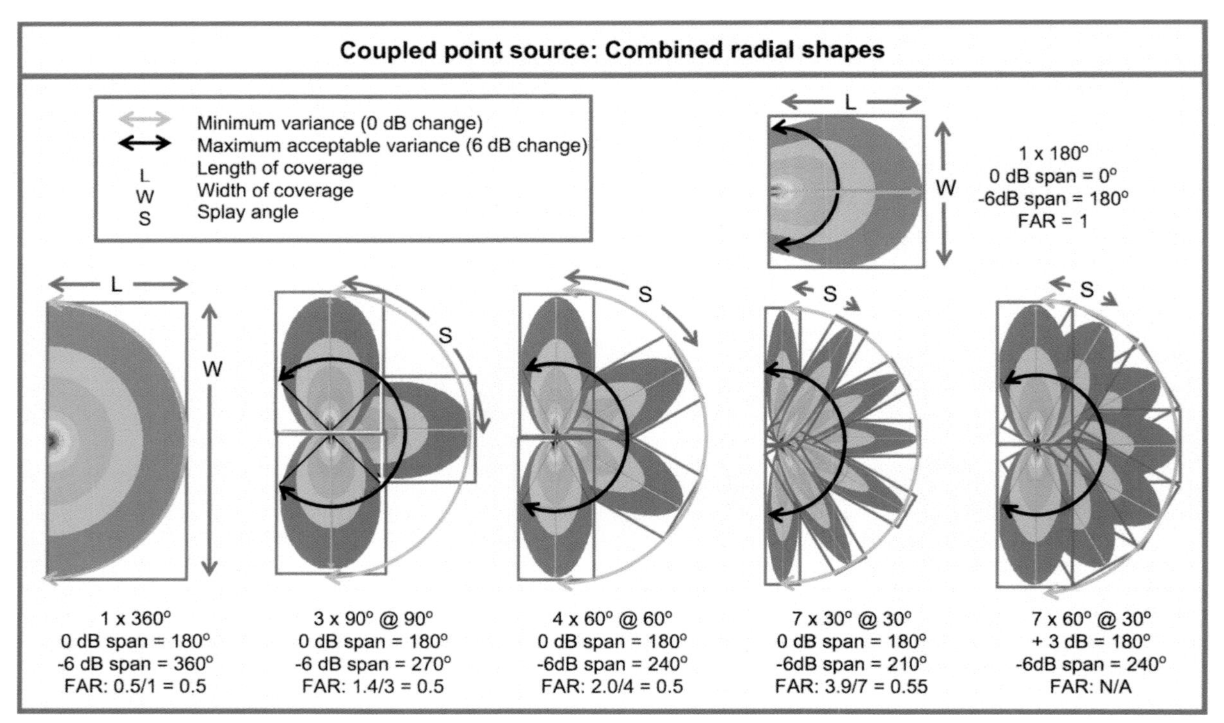

FIGURE 9.9
Quantity effects on the combined aspect ratio for the symmetric coupled point source

minimum-level variance is created when the unity splay angle is used. The minimum-variance line runs between ONAX of the outermost elements. The line of maximum acceptable variance extends beyond the last elements until the -6 dB point is reached. An array for a given angular span of minimum variance can comprise a small number of wide elements or a large number of narrow elements. The latter scenario creates a higher ratio of coverage within the minimum-variance line compared with the outside edges framed within the maximum acceptable span.

An important feature here is the difference between the minimum-variance line (0 dB) and the maximum acceptable variance line (0 to -6 dB). The upper-right portion of Fig. 9.9 shows a single 180° speaker, and its standard square aspect ratio shape. The radial line of maximum acceptable variance spans 180°, and yet the radial line for 0 dB spans 0°. Why? Because a radial arc drawn with the speaker at the center would trace a line that immediately shows level decline as it leaves on axis until the -6 dB point is reached. Compare this with the 360° pattern (lower left), which has an FAR of 0.5. This speaker has a span of 180° on the 0 dB line and 360° on the 0 to -6 dB line. Compare and contrast the two singles with the four combined arrays. The 360° single and the four arrays all stay at 0 dB for a span of 180° (the minimum-variance line) but differ in how far they go before falling to -6 dB (the maximum acceptable variance span). The lesson here is that an array filled with narrow elements has a sharper edge than one comprised of fewer wide elements. The "edge" is revealed as the difference between the minimum-variance and maximum-acceptable-variance lines. If we want 180° of 0 dB level variance we will need to create an array with a 180° span between the on-axis points of the outermost elements. We can do this with three speakers at 90° (0 to -6 dB = 270°), 181 speakers at 1° (0 to -6 dB = 181°) or anything in between. If we are concerned with leakage we will find advantage in keeping the individual elements narrow. An additional note concerns the last scenario, which shows overlap between the elements. Although the overlap will add power capability to the front of the array and may cause increased ripple variance, this will not change the ratio of the minimum-level variance and maximum acceptable variance spans.

The combined FAR for our example point source arrays settles around 0.5, which is the FAR value of a 360° individual element. This is no coincidence. The array's radial spread creates an equal-level arc with the same shape as the 360° single element. The symmetric coupled point source is essentially a sectional version of the 360° shape that can be filled as much as desired.

9.2.2.1 SYMMETRIC COUPLED POINT SOURCE

The symmetric coupled point source can provide curved lines of equal level radiating outward from the virtual point source. The combined response strongly resembles the radial spread of the individual FAR patterns as long as combing is minimal (Fig. 9.10). The correlation to the simplified icons is best with low angular overlap (isolation) and low displacement (coupling), and worst with high overlap and high displacement (combing). The left panels show a second-order array with minimum overlap.

The symmetric coupled point source fills an arc. Two modes of overlap behavior characterize the symmetric point source array over quantity. The 100% overlap model is, of course, a line source, (not a point source). In any case it is the ultimate extreme limit, and its behavior is characterized by power and FAR multiplication. The other extreme is unity splay, which is characterized as coverage angle multiplication. This has the least on-axis power addition, but don't forget that we are spreading power addition across the arc. The middle-ground behavior is the partial overlap configuration, which is characterized as splay angle multiplication. This is the middle ground in on-axis power as well. Note that coverage angle and splay angle are one and the same at unity splay, but not at others.

Angular overlap effects on combined coverage angle

1. 0–95% overlap: combined coverage = (quantity) × (splay angle between elements).
2. 100% overlap: combined coverage = (quantity) × (FAR).

What are our options if we need 40° of coverage? There are countless ways to create a 40° pattern. The following example illustrates the role of overlap in the computation of combined coverage angle.

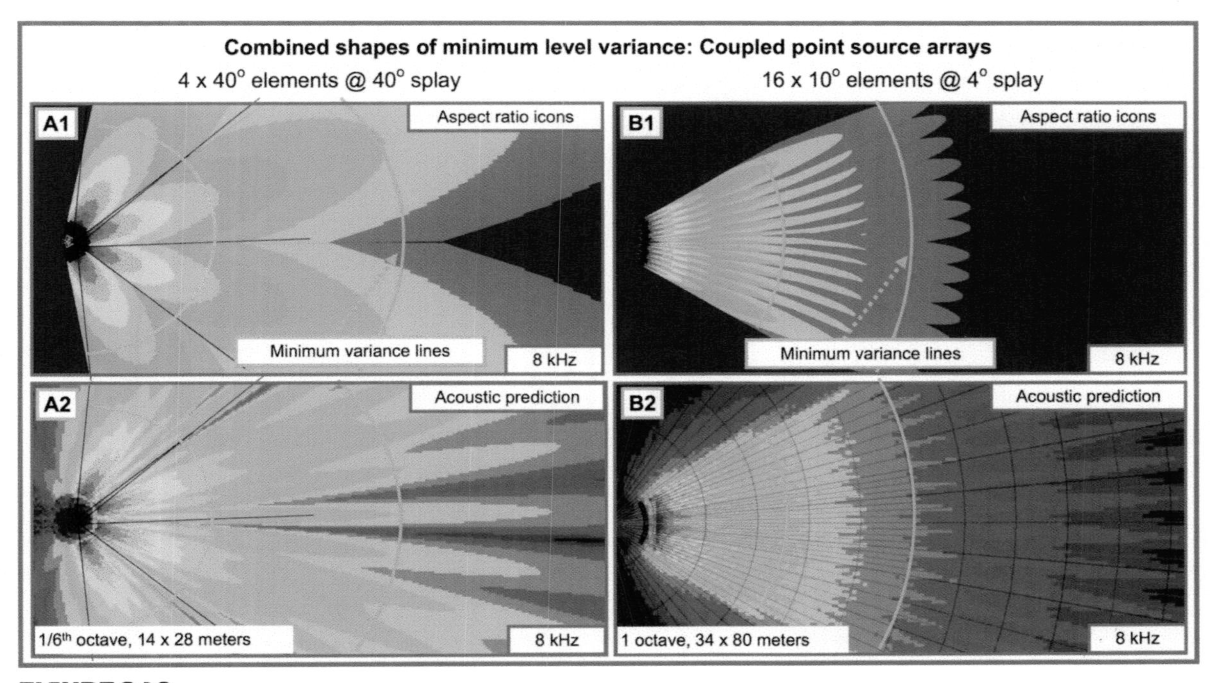

FIGURE 9.10
Minimum level variance shapes for the symmetric coupled point source. Left: constant speaker order, unity splay angle and constant level. Right: constant speaker order, 60% overlap splay angle and constant level.

Combined coverage angle of 40°

- 1 × 40° speaker (FAR 3).
- 2 × 20° speaker @20° splay: 0% overlap (coverage of 20° × 2 = 40°).
- 2 × 27° speaker @20° splay: 25% overlap (splay of 20° × 2 = 40°).
- 2 × 40° speaker @20° splay: 50% overlap (splay of 20° × 2 = 40°).
- 4 × 40° speaker @10° splay: 75% overlap (splay of 10° × 4 = 40°).
- 8 × 20° speaker @5° splay: 75% overlap (splay of 5° × 8 = 40°).
- 2 × 80° speaker @0° splay: 100% overlap (FAR 1.5 × 2 = 3).

There is always a tradeoff of power and ripple variance. Displacement must be kept small when overlap is high. The third-order speaker is our best choice in high-overlap designs. Low-overlap designs favor the first- and second-order plateau beamwidth with its extended isolation zone.

9.2.2.2 ASYMMETRIC COUPLED POINT SOURCE

Differing drive levels create an asymmetric coupled point source, bending the radial shape into an elliptical or diagonal contour. The aspect ratio icons scale with the level change allowing us to see the asymmetric shaping. A level reduction doesn't change a speaker's coverage angle but it does change its *shape*: It gets smaller (Fig. 9.11). A 6 dB loss reduces the icon to half-scale, which reveals its role in context to louder elements. An example asymmetric point source is shown in Fig. 9.12. Notice in panel "B" that the array configuration has highly overlapped coverage angles in the on-axis area of the 10° speaker. Once level is factored in we see that the inverse is true: This is actually the most isolated area, a point made clear by the scaled FAR icons and the combined prediction. It's easy to get worried about the interaction of wide speakers such as downfills, frontfills and delays when we view their overlapping radial coverage lines. It becomes clear they are no threat once level is factored in with FAR scaling.

Compensated unity splay angle

The methods used to create the asymmetric point source array are shown in two scenarios in Fig. 9.13. The left panels show a log-level taper and constant splay angle (50% overlap) among matched components. Such an array aims the on-axis point of each succeeding element at the -6 dB point of the one above. Each layer is

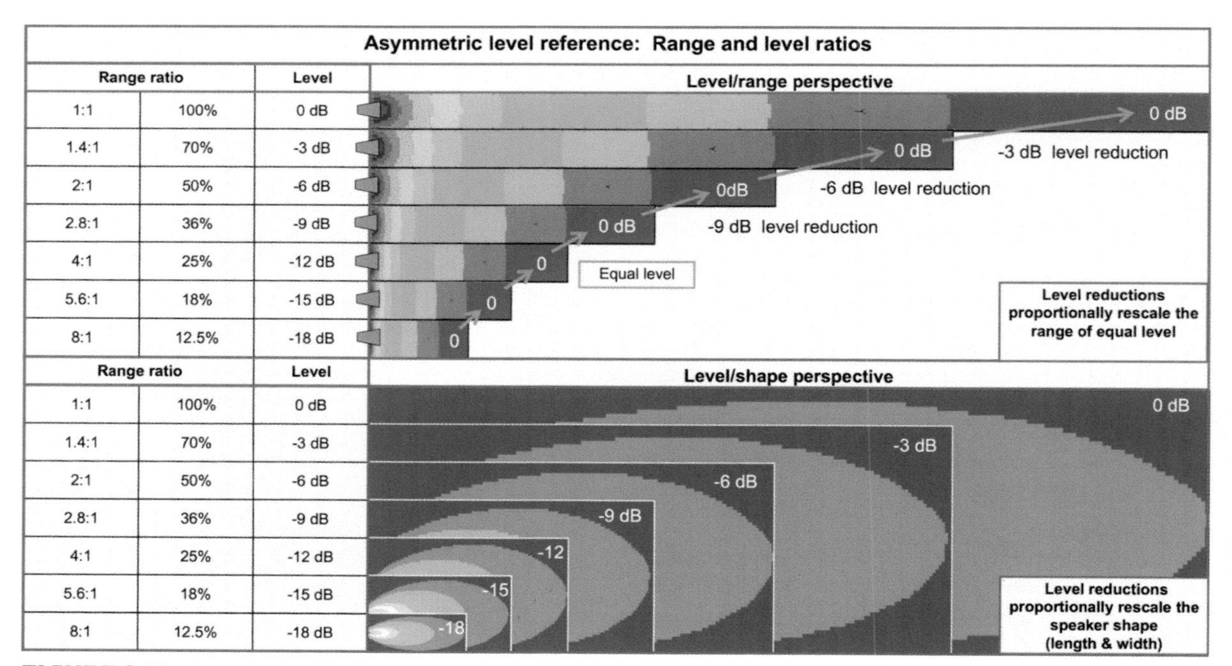

FIGURE 9.11
Level/range ratios for sources with matched origins, including the FAR scaling

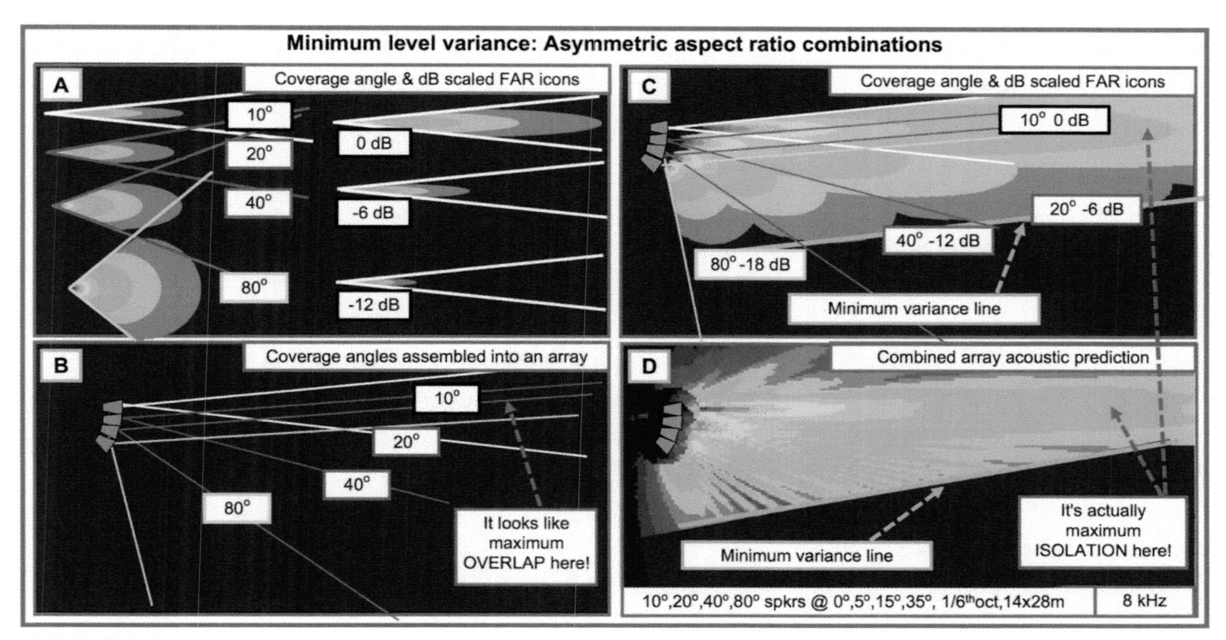

FIGURE 9.12
Aspect ratio icons scaled with level in their individual and combined shapes

successively tapered by -6 dB resulting in a unity XOVR in front of each lower speaker. Offsetting the level taper and splay to achieve a unity XOVR is termed the "compensated unity splay angle." The maximum amount of layer separation is used here (6 dB). The result is a curved minimum-variance region that continues indefinitely over distance. The right panel shows the same principles applied with unmatched elements that double their coverage with each layer. The compensated unity splay angle is applied again but each layer is separated by a larger splay angle in order to preserve the relationship of the on-axis aim point to the -6 dB edge of the unit above. The result is a diagonal minimum-variance line linking XOVR to XOVR, the product of the complementary asymmetry of changing splay and coverage angles.

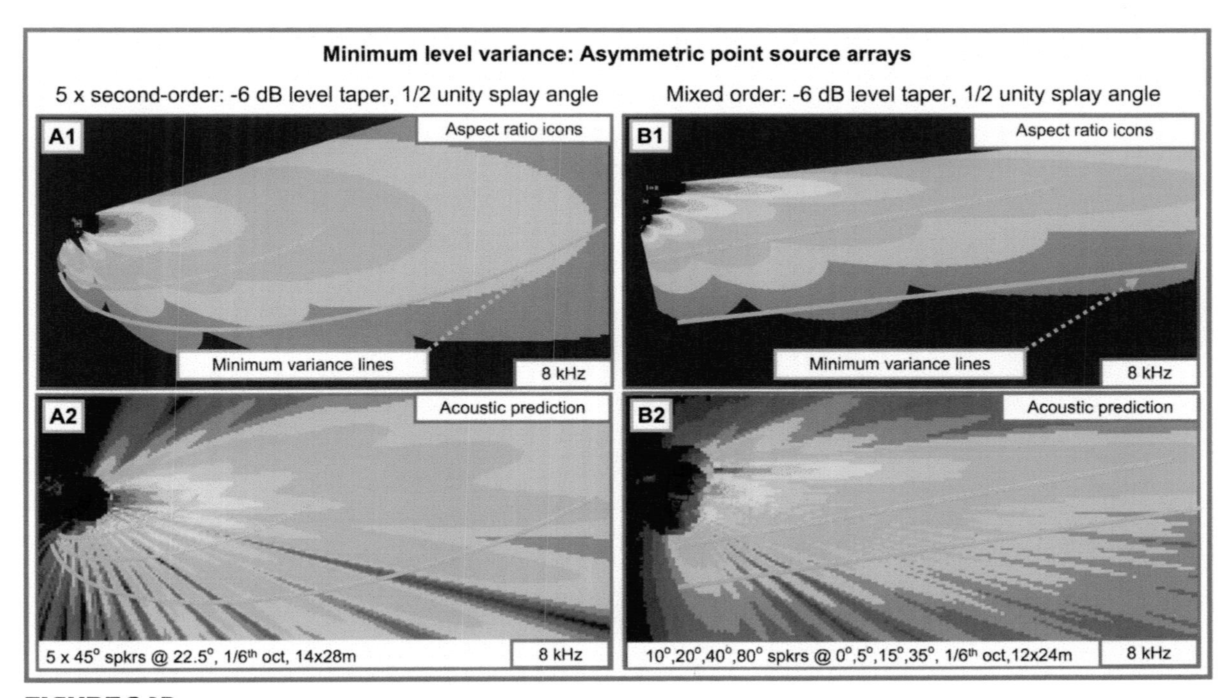

FIGURE 9.13
Minimum level variance shapes for the asymmetric coupled point source. Left: constant speaker order and splay angle, with tapered level. Right: mixed speaker order with splay angle tapered to provide a constant overlap ratio, with tapered level.

The asymmetric coupled point source is chosen for the same reasons as the symmetric point source but conforms to a different shape. The power addition of the asymmetric point source is self-scaling because the entire rationale for this array is that it fits into a variable distance shape. The levels are tapered as appropriate to scale the array to the distances presented. The unity splay angle must then be compensated as appropriate for the level offset factor.

There is no "correct" element with which to start the asymmetric coupled point source. We must start with something and then the process begins as we see what is left. After the next piece is added we re-evaluate again until the shape is filled. The process can be restarted until a satisfactory assemblage of puzzle pieces is found.

It is simple to find the unity splay in a symmetric array. A pair of 30° speakers will be splayed 30° apart. We can just add them and divide by two, as long as the levels are matched.

$$(\text{Coverage}^1 + \text{Coverage}^2)/2 = \text{unity splay}$$

$$(30° + 30°)/2 = 30°$$

The spatial crossover is at the geometric and level mid-point: 15° off axis to either element. What about if we want to merge a 30° and 60° speaker? The same equation applies, as long as the levels are matched.

$$(30° + 60°)/2 = 45°$$

The elements meet at the -6 dB edge of both elements: 15° off axis from the 30° element and 30° off axis from the 60° unit. The spatial crossover is the level center, but not the geometric center.

The equation must be modified whenever levels are offset (Figs 9.14 and 9.15). There is no best splay angle between two speakers, until we know the distance/level relationship. If we take two matched elements and turn one down 6 dB, the standard unity splay angle will not provide unity results. The geometric center finds -6 dB from one element and -12 dB from the other. What is the unity splay angle then? A change of 6 dB is a level difference of 50%. The splay angle must be adjusted by the same ratio to return to unity performance at XOVR. A reduction of 6 dB shrinks the speaker's range in half. That is the decisive number.

The compensated unity splay equation

$$((\text{Coverage}^1 + \text{Coverage}^2)/2) \times (\text{Range}^2/\text{Range}^1) = \text{compensated unity splay}^*$$

*assumes that levels are set in proportion to distance

Here is an example of two 30° speakers, with one covering half the distance of the other (-6 dB)

$$((30° + 30°)/2) \times (0.5/1) = \text{compensated unity splay}$$

$$(60°)/2) \times (0.5) = \text{compensated unity splay}$$

$$0° \times 0.5 = 15°$$

Next we join a 30° speaker with a 60° element that is covering 70% of the range (-3 dB).

$$((30°+ 60°)/2) \times (0.7/1) = \text{compensated unity splay}$$

$$((90°)/2) \times (0.7) = \text{compensated unity splay}$$

$$45° \times 0.7 = 31.5°$$

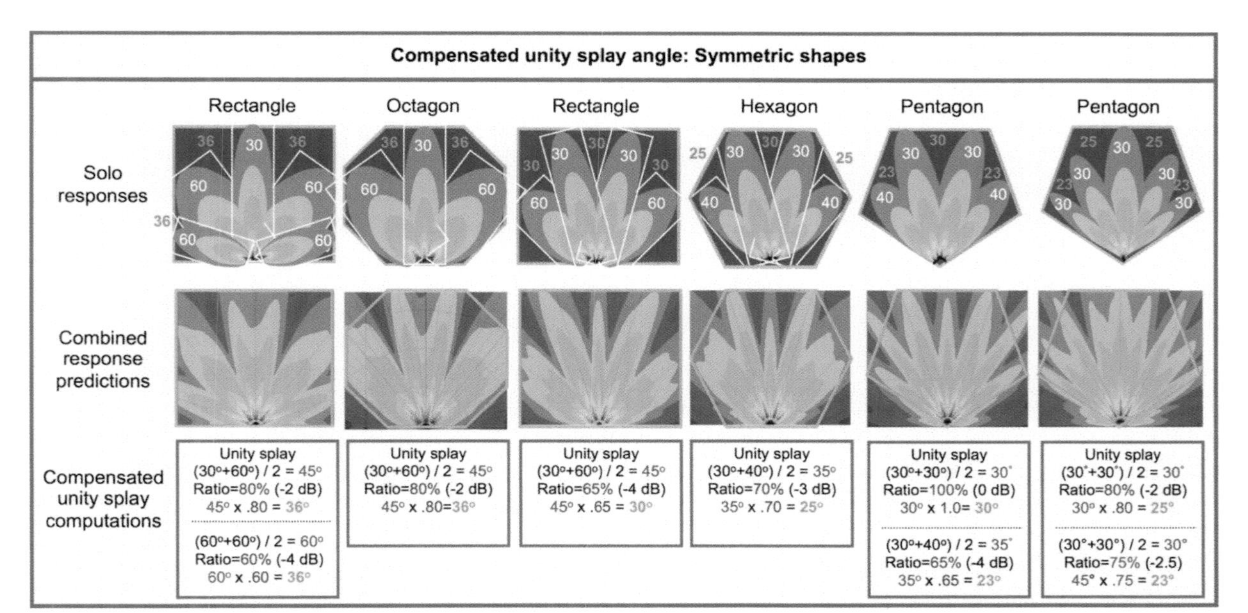

FIGURE 9.14
Compensated unity splay angle examples for typical horizontal plane shapes

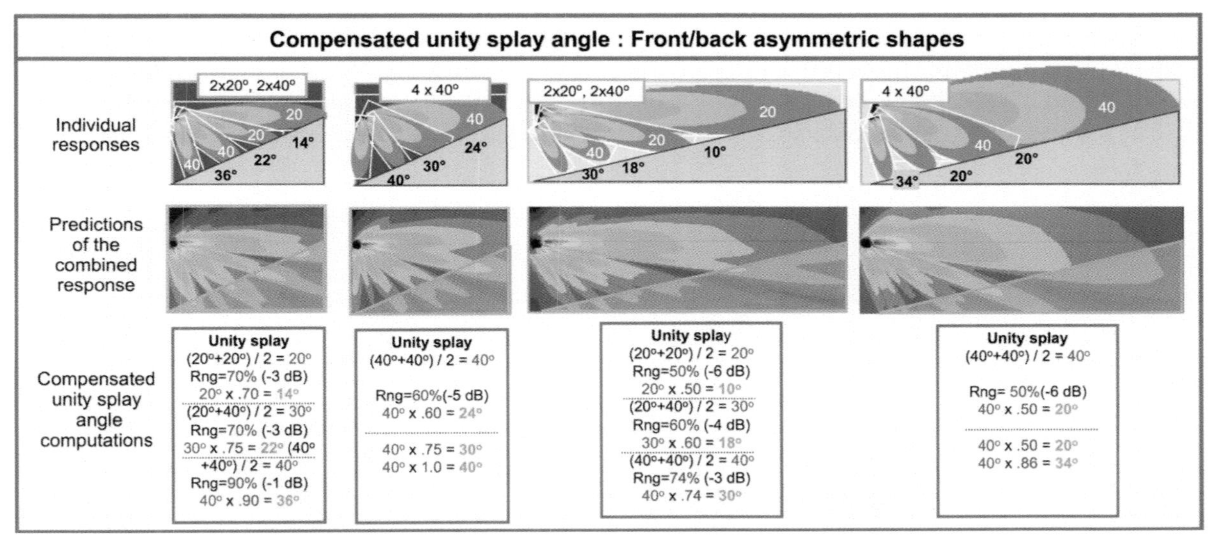

FIGURE 9.15
Compensated unity splay angle examples for typical vertical plane shapes

9.3 LEVEL VARIANCE OF UNCOUPLED ARRAYS

The minimum-variance behavior of uncoupled arrays differs fundamentally from the coupled arrays. The differentiating factor is range. The coupled arrays set their MV shape and hold it over an unlimited range, whereas the uncoupled arrays are in constant change. FAR icons help to visualize the changing shape over distance and we will again compare them to the combined predicted responses.

The advantage of the FAR approach becomes clear when we need to cover a shape that's wider than deep (an aspect ratio of <0.5). Such a shape requires a speaker with a coverage angle beyond 360°, a challenge that only marketing departments can overcome. The best way to create overly wide coverage shapes is to spread multiple sources, i.e. uncoupled arrays. The coverage shape must be characterized as depth by width, which makes the FAR shape the best tool for uncoupled array characterization.

9.3.1 Uncoupled line source

9.3.1.1 SYMMETRIC UNCOUPLED LINE SOURCE

The combined shape of the uncoupled line source is a lateral extension characterized as side-by-side stacking of the individual FAR shapes. The combined FAR is reduced as the width multiplies for a given depth. The combined FAR is found by dividing the single-element FAR by the number of elements, e.g. a line of four 90° elements will spread to a combined FAR of 0.35 (1.4/4). The lowest possible FAR for a single element is 0.5 (360°). This provides the base shape for the example scenarios of Fig. 9.16. Three sequential element quantity doublings (while halving the FAR) are shown. Each combines to approximately the same FAR as the single 360° element (0.5).

We are connecting the lateral width shapes (section 3.10.5.5) into a continuous minimum-variance line. The mid-point depth of the FAR shape connects the on-axis point of each element through the spatial crossover of the adjoining elements until the last element is reached and the level falls to the maximum acceptable variance (-6 dB). The uncoupled line source is more accurately characterized as a combined shape (depth/width) than a coverage angle. Our combined example fills the same FAR shape as the single 360° speaker, and fills in greater amounts of the rectangle as the slices become smaller. The outer edges sharpen as elements narrow, as seen previously in the coupled point source. In practice this is rarely a desired trait for uncoupled arrays, which almost exclusively use first-order elements (≥60°). Narrow elements must be closely spaced and therefore may function as coupled arrays in the lower frequencies, which would greatly complicate the coverage picture.

It is reasonable to ponder how the coupled and uncoupled versions of the line source array could provide such different responses. Coupled lines *extend the length* of the aspect ratio whereas the uncoupled versions *extend the width*. Coupling multiplies, whereas uncoupling divides. The difference is attributed to their respective positions in the parallel pyramid. The fully coupled response is the top of the pyramid, whereas the fully uncoupled is the bottom. In reality, the uncoupled array is simply a coupled array waiting to happen. We just have to go low enough down in frequency to find it. (Recall back in Fig. 9.7 where the third-order speaker coupled while the second-order combed.) The array is fully coupled once we move far enough forward for *all* elements to overlap. From this

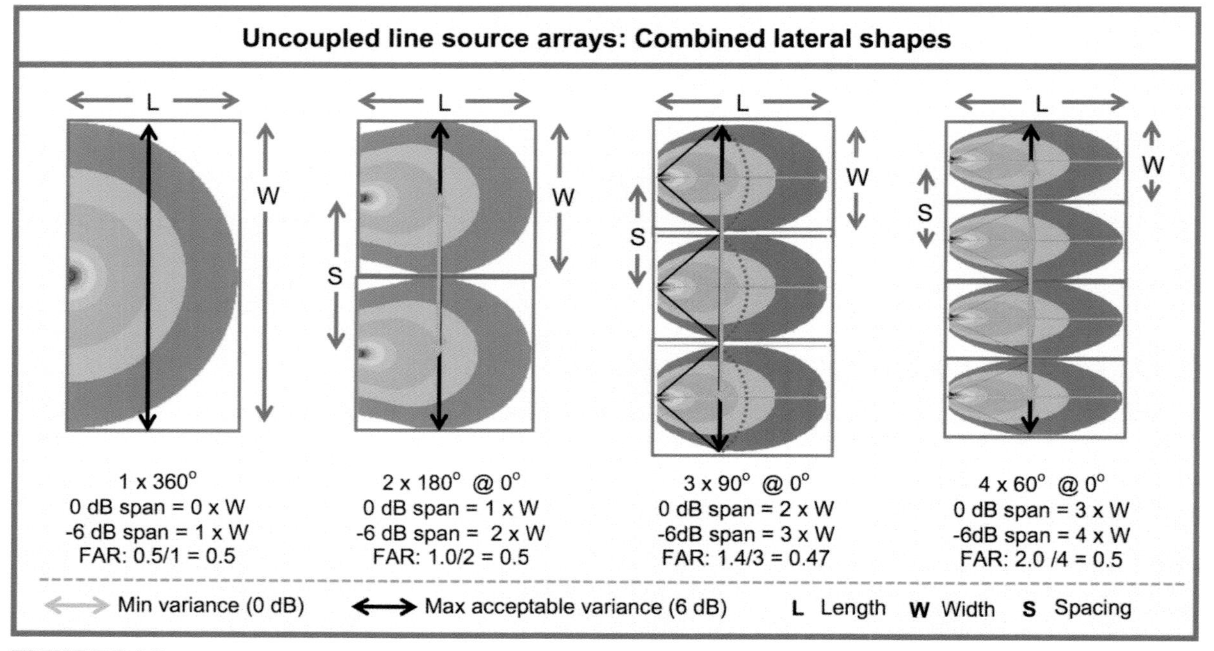

FIGURE 9.16
Minimum level variance shapes for the symmetric uncoupled line source array. The combined aspect ratio of an uncoupled line source matches the shape of the individual element aspect ratios placed in a lateral line. Note the contrast to the coupled line source multiplication shown previously in Fig. 9.7.

point on, it's no longer described as a lateral shape (coverage width) but as a rectangular FAR shape with a defined coverage angle.

A coupled array, conversely, is an uncoupled array that's already congealed. Even the most compact array imaginable has some point in its near field where we've not yet reached the unity class crossover. It is uncoupled at that point (defined by width, not angle). The keys are wavelength and displacement. The transition point from coupling to uncoupling is wavelength dependent (and therefore frequency dependent).

There are several reasons to split this hair. The first is that the bottom two uncoupled floors of the parallel pyramid are among the minimum-variance shapes we commonly use in our designs (frontfills, underbalcony delays, etc.). Second, the transition zone between these two worlds is a poorly understood area in modern speaker array design. Long, coupled line source arrays and hybrid "J" arrays are often designed so that large amounts of the listening area are inside the highly volatile and variant space between the uncoupled and coupled worlds. It is very challenging to characterize array behavior in the metamorphic zone between the opposing forces of isolation zone spreading (at the bottom) and coupling zone beam concentration (at the top). The array cannot be defined as either a constant width or constant angle in this transition zone. If it can't be defined, our chances of finding a minimum-variance alignment strategy are poor. This is at the heart of why we see "configuration dependent" on the specifications for line array coverage.

Before we move on let's review the coupled line source shown previously in Fig. 9.7. The reason the aspect ratio icons didn't match the predicted response was because we used the uncoupled method (icon spreading) rather than the coupled method (icon stacking). The predicted response, meanwhile, is working its way up the pyramid.

Determining the depth limit of minimum variance for such arrays is a simple matter because it is directly related to the aspect ratio. Refer to Fig. 9.17. The minimum-variance coverage zone begins at the mid-point depth of the aspect ratio shape (the unity line, where the coverage angles meet and close the gap) and ends at the full depth (the limit line, where we begin to reach three arrivals at the listeners). Beyond the limit line listeners will have three or more arrivals resulting in high ripple variance.

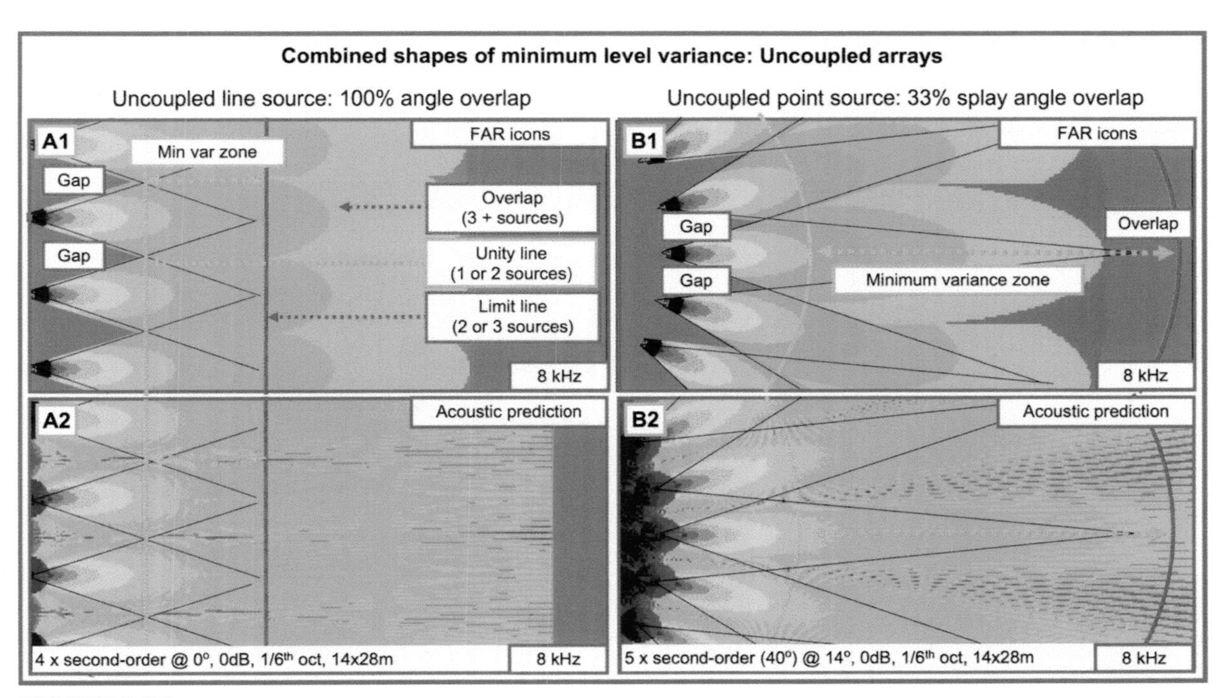

FIGURE 9.17
Minimum level variance shapes for the symmetric uncoupled line and point source arrays. Left: four-element, second-order speaker array with 0° splay angle, matched line of origin, constant lateral spacing and levels. Right: a five-element, second-order speaker array with 14° splay angle, constant lateral spacing and levels. Note the depth extension of the minimum-variance zone.

9.3.1.2 ASYMMETRIC UNCOUPLED LINE SOURCE

Minimum-level variance can be spread asymmetrically over a limited area with uncoupled sources if level, position and orientation are carefully coordinated. The principle involved is one of offsetting effects, in this case distance-related loss and electronic gain. Consider a race where all runners share the same starting point. The faster runners will always lead because they run a proportionally larger distance over time. Now let's handicap the race with staggered starting positions to give the slower runners a proportional head start. The slower runner's initial lead immediately begins to fade, followed by one moment where everyone comes together as the head start compensation has elapsed. From this point the race is effectively restarted and the fastest runners plow ahead.

Compensated unity spacing

This is the nature of interaction between sound sources at unequal drive levels: The louder source always wins. If they are to be made equal for any position in front of the origin of the louder speaker, we must handicap the race by moving the quiet speaker forward by an amount proportional to its level offset (Fig. 9.18).

Two exaggerated examples of offsetting distance and level (Fig. 9.19) serve to illustrate the mechanism. Individual matched-color FAR shapes within the combined response indicate areas of high isolation and minimum level variance. The first scenario shows a logarithmically spaced speaker series with log-tapered levels, i.e. the distance from each sequential speaker is half the distance and half the level (-6 dB) of the previous. The compensating offsets and levels create a diagonal equal level contour line angling up and away from the louder speakers at the unity crossover points. The line remains straight as we move farther but changes its tilt angle with distance (the range-dependent shaping described earlier). The contour can be custom curved with different ratios of level taper and spacing.

An alternative method (also Fig. 9.19) creates a straight contour by log-staggering the starting points and tapering levels proportionally. The levels match to create a vertical unity crossover line. Staggered origins create different doubling distances for each source. As quickly as the speakers combine to create a straight line, the contour begins a tilt in favor of the long-distance speakers. Again the uncoupled arrays can only maintain a given shape for a limited range. There are gap areas prior to the unity range and the longer-range (louder) speakers dominate the soundscape at longer distances.

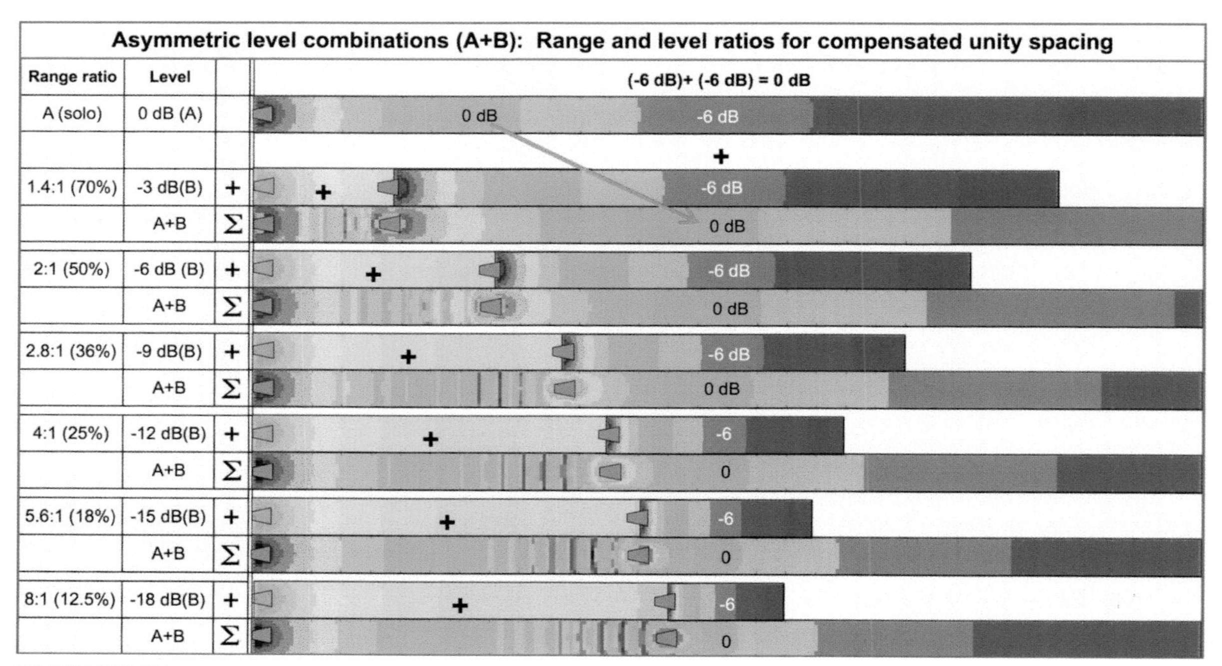

FIGURE 9.18
Level/range ratios for sources with unmatched origins, including the summed levels

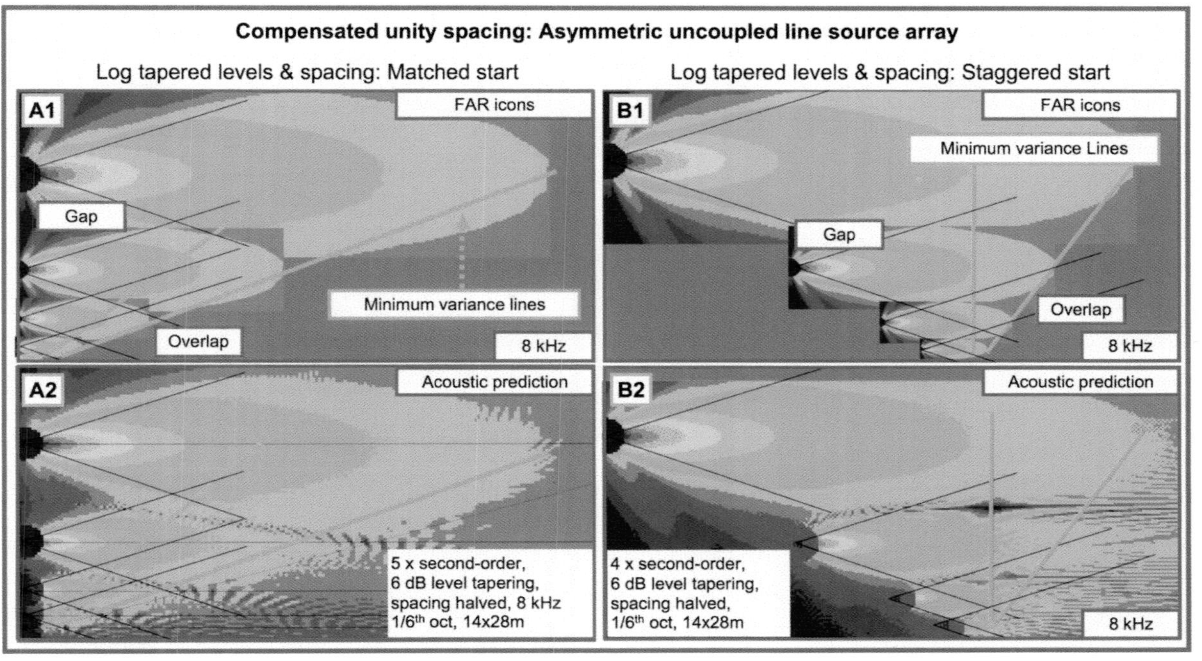

FIGURE 9.19
Minimum level variance shapes for the asymmetric uncoupled line source. Left: five-element, second-order array with 0° splay angle, matched line of origin, compensated unity spacing and levels. Right: four-element second-order array with 0° splay angle, log-staggered line of origin, compensated unity spacing and level.

9.3.2 Uncoupled point source

The uncoupled point source uses both displacement and angular isolation, which greatly expand the usable range. The angular separation moves the unity line proportionally outward but has a stronger effect on the limit line (the angular overlap of three sources). The displacement and degree of angular isolation between adjacent pairs yield twice that for the next element. The uncoupled point source is often chosen over the uncoupled line source for this reason.

9.3.3 Uncoupled point destination

9.3.3.1 SYMMETRIC UNCOUPLED POINT DESTINATION

The uncoupled point destination array places isolation mechanisms in opposition (Fig. 9.20). This array isolates by outward displacement and reverses isolation (overlaps) by inward angle. This is overlap without coupling so there is little upside for the addition. All arrays meet at their edges somewhere but point destinations are the only type where the ONAX points cross. In short, we've been "double-crossed," and that's never a good thing. The on-axis overlap limits the usable depth to around half the distance to the ONAX crossing point. Displacement isolation dominates when speakers are aimed minimally inward, resembling a range-reduced uncoupled line source. Angular overlap becomes dominant when the inward angle is high, which restricts the usable range to the near field of each element. An inward angle of 45° could be considered the break point between displacement and angular dominance. The isolated near-field area closely resembles the aspect ratio icon in the symmetric version until combing dominates as the ONAX crossing approaches. In short, this array cannot be used as an array once we pass 45° of inward tilt, but rather as individual responses with a usable area located between the individual elements and the train wreck overlap zone. The most symmetric versions are infill arrays and sidefill monitors.

9.3.3.2 ASYMMETRIC UNCOUPLED POINT DESTINATION

The right panel of 9.20 shows the asymmetric version, recognizable as the vertical plane of the "delay speaker." Assuming the systems are synchronized, we can observe the forward-range extension provided by the delay speaker.

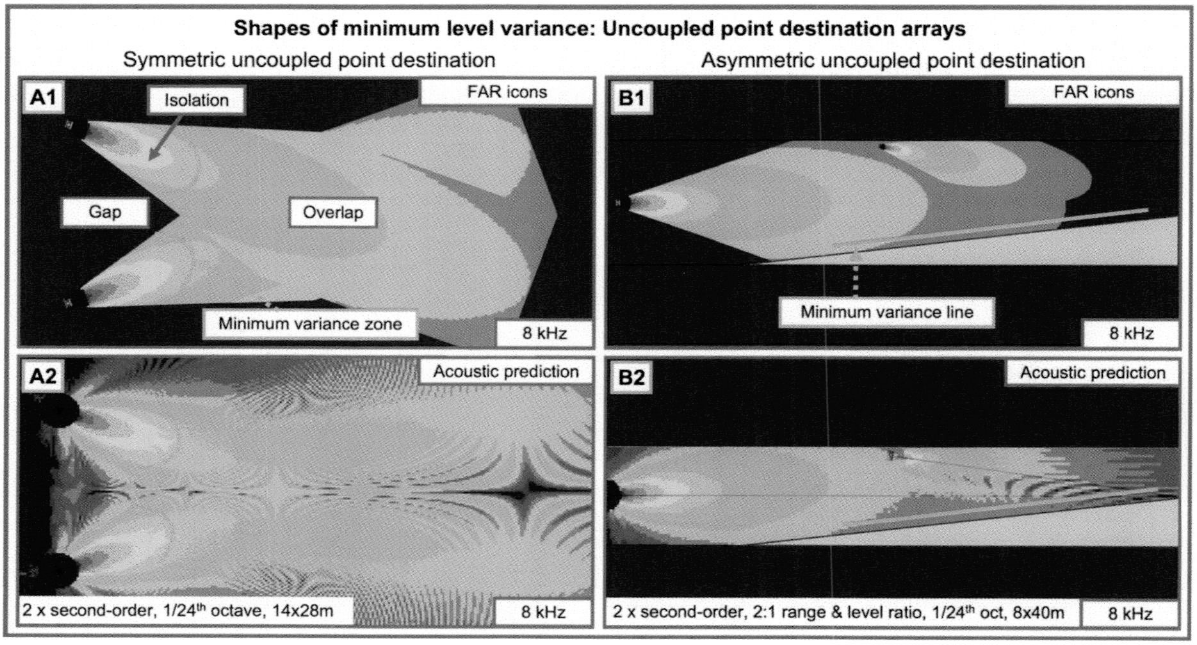

FIGURE 9.20
Minimum-level variance shapes for the uncoupled point destination array. Left: symmetric version. Right: asymmetric version.

The combined levels reduce the loss rate over distance, thereby creating a forward line of equal level on axis to the elements. This is the only means of level extension that does not rely on the offsetting effects of distance and axial loss rates. When used in combination with those effects, the loss rate can be reduced to the lowest amount practical. Here the on-axis responses combine for a 6 dB level increase. The locations before the meeting point are both closer and more off axis to both elements, thereby providing a level stalemate. The range of forward level extension depends upon the distance/level ratio (previously shown in Fig. 9.18).

9.4 SPECTRAL VARIANCE OF COUPLED ARRAYS

Minimum spectral variance can be summed up in two words: flat beamwidth. The goal is combination of speakers that create the same shape over frequency. If the elements already have a constant coverage shape, i.e. flat beamwidth, we can repeat the minimum-level variance strategies albeit with a 600:1 range in wavelengths to consider.

The reality is that there are few (if any) speakers with a uniform beamwidth over their full range. We must assume an overly wide LF range, with the question being shapes of the MF and HF ranges. This returns us to the two standard beamwidth families: plateau and proportional. Combined minimum spectral variance is accomplished by compensating for the individual shape variance with complementary amounts of isolation and overlap. We use overlap (coupling) to narrow the overly wide ranges and isolation to widen the narrow ranges. Plateau beamwidth elements can be combined with isolation dominant in the plateau range and increasing overlap below. The plateau widens and the LF range narrows to flatten the combined shape, achieving minimum variance with a small quantity of elements.

Minimum spectral variance with plateau beamwidth elements

- Isolation zone widens the plateau range (maximum 2×).
- Gradually increasing overlap narrows the response below the plateau (maximum 0.5×).
- Combing immunity rises with isolation but is offset with frequency.
- Combing risk rises with overlap but is offset by increasing wavelength.
- Increasing quantity incrementally flattens the combined beamwidth.

Proportional beamwidth elements have a long way to go and will need quantity to get there. The ratio of isolation to overlap must be frequency proportional to compensate for the gradual narrowing of the elements. Isolation increases with frequency providing a gradual widening force. Overlap increases with wavelength providing a gradual narrowing force. The combined beamwidth will flatten when the overlap and isolation forces are equal and opposite to the original shape of the element.

Minimum spectral variance with proportional beamwidth elements

- Isolation zone widens the highest frequency range (maximum 2×).
- Isolated frequency range expands proportionally downward with quantity.
- Gradually increasing overlap narrows the response below the isolated range (maximum 0.5×).
- Combing immunity rises with isolation but is offset with frequency.
- Combing risk rises with overlap but is offset by increasing wavelength.
- Increasing quantity incrementally flattens the combined beamwidth.

We have two tools with which to move forward. The dominant mechanism for the plateau speaker is isolation, which allows it to be used in both coupled and uncoupled arrays. Overlap dominates the proportional beamwidth speakers, which limits them to coupled arrays. Isolation spreads the power, whereas overlap concentrates it.

Arrays can be characterized as combinations with five possible outcomes: beam narrowing (coupling zone overlap), beam spreading (isolation zone), a stalemate between overlap and isolation, beam scattering (combing zone) and finally beam cancellation (applied to subwoofers in the next chapter). Our goals are speaker array methods that lead to minimum spectral variance. We look to create a series of equal-level contours over frequency that lay over each other like floors of a building. When the response becomes spectrally tilted, it will be tilted everywhere. This is to dream the impossible dream, but some approaches will get us a whole lot closer than others.

Our investigation will once again use the familiar pattern of isolating the main parameters and viewing their contribution to the overall response by a series of steps such as successive doublings.

Four coupled array types are discussed: the line source, symmetric point source, asymmetric point source and the hybrid line/point combination, also known as the "J" array. Each differs in its control of beam concentration and spreading. Two will pass the minimum-variance test, and become candidates for optimized design.

9.4.1 Coupled line source

The coupled line source contains an unlimited quantity of speakers at the most limited number of angles: 1. It's all overlap with no isolation and contains no mechanism for widening, only narrowing. This is an inherent trap. A coupled line source can achieve a flat combined beamwidth width as long as the elements are flat to begin with. The combined result is a narrowed version of the original at all frequencies. This is a unicorn hunt because this element does not exist in a form capable of maintaining coupled response (within 0.5 λ) over the full range. Nonetheless, we will go through the exercise to illustrate the spectral effects of the coupling zone in its purest form.

9.4.1.1 ELEMENTS

The first piece of business is to select the best element to use as a building block for a minimum spectral variance coupled line source. There are no good choices, but some are less bad than others. We have no avenue to spread the coverage without splay angle or specialty processing such as beam steering. This leaves us with pattern narrowing at all frequencies and looming ripple variance due to displacement. A flat beamwidth element will run into ripple variance trouble with rising frequency and quantity. We might be able to survive a combination of two or three

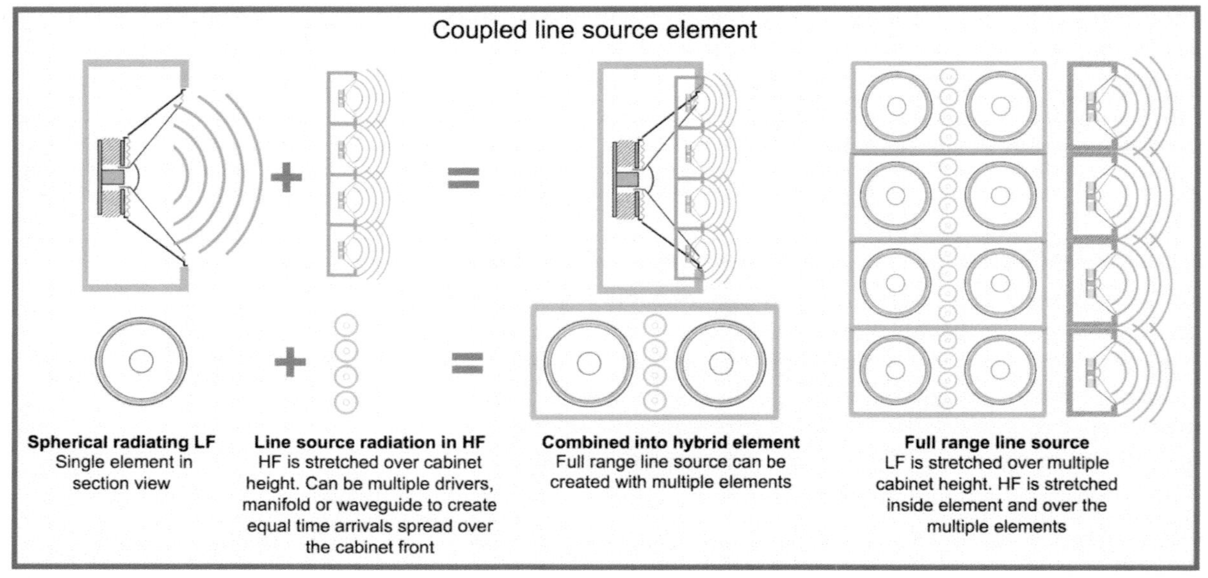

FIGURE 9.21
Element choice for the coupled line source

elements but displacement will soon catch up to the small wavelengths. Proportional beamwidth elements have greater ripple immunity but lack a mechanism to change their ever-narrowing ways.

The worst-case scenario is a large-format first-order element: high overlap and large displacement, but don't think you won't see it just because it's a totally stupid idea. Flat beamwidth with 100% angular overlap has a simple math formula: frequency proportional combing. Smaller format and higher order reduce the symptoms but do nothing to alleviate the disease. Proportional beamwidth elements under the same conditions also have a simple formula: quantity proportional narrowing.

Nonetheless, we will go forward with the element we are most likely to encounter in the wild: the third-order proportional. There is much to learn here that will apply to the other array types (Fig. 9.21).

9.4.1.2 LINE LENGTH

One question sound engineers love to ask each other is "How long is yours?" We are talking about line array length, of course, a major force in LF directional control.

Recall that the beamwidth of a single speaker approaches 90° when the transmitted wavelength equals the radiating cone size. Lower frequencies are wider than 90° and higher ones are narrower until the wavelengths are so small that lobing scatters the pattern. A coupled line source also has a size-related directivity milestone: Coverage is 72° when the transmitted λ equals the line length, e.g. an array 2 m long will be 72° at 170 Hz. The coverage widens below and narrows proportionally above. The narrowing is also frequency/wavelength-limited by lobing, in this case by the combing interaction caused by the element spacing.

Some notable conditions are attached to the 1 λ = 72° standard coverage. Line-length equations assume omnidirectional sources, so the coverage numbers will shrink with directional elements. We also need enough quantity for the line shape to be fairly full of speakers. Think of the 1 λ length as a spread of 360° of phase offset over the elements. With just two elements, the sources are a full λ apart, causing a huge side lobe (0° + 360°). Three elements space them at 180° intervals, which should set off alarm bells. Five elements bring the spacing to 90° intervals, which can achieve the orderly chaos needed to cancel the side radiation. Beyond this point the quantity has less impact on the cutoff frequency. Line lengths of 3 meters will have the same cutoff frequency (113 Hz) whether we fill it with eight, twelve or sixteen elements. Line lengths of 1.35 m, 2.7 m and 5.4 m correspond to cutoff frequencies of 250 Hz, 125 Hz and 63 Hz respectively.

Primary features of line length

- Standard attributes apply to omnidirectional sources.
- Nominal coverage is 72° when line length = 1 λ, 36° when length = 2 λ and so on.
- Directional elements reduce the nominal coverage proportionally.
- Line must be effectively continuous (sufficient density to maintain coupling).

Angular reduction is proportional at frequencies above the 1 λ length, i.e. we get 36° when we double the length, or double the frequency. The narrowing continues with successive doublings. Simple proportionality does not hold below the 1 λ frequency as the coverage approaches 360°. This is similar to how the coverage angle and FAR shape were proportional for narrow angles but not so for wide angles (Fig. 3.40). Line length/quantity doubling causes angular halving (the FAR multiplication described in section 9.2 above).

It's easy to exaggerate the ramifications of line length, as also often happens with coverage angle. The 1 λ frequency is a convenient reference point but not something that stands out like a misaligned crossover. It's not as if 72° is the world's greatest coverage angle. The frequencies below this marker widen gradually, not suddenly, so don't be disappointed when adding that sixteenth box reduces the coverage all the way to 68°.

Line length is decisive only in the very low end of these arrays, where omnidirectional elements are otherwise unsteered. As frequency rises and elements become directional, the line length extends the uncoupled/undefined area of the parallel pyramid. The extension varies with frequency, becoming steadily longer in the HF. This results in widespread spectral variance over angle and over distance.

It is often believed that line length is a property exclusively relevant to the coupled line source (all boxes at 0°). The coupled point source also has length, albeit in curved form. A few degrees of splay between boxes is negligible to omnidirectional elements, so the difference between a line and point source in the relevant range is not worth spending any time to analyze.

We can summarize the line-length behavior by the Goldilocks method: too big, too small and just right. We have less than full control when λ > line length, extra control when λ < line length and "just right" control when λ = line length (Fig. 9.22).

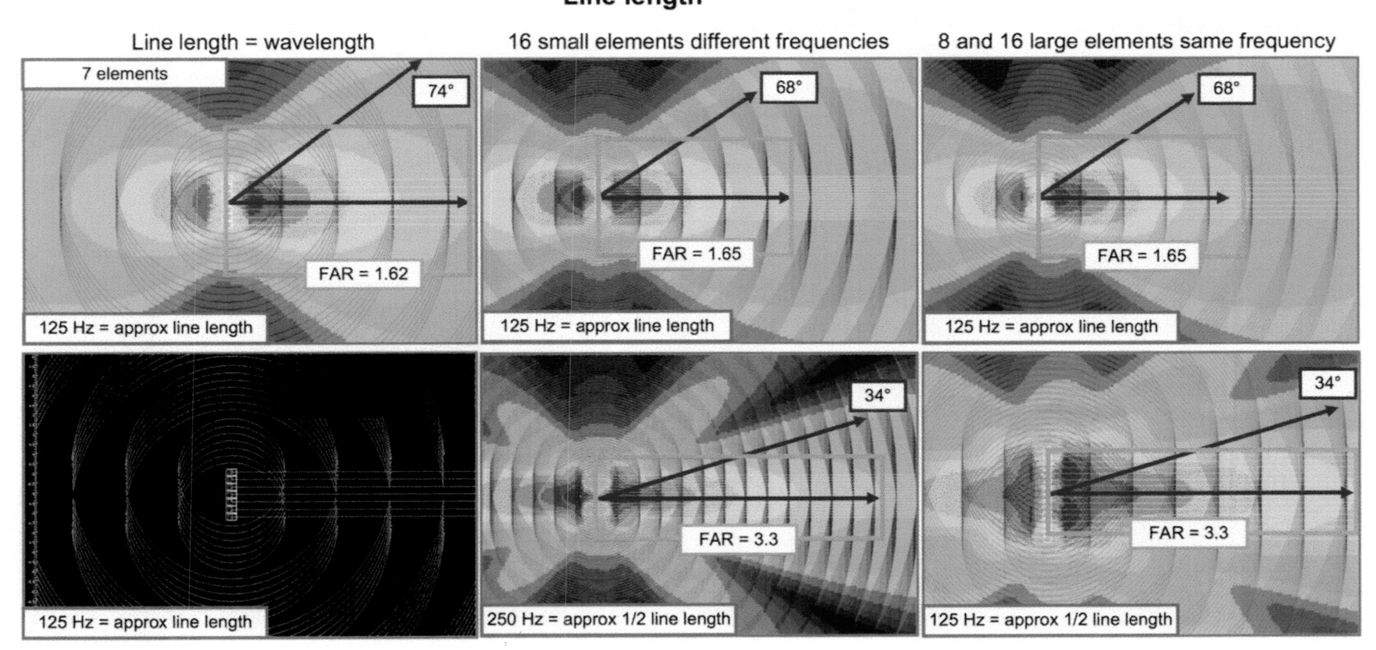

FIGURE 9.22
The effects of line length in the coupled line source, length is 2.76 meter (1 λ @125 Hz). The circles are sized to the wavelength of the frequency shown (lower left). Notice the front lobe follows the confluence of the circles and cancellation occurs in places where the circles diverge. The coverage angle is around 70° regardless of quantity when the line length equals the wavelength (upper panels). The coverage angle is reduced by half when λ equals half the line length (lower center and right panels).

9.4.1.3 ELEMENT SPACING

There is no theoretical limit to line length and its endless narrowing. The physical limit is reached when the element displacement is large enough to uncouple the array and lobe. This is a classic "catch 22" scenario. We can narrow the array forever if we have an infinitely small displacement. On the other hand, there won't be any narrowing at all without displacement between the elements. Displacement is the steering mechanism.

Primary features of element spacing

- Displacement provides phase offset in the uncoupling plane (the steering mechanism).
- Displacement should not exceed 2/3 λ. A displacement of 1 λ produces fatal side lobes.
- Low displacement and quantity minimize steering (high amounts maximize steering).

Recall the spatial icon for summation (Fig. 4.16), which contains two primary features: the coupling and uncoupling lines. The line source concentrates its forward beam onto the coupling line while physically spreading along the uncoupling line. The coupling line has the least displacement/time offset between elements and the uncoupling line has the most. This causes maximum forward extension along the coupling line and lesser amounts along the uncoupling line. The sideways coupling will be substantially less than the forward coupling as long as the total phase offset between the first and last elements is <360°. A constant displacement between elements yields evenly spaced intervals of phase offset in the uncoupling plane. For example: ten elements spread at 36° form a perfect picket fence of arrivals in the uncoupling plane, none of which fall on top of each other (from 0° to 324°). An eleventh box would arrive at 360°, which adds constructively to the first box because it is 1 λ behind. This is the birth of a side lobe, a wart that will grow as more elements are added. Additional boxes from here forward will add along both the coupling and uncoupling lines. The difference is that everybody is on the lead lap on the coupling line and they are lapping each other on the side. The front lobe still has maximum addition but there's a storm brewing on the sides.

The lapping accrues faster as frequency rises. Eventually we reach the end game, which is when λ = displacement. This harmonic convergence places elements at 360° intervals along the uncoupling line. Every element adds constructively to the next (at that frequency) sending a powerful beam to the side. We can't stop this process, but we can reduce its effects. We need to control the amplitude when we can't win with phase, thereby keeping the side lobes small enough to minimize damage.

So it's a race. We can keep the elements wide and couple their power as long as we can keep them close (in wavelength). We have two choices as displacement approaches λ: Narrow the element (isolate) or cross it over to another element with smaller displacement, which restarts the race (Fig. 9.23).

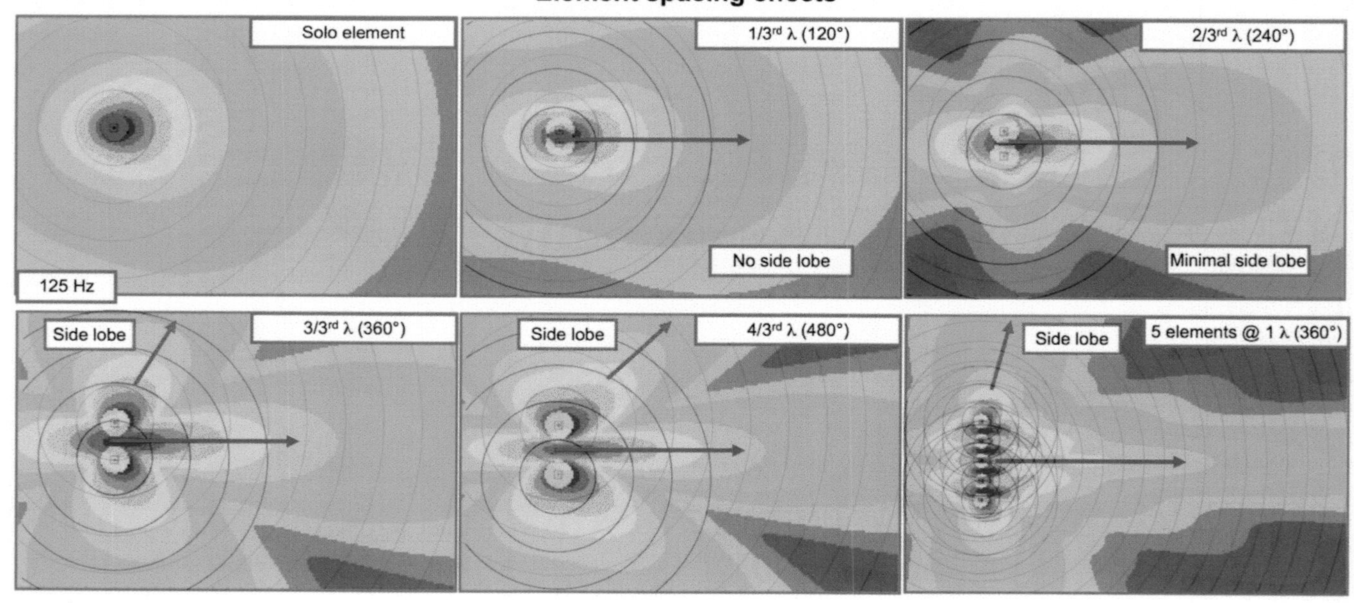

FIGURE 9.23
Element spacing effects for the coupled line source. Spacings from ⅓, 4⁄3 wavelength are shown. The side lobe becomes strong at spacings beyond ⅔ λ.

9.4.1.4 QUANTITY EFFECTS

Coupling and displacement team up to create narrowing. The potential counterforces to this are isolation by splay angle and/or uncoupling, neither of which are present in the coupled line source. The futility of attempting to use quantity to reduce the spectral variance of proportional beamwidth speakers is shown in Fig. 9.24. Successive quantity doublings definitely narrow the combined beamwidth, but they do so for all frequencies. It's an endless cycle of division, with each additional element asymptotically narrowing the entire response, i.e. approaching, but never quite reaching, 0° coverage. The chart's vertical scaling (in degrees) gives the illusion of reduced spectral variance with quantity, i.e. the sixteen-element beamwidth is *visually* more horizontal than lower quantities. This is most certainly *not* the beamwidth plateau described earlier. The HF beamwidth is only changing a few degrees/octave, but such angular changes are huge when they are the difference between 4° and 2°. Coverage angle over frequency remains *proportionally* equivalent in all cases, i.e. if it's 12× wider at 1 kHz than 8 kHz for a single box, then it's 12× wider with sixteen boxes.

The illusion breaks down when we rescale the same array's coverage into FAR over frequency (Fig. 9.25). FAR scaling reveals the beam concentration of the parallel pyramid. The beam continuously narrows as quantity increases and the shape of the coverage extends forward toward infinity. The coupled line source's LF range will never get any closer to its MF and HF shapes no matter what the quantity. A beamwidth change of 1° appears negligible in the vertical scaling. But a 1° change between 90° and 91° is a very different matter than between 1° and 2°! The FAR scaling reveals this.

The forward distance required to reach a fully coupled response is inversely proportional to frequency. The distance for each element to meet its closest neighbor extends as the FAR shape narrows. A fully coupled pyramid could be 32× further at 16 kHz than 125 Hz. Spectral tilt varies over distance because nearby listeners are in coverage of only a few HF drivers (and all the LF). Far-field listeners hear all the HF and LF.

A set of predictions with three successive quantity doublings is shown in Fig. 9.26. Let's look at how this stacks up on our scorecard. The coverage pattern shapes for each range will lie on top of each other if minimum spectral variance has been achieved. In our case the coverage narrows with frequency.

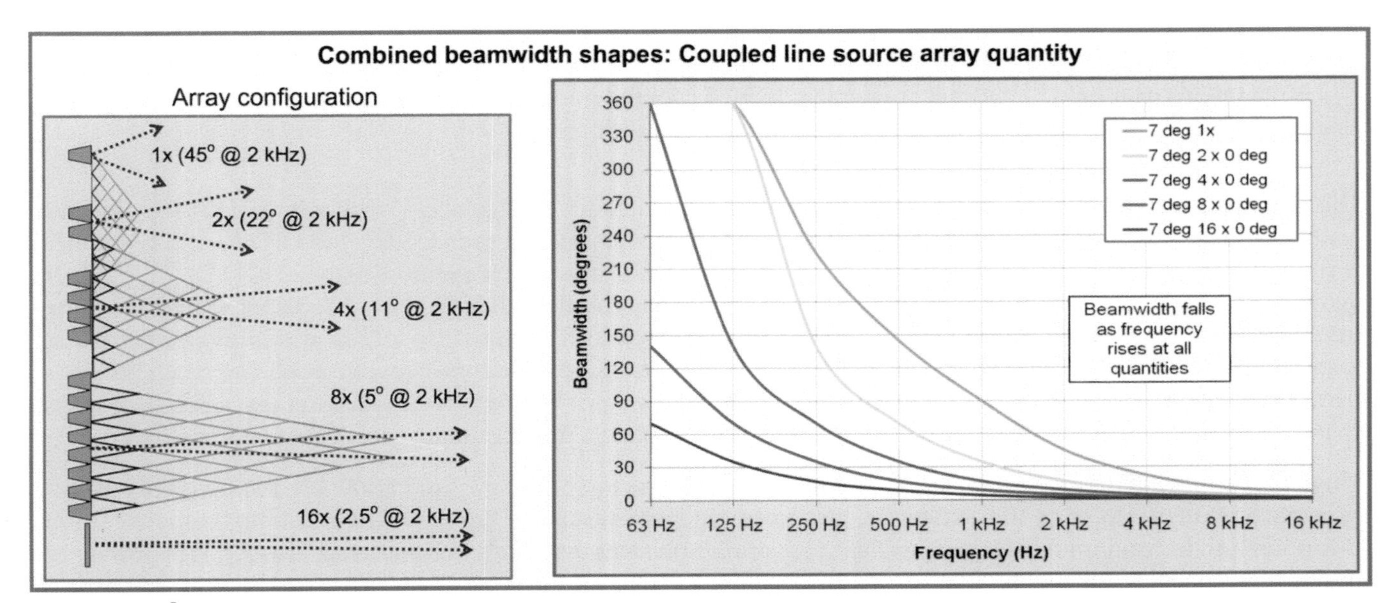

FIGURE 9.24
Beamwidth vs. frequency plots for the third-order coupled line source

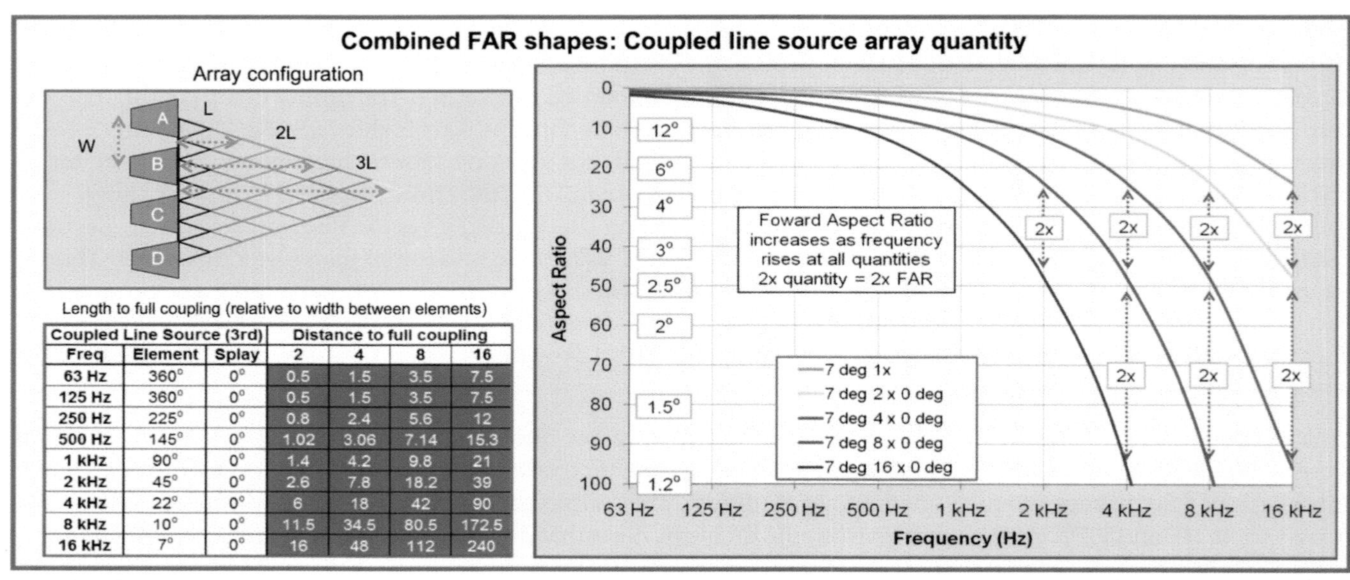

Coupled Line Source (3rd)			Distance to full coupling			
Freq	Element	Splay	2	4	8	16
63 Hz	360°	0°	0.5	1.5	3.5	7.5
125 Hz	360°	0°	0.5	1.5	3.5	7.5
250 Hz	225°	0°	0.8	2.4	5.6	12
500 Hz	145°	0°	1.02	3.06	7.14	15.3
1 kHz	90°	0°	1.4	4.2	9.8	21
2 kHz	45°	0°	2.6	7.8	18.2	39
4 kHz	22°	0°	6	18	42	90
8 kHz	10°	0°	11.5	34.5	80.5	172.5
16 kHz	7°	0°	16	48	112	240

FIGURE 9.25
FAR vs. frequency plots for the third-order coupled line source

Quantity effects on the coupled line source (Fig. 9.26)

- Coverage pattern narrowing is proportional to quantity (50% reduction/doubling).
- The ratio of spectral variance is unchanged (all frequencies are narrowed).
- Beam coupling (100% overlap in the parallel pyramid) is the primary force.
- Larger quantities *appear* to reduce spectral variance because the viewing range does not extend far enough to show full coupling in the higher frequencies. The lower ranges run the full course of the pyramid while the higher ranges have not yet reached the summit.

Let's see if we can find a minimum spectral variance line. A lateral minimum-*level* variance line runs across the near-field response of the array (Fig. 9.27). This is the bottom level of the parallel pyramid, the unity line for the uncoupled line source as described previously in section 4.4.2 (and soon in section 9.5). This is not a minimum spectral variance line, however, because its location changes drastically over frequency. This difference can easily be a 20:1 factor in an array of proportional beamwidth elements!

Three stages of combination in the coupled line source

- Uncoupled (near field): Coverage is laterally defined over the line length.
- Semi-coupled (mid-field): Coverage morphs between definitions.
- Coupled (far field): coverage is angularly defined.

The near-field/uncoupled approach has been eliminated as a contender. Let's go deep. We'll wait until the array is fully coupled and takes the shape of a single speaker. We are looking for an MV line from ONAX_{FAR} to $\text{OFFAX}_{\text{NEAR}}$. The problem is that ONAX_{FAR} at 100 Hz is a few meters from the speakers while at 8 kHz it should be labeled $\text{ONAX}_{\text{EXTREMELY-FAR}}$. It takes 20× farther for the HF to reach the top of the pyramid (fully coupled) than the LF, so we are up against the same impossible challenge. ONAX keeps moving deeper with frequency. Its relationship to OFFAX shows an ever-narrowing angular response, just like the original element (surprise!). The metamorphic zone between won't fare much better because its length and width rescale over frequency. There is no single ONAX or OFFAX location that can be deemed as spectral representative for the array. So where's the best spot to place a mic for tuning this array? I have no idea.

There is one location that can serve as a reference milestone: the top of the pyramid at the highest frequency. All speakers are fully coupled at all frequencies. Let's use eight elements as our example. The combined response from this point and beyond is effectively equivalent to a single array element that's been coverage divided and power multiplied. The response will be flat (if it started that way) except for the effects of air loss (HF) and reflections. Standard inverse square law loss and angularly defined propagation are here to stay.

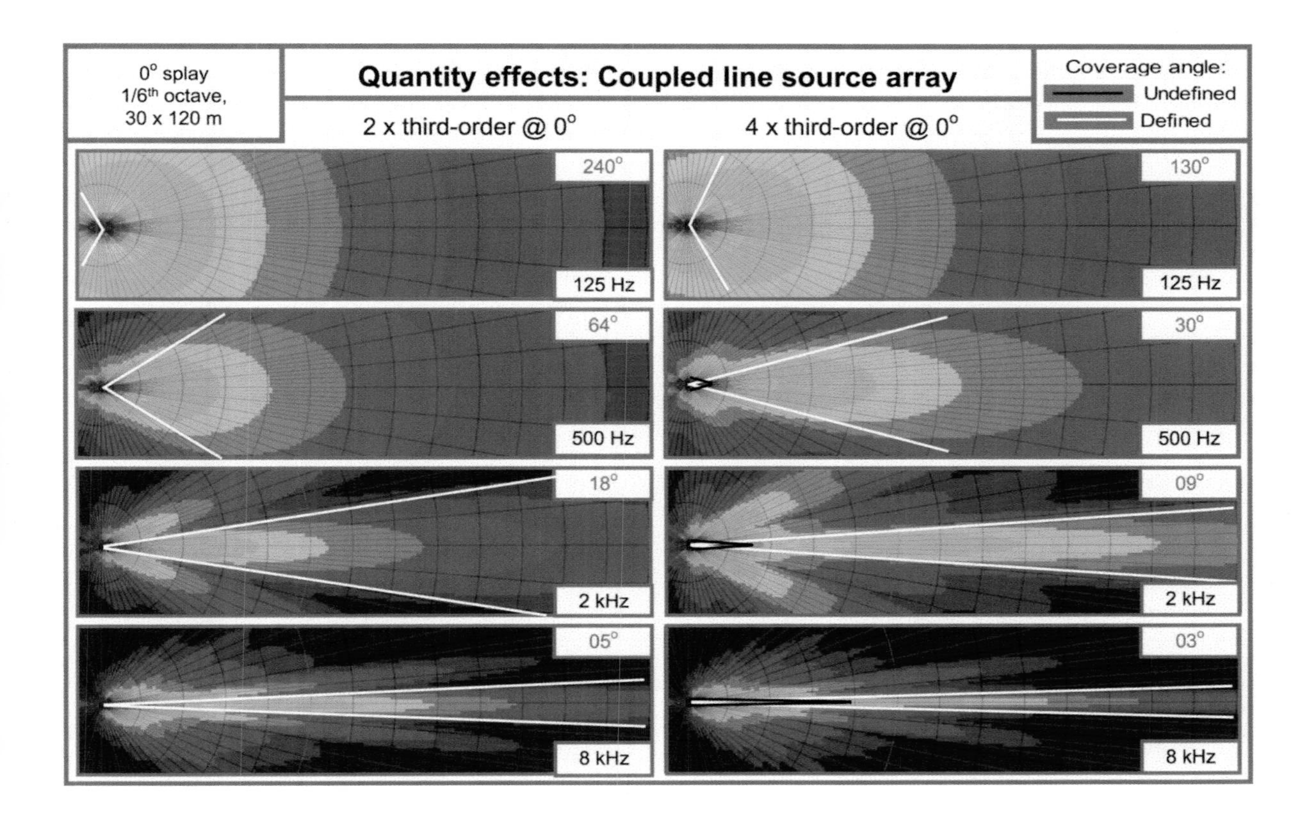

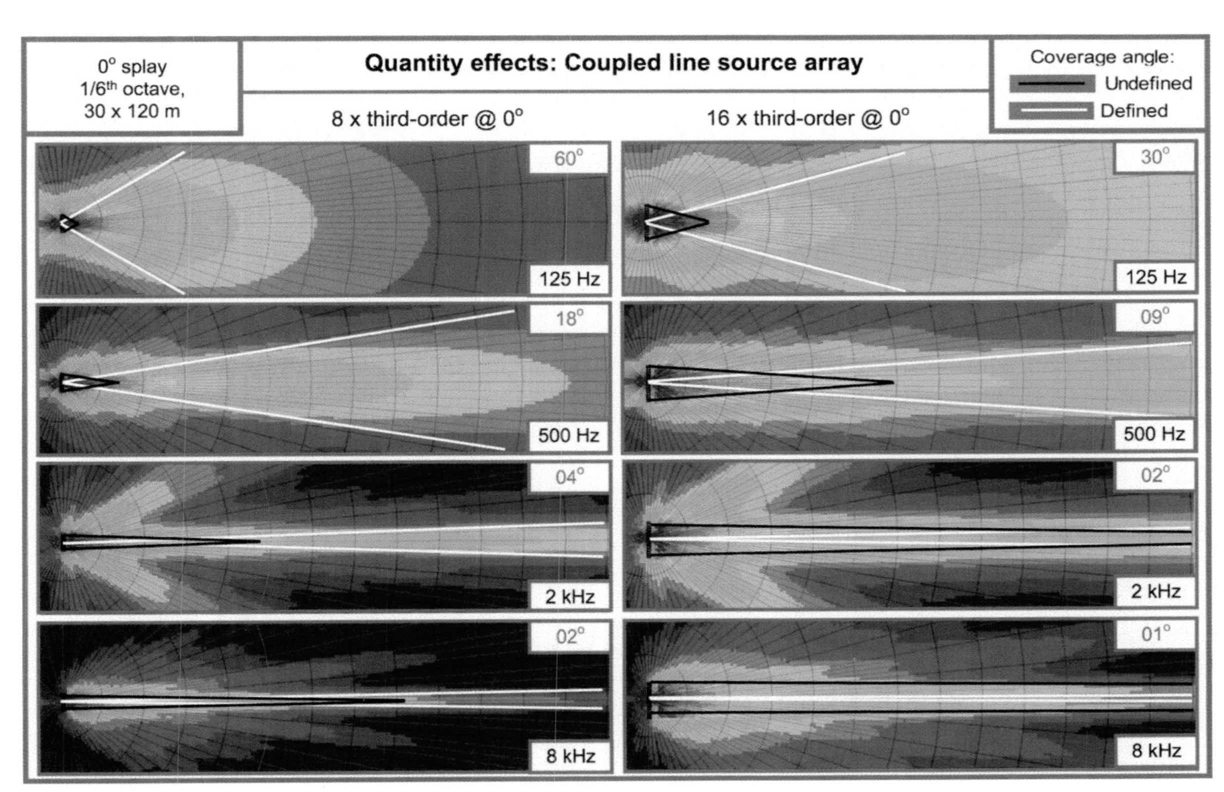

FIGURE 9.26A, B

Quantity effects vs. frequency for the coupled line source

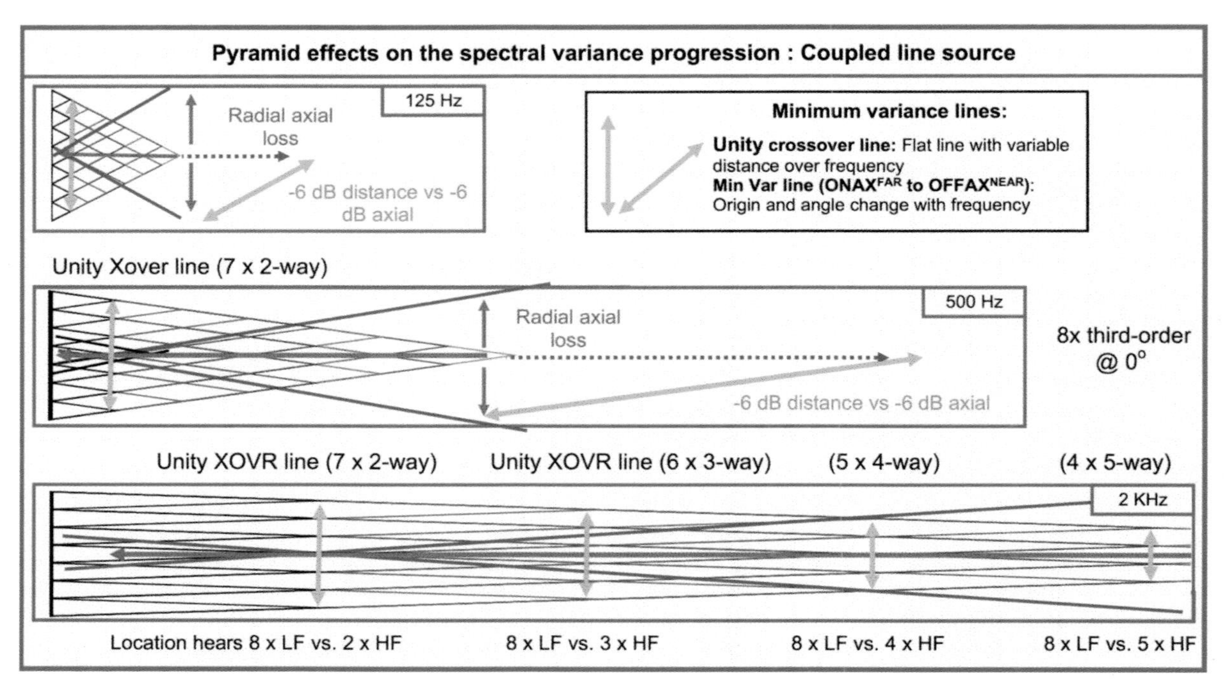

FIGURE 9.27
Parallel pyramid effects on spectral variance for the coupled line source

Let's call this "flat" and move inward toward the array. The frequency response begins to tilt in favor of the lows (pink shift) because each step forward reduces the quantity of elements at the high end, from eight to six, four and finally two. We are in the coverage of only two HF elements but still might be hearing four MF and eight LF drivers. The result is increasing spectral tilting as we approach the array. Conclusion: high spectral variance over distance.

Spectral variance is at least slightly reduced between the reference point and the near field by air absorption (small consolation, perhaps). The pyramid top, the most distant point, has the most HF air loss and the least axial loss. Moving closer decreases the air loss, but fewer speakers are on axis. The results are application- and weather-dependent.

Let's return to the reference point and move radially. The same for axial reduction applies as we lose HF elements at a faster rate than the MF and LF. Spectral tilting increases as we move off axis. Conclusion: high spectral variance over angle.

It's often claimed that coupled line source behavior mimics a single speaker. It's a speaker that takes a long distance to find its shape! This "speaker" can only maintain a minimum-variance line for a limited frequency range and distance. Listeners and optimization engineers will likely find themselves in the metamorphic transition zone between dimensional and angularly defined radiation. Listeners don't care, but the distinction is relevant to optimization, because our strategies require a definable and stable coverage pattern. A system in transition can only be tuned for a point in space. These systems can be made loud and powerful. The modern "line of sound" is a huge improvement from the old-style "wall of sound" because the interference is limited in one plane and greatly reduced in the other. But we need not settle so quickly for power over minimum variance. The modern third-order speaker can still fulfill that promise, as we will soon see.

9.4.1.5 ASYMMETRIC LEVEL EFFECTS

There is only one option for asymmetry in the coupled line source array: variable level (Fig. 9.28). The elements charged with the closest coverage can be level-tapered to compensate for proximity. This is a misnomer because there is no actual separation into near and far coverage. The results of such level tapering are, not surprisingly, largely ineffective. The strongest effects are found in the near field (and the HF range) where uncoupled behavior is the dominant mechanism. Level tapering of the HF range will tilt the unity line in the manner of the asymmetric

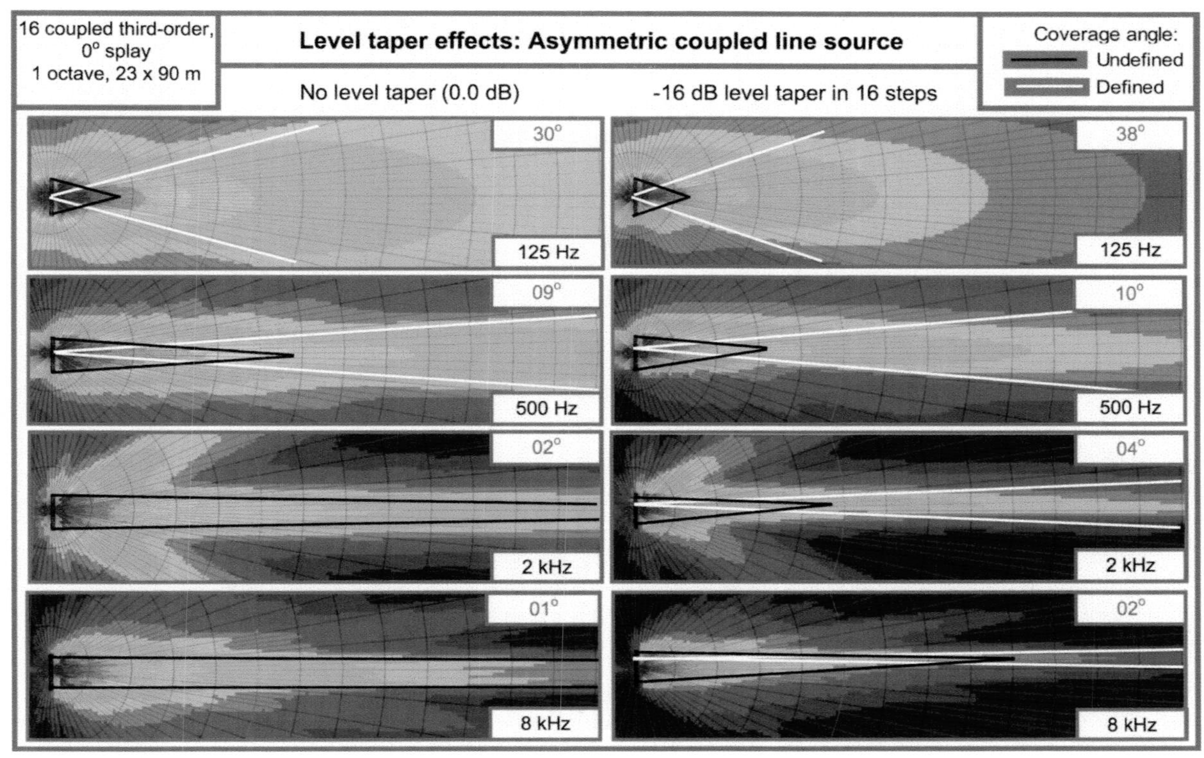

FIGURE 9.28
Asymmetric level effects vs. frequency for the coupled line source array

uncoupled line source (section 9.5.1.2). Coupling dominates as frequency falls and the result is little more than a minor re-centering (and slight widening) of the forward beam (and corresponding loss of LF power capability). Level tapering inherently reduces maximum SPL capability but can be worthwhile in cases where it can effectively reshape the response toward minimum variance. Amplitude-driven shaping is ineffective without isolation, which makes full-spectrum level tapering a fool's errand for the coupled line source. Asymmetric level adjustment should be reserved for the HF in the near field. Even so, such adjustments will have minimal benefit. Covering listeners at near and far distances with a pile of speakers set at 0° is a massive design fail for which HF-level tapering will be little more than a Band-aid®.

9.4.2 Coupled point source

The coupled point source is a hybrid of angular overlap and isolation. It contains mechanisms for widening and narrowing, which can be simultaneously applied. A coupled point source can achieve a flat combined beamwidth when elements are flat to begin with or proportionally narrowing with frequency.

The coupled point source can contain an unlimited quantity of speakers but is limited to a combined beamwidth of no more than 360° (duh!). The individual elements, splay, quantity and level asymmetry all have roles to play. We again isolate each mechanism in search of minimum spectral variance.

9.4.2.1 ELEMENTS

All of our standard speaker elements are coupled point source candidates. The first- and second-order flat beamwidth types are used for radial extension. They function well when angular isolation is dominant. Proportional beamwidth elements are preferred for pattern narrowing and highly asymmetric shaping. These can also widen the response, but require substantially more quantity than the first- and second-order elements.

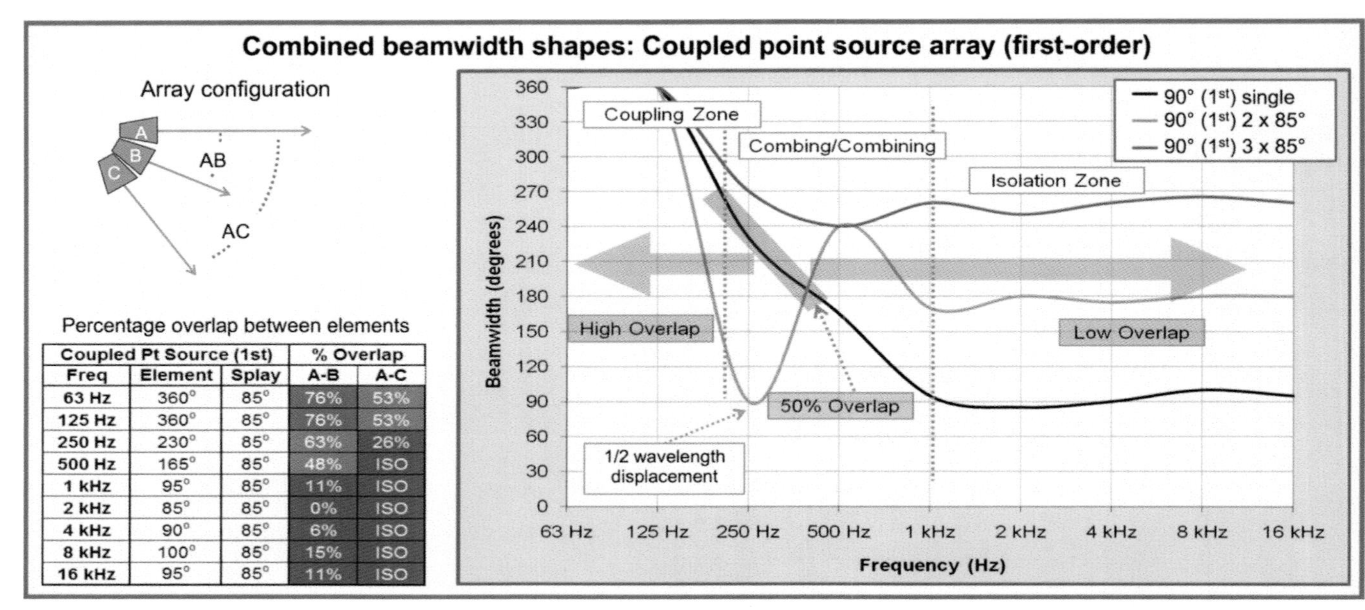

Coupled Pt Source (1st)			% Overlap	
Freq	Element	Splay	A-B	A-C
63 Hz	360°	85°	76%	53%
125 Hz	360°	85°	76%	53%
250 Hz	230°	85°	63%	26%
500 Hz	165°	85°	48%	ISO
1 kHz	95°	85°	11%	ISO
2 kHz	85°	85°	0%	ISO
4 kHz	90°	85°	6%	ISO
8 kHz	100°	85°	15%	ISO
16 kHz	95°	85°	11%	ISO

FIGURE 9.29
Beamwidth plots for the first-order coupled point source, unity splay and level

9.4.2.2 BALANCING EFFECTS OF ANGULAR OVERLAP AND ISOLATION

The beamwidth behavior of a first-order symmetric coupled point source array is shown in Fig. 9.29. The single unit is compared with quantities of two and three units arrayed at the unity splay angle. The range above 1 kHz shows radial coverage expansion by beam spreading (isolation zone summation). The combined beamwidth above 1 kHz is a simple multiple of the individual elements. The three-element array has a constant beamwidth of around 260° for six octaves. This is a minimum spectral variance array.

These configurations share a common feature present in the point source arrays. The behavior can be seen as split along the 50% overlap line. Isolation zone spreading dominates below the 50% line and coupling zone beam concentration dominates above (along with possible combing). It's easy to spot the 50/50 line: the point where the combined coverage equals the individual element angle.

Fig. 9.29 also contains a table of the individual overlap percentage relationships among multiple array elements. The three-element configuration reveals push and pull at 250 Hz, which tends toward coupling between adjacent elements (A–B and B–C) and isolation between the outers (A–C). This push–pull relationship becomes very important as element quantity rises.

Note re. Fig. 9.29: The 250 Hz to 500 Hz range merits a brief explanation. There is enough angular isolation above 1 kHz to characterize the beamwidth without physical scaling, i.e. the elements could be large or small, and their displacement scaled accordingly. Displacement is a defining factor below 1 kHz as angular overlap rises steadily. This two-element array has 1 λ displacement (0.69 m) at 500 Hz (and 0.5 λ at 250 Hz). This type of beamwidth variance is found in any two-element array as the frequency falls under 1 λ displacement, with the narrowest point being reached at 0.5 λ (see section 4.2.2.3). Element size affects the frequency range, with larger displacements moving the range downward. Complexity rises as elements are added because each has λ displacement relationships to every other. Readers should view the interactions below the isolated frequency range as trends that will adjust their frequency ranges with the physical size of the elements. In short, we can say with some certainty that a coupled pair of first-order speakers splayed at 90° will narrow (at 0.5 λ) but not necessarily at 250 Hz.

These principles also apply for the second-order system (Fig. 9.30). The physical size and driver complement is the same as the first-order elements (Fig. 9.29), so the disparities can be attributed to HF horn differences and splay angle. The isolation zone range is now limited to 4 kHz and up where it spreads by increments of the 40° unity splay angle.

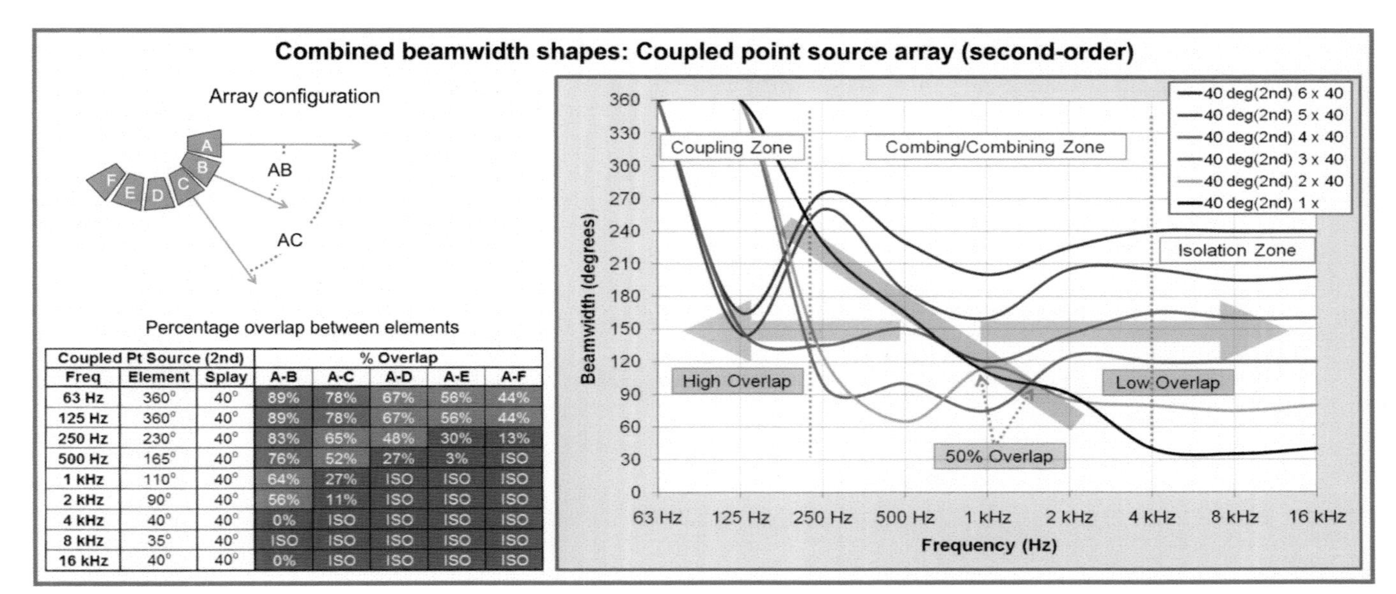

Coupled Pt Source (2nd)			% Overlap				
Freq	Element	Splay	A-B	A-C	A-D	A-E	A-F
63 Hz	360°	40°	89%	78%	67%	56%	44%
125 Hz	360°	40°	89%	78%	67%	56%	44%
250 Hz	230°	40°	83%	65%	48%	30%	13%
500 Hz	165°	40°	76%	52%	27%	3%	ISO
1 kHz	110°	40°	64%	27%	ISO	ISO	ISO
2 kHz	90°	40°	56%	11%	ISO	ISO	ISO
4 kHz	40°	40°	0%	ISO	ISO	ISO	ISO
8 kHz	35°	40°	ISO	ISO	ISO	ISO	ISO
16 kHz	40°	40°	0%	ISO	ISO	ISO	ISO

FIGURE 9.30
Beamwidth plots for the second-order coupled point source, unity splay angle and level

There are several noteworthy trends below 4 kHz where the element quantity has a strong effect on spectral variance. The first is in the two-element scenario, between 500 Hz and 1 kHz. This characteristic signature (1 λ/0.5 λ) is an octave above the two-element first-order array (Fig. 9.29). The physical displacement is halved due to the smaller splay angle (closer at the front).

The next trend is toward coupling zone beam concentration. The combined angle at 125 Hz and below is consistently narrower than its 360° solo response. The beam narrows as quantity rises: the unmistakable workings of beam concentration.

The 100% overlap sequence of the parallel pyramid no longer applies. Instead there is a complex sequence of differing overlap percentages between elements over frequency. Outermost elements are the most isolated, putting the brakes on the beam multiplier and eventually moving toward spreading. This is evident in the 250 Hz response, which widens in the large quantities (five and six elements).

The range between 200 Hz and 4 kHz is a battleground between two competing and offsetting factors: angular isolation and wavelength displacement. The physical realities of multiple element clusters include layered λ displacements as the number of elements increases. Each element is displaced the same amount from adjacent units but the interaction continues between separated units as well, albeit reduced by angular isolation. This layering of interactions creates the complex weave found here. As a quick example consider a five-element array with a 0.5 λ displacement. The center cabinet is 0.5 λ from its adjacent neighbors, but 1 λ from the outside units. The outermost units are 2 λ apart from each other. The interactions are strong as long as the elements have a wide pattern in the given frequency range.

The 250 Hz to 500 Hz range provides a glimpse into the competing factors. The combined beamwidth is narrower than a single unit with quantities of two to four elements (beam concentration is dominant). The angular spread is so wide with five and six units (200° to 240°) that the outermost elements are isolated enough for beam spreading, even while the central elements are dominated by beam concentration. The combined result is widened slightly beyond the single element.

An alternative second-order system uses large-format horn-loaded systems with 30° nominal pattern above 2 kHz (Fig. 9.31). The horn-loaded low driver reduces the LF beamwidth, which extends the isolated frequency range beyond the front-loaded second-order speaker in spite of the reduced splay. The result is noticeably more uniform beamwidth over quantity. The beam-spreading and beam-concentration forces push the response in opposite directions as quantity increases. The HF response is continually widened, whereas the LF response is narrowed, resulting in a more consistent combined beamwidth than the single elements. The mid-range area sees both concentration (quantities <3) and spreading (quantities >3).

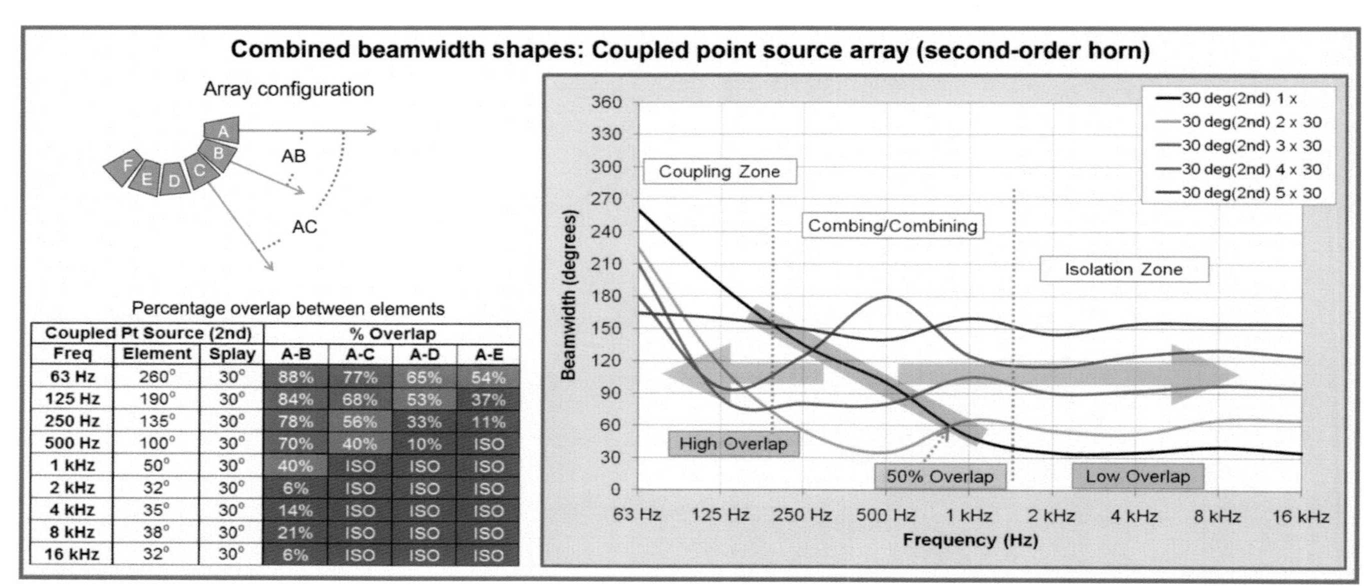

Coupled Pt Source (2nd)			% Overlap			
Freq	Element	Splay	A-B	A-C	A-D	A-E
63 Hz	260°	30°	88%	77%	65%	54%
125 Hz	190°	30°	84%	68%	53%	37%
250 Hz	135°	30°	78%	56%	33%	11%
500 Hz	100°	30°	70%	40%	10%	ISO
1 kHz	50°	30°	40%	ISO	ISO	ISO
2 kHz	32°	30°	6%	ISO	ISO	ISO
4 kHz	35°	30°	14%	ISO	ISO	ISO
8 kHz	38°	30°	21%	ISO	ISO	ISO
16 kHz	32°	30°	6%	ISO	ISO	ISO

FIGURE 9.31
Beamwidth plots for the horn-loaded second-order coupled point source, unity splay and level

The five-element array is a notable example of minimum spectral variance. The offsetting array factors have met in the 160° range over the entire frequency range of the device.

We next examine the effect of angular overlap for a fixed quantity of elements. In this case (Fig. 9.32) we will reuse the five-element horn-loaded second-order array just featured. The effects of increasing angular overlap are compared with the response using the unity splay angle of 30°. A splay angle of 22.5° gives us 75% isolation and 25% overlap. The combined result approximates a 25% reduction in beamwidth over the full range. The overlap side effects aren't seen in the beamwidth: increased ripple variance (-) and increased power addition (+).

An undesirable effect arises at 50% overlap (15° splay). The 125 Hz range becomes over-steered due to a mix too heavy on the overlap side. The solo element is 180°, which leaves adjacent units at 15° splay with over 90% overlap. The combination leaves the lower range squeezed to as little as 60° while the plateau range above 2 kHz holds at 90°. Everybody loves LF directional control but you can have too much of a good thing. The outermost 10° (per side) of coverage has full HF extension while the LF range is down more than 6 dB. The beamwidth of an array (or a single element) should never be narrower in the mids and lows than the HF range unless you want it to sound like a telephone. A highly overlapped array of flat beamwidth elements also features increased ripple variance. But at least it's really loud! This is an array capable of making us leave the room.

Finally we reach 100% overlap with a 0° splay angle, a point source to line source conversion. The parallel pyramid collapses the beamwidth as expected. Narrowing continues until displacement breaks down the beam concentration and combing/lobing scatters the pattern (in this case the break point is around 4 kHz). The combined beamwidth in the scatter zone resembles the individual element, albeit with massive ripple variance. The beamwidth expands by a 10:1 factor between 4 and 8 kHz, which removes all hope of minimum-variance performance for this array. It is noteworthy that highly overlapped flat beamwidth arrays were the most common type for concert sound before the advent of third-order proportional beamwidth speakers in the 1990s. Power, yes. Minimum variance? No.

Arrays of third-order speakers (Fig. 9.33) use splay angles that are a small fraction of those used previously for first- and second-order systems. The frequency range where the unity splay angle holds is the lowest ever: practically none. The response of a single element is compared with successive doublings up to eight units. There are several important trends. We begin with the beam-spreading behavior at 8 kHz. Notice that the single unit response (15°) is wider than the array of two (8°), and equal to that of four. This is consistent with the fact that the chosen

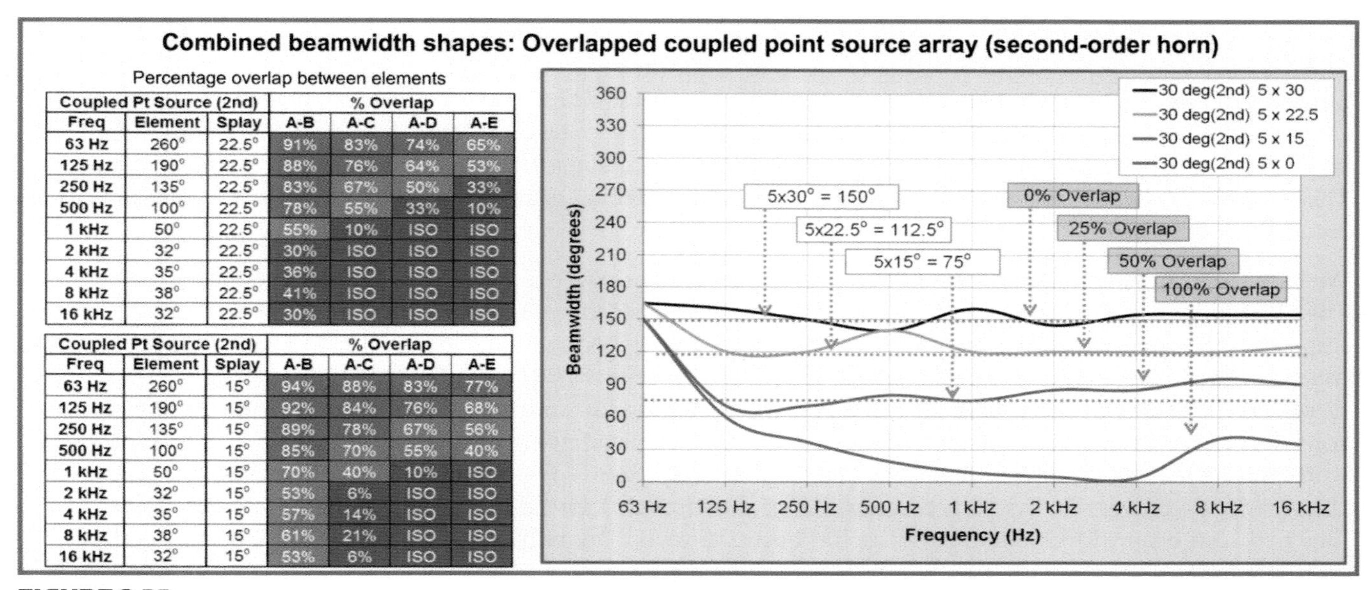

Combined beamwidth shapes: Overlapped coupled point source array (second-order horn)

Percentage overlap between elements

Coupled Pt Source (2nd)			% Overlap			
Freq	Element	Splay	A-B	A-C	A-D	A-E
63 Hz	260°	22.5°	91%	83%	74%	65%
125 Hz	190°	22.5°	88%	76%	64%	53%
250 Hz	135°	22.5°	83%	67%	50%	33%
500 Hz	100°	22.5°	78%	55%	33%	10%
1 kHz	50°	22.5°	55%	10%	ISO	ISO
2 kHz	32°	22.5°	30%	ISO	ISO	ISO
4 kHz	35°	22.5°	36%	ISO	ISO	ISO
8 kHz	38°	22.5°	41%	ISO	ISO	ISO
16 kHz	32°	22.5°	30%	ISO	ISO	ISO

Coupled Pt Source (2nd)			% Overlap			
Freq	Element	Splay	A-B	A-C	A-D	A-E
63 Hz	260°	15°	94%	88%	83%	77%
125 Hz	190°	15°	92%	84%	76%	68%
250 Hz	135°	15°	89%	78%	67%	56%
500 Hz	100°	15°	85%	70%	55%	40%
1 kHz	50°	15°	70%	40%	10%	ISO
2 kHz	32°	15°	53%	6%	ISO	ISO
4 kHz	35°	15°	57%	14%	ISO	ISO
8 kHz	38°	15°	61%	21%	ISO	ISO
16 kHz	32°	15°	53%	6%	ISO	ISO

FIGURE 9.32
Beamwidth plots for a five-element, horn-loaded, second-order coupled point source at various splay angles and unity level

angle (the manufacturer's maximum splay with the standard rigging frames) is less than the unity splay angle. The outermost units overlap less and the beam begins to spread as quantity increases. Overlap becomes stronger and beam concentration becomes increasingly dominant as frequency falls. The result is a gradual flattening of the combined beamwidth, in sharp contrast to the steady downward slope of the single element. This is the key design feature of the third-order speaker. A point source array comprising steadily narrowing elements will cause steadily offsetting effects among the beam behaviors. For a given splay angle, as the individual beams narrow (frequency rising) the combined beam spreads due to isolation. As the individual beams widen (frequency falling) the combined beam concentrates due to overlap. The result is a push–pull effect that flattens the beamwidth over larger frequency ranges as the quantity increases. Observation of the response over quantity reveals an approximate doubling of the flattened area with each doubling of quantity. Two units yields flattening down to 2 kHz, four units extend it down to 1 kHz and eight units flatten the beamwidth all the way to 500 Hz.

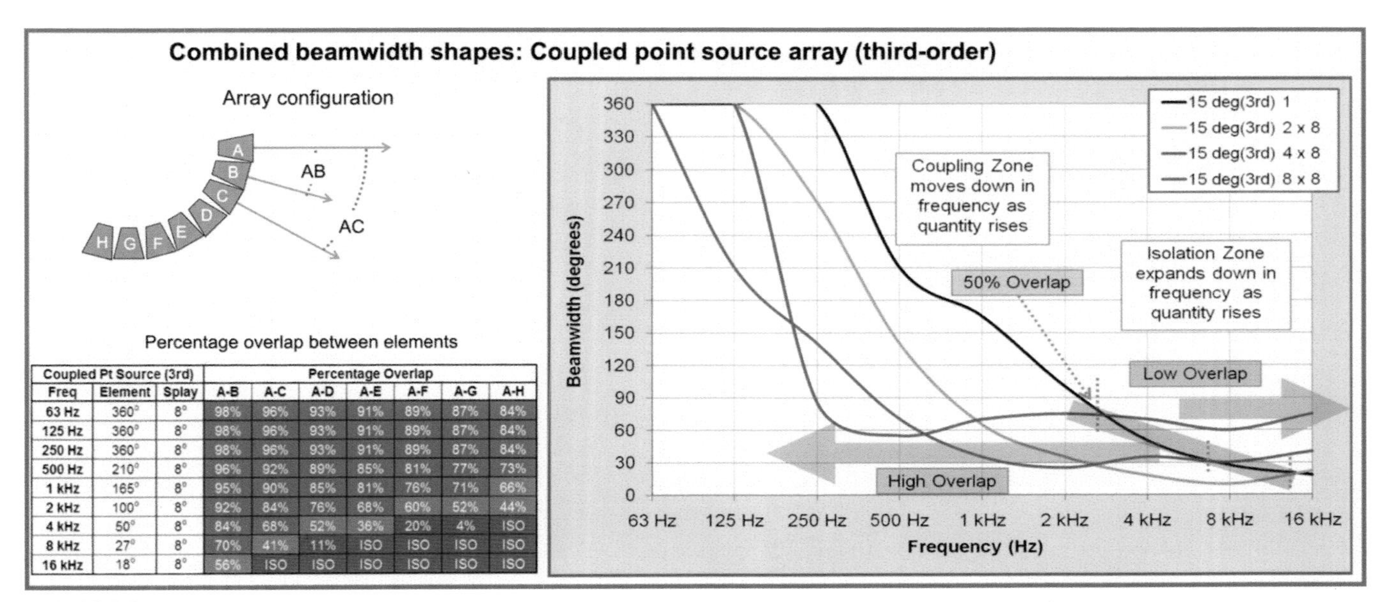

Combined beamwidth shapes: Coupled point source array (third-order)

Percentage overlap between elements

Coupled Pt Source (3rd)			Percentage Overlap						
Freq	Element	Splay	A-B	A-C	A-D	A-E	A-F	A-G	A-H
63 Hz	360°	8°	98%	96%	93%	91%	89%	87%	84%
125 Hz	360°	8°	98%	96%	93%	91%	89%	87%	84%
250 Hz	360°	8°	98%	96%	93%	91%	89%	87%	84%
500 Hz	210°	8°	96%	92%	89%	85%	81%	77%	73%
1 kHz	165°	8°	95%	90%	85%	81%	76%	71%	66%
2 kHz	100°	8°	92%	84%	76%	68%	60%	52%	44%
4 kHz	50°	8°	84%	68%	52%	36%	20%	4%	ISO
8 kHz	27°	8°	70%	41%	11%	ISO	ISO	ISO	ISO
16 kHz	18°	8°	56%	ISO	ISO	ISO	ISO	ISO	ISO

FIGURE 9.33
Beamwidth plots for the third-order symmetric coupled point source. Splay angle is 50% overlap, unity level.

9.4.2.3 SPLAY ANGLE EFFECTS OVER FREQUENCY (SYMMETRIC)

The presence of an angle, even a small one, opens an opportunity for reduced spectral variance. Even small amounts of angular isolation will begin reversing the relentless pattern narrowing. Fig. 9.34 shows sixteen third-order speaker elements at unity level with three successive splay angle doublings. A gradual emergence is seen from the familiar parallel beam concentration behavior toward a new dominant factor: radial beam spreading. These two behaviors are the governors of the low- and high-frequency ranges respectively, and they will meet at some point in between.

We begin with the smaller angle increment of 0.5° per element. The combined angle spread is 8° (16 × 0.5°). The LF responses (125 Hz and 500 Hz) are virtually unchanged from the coupled line source at 30° and 9° respectively (see Fig. 9.26B for reference). This is not surprising because a combined spread of 8° will not provide much counterforce to the overlapping effects of sixteen elements with 300° individual patterns. The HF response (2 kHz and 8 kHz) shows a substantive change, expanding from 2° and 1° respectively to 8° each. The pyramid convergence has been slowed by the introduction of splay angle allowing the expansion of the beam to the *angular* width, rather than the *line* width. Note that the angular shape holds over distance in contrast to the line source's changing pattern over distance. It is no coincidence that the splay spread of 8° and combined coverage angle are matched. Our beamwidth flattening device has begun its work. This will be the typical response when isolation zone summation is dominant.

The isolation zone signature dominates the HF ranges as the splay angle doubles to 1°. The 16° spread is clearly defined as an angular area with sharp and distinct edges. The 500 Hz response is showing early signs of push and pull as it widens slightly (from 9° to 12°).

A coverage shape created by radial isolation is distinct from one created by beam coupling even if both are described by the same coverage angle. The distinction is often missed because coverage angle is defined by comparing on axis to the -6 dB point. The radially isolated shape holds its 0 dB value over the arc and then finally drops sharply to -6 dB and more. The beam-coupled version begins its decline as soon as the on-axis point is left.

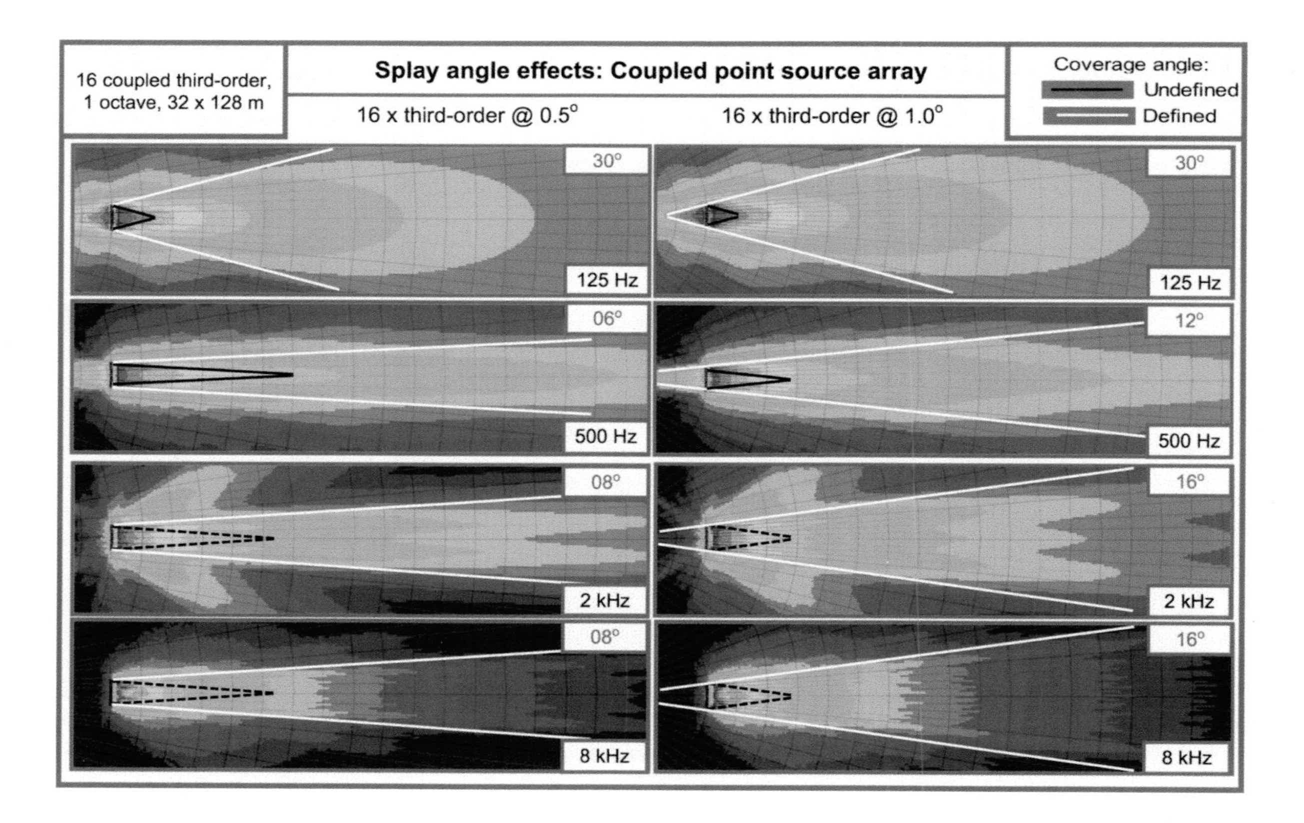

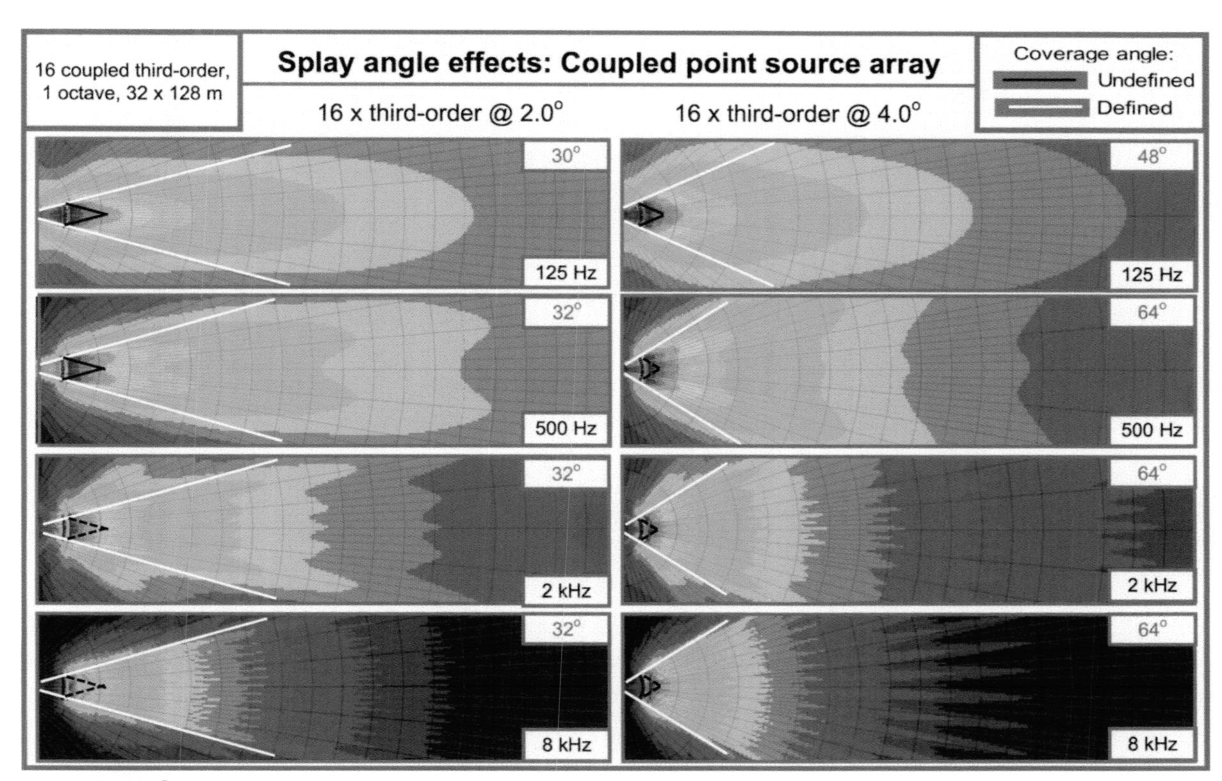

FIGURE 9.34A, B
Splay angle effects vs. frequency for the coupled point source

This is a rectangular shape similar to a single-speaker response. Only the 0 dB and -6 dB points can be expected to match when such pattern shapes are compared.

Let's double the angle again. Now we have 2° splays creating a full spread of 32° (Fig. 9.34B). The 500 Hz response is now finding itself in the crossroads, while the upper ranges show highly developed spreading. We have reached a significant milestone: The coverage pattern over this 8-octave span is nearly perfectly matched. All four responses show -6 dB points approximately 32° apart. The final doubling (4°) leaves only 125 Hz in beam concentration mode. We have actually gone too far. Coverage from 500 Hz and up has spread to the 64° splay but the 125 Hz range has only just begun to put the brakes on its beaming. Its 48° beam leaves the outer coverage areas lacking in LF response while the mids and high are still going strong. Those that can't get enough of line length might consider that there can be too much of a good thing. If we have subwoofers to cover down there, however, this point may be academic.

Let's evaluate these four arrays by our minimum-variance criteria. They all show lower spectral variance than we found in the coupled line source arrays. Radial spreading stemmed the tide of endless narrowing found previously when all elements had 100% overlap at all frequencies. The lowest spectral variance is found in the 32° and 64° arrays. An added bonus is that these will have the lowest ripple variance due to reduced overlap. We have found a minimum-variance array configuration: the symmetrical point source.

This basic framework is applicable to all speaker orders, not just third-order speakers (Fig. 9.35). Here we see four different recipes for 100° coverage, ranging from a pair of 50° speakers to a quintet of 20° elements. In all cases the HF response falls in the 100° due to radial isolation. In general we are able to maintain a more consistent beamwidth with increased quantity.

9.4.2.3 ASYMMETRIC LEVEL EFFECTS

We now investigate the effects of level asymmetry. Refer to Fig. 9.36. Our example thirty-two-unit array has a constant 1° splay angle. The level is continuously tapered so that each element receives a successively lower drive level. Disclaimer: Kids, don't try this at home. Such a drastic level taper is not recommended practice. It's used here to illustrate the trends.

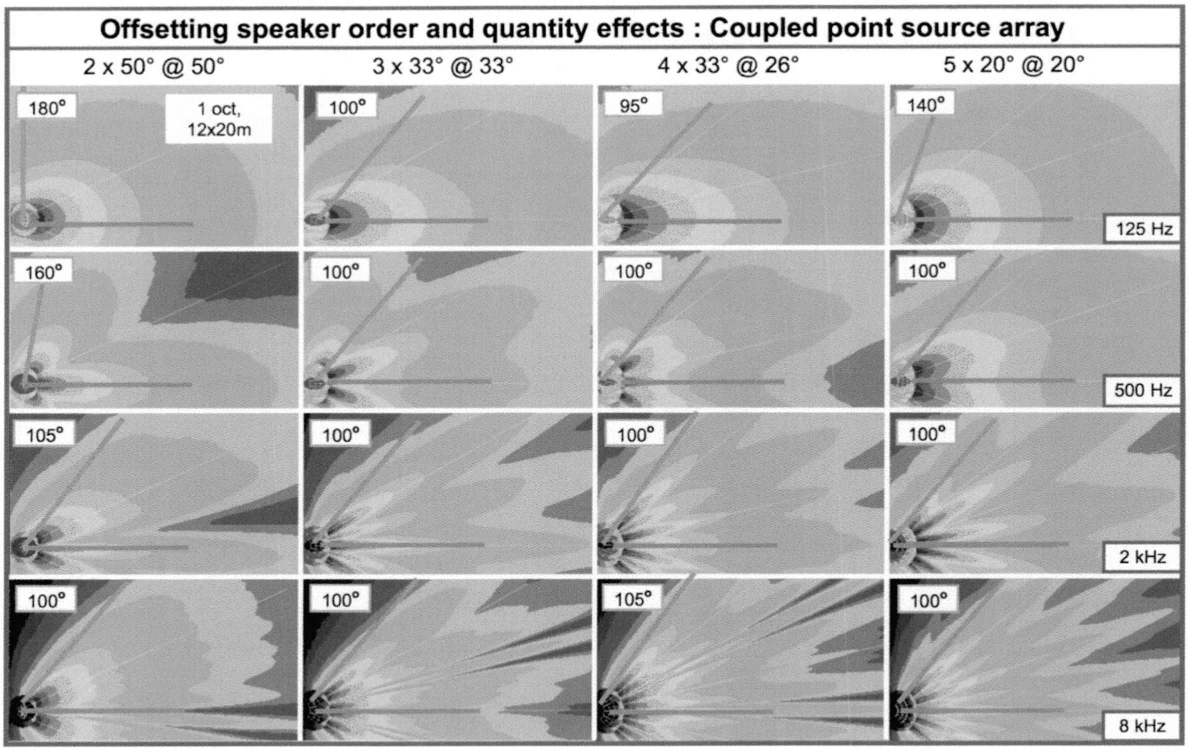

Offsetting speaker order and quantity effects : Coupled point source array

2 x 50° @ 50° | 3 x 33° @ 33° | 4 x 33° @ 26° | 5 x 20° @ 20°

1 oct, 12x20m

125 Hz: 180° | 100° | 95° | 140°

500 Hz: 160° | 100° | 100° | 100°

2 kHz: 105° | 100° | 100° | 100°

8 kHz: 100° | 100° | 105° | 100°

FIGURE 9.35
Four different coupled point source scenarios create a 100° of radial coverage, ranging from small quantities of wider elements to larger quantities of narrower elements. The 2 × 50° option (left) shows the highest variance over frequency because the LF region stays wider than the HF. Note that the element in the middle two scenarios uses a horn-loaded LF driver, which helps preserve minimum spectral variance with relatively small quantities. The 5 × 20° reverts to front-loaded LF but achieves a flattened beamwidth by coupling the overlap of the five LF drivers.

Characterization of an asymmetric shape presents a challenge to our traditional thinking. Where is "on axis" of an asymmetric coverage pattern? What do we do with off-axis points that fall at different angles/distances from the "center"? We will define center as the point of highest level. Off axis is calculated from there, and will have two individual angles (the edges) and one combined angle (the radial spread). That is simple enough, but what do we do when different frequency ranges have different centers?

Let's reflect for a moment about our lessons from the symmetric version. The centers were matched over frequency but the edges differed. We see sculpted radial edges in the isolated range and one big rounded beam in the coupled zone. The range between moves gradually between the shapes. The HF range plays the leading part because its malleable edges are the easiest to conform to an asymmetric target shape. Our goal will be to extend the same picture down in frequency.

Asymmetric array evaluation

- Center (on axis): the angular orientation of the longest distance/loudest point.
- Top/bottom (off axis): -6 dB points along the target shape, above and below center.
- Total (TTL): the combined angular spread from top to bottom.
- Symmetry percentage: the ratio of angular coverage above and below center.

Constant coverage angle over frequency is no longer enough. We must know where the lows and highs are centered, and how well they match our asymmetric target.

Our first array has a constant symmetric splay angle (1°) and asymmetric level taper of 8 dB in thirty-one successive ¼ dB steps. The HF ranges are sculpted into a diagonal MV line in scale with our expectations regarding level and distance (review Fig. 9.11). The farthest point (on axis) in the 8 kHz range is at +3°. The coverage extends just 2° above this point and 31° below for a combined total of 33°. The line of minimum variance is 2.5× farther to the top than the bottom (equivalent to 8 dB). Obviously this is highly asymmetric, and yields a result of 6% symmetry. The beam "center" in the 2 kHz response is 2° lower, but the top and bottom angles remain unchanged. This is the start of a trend toward lower centering and increased symmetry (14%) as frequency falls. The center has moved down to -4° at 500 Hz. The geometric center of the array, marked by a blue dotted line, is at -12.5° (the top box is aimed at +4°). The geometric center tells us where the beam center *would* have been if the array were symmetric. This serves as a reference point as to how far we have moved the array from its symmetric tendencies. The 500 Hz beam center has been lifted 8° by the level tapering but still falls below the HF responses. The LF (125 Hz) maintains nearly the same coverage angle as the other ranges but is aimed far down (-12°) and is 93% symmetric. What does this all mean? Here is a simple summary: The center at 8 kHz is off axis at 500 Hz and 125 Hz. The level tapering bends the HF response but barely moves the LF responses. It's a start.

Let's investigate more by doubling the level taper to 16 dB (Fig. 9.36, right side). The sculpted HF bends farther outward with increasing level asymmetry. The minimum-variance line falls to a 15° slope. This would be a better fit for a lower rake in the audience seating (vertical plane). A steering control is revealed that allows us to form-fit the response over tilted surfaces with various rake angles. The steering is highly effective when angularly isolated (HF) and less so when coupled (LF). Symmetry has decreased in all frequency ranges, but this does little to reduce the disparity between the LF and HF patterns. The longest beam of HF coverage is still off axis to the LF coverage.

It is widely reported that the LF range is less "controlled" as a result of level tapering. This depends on your definition of "controlled." With matched levels the beam is the most symmetric but not necessarily the narrowest, the property typically synonymous with "control." It does us no good to evaluate coverage angle in symmetric terms when we have an asymmetric shape to fill. Level tapering can increase "control" if we consider it a measure of how well the coverage pattern fits into the shape. If we remove the level taper for this array, the increase in "control" comes at the cost of pushing the LF beam center farther downward to the geometric center (-12.5°). Level tapering's loss of "control" comes with the bonus of steering the beam in the direction of the minimum-variance frequency response shape. Large-scale LF-level tapering is a waste of power resources and therefore not widely practiced. We will see later that asymmetric phase steering will be more effective and efficient. For now we will let it suffice to show that the HF and LF require different steering mechanisms.

It is worthwhile to ponder the question of priorities at this juncture. Let's assume that we cannot always achieve minimum spectral variance over the full range. Which range would be the best to focus on and which to let go? We know already that the HF range will be easiest to control, whereas the LF range will usually be overlapped and therefore more challenging. We also know that efforts to control the LF range by beam concentration may backfire on us by over-steering the upper ranges. The answer to our prioritization lies in the room. The speaker/room summation will be a far more dominant factor in the LF range than the HF (unless we have a *really* bad room). We must expect that the low end will be strongly affected. It is still possible to get usable LF energy from the speaker/room summation even when our shape is not a match there. The room's contributions will be too late beyond the LF band to provide much usable addition, without costly combing. Therefore we must prioritize downward with frequency. First priority is to match the high mids to the highs. Then we add the low mids and break out the champagne if we make it all the way down.

The final consideration for this array is ripple variance. Spatial crossover areas can exhibit substantial ripple, as we know from our in-depth study of summation. There are thirty-one highly overlapped asymmetric spatial crossovers in this case, each of which has some volatility. On the other hand, there are two significant factors that reduce the ripple:

1. The angular splay and level tapering reduce the amount of multi-element overlap found in the symmetric coupled line source array. Isolation increases along with displacement between non-adjacent elements.
2. Time offsets can be kept low (provided the boxes are small and tightly configured) due to the minimal displacements between the overlapping sources.

We can conclude that the level-tapered asymmetric coupled point source has potential for minimum-variance performance in all three categories: level, spectral and ripple. (Just don't try it with 16 dB of taper.)

9.4.2.4 LEVEL INCREMENT EFFECTS

How much effect does the number of level taper increments have on the outcome? If we get 16 dB of level taper, does it matter whether it's done in thirty-two steps or two? We can immediately guess that there will be some effect, but how much and at what frequency? The answer is as incremental as the question. The finer we slice the response the smaller will be the tatters on the sculpted edges. The example shown in Fig. 9.37 shows the response with thirty-two (left) and four increments (right) respectively. The differences are confined almost entirely to the isolated HF range. This is not surprising because isolation and individualized response control go hand in hand. The four-step scenario leaves telltale fingerprints in the 8 kHz response. The key concept here is that incremental-level changes are a direct cause of spectral variance. The HF shape becomes unmatched to the lower-frequency ranges whose edges are smoothed over by higher overlap percentage.

The squaring off in the HF shape will rise with isolation. We can expect the break points to become more obvious and wider in range if the splay angle were opened and isolation increased. A first-order array requires the lowest ratio of elements to increments whereas a third-order system is the most capable of grouping multiple elements on a single channel of processing. There is a practical tradeoff between level increments and spectral variance. Fine slicing has signal processing and amplification costs, more optimization steps and increased opportunity for wiring errors. The coarse approach limits our ability to fine-tune the shape.

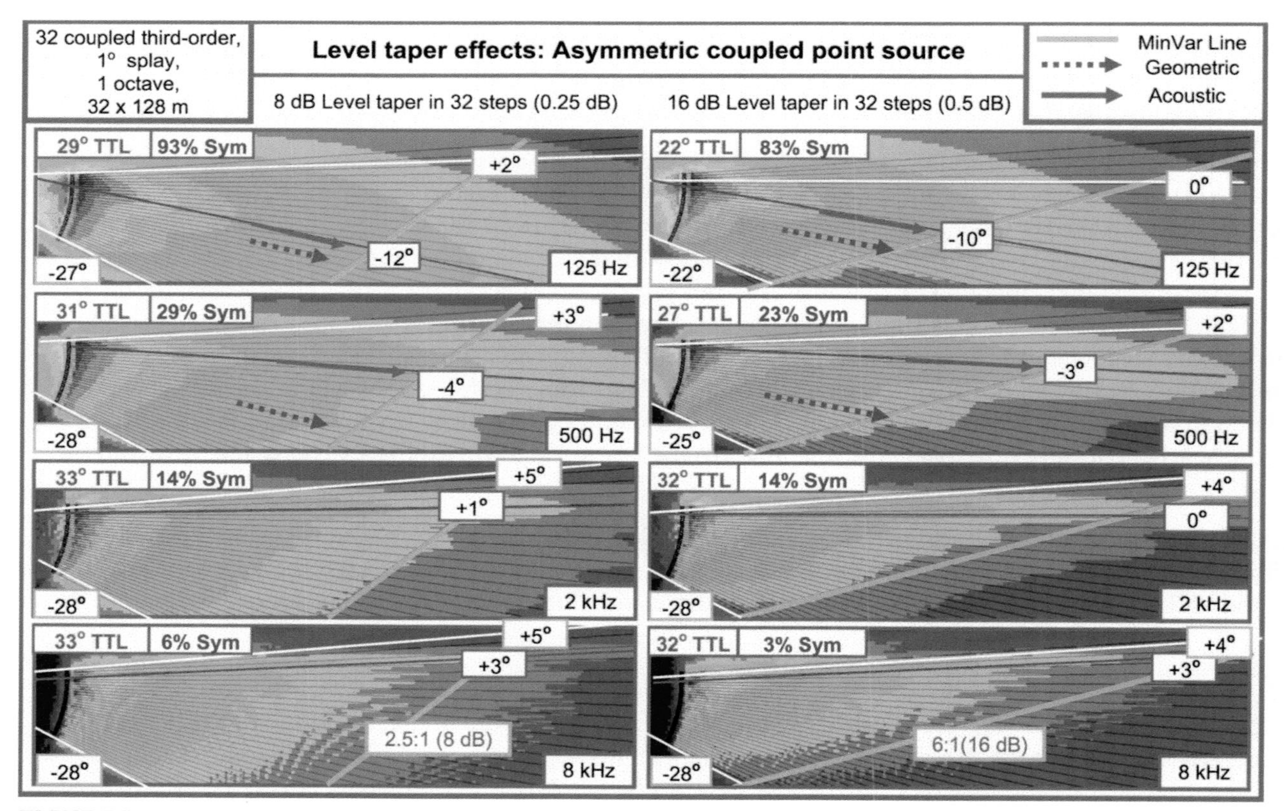

FIGURE 9.36
Level asymmetry effects vs. frequency for the coupled point source

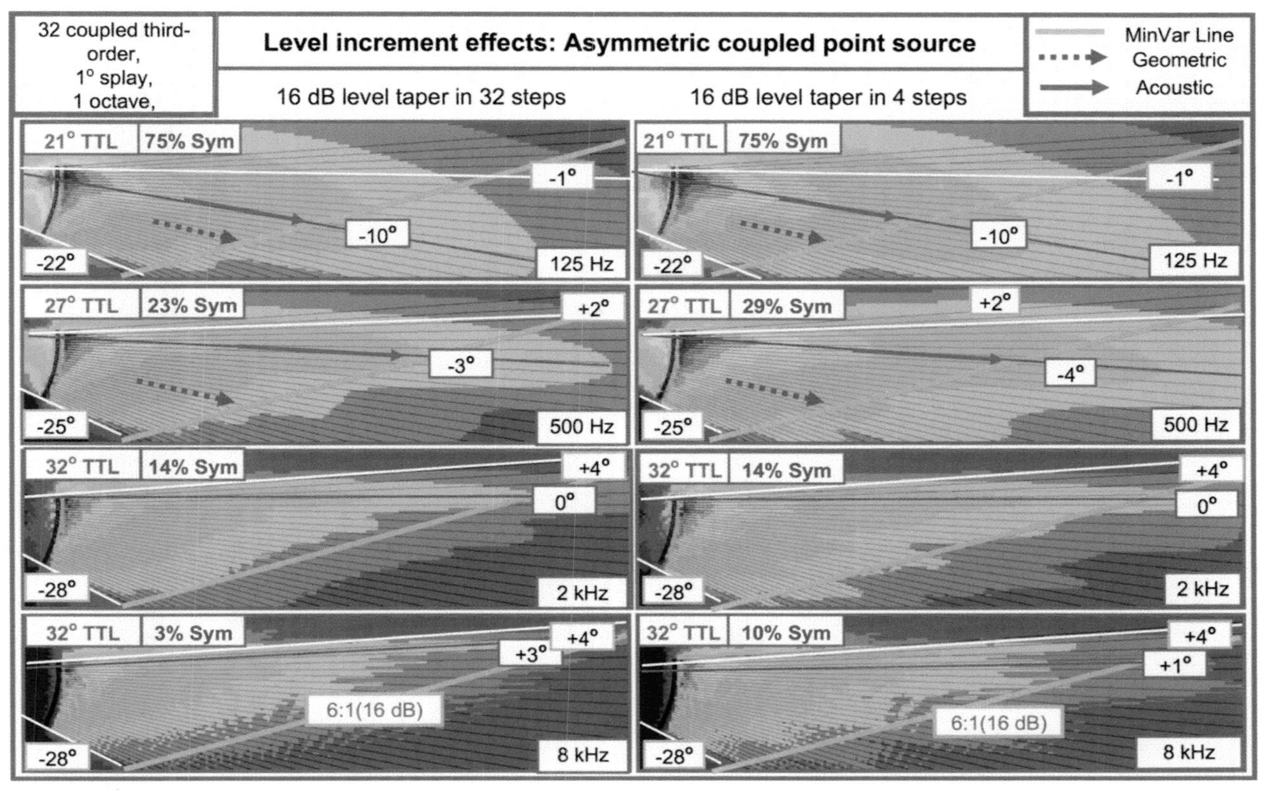

FIGURE 9.37
Effect of number of level increments vs. frequency for the asymmetrical coupled point source

9.4.2.5 SPEAKER ORDER EFFECTS

We can fill a 60° symmetric radial shape with one 60° speaker or an array of sixty speakers splayed at 1°. That's extreme, but it short-circuits the discussion of speaker order for the coupled point source. The calculus changes significantly once we add asymmetry to the equation. Any order of speaker can fill a shape with a small range ratio but only high-order speakers can handle the extreme ratios. Simply put, narrow speakers can make a segmented copy of a wide speaker but the low-order elements cannot return the favor. Nonetheless there is enough shaping capability in lower-order speakers to make them a viable design option for the asymmetric coupled point source.

Let's explore this by viewing second- and third-order arrays with nearly matched parameters: length, splay angle spread and level tapering (Fig. 9.38). The target shape has a 3:1 range ratio (10 dB) so we are not going easy on the second-order system. The element coverage angle and quantity differ, of course, with the third-order arrays having a larger quantity of narrower elements. The responses reveal substantial areas of similarity and some important differences. The similarities suggest that array configuration is as important as the nature of the individual elements. Matched array configurations comprised of different elements have greater similarity than matched elements configured differently.

The four frequency range pairings show the same basic shape. The differences between the centers, tops and bottoms are statistically insignificant in all ranges except 500 Hz, where the second-order system shows a strong down-lobe. The second-order system has more HF leakage above the array top (due to the wider elements) and more ripple variance in the coverage area (wider angles, higher displacement).

What can we conclude from this? Minimum spectral variance can be achieved with this configuration regardless of speaker order, provided the offsetting quantities are applied. The third-order approach has the potential advantage of increased power and asymmetric shaping from increased quantities of overlapped elements.

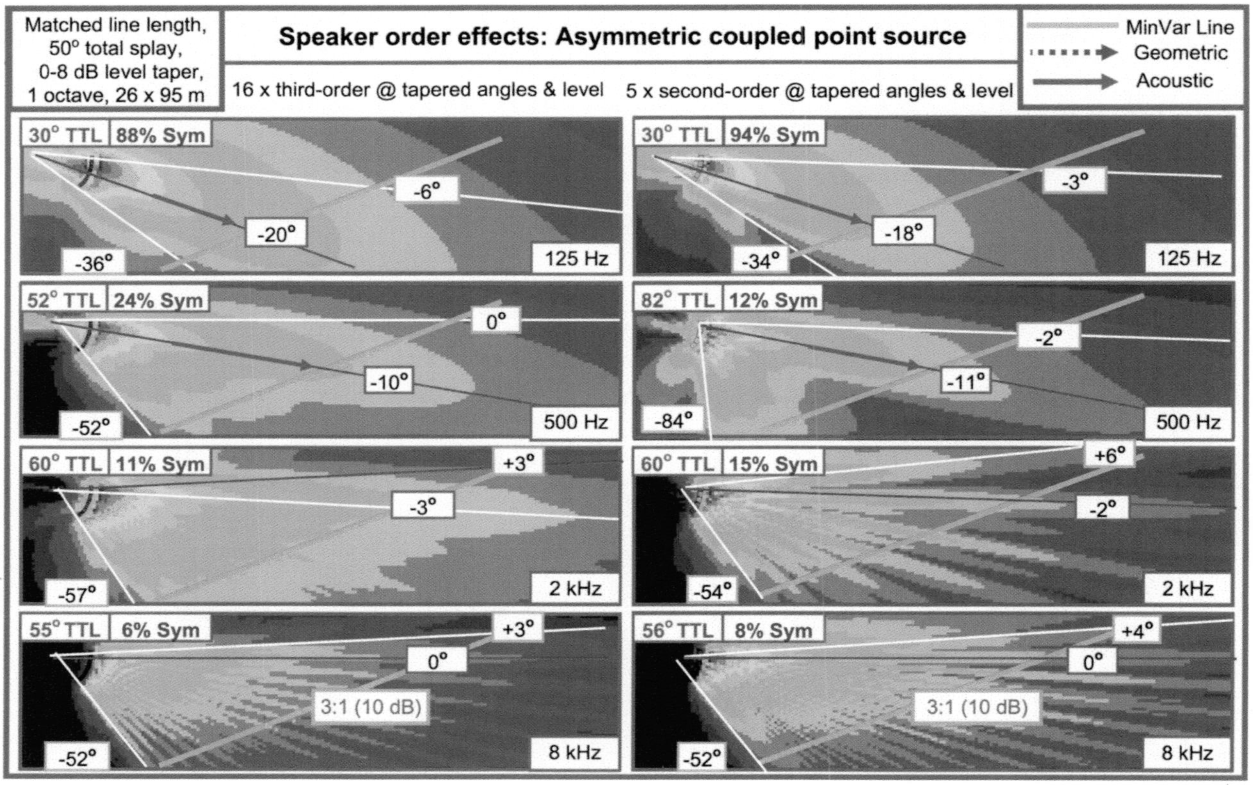

FIGURE 9.38
Speaker order effects vs. frequency for the asymmetric coupled point source

9.4.2.6 COMBINED ASYMMETRIC SPLAY, LEVEL AND ALL-PASS DELAY EFFECTS

There are three primary asymmetric steering mechanisms for the diagonal minimum-variance line: level, angle and delay. The three mechanisms can be combined to produce shaping effects exceeding their individual efforts. The effects of level tapering and asymmetric angles are shown separately and combined in Fig. 9.39A. The centers, tops and minimum-variance lines are closely matched above 125 Hz. (The bottoms are not matched because the angular spread differs.) The combination of these matched slopes has a doubling effect (the left panels of Fig. 9.39B). The combination of 6 dB level taper and a 6:1 ratio of angular asymmetry creates a shape similar to the 16 dB taper seen in Fig. 9.36, yet with greater LF shape conformity. This combination has the lowest spectral variance we have yet seen, with the LF response tilted down only a few degrees.

The final combination in this series includes delay asymmetry. This is not delay in the traditional sense, but rather a frequency-selective delay that will steer the LF range upwards without affecting the HF range. The upper cabinets receive the most delay, tipping their focus upward. The delay is inversely proportional to frequency. The 125 Hz range of the uppermost element has the largest delay (3.2ms) whereas its 500 Hz area has ¼ of this amount (0.8 ms). Delays decrease proportionally as we move down the stack. This array has the lowest spectral variance yet again, varying only a single degree in its center orientation.

9.4.2.7 SCALAR EFFECTS

Will the same coverage shape emerge if we shrink the array in half but maintain the same overall angle and level relationships from top to bottom? The answer is a qualified yes. Coverage angle expansion in the LF range is assured (reduced line length), but the HF range behavior is worth looking at for a few rounds of scalar doubling (Fig. 9.40). We can rescale by simultaneously doubling and halving parameters. When quantity is halved, the splay angle is doubled. The modeled space is also halved so that all dimensional relationships are preserved. We create the same asymmetric shape with half the array in half the space.

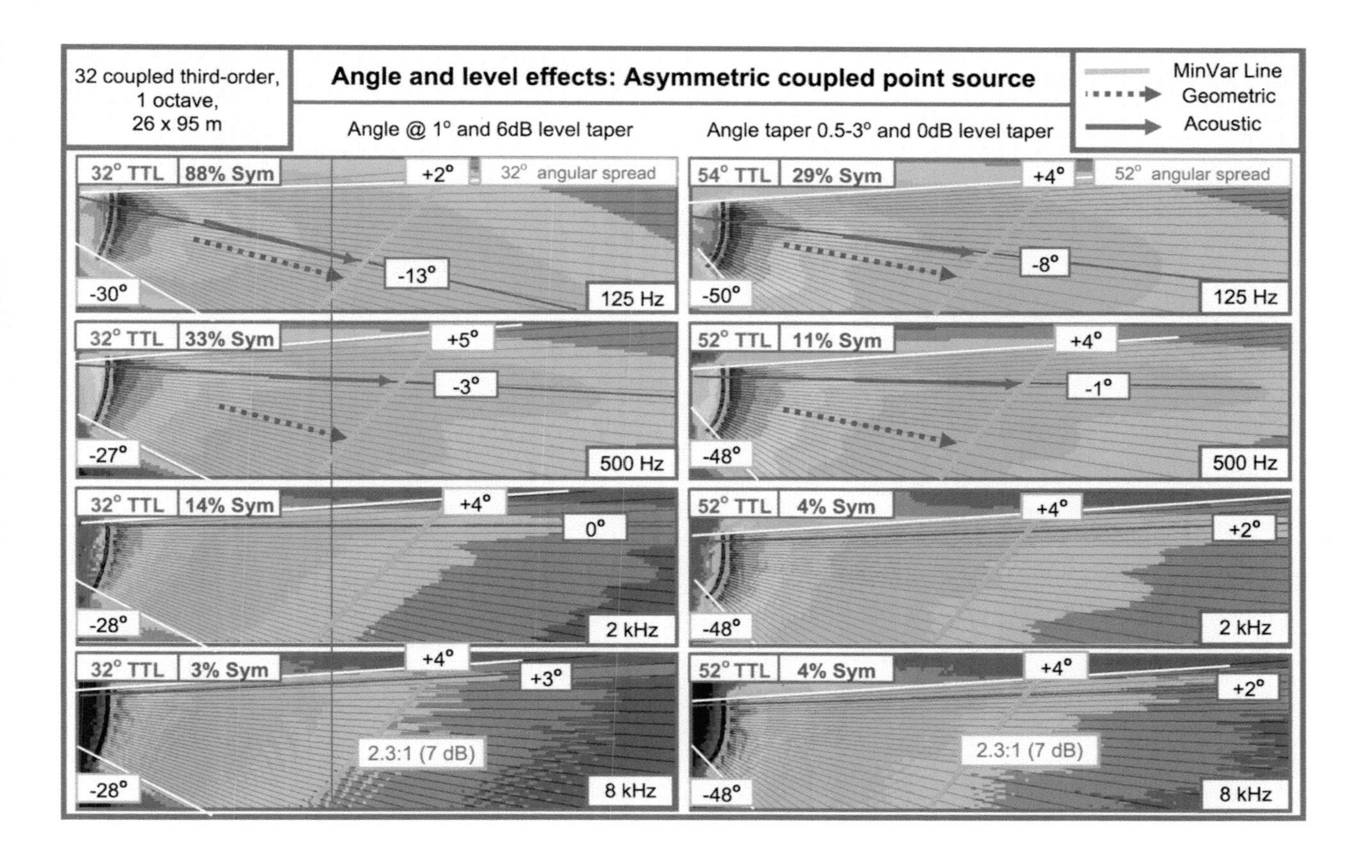

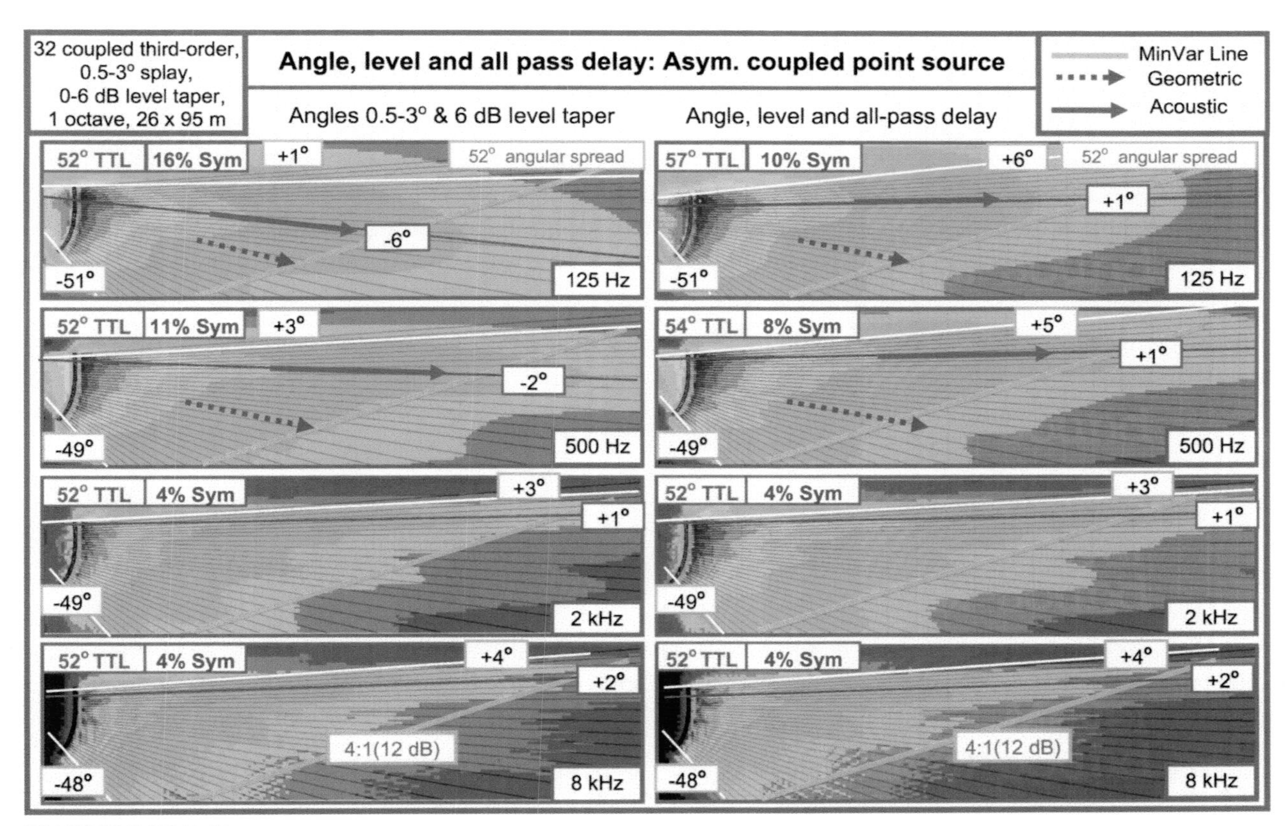

FIGURE 9.39A, B
Separate and combined asymmetric level and angle effects vs. frequency for the coupled point source

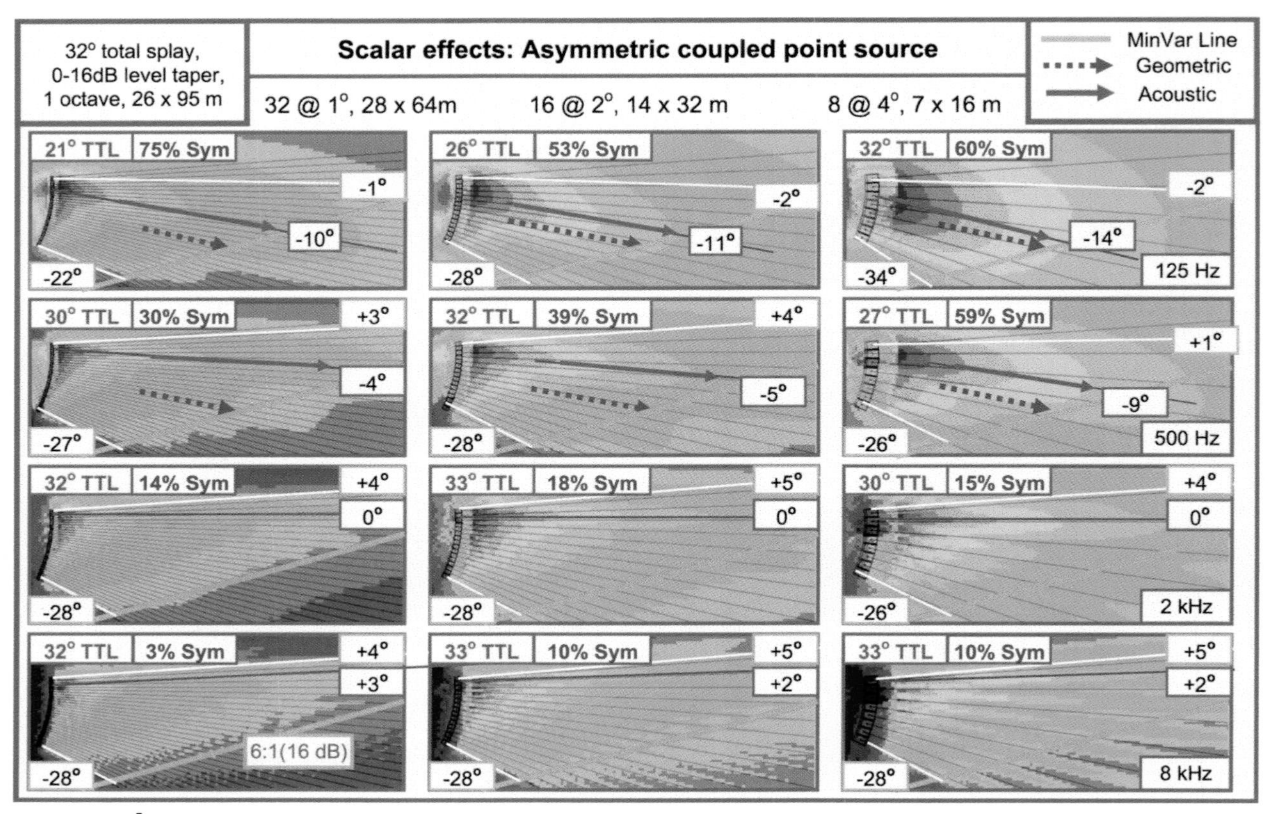

FIGURE 9.40
Scalar effects vs. frequency for the asymmetric coupled point source

The coverage shape in each is functionally equivalent above 2 kHz. The shaping behavior of isolated array elements is scalable. All show similar spreading behavior over the 32° angle and 8 dB level taper. This has its limits. We will gap in the HF if we splay the elements beyond their limits. This is evident in the small-scale 8 kHz response, which could not stand another doubling. Another downscaling could be done with the same quantity of smaller boxes. The range ratio also affects scalability. Highly asymmetric shapes require quantity. We can better preserve shape with the same quantity of smaller boxes of similar splay than halving the big box and doubling splay.

Scalar quantity is primarily a power capability and low-frequency steering issue as long we have enough quantity to cover the range ratio.

9.4.3 Hybrid line/point source

Multiple unit arrays can be configured as a line source/point source hybrid. This is common practice in many third-order line array applications. Most modern system designs do not place all of the array elements in a straight vertical line. At the very least designers feel compelled to tilt a few enclosures at the bottom of the array toward the early row of seats.

The hybrid array is inherently asymmetric. The upper section is a symmetric coupled line source, whereas the lower is a coupled point source. The key parameter is the transition point between the arrays, inherently an asymmetric spatial crossover. Relative quantities and ratio of asymmetry of the two array types affect the crossover behavior.

Another series of doubling scenarios isolates the effects. We will modify the line source/point source allocation of the sixteen elements while maintaining a consistent angular spread. The line source section is presented only in symmetric form, because we have previously shown the futility of level tapering it (see Fig. 9.28). For brevity the point source section is only shown in symmetric form.

The doublings of line/point resource allocation vary from 2/14, 4/12 to 8/8 for the sixteen-element array (Fig. 9.41). The 8/8 scenario is the most telling. The HF ranges clearly show two array types divided into separate worlds. The line source elements show beam concentration narrowing, while point source units perform radial beam spreading. The level changes are so severe that four distinct coverage edges are found (a top and bottom for each and a gap in the middle). The combined shape looks like a gun pointed at the rear of the hall (and probably sounds like it too). This is a most unlikely shape for an audience plane.

The two array types don't mix because line source beam concentration steers the energy away the point source spreading. The coupling increases the on-axis level, narrowing the beam enough to take it out of reach of the spreading components. Energy from the radially isolated elements is unable to merge with the slope of the parallel pyramid. The schism heals as frequency falls and overlap dominates both array types.

The hybrid attempts to mix two shapes that can't stay married over the long haul: a variable pyramid and a fixed arc (see Fig. 9.42). Recall that the coupled line source retains the proportional beamwidth of its elements, whereas the coupled point source flattens it out. What's the best splay angle to connect a speaker that changes its shape over frequency to one that doesn't? I have no idea.

There is an additional complication. The coupled line source has to pass through its metamorphic stage before becoming a single proportional beamwidth speaker. The semi-coupled line source adds a lateral extension to the fully coupled point source, forming a combined "J" shape. It's a short-term marriage that grows apart as the line source continues to evolve toward complete coupling. The line source shape is not stable over distance, whereas the point source is. Shapes stabilize once the pyramid top has been reached at all frequencies, but the line source's finished shape is not capable of joining its point source neighbor below.

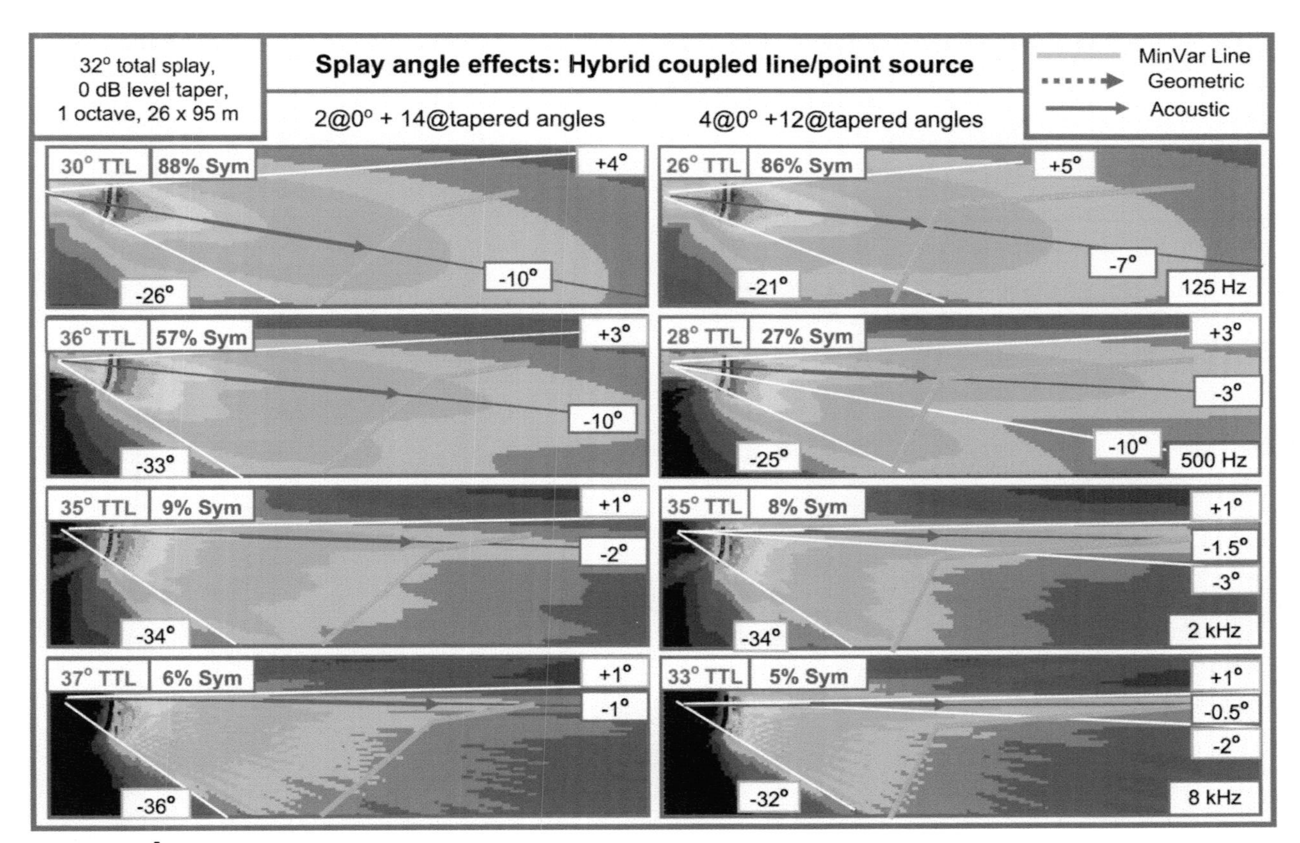

FIGURE 9.41 A
Asymmetric splay angle effects vs. frequency for the hybrid coupled line/point source

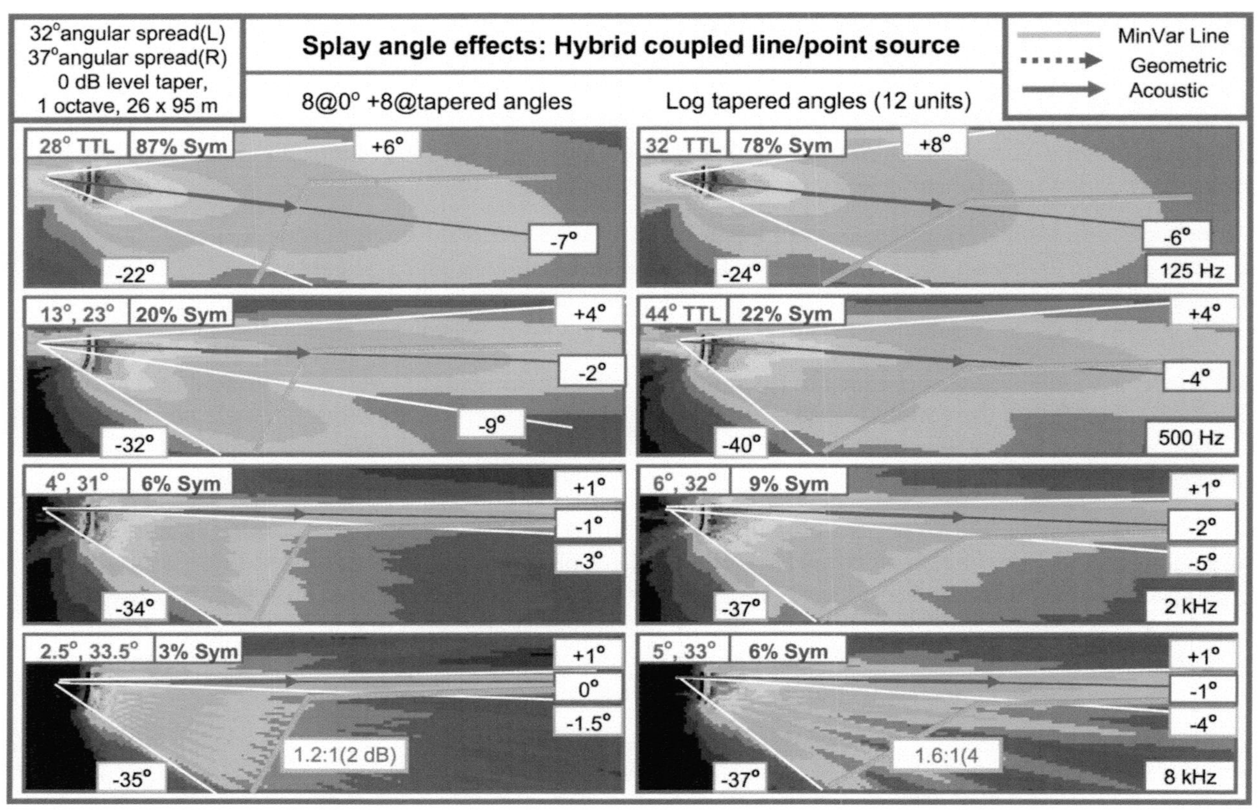

FIGURE 9.41B
Asymmetric splay angle effects vs. frequency for the hybrid coupled line/point source

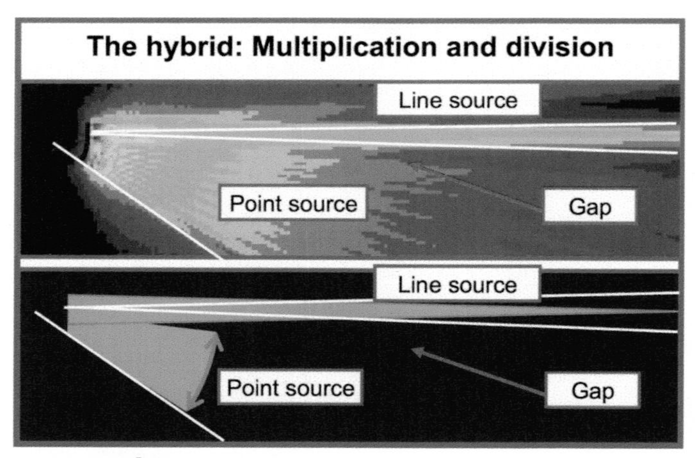

FIGURE 9.42
Coverage shapes for the components of the hybrid coupled line/point source

The 4/12 allocation reduces the on-axis addition and lessens the line source narrowing. The HF responses show a slight thickening in the merger area between the beam concentrating and spreading zones, but the main trends remain. A loaded gun is still aimed at the audience. The gun barrel persists even for the 2/14 allocation.

An additional scenario is a "log" angle taper, recommended by a speaker manufacturer. This is actually a form of the asymmetrical point source, because no angles are 0°. The initial angles are so minute (0.2°) that it looks more like the hybrid arrays than the asymmetrical point sources covered earlier. The gun barrel remains in view for the HF responses.

9.5 SPECTRAL VARIANCE OF UNCOUPLED ARRAYS

Our study of coupled arrays revealed the challenges to creating consistent level contours over frequency. This is a "walk in the park" compared with the hazards of uncoupled arrays. Ripple variance now becomes the range-limiting force due to the large-scale displacement of uncoupled elements. The frequency-dependent spatial properties of summation play a major role in how we approach minimum spectral variance in uncoupled arrays. When substantial displacement and overlap are found together, all hope of minimum spectral variance goes with it. We can't run through all possible iterations, but instead focus on directions where limited success is possible.

Uncoupled arrays can create a reasonably consistent shape in three directions: lateral, radial and forward. Example applications will reveal the trends in these directions.

9.5.1 Uncoupled line source

9.5.1.1 SYMMETRIC UNCOUPLED LINE SOURCE

We begin with the symmetric uncoupled line source (Fig. 9.43). The minimum-variance area runs from 50% to 100% of the FAR shape as discussed previously (section 8.6.2). Coverage depth is limited by ripple variance because there is no angular isolation. Success is measured by the consistency of the coverage start and stop points (the unity and limit lines). Therefore, a consistent combined shape requires a consistent element shape, which favors the first-order speaker. As speaker order increases, the start and stop areas move forward only in the highs, leaving the mid-range and lows behind. As spacing widens, the start and stop points will scale proportionately. Frequency does not scale so the ripple variance will differ with displacement. The effects of these variables were shown previously in the figures in Chapter 4 and for brevity are omitted here.

9.5.1.2 ASYMMETRIC UNCOUPLED LINE SOURCE

We move forward with an example of shape-shifting with the uncoupled line source (Figs 9.44 and 9.45). There are an infinite number of ways to mix these three variables: relative level, spacing and speaker order. For brevity we will recycle the two scenarios we used to illustrate level variance in Fig. 9.19, now adding spectral variance information to these compensated unity spacing configurations.

The first scenario aligns the starting points, creating a diagonal unity line, while the second staggers them to create a straight line (in one location). The HF responses hold discernible shapes but chaos reigns below that. Isolation by level is sufficient in the HF range because of the directional control there. The lack of isolation in the lows leaves the attenuated speakers at the mercy of the late arrivals from the dominant sources. We have three main choices here: surrender the LF custody to the dominant elements (i.e. roll off the low end in the small units), use elements with directional control in the LF (this becomes more difficult with the smaller speakers) or live with the fact that the LF range is in the combing zone. The convergence of high displacement, high overlap, asymmetric level and low directivity creates an impossible puzzle, which no delay strategy can solve. This is not to say we

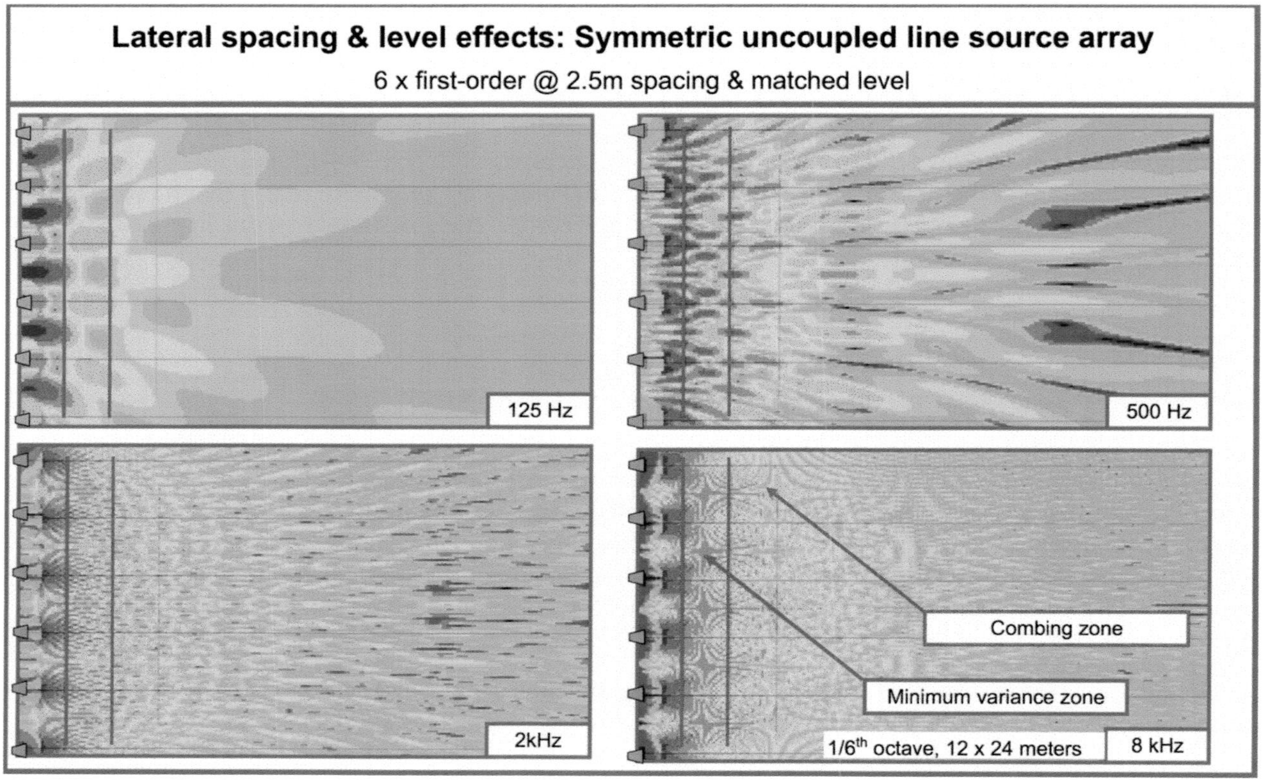

Lateral spacing & level effects: Symmetric uncoupled line source array
6 x first-order @ 2.5m spacing & matched level

FIGURE 9.43
Coverage shape vs. frequency for the uncoupled line source

should never use the asymmetric uncoupled line source, but simply to keep our expectations realistic about what we can do with it.

The staggered scenario faces the same challenges and fares no better. Isolation and combing patterns populate the same locations in the spectrum but different locations in the room.

More directionality looks like a good solution here, because the problems are created by overlap. This is true but not in the expected way. If we changed over to second- or third-order speakers we would likely make the problems worse, because the issue is LF control, not HF. The ideal element for this array has a flat beamwidth for its entire range. This would give us the same level of isolation over the whole spectrum and create the maximum uniformity of combined shape. High-order elements are the least likely to have an extended flat beamwidth. It's possible but will take some size (e.g. horns, parabolic dish or column arrays).

It would all be well and good to conclude that use of such arrays should be abandoned in favor of the powerful and low-variance coupled arrays. But we can't. The asymmetric uncoupled line and point source come into play when we see the big picture. Step back, look at these figures and replace the individual elements with coupled arrays. At the top are the upper mains on the top of the proscenium side wall. Next are the lower mains that are flown three meters off the deck. Then we have the front fills. The combination of these coupled arrays will behave as the asymmetric uncoupled arrays found here. The big picture is a weave of the little ones.

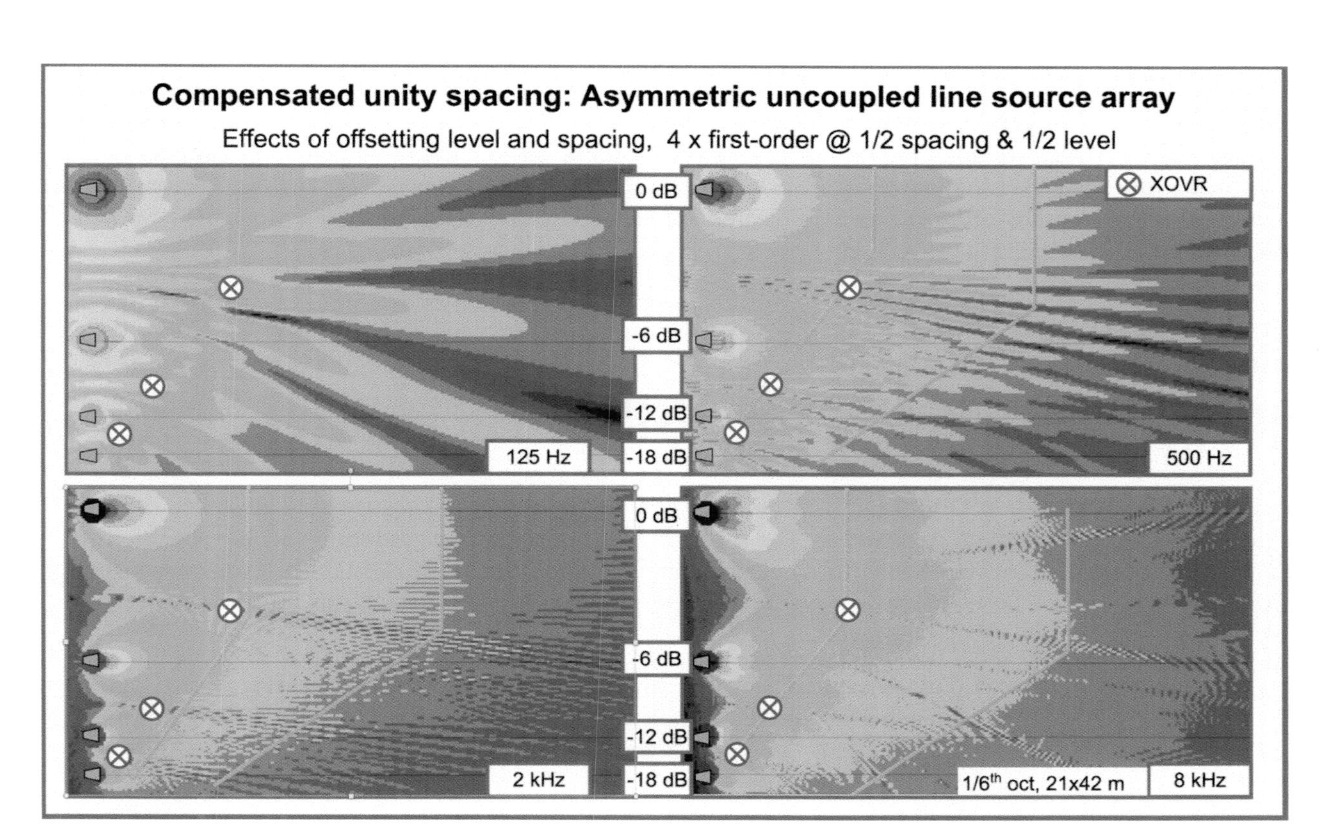

FIGURE 9.44
Compensated unity spacing effects vs. frequency for the asymmetric uncoupled line source (matched origin)

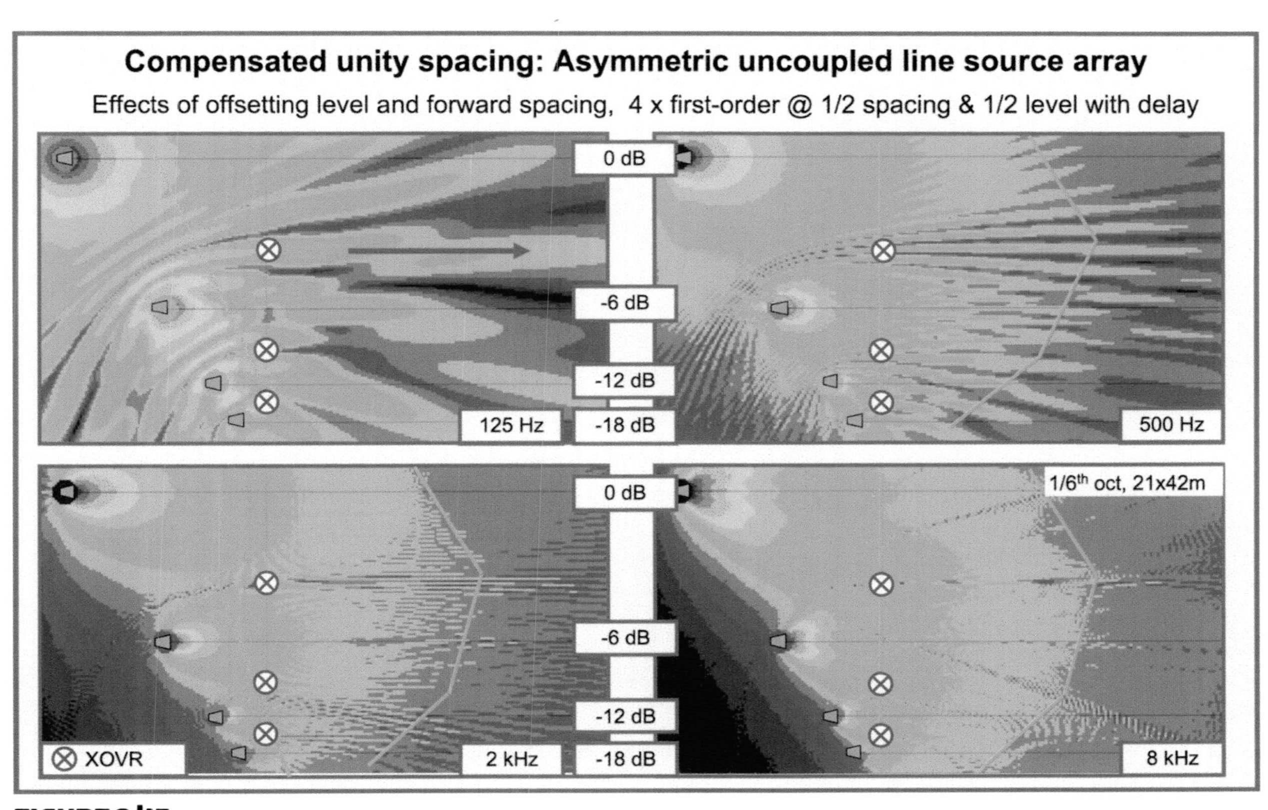

FIGURE 9.45
Compensated unity spacing effects vs. frequency for the asymmetric uncoupled line source (unmatched origin)

9.5.2 Uncoupled point source

The uncoupled point source array has an arc of minimum level variance whose start and stop points are determined by displacement and angular overlap. The response of a second-order system is shown in Fig. 9.46. The enclosed area shows high conformity over frequency, albeit not immune to ripple variance as frequency falls. An asymmetric version can be made in a variety of forms: level, angle and speaker order. The trends of these actions are largely consistent with those just discussed.

9.5.3 Uncoupled point destination

The combination of two sources that are forward displaced and operated at different levels has a limited depth of interaction (e.g. a main system and a delay). The decisive factors are the distance and level ratios. The two sources will have matched levels at some forward distance, the spatial crossover point. Both sources will lose level as they continue forward, but at different rates. The losses are asymmetric because they are based on the individual doubling distances from the sources. The more distant main speaker gradually becomes dominant in level after the crossover point. The transition area size depends upon the range ratio to the crossover meeting point. The affected area's size is inversely proportional to the range/level ratio, shown previously in Fig. 9.19.

The spatial crossover point is created at the same distance in all cases of our example scenario (Fig. 9.47). Range and level ratios are offset to create a matched combined response at XOVR. The combined level at XOVR, not coincidentally, is matched to the mid-point depth of coverage from the primary source (the typical ONAX A).

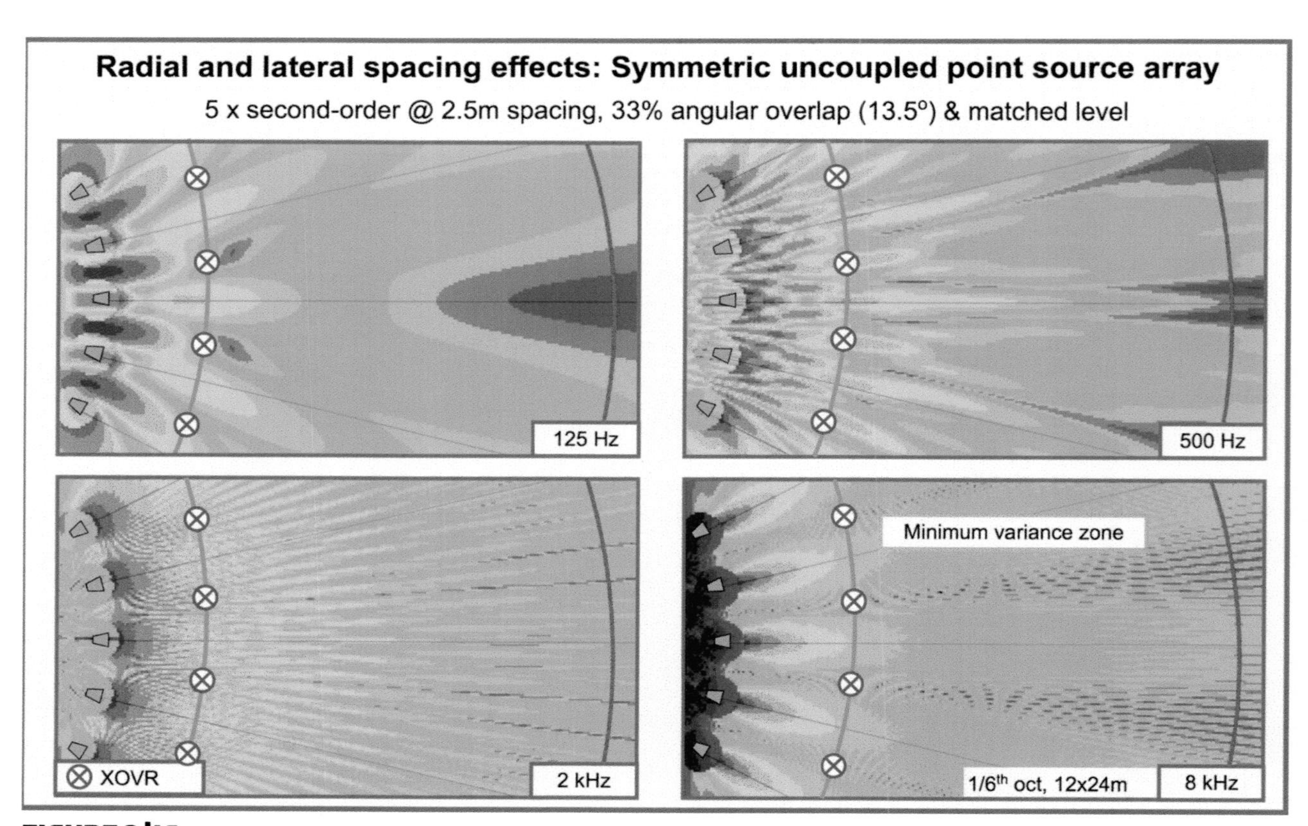

FIGURE 9.46
Coverage shape vs. frequency for the uncoupled point source

Let's begin with a range ratio of 2:1. This range ratio from a single source would create 6 dB of level difference by the inverse square law. Attenuating the forward speaker by 6 dB duplicates this. They both arrive at the meeting point at equal level and combine to restore us to 0 dB. The loss rate over distance beyond XOVR is greater for the forward speaker, so it gradually drops out of the combination equation as we move onward. The remaining iterations in our example move in doubling distance ratios and 6 dB incremental-level reductions. The result is a constant level at XOVR but a steady reduction of the interaction range both before and after the spatial crossover. High ratios decrease the ill effects of "backflow" from the remote speaker, but also reduce the range of forward extension.

The level in the spatial crossover area is raised by 6 dB from the baseline in all cases, which is the sonic equivalent of moving the listeners forward to the center of the room. The 2:1 scenario is able to substantially arrest the level loss at the rear of the room but this comes at the cost of ripple variance near the delay speaker. The ripple is highest between the sources, especially near the back of the delay speaker. This is due to the time offset between the sources, which are set to be phase aligned at the spatial crossover. This "backflow" area behind the delay speaker is a recognizable feature of our triangulation discussion: It is on the straight line of maximum ripple variance between two sources. The 4:1 (-12 dB) and 8:1 (-18 dB) scenarios have less forward addition and proportionally less backflow.

A secondary factor is the angular orientation of the speakers. Figure 9.48 shows the effects of various angular orientations with the same distance ratio. Each angular orientation creates a different line of equal time, in contrast to lines of equal level created in the previous exercise. The delay is synchronized at XOVR and the equal time contours are marked "sync" on the drawing. This line indicates the locations that remain equi-temporal between the speakers. Additional lines show incremental time offsets of 1 ms up to 5 ms apart. The 5 ms limit indicates

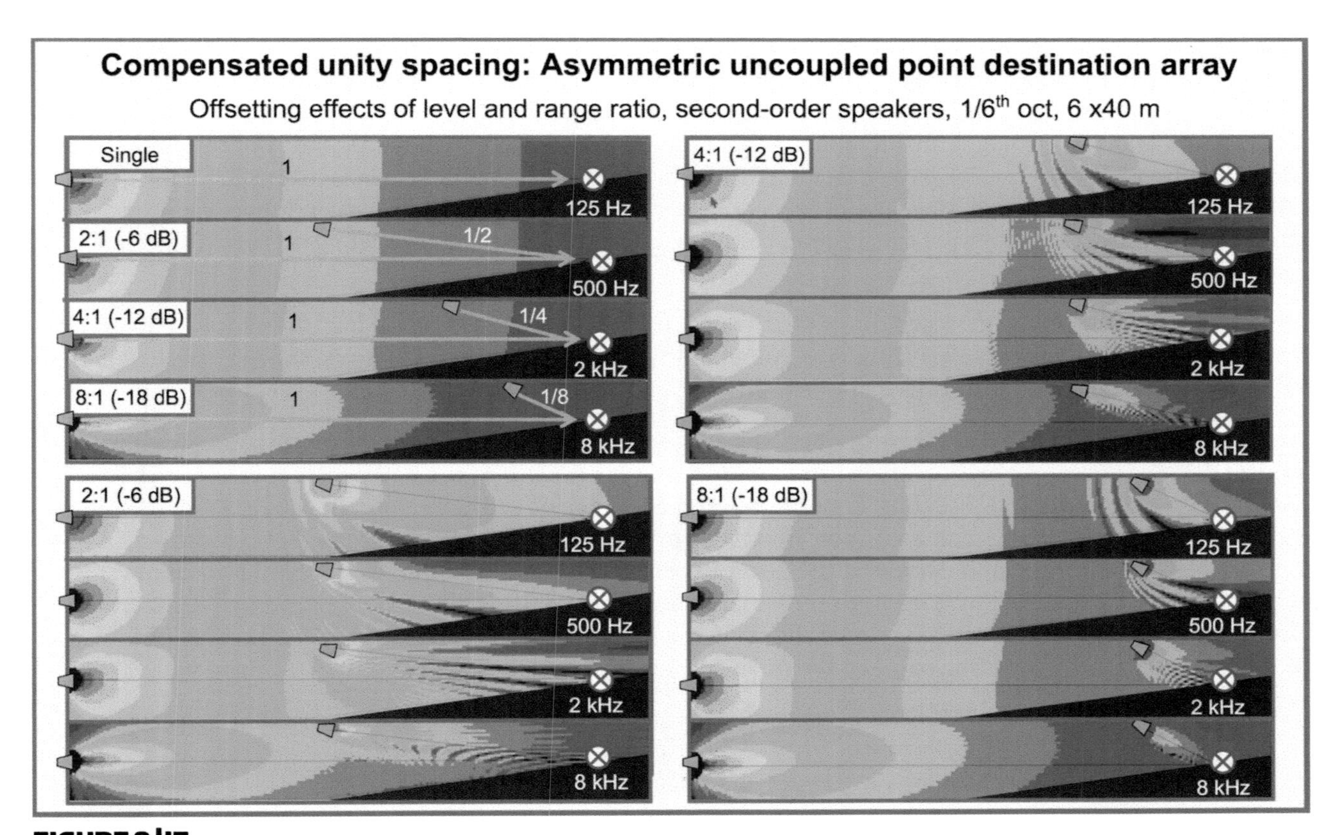

FIGURE 9.47
Level and range ratios applied to the asymmetric uncoupled point destination

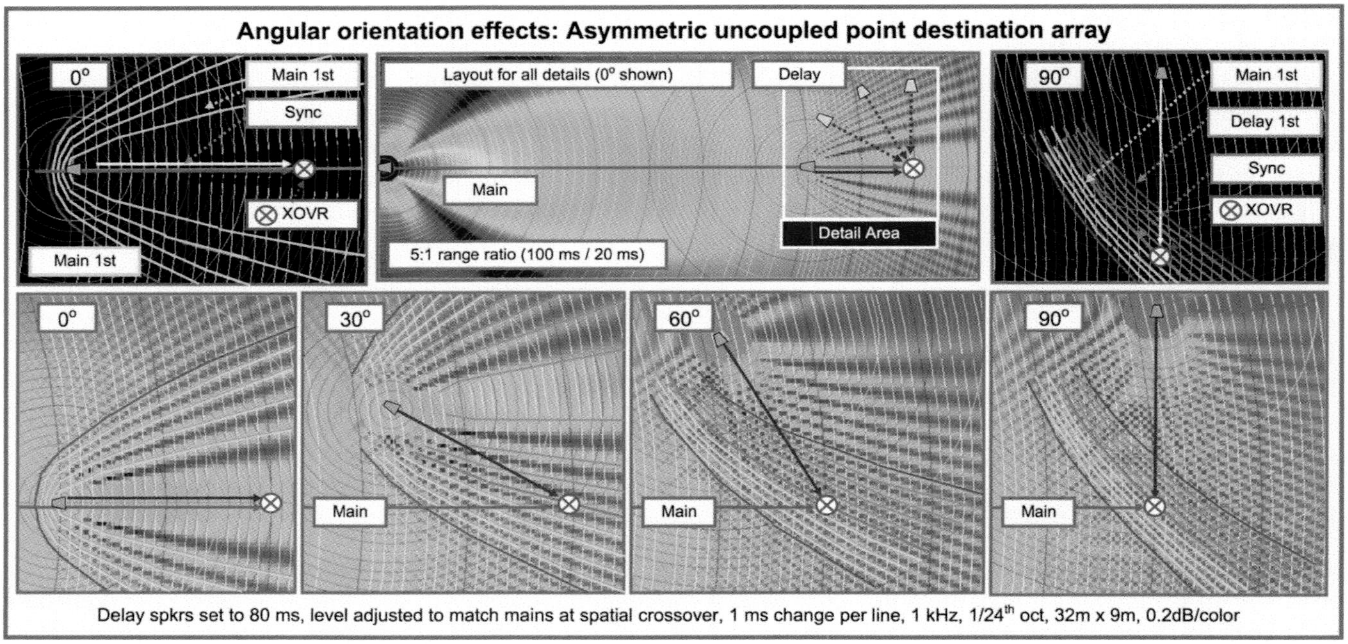

Angular orientation effects: Asymmetric uncoupled point destination array

FIGURE 9.48
Angular orientation effects applied to the asymmetric uncoupled point destination

a range where the comb filtering effects would extend down to 100 Hz, low enough to degrade the entire working range of most delays.

The delay speakers lead the mains in some places and lag in others. The first notable trend is that rising angular offset decreases the usable area (inside the 5 ms limits). A second trend line tracks the relationship of time and level offset. Note that the delay speaker is ahead of the mains in the area beyond XOVR. Recall that the doubling distance loss rates favor the main speaker. We have offsetting factors in the battle for sonic image (sections 5.5.4.2 and 8.4). The image stays constant as we move deeper due to the offsetting factors. The main leads in time as we move forward of XOVR and the delay leads in level (until we move off axis). The competing offsets again create an image stalemate.

The mid-range plots show how increased directionality over frequency decreases in the backflow zone between the mains and delay. The interaction range shrinks as vulnerability to ripple rises (directivity vs. shorter wavelengths), another pair of offsetting effects.

Let's return to Fig. 9.48 and focus on how the changing range ratio alters the angular relationship between the sources. The delay's inward orientation angle increases as it moves deeper into the hall, which increases ripple variance (see Figs 4.52 and 4.53). Backflow is negligible and the range of addition is small by 8 kHz where isolation has reached its maximum. Such small additions, however, are progress toward our goals. We have improved the front/back level uniformity and decreased the direct to reverberant ratio (another form of reduced ripple variance) in the rear. The dominant force in ripple variance at the rear of the hall would be room reflections, and our refreshed direct sound, even with the ripple shown, has a high prospect for overall ripple reduction. This also reduces spectral variance by restoring some of the main speaker's HF air loss.

9.6 MINIMUM-VARIANCE INVENTORY

We can now take stock and sort the various elements and array types. This will prepare us to choose the best available options for our designs.

9.6.1 Summary of properties

APPLICATIONS FOR FIRST-ORDER SPEAKERS

- Most suitable for single-element applications with a low range ratio.
- Least suitable for high percentage angular overlap.
- Least suitable for power coupling.
- Beamwidth must not reverse direction as frequency rises. Coverage must remain constant (not widen at higher frequencies) once the plateau section has been reached.
- Should be arrayed at or near the unity splay angle of the beamwidth plateau.

APPLICATIONS FOR SECOND-ORDER SPEAKERS

- Most suitable for single-element applications with a medium-range ratio.
- Moderately suitable for power combination.
- No beamwidth reversals (as above for the first order).
- Should be arrayed at or near the unity splay angle of the beamwidth plateau.

APPLICATIONS FOR THIRD-ORDER SPEAKERS (PROPORTIONAL BEAMWIDTH)

- Least suitable for single-element applications.
- Most suitable for high-percentage angular overlap.
- Most suitable for power combination.
- Proportionally narrowing beamwidth slope with no reversals as frequency rises.
- Should be splayed at an angle greater than 0°.

POWER CAPABILITY VS. VARIANCE CONSIDERATIONS

- Coupled arrays can increase power capability beyond a single element while maintaining minimum variance over long distances.
- Coupled point sources have the potential to maximize power addition while maintaining minimum variance over a wide-frequency span and spatial area.
- Coupled line sources can provide unlimited power but are incapable of minimum spectral variance.
- Asymmetric level tapering can reduce variance over an asymmetric space, with a tradeoff in decreased overall power capability at the geometric center of the array.
- Uncoupled arrays can increase distributed power capability beyond a single element. Minimum variance can only be maintained over a limited range.

PRIMARY MINIMUM-VARIANCE COVERAGE SHAPES

- **Rectangular (FAR version)**: as found in the individual element.
- **Radial**: as found in the symmetric point source. Coupled version is range-limited only by propagation loss and spectral variance from air loss (HF). Uncoupled version is range-limited by overlap-induced ripple variance.
- **Lateral**: as found in the uncoupled line source.
- **Diagonal**: as found in the asymmetric coupled point source.
- **Forward**: as found in the asymmetric uncoupled point destination (delay systems).

9.6.2 The minimum-variance menu

The building blocks for our systems and subsystems have been tallied. We have a menu of choices to fill the acoustic space. The first choice, our appetizer, is the single speaker.

MINIMUM-VARIANCE PRINCIPLES FOR SINGLE SPEAKERS

- **Minimize range ratio**: Position speakers to minimize range ratio in at least one plane. This may be limited by sonic image concerns (e.g. a high center cluster) or sightlines (e.g. trying to lower a center cluster). Frontfills are another case of high-range ratio due to sightlines. Note: This carries forward for arrays as well.
- **Use low-order speakers**: The lower spectral variance of first- and second-order speakers makes them the preferred choice for solo applications. Third-order speakers can only be used as soloists when the coverage is virtually two-dimensional (e.g. a frontfill aimed just above a flat listener plane).
- **Complementary symmetry**: Match the properties of symmetry (symmetric orientation for symmetric shapes, asymmetric orientation for asymmetric shapes).
- **Shape matching (horizontal)**: Match the speaker shape (FAR) to the audience shape.
- **Shape matching (vertical)**: Match the speaker's MV line to the listener plane.
- **Limitations**: Use an array instead of a single element when the range ratio exceeds 2:1 or if the coverage spills too much onto the room surfaces.

The main course is coupled arrays. This will fill up the bulk of the space. The menu is shown in Fig. 9.49.

MINIMUM-VARIANCE PRINCIPLES FOR POINT SOURCE ARRAYS

- **Fill the radial shape**: The array must fill the arc angle of the audience shape. The number of elements required will depend upon the element speaker order. Low-order speakers must have minimal overlap. High-order systems can be overlapped for maximum power addition.
- **Flatten the beamwidth**: Use isolation to shape the HF and overlap to shape the LF.
- **Minimize displacement**: Keep elements as close as possible. Highly overlapped HF is the most critical displacement issue.
- **Compensate the range ratio**: Power allocation and directional control should be reflected in the range ratio (e.g. we need to send 6 dB more to the back than the front of a space with a 2:1 range ratio). Different models, percentage overlap, asymmetric level or all of the above can achieve this.

Dessert is now served: uncoupled arrays. These follow the main course well as long as we keep them relatively small. Uncoupled arrays can combine single speakers or previously combined arrays but they can only maintain minimum variance over a limited range. Therefore they are typically comprised of first- or second-order elements (or a coupled array that resembles a first- or second-order element). The menu is also shown in Fig. 9.49.

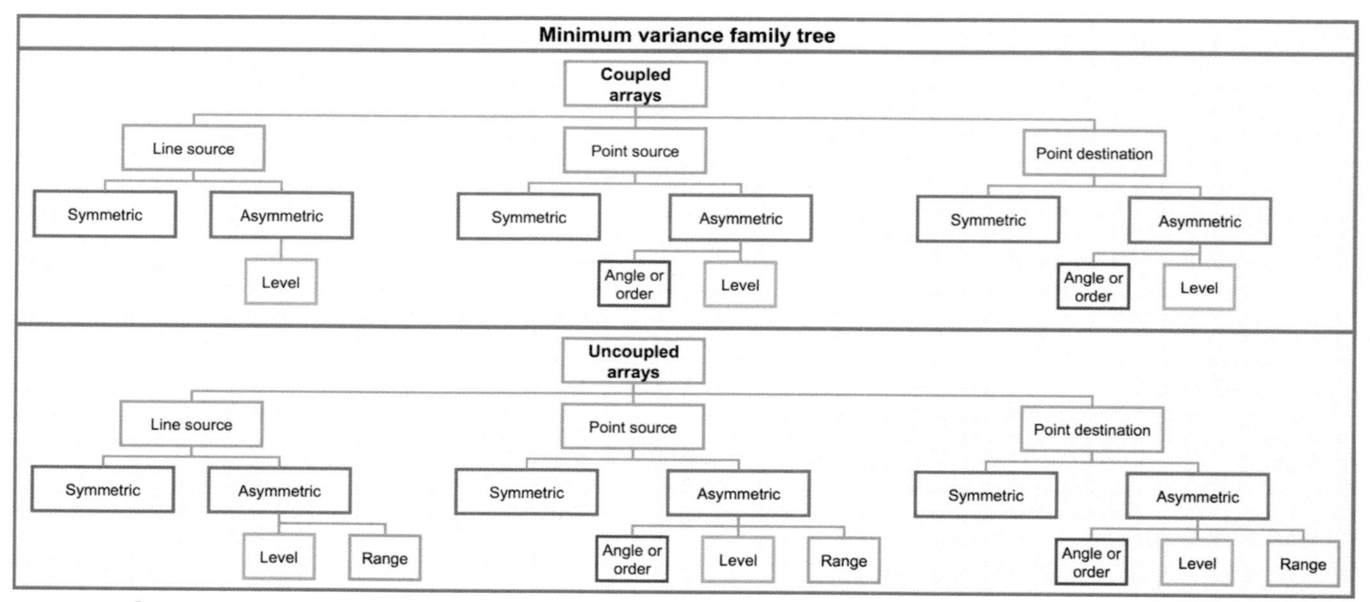

FIGURE 9.49
The minimum-variance family tree

MINIMUM-VARIANCE PRINCIPLES FOR UNCOUPLED ARRAYS

- **Isolate**: Achieve isolation by angular offset, displacement, or both. Exception: delays.
- **Set the range**: Uncoupled array depth is initiated (and ended) by overlap. Before overlap, the response is essentially a single element. Overlap of adjacent elements begins the array coverage (unity line) and overlap of non-adjacent ends it (limit line).
- **Scale the level**: Uncoupled arrays can be scaled by relative level to appropriate depth.
- **Use low-order elements**: Changes in beamwidth result in frequency-dependent operating depth (the MF may be past the limit line before the HF has reached the unity line). Low-order flat beamwidth speakers ensure minimum variance in operational depth.

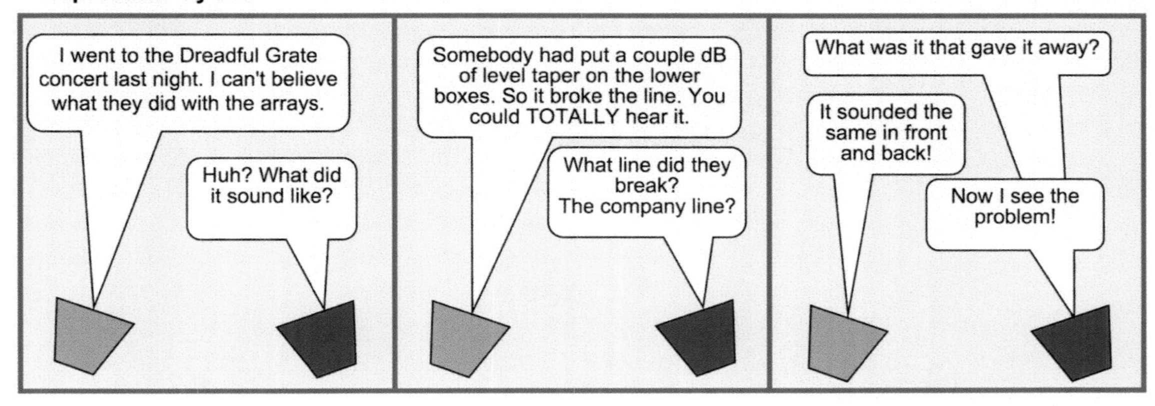

CHAPTER 10

Cancellation

> *cancel v. obliterate, cross out; annul, make void, abolish, countermand, neutralize.*
>
> **Concise Oxford Dictionary**

Many forms of philosophy emphasize the ultimate balance in the universe. The ancient Eastern concept of Yin and Yang illustrates the balance between the light side and the dark side. Modern Western philosophy does the same with Luke Skywalker and Darth Vader. This chapter explores the dark side of speaker summation: cancellation. This dark force can be used to actively steer sound away from some selected areas.

We have studied the light side extensively. The confluence of unity level and phase produces the maximum brightening (+6 dB). Increasing level offset (isolation) reduces the intensity gradually. By contrast (literally), the confluence of unity level and anti-phase (180°) produces the maximum darkening (-inf dB). Increasing level offset (isolation) also reduces the intensity of effect, but this time it's not gradual. Small amounts of isolation *strongly* reduce the cancellation effect. Phase offset has similarly asymmetric effects on light and dark. Rising offset on the light side causes gradual dimming. The power of darkness diminishes rapidly with even small offsets. Perfect discipline of level and phase is required to maintain the dark side. Now we know why Darth Vader was such a hard ass.

The coverage patterns of individual LF elements are so wide that angular level offset ranges from little to none. There is only so far we can splay speakers apart before coming back around again. Isolation is assumed to be the minority player for all coupled configurations, bringing phase to the forefront as the steering mechanism. Our summation zones are reduced to Yin and Yang: coupling and cancellation. Phase is the judge. The coverage shape is created by steering the phase addition and subtraction.

The cancellation zone is the back side of the phase wheel (phase offset between 120° and 180°). Strong cancellations (>6 dB) are reserved for phase offsets >150°. The most well-known cancellation mechanism is polarity reversal between adjacent elements. Such reversals don't necessarily reduce the sound level in the room, but they certainly move it around. Polarity reversals create a frequency-independent phase offset of 180° at an otherwise symmetric spatial crossover, which splits the beam sideways into a figure eight.

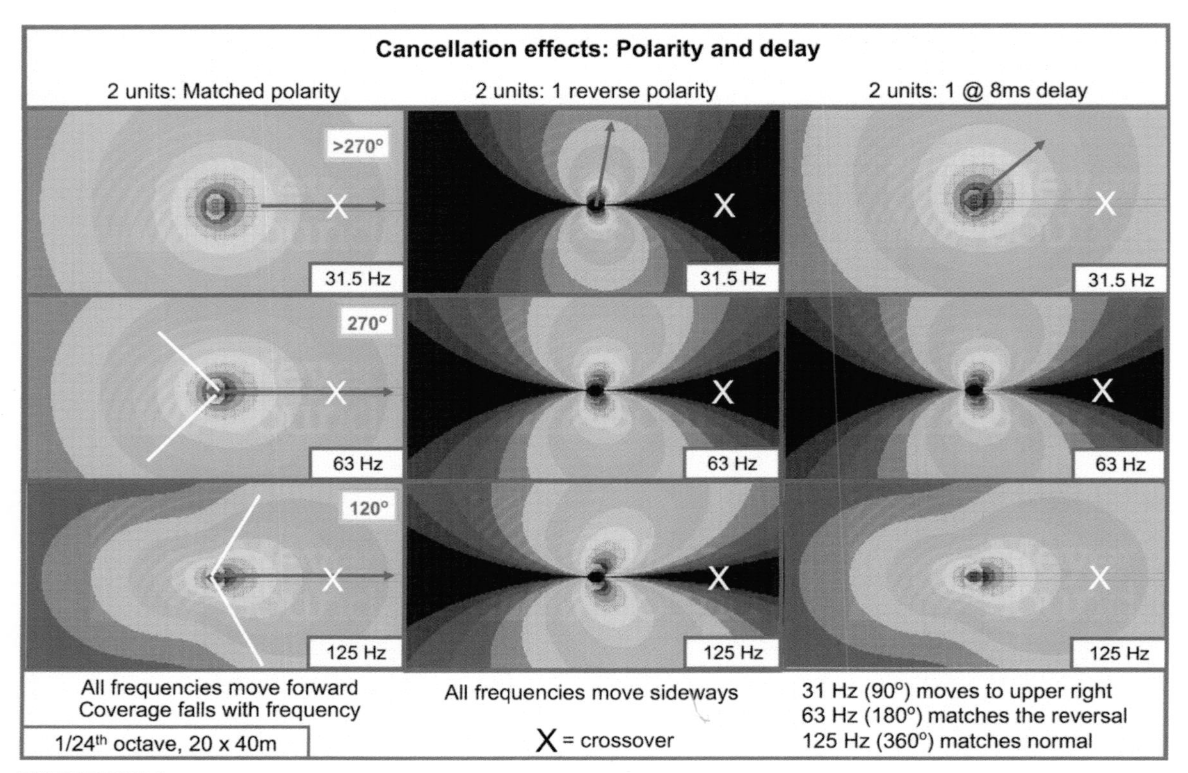

FIGURE 10.1
Comparison of a two-element subwoofer array with normal polarity, reverse polarity and 8 ms delay between the elements

Delay offset at a spatial crossover can mimic a polarity reversal, but only at one frequency. Delay offset is frequency-dependent phase steering, whereas polarity is a full-range control. We begin by comparing polarity reversal and delay offset in a two-element array (Fig. 10.1). The polarity reversal vacates the crossover area at all frequencies. By contrast, a delay of 8 ms has three distinct frequency-dependent effects. A beam is steered toward the delayed element's side at 31 Hz (λ, ¼ λ offset), identical to the polarity reversal at 63 Hz (180°, ½ λ) and the same as the original combination at 125 Hz (360°, 1 λ). Meet the players (in order): beam steering, cancellation and coupling. Let's put them to work.

10.1 SUBWOOFER ARRAYS

The normal placement rules for full-range speakers do not apply to subwoofer arrays. Two unique features create this opportunity: (a) separate enclosures with limited frequency range; (b) large wavelengths able to diffract around neighboring objects, most notably other subwoofers. No sane person places a full-range speaker facing directly into the back of another such speaker. We'd be crazy *not* to consider this option for subarrays.

Individual subwoofers come in two basic flavors, omnidirectional and cardioid. The omnidirectional version, like its microphone counterpart, is not truly omni and narrows with frequency. Commercially available cardioid versions use phase offset to combine front and rear drivers to cancel behind and couple forward.

It's always best to meet the individual elements, before combining (Fig. 10.2). Our representative element would appear to have a flat beamwidth of 360° over its 31 Hz to 125 Hz range, if evaluated by the standard method (OFFAX = -6 dB). This should ring some alarm bells because we know the beamwidth of cone drivers naturally narrows with frequency (section 2.7.3.1). This shows the vulnerability of angular specification for overly wide devices. The coverage angle of 360° is the loosest spec in all of audio. There's a ±6 dB range of coverage patterns that meets this spec. Does it matter to you if the level at the rear is the same as the front, -6 dB or +6 dB? All qualify as 360°. Wait, there's more! How about 0 dB front and back and +6 dB on the sides? Still 360° on the spec sheet. We won't miss any of these differences if we follow the 0 dB contour instead of classifying the

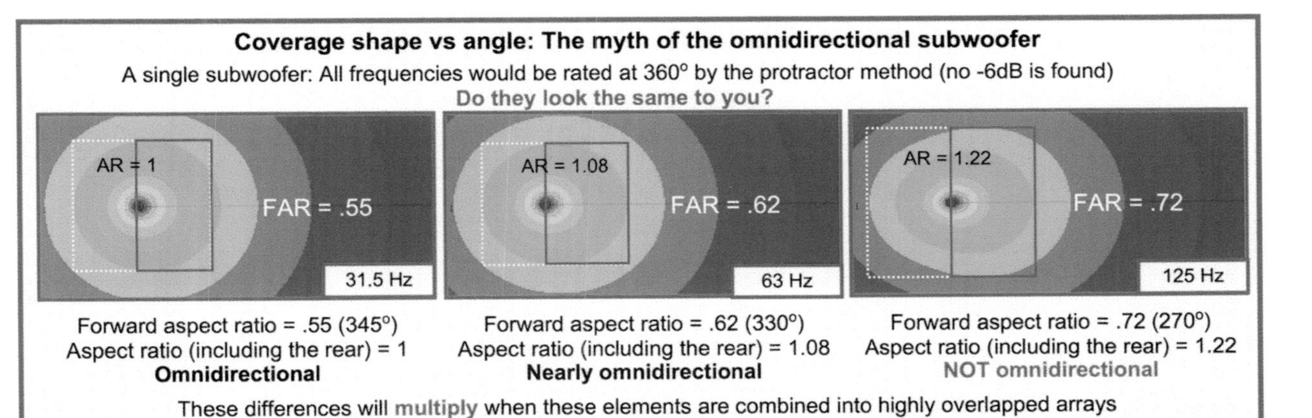

FIGURE 10.2
The directional characteristics of "omnidirectional" subwoofers

shape radially. The aspect ratio shape (the rectangulation of the 0 dB contour) sees it all. We've previously used a forward-thinking characterization (FAR), but now the back side is an equally important part of the shape. We include front/back and side/side in the total aspect ratio characterization. The shapes of truly omnidirectional 31 Hz and quasi-omni 125 Hz don't appear significantly different at first glance but become clear as quantity increases.

Directional arrays can be created with combinations of omni elements. There are, as usual, various options available and we will, as usual, isolate their effects and see where it leads. We'll focus on coupled arrays, because uncoupled subwoofer arrays are degraded by ripple variance. There is simply too much overlap unless the uncoupled elements are cardioid.

Coupled subwoofer array options

1. Lateral arrays: the coupled line source.
2. Radial extension: the coupled point source.
3. Forward arrays: end-fire arrays, dual element in-line, gradient.

Before we go any further I want to mention an excellent resource to aid your understanding of subwoofer arrays: SAD (subwoofer array designer) by Merlijn Van Veen (www.merlijnvanveen.nl).

10.1.1 Coupled and uncoupled line source

10.1.1.1 QUANTITY AND SPACING EFFECTS

The coupled line source is first, as usual, and we'll use the familiar doubling technique to expose the trends (Fig. 10.3). Progressive narrowing over frequency becomes apparent even in small quantities. An interesting convergence appears as quantity doubling yields the same coverage at half the frequency. The parallel pyramid is back: full-range narrowing that preserves (and makes obvious) the spectral variance between 31 Hz and 125 Hz.

Element spacing has a similar effect (Fig. 10.4). Array length is held constant while successively smaller quantities are spread apart to fill it. There are limits to this, because the wavelengths we are transmitting don't rescale. The wide spacing scenario (3.2 m) shows evidence of combing zone summation, which would worsen as frequency and/or displacement rise. An important trend is seen in the progressively expanding base of the parallel pyramid. The coupling multiplication is reduced proportionally as displacement rises, i.e. we are in the process of uncoupling. Notice that element quantities of 4, 8 and 16 maintain a nearly constant coverage angle for a given frequency. Still no progress, however, toward reducing spectral variance.

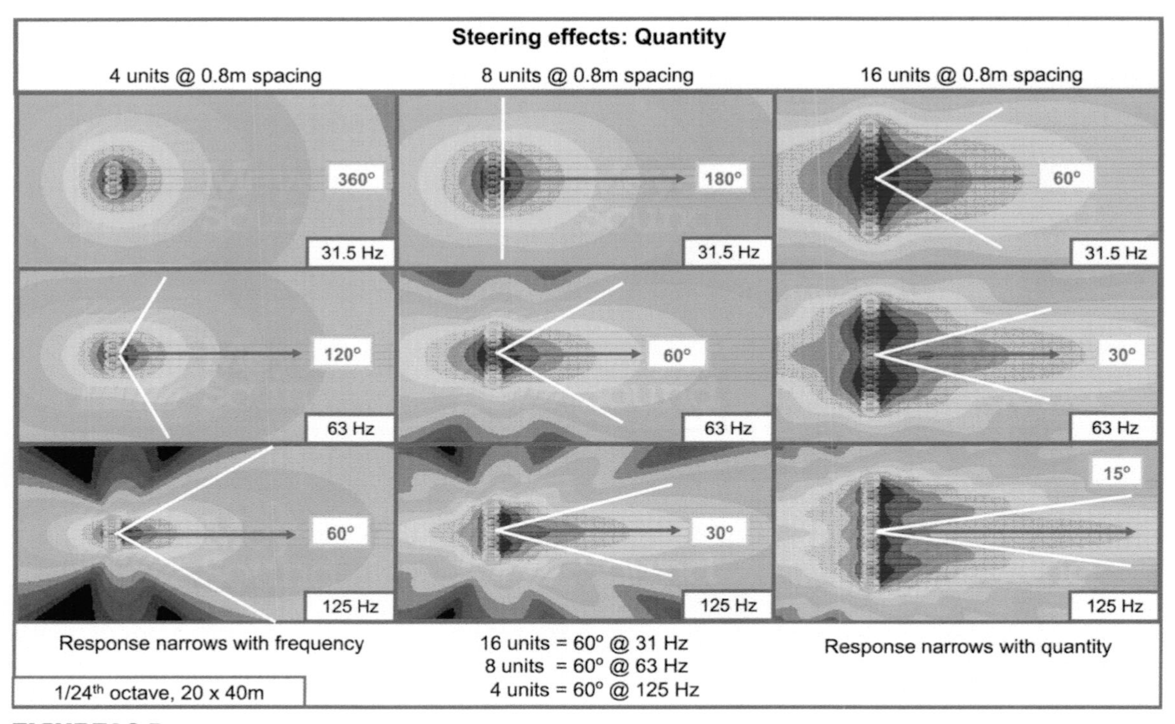

FIGURE 10.3
Quantity effects at fixed spacing for subwoofer coupled line source arrays

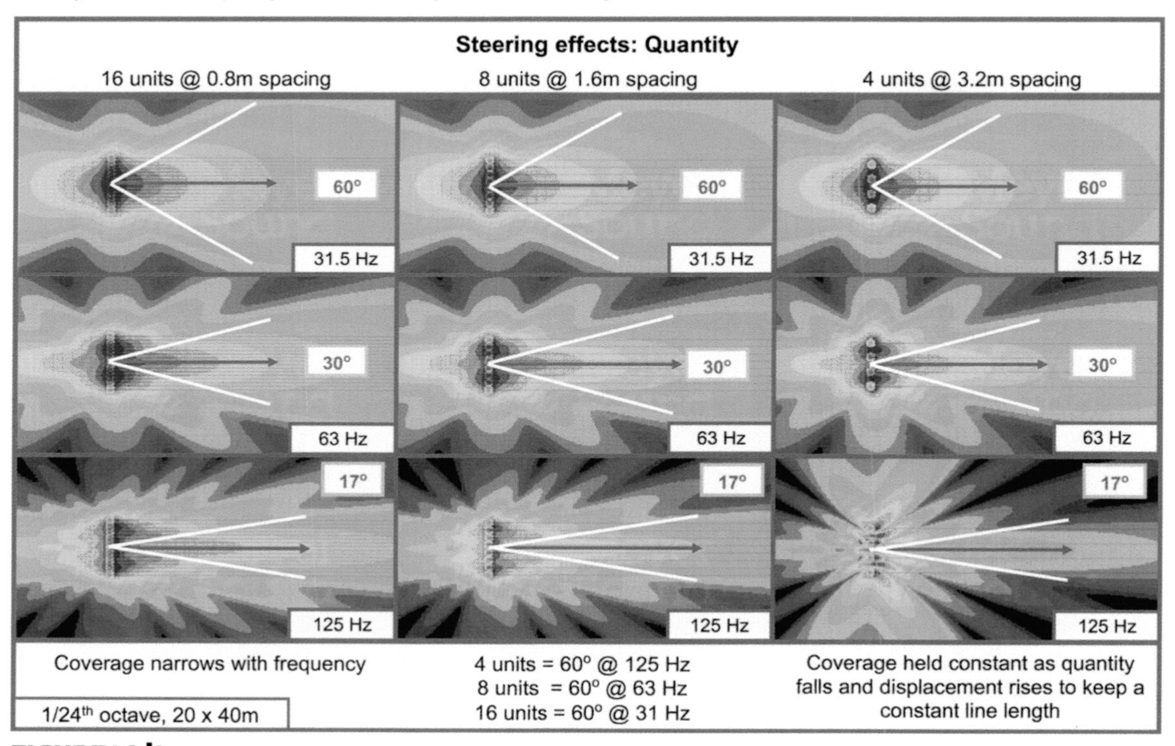

FIGURE 10.4
Offsetting spacing and quantity for subwoofer coupled line source arrays

10.1.1.2 BEAM STEERING BY DELAY

The bass response peak at the center of L/R systems is so well known to concert sound engineers that it has its own name: power alley. We know it here as the coupling zone, spatial crossover XLR. Widely spaced L/R systems have small alleys and vice versa. The width shrinks with frequency and then the combing begins. It's an overlap crossover that can't stay in time as we move off center. Can we do anything about it? We can steer the array physically by splaying the left and right channels apart. Alternatively we can steer the pattern electronically.

The subwoofer's limited spectrum enables simple beam steering, i.e. asymmetric delay between elements. "Simple" beam steering means a single delay value for an entire element: fixed time offset (and therefore variable phase). Complex beam steering uses frequency-dependent delay (e.g. all-pass and FIR filters), which can give variable time offset and fixed phase offset (or other options). Simple steering is inherently range limited (subwoofers only) but easier to comprehend and implement. Complex steering can be applied to subs or full-range systems, but requires adult supervision.

Let's move forward with simple beam steering, i.e. asymmetric time offset. There will be a 4× multiplication of effect for a subwoofer operating over a 30–120 Hz range. The multiplier reduces to 3× if we cut the subs off at 90 Hz. Restricting the range makes for more consistent effect over the sub as a whole. Let's see if we can put this to practical use.

Let's first establish some beam steering goals. First we'll try moving the coverage pattern off center. Then we'll try to reduce spectral variance, i.e. achieve a constant beamwidth.

We begin with a symmetric coupled line source, and the standard parallel pyramid comes into view (Fig. 10.5). The pyramid height rises with frequency, so the 125 Hz peak is the highest and narrowest. All frequencies share the same dead-center axial orientation. Now let's linearly delay the elements in one direction. The pyramid's equal timing lines moved sideways, but the equal-level lines stayed in place. The beam bends toward the synchronous time center unless the level side provides a resistive force, i.e. isolation. The extent of the movement will be proportional to the frequency and the time offset. The 30 Hz peak moves only 25% as far off center as the 125 Hz range. It is a lot easier to push a dog than an elephant. A log delay taper (Fig. 10.5) is more effective at maintaining a constant angular change. The effect is a half-conversion of the line source into a point source. The delay creates the arrival times of the point source (albeit with the amplitude aspects of the line source). The result is a partial steering outward.

Delay-induced beam steering is essentially a willfully misaligned spatial crossover. The 0 ms time offset location is moved off the 0 dB level offset location. We see we can bend the beam, but we've made no progress toward flattening the beamwidth yet.

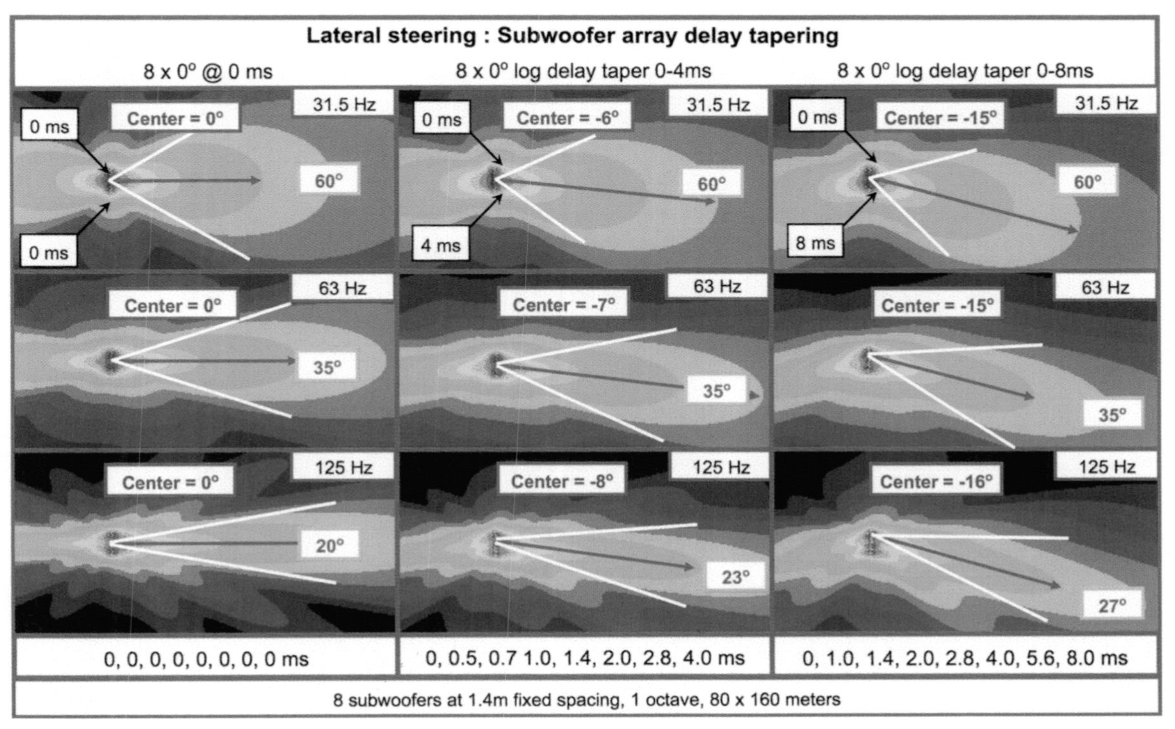

FIGURE 10.5
Lateral steering of the coupled line source with delay tapering

The next step is a double steer. Half the array steers north, while half steers south. We can now take advantage of the fact that the higher frequencies are more steerable. Double steering pulls the 100 Hz range outward (in both directions) more strongly than 30 Hz. Recall that 100 Hz was narrower than 30 Hz. The double steering reduces spectral variance by making the 100 Hz stretch look more like the 30 Hz response.

10.1.1.3 BEAM STEERING BY LEVEL TAPERING

We can consider the possibility of level tapering. This is a revival of the asymmetric coupled line source that proved ineffective for full-range systems (section 9.4.1.5). Can this work better for subwoofers? It's actually even less effective because there is more overlap and less isolation in the subwoofer range. Level tapering without isolation does nothing but reduce headroom and widen the coverage (slightly). We are trying to steer the *Titanic* with a rudder that's way too small, and we all know how that turned out. We can bring this down to a simple math equation: level tapering*(% isolation) = potential benefit.

10.1.2 Coupled and uncoupled point source

Add splay and we have a coupled point source (Fig. 10.6). Angular isolation was the most effective means of reducing spectral variance in full-range systems. Can it work for LF devices?

Let's take a moment to consider why this is not common practice. It can be very difficult in practice to find space to curve an array. Stage fronts are flat, the security perimeter is flat, and on it goes. We'll have to make a strong case to get such valuable real estate set aside for us. Second, there is widespread belief that subwoofer responses are *de facto* "too wide," which is often overcompensated by excessive narrowing. It takes a lot of subs to get 30 Hz to squeeze exactly into the room, but 100 Hz is very prone to over-steering. The spectral variance of proportional beamwidth strikes again. If we splay the array outward, a level reduction at the center is a certainty (and a step in the right direction toward reducing power alley). The third factor comes from the SPL Preservation Society. Anything that steers energy away from front of house (FOH) will reduce the dB SPL (at FOH). Enough said.

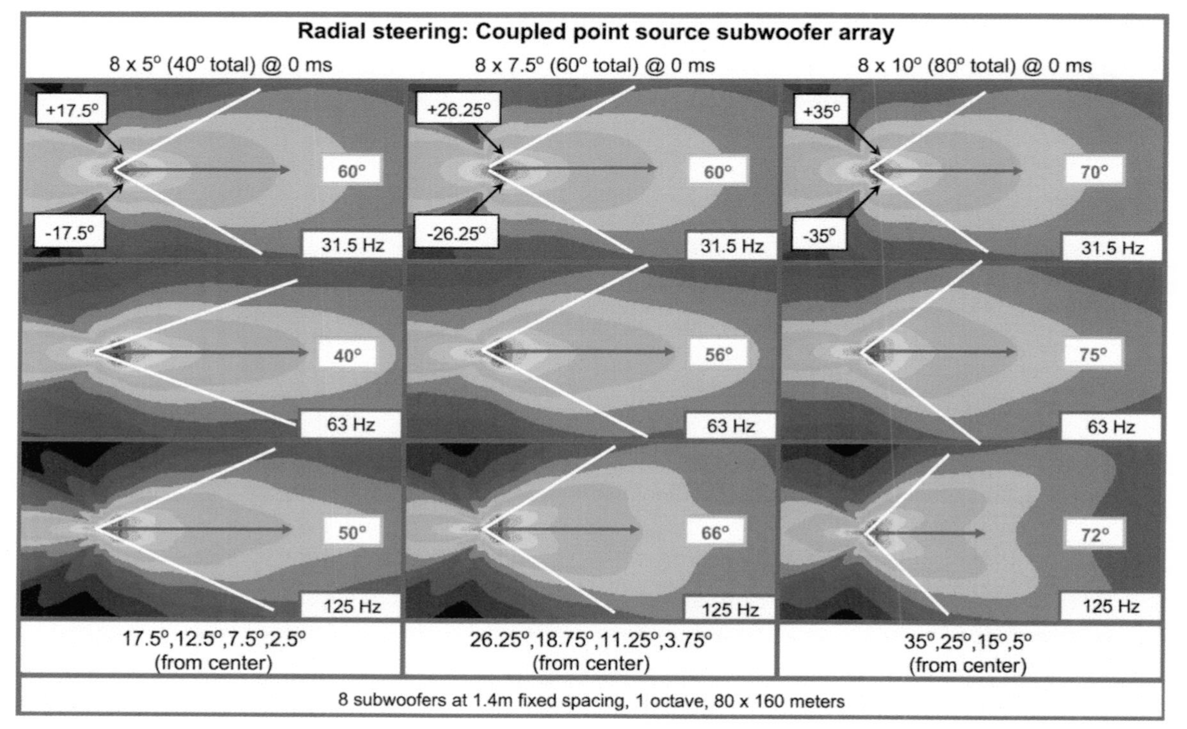

FIGURE 10.6
Radial steering of the coupled point source

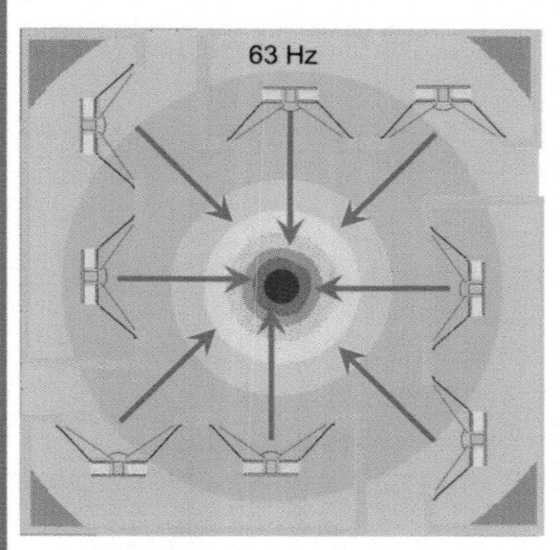

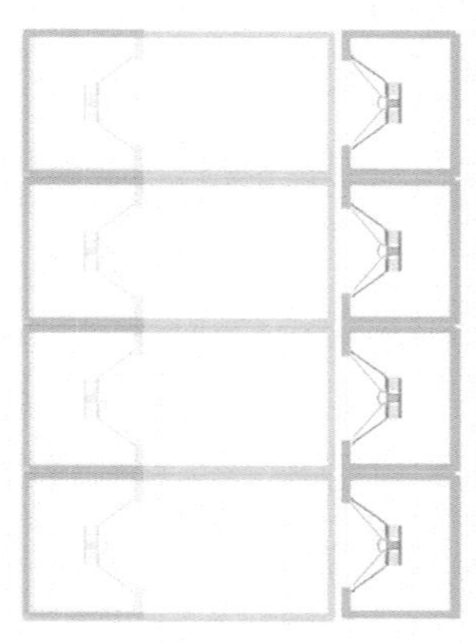

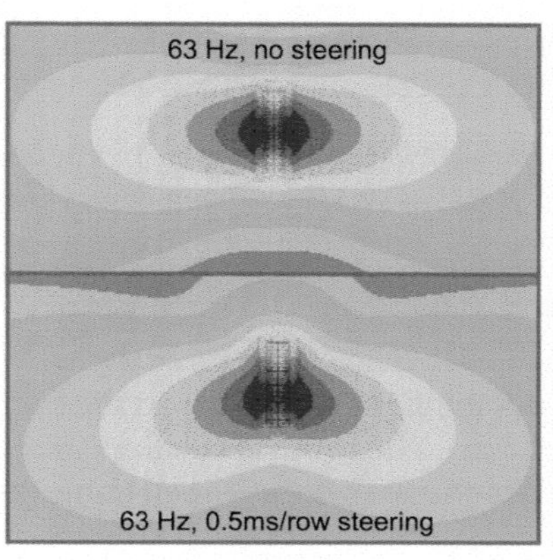

FIGURE 10.7
TM array

10.1.3 TM array

The TM array is a unique solution to the challenge of large-scale "in the round" applications. The TM array has many unique features, not the least of which is that it was not invented by Harry Olson before you were born. It was Thomas Mundorf who innovated the concept for the Metallica tour (Fig. 10.7). The design goal is omnidirectional coverage in the horizontal plane and a controlled beam in the vertical. The horizontal configuration is four boxes facing into each other (the ultimate symmetric coupled point destination). The principle is zero displacement in the horizontal plane, and can only work in the LF range where the physical obstruction of the speakers to each other is not destructive. A coupled line source is used in the vertical plane. It must be long enough, and high enough, to provide enough vertical steering to get over the stage (in the case of Metallica the drummer). The vertical beam can be steered downward by the beam steering described above. The TM array is the minimum spectral and ripple variance leader in the horizontal plane.

10.2 CARDIOID SUBWOOFER ARRAYS

The reasons for choosing cardioid arrays are obvious. Rarely do we suffer from insufficient back lobe! Reasons for *not* choosing them, not so much. There are benefits, but also costs.

Cardioid subwoofer considerations

- Reduce stage leakage: Typical configurations can yield >20 dB front/back ratios.
- Improved D/R ratio: Rear/side control reduces early house reflections.
- Pattern optimization: Steering reduces horizontal coverage (not just rear).
- Efficiency loss: reduced maximum SPL (compared with all subwoofers as a block).
- Cost/practical issues: extra space required, special rigging, etc.
- May be ineffective due to local acoustics, e.g. under stage, recessed in a wall.
- Not always applicable: Why cancel the rear if we're against a wall?
- Time stretch: Delayed elements arrive behind the front (gradient).

Two cardioid configurations are in common use: end-fire and gradient (2-element in-line and inverted stack versions). The end-fire is front steered: coherent phase on the front side and random in the rear. Gradients are rear steered: phase matched (and polarity inverted) on the backside, and quasi-coupled in the front.

Cardioid steering is an active process that requires linear performance to maintain its pattern. The pattern will be dynamically re-steered if some elements limit before others. Therefore matched elements at unity level are recommended for all configurations. It can be tempting to taper levels to improve performance at a particular location but this can trigger unexpected steering at show time.

Cardioid subwoofer arrays can be used as elements of the linear and radial extension arrays described above, even the beam-steered versions. It looks like a graveyard, but it works!

10.2.1 End-fire

We can use the same language to describe the end-fire array and the most infamous sound engineer haircut, the mullet: business on the front side, party in the back. Elements form a forward line with predetermined spacing (typically around 1 m). The forward speakers are incrementally delayed to sync in front. A four-box array with 1 m spacing would delay the 2nd box by 1 meter, the 3rd box by 2 meters and the 4th by 3 meters. It sounds silly to delay a box 3 meters but that is exactly correct (it's 8.82 ms, the propagation time for 3 meters). All elements arrive at the front at 3 meters (8.82 ms), but each uses a unique mix of acoustic and electronic distance. The rearmost box is 3 m of acoustic, the front is 3 m of electronic and the middle two are mixes of 2 + 1 and 1 + 2 respectively.

What's happening in the back room? Everybody's out of it. The arrivals are equivalent to path totals of 0, 2, 4 and 6 meters. The forward and rearward acoustic paths are the same, but the timing chain that sync'd us in front puts us 2× out of sync in the back. The front speaker is delayed 3 meters and then has to travel another 3 to get back the rear speaker. The timing back there is 0 ms, 5.94 ms, 11.76 ms and 17.64 ms. It's a party all right and the guests are staggering all over the place.

Let's digress for a moment to clarify rearward radiation from speakers. This is vital to grasp the end-fire concept and often misunderstood. It's easy to think that rearward propagation is opposite in pressure direction (polarity) from the front. This is true of a fan (high pressure in front, low pressure in the rear) but not for speakers. Waves emanating from the rear are the same polarity as the front, just a little later in time (assuming drivers in the front). There was one speaker model that blew everybody away because it had the propagation properties of a speaker *and* a fan. It was used in a famous advertisement for Maxell, but sadly I have not been able to find one to measure.

Fig. 10.8 shows the 1-meter, 4-element end-fire array described above. The timings for each element, and their resulting phase positions, are shown along with the physical model. There is zero time offset and zero phase offset

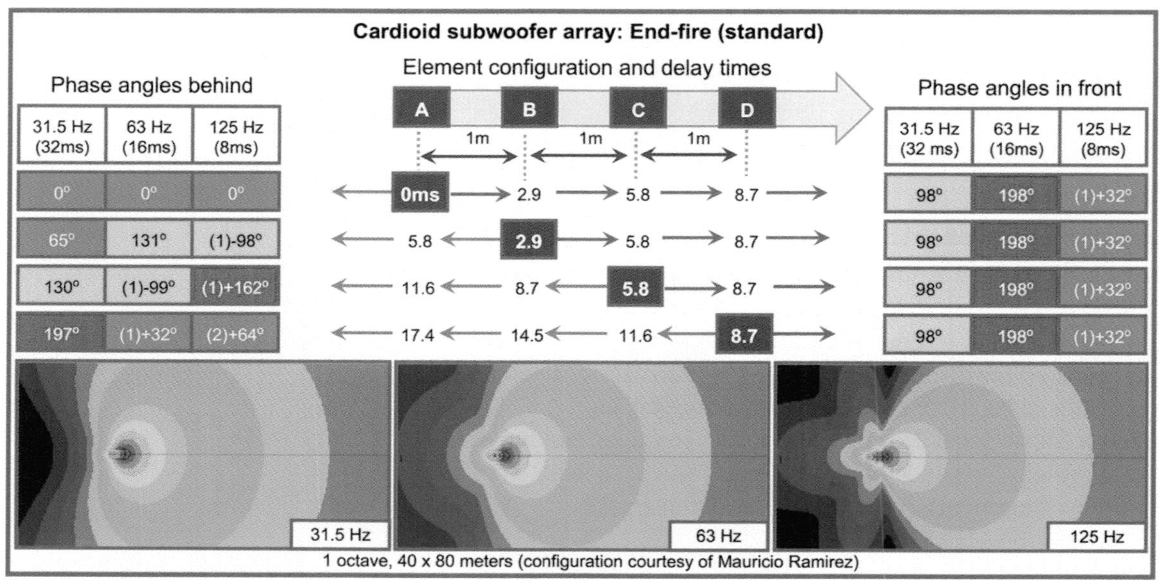

FIGURE 10.8
The standard end-fire array with physical model, timing chain and coverage pattern

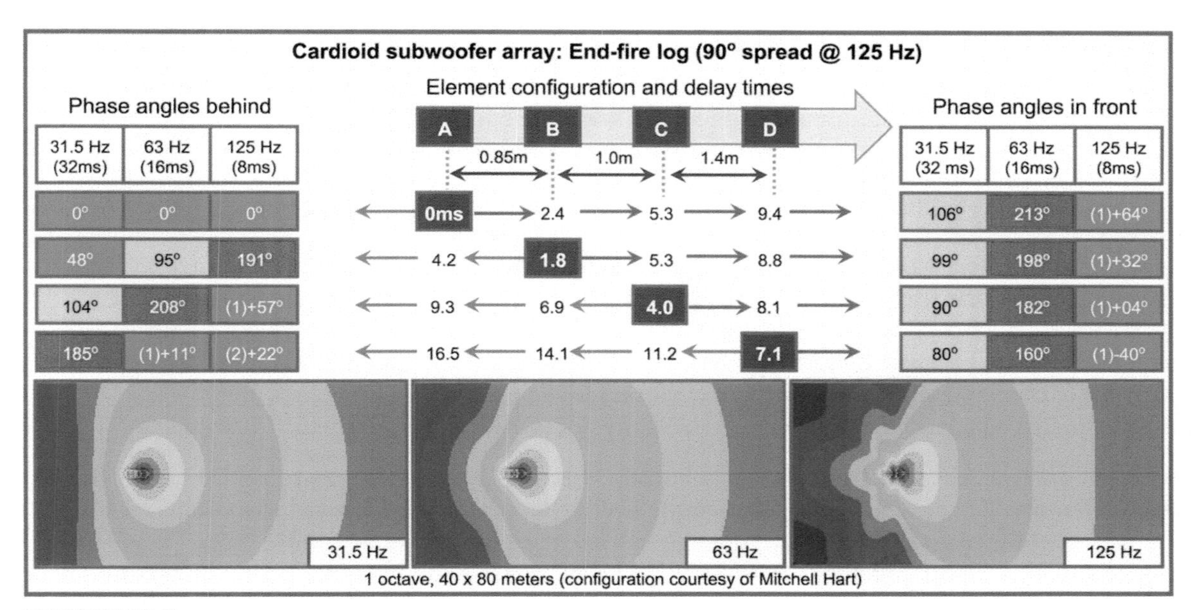

FIGURE 10.9
A log version of the end-fire array with 90° relative phase spread at 125 Hz

in front. It's their phase matching that's critical here, not the absolute numbers. In practice the timings may be slightly larger than a straight line of sound propagation would suggest. That's because it's not a straight line! The wavefronts have to wrap around the cabinet(s) in front of them before they break into the clear. This is another case where size matters, this time the size of the speaker enclosures. We can start with these design values, but the optimization values should be found in the field by observing the phase response and matching them up.

The situation at the rear is staggered time offset, which creates a thoroughly scrambled mix of phase offsets over frequency. The front is exclusively full-range coupling zone summation while the rear is a punch bowl full of every summation zone for every frequency. Party on Garth!

An alternate version can provide a wider horizontal spread by altering the timing (Fig. 10.9). The sync location is moved off to the side, which evens out the coupling effect across the front. There is 90° of accumulated phase offset in the front, which flattens the front a bit without raising combing concerns.

The end-fire is a range-limited array configuration, but it's not what you think. It gets better with age. It gains efficiency as we get farther away because the doubling distance effects become less significant. Maximum coupling and cancellation require matched amplitude. Our speakers are spread over a 3 meter depth, so it's going to take some distance before we get the full effect. Close range knocks a little bit off the coupling, but it takes a lot off the cancellation. Our designs should consider how quickly we need a fully matured array. The lead singer won't be too impressed with the cancellation if he's only a meter behind the last box. The end-fire array eventually approaches 100% efficiency, but the maximum SPL of a standard coupled line source would be greater.

Spectral variance of an end-fire array is a break-even with its elements. The pattern narrows at all frequencies, so the 100 Hz range will remain proportionally narrower than 30 Hz.

Four elements is the most common end-fire quantity because it is effective and reasonably practical. Economizing to three units sharply reduces the randomization in the rear, leading to frequency-dependent reduction. Never end-fire with just two elements. It's a one-note-wonder on the back side. Use the gradient in-line instead (same physical, different settings). We don't have to stop at four, bearing in mind that the horizontal pattern narrows with quantity. Get crazy! RF antennas will end-fire 10+ deep.

The end-fire is the superior cardioid steering configuration for its high efficiency and minimum phase offset on the front side. The reasons to choose other configurations are usually practical (real estate, rigging, etc.) rather than performance. Harry Olson introduced the audio world to the end-fire array, but it was the antenna folks that got the party started.

10.2.2 Gradient (in-line)

The 2-element in-line technique is a smaller-scale version with large-scale results. It's not half an end-fire. It is synchronized going rearward, where the end-fire goes the other way. Some folks call it a "reverse end-fire" but there is a very important distinction: the end-fire won't work with only 2 elements and the gradient only works *with* 2. You can use a million speakers, but it's always a 2-element configuration (AB). The end-fire can be a million elements (ABCD . . .).

The gradient's operating principle is to sync by delay and cancel by polarity. It's two forward-facing speakers spaced 1 meter or less apart. Delay the rear speaker by the distance between them, add polarity reversal and serve. The front/back ratio can exceed 20 dB over the full operating frequency range. This can be more effective than an end-fire on the cancellation side because the tuning is based on time offset rather than scrambled phase. A gradient requires far less real estate than an end-fire and matures at closer range because it only spans 1 meter. These practical benefits are weighed against the side effects on the front side. Or should I call them "front effects"?

We've delayed a speaker that was already late and then polarity reversed it. What could go wrong? Let's use 1 meter spacing and take an inventory. The delay is 3 ms (I turned the heat up to get the sound speed to a round number) so the rear speaker arrives at the front side 6 ms late, the equivalent time offset to 1 λ at 166 Hz. 83 Hz is ½ λ so it looks like it's going to cancel, but wait, polarity reversal saves the day! That's 180° (time) and 180° (polarity) so we are on the coupling side. Meanwhile 166 Hz is going to cancel because it's 360° (time) and 180° (polarity). Our subwoofer is already finished by 100 Hz so we escape (if not, we'll need to reduce the spacing and try again). Our final stop is 42 Hz, which has 90° (time) and 180° (polarity). We've reached the frequency of polarity indifference, because +90° is the same as 90°. The polarity reversal helps us between 42 Hz and 83 Hz. Below 42 Hz it hurts as it brings the back lobes more in phase (we are trying to cancel, remember). The 90° phase offset allows us to net just 3 dB addition in front (instead of the 6 dB we would get with a standard coupling). All told we get a frequency-dependent coupling efficiency in the front in exchange for broadband cancellation in the rear. The front side sees gains of 3 dB or more for about 1.5 octaves (in this case centered at 83 Hz). The efficiency loss will diminish over distance, but its frequency-dependent aspects will remain because the phase offset is constant over distance.

There is another side effect to this staggered timing. There is still 6 ms of time offset between two speakers running at equal level. We tricked the rear speaker into not combing in front but it's still 6 ms late, which stretches the time response. An impulse reproduced through this system will be *de facto* stretched in time, making the LF response

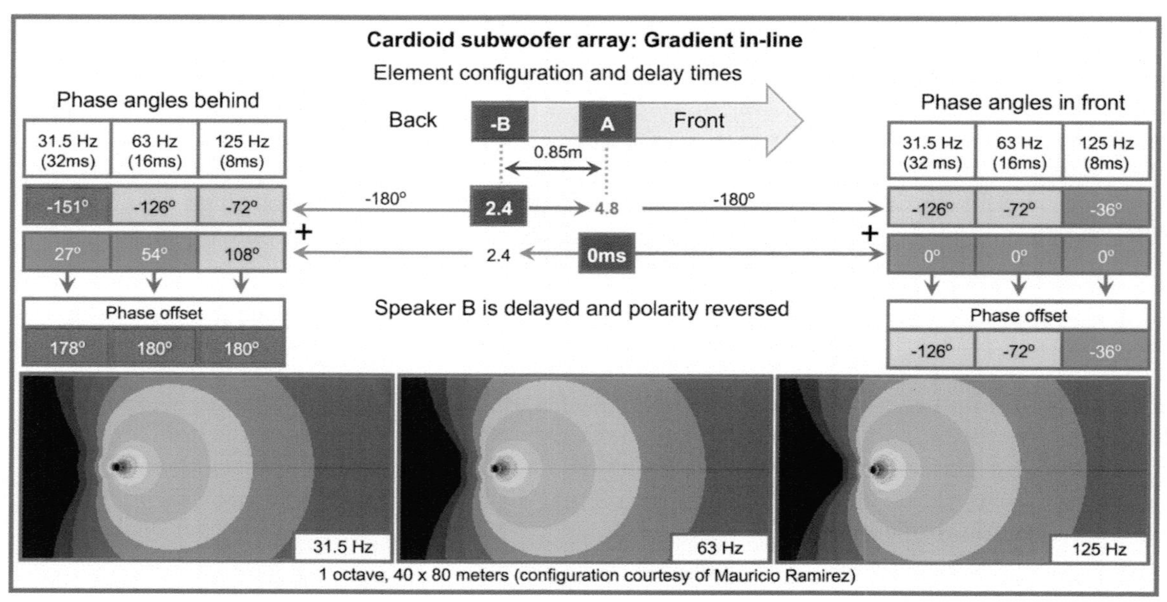

FIGURE 10.10
The gradient in-line cardioid configuration

less "tight." This may be a worthwhile tradeoff because the reflections it reduces might be as strong as the one we inherently create here. This is a classic TANSTAAFL choice.

This is a minimum spectral variance configuration. The pattern is almost identical over frequency. The element coverage angle widens in the LF, which enables substantial cancellation on the sides. The sides are less affected as the element pattern narrows and the factors offset beautifully. The lowest/widest frequencies are steered more while the highest/narrowest are steered less, which nets a balanced cardioid pattern.

The gradient method was originated by, yeah you got it, Harry Olson (Fig. 10.10).

10.2.3 Gradient (inverted stack)

The inverted stack works under the same principles as the 2-element in-line, but with two twists that give it a substantial advantage. The inverted stack has it all: time offset, polarity reversal and level offset (the new ingredient). The physical configuration (typical) is two speakers facing forward with a rear-facing unit in the middle. Why turn a 360° speaker around? Because it's one of those -6 dB at the rear-type of 360° speakers.

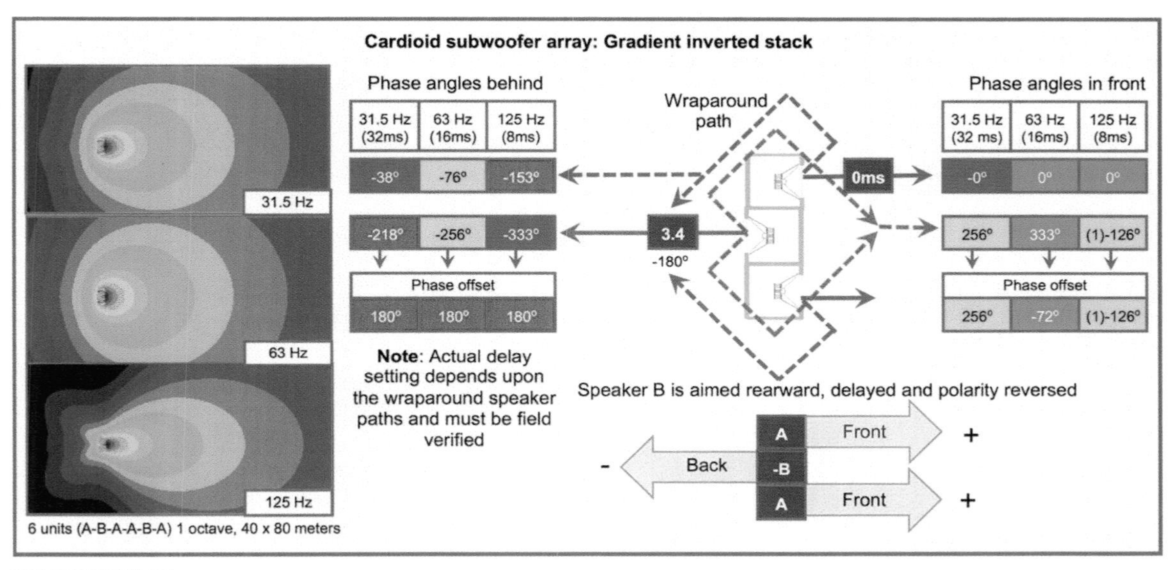

FIGURE 10.11
The gradient inverted stack cardioid configuration

End fire: standard (narrow)
Fixed increment spacing (1m). All elements are synchronized at ONAX (0 ms time offset). The same timing increment is maintained at all quantities (2.94 ms).

Channel #	3 box	4 box	5 box	6 box	7 box	8 box	
1	0	0	0	0	0	0	
2	2.94	2.94	2.94	2.94	2.94	2.94	Increment
3	5.88	5.88	5.88	5.88	5.88	5.88	2x
4		8.82	8.82	8.82	8.82	8.82	3x
5			11.76	11.76	11.76	11.76	4x
6				14.7	14.7	14.7	5x
7					17.64	17.64	6x
8						20.58	7x

End fire: staggered (wide)
Fixed increment spacing (1m). Timings are staggered to accumulate 60° total phase offset at 100 Hz @ ONAX (1.8 ms time offset).
Timing increment is quantity dependent but always totals 1.8 ms time offset at ONAX

Channel #	3 box	4 box	5 box	6 box	7 box	8 box	
1	0	0	0	0	0	0	
2	2.04	2.34	2.49	2.58	2.64	2.683	Increment
3	4.08	4.68	4.98	5.16	5.28	5.366	2x
4		7.02	7.47	7.74	7.92	8.049	3x
5			9.96	10.32	10.56	10.73	4x
6				12.9	13.2	13.42	5x
7					15.84	16.1	6x
8						18.78	7x

Gradient arrays:

In-line:
A: Polarity normal, 0ms delay
B: Polarity inverted, 2.94ms x displacement

0.75 m: B delay = 2.0 ms
0.85 m: B delay = 2.5 ms
1.0 m: B delay = 2.94 ms
1.25 m: B delay = 3.7 ms
1.5 m: B delay = 4.4 ms

Inverted stack:
A: Polarity normal, 0ms delay
B: Polarity inverted, 2.94ms x wraparound path

FIGURE 10.12
Cardioid master reference table

The forward speakers are 0 dB in the front, so the pair add to +6 dB (1 + 1 = 2). They are each -6 dB in the rear, so they add to 0 dB (0.5 + 0.5 = 1). The rear-facing speaker is -6 dB in the front and 0 dB in the rear. We now have 0 dB + 0 dB in the rear. The timing sequence is done exactly the same as the in-line array and the rear cancellation yields 0 dB (0°) + 0 dB (180°) = -inf (1+ (1*-1)) = 0. There's a difference on the front side. The rearward speaker is 1 against 2 and it's backwards so it loses another 6 dB. It's +6 dB vs. -6 dB in front (2 + 0.5 = 2.5). The time-stretching part of the equation moved from 0 dB (in-line) to -12 dB (gradient), which means the tightness we gain from controlling the pattern far outweighs the tightness we lose from a latecomer that's 12 dB down.

The gradient's time offset comes from the displacement between the forward and rearward drivers. The displacement must net at least 3 ms to prevent the back side from cancelling below 30 Hz in front (the phase offset would exceed 120°). The in-line configuration must be kept close enough to escape trouble at the top. The gradient must be spaced far enough to prevent trouble at the bottom. The propagation paths physically wrap around the sides of the box to reach their front and back meeting points. Deep/wide boxes are range limited at the top, but have more coupling at the bottom. Shallow/skinny boxes are the opposite.

The optimization process uses the phase response to maximize the cancellation. This is also a minimum spectral variance configuration for the same reasons as the in-line above. The final note about the gradient is that it's extremely practical to stack or fly, which adds to its allure all the more (Fig. 10.11). A reference chart for typical configurations of the end-fire and gradient in-line arrays is found in Fig. 10.12.

CHAPTER 11

Specification

The sound design process cannot begin until we receive answers to some specific questions. What are you going to use a sound system for? What is the room shape? Where is the stage? Where can we put speakers? The list goes on forever but we can focus on some of the most important ones.

specification n. detailed description of construction, workmanship, materials, etc. of work (to be) undertaken by architect, engineer, etc.
specify v.t. name expressly, mention definitely (items, details, ingredients, etc.).
Concise Oxford Dictionary

- Budget: Do you have any money to spend? This sets the bar on quality, power and complexity.
- Power scale: Do we want loud, stupid loud or insanely loud?
- Spectral allocation: Voice, concert, musical theater, DJ, worship? Translation: How many subs?
- The room: Was it designed by someone who understands loudspeaker systems?
- Locations: Where are the performers and the audience? Where can we put speakers?
- Architectural police: You *seriously* want to hide the speakers in a soffit behind a scrim?
- Channels: Left, right, vocal surrounds, effects?
- Mains: How many main systems do we need? L/R pair or more? (Don't tell me less.)
- Fills: What areas need to be covered with fills?

That's the tip of the iceberg, but it's enough to get us started.

The sound design process is always one of compromise, but that does not mean we need to compromise our principles. There is no job with unlimited budget and placement options. We are the sound experts. We have a duty to inform others about how their design decisions affect the sound quality. It's up to us to inform them that hiding the speakers in soffits is the most expensive downgrade available. Then we can reach a suitable compromise or creative solution based on the weight of all factors. Our decisions have both positive and negative effects. How can we make these choices? Let's take some guidance from others.

TANSTAAFL

Science fiction author Robert Heinlein penned TANSTAAFL in his 1966 novel *The Moon Is a Harsh Mistress*. This term (pronounced "t-ahn-st-ah-ful") is an acronym for "There ain't no such thing as a free lunch." The concept is applicable to decision-making in all walks of life. Every decision comes with a cost, some form of tradeoff. A simple illustration: two overlapping speakers in a point source array. Increased overlap yields higher power capability but raises the ripple variance. Decreased overlap yields less power addition and minimizes ripple variance. The free-lunch option (maximum addition with minimum variance) is *not* on the menu. Take the power option for AC/DC and opt for minimum variance for *Les Mis*.

Sales and marketing departments are trained to suppress our consideration of TANSTAAFL. Each day an assortment of free offers is presented to us. All have strings attached. If it seems too good to be true, it is.

Acoustic triage

The medical community uses a resource allocation and prioritization system under emergency conditions known as "triage." Resources are distributed based on a combination of probability of success and degree of need. Triage prevents excessive resource allocation for lost causes, and prioritizes those with minor problems to the waiting room. It's pointless to use the entire blood bank for a single doomed patient when this limited resource could save hundreds who have a fighting chance. It would be equally irresponsible to treat those with bruised toes ahead of those in critical, but recoverable, condition.

Our acoustic triage model identifies areas that require special care. We allocate resources in search of solutions for one area that don't create problems for others. We use triage to help our decision-making when conflicting solutions arise. An example: A small distant balcony area needs coverage. The main cluster can be aimed upward, resulting in excellent imaging for the balcony seats. Did I mention the 10 meters of mirrored wall directly above the top row? Optimal imaging for the upper seats results in strong-strong reflections-reflections for everybody else. Triage principles direct us to abandon this strategy and prioritize clarity for many over image for a few. Overbalcony delay speakers can cover this area with minimal risk to others.

The principles of minimum variance, TANSTAAFL and triage are the philosophical underpinnings of our specification strategies. They combine to create an approach that seeks the maximum benefit for the majority, is realistic about the tradeoffs required and is committed to allocation of the resources to the areas where tangible benefits can be assured.

11.1 SYSTEM SUBDIVISION

Transmission subdivision begins with channels and passes through arrays and single speakers to the listener. Channels are waveform streams and the speakers are the delivery system.

11.1.1 Channels

The left channel is a unique waveform stream sent to all of the "left channel" speakers. Members of the left channel maintain a correlated stable summation relationship with each other. The same goes for all right channel speakers, but not so for the relationship *between* left and right. Their relationship is uncorrelated and subject to change without notice. Therefore we design and optimize our systems on a per/channel basis. Everything left links to left and so on. The relationships between left and right, music and vocal systems, side and rear surrounds are placed under artistic control.

We need to agree on the macro allocation of channels and design the system around that. Coverage of the space is divided between the mains and fills carrying the same channel. This serves as a guideline but is not a rigid rule. The frontfill, for example, could be linked to the mains even if it receives a matrix feed that contains less than the full mix. The system is designed for minimum-variance transmission for correlated signals, even if the actual mix deviates from this. The modern sound system has lots of opportunities for cross-pollination in the matrix. The system is calibrated to achieve minimum variance with a unity matrix.

Channels can play an infinite number of roles but the usual suspects are pretty straightforward.

TYPICAL CHANNEL TYPES

- **Main**: left/right or mono mix of the complete show.
- **Voice**: separate channel for voice only. Typical of musical theater systems.
- **Music**: left/right system for music only (no voice). Musical theater again.
- **Subwoofer**: the subwoofers driven as an independent channel, because . . . subwoofer.
- **LFE (low-frequency effect)**: the cinema name for subwoofer, the "point one" in 5.1.
- **Surround**: sound image sourcing on the sides, rear and overhead and spatial immersion.
- **Special**: effect sources to localize a sound cue to a particular location.

11.1.2 Systems and subsystems

A channel stream can be divided into subsystems to manage different speakers within the transmission chain. The purpose of subdivision is to tailor the response to create the desired coverage shape. Simple shapes covered by a single system require no subdivision. More complex shapes require subdivision to customize the fit. It's not always necessary to subdivide to the component level. The additional expense of subdivision is not required when all elements are identically driven. The subdivision decision rests on symmetry. Asymmetric speaker groupings require subdivision, whereas symmetric ones don't.

System subdivision concerns four differentiation layers: speaker type, level, delay and equalization. Processor function subdivision is mandatory when different speaker models are used. Subdivision has merit when matched speakers need to cover different shapes or have asymmetric splay or spacing (Fig. 11.1).

Subdivision is a countermeasure to asymmetry and complexity. How much asymmetry is required before subdivision is merited? Let's be practical and apply TANSTAAFL. Every subdivision costs material, installation, ongoing maintenance, calibration time and opens up the door for human error. A range ratio of 1.4:1 (3 dB) is a practical threshold for subdivision decisions. We will again use alphabetical hierarchy to denote subdivision levels. Two matched speakers covering different depths would form an AB subdivision. A three-element combination could use AAA, AAB or ABC depending on the range ratios faced.

Once level asymmetry is introduced, delay and equalization follow suit. We must assume a speaker driven at reduced level is operating at closer range. Why else reduce it? Such elements are EQ'd separately because they encounter different local acoustical environments. Level asymmetry forces the spatial crossover off the equidistant geometric center to a point closer to the lower-level element. Delay can be used to phase align the crossover, in

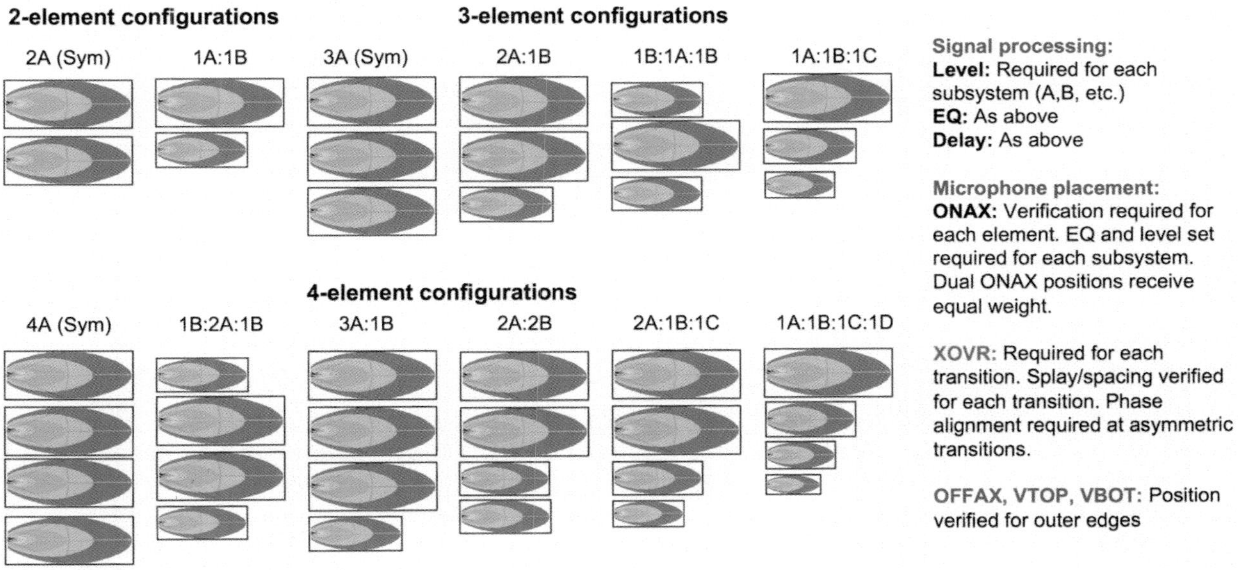

FIGURE 11.1
System subdivision strategy detailing the number of processing channels and calibration mic positions

proportion to the level offset and displacement. Such delay may be deemed impractical with small-level offsets and/or when the speaker geometry maintains extremely low displacement (such as third-order proportional beamwidth units). In such cases the potential benefits may not be worth the expense, trouble and (most important) risk of error.

11.1.3 Subsystem relationships

Let's divide the subsystem family tree into three principal parts: mains, fills and low-frequency systems (subwoofers). Each has sibling relationships and also to the larger groups. Members of a main array are a tightly knit family. They stick together through their entire range to make a composite main element. Main systems such as left and right relate on equal terms. Frontfills and underbalcony delay elements also relate as peers but they are an extended family spread over the space. They also join the mains, but not as equals. Mains are the dominant players over the various fill types. The LF devices can be left on their own because there is safety in numbers. They do better in packs (coupled arrays), which affects our strategies for combining them with the mains.

Every relationship between speaker elements can be characterized as one of the previously studied arrays. We have elements within arrays, and arrays as elements within arrays. For example, a coupled point source main and uncoupled point source frontfill can join together as an uncoupled point destination.

SUBSYSTEM FAMILIES

- **Mains**: principal sources to cover the majority of the room. Single speakers or coupled arrays.
- **Coupled fills**: radial extenders coupled to the mains (sidefills, downfills and rearfill).
- **Uncoupled fills**: auxiliary sources to cover leftover areas. Single speakers or uncoupled arrays.
- **Subwoofer**: spectral extension to the mains. Coupled arrays.

11.2 POWER SCALING

We need to set a scale for the system level. Are we doing rock and roll in an arena or a modern worship service? Sorry, those are both at the same level. Try again. Is it poetry readings or electronic dance music? We set the size from the top down: mains to fills and mains to subs. We are applying the principle of the analog transmission pipeline. We need to scale the pipeline for the program material and for its relative distance.

11.2.1 Main systems

How much power do we need? The answer is more. We have previously classified the speaker systems by power (section 2.7.6). Here we determined four general orders of magnitude for systems in 10 dB steps (Class 4 = 140 dB SPL @1 m, Class 3 = 130 dB SPL and so on). We also discussed the expected SPL levels for various program materials (Fig. 2.24).

This leads to a fundamental choice in the system design: the power scale of main system elements. There are no exact answers here so we have to make do with guidelines based around "typical" program material and venue size. Speaker element classes are in 10 dB steps, which correspond to 10× multiples of power (recall that power is a 10 Log_{10} formulation). We can scale crowd size for equivalent acoustic power (Fig. 11.2). A given program material needs a certain class of element for a given venue size, which rescales from there with different venues. For example, standard pop in a 5000-seat venue would do well with Class 4 elements but could use Class 3 for a 500-seat venue.

The coverage needs of the main system can play a role in the element selection. Let's say our budget will allow 8/side speakers of one type or 12/side of another. The former has 3 dB more SPL capability and 3 dB more per/element cost. Each group will generate the same SPL and debt payment. The best choice depends on coverage spread and the range ratio. Quantity is our friend when we need to cover a highly asymmetric and/or wide shape. The advantage moves to fewer higher-power elements when the spread is small and/or the range ratio is low. We will dig deeper into this soon. For now, we have a basic plan to choose a main speaker element.

Main system speaker element power scale reference					
	Speaker class	Class 4	Class 3	Class 2	Class 1
Program material	Sound levels	140+ dB SPL	130+ dB SPL	120+ dB SPL	110+ dB SPL
DJ/EDM	130+ dB SPL	10+ people	1+ people	Headphones	Hearing aid
High level pop/metal/rock/hip-hop	120+ dB SPL	100+	10+	1+	Headphones
Standard pop	110+ dB SPL	1000+	100+	10+	1+
Soft pop/Jazz/musical/traditional	100+ dB SPL	10,000+	1000+	100+	10+
Announce	90+ dB SPL	100,000+	10,000+	1000+	100+

FIGURE 11.2
Power scaling recommendations for main systems. Table is arranged by program material for audience sizes ranging from 1 to 100,000.

Fill system speaker element power scale reference										
Main/fill Range ratio		Class 4		Class 3		Class 2		Class 1		Main class
		>140	140	135	130	125	120	115	110	
1.8:1	5 dB	140	135	130	125	120	115	110		
3:1	10 dB	135	130	125	120	115	110			
6:1	15 dB	130	125	120	115	110				Fill class
10:1	20 dB	125	120	115	110					
18:1	25 dB	120	115	110						
30:1	30 dB	115	110							
60:1	35 dB	110								

FIGURE 11.3
Power scaling recommendations for fill systems relative to the mains. Axial differences are not included here.

11.2.2 Fill systems

We can move on to fill speakers once we have established the power scale for the mains. The smaller fills need to keep up with the mains. They do it by operating at a shorter range and/or by letting the mains do half the work. The range ratio gives us the answer. If the fill speakers are 10 dB closer (range ratio = 3), we can safely use a model that's 10 dB less powerful (e.g. a Class 4 mains with a Class 3 fill). Higher-range ratios allow for larger drops in power and so on (Fig. 11.3).

SPL calculations for combined arrays in the far field are very complicated. SPL ratings represent the entire frequency range with a single number. Such simplification serves us poorly when comparing two very different animals, e.g. off axis, coupled mains against on axis, uncoupled fill. Such frequency responses are highly unmatched (mains are heavily pink-shifted while the fills are flat). The fills will never keep up with the mains LF response (and we don't need them to). It's a merger of equals only in the isolated HF range, while the fills happily freeload off the mains in the ranges below.

We can ease the mental burden by focusing on the mains/fill meeting point: XOVR. It is *here* that the isolated range of the fills must keep up with the mains. It comes down to a one-on-one relationship, e.g. the bottom element of the mains and a frontfill. The calculation factors reduce to three items: difference in SPL capability, range and axial orientation. Let's start with max SPL and range first and add the axial part later.

If the main element has twice the SPL (+6 dB) and has to go twice the distance (-6 dB), then we are all even (as long as both speakers are on axis at XOVR).

MAINS AND FILL POWER SCALING EXAMPLE

1. Mains element (A) maximum SPL is 132 dB SPL @1 m (Class 3).
2. Propagation distances to XOVR AB are 12 m (main) and 3 m (fill), i.e. range ratio = 4:1 (12 dB).
3. Fill could be a Class 2 speaker (120 dB SPL) because 132 - 12 = 120.

The third factor, axial orientation, plays out differently with each pairing. No additional calculations are required if the mains and fills have axial equality (ONAX–ONAX or OFFAX–OFFAX). Main and underbalcony speaker aim into the same area, an ONAX–ONAX pairing. Main + coupled sidefill combinations target different areas but join their radial edges (OFFAX–OFFAX). The main + frontfill combination joins the main's off-axis bottom edge (VBOT) to the frontfill's ONAX. This gives the frontfills a 6 dB advantage (maximum), which is added to the range ratio used for power scaling.

MAIN AND FRONTFILL POWER SCALING EXAMPLE (INCLUDES AXIAL EFFECTS)

1. Mains element (A) maximum SPL is 138 dB SPL @1 m (Class 3).
2. Propagation distances to XOVR AB are 12 m (main) and 3 m (fill), i.e. range ratio = 4:1 (12 dB).
3. Add the frontfill's 6 dB axial advantage (ONAX–OFFAX) to make the total 18 dB.
4. Frontfill could be a Class 2 speaker (120 dB SPL) because 138 - 18 = 120.

Three main possibilities are found for main/fill pairings: ONAX–ONAX, OFFAX–OFFAX and ONAX–OFFAX. Some typical configurations are shown below.

LEVEL SCALING EFFECTS OF AXIAL ORIENTATION OF MAINS AND FILLS (TYPICAL)

- Main + frontfill: OFFAX main (vert) and ONAX frontfill. ≤6 dB advantage for frontfill.
- Main + centerfill: OFFAX main (hor) and ONAX centerfill. ≤6 dB advantage for centerfill.
- Main + downfill: OFFAX main (vert) and OFFAX (vert) downfill. No advantage.
- Main + sidefill: OFFAX main (hor) and OFFAX (hor) sidefill. No advantage.
- Main + infill: OFFAX main (hor) and OFFAX (hor) infill. No advantage.
- Main + delay: ONAX main and ONAX delay. No advantage.

11.2.3 Low-frequency systems (subwoofers)

The LF power scaling must keep parity with (or exceed) the mains. This is as much about quantity as the particular element. Small main systems in small spaces can use baby subwoofers with 10″ and 12″ drivers, but the majority of applications select from a pool of 2 × 15″ and 2 × 18″ elements. This is not to say that there is equivalence between Brand X and Brand Y or even Model 2–18 and Model 18–2 by the same maker. Linearity, maximum levels, frequency response and harmonic distortion are all over the map. You must conduct your own evaluations. The choice between 15″ and 18″ seems to be driven more by program material and acoustics than by SPL. I've been told that 15″ drivers are "punchier" and have more . . . (hits self on chest). Such descriptions may be realistic, but I've never been able to verify these qualities in a prediction plot or analyzer display. In general, smaller drivers don't go quite as low as the larger ones, which may be desirable for certain music and/or when the LF reverberation is long.

We previously evaluated fill system scaling on a one-to-one relationship with a main element at XOVR. The mains + LF junction is the polar opposite: All the mains join all the subs in all the room. Conventional SPL data is a very poor fit for comparison here. SPL is a single-number answer to a spread spectrum, and these two devices are spread over vastly different spectrums. They only overlap for one octave (at most) and far less for lovers of double black diamond crossover slopes. We need the subs to keep up with the mains at the crossover frequency and take it down from there.

The MAPP program calculates the spectrum of main and subs at their maximum SPL capability. This can be used to evaluate the LF to main ratio in a familiar three-step process: look at A (main) solo, B (sub) solo and A + B. The combined response in the LF range should comfortably exceed the base level for the midrange (+6 to 10 dB). An example is shown in Fig. 11.4.

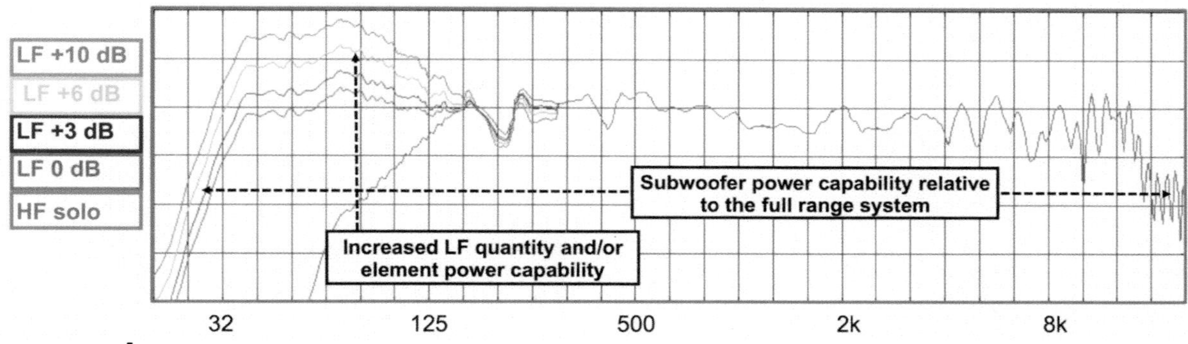

FIGURE 11.4
Power scaling example for an LF system relative to the mains

11.3 SINGLE SPEAKER COVERAGE AND AIM

Our speaker has scale. Now it needs shape. How wide do we need? This depends on our target shape and our perspective toward it. Where do we aim it? Same answer.

11.3.1 Horizontal coverage angle

Let's start with a speaker on the centerline of a rectangular room (Fig. 11.5). There are three logical break points from which to evaluate coverage: start, middle and end. If we specify a speaker wide enough to cover the earliest rows, it's too wide for every other one. This provides minimum level variance for all rows but steadily rising ripple variance as we go back (if the room happens to have side walls). TANSTAAFL and triage. The opposite extreme is found by using the last row as the coverage target. We get just enough there and not enough for every other row, a steady rise in level variance as we move forward. Ripple variance is minimized, but tell that to the half of the audience without coverage. The choices so far are perfect front with over coverage or perfect rear with under-coverage.

We can even out the under- and over-coverage errors by designing around the mid-point width. The minimum-variance coverage angle is found by matching the speaker's FAR value to the depth/width ratio of the shape. The result is 6 dB level and spectral variance across the mid-point depth of the room. The front half is under-covered and back half over-covered in equal proportion when the target is a simple rectangle. Notice how the area of the FAR shape is equal to the area of the room shape (the missing coverage in front is exactly the same shape as the overage in the rear). Ripple risk rises in the rear while gap risk rises in front. The front corner gaps can be reduced by raising up the speaker (which widens the effective coverage width by the lateral shape). Alternatively we can add fill speakers to plug the front corner gaps. Both the underage and overage errors are reduced if the room has expanding splay walls.

HORIZONTAL COVERAGE ANGLE DETERMINATION (SINGLE SPEAKER)

- Find the front/back depth along the centerline (speaker to last row).
- Find the width along the mid-point depth.
- Depth/width is the FAR. Convert FAR to angle (again Fig. 11.5).

11.3.1.1 FRONT/BACK ASYMMETRIC SHAPES

Not all listening spaces are rectangles. Even so, the best fit for coverage angle links the mid-point depth and width. Horizontally symmetric shapes with matched depth and width will need the same coverage angle despite differences

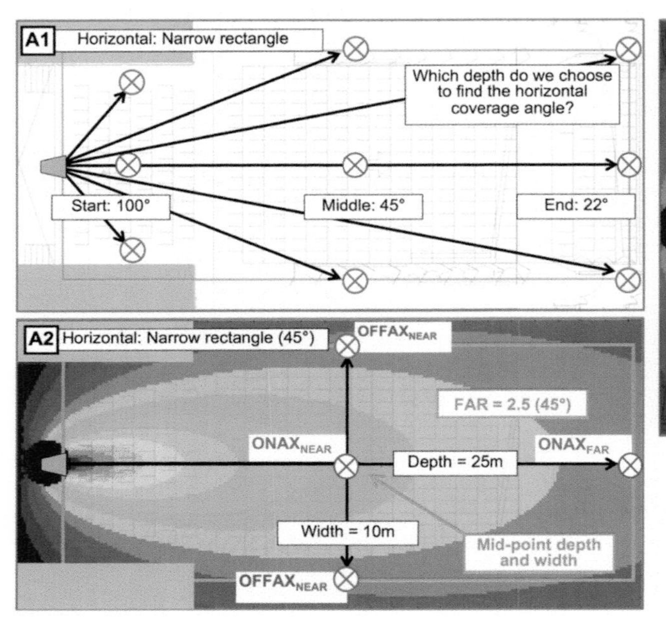

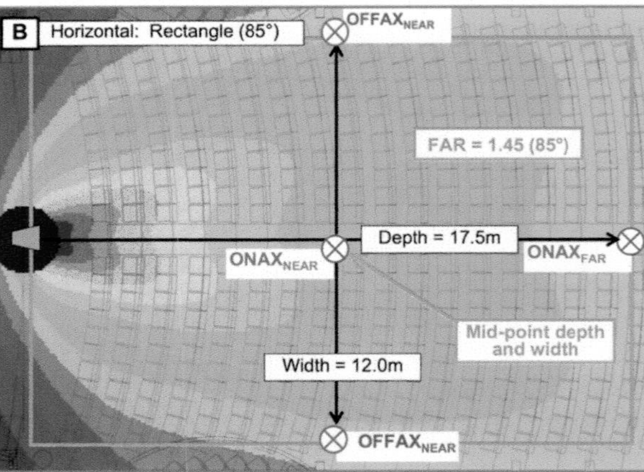

FIGURE 11.5
Single-speaker coverage in the horizontal plane (fully symmetric). (A1) The required coverage angle for this narrow rectangle varies from 100° to 22° depending on the depth used for calculation. (A2) The FAR method is applied to the same shape to find the average coverage angle of 45°. (B) The FAR method is applied to a wider rectangle.

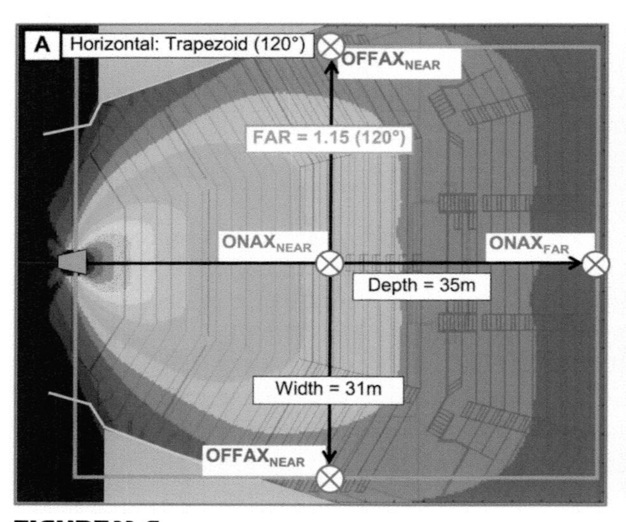

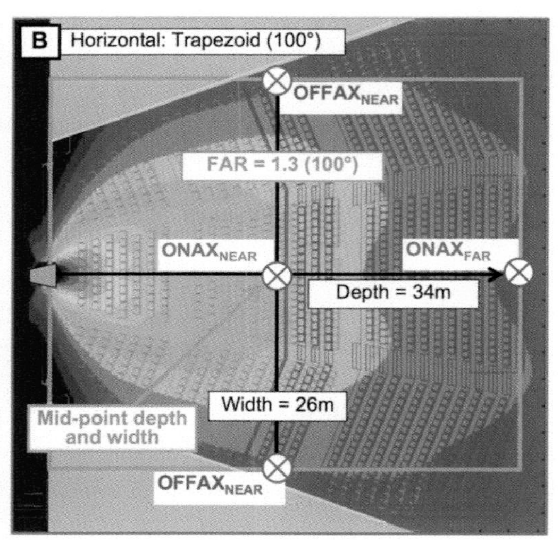

FIGURE 11.6
Single-speaker coverage in the horizontal plane (front/back asymmetric). (A) 120° trapezoidal shape with mild asymmetry. (B) 100° trapezoidal shape where the rectangulation is clearly wider in front and narrower in back.

in details. Many performance spaces are wider in back than front. We only have one speaker to cover this shape so we can't sweat the details. Some typical shapes are shown in Fig. 11.6.

11.3.1.2 LEFT/RIGHT ASYMMETRY

Left/right asymmetry adds another layer to the search for the middle/middle. Examples include an off-center speaker location into a symmetric shape, or a centered speaker into an asymmetric space. The concept is still the same: Aim the speaker through the middle of the middle and find the FAR from the depth/width ratio. The key to maintaining minimum-level variance is to match symmetry with symmetry, or to compensate asymmetry with a complementary asymmetry. The speaker placed in the corner of a rectangle is aimed at the opposite corner to maintain symmetric balance.

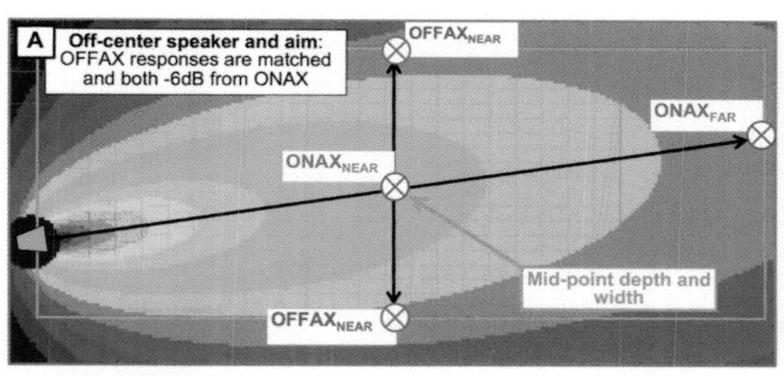

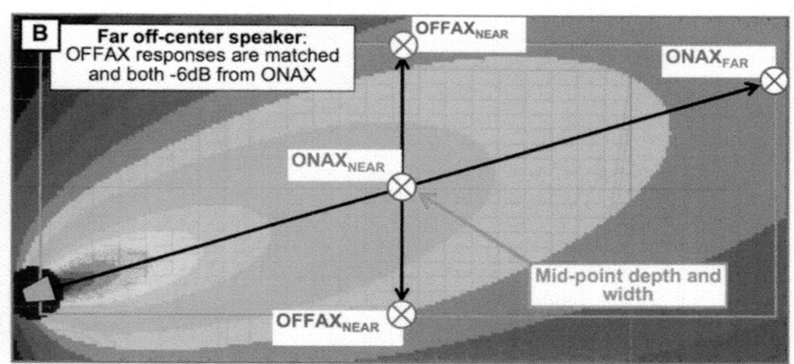

FIGURE 11.7
Horizontal aim from an off-center location in a symmetric shape

11.3.2 Horizontal aim

The shape is defined as depth (from speaker to last seat) and width (outermost seat to seat at the mid-point depth). Aim target is defined as the middle/middle (middle seat at the mid-point depth). A centered speaker obviously aims along the front/back centerline, which includes the mid-point center. Off-center speakers aim through the mid-point center to offset their asymmetric orientation (Fig. 11.7).

11.3.3 Vertical coverage angle

There's no point in discussing symmetry for vertical plane coverage until they start selling seats on the ceiling. The vertical plane is the inherently asymmetric plane and it's only one seat deep. It's a single line of coverage, not depth and width. Coverage angle is determined by the difference between the uppermost seat (VTOP) and the lowest/nearest (VBOT). Minimum-variance coverage angle specifications incorporate both the angular and range differences between these locations.

Let's use 50° as our example angular spread from VTOP to VBOT (Fig. 11.8). Our job is done if the two locations are equidistant (range ratio = 1). A centered 50° speaker fills the symmetric shape from top to center to bottom with -6 dB, 0 dB and -6 dB respectively. The situation changes radically when the range ratio rises to 2:1. VTOP is now twice as far, so we lose another 6 dB in transit. The response is -12 dB at VTOP and still -6 dB at VBOT. We can get 6 dB back at VTOP if we aim the speaker upward, but then we fall below -6 dB at VBOT. We need twice the speaker (100°) aimed directly at VTOP to maintain minimum variance. It's no coincidence that range ratio doubling is compensated by coverage angle doubling. The minimum-variance coverage angle for the vertical plane is found by multiplying the angular spread by the range ratio.

We have just met the outermost extremes, range ratios of 1:1 and 2:1 yielding coverage angles of 50° and 100°. A range ratio of 1.1:1 (+1 dB) moves the minimum coverage angle upward by a factor of 1.1 (+10%) to 55°. The speaker is aimed above the vertical mid-point, therefore extra coverage is needed to reach the bottom. Speaker coverage angle rises proportionally with range ratio until the limit is reached at 2:1 (100° speaker aiming at VTOP). Both level and spectral variance are minimized while the potential for ripple variance from surfaces above the target is maximized. A single-speaker approach should be abandoned in favor of an array if the ripple variance is too great or range ratio exceeds 2:1.

It may help to visualize this as a waste/efficiency model. Filling a 1:1 shape allows for maximum efficiency. At the other extreme is the 2:1 shape, which requires twice as much speaker as we use for actual coverage. That 50% waste product is going to cost us ripple variance etc. It's time to make an AB array when the waste gets too high.

11.3.4 Vertical aim

Aim is found by the range ratio compensated coverage method (Fig. 11.9). Let's standardize VTOP as 0° (relative). We will use 50° as an example coverage target, so the range is from 0° to -50° (mid-point of -25°). A 1:1 range ratio leaves the aim point at the vertical center (-25°). Aim point rises proportionally toward VTOP with range ratio. A range ratio of 1.1 (1 dB) moves the aim point upward by a factor of 1.1 (+10%) to -22.5°. Increasing the ratio raises the speaker farther until the limit is reached at 2:1 (speaker is aiming at 0°, the top of the target).

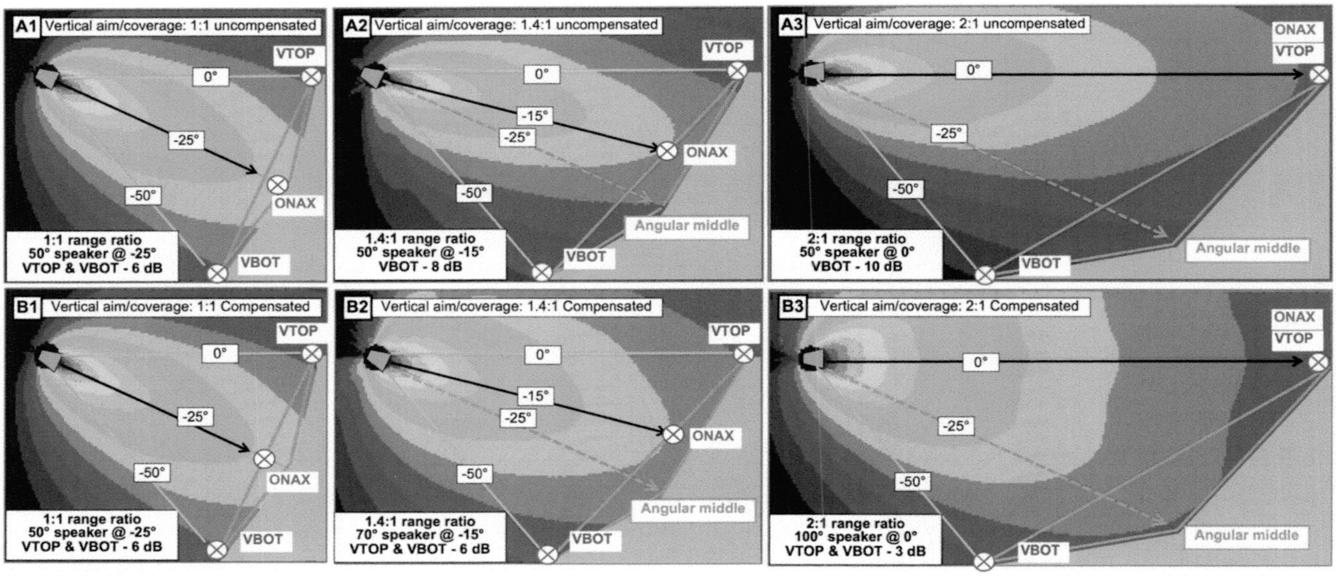

FIGURE 11.8
Compensated unity aim in the vertical plane. (A) The speaker is aimed to compensate for three levels of asymmetry but the coverage angle is uncompensated, resulting in up to 10 dB variance for the 2:1 shape. (B) Both aim and coverage angle are compensated for the three shapes, reducing variance to around 3 dB for the 2:1 shape.

Range ratio (VTOP/VBOT)	Target coverage angle (VTOP-VBOT)							
	20°		30°		40°		50°	
	Speaker		Speaker		Speaker		Speaker	
	Coverage	Aim below VTOP	Coverage	Aim below VTOP	Coverage	Aim below VTOP	Coverage	Aim below VTOP
1:1 (0 dB)	20°	-10°	30°	-15°	40°	-20°	50°	-25°
1.1:1 (1 dB)	22°	-9°	34°	-13°	45°	-18°	56°	-22°
1.25:1 (2 dB)	25°	-7°	38°	-11°	50°	-15°	63°	-19°
1.4:1 (3 dB)	29°	-6°	42°	-9°	57°	-12°	71°	-15°
1.6:1 (4 dB)	32°	-4°	47°	-6°	63°	-8°	79°	-10°
1.8:1 (5 dB)	36°	-2°	53°	-3°	71°	-4°	89°	-6°
2:1 (6dB)	40°	-0°	60°	-0°	80°	-0°	100°	-0°

Range ratio (VTOP/VBOT)	Target coverage angle (VTOP-VBOT)							
	60°		70°		80°		90°	
	Speaker		Speaker		Speaker		Speaker	
	Coverage	Aim below VTOP	Coverage	Aim below VTOP	Coverage	Aim below VTOP	Coverage	Aim below VTOP
1:1 (0 dB)	60°	-30°	70°	-35°	80°	-40°	90°	-45°
1.1:1 (1 dB)	67°	-26°	79°	-31°	90°	-35°	101°	-40°
1.25:1 (2 dB)	75°	-22°	88°	-26°	101°	-30°	113°	-33°
1.4:1 (3 dB)	85°	-18°	99°	-21°	113°	-23°	127°	-26°
1.6:1 (4 dB)	95°	-13°	111°	-15°	127°	-17°		
1.8:1 (5 dB)	107°	-7°	125°	-8°				
2:1 (6dB)	120°	-0°						

FIGURE 11.9
Vertical aim and speaker coverage reference chart for target angles between 20–90°. The chart shows the range ratio compensated aim (down from VTOP) and speaker coverage angle (compensated) required to fill the shape.

11.3.5 Horizontal/vertical coverage width

The width of effective coverage across the room for a given speaker is a function of both the vertical and horizontal planes (Fig. 11.10). We just determined the horizontal coverage angle and now we can see whether we need fill speakers in the front.

EFFECTIVE COVERAGE WIDTH DETERMINATION

1. Section view: Find the range to the closest seat (coverage start).
2. Find the lateral coverage width for that range from the lateral aspect ratio (Fig. 3.41).
3. Plan view: Draw a line of the determined coverage width along the closest row.

Fill speakers are needed if the coverage width line does not reach across the shape.

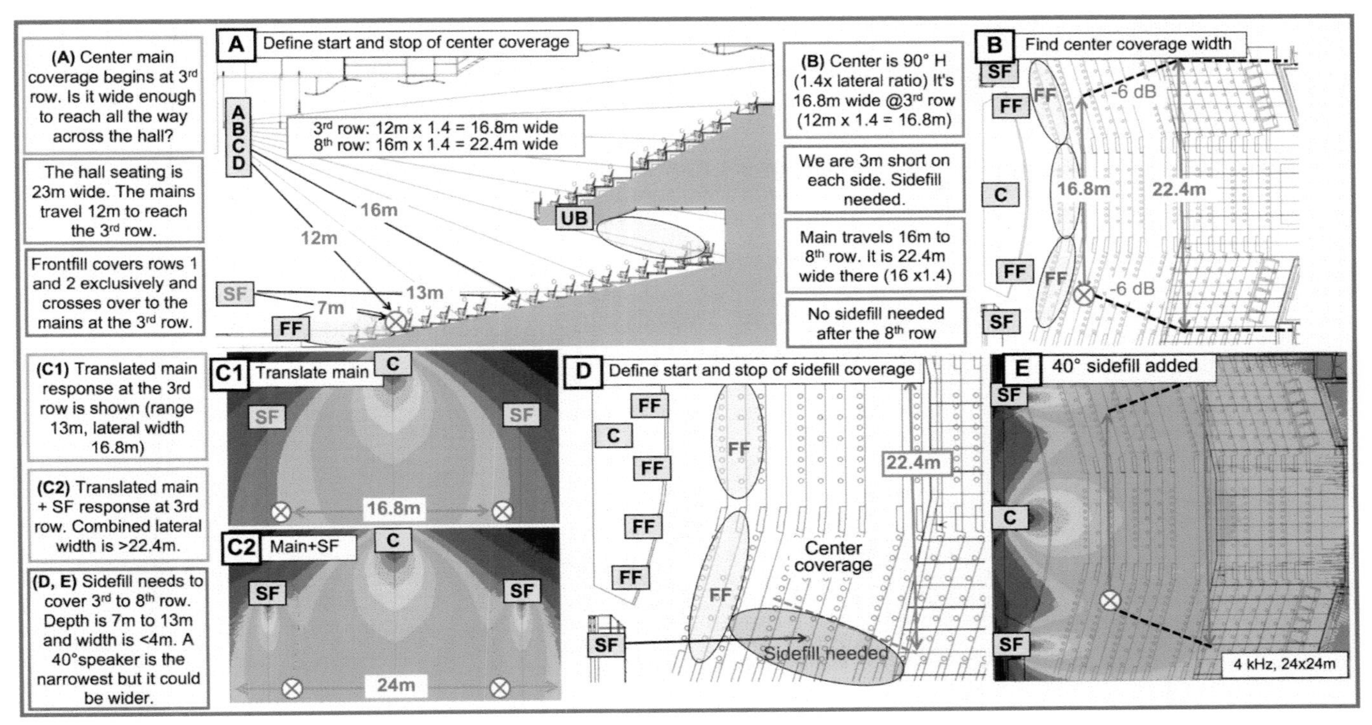

FIGURE 11.10
Example application for coverage width determination. The question is whether the center main can cover the whole width or if it needs sidefills. (A) Start of coverage is found and width is calculated. (B) Width is not long enough and areas needing fill are defined. (C) The required width and depth of the sidefill is found. (D and E) Sidefills are added.

11.4 COUPLED ARRAY COVERAGE, AIM AND SPLAY

11.4.1 Coupled point source: horizontal plane

We can look at filling these same shapes with more than a single speaker. Coverage subdivision allows for greater flexibility and detail. The coupled point source is the array of choice here. Its inverted form, the coupled point destination, is specified only under duress because of some physical requirement.

The use of multiple elements makes coverage angle and aim become interdependent variables. The symmetric coupled point source should be aimed at the middle of a symmetric coverage shape. If the coverage shape is asymmetric then the array should be aimed at . . . no wait, the array should be asymmetric. Once we've gone asymmetric we have aim *points*, which are analyzed on a case-by-case basis depending on levels, order and splay.

11.4.1.1 SYMMETRIC

Symmetric versions subdivide evenly. An example 80° horizontal coverage target (FAR = 1.55) can be covered with 2 × 40° elements, 40 × 2° elements or any combination between that multiplies to 80°. Symmetric subdivision transforms the coverage shape from the single-speaker rectangle to the radial fan (in the HF at least). The fully symmetric coupled point source is a one-trick pony. It improves the fit into radial fan-shaped rooms, but worsens it for rectangular rooms. The symmetric radial array spreads an MV line across the middle depth. This is ideal for a fan-shaped room because it effectively spreads an MV line at all depths. Rectangular rooms face the under/over coverage tradeoff we previously solved for a solo with the mid-point compromise. An 80° symmetric point source has the same area of under-coverage as the single, but a lot more over-coverage. The single speaker did not point at the side walls. Many, if not all, of the elements of the coupled point source are aimed at the side walls, substantially raising the ripple risk. This does not eliminate the coupled point source for rectangular horizontal shapes. We just need to use the asymmetric version.

The combined coverage angle can be approximated as splay angle × quantity. The approximation is closest when angular isolation is dominant or when the quantities are high enough to force the response into the shape enclosed by the total angular splay (Figs 11.11 and 11.12).

Horizontal coverage angle determination (symmetric)

- Find the mid-point depth along the centerline (mid-point from speaker to last row).
- Find the radial angle that best follows the mid-point depth from edge to edge. This can be used for a rectangle, a fan-shaped room or even a complete circle.
- Divide the coverage angle by element quantity (or multiply the elements until they fill the coverage).

11.4.1.2 ASYMMETRIC

We can customize the shape with asymmetric levels or elements. We'll use the compensated unity splay angle as the shaping agent. There is no "correct" element with which to start. The process begins with the "A" element, the speaker in charge of the largest area. The next element covers the majority of the remainder and proceeds from there until the shape is filled or other systems take over.

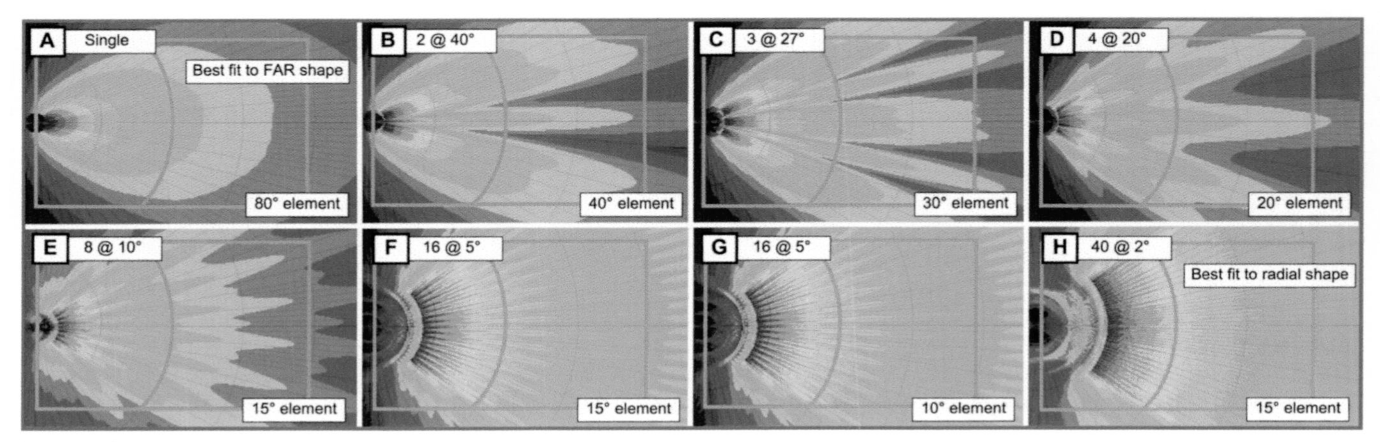

FIGURE 11.11
Symmetric horizontal coverage examples. A single 80° element (A) is compared with various arrays that create 80° of radial coverage. (B–H) Arrays ranging from 2 elements @40° to 40 elements @2° are shown. Both the FAR shape and radial minimum variance lines are shown to emphasize the transformation from single speaker (rectangular) to coupled point source (radial shaping). Rising quantity and overlap along with falling element coverage angle increase the uniformity along the radial line and sharpen the edges.

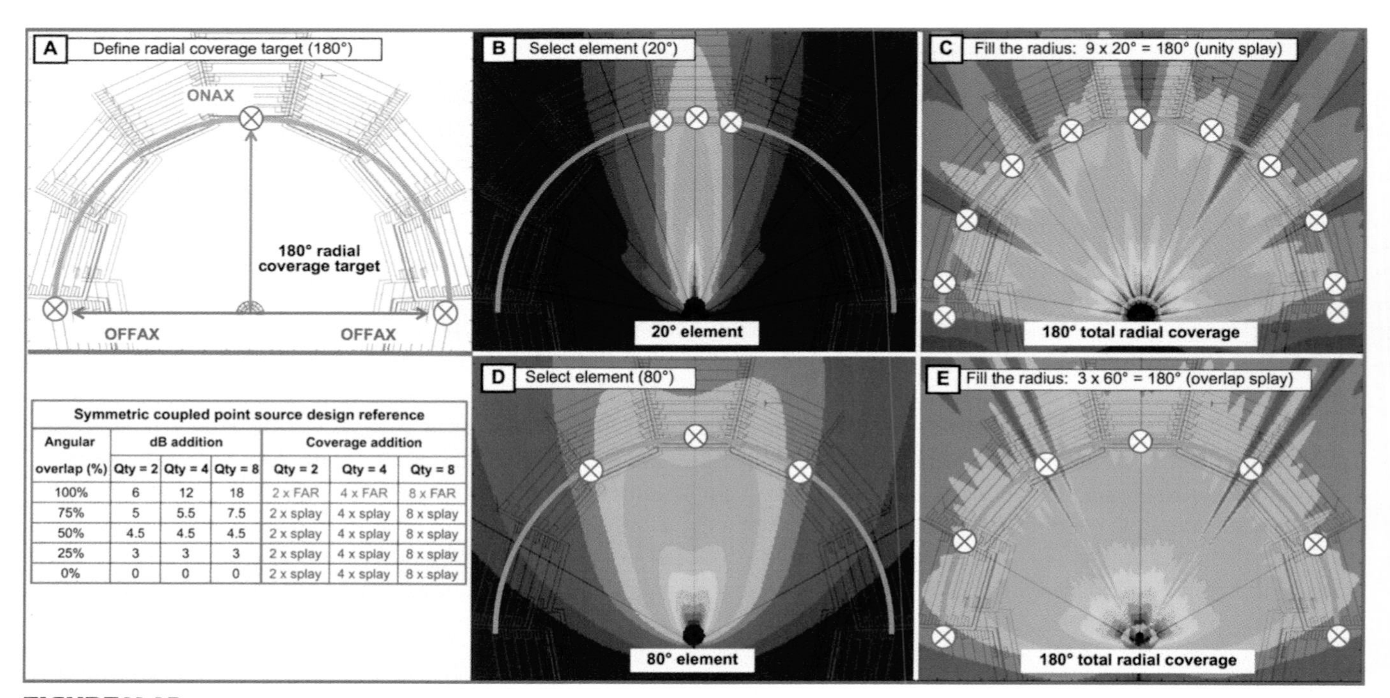

Symmetric coupled point source design reference

Angular	dB addition			Coverage addition		
overlap (%)	Qty = 2	Qty = 4	Qty = 8	Qty = 2	Qty = 4	Qty = 8
100%	6	12	18	2 x FAR	4 x FAR	8 x FAR
75%	5	5.5	7.5	2 x splay	4 x splay	8 x splay
50%	4.5	4.5	4.5	2 x splay	4 x splay	8 x splay
25%	3	3	3	2 x splay	4 x splay	8 x splay
0%	0	0	0	2 x splay	4 x splay	8 x splay

FIGURE 11.12
Example application for the symmetric coupled point source design procedure: (A) coverage target is defined, (B, C) the design process for a narrow element, (D, E) the same process using wider elements

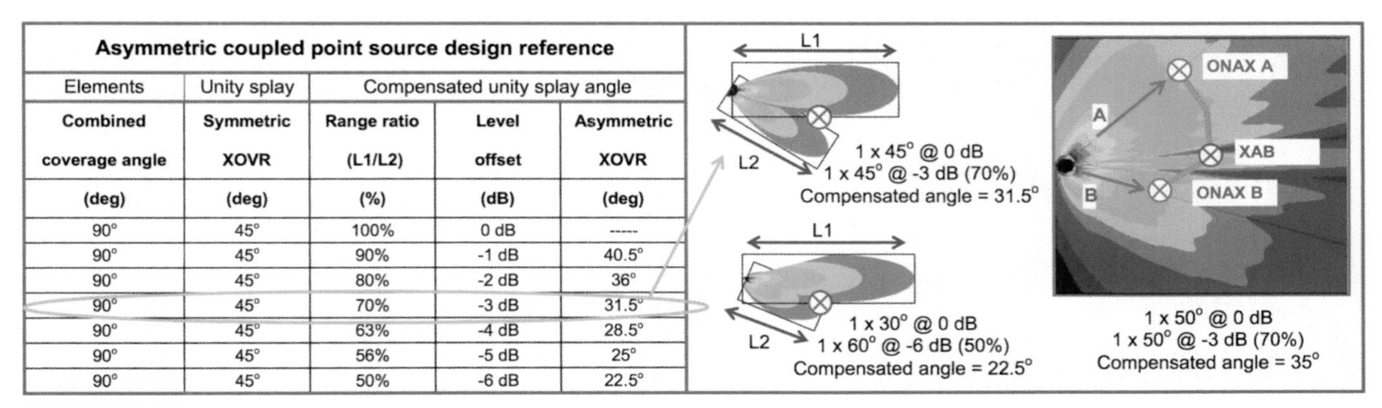

Asymmetric coupled point source design reference				
Elements	Unity splay	Compensated unity splay angle		
Combined coverage angle (deg)	Symmetric XOVR (deg)	Range ratio (L1/L2) (%)	Level offset (dB)	Asymmetric XOVR (deg)
90°	45°	100%	0 dB	-----
90°	45°	90%	-1 dB	40.5°
90°	45°	80%	-2 dB	36°
90°	45°	70%	-3 dB	31.5°
90°	45°	63%	-4 dB	28.5°
90°	45°	56%	-5 dB	25°
90°	45°	50%	-6 dB	22.5°

FIGURE 11.13
Asymmetric coupled point source compensated unity splay design reference

Asymmetric coupled point source design procedure (Fig. 11.13)

1. Coverage shape is defined by angular spread and range ratio. Ratio ≥2:1 is assumed.
2. Aim A toward the farthest area. A should have the narrowest coverage angle in the array.
3. Select a coverage angle for A that fits the local area.
4. Estimate location for XOVR AB (the transition into the next element). Start with the unity splay location and compensate for shorter ranges if applicable. The B element continues coverage from here.
5. Select coverage angle of B as needed to continue (or finish) coverage. Calculate the compensated unity splay if the range for B < A ((coverage of A + B)/2* range ratio).
6. Position B at the compensated angle and appropriately scaled range. Assess and fine tune as required.
7. Continue process with the third element (C) and so on until the shape is filled.

One additional consideration is the physical size of the array. The boxes take up space. An array gets closer to people as it gets larger, which affects both the range ratio and the coverage target. Doubling the element size, or quantity, will require a revisit to the design, a classic "chicken and egg" problem. All I can suggest is to be as realistic as possible during the design process about the physical size and reassess as needed when things change a great deal.

11.4.2 Coupled point source: vertical plane (constant beamwidth elements)

The asymmetric coupled point source lays a line of coverage over the vertical plane. Range ratio and angular spread between VTOP and VBOT remain the primary design factors. The process is similar to the horizontal plane, but typically with higher-range ratios and narrower elements. Range ratio, our asymmetry gauge, provides a convenient indicator of when to subdivide. Just look at the number. A 2:1 ratio needs two sections (AB), 3:1 needs three (ABC) and so on (Fig. 11.14).

The process is top/down, with the narrowest element covering the longest distance and the widest covering the shortest. An AB array will have a single partition located well past the mid-point depth. Putting the break around 2/3 depth allocates the resources fairly evenly because the upper systems have farther distance to cover. The coverage angle of the A element should also be narrower than B, also helping to level out the resource allocation. Recall that we needed a 100° speaker to cover a 50° angular spread with a 2:1 range ratio. An AB array can cover this same shape with a 75° of speaker (A = 25° and B = 50°). The AB approach doubles the number of devices and reduces the waste, both of which result in more SPL at the seats and proportionally less on the ceiling (in this case just 12.5° above VTOP instead of 50°) We can subdivide further, use smaller elements and trim down even more of the overage.

Higher range ratios require more subdivision. There are more partitions and a wider range of individual element angles, but the principle remains the same. Element A has the longest throw, narrowest coverage and smallest depth. Then comes B, then comes C and so on.

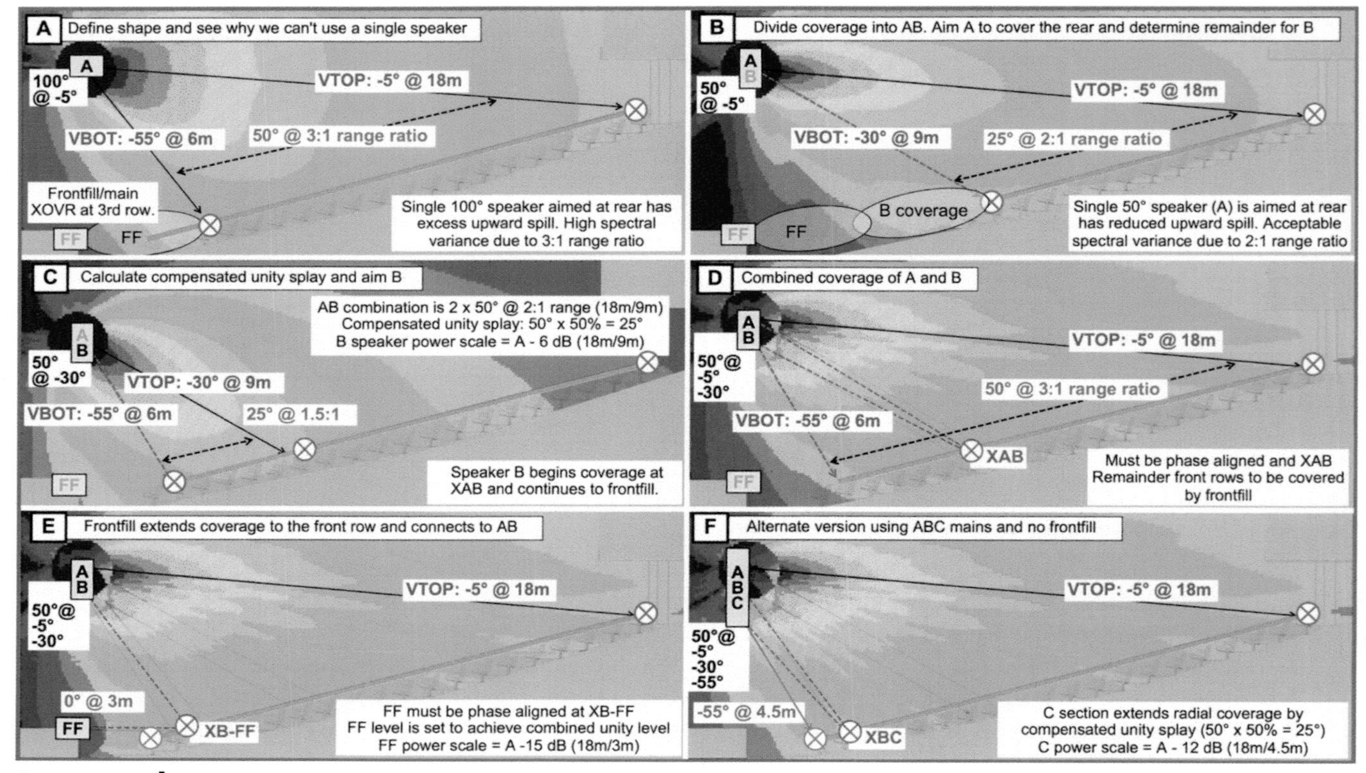

FIGURE 11.14
Example application for AB and ABC vertical combinations of the asymmetric coupled point source

11.4.3 Coupled point source: vertical plane (proportional beamwidth elements)

We don't have to stop at ABC. Proportional beamwidth elements allow us to run the entire alphabet (if we want to). These elements provide increased flexibility in shaping because we can custom mix angular overlap and isolation to our needs, even in the face of very high-range ratios. The target coverage shape is characterized by its angular spread and range ratio, e.g. 20° of coverage with a range ratio of 2:1 (6 dB). We can fill shape with speakers, dividing the coverage into finer slices as quantity rises. There is a minimum number required, which depends on the element and target shape. The maximum is limited only by budget and rigging standards.

We can find the absolute minimum quantity required once we know (a) the element maximum splay, (b) target angle and (c) range ratio. Let's start by filling a simple 60° shape with a 1:1 range ratio. Our element is 10° so we need at least six boxes (6 × 10° = 60°). Take note of the fact that the "average" splay angle for this array is 10°. That seems silly now but its importance will become clear soon. We can add boxes and reduce the splay proportionally (e.g. 12 @5°, 15 @4° and so on as long they multiply to 60°). The average splay angle is falling as quantity rises. The design process is simple with a 1:1 range ratio: quantity × splay = coverage angle.

Now let's double the range ratio (2:1, 6 dB) and see how this affects the minimum box count for our 60° example. We get the array to follow the 2:1 shape by having more overlap in the upper boxes (i.e. reduced splay). We can't get to the bottom with six boxes now because we can't splay beyond the element limit of 10° without gapping. How many more do we need? The answer is eight boxes, which will require some explanation.

We filled the 1:1 shape with matched elements at matched level and splay. The 2:1 shape can be made from matched elements, matched level and unmatched splay. We need to see at least a 1:2 ratio between the narrowest and widest splay angles. If we stick with six boxes we still have an average splay of 10° with narrower splays above and wider below. We could solve this with angles ranging from 7.5° (top), 10° (middle) and 15° (bottom) except for one small problem: Our 10° element leaves gaps in coverage. The maximum splay at the bottom is 10° and we can achieve our desired splay ratio with 5° at the top. The average splay, then, is 7.5°, which gives the eight-box count (60°/7.5 = 8). The intermediate splays fill in between the milestone values, e.g. 5°–5°–5°–7.5°–10°–10°–10° or 5°–6°–7.5°–7.5°–9°–9°–10. We can add more boxes and keep the same shape by maintaining the same top/average/bottom ratios.

This doesn't mean a 2:1 range ratio always requires at least eight boxes. If we reduce the target angle to 30° we need three boxes at 1:1 (10°–10°) and four boxes at 2:1 (5°–7.5°–10°). A target angle of 15° needs two boxes at 1:1 (7.5°) and three for 2:1 (3.75°–7.5°). Three rules are followed in all cases here: (a) quantity × average splay = coverage angle, (b) splay ratio is the inverse of range ratio and (c) the widest splay does not exceed the element limit.

Let's return to the 60° target and extend the range ratio to 3:1 (10 dB). We need a splay ratio under 1:3. A minimum splay of 3.3°, maximum 10° and average 6.6° satisfies our needs with a box count of ten (60°/6.6°). Reducing the coverage target angle by half and the box count does the same. Reduce the element maximum by half (to 5° maximum splay) and the box count doubles.

The saga continues with other coverage targets, range ratios and element limits, but this is the basic outline. Note that this represents the minimum box count to create these shapes without help from level tapering. Results in the field will vary from model to model depending upon the quality of the beamwidth shaping in the elements.

It is not necessary to drive every element individually. The array can be broken into sections (ABC etc.) that contain multiple elements. The channel quantity should at least match the range ratio (rounded upward). As a general rule, the upper sections should contain the most elements and gradually fewer as we go down. This mimics the ABC approach we established for single elements and allows for more level tapering at the bottom if needed. The choices for where to segment the array will be strongly influenced by the room geometry, such as a balcony or a change in the seating rake.

We have now constructed a segmented asymmetric point source (elements ABC etc.) out of unlimited quantities of single-speaker sub-elements, which puts it into the realm of a practical, tunable system (Fig. 11.15).

11.4.4 Asymmetric-composite coupled point source: Vertical plane

At the end of the day we are going to need to place mics in the hall and optimize the array. We have previously established a systematic approach with the key locations: ONAX, OFFAX and XOVR. These locations are easily found for single elements. Things get more complex when the element "A" is a composite made up of four speakers. Finding the center of a composite is not as simple, especially if there are three different splay angles between the four boxes. Let's look a bit deeper.

11.4.4.1 ONAX AND ONAX-ISH

On axis is important because it's the key reference point for predictable spatial behavior. We literally cannot find off axis if we haven't found on axis first. A splayed pair (a two-element composite) has three on-axis points (one for each box and the center of the pair). We create the composite using the individual element on axis locations. The composite center then becomes the reference for aiming the combined assembly and combining it with other elements. The

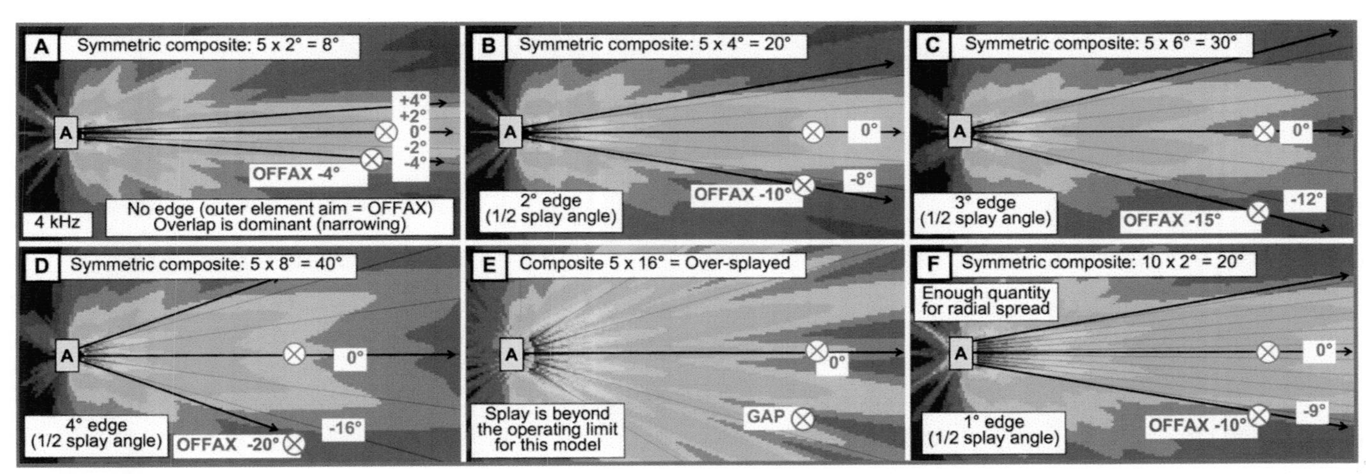

FIGURE 11.15
Composite point source element examples. (A) High overlap and low quantity can create a narrowed composite, i.e. 5 × 2° would be expected to create 10° but instead creates 8°. (B–C–D) Progressively wider splays proportionally expand the composite coverage angle and soften the edges. (E) Excessive splay leads to gaps within the composite. (F) Increased quantity adds enough isolation to overcome the narrowing of highly overlapped elements, i.e. 10 × 2° = 20° (compare with panel A).

composite's ONAX location is clearly identifiable as long as all the relationships inside it are symmetrical. We can keep adding elements but the center remains the center for the full frequency range and everything is referenced from there. ONAX is king of the hill with the familiar progression toward OFFAX in either direction. The center cannot hold once asymmetry in level, spacing, splay or phase is introduced. The asymmetry initiates a steering force that moves things off center. The greater the asymmetry, the stronger the steering. The situation is further complicated by the fact that the steering is likely frequency dependent, i.e. the lows, mids and highs are moved by different amounts (and to different places). Our central area has changed from ONAX to ONAX-ish (Fig. 11.16).

Let's consider a few scenarios to bring the point home. Would you be comfortable driving two different speaker models with the same processor channel? Of course not. We all accept that different speakers require different tunings. How about a single matched pair? Fine. A matched trio? Still fine. What exactly is a composite of five matched boxes with unmatched splay angles? It's kind of matched and most definitely not. Where is the middle of a composite with 1°–2°–3°–4° splays? We can find ONAX (for all frequencies) of a symmetric composite with 5 × 2°. Not so for the previous. The HF is steered more upward with the tight angles and the LF doesn't notice the difference between any of the splays and heads to the middle. We are back to ONAX-ish again.

An ONAX-ish location doesn't make optimization impossible. It just makes it more challenging, because there is less certainty that the mic position is the best representative for the composite response. Asymmetric composite elements have approximated centers providing approximated answers. The more asymmetric the more approximate (and margin for error). Tuning work is difficult enough already without wondering if we are in the middle or fringe of the element. We want the clearest viewpoint because a processing channel affects the entire composite. A symmetric composite has a 100% identifiable center (the loudest, flattest location), and a predictable progression of responses from there. We can place mics knowing A, B, C and D are directly in our sights and then combine them to uniformity. We are building an asymmetric coupled point source macro array out of symmetric elements (just as we would with single speakers).

11.4.4.2 DEFINING THE COMPOSITE ELEMENT

A symmetric composite can be defined by quantity × splay = composite angle, e.g. 6 × 10° = 60°. We can describe the eight-element AB array from earlier as A: 4 × 5° (20°) over B: 4 ×10° (40°). Notice that A + B adds up to 60°. Also note that 7.5° (the average angle in the middle) is missing in action. The seven actual splay angles are 5°–5°–5°–7.5°–10° –10°–10°. Remember that splay angles don't make sound. Speakers do (there are eight of them). We hear four elements splayed at 5° when we turn on the A section. Think of each speaker as a 5° element, i.e. ±2.5° from center. Their combined 20° coverage results from 3 × 5° splays and 2.5° above and below the outer elements. The B section gets it 40° from 3 × 10° splays and a 5° remainder on the outer edges. The transition splay angle (7.5°) connects the bottom of A (2.5° underneath) and the top of B (5° overhead). All told the array spans 60° of coverage from +2.5° to -57.5°. You might wonder how the same speaker could transform from ±2.5° to ±5°. It can't as a soloist but it can't resist in the face of all the overlap with its neighbors. It's array groupthink. Overlap sharpens the edges

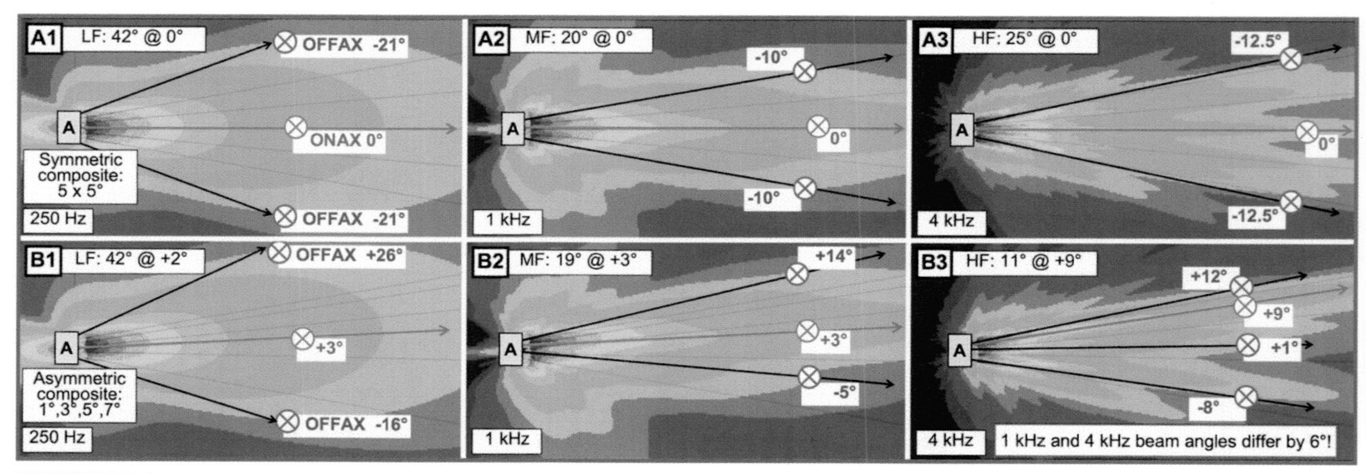

FIGURE 11.16
Symmetric and asymmetric composite elements. Notice that the symmetric version (A) remains centered. The asymmetric version has a moving center and highly unpredictable sides.

(recall Figs 11.11 and 11.15). Finally notice that the center of A (-7.5°) is 30° away from the center of B (-37.5°). I guess it's just a coincidence that this is exactly the unity splay angle for pairing a single 20° and 40° speaker.

Asymmetric composite coupled point source design procedure (Figs.11.17 to 11.18)

1. Define target top and bottom (VTOP and VBOT) as angle and range (e.g. -5° @20 m).
2. Define the target shape from VTOP to VBOT as coverage angle and range ratio (e.g. 30° @2:1).
3. Define maximum splay (depends on element) and minimum (max/range ratio).
4. Calculate the average splay angle ((max + min)/2), e.g. (10° + 5°)/2 = 7.5°.
5. Calculate minimum element quantity (target coverage angle/average splay).
6. Subdivide the array into composite segments with splay ratios scaled to the shape.

11.4.4.3 PREVENTING AND INDUCING SHOCK

Splay angle asymmetry is the main ingredient of this type of array, but too much, too quick can shock the system. It is important to remember that we are combining sets of proportionally narrowing elements. Differences of a few degrees are a big deal to the highest frequencies and then provide proportionally less impact as we go down. The key to splay angle change is to think proportionally, i.e. in ratios. Simply put, a 3° change can be major or minor, depending on what it changes *from*. Transitioning from 1° to 4° splay is major, whereas moving from 9° to 12° may be barely detectible. The former changes by a factor of 3.0, whereas the latter only differs by 0.3. We are seeking to make a flattened beamwidth from proportional elements. If the transition is too rapid

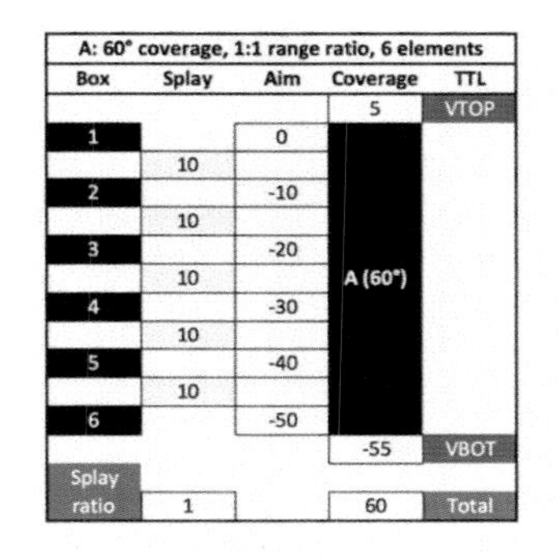

A: 60° coverage, 1:1 range ratio, 6 elements

Box	Splay	Aim	Coverage	TTL
			5	VTOP
1		0		
	10			
2		-10		
	10			
3		-20		
	10		A (60°)	
4		-30		
	10			
5		-40		
	10			
6		-50		
			-55	VBOT
Splay ratio	1		60	Total

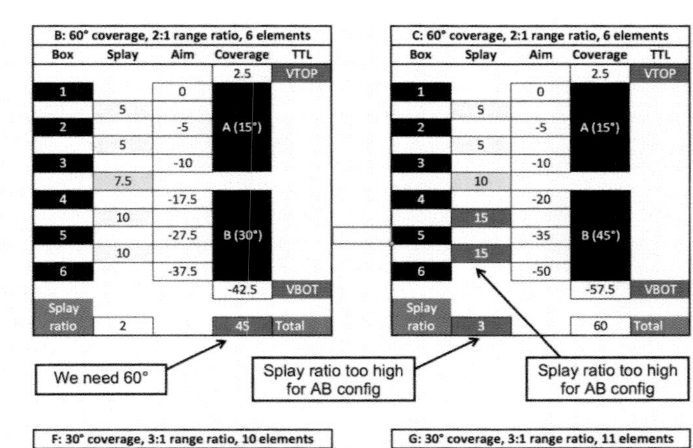

B: 60° coverage, 2:1 range ratio, 6 elements

Box	Splay	Aim	Coverage	TTL
			2.5	VTOP
1		0		
	5			
2		-5	A (15°)	
	5			
3		-10		
	7.5			
4		-17.5		
	10			
5		-27.5	B (30°)	
	10			
6		-37.5		
			-42.5	VBOT
Splay ratio	2		45	Total

C: 60° coverage, 2:1 range ratio, 6 elements

Box	Splay	Aim	Coverage	TTL
			2.5	VTOP
1		0		
	5			
2		-5	A (15°)	
	5			
3		-10		
	10			
4		-20		
	15			
5		-35	B (45°)	
	15			
6		-50		
			-57.5	VBOT
Splay ratio	3		60	Total

D: 60° coverage, 2:1 range ratio, 8 elements

Box	Splay	Aim	Coverage	TTL
			2.5	VTOP
1		0		
	5			
2		-5		
	5		A (20°)	
3		-10		
	5			
4		-15		
	7.5			
5		-22.5		
	10			
6		-32.5		
	10		B (40°	
7		-42.5		
	10			
8		-52.5		
			-57.5	VBOT
Splay ratio	2		60	Total

E: 30° coverage, 3:1 range ratio, 9 elements

Box	Splay	Aim	Coverage	TTL
			0.75	VTOP
1		0		
	1.5			
2		-1.5	A (4.5°)	
	1.5			
3		-3		
	2			
4		-5		
	3			
5		-8	B (9°)	
	3			
6		-11		
	4.5			
7		-15.5		
	5.5			
8		-21	C (16.5°)	
	5.5			
9		-26.5		
			-29.25	VBOT
Splay ratio	3.7		30	Total

F: 30° coverage, 3:1 range ratio, 10 elements

Box	Splay	Aim	Coverage	TTL
			0.75	VTOP
1		0	A (6°)	
	1.5			
2		-1.5		
	1.5			
3		-3		
	1.5			
4		-4.5		
	2.25			
5		-6.75	B (9°)	
	3			
6		-9.75		
	3			
7		-12.75		
	4			
8		-16.75	C (15°)	
	5			
9		-21.75		
	5			
10		-26.75		
			-29.25	VBOT
Splay ratio	3.3		30	Total

G: 30° coverage, 3:1 range ratio, 11 elements

Box	Splay	Aim	Coverage	TTL
			0.75	VTOP
1		0		
	1.5			
2		-1.5		
	1.5		A (4°)	
3		-3		
	1.5			
4		-4.5		
	1.75			
5		-6.25		
	2			
6		-8.25	B (6°)	
	2			
7		-10.25		
	2.5			
8		-12.75		
	3		C (6°)	
9		-15.75		
	4.5			
10		-20.25		
	6		D (12°)	
11		-26.25		
			-29.25	VBOT
Splay ratio	4.0		30	Total

H: 30° coverage, 3:1 range ratio, 12 elements

Box	Splay	Aim	Coverage	TTL
			0.5	VTOP
1		0		
	1			
2		-1		
	1		A (4°)	
3		-2		
	1			
4		-3		
	1.5			
5		-4.5		
	2			
6		-6.5	B (6°)	
	2			
7		-8.5		
	2.5			
8		-11		
	3			
9		-14	C (9°)	
	3			
10		-17		
	4.5			
11		-21.5		
	5.5		D (11°)	
12		-27		
			-29.75	VBOT
Splay ratio	5.5		30.25	Total

FIGURE 11.17
Composite builder examples for showing 60° and 30° total coverage of 1:1 to 3:1 range ratios. Element limit is given as 10°. (A) Fully symmetric, minimum quantity. (B) Failed due to insufficient quantity to cover the target coverage angle. (C) Failed due to over-splay between elements (15° splay for a 10° element) and shock (splay ratio between adjacent composites exceeds 2:1). (D) Minimum quantity for 2:1 range ratio at 60°. (E–H) Examples of 30° coverage at 3:1 range ratio with various quantities.

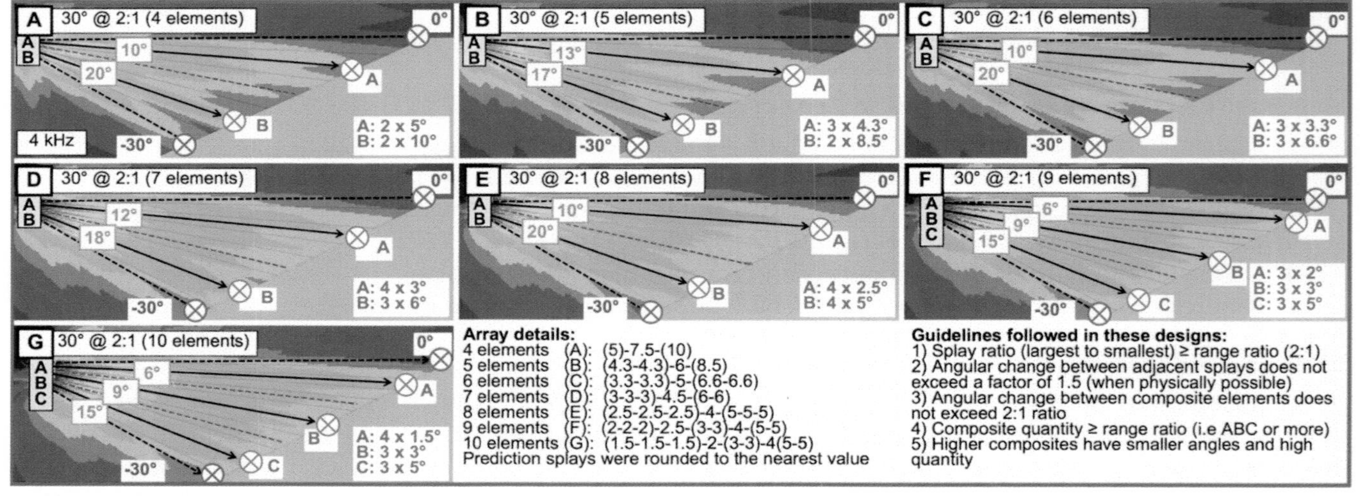

FIGURE 11.18A
Composite combination examples to create 30° coverage at 2:1 range ratio with various element quantities

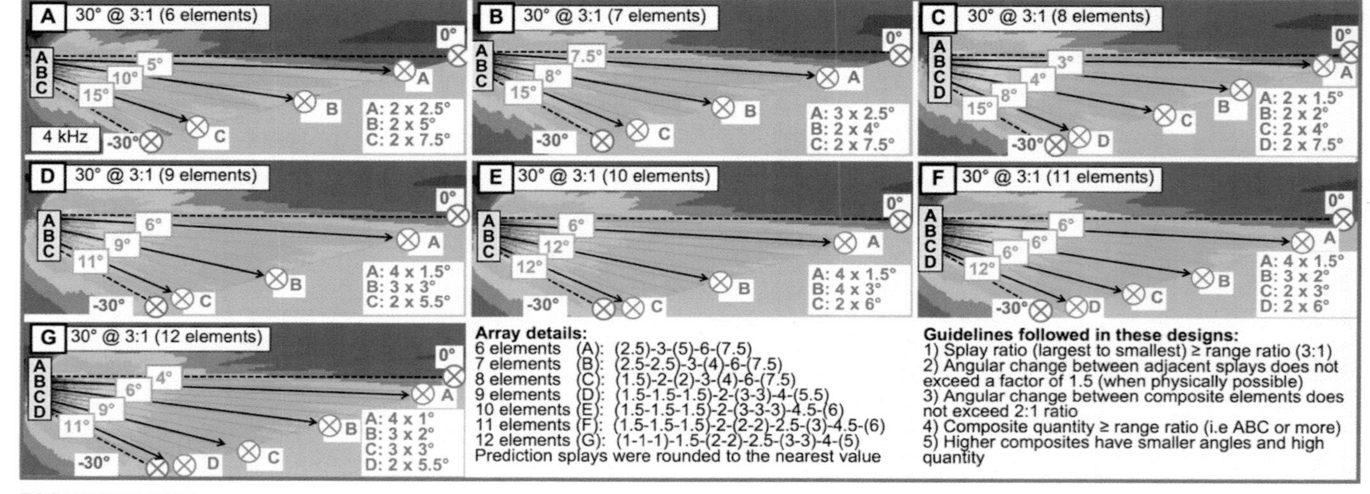

FIGURE 11.18B
Composite combination examples to create 30° coverage at 3:1 range ratio with various element quantities

we will "shock" the combined response with oversteered HF and with lower ranges that fail to follow. This can be prevented by keeping the transitions gradual (as ratios). Let's set a goal of never exceeding a 2:1 change and see how that plays out. From a 2° splay we can go down to 1° or up to 4°. From 4° we can go up to 8°. Those are ratio changes and represent, in my experience, the absolute upper limit. My personal approach favors limiting changes to

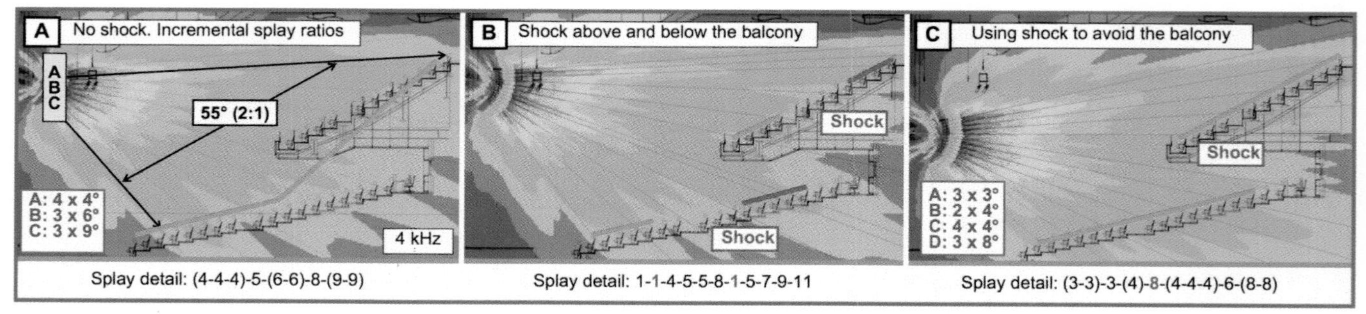

FIGURE 11.19
Shock induction and prevention. Incremental splay ratio changes will minimize risk of shock. Large changes can induce shock for good (balcony avoidance) or bad (hot spots and gaps).

≤1.5:1 for adjacent splays and 2:1 changes for composite splays. The following sequence illustrates this approach for an eight-element ABC array: 2°–2°–3°–4°–4°–6°–8°. That's 3 × 2° (6°), 3 × 4° (12°), 2 × 8° (16°). No adjacent splay change exceeds 1.5:1 and the composite to composite increments move at a 2:1 rate from 2°–4°–8°. You will see variations of this approach over and over again as the example arrays are shown here (Fig. 11.19).

On the other hand, we can choose to purposefully induce shock in order to gap the response and avoid balcony fronts and other undesirables. Once again it is the change in splay ratio (both above and below) that is decisive.

11.5 UNCOUPLED ARRAY COVERAGE, SPACING AND SPLAY

Uncoupled arrays are range limited regardless of propagation plane. Therefore the specification process does not differ between the planes in the room, but rather the planes of the arrays. We focus on two coverage line milestones: unity and limit (Figs 11.20 and 11.21). Note: Daniel Lundberg created a program that performs these calculations. It's available at www.lundbergsound.com.

11.5.1 Uncoupled line source (symmetric)

Used for multiple mains, overheads, surrounds and various fills (frontfill, underbalcony, etc.).

SYMMETRIC UNCOUPLED LINE SOURCE DESIGN PROCEDURE

1. The target shape is a straight line of coverage with matched starting depth (the unity line).
2. Define range to start of coverage: unity line depth, e.g. 3 m.
3. Define spacing: Element lateral aspect ratio × unity depth, e.g. 1.25 (80°) × 3 m = 4 m spacing.
4. Define quantity: Total coverage length/spacing, e.g. 14 m length/4 m spacing = quantity of 4.
5. Define limit depth: Limit depth for an uncoupled line source = 2× the unity depth.

Symmetric uncoupled line source design reference

Element		Known unity line (D)		Known spacing (S)	
Coverage (deg)	Aspect ratio FAR	Limit depth D^{LIM} = 2 x D^{UNI}	Spacing S= D^{LIM}/FAR	Unity depth D^{UNI} = 1/2S x FAR	Limit depth D^{LIM} = S x FAR
180°	1.00	2x D unity	2.00	0.50	1.00
150°	1.04	2x D unity	1.92	0.52	1.04
120°	1.15	2x D unity	1.74	0.58	1.15
90°	1.41	2x D unity	1.42	0.71	1.41
80°	1.56	2x D unity	1.28	0.78	1.56
70°	1.74	2x D unity	1.15	0.87	1.74
60°	2.0	2x D unity	1.00	1.00	2.00
50°	2.3	2x D unity	0.87	1.15	2.30
45°	2.6	2x D unity	0.77	1.31	2.61
40°	3.0	2x D unity	0.67	1.50	3.00

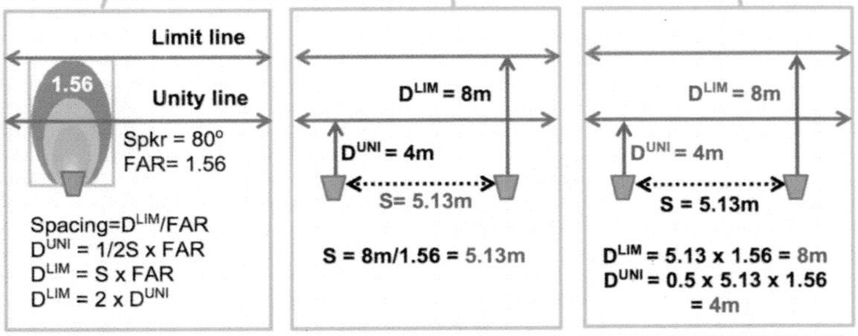

Limit line
1.56
Unity line
Spkr = 80°
FAR= 1.56
Spacing=D^{LIM}/FAR
D^{UNI} = 1/2S x FAR
D^{LIM} = S x FAR
D^{LIM} = 2 x D^{UNI}

D^{LIM} = 8m
D^{UNI} = 4m
S= 5.13m
S = 8m/1.56 = 5.13m

D^{LIM} = 8m
D^{UNI} = 4m
S = 5.13m
D^{LIM} = 5.13 x 1.56 = 8m
D^{UNI} = 0.5 x 5.13 x 1.56 = 4m

FIGURE 11.20
Symmetric uncoupled line source design reference

11.5.2 Uncoupled line source (asymmetric)

Used for multiple mains, overheads, surrounds and various fills (frontfill, underbalcony, etc.).

ASYMMETRIC UNCOUPLED LINE SOURCE DESIGN PROCEDURE

1. The target shape is a bendable line of coverage with variable starting depths (the unity line).
2. Select the longest-range element (the highest-order element if different models are used).
3. Calculate the element coverage by its lateral width. This is the unity spacing between elements.
4. The lateral width lines of attenuated elements must be scaled down proportionally (compensated unity).
5. Place additional elements at the compensated unity spacing until the line length is filled.
6. The limit line is found at twice the various unity line ranges.

11.5.3 Uncoupled point source (symmetric)

Used for multiple mains and various fills (frontfill, underbalcony, sidefill, rearfill, etc.).

SYMMETRIC UNCOUPLED POINT SOURCE DESIGN PROCEDURE (FIG. 11.22)

1. The target shape is an arc of a given radius with matched starting depth (the unity line).
2. Select an element and place it in the central area of the arc.
3. Calculate the element coverage by its lateral width. This is the unity spacing between elements.
4. Space/splay additional elements so that the lateral width lines connect radially until the arc is filled.
5. The limit line range is highly variable. Low angular overlap extends the limit line.

11.5.4 Uncoupled point source (asymmetric)

Used for multiple mains and various fills (frontfill, underbalcony, sidefill, rearfill, etc.).

ASYMMETRIC UNCOUPLED POINT SOURCE DESIGN PROCEDURE

1. The target shape is an elliptical segment with a given radius with unmatched starting depth (unity line).
2. Select an element and place it in the deepest area of the arc.
3. Calculate the element coverage by its lateral width. This is the unity spacing between elements.
4. The lateral width lines of attenuated elements must be scaled down proportionally (compensated unity).
5. Space/splay additional elements so that the lateral width lines connect radially until the shape is filled.
6. The limit line range is highly variable. Low angular overlap extends the limit line.

11.5.5 Uncoupled point destination (symmetric)

Standard applications: infills, monitor sidefills.

SYMMETRIC UNCOUPLED POINT DESTINATION DESIGN PROCEDURE

1. Target shape resembles a pair of rectangles connected with a hinge, or a boomerang. For real.
2. Unity line is at half the distance to the location where the ONAX aim points cross.
3. Select an element for its ability to cover its local area as a soloist.
4. Calculate the element coverage by its lateral width.
5. The inner edges of the lateral shape must meet at center (unless covered by others such as frontfill).
6. The limit line is the point where the ONAX aim points cross.

11.5.6 Uncoupled point destination (asymmetric)

Standard applications: combinations of arrays (mains + delays, mains + frontfill, etc.).

A1 Find frontfill start: 2.5 m to 1st row
D_{UNITY} = 2.5m
D_{LIMIT} = 5.0m
FF
2.5m
2.5m

B1 Find U-Balc start: 3.5 m to 1st row
D_{UNITY} = 3.25m
D_{LIMIT} = 6.5m
UB
6.5m
3.25m

Note: The start of coverage is found by triangulation. In this case the element vertical coverage is 50°. The coverage starts at -25° from on axis (the vertical coverage edge).

A2 Calculate frontfill spacing: LAR x D_{UNITY}
FF
3.5m
Spacing for a 2.5m unity line with 90° elements:
1.4 x 2.5m = 3.5m
FF
3.5m
FF
2.5m
2.5m
Limit for a 2.5m unity line (0° splay):
2 x 2.5m = 5.0m
3.5m
FF
90° @ 0°
LAR = 1.4

B2 Calculate U-Balc spacing: LAR x D_{UNITY}
UB
4.0m
UB
4.0m
UB
3.25m
3.25m
4.0m
UB
80° @ 0°
LAR = 1.25

C Overhead speaker spacing: Unity line only
OH
OH
80° @ 0°
LAR = 1.25
3.2m
3.2m
4.0m
Spacing for a 2.5m unity line with 90° elements:
1.25 x 3.2m = 4.0m
Limit line is underground so we don't care!

FIGURE 11.21
Symmetric uncoupled line source design procedure. Example frontfill and underbalcony applications.

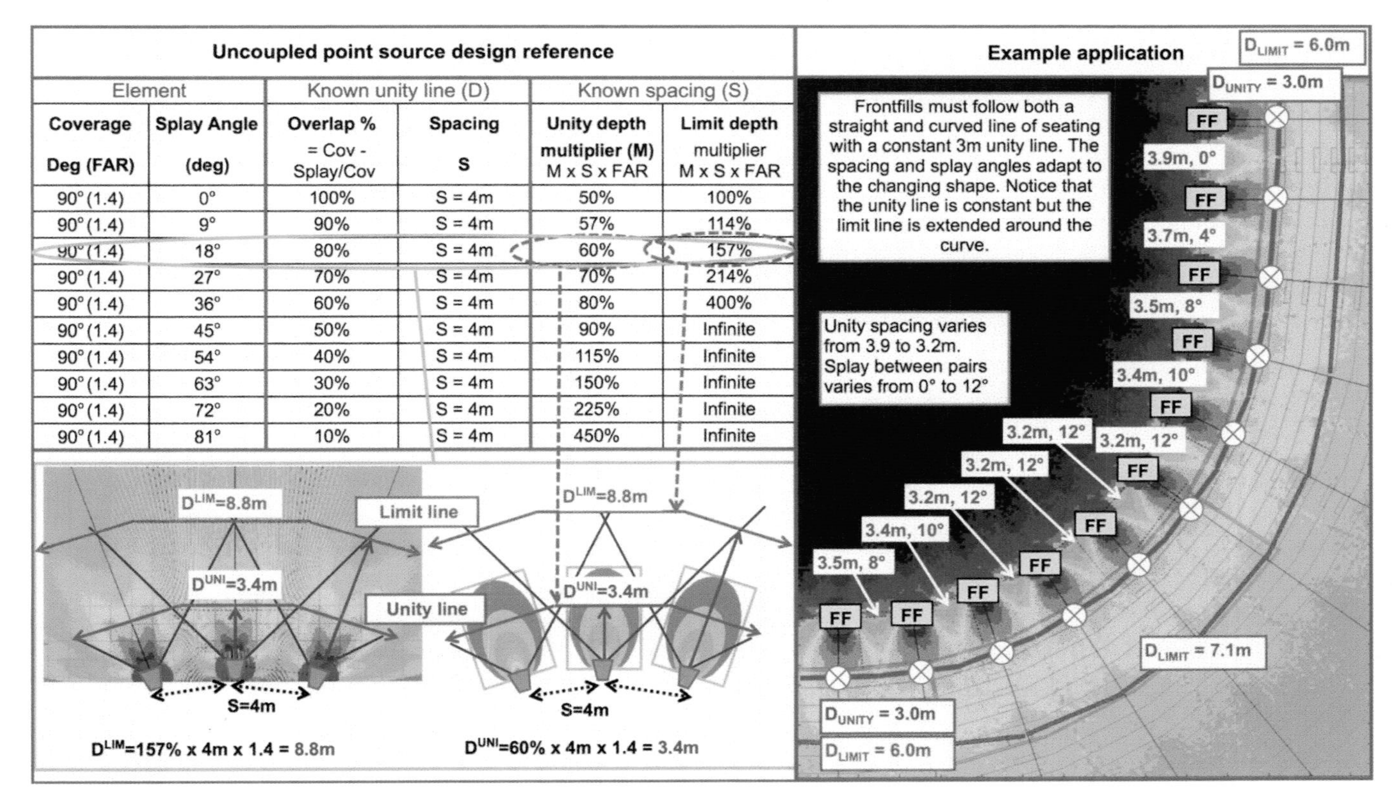

Uncoupled point source design reference

Element		Known unity line (D)		Known spacing (S)	
Coverage Deg (FAR)	Splay Angle (deg)	Overlap % = Cov - Splay/Cov	Spacing S	Unity depth multiplier (M) M x S x FAR	Limit depth multiplier M x S x FAR
90° (1.4)	0°	100%	S = 4m	50%	100%
90° (1.4)	9°	90%	S = 4m	57%	114%
90° (1.4)	18°	80%	S = 4m	60%	157%
90° (1.4)	27°	70%	S = 4m	70%	214%
90° (1.4)	36°	60%	S = 4m	80%	400%
90° (1.4)	45°	50%	S = 4m	90%	Infinite
90° (1.4)	54°	40%	S = 4m	115%	Infinite
90° (1.4)	63°	30%	S = 4m	150%	Infinite
90° (1.4)	72°	20%	S = 4m	225%	Infinite
90° (1.4)	81°	10%	S = 4m	450%	Infinite

FIGURE 11.22
Uncoupled point source design reference with example frontfill application

ASYMMETRIC UNCOUPLED POINT DESTINATION DESIGN PROCEDURE

1. The target is an overlap region of main and fills. Its shape is already pre-formed by the elements.
2. The unity line is XOVR main–fill. Its location depends on the fill level in the overlap region.
3. Aim the fill system to maximize forward coverage extension while remaining linked to the main.
4. The fill system must have enough coverage and power scale to match the main at XOVR.

11.6 MAIN SYSTEMS

Main systems range from multi-element stadium arrays to single ceiling speakers and every size and quantity between. If it's at the top of the food chain in the venue, it's the main.

11.6.1 Center main (C)

Center main systems are the principal element for mono systems and the voice channel for L/C/R systems. The coverage requirements are the same in both cases but the power scale changes (the mono main needs more power because it will have music *and* voice).

STANDARD FEATURES OF THE CENTER MAIN

- Elements: (H) first, second order, (V) mostly third or second order. First order only as a soloist.
- Array type: (H) solo, coupled point source, (V) asymmetric (composite) coupled point source.
- Coverage range: unlimited.
- Sonic image: centered, but usually high because of stage sightline limits and gain before feedback.
- Timing: arrives very late to the floor. High location requires anti-delay to sync to analog live sources.
- Range ratio: low(er). High location keeps the closest listeners fairly far away (compared with L/R).
- XOVR to frontfill: reduces image distortion in early rows and leakage onto stage.
- XOVR to sidefill: typically L/R deck speakers to fill coverage gaps in the near side areas.
- XOVR to underbalcony delays: High main position means we often lose sightline under the balcony.
- Form before function: Architects want to hide our unsightly speakers. Don't fence me in!
- Not popular: Nobody wants to mix on them. Specify only when the L/R option is unavailable.

The mono center main is your grandfather's array. There is little arguing that it's optimized for maximum vocal intelligibility. This may be the best choice for voice-only applications, but not necessarily. Intelligibility is a very high priority but feedback always comes before it. And we are in big trouble if feedback comes before we have enough gain. A high center cluster with less-than-stellar directivity poses a clear and present danger to acoustic gain. Our triage rating can evaluate the L/R signal degradation against the potential for higher gain and rental charges when the band arrives.

TYPICAL REASONS TO SPECIFY A CENTER MAIN

- Part of an L/C/R system.
- 360° seating or extreme wraparound balconies that leave no place for an L/R system.
- Heritage hall that won't allow L/R hang points.
- Someone has fond memories of Altec speakers.

The height of a center system is sometimes an open variable. In most cases our answer is "as low as possible," usually limited by follow spot sightlines or fashionistas. I suppose it's possible for a center cluster to be too low, but I don't think I've ever seen it. This leaves us to specify coverage, aim, splay and power scale. Refer to Fig. 11.23 as we step through the design process.

11.6.1.1 HORIZONTAL AIM

1. Multiply the compensated unity FAR by the minimum-variance combing zone ratio.
2. Aim it at center.

11.6.1.2 HORIZONTAL COVERAGE ANGLE

Use the single-speaker coverage method (section 11.3.1) for any single-wide horizontal configuration (e.g. stripe of line array speakers). Use the symmetric or asymmetric point source methods (section 11.4.1) for horizontal arrays.

Coverage is analyzed at mid-point depth. Variance between ONAX, OFFAX should be <6 dB. If variance is >6 dB then specify a wider main or supplement with sidefill.

11.6.1.3 VERTICAL AIM AND COVERAGE

Use the single-speaker (section 11.3.3), asymmetric point source (section 11.4.3) or composite point source method (section 11.4.4).

11.6.1.4 HORIZONTAL COVERAGE WIDTH

We will need to calculate the coverage width on the floor under the array to see how many seats must be covered by the sidefill deck system. Use the lateral coverage width method (section 11.3.5) from the bottom of the array to the frontfill/main XOVR.

11.6.2 Left/right mains (L/R)

Room shape plays a small part in the choice of L/R but a big part in who hears both of them. The basic rectangle is the most friendly and the wide fan, the least. Wide fans often require sidefill, centerfill and/or infills to extend coverage and close gaps.

STANDARD FEATURES OF THE LEFT/RIGHT MAINS

- Elements, array type and coverage range: Same as the mono main.
- Sonic image: Localizes strongly to each side due to precedence effect (except for a small center area).
- Timing: Can arrive early to the floor if hung low. Add delay to sync to analog live sources.
- Range ratio: variable. Can be high if mains are low or low(er) if they are hung high(er). Really.
- XOVR to frontfill: Same as center main above.
- XOVR to sidefill: Can be flown (coupled) or deck speakers (uncoupled), if needed to extend coverage.
- XOVR to centerfill: Flown above center stage to fill the gap between L/R and frontfill.
- XOVR to underbalcony delays: Low main position can ease the need for underbalcony systems.
- Popular: Nobody gets fired because they had the crazy idea of spec'ing L/R.

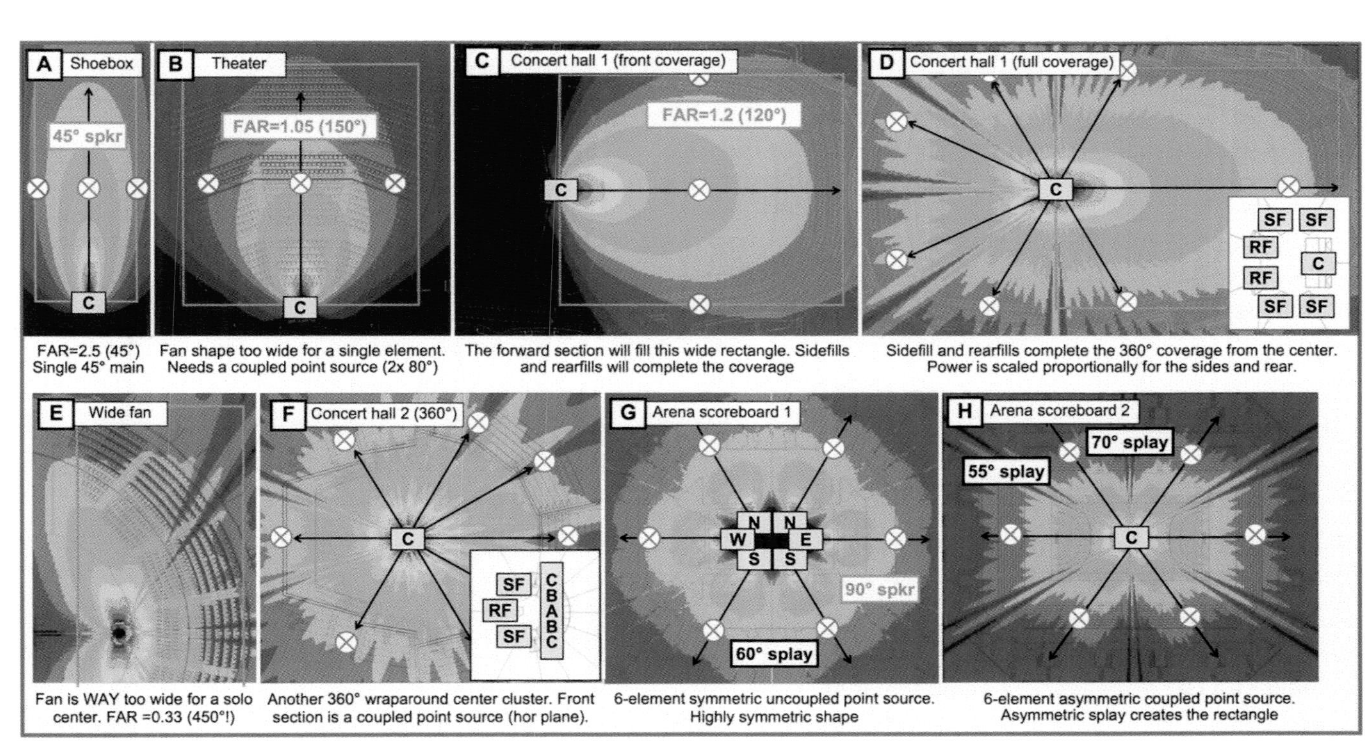

FIGURE 11.23
Center main system design examples for the horizontal plane

11.6.2.1 MAINS HEIGHT

Left/right mains have a wider range of height options than the center. Proscenium stages present the opportunity to move them lower. Therefore this is the time to discuss how high to hang the mains.

The mains height has a range of design implications. There are tradeoffs at every turn. Low mains have advantages in sonic image, but potentially high range ratios that make it loud in front. High mains have image issues and can be late to the party, but have a big advantage in the level and spectral variance categories. Gain before feedback can go either way. In the ancient days of sound reinforcement our choices were image-killing center clusters or ear-killing L/R mains on the deck. Modern systems can pick and choose the optimal compromise altitude.

The room shape plays a part, of course. Finding the middle ground in this case means finding the middle height. We seek to spread the vertical coverage evenly. It's not as simple as shooting from the middle height because the upper-level seats are very often farther than the lower-level ones. This leads us upward, which turns out to be a safer and more reliable position to go for the long throw.

Let's compare the two extremes before we work our way to the middle: speaker at the height of the top row versus speaker on the deck. Listeners below versus above the speakers. The high main option has lower-range ratio, better defense against ceiling reflections, worse imaging, is late on the floor and risks spilling onto the stage. The low main (no, it's not for lunch) has to kill the floor to get to the back, seriously risks lighting up the ceiling (if it can actually cover up there), but has great imaging.

Smiles and frowns

A speaker with a 40° vertical pattern can give or take a few degrees of its pattern without getting emotional. The same cannot be said about today's super-directional proportional beamwidth speakers. If we change them a few degrees they get happy or sad. Their patterns can either frown or smile, depending on their orientation to the shape. This is 3-D geometry so let's take it step-by-step (Fig. 11.24).

The vertical ONAX aim of a speaker is flat across the horizontal plane (unless it's a wacko horn). Put a speaker on a turntable and spin it, and ONAX circles the room at the speaker height (if the speaker is not angled up or down). That's a big "if" there. An upward-angled speaker will circle the room above the speaker height (the easy part) but it may be at different heights all over the room (the tricky part). A speaker in the center of a circular room draws 360° at the same height for a given tilt. But the height is going to vary if the speaker is not centered, and/or the walls are different distances from it. The aim angle is constant but the height above the speaker accrues with distance (e.g. 1 m rise for every 10 m distance). An up-aimed speaker that hits seats at 10 and 12 m ranges arrives at heights of 1 and 1.2 m. How does that happen? The left main is closer to the leftmost seat in the last row than the middle seat. The seats are the same height above the speaker but not the same *angle* above the speaker. We now see that ONAX is a constant angle but not a constant elevation. Smile. Increased up-tilt and and/or horizontal range ratio make the smile even bigger. Down-tilt makes it frown and a flat orientation leaves a blank stare.

This has substantial practical implications for speaker aiming, whether the intent is coverage or gapping (e.g. avoiding the balcony). We might not hit what we aim for, and we might not miss what we hope to. A center speaker can be up-tilted into a fan-shaped room (or balcony) and maintain a constant ONAX on the listener plane. ONAX rises if the back wall flattens. L/R mains are virtually assured of smiling when up-tilted, unless the back wall is convex (i.e. deeper on the sides than the center). Yes. That would be crazy.

Balcony avoidance strategies have the same issues; it's just that we are aiming XOVR (the gap) at the balcony. Our balcony avoidance becomes audience avoidance if the gap frowns or smiles.

Speaker height effects (smiles and frowns)

- Flat: Speaker is on the listener plane, regardless of depth.
- Smile: Speaker is below the listener plane. Listeners are at different depths (from speaker point-of-view).

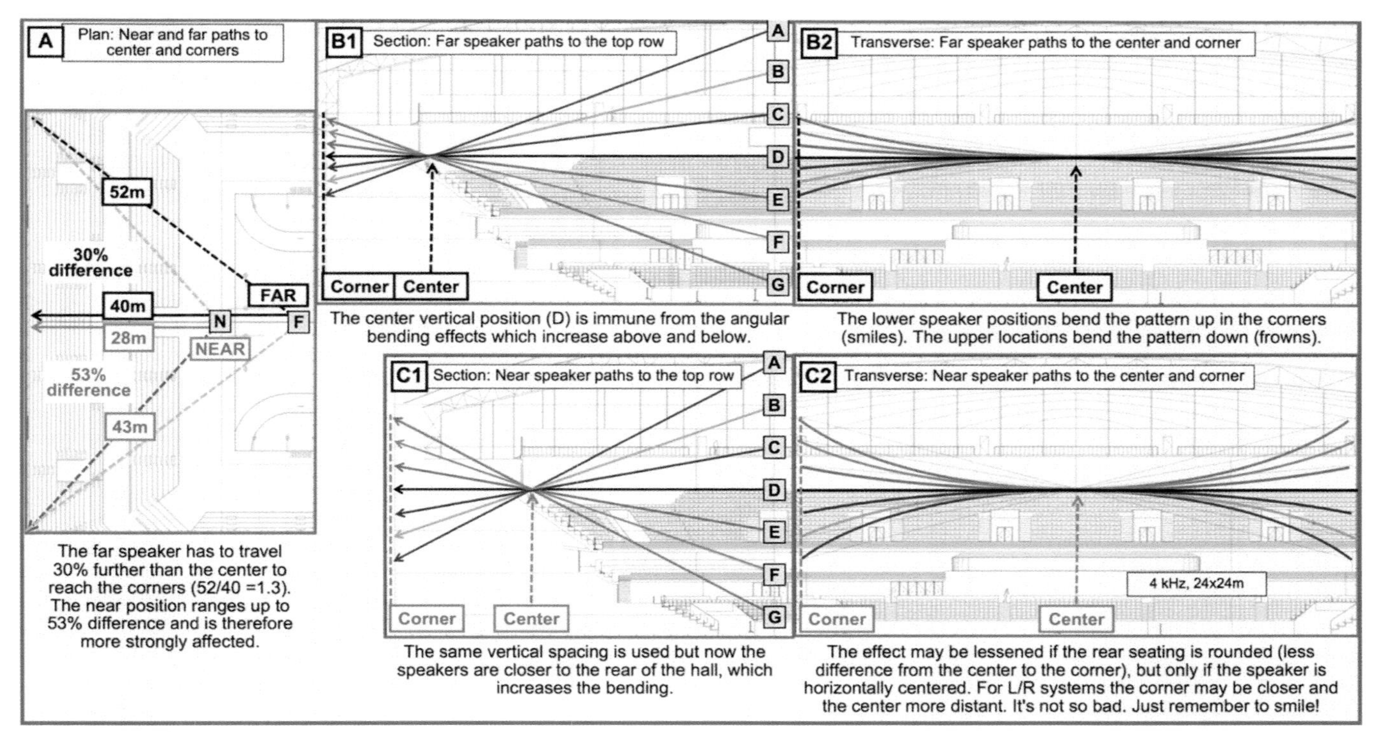

FIGURE 11.24
Smiles and frowns. Pattern bending due to differences in path length.

Our designs need to see this coming. It's far too easy to specify surgical-level shaping on a 2-D section view that will end up smiling its way to places unknown in the venue. We can't stop it, but sometimes we can minimize it by limiting the tilt on speakers throwing the longest distances (with the narrowest angles).

Mains height has the strongest effect on the bending errors. The smile/frown geometry is symmetric, but the risks are not. A smile above the last row poses more danger than some frowning down in front. The rear is where a degree or two matters due to the long throw and narrow speakers. Minimizing the mains up-tilt reduces the risk of over- or under-coverage. This potential benefit can be factored into cluster height decisions. Coupled main systems that want to gap around the balcony should seek a height that places the gap in the array around the balcony height. This ensures that the gap stays on the balcony. Gapping will be less effective (and potentially degrading) if the gap angle has to tilt up or down to the target.

Prove it yourself:

1. Take a laser pointer and aim it flat onto a wall. Swivel side to side. The pattern is a flat line.
2. Return to center. Aim up at an identifiable line, such as the wall/ceiling joint. Swivel again, being careful to maintain the same angle on the laser. Notice that light does not stay on the joint as you swivel horizontally. It smiles upward. It's not you, dear. It's geometry.
3. Next aim to a line below you and repeat the process. You made it frown now, you meanie.

11.6.2.2 HORIZONTAL AIM

Left covers left and right covers right. We have previously discussed the sad reality of how well stereo works in concert applications (section 5.4). Therefore, we will bisect the room and aim each main through its respective middle-middle. This approach creates a balanced ratio of under and over-coverage from front to back. In this case there are approximately equal amounts of actual wall reflection and virtual wall reflection (the leakage from the other side). The sources of horizontal ripple variance are balanced.

There is a justifiable fear of excess horizontal coverage splashing onto the side walls. This can lead designers of L/R systems to turn inward to avoid them. TANSTAAFL shows us that we just added certain inward leakage in exchange for (potential) outward leakage, i.e. the wall reflection. We must ask the question: Which is worse? It's a real reflection against a faux-flection (the late arrival from the opposite side). Let's tally things up. There is HF air loss in all paths so that evens out. Is the side wall absorptive at all? Will we hear its reflection directly or have to wait until it comes off the back wall? We have little to fear if the HF is absorbed and we don't hear it until it's come off the back wall. The "wall" for the faux-flection is the centerline of the room. It's a perfect mirror, absorbing nothing. The more we turn L and R inward, the more we are sending on-axis HF into the mirror. As I said before: potential from the wall versus certainty from the opposite side? Neither wall is going away. The middle-middle approach minimizes the horizontal level/spectral variance and balances the horizontal sources of ripple variance (Fig. 11.25).

11.6.2.3 HORIZONTAL COVERAGE ANGLE

Coverage is analyzed on the bisected space (as above). Left: Use the single speaker coverage method (11.3.1) for a single stripe of line array speakers (first-order). Use the symmetric or asymmetric point source methods (11.4.1) for horizontal arrays. Coverage is analyzed at mid-point depth. Variance between OFFAX L, ONAX L and XOVR LR should be <6 dB. If greater than 6 dB, change to wider main or supplement with side and/or centerfill.

11.6.2.4 VERTICAL AIM

Use the same methods as center main above. Expect higher range ratio if the L/R mains are lower than the center.

11.6.2.5 VERTICAL COVERAGE

Use the same methods as for center main.

11.6.2.6 HORIZONTAL COVERAGE WIDTH

The lateral coverage width at the vertical bottom of the main (VBOT) must be evaluated (11.3.5) to determine whether centerfills and/or sidefills are required to cover nearby seating beyond the frontfill. Add centerfill (or sidefill) if coverage width does not reach center (or side).

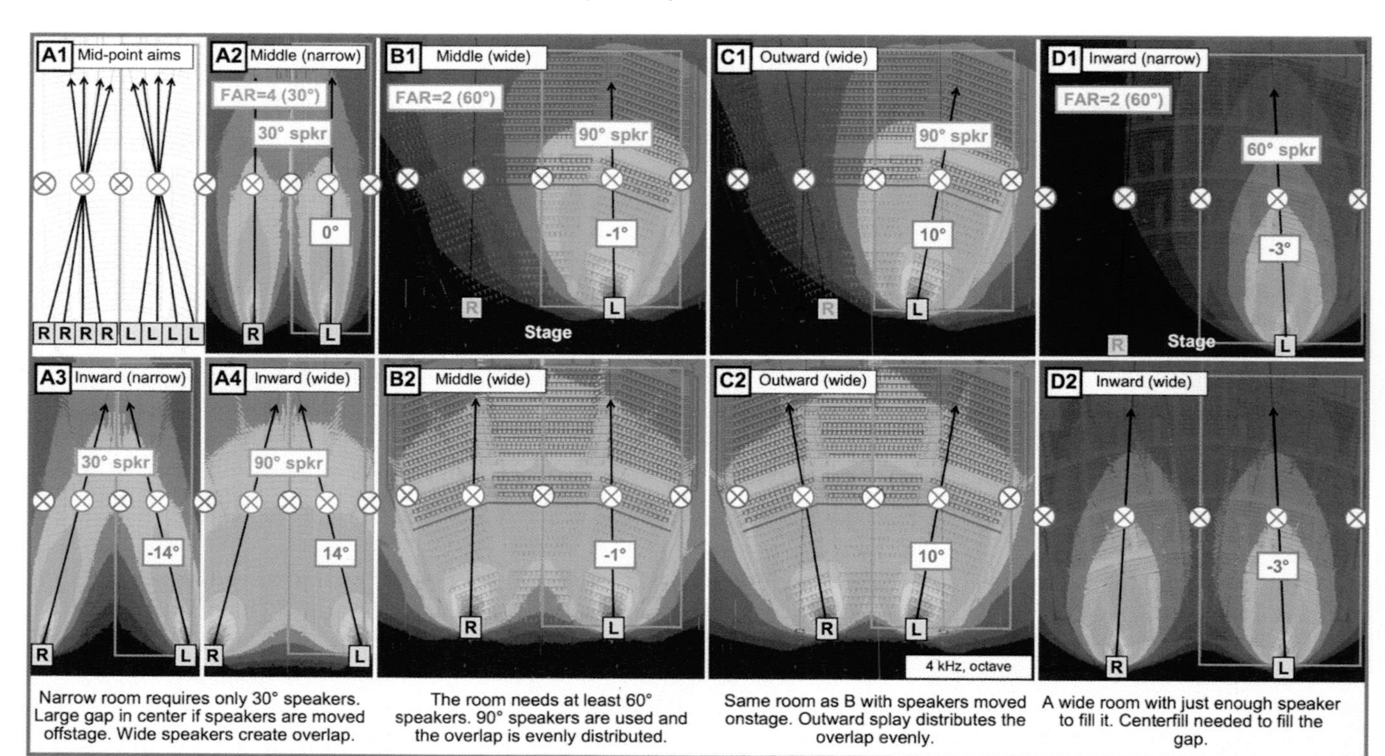

FIGURE 11.25
Left/right (L/R) main system design examples for the horizontal plane

11.6.3 Left/center/right mains (L/C/R)

L/C/R system design is exactly what the name implies: left, center and right. Combine the previous design specifications to make a multichannel system, presumably separating music and voice. The common mistake in L/C/R designs is a center channel with incomplete coverage (such as center not wide enough or no deck sidefills). Engineers then add voice to the L/R system, which creates massive ripple variance and things go downhill from there. No worries. Louder will fix it!

Mixing on multichannel systems is orders of magnitude more complicated and L/C/R is no exception. Some channels are mixed electrically in the console and others acoustically in the room, and still others both. Stage sources can add yet another layer. The acoustical interaction of correlated signal is complicated enough already but at least it's stable and predictable. Multichannel systems have spatial interactions that change on a channel-by-channel, moment-by-moment basis over the room. Engineers/designers need time to walk the hall and explore the spatial matrix. No mix position is a perfect representative of the whole room for any system, even a monaural one, but the translation primarily involves spectral variance. The mix position for a multichannel system must represent both the channel balance as well as their respective spectrums. If center is down at the mix position the vocals are going to be hot elsewhere. Multichannel is practical and advantageous for ongoing long-term productions, where tech and rehearsal periods are sufficient to dial it in. Visiting pop music engineers don't have time to adapt to such spatial unknowns, particularly for the critical voice transmission. They expect to mix *in* the console, i.e. L/R.

L/C/R systems present a terrible temptation to L/R-minded touring engineers. It's difficult for any engineer to resist the opportunity for more power, so they may be tempted to send an L/R summation into the center system. Massive ripple variance is the result.

Room shape plays a big part in the viability of L/C/R. The rectangle is the most friendly and the wide fan the least. Fan-shaped rooms often require a doublewide center array, driving up costs and leaving only a tiny minority of seats hearing L/C and R. It's time to rethink your L/C/R dreams if you find all main systems needing radial extension. Such systems are too spread apart to mix well acoustically. Let's place the violins offstage left, the singer in the center catwalk and brass offstage right. Now make it sound good.

My personal policy is that L/C/R systems are for consenting adults only, i.e. venues or productions that will be operated by the same personnel each night. Musical theater, dedicated showrooms (like Vegas), theme parks and houses of worship are prime candidates. The local roadhouse, not (Fig. 11.26).

Standard features of the L/C/R mains:

- Multichannel strategy: L/R for music and center for voice is the most common.
- Requires familiarity: Mix is both electrical and acoustic, adding spatial complications.
- Full coverage center required: Cover the whole room or cover your ears.

11.6.4 Multi-mains

We can expand to three mains when left and right is simply not enough coverage. They might be labeled left, center and right, but it's multi-mains (mono) not L/C/R (multichannel). The difference is function, not form. The three channels cover their respective areas as equals, with crossover transitions in between. This is often found in fan-shaped houses of worship with a stage that pushes deep into the house. There is no reason to stop at three. I've done parade routes with over one hundred mains.

Multi-mains are the macro version of the uncoupled line, or point source. The elements may be singles or coupled point source arrays. It's easy to visualize numerous examples of multi-mains in the horizontal plane, but there are a few applications in the vertical plane. They all involve balconies, which can split the mains into upper and lower systems of comparable size. The upper and lower element can be singles or more in either plane and combine as an uncoupled line or point source in the vertical plane. Yes, it's possible to have a symmetric uncoupled line source of asymmetric coupled point sources (Fig. 11.27).

STANDARD FEATURES OF MULTI-MAINS

- Mono or repeating L/R strategy: Either way the systems are designed around the mono model.
- Range limited: Uncoupled macro elements means coverage is optimized between the unity and limit lines.

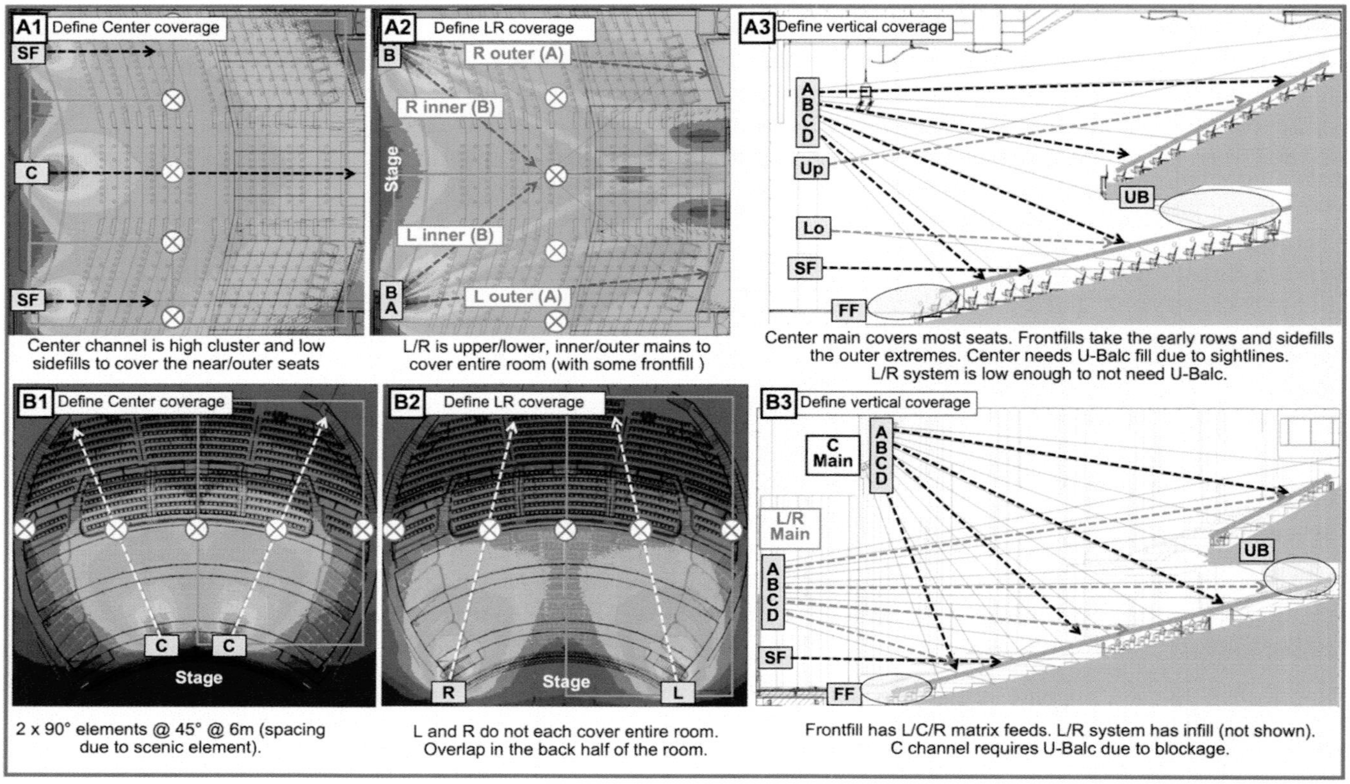

A1 Define Center coverage
SF
C
SF
Center channel is high cluster and low sidefills to cover the near/outer seats

A2 Define LR coverage
B
R outer (A)
R inner (B)
Stage
L inner (B)
B
A
L outer (A)
L/R is upper/lower, inner/outer mains to cover entire room (with some frontfill)

A3 Define vertical coverage
A
B
C
D
Up
UB
Lo
SF
FF
Center main covers most seats. Frontfills take the early rows and sidefills the outer extremes. Center needs U-Balc fill due to sightlines. L/R system is low enough to not need U-Balc.

B1 Define Center coverage
C
C
Stage
2 x 90° elements @ 45° @ 6m (spacing due to scenic element).

B2 Define LR coverage
R
Stage
L
L and R do not each cover entire room. Overlap in the back half of the room.

B3 Define vertical coverage
C Main
A
B
C
D
L/R Main
A
B
C
D
UB
SF
FF
Frontfill has L/C/R matrix feeds. L/R system has infill (not shown). C channel requires U-Balc due to blockage.

FIGURE 11.26
Left/center/right (L/C/R) main system design examples (horizontal and vertical planes)

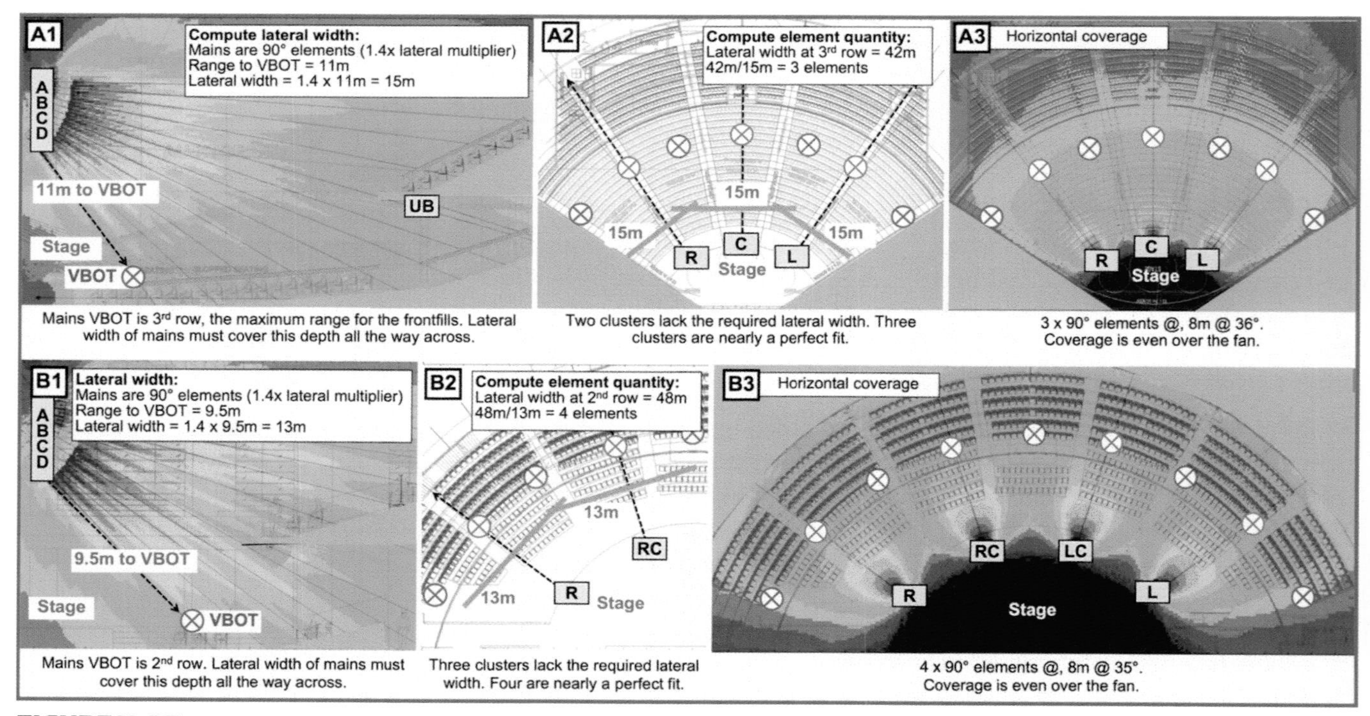

FIGURE 11.27
Multi-main design examples focusing on the required quantity of main clusters

11.6.5 Upper/lower mains (UpLo)

Multi-mains can go vertical as well, although it's unlikely that we'll expand beyond two elements. Vertical expansion is also driven by room shape, in this case a very specific room shape: the balcony. Everybody loves breaking the mains into left and right, but upper and lower, not so much. This section details the tradeoffs involved in battling the balcony.

We are talking about L/R mains. There's no such thing as a lower center. Center mains solve balcony coverage with L/R sidefills near the deck and lots of delays. The question for the left main is when to surrender to uncoupling. We all want the band to stay together, but sometimes the members need to go their separate ways.

11.6.5.1 SINGLE AND DOUBLE SLOPES

The typical listener plane is a slope rising with distance. We solve this shape with the asymmetric point source. The simplest listener plane is a constant slope, a consistent rise over distance. We commonly encounter more complexity, with a steeper slope in the rear than front. The coupled point source can adapt to this shape by complementary asymmetry, even with very substantial differences in rate of rise (Fig. 11.28).

Balconies add a second listener plane, which is where the trouble starts. We are now double-sloped, a shape that calls out for uncoupling. There are two primary strategies: Treat the shape as a single complex slope (and stay coupled) or treat them as distinct slopes and solve them separately (uncoupled).

We can keep the main array coupled by plowing a line of best fit through the balcony details. The downsides are level variance (balcony front will be louder) and ripple variance (balcony front reflection). The upside is an extended frequency range in the coupling zone by keeping all sources close. A deep balcony increases the level variance. A tall, reflective balcony front increases the ripple. That's what we're up against. How deep is too deep? How do we know when the front will give us trouble? We'll be able to wrap our heads around balcony acoustics once we see how acoustics wraps itself around balconies.

The double slope has four coverage target milestones: $VTOP_1$, $VBOT_1$, $VTOP_2$ and $VBOT_2$. Each has a unique angle and range relative to the mains, and each pair has a unique angular spread and range ratio relative to each other. It's the inner pair's relationship ($VBOT_1$–$VTOP_2$) that has the twist. Let's plug in some numbers and see the results (Fig. 11.28).

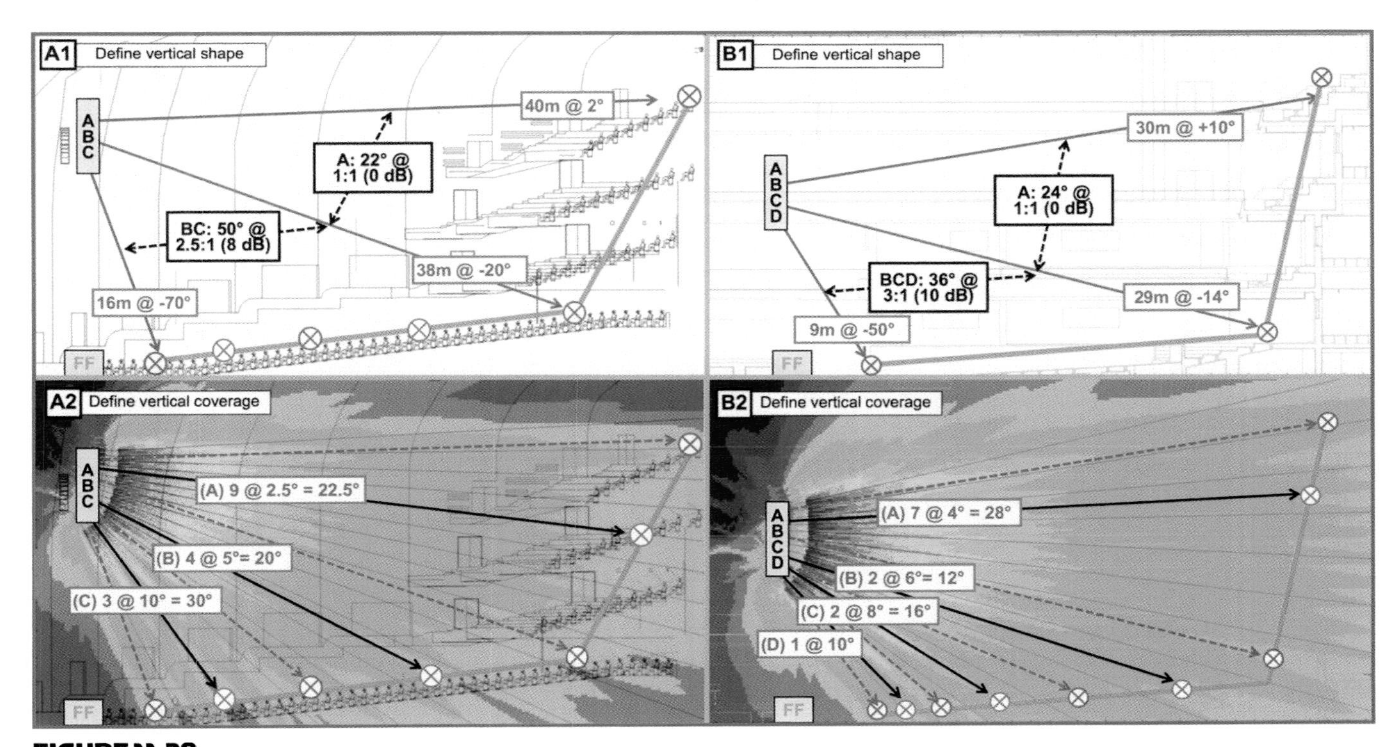

FIGURE 11.28

Example applications with double-sloped vertical shapes. Multiple shallow balconies present a nearly symmetric (1:1) range ratio whereas the floor is highly asymmetric (>2:1). Coverage can be divided between the two slopes.

We start with a pair of similar slopes, stacked directly on top of each other (Fig. 11.29 (B)). Each has a 20° spread and 2:1 range ratio. Our speaker is in the middle so it covers from $VTOP_1$ (+20°) to $VBOT_2$ (-20°), with a 2.8:1 range ratio (9 dB). In between are $VBOT_1$ (+0°) to $VTOP_2$ (-0°), which has a 1.4:1 range ratio. We know how to solve a 6 dB range ratio spread over a 40° angular spread. How do we solve a 3 dB change that happens in a 0° spread? OK. It can't be 0° because the balcony has to be thick enough to hold people, but it can be very, very small.

11.6.5.2 RETURN RATIO

Listen up main. Stick together. Here is our mission: Go deep then gradually come closer for 20° and then instantly go deep again and repeat. If you are not convinced yet that this is mission impossible then add range ratio until you surrender. A wider balcony front gives more angle to work with, but with friends like this, who needs enemies.

Let's make a single modification to the shape and do the exercise again (Fig. 11.29 (A)). Angle the lower floor upwards so its top aligns with the upper floor's bottom. What's different? $VBOT_1$ and $VTOP_2$ are still both at 0° but they now have a 1:1 range ratio. There is no longer a zig-zag in the middle. We would surely cover this with a single main. It's also not a balcony any more but it reveals the mechanism, the return ratio, the primary indicator for splitting the array. Every inch we slide the upper floor forward increases the discontinuity between $VBOT_1$ and $VTOP_2$. Such sharp turns in coverage require angular isolation and we don't have it. Return ratio (in dB) quantifies the level difference the balcony forces us to overcome. We can saw a line of best fit through a shallow balcony with a small return ratio and keep the array together. Return ratios of 6 dB or more cannot be smoothed over (Fig. 11.29 (C)).

11.6.5.3 SECONDARY OPTIONS FOR UPPER/LOWER MAINS

There are still options short of breaking up. We can outsource coverage to others, specifically the underbalcony area, which can be covered by delays. The area covered by the delay is taken off the custody requirements of the main. VTOP2 moves closer, reducing the range ratio and opening up angle, a double bonus. If the delays can bring the return ratio in bounds we may be able to tough it out as a single main (Fig. 11.29 (D)).

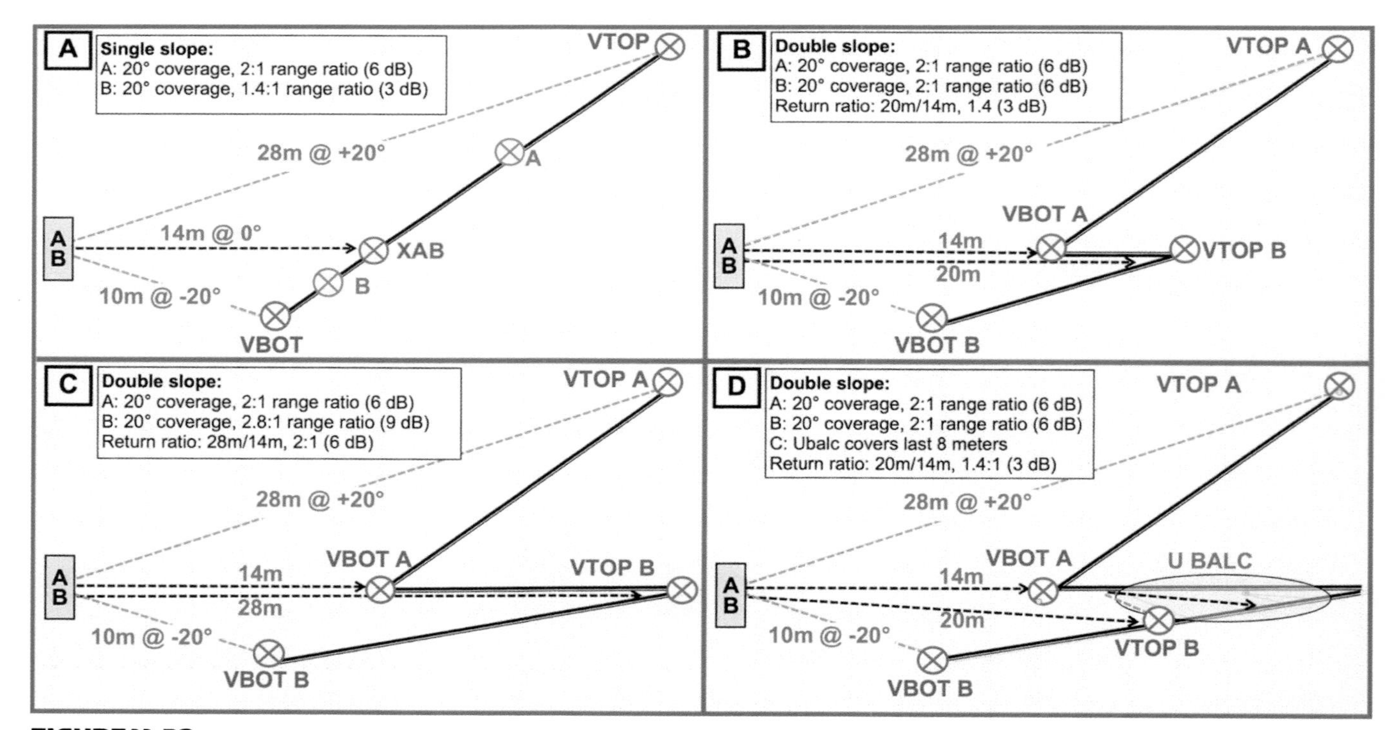

FIGURE 11.29

Upper/lower decision examples. In all cases there is 20° of coverage required above and below the speaker location. (A) Single main can cover the continuous slope. (B) The 3 dB return ratio is low enough that splitting is not required. (C) The 6 dB return ratio indicates splitting is best. (D) Underbalcony speakers reduce the return ratio to 3 dB (no splitting required).

The mains height also plays a role. We have looked at mains in the middle. Going upward reduces the angular spread between $VBOT_1$ and $VTOP_2$ (as if it wasn't small enough already). Going higher leads to occultation (the blocking of the sightline to the speaker) underneath, which reduces return ratio by coverage reduction. Delays have moved from optional to mandatory. Occultation seriously downgrades the underbalcony area and should not be considered fair trade for return ratio gains. Moving the mains under the balcony line ensure sightlines to the back and opens up the angular spread (the mains can see the underbalcony ceiling now). Return ratio shows no improvement and balcony coverage will become more challenging. The upper-level slope is flattening (from the mains POV). The upper level needs more severe shaping due to reduced angle and rising range ratio. Such severe asymmetry is difficult with a single slope. Asking the mains to do that upstairs *and* on the floor is a very tall order.

There is one more height-related consideration: the coverage pattern bending we previously termed "smiles and frowns" (recall Fig. 11.24). This is relevant to high and low orientation to balcony fronts (and our attempts to avoid them). Only a centered main can be precisely steered to overcome a high return ratio. Upper or lower positions cause the coverage transition to appear at different heights across the room.

Here is the consolation for this exercise. All is not lost when we raise the white flag and divide the array into upper and lower sections. Instead of trying to make one array do something it hates to do (double sloping) we get two arrays being what they love to be: asymmetric coupled point sources drawing a single slope. The ripple variance in the low end might be a very small price for this payoff.

The design process is basically a two-layer cake. Upper and lower are separately analyzed and comparably power scaled. In some cases it's possible to move the upper main deeper into the house (because it starts at the balcony). This is free money in terms of power and signal/noise ratio, as long as the image is not compromised (Fig. 11.30).

11.6.5.4 BALCONY FRONTS

There's an old expression to sum up the sound engineer's perspective on this: The only good balcony front is a dead balcony front. Balcophobia is a serious malady in the sound community. Designers go to great lengths to avoid what acousticians go to great lengths to install: lively balcony fronts. We've all been burned by this, so it's worth a few paragraphs to put things in proper perspective. Some balconies are poisonous whereas others are harmless, but many of us run from both kinds.

Bad balconies are tall, lively, featureless (single angle, not diffuse). Glass and steel: bad. They have bad angles with respect to our speakers, sending sound back on stage or onto paying patrons. Worst is the flat, curved balcony in a fan-shaped room with the stage as its focal point. Been there, done that.

Good balconies are short, dead, diffuse, filled with lighting gear, multi-angled and inclined to send our sound harmlessly into the open air such as the ceiling. Size up the live surfaces of your local balcony. It's harmless below 500 Hz when it's less than 0.5 m tall. The key is not to hurt the design over avoiding something that won't hurt you.

Standard features of the upper/lower mains

- Application: separate coverage of over and underbalcony areas. Horizontal design is same as L/R.
- Elements: (V) coupled point source (large), second order (small).
- Array type: (V) uncoupled line or point source.
- Coverage range: $VTOP_1$ to $ONAX_1$ to $VBOT_1$ to $VTOP_2$ to $ONAX_2$ to $VBOT_2$.
- Sonic image: can be generally low for both levels.

11.6.6 Relay mains

Relay mains are satellite power boosters for forward extension. They are powerful enough to take over from the mains, hence the term relay. There is no limit to the quantity or scale of relay mains. The limiting factor is the amount of destructive interference between all the various mains and subsystems.

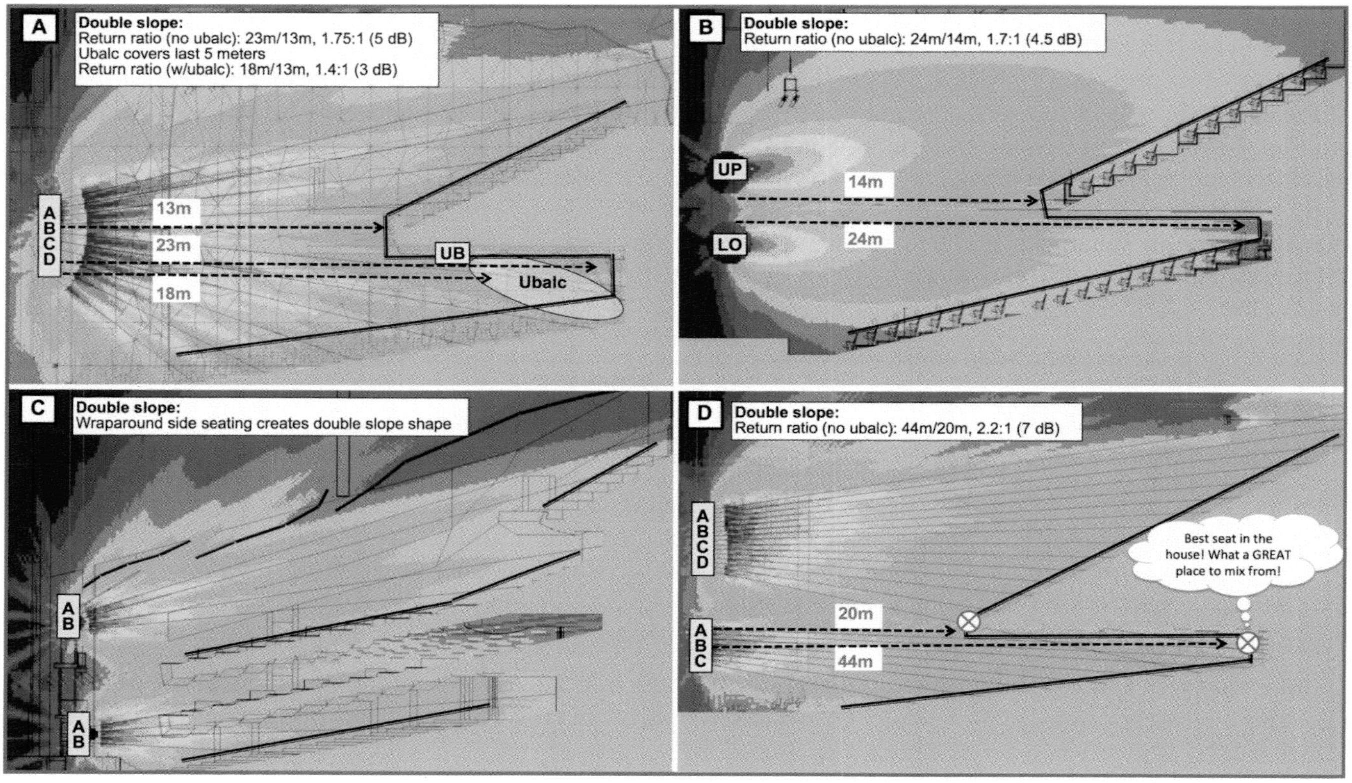

FIGURE 11.30
Upper/lower-main design examples. (A) U-balc reduces the return ratio enough to use a single main. (B–D) Upper lower mains are used.

The macro shape of relay mains is the same as delay fills (section 11.7.2), an uncoupled point source or line source in the horizontal plane. Maximum range is the top priority of relay mains, so the uncoupled point source is preferred for its limit line extension.

Relay mains will ideally reconcile back to the stage mains as "spokes on a wheel." Additional sets of relay towers will continue forward as spokes with the stage mains as the common hub.

Backflow is the limiting issue. Delay offsets between the stage mains and relay mains rise at the maximum rate behind the relay towers. There is no better application for cardioid low-frequency arrays and all manner of beamwidth control.

ELEMENTS

Height and directional control in both planes are required to achieve the maximum B4MBNW4U ratio. Not familiar with this metric? It's the "better for me but not worse for you" index. The AES standards committee is working on the details but I think it speaks for itself. Height gives us a running start by keeping the closest listeners (and fastest doubling distances) out of the coverage. Vertical control extends the coverage start and horizontal control extends the range to the stop. It's easy to see that more is better for height and vertical control, within reason. We need more towers if we narrow the horizontal, whereas overly wide towers drive up the annoyance factor with overlapping arrivals that are 10s or 100s of milliseconds late. The point of diminishing returns is found by 90°, which means a typical single element is suitable, but doublewide arrays are trouble. We have only one delay setting available for these satellites, which means we are destined to fall out of time in the horizontal plane. The wider we go, the sooner the delay offsets tip the balance in favor of annoyance.

The rearward path has the fastest rate of time offset change. Anything we can do to dampen the back lobe is a plus. Acoustic absorption, cardioid steering and active noise cancelling are all options worth exploring.

FORWARD SPACING

Relays are power-boosting systems. The stage mains start to sag and Tower 1 pumps it up. The cycle repeats at Tower 2 but there's a twist. The later links don't have to go it alone. They are riding on the backs of the earlier mains, which have gone a longer way. This cuts both ways. Signal from the stage main has fallen the most, but its rate of loss is the slowest. This is a real-world application of staggered starts and their different doubling distances (Fig. 9.18). We don't seek unity line in this application. We are boosting. The residual level of the earlier systems allows us to taper the power scaling for relays as we move outward. We won't bring it down to unity scaling (like our fill systems) but we won't make every tower a clone of the stage. Dropping 3–6 dB per relay link takes advantage of the piggyback opportunity and reduces the risk of backflow.

It is a ratio proportional scaling, of course. Each tower is scaled down in level and distance between in proportional measure. For example, a four-link relay chain downscaling in 3 dB increments can be spaced at 70, 49 and 35 meters (Fig. 11.31). The 6 dB alternative would be 70, 35 and 18 meters. The former has more boost and range, and the latter has less backflow.

This is one area of system design where we must get specific about scale. We could hypothetically use big relays in big spaces and small relays in small spaces, but in practice we avoid relays in small space (by using big systems). In really big spaces, especially outdoors, the relay is mandatory.

A large line array of Class 4 power-scaled elements should be viable up to 75 or 100 meters. We are at the mercy of air loss and weather. The weather is a liability in two ways: acoustical degradation and the safety limits of our speaker tower height. Sadly the latter issue has had devastating consequences when limits are exceeded.

The relay towers (or crane lifts) will typically have lower height and weight limits than the mains. This is fine because the relays will be power scaled under the mains anyway (in combined quantity if not individual elements).

Standard features of the relay mains

- Application: Power boosting forward extension.
- Elements: (H) first, second order, (V) coupled point source (large), second order (small).
- Array type: (H) uncoupled line or point source, (V) uncoupled point destination with stage mains.
- Coverage range: unity line (ONAX to XOVR to ONAX) to limit line (triple speaker coverage or more).
- Sonic image: You're a giant tower in plain view. Just add 10 msec of extra delay and you'll disappear.

11.6.7 Overhead mains (ceiling speakers)

It seems ridiculous to call a ceiling speaker a main system, but tell that to engineers in hotels, convention centers, theme parks, restaurants and retail. They are a bigger market than concerts, by the way.

Overhead mains are typically a matrix of uncoupled line sources. A hallway might use a single line, whereas an open room needs a two-dimensional matrix of line sources. What should we call it, an uncoupled waffle source? A room with flat floor and ceiling can be designed with equally spaced and power-scaled elements. The spacing is found by the uncoupled line source calculator (Fig. 11.20) for sitting (or standing) head height.

It is mandatory to use the lateral-width shape for overhead system spacing. The listener plane is a flat line of coverage, not radial unless your audience is in a bowl. The lateral width is derated from the radial width (e.g. a 90° radial coverage yields 70° of lateral). We can use the derated lateral values and draw coverage lines from the speakers to the listener plane. Just stack the triangles side-by-side until the space is full.

The spacing and power scaling must adapt to the depth changes in room with sloping floors or ceilings, balconies, etc. The example system shown in Fig. 11.32 is the real-world application of the asymmetric uncoupled line source with unity-compensated spacing shown back in Fig. 9.19.

Standard features of overhead mains

- Element: first order only except for very high ceilings.
- Array: uncoupled line source.
- Range: listener head height.

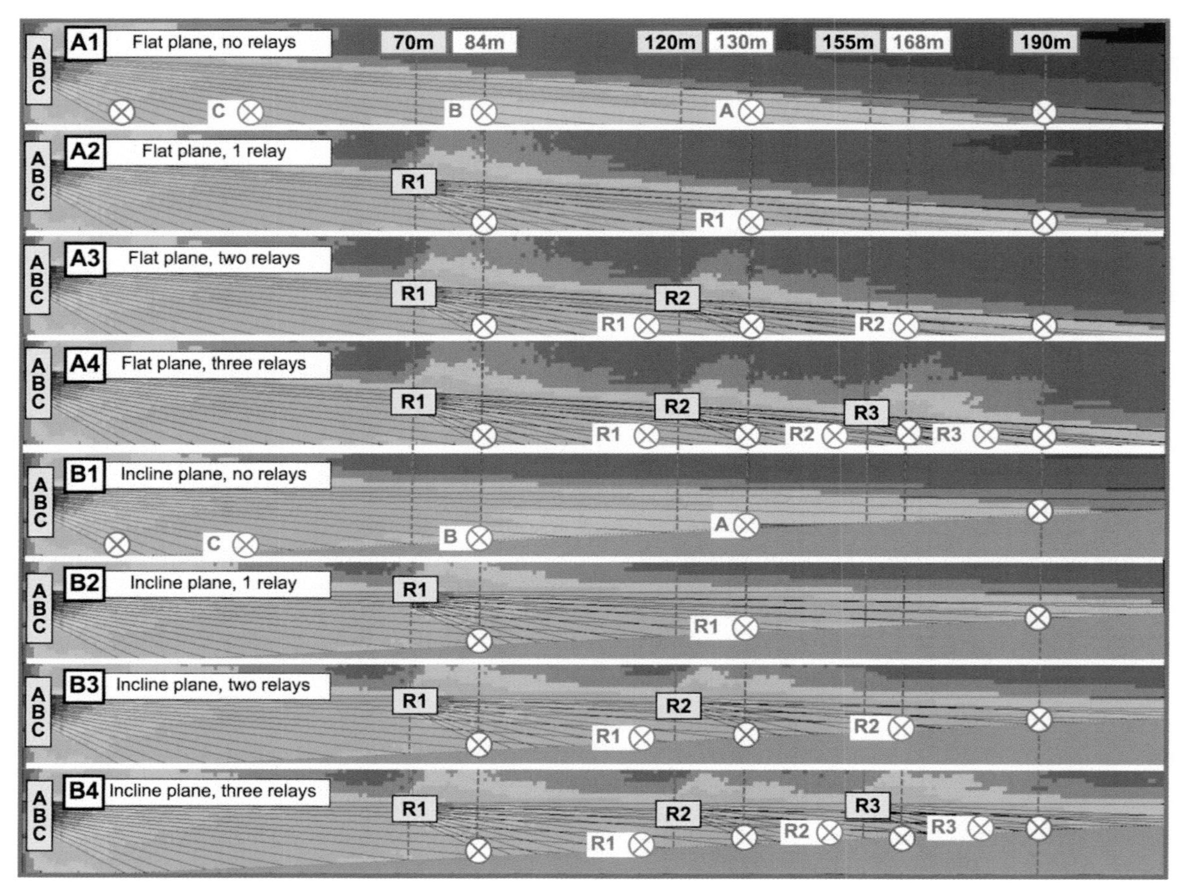

FIGURE 11.31
Relay mains design examples. The range between successive relays falls incrementally

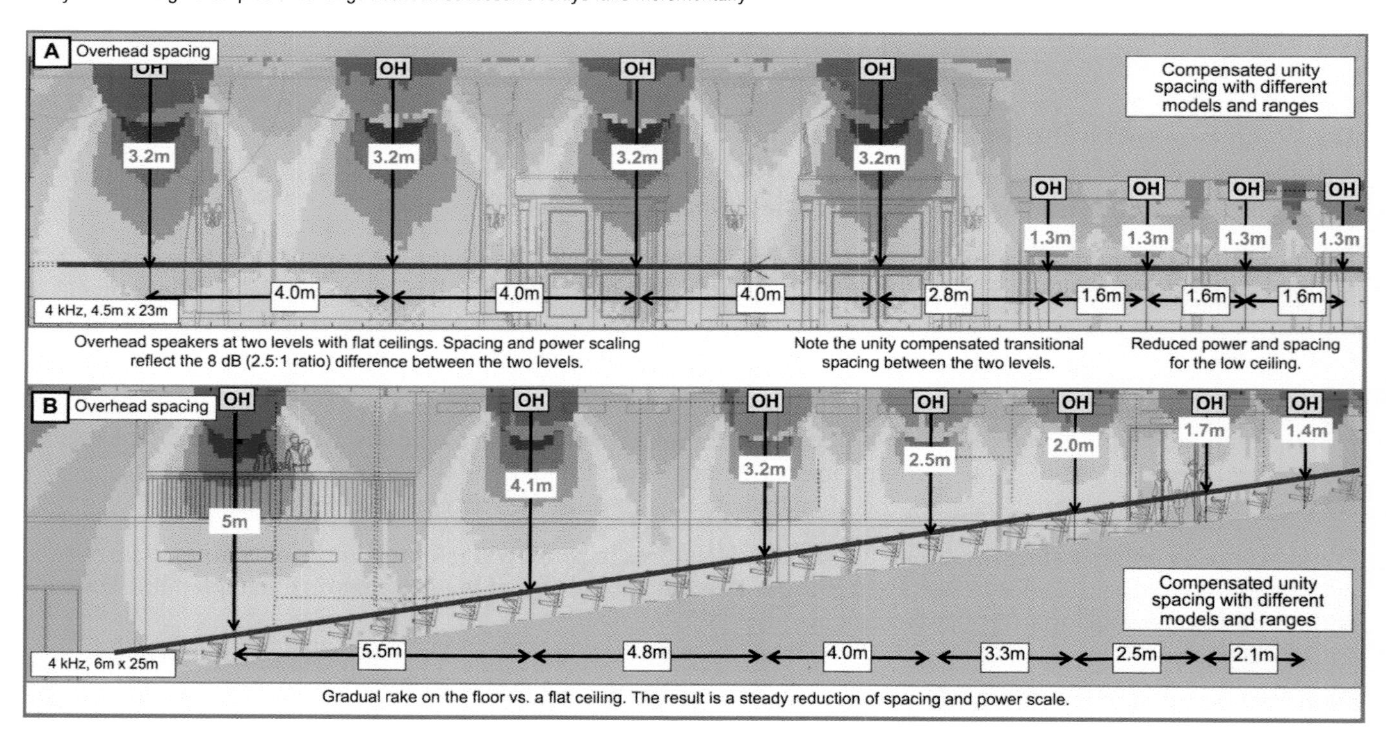

FIGURE 11.32
Overhead main design examples. Spacing and power scaling adapt to the ceiling height.

OVERHEAD MAINS SPECIFICATION

- Spacing: compensated unity spacing using the lateral aspect ratio reference (Fig. 3.41)
- Power scaling: application dependent but rarely used for high-power systems.

11.7 FILL SYSTEMS

Fill systems supplement the mains. They have less power, cover fewer people and yield to the needs of the mains. Where do we draw the line between mains and fills? If I turn the lower speaker in my main array down 1 dB is it now a downfill? Hardly. Recall that the range ratio breaks for main-array subdivision were 6 dB, so this gives us a good guideline. We call it mains and sidefills when our side array only goes half as far as the front array. Likewise for a downfill or rearfill. All of the other fill types are uncoupled from the mains, so no incremental differentiation is required. Frontfills are fills no matter what you try to do with them.

We specify these systems to fill the gaps. We define their start and stop points, which gives us the range info we need to power scale them to keep up with the mains. There are lots of ways to fill the shapes. We will select the element coverage angle and quantity to get the combined shape we need.

11.7.1 Frontfill (FF)

Frontfills are largely one-dimensional, making them the easiest fills to specify. The horizontal plane is where the work is done. The mixture of element horizontal angle, spacing and splay determines the start and stop of optimal coverage. There is only a single vertical plane element and its coverage angle barely matters. The speaker is aimed point blank into the faces of listeners on a flat plane. It's hard to miss as long we don't go too narrow. There is little to gain and much to lose from narrow vertical frontfills. The vertical spill from wide frontfills goes harmlessly into the laps of the front row and the air above. We should be OK as long as the speakers can see the last row of seats they cover. Speakers placed so low that they can't see past the first row are known as footfills.

STANDARD FEATURES OF FRONTFILL SYSTEMS

- Application: coverage and image source for seating on the stage perimeter.
- Elements: (H) first, second order, (V) all orders.
- Array type: (H) solo speaker, uncoupled line or point source, (V) solo speaker.
- Coverage range: unity line (ONAX to XOVR to ONAX) to limit line (triple speaker coverage or more).

We don't really ever *need* frontfills, but they are so nice to have. The mains could cover the front but there are leakage risks. Frontfills are a coverage relief valve, providing a buffer zone between the mains and the open mics on stage. They also serve as image sources toward the stage. The default answer is yes, we need frontfills unless both coverage and imaging are handled by others, e.g. stage sources, actors, etc. The frontfills are redundant when abundant stage sound leaves us unconcerned about gain and image in front (Fig. 11.33).

FRONTFILL (FF) SPECIFICATION SUMMARY

- Frontfill need: Yes, except for small venues with light reinforcement and high ratios of stage sound.
- Coverage start: unity line at first row.
- Horizontal width: OFFAX edge at first-row last seat. Sooner if other low sources cover outer areas.
- Horizontal spacing/splay: unity line connecting the lateral width. Uncoupled array calculator (Fig. 11.20).
- Power scaling: Find range ratio between mains (-6 dB) and FF at the coverage start, and derate the FF.
- Vertical aim: Aim at head height of listeners at the limit line depth.
- Vertical coverage: only a few degrees needed. Minimal benefit (and real risk) to using narrow elements.

There is an alternative frontfill solution for stages too low to give us the required sightline. Go up: frontfills on the ceiling aiming down at the front rows. This option applies to venues with low ceiling and low stages. The horizontal coverage/spacing is calculated the same way and vertical aiming is again at the limit line.

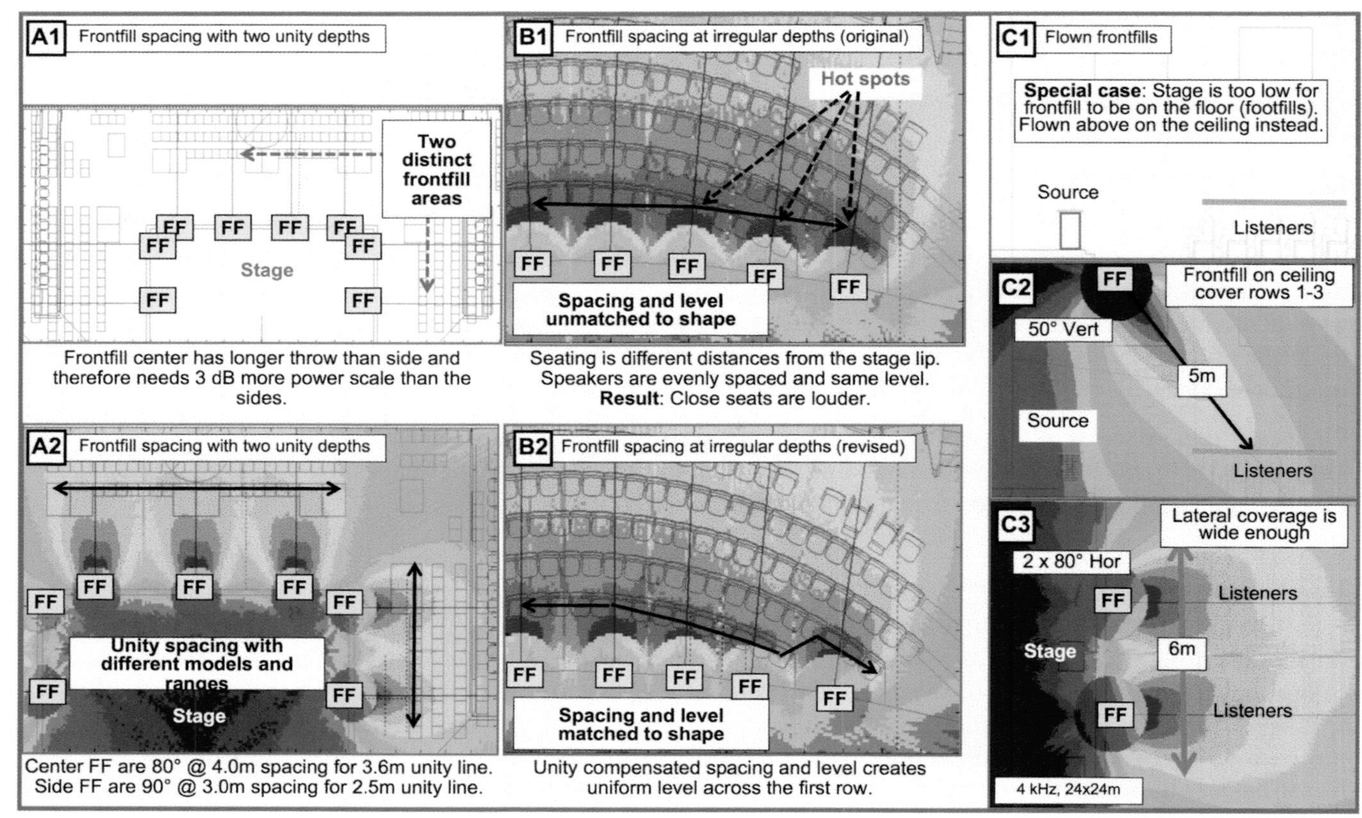

FIGURE 11.33
Frontfill design examples. Spacing and power scaling adapt to the starting depth.

11.7.2 Underbalcony (UB) and overbalcony (OB) fill (delays)

There are innumerable possibilities for "delay" speakers, just add delay. The default meaning for "delays" in our industry (and this section) is a row of under- or overbalcony fills. The standard configuration is like frontfills moved into the house and attached to the ceiling. This adds the vertical plane into the equation. The unity and limit lines are still set by the mix of element angle, spacing and splay, but the height of the speakers above the listeners adds triangulation into the range ratio.

11.7.2.1 NEEDS ASSESSMENT

Do we really need these delays? The answer is easy if we can't see the mains. Otherwise it's a very gray area. It helps to remember the primary challenge for underbalcony areas: strong early reflections from multiple surfaces. Low-clearance underbalcony spaces will have stronger and earlier reflections than those with more air space overhead. Clearance is a key factor for evaluating delay necessity.

Changing the direct sound transmission distance minimally affects the timing of reflections from nearby underbalcony surfaces. By contrast, transmission distance strongly affects the direct/reflected-level relationships there. Propagation loss doubles with distance, including the reflection paths. More distant sources have longer doubling rates, so their reflections don't lose as much level. The loss rate for reflections from close sources is much greater (the reason why delay speakers improve D/R ratio here). Sources close to the balcony front have more of the positive qualities that make underbalcony delays work. A distant source has a D/R ratio disadvantage underneath, even though the surfaces are the same.

It is often mistakenly believed that underbalcony areas require restoration for HF loss. And yet direct sound transmission under a balcony is no different than outdoors (unless the path is blocked). If being under a balcony changed the direct sound, then why wouldn't over the balcony be the same? It's actually LF range buildup and combing from strong early reflections. The HF is the *least* affected, not the most. The overall effect is a pink-shifted frequency response, hence the *perception* of HF loss. This can be fixed by a jackhammer removing the balcony.

All of this discussion holds for overbalcony spaces as well and helps to evaluate delay for the uppermost areas of single-slope spaces. This is relevant wherever there's a roof.

11.7.2.2 CLEARANCE AND RETURN

Two key variables have been revealed: underbalcony clearance and mains transmission range. High clearance means the mains can do it alone unless they are extremely far. Low clearance means trouble and we'll need to bring the mains in close to overcome the damage.

Our second round of balcony battles involves ratios as well. The primary indicator is the clearance ratio, the shape of the underbalcony air space (height/depth). The return ratio, which helped decide when to split the mains (section 11.6.5.) returns in a secondary role here. The balcony range ratio is the difference between covered and uncovered transmission lengths. We need two lengths and one height: (a) main to balcony front, (b) balcony front to last row and (c) the average clearance above our ears under the balcony.

THE RATIOS DRIVING DELAY NECESSITY

- Clearance ratio: height/depth of the air under the balcony (head height to ceiling/$VBOT_1$–$VTOP_2$).
- Return ratio: main to last seat/main to balcony front (M–$VTOP_2$/M–$VBOT_1$).

Even a shallow underbalcony area needs delays if the clearance is inches above our heads. We literally sense the need for more air under a low ceiling. We can go deeper without a delay if we get more breathing room. This is the clearance ratio: vertical air space to depth. Take a meter off the floor–ceiling height and you've got the seated head clearance. Underbalcony areas with clearance ratios below 50% (height/depth) are flagged as contenders for delays, sending us to the second round of qualification (the return ratio). We will need a final score above 50% to be exempt from delays. The range ratio raises the clearance score; the question is will it be enough to reach the combined threshold of 50%?

We can reframe our decision as "can the main overcome the clearance challenge?" We look at the underbalcony area from the main's perspective to assess the chances. The return ratio, which links inversely to D/R ratios in the underbalcony area tells us a lot. A main speaker parked near the balcony front has a D/R ratio advantage over a distant source. It can penetrate the underbalcony space more effectively than one that's traveled further.

These two ratios combine to give us a composite quality score for the main system with a threshold at 50%. Note that this index is statistically based. "Needs delays!" is not likely to show up in our design data, but sure makes itself clear during optimization. The quality threshold is derived from some 100 case studies analyzed for their delay decision and the optimization field results. The delay decision results are sorted into three categories: needed and specified, not needed but specified, and needed but not specified (in best to worst order).

COMPOSITE SCORE RESULTS OF THE CASE STUDIES

- <50%: Delays should be specified
- 50–70%: Delays may be merited depending on tertiary factors (below)
- >70%: Delays should not be specified (may end up turned off during optimization).

The custody of the underbalcony area is reallocated once delays are added. The area covered by the delays is removed from the mains scope of work and the quality score is recalculated. The underbalcony depth is effectively reduced, raising both ratios, hopefully combining to exceed the 50% threshold. If not, a second set of delays must be considered.

The composite score also gives insight about delay placement and power scale. A score just under 50% indicates to us that we should place the delays at a depth appropriate to cover the last few rows. The answer "yes delays" does not mean the entire underbalcony area needs coverage from speakers mounted on the balcony, please! This all too common practice degrades seats that didn't need help and is ineffective where needed most.

It is worth noting the return ratio's role in both balcony decisions (high return ratios lean us toward splitting the coverage in either case). Mains with a short throw to the balcony are pushed toward an upper/lower split (section 11.6.5), whereas distant mains are pushed toward adding delays.

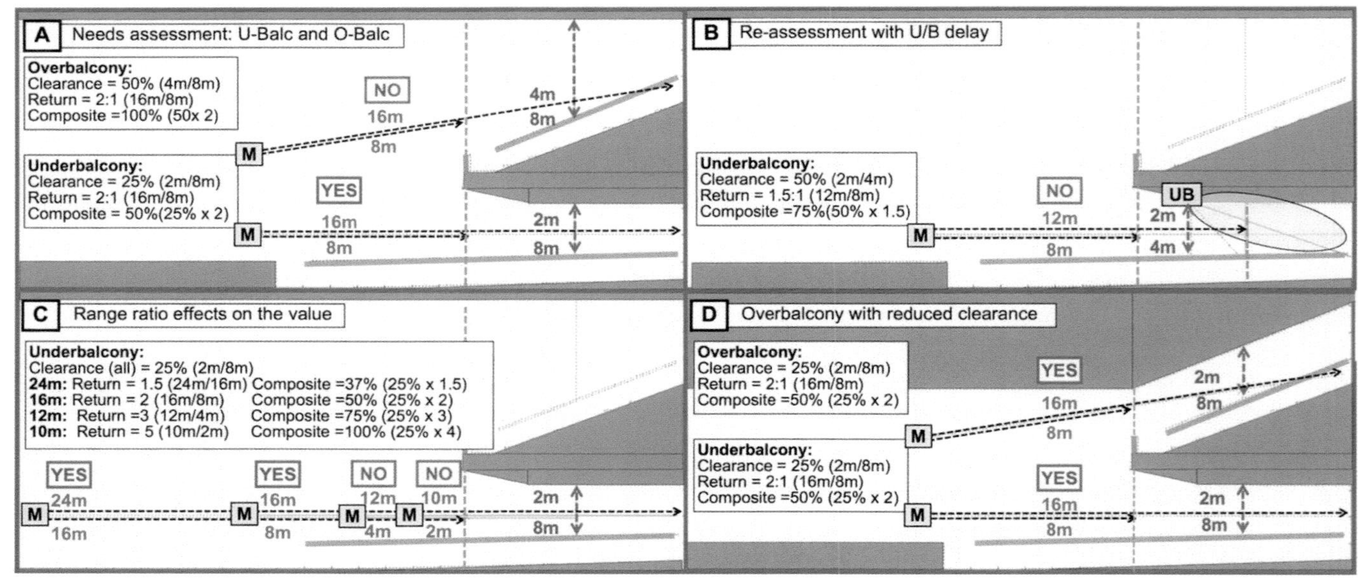

FIGURE 11.34
Needs assessment for balcony fills. A variety of scenarios show the process of determining whether or not fills are needed over and under a balcony.

11.7.2.3 TERTIARY FACTORS AFFECTING DELAY SYSTEMS

We can also add some short subjects that influence the decision to lesser extents.

Tertiary considerations for composite scores in the grey zone (50–70%)

- Main directivity: A highly controlled main improves the quality by reduced reflections.
- Main height/aim: The most favorable orientation is a flat main skimming under the balcony ceiling.
- Upper/lower mains: Split mains often have more favorable orientation to the underbalcony area.
- Room acoustics: Strong room reverberation weighs in favor of adding delays (under and over).
- Overhead angle: Lively back wall, down-angled balcony ceiling weigh in favor of adding delays.
- Program material: Voice transmission weighs in favor of delays more than music (e.g. LCR systems).

11.7.2.4 FIELD EXAMPLES

Our first room (Fig. 11.34) has a single balcony with 2 m × 8 m air space. Delays are very likely with such a low clearance ratio (25%). We'll start with the mains placed 8 m from the balcony front (Fig. 11.34 (A)). This yields a 2:1 return ratio (16 m/8 m). Multiply the ratios to get the quality composite, 0.25 × 2 = 0.5 (50%). We'll add delays and recalculate the score based on the depth remaining in the main's custody (Fig. 11.34 (B)). Only the first 4 m are left to the mains, which improves the clearance ratio to 50% (2 m/4 m). The return ratio falls to 1.5 (12 m/8 m). The composite is 75% (0.5 × 1.5) so we know we don't need a second set of delays at the balcony front.

Let's reset and illustrate the other mechanism with another scenario. We'll increase the return ratio by moving the main closer to the balcony (Fig. 11.34 (C)). Halving the distance to 4 m increases the return ratio to 3:1 (12 m/4 m). This brings the composite to 75% again (.25 × 3), but this time without delays. The mains can overcome an unfavorable clearance ratio, but they have to be in your face to do it. The closer we get, the more we are taking on the same underbalcony perspective as a delay.

Let's reset again and move the speaker farther away, doubling the distances to the balcony front. Delays are a certainty now; the only question is a second ring. Range ratio falls as the mains move further away, reducing their penetration depth under the balcony. Our example hits a tipping point at around 36 meters, and a forward set of underbalcony delays would be helpful at the very front of the balcony.

Return again to the original placement and consider what would happen if we double the clearance under the balcony to 4 m. Clearance doubles to 50% raising the composite score to 100% (0.5 × 2), which makes the specification a no-brainer. We could move the mains out to 28 meters before we enter the gray zone under 70%.

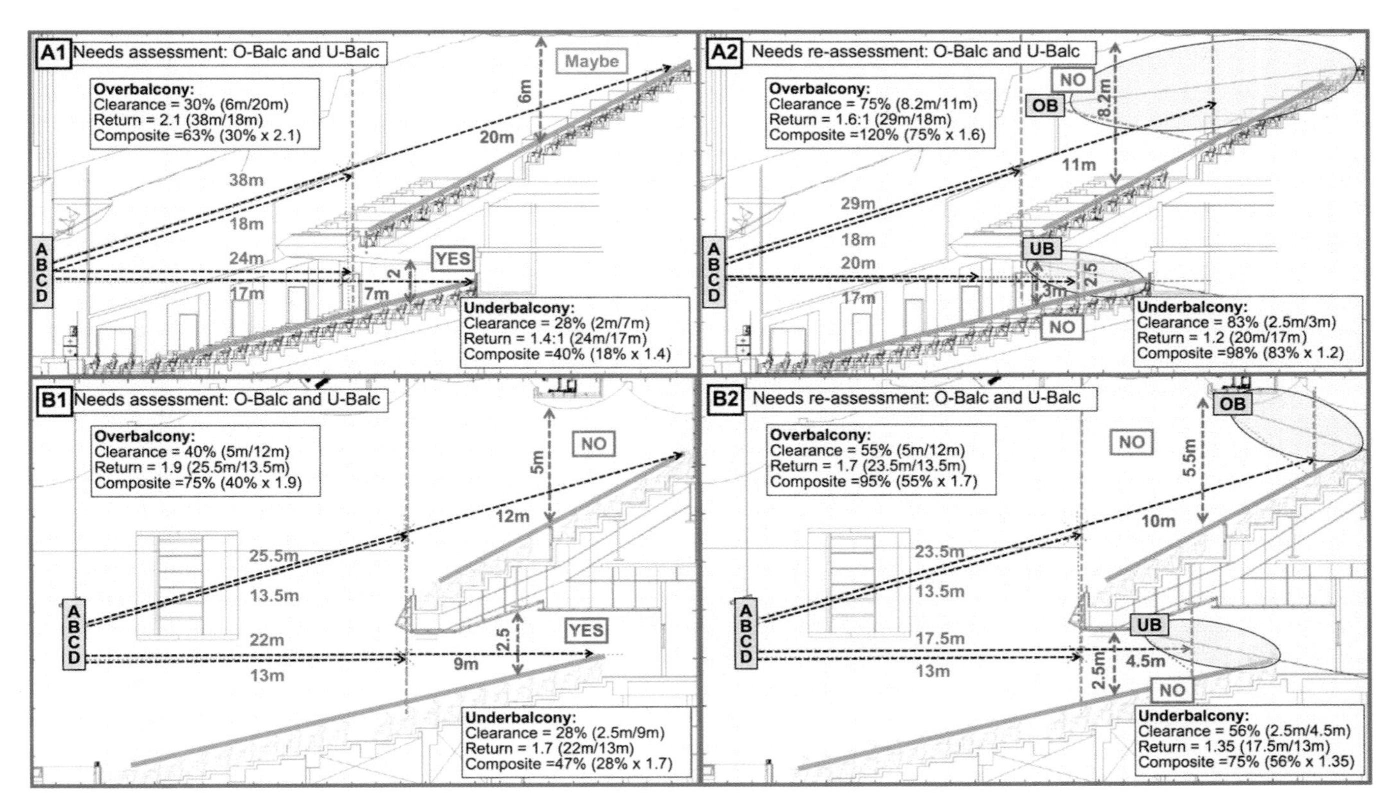

FIGURE 11.35
Needs assessment for under/overbalcony fills for two sample halls. The rooms are evaluated before and after delays have been added.

This brings up an interesting aspect of our example application. The overbalcony area is the same depth and twice the height of the underbalcony. We can (and should) apply the same equations to the upside as the downside. As the saying goes "One man's ceiling is another man's floor." The clearance has doubled to 50%, raising the composite score to 100% (0.5 × 2), an easy decision. We could move the mains out to 28 meters before we enter the gray zone under 70% for the overbalcony area. The exact same thing would happen if we doubled the clearance *under* the balcony.

Fig. 11.34 (D) shows a modified ceiling structure for the second floor that drops the average height to 2 m, making it functionally identical to the underbalcony area. Don't think it hasn't been done just because it's a bad idea!

The next hall is also two floors with a deep overbalcony (Fig. 11.35, upper). The composite scores for both the overbalcony (63%) and underbalcony (40%) areas call for delays. The low clearance above leaves us with a lot to cover, raising concerns about getting too loud in the closer rows. The overbalcony speakers are power scaled for much longer range than the underbalcony. The front half of the balcony now has a composite score of 120% so a second set is not needed. A third hall (Fig. 11.35, lower) has an easy call for the underbalcony system (47% composite) but the overbalcony area was just over the line at 75%. A marginal case like this can opt for a small-scale delay system to cover the last rows.

11.7.2.5 FEATURES AND SPECIFICATIONS

The horizontal spacing and splay of the delays is just like the frontfills except that the unity line length is triangulated in the vertical plane. The process is about placement and spacing once we have assessed the depth where coverage must start. The delays are placed ahead of the unity line so they can be aimed favorably to reach the rear. Keeping the unity line at <25° under the delay is recommended because the speaker will likely need to be twice the coverage angle to fill the shape (many delay areas have range ratios of 2:1). Space the delays so that the main–delay unity line (depth) and the delay–delay unity line (width) are in the same row (uncoupled line or point source spacing calculator). Power scaling is a merger of ONAX responses, so derate the mains by the range ratio between mains and delays at XOVR (Fig. 11.36).

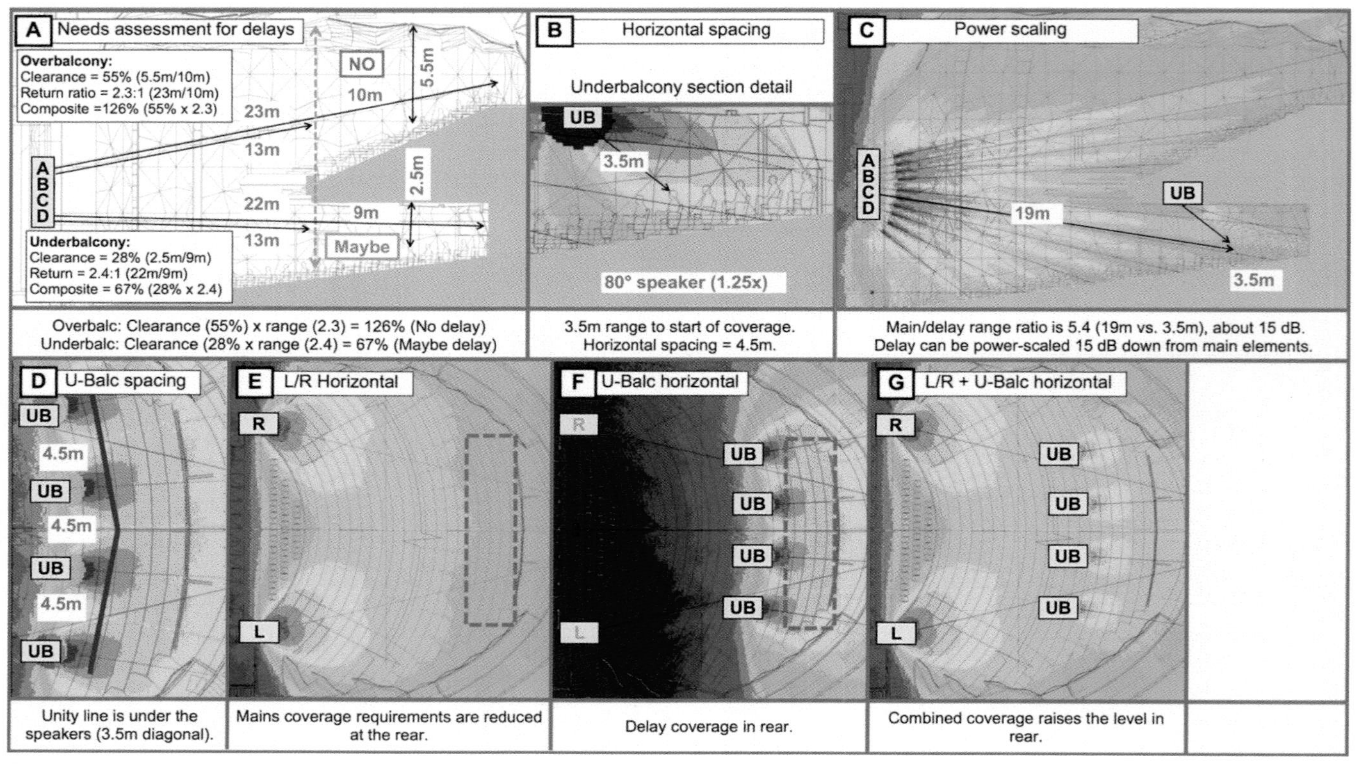

FIGURE 11.36
Example design process for underbalcony delays

Standard features of delay systems

- Application: forward extension and direct/reflected ratio improvement.
- Elements: (H) first, second order, (V) all orders (high order used when placement is too far ahead).
- Array type: (H) solo speaker, uncoupled line or point source, (V) solo speaker.
- Coverage range: Starts at horizontal unity line (if within vertical coverage). Finish at the limit line.
- Sonic image: vertical distortion if level and/or location are too high. Horizontal image is favorable.

Delay specification summary

- Needs assessment: I think we covered that.
- Coverage start: the first row that is both within the vertical pattern and horizontal unity line.
- Horizontal width: OFFAX edge at unity line last seat.
- Horizontal spacing/splay: unity line connecting the lateral width. Uncoupled array calculator (Fig. 11.20).
- Power scaling: Max SPL_D = Max SPL_{MAIN} - ($Range_{MAIN/D}$). Calculate at XOVR MN–D.
- Vertical aim: Aim at head height of listeners at the limit line depth.
- Vertical coverage: Must reach down to listeners at the unity line (but not too much before that).

11.7.3 Coupled sidefill (SF)

The two general categories for sidefill (coupled and uncoupled) are distinct enough to require separate design strategies. The coupled version is simply a radial extension of the mains. It's functionally equivalent to adding lower-level horizontal elements to an asymmetric point source array. Consider it a mini-main. This is a staple for thrust stages and arenas, where the side seating area is a major portion of the space but shallower than the front area. The coupled sidefill's role is clear-cut with single element mains in the horizontal plane (such as a line array). The compensated unity splay angle is used, yielding a consistent shape over distance. The power scale for a coupled sidefill will be close to the mains, because 6 dB (2:1) is the maximum range ratio allowed between adjacent coupled elements. The vertical plane is handled as we would a main speaker (or array), adapted for its local vertical shape.

In some cases the sidefills end up far enough away from the mains that we would classify them as uncoupled. Our design will need to incorporate the start of coverage etc., as with all uncoupled arrays. The design is functionally equivalent to multi-mains (section 11.6.4) and the distinction is unimportant. We will save the term "uncoupled sidefill" for subsystems covering only the near outside seating.

Room shape plays a big part in evaluating the need for sidefills. Wide fans, arenas and stadiums will likely need them. Proscenium stages in a shoebox house are the least likely.

STANDARD FEATURES OF COUPLED SIDEFILL SYSTEMS

- Application: radial extension of the mains where substantial coverage is required.
- Elements: (H) first, second order, (V) all orders.
- Array type: (H) coupled point source combined with mains, (V) solo speaker, coupled point source.
- Coverage range: unity line (XOVR main–SF to ONAX SF to OFFAX SF).

11.7.4 Uncoupled sidefill (SF)

Uncoupled sidefills plug outside gaps in the early seating areas. It's best to locate them near the stage level so they can be run just loud enough to fill the gap and yield the rest of the floor to the mains. They join the mains as an asymmetric uncoupled point source to provide a mix of outward radial and downward vertical extension. Common versions include solo speakers or small arrays near the stage deck. They fill the nearby side areas regardless of whether the mains are missing vertically, horizontally or both.

The uncoupled sidefill is a mandatory component for center main systems (and the center channel of an L/C/R main). Another application is to help L/R systems in wide rooms or hung high.

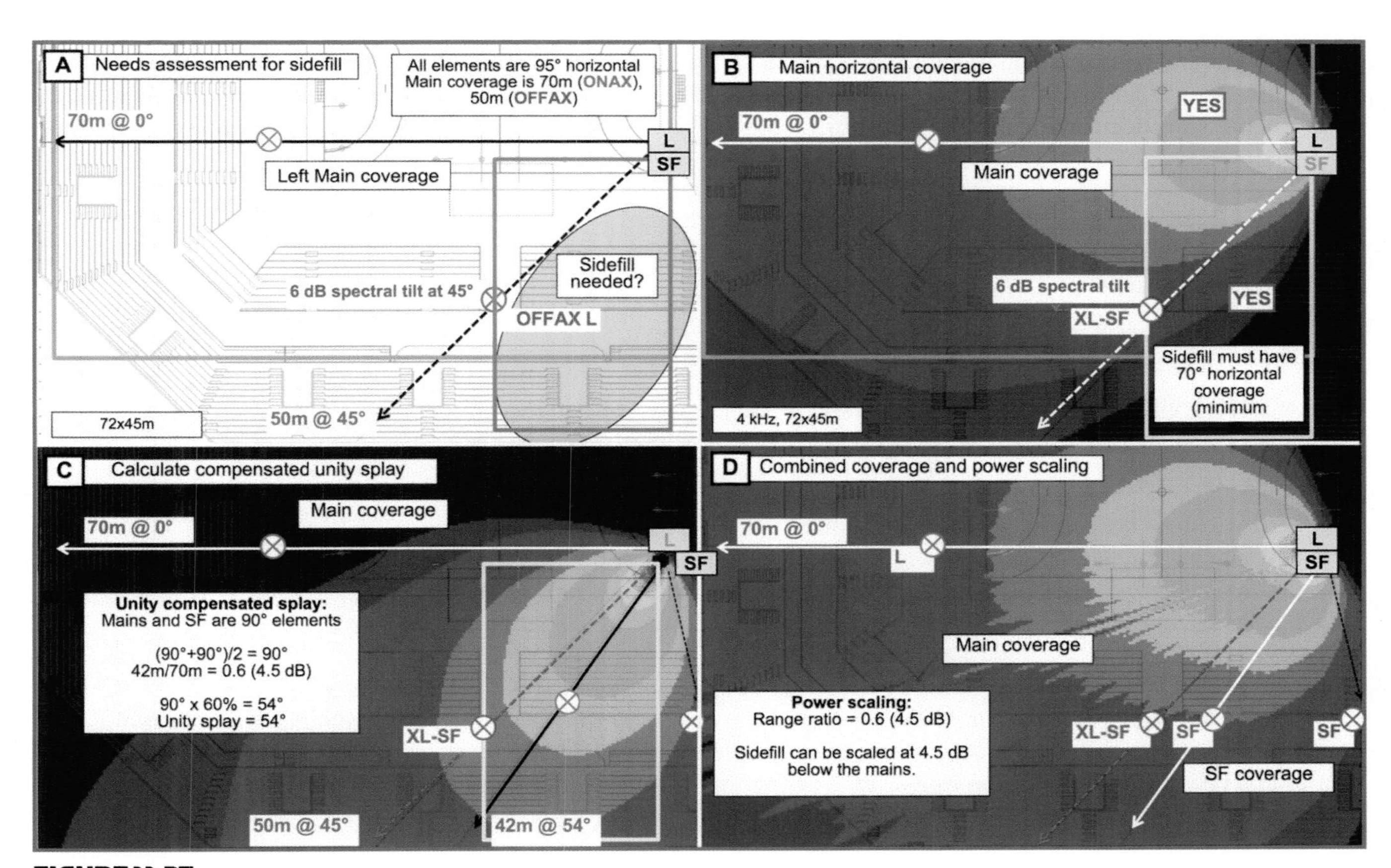

FIGURE 11.37
Example design process for coupled sidefill

STANDARD FEATURES OF UNCOUPLED SIDEFILL SYSTEMS

- Application: Coverage and image source for outside seating near the stage (beyond the frontfill).
- Elements: (H) first-, second-order, (V) first-, second-order.
- Array type: (H) solo speaker, uncoupled line or point source, (V) solo speaker.
- Coverage range: XOVR Main-SF to ONAX SF to OFFAX SF.
- Sonic image: Located on side of stage, so it can (hopefully) be placed fairly low.

DO WE REALLY NEED THESE SIDEFILLS?

Somebody is going to ask this question so let's find our justification. The decisive parameter is coverage width on the post-frontfill area of the floor. This is also a common theme for infill, centerfill and rearfill so it's worth our time to explore it in depth. We need to know if the mains make to the outermost seats near the front. There's good reason for the mains not to cover these seats: a nearby wall at a really bad angle for us. The precise location of the uncovered area depends on the frontfill coverage depth. Let's call it the third row. If the mains can't take it from here, then it's time to call for backup: sidefills.

How wide are mains at the frontfill finish line? We know the coverage pattern is a factor but the bigger one is height. We won't need sidefills if the mains are a mile high (even if they have a 1° coverage pattern). Lower, closer, narrower mains in a wide hall need more sidefill help than higher, farther, wider mains in a narrow hall. The mains lateral width must extend to the outermost seat in the third row (or call for help).

SIDEFILL NEED ASSESSMENT (START)

- Prerequisite: Mains are already aimed. Frontfills will cover to the third row for our example.
- Section: Find the main's range to XOVR M–FF (e.g. bottom of main to heads in the third row = 6 m).
- Refer to Fig. 3.41 for the main speakers' lateral width ratio (e.g. 90° speaker = 1.4).
- Multiply the range by the lateral ratio. (e.g. 6 m*1.4 = 8.4 m). Mains are 8.4 m wide in the third row.
- Plan: Draw a perpendicular 8.4 m line through XOVR M–FF. This is the mains coverage width.
- Sidefill is needed for seats outside the line (e.g. 4 m short), infill or centerfill for the inward side.

We now know we need four meters of sidefill. We can determine coverage angle by reverse engineering from the lateral width and range. We can get 4 m of lateral width from a 60° speaker at 4 m or an 80° speaker at 3.2 m. Other combinations will work as well. Overage is a safer bet than trying to be surgical here. Horizontal aiming follows the single-speaker method: middle/middle.

The horizontal plane is done. On to the vertical aim. The missing piece is how much further the sidefill needs to go. Is it one row or twenty? We find the answer by looking at the lateral width of the mains again. We work our way back until the mains don't need our help any more.

SIDEFILL NEED ASSESSMENT (STOP)

- Section: Move back from our starting depth a few rows and reassess the range to the mains.
- Re-calculate the lateral coverage width and plot the length at the new depth on the plan view.
- Bracket the depth until the speaker coverage width equals the seating coverage width (e.g. ninth row).
- This is the limit depth. Sidefill covers from the third-row seats 20–30 and finishes at the ninth row.

Vertical coverage/aim can be evaluated by the single-speaker method (11.3.2) or save your brain and aim it around the top row of the SF coverage. The sidefill is easing its way under the mains so losing level over distance is a desirable quality for the sidefill. This is a fill speaker and we don't need to get crazy about an exact vertical and horizontal pattern fit. It's better to be overly wide than narrow. We can tolerate some angular overlap as long as we keep the level and timing under control.

This is a lot to go through over a stinking little sidefill. Thankfully this process will be recycled for the remainder of the fill systems. We have the theme. The rest are variations.

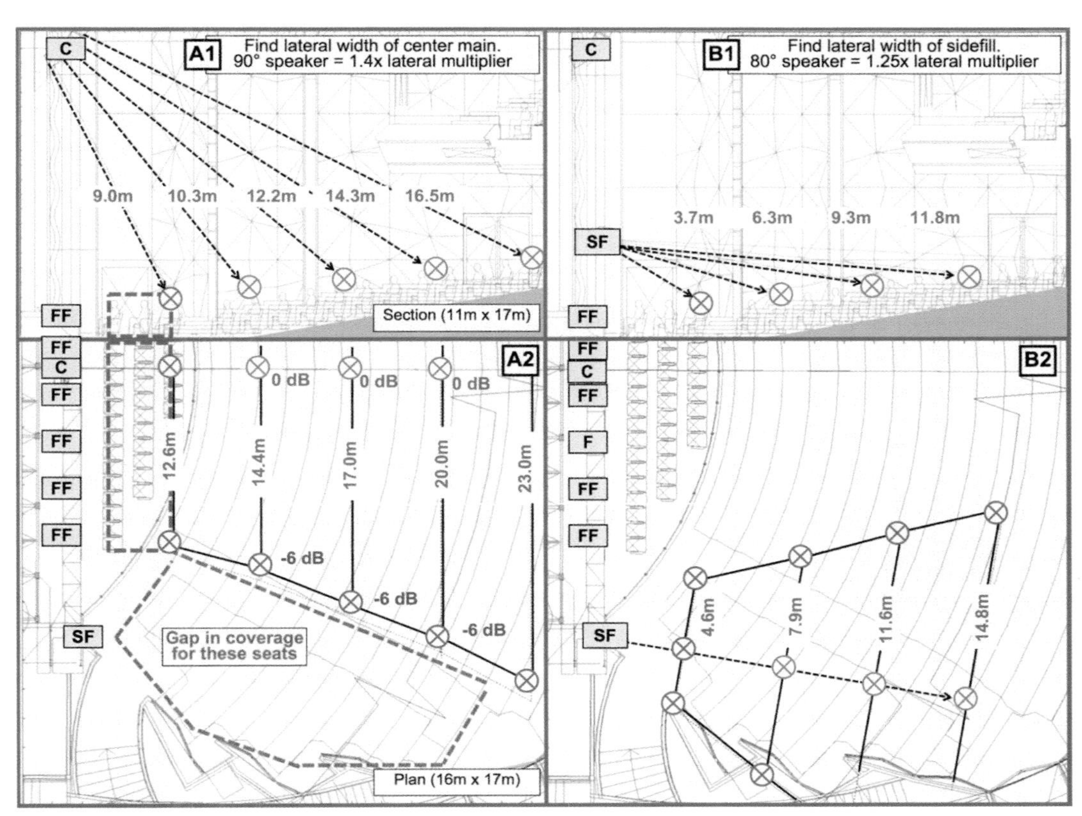

FIGURE 11.38
Example design process for uncoupled sidefill: needs assessment

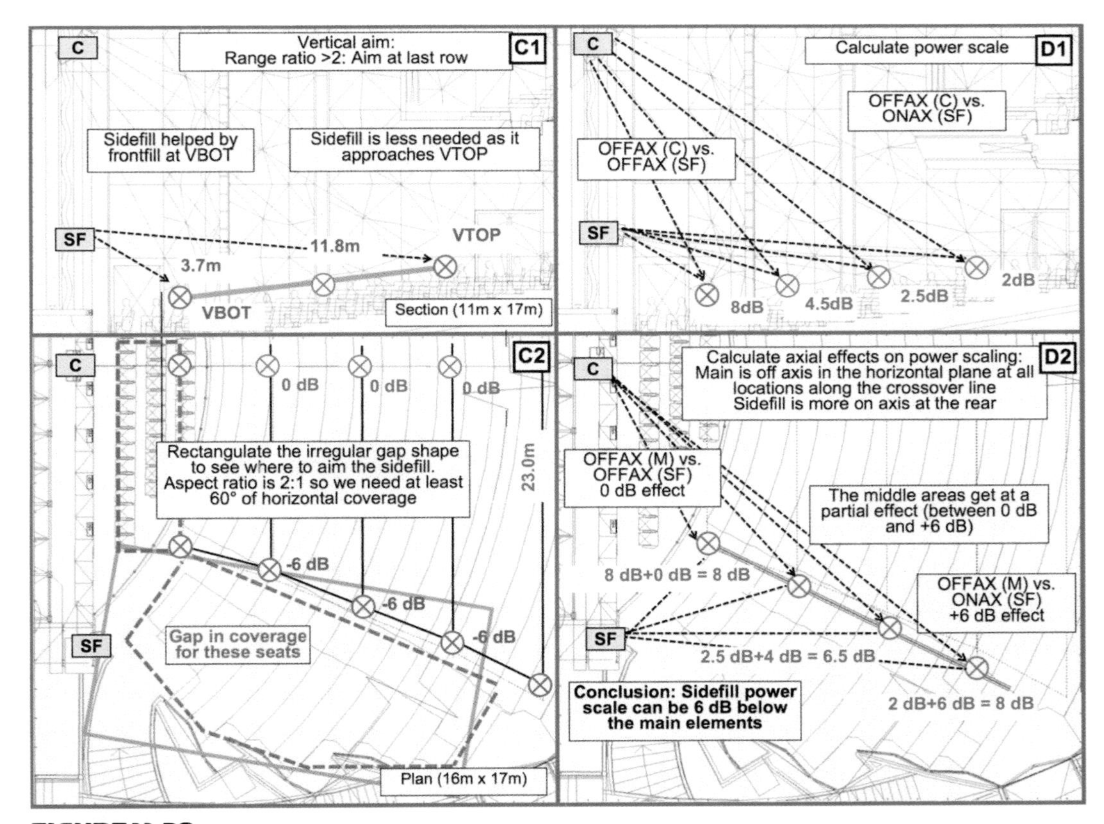

FIGURE 11.39
Example design process for uncoupled sidefill: aim and power scaling

UNCOUPLED SIDE FILL (SF) SPECIFICATION SUMMARY

- Need assessment: Yes, if the L/R mains don't reach the near outer area by the frontfill limit line.
- Horizontal width: lateral width calculation at start of coverage (range*lateral ratio).
- Horizontal aim: middle of its coverage.
- Vertical aim/coverage: compensated unity coverage and aim (Fig. 11.9).
- Power scaling: Max SPL_{SF} = Max SPL_{MAIN} – ($Range_{MAIN/SF}$). Calculate at XOVR MN–SF.

11.7.5 Centerfill (CF)

A centerfill for L/R mains is an inverted version of the uncoupled sidefills for a center main. The center plugs the inside gap of an L/R array while the sidefill plugs the outside gap for the center array. The center gap shape is a triangle that's widest at the bottom and tops out at the unity XOVR of left and right. The design process is very similar to the sidefill so we can simply describe this in macro steps and add only the relevant details.

STANDARD FEATURES OF CENTERFILL SYSTEMS

- Application: gap filler between L/R and FF systems. Sonic image stabilization (toward center).
- Elements: (H) first, second order, (V) all orders.
- Array type: (H) solo speaker or point source, (V) same.
- Coverage range: from post-frontfill center area to unity XOVR LR at center.

CENTER FILL (CF) SPECIFICATION SUMMARY

- Need assessment: Yes, if the L/R mains don't reach the center before the frontfill limit line.
- Horizontal width: lateral width calculation at start of coverage (range × lateral coverage width multiplier).
- Horizontal aim: middle of its coverage (and the room).
- Vertical aim/coverage: compensated unity coverage and aim (Fig. 11.9).
- Power scaling: Max SPL_{CF} = Max SPL_{MAIN} – ($Range_{MAIN/CF}$). Calculate at XOVR MN–CF.

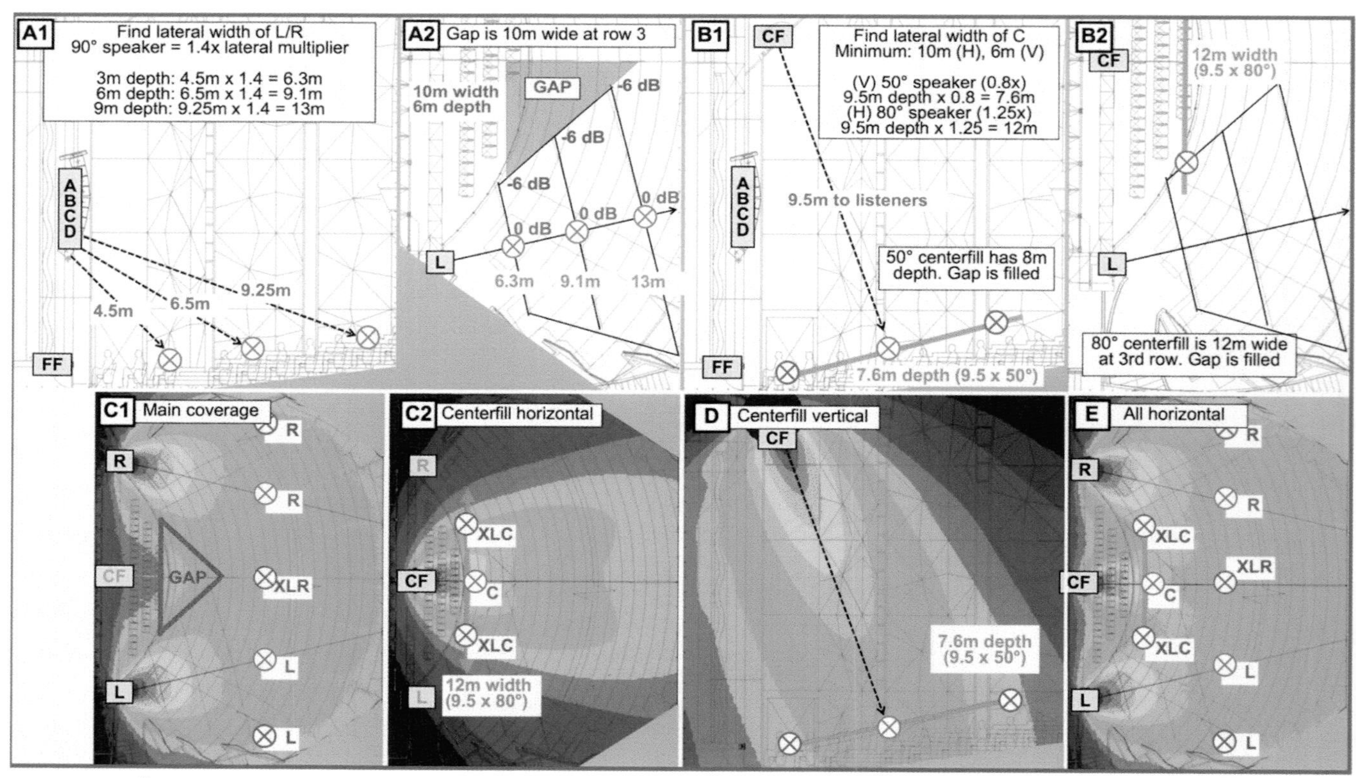

FIGURE 11.40

Design example for centerfill. (A) Needs assessment, finding the gap between the L/R mains. (B) Defining coverage of the centerfill (C) solo and combined coverage.

11.7.6 Infill (IF)

Sidefills are sometimes known as "outfills." This provides a strong hint that infills are another version of inverted sidefill. Infills are the down low version of center gap fillers (center is the high version). Infills are often a choice of last resort because of two strong disadvantages: extreme horizontal sonic image variance and high amounts of ripple variance. Let's illustrate this with an example. A sixth-row seat is near (but not exactly at) center. It's in the dead zone between FF and the L/R mains. Option A is centerfill. Option B: infills. The centerfill option adds a single source to the area capable of filling in the direct sound and centrally stabilizing the image. There is sure to be some ripple but we can minimize it with aim, directional control, delay and level. By contrast, the infill option adds two sources that can only sync to each other at the exact center. Just off center they are barely different in level and majorly different in time over a large part of the area. This is a point destination ripple variance disaster that consoles itself by remembering, "at least I'm not a monitor sidefill." And for bonus points it pulls the image hard outward to the sides. You've been warned, but sometimes you gotta do what you gotta do.

Infills are so unlikely with a center main that we can move straight to the L/R config. They are inward radial extenders to fill the center gap. The design process is again close to the uncoupled sidefill so we will go through the macro steps and add relevant details.

STANDARD FEATURES OF INFILL SYSTEMS

- Application: gap filler between L/R (or center) and FF systems.
- Elements: (H) first, second-order, (V) usually second order but all can be used.
- Array type: (H) symmetric uncoupled point destination, (V) solo speaker or point source.
- Coverage range: from post-frontfill center area to unity XOVR LR at center.
- Sonic image: vertically favorable but strong outward horizontal image distortion is common.

INFILL (IF) SPECIFICATION SUMMARY

- Need assessment: Yes, if the mains don't reach the center by the frontfill limit line.
- Horizontal width: lateral width calculated at start of coverage (range × lateral coverage width multiplier).
- Horizontal and vertical aim: typically the room center, last row of coverage before mains gap closes.
- Vertical coverage: coverage angle must reach down to start of coverage, typically XOVR FF–IF.
- Power scaling: Max SPL_{IF} = Max SPL_{MAIN} - ($Range_{MAIN/IF}$). Calculate at XOVR MN–IF.
- Less is less: The more they are turned inward, the more combing we get.

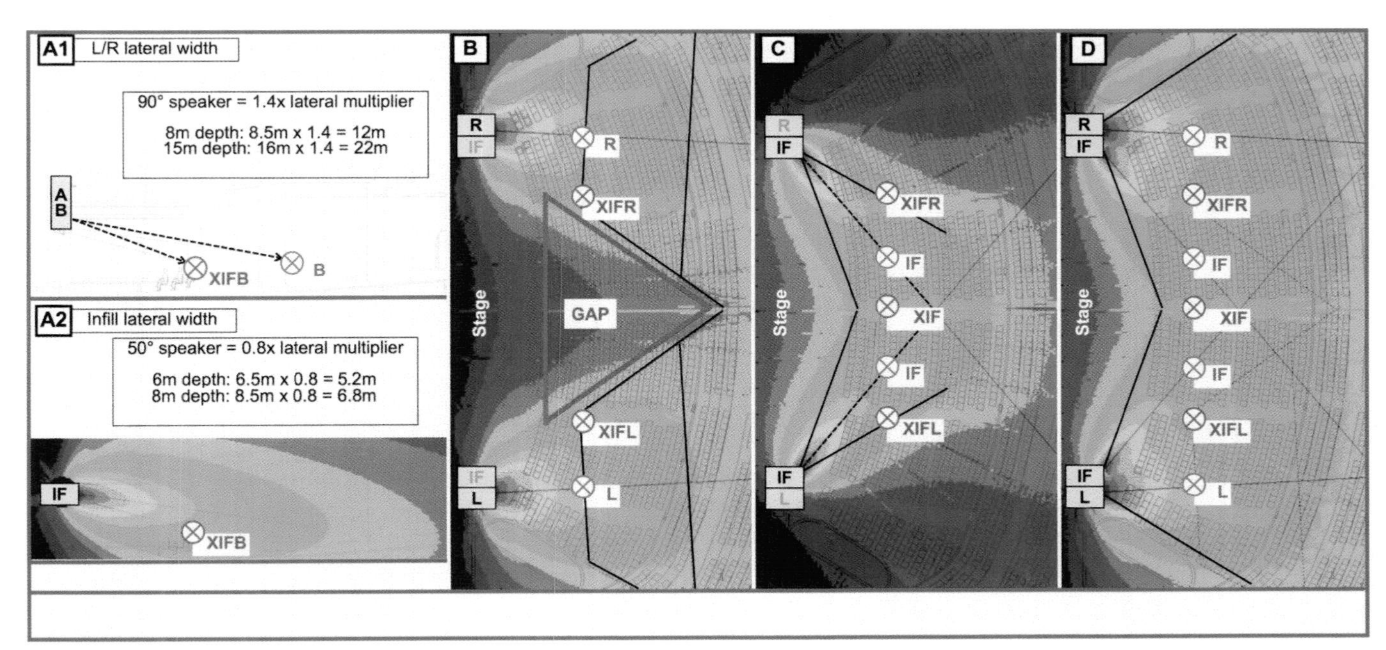

FIGURE 11.41
Infill design example: (A) needs assessment; (B) defining the gap between the L/R main; (C) infill solo coverage; (D) combined coverage

11.7.7 Rearfill (RF)

Rearfill are rarely specified because nobody wants to watch a performer's butt. Ok, some people do, but rearfills are still rare. Rearfills are a fill fill, i.e. they don't connect to the mains (they radially extend the sidefills that radially extended the mains). Their macro role is an uncoupled point source in the horizontal plane to complete the 360° coverage. More than a single rearfill may be needed to fill the horizontal shape. A point source or uncoupled line source array are common solutions, depending on coverage shape and physical options. Vertically they are singles or couple point sources. Rearfill system design can't begin until mains and sides are complete and we see what's left.

STANDARD FEATURES OF REARFILL SYSTEMS

- Application: coverage extension to seating behind the stage. Radial extension of sidefills.
- Elements: (H) first, second order, (V) usually second order but all can be used.
- Array type: (H) symmetric uncoupled point source, (V) solo speaker or point source.
- Coverage range: from XOVR SF–RF to ONAX RF to XOVR SF–RF (opposite side).
- Leakage challenges: mains back lobe, house reflections, stage monitors, stage sources.
- Feedback risk: Location choice must consider ramifications of firing across/over the stage.
- Limit the range: Keep the level down by minimizing the throw distance.
- Timing: Rearfills must arrive early enough to sync to stage sources and leakage.

Some rearfill areas have frontfill speakers. This is a terminology issue, not a typo. What should we call speakers in front of the audience that's at the rear of the stage? If fill speakers are on the stage lip, in the stage front or on stage stands firing up into the faces of the audience we will call them frontfills, OK?

Location is important for all speakers but none more than the rearfill. We can hang close to the mains and stay coupled, or move closer to the rear and uncouple. Coupling is usually the best option under normal circumstances, but rearfills are far from normal. Let's not forget that there's a show going on behind the mains. Firing speakers over their heads is risky business.

REARFILL LOCATION CONSIDERATIONS (COUPLED IS NEAR THE MAINS, UNCOUPLED IS NEAR THE REAR SEATING)

- Coupled: maximum coupling, minimum range ratio/angular spread. Maximum throw, late arrival.
- Uncoupled: minimum coupling, maximum range ratio/angular spread. Minimum throw, on-time arrival.

The coupled approach can use a narrower speaker but it has to be run louder, a break-even trade at best. Coupling requires sync'ing to the main (which were delayed to help the forward transmission). The rearfill arrives so far behind the direct sound (and stage monitors) that it's more like a rear wall reflection than a PA. The audience area behind the stage is polluted with back lobes from the mains, leakage from stage monitors, and the direct sound of the instruments. We want to add as little to the mess as possible, so keeping the level down is the number one priority. Uncoupling and getting as close as possible keeps the level down and gets us closer in time.

REARFILL (RF) SPECIFICATION SUMMARY

- Needs assessment: Yes, if there are no frontfills and the sidefills leave a coverage gap in the rear.
- Horizontal width (solo or uncoupled): lateral width calculated at coverage start (range*lateral ratio).
- Horizontal array splay (coupled): compensated unity splay angle.
- Horizontal aim: single speaker (solo or uncoupled), centered (coupled).
- Vertical aim and coverage: solo speaker method. Coverage angle must reach both VTOP and VBOT.
- Power scaling: Max SPL_{RF} = Max SPL_{SF} - ($Range_{SF\text{-}RF}$). Calculate at XOVR SF–RF.

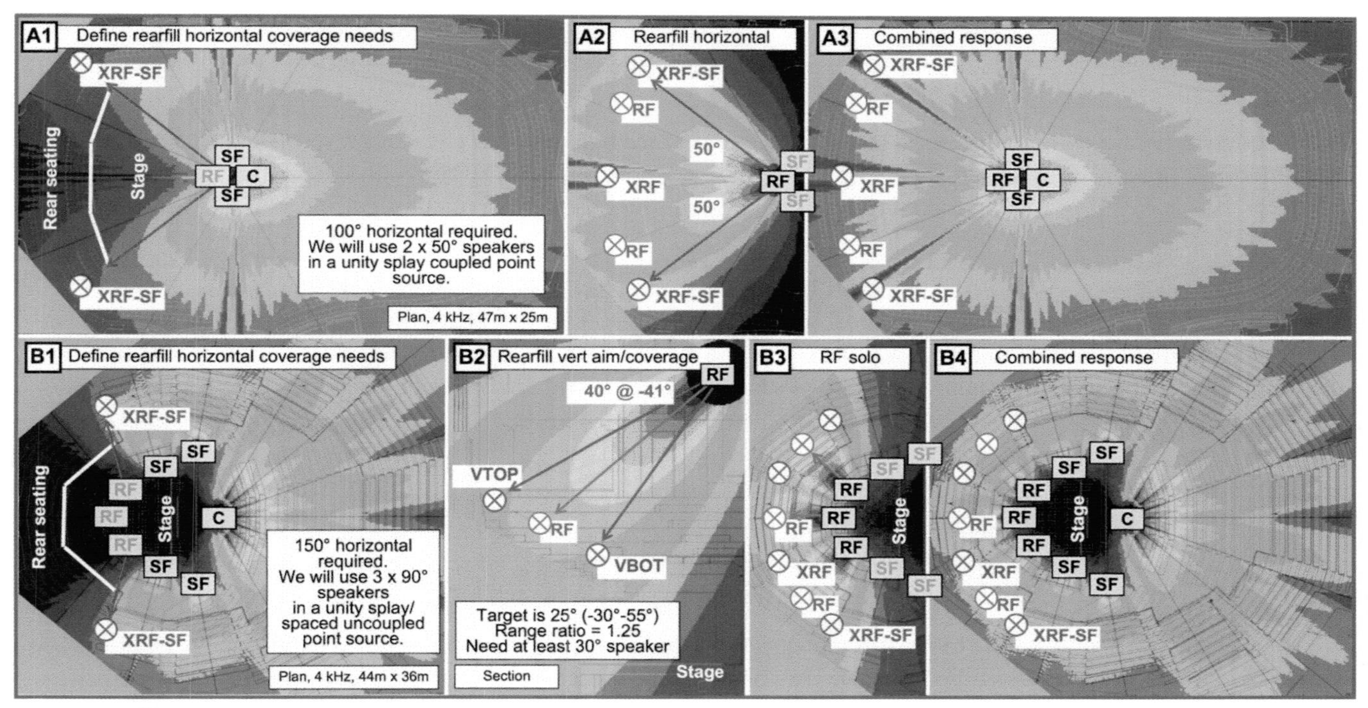

FIGURE 11.42
Design examples for rearfill: (A) coupled point source example (B) uncoupled point source example

11.7.8 Shadow fill

Some seating areas lack line of sound to the speaker system. Such shadowed areas are most common under balconies, especially the multi-level wraparound types. They still sell these seats so we have to cover them, often making them the most expensive seats in the house (for sound). LF leakage diffracting around the blockage is not usable acoustic power, so the shadow fill must be power scaled to go it alone. This contrasts to a standard underbalcony fill, which shares power with the mains. Shadow fills require 6 dB more power scale than a comparably ranged underbalcony or sidefill. Sonic image distortion is a lost cause that can't be solved by clever delay settings. The image is guaranteed to be at the shadow speaker because it is the sole source of HF information, the dominant range for localization. The best we can do is place the speaker between the audience and stage, thereby imaging in its general direction. Shadow fills coverage shapes are totally application dependent, so we can only issue general guidelines here.

SHADOW FILL

- Application: seating areas without line of sight to mains or equivalent fill system.
- Elements: application dependent but generally first or second order.
- Array type: application dependent but generally solo or uncoupled arrays.
- Range: Must cover all occulted seating (and minimally disturb others).
- Power scaling: Mid-depth shadow fill is equal to mid-depth main.
- Location tips: Place for best sonic image and minimum operating range (to minimize leakage).

11.8 EFFECTS SYSTEMS (FX)

Our main and fill design strategies are often concerned with sonic image control, with the common goal of keeping the listener's attention away from the speakers. Effects systems are even more concerned with sonic image, but in the exact opposite way. The effect system approach is "Hey! Look over here!" They intentionally bring the attention in their direction. FX systems are designed as a confederacy of loosely affiliated channels, each with their own roles. The best-case scenario for an FX speaker is that everyone perceives its role the same way, e.g. the left rear surround images to the left and rear of everyone. This is easier said than done.

Side, rear and overhead surrounds are the most common effect systems. This would create the "Point Destination of Doom" if they were all driven with the same signal. Sane people drive them with unique channels, usually as soloists or groups of uncoupled line sources.

Independence always comes with responsibility. Surrounds are independent channels responsible for covering the entire room. L/R surrounds are not like the L/R mains. Each surround covers both sides, near and far. The challenge for surrounds is covering an extended depth with a range-limited array (the uncoupled line source). TANSTAAFL and triage will be called upon shortly.

It's difficult for FX speakers to do their job without annoying someone, somewhere. This doesn't sound very scientific but it's a real design factor. It's mostly about range ratio. Our challenge is to get the attention of folks on the opposite side without giving heart attacks to the nearby listeners. FX speaker range ratios often exceed anything we find in our main arrays. Visualize rear surrounds placed just above the heads in the twenty-fourth row trying to reach the twelfth row. Keep going. You're only half way there.

11.8.1 Side surround (SS)

Five to six parameters are typically in play: element coverage (H and V), height, horizontal spacing and vertical and/or horizontal aim. If the speaker mounting is only adjustable in one plane then it is usually best to choose the vertical. Dedicated surround enclosures usually have a downward tilt built in, which hopefully will have the correct angle.

We need to establish a unity line to set the spacing. We used the first row for the frontfills, but they were at ground level. The surrounds will be raised up, making it seem more like our underbalcony fills (at first glance). Our underbalcony systems don't have to cover the nearest rows under them (the mains handle that) but the side surrounds don't have that luxury. So our coverage target starts nearby like the frontfills, from a position like the U/B fills and with the range extension of a main system. No problem. Low surrounds need higher quantity and have shorter range. High surrounds needs smaller quantity and have longer range, but if we go too high they lose their identity as side surrounds, becoming more like overhead surrounds. TANSTAAFL. We might need 10 or 100 side surrounds depending our approach. Compromise height/spacing provides the best balance between identity loss, gaps and excess coverage. We have been through this before on the main system horizontal aiming (11.3.2). Let us adapt the compromise for the surrounds.

The uncoupled line source has three distinct zones of coverage (by depth): pre-unity (coverage and gaps), minimum variance (unity line to limit line) and post-limit (overlap and more overlap). The first two zones are of equal length, and the third is unlimited. Surrounds are standard in cinemas so let's characterize the zones as movie roles: pre (good-bad), MV (good-good) and post (bad-bad). We can even out the good/bad distribution by dividing the room width into thirds. A unity line at ⅓ width creates a limit line at 2/3 width, an even distribution of the good, the bad and the ugly.

Height plays a secondary role that aids distribution. Raising the speakers up on the wall triangulates the path from speaker to listeners. This puts more of the near listeners inside the horizontal coverage, with negligible difference to the opposite side. This can move the unity line significantly closer to the speaker side without much effect on the limit-line position. Height expands the good, without adding much bad.

Three factors limit our choice of height: the ceiling, the vertical coverage and being perceived as overhead. We'll cover the ceiling issue later. There is no bright line between perceiving an image as lateral or overhead, but we can still use some logic and basic geometry to draw a gray one: 45°. Below this is more lateral and above is more overhead. Therefore the near-side seats may perceive the sides as ambiguously overhead-ish if we go too high.

Let's aim the vertical. We already know that a range ratio of 2:1 (or higher) obliges us to aim the speaker at the farthest seat. Ratios >2 are assured here so no calculation is required. Aim it at the farthest seat on the opposite side. We have known ONAX and known unity line (⅓ of the distance to ONAX). The angular spread between ONAX and the limit line is half our required coverage. An example: ONAX @-10°, limit line @-35°. That's 25° from ONAX to OFFAX (the limit line) so we need at least a 50° speaker.

Spacing, the final frontier. We know the vector distance from the speaker to the unity line. That number is used to calculate the lateral width. Example: 90° element (H) with 3 m path to the unity line—horizontal spacing is 4.2 m (3 m × 1.4 = 4.2 m).

The above strategy evenly divides the under- and over-coverage. Surround signals are generally more tolerant of ripple variance so it may be advantageous to tilt the game in favor of less gapping. It takes a lot of combing to degrade our experience of a cannon blast. The pre-unity area can be shortened (and post-limit area extended) by wider elements, closer spacing or lower height.

Power scaling for surrounds is content and application based. Cirque du Soleil and other heavily spatialized megashows require power scaling within 6–10 dB of the main systems. Typical musical theater surrounds are 15–20 dB below the mains. Cinema surrounds must follow SMPTE standards based on their summed response but generally fall 15–20 dB below the mains on a solo basis.

STANDARD FEATURES OF LATERAL SURROUND SYSTEMS (SIDE AND REAR)

- Application: intentional sources for sonic image localization away from stage.
- Elements: (H) first order, (V) first or second order (preferred for getting better coverage to center).
- Array type: (H) uncoupled line source, (V) solo.
- Range: all seating (and minimally disturb nearby listeners).
- Power scaling: application dependent but commonly 15–20 dB below mains.

11.8.2 Rear surround (RS)

Rear surrounds in a room with a high ceiling and no balcony follow the same design guidelines as the sides. Raked seating, low ceilings and balconies require a different approach. Recall how raked seating presents a diagonal target to the main system. VTOP is high and far and VBOT is near and low. We aim toward the back to counter the rake. Rear surrounds come from the opposite direction and see the rake from a totally different angle, literally. It looks like a flat floor from the speaker perspective because we have to aim down to the front while skimming over the rear seating. We have much more of a "frontfill in the back" perspective unless the ceiling is high in the back. Level variance: strike one.

Low ceilings, common at the rear of a raked room, compound the issue by forcing the speaker placement to be barely above standing height. This yields insanely high-range ratios that require a laser speaker to counter. We can

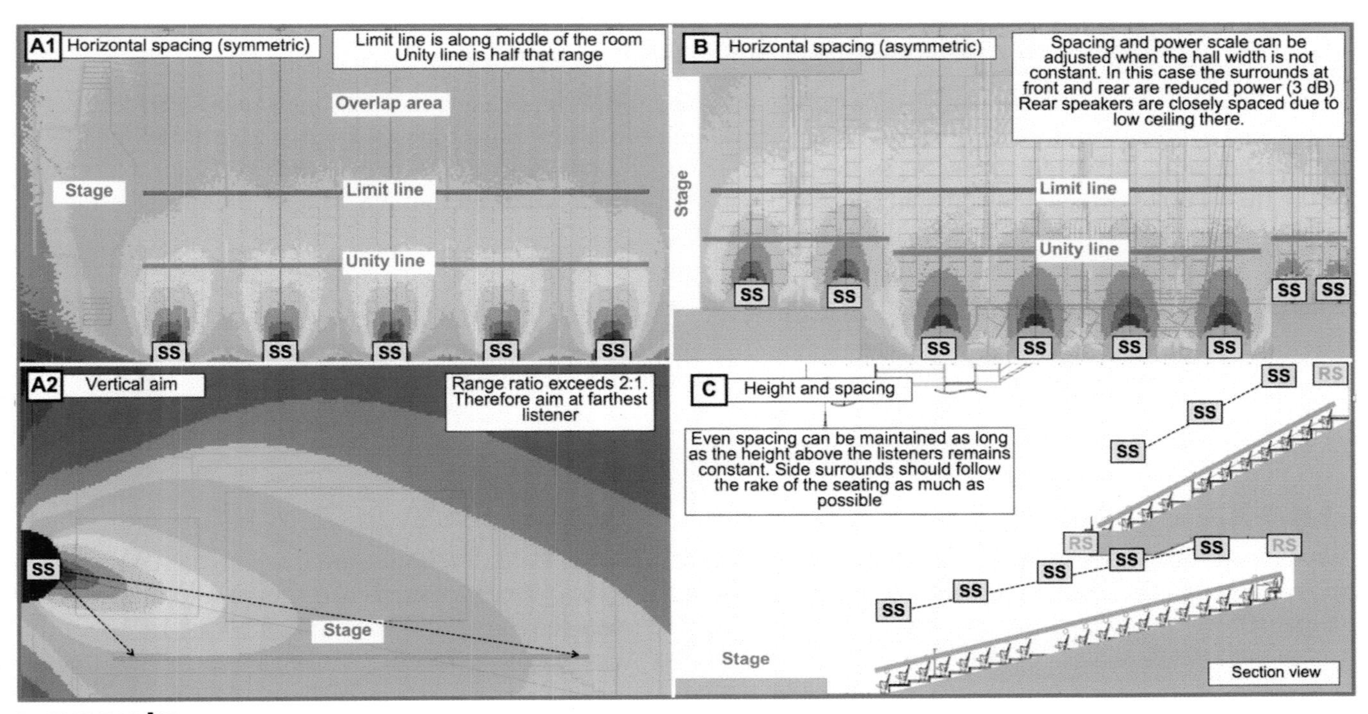

FIGURE 11.43
Design examples for side surrounds. (A) Single flat floor vertical plane and rectangular horizontal plane. (B) Asymmetric width in the horizontal plane. (C) Two-floor vertical with balcony and raked seating.

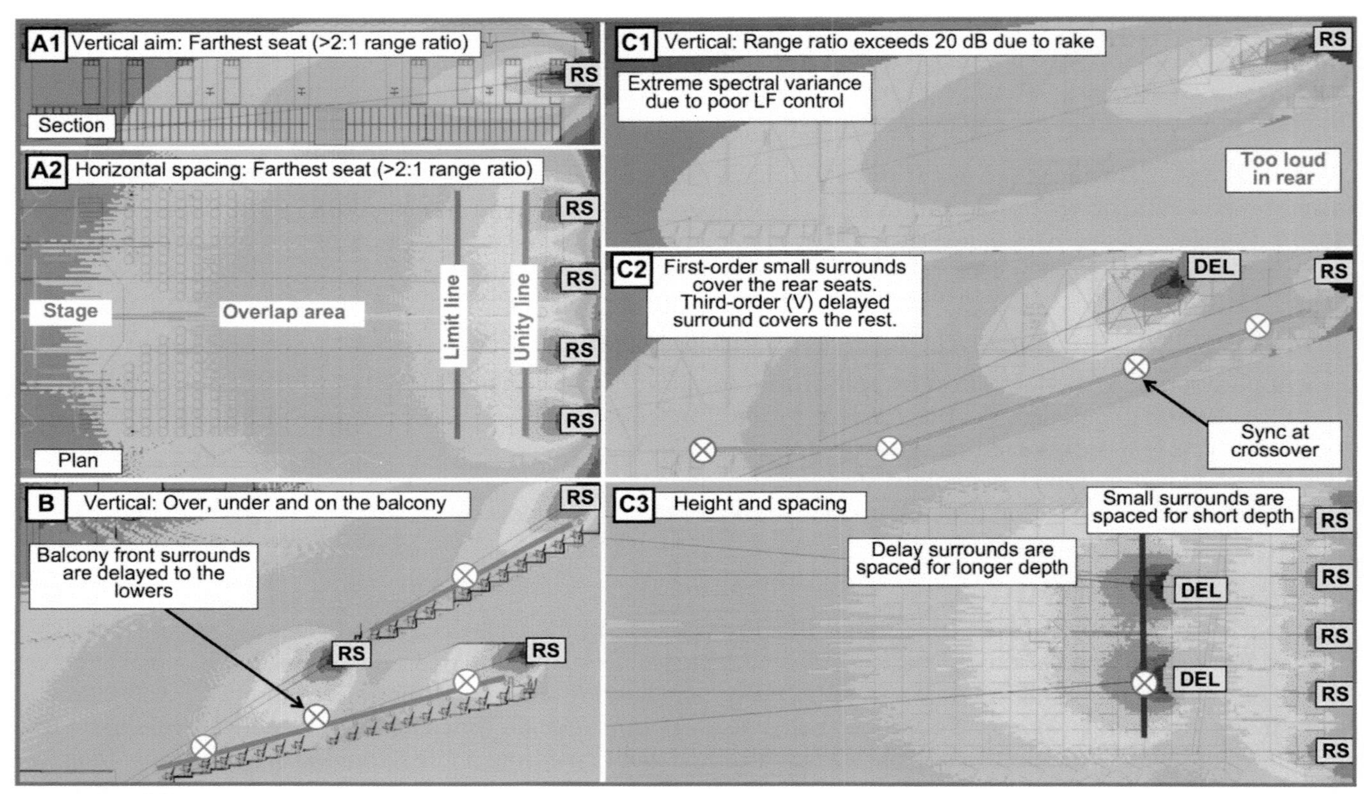

FIGURE 11.44
Design examples for rear surrounds. (A) Single flat floor vertical plane and rectangular horizontal plane. (B) Two-floor vertical with balcony delay. (C) Extreme range ratio is reduced by a delay system.

use third-order speakers, but they won't control anything below the HF because they are solo units in the vertical plane. Spectral variance: strike two. Ripple variance off the ceiling: strike three.

The worst-case scenarios are underbalconies where the rear surrounds are like a caged animal. Delays can come to the rescue, placed on the ceiling ahead of the rear wall units and/or on the balcony front. It's worth noting that these delays may cover longer ranges than the "main" rear surrounds (the ones they will sync to). This is a rare case of the fill leading the main. Balcony rail surrounds will typically throw longer than the units at the rear of either level, and therefore will be power scaled up proportionally.

11.8.3 Overhead surround (OH)

Overhead surrounds are functionally equivalent to the overhead mains described previously (11.6.7). The relationship of element coverage, spacing and range apply equally. The FX power scaling will likely be less than a mains application, but this is totally application dependent. *Peter Pan*'s Tinkerbell moves around a lot in the air. *Miss Saigon*'s helicopter moves a lot of air around.

11.8.4 Special FX

What do we call a Class 4 power-rated, five-element point source array inside a robotic prop that's used for only one minute for one voice? Ursula the Sea Witch, of course (Tokyo Disney Sea). That's a special effect source. It ensures every kid in this 360° surround theater knows exactly who's talking. There are a million unique special FX applications, but one common aspect to their specification process: each is a soloist that must cover as much of the space as possible/practical.

11.9 LOW-FREQUENCY SYSTEMS (LF)

Low-frequency systems lack isolation from each other or the room. The interactions are so spatially variant that we are faced with triage and TANSTAAFL as standard operating procedure here. We have previously discussed power scaling to the mains. This section focuses on the macro placement options. There are three places we are likely to see flown LF mains: near the L/R mains, above center stage or above the drummer (TM array for 360° arena applications).

11.9.1 Flown LF mains (L/R)

STANDARD FEATURES OF FLOWN LF MAIN SYSTEMS (L/R)

- Minimum-level variance: Audience is out of the close range (reduced range ratio).
- Maximum coupling with mains: Proximity to mains minimizes combing.
- Back lobe/side lobe cancelling options: Phase steering is possible when coupled close to L/R mains.
- Minimum floor coupling: Sub-floor-woofers (the free ones we get from a ground stack) are uncoupled from the flown LF.
- LF center coupling nose in the L/R center: Commonly known as power alley. The nose narrows with frequency.
- Spectrally continuous sonic image: LF range is linked to MF and HF range because LF and main locations are close.

11.9.2 Flown LF mains (C)

STANDARD FEATURES OF FLOWN LF MAIN SYSTEMS (CENTER)

- Cardioid array is highly recommended: Stage leakage will be excessive without cardioid configuration.
- Minimum variance: Location presents a low-range ratio, maximum coupling and minimum ripple.
- Spectrally semi-continuous sonic image: LF is vertically linked to mains but horizontally separated. Most LF material in the L/R channels is panned center so this is unlikely to be a detectible spectral sonic image distortion.

11.9.3 Flown LF mains (TM array)

STANDARD FEATURES OF FLOWN LF MAIN SYSTEMS (TM ARRAY)

- One hit wonder: It's "the bomb" for 360° center stage coverage. It bombs for all other applications.
- No short people: Must be at least 2 meters tall to provide sufficient vertical control.
- Minimum variance: No combing and no spectral variance over the 360° shape.
- Spectrally non-continuous sonic image: unlinked to mains (V and H). Low/near seats may sense disconnection.

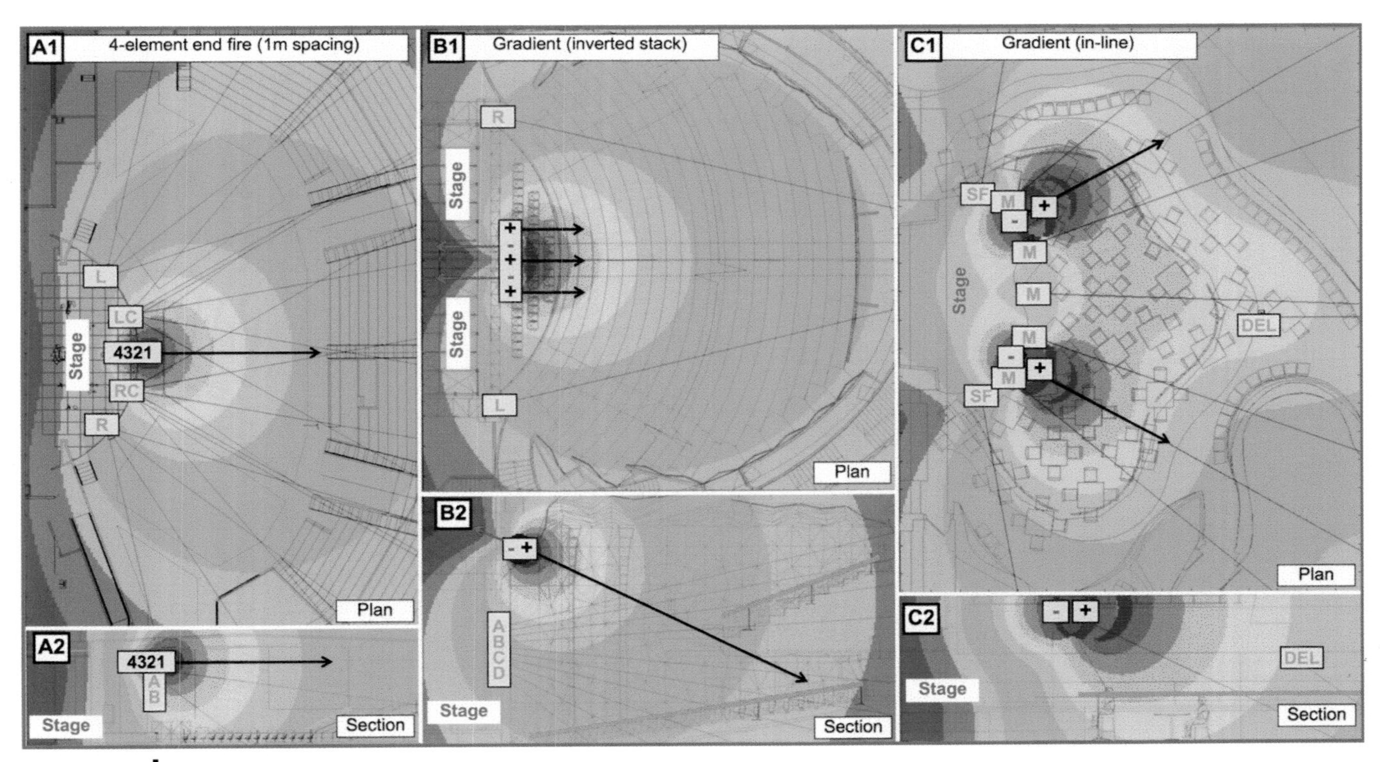

FIGURE 11.45
Design examples for flown LF systems. (A) End-fire. (B) Gradient (inverted stack). (C) Gradient (in-line).

11.9.4 Deck LF (L/R)

Stacked subs on left and right need no introduction. We know where they live.

STANDARD FEATURES OF DECK L/R SUBWOOFER SYSTEMS

- Cardioid recommended: stage leakage can otherwise be excessive.
- Maximum floor coupling: All elements are close to the floor.
- High level variance: Audience members are close on the first floor and far away in the balcony.

11.9.5 Deck LF (C)

LF devices can be spread across the stage front. These are buried under the stage in permanent installs, and often in front of the stage for touring rigs. They must be kept low profile, which limits vertical coverage options. Units can be spaced up to 2 meters apart. Cardioid steering can help minimize stage leakage. Speakers placed under the stage may respond unexpectedly to cardioid steering, i.e. they may not work or be less effective. Solid decks may provide enough isolation to reduce the need for steering (while also making it harder to achieve). Deck subs will couple in the room center and then begin a steady combing

STANDARD FEATURES OF DECK LF (CENTER)

- Multiple drive channels: recommended for cardioid control and beam steering.
- Cardioid recommended: stage leakage can otherwise be excessive.
- Horizontal beaming: prone to excessive narrowing unless beam steering is implemented.
- Maximum floor coupling: All elements are close to the floor.
- Sonic image: linked to stage (V and H). Separated from L/R mains (V and H).
- Keep CPR paddles charged: very loud in areas where people are sure to crowd in close.

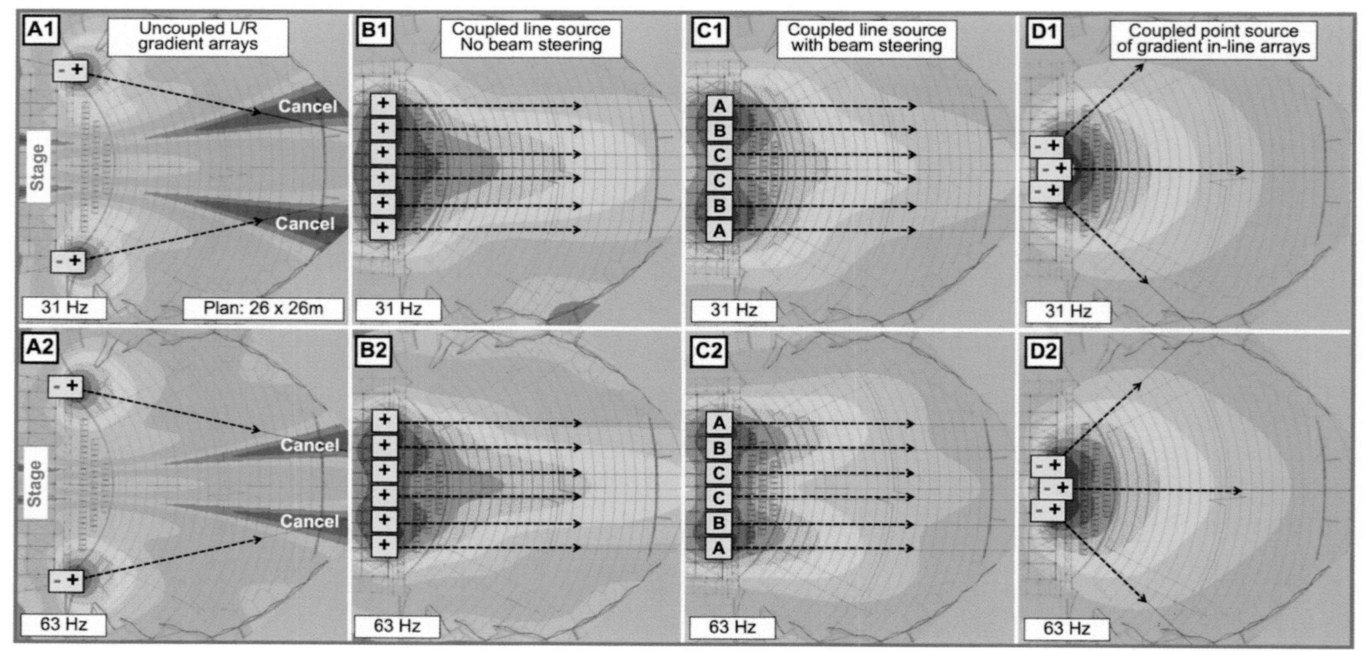

FIGURE 11.46
Design examples for LF deck systems. (A) Uncoupled L/R gradient. (B) Un-steered coupled line source. (C) Steered coupled line source. (D) Coupled point source of gradient elements.

FIGURE 11.47
Design examples for delayed LF systems. (A) No delay. (B) Delayed omnidirectional LF. (C) Delayed gradient. (D) Delayed end-fire.

11.9.6 Delayed LF

LF delays enable forward extension to areas that would otherwise lack the impact of high D/R ratios in that range. The limiting factor is backflow, which can be best managed by operating at low level and/or using cardioid steering.

STANDARD FEATURES OF LF DELAYS

- Cardioid mandatory: Must be cardioid configuration or stage leakage will be excessive.
- Minimum variance: triple bonus: high-range ratio, single beam to fit room and minimum ripple.
- Sonic image: linked to mains (V). Separated (H). It's moved to center but few will complain.

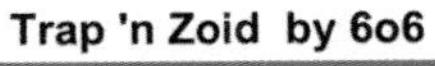

PART III

Optimization

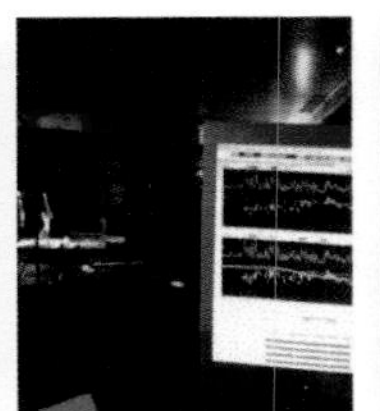
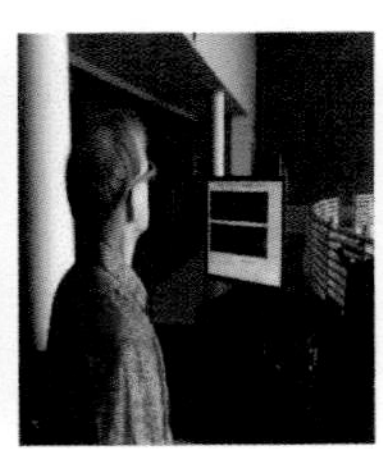

CHAPTER 12

Examination

It's time to put our sound design to the test. There are lots of questions we need answered. Was the system installed per our specification? Were the speakers hung correctly? Are they wired right? Is signal flowing to the right places? We will also question our design. We either prove the design or improve the design as needed.

The examination process requires objective measurement tools. Without examination, superstition and science are on the same footing. Examination puts an end to circular discussions about what might happen if we change something and or what might be the cause of some particular problem. The theories can be tested, the questions resolved and we can move forward having learned something.

> **examine** *n.* 1. *detailed inspection (of, into).* 2. *testing of knowledge or ability.*
> **examine** *v.t. inquire into nature, condition etc., of, test by examination, theory, statement; formally question.*
> Concise Oxford Dictionary

Examination tools take a variety of forms, from the simple physical tools used to measure a speaker angle to the complex audio analyzer. Each has a role to play in the big picture. The most challenging to us is the audio analyzer, the diagnostic tool used to monitor the variance in the electronic and acoustic signals as they move through the transmission path to the listener. A key role of the analyzer is to help us understand what we hear. These tools inform the designer and optimization engineer about the extent of level, spectral and ripple variance over the space and tell us how well the original signal has been preserved.

12.1 PHYSICAL MEASUREMENT TOOLS

12.1.1 Inclinometer

An inclinometer uses gravity to measure vertical angle. It's the standard speaker focus tool, available in mechanical (analog) and digital models. The iPhone app is the cheapest and most expensive. It costs only $1 until you drop it from the top of the cluster.

You already know how and why to use an inclinometer. Here a few pointers just in case.

- Make sure you are on a flat side of the speaker. Beware the trapezoid!
- Make sure you are on a flat side of the inclinometer. Beware the iPhone case!

12.1.2 Laser

The laser is another speaker focus tool in common practice. Accurate and reliable versions are widely available. A laser placed on the speaker or flying grid shows us exactly where it's pointed. We can locate ONAX in the house before even sending signal.

Once again there is no need for detailed instructions here, just a few pointers.

- Make sure you are on a flat side of the speaker. Beware the trapezoid!
- Presentation laser pointers are not accurate enough. If it looks like a pen, don't use it.
- A laser placed above a speaker (e.g. on the grid) lights a position above the actual speaker aim by a fixed vertical offset (in meters, not degrees).
- Same as above for a laser placed on the side of a speaker for the horizontal plane.

12.1.3 Disto

The laser distance or range finder sends a beam and reads its reflection. It can determine the distance traveled with high accuracy. I don't recommend this for measurement of delay times. (We have better tools for this.)

12.1.4 Thermometer/hygrometer

Sound speed is temperature dependent. A thermometer can tell you whether your system is running fast or slow today. We don't optimize systems for a particular temperature so this tool doesn't play a major part in the process. Temperature changes are a larger concern for ongoing operations such as festivals and long-running shows or venues with inadequate heating, ventilation, and air conditioning (HVAC) systems. Large-scale temperature shifts during performance can be dynamically compensated by adjusting delay times in proportion to the change.

Humidity can also be worth monitoring with a hygrometer for the same reasons because it affects HF air loss. Humidity readings can be observed and changes in the high-frequency response can be expected and compensated.

12.2 MEASUREMENT MICROPHONES

The microphone is our surrogate ear. Its faithfulness to our ear's response would seem to be the defining factor. The reality, however, is not that simple.

12.2.1 Comparing microphones to our ears

Microphones differ from our aural system in very distinct ways. Stage applications have strong factors that favor microphones far different from our ears. First, our ears are a binaural system separated by our head. It takes a pair of mics placed on an artificial head shape to recreate this experience, a practice known as "dummy head" or binaural recording. Binaural recordings are fascinating, but must be played back on an analogous speaker system: headphones. It works great for virtual-reality applications and some forms of acoustic research, but is not applicable to system optimization.

Another approach for simulating human hearing is a stereo recording pair. Mics are placed in close proximity and splayed apart. The individual mics are more directional than our individual ears but it creates a familiar experience. The mics must be placed in the far field to allow the stage sources to mix together. This is fine for recording but used only sparingly for sound reinforcement due to high leakage and low acoustic gain. Mic placements that mimic our hearing are out for the stage. What will work?

Directional control is a top priority for sound reinforcement stage mics. Stage sources and the main speakers will find their way into all open mics, which adds combing and coloration to our signal. There are several defenses against the combing. Close mic'ing with directional mics yields isolation. This creates a false perspective (section 5.5.4), such as transporting the listener inside the kick drum. The mix console is in charge of restoring a realistic perspective, typically through EQ.

A second defense is physical isolation on stage with baffles, drummers in fish-bowls or asking the musicians to confine their levels to the local area. All musicians want to hear themselves and *some* wish to hear others. This presents an additional complication: stage monitors. These sound sources introduce an additional leakage path that leaves us wanting to maximize microphone directional control.

A pop music group stage is an acoustical combat zone. Our countermeasure begins with maximum isolation and a plan to reassemble the pieces in the mixer, i.e. to fix it in the mix. Microphones that simulate human hearing are rarely up to this task.

12.2.2 Measurement microphone specifications

The mics used for system optimization are special-case recording models. There is no need for isolation. We are passively monitoring the transmission at various locations, not adding signal to the stream. Our listening experience is affected by both direct sound and the reverberant field, and our monitoring should reflect that. An omnidirectional mic, aimed in the direction of the sound source, is a reasonable approximation to a single ear's response. Both the pinna and a typical omnidirectional mic have a relatively small amount of front/back directionality in the HF range. Cardioid mics, by contrast, are far more directional than the ear and change their low-frequency response over distance (the proximity effect). A single omnidirectional mic provides no localization clues but the summation frequency response is a close approximation to the perceived tonal response. Therefore, our surrogate ears for transmission monitoring will be single omnidirectional mics placed at strategic positions.

Mics used for system optimization are classified as "free field" type. They yield a flat response on axis and slightly rolled off HF response as frequency rises. The alternate type of measurement mic is the "random incidence" type (also called "diffuse field"), which is designed for characterization of totally incoherent sources such as HVAC, environmental noise, etc. These will show a huge HF peak on axis and are not usable for system optimization.

Performance criteria for measurement mics goes beyond extremely flat frequency response alone. We need low distortion, and dynamic capability sufficient to measure the system transient peaks without clipping. The self-noise must be below the ambient noise level in our space. Moving lights have a nice way of making microphone self-noise academic. Another key parameter is stability over time and temperature. Cheap $20 measurement mics might match the $2000 models on Tuesday morning, but that does not ensure they will in the evening.

Transmission monitoring requires the mic to be moved around the room. Better yet, move multiple mics around. This makes mic matching a matter of paramount importance. Mics don't need to be matched to set delay times, but every other aspect of system optimization requires it. Mics should have matched sensitivity and frequency response. Most analysis systems can compensate for sensitivity offsets via software. This allows us to merge our mic kit on a job site and increase quantity. Some systems allow for software compensation of frequency response deviations. This should be undertaken with some caution as only the on-axis field will be factored into the correction. Ninety percent of the action in matching high-quality mics is above 4 kHz. Mics that diverge significantly below that raise concerns. In the end our choice is to use matched mic sets or put on our walking shoes (or both).

MEASUREMENT MIC SPECIFICATIONS

- **Frequency response**: 27 Hz to 18 kHz ±1.5 dB.
- **Coverage pattern**: omnidirectional.
- **Type**: free field (not diffuse field/random incidence).
- **THD**: <1%.
- **Max SPL without overload**: >140 dB SPL.
- **Self-noise**: <30 dB SPL (A weighted).
- **Sensitivity matching**: ±0.5 dB SPL (1 kHz).
- **Frequency response matching**: ±0.5 dB <4 kHz and ±1.5 dB above.

12.3 SIMPLE AUDIO MEASUREMENT TOOLS

12.3.1 Volt/ohm meter (VOM)

Volt/ohm meters (VOMs) are a standard part of the audio toolkit. We typically use it more for AC mains (power) than L/R mains (audio). The VOM provides AC and DC voltage testing, continuity and short circuit detection. AC voltage and current measurements can be average measurements calibrated as RMS, or true RMS (root mean square). The VOM's principal role is in the pre-verification stage of the system.

12.3.2 Polarity tester

The polarity tester, also known as the "phase popper," is a combination pulse generator and receiver. The generator drives the line and the receiver decodes the electrical (or acoustical) signal at the other end. The electrical readings should be reliable as long the response is flat over the full frequency range. Phase poppers are easily misled when measuring something that's not full bandwidth and/or has frequency-dependent delay. That eliminates all loudspeakers, so we can sum this up very simply: Use polarity testers for cables (only). Use batteries and FFT analyzers for speaker polarity testing.

12.3.3 Listen box

This is an audible audio test device, a battery-powered miniature amplifier and speaker that listens to the signal anywhere in the path. The high-input impedance minimizes loading. It's a fast and efficient tool useful in the pre-verification stage for locating where our signal is, and isn't.

12.3.4 Impedance tester

The impedance tester is an advanced tool for testing installed audio cables. It measures the AC impedance rather than just the DC resistance in the line. Its major advantage over the VOM is its ability to troubleshoot an installed cable with speakers, transformers and capacitors in line. The impedance tester can see the transformer impedance and give an accurate reading. DC blocking capacitors installed for HF driver protection will appear as an open circuit to a VOM, potentially misleading us to conclude the driver is blown. An impedance meter, however, will see through the capacitor to the driver. An impedance tester is highly recommended for 70 volt systems.

12.3.5 Oscilloscope

An oscilloscope is a waveform analysis device that displays voltage over time. Both axes are independently adjustable, allowing for the monitoring of virtually any electrical signal. DC voltage, line voltage and audio signals can be viewed. They are not often used for system calibration, but are a useful troubleshooting and verification tool. The oscilloscope tracks the peak-to-peak response of the waveform at any stage including the amplifier outputs. In the modern era this function is carried out by digital information networks that simultaneously monitor the multiple amplifier outputs.

12.3.6 Sound level meter (SPL meter)

SPL meters integrate level over frequency and time into a single dB SPL reading. SPL subunits correspond to user settings for frequency weighting (A, B, C and Z) and time constants (fast, slow, continuous, peak). SPL meters are operational rather than optimization tools. There are no calibration parameters for which the SPL meter is not out-performed by others (mostly the FFT analyzer). Calibration decisions are made by analyzing frequencies individually (the FFT) rather than integrated collectively (the SPL meter). The sound level meter is, however, the foremost tool for mix engineer bragging rights and outdoor concert law enforcement.

12.3.7 Real-time analyzer (RTA)

The standard RTA is a ⅓ octave audio analyzer. It has been used to tune tens of thousands of sound systems over the years. All of them in the past. The RTA is DOA. It is barely worth writing about so I will be brief. The RTA was the industry-standard audio analyzer back in the 1980s. Tuning a sound system was a one-step process: equalization.

The procedure had two steps: equalization while looking at the RTA, followed by equalization while not looking at the RTA, a.k.a. by ear. The process amounts to an internal argument between the RTA and its operator's ears.

We do things differently now. We *optimize* systems. Equalization is only one part of a process that requires data that the RTA can never supply us with. The RTA cannot give us clear enough information to identify reflections, aim speakers, set splay angles, set levels, delays. It can't tell time, compute phase or even polarity. It doesn't know direct from reverb, program from forklift. Its resolution is so coarse it can't even do equalization well. It's not even cheap any more!

There's no good reason to buy a stand-alone RTA. Modern FFT analyzers duplicate the RTA functions and also have the actual features we need to do a complete optimization. The number of things the RTA can do better than a modern FFT analyzer is zero.

12.4 FAST FOURIER TRANSFORM ANALYZER (FFT)

It wasn't long ago that the tuning of sound systems was seen almost exclusively in artistic terms. Engineers would play a tape or talk into a mic and walk around and make the important decisions about how the system was aimed, relative levels, crossover points, timing, equalization and more. Systems were admittedly not as complex as now. Or were they? It's true that most systems had fewer channels, subsystems and signal processing than currently, but in other ways they were much more complex. The modern signal path is often a one-stop flight from mix console to full-range powered speakers with a layover at the signal processor. In the old days it was like going go coast to coast on Southwest Airlines. Console to equalizer to delay line to crossover to limiter to amplifier and finally to the speaker. And that's just one path to one high driver! It is the simplification and standardization of modern signal processing and speakers that make it easy for artistic sound designers to immerse themselves in soundscapes of ever-greater complexity. If the components of these super-complex systems could not be optimized to work together in a predictable way, few designers would risk their shows on wayward subsystems that have creative ideas of their own. The modern audio analyzer is the tool of choice for reversing entropy in the sound system.

Analyzers have been out there for a long time, of course. But in the past they were crude tools, capable of only limited duties. The sound systems themselves were crude as well, and in many ways they have grown up together. The sound system moved out of the garage and the analyzer out of the laboratory in the 1980s, and they have been on the road together ever since. An analyzer used to weigh 100 pounds and cost more than a mix console. Now they weigh nothing and cost less than a microphone. Everybody has one, but there are lots of us that don't really know what's going on inside of these tools and what the data really means. Our chances of making good tuning decisions increases greatly if we understand the perspective of the modern audio analyzer. This section is a look under the hood. We won't dig deep enough to ready you to design your own analyzer, but with the goal of making you a better driver.

12.4.1 Fast and Fourier

The FFT analyzer is the industry-standard optimization tool. First let's reverse engineer the analyzer's name. The "Transform" is the conversion of a sampled waveform from amplitude vs. time to a series of amplitude and phase values over frequency. The simple expression for this is a transformation from the time domain to the frequency domain, e.g. a continuous stream of music goes in and is separated into slices by frequency.

The Fourier Transform (FT) is the practical application of the Fourier theorem, named for eighteenth-century mathematician Jean-Baptiste Joseph Fourier. The Fourier theorem goes both directions: time domain to frequency domain and back again. No waveforms were injured in the making of this transform (in its purely theoretical form). This was shown back in Fig. 1.1.

The "Fast" part is where things get tricky. The Fourier theorem requires infinite iterations to be infinitely accurate. We have a show at 8:00 p.m. We can't wait forever for the most correct answer so we end the calculation when we have sufficient accuracy. That's the "Fast" part which moves the transform from theoretical perfection into the practical world and into an algorithm we can program into a computer. Computers *hate* infinity!

The FFT analyzer gives us amplitude and phase over frequency in more detail than we can ever use. Great. Unfortunately, the raw data is marginally usable in the practical world of optimization. We had to build in a lot of special features to adapt this tool for our trade. Here are a few examples of the limitations of the basic single-channel FFT analyzer. The linear frequency response is mismatched to our log hearing system. A known (and precisely constructed) input signal is required. The phase response is like a clock with just a second hand. The list goes on.

How do we get around this? We throw money at it. We stack up lots of analyzers in series and parallel to make a composite system that suits us. The standard frequency and phase responses of modern analyzers appear as single lines but are actually derived from the computations of around sixteen FFT analyzer modules. Details to follow.

12.4.2 Transform

The basic transform flows from sampled waveform to time record to real and imaginary numbers to magnitude and phase over frequency. First item of business: The real and imaginary numbers are on a "need to know" basis. We don't need to understand this to tune sound systems. This leaves us with two transitions: input to time record and then onto the frequency domain. The sampled input waveform runs by the same rules as the digital audio we deal with every day, i.e. we sample at >2× the highest frequency we want to use. We will use 50 kHz as an example to get a usable 22 kHz bandwidth.

The 50 kHz sample frequency yields time increments of 0.02 ms. Let's start the counting game. A set of 250 samples loads a 5 ms stream of data into the buffer (250 × 0.02 ms). The 5 ms sample period enables one cycle of 200 Hz to exactly fit into the buffer. The time/frequency domain transform begins its interrogation of the buffered data.

Let's listen in on the conversation. "Did anyone here complete exactly one cycle? If so, (a) how big are you? and (b) what part of the phase cycle were you in the middle of the time buffer?" The former tells us the magnitude at 200 Hz, and the latter tells us the phase. The next step is to ask if anybody completed two cycles, which gives us the status report on 400 Hz. The process continues for as long as we like (until we reach the highest frequency allowed by the sample rate). In short, a 5 ms sample slices the spectrum into 200 Hz (linearly spaced) increments (Figures 12.1 and 12.2). The linear spacing provides uneven resolution to our log hearing (1 octave between 200Hz and 400 Hz, ½ octave between 400 Hz and 600 Hz, ⅓ octave between 600 Hz and 800 Hz, and so on).

12.4.2.1 INFINITY AND BEYOND

The first challenge is linear data in a log world (the one in our heads). The second is subtler. What about frequencies that aren't integer multiples of 200 Hz? How do we count 300 Hz? Do we spread its amplitude across the 200 Hz and 400 Hz bins? The phase value for a signal that completed 1.5 cycles won't be the same at the beginning and ending of the time record. Which is right? What can we do? We can slice it into smaller increments by capturing a longer waveform. Ten milliseconds lets us count in 100 Hz increments. We can see 300 Hz now, but what about 350 Hz? There can always be something in between no matter how fine we slice it. Infinite resolution requires an infinite time record (the FT's infinity challenge discussed above).

The price for finite measurement is assuming it matches what we *would* see if we kept measuring. The sample is treated as if it "always was, and always will be." What we really need is confidence that it accurately represents right now and the foreseeable future. Good news! It can.

Let's look at this challenge with an analogy. If our time sample were the complete song "Stairway to Heaven" we would be operating under the assumption it has been played continuously back-to-back *forever*. Yes, I know it *feels* that way. But the loop must have the *entire* song, so that the quiet, whiny ending meets the quiet, whiny beginning. If the time record cut off before the song ended (while Jimmy Page was still wailing with his amp set to 11), the restart would be abrupt and we would know the song does not remain the same. Yes, that is a long analogy, but the alternative is integral math equations.

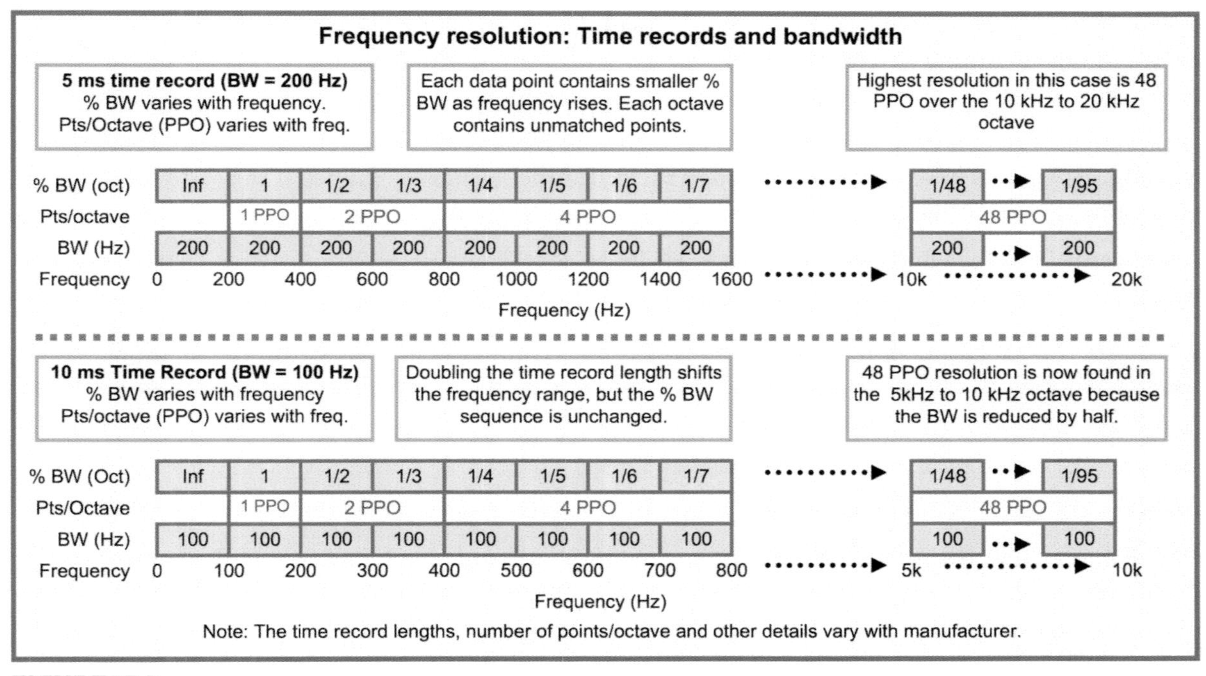

FIGURE 12.1
The relationship between the time record length and frequency resolution in the FFT analyzer

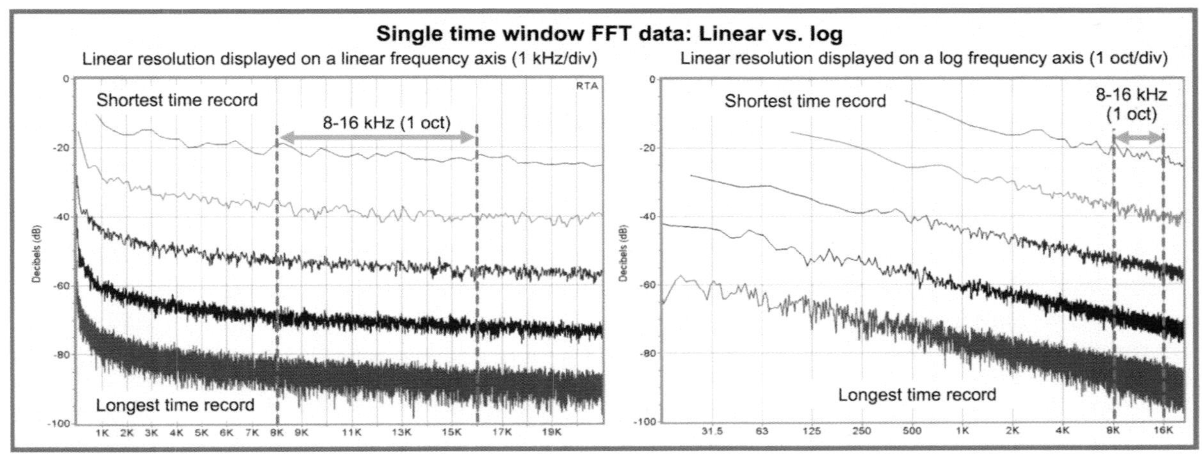

FIGURE 12.2
Linear vs. log frequency resolution: A single FFT response computes the data with linear frequency spacing, displayed here with both linear and log frequency scaling. The linear/linear display (left) shows equal spacing between frequency bins. The linear/log-spaced version shows wide LF spacing and progressively smaller spacing with frequency. Linear frequency resolution depends upon time record length, shown here in sequential quadrupling. Note the poor LF resolution (short time record) and the excessive HF resolution (longest time record) (courtesy of SIA-SMAART).

12.4.2.2 WINDOW FUNCTIONS

This is where the middle section, the time record, comes into play. The time record is a modified version of the raw sampled waveform, making it ready for the transform. We can't place limits on program material and we don't have all day. The time record must give reasonably valid representations of all frequencies within the bandwidth. We can give up perfection for the frequencies that are integer multiples to gain equality for all.

The remedy is a time "window," a mathematical simulation of a gain control between the buffered waveform and our second stage: the time record. The time record is treated as if it's a sample of an infinitely repeating sequence. This construct only holds up if the end of the time record matches the beginning. The time record must be circular, like a tape loop. Audio engineers can visualize the window function as a VCA or time-triggered gate. Let's resume with our example 5 ms time record. It's closed at the start (0 ms) and later begins to open. The window eventually opens fully and then closes when 5 ms has elapsed. The final product is a modified version of the original waveform with amplitude weighting that favors the middle over the beginning and ending.

How does this help? First, we are assured that our Fourier transform assumption (the time record can be looped) has been satisfied. Second, we can see there are certain to be costs to modifying the waveform. The costs vary by program material. Sine waves have distortion added to them. Transients may be ignored if they arrive at the beginning or end among other things.

There are different versions of the window functions and each has its own favorite source material. (None are fans of Zeppelin though.) All windows close at the beginning, open fully in the middle and close at the end. It's like an envelope follower where window type has unique rise and fall slopes. The Hann and Blackman-Harris windows are often used for random sources (noise or music), whereas the Flattop window is favored for sine wave testing. How big are these errors? Most are 40–60 dB down from a full-scale signal. Perfect? No. End of the world? Not sure. We'd have to measure to infinity to know.

12.4.2.3 FINDING PHASE

We now have a full set of linear frequency response data, with amplitude and phase. The amplitude part is straightforward. Bigger is bigger. The phase part, not so much. The phase value we've acquired is just a position on a circle relative to . . . um, something. A steady sine wave will have a repeating sequence of phase values, but random signals will have random phase. Simply put, we can't do anything with phase data until we have a time reference for comparison, which is to say *relative* phase. A single-channel analyzer can stake out a fixed time reference. A dual-channel analyzer uses one channel as the reference for the other, which allows us to accurately track relative phase with known or random sources such as music. This is a key benefit to the dual-channel analyzer.

12.5 FIXED POINTS/OCTAVE (CONSTANT Q TRANSFORM)

The ultimate analyzer would acquire high-resolution data in complex form with a display that's intuitive to our ears. It could be derived from linear data (including phase), but displays a constant frequency resolution per octave: a lin/log hybrid. One option is to simply stretch and squeeze the frequency response display to make the linear data fit in a log display. Visualize a Slinky. This is a video solution for an audio problem. The challenge is resolution, not display: very low resolution in the LF range and very high resolution in the HF. Our 5 ms example has frequency data points every 200 Hz. That's 0 Hz, 200 Hz, 400 Hz . . . 19,600, 19,800, 20,000. Notice that we skipped over the subwoofers? On the other hand we have fifty slices in the top octave (10k–20 kHz).

A longer time record increases LF resolution (a good thing) and HF resolution (too much of a good thing). This is yet another battle with infinity. Instead let's try a compromise approach: Take both short *and* long time windows. We'll take the best and leave the rest. We take multiple FFT measurements in parallel and patch the responses together into a single "quasi-log" display. Separate time records are taken, each sized to target a particular octave.

The scheme is a sequential doubling of time records (5 ms, 10 ms, 20 ms and so on) (Fig. 12.3). Visualize a series of cameras with different shutter speeds operated by a single trigger. Each doubling halves the sample frequency resulting in a matched number of data points an octave lower. We'll use only a single octave of each, in our example case 48 points/octave. The chosen octave from each band is spliced together to make a composite full-range response with consistent per octave resolution. The linear resolution within each octave is stretched to create a log display. The lin/log stretching of a single octave is barely detectible (especially compared with the full-range stretching we'd get if we used only a single time record). The process can continue infinitely but reaches a practical limit due to the expanding time record length.

The fixed points/octave (PPO) measurement window includes a frequency/wavelength proportional ratio of direct sound and late arrivals. A base resolution of 24 PPO includes the direct signal and late arrivals within a 24 λ duration. This makes it easy to remember: 24 PPO = 24 λ. Late arrivals (reflections or other summations) beyond this limit are seen as uncorrelated to the original signal, i.e. noise. Therefore, we are seeing a relatively constant ratio of direct sound to early reflections over frequency. Contrast this with any single fixed time record, where we would

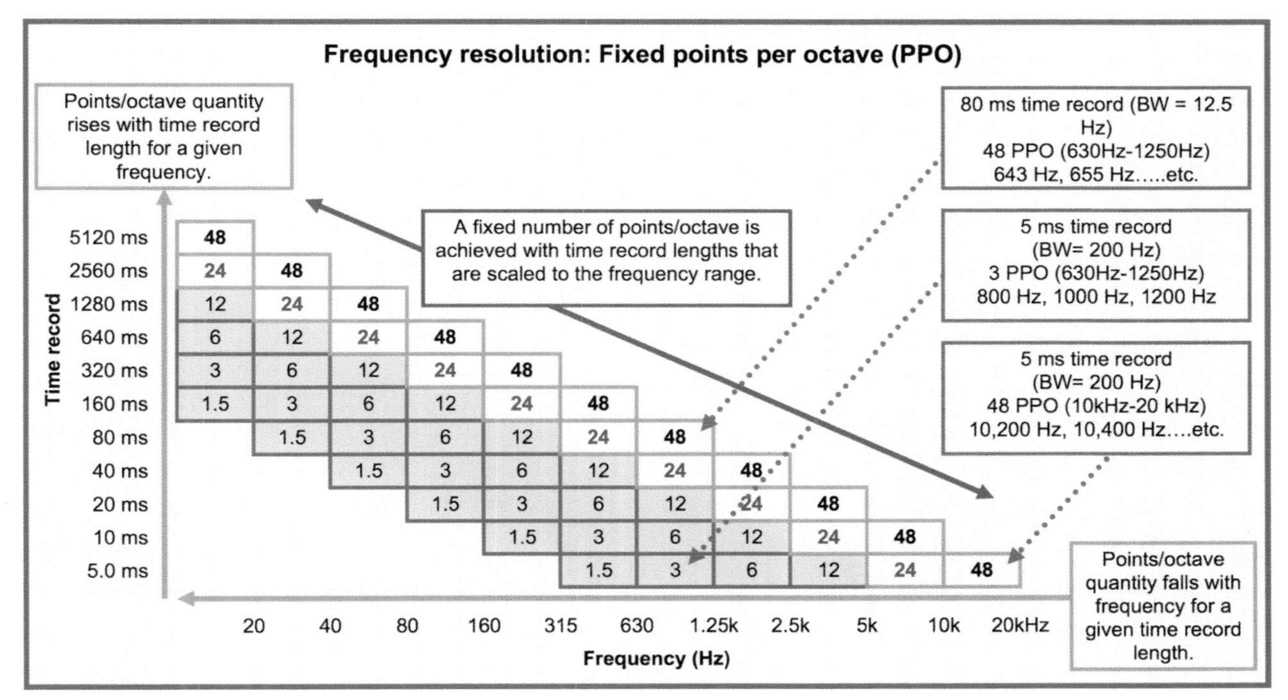

FIGURE 12.3
The fixed points/octave (PPO) also known as the Constant Q transform

see a different number of wavelengths for every frequency (more HF than LF). Consider an analyzer with a single time record length of 24 ms. That's a 24 λ window at 1 kHz, which corresponds to the edge between perceived spaciousness and separation (section 5.3). It also includes 240 λ at 10 kHz (perceived as an echo) and only 2 λ at 80 Hz (perceived as strong tonal variation). Our ideal analyzer shows tonal, spatial and echo perception equally over frequency. This requires the multiple time records of the fixed PPO transform.

How much frequency resolution is enough? The answer depends on the question we want the analyzer to answer. Splay angle questions require far less resolution than EQ questions. Excess resolution is better than not enough, so we'll focus on the minimum needed for a particular decision.

Equalization and comb filter identification require the highest resolution. The tonal/spatial/separation perception zones come into play in these decisions. We have seen that 1/6 octave resolution is regarded as the "critical bandwidth" for perception of tonal variation. We also saw that frequency response ripple beyond 1/24 octave resolution (24 λ) is perceived as separated rather than spacious or tonally modified. Therefore resolutions of 1/24 to 1/48 octave will be sufficient for our primary needs in the frequency domain (see Fig. 5.10)

Recommended frequency resolution for specific tasks

- **Noise floor**: low resolution (⅓) OK for general use.
- **Hum and oscillation**: high resolution (1/24) to identify individual frequencies.
- **Ripple identification**: high resolution (1/24) required to see details.
- **Equalization**: high resolution (1/24) best for identification of center frequency, bandwidth and level of peaks and dips to be equalized.
- **Phase alignment**: high resolution (1/24) required to monitor best-case coupling.
- **Level setting**: medium resolution (1/6) enough to identify 0 dB point.
- **OFFAX identification**: medium resolution (1/6) enough to identify -6 dB point.
- **XOVR identification**: medium resolution (1/6) enough to identify -6 dB point.

We must define three parameters to characterize a peak or dip: center frequency (F_c), maximum deviation (±) and percentage bandwidth (BW). This requires at least three frequency data points: F_c and those where the response has returned to unity (above and below). Additional resolution fills in the shape between these points.

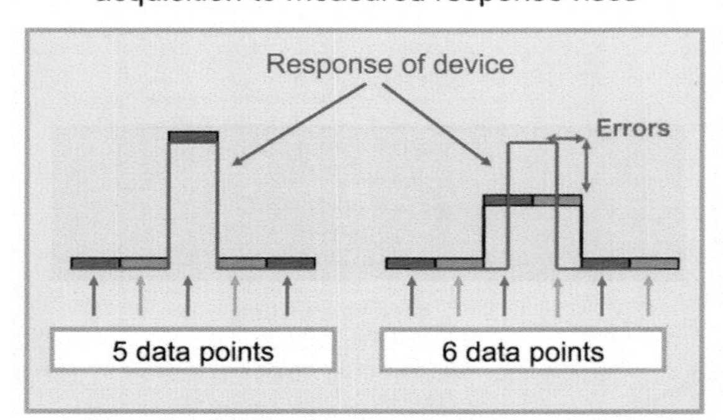

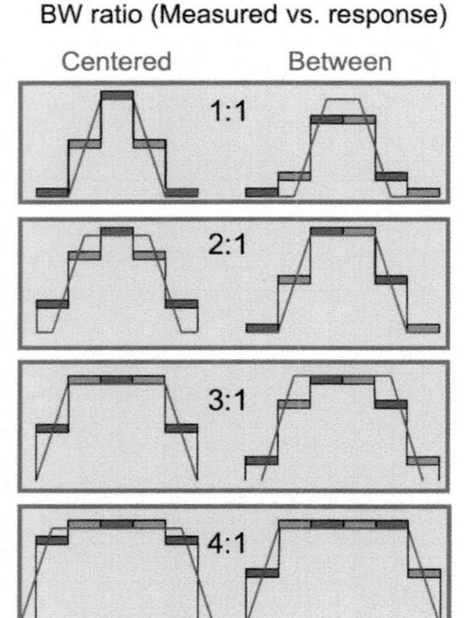

FIGURE 12.4
Bin/accuracy issues

The analyzer divides the spectrum into data points, "bins." Each bin has an F_c and effective BW that connects to the adjacent bin at unity gain. The F_c and BW of a measured device's peak or dip may (or may not) match the bin's F_c and/or BW. These two open variables create a range of possibilities. F_c can match one bin or fall somewhere between two. The device's BW might match the bin's BW or be narrower or wider.

A sequence of scenarios with varying ratios of analyzer resolution vs. a device's variations is shown in Fig. 12.4. We might get lucky and match (good accuracy) or fall between the cracks (errors) when our analyzer resolution equals the bandwidth variations of the measured device (1:1 relationship). The error window shrinks as the ratio goes up. A 4:1 ratio of "over sampling" reduces the error rate enough to characterize F_c and BW without concern about whether they are on a bin center. Therefore a 1/24 octave analyzer ensures enough accuracy to perform equalization within the 1/6 octave tonal perception zone (4:1 oversampling* 1/6 octave = 1/24 octave).

12.6 SINGLE-AND DUAL-CHANNEL ANALYSIS

12.6.1 Single-channel spectral analysis

Single-channel analyzers compare to a fixed standard (Fig. 12.5). The unknown is referenced to the known (some standard value). Single-channel measurements, referred to here as **spectrum**, require a single unknown: the output response. This is compared with an internal reference standard such as 0 volts or 0 dB SPL.

SINGLE CHANNEL ANALYZER PRIMARY APPLICATIONS

1. Monitor source frequency content, level and dynamic range.
2. Total harmonic distortion (THD).
3. Maximum input and output capability.
4. Hum and noise floor.

The above verification procedures are detailed step-by-step in Chapter 13.

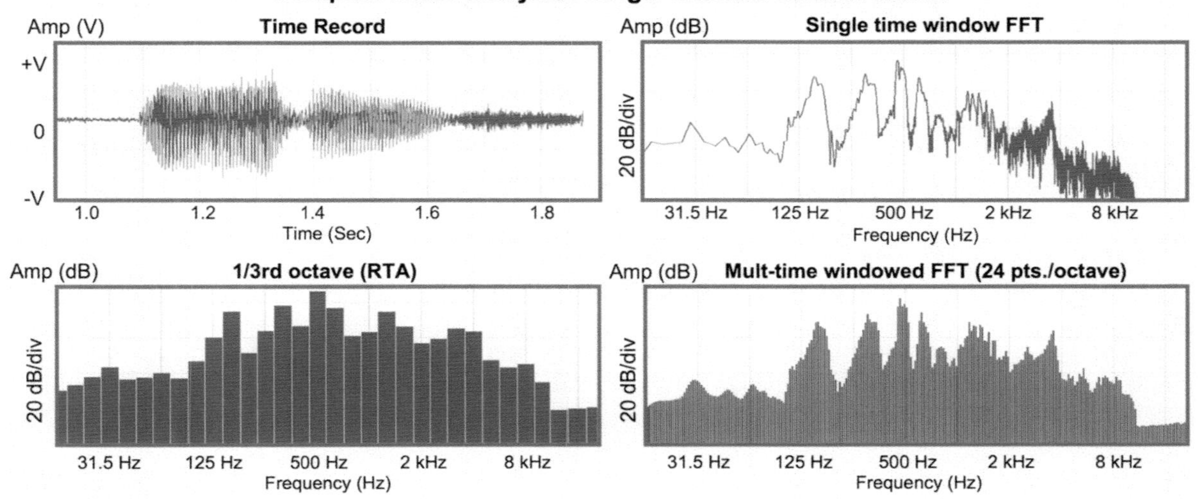

FIGURE 12.5
Single-channel spectrum of a voice measured with a multiple time-windowed, fixed points per octave FFT analyzer (24 PPO) (courtesy of SIA-SMAART)

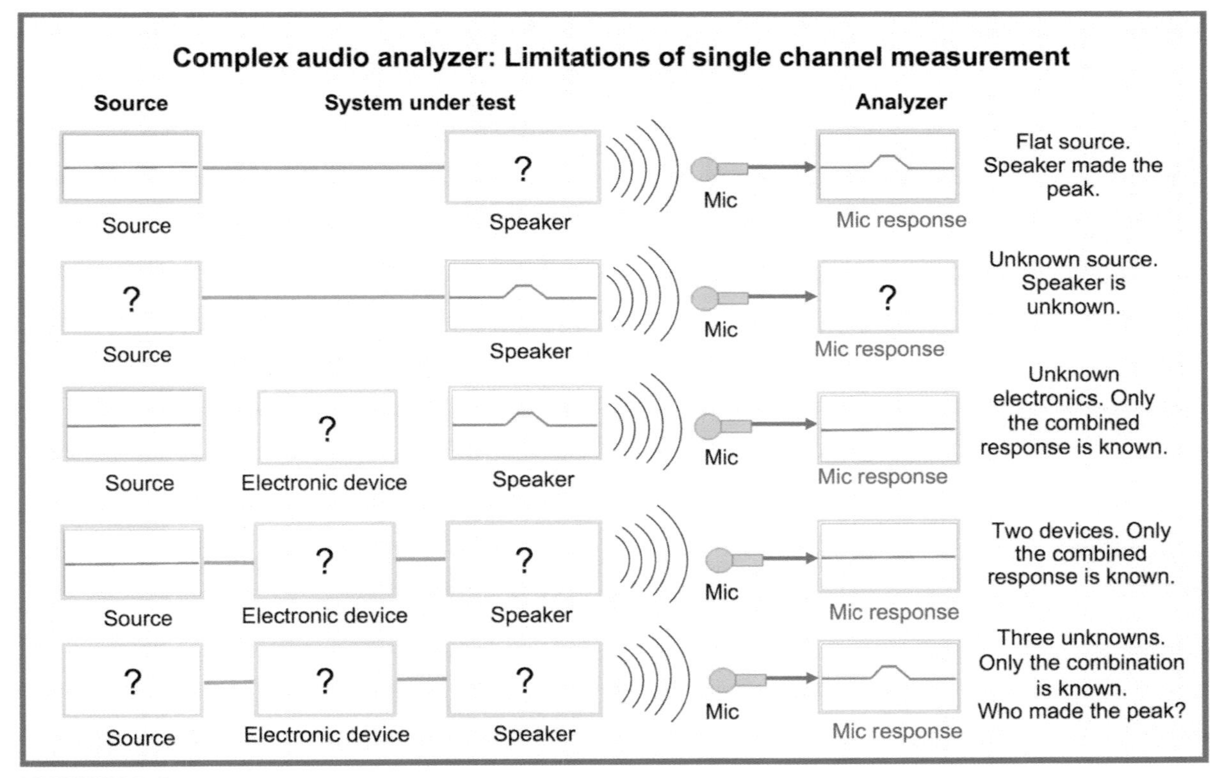

FIGURE 12.6
Limitations of single-point audio measurements

Single-channel measurements have limited applications because they require a known signal (Fig. 12.6). Music flowing through an equalizer is seen by the analyzer as the music *and* the equalizer. What part is the music, and what part is the equalizer? The equation has two unknowns: input (the music) vs. output (the music and the equalizer). This is fine if we only want to see the spectrum flowing through the system. It's of limited use in deciding what to do about it. This is observation, not optimization.

Single-channel frequency response measurements use pink noise or other source signals known to have a flat spectrum. THD measurements use a low-distortion sine wave, whereas noise floor measurements require no source at all.

A second limitation arises with multiple components in series, i.e. every sound system. The known source drives the first input and we don't see it again until the last output. A single-point measurement system can't discern the individual effects of the various components without breaking the signal chain or making assumptions. We know something happened but we don't know which device(s) did it. The first device passes an unknown output to everybody downstream. The second device has unknown input *and* output. Bzzt! Thanks for playing.

12.6.2 Dual-channel analysis (transfer function measurement)

We have gone to a lot of trouble to create a high-resolution frequency response. It's essentially a high-definition real-time analyzer until we add a second channel. Then it's a whole new ball game. The RTA doesn't know processor from speaker from room, our show from a forklift, or direct sound from recycled reflection. By contrast, the dual-channel FFT analyzer tells us whether it knows the answer or is guessing via the coherence function. This won't assure good decisions, but surely ups the probability. The RTA only answers amplitude, leading users to believe that EQ is the only question. The modern dual-channel FFT guides us to see sound system challenges in their full complexity. Better diagnosis makes for better treatment. Speaker aiming, splay, spacing, delay, crossover alignment and level tapering are all aided by the information provided by transfer function measurement. We can tell time, detect distortion, compression and changes in sound speed while the band plays and audience dances (Fig. 12.7).

The term "transfer function" describes the behavior of linear devices in the signal path. Devices can be passive, such as a wire, or active, such as an equalizer. Transfer function analysis sees whatever goes on between the input and output of a device. The difference between the gozinta and the gozoutta.

The math of the transfer function is complicated under the hood but intuitive on the exterior. For level analysis it's modeled as division: output over input. For temporal analysis (time and phase) it's subtraction: output minus input.

Let's start simple. We have a +6 dB transfer function amplitude ratio when a 1 V input generates a 2 V output (out/in = 2:1). We have a transfer function time of 1 ms when a 1 ms input generates a 2 ms output (output minus input = 2–1 = 1 ms). A linear device maintains a constant transfer level and transfer time, and will have predictable output behavior *relative* to its input. We can expect an output of 8 V @11 ms when we put in 4 V @10 ms. This operation takes place 48 times per/octave to give us transfer amplitude and transfer time (phase) at every frequency.

The beauty of transfer function analysis is portability. Pick any starting and stopping point in the signal path and see the difference in level and time. We can span from console input to the sound at the last row, a single device

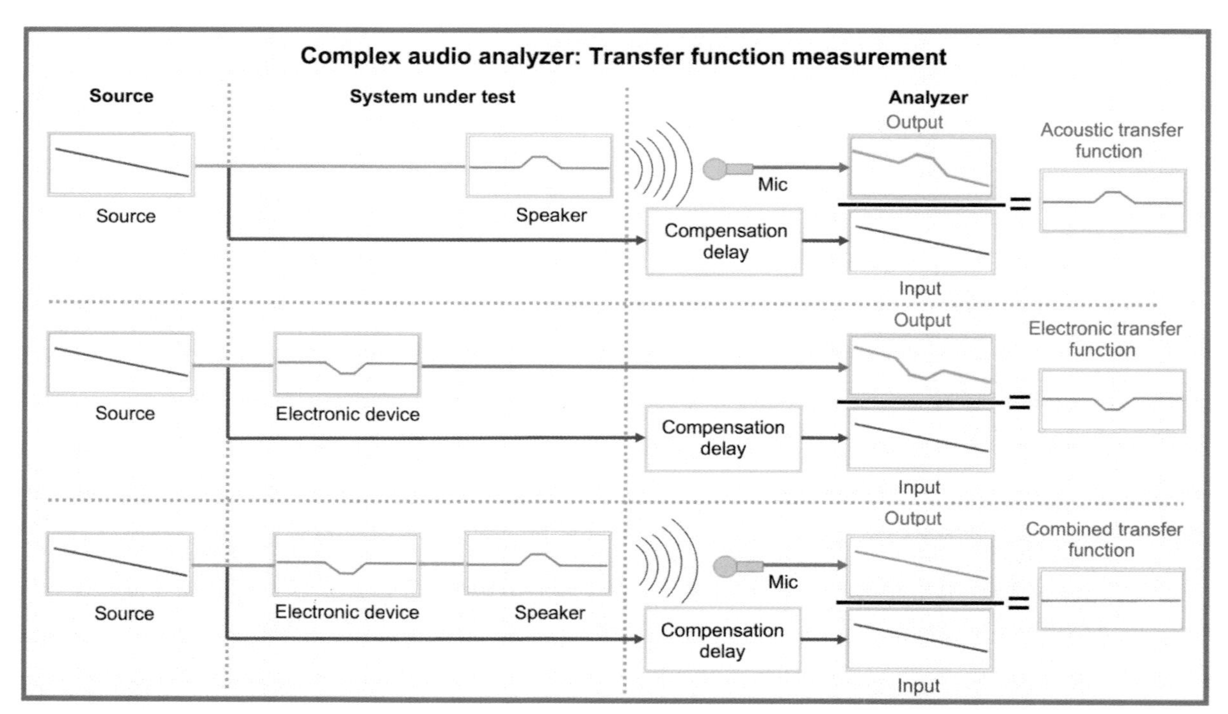

FIGURE 12.7
Two-point transfer function measurement flow block. The acoustic, electronic or combined response can be found between the two measurement access points.

or a single resistor in a circuit. It's physically two parallel probe points (in and out) "Y'd" into the analyzer. Signal flows through the system as normal and we tap off a copy for analysis.

Let's put the concept of linear relationship to bed by some examples. One kilohertz alone goes in, 1 kHz alone goes out: linear. One kilohertz goes in, 1 kHz and 2 kHz show up at the output: non-linear (harmonic distortion). Pink noise in and pink noise out: linear. Punk noise in and (the same) punk noise out: linear. Good jazz in and Kenny G out: non-linear. We can get a valid transfer function using a random source because it's not random to us. We have a copy of the input. Even the best system we measure has some non-linear behavior. It has a noise floor and some distortion. If a system has too much noise, or too much distortion, the transfer function data is not stable, reliable or repeatable.

The advantage to source independence, i.e. being able to use random signals such as music, is extremely obvious, especially when the lights go down. There is a catch though: the two channels must be synchronized when they arrive at the analyzer inputs. The relationship between matched signals becomes (for lack of a better word) random, when they are out of sync. Any transfer function with a speaker involved is going to be out of sync because acoustic propagation always loses the race. This is handled by placing a delay line in series with the analyzer inputs and compensating for any differences in arrival times. The delay affects our analysis but is not part of the sound system.

We can illustrate the challenge with a few variable sequences. We put "aaaaaaaa" in and get "**aaaaaaaa**" out. Analysis reveals that the device amplifies the signal to bold-face type. Easy. It wouldn't matter if we sampled only one "a" or all eight. And it wouldn't matter if we compared the first four output letters with the middle four input. We can fall out of sync and not care, as long as we have an endlessly repeating sequence.

Change the input sequence to "abcdefgh" and we get "**abcdefgh.**" It still doesn't matter if we sample one letter or all letters, as long as we are in sync and compare the same letter. If we are *not* in sync then "cdef" at the input could yield "**abcd**" at the output. Now our analyzer is confused about the bold-faced liar it's measuring. So it is with music, at least the kind that is not an endless repeating sequence.

The transfer function of a hypothetically perfect transmission system would be zero. Zero change in level, zero delay, and zero noise at all frequencies. Actual devices have some deviation in transfer level and time, and add noise. Transfer function measurement detects the changes and displays them in various ways. Recall how the single-channel spectrum related the output to a fixed standard. Transfer function measurement is relative to a variable standard (the source signal) and therefore uses relative terms: relative amplitude, relative phase, relative time and S/N ratio.

Source signal at the input channel (often termed the reference) is designated as "known" and the output signal as "unknown" (often termed the measurement). This source becomes the standard used for comparison whether it's music, speech or random noise. This should help clarify how transfer functions distinguish between "signal" and "noise." When pink noise is our source, we call it "signal" (pink signal?). When unrelated signal (any color) appears at the output, we call it "noise."

The source must contain energy at a given frequency for us to make conclusions about it. We know nothing about 1 kHz until it's sent in and we see what comes out. A full-range signal is required but it doesn't have to happen all at once. Data can be averaged over time, allowing the use of less dense input signals.

Certain conditions must be met to obtain a valid transfer function measurement. They can be approached, but never perfectly met, so we will view them in practical terms.

CONDITIONS FOR A VALID TRANSFER FUNCTION

1. **Stability**: Device under test (DUT) must have stable transfer level and transfer time during an individual sample period. For example, we can't measure a speaker on a bullet train (Herlufsen, 1984, p. 25).
2. **Time invariance**: DUT must not change its response over time. This refers to the averaging period of multiple samples. A speaker's response will change with temperature, but, if the weather holds over the averaging time, the measurement is valid for that period (Herlufsen, 1984, p. 25).
3. **Linearity**: DUT must be linear. Output must be proportional to the input. We can have gain or loss, filters, delay and more but they have to be stable. Clipping, distortion, compression, limiting, companding and any dynamic processing are non-linear (Herlufsen, 1984, p. 25).

We'll never make a perfectly valid transfer function but we can do plenty with our imperfect data. The environment where loudspeakers live is a very dense jungle and transfer function measurement has proven effective at capturing the response and taming it. Various aspects make this environment hostile. There is noise, lots of noise. There is non-linearity and there are changing weather conditions. The most hostile of all, however, is an audience being subjected to pink noise. (Use Pink's music instead).

KEY FEATURES OF THE DUAL-CHANNEL FFT

1. **Source independence**: Able to use program material as the source. Analysis continues with audience present. Can you think of an application for that?
2. **Non-intrusive access**: Two measurement points (input and output) can be fed from any two points in the system without interrupting the signal path.
3. **Complex frequency response**: Dual-channel method provides relative level, relative phase, relative time and signal relative to noise data.
4. **Best fit for non-linear data**: Non-linear response is detected and shown in coherence trace. Analyzer has a copy of the source signal to compare with the output.
5. **Noise immunity**: Able to identify noise (as the non-linear behavior above). Minimizes errors in the transfer response by averaging.

12.7 TRANSFER FUNCTION AMPLITUDE

The relative level between output and input is commonly expressed in dB. Transfer function amplitude is simplistically modeled as division:

Output/Input = Transfer function amplitude

Gain is shown when output > input. Loss/attenuation is shown when output < input. Transfer function amplitude is source independent. The source drive level or frequency content will not affect the outcome of the measurement, provided it is sufficient to rise above the noise floor and below clipping. Response will not change as long as the output/ratio is constant regardless of input signal.

12.8 TRANSFER FUNCTION PHASE

A frequency response is a series of solitary amplitude and phase values. A single amplitude value needs no further explanation. Unity is unity. A gain of 6 dB is exactly what you think. Phase is different. We can't conclude anything from a solitary phase value. The technique for reading phase response is literally best described as "connect the dots."

Let's begin with the simplest one: 0°. It seems like 0° should correlate to unity time (no time difference between input and output). Not necessarily. Unity time could be a reading of 180° if there is a polarity reversal between the measured channels. A lot of different things can lead to a 0° reading. Let's see how many we can come up with in 60 seconds.

Six degrees of separation (0° version)

- Unity time (0°) + normal polarity (0°) = 0°.
- 1 λ behind (360°) + normal polarity (0°) = 0°.
- 1 λ ahead (360°) + normal polarity (0°) = 0°.
- 2 λ behind (720°) + normal polarity (0°) = 0°.
- 0.5 λ behind (180°) + reverse polarity (180°) = 0°.
- 1.5 λ behind (540°) + reverse polarity (180°) = 0°.

All of these possibilities are real. The single value does not provide us the *context* we need to discriminate between them. It's just position on a circle. It's like receiving the answer "22 seconds" to the question "What time is it?" It's true, but not helpful. We need the minute, the hour, the day and the year to put "22 seconds" in context.

Finding the phase of a second frequency enables us to link two values. The connecting line is the phase slope, which can be decoded into time by discerning the rate of phase shift over frequency. Let's break it down. First we find the slope direction, read left to right on the frequency axis. A downward slope is delay (output after the input). This is normal and intuitive. An upward slope indicates anti-delay (output before the input). This is also normal and extremely unintuitive. There are two places you are likely to encounter this: any time we have an internal delay in our analyzer and with filters in our measured circuit. The first is easy. We use an internal compensation delay to sync up output and input. The phase slope rises if we put in too much delay. We see this when the temperature rises in a room, because the speed of sound accelerates and gets to our mics quicker. Wrapping our heads around anti-delay in the filter scenario requires a PhD in filter theory and even then it's controversial. Suffice to say that our rendering of the phase in filters is a simplification of a more complex reality. Nonetheless, you will see both a rising slope and falling slope when you view the phase response of a filter.

Decoding the slope into timing information comes next. It's linear, so the slope's rate of change is clocked in Hz, not octaves. It's useless to analyze the entire 20 kHz bandwidth in one shot, so we'll break the spectrum down into digestible pieces. Let's choose a range of frequencies and start. A shift of 90° between 250 Hz and 500 Hz decodes to 1 ms of delay. Why? It moved 90° (1/4 λ) in a 250 Hz span (500–250). At that rate it would shift 180° (1/2 λ in 500 Hz and 360° (1 λ in a 1 kHz span). The rate of change is 360° per 1000 Hz (i.e. 1 msec) no matter what frequencies start and stop the slope.

A steeper slope (more phase shift over the same frequency span) indicates more time. A constant phase shift rate indicates frequency-independent delay. A variable rate of change indicates frequency-dependent delay. The display can be confusing because we have the workings of linear time on a log frequency axis. It helps to remember that frequency-independent delay appears as an ever-steepening phase slope. Increasingly bent phase slope (from LF to HF) is flat time. A constant phase slope (from LF to HF) is bent time.

Factors that provide context for reading phase

- **Frequency**: Tells us the time period for the given range.
- **Phase slope rate**: Phase shift vs. frequency decodes to time (phase delay).
- **Phase slope direction**: Tells us who came first (input or output).

12.8.1 Polarity and relative phase

Let's separate two concepts that get mixed up together and complicate matters: polarity and relative phase. Polarity is a frequency-independent directional indicator. Positive polarity means the input and output signals track together in the same direction. Polarity has a phase component but no delay component. There is either 180° of phase shift at all frequencies, or none. We can run an auto race clockwise or counterclockwise without changing the outcome. This is not to say polarity is irrelevant. Speakers and racecars need to run in the same direction or risk head-on collisions.

Relative phase is frequency dependent. The quantity "90° phase" has no meaning without frequency.

12.8.2 Wraparound

The phase response over frequency display contains an unexpected and potentially confusing visual feature that requires explanation. The vertical axis is a straight-line representation of a circular (0–360°) function. We need to wrap our head around how it displays phase shifts greater than 360°. It wraps it around from the top to the bottom connecting 360° to 0° just like your clock connects 60 seconds to 0.

The source of the confusion is that 0° and 360° appear to be the same points when viewed without context. The 360° reading is, however, one wavelength late (or early). Our previous study of summation told us the important distinction between being synchronous and offset by one cycle. The wraparound line differentiates the otherwise identical radial phase angle positions such as 10° and 370°. The wraparound provides *context* to the radial angle. A phase response over frequency that contains no wraparounds has confined all of its time offsets to within a single wavelength at all frequencies. A response with multiple wraparounds has some ranges that are more than 1 λ behind others. The conversion from radial phase to phase over frequency is illustrated in Fig. 12.8.

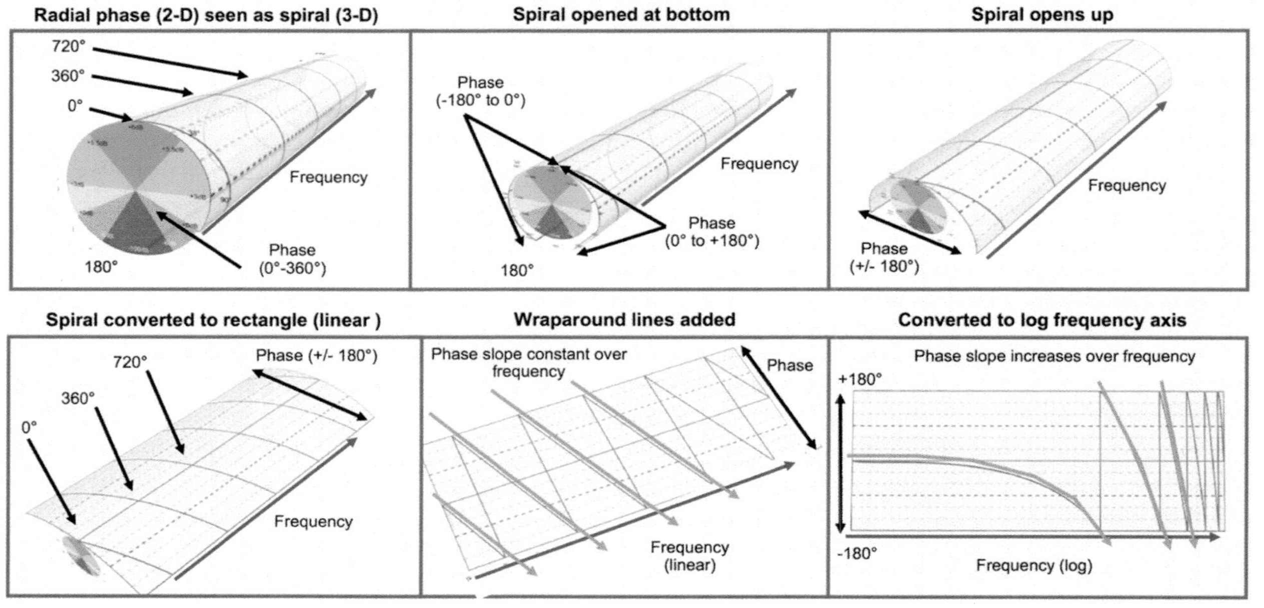

FIGURE 12.8
Phase response "unwound" (3-D artwork provided by Greg Linhares, courtesy of Meyer Sound)

The most common phase vs. frequency display has 0° at center and ±180° at the edges. The wraparound line jumps up when the response reaches -180° and connects to the next point, +179° (the same radial angle as -181°). The wraparound vertical line is a visual artifact and not indicative of phase discontinuity in the measured device.

The phase response is a lin/log merger zone. Its operation is time based and, therefore, linear. Its audible effects are logarithmic. Let's bring the two worlds together. Frequency-independent delay creates perfectly even spacing between the wraparounds. Viewing this on a linear frequency scale makes the linear nature of the wraparounds crystal clear (Fig. 12.8). Wraparounds accrue with each multiple of the base frequency ($F = 1/T$). The phase slope stays constant, because the delay is also constant. The downside is that it does not resemble how we hear it.

This contrasts with the log representation, which shows the wraparound rate and phase slope rising with frequency. We'll have to learn how to read phase slope the way we hear it, i.e. log, even though it is more difficult visually.

12.8.3 Phase delay

Flat phase over frequency comes in two principal forms: 0° and 180°. Both have zero phase delay over frequency. Delay causes the phase to take on a slope, which we can decode into phase delay. The decoding formula is another variation of $T = 1/F$, which we have encountered before.

The phase delay formula is:

$$T = \frac{\frac{\text{Phase}_{HF} - \text{Phase}_{LF}}{360}}{\text{Freq}_{HF} - \text{Freq}_{LF}}$$

where T is the phase delay in seconds, Phase_{HF} is the phase angle at the highest frequency in degrees, Phase_{LF} is the phase angle at the lowest frequency in degrees, Freq_{HF} is the highest frequency, and Freq_{LF} is the lowest frequency.

This formula can be applied to any range of frequencies, and yields the *average* amount of phase delay over the selected range. It's a good idea to limit the frequency span for phase delay computations, especially for loudspeakers, which vary widely over frequency. This expression can be simplified to phase shift over a given frequency span:

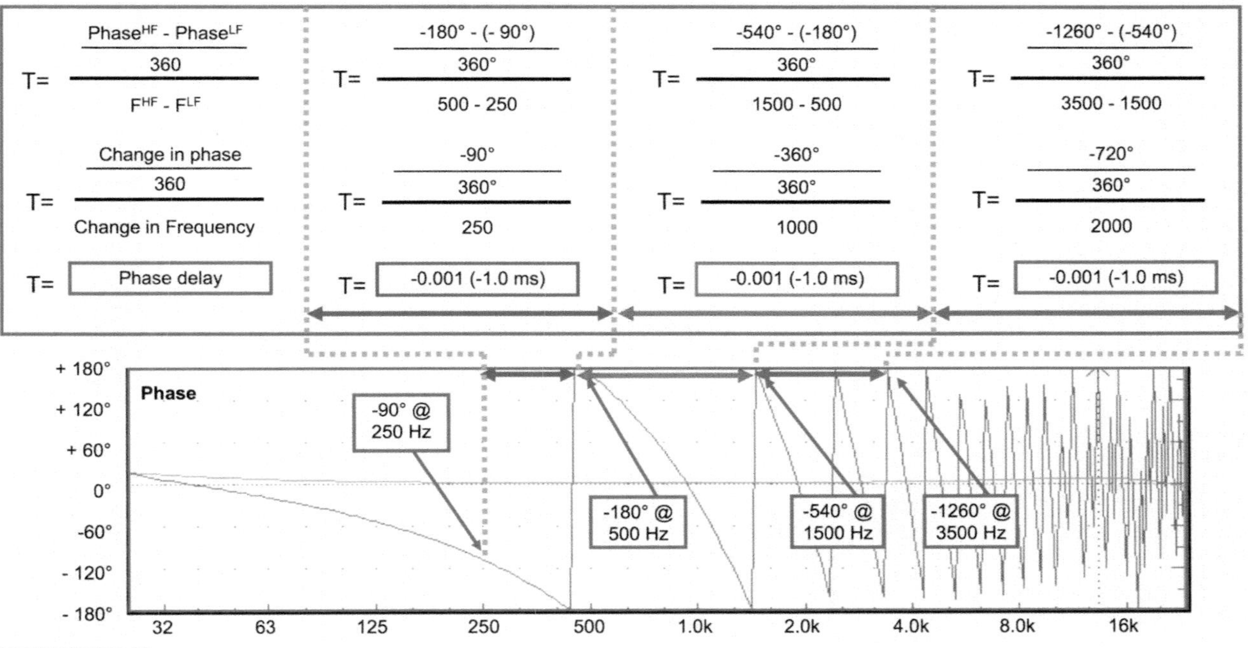

FIGURE 12.9
Example application of the phase delay formula for an electronic device with a fixed 1 ms delay over all frequencies

$$T = \frac{\frac{\text{Phase change}}{360}}{\text{Frequency change}}$$

An example application of the formula indicates 1 ms of phase delay at all frequencies (Fig. 12.9).

12.8.4 Phase slope

Frequency-independent delay creates constantly increasing slope over frequency (log display is assumed from now on). Each succeeding octave has double the number of wraparounds, because it's double the frequency span. Each succeeding wraparound will be incrementally steeper than the previous, e.g. if the slope at the first wraparound is x degrees, then the next wrap will be $2x$ degrees, followed by $3x$, $4x$ and so on.

Unfortunately it is not possible to relate the slope angle to a particular phase angle, because the vertical to horizontal ratios of a graph are variable. We can't just see a 45° slope at 1 kHz and decode that as 1 ms. The graph could be short and wide, or tall and narrow, so the slope can be Photoshop manipulated. This does not eliminate its usefulness. The slope angle indicates the number of wavelengths of delay, rather than the time. The same slope angle at different frequencies indicates the same number of wavelengths offset.

SUMMARY OF PHASE SLOPE PROPERTIES

- **For a given frequency**: Slope angle is proportional to wavelengths of delay.
- **For a given slope angle**: Phase delay is inversely proportional to frequency.
- **For a given phase delay**: Slope angle is proportional to frequency.

The phase slope informs us of response trends (Fig. 12.10). This has practical application when combining speakers with unmatched phase delay over frequency. This is everything we need to identify phase delay at any frequency.

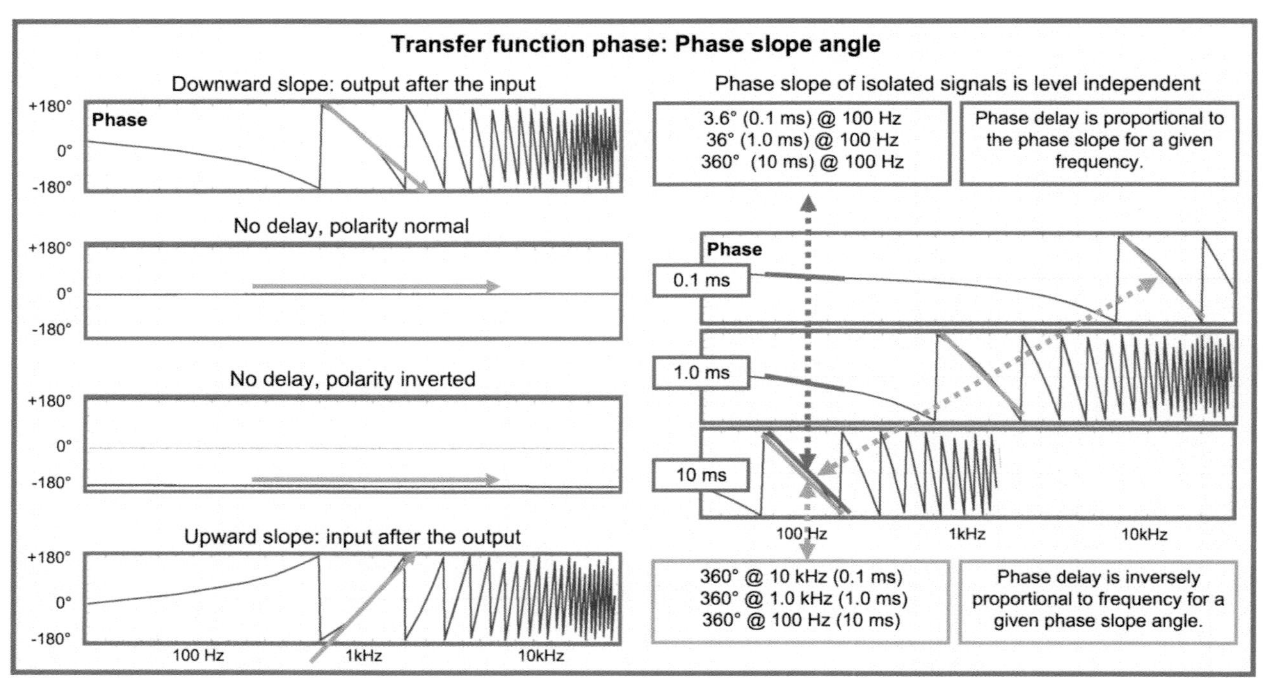

FIGURE 12.10
Reading the phase slope

12.8.5 Frequency-dependent delay

Many audio devices have frequency-dependent delay (phase delay over frequency). For example: everything. Even a cable has phase delay due to capacitance, inductance and length. It's a safe bet that a device with non-flat frequency response within the measured range will have frequency-dependent phase delay. In addition there can be phase delay effects from filters we can't see within the passband (transient intermodulation filters (TIM) and anti-aliasing filters above and AC coupling circuits below). Frequency divider filters, equalization boost and cut (analog or digital) also add frequency-dependent phase delay. The most dramatic case is the loudspeaker. A well-designed "phase corrected" system can have extended ranges of minimal phase shift, while "sales corrected" versions will have small ranges and wide variance. Both will fall behind in the LF range.

12.8.6 Filter phase effects

Phase response is affected by filter center frequency, slope, topology and magnitude.

PHASE DELAY SUMMARY FOR EQUALIZATION FILTERS

- Phase delay is inversely proportional to the center frequency.
- Phase delay is inversely proportional to the bandwidth.
- Each filter topology has specific phase delay characteristics.

A filter centered at 1 kHz will have twice the phase delay as one at 2 kHz. A narrow filter has more phase delay (over a smaller frequency range) than a wide filter (over a larger range) for a given frequency. See Fig. 12.11.

12.8.7 Frequency (spectral) divider phase effects

Phase response is affected by filter corner frequency, topology and filter slope (order).

PHASE DELAY SUMMARY FOR LPF AND HPF FILTERS

- Phase delay is inversely proportional to the corner frequency.
- Phase delay is inversely proportional to filter order (slope).
- Each filter topology has specific phase delay characteristics.

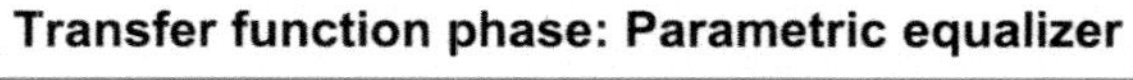

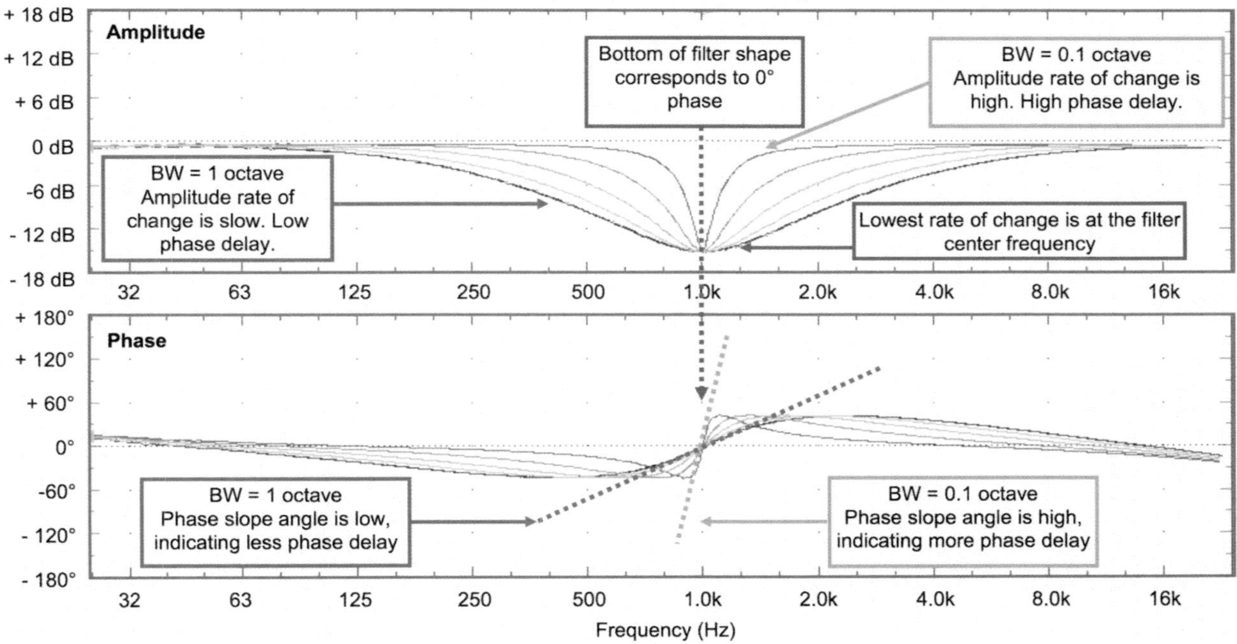

FIGURE 12.11
Phase slopes for example filters used in parametric equalizers

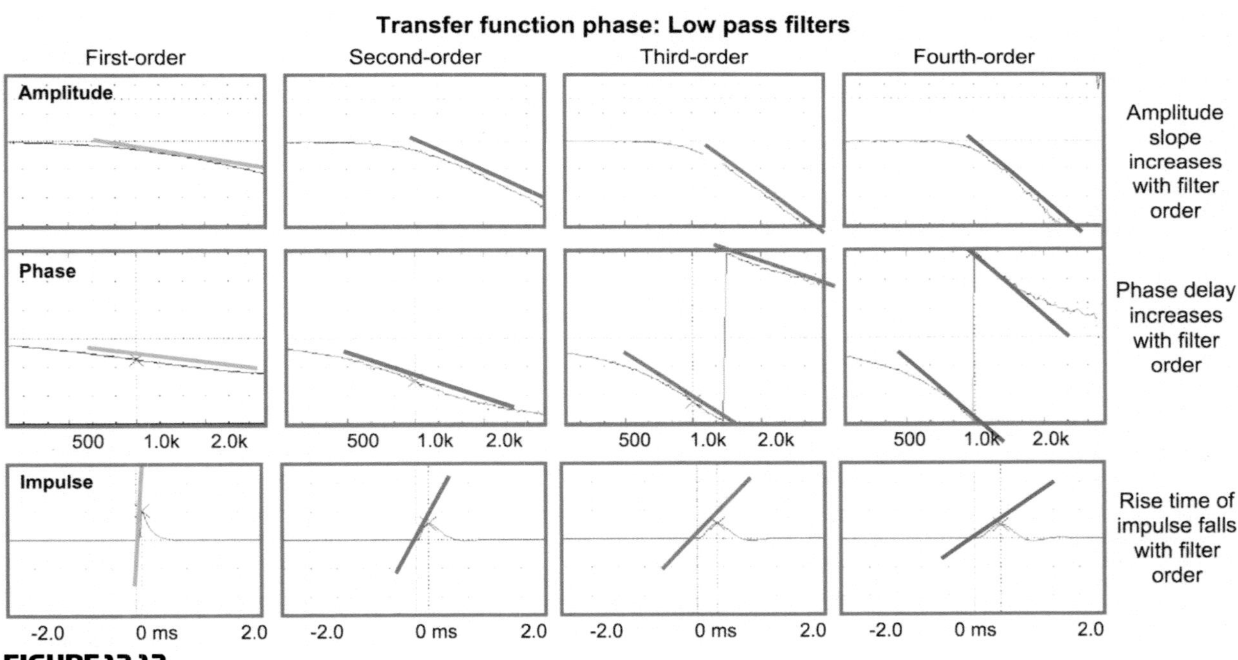

FIGURE 12.12
Phase slopes for example low-pass filters used in spectral dividers

A filter that turns at 1 kHz will have twice the phase delay as one that turns at 2 kHz.

A steep filter slope creates more phase delay (over a smaller portion of the passband) than a wide filter (over a larger portion) for a given corner frequency. See Fig. 12.12.

12.8.8 Loudspeaker phase effects

Loudspeakers are put to the ultimate challenge: mechanical devices that must transmit wavelengths varying in size by a 600:1 factor. Producing all these frequencies at the same level and same time is a seemingly impossible task for a single speaker. Our desire for high power, directional control and low distortion creates the need for multi-way systems and the mechanics become even more complex. The inevitable result is a loudspeaker response with frequency-dependent delay.

CAUSES OF FREQUENCY-DEPENDENT DELAY IN LOUDSPEAKERS

1. Different radiation modes of individual speaker over frequency.
2. Mechanical displacement in multi-way systems.
3. Crossover irregularities in multi-way systems.

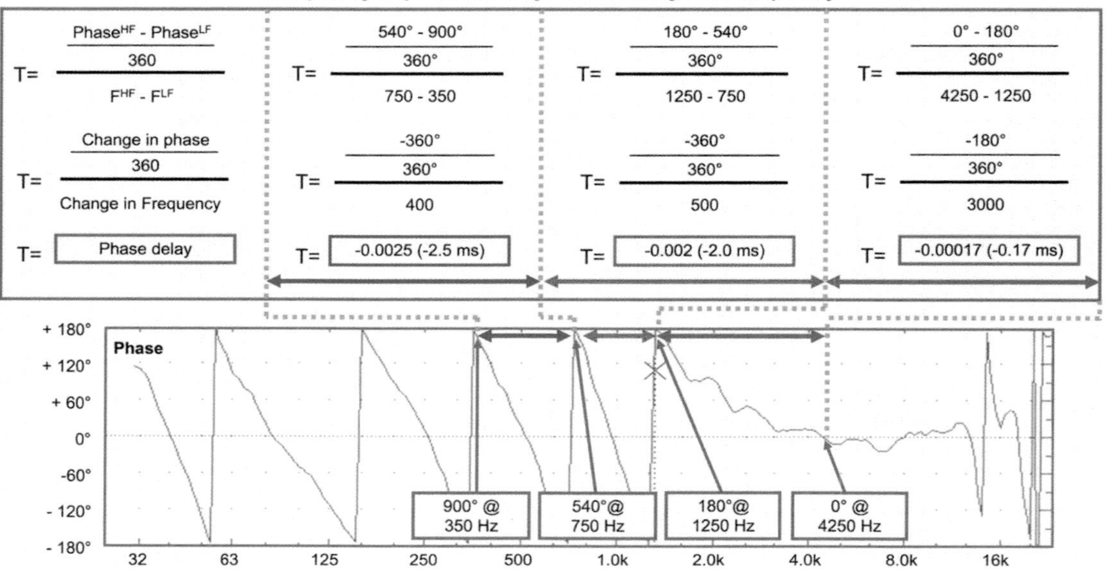

FIGURE 12.13
Phase response of an example loudspeaker with variable phase delay over frequency

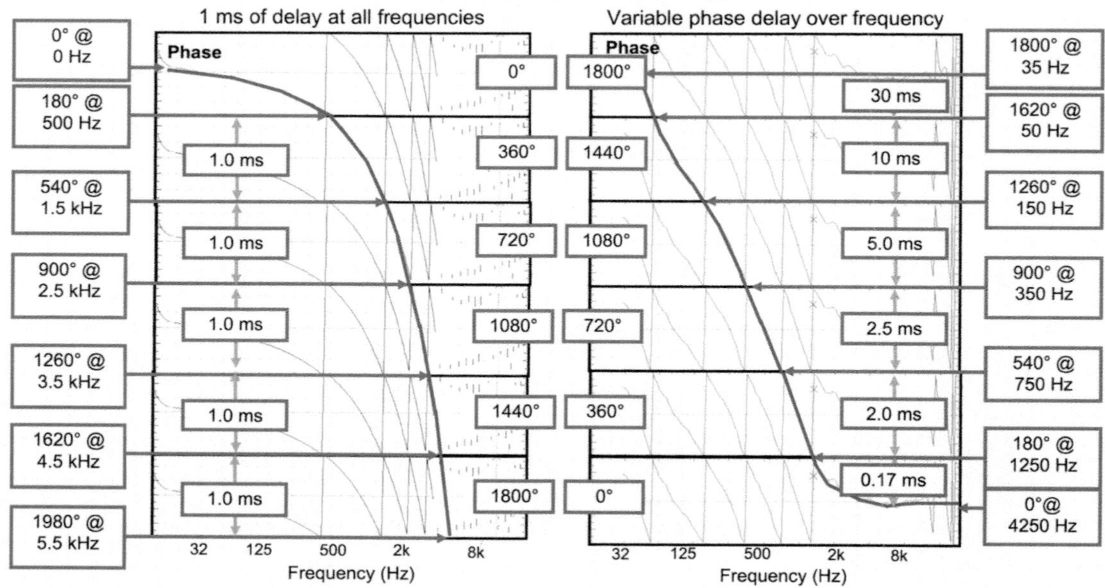

FIGURE 12.14
Unwrapped phase responses over frequency for the example electronic device (see Fig. 12.9) and example speaker (see Fig. 12.13). The unwrapped representations give a wider context to our view of the phase slope over frequency. The contrast between fixed delay and frequency-dependent delay is clearly seen. All the ghosted phase responses shown in each panel are the same. They are stacked and the trace is unwrapped by connecting the bottom of the first "wrap" to the top of the second.

It is virtually assured that a loudspeaker's highest frequencies will lead its lowest. This is due to the nature of the different modes of radiation of speakers. Sound radiates from the speaker with a piston-type motion when λ is smaller than the speaker diameter. A well-designed speaker system is capable of creating an even phase response over frequency over this range. The radiation style changes when λ exceeds the piston size. One of the artifacts of this is phase delay, which increases as the wavelength/speaker rises. This relationship is scalar, so even mid-range frequencies such as 1 kHz are large compared with an HF driver less than half its size. For subwoofers, the entire operating range contains wavelengths that are huge compared with the radiating devices. As a result, subwoofer phase response will show a steady increase of phase delay as we go down.

Three factors play parts in the phase response of multi-way systems: physical driver displacement, the electronic response and the motion factors described above (which act on the two drivers differently due to their different diameters). These factors all come home to roost in the combination at the acoustical crossover.

It will suffice for now to observe the phase delay without venturing further into its causes. The phase delay formulas are applied to an example multi-way speaker to illustrate the decoding process for frequency-dependent delay (Figs 12.13 and 12.14).

12.8.9 Summation phase effects

An important distinction must be made between wraparound in an isolated speaker (indicative of phase delay) and a similar version found in a summed response (indicative of time offset arrivals). We have discussed how summation modifies the amplitude response in detail, but barely touched upon the look of the summed phase response. In the same way that the summed amplitude is affected by relative level and phase, so it shall be with phase. The summed phase value will be unchanged, regardless of relative level when the phase responses are matched (where else could it be?). The combined phase value will be found *between* the two individuals when the values are offset. It's at the mid-point when levels are equal. It's closer to the phase value of the dominant source when they are unequal. The maximum disparity between the individuals (180°) causes the summed response to resemble a wraparound, but in this case represents a real acoustic result, rather than a visual artifact of our analyzer. The wraparound and summation phase disturbances can both be seen in Fig. 12.15.

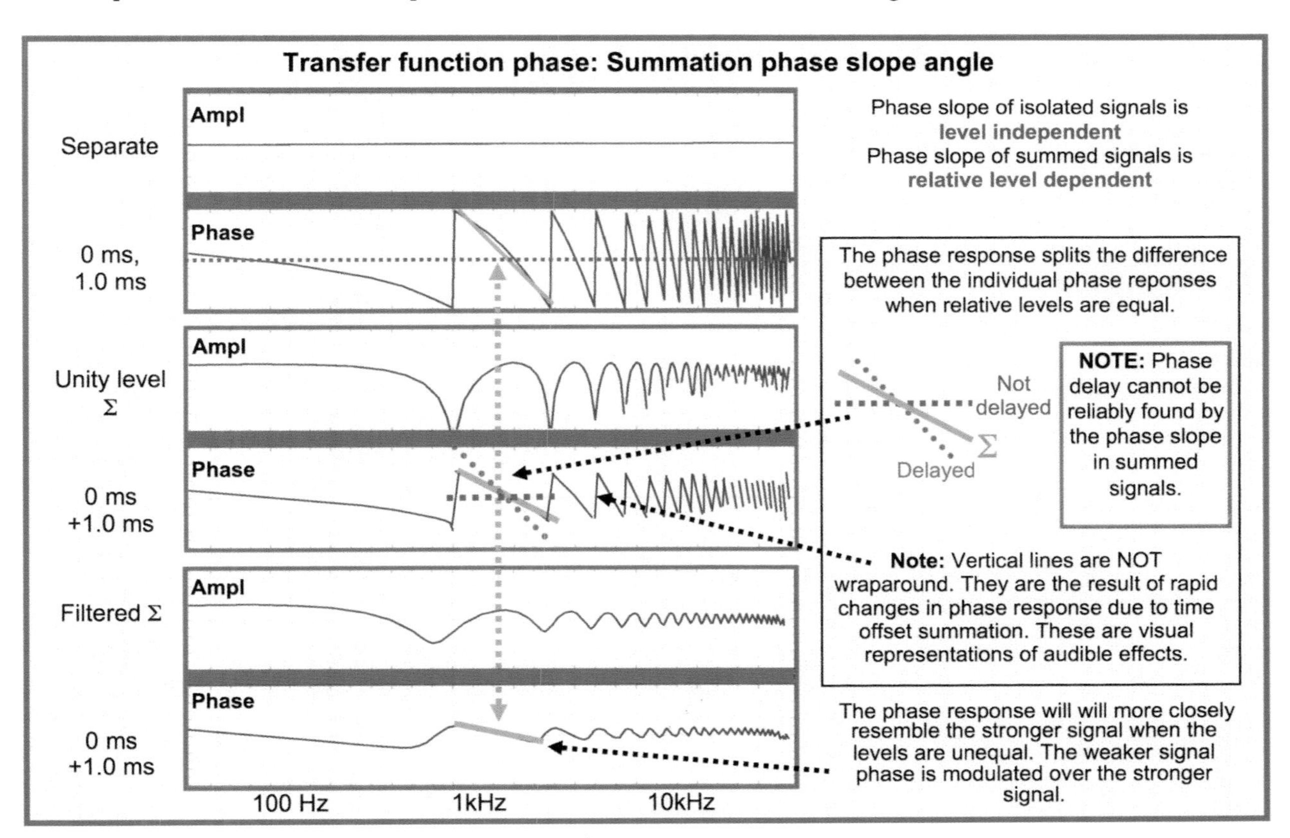

FIGURE 12.15
Summation effects on the phase slope

The summation phase slope is a compromise value between two (or more) oppositional parties. This affects how we apply the phase delay formula to convert slope to delay time. In short, the phase delay formula only works on isolated responses. The compromise slope is potentially misleading. The reduced phase slope angle could be misinterpreted as reduced phase delay, when it is actually the simultaneous presence of conflicting values. Evidence of the conflict remains on the screen in the form of the comb filtering in the amplitude response, and the ragged edges in the phase response where the phase responses have reached their maximum conflicts. In cases where substantial summation is occurring, we can no longer decode phase slope angle directly to phase delay. The combined phase slope does reveal the presence of the component parts, which are superimposed over each other. This is why our references to summation phase include the individual time components (see Figs 4.13 and 4.14). The compromise phase value migrates toward the dominant party. Once again, size matters. The lower-level signal causes slope changes at a constant rate over frequency (the comb frequency), because this is determined by the time offset. The combined slope shows the phase response of the low-level contributor as modulations above and below the dominant phase trace.

Some analyzers expand the phase vertical scale up to thousands of degrees to eliminate wraparound, achieving a response similar to the unwrapped scenarios shown in Fig. 12.14. Such displays can be informative but also error prone when the phase data has strong summations (reflections or other speakers). Decoding summed phase response is challenging even for the most experienced and usually requires context (such as seeing a second speaker path).

12.9 SIGNAL AVERAGING

The acoustic environment is hostile to measurement. Any single acoustic measurement is prone to error from ambient noise, reverberation, weather changes and more. Averaging improves our accuracy by providing statistical confirmation of the signal and statistical dilution of the noise. A noise-free signal receives no benefit from averaging. The signal is the signal and there is nothing to average out. Noise adds summation layers on top of the signal, which add (or subtract) to/from the signal. Uncorrelated noise adds randomly to the signal (unstable summation). We average in the signal and average out the noise when the multiple samples are taken together.

Averaging is statistical so we can examine the numbers to see the range of differences between them, i.e. the deviation. This provides a measure of stability that helps us ascertain the data quality. Deviations are large when the noise is strong compared with the signal because the noise dominates the response. Increasing the number of averages improves our statistical accuracy. Measurements in noisy environments benefit from higher numbers of averages. The summation effects of the noise are statistically reduced in our computation, but still present in the room (an important distinction). The measure of deviation is coherence, which will be covered shortly.

12.9.1 Averaging types

There are many different equations for data averaging, but we won't get a clearer picture by averaging then together. The schemes can be divided into a few main categories: the waveform math, the sample weighting and how the current numbers are maintained (Fig. 12.16).

There are two primary math options: RMS and vector. RMS (root-mean-square) averaging is used for single-channel spectrum RTA-type simulations and for impulse response averaging. RMS averaging is suitable for averaging of random signals. It's relatively poor at discriminating random from correlated signals, so not well suited for most transfer function measurements. Vector averaging, as the name implies, uses vector values between the amplitude and phase (real and imaginary) aspects of the signal. Vector averaging is highly sensitive to variations in either amplitude or phase (and wind), which gives it preference for transfer function applications.

Weighting schemes (not to be confused with A weighting etc. for SPL) fall roughly into two types: weighted or unweighted. The term refers to the proportional value each sample contributes to the average. Unweighted averaging treats all samples equally (the norm unless noted). Weighted schemes give more votes to certain samples. Exponential averaging gives younger samples (exponentially) higher weighting than older ones, making it quicker to react to dynamic changes. Unweighted averaging is slower to react because it places history on par with the present. Exponential has less stability but higher speed than unweighted, for a given sample number.

Signal averaging: Speed and stability

FIFO (first in, first out)

FIFO averaging continually updates. New data arrives that is higher in value than all previous(our example sample # 1), causing the average to rise. The average updates fastest when the number of averages is low. Stability rises as more averages are used.

Sample #	1	2	3	4	5	6	7	8
Value	120	115	110	105	100	95	90	85
Weighting	1	1	1	1	1	1	1	1
Weighted value	120	115	110	105	100	95	90	85
Avg value (8)	102.5 (820 / 8)							
Avg value (4)	112.5 (450 / 4)							
Avg value (2)	117.5							

Accumulate

The average is kept as a single lump value of unlimited accumulations. This scheme has variable speed and stability. It is fast initially and unstable, then becomes slow, stable and increasingly immune to change.

Sample #	1	2-8
Value	120	700
Weighting	1	7
Weighted value	120	100
Avg value (All)	102.5 (820 / 8)	

Weighted (most recent has highest value)

The new sample has the highest weighting. The average value moves rapidly toward the new value. Weighted averages are the fastest acting, but least stable.

Sample #	1	2	3	4	5	6	7	8
Value	120	115	110	105	100	95	90	85
Weighting	1	1/2	1/3	1/4	1/5	1/6	1/7	1/8
Weighted Value	120	58	36	26	20	16	13	10
Avg value (8)	110 (300 / 2.7)							
Avg value (4)	115.4 (240 / 2.1)							
Avg value (2)	118.3 (178/1.5)							

FIGURE 12.16
Averaging types

The final parameter is the accommodation of old samples. Some averagers hold on to them, whereas others throw them away. An accumulating averager adds each new sample to the combined result on an ongoing basis. The first two samples are added together and divided by two. The third sample piles on and the average is derived from the trio. This continues forever like Star Wars movies. Accumulators have the advantage (and disadvantage) of maximum long-term stability. The advantage is noise immunity. The down side is holding on to outdated information (the speaker is already in the truck). The accumulator is a hoarder. It never parts with anything it collects. That's fine as long as nothing ever changes. If it does we need to flush it and restart the accumulation.

A second scheme is a fixed number of averages and automatic restart, like a windshield wiper in light rain. The response builds up for a while and then, swish, restart. This has its place when changes are expected, such as setting an equalizer or delay line.

A third scheme is the "first in, first out" (FIFO) style. The data flows through a sample pipeline of fixed length and new data is flushed through it sequentially. New data pushes out old data so there is never a need for restarting. The average is always up to date with only the most recent samples. It's CNN averaging: everything is "Breaking News!"

12.9.2 Speed vs. stability

The number of averages is user selectable. Higher numbers give greater stability, but slow the reaction time to changes. The best choice is application dependent. We use higher numbers in noisy environments (such as acoustic measurements), sacrificing speed for stability. Electronic measurements are low noise, allowing for lower numbers of averages without significant loss of stability.

12.9.3 Amplitude thresholding

We can't assume every frequency in the input signal will be at the full level all the time, even though reggae musicians set this as a goal. Maximum bass at all frequencies, mon. It's easier to make accurate measurements with strong signals than weak ones near the noise floor. Each sample of program material has some strong and some weak frequencies. We are not in such a hurry that we need to use every piece of data. We can be selective on a sample-by-sample, frequency-by-frequency basis. The mechanism is amplitude thresholding, which operates in principle like the noise gates on drum mics. Each sample is screened for level over frequency. We will send 1 kHz to the averager if it's strong. We can wait for the next bus at 2 kHz if it's weak. We restrict the averager to working with pre-approved samples (ones strong enough to get good data). It's hard enough already to get good signal to noise in the world of acoustics. We don't need to make this more difficult by using data that's lost in the noise.

Note: Not all dual-channel FFT analyzers utilize this capability. Those who do use it enjoy a higher stability and noise immunity in the transfer function.

12.10 TRANSFER FUNCTION COHERENCE

Coherence is the answer to the question: "Hey analyzer, do you have any idea what you're talking about?" Coherence values range from 1 (yes, I mean it) to 0 (I am just making it up). Educated guess is in the middle. The real math is beyond our scope but we can comprehend the principles pretty easily. Coherence is statistically derived, i.e. it requires an average of multiple samples. We are looking for agreement between samples. We have confidence in the averaged value when all provide the same transfer function values, and lose confidence when answers are all over the map. Why would there be disagreement? Noise. Any transfer measurement contains a noise component, i.e. uncorrelated signal at the output that wasn't in the original input. The question is how much. The answer is coherence. The S/N ratio should be very high in a line-level device. Everything moves in noise's favor in the acoustic world.

Coherence is evaluated on a frequency-by-frequency basis, so we can see which ranges are faring better than others. We use coherence to make decisions, and most importantly to not make decisions. Very high coherence tells us we can make adjustments with confidence that they will have a predictable effect. Low coherence means that there is more going on here than meets the ear. We could find ourselves turning knobs on the EQ and seeing little or no improvement. Adjustments such as adding absorption, adjusting splay angle and setting delays can produce great improvements in coherence. Many audio engineers will differ on whether a particular EQ setting is an improvement. Coherence, on the other hand, correlates highly with quality of experience. If we make an adjustment that improves coherence there is usually widespread agreement.

12.10.1 COHERENCE DEFINED

Coherence is a statistical value, derived from the deviations between the amplitude and phase values in the averager. Deviations are high and coherence is low when samples contain extraneous noise. Coherence is calculated for each frequency bin on a scale of 0 to1 (1 being perfect stability with no noise). Fixed linear differences, such as voltage gain, do not degrade coherence (because things remain stable). By contrast, delay between the measurement input and output causes the two time records to be drawn from different waveforms. The differences (even slight ones) reduce the transfer function stability, which degrades coherence (in the measurement, not the system). A completely uncorrelated relationship between the two measured channels reduces the coherence value to 0. The relationship is termed **causal**, i.e. the output response was *caused* by the input when the output and input are linearly linked. It is termed **non-causal** when the output signal is unlinked to the input.

Coherence is a measure of the causal output signal strength relative to the non-causal output noise component. We can attribute coherence loss to the measured device (not analysis error) once we compensate for any time offset between the measured signals. Coherence is 1 when the output is all signal and no noise. It's 0.5 (50%) when signal and noise are equal and 0 (0%) when the output is all noise and no signal (Fig. 12.17).

Let's put on some music to illustrate. A passage is first heard loud and then soft. We recognize the music as changed, yet fully correlated (coherence = 1). Listen to the first 10 seconds again. Then start 5 seconds in and play it again for 10 seconds. The musical passages are half-related (coherence = 0.5). Finally, listen to some good music and then Barry Manilow. These two forms of music *should* be 100% uncorrelated (coherence = 0).

COHERENCE EXPRESSED IN ANALOGOUS TERMS

- **Data quality index:** The analyzer's confidence in the amplitude and phase values.
- **Signal-to-noise ratio:** Noise degrades the coherence value.
- **Stability indicator:** indicative of the stability/repeatability of the measurement.

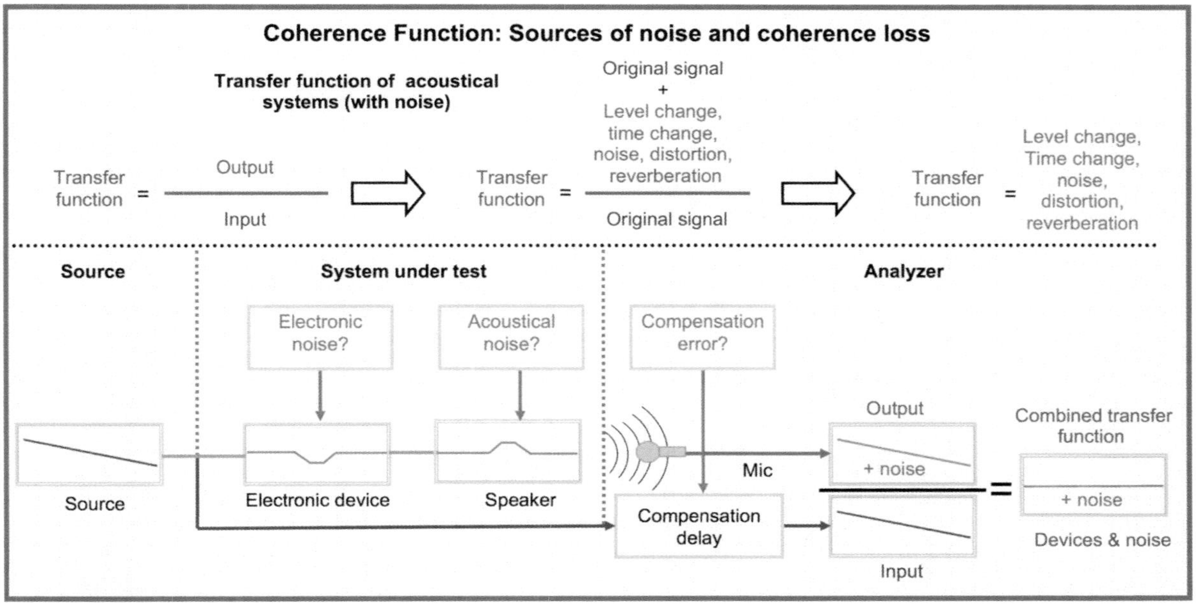

Coherence Function: Sources of noise and coherence loss

Transfer function of acoustical systems (with noise)

Transfer function = Output / Input

Transfer function = Original signal + Level change, time change, noise, distortion, reverberation / Original signal

Transfer function = Level change, Time change, noise, distortion, reverberation

Source | System under test | Analyzer

Electronic noise?
Acoustical noise?
Compensation error?
Source
Electronic device
Speaker
Mic
Output
+ noise
Input
Compensation delay
Combined transfer function
+ noise
Devices & noise

FIGURE 12.17
Flow block diagram of transfer function measurement with noise sources

We can look at a frequency response and assess quality on two levels: measurement quality and system under test quality. Low coherence results from inconsistent results. Something is causing it. Either we are measuring it wrong or it's in the system (we need to know which). Coherence requires averaging, and looks for discrepancies between the individual responses that comprise the average. Perfect coherence connotes perfect agreement between all samples in the averager. Coherence drops when samples differ.

12.10.2 Factors affecting coherence

How does the coherence function detect noise in our measurements? As our analyzer sees it, there are two possible kinds of output signals: causal signals correlated to the input and non-causal. Invited guests and party crashers. We know who was invited because we have a complete guest list: the input time record. We recognize the output waveforms on the invitation list (and those that weren't). We don't have bouncers to keep them out of the room, but we can identify them and monitor their effects. Let's measure a speaker in a room (Fig. 12.18).

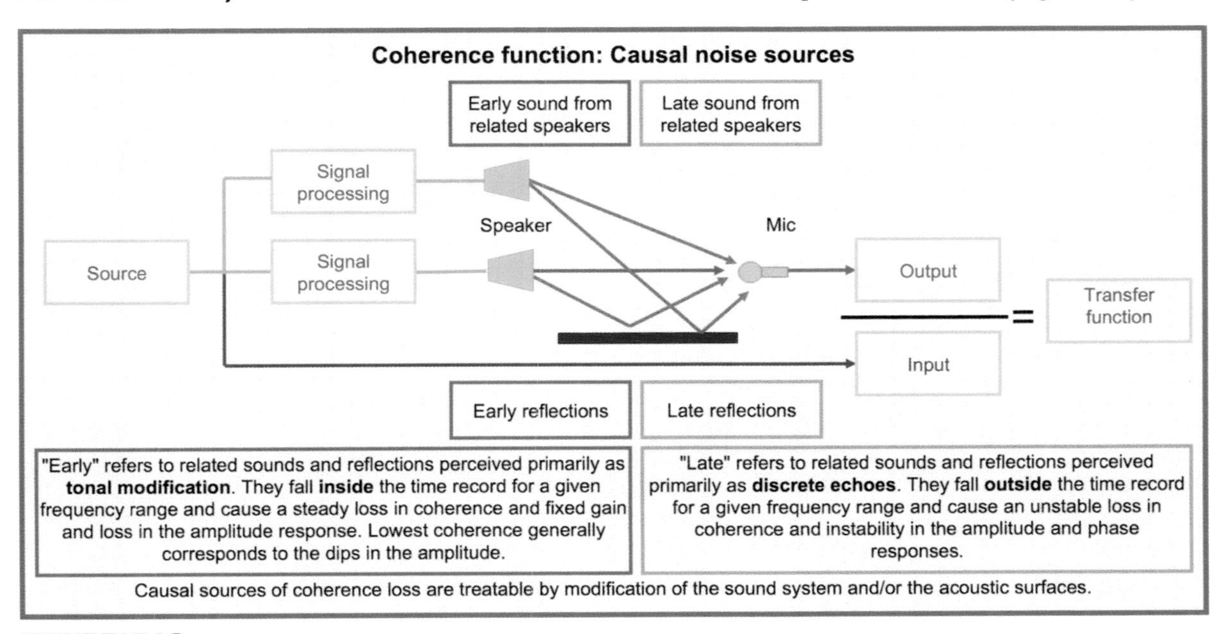

FIGURE 12.18
The coherence function and causal noise

CAUSAL SIGNALS AT THE OUTPUT INCLUDE

- Source signal.
- Copies of the source signal from other speakers.
- Copies of the source signal from reflections.

The source signal is recognized at the output. Copies include secondary sources such as reflections or additional speakers driven by the same source. We must be able to differentiate between causal and non-causal signals, because the optimization strategies differ markedly for each. The coherence factor of a causal signal will remain consistent over a series of averages, as will amplitude and phase. This is indicative of a stable relationship between the output and input signals and a stable S/N ratio. Strong combing makes a series of high and low coherence values that track the peaks (stable and high) and dips (consistently low). The stability of the coherence validates that the peaks and dips result from causal summation, which helps to guide us toward applicable solutions (level, delay, splay, absorption, etc.).

Non-causal interactions create amplitude and phase instability, which degrades coherence. Averaging helps to steady the data statistically for measurement (but not acoustically). The most effective optimization strategies separate the signal (causal) from the noise (non-causal) because different solutions apply. First minimize the noise instability before undertaking strategies such as equalization on the stable remainder. An illustrative example is speaker focus angle and acoustic treatment. We can reduce noise by aiming the speaker correctly and adding absorption. The coherence value tells us if we are making progress toward noise reduction before we equalize the stable remainder.

The end result is a frequency response with causal and non-causal elements. The coherence values reflect their mix proportions; stable values show strong correlated presence. The data will never stabilize sufficiently for us to obtain a definitive response of the system, if the uncorrelated signal is too large. This sends a clear message that some major work needs to be done *before* equalization is attempted.

NON-CAUSAL SOURCES (INCLUDE BUT ARE NOT LIMITED TO)

- Distortion, hum, noise and leakage in the sound system.
- Moving lights, audience participation, HVAC, forklifts and grinders in the room.
- Late-arriving echoes (from the fixed PPO analyzer point of view).
- Late-arriving sound from other speakers (from the fixed PPO analyzer point of view).

Items on the first line have infiltrated the sound system but are not part of our intended transmission. Coherence reveals them as intruders. The next line contains aliens invading our acoustic space with no relationship to the original signal. Their effects are seen as random deviations (modulations) above and below the correlated stable frequency response. One common aspect so far is that EQ will be ineffective. The last two items are a special case, because the signals are late copies of the original. Exactly when is late? This answer requires revisiting the fixed PPO (Constant Q) transform.

12.10.3 Direct, early and late arrivals

The composite fixed PPO response is made up of multiple time records ranging from a few milliseconds to over half a second. The hall's acoustical properties are selectively time windowed so that a consistent ratio of direct to delayed wavelengths is classified as causal, and later arrivals are classified as noise. Fixed PPO means, by definition, a fixed number of late wavelengths fit in the time window. Recall how the fixed PPO transform is rooted in our tonal, spatial and echo perception thresholds. Direct sound and stable summations of early causal signals (the sources of tonal perception), are the prime candidates for equalization. Their features appear clear and stable on the analyzer. Signals perceived as echoes create instability in the analyzer. Their features appear fuzzy and unstable. There is strong evidence that the EQ solutions will fail when the frequency response target curve is unstable. The display guides us toward superior solutions such as acoustic treatment, speaker focus and phase alignment.

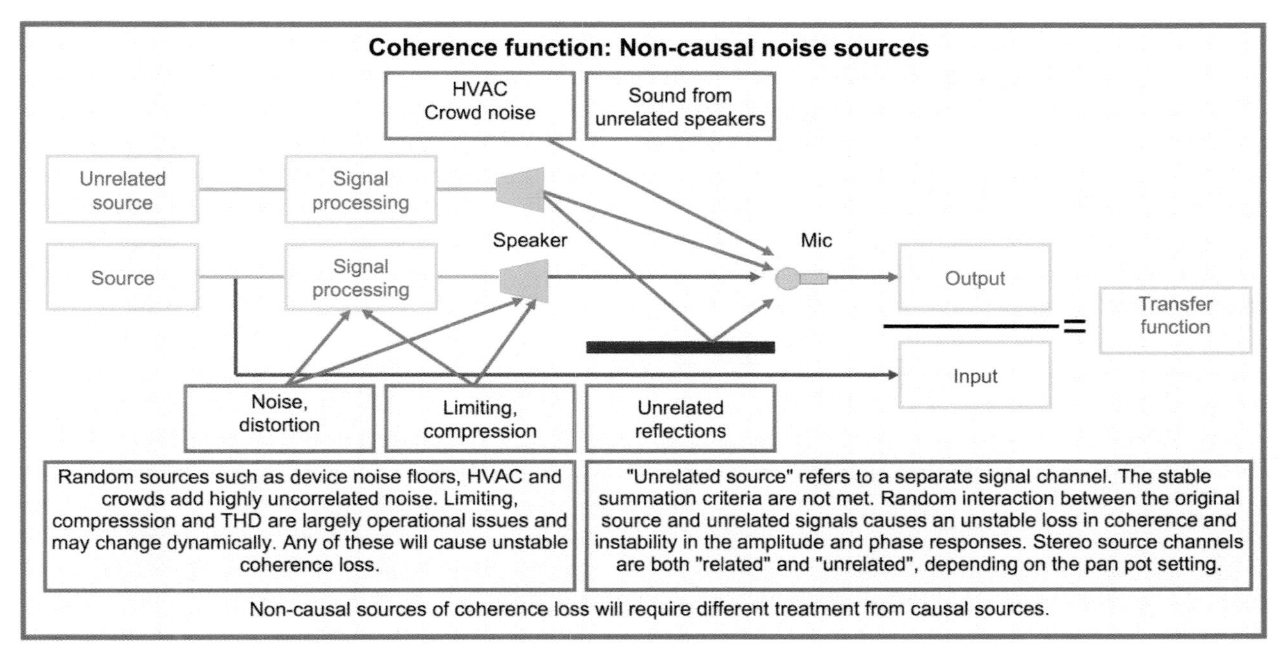

FIGURE 12.19
The coherence function and non-causal noise

This also applies to correlated signals arriving from other speakers. Look around the room and consider the remaining treatment options. All options remain open for speakers within a coupled array and nearby reflections. EQ options decrease with frequency (and time) for late speaker arrivals or reflective surfaces. Low frequencies and short reflections remain in the EQ game while high frequencies and long reflections create overly fine combing. The fixed PPO high-resolution frequency response display emphasizes the area of practical equalization.

A stable response indicates all optimization options remain open to us: EQ, level, delay, splay, spacing and acoustic treatment. An unstable response, caused by late reflections and other speakers, is beyond the equalization horizon, but all other options still apply. Unrelated non-causal noise sources require solutions outside the sound system, such as acoustic isolation, or taking away the forklift keys.

Let's add one more level of detail: the distinction between causal data within the time record, and that which falls outside. Direct sound is the former. A single reflection can arrive within the time window, outside it or straddle both (recall the multiple time record lengths that comprise the fixed PPO frequency response). The reflection is recognized as causally related if it arrives within the same time record as the direct signal. The reflection is, after all, the child of the direct sound so they are definitely related. Reflections arriving after the time record closes are disowned by the analyzer, which restarts the process from scratch with no knowledge of previous events. Reflections still bouncing around the room from previous signals become instant strangers to the analyzer and are treated as noise (Fig. 12.19). The threshold is frequency dependent. A 100 ms reflection, for example, falls far outside the short HF time records and well inside the long LF time records. The analyzer sees the HF portion as non-causal (outside of the time window) and the LF portion as causal (inside the window).

12.10.4 Instability in the causal signal

There is no guarantee that the causal signal will not change, e.g. an EQ can be adjusted while being measured. The output signal is recognized as causal, but changing relative to the input signal. The result is coherence loss, even though the device is operating normally. Coherence is based on the range of variation between the samples that make up the averaged value. Coherence falls as variation rises, whether due to internal dynamic changes or external noise. Coherence will rise again once the response stabilizes. Simply put, coherence hates instability, no matter where it comes from.

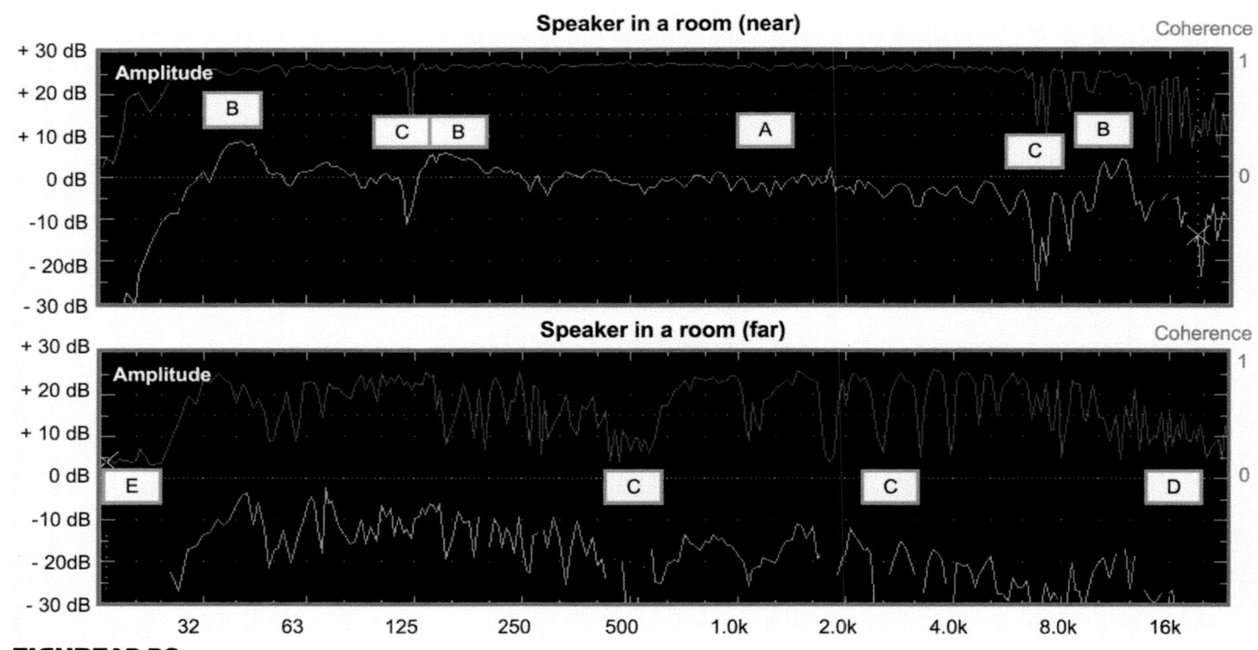

FIGURE 12.20
Field example of the coherence function.

BASIC TRENDS OF THE COHERENCE TRACE (FIG. 12.20)

- Temporary drop when the system has dynamic changes.
- Permanent stable response (high and low) with causal ripple variance.
- Unstable loss when non-causal noise is added to the output signal.

12.11 TRANSFER FUNCTION IMPULSE RESPONSE

We have seen how the phase response was plainly viewable in the frequency domain, and encrypted in the time domain. The impulse response polarity reverses this paradigm: straightforward time domain and encrypted frequency response. An impulse response and transfer function frequency response derived from the same time record contain the same information, albeit from a vastly different perspective. The time domain representation of the impulse separates the direct sound and individual reflections, making them easily identifiable. The impulse response displays the arrivals in the order received and gives away their positions by the timing. This contrasts with the frequency response, which shows the effects of reflections superimposed on the direct sound response (combing). The timing sequence, relative strength and polarity of each arrival can be found.

Impulse response is a wonder of simplicity to put to practical use. It shows direct sound and reflection arrivals in units we can easily understand: milliseconds. It is much easier to read the impulse response than to understand how it is derived. Under the hood we go again, with the "as simple as possible without getting a mathematician upset" approach.

The impulse response is a mathematical construction of a hypothetical experiment we don't have to actually perform. Nice! It's the answer to the question: "What would an oscilloscope show us if we put a single perfect impulse into the system?" Pop goes the woofer and we wait for its arrival at a mic, and then stick around to see it arrive again, and again off the floor and other surfaces. That's the easy part. Now the questions begin. What is a "perfect" impulse? How can we see the impulse response on the analyzer when we don't hear one coming out of the speaker?

Mathematicians don't throw words like "perfect" around lightly. You win a prize if you guessed this is another infinity construction. An infinitely steep transient into a system with an infinite frequency range makes an infinitely tall impulse, rising from, and returning to zero in (you guessed it) an infinitely small time period. In the practical world we eliminate infinity and limit the range of interest (e.g. up to 20 kHz) so the pulse slope, height and width become finite. A single perfect impulse is an audio stimulus signal approximated by the following recipe: all frequencies, equal level, in phase, minimum duration. Visualize a one-lap race for all frequencies with everyone ready to run once the starter pistol fires. In fact, acousticians use starter pistols as impulse generators for room analysis! A system with a perfectly flat amplitude and phase response over frequency will return an impulse that rises and returns in the minimum time with no additional features such as overshoots, undershoots or ringing. That's the only way we can ever see a perfect impulse response. Any peaks or dips in the amplitude and/or differences in phase slope will disfigure the impulse in characteristic ways that are very revealing for us.

Time offset between the output and input shifts the *x*-axis location of the impulse. Delay moves it to the right (output is after the input) and anti-delay to the left (output precedes the input). We know the output cannot precede the input in physical reality but our analyzer is just responding to two signals, which can come from anywhere. We have internal delay in our analyzer so this is not an exotic occurrence.

Let's see what the impulse response reveals. The standard features are a dead zone before the arrival, the shape of the impulse and any secondary arrivals that follow. The dead zone is the transfer time, i.e. latency or acoustic propagation. The analyzer finds this by recognizing the content of the input and output signals as similar, but offset in time. The next thing we notice is the orientation of the impulse, which reveals the polarity (positive or negative). The third feature is the rate of the pulse's rise and fall. A steep vertical line with no ringing indicates flat amplitude and phase. A rounded impulse indicates the system has HF loss and or frequency-dependent phase shift. Ringing (before or after) is the hallmark of filtering (electronic or acoustic). Reflections are extra copies arriving behind the direct signal. These are recognized as the children of the input and shown as secondary transient peaks lining up behind the first arrival (Fig. 12.21).

How does our analyzer give us a display of transient peaks that look like snare drum hits when we put in the music of 101 strings? This is where the math magic comes in. Recall that the original FFT transform took our time domain signal and converted it to the frequency domain. Then we took two channels to create a transfer function phase and amplitude over frequency. This impulse response is the result of a second generation of FFT calculations, in effect an FFT of the amplitude and phase over frequency, which converts us back to the time domain, termed the Inverse

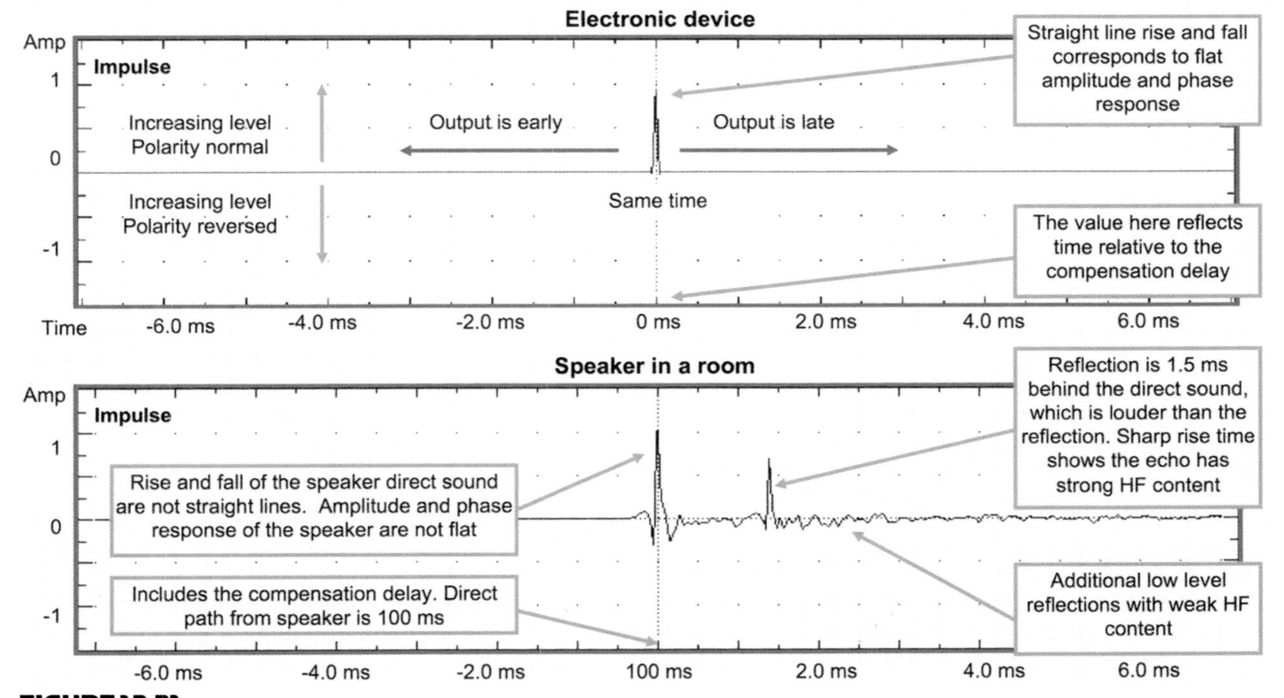

FIGURE 12.21
Impulse response

FourierTransform (IFT). The IFT result is a simulation of the output waveform that would result from a perfect impulse passing through the measured system. Because the transfer function amplitude and phase are source independent, we are able to build our impulse simulation by listening to pop music singles (not just the sound of a single pop).

We have previously discussed the circular relationship between the time domain and frequency domain. Therefore a frequency response and impulse response derived from the same data are interchangeable. This is not applicable in our case because we obtain our frequency and time domain responses differently. Our frequency response is a quasi-log composite of multiple linear time records of different lengths. It's a "time composite" as well as frequency composite. We can't convert that into a quasi-log 48 PPO composite impulse response. Our actual impulse response comes from a single linear time record. We can convert that back to a frequency response, but it's linear, so why bother? We use what's best for each domain, even if it means the data is not interchangeable.

Six answers found in the impulse response

1. **Relative time**: transfer time through the system.
2. **Relative level**: transfer level, heavily weighted toward HF (linear frequency data).
3. **Polarity**: any driver viewed individually or HF drivers in a multi-way.
4. **Frequency-dependent delay**: yes/no is discernible. Beyond that is Jedi level only.
5. **HF roll-off**: yes/no is discernible. Rounded impulse signifies roll-off.
6. **Secondary arrivals**: answers 1–5 for reflections, sound from other speakers.

The arrival times of reflections or secondary sources are indicated by horizontal location. The vertical scale has a polarity plot twist. Unity gain is 1 if the polarity is normal and -1 if the polarity is inverted. Consider what happens when gain and polarity both change. It's simple to visualize that +6 dB, polarity normal, gives a linear gain value of +2. But what is +6 dB, polarity inverted? That would be -2, of course, which can be confused with -6 dB, which it's clearly not (-6 dB is 0.5). The directional component (polarity) is independent of level, which is why the vertical scale is linear.

Summary of impulse response outcomes

- Horizontal center: channels sync'd (normalized to internal compensation delay).
- Left of center: output is early by the amount of time offset.
- Right of center: output is late by the amount of time offset.
- Amplitude level is 1 or -1: unity gain.
- Amplitude level is between -1 and 1: loss in level.
- Amplitude level is more than 1 or less than -1: gain in level.
- Vertical line goes up: polarity normal.
- Vertical line goes down: polarity inverted.
- Straight vertical line: no phase delay (frequency-independent delay).
- Stretched/distorted vertical line: phase delay (frequency-dependent delay).
- Ringing behind impulse: phase delay from a filter (frequency-dependent delay).
- Steep vertical rise and fall: flat full range frequency response.
- Reduced height and dampened vertical rise and fall: HF roll-off.

These features are common to many impulse response renders. Additional computations can create a variety of analyzer-specific enhancements. The linear amplitude scale is not optimally suited for ease of identification and characterization of echoes, a weakness of the standard impulse response. The Hilbert Transform is a popular modification of the impulse response (Fig. 12.22). This is very simplistically described as taking the absolute values of the linear impulse response and logging the vertical scale. The negative half of the impulse response is folded upward and added to the positive, which eases reflection identification. Level is in dB (makes life easier) but polarity is ambiguous unless otherwise noted in the display (makes life harder).

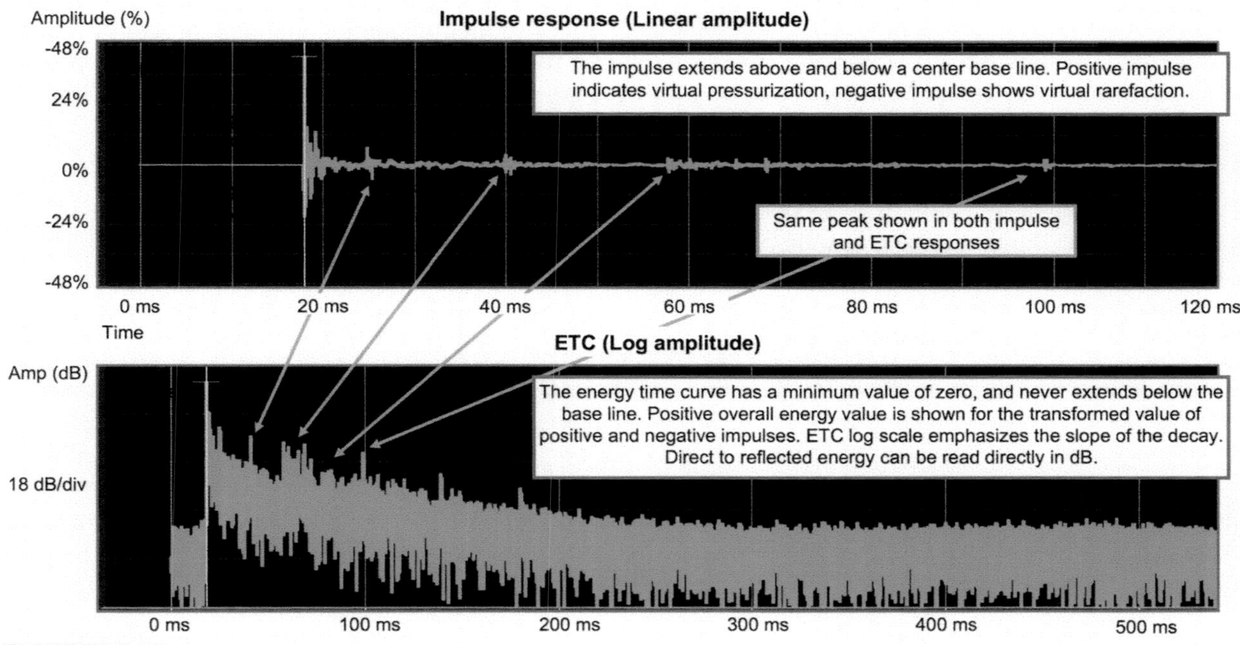

FIGURE 12.22
Comparison of log and linear impulse response (courtesy of SIA-SMAART)

12.12 PUTTING THE ANALYZER TO WORK

Transfer function amplitude answers the question of gain or loss both as a whole and over frequency. Phase answers the question of what frequency arrived when. Impulse response precisely identifies direct sound and reflection timings. Coherence alerts us when our data is contaminated by noise. That's a lot of information we can put to practical use.

Impulse response is the tool for setting delays and identifying surfaces causing the most trouble. Hopefully we are allowed to treat them. Phase response is key for joining speakers together at crossover: subs to mains, MF to HF and more. Phase response also reveals the compatibility between different speaker models. Maybe they meet at the highs but what about the mids? There are specialized phase filters that can help with this.

Transfer function amplitude helps find the best aim, splay angle and spacing. We measure the level on axis and then compare it with the level at the other key locations. For a single speaker we compare the on-axis response to the outer edges of coverage and move the speaker until minimum differences are found. For splay angle we compare the individual on-axis levels with that at the intersection between the speakers. The angle can be adjusted to minimize the difference. Spacing is done much the same way. Move the speakers apart until the level at the mid-point matches the level on axis to the individuals.

Equalization uses the coherence function to select the best candidates and which are better off trying other solutions. Transfer function amplitude then shows us the peaks and dips that need treatment and how well they have been equalized.

The features presented here barely scratch the surface of the computational power available in the dual-channel FFT, much less the full gamut of modern acoustical analyzers. There are Nyquist plots, cepstrum responses, Wigner distributions, time spectrographs, the modulation transfer function, intensity computations, RASTI, STI II and on and on it goes. Each contains unique information and presentation. There is no technical reason to exclude them from discussion here. The math behind them is likely to be every bit as sound as those we have covered in detail. My conclusion as a practitioner of system optimization for over thirty years is that we don't need them to make our decisions and it's complicated enough already. The basic functions described provide enough answers to set the equalizer, delay, level, speaker focus and where to put the fiberglass. Ours is a practical trade, not an R&D foundation. The optimization stage is driven by these basic FFT analyzer functions: single-channel spectrum, transfer function amplitude, phase, coherence and impulse response.

12.12.1 Other complex signal analyzers

The fixed PPO (Constant Q) dual-channel FFT owns the market for system optimization. It's the only complex signal analyzer used on tour. A few alternative platforms remain in use for permanent installations although their market share continues to decline. The industry has standardized around the FFT but it was not always so.

The most significant alternative is the "calculated semi-anechoic response" family. These analyzers strive to capture the direct sound response stripped of the reflection summation effects. The reflection effects are suppressed by filtering in the frequency or time domain. The response has far greater noise immunity than the FFT, because the reflections are removed (in the analyzer, not the room). The extent of reflection suppression is adjustable, allowing users to dial in as much (or as little) of the room as they like.

The original concept was developed by Richard Heyser as time-delay spectrometry (TDS) and manufactured as TEF™. The basic scheme is this: A sine wave sweeps upward at a constant linear rate. A tracking band pass filter sweeps at the same rate as the source, which sets up the following situation: The analyzer only sends (and receives) one frequency at a time. Both are moving targets. We line them up by delaying the filter by the transit time to the mic. When a reflection from one frequency arrives at the mic, it encounters a filter that favors the direct sound of a higher frequency that's just now arriving. The direct sound is allowed in. Latecomers will not be seated. Any reflection slower than the sweep rate is suppressed. Faster sweeps suppress more of the room but at a cost: resolution. As resolution falls we lose more than just the peaks and dips caused by reflections. The direct path gets smoothed over as well (including EQ filters).

The maximum-length sequence sound analyzer (MLSSA) captures a simulated anechoic response with a time domain capture: a time-filtered impulse response. The reversible FFT can convert time to frequency and vice versa. MLSSA obtains an impulse response from a known periodic source containing all frequencies in a calculated non-repeating pattern (the maximum length sequence). The impulse is truncated, i.e. all values past a selected amount of time are set to zero. Say goodbye to the reflections! Recall the interrelated nature of the frequency and time domain. An echo in the time domain shows up as combing in the frequency domain. Suppressing combing reduces the peaks and dips in the frequency domains and also shrinks the magnitude of the reflections shown in the time domain. That's what we'd see if we covered the wall with drapes. If we truncate the echo out of the impulse response, the

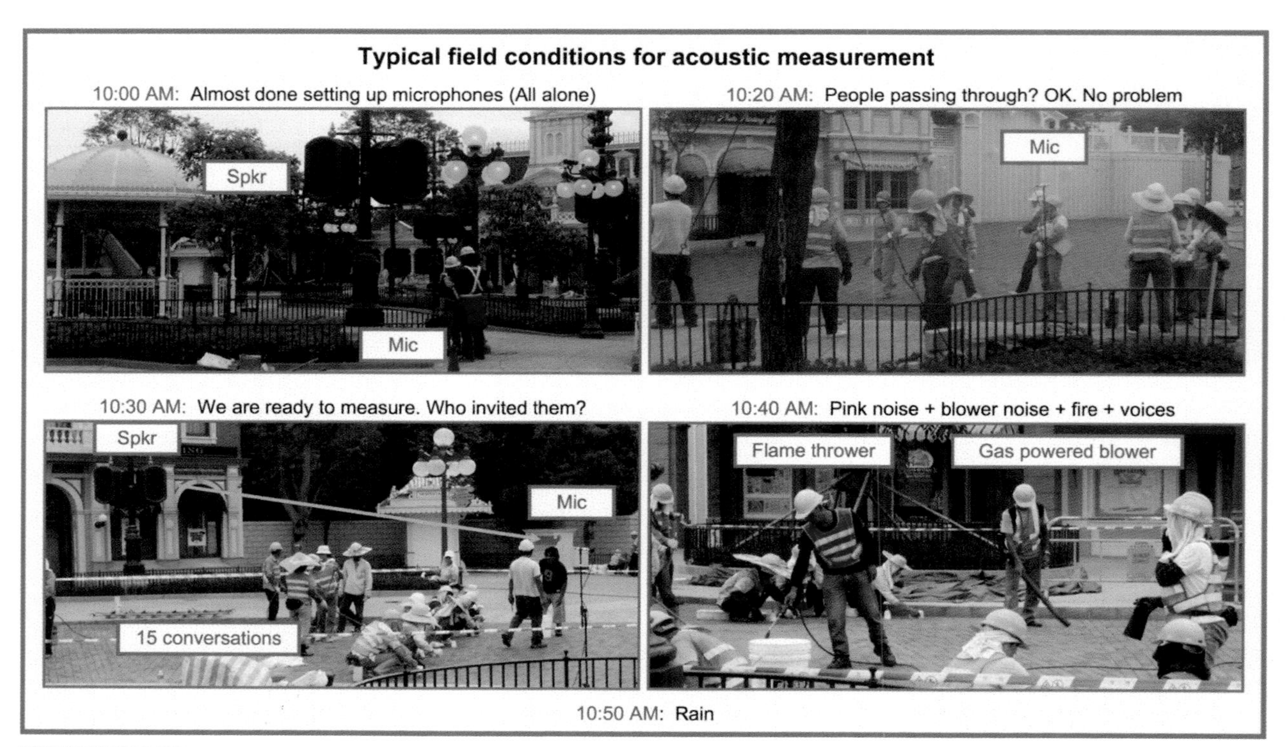

FIGURE 12.23
When we began our setup there was no one in sight and perfect quiet. By the time we were ready for our first measurements we had dozens of workers sealing the grout, a flame thrower to pre-heat the bricks, a gas-powered blower to clean out the dirt and a brick-cutting tool all joining the measurements. This is an illustrative example of "non-causal" coherence degradation.

frequency response derived from this will no longer show combing. It looks like we treated the wall (and costs less). The cost is frequency resolution (again). The more time we remove from the impulse the lower the resolution.

These systems can display data that looks far superior to the high-resolution fixed-point FFT (because it's way lower resolution). The data will look closer to the FFT if we slow down the sweep or truncate less of the impulse of these systems. But then we're stuck with linear frequency resolution (which us FFT folks haven't had to see since 1991).

There has been much debate over how much of the room to leave in the measurement. The linear approach (TEF and MLSSA) cuts a single line in time. The fixed PPO FFT approach is consistent with our log proportional tonal response (section 5.3). The market has spoken, so there is little need to carry on the great debate. There is one last thing to say about it: The only decisions at stake between these platforms are the EQ curve and identification of reflection problems. All other optimization operations (speaker aiming, splay, spacing, delay and level setting) are essentially seen the same by all (if you know how to look for them). The EQ will look different and one of the systems will make sure you see the problems created by the balcony front.

12.12.2 Analysis systems

The perfect analyzer alone does not an optimization tool make. A dedicated optimization system contains everything needed to access the sound system without interrupting the signal flow, injecting hum or noise. We need access to both line-level signals and microphones for comparison and must have delay compensation for latency and acoustic propagation delays.

KEY FEATURES OF THE ANALYSIS SYSTEM

- 24–48 point/octave dual channel FFT.
- Spectrum vs. frequency.
- Transfer function amplitude vs. frequency.
- Transfer function phase vs. frequency.
- Transfer function coherence vs. frequency.
- Impulse response.
- Line-level interface to accept electronic input signal (pre-processor).
- Line-level interface to accept electronic output (post-processor).
- Microphone preamp interface.
- Internal delay line for electronic signals with .02 ms minimum resolution.
- Omnidirectional high-grade linear reference microphone.

Examination provides knowledge of our system but no disease has ever been cured by diagnosis alone. Treatment plans and proof of their effect comprise the remainder of this book.

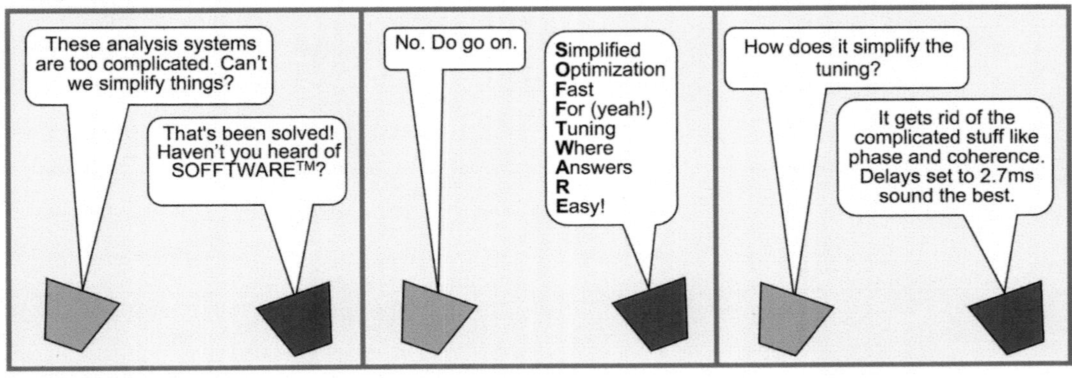

REFERENCE

Herlufsen, H. (1984), *Dual Channel FFT Analysis (Part I)*, Bruel & Kjaer.

CHAPTER 13

Verification

The end product of the optimized design is speakers in position, acoustic treatment in place and equalizer, delay and level settings programmed into signal processors. Wouldn't it be depressing and embarrassing to learn we had done all the smart stuff and left in the stupid? We equalized speakers that had polarity reversals or other unexpected "features" such as the scenery painter's cardboard taped over its front. (Yes, this really happened.) Such discoveries are business as usual when found early in the process and a huge loss of face and time when found near the conclusion. If discovered by someone else after we've signed off on the job, we look like utter fools (deservedly). Our entire system calibration crashes like a house of cards over false assumptions that should have been discovered before a single processor setting was applied.

The verification stage ensures these types of challenges are sorted out in advance through a series of routine tests of system components and interconnection wiring. The verification process often reveals no problems with the system, which is boring but reassuring. Skipping the verification process ensures you will find no problems (because you're not looking). This can make the calibration stage much more exciting! Personally, I would sooner leave a system verified and un-calibrated, than calibrated and unverified.

> ***verification*** *v.t.* *1. a confirmation of truth or authority. 2. a formal assertion of validity. 3. the process of determining whether or not a set of established requirements are fulfilled. 4. establish the truth or correctness of by examination or demonstration. Synonymous with: confirmation, check, or substantiation.*
>
> **Concise Oxford Dictionary**

Let's not get carried away. Our task is to ready the system for calibration, not to run a specification proof of every piece of gear. The assumption is we use the same primary measurement tool for verification and calibration: the dual-channel FFT analyzer.

13.1 TEST STRUCTURE

This is a test. I repeat. This is only a test.

Verification tests are not open ended, or philosophical. We want specific answers to specific questions. We structure the questions to obtain reliable answers. The question/answer structure is the test procedure.

Test procedure structure

- **Subject**: What do we want to know (e.g. polarity or maximum SPL)?
- **Procedure**: What tools and steps are required (FFT analyzer phase, SPL meter)?
- **Units**: How is it quantified (e.g. normal or inverted, 112 dB SPL)?
- **Standards**: What is expected and/or acceptable (e.g. normal, 110 dB SPL)?
- **Result**: What is the outcome (e.g. pass, pass)?

The device under test (DUT) is the physical object we are measuring, whether a single cable, speaker, electronic component or the entire transmission chain. The test signal is the source, which is often a specific known signal.

13.1.1 Testing stages

The transmission path spans from stage source to listener with any number of routes between. The verification tests can examine any single component or the entire transmission path. This differs from calibration, which is concerned only with the back half of the transmission path (post-console outputs). Simply put, it pays to verify the upstream signal path so that we can be sure there are not major problems that will compromise the end product.

FOUR STAGES OF VERIFICATION

- Eyes and ears: The common sense checking we can do without an analyzer.
- Self-verification: Test the analysis system to ensure it measures accurately.
- Pre-verification: Check the system before calibration (includes upstream and down).
- Post-verification: Check the system after calibration.

13.1.1.1 EYES AND EARS

We can do lots of work before we turn on the analyzer. Ears, eyes and simple tools such as the laser and inclinometer can expose big and small problems in short order. Visual inspection of speakers, racks and a quick channel-by-channel listen can find things right away that can be fixed while we move on to other items. If you can't see the speakers (behind scrims etc.) you absolutely must go to the catwalk or open the soffit and meet it.

Eyes and ears verification examples

- Gross continuity: Is the left channel driving the left side?
- Fine continuity: Is the amplifier driving the correct driver?
- Gross defect: Why does the third surround speaker sound totally different?
- User settable parameters: Does the horn rotate? Is there a switch on the back?
- Hum and noise: Do we need an analyzer if we can already hear it?
- Aim: Is the speaker aimed correctly (laser or inclinometer can help)?
- Occultation: Is anything blocking the speaker (I-beam) or its path (balcony)?
- Future problem: How long before that unsecured connector falls out?

13.1.1.2 ANALYSIS SYSTEM SELF-TEST

Analyzer self-verification ensures that problems we discover are actually in the sound system, not the diagnostic tool. Our test signals will need verification. Single-channel measurements need a high-quality pure sine wave. Transfer measurements need a full range (not necessarily flat) source. The modern analysis system may be turn-key or assembled from software programs and purpose built or custom hardware. Self-verification is largely the same in either case.

Analysis system self-verification

- Analyzer noise floor: Must be lower than the DUT.
- Analyzer and source frequency response: Must be flatter than DUT.
- Sine tone: source and analyzer THD: Must be lower than the DUT.
- Maximum-level capability: Must be higher than the DUT. No compression.
- Latency: stable and matched (dual channel).

- Microphone: known sensitivity, flat, linear and stable (no compression).
- Microphones: matched sensitivity and response (multiple mics).

We cannot assume that a source is flat but we have to start somewhere. Pink noise and sine tone sources are available in hardware formats and as .wav files. Unless self-verification shows problems we can assume the sources to be functional for our purposes. Begin with an electronic test, which should show up any source problems.

The acoustic side includes a microphone, which is extremely hard to independently verify in absolute terms. A calibrator (piston phone) can verify the sensitivity but we are vulnerable in regards to its frequency response. Microphone verification is mostly in relative terms, seeking to match responses so we can place multiple mics around the room and discern differences in the system response (Fig. 13.1).

13.1.1.3 SOUND SYSTEM PRE-VERIFICATION

System pre-verification investigates whether equipment was installed as specified. This includes wiring, continuity, speaker positions and more. The downstream half (console outputs and onward) includes the components we measure during calibration. The vulnerability of only looking downstream is that the mix console may be sending a defective signal (which we use as the reference). Upstream verification uses an earlier reference (our own noise source) and checks up to the transition point: the console outputs. The poster child for upstream verification is reverse polarity of left or right. Downstream each channel checks out as matched, but the reversal would be detected in the upstream measurement. Upstream verification keeps us from taking the blame for accurately transmitting a defective signal.

The calibration process includes everything from the console outputs to the speaker. Verification of these components is mandatory. We may also choose to verify upstream components, such as the console, outboard gear, stage microphones, etc.

13.1.1.4 SOUND SYSTEM POST-VERIFICATION

The latter stage, post-verification, consists primarily of symmetry checking the calibrated system. For example, a simple stereo system is initially fully pre-verified and calibrated on one side and then post-verified on the other. Post-verification can continue indefinitely. A permanently installed system can be continuously post-verified to ensure it maintains its

Verification Test Reference

	Examination tool	Verification role
Physical tools	**Inclinometer**	Determine vertical focus angle for speaker and/or vertical splay angle between speakers.
	Protractor	Determine splay angle between speakers or a speaker and a surface.
	Disto	Measures distances to surfaces,cluster height cluster etc. Verification of proper installation and symmetry.
	Laser pointer	Determine focus angle for the speaker. Especially useful for post calibration verification of symmetry.
	Thermometer	Establish baseline and monitor changes in variable temperature environments.
	Hygrometer	Establish baseline and monitor changes in variable humidity environments.
Simple audio tools	**VOM**	Continuity testing. High voltage testing such as amplifier outputs and line voltage. Essential general purpose tool.
	Polarity Tester	Cable testing
	Listen Box	Continuity testing.Signal routing. Hum and distortion detection.
	Impedance tester	Verify the wiring of speaker lines and the presence of speakers on the line.
	Oscilloscope	Misc. signal tests. Amp outputs, high voltage testing. Detection of DC and/or oscillations beyond the audio band.
	Sound Level Meter	Full or partial bandwidth SPL verification, microphone calibration. General and weighted SPL measurements
	Mic calibrator	SPL calibration of microphone sensitivity
	RTA	Doorstop
Complex tools	**Ears**	Advance troubleshooting. Signal routing, continuity. THD, noise & freq response. Essential general purpose tool.
	Eyes	Advance troubleshooting. Detection of path obstruction, symmetry errors, noise sources, smoke.
	Dual Ch. FFT analyzer	Electronic and acoustic signal test. Details to follow.

FIGURE 13.1
Verification test reference

originally calibrated response. Most post-calibration verification consists of "OK, next!" as we check left versus right and move through sixteen surround speakers, but the work has its reward when we find the one that falls out of line.

Post-calibration verification measurement examples

- Do symmetrically matched speakers have symmetrically matched response?
- Were copy/paste processor parameters programmed correctly?
- Was this speaker or microphone damaged during the course of the tour?
- Is the system response the same as six months ago?

13.1.2 Access points

The standard transmission signal path flows through four major milestones: source output, console output, processing output and speaker output (measurement microphone input). The upstream side ends at the mix console outputs or equivalent. This is the location where the finished art form is put on the loading dock for transport, our electronic reference signal. Signal flows through the processor and then on the speakers. We monitor the processor output so we can see what we are doing to the signal heading to the speaker. I don't need to tell you what the mic does (Fig. 13.2).

Any device can be taken off-line for verification if access is not provided. Probably not a good idea during the show. Key verification parameters are shown in the reference chart. Digital audio systems also need to provide us with access points so that we can see what they're doing. These may be furnished as dedicated channels of converted analog or raw digital (as long as our analyzer can read it). If they are not provided we will have to get creative to find a way to at least get a clone/copy of what it's doing through a borrowed channel or such.

13.1.2.1 PHYSICAL ACCESS

Speaker position, splay angle and focus angle can be verified by the standard physical measurement tools. A tape measure confirms height and an inclinometer the vertical angle. Horizontal and vertical aim are aided by lasers. The final focus positions will be determined by acoustic performance as part of the calibration process but the physical tools are extremely useful for symmetry verification after one side has been calibrated.

13.1.3 Test setup

The four general-purpose verification setups are single and dual channel (electronic and acoustic). Setup 1 (Fig. 13.3) is single-channel measurement of the DUT (or series of DUTs) with known source signal. Setup 2 (Fig. 13.4) is a

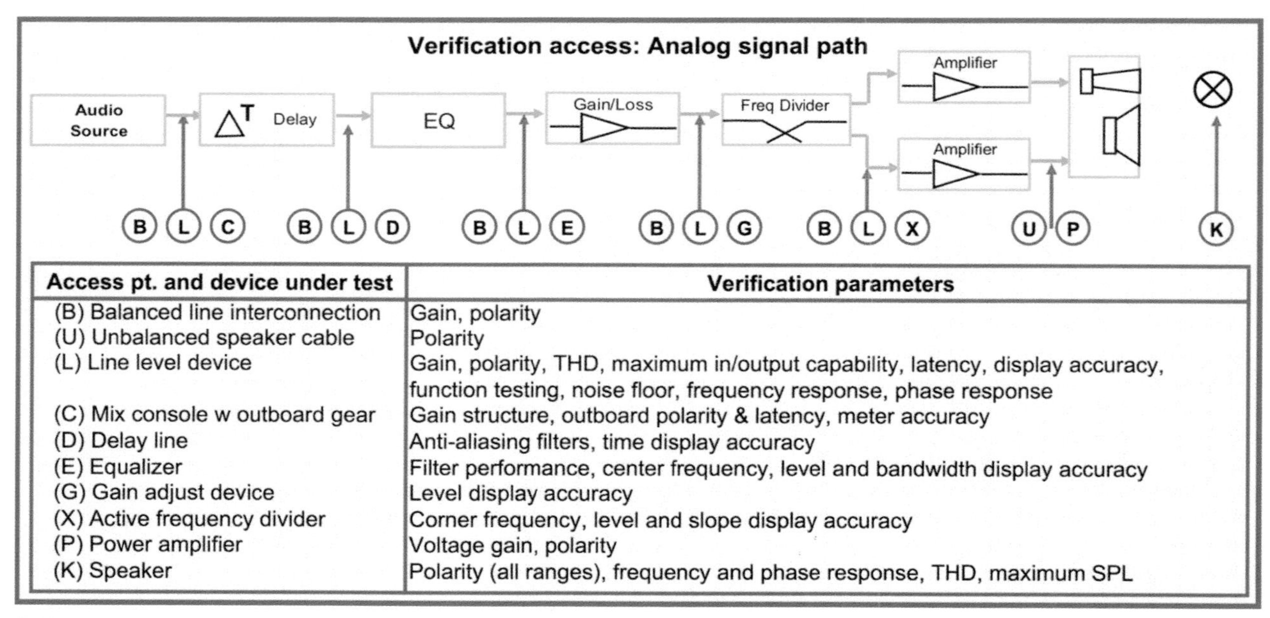

Access pt. and device under test	Verification parameters
(B) Balanced line interconnection	Gain, polarity
(U) Unbalanced speaker cable	Polarity
(L) Line level device	Gain, polarity, THD, maximum in/output capability, latency, display accuracy, function testing, noise floor, frequency response, phase response
(C) Mix console w outboard gear	Gain structure, outboard polarity & latency, meter accuracy
(D) Delay line	Anti-aliasing filters, time display accuracy
(E) Equalizer	Filter performance, center frequency, level and bandwidth display accuracy
(G) Gain adjust device	Level display accuracy
(X) Active frequency divider	Corner frequency, level and slope display accuracy
(P) Power amplifier	Voltage gain, polarity
(K) Speaker	Polarity (all ranges), frequency and phase response, THD, maximum SPL

FIGURE 13.2
Access points for verification and the parameters to be tested

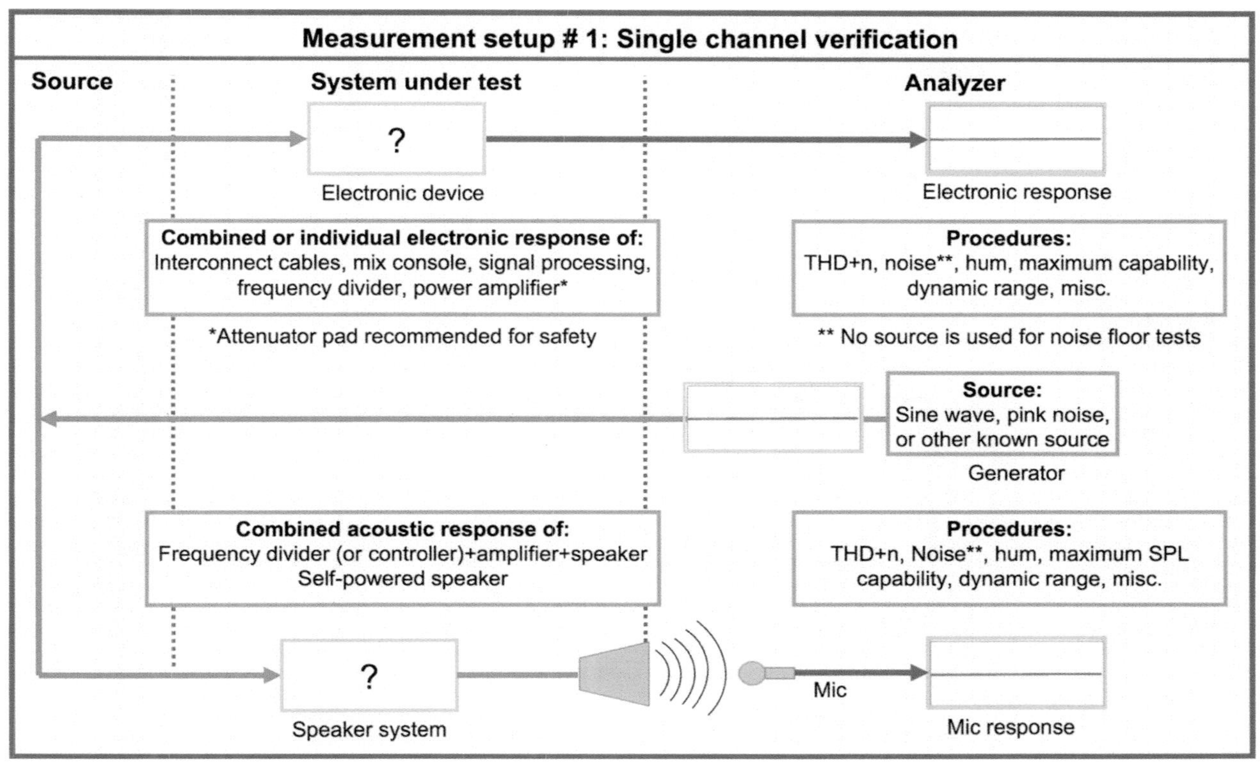

FIGURE 13.3
Verification test setup 1: flow block of the test setup for single-channel verification procedures

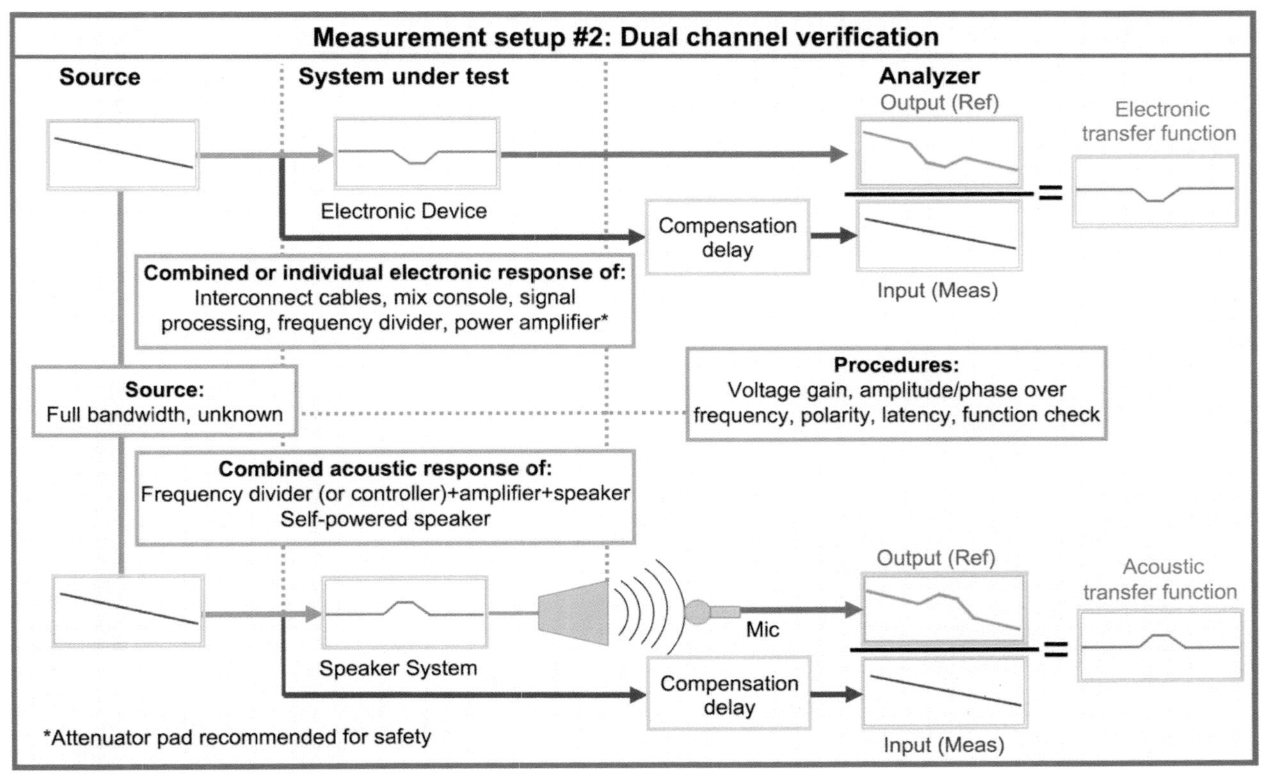

FIGURE 13.4
Verification test setup 2: flow block of the test setup for dual-channel (transfer function) verification procedures

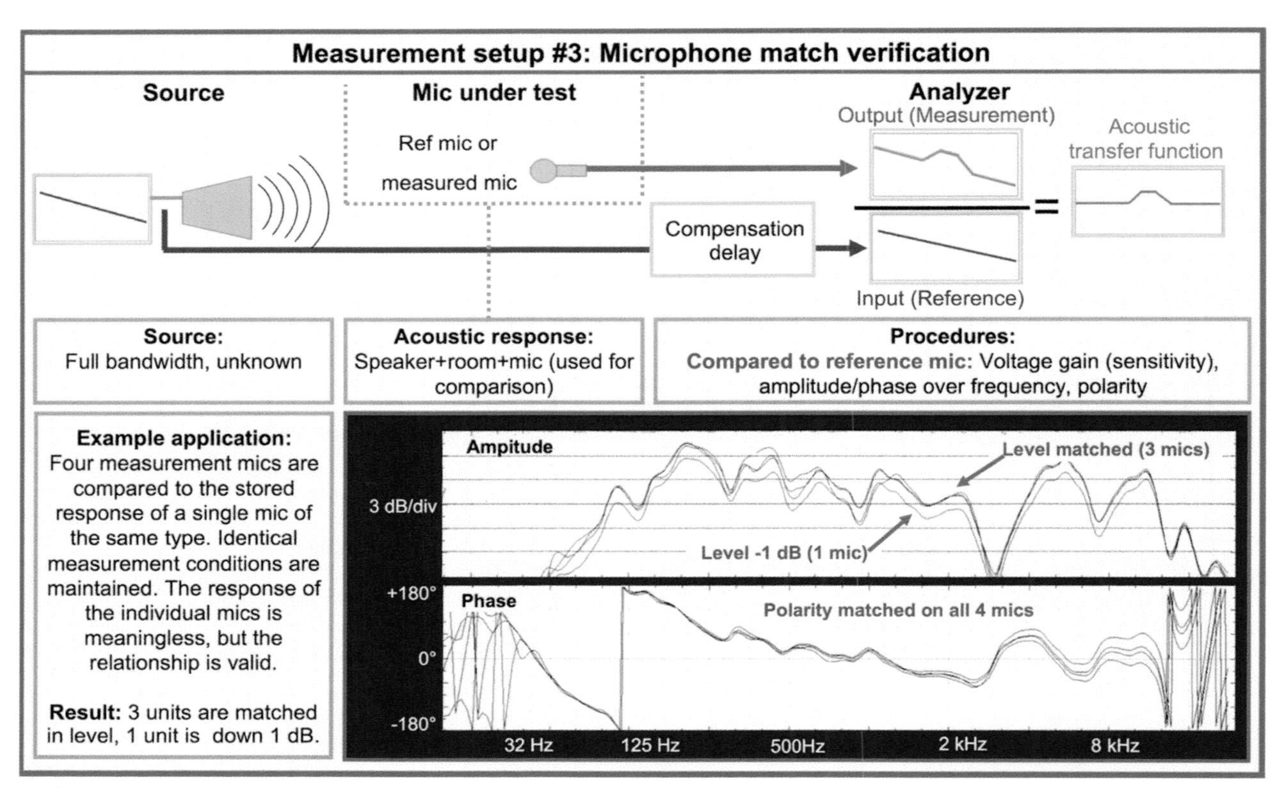

FIGURE 13.5
Verification test setup 3: setup flow block and example application for microphone level and frequency response matching. Consistent microphone placement is required for accurate results.

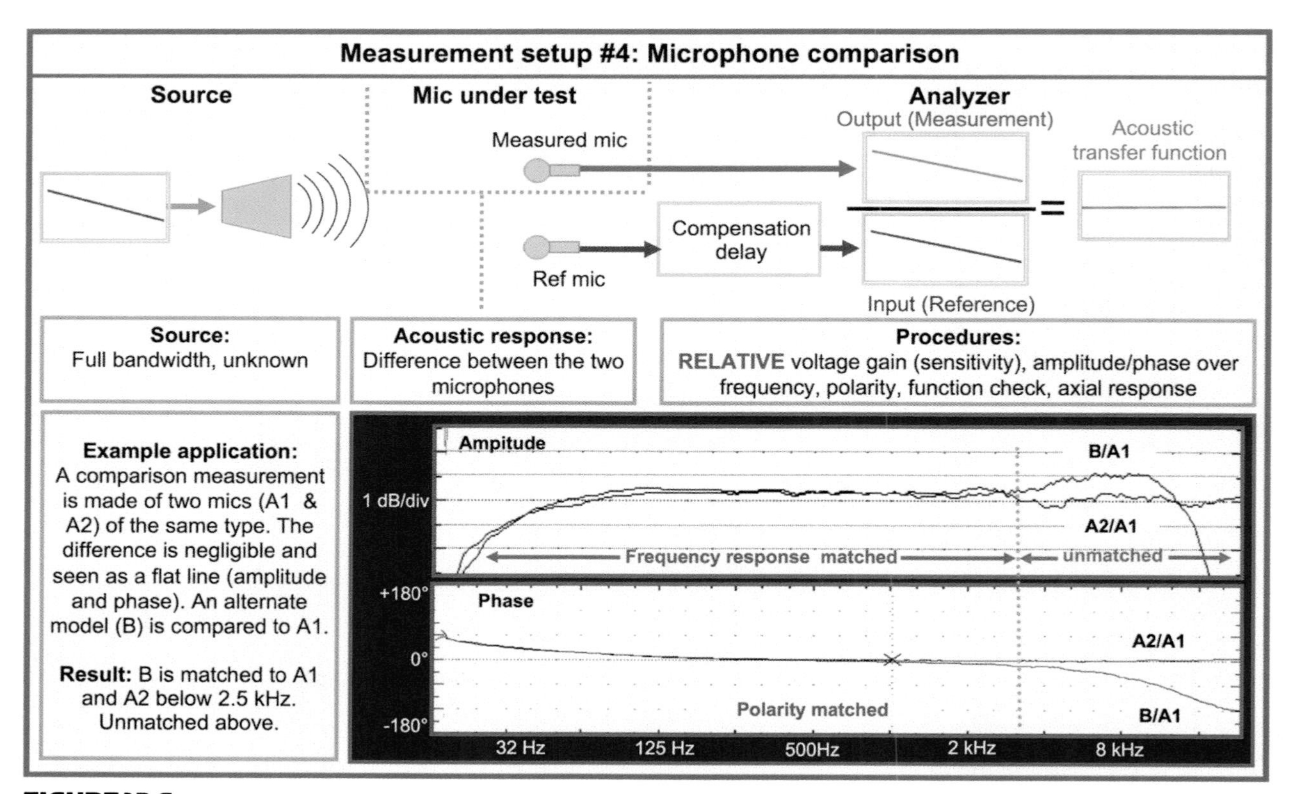

FIGURE 13.6
Verification test setup 4: setup flow block and example application for microphone response comparison

transfer function measurement scenario that spans any number of devices between the two measurement access points. There are also three additional specialized transfer function setups that provide verification of the analyzer and microphones (Figs 13.5 and 13.6). Refer to these flow block diagrams for the test procedures outlined in this chapter.

Verification is performed on individual components and on the system as a whole. The component tests tell us about a particular device, but not how it behaves when interconnected to others. The fully assembled system is the final test because it must cleanly drive the amplifiers to full power.

13.2 VERIFICATION CATEGORIES

13.2.1 Noise vs. frequency

No rock concert or sporting event would be complete without a performer shouting "Let's make some noise!" Our sound system needs no such encouragement. The upper limit of dynamic range is the *noise* the performers are making. The lower limit is the noise we didn't invite to the party. Hopefully these can be kept far apart.

Noise floor measurements should be conducted with devices operating at the expected/normal voltage gains (e.g. unity gain). The DUT input should be loaded at the source impedance it will see in the system. Mute the input of the driving device if testing in line, or terminate it with resistors that match the standard source impedance for off-line testing. In-line verification is preferable because the primary concern is the as-built system rather than R&D of solo elements on a bench.

Noise floor (and maximum capability, described below) are affected by gain settings on multi-stage devices (those with input *and* output level controls), even when they combine to a net unity gain. Early-stage gain followed by late-stage loss may have a different noise floor than unity/unity or loss followed by gain. Therefore much can be learned from iterative changes in the internal stage gains (Fig. 13.7).

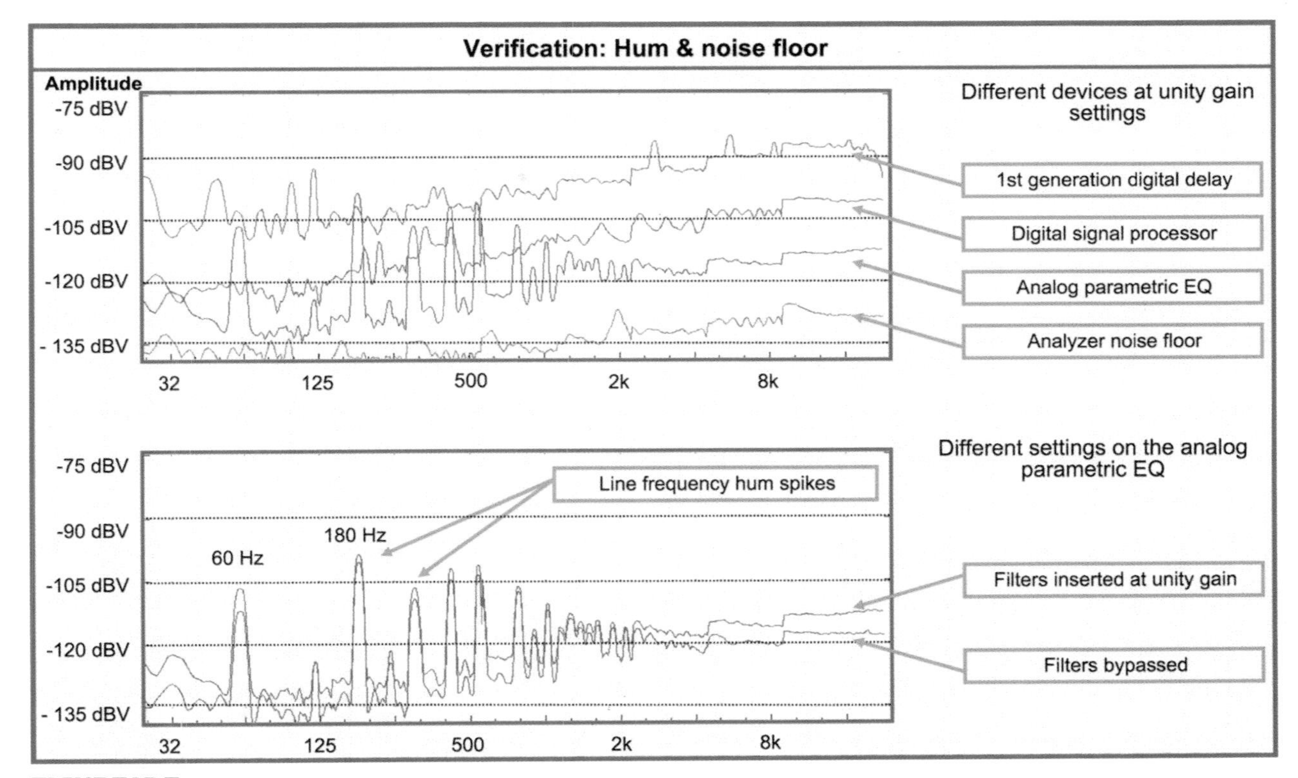

FIGURE 13.7
Example application of the hum and noise verification procedure

13.2.2 Total harmonic distortion + noise (THD + *n*)

Total Harmonic Distortion (THD) measurements are conducted by driving the DUT with a pure sine wave (the fundamental). The presence of added harmonics in the output indicates distortion. THD is calculated by comparing the fundamental level to its harmonic multiples (2×, 3×, 4×, etc.). THD + *n* includes non-harmonic frequencies (noise and other types of distortion products). THD + *n* is a more thorough measure and is generally recommended (particularly for digital). The fundamental level must be well above the noise floor to immunize the measurement from having the harmonics masked under the noise. It is recommended to drive the system in the nominal operating range (such as 0 dBV for a line-level device).

The %THD calculation is derived from the integration of the harmonic series levels to a single value (hence the terms "total harmonic"). A reading of 100% THD indicates the combined harmonic levels equal the fundamental. Ten percent indicates the integrated harmonics are 1/10 the fundamental (-20 dB). For example, the fundamental is 1 V (-20 dBV) @1kHz and harmonics series equals 0.1 V (-20 dBV). The addition is done by the "square root of the sum of the squares" method:

$$\%THD = SqR\ (A^2 + B^2 + C^2 \ldots)/Fundamental$$

Fortunately many analyzers perform this calculation for us, so we won't need to figure out the square root function on our cell phone. THD can be approximated as the ratio between the fundamental and the strongest harmonic, if the analyzer doesn't compute THD directly. Each 20 dB of harmonic-level reduction corresponds to a decimal point change in distortion level. This provides a quick reference for approximation: -20 dB = 10%, -40 dB = 1%, -60 dB = 0.1%, -80 dB = 0.01%.

Any sine wave source signal has distortion of its own, which sets a lower limit to our measurement. It's best to pre-test the generator's THD directly before attributing distortion to a DUT, another facet of analyzer self-verification.

Note: It is possible to receive misleading readings in THD + *n* measurements using FFT analyzers. The FFT window needs to be optimized for a sine wave input signal to minimize leakage in the computation. Flattop window is recommended unless otherwise specified by the analyzer manufacturer (Fig. 13.8).

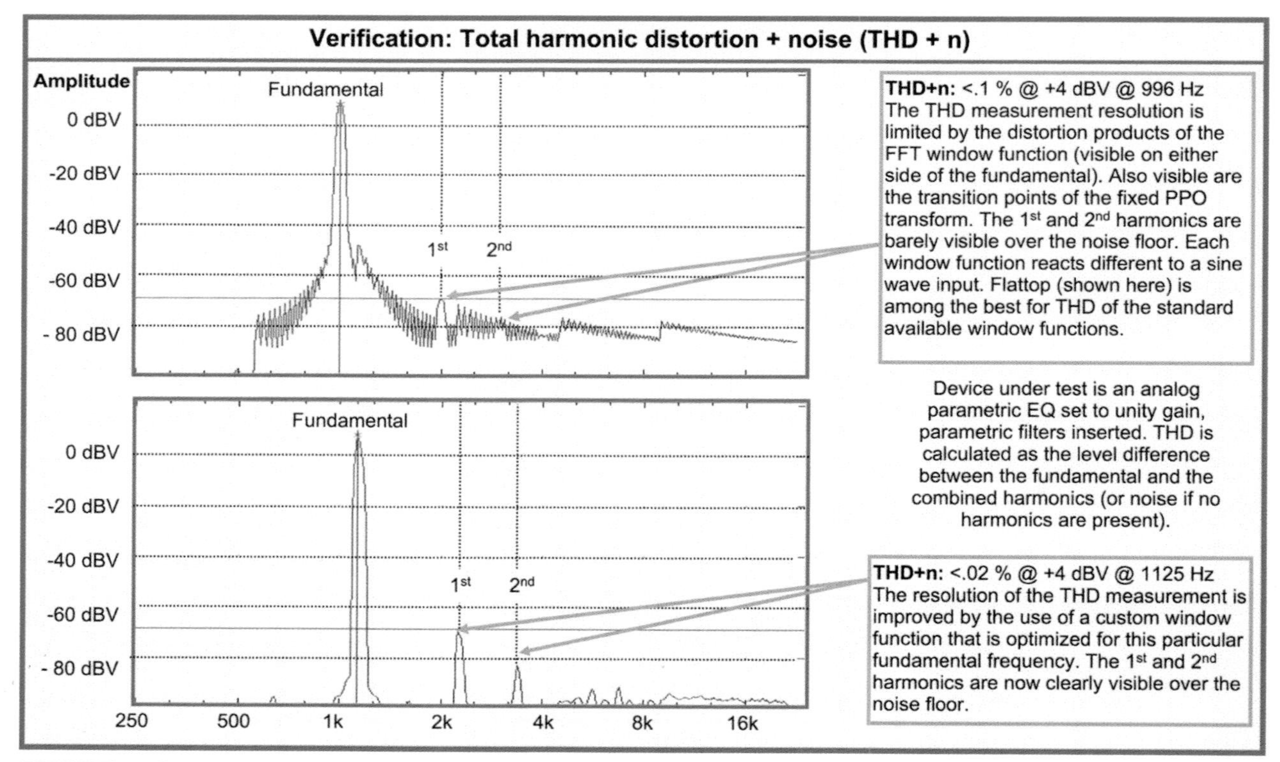

FIGURE 13.8
Example application of the THD + *n* verification procedure

13.2.3 Maximum input/output capability vs. frequency

Maximum input/output (I/O) capability is the dynamic range's linear upper limit (the clip point or onset of limiting). An analog electronic, or digital device with a flat frequency response, should be consistent over frequency, so a frequency-by-frequency test is not warranted unless a problem is suspected.

Speakers are mechanical devices and therefore require much greater scrutiny over their frequency range. Capability limits are highly frequency dependent, the onset is much more gradual and distortion rises gradually as the maximum is approached. A speaker's maximum SPL rating for a given frequency holds for *only* that frequency. Furthermore, the dB SPL readings for a single frequency will not correspond to the manufacturer's stated specifications, which are typically integrated over the full range of the device. Don't be surprised if a speaker rated at 130 dB SPL (a full bandwidth specification) can only reach 110 dB SPL with a sine wave.

SINGLE AND MULTI-STAGE GAIN DEVICES

Recall the discussions of multiple gain stages in the analog and digital pipelines (section 3.1). Net unity gain is possible with offsetting levels at various stages (e.g. input +20 dB/output -20 dB). This may (or may not) affect the dynamic range, reducing the maximum (inter-stage clipping), raising the noise floor, or both. Combinations with the widest dynamic range are found by comparing various inter-stage gain configurations.

A single gain stage device is simple. Set it to unity and measure the minimum (noise) and maximum (clipping). Multi-stage devices are also measured at net unity but should be run through push/pull iterations of various stages to feel our way around the pipeline plumbing. Multi-stage analog devices (e.g. mix consoles) may have misleading silkscreen markings (sometimes 0 is just a pretty name). The goal is to find signal path constrictions and reveal inter-stage overload conditions that don't show up in the meters.

Dynamic range as a whole is limited by the system's weakest link. We can't fit a 120 dB signal through a 100 dB device, barring some form of compression and expansion (like Dolby™), we either lose 20 dB of headroom at the top, gain 20 dB of noise at the bottom, or spread the loss between. The dynamic range gap between the analog and digital audio domains has closed but we must be vigilant about gain structure at the transitions between these domains.

Two ways to unity gain through a device

1. Input unity, output unity: standard for devices with high dynamic range. Result is the least amount of change of maximum in/out capability, and noise.
2. Input +20 dB, output -20 dB: found in low dynamic range devices. Result is 20 dB loss of maximum in/out capability, and minimal addition of noise.

The first scenario is true unity gain. The second is pseudo-unity gain. Input boost was popular in 16- and 18-bit digital device era, when analog still ruled the dynamic range by 20 dB. The digital devices essentially deleted the top 20 dB of analog capability in order to match the digital and analog noise floors. The real goal was noise reduction (the noise of complaining engineers). The lost headroom can become serious during operation, if the processing is unable to drive the amplifiers to full power.

Gain structure management is mostly an operational issue. We can aid the process by finding the upper and lower range limits of each of the components and their interconnection as a whole. This provides the highest assurance that we can pass the widest-range signal through the system from start to finish (Fig. 13.9).

13.2.4 Dynamic range vs. frequency

Dynamic range is derived from the results of the noise and maximum input/output (I/O) capability tests. The difference between the two values is the dynamic range. One important consideration is the multi-stage gain settings. They must be the same for both the upper and lower dynamic range limits. There is no point in gaming the processor to get the best measurements of the noise floor (gain cranked down) if we can't use the same settings to get the maximum output. In other words, if our settings reduce the noise by 10 dB and reduce the maximum capability by 10 dB, then we are right where we started. The most important thing is to find the settings that allow

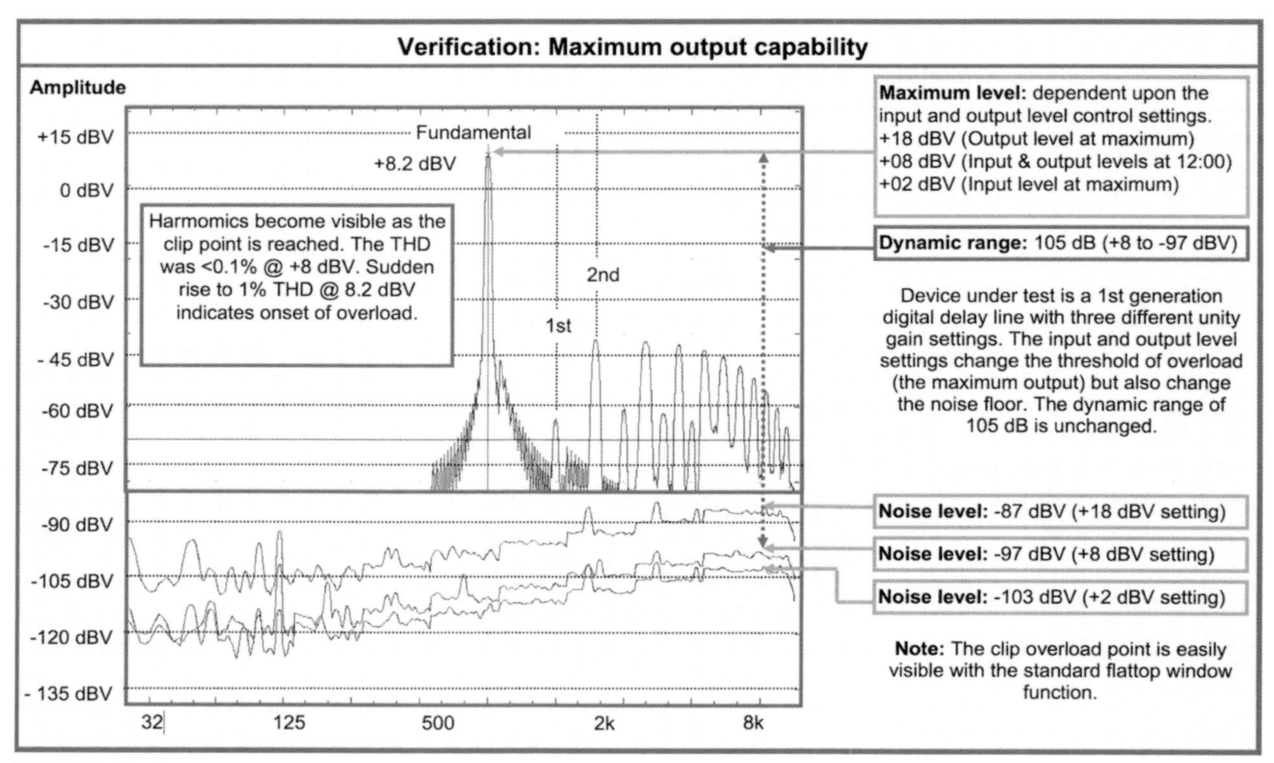

FIGURE 13.9
Example application of the maximum input/output capability verification procedure

the greatest dynamic range. The dynamic range test requires that we measure noise floor and maximum capability at the same processor settings (Fig. 13.10).

13.2.5 Latency

Latency is a fact of life for audio transmission. It is so small in the analog world that we barely consider it. It's there, however, in transformers and other transducers. Digital systems have latencies all over the map. A modern signal path picks up latencies like hitting tollbooths on the turnpike: the A/D at the mic splitter, the optical send from stage, the mix console, the plug-in effects, the system DSP, network to the amplifiers and more. Every one of these devices says 0 ms on their user interface and they are all lying. It is up to us to sort out the bus schedule because we need to know every place where time slips. We need to know all the fixed latencies and the chameleons that change their timing as they see fit. We need to know the bottom-line number for several reasons: to know how long we will have to wait before a signal leaves our speakers, and to anticipate any opportunity where related signals with different latencies might be summed.

The most important latency-related division is between fixed and variable devices. Fixed-latency devices are straightforward. Measure the latency once and enact a strategy for dealing with any summed time-offset paths. Variable-latency devices require constant vigilance. Such devices usually hold a consistent value until something is modified, e.g. recompiling, reallocating network bandwidth, specialized filters, etc. Systems with variable-latency devices require ongoing verification of the timing chain. Some units are multiple-variable latency devices, i.e. latencies on a per/signal path basis. They operate like a restaurant that serves whatever is ready at the moment, rather than waiting for the fries to be ready and served with the burger. These devices usually have a user-selectable override of the multiple-variable latency function, allowing for all channels to wait for the slowest one. Selecting matched latency takes more DSP memory resources but saves memory resources on the human side.

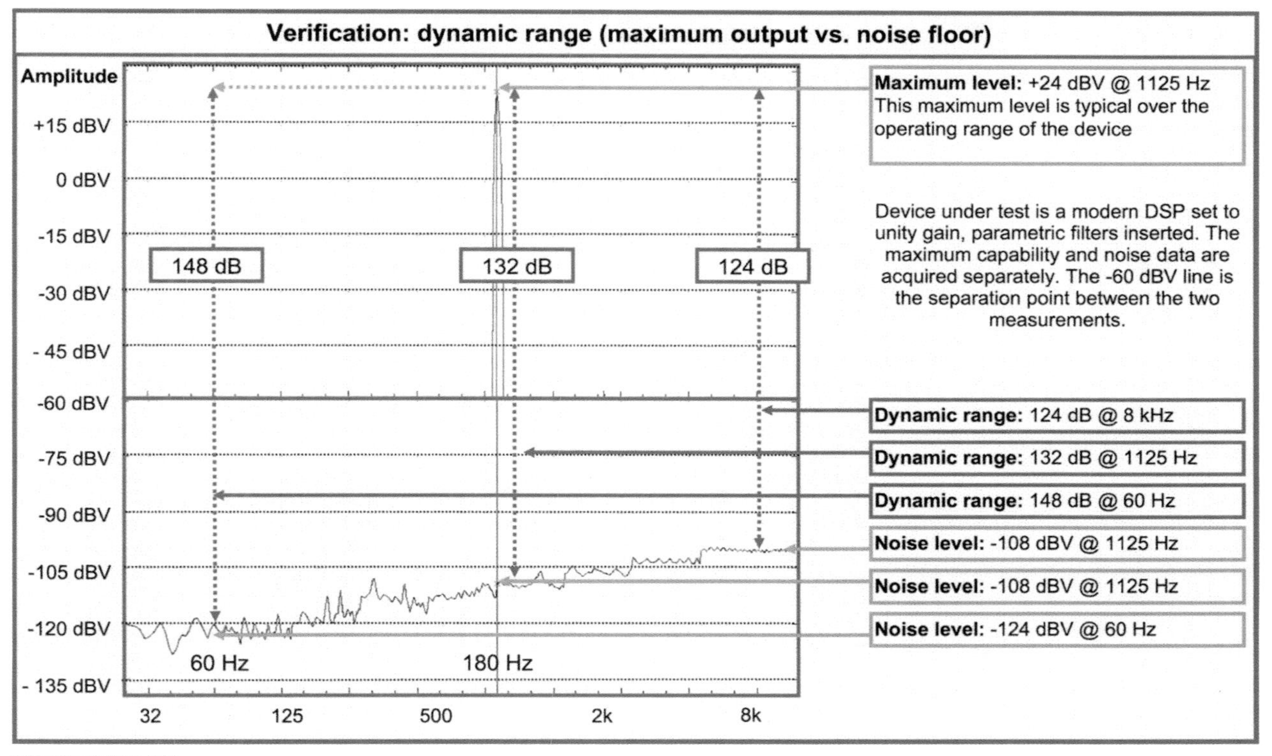

FIGURE 13.10
Example application of the dynamic range verification procedure

Networked audio opens up many new options for latency offsets. These can accumulate through differences in the network transmission paths. Eventually the signal must leave the network and re-emerge as audio. We cannot assume that networked multichannel systems have latency-matched outputs.

Digital and networked systems should be monitored for transmission stability, i.e. freedom from clocking errors, drift, jitter, etc. Latency measurements can be monitored over a period of time to ensure stability.

Latency should be measured first with default DUT settings (unity gain, filters bypassed, etc.) Then change the settings to see if the latencies are stable or variable (Fig. 13.11).

FIXED AND VARIABLE LATENCY CONSIDERATIONS

1. Models with **fixed and matched** latencies.
2. Models with **fixed and unmatched** latencies.
3. Models with **variable and (sometimes) matched** latencies.
4. Models with the **matched internal** latency for **each channel**.
5. Models with **variable** latency for **each channel (due to user settings)**.
6. Models with **different** latency for in-network and out-of-network signals.
7. Signal paths that travel through **multiple** devices in **series**.

13.2.6 Polarity

Polarity is mostly a "go, no-go" test. For electronics there is nothing to it. Impulse response or phase tell it all. Individual loudspeaker drivers are gray zones, especially after they go through the LPF and/of HPF filters. The best speaker polarity tester is a battery but sometimes we can't get a rigger to rappel down, open up the cabinet and get to the terminals. The impulse response is fairly reliable for HF drivers but less so for LF drivers. The impulse can be used only for the HF driver in two-way speakers. Polarity verification for loudspeakers is best done by analysis of the phase trends, but again this is more clear-cut for drivers alone than in multi-way speakers. The polarity of an

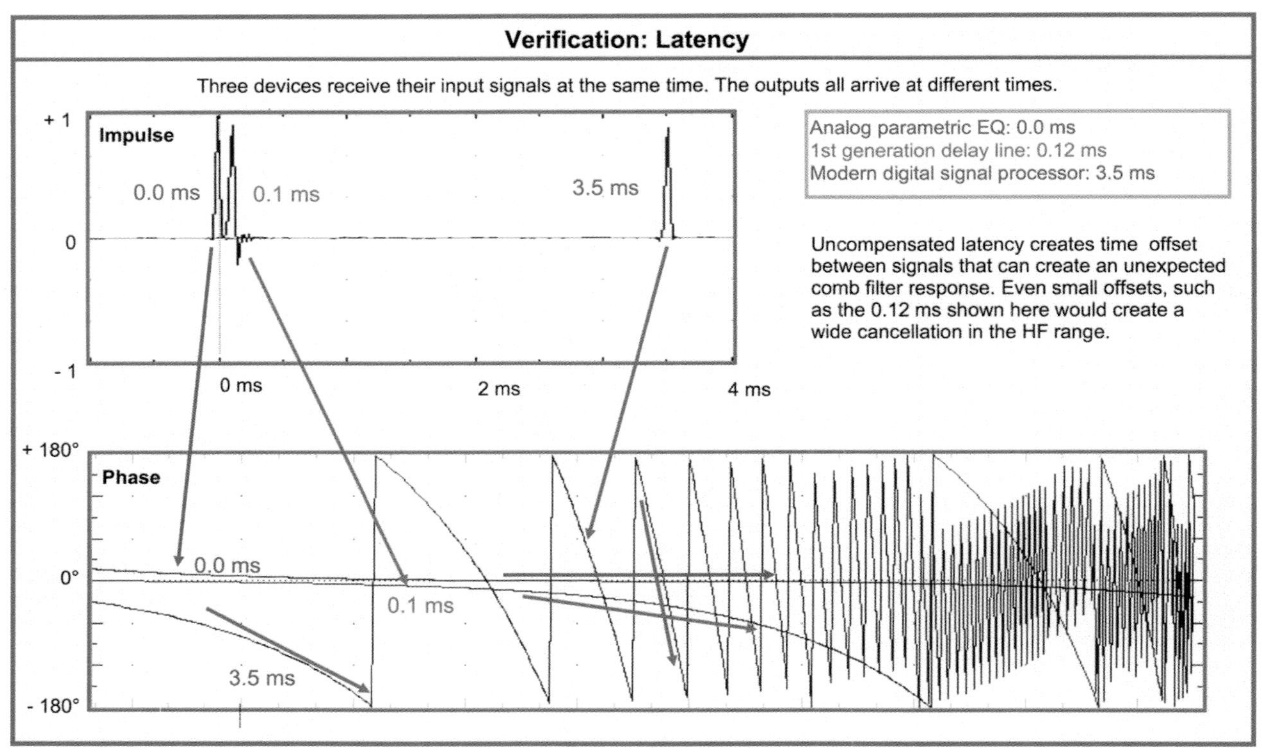

FIGURE 13.11
Example application of the latency verification procedure

LF driver is really an open variable that can change (and still be optimal) with the spectral crossover strategy, filters, etc. The most important verification aspect is matching, i.e. does this speaker match its brother, is it compatible with its cousin, etc.?

13.2.7 Amplitude vs. frequency

The electronic version of this test is so straightforward that we immediately move on to the acoustic version. It's difficult to make conclusions about the frequency response of a speaker system due to ripple variance (unless in an anechoic chamber). Even extremely close measurements have floor reflections and other noise in the data. The key is to make sure the reflections don't fall in places that lead to false conclusions.

We can discern trends in HF and LF roll-off in spite of local ripple variance (unless they happen to fall right on the roll-off point). We can see a spectral crossover, though not as clearly as can be done under controlled anechoic conditions. The search for the optimized spectral crossover is a good example of TANSTAAFL in action. As we get closer to the speaker, we gain immunity from the ripple variance. At the same time, our perspective of the spectral crossover becomes increasingly near-sighted and we can make crossover alignment recommendations that will serve us poorly in the far field.

Our most effective work will be in the form of comparison. As long as we reduce the question to one of difference, rather than absolute value, we have leveled the verification playing field (Figs 13.12 to 13.15).

FREQUENCY RESPONSE COMPARISON EXAMPLES

- Do matched models have matched polarity and drive level for component drivers?
- Do symmetrically matched speakers have symmetrically matched responses?
- Are two different speaker models phase compatible over their shared range?
- Are two drivers phase compatible at the spectral crossover?

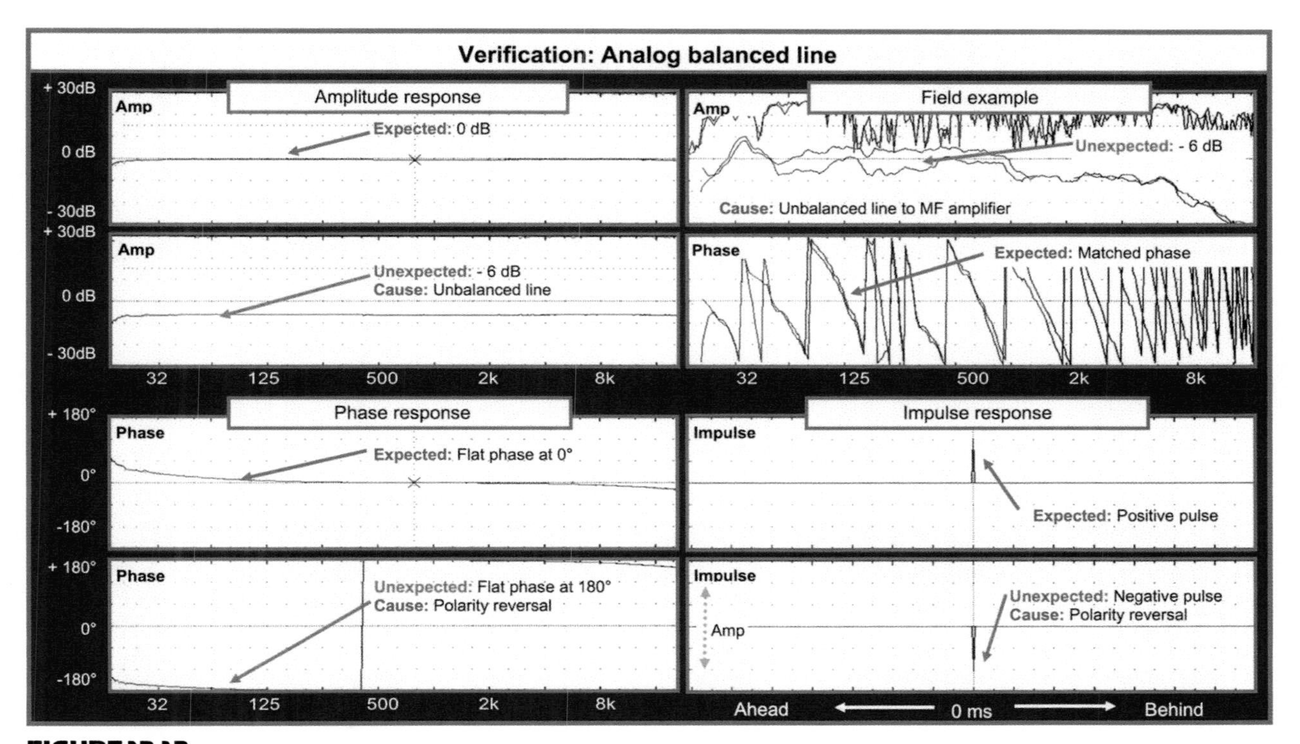

FIGURE 13.12
Example application of the electronic device polarity and level verification procedures

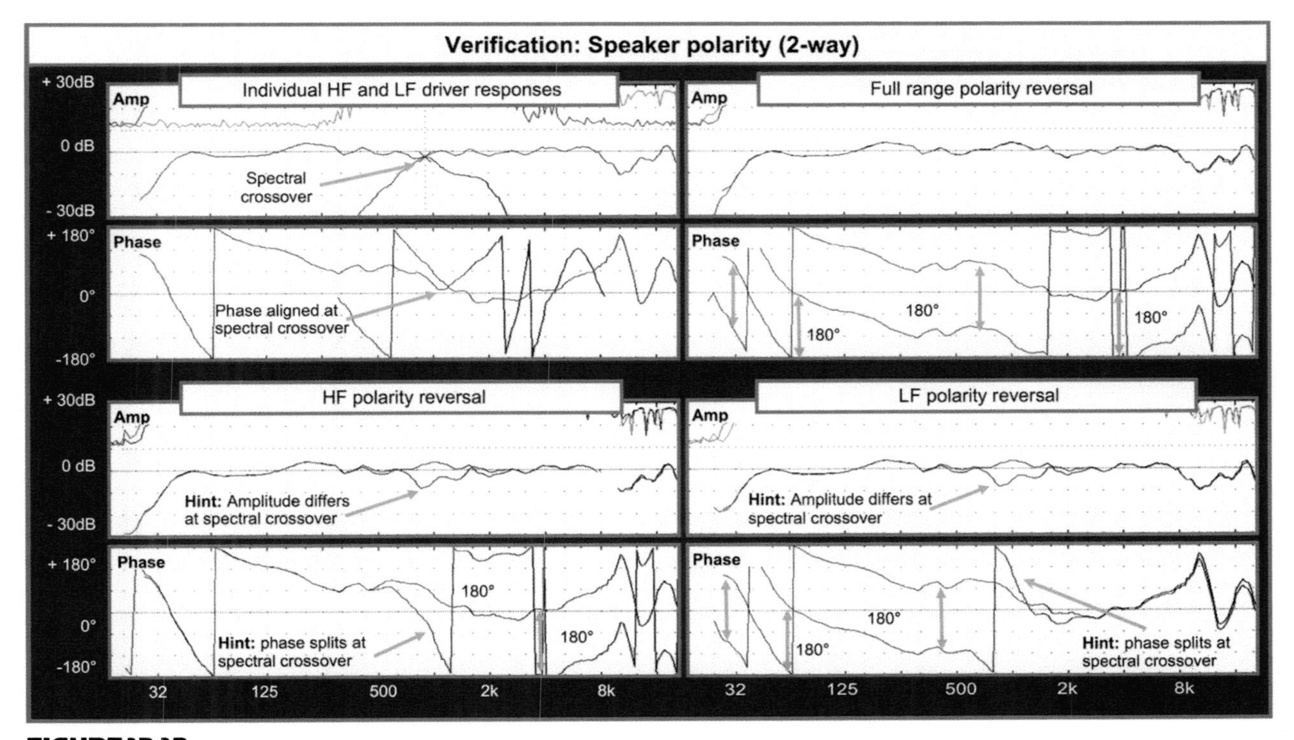

FIGURE 13.13
Example application of the speaker polarity verification procedure

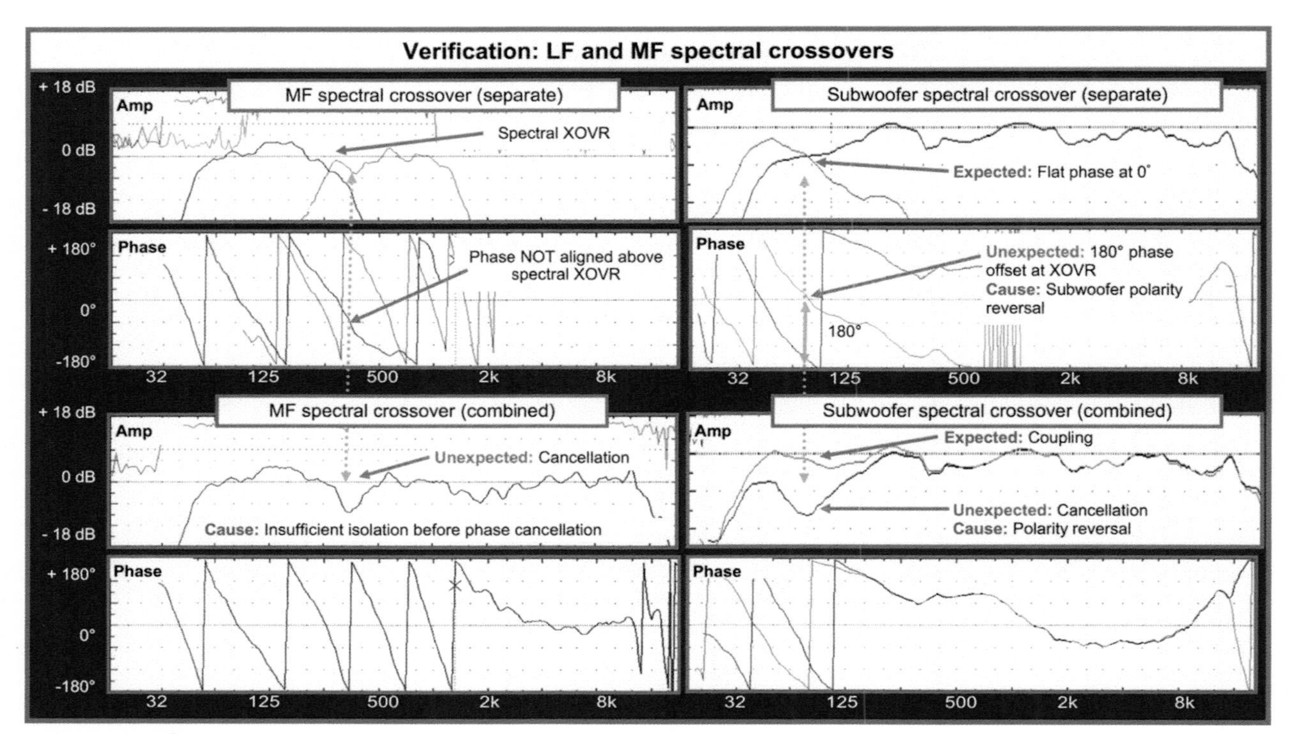

FIGURE 13.14
Example application of the amplitude and phase response verification procedures for spectral crossovers

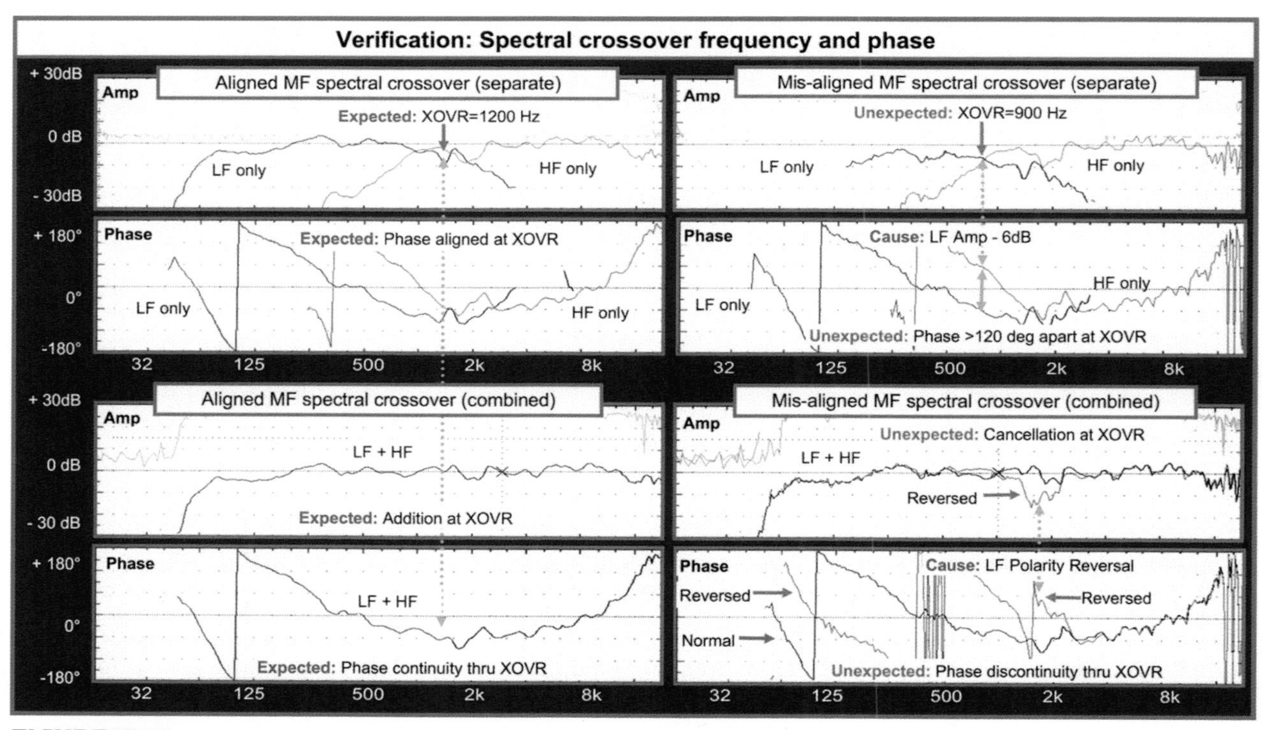

FIGURE 13.15
Example application of the amplitude and phase response verification procedures for spectral crossovers

13.2.8 Phase vs. frequency

13.2.8.1 ELECTRONIC DEVICES

There shouldn't be much to see in the phase response of an electronic DUT with a flat frequency response. Expect 0° over the full range except at the LF and possibly HF extremes. Phase shift in the LF extreme comes from AC coupling and should be restricted to minimal amounts within the passband. It's a different story at the HF extreme. It may be ultrasonic or anti-aliasing filtering in the DUT or it could be an artifact of the analyzer's time resolution limit. If the DUT latency falls somewhere between the compensation delay increments, there will be a phase delay remainder in the measurement. It's a time resolution issue similar to the bandwidth resolution (see Fig. 12.4). For example, what happens when the DUT has a latency that falls at the mid-point between increments, such as 10 µsec when the analyzer's internal delay has 20 µsec (0.02 ms) increments, then? The analyzer sees 10 µsec of phase delay and charts the response accordingly. That's 72° @20 kHz, so it won't go unnoticed. What can we do about it? Some analyzers allow for sub-incremental correction, some don't. The analyzer's phase judgment can't exceed the time record increments.

Phase shift resolution for a given time compensation increment

- 20 µs: ±36° @ 10 kHz, ±72° @20 kHz.
- 10 µs: ±18° @ 10 kHz, ±36° @20 kHz.

13.2.8.2 SPEAKERS

We don't expect a speaker's phase response to be flat over frequency. This is verification (not calibration), so we are looking for installation errors, not fine-tuning. Mostly we are looking for phase matching between same speakers and compatibility between different models. Compatibility issues can be addressed now or later during calibration (Fig. 13.16).

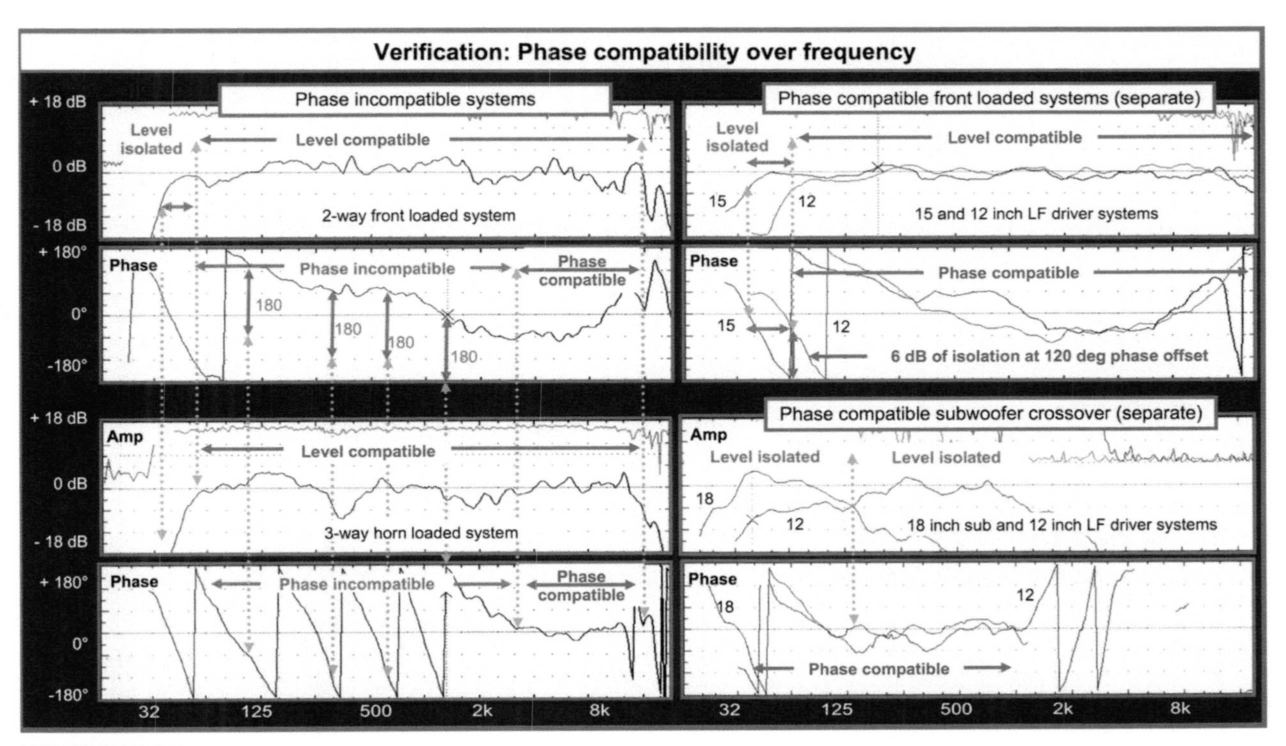

FIGURE 13.16
Example application of the amplitude and phase response verification procedures for speaker system range and compatibility

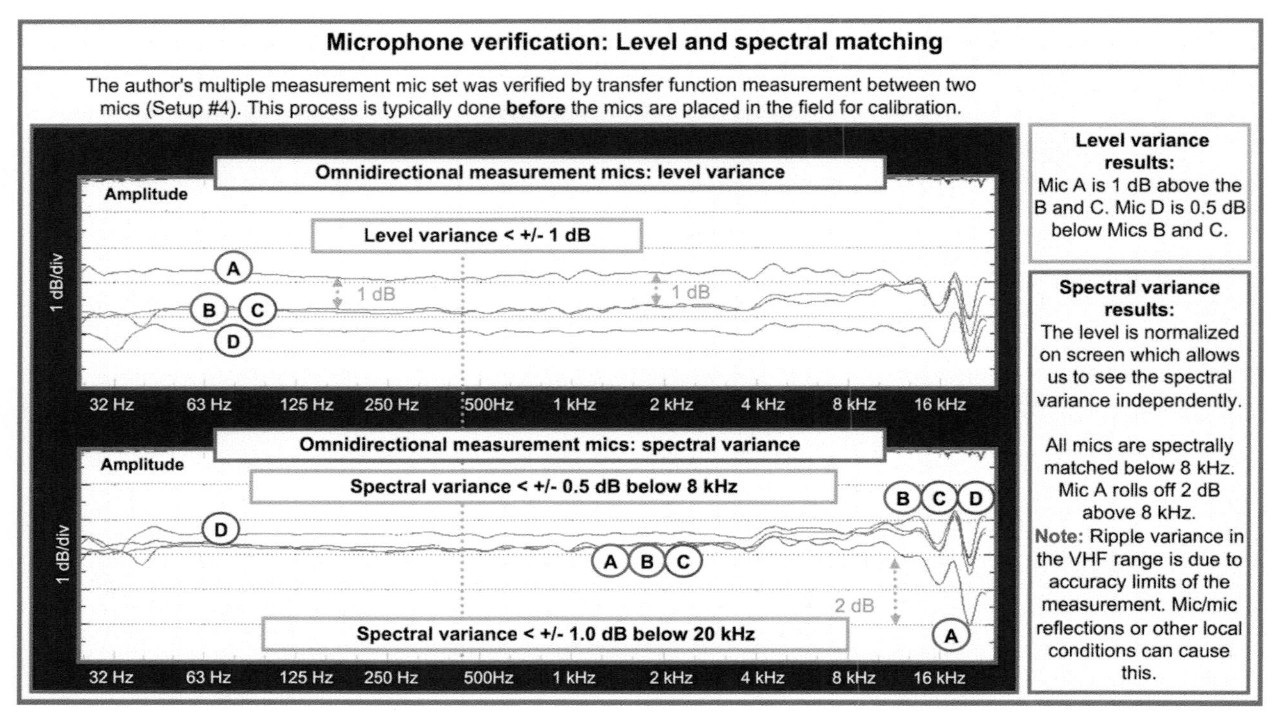

FIGURE 13.17
Field example of microphone matching procedure

13.2.9 Microphone verification

Microphone testing requires a special setup configuration. We need a known flat acoustic source to characterize a microphone response, which would require a known flat microphone to verify its response. This is a circular argument where the Bureau of Standards has the final say. This is why we choose measurement mics very carefully, based upon the credibility of the manufacturers.

In practice we can only test mics in relative terms. The comparison can be done serially (measure one mic, replace it and compare) or in parallel (a transfer function comparison between two microphones in a matched acoustic field) (Fig. 13.17).

13.2.9.1 MIC VS. MIC TRANSFER FUNCTION

Note: First-order source speakers make it easier for the mics to evenly share the same sound field. Place the mics as close as possible to the speaker to reduce the room/speaker summation effects. The displacement plane of the microphones must be opposite the driver displacement plane in a two-way speaker (i.e. if a speaker is HF over LF then the mics are displaced horizontally).

13.2.9.2 MICROPHONE POLAR RESPONSE

For on-axis characterization, both mics are placed on axis to the source. To characterize the axial response of a mic, the mic under test is rotated in place, keeping its diaphragm in approximately the same plane. Precision is limited by reflections between the microphones, and small differences in path lengths of local reflections. Either can cause frequency response ripple. These measurements clearly show the axial trends such as the off-axis spectral tilt of an omnidirectional mic. Axial analysis of a cardioid microphone can help optimize stage monitor placement for maximum rejection (Fig 13.18).

13.3 ANALYZER SELF-VERIFICATION PROCEDURES

Analyzer self-test: hum and noise over frequency		
Source: None	**Units**: volts (dBV, dBu)	**Setup**: #0A (self)
Resolution: > = 1/24	**FFT Window**: Hn/Flt	**Averaging**: 8+
1: Connect to signal generator. Mute generator.		
2: Measure DUT hum/noise floor with 1 ch. FFT.		
3: Result specified as V (or dBV, dBu) @F, e.g. -90 dBV @10 kHz.		
4: Store trace of hum and noise levels for reference.		
PASS/FAIL: Analyzer hum and noise levels set limit to what we can measure		

Analyzer self test: THD + *n* over frequency		
Source: Pure sine tone	**Units**: %THD + *n*	**Setup**: #0A (self)
Resolution: > = 1/24	**FFT window**: flattop	**Averaging**: any
1: Drive DUT at nominal level with sine wave. 1 V (0 dBV) @1 kHz is typical.		
2: Measure the fundamental frequency level and mark as reference.		
3: Measure harmonic level relative to fundamental. Estimate THD (-20 dB = 10%, -40 dB = 1%, -60 dB = 0.1%, -80 dB = 0.01%).		
4: Outcome specified as %THD @V @F, e.g. 0.01 %THD @0 dBV @1 kHz.		
5: Store trace of THD levels for reference.		
PASS/FAIL: analyzer THD levels set limit to what we can measure		

Analyzer self-test: maximum I/O capability over frequency		
Source: pure sine tone	**Units**: volts (dBV/dBu)	**Setup**: #0A (self)
Resolution: > = 1/24	**FFT window**: flattop	**Averaging**: any
1: Raise sine wave source level (1 kHz typical) until clipping/limiting onset.		
2A: (Analog) Result spec'd as max V (dBV/dBu) @F, e.g. +20 dBV @1 kHz.		
2D: (Digital) Result spec'd as dBFS @F, e.g. +20 dBV = 0 dBFS @1kHz.		
3: Store trace of max capability levels for reference.		
PASS/FAIL: analyzer maximum capability levels set limit to what we can measure		

Analyzer self-test: latency		
Source: noise or music	**Units**: T (ms)	**Setup**: #0A (self)
Resolution: .02ms or better	**FFT Window**: Hann	**Averaging**: any
1: Drive DUT at nominal level with broadband signal.		
2: View impulse response and compensate I/O time offset with analyzer delay.		
3: Verify flat phase (therefore latency is frequency independent).		
4: Result specified as time (*x* ms), e.g. 2.5 ms.		
FAIL: instability in latency, latency offset between summed signals		

Analyzer self-test: transfer function		
Source: noise or music	**Units**: dB, phase°	**Setup**: #0A (self)
Resolution: > = 1/24	**FFT Window**: any	**Averaging**: any
1: Send single source to all inputs ("Y" cord): console, processor, mic.		
2: Observe matching of each individual channel.		

3: Observe matching of each transfer function pairing (0 dB and 0°).		
4: Variation result specified as *x* dB, *x* phase°. Should be <±0.2 dB, <15° @20 kHz.		
PASS/FAIL: amplitude should be flat within ±0.2		
PASS/FAIL: phase should be flat within ±15°		
FAIL: polarity inversion		
FAIL: Frequency response errors or instability		

Analyzer self-test: multiple measurement microphone match		
Source: Pink noise	**Units**: volts (dB, phase)	**Setup**: #3
Resolution: > = 1/24	**FFT window**: flattop	**Averaging**: any
1: Assumption: 1 mic is "normal" and used as reference.		
2: ONAX of mic is aimed at ONAX of speaker no closer than 2 m.		
3: Store and recall XFR amplitude and phase of reference mic.		
4: Replace reference mic with DUT mic in same place and compare responses.		
5: Sensitivity result is the difference in overall level between mics.		
6: Variation result is the difference in frequency between mics.		
7: Polarity/phase result is the difference in polarity/phase between mics (+ or-).		
PASS/FAIL: sensitivity matched within ±0.25 dB (can be matched in software?)		
PASS/FAIL: amplitude should be flat within ±1 dB (can be matched in software?)		
PASS/FAIL: phase should be flat within ±30° (can be matched in software?)		
FAIL: polarity inversion		
FAIL: instability in sensitivity or frequency response		

13.4 ELECTRONIC VERIFICATION PROCEDURES

Test: universal properties (unless otherwise specified)
Dual channel: analyzer measures transfer function between DUT input and output
Single channel: analyzer measures DUT output spectrum (amp vs. frequency)
Optimize analyzer input gains to the maximum level short of overload
Use impulse response to set analyzer delay to sync dual-channel measurements
Multi-gain stage DUTs should be run through various gain iterations of relative levels

Test: noise over frequency		
Source: none	**Units**: volts (dBV, dBu)	**Setup**: #1A (single)
Resolution: > = 1/3	**FFT window**: Hann	**Averaging**: 8+
1A: Analog: (In-line) mute driving device. (Off-line) terminate input at source Z.		
1D: Digital: Mute driving device or disconnect input.		
2: Measure DUT noise floor with 1 ch. FFT.		
3: Outcome specified as V (or dBV, dBu) @F, e.g. -90 dBV @10 kHz.		
PASS/FAIL: electronic device noise should be <-90 dBV.		

Test: hum over frequency		
Source: none	**Units**: volts (dBV/dBu)	**Setup**: #1A (single)
Resolution: > = 1/24	**FFT window**: flattop	**Averaging**: any
1A: Analog: (In-line) Mute driving device. (Off-line) Terminate input at source Z.		
1D: Digital: Mute driving device or disconnect input.		

2: Measure DUT hum spectrum with 1 ch. FFT.		
3: Outcome specified as V (or dBV, dBu) @F, e.g. -90 dBV @50 Hz.		
PASS/FAIL: electronic device hum should be <-90 dBV		

Test: THD + *n* over frequency		
Source: pure sine tone	**Units**: %THD + *n*	**Setup**: #1B (single)
Resolution: > = 1/24	**FFT window**: flattop	**Averaging**: any
1: Drive DUT at nominal level with sine wave. 1 V (0 dBV) @1 kHz is typical.		
2: Measure the fundamental frequency level and mark as reference.		
3: Measure harmonic level relative to fundamental. Estimate THD (-20 dB = 10%, -40 dB = 1%, -60 dB = 0.1%, -80 dB = 0.01%).		
4: Outcome specified as %THD @V @F, e.g. 0.01%THD @0 dBV @1 kHz.		
PASS/FAIL: electronic device THD + *n* should be <.05%		

Test: maximum I/O capability over frequency		
Source: pure sine tone	**Units**: V, dBV/dBu	**Setup**: #1B (single)
Resolution: > = 1/24	**FFT window**: flattop	**Averaging**: any
1: Raise sine wave source level (1 kHz typical) until clipping/limiting onset.		
2A: (Analog) Result spec'd as max V(dBV/dBu) @F, e.g. +20 dBV@1 kHz.		
2D: (Digital) Result spec'd as dBFS @F, e.g. +20 dBV = 0 dBFS @1 kHz.		
PASS/FAIL: electronic device maximum capability should be >+16 dBV		

Test: dynamic range over frequency		
Source: N/A	**Units**: dB	**Setup**: #1B (single)
Resolution: > = 1/24	**FFT window**: N/A	**Averaging**: N/A
1: View hum/noise results and find worst-case value @F, e.g. -100 @1 kHz.		
2: View results of maximum capability test @F, e.g. +20 dBV @1 kHz.		
3: Result spec'd as range from noise to max dB @F, e.g. 120 dB @1 kHz.		
PASS/FAIL: electronic device dynamic range should be >110 dB		

Test: latency		
Source: noise or music	**Units**: T (ms)	**Setup**: #2A (dual)
Resolution: 0.02 ms or better	**FFT window**: Hann	**Averaging**: any
1: Drive DUT at nominal level with broadband signal.		
2: View impulse response and compensate I/O time offset with analyzer delay.		
3: Verify flat phase (therefore latency is frequency independent).		
4: Result specified as time (*x* ms), e.g. 2.5 ms.		
PASS/FAIL: latency budget is application dependent—stage monitors <10 ms		
PASS/FAIL: latency budget is application dependent—concert mains <20 ms		
FAIL: instability in latency, latency offest between summed signals		

Test: polarity		
Source: noise or music	**Units**: + or –	**Setup**: #2A (dual)
Resolution: > = 1/24	**FFT window**: Hann	**Averaging**: Any
1: Drive DUT at nominal level with broadband signal.		
2: XFR function measurement between input and output of DUT.		

3: View impulse response and compensate I/O time offset with analyzer delay.		
4: Result specified as + (impulse upward) or - (impulse downward).		
5: Verify mostly flat phase along 0° or 180° line within DUT operating range.		
6: Polarity result specified as + (phase at 0°) or - (phase @180°).		
PASS: polarity consistently adheres to pin 2 hot standard		
FAIL: inconsistent or non-standard polarity		
Test: amplitude response over frequency (gain, range and variance)		
Source: noise or music	**Units**: ±dB @F	**Setup**: #2A (dual)
Resolution: > = 1/24	**FFT window**: Hann	**Averaging**: any
1: Drive DUT at nominal level with broadband signal.		
3: Gain result specified for nominal value in dB (+ or -).		
4: Range result specified from F Lo to F Hi within 0 to -3 dB.		
5: Variance result specified as range from ±dB within nominal 0 dB passband.		
PASS/FAIL: electronic device should be flat ±0.25 dB 20–20 kHz		
FAIL: discrepancy between user interface and actual gain or features		
Test: phase response over frequency (range and variance)		
Source: noise or music	**Units**: phase°, T (ms) @F	**Setup**: #2A (dual)
Resolution: > = 1/24	**FFT window**: Hann	**Averaging**: any
1: Drive DUT at nominal level with broadband signal.		
2: XFR function phase measurement between input and output of DUT(s).		
3: Range result specified as phase deviation in passband (or specified frequency range).		
4: Variance result specified as range from ± degrees within nominal passband.		
PASS/FAIL: electronic device phase should be flat ±30° dB 20–20 kHz		
Test: phase delay over frequency		
Source: noise or music	**Units**: T (ms) @F span	**Setup**: #2A (dual)
Resolution: > = 1/24	**FFT window**: Hann	**Averaging**: any
1: Drive DUT at nominal level with broadband signal. Compensate latency.		
2: XFR function phase measurement between input and output of DUT(s).		
3: Select frequency range of interest. Measure phase shift between F Hi and F Lo.		
4: Apply phase delay formula (section 12.8.3) to convert to time.		
5: Variance result specified as T (ms) within the specified frequency range.		
PASS/FAIL: latency budget is application dependent—stage monitors <10 ms		
PASS/FAIL: latency budget is application dependent—concert mains <20 ms		
FAIL: instability in latency, latency offset between summed signals		
Test: compression (onset)		
Source: sine	**Units**: ±dB @F	**Setup**: #2A (dual)
Resolution: > = 1/24	**FFT window**: Hann	**Averaging**: any
1: Drive DUT with low-level sine wave signal (F), e.g. -10 dBV @1 kHz.		
2: XFR function amplitude measurement between input and output of DUT(s).		
3: Store and recall XFR amplitude value @F, e.g. 0 dB@1 kHz.		
4: Increase drive level until XFR amplitude value begins to fall (limiting onset).		

5: View output (single channel) to find onset level.		
6: Compression onset-level result is specified in V, dBV or dBu.		
PASS/FAIL: application dependent. Depends on intended onset threshold		
FAIL: onset too early (or late), in wrong frequency band, excess THD		
Test: jitter and clocking errors (range and variance)		
Source: noise or music	**Units**: cycles, phase°	**Setup**: #2A (dual)
Resolution: > = 1/24	**FFT window**: Hann	**Averaging**: any
1: Drive DUT at nominal level with broadband signal. Compensate latency.		
2: XFR function phase measurement between input and output of DUT(s).		
3: Store and recall XFR phase trace as time/stability reference.		
4: Phase trace should be time invariant unless there is clock drift or jitter.		
5: Result specified as range and frequency of phase/clock cycle deviation.		
PASS/FAIL: stability with ½ clock cycle in the phase response (pass) or beyond (fail)		

13.5 ACOUSTIC VERIFICATION PROCEDURES

Test: noise over frequency		
Source: none	**Units**: volts (dB SPL)	**Setup**: #1B (single ch.)
Resolution: > = ⅓	**FFT window**: Hann	**Averaging**: 8 or more
1: Mute speaker to acquire baseline of noise in the room.		
2: Measure, store and recall ambient noise floor with 1 ch. FFT.		
3: Unmute speaker(s) and measure noise floor with 1 ch. FFT.		
4: Result specified as dB NC (ambient) and dB SPL A weighted (speaker system).		
PASS/FAIL: speaker noise should be < ambient noise in listening area (at all frequencies)		
Test: hum over frequency		
Source: none	**Units**: volts (dB SPL)	**Setup**: #1B (single ch.)
Resolution: > = 1/24	**FFT window**: flattop	**Averaging**: optional
1: Mute speaker to acquire baseline of noise in the room.		
2: Measure, store and recall ambient noise floor with 1 ch. FFT.		
3: Unmute speaker(s) and measure hum spectrum with 1 ch. FFT.		
4: Result specified as dB NC (ambient) and dB SPL @F (speaker system).		
PASS/FAIL: speaker hum should be < ambient noise in listening area (at all frequencies)		
Test: THD + *n* over frequency		
Source: pure sine tone	**Units**: %THD + *n*	**Setup**: #1B (single ch.)
Resolution: > = 1/24	**FFT window**: flattop	**Averaging**: optional
1: Drive DUT at level >20 dB below rated max, e.g.100 dB SPL for 124 dB spkr.		
2: Measure the fundamental frequency level and mark as reference.		
3: Measure harmonic level relative to fundamental. Estimate THD (-20 dB = 10%, -40 dB = 1%, -60 dB = 0.1%, -80 dB = 0.01%).		
4: Result spec'd as %THD @dB SPL @F, e.g. 1%THD @100 dB SPL @1 kHz.		
PASS/FAIL: speaker THD + *n* should be <1% in the range between 250 Hz–5 kHz.		

Test: maximum SPL capability over frequency		
Source: pure sine tone	**Units**: volts (dB SPL)	**Setup**: #1B (single ch.)
Resolution: > = 1/24	**FFT window**: flattop	**Averaging**: optional
1: Drive DUT at low level (>20 dB below max rating), e.g. 80 dB SPL @1 kHz.		
2: XFR amplitude. Store and recall trace for reference value @F, e.g. 0 dB @1 kHz.		
3: Increase drive level until XFR amplitude value begins to fall (limiting onset).		
4: View output (single channel) to find onset level.		
5: Prorate SPL value by distance to obtain 1 m equivalent (if desired).		
6: Result is specified as max dB SPL @F @m, e.g. 120 dB SPL @1 kHz@1 m.		
PASS/FAIL: speaker max capability @f should be within 20 dB of overall rating		

Test: maximum SPL capability (full range)		
Source: pink noise	**Units**: SPL (pk, cont, A, C, Z)	**Setup**: #1B (single ch.)
Resolution: > = 1/24	**FFT window**: flattop	**Averaging**: optional
1: Follow steps 1 to 5 of above test but use full-range source.		
2: Result spec'd as max dBSPL (pk, cont, weight) @m, e.g. 120 dB SPL pk "Z" @1 m.		
PASS/FAIL: speaker max capability should be within 3 dB of overall rating		

Test: polarity (single driver)		
Source: noise or music	**Units**: + or -	**Setup**: #2B (dual ch.)
Resolution: > = 1/24	**FFT window**: Hann	**Averaging**: optional
1: Drive DUT at nominal level with broadband signal.		
2: XFR function measurement between input and output of DUT.		
3: View impulse response and compensate I/O time offset with analyzer delay.		
4: Outcome specified as + (impulse upward) or - (impulse downward).		
5: Verify relatively flat phase along 0° or 180° line within speaker's operating range.		
6: Polarity result specified as + (phase at 0°) or - (phase at 180°).		
PASS: polarity consistently adheres to pin 2 hot standard (+)		
FAIL: inconsistent or non-standard polarity		

Test: polarity (two-way speaker)		
Source: noise or music	**Units**: + or -	**Setup**: #2B (dual ch.)
Resolution: > = 1/24	**FFT window**: Hann	**Averaging**: optional
1: Determine HF driver polarity by single-driver method above.		
2: XFR function phase measurement between input and output of DUT.		
3: Store and recall solo HF driver amplitude and phase response.		
4: Mute HF and measure solo LF. Compare phase responses in the XOVR region.		
5: Adjust delay and/or polarity for best phase correlation between HF and LF.		
6: Result specified by maximum coupling and minimum phase delay through XOVR.		
PASS: XOVR summation approximates unity		
FAIL: XOVR summation <-3 dB		

Test: amplitude response over frequency (gain, range and variance)		
Source: noise or music	**Units:** ±dB @F	**Setup:** #2B (dual ch.)
Resolution: > = 1/24	**FFT window:** Hann	**Averaging:** Optional
1: Drive DUT at nominal level with broadband signal. Compensate latency.		
2: XFR function amplitude measurement between input and output of DUT(s).		
3: Range result specified from F Lo to F Hi within 0 to -6 dB.		
4: Variance result specified as range from ±dB within nominal 0 dB passband.		
PASS/FAIL: (Range) Speaker should extend over published range (within ⅓ oct)		
PASS/FAIL: (Variance) Speaker nominally covers published range (beware of reflections)		

13.6 ADDITIONAL VERIFICATION OPTIONS

There is no limit to the level of verification we can perform on the system. It's not practical or necessary to perform a full set of verifications each night for ongoing operations such as a repeating show or touring system. We can move safely forward with a reduced verification menu when the system has only been power cycled or reconnected, rather than rewired, since yesterday. We can always retreat to verification procedures to locate problems found in the calibration stage.

A thorough verification stage provides the solid foundation required for the critical calibration decisions. It's tempting to view the available time and conclude it would be better spent doing calibration rather than verification. Do so at your peril. There is no glory in a great verification, but plenty of shame in a bad one.

In the end we each arrive at a verification level we can live with. For touring systems, a thorough verification in the shop reduces the requirements on the road. A permanent install has the highest requirements because of how many disciplines are involved, and the long-term impacts of any errors. Nobody wants to be the one who missed

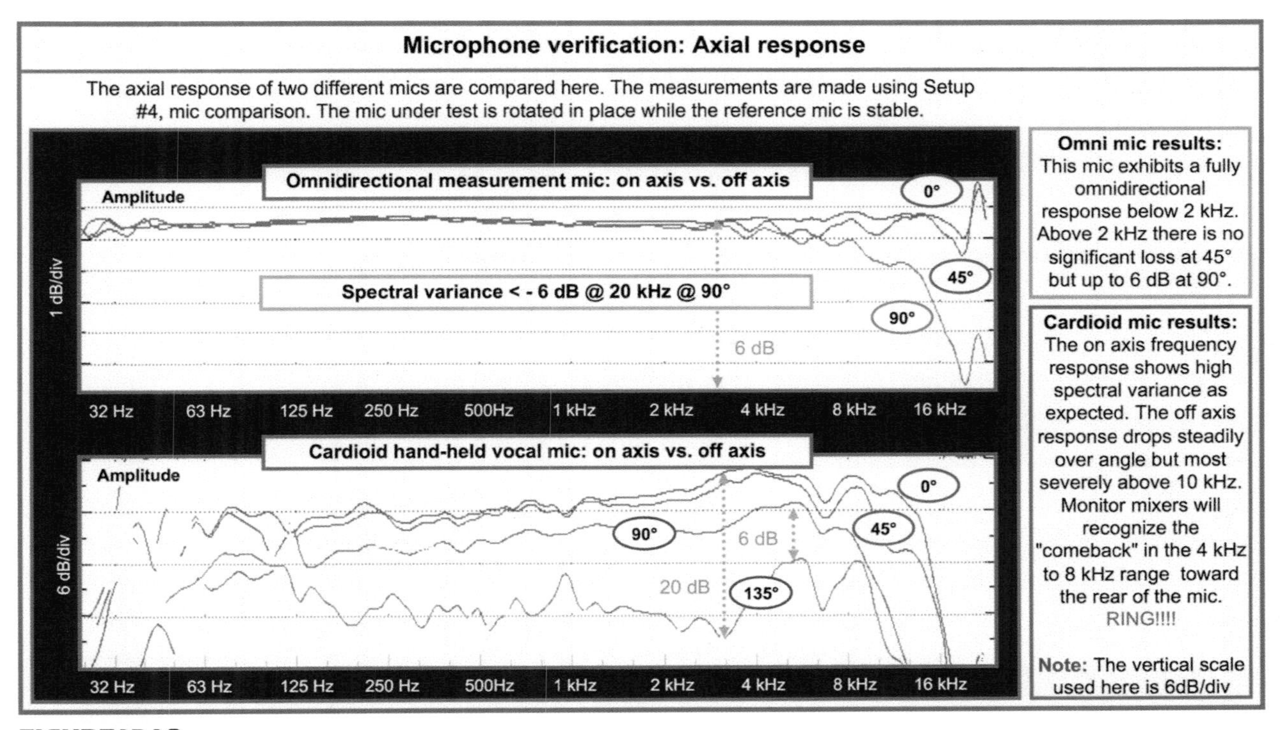

FIGURE 13.18
Field example of on- and off-axis microphone responses

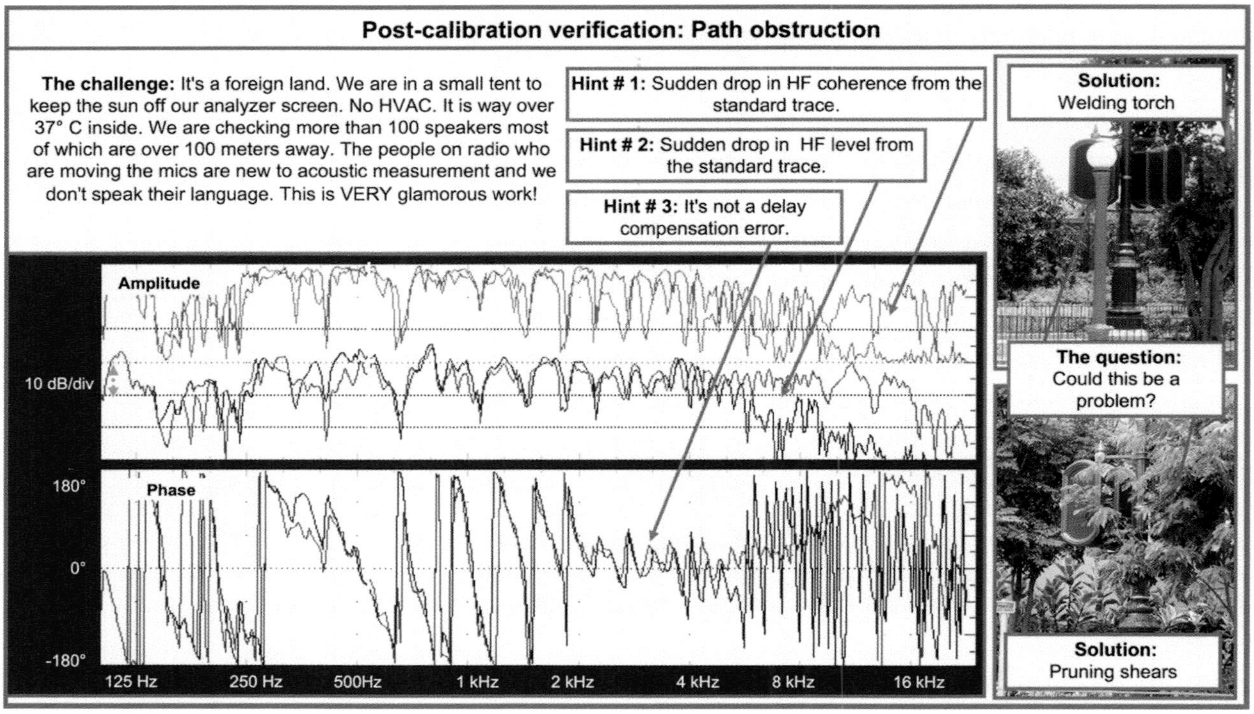

Post-calibration verification: Path obstruction

The challenge: It's a foreign land. We are in a small tent to keep the sun off our analyzer screen. No HVAC. It is way over 37° C inside. We are checking more than 100 speakers most of which are over 100 meters away. The people on radio who are moving the mics are new to acoustic measurement and we don't speak their language. This is VERY glamorous work!

Hint # 1: Sudden drop in HF coherence from the standard trace.

Hint # 2: Sudden drop in HF level from the standard trace.

Hint # 3: It's not a delay compensation error.

Amplitude
10 dB/div
Phase
180°
0°
-180°
125 Hz
250 Hz
500Hz
1 kHz
2 kHz
4 kHz
8 kHz
16 kHz

Solution: Welding torch

The question: Could this be a problem?

Solution: Pruning shears

FIGURE 13.19
Example application of the post-calibration verification

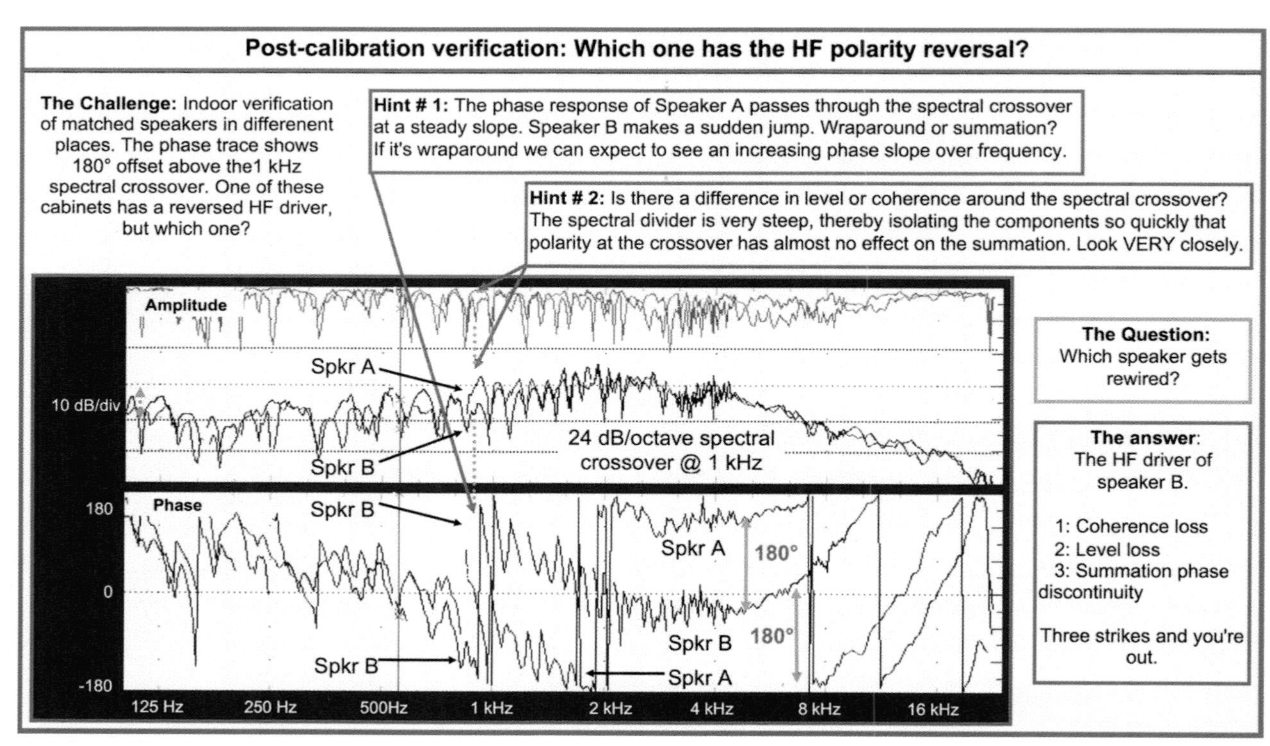

FIGURE 13.20
Example application of the post-calibration verification

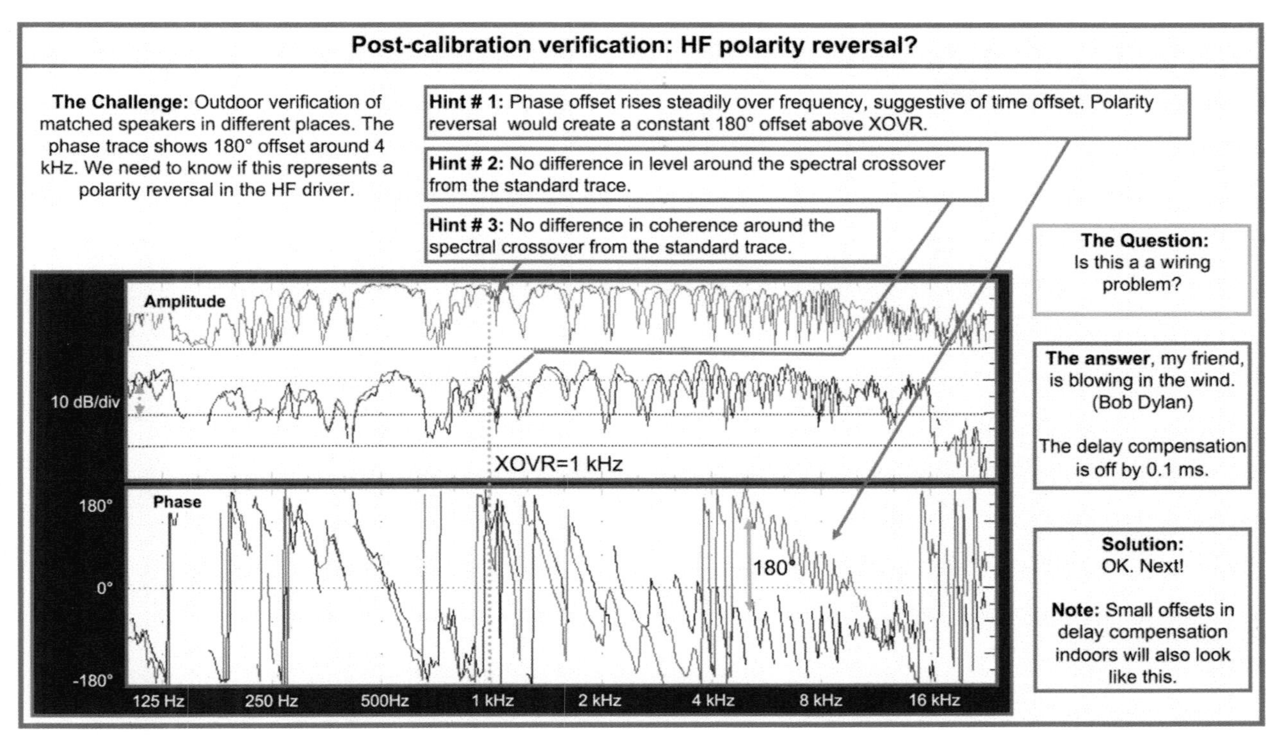

FIGURE 13.21
Example application of the post-calibration verification

the polarity reversal that gets discovered years later. Some examples of post-calibration verification in the field are found in Figs 13.19 to 13.21.

Verification was explained to me like this on my first day in professional audio: "Assumption is the mother of f@$%-up" (Tony Griffin, 1976).

CHAPTER 14

Calibration

> *calibrate v.t. find calibre of; calculate irregularities of before graduating it; graduate with allowance for irregularities; correlate readings (of instrument etc.) with a standard.*
>
> **Concise Oxford Dictionary**

The system is verified, speaker positions are roughed in and we are prepared to set signal processor parameters. The stage is set to complete the optimization process. Calibration operations progress from simple to complex. Each verified subsystem is individually calibrated and combined into larger subsystems, where secondary adjustments are made to compensate for the summation effects. The process is complete when all related subsystems are grouped into an aggregate combined system with minimum response variance throughout the listening space.

The calibration process, like verification, is a series of controlled procedures designed to give specific answers such as speaker position, delay time, equalization and level setting. Calibration answers are never as cut and dried as those found in verification. There is not a one-size-fits-all, step-by-step set of procedures for all system designs. Each application has unique combinations of speakers, processing and room acoustics. We need an adaptable cookbook ready for the hundreds of contingencies that arise. This section contains the standard procedures for calibration and enough support and explanation to provide a firm but flexible foundation to adapt them to each application.

Calibration is the process of proofing and fine-tuning. Parts of the design must be proven (such as aiming and splay angles) and others require on-site adjustment (such as delay and EQ). We expect an EQ to be properly wired (verification) but don't expect it have our desired settings when unpacked (calibration). Speaker placement is a variation on this theme. Verification: checking the initial placement/aim against the drawings. Calibration: finding the *best* placement/aim.

Calibration goals of uniformity, efficiency, clarity and plausibility are fairly universal. They are not unique or original to this text or to dual-channel FFT analysis. Calibration's challenges are also universal. We must control each speaker's direct sound distribution and its summation with others. We use summation to maximize the power addition and directional control, and seek to minimize its combing effects. Dynamic conditions require ongoing monitoring and active calibration to maintain consistency over time.

Calibration inevitably boils down to a series of decisions and signal processing settings. A monkey can place the speakers, turn the knobs on the processor and somebody can still mix a show on it. The monkey's settings are unlikely to create uniform response over the space. It takes very evolved techniques to create uniformity. We are well on the way if we have adhered to the minimum-variance principles during the design and verification stages. Calibration is the final push in the process of achieving uniformity.

Calibration goals

- **Minimum variance**: level, spectral and ripple (same sound everywhere).
- **Maximum coherence**: highest intelligibility, direct/reverberant ratio, clarity, etc.
- **Maximum efficiency**: full use of coupling capability for maximum power addition.
- **Sonic image control**: the sound image appears where we want it.

Calibration challenges

- **Direct sound distribution**: even level of direct sound everywhere.
- **Speaker/speaker summation**: maximum coupling and minimum combing.
- **Speaker/room summation**: minimize reflections except constructive LF coupling.
- **Dynamic conditions**: wind, changing humidity, temperature and room empty to full.

Calibration strategies

- **Philosophy**: guiding principles for gray-area decisions (choosing the winner when all win–win options have been exhausted).
- **Data access**: the critical transmission path probe points required to make informed decisions—console out, signal processor out and the speaker's response in the room.
- **System subdivision**: appropriate separation of the signal path into the optimal number of processing channels and speaker subsystems. We need enough flexibility to independently set parameters such as equalization, level, delay, etc., without wasting money on redundant channels.
- **Methodology**: a set of procedures, playbook or roadmap. The methods for reaching the goals can be reduced to a series of specific tests for finding the relevant answers.
- **Context**: data for any given point is neither good nor bad in its own right. Context gives us expectations, an adjustable standard to judge the result. Is this the *expected* response? For example, extensive pink shift is expected at OFFAX, but unexpected at ONAX.

Calibration decisions

- **Single speaker (or system) aim**: optimal vertical and horizontal focus.
- **Array splay angle**: optimal angular orientation between elements.
- **Array spacing**: optimal spacing between elements.
- **Acoustical modification**: Identify and treat acoustic problems (i.e. reflections).
- **Level setting**: Scale the relative levels to create minimum-level variance.
- **Delay setting**: Phase align crossovers to minimize ripple variance.
- **Solo EQ**: Minimize spectral variance with a matched standard response.
- **Combined EQ**: Compensate for LF coupling to minimize spectral variance.

Calibration subdivision

- **Source:** Everything before the speaker system processing. The art side of the system.
- **Processor:** The signal processing used for calibrating the sound system.
- **Speaker/room:** The sound system in its acoustic environment

Calibration test reference	
Calibration tool	**Calibration role**
Physical tools	
Inclinometer, protractor, laser pointer	Speaker focus fine tuning
Thermometer, hygrometer	Dynamic environmental monitoring
Disto, tape measure, meter stick	Subwoofer spacing, cluster trim height etc.
Simple audio tools	
VOM, polarity or impedance tester, listen box, o-scope	Off line troubleshooting for discoveries
SPL meter	Bragging rights, police action
Complex tools	
Ears, eyes, brain	Not sure, but they are already paid for
Dual-channel FFT analyzer	Electronic and acoustic signal test

FIGURE 14.1
Calibration test reference

14.1 APPROACHES TO CALIBRATION

We interrupt this exercise in objectivity and science with a philosophical discussion. Calibration includes many extremely gray, "lesser evil" decisions. Verification, by contrast, had clear-cut answers. We fix a speaker with reverse polarity or distortion, knowing everybody benefits. Calibration adjustments that help one location often degrade it at others. Sadly, this is virtually assured with EQ, level and delay setting, the cornerstones of calibration. This presents an ethical dilemma: How do we decide the winners and losers? We strive toward universal solutions but inevitably the win–win options are exhausted and we face win/break-even/lose. Let's interpret our ethical dilemma in socio-political terms and see which model provides the best decision-making directives.

Anarchy

The anarchy political model is structured on a total lack of structure. Governing authority is absent and every individual is essentially self-governing, with no common goals. Whoever grabs the controls can calibrate the system for their location, which just *might* happen to be the mix position. There's no need for an acoustical analyzer's objectivity because it's every man for himself. This would be comical if not so often true.

Monarchy

In this model, a single party makes decisions without having to answer to facts and science, and little or no regard for its effects outside the royal circle. The mix position is the castle and the inner circle includes the mix engineer, band manager and court jesters. The mix area is monitored with the finest analyzer available and calibrations can ensure the maximum power concentration there. A regal decree states that all seats in the house benefit from concentrating all resources at the castle.

Capitalism

Capitalism promotes the idea that quality should be commensurate with price (the "money" seats over the "cheap" seats). It's easy because the expensive seats are closer, have higher levels and direct-to-reverberant ratios. The "cheap" seats are disadvantaged in both categories but there is a whole lot more of them (which add up to a lot of money). Distant seats will almost always be at a sonic and visual disadvantage but we don't have to settle for widespread input inequality. A more uniform sound redistribution strategy can reduce inequality without compromising the highly advantaged seats.

Democracy

In the democratic model each seating area is given equal representation. When decisions benefit one seating area above the next, the effects are evaluated on a majority basis. Two principal factors to consider: quantity affected (large or small) and symmetry of effect (positive vs. negative). We go forward if quantities are equal and positive effects outweigh the negative. The majority rules when effects are equal. Otherwise we use the "triage" method. We don't give up a 3 dB improvement for 1000 seats because it creates a 20 dB hole in ten. Conversely, we don't make ten seats perfect if that screws up 1000 (see monarchy above). We view the population asymmetry vs. that of positive and negative effects.

Democracy seems the best model for design and optimization strategies. To implement this requires us to do more than measure one position and issue proclamations. We must survey every population sector. It's not practical to measure all 12,000 seats in an arena. Certain seats are carefully selected to be local area representatives. We'll meet them later.

TANSTAAFL and triage

TANSTAAFL ("There ain't no such thing as a free lunch") and the decision-making structure of acoustic triage, introduced in Chapter 11, are equally applicable to calibration. TANSTAAFL comes into play at each calibration decision: There are no effects without side effects. A setting that improves one location *must* affect others. Perhaps it's a big upside/small downside, vice versa or a stalemate. It's very tempting to delude ourselves that a single-position solution has a global benefit. TANSTAAFL underscores the need to measure in multiple locations to monitor changes. Remember: If something seems too good to be true, it probably is.

The solution for one area may simply be transporting the problem to a new location, which sounds like a break-even proposition but not necessarily. Taking out the garbage does not eliminate it, but moves it to a far preferable location. Solutions for highly populated areas may outweigh the minority area side effects. For example, the high ripple variance near a spatial crossover can be placed down the length of an aisle. We move the problem to the ushers instead of paying customers. TANSTAAFL keeps us on the lookout for side effects. Acoustic triage helps us decide what to do about them.

14.2 MEASUREMENT ACCESS

14.2.1 Access points: console/processor/mic

The signal flows serially through three distinct sections: source, signal processing and the speaker system in the room. Signal is created at the source, the processing manages it and the speaker system delivers it. We need an access point at the output of each section.

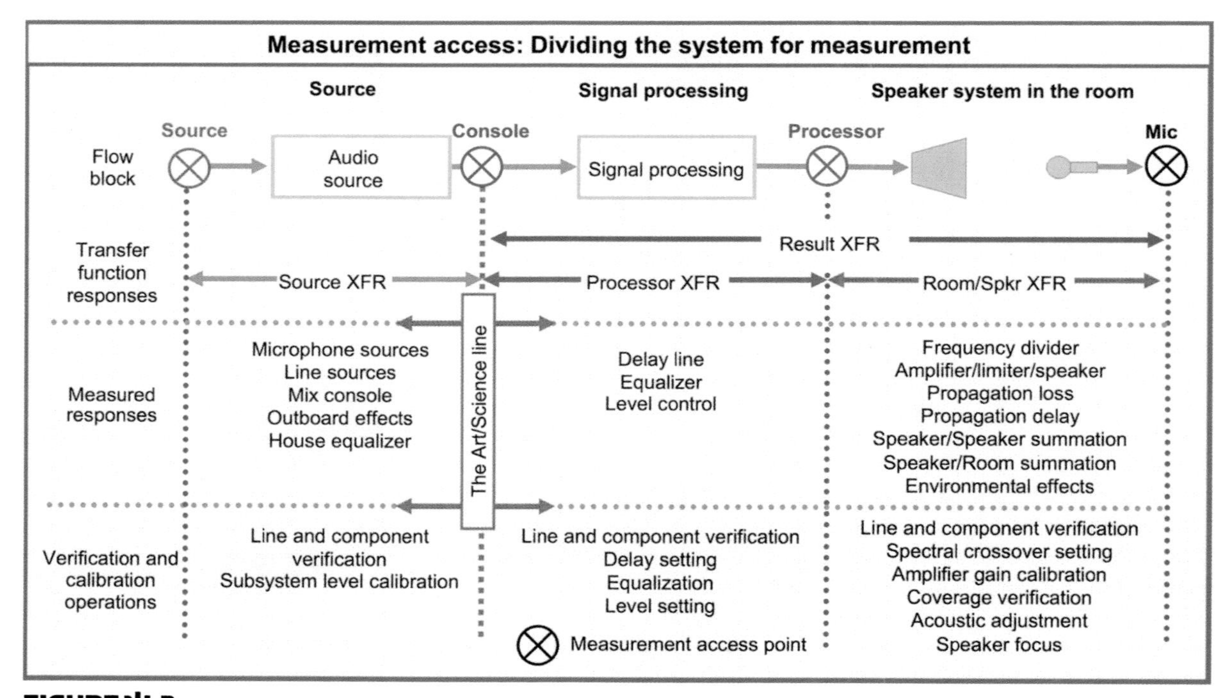

FIGURE 14.2
Flow block of the electronic and acoustic measurement access points (console, processor and microphone) and the three transfer function responses (room/processor/result)

Access points for measurement

- **Console output**: the art/science reference standard. Connection point is post-matrix and pre-speaker system processor. Parallel feeds are sent into the analyzer and processor.
- **Processor Output**: dual-use test point—processor output and speaker system reference signal. Connection point is post-processor and pre-crossover. Reference signals must be full range. Parallel feeds are sent into the analyzer and speaker system.
- **Microphone**: The surrogate ear at the end of the measurement chain. Used to monitor the speaker in the room. Can be compared to console output or processor out.

14.2.2 Three transfer functions: room/processor/result

The three access points yield three distinct two-point transfer function results. Their roles in verification and calibration are outlined in Fig. 14.3.

THE THREE TRANSFER FUNCTIONS

1. **Room/speaker**: processor output (speaker system in) vs. mic (speaker system out).
2. **Processor (EQ)**: source (processor in) vs. processor output.
3. **Result**: source vs. mic (speaker system out).

14.2.3 Alternative access options

The best-case scenario is direct physical access to the processor input and output connections. Analog transmission allows a parallel split but not so in digital. Next best is a substitute test point that delivers a copy of the desired signal, such as an unused output channel, network port or other surrogate. Naturally we want this to track as closely as possible to the signal flowing into the speaker system.

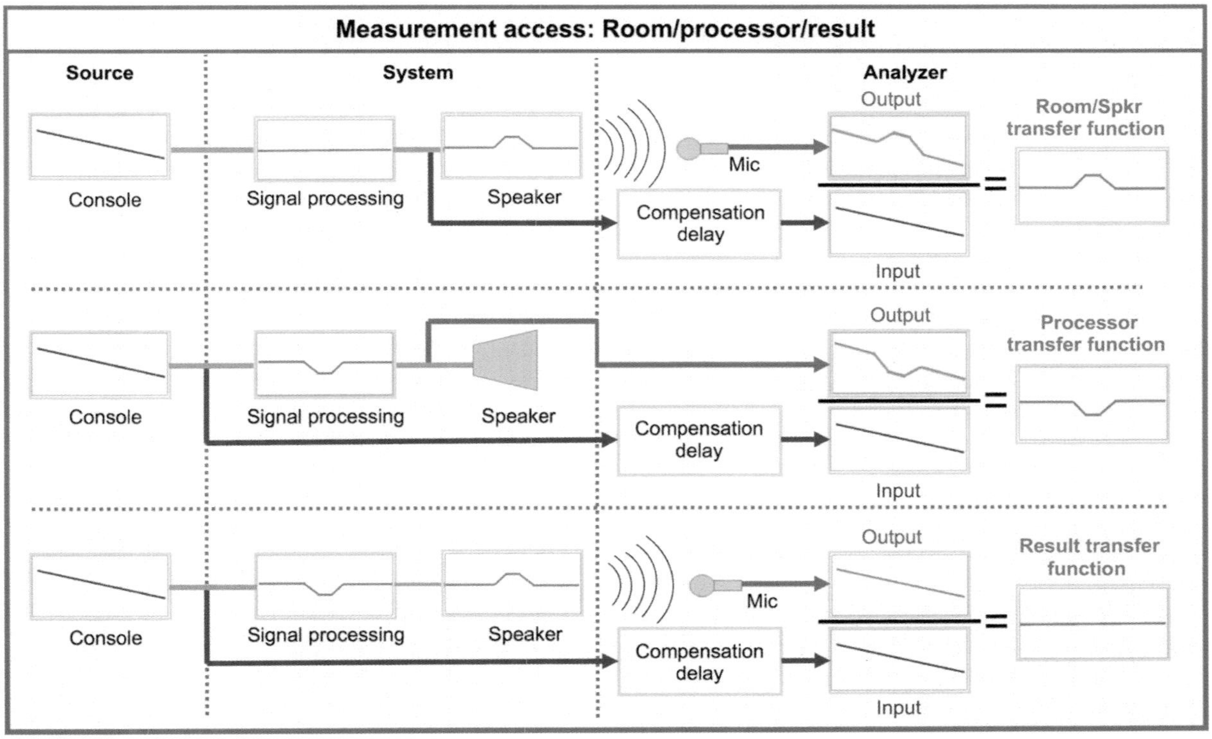

FIGURE 14.3
Flow block of measurement access points of the three transfer functions and their roles in the equalization process

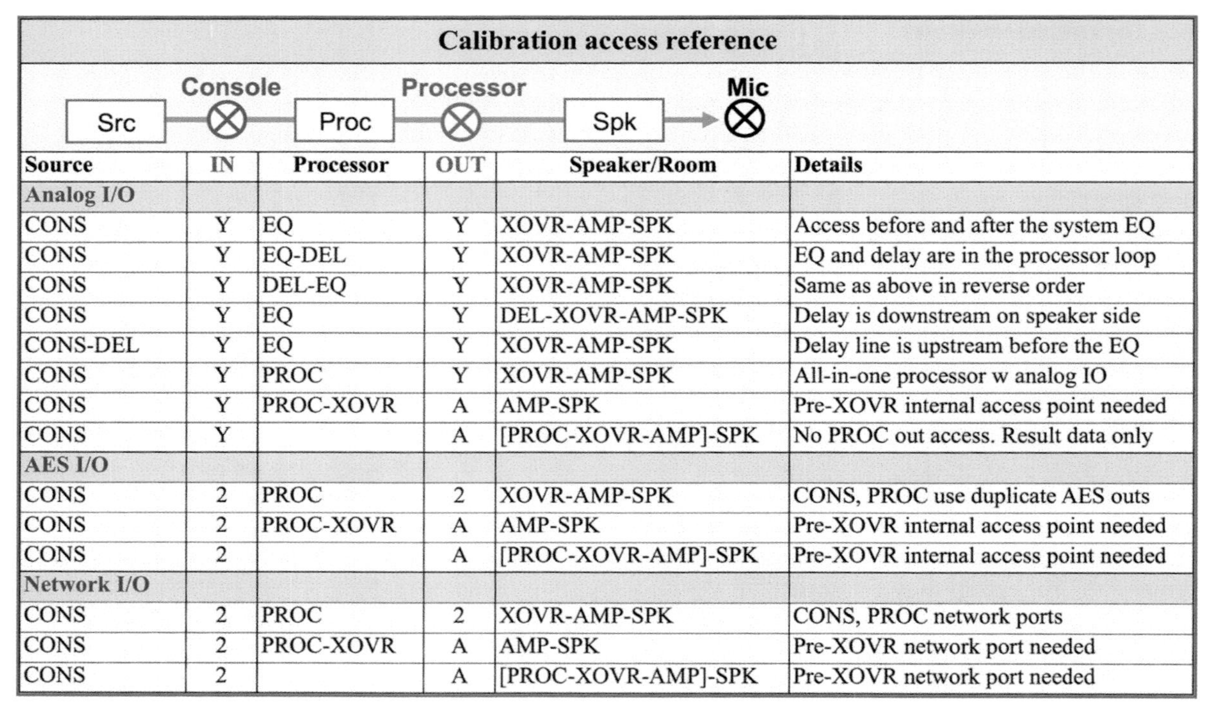

Calibration access reference

Console, Processor, Mic

Src → Proc → Spk → Mic

Source	IN	Processor	OUT	Speaker/Room	Details
Analog I/O					
CONS	Y	EQ	Y	XOVR-AMP-SPK	Access before and after the system EQ
CONS	Y	EQ-DEL	Y	XOVR-AMP-SPK	EQ and delay are in the processor loop
CONS	Y	DEL-EQ	Y	XOVR-AMP-SPK	Same as above in reverse order
CONS	Y	EQ	Y	DEL-XOVR-AMP-SPK	Delay is downstream on speaker side
CONS-DEL	Y	EQ	Y	XOVR-AMP-SPK	Delay line is upstream before the EQ
CONS	Y	PROC	Y	XOVR-AMP-SPK	All-in-one processor w analog IO
CONS	Y	PROC-XOVR	A	AMP-SPK	Pre-XOVR internal access point needed
CONS	Y		A	[PROC-XOVR-AMP]-SPK	No PROC out access. Result data only
AES I/O					
CONS	2	PROC	2	XOVR-AMP-SPK	CONS, PROC use duplicate AES outs
CONS	2	PROC-XOVR	A	AMP-SPK	Pre-XOVR internal access point needed
CONS	2		A	[PROC-XOVR-AMP]-SPK	Pre-XOVR internal access point needed
Network I/O					
CONS	2	PROC	2	XOVR-AMP-SPK	CONS, PROC network ports
CONS	2	PROC-XOVR	A	AMP-SPK	Pre-XOVR network port needed
CONS	2		A	[PROC-XOVR-AMP]-SPK	Pre-XOVR network port needed

FIGURE 14.4
Measurement access points for various system configurations. Access to processor input and processor output measurement points are indicated by "Y" (parallel analog), "2" duplicate AES output/network port or "A" dedicated pre-crossover access point.

14.3 MICROPHONE PLACEMENT

Let's begin the hunt for the perfect mic placement. It will be harder to find than a unicorn (at least we know what a unicorn looks like). What would a perfect mic placement even look like? It would be one that accurately represents the system response over the whole room. But one position can only speak for all when the sound is the same everywhere (so every position is perfect). Even the sound of a perfect speaker in a perfect room differs at almost every location (so any mic position is imperfect). Maybe it doesn't matter. It does, because the *differences* over location aren't random. They follow patterns. Mic positions sort out the patterns and connect them together. Each mic position has specific roles to play with purposeful (not random) placement. There are best locations for equalization, best for speaker positioning and best for delay setting, but they are not the same.

The approach proposed here uses individual responses with clearly predefined context. Instead of seeking to find a common average response, we seek out the differences between the key locations that characterize the expected system behavior. The mic positions are at the critical milestones of the standard variance and summation progressions. We can interpolate the response between the milestones based on the known progressions.

14.3.1 Placement classification

There are six classifications of calibration mic positions (Figure 14.5). These positions have specific locations and roles in the calibration procedures.

- ONAX: located "on axis" to the speaker. ONAX is used for equalization, level setting, architectural modification and speaker position. ONAX is found at the point of maximum isolation from neighboring speaker elements. The mic will be most literally "on axis" to the element when the speaker has a symmetrical orientation to the space. When asymmetrically oriented, the ONAX mic is located at the mid-point between the coverage edges, rather than the speaker's on-axis focus point. ONAX locations for arrays are on a per-element basis. Spatial averaging techniques can be used for multiple positions inside the general ONAX area, or for multiple ONAX locations in a symmetrical array. In order to be classified as ONAX, a mic position needs significant isolation between its target and other elements in the array.

- **OFFAX**: located at the intended horizontal coverage edge and/or "off axis" to the speaker. The position is defined by the listening space shape (e.g. the last seat in the row), not the speaker. This may or may not be at the actual edge of the speaker's coverage. OFFAX positions are analyzed in relation to ONAX data. Our OFFAX goal is variation of <6 dB (the maximum acceptable variance) from the ONAX response.
- **VTOP (vertical top)**: same as OFFAX but for the highest seats in the vertical plane.
- **VBOT (vertical bottom)**: the lowest seats in the vertical plane.
- **XOVR**: located at the spatial crossover. XOVR is on the geometric centerline of symmetric arrays, but closer to the low-level speaker in asymmetric arrays.
- **SYM**: location symmetrically opposite to an element elsewhere in the system. Primary role is symmetry verification rather than calibration. Symmetric positions are usually copies of ONAX positions.

MEASUREMENT MIC CLASS BY TASK

- **Solo equalization**: ONAX position with option of local spatial averaging.
- **Combined equalization AB**: ONAX A and ONAX B (compare with solo responses).
- **Speaker aim (hor)**: ONAX compared with OFFAX (L and R if asymmetric).
- **Speaker aim (vert)**: ONAX compared with VTOP and VBOT.
- **Speaker splay angle**: ONAX compared with XOVR.
- **Speaker spacing**: ONAX compared with XOVR.
- **Level setting AB**: ONAX B (compared on ONAX A).
- **Delay setting AB**: XOVR AB (compare arrivals of A and B).
- **Architectural**: any position (wherever problems arise).

MEASUREMENT TASK BY MIC CLASS

- **ONAX**: solo EQ, combined EQ, level, aim, splay, spacing.
- **OFFAX, VTOP, VBOT**: aim.
- **XOVR**: splay, spacing, delay.

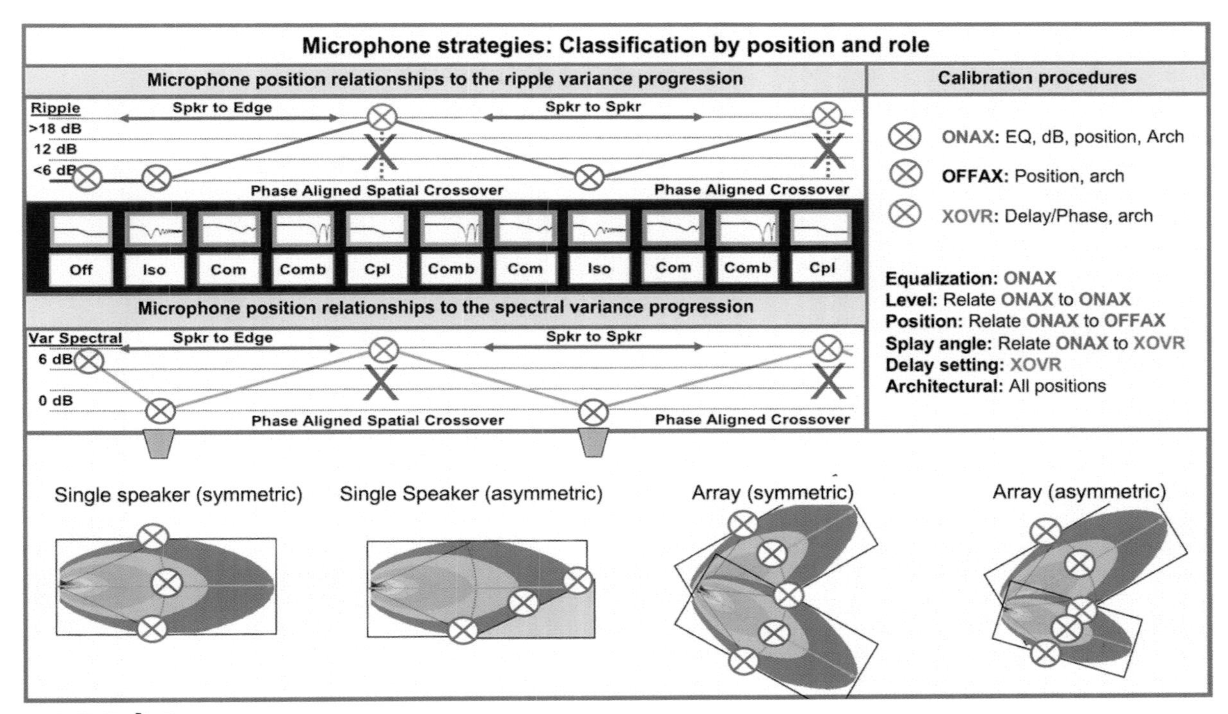

FIGURE 14.5
Mic position classifications and roles

MIC POSITION CLASSES TO CHARACTERIZE A SINGLE SPEAKER (A)

- **OFFAX$_L$**: horizontal left edge.
- **ONAX**: mid-point of coverage (horizontal and vertical).
- **OFFAX$_R$**: horizontal right edge (if asymmetric).
- **VTOP**: vertical top.
- **VBOT**: vertical bottom

MIC POSITION CLASSES TO CHARACTERIZE A HORIZONTAL SPEAKER ARRAY (AB)

- **OFFAX A**: horizontal outer edge of A (opposite of B).
- **ONAX A**: mid-point of A speaker coverage (horizontal and vertical).
- **XAB**: equal level point between A and B coverage (between ONAX A and B).
- **ONAX B**: mid-point of B speaker coverage (horizontal and vertical).
- **OFFAX B**: horizontal outer edge of B (opposite of A).
- **VTOP**: vertical top.
- **VBOT**: vertical bottom.

Note: For a vertical version the XAB mic goes between ONAX A and B in the vertical plane.

14.3.2 Placement strategies

14.3.2.1 SINGLE SPEAKER

Symmetric orientation

The ONAX position is mid-point width and depth. OFFAX positions are along the coverage shape edges (last seats or crossover to another system) at mid-point depth (Fig. 14.6).

Asymmetric orientation

The asymmetric orientation is typical of vertical applications. The VTOP and VBOT positions bracket the top and bottom of the coverage shape. The ONAX position is closer to VTOP, depending on the amount of asymmetry. ONAX will be close to center if the range ratio is small and will move up as the range ratio rises.

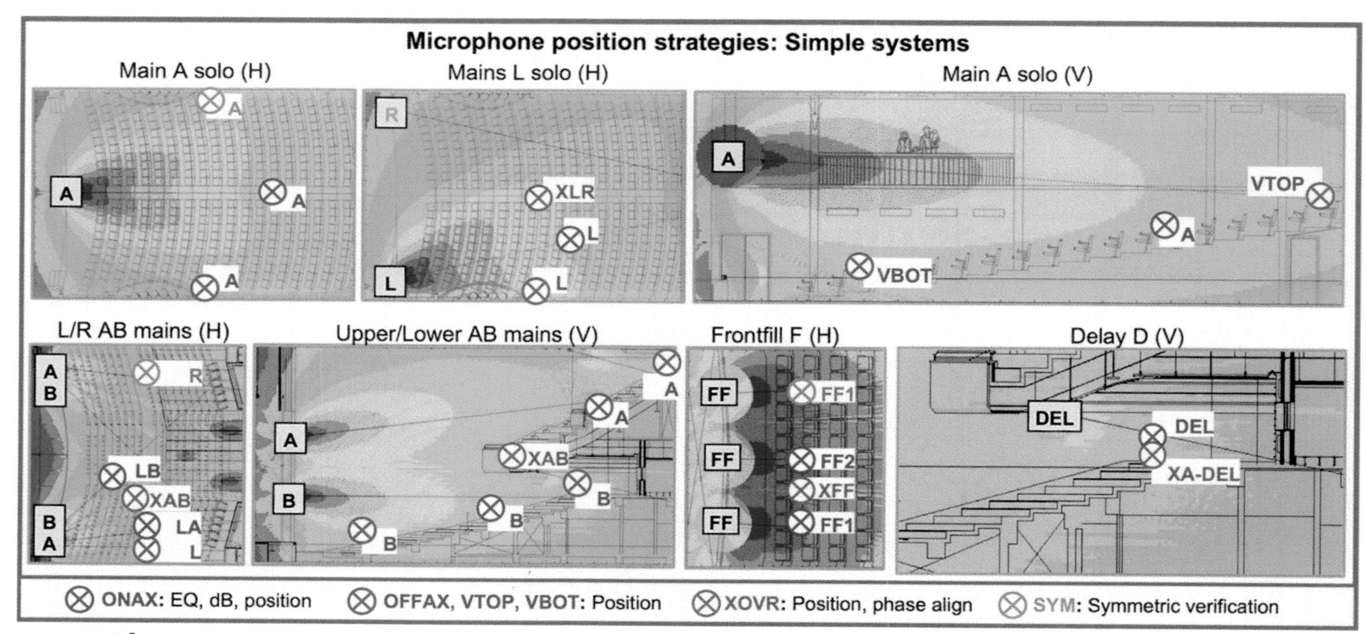

FIGURE 14.6
Mic position strategies for simple systems

14.3.2.2 COUPLED ARRAYS

Line source (symmetric)

The coupled line source has no angular or displacement isolation. ONAX A_1, ONAX A_2 and XA$_1$A$_2$ are essentially merged, reducing placements to ONAX AA and OFFAX AA.

Point source

A two-element symmetric version (A_1A_2) uses a 3++ mic spread: OFFAX A_1, ONAX A_1, XAA, SYM, SYM (ONAX A2 and OFFA2 are normally just symmetrical opposites). Expansion to three elements ($A_{1\text{-}1}$, A_2, $A_{1\text{-}2}$) adds the ONAX A2 position. The four-unit version ($A_{1\text{-}1}$, $A_{2\text{-}1}$, $A_{2\text{-}2}$, $A_{1\text{-}2}$) adds another SYM position and on it goes.

A two-element asymmetric version (AB) uses a five-mic spread: OFFAX A, ONAX A, XAB, ONAX B, OFFAX B. Each incremental expansion requires two more positions, e.g. an ABC array converts OFFAX B to XBC and adds ONAX C and OFFAX C (Fig. 14.7).

Point destination

Mic position strategies for the coupled version are identical to the coupled point source.

14.3.2.3 UNCOUPLED ARRAYS

Line source

The element/mic relationship follows the coupled point source. Speakers and mics are arrayed in a line (also Fig. 14.7). ONAX is in front of the speaker at the mid-point depth (unity line). The XOVR location is the geometric mid-point in symmetric systems and closer to the quiet element (B) when asymmetric. OFFAX positions verify the array has reached its intended extension. SYM mics verify remaining element at their ONAX locations.

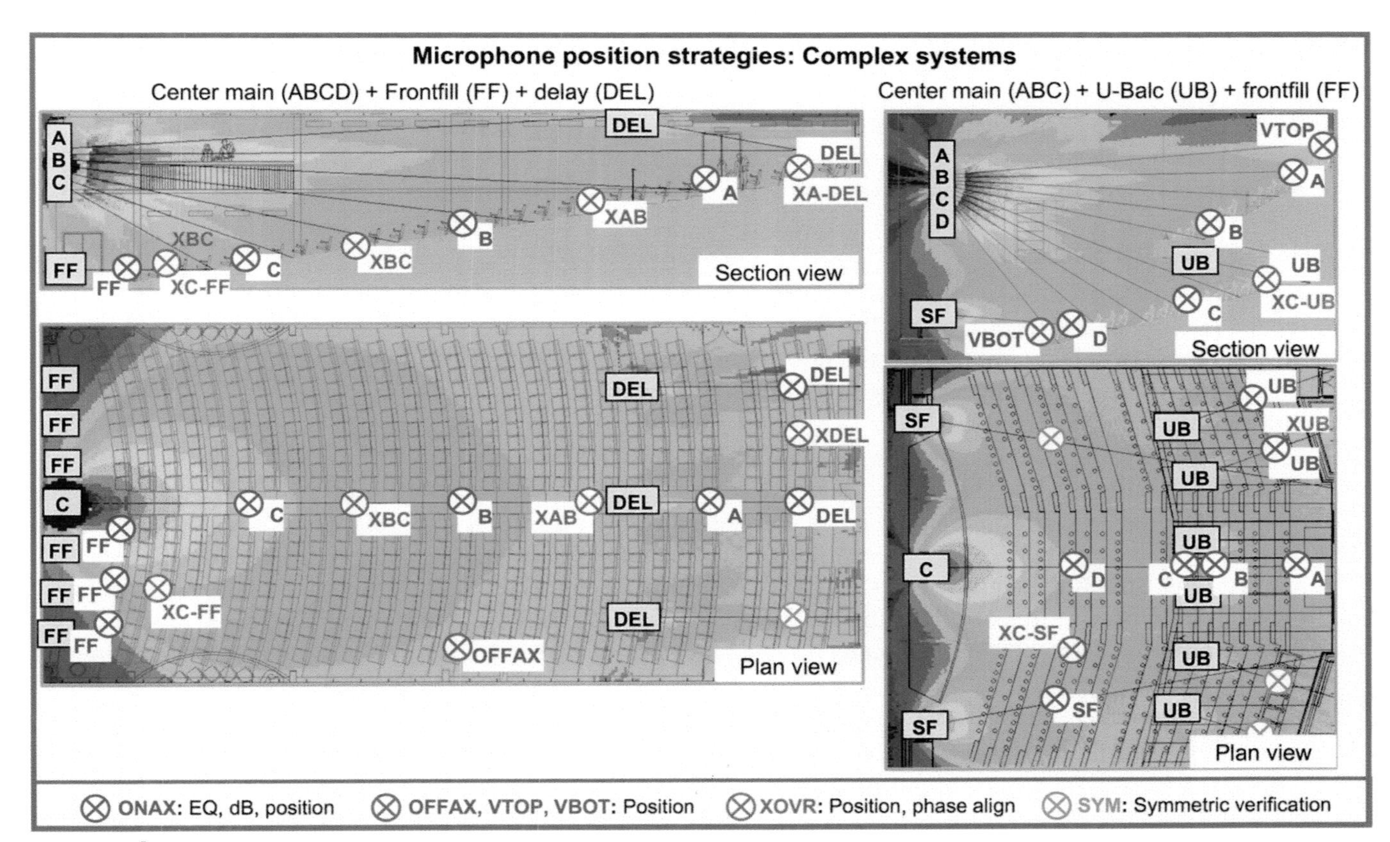

FIGURE 14.7
Mic position strategies for complex systems

Point source

The uncoupled point source is a hybrid of the coupled point source and uncoupled line source. So are the mic locations. XOVR locations in the symmetric version use the mid-point angular relationship (like the coupled point source) and the mid-point depth (like the uncoupled line source). The XOVR mic migrates toward the quiet source (B) in the asymmetric version. ONAX B and OFFAX B coverage (and mic) depths are rescaled in accordance with their level relationship to A.

Point destination

The crossing patterns create another merger of ONAX and XOVR positions. The key to understanding the symmetric version is to see the pattern cross point as the end of coverage (not the middle). Therefore the mid-point depth (where we place ONAX) is half way to the cross point. This location has enough level isolation to function as an equalizable ONAX. XOVR is found on a line midway between the two ONAX locations. OFFAX is on the outer edge at mid-point depth.

The asymmetric version of the point destination requires an alternate strategy. If the B speaker is much smaller scale (e.g. a delay or minor fill speaker) it will not have the privilege of isolation at any location. ONAX B and XOVR AB are one and the same. Level, EQ and delay are all set from this location. OFFAX is somewhat irrelevant here because there is another speaker (A) that will take over after the B speaker fades out.

14.3.3 Placement details

MEASUREMENT MIC AIMING CONSIDERATIONS

- **Point mics at the speakers**: Measurement mics are *mostly* omnidirectional. We don't have to be exact (±30° has almost no effect). Don't point them at 90° off axis to the source (-6 dB HF loss).
- **Don't point mics at the ceiling**: The random incidence approach used by acousticians to sample the room or noise analysis (like HVAC systems). NEVER recommended. Your PA is coherent (hopefully). Point the mic at the speakers (free-field approach).
- **Hanging mics off the balcony rail**: A measure of last resort used when optimizing with an audience present. Best, if possible, to keep the mic aimed at the speakers (not twisting around in circles). The lesser alternative is hanging straight down. They are 90° off axis so assume the VHF is down 6 dB.
- **Spatial crossovers**: point the mic *between* the speakers (e.g. frontfill and mains).

MICROPHONE LISTENING POSITION HEIGHT CONSIDERATIONS

- **Floor reflections**: All measurement mics receive floor reflections (unless there is no floor) and wall and ceiling reflections (unless we are outdoors). The floor bounce vs. microphone placement question concerns which position most closely resembles how the floor reflection will be experienced by listeners during the show. Is it closer to the sitting, standing or laying on the floor placement?
- **Sitting head height**: Mic placed at ear height seems like the obvious choice. Placing the mic very close to a seat back can put a strong reflection in the data that disappears when someone sits there. It's a minor issue in a soft-seater but significant with hard backs. Preferred in applications where standing height would be too high for an accurate reading. (e.g. frontfills where standing height would be off the vertical axis) (Fig. 14.8).
- **Standing head height**: mic placed at standing height is suitable and practical for many standard applications. Standing height reduces seat back reflection strength. Vertical angle errors can be assessed on a case-by-case basis (e.g. frontfill or top row). Vertical angle can be compensated in raked seating by placing the mic one row below the vertical target, i.e. standing height in the closer row is a match for sitting height one row back.
- **Ground plane**: microphone is placed on the floor (the speaker/room spatial crossover). The floor reflection couples in phase so there's no combing (theoretically). In practice it's difficult to prevent VHF combing. The response is +6 dB over a standing/sitting height mic (important when comparing to other types). Ground plane is preferable when sitting/standing positions are heavily compromised by floor reflections that won't be present (or will be significantly different during the show), e.g. an empty flat arena floor with no seating or audience (yet).

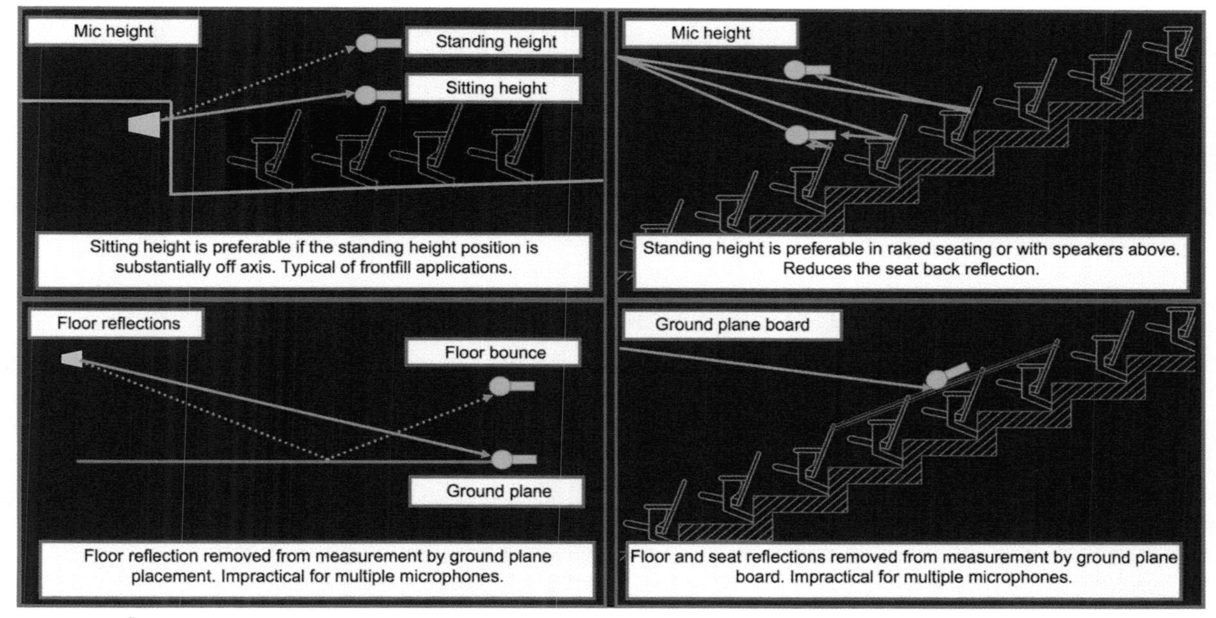

FIGURE 14.8
Mic placement details

14.3.4 Spatial averaging

There are various ways to average the response of multiple microphones or placements. This is common practice for acoustical studies, which are statistically based on hundreds of reflection paths or for random interactions such as HVAC noise. Spatial averaging is of limited use in optimization because very few decisions are statistically based. The closest fit is EQ, which can use spatial averaging to broaden its view. The goal is to equalize the common trends in a speaker system's coverage area rather than just a single spot. A spatially averaged response becomes the majority voice for the area.

Spatial averaging raises confidence about EQ decisions but does nothing to improve the spectral uniformity within its coverage zone. In the end a single EQ is applied whether it comes from a single measurement, 1000 spatially averaging points or Gaucho. The other optimization operations (aiming, splay, delay, level and architectural modification) are based on specific arrivals and locations, not on averages. (Fig. 14.9)

SPATIAL AVERAGING TECHNIQUES

- **Summed mics**: two mics into a "Y" cord (or actively summed). Don't EVER do this! There will be time offset between the mics that will create combing that nobody can hear and has nothing to do with the sound system. Other than that it's fine.
- **Moving mics**: Moving measurement mics sample the room like chlorine in a swimming pool. There's no phase response or coherence and we can't tell direct sound from reverb. Used for measuring random sources (hopefully your PA is not). NEVER recommended.
- **Multiplexed mics**: Switcher cycles between mics while the analyzer continuously measures to get a spatial average. All the same problems as a mic in motion. Sound is not spread by osmosis. NEVER recommended.
- **Mathematical trace averaging**: Multiple traces can be mathematically averaged with some intelligent management (e.g. weighting in favor of the data with high coherence). This can give usable data for equalization provided the mic placements are within the ONAX coverage area. Recommended for EQ only.
- **Optical averaging**: multiple traces are overlaid on the screen and trends detected. Data analysis is done visually rather than computationally (e.g. coherence weighting etc.). Keeping each trace in its original state allows for location-based weighting (e.g. disregard the LF bump at mic 4 because it's right next to a wall). Recommended for EQ only. Note: Math and optical averaging can be applied to the same data.
- **Star pattern**: A standard pattern of mic placement for spatial averaging (originated by Roger Gans). Begins at ONAX and moves short distances horizontally and vertically from there.

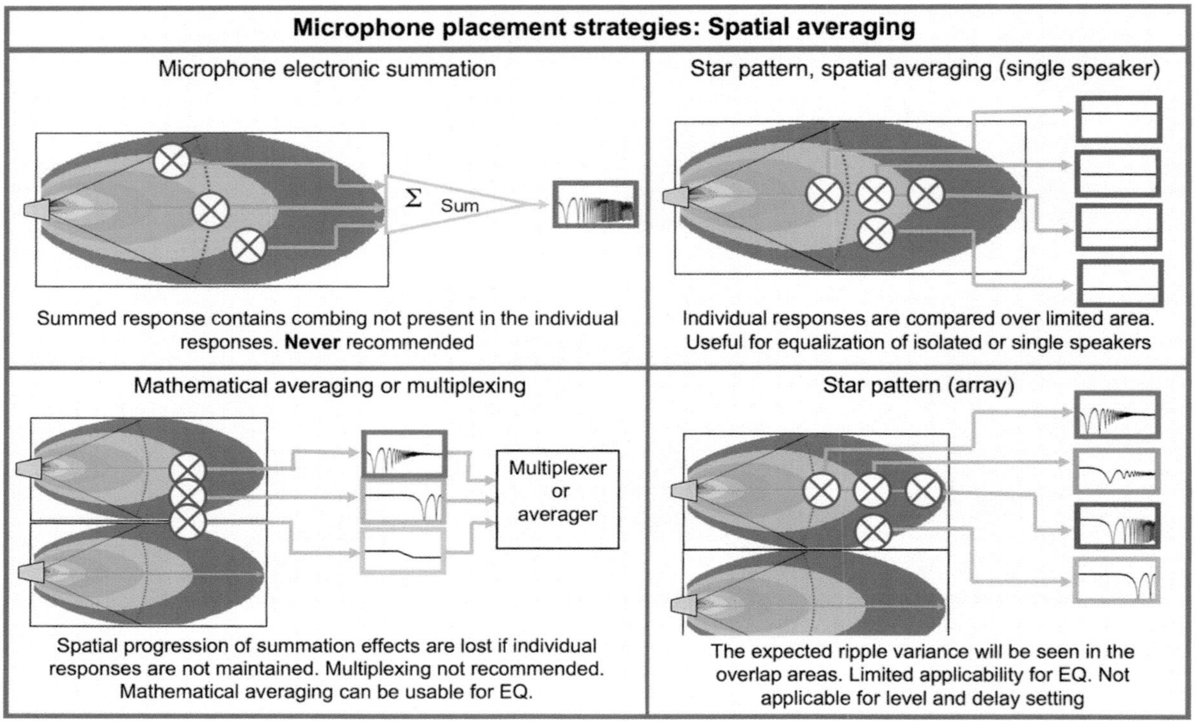

Microphone placement strategies: Spatial averaging
Microphone electronic summation
Σ Sum
Summed response contains combing not present in the individual responses. Never recommended
Star pattern, spatial averaging (single speaker)
Individual responses are compared over limited area. Useful for equalization of isolated or single speakers
Mathematical averaging or multiplexing
Multiplexer or averager
Spatial progression of summation effects are lost if individual responses are not maintained. Multiplexing not recommended. Mathematical averaging can be usable for EQ.
Star pattern (array)
The expected ripple variance will be seen in the overlap areas. Limited applicability for EQ. Not applicable for level and delay setting

FIGURE 14.9
Spatial averaging strategies

Amplitude (dB)
30
20
10
0
-10
-20
-30
16
31
63
125
250
500
1000
2000
4000
8000
16000
Frequency (Hz)
Phase (deg)
180
120
60
0
-60
-120
-180

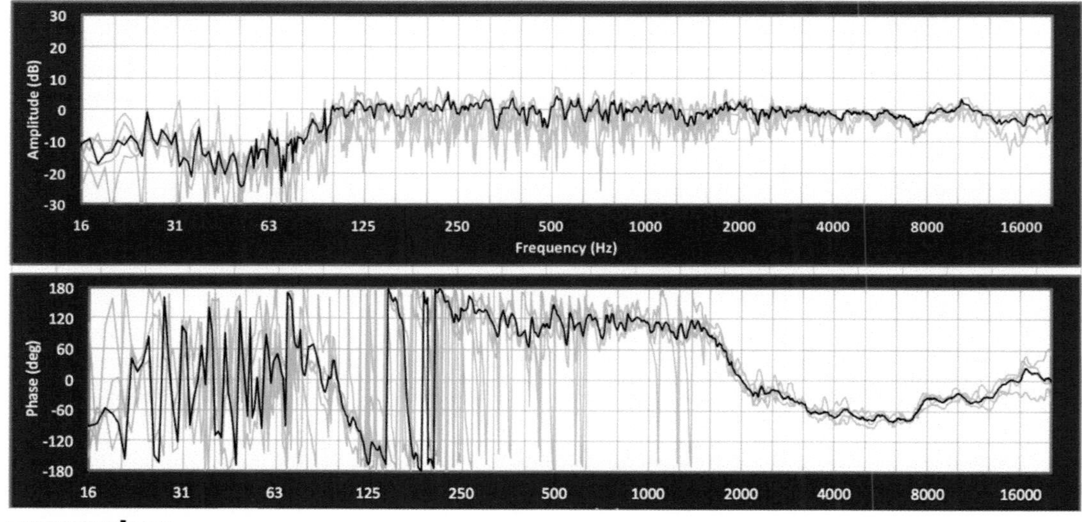

FIGURE 14.10
Mathematical spatial averaging

14.3.4.1 MATHEMATICAL TRACE AVERAGING

There are many approaches to mathematical averaging of frequency responses. Check your analyzer for details on its methods. Only the broad strokes are outlined here.

An average can be found by adding individual levels and dividing. The math is linear and result shown as log. For example, values of +6 dB and 0 dB are averaged as (2 + 1)/2 = 1.5 (+3.5 dB). The average is the linear mid-point, but the log result is closer to the higher-level signal, e.g. +6 dB and -20 dB don't average to -7 dB (13 dB below and above the samples). The linear equivalent is (2 + 0.1)/2 = 1.05 (+0.5 dB) (5.5 dB below and 20.5 dB above the samples). The average changes just 0.5 dB if the dip is deepened to -40 dB (2 + 0.01)/2 = 1.005 (0 dB). The difference between the louder sample (+6 dB) and the average (0 dB) is obviously significant. Studying summation revealed that 20–40 dB dips are likely to stay down in only a small area, whereas 6 dB peaks may spread over a wide area. Studying perception revealed greater tonal sensitivity to wide peaks over narrow dips. Therefore we should be wary of accepting 0 dB as the best representative here. When samples agree, the averaging builds confidence. When samples differ, the average is suspect. There's safety in numbers when math averaging is used: get a lot of samples (Fig. 14.10).

14.3.4.2 OPTICAL TRACE AVERAGING

The upper side of a frequency response curve, the peaks and unaffected areas, dominate the perceived tonal character (section 5.3). The tonal envelope is the primary shape used for position, level and EQ adjustments. Multiple measurements around ONAX can be overlaid to create a layered spatial average for equalization. The tonal envelope trends are found by eye, hence "optical averaging." Null structures are ignored for the moment and the focus placed on the most dominant tonal features. Be sure to normalize the individual traces to level the viewing field (i.e. push up traces from further distances) (Figs 14.11 and 14.12).

14.3.4.3 LIMITATIONS OF SPATIAL AVERAGING

All this begs the question of how much spatial averaging is advisable or appropriate. A practical number of positions (four, eight, sixteen) are statistically insignificant. Wavelength plays a part as well. Low frequencies will change very little from mic to mic compared with highs. Mics a meter apart are worlds away at 10 kHz, but not at 100 Hz. Would it be better to change the splay angle or add some delay between the array elements, if eight samples averaged out to a +6 dB octave-wide peak at 2 kHz? That's a nonsensical question because all we can conclude from spatial averaging is *equalization*. The data that drives the other procedures is not statistical. It's found in specific locations, not general or averaged ones.

Mathematical spatial averaging combines multiple measurements into a single value but does not necessarily indicate the variance *between* the positions. Standard deviation equations can find the variation range, but this is

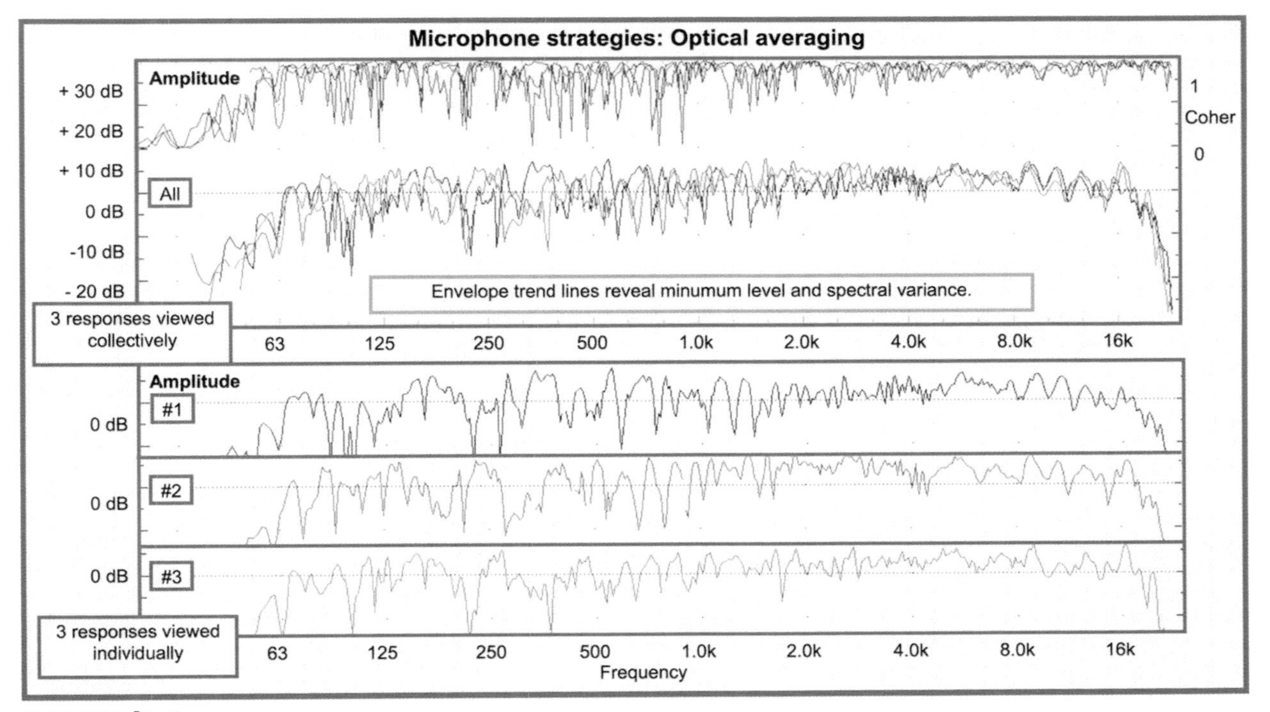

FIGURE 14.11
Optical spatial averaging

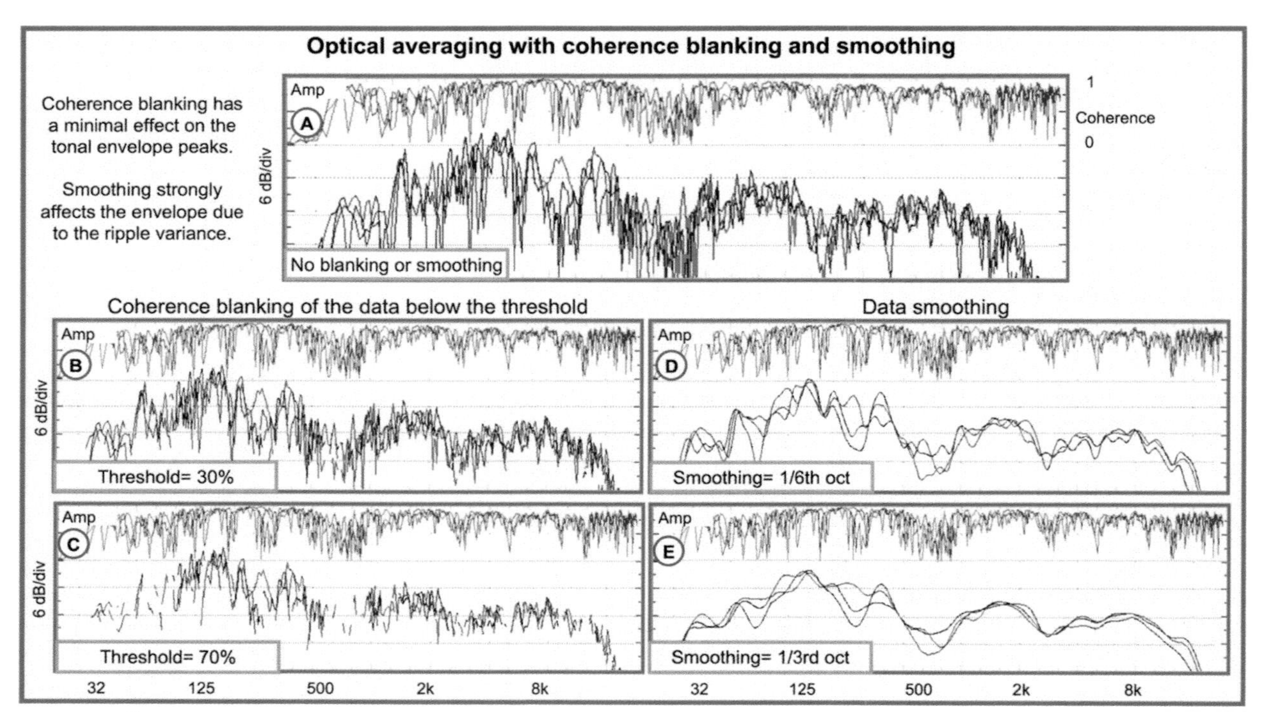

FIGURE 14.12
Optical spatial averaging with coherence blanking and smoothing

very tough with high-resolution data in our noisy reflective environment. We still don't know who's hot and who's not without going back to the individual traces. The optical method leaves the data intact at each position, thereby preserving the individual circumstances.

Spatial averaging is effective at finding the "average" tonal response so we can apply an "average" equalization for the area. We must be careful not to dilute the sample. It is recommended to stay well within the speaker's coverage limits (OFFAX data drags down the HF response) and be vigilant about the summation zone progression (combing zone randomizes the data).

14.3.4.4 MICROPHONE PLACEMENT IN CONTEXT

Learning to read data in context is an acquired skill. More than thirty years later, I am still working on it. What should the data look like here? Is this the expected response? High-resolution data is challenging because there is so much detail that everything looks terrible. The question is whether it's normal terrible or bad terrible. Context gives rise to expectations, which can be met, exceeded or failed. Most contextual information comes from comparison (Figs 14.13 to 14.15).

Mic placement context examples

- **ONAX area**: expected to be the speaker's best-case scenario. If other areas look better then context tells us to investigate immediately. Possibly our mic position is not correct or the speaker is not aimed where we think, or something is in the way.
- **Distance**: How far away are we? HF loss at 50 m is expected, but if it's down at 5 m we need to investigate (aimed wrong, blockage, bad driver, etc.). A fight for coherence is expected at 50 m, but not at 5 m.
- **Room acoustics**: Are we in a super-reverberant symphony hall? A glass and cement hockey arena? Is it windy? Any of these conditions lowers expectations.

- **Local conditions**: Is there line of sight to the speaker? Are we near a strong local reflection source? Could the bald guy sitting ahead of the mic be a reflection source? Yes (true story, Osaka, Japan, 1989). Investigate the local conditions for contextual clues before taking action. Don't EQ the system for the bald guy.
- **Array type**: We expect lots of pink shift in a twelve-element coupled array. Low-frequency overlap should cause the spectrum to substantially tilt. A flat combined response might seem like cause for celebration: "Look mom, no EQ!" Because tilt was expected we should investigate whether the individual speakers were not flat to start with, or begin searching for LF driver polarity reversals.
- **Inverse square law**: Its repeal is unlikely despite marketing claims to the contrary. If we don't see the expected drop of 6 dB/doubling then look for contextual clues. Did we move more off axis (or more on axis) as we moved back? Are we in the near-field of an array (parallel pyramid)?
- **Off axis**: It's simple to measure on axis and declare a speaker's placement as optimal. All we actually know is that sound arrived at the mic. The on-axis data alone has no positional context. We only learn whether the aim is correct or whether or not we have enough coverage by comparing the ONAX response to OFFAX or VTOP or VBOT (the positions that create a context for ONAX).
- **Spatial crossover area**: Is an octave-wide 20 dB dip at 5 kHz a problem? Perhaps if found near ONAX. It's right on schedule when found a few inches from XOVR.
- **Stereo symmetry**: We measured the left side. We expect the right side to match.
- **Audience presence**: We measured and stored the empty room response. Now the audience is in, the band is hot and so is the room. Do the response changes we see make sense relative to the actual physical changes? Are they even *possible*, or are we being tricked by leakage? Carefully consider this before turning a knob.

Bring everything to the table: common sense, auditory and visual clues, past experience with the components, the contractor or the hall.

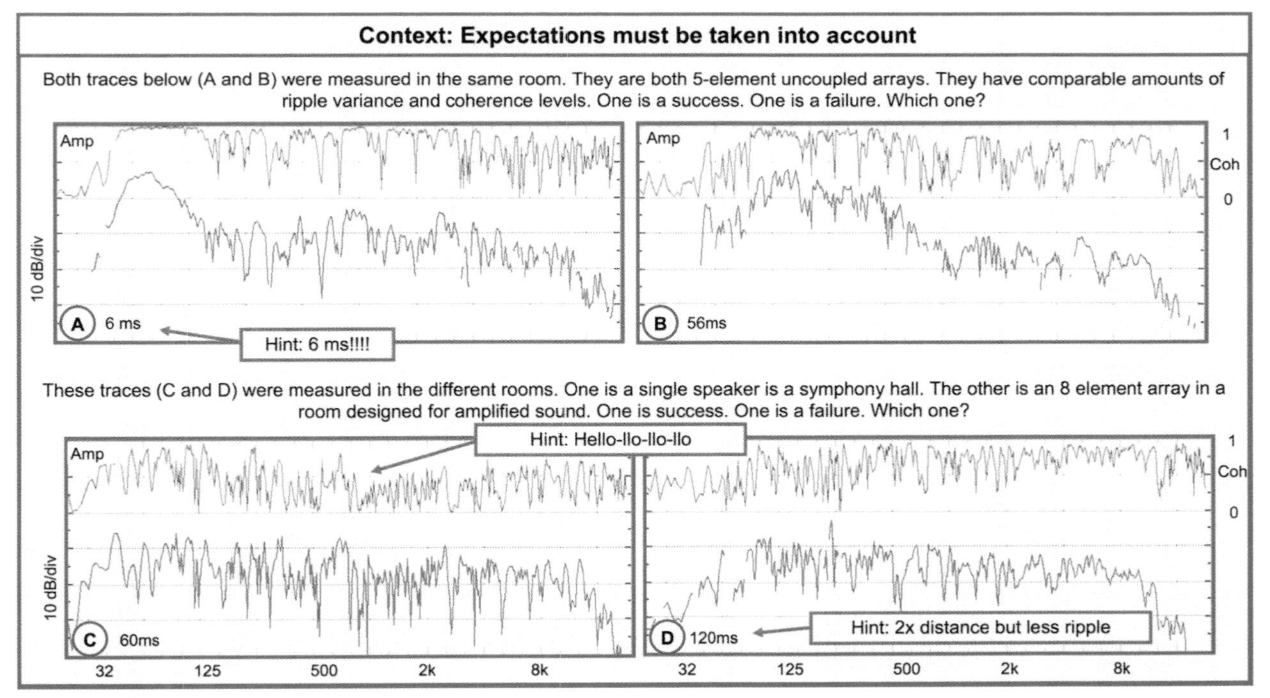

FIGURE 14.13
Similar responses with different expectations. We expect the quality at a short range to be high. Here that is not the case. The short-range measurements reveal excessive overlap in an uncoupled line source array.

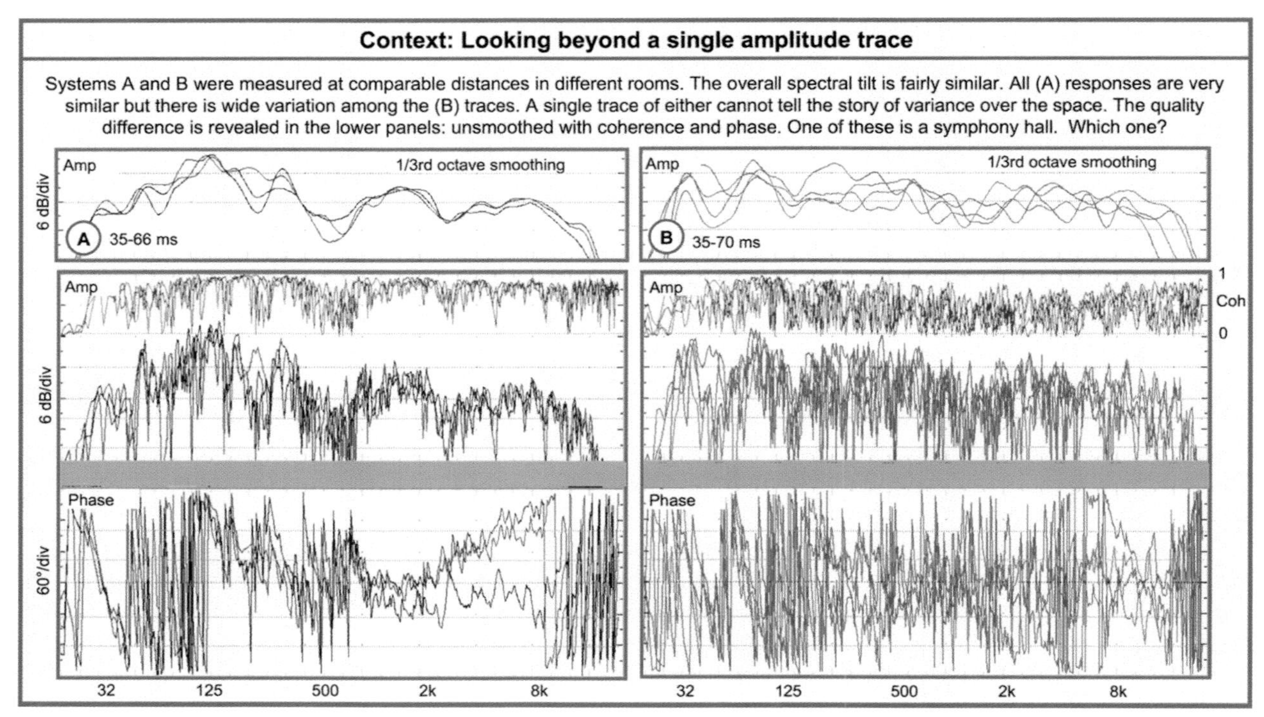

FIGURE 14.14
An illustration of how looking at a smoothed amplitude trace leaves us without the clues provided by other measurements such as coherence and phase

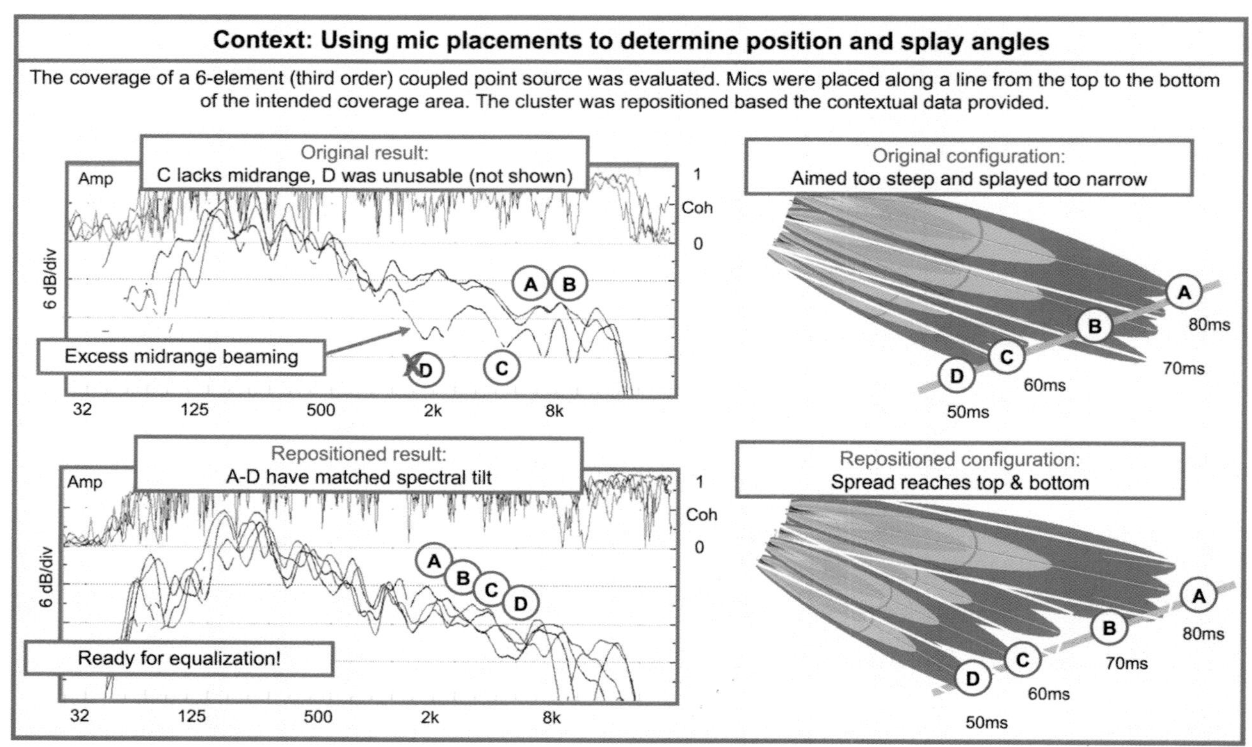

FIGURE 14.15
Using a contextual mic placement strategy to determine speaker position

14.4 PROCEDURES

14.4.1 Acoustic treatment

The calibration process may reveal troublesome acoustic features. If we're lucky we might be able to do something. We won't do statistical acoustical studies or reduce the reverb time by some number. This is about identifying a particular surface and its relationship to a particular speaker. It might be a reflecting surface near the speaker, the balcony front or the back wall. Any mic location could reveal a treatable acoustic problem: strong early reflections. Surfaces near the speaker can benefit from even small amounts of absorptive material to dampen the HF reflections. More distant surfaces require larger areas of absorption because the sound from speakers has spread. The offending surface can be spotted in the impulse response and/or comb filter signature in the frequency response.

Calibration seldom includes enough acoustic treatment to make a statistically significant reverberation time change. That's another scope of work entirely. Adding a few panels, drapes or other soft goods can have a large effect on some key reflections but are unlikely to reduce the overall tail. Our acoustical work is typically small scale, extremely cheap and doable without heritage committee approval or fire safety studies.

ACOUSTICAL TREATMENT SUMMARY

- **Identify problem**: Encounter strong reflection(s) in impulse and/or frequency response.
- **Localize source**: Identify the surface/speaker relationship, size of affected area.
- **Treatment**: Add absorption or otherwise treat the surface.
- **Observe result**: Compare the treated response to the untreated original.

14.4.2 Level setting

Levels are set to match each ONAX position to ONAX A, the gold reference standard. A minimum level variance line connects ONAX positions through the unity class spatial crossover (XOVR). Level setting matches the isolation zones, while proper splay and spacing match the shared areas. Level setting takes a proactive rather than reactive role. Speakers are brought into level compliance with the listening area, rather than the opposite. The process is fully detailed in Procedure 14.12 (Fig. 14.41).

LEVEL-SETTING PROCEDURE SUMMARY

- **Set GOLD reference**: Equalize A solo at ONAX A and store as GOLD A reference.
- **Set level**: Equalize B solo at ONAX B and level set to match GOLD A.

14.4.3 Single speaker aim

Speaker aim seeks to minimize level, spectral and ripple variance in the key relationship between the speaker and room. A single speaker is aimed as a soloist in the horizontal and vertical planes. Arrays are sometimes aimed in the same way, in at least one plane (e.g. a modern line array aims as a single speaker in the horizontal plane and as an array in the vertical). The aiming of symmetric arrays is functionally equivalent to a single (multi-cell) speaker. The mic position strategy begins at ONAX and brackets the outer edges of each plane. Speaker position adjustment is complete when the relationship between ONAX, OFFAX, VTOP and VBOT positions reaches minimum variance. The process is fully detailed in Procedures 14.1 to 14.5 (Figs 14.31 to 14.34).

SOLO SPEAKER AIM PROCEDURE SUMMARY

- **Normalize ONAX level**: Store response of solo speaker as reference.
- **Find OFFAX relationships**: Compare both OFFAX responses to ONAX.
- **Adjust horizontal**: Find aim with minimum variance between OFFAX–ONAX–OFFAX.
- **Find vertical relationships**: Compare VTOP and VBOT responses to ONAX.
- **Adjust vertical**: Find aim with minimum variance between VTOP–ONAX–VBOT.

14.4.4 Splay/spacing adjustment

Splay angle and spacing adjustment seek to minimize level, spectral and ripple variance in the key relationships between the speakers in an array. The mic position strategy begins at ONAX A and spans to ONAX B through XAB. Splay/spacing adjustment is complete when the relationship between ONAX A, XAB and ONAX B reaches minimum variance. The process is fully detailed in Procedures 14.6 to 14.8 (Figs 14.35 to 14.38).

ARRAY SPLAY/SPACING PROCEDURE SUMMARY

- **ONAX A level**: Store equalized response of A solo as reference.
- **ONAX B level**: Set B solo level at ONAX B to match ONAX A as above (equalized).
- **Find XAB**: Find location where A solo response is -6 dB from ONAX A reference.
- **Adjust splay/spacing**: Splay/space B solo speaker until response matches A at XAB.
- **Combine**: Measure A + B at XAB (set delay if necessary). Should match ONAX A.
- **Verify**: Measure A + B at ONAX A, XAB and ONAX B. All should match.

14.4.5 Equalization

Equalization has a unique emotional position in the calibration landscape. It's a subject near and dear to the mix engineer due to its key role in the perceived tonal response. All calibration parameters play important roles in the mix position tone, but none are scrutinized as closely as EQ. It's understandable because the mix console has 500 EQ knobs of its own and zero for adjusting splay angle. Engineers can use their input channel EQ for fine-tuning rather than gross compensation if the system is tuned with the tone they expect.

There are almost innumerable EQ stages in the signal chain. The end result is the accumulation of all their effects.

MAJOR EQ STAGES

- **Input channel EQ**: artistic adjustment of individual channels.
- **Subgroup EQ**: artistic adjustment of summed channel groups.
- **Mix console output channel EQ**: artistic adjustment of complete mixes.
- **Master EQ**: artistic/objective tonal shaping to the target curve.
- **System EQ**: objective system optimization for minimum spectral variance.
- **Speaker system internal processing EQ**: objective manufacturer settings.

There is a trickle-down aspect to the EQ chain. Errors left at the back end of the chain need to be compensated by a previous stage (e.g. peaks left in the speaker response by the manufacturer's internal settings must be fixed in the system EQ). Problems left in the system EQ are treated in the master EQ upstream. Unfortunately the upstream stage repairs are less effective than fixing the problem within its own stage. The farther the problem gets kicked upstream the more detached the solutions become from the causes. The ultimate example is an error in the manufacturer's internal settings (last stage) having to be corrected on 96 input EQs (first stage). The ultimate goal is to keep each EQ stage focused on treating the local conditions under its control. Optimization focuses on the internal manufacturer EQ (verification) and the system output EQ (calibration). Matching the target EQ curve can be done at the master EQ or within the system EQ. These stages set the tonal envelope for the system, leaving the upstream stages to concentrate on the individual input channels and their interactions with others in their mix.

There are some grand old sayings about EQ: "the best EQ is the least EQ" and "he who EQ least, EQ best." Everybody acts like they believe this and then the next subject is how this console/processor/plug-in/whatever is so awesome because it has these incredible EQ filters. So which is it folks? My adaptation of the old sayings are these: "the best EQ is the best EQ" and "he who EQ best, EQ best."

14.4.5.1 EQUALIZATION IN PERSPECTIVE

Equalization is the easiest and the hardest part of tuning sound systems. Locating peaks and dips and bringing them into submission is child's play compared with making sure the speakers are aimed right, splayed right, spaced right, delayed and level set for even coverage. The hard part about equalization is the guessing game. Do they want it flat? If not flat, then what? Cinema has an X Curve. Everything else in the industry uses the "read my mind" curve. I recently attended a presentation where I learned the answer to the question I have been grappling with for thirty years: a 12 dB slope from lows to highs sounds best. Period. My mind flashed back through a thousand tunings and concluded that +12 dB is indeed the average between my clients that like it flat and the ones that go for 24 dB extra low end. We are within ±12 dB of a one-size-fits-all answer.

Where does this all come from? Why does "flat" wear a white hat on your console, amplifier, cables and even individual speakers and a black hat as soon as these components are fully assembled? How comfortable are you with me inserting a cable with a 12 dB LF to HF slope on a flat system if I can get you the same frequency response as a flat cable on a tilted speaker system?

Where does the urge to tilt come from and why do we save it for last? It's a product of the natural world and how we experience sound sources over distance. Frequency response tilt is a primary factor in how we distance range a source. The guessing game of how much tilt to put in a system is the artistic decision of how close we want the audience to feel to the sound system (and the signals we are sending through it).

The "equal" in EQ

The term "equalization" implies that we are compensating for the inequality in the speaker response with inequality in the electronic filter response. Sorry Mom: Two wrongs *do* make a right. A peak in the speaker gets a dip filter of the same center frequency, bandwidth and dB level. The combination of speaker/room and filters is flat, i.e. "equalized." Saying that you equalized it "flat" is redundant. If it's not flat it's not equal, and if it's not equal it is certainly *not* equalized. Our industry is fast and loose with terminology and this is no exception. It is rather amusing to note that equalizing something to equality (i.e. flat) is considered an aberration. Everybody knows the last thing we want to do is put too much equality in our equalization. It's like fish that's too fishy.

Fear of flatting is a common phobia in speaker treatment but not so in other parts of the system. For example, if you saw the HF rolled off 1 dB in the electronic signal path would you leave it that way or equalize it (i.e. make it flat)? It's difficult to argue against flat in this case but this flies out the window once sound gets into the air.

Sonic image range

Our perception of sonic image is multifaceted. We immediately think of vertical and horizontal localization but there is a third dimension: depth. That's where EQ fits in. The frequency response of any sound source transmitting through the atmosphere of planet Earth changes over distance. The response tilts in favor of lows over highs. The reasons are the (A) atmosphere (loss of HF response in the air) and (B) planet Earth (reflections off the ground and other surfaces favoring the LF response). A "normal" violin sound in the near field is very different than far away. If it only got quieter over distance we could easily fool our sonic range finder, but there's more to it. Mezzo forte sounds louder, but not closer than pianissimo. The "normal" near response has more HF (less air) and flatter LF (less room) compared with a level-normalized midrange. It's interesting that when people ask for "more air" in their mix they are looking for VHF extension, which is exactly what adding real air to the transmission takes away.

Perceived sonic range is a combination of spectral tilt and direct/reverberant ratio clues. Spectral tilt in favor of the lows and/or rising reverberation expands the perceived depth. The on-axis response of a distant source (compared with its near-field response) will be unequalized in favor of the lows (perceived as far away). The near-field off-axis response of the same source will also be unequalized (due to being off axis). It is perceived as far away even when close (due to the HF loss). A natural sound source can be perceived at its actual depth or further away (off axis or strongly reverberant).

Loudspeakers have the unique capability of sonic range ventriloquism. Unlike the violin they can be perceived as closer than reality. A sound system that's "in your face" is perceived as closer sonically than visually. This is accomplished by equalization and reverberation control. A very flat and directionally controlled speaker is more likely to create an up close and personal experience than a heavily tilted wide one (Fig. 14.16).

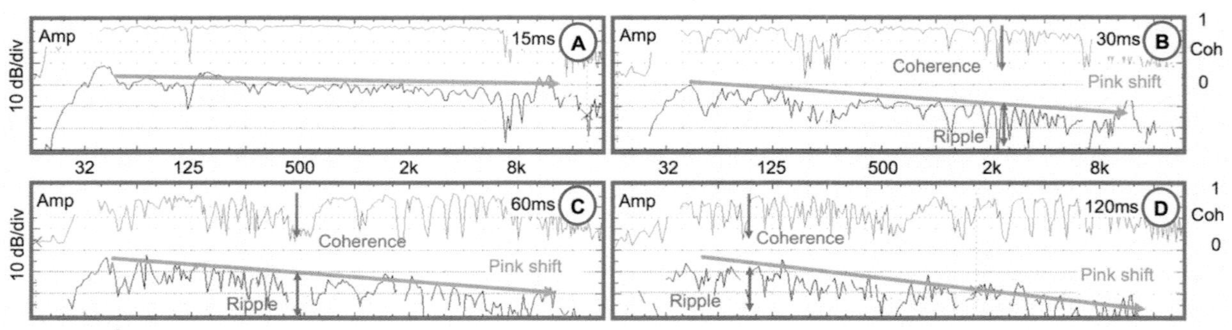

FIGURE 14.16
EQ sonic perspective

Linking equalization and sonic image depth

There's no image depth perspective for a length of wire, and hence no objections to equalizing any losses it accrues and returning its response to flat. Speakers always have an image perspective, as do the input signals they are transmitting. Let's compare two paths to the listener: direct sound from a natural instrument (e.g. violin) versus through a mic, console and speaker. Our example direct path is 10 m. The mic is placed 1 m from the violin and the speaker is 4 m from the listener (a total path of 5 m). If the mic and speaker are flat then the amplified sound will *not* match the direct sound. This might seem strange at first but remember that the direct sound has 10 m of HF air loss and LF room reflections whereas the amplified path has 5 m. The amplified path moves the sonic depth perspective to a range closer than the 10 m direct path. The question is how much closer.

Let's look at this in two parts: HF (air loss) and LF (room reflection addition). The air loss portion is simple: More air equals more loss so the flat mic/console/speaker chain delivers an HF sonic range of 5 m. The LF portion is more complex because the directionality of the mic and speaker plays a role. Increased directionality reduces the reflections going into the mic and being added to the speaker transmission. Directionality reduces the upward LF tilt, thereby shrinking the image range to even less than the 5 m of actual path. An omnidirectional natural source reinforced with highly directional mics and speakers will experience the highest amount of depth reduction.

We can make the amplified path match the direct path with equalization. We can reduce the LF addition and restore the HF losses until the direct and amplified paths match. We are tilting the 5 m amplified path in favor of the lows to mimic the 10 m direct path. Now we have a 10 m sonic image depth and the ability to add reinforcement gain as desired. If we have tuned our system with a matched frequency response over the room then we have approximately placed all listeners at a sonic depth of 10 m. The sonic perspective can be made closer by reducing the tilt, or pushed deeper by tilting more.

On stage we use close mic'ing to isolate the channels for control, an overly close sonic image perspective. Returning the sonic perspective to something normal requires adding tilt into the mic response, console or speakers. Mics are all over the stage at different distances and axial orientations, and there are also perspective/free direct feeds (e.g. keyboard or bass). Therefore the speakers tend to carry most of the sonic range load.

The target curve

What's the optimal sonic depth? That's the million dollar question of course, because the answer gives us a target curve for equalization. In practice we're likely to encounter huge differences in visual depth between performers and near and far listeners. A mid-sized concert might have listeners ranging from 4–40 m and easily 80 m for an arena. We want to reduce the sonic perspective and bring the audience closer to the performers. But how close? A 50% reduction? Bring everybody on stage? Inside the kick drum?

If we close-mic the instruments and make the system flat from DC to light we are reducing the depth to an uncomfortable and unnatural perspective. I love Bonnie Raitt but I don't want her singing direct into my ear like I'm a hand-held. If plausibility is important (such as musical theatre), then we need to limit the reduction to

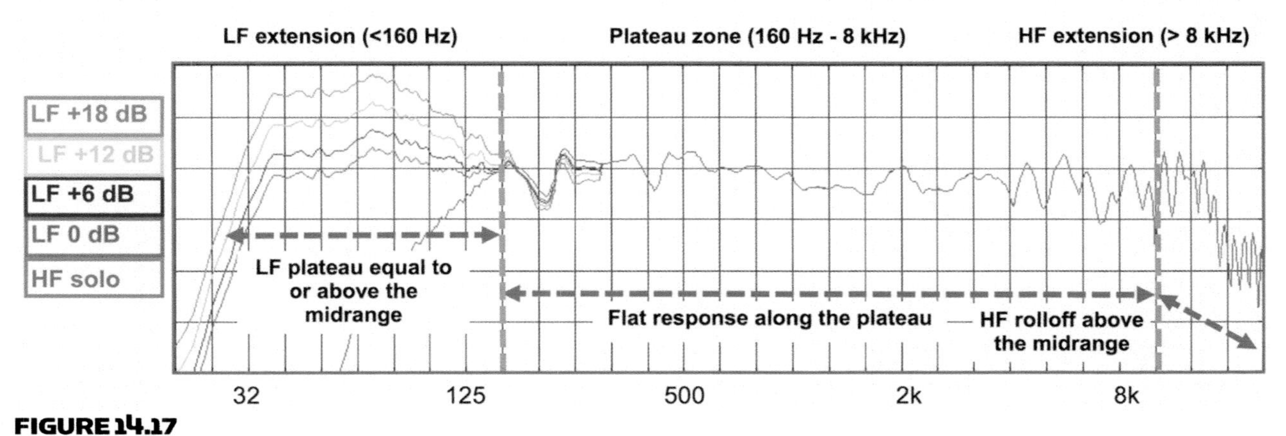

FIGURE 14.17
EQ perspective/target curve

something short of detectibly unnatural. Shrinking the distance in half may be enough to bring people in without awareness of the speakers. As the venue scale rises, the amount of tilt required to maintain a median sonic distance goes up.

We are a long way toward a scalable and consistent sonic perspective if a constant frequency response is maintained throughout the room. Spectral uniformity keeps the perceived depth constant and the secondary mechanism, direct/reverberant ratio, now moves into play. As we move closer to the sound system (and stage) the D/R ratio should rise, which correlates to reduced depth. The D/R naturally falls in the rear and the perceived depth increases.

Let's tune the system to a desired target curve at the room's mid-point depth and chart the sonic depth variance from there. The D/R ratio rises as we move forward (sonically closer) and the MF/HF response stays the same (sonically constant). The LF probably rises closer in, a fact of life in PA world (sonically farther). We have a bit of a stalemate in perspective. We lose D/R ratio moving back in the hall (sonically farther), but keep the same LF/MF/HF response (sonically constant) until we reach the extreme rear where the LF rises up (sonically farther). We are moving the image away but at a lower rate than natural sound (because we are keeping the HF response fairly constant).

Everyone has expectations in their head of target curves, but they are not as simple as following a graph. I have done work with the same band, same mixer and worked closely to achieve their target curve/sonic depth and stored it in the analyzer memory. A night later in a new city we tuned the same system to the same target curve but did not end up there. The hall was much more reverberant and the mix engineer wanted less LF range up-tilt. My conclusion: He wanted the same sonic perspective. More tilt, less reverb (hall 1) and less tilt, more reverb (hall 2) came out to the same sonic depth.

None of this holds up over the space unless the system is calibrated for minimum spectral variance (the real job of calibration). There are very advanced tools for gently shaping the response into the desired target shape. I'll give a shot at it, but if my guess doesn't feel right to the client, then we'll rock it up (or down) until the desired effect is achieved (Fig. 14.17).

14.4.5.2 THE ROLE OF EQUALIZATION

The equalizer's principal roles

- **Global sonic range perspective**: Control of the overall spectral tilt of the system, the target curve, subject to the client's artistic discretion. This is a global parameter.
- **Minimum spectral variance**: spectral tilt matching subsystems to minimize differences throughout the listening space. Different filter settings are applied on a subsystem-by-subsystem basis to match the target curve for a unified global effect.

- **Minimum ripple variance (speaker/room)**: Reduce peak/dip variations within each subsystem driven by a particular equalizer.
- **Minimum ripple variance (speaker/speaker)**: Reduce peak/dip variations within each subsystem driven by a particular equalizer.

14.4.5.3 LIMITATIONS OF EQUALIZATION

Equalization affects a speaker system's entire coverage the same way. It's a global solution that's most effective when facing global challenges (such as spectral tilt). It's least effective when facing widespread local differences (such as ripple variance).

The effective range of equalization filters

- **Spectral tilt reduction/increase**: practically unlimited.
- **Spectral variance in a single speaker**: Filters can correct for spectral variance in the on-axis response but cannot affect the spectral variance over the space of a single device. That is governed by the speaker's beamwidth and its orientation to the space. Simply put: EQ cannot change a speaker's coverage pattern.
- **Spectral variance between two devices**: can be minimized by applying separate filters, which result in matched spectral tilts.
- **Spatial area and frequency**: The effective spatial area for ripple variance reduction is inversely proportional to frequency.
- **Spatial area and bandwidth**: The effective spatial area for ripple variance reduction is proportional to bandwidth.

An example system helps see the limitations in practice: eight speakers and eight five-band parametric equalizers. We can EQ the entire system at the mix position as one channel, precisely minimizing minute ripple variance with forty narrow bandwidth filters. We've created a perfect ripple variance decoder for exactly one position (until the weather changes). Unfortunately our ripple decoder key doesn't fit for any other locations and increases the variance everywhere else. We've applied a global "solution" to a local problem. TANSTAAFL: The fix for the mix costs everyone else. Acoustic triage: We used up the entire blood bank for one patient. In the end, all this effort yields very little. The mix position is in a feedback loop that makes it artistically self-calibrating. If the tone is not right, the mix engineer adjusts the mix until it is. It's a servo loop that affects 10,000 people but gets feedback from only one. Global system EQ is primarily for the mix engineer's convenience, to reduce the EQ required on the individual channels. The primary optimization issue, the *difference* between the mix area and the other seats, is unchanged by the presence of forty filters (or none at all). A system with only a single EQ (or discrete stereo pair) has only two paths to minimum spectral and ripple variance: speaker position and acoustic treatment. A key that reduces differences over the space is the one we should be seeking.

Let's repatch and give each speaker a dedicated five-band EQ. The filters are used to minimize the response *differences* over the space. Each EQ addresses the strongest room/speaker summation trends in their local area. The coupled component of the combined system response can then be jointly equalized. If the arrays have been properly aimed, splayed and spaced we are well on our way to minimum variance.

14.4.5.4 SOLO EQ, SINGLE SYSTEM EQ (A, B, . . .)

Equalization proceeds in two stages: solo (A or B) and combined (A + B). Solo EQ corrects each subsystem in its local environment. This includes speaker/speaker coupling within the subsystem, speaker/room interaction and HF air loss. The relationship between acoustical effects and the equalizer is complementary at a 1:1 ratio, e.g. an octave-wide 6 dB peak at 200 Hz is equalized by a 6 dB cut of the same width and center frequency (Fig. 14.18).

The solo EQ process

- **Establish a spectral target curve**: This can be a verifiable standard (such as the cinema X Curve), a specific request from a client (such as "make it like last night in Toronto") or a generalized guess about what the average engineer will expect when they encounter the hall/system (e.g. flat, +6 dB LF, +12 dB LF, etc.).

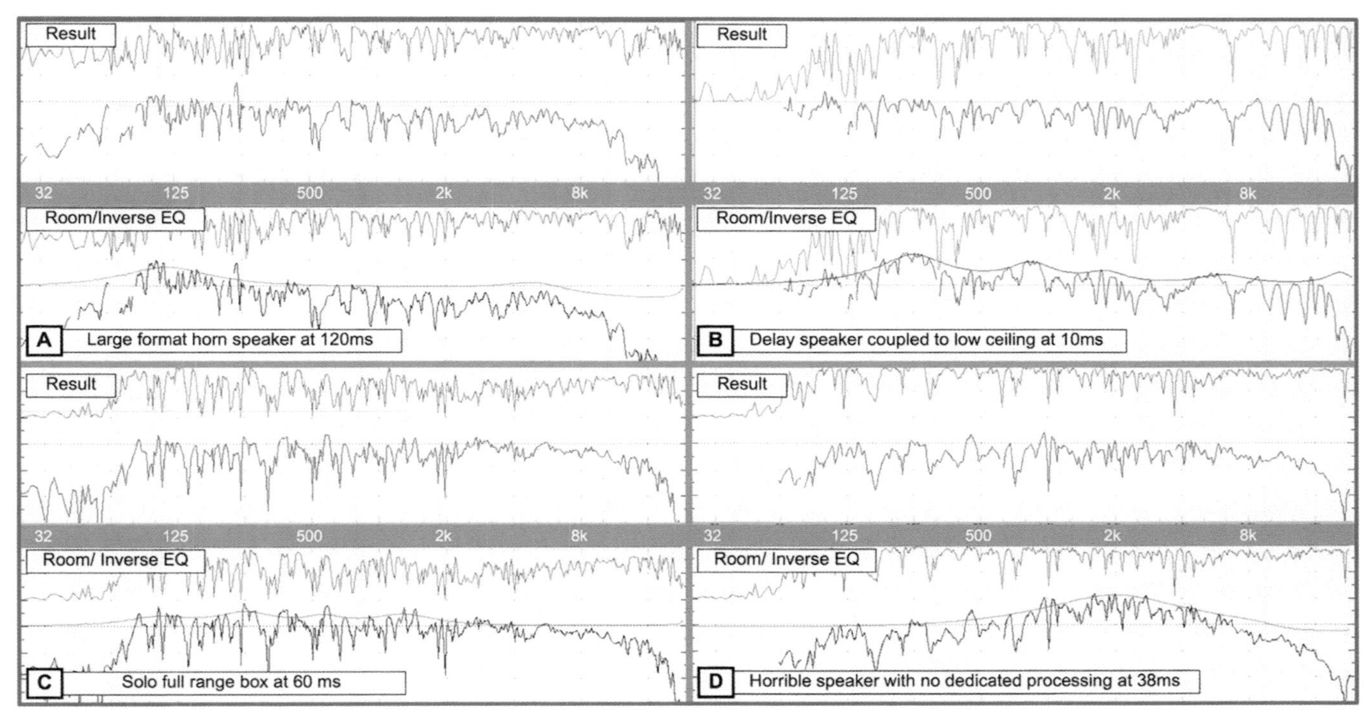

FIGURE 14.18
Solo complementary EQ examples

- **Find representative response (pre-EQ)**: Use solo ONAX or solo spatial averaging. We are looking for the strongest tonal modifiers in the response (such as wide peaks over wide areas, rather than microscopic details that affect only one seat).
- **Apply complementary EQ**: Create an inverse of the representative response with filters (matched level and bandwidth, and equal but opposite level). Peaks are met with dips and vice versa until the representative response matches the target curve.

14.4.5.5 SYMMETRIC COMBINED EQ ($A_1 + A_2$)

Combined EQ corrects for the coupling of previously equalized subsystems. In practice this should primarily be focused on the LF range, where coupling zone summation is maintained between the three locations: ONAX A_1, XAA and ONAX A_2. Combined EQ for a symmetric pair is complementary (like solo EQ), e.g. 3 dB of acoustical coupling is met by a 3 dB cut in *both* equalizer channels. Maximum combination of a symmetric pair (6 dB) can only occur when the patterns are 100% overlapped, i.e. the lobe from the opposite side is equal to the one on the local side. As overlap falls the coupling decreases as does the required EQ compensation. Combined EQ for a symmetric pair is easy in practice. Compare the stored ONAX A_1 solo response to the combined response and quantify the addition (e.g. +6 dB below 150 Hz and +3 dB from 400 Hz to 150 Hz). Apply a matched complementary EQ to *both* channels. It helps to analyze the symmetric interaction to prepare us for the asymmetric version, which is not at all straightforward.

Consider three locations (two unique): ONAX A_1, XAA and ONAX A_2. There are six paths (three unique): each speaker to each location. The responses at XAA are guaranteed to match no matter what the directional pattern is because it is the geometric center of a symmetric pair (otherwise you need to fix the crossover). The two paths to the ONAX A_1 position are *not* guaranteed to match. If the speakers are perfectly omnidirectional then the two paths will be nearly identical (only the physical displacement differs). Level differences between the paths rise with directionality (Fig. 14.19).

Symmetric combined system EQ locations

- **XAA:** A_1 and A_2 solo responses are assumed matched (0 dB, 0° relative). The combined total here is +6 dB above the solo responses. The combined spectrum is the same as the solo spectrums (just louder) so there is nothing new to EQ here.

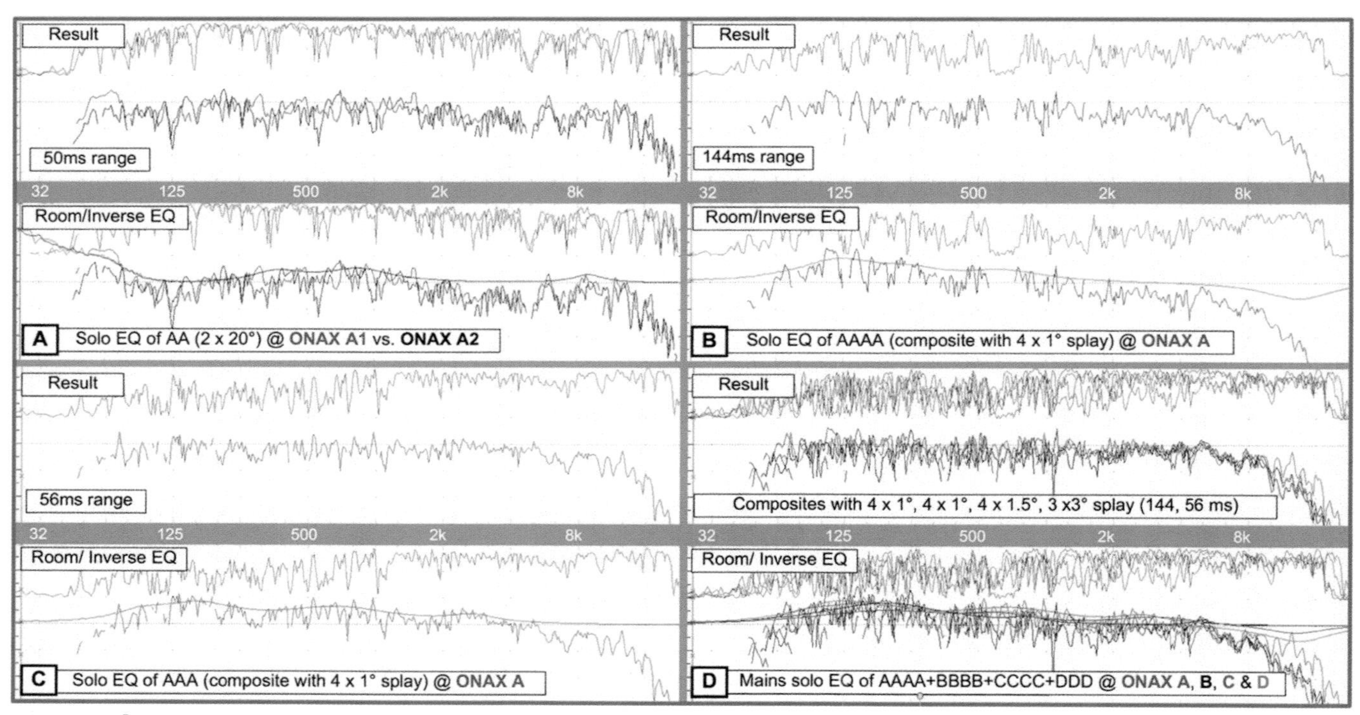

FIGURE 14.19
Coupled symmetric complementary EQ examples

- **ONAX A_1:** The solo EQ'd response of A_1 becomes the 0 dB, 0° reference point for the combination. Solo response of A_2 @ONAX A_1 is an open variable, which can approach, but never achieve, perfect matching to A_1 (A_2 is displaced by some distance, creating some level and phase offset). The combined response can approach +6 dB (omnidirectional speaker with minimal level and phase offset) or much less (directional speaker with substantial level and/or phase offset).

In this example, a 3 dB addition occurs in three scenarios (using phase multiplier (4.1.5.1)):

- A_2 arrives at -8 dB, 0° ((1 + 0.4)*1) = 1.4 (+3 dB).
- A_2 arrives at 0 dB, 90° ((1 + 1)*0.7) = 1.4.
- A_2 arrives at -5 dB, 45° ((1 + 0.55)*.91 = 1.4).

The same EQ is applied in all cases: 3 dB cut in both channels.

- **ONAX A_2:** the symmetric opposite of ONAX A_1.

The LF range will typically have more overlap and lower phase offset than the MF and HF ranges. Therefore the applied EQ will have the most reduction in the LF range.

Note: A fully symmetric pair, trio, etc. can be driven by a single processing channel because it's expected to have identical processor settings.

14.4.5.6 Asymmetric combined EQ (A + B)

Combined EQ for an asymmetric pairing is a complex challenge and a positive outcome is not assured. We start with two equalized solo responses, each level scaled for its own zone size. We hope to combine them into a single asymmetric shape that preserves the uniformity we had as soloists. The unequalized combination raises the LF response through coupling zone summation that is not evenly shared between the two parties, which changes the spectral and spatial shapes. Combined EQ, applied asymmetrically, seeks to reduce the variance of the raw combination. We are actually changing the (combined) polar pattern with EQ by applying asymmetric levels over frequency. Combined EQ might fully or partially compensate the spatial and spectral variance of the raw AB pair. It might only break even. We want to make sure we don't make it worse.

Three factors dominate: range ratio, coverage pattern overlap and phase offset. The outcome has two parts for a given frequency: the amount of level change at ONAX A and B and the difference between them (the asymmetry).

Range ratio affects the distribution of leakage from speaker A into the ONAX B area and vice versa. The solo responses at each location are level matched (0 dB), but the leakage (side lobe, down lobe, etc.) into the neighbor's coverage differs. Speaker A is louder so it leaks more into the B coverage area than vice versa (a small difference when the range ratio is low, but can be up to 6 dB with a range ratio of 2:1).

The overlap/isolation between the systems is evaluated on a continuum of level offset (the difference between what A spills into B's area and vice versa). If we have lots of isolation (speaker A level is low at ONAX B) then there is little we need to do. If we lack isolation (speaker A level is high at ONAX B) then there is little we *can* do.

Example application of range ratio and overlap/isolation

- **Range ratio = 2:1 (6 dB)**: ONAX B is ½ the range of ONAX A.
- **Isolation = (6 dB)**: Each speaker's coverage is -6 dB (by angle) at the opposite ONAX.
- **Level settings**: Speaker A's drive level is set to 0 dB, speaker B to -6 dB.
- **Solo responses at ONAX A**: 0 dB from speaker A and -12 dB leakage from speaker B (-6 dB level, 0 dB from the standard range and -6 dB from isolation).
- **Solo responses at ONAX B**: 0 dB from speaker B and 0 dB leakage from speaker A (0 dB level, +6 dB from the closer range and -6 dB from the isolation).
- **Combined response at ONAX A**: 0 dB + (-12 dB) = +2 dB (1 + 0.25 = 1.25).
- **Combined response at ONAX B**: 0 dB + 0 dB = +6 dB (1 + 1 = 2).
- **Net variance between ONAX A and B (pre-EQ)**: 4 dB.
- **Settings to restore ONAX A and B to 0 dB**: A = -0.1 dB, B = -40 dB (DON'T!!!).

The above example shows how we would need a 40 dB difference in filter settings to overcome a net variance of 4 dB between ONAX A and B. This is a case where the side effects (loss of coupling power capability) far outweigh the benefits. In this case we don't have nearly enough isolation to overcome such a large range ratio, but there are many cases where we can. We draw the limit at 6 dB of additional cut for the B speaker. The combination of two systems can't add more than 6 dB so we should not ask our filters to cut more than that.

Let's take an inventory of the big picture now.

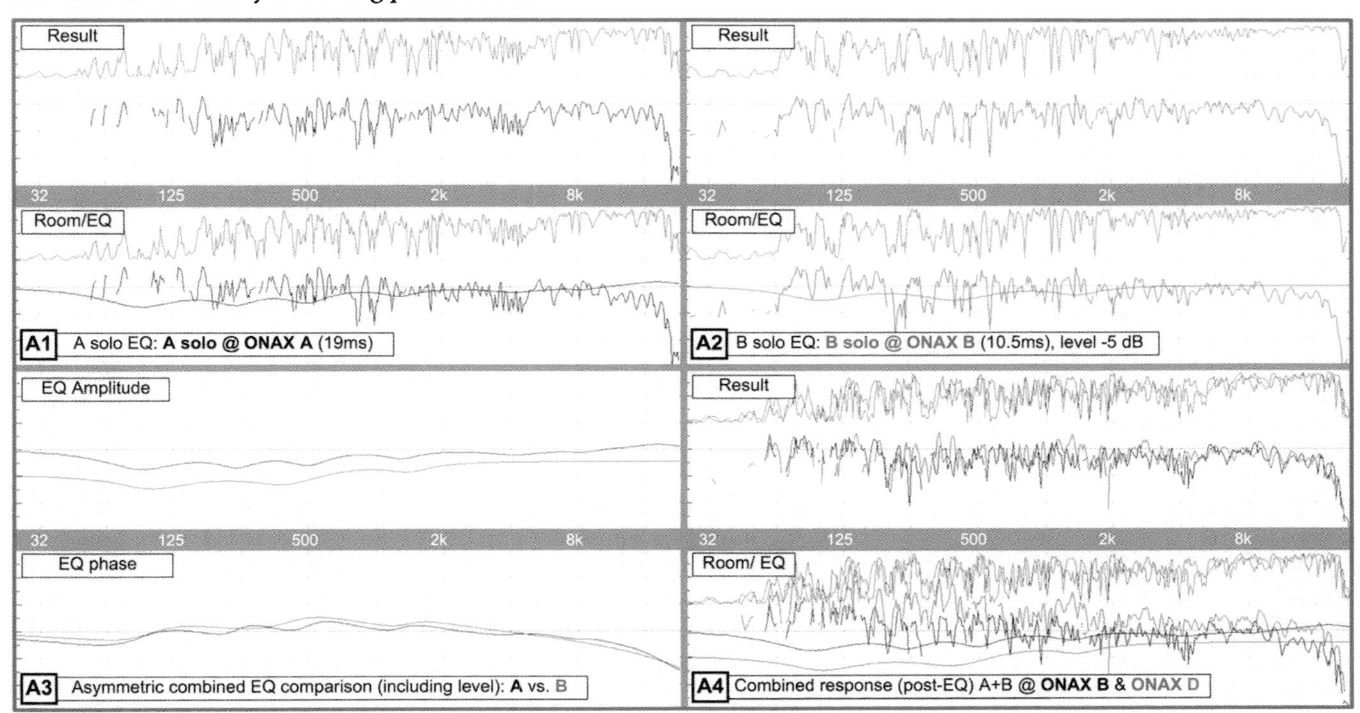

FIGURE 14.20
Coupled asymmetric complementary EQ examples

Combined EQ in asymmetric arrays (AB)

- **Primary goal**: minimum variance between ONAX A and B after combination.
- **Primary challenge**: combined level at ONAX B is higher than ONAX A.
- **Primary technique**: Set filters for A and B at different levels (B is reduced more).
- **Determining factors**: overlap/isolation, range ratio and phase offset.
- **Limiting factor**: excess level offset between A and B (6 dB maximum).
- **Outcome (+)**: minimum variance or variance reduction (compared with pre-EQ).
- **Outcome (-)**: no variance reduction (compared with pre-EQ) or excess power loss.
- **Highly overlapped**: reduces independent control. More difficult to reduce variance, especially at high range ratios. No reduction is possible with 100% overlap.
- **Highly isolated**: increases independent control. Minimum variance is achievable, even at high-range ratios.
- **Low-range ratio**: minimally asymmetric behavior (almost symmetric). Small pre-EQ variation. Minimal level differences required between A and B filter settings.
- **High-range ratio**: highly asymmetric behavior. Large pre-EQ variation. Large level differences required between A and B filter settings.
- **Phase offset**: rising offset decreases coupling (more in the B area than A due to level imbalance). Can reduce the pre-EQ variance by bringing combined B level closer to A, which reduces the amount of level asymmetry required to compensate a given range ratio and overlap. If the phase offset is >120° then combined EQ should not be used.
- **Best-case MV scenario**: low overlap, low range ratio and low phase offset.
- **Worst-case MV scenario**: high overlap, high range ratio and high phase offset.
- **Minimum benefit**: combined response at ONAX A matches A solo (target curve).
- **Maximum benefit**: combined responses at ONAX A and ONAX B match the solo reference.

Let's start with one extreme: no isolation. If the speaker is 100% omnidirectional then it's 0 dB down at all angles. Splay angle doesn't matter if the two patterns are circles. There is no isolation, so the closer you are, the louder it will be. It will be louder at the B speaker in direct scaling with the range ratio. Filters on either speaker will bring down the levels at both (by the same amount). The only thing to be done is to restore the response at ONAX A to the target curve (and let B go above that). See Fig. 14.21.

Let's go to an opposite extreme: high isolation (12 dB). The response can be restored to the original target curve in both locations at all range ratios with <6 dB difference. The full range of filter-level settings over range ratio and isolation is found in Fig. 14.22.

Phase offset is the final component. The speaker elements are displaced and so the leakage path arrives late. This decreases the efficiency of the coupling, most notably on the ONAX B side because the levels are closer. The result can be less than full addition on the B side. This can reduce the net variance between ONAX A and B, and thereby reduce the cut levels required. A small detail is that the phase offset is not necessarily symmetric. If the B speaker has been delayed to optimize at XAB then the phase offset will be less at ONAX B and more at ONAX A. This reverses the net variance trend in the opposite direction: better addition at B (where we don't need it). The likelihood of phase offset asymmetry rises with range ratio.

In the end we have to measure at ONAX A and B and start to turn knobs. But this is what is going on behind the scenes (Fig. 14.22).

100% Overlap (0 dB)	@ ONAX A	A	B	@ ONAX B	VARIANCE
A solo (Lvl = 0 dB)	+0.0 dB			+6.0 dB	
B solo (Lvl = -6dB)	-6.0 dB	0 dB	-6 dB	+0.0 dB	
A+B	+3.5 dB			+9.5 dB	6 dB

FIGURE 14.21
AB combination with 100% overlap and 2:1 range ratio

1.4:1 Range Ratio (B range is 3 dB shorter than A) Results with symmetric EQ					
100% Overlap (0 dB)	@ ONAX A	A	B	@ ONAX B	VARIANCE
A solo (Lvl = 0 dB)	+0.0 dB			+3.0 dB	
B solo (Lvl = -3dB)	-3.0 dB	0 dB	-3 dB	+0.0 dB	
A+B	+4.5 dB			+7.5 dB	3 dB
70% Overlap (-3dB)	@ ONAX A	A	B	@ ONAX B	VARIANCE
A solo (Lvl = 0 dB)	+0.0 dB			+0.0 dB	
B solo (Lvl = -3dB)	-6.0 dB	0 dB	-3 dB	+0.0 dB	
A+B	+3.5 dB			+6.0 dB	2.5 dB
50% Overlap (-6 dB)	@ ONAX A	A	B	@ ONAX B	VARIANCE
A solo (Lvl = 0 dB)	+0.0 dB			-3.0 dB	
B solo (Lvl = -3dB)	-9.0 dB	0 dB	-3 dB	+0.0 dB	
A+B	+2.5 dB			+4.5 dB	2 dB
35% Overlap (-9 dB)	@ ONAX A	A	B	@ ONAX B	VARIANCE
A solo (Lvl = 0 dB)	+0.0 dB			-6.0 dB	
B solo (Lvl = -3dB)	-12.0 dB	0 dB	-3 dB	+0.0 dB	
A+B	+2.0 dB			+3.5 dB	1.5 dB
25% Overlap (-12 dB)	@ ONAX A	A	B	@ ONAX B	VARIANCE
A solo (Lvl = 0 dB)	+0.0 dB			-9.0 dB	
B solo (Lvl = -3dB)	-15.0 dB	0 dB	-3 dB	+0.0 dB	
A+B	+1.4 dB			+2.6 dB	1.2 dB

2:1 Range Ratio (B range is 6 dB shorter than A) Results with symmetric EQ					
100% Overlap (0 dB)	@ ONAX A	A	B	@ ONAX B	VARIANCE
A solo (Lvl = 0 dB)	+0.0 dB			+6.0 dB	
B solo (Lvl = -6dB)	-6.0 dB	0 dB	-6 dB	+0.0 dB	
A+B	+3.5 dB			+9.5 dB	6 dB
70% Overlap (-3dB)	@ ONAX A	A	B	@ ONAX B	VARIANCE
A solo (Lvl = 0 dB)	+0.0 dB			+3.0 dB	
B solo (Lvl = -6dB)	-9.0 dB	0 dB	-6 dB	+0.0 dB	
A+B	+2.5 dB			+7.5 dB	5 dB
50% Overlap (-6 dB)	@ ONAX A	A	B	@ ONAX B	VARIANCE
A solo (Lvl = 0 dB)	+0.0 dB			+0.0 dB	
B solo (Lvl = -6dB)	-12.0 dB	0 dB	-6 dB	+0.0 dB	
A+B	+2.0 dB			+6.0 dB	4 dB
35% Overlap (-9 dB)	@ ONAX A	A	B	@ ONAX B	VARIANCE
A solo (Lvl = 0 dB)	+0.0 dB			-3.0 dB	
B solo (Lvl = -6dB)	-15.0 dB	0 dB	-6 dB	+0.0 dB	
A+B	+1.5 dB			+4.5 dB	3 dB
25% Overlap (-12 dB)	@ ONAX A	A	B	@ ONAX B	VARIANCE
A solo (Lvl = 0 dB)	+0.0 dB			-6.0 dB	
B solo (Lvl = -6dB)	-18.0 dB	0 dB	-6 dB	+0.0 dB	
A+B	+1.1 dB			+3.5 dB	2.4 dB

1.4:1 Range Ratio (B range is 3 dB shorter than A) Results with asymmetric EQ					
100% Overlap (0 dB)	@ ONAX A	A	B	@ ONAX B	VARIANCE
A solo (Lvl = 0 dB)	+0.0 dB			+3.0 dB	
B solo (Lvl = -3dB)	-3.0 dB	0 dB	-3 dB	+0.0 dB	
A+B	+4.5 dB			+7.5 dB	3 dB
70% Overlap (-3dB)	@ ONAX A	A	B	@ ONAX B	VARIANCE
A solo (Lvl = 0 dB)	+0.0 dB			+0.0 dB	
B solo (Lvl = -3dB)	-9.0 dB	0 dB	-6 dB	-3.0 dB	
A+B	+2.5 dB			+4.5 dB	2.0 dB
50% Overlap (-6 dB)	@ ONAX A	A	B	@ ONAX B	VARIANCE
A solo (Lvl = 0 dB)	+0.0 dB			-3.0 dB	
B solo (Lvl = -3dB)	-15.0 dB	0 dB	-9 dB	-6.0 dB	
A+B	+1.5 dB			+1.5 dB	0 dB
35% Overlap (-9 dB)	@ ONAX A	A	B	@ ONAX B	VARIANCE
A solo (Lvl = 0 dB)	+0.0 dB			-6.0 dB	
B solo (Lvl = -3dB)	-16.0 dB	0 dB	-7 dB	-4.0 dB	
A+B	+1.3 dB			+0.0 dB	1.0 dB
25% Overlap (-12 dB)	@ ONAX A	A	B	@ ONAX B	VARIANCE
A solo (Lvl = 0 dB)	+0.0 dB			-9.0 dB	
B solo (Lvl = -3dB)	-13.0 dB	0 dB	-4 dB	-1.0 dB	
A+B	+1.7 dB			-1.8 dB	0.1 dB

2:1 Range Ratio (B range is 6 dB shorter than A) Results with asymmetric EQ					
100% Overlap (0 dB)	@ ONAX A	A	B	@ ONAX B	VARIANCE
A solo (Lvl = 0 dB)	+0.0 dB			+6.0 dB	
B solo (Lvl = -6dB)	-12.0 dB	0 dB	-12 dB	-6.0 dB	
A+B	+2.0 dB			+8.0dB	6 dB
70% Overlap (-3dB)	@ ONAX A	A	B	@ ONAX B	VARIANCE
A solo (Lvl = 0 dB)	+0.0 dB			+3.0 dB	
B solo (Lvl = -6dB)	-12.0 dB	0 dB	-9 dB	-3.0 dB	
A+B	+2.0 dB			+6.05 dB	4 dB
50% Overlap (-6 dB)	@ ONAX A	A	B	@ ONAX B	VARIANCE
A solo (Lvl = 0 dB)	+0.0 dB			+0.0 dB	
B solo (Lvl = -6dB)	-18.0 dB	0 dB	-12 dB	-6.0 dB	
A+B	+1.0 dB			+3.5 dB	2.5 dB
35% Overlap (-9 dB)	@ ONAX A	A	B	@ ONAX B	VARIANCE
A solo (Lvl = 0 dB)	+0.0 dB			-3.0 dB	
B solo (Lvl = -6dB)	-21.0 dB	0 dB	-12 dB	-6.0 dB	
A+B	+1.0 dB			+1.5 dB	0.5 dB
25% Overlap (-12 dB)	@ ONAX A	A	B	@ ONAX B	VARIANCE
A solo (Lvl = 0 dB)	+0.0 dB			-6.0 dB	
B solo (Lvl = -6dB)	-24.0 dB	0 dB	-12 dB	-6.0 dB	
A+B	+0.5 dB			+0 dB	0.5 dB

FIGURE 14.22
Coupled asymmetric complementary EQ scenarios

Limitations to asymmetric combined EQ

There are tangible risks to offsetting levels and EQ within a coupled speaker array. Gentle slopes and gradual level transitions will reduce the risk whereas steep slopes and transitions will increase it. The phase aspect has been previously discussed but the dynamic range effects merit discussion. Asymmetric EQ and level taper will reduce the dynamic range and potentially cause dynamic shifting, i.e. some parts of the system may reach limiting before others. This can be a classic TANSTAAFL tradeoff between maximum uniformity, stability and headroom. Detectible instability in the frequency and spatial response are unacceptable outcomes. In short, dynamic uniformity is a vital part of level and spatial uniformity. Level taper effects on dynamic range were discussed previously (Fig. 4.7), which can be used as a reference point. This is one area where the program material and usage are extremely relevant. If the system is to be operated at its absolute maximum at all times (e.g. a heavy metal music festival), then front/back level uniformity must yield to maximum dynamic range and stability, i.e. level tapering and asymmetric EQ must be absolutely minimized. By contrast, if the system will be operated with ample headroom under controlled conditions (e.g. musical theater), then asymmetric EQ and level tapering become viable options to maximize uniformity. This decision must be evaluated in the field based on risk and return. My personal approach, which will be shown in Chapter 15, is generally to compromise, using minimal asymmetric EQ and leveling to reduce, but not necessarily fully eliminate, the level increase in the front of the room. Risks are relatively low if the asymmetry can be kept low through a large majority of the array. Increased asymmetry at the bottom has more potential benefit and is less likely to cause audible instability to the majority shareholders above them. Bear in mind, however, that much of my work is in applications that highly prioritize uniformity and high fidelity, which leaves me with more room for asymmetry than those who optimize systems that absolutely must prioritize for maximum stability while delivering bone-crushing power.

Real-world loudspeakers will have variable amounts of overlap over frequency. An array of constant beamwidth models will have an upper range with constant overlap and a lower range with increasing (and variable) overlap. The effectivity of asymmetric EQ techniques therefore varies over frequency, with the HF range enjoying the most effect for the lowest side effects and risks.

An array of proportional beamwidth models has proportional overlap, i.e. it rises gradually and steadily with wavelength. Combined EQ can be approached here by segmenting the spectral range into isolated (HF) and overlapped (LF) regions. EQ and level setting for the isolated HF is carried out by the ABC approach seen throughout here. By contrast, the LF region is approached as a single block, with all EQ and level settings matched (i.e. symmetric). This approach will likely yield less uniformity than the fully asymmetric approach but eliminates the dynamic risks inherent in asymmetric EQ.

Combined asymmetric equalization procedure summary

- **Establish target curve**: determine desired response.
- **Find representative response (pre-EQ)**: Use solo ONAX or solo spatial averaging.
- **Apply solo EQ**: Filter to shape the representative response to match target curve.
- **Combine A + B**: Observe response changes at A, B and XAB.
- **Apply combined EQ**: Filter to shape the representative response to match target curve.
- **Asymmetric filter levels**: Observe uniformity of A vs. B and compensate if needed.

14.4.6 Delay setting

Measure the time offset between a main and delay speaker, and type the number into the delay line. The operation is simple but the underlying strategy is more complex.

There's a continuum of relationships between delays and their references. On one side is the stationary, fully correlated reference (e.g. mains and delay fed the same signal). On the other is referencing to a moving target and semi-correlated source, such as an actor on stage. In between is stationary/semi-correlated, such as a fill speaker fed a different mix than its reference (frontfill on an aux bus or mono summed delays). An inverted version of stationary/semi-correlated is delaying the mains to the Marshall stack.

Stationary/correlated gives us the most control, and the relationship is fixed and stable over the space pending a change in the weather (literally). Delaying to moving actors is a tradeoff of realistic imaging vs. consistent response over time and space. Delaying to the back line is surrendering to the reality of "if you can't beat 'em, join 'em."

The relationship between stage sources and our sound system fails the stable summation criteria (section 4.1.1.1). The waveforms are only partially correlated because the referenced stage sound doesn't match the full mix in the speakers. The level relationship varies with changes in the mix and timing changes as actors move around the stage.

Synchronization to stage sources should be undertaken with caution. Only the absolute minimum number of subsystems (those closest to the stage) should be "aligned" this way. All remaining systems should be synchronized to the closer speaker systems (see Fig. 14.23).

14.4.6.1 PRECEDENCE EFFECT

It is almost certain that someone will bring up the precedence effect (a.k.a. the Haas effect) as we go about setting delays. There is a pervasive practice of purposeful mis-calibration of delay settings for the intended benefit of superior sonic imaging attributed to the precedence effect (section 5.2.4.2). The strategy is this: Find the delay time

Time and level reference sources for delay setting			
Source	Waveform correlation	Time relationship	Level relationship
Stage actor or singer	Semi-correlated	Variable	Variable
Pit orchestra	Semi-correlated	Variable	Variable
Drum kit, acoustic source	Semi-correlated	Fixed	Variable
Guitar backline	Semi-correlated	Fixed	Variable
Subwoofers on aux send	Semi-correlated	Fixed	Variable
Stage monitors	Semi-correlated	Fixed	Variable
Other channels (e.g. L vs. R)	Semi-correlated	Fixed	Variable
FX speakers, surrounds, etc.	Uncorrelated	Fixed	Variable
Subsystems on same channel	Correlated	Fixed	Fixed

FIGURE 14.23
Delay reference points

and then add 5, 10, 15, 20, some number of ms, of extra delay to move the image away from the delay speaker. Image is not everything, so we should be aware of the side effects.

Precedence effect side effects and considerations

- The precedence effect is a binaural function applicable to the horizontal plane only.
- Time offset increases ripple variance, decreases coherence and creates the *perception* of increased reflections. The delay speaker's job is to decrease (not increase) ripple variance.
- Time offset decreases power addition, leading to raised delay speaker level settings than otherwise needed.
- Main and delay are phase aligned *somewhere* and equal level *somewhere else*. We just don't know where. We only know where they're definitely *not* aligned.

There is no doubt that the precedence effect can move the image down. There is also no doubt that setting delays to arrive late can bring the quality down. The time offset side of sonic image perception has a very short shelf life (less than 10 ms) and only works in the horizontal plane. The less famous partner in the sonic image equation is level offset, which has a wider range of action, works in both planes and minimizes combing (rather than adding it). In short, we can *mostly* win the localization game with proper levels at minimal cost in side effects. Coupling, combing and isolation zone behaviors follow the standard progression. It's not glamorous, but maximizes intelligibility and uniformity. Adding precedence delay provides an almost magical image improvement in some seats but removes the possibility of coupling summation and ensures widespread combing. Adding this faux-reflection to the underbalcony area probably won't enhance intelligibility. The lack of coupling and intelligibility loss can lead to a perfectly logical (and perfectly illogical) outcome: turning up the delay because it doesn't seem loud enough and we can't understand the words. Or perhaps we could set the timing back to synchronous and turn it down? This is a classic case of TANSTAAFL. You can't get something for nothing. Conclusion: Proceed with caution.

The case for the phase-aligned crossover at ONAX D (delay) is advanced by its substantial sonic image stability, a combination of level and time offset. Equal level and equal time centers the image between the main and delay. Leading in one category and trailing in the other (in the right proportions) can also center the image. It's obvious that we can perfectly synchronize the mains and underbalcony delays in only one or two spots, but this is also true for the level portion (they can only stay perfectly equal in level in one or two spots). Time offsets creep in by differences in the triangulation of the paths. Level offsets arise by the ratio of the path lengths (the inverse square law losses). The image falls between the main and delay if we match their time and level at ONAX D (0 ms/0 dB). The delay speaker drops level more rapidly as we move rearward (shorter doubling distance) but moves ahead in time (triangulation reduction). Leading in time and lagging in level: image in the middle. The inverse is true for the rows in front of ONAX D. The nearby delay speaker rises in level faster than the distant mains (shorter doubling distance) but arrives later (triangulation). Leading in level and lagging in time: image in the middle. Will the image always be in the middle? This is situation dependent but the trend line is favorable for the best overall image stability (Fig. 14.24).

When clients ask me to add extra delay for precedence effect enhancement, I have proposed the following: Listen to the delays set up as a phase-aligned crossover and then tell me how much extra delay they think it needs to get good imaging. You win a prize if you guessed the usual answer is "leave it alone."

14.4.6.2 DELAY DILEMMAS

The timing relationship between two speakers (A and B) has only one answer at each location (A = B @XAB). Adding a third speaker complicates the timing because it is almost assured that each location has one right answer (and one wrong answer).

Timing relationships of three systems

- **A = B ≠ C @XAB**: A and B are synchronized and C is not.
- **B = C ≠ A @XBC**: B and C are synchronized and A is not.
- **A = C ≠ B @XAC**: A and C are synchronized and B is not.
- **A = C = B @XABC**: A, B and C are synchronized. Can only happen in one location.
- **A ≠ B_1 = B_2 @XB_1B_2**: B_1 and B_2 are synchronized and A is not.
- **A = B_1 ≠ B_2 @XAB_1**: A and B_1 are synchronized and B_2 is not.

Practical examples of three-way delay dilemmas

- **Left (L) to center (C) to right (R)**: It's a given that no one in their right mind is going to put left and right out of sync in the middle so L must equal R at XLR. Center synchronization is the question. If we sync C with L and R at center then we have L = C = R @XLCR and L ≠ C ≠ R at every other location. If we sync C with L at XLC (between ONAX L and ONAX C) then we have the following:

L = C, R ≠ C, L ≠ R @XLC L ≠ C, R ≠ C, L = R @XLR L ≠ C, R = C, L ≠ R @XRC

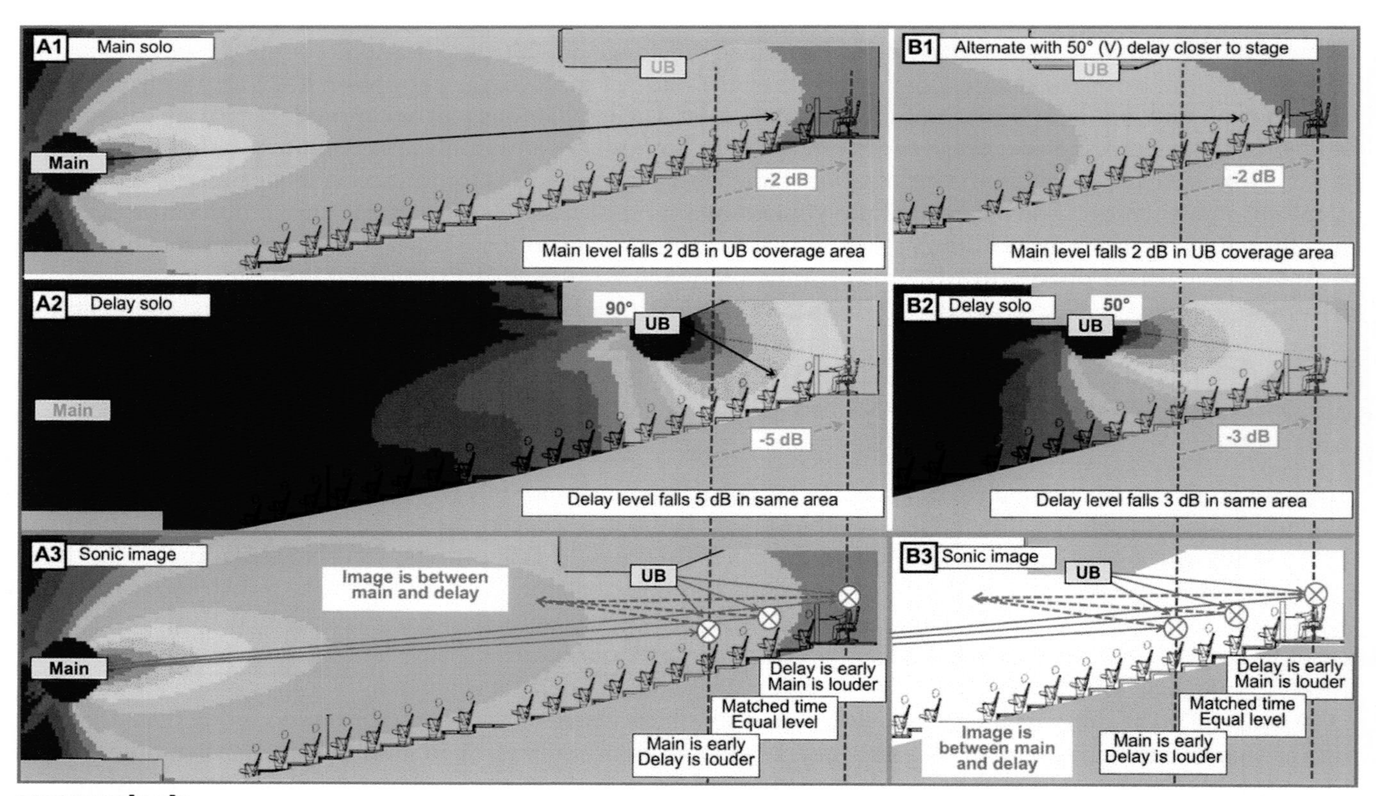

FIGURE 14.24
Precedence effect tradeoffs

In the first scenario there is a single-point triple synchronicity and every other location has none. In the second scenario there is one synchronous pair at left/center, center and right/center.

- **Main (A) to frontfill (B_1) to frontfill (B_2)**: If we sync each frontfill to the mains then the frontfills don't sync to each other ($A = B_1$, $A = B_2$, $B_1 \neq B_2$). If we keep the frontfills together then they can't both sync to the mains ($A = B_1$, $A \neq B_2$, $B_1 = B_2$).
- **Main (A) to underbalcony (B_1) to underbalcony (B_2)**: same as frontfills.
- **Main (A) to infill (B) to frontfill (C)**: This is a 3-D pretzel. Mains coming down from above, infills coming in from the side and frontfills shooting straight ahead. Infills are found on left and right so they will meet each other in the middle. Best not to sync to the mains in the middle (see the LCR dilemma above) but rather use an off-center mid-point. The frontfills can meet the mains in the center, and the infills nearer the sides, but then we require them to be out of sync to each other. Another layer can add to this fun: a live stage source. Decisions, decisions.
- **Stage source (S) to main (A) to frontfill (B)**: This has two wild variables— The stage source is movable and has an unstable level relationship to the speakers. The frontfill is the dilemma: Do we sync them to the mains or the stage source? If the mains are centered overhead the frontfill can be delayed to them and never worry about being ahead of the actors. The difference in frontfill timings to the mains and a centered stage source would be very different. In that case the stage source is usually chosen with L/R mains.

Such delay dilemmas are encountered once we move beyond the single-system stage of system optimization. Setting all eight underbalcony speakers (B_1 to B_8) to the same delay time creates a series of phase-aligned spatial crossovers. We'll need four different times to individually sync them to the mains. It would be great if there were a clear way to choose whether to sync main to delay ($A = B_1$, $A = B_2$, . . .) or underbalcony to underbalcony ($B_1 = B_2 = B_3$, . . .). There is! Democracy (the number of people affected). It's right there in our lettering system. A is the biggest system. The A + B overlap area is larger than the $B_1 + B_2$ overlap area. Every person listening to the underbalcony delays also hears the mains (A + B). Only a small minority hear $B_1 + B_2$ and they are actually hearing $A + B_1 + B_2$ so sync'ing to A is still the priority even there (Fig. 14.25).

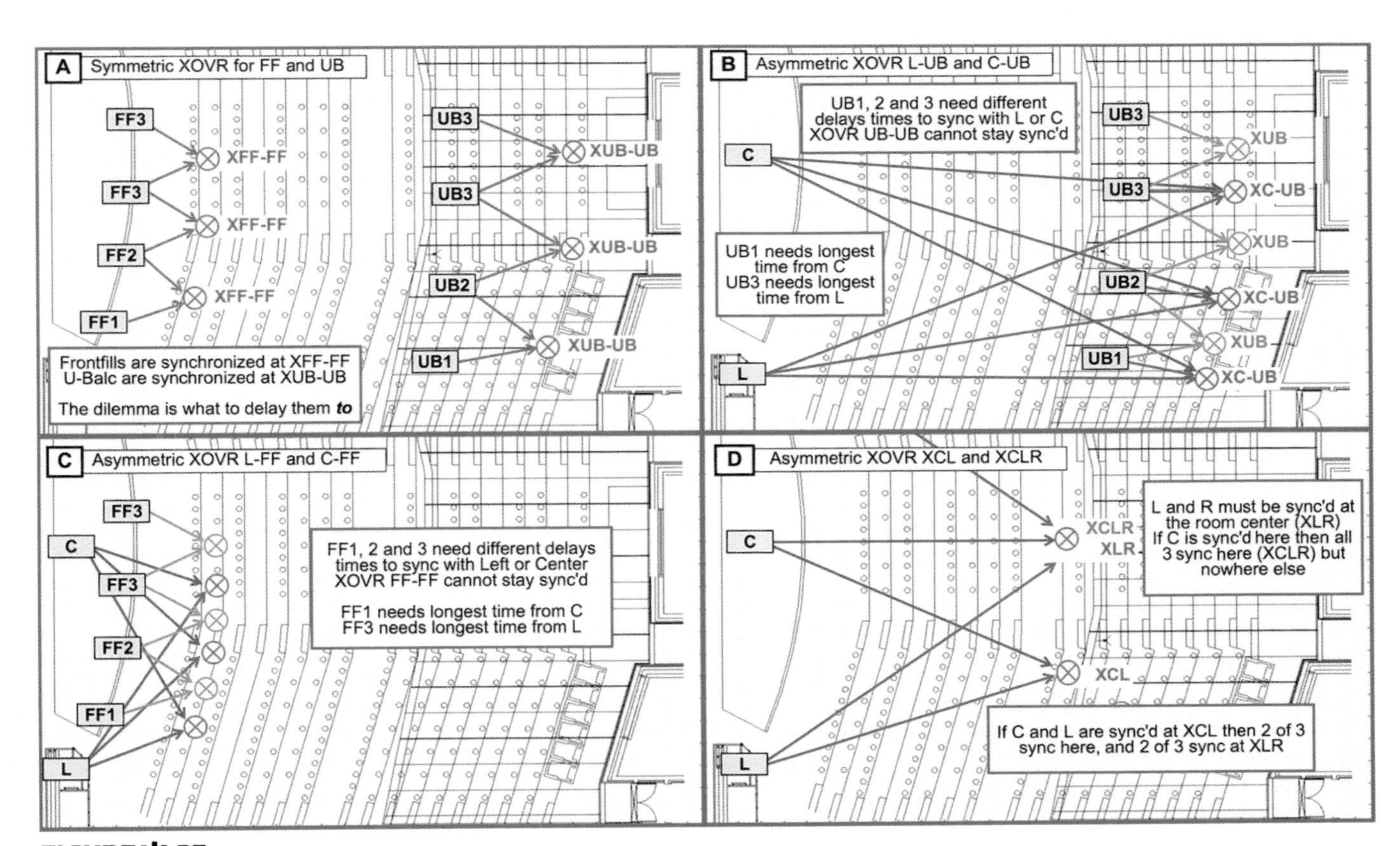

FIGURE 14.25
Delay dilemmas for an LCR system with frontfills and underbalcony delays

14.5 ORDER OF OPERATIONS

Now we organize the individual calibration procedures into a game plan, an order of operations that begins with each element in each subsystem until all are assembled. The order proceeds by the democracy principle: The longest throw system (A) is assumed to cover the largest number of people, while B is next and so on.

Once the individual subsystems have been calibrated we can begin combining them together. Our hope is the pieces will fit together like a puzzle. If not, we may have to go back and do some rework on the subsystems.

Operations to be completed on individual systems before combination with others

1. Speaker focus as found by the ONAX, OFFAX, VTOP, VBOT and XOVR positions.
2. Initial level and solo equalization setting at the ONAX position(s).
3. Acoustic treatment for the local area (any mic position).

The combination process includes

1. Delay setting at the XOVR position.
2. Speaker position adjustment (splay or spacing) to minimize variance between ONAX and XOVR.
3. Combined EQ at the ONAX positions. The order of combination operations can be reduced to a simple math equation derived from the ABC system hierarchy.

Example of system combinations as an equation

$$(((A_1 + A_2) + (B_1 + B_2)) + (C_1 + C_2))$$

The first operations are the combinations of symmetric elements within a given system, such as $(A_1 + A_2)$. Next combine asymmetric systems in hierarchical order: As to Bs and then add the C and onward to the running total.

The priorities for calibration order of operations

1. **Coupled elements**: Coupled symmetric elements are internally combined $(A_1 + A_2)$ before external combination (+B). For example, elements of a symmetric coupled point source main cluster are combined before adding sidefills $((A_1 + A_2) + B)$.
2. **Uncoupled elements**: Uncoupled symmetric elements are internally combined $(B_1 + B_2)$ before external combination (+A). For example, elements of a symmetric uncoupled line source frontfill are combined before adding them with the main cluster $((B_1 + B_2) + A)$.
3. **Coupled subsystems**: Coupled asymmetric subsystems are combined (A + B) before external combination (+C). For example, the main cluster downfill first joins the upper mains before combining with the uncoupled frontfills ((A + B) + C).
4. **Distance hierarchy**: The system that covers the longest distance (A) takes priority over subsystems that cover closer areas (B, C, etc.).

Example order of operations equation

(((main upper inner + main upper outer) + (main lower inner + main lower outer)) + (frontfill inner + outer))

Calibration strategy reference charts are found in Figs 14.26 to 14.28. Each chart contains element quantity, microphone position class, their functions and the signal-processing role during calibration. The charts show representative variations of speaker model, desired throw distance and spacing. The spacing may be angular and/or lateral, depending upon the array type. The series doesn't cover every possible scenario, but the iterations show the logic for any array combination encountered in the field. Symmetric arrays require the minimum number of mic positions and signal-processing channels. They also have the maximum number of SYM mic positions, which are the simplest of the operations. By contrast all forms of asymmetry (model, distance or spacing) require dedicated mic positions and separate processing channels.

Calibration strategies: 1-element										
Layout	Speaker				Microphone		Processing			
	Element	Model	Distance	Spacing	Type	Function	Channel	EQ	Level	Delay
					OFFAX A1	Position				
	1	A	a		ONAX A	Calibrate	A	EQ	0 dB	
					OFFAX A2	Position				

Order of operations					
Step #	Speaker		Microphone		Procedures
	Speaker	Mode	Type	Function	
1	A	Single system	ONAX A	Position (aim)	Set gold reference level
2	A	Single system	OFFAX A1 & A2	Position (aim)	Adjust for minimum variance between OFFAX & ONAX
3	A	Single system	ONAX A	Calibrate	A solo EQ

FIGURE 14.26
Calibration strategies for a single-element configuration asymmetric orientation is shown. Symmetric version requires only a single OFFAX measurement.

Calibration strategies: 2-element array: Symmetric (2A)										
Layout	Speaker				Microphone		Processing			
	Element	Model	Distance	Spacing	Type	Function	Channel	EQ	Level	Delay
					OFFAX A	Position (aim)				
	1	A	a		ONAX A1	Calibrate		EQ	0 dB	
				AA	XOVER AA	Pos (splay/space)	A			
	2	A	a		ONAX A2	Symmetric		EQ	0 dB	
					OFFAX A	Symmetric				

Calibration strategies: 3-element array: Symmetric (3A)										
Layout	Speaker				Microphone		Processing			
	Element	Model	Distance	Spacing	Type	Function	Channel	EQ	Level	Delay
					OFFAX A	Position (aim)				
	1	A	a		ONAX A2	Calibrate		EQ	0 dB	
				AA	XOVR AA	Pos (splay/space)				
	2	A	a		ONAX A1	Calibrate	A	EQ	0 dB	
				AA	XOVR AA	Symmetric				
	3	A	a		ONAX A2	Symmetric		EQ	0 dB	
					OFFAX A	Symmetric				

Order of operations					
Step #	Speaker		Microphone		Procedures
	Speaker(s)	Mode	Type	Function	
1	A	Single system	ONAX A1	Position (aim)	Set gold reference level
2	A	Single system	XOVR AA	Pos (splay/space)	Set for min. variance @ ONAX A1, XOVR AA, ONAX A2
3	A	Single system	OFFAX A	Position (aim)	Set for min. variance between OFFAX A & ONAX A1
4	A	Single system	ONAX A1 & A2	Calibrate	A solo EQ based on both ONAX positions

FIGURE 14.27
Calibration strategies for fully symmetric systems. The approach for higher quantities can be interpolated from the trends shown here.

Calibration strategies: 3-element array: Asymmetrical model, distance and spacing mix (1A:1B:1C)										
Layout	Speaker				Microphone		Processing			
	Element	Model	Distance	Spacing	Type	Function	Channel	EQ	Level	Delay
					OFFAX A	Position (aim)				
	1	A	a		ONAX A	Calibrate	A	EQ	0 dB	
				AB	XOVR AB	Phase align, position				0 ms
	2	A (or B)	b		ONAX B	Calibrate	B	EQ	0 dB	
				BC	XOVR BC	Phase align, position				0 ms
	3	B (or C)	c		ONAX C	Calibrate	C	EQ	0 dB	
					OFFAX C	Position				

Order of operations					
Step #	Speaker		Microphone		Procedures
	Speaker(s)	Mode	Type	Function	
1	A	Single system	ONAX A	Position (aim)	Set gold reference level
2	A	Single system	OFFAX A	Position (aim)	Adjust for minimum variance @ OFFAX A & ONAX A
3	A	Single system	ONAX A	Calibrate	A solo EQ
4	B	Single system	ONAX B	Calibrate	Set B level to match gold reference A
5	B	Single system	ONAX B	Calibrate	B solo EQ
6	A+B	Combined system	XOVR AB	Pos (splay/space)	Set for min. variance @ ONAX A, XOVR AB, ONAX B
7	A+B	Combined system	XOVR AB	Phase align	Set B delay to match A @ XOVR AB
8	A+B	Combined system	ONAX A & ONAX B	Calibrate	A+B combined EQ
9	C	Single system	ONAX C	Calibrate	Set C level to match gold reference A
10	C	Single system	ONAX C	Calibrate	C solo EQ
11	(A+B)+C	Combined system	XOVR BC	Pos (splay/space)	Set for min. variance @ ONAX B, XOVR BC, ONAX C
12	(A+B)+C	Combined system	XOVR BC	Phase align	Set C delay to match B @ XOVR BC
13	(A+B)+C	Combined system	ONAX A, B, C	Calibrate	(A+B)+C combined EQ
14	(A+B)+C	Combined system	OFFAX C	Position	Verify coverage edge C. If fail then D speaker needed

FIGURE 14.28
Calibration subdivision strategies for three elements in the ABC configuration. The sequence can be continued indefinitely.

14.6 PRACTICAL APPLICATIONS

The calibration of even the most complex sound system design can be broken down into the series of procedures and order of operations. Step by step, subsystem by subsystem, the pieces are stitched together like quilt sections into a single fabric. Figs 14.29 and 14.30 show representative examples of the process.

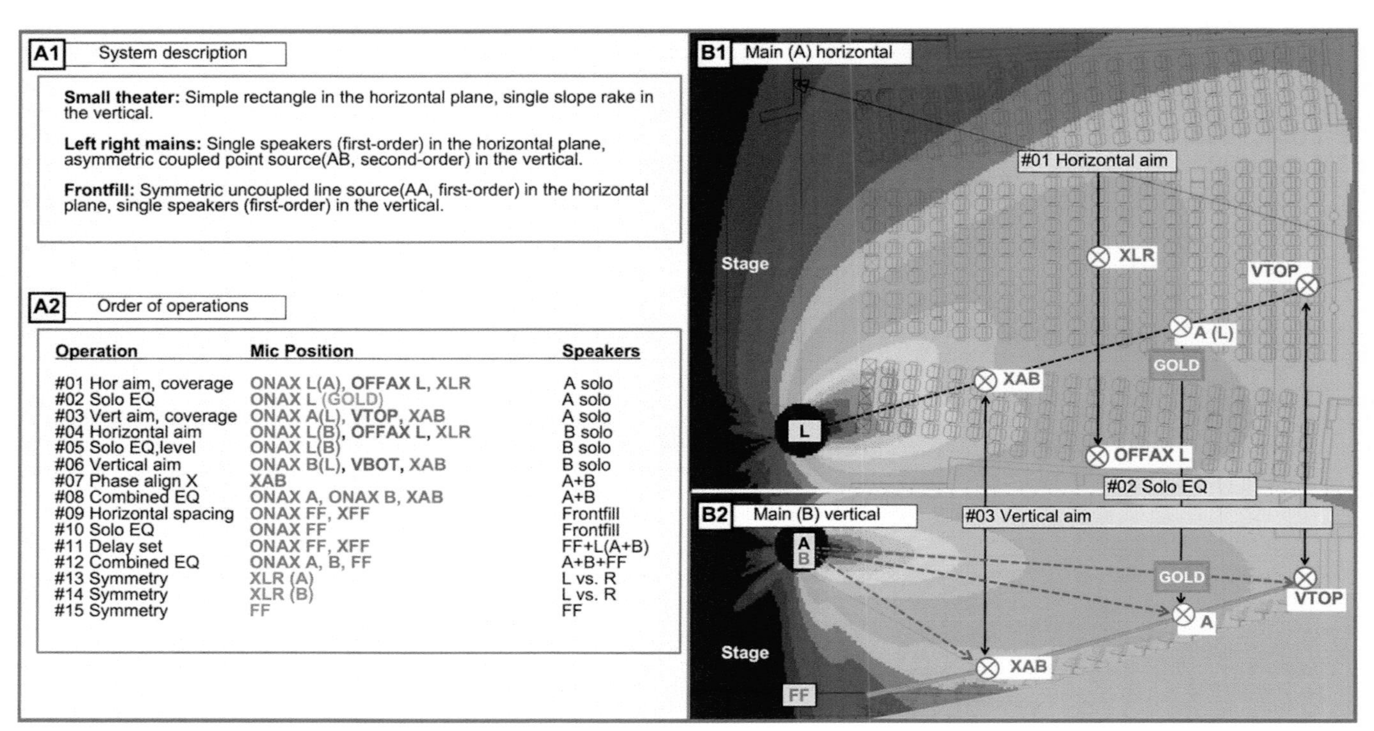

Operation	Mic Position	Speakers
#01 Hor aim, coverage	ONAX L(A), OFFAX L, XLR	A solo
#02 Solo EQ	ONAX L (GOLD)	A solo
#03 Vert aim, coverage	ONAX A(L), VTOP, XAB	A solo
#04 Horizontal aim	ONAX L(B), OFFAX L, XLR	B solo
#05 Solo EQ,level	ONAX L(B)	B solo
#06 Vertical aim	ONAX B(L), VBOT, XAB	B solo
#07 Phase align X	XAB	A+B
#08 Combined EQ	ONAX A, ONAX B, XAB	A+B
#09 Horizontal spacing	ONAX FF, XFF	Frontfill
#10 Solo EQ	ONAX FF	Frontfill
#11 Delay set	ONAX FF, XFF	FF+L(A+B)
#12 Combined EQ	ONAX A, B, FF	A+B+FF
#13 Symmetry	XLR (A)	L vs. R
#14 Symmetry	XLR (B)	L vs. R
#15 Symmetry	FF	FF

FIGURE 14.29
Order of operations applied to an example system

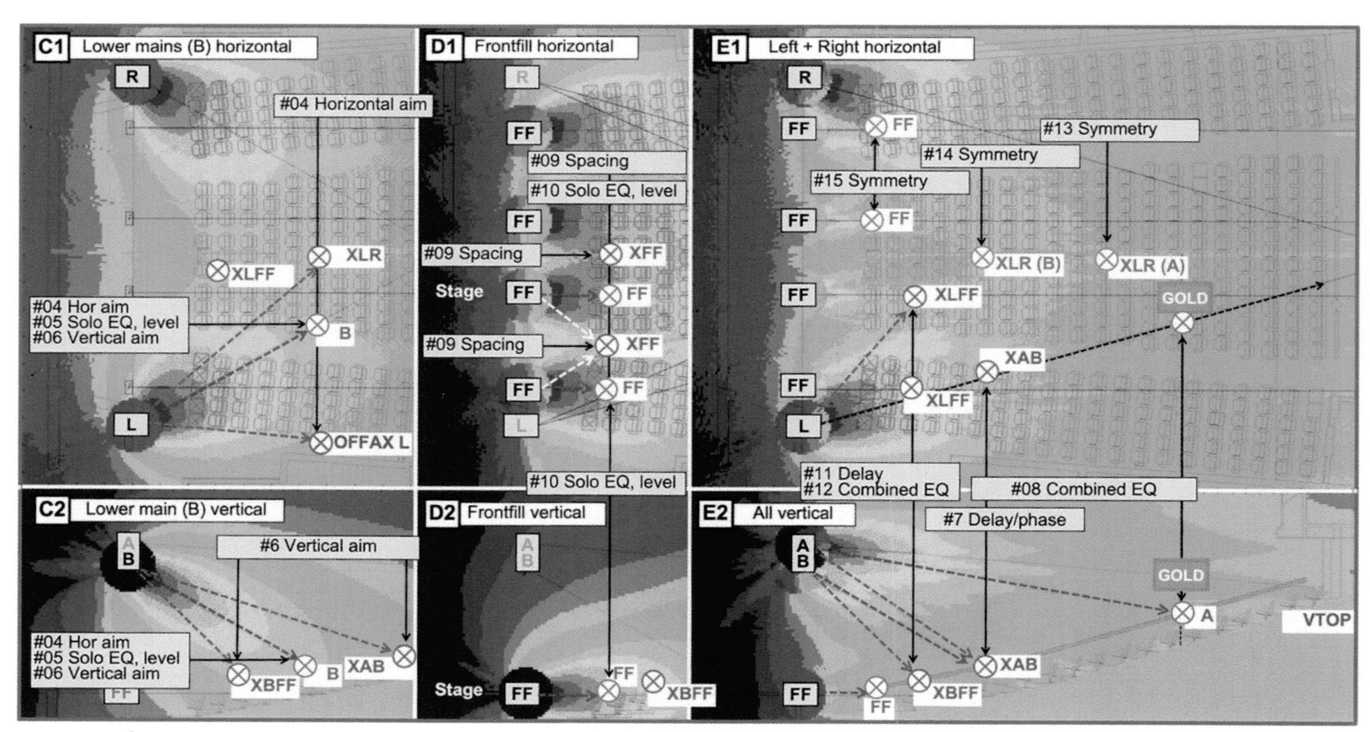

FIGURE 14.30
Order of operations applied to an example system continued

14.7 CALIBRATION PROCEDURE COOKBOOK

Procedure 14.1: Aiming a solo element (horizontal)

Goal: Solo speaker aimed for minimum variance in the horizontal plane (e.g. solo main, uncoupled solo element) (Fig. 14.31).

Procedure 14.1: Aim a single speaker (horizontal)
Speaker elements: a solo
Mic #1 ($OFFAX_1$): last seat on the side (mid-point depth)
Mic #2 (ONAX A): middle seat (mid-point depth)
Mic #3 ($OFFAX_2$) (optional if symmetric): last seat opposite side (mid-point depth)
Procedure: 1. Define the intended horizontal coverage limits and depth, and place mics as above. **Note**: Do not count the distance *before* coverage starts when calculating average coverage depth (e.g. the front rows that are underneath the vertical coverage). If covering rows 3–23, the middle row is 13. 2. @ONAX: solo EQ (not required, but helps finding the off-axis edge). 3. Compare the ONAX response to $OFFAX_1$ and $OFFAX_2$. 4. Aim speaker to minimize variance between ONAX, $OFFAX_1$ and $OFFAX_2$.
Outcomes and actions:
If $OFFAX_1$ > $OFFAX_2$ then aim speaker more toward $OFFAX_2$
If $OFFAX_1$ < $OFFAX_2$ then aim speaker more toward $OFFAX_1$
PASS/FAIL: Fill speaker required if $OFFAX_1$–ONAX–$OFFAX_2$ variance >6 dB

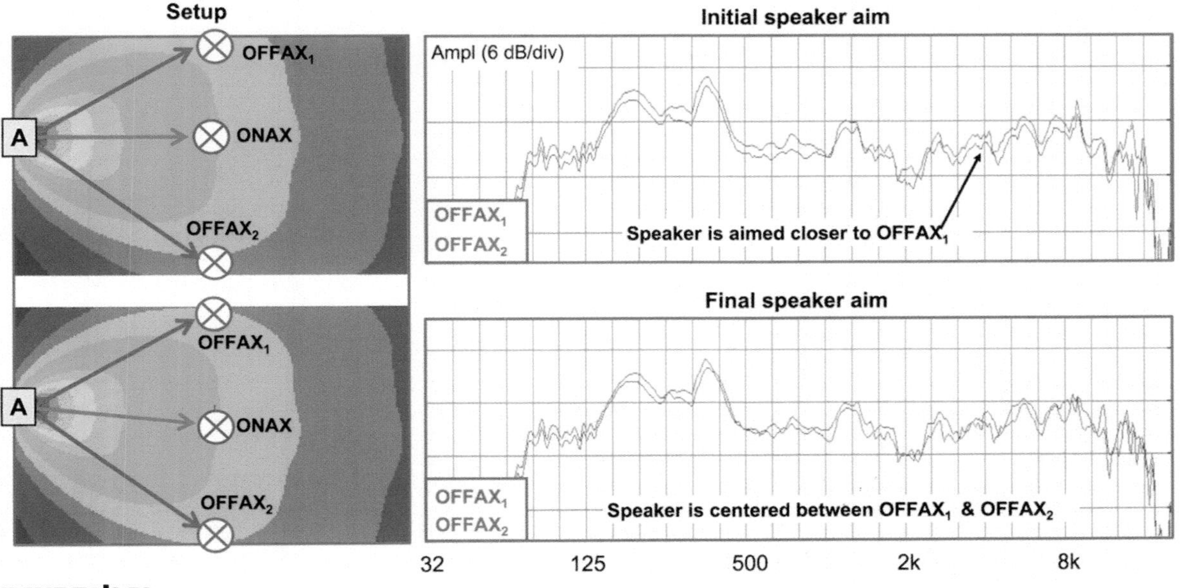

FIGURE 14.31
Single speaker aim (horizontal)

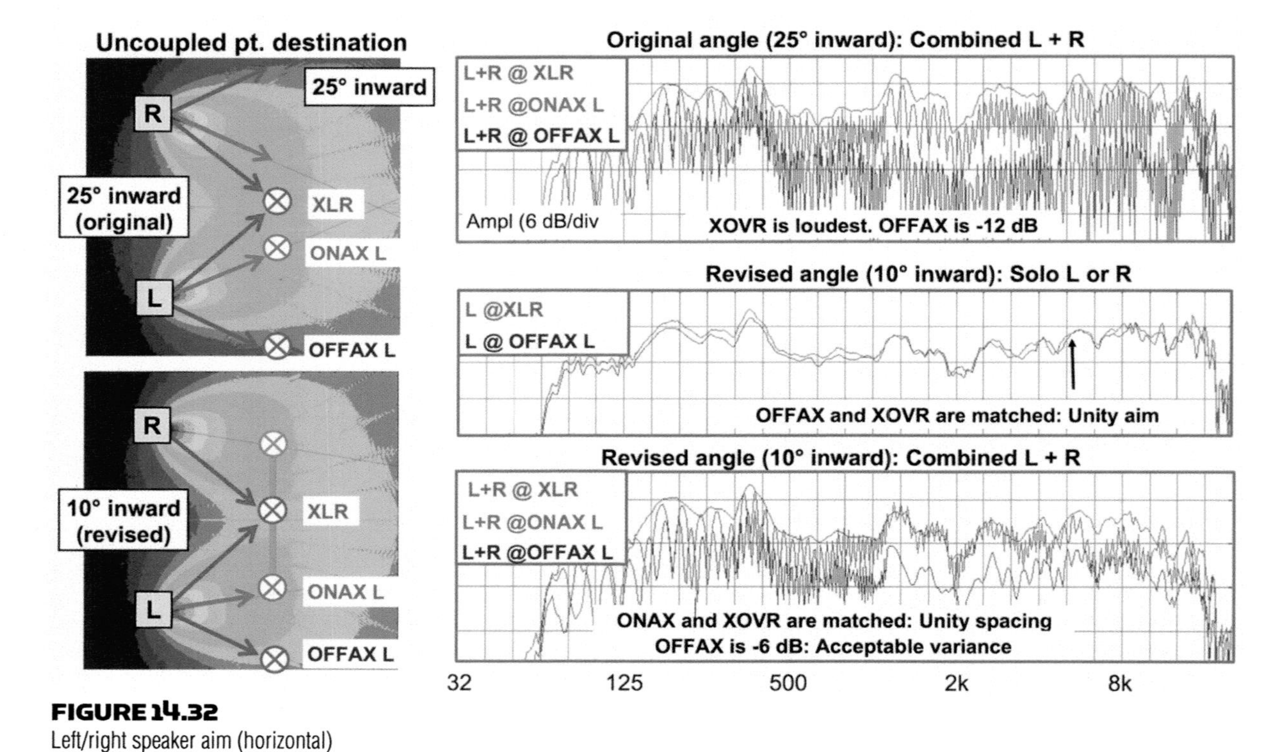

FIGURE 14.32
Left/right speaker aim (horizontal)

Procedure 14.2: Aiming the left (or right) mains (horizontal)

Goal: Left main is aimed for minimum variance in the horizontal plane (same for right).

The procedure is similar to procedure 14.1 (aiming a single speaker) but the room center is considered a virtual wall (Fig. 14.32).

Procedure 14.2: Aim the left main speaker (horizontal)
Speaker elements: left solo
Mic #1 (OFFAX L): last seat on the left side (mid-point depth)
Mic #2 (ONAX L): middle seat between center and outermost left (mid-point depth)
Mic #3 (XLR): room center (mid-point depth). XOVR for left/right
Procedure: 1. Define L/R mains coverage depth (same method as 14.1) and place mics. 2. @ONAX L: solo EQ (not required, but helps finding the off-axis edge). 3. Compare the ONAX L response to OFFAX L and XLR. 4. Aim speaker to minimize variance between ONAX L, OFFAX L and XLR.
Outcomes and actions:
If OFFAX L > XLR then aim speaker toward XLR (inward)
If OFFAX L < XLR then aim speaker toward OFFAX L (outward)
PASS/FAIL: Fill speaker required if OFFAXL–ONAX–XLR variance >6 dB

Procedure 14.3: Aiming a solo element (vertical)

Goal: Solo speaker vertically aimed for minimum variance (e.g. solo main, surround) (Fig. 14.33).

Procedure 14.3: Aim a single speaker (vertical)
Speaker elements: a solo
Mic #1: (VTOP), uppermost coverage area (on axis horizontally)
Mic #2: (ONAX A), mid-point depth (on axis horizontally)
Mic #3: (VBOT), bottom of coverage (on axis horizontally)
Procedure: 1. Define the intended vertical coverage limits and place mics (e.g. from the third-row frontfill XOVR to the twenty-third-row overbalcony delay XOVR). 2. @ONAX: solo EQ (optional, but eases evaluation of other responses). 3. Compare the ONAX response to VTOP and VBOT. 4. Aim speaker to minimize variance between ONAX, VTOP and VBOT.
Outcomes and actions:
If VTOP > VBOT then aim speaker down
If VTOP < VBOT then aim speaker up
PASS/FAIL: Fill speaker required if VTOP–ONAX–VBOT variance >6 dB

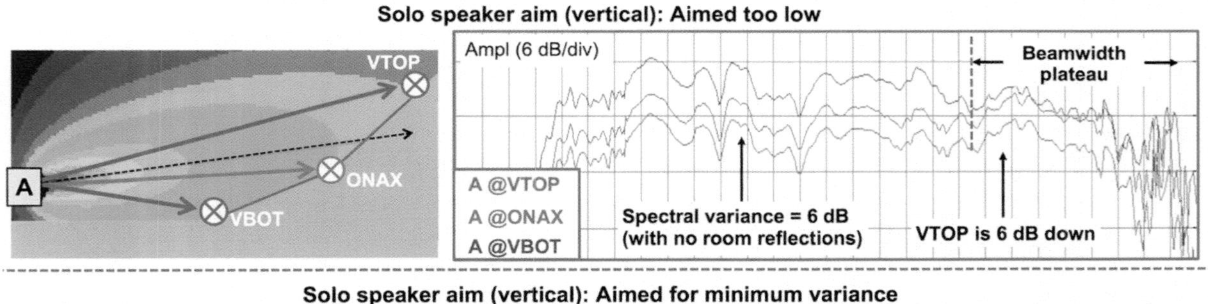

Solo speaker aim (vertical): Aimed for minimum variance

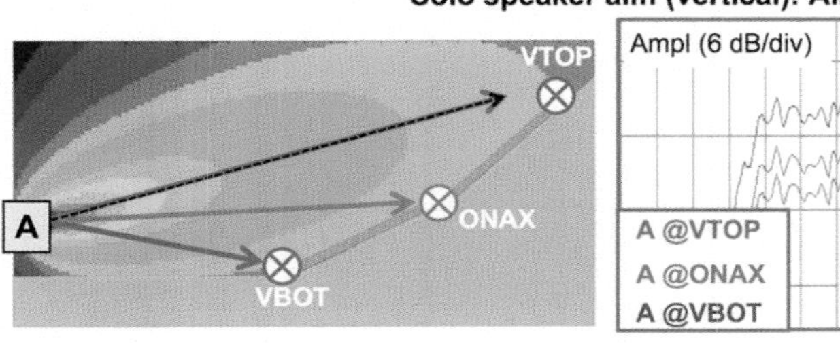

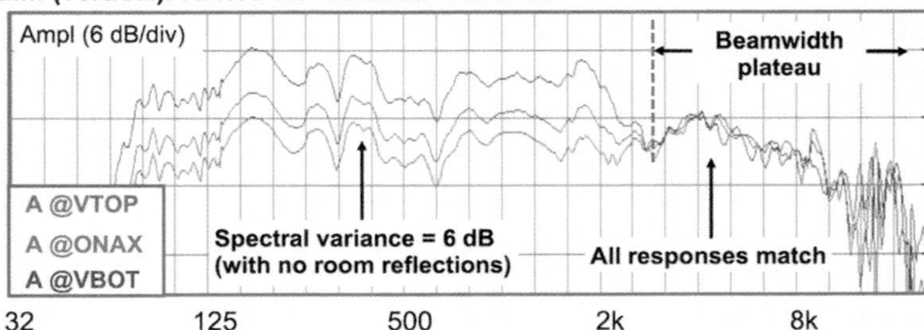

FIGURE 14.33
Single speaker aim (vertical)

Procedure 14.4: Aiming the top element (A) of an AB array (vertical)

Goal: Upper speaker of a coupled point source array is vertically aimed for minimum variance in the upper portion and preparation for connection to the B speaker below it (Fig. 14.34).

Procedure 14.4: Aim the top element of an AB array (vertical)
Speaker elements: speaker A of array AB, or ABC, etc.
Mic #1: (VTOP), uppermost coverage area (on axis horizontally)
Mic #2: (ONAX A), mid-point depth (on axis horizontally)
Mic #3: (XAB), bottom of A coverage, future connection point to B
Procedure: 1. Define the uppermost vertical coverage limit and place mics. 2. @ONAX A: solo EQ (optional, but eases evaluation of other responses). 3. Compare the ONAX response to VTOP. 4. Aim speaker to minimize variance between ONAX and VTOP. 5. Move XAB mic until -6 dB point is found. Will be used for aiming B.
Outcomes and actions:
If VTOP > ONAX A then aim speaker down
If VTOP = ONAX A then minimum variance has been achieved
PASS/FAIL: Delay fill speaker required in rear if VTOP–ONAX variance >6 dB
PASS/FAIL: If ripple variance is too strong from above then maximize ONAX A–VTOP variance (6 dB) by aiming speaker down

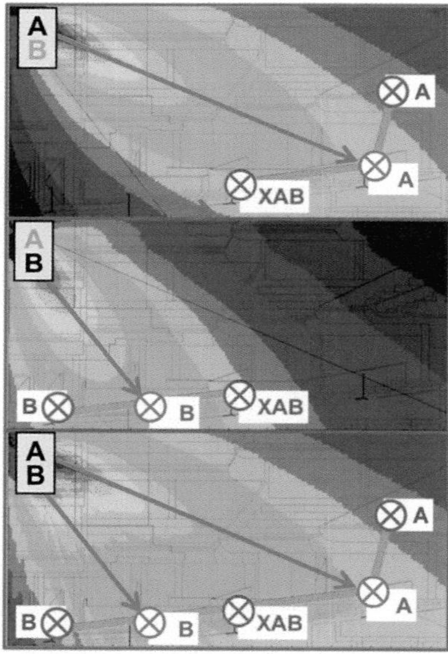

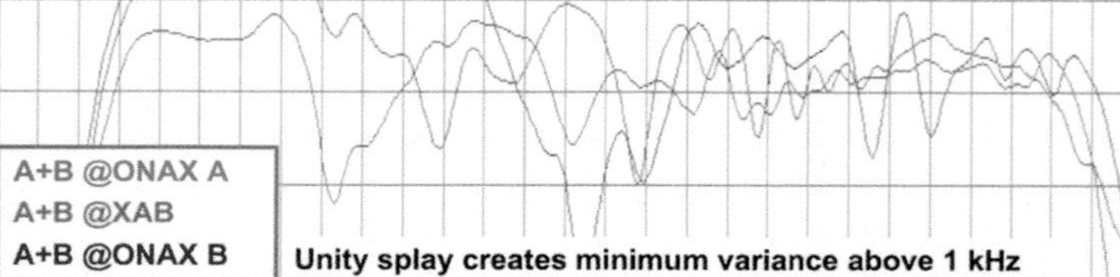

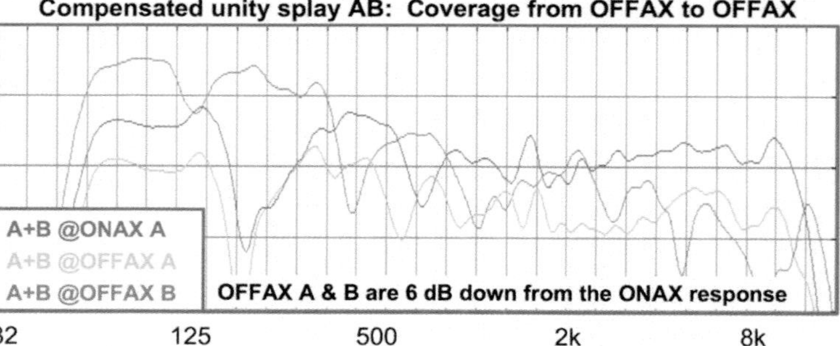

FIGURE 14.34
Compensated unity splay (AB vertical)

Procedure 14.5: Find the unity splay angle (AA symmetric)

Goal: Find the unity splay angle for minimum variance of a symmetric pair, trio, etc. Array elements cover the same depth and are driven at the same level (an $A_1 + A_2$ combination). The procedure assumes the horizontal plane but works for either. The goal is a radial minimum variation line between ONAX A_1, XAA and ONAX A_2.

It's not always practical to solo the individual elements because they may be wired in parallel etc. In such cases the responses are only seen in combined form. Both are covered here (Fig. 14.35).

Procedure 14.5: Find unity splay angle (AA symmetric)
Speaker elements: A_1 and A_2 (matched) speakers
Mic #1 (ONAX A_1): on axis to A_1 at mid-point depth
Mic #2 (XAA): radial mid-point between A_1 and A_2 (equidistant to ONAX A_1)
Mic #3 (ONAX A_2): on axis to A_2 at mid-point depth (can be a SYM verification)
Procedure (if individual elements can be muted): 1. Drive A_1 solo. 2. @ONAX A_1: A_1 solo EQ (optional, but eases evaluation). 3. @XAA: Move mic until A_1 solo @XAA is -6 dB from A_1 solo @ONAX A_1. 4. @XAA: Mute A_1 and drive A_2 solo. 5. @XAA: Adjust splay until A_2 solo = A_1 solo @XAA (-6 dB re. ONAX A_1). **Procedure (if individual muting is not possible)**: 1. All speakers driven. Place mics as described above. 2. @ONAX A_1: solo EQ A_1A_2 (speakers are combined but 1 channel of EQ). 3. @XAA: Compare speakers A_1A_2 to A_1A_2 @ONAX A_1. 4. @XAA: Adjust splay until A_1A_2 @XAA = A_1A_2 @ONAX A_1. Note: Gross adjustments may require updating mic positions. Keep XAA, ONAX A_1 and A_2 on their centers to maintain accuracy.
Outcomes and actions:
If XAA > ONAX A then increase splay
If XAA < ONAX A then decrease splay
PASS/FAIL: If splay is too narrow but overall array is too wide and hits reflective walls, then compromise is required (a ripple variance tradeoff). Reduce splay until the array overlap is proportional to room/speaker overlap
PASS/FAIL: If splay is unity but overall array is too narrow (OFFAX A is down more then 6 dB) then fills are required

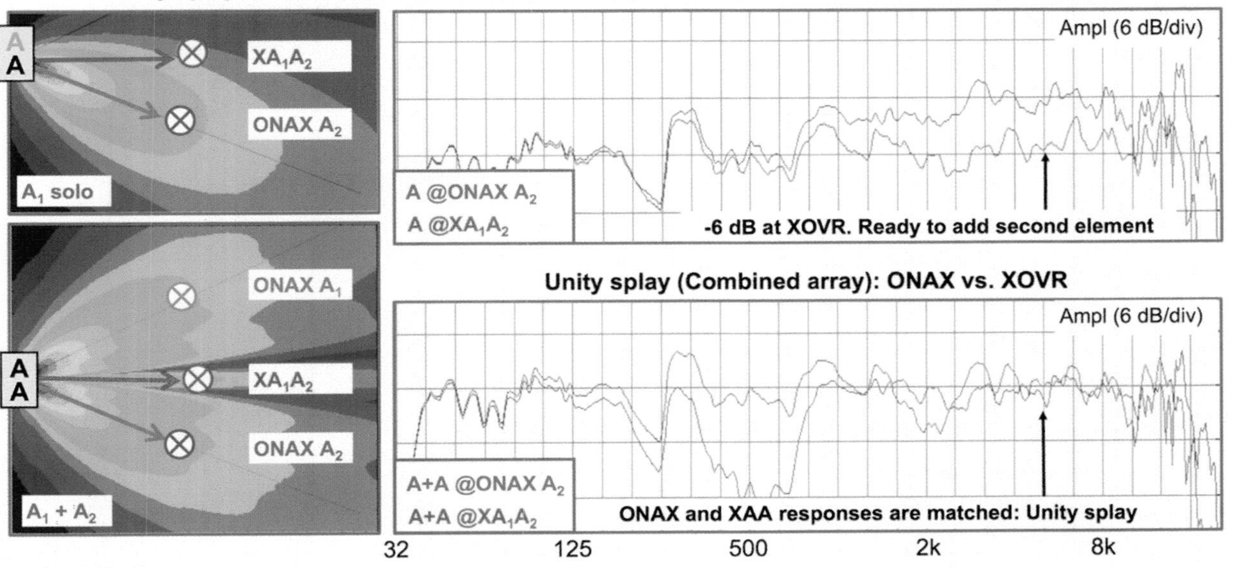

FIGURE 14.35
Unity splay (AA symmetric)

Procedure 14.6: Find the unity splay angle (AB asymmetric)

Goal: Find the compensated unity splay angle for an asymmetric pair, trio, etc. Array elements cover different depths and are driven at different levels (A + B combination). The procedure assumes the horizontal plane but works for either. The splay angle goal is a minimum variation line between ONAX A, XAB and ONAX B (Fig. 14.36).

Procedure 14.6: Find unity splay angle (AB asymmetric)
Speaker elements: A and B (A is the longer throw system)
Processing: channels A and B for EQ, level and delay
Mic #1 (ONAX A): on axis to A at mid-point depth
Mic #2 (XAB): mid-point between A and B (exact location TBD)
Mic #3 (ONAX B): on axis to B at mid-point depth of its coverage
Procedure (if individual elements can be muted): 1. Prerequisite: A and B are level set and EQ'd to match in their zones. 2. @ONAX A: Measure A solo for reference. 3. @XAB: Compare A solo here to A solo @ONAX A. 4. @XAB: Move mic until A solo response = -6 dB. 5. @XAB: Mute speaker A and drive speaker B solo. 6. @XAB: Adjust splay until B = A solo response (both -6 dB re. ONAX A). 7. @XAB: Delay the B speaker (if needed) to phase align B to A.
Outcomes and actions:
PASS: If XAB = ONAX A then splay is optimized
FAIL: If XAB > ONAX A (or B) then increase splay. If less then decrease splay
PASS/FAIL: Ripple variance tradeoffs as described above in the symmetric version

Compensated unity splay AB: B is -3 dB (2 x 80° @ 56°)

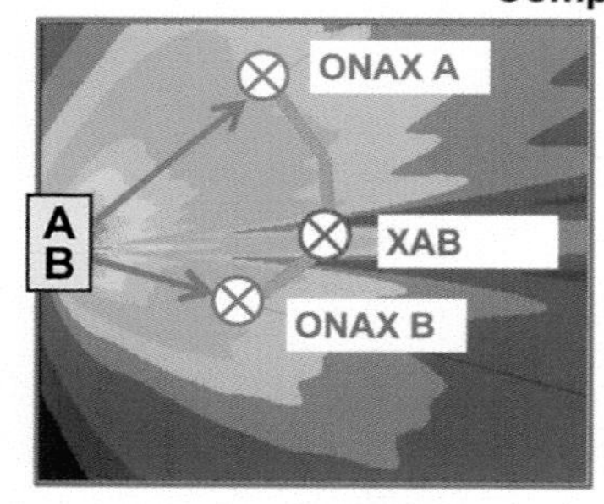

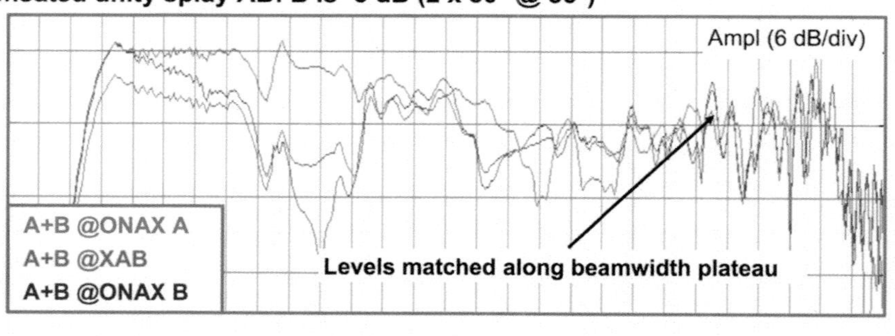

Uncompensated unity splay AB: B is -3 dB (2 x 80° @ 80°)

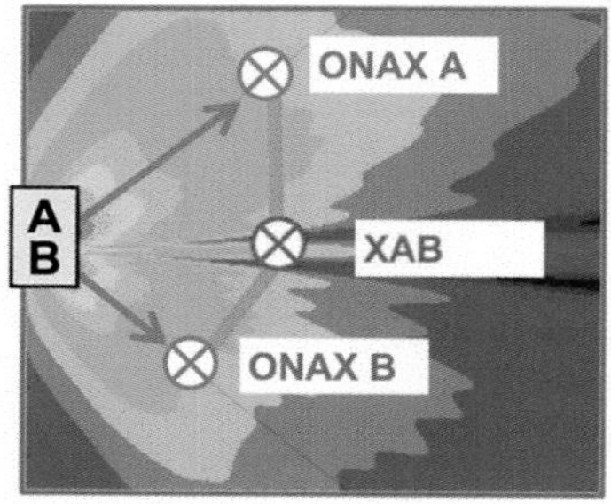

Note: XAB mic moves closer

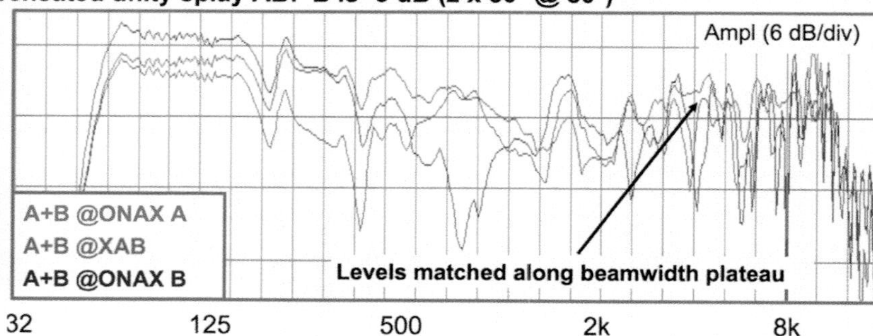

FIGURE 14.36
Compensated unity splay (AB horizontal)

Procedure 14.7: Unity gain uncoupled array spacing (AA symmetric)

Goal: Find the unity spacing for minimum variance of a symmetric pair, trio, etc. Array elements cover the same depth and are driven at the same level (an $A_1 + A_2$ combination). The procedure assumes the horizontal plane but works for either. The goal is a linear minimum variation line between ONAX A_1, XAA and ONAX A_2.

This procedure is functionally analogous to the symmetric unity splay (procedure 14.5), with linear spacing providing the isolation instead of splay angle. Differences are shown here for brevity (Fig. 14.37).

Procedure 14.7: Find unity spacing (AA symmetric)
Speaker elements: A_1 and A_2 (matched) speakers
Mic #1 (ONAX A_1): on axis to A_1 at intended unity coverage line
Mic #2 (XAA): linear mid-point between A_1 and A_2 (unity line depth)
Mic #3 (ONAX A_2): on axis to A_2 at unity line depth (can be a SYM verification)
Procedure (if individual elements can be muted): 1. Follow steps 1–5 of procedure 14.5 (unity splay AA, solo muting). 2. @XAA: Adjust spacing until $A_2 = A_1$ solo @XAA (-6 dB re. ONAX A_1). Procedure (if individual muting is not possible): 1. Follow steps 1–3 of procedure 14.5 (unity splay AA, no solo muting). 2. @XAA: Adjust spacing until $A_1 + A_2$ @XAA = $A_1 + A_2$ @ONAX A_1. **Note**: Gross adjustments require updated mic positions. Keep XAA, ONAX A_1 and A_2 on their centers for best accuracy.
Outcomes and actions:
PASS: If XAA = ONAX A then spacing is optimized
FAIL: If XAA > ONAX A then increase spacing. If < ONAX A then decrease spacing
PASS/FAIL: If spacing is unity but overall array is too narrow (OFFAX A is down more then 6 dB) then more elements are required

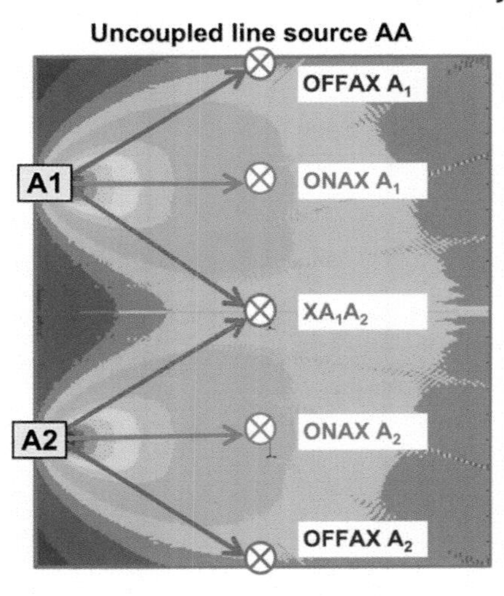

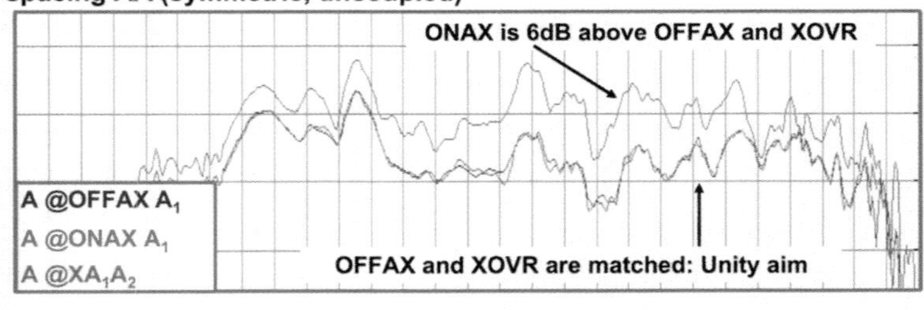

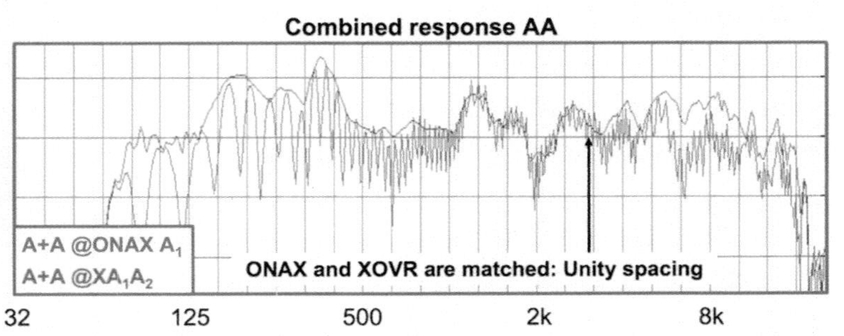

FIGURE 14.37
Unity spacing for the symmetric uncoupled line source (AA)

Procedure 14.8: Find the compensated unity spacing (AB asymmetric)

Goal: Find the compensated unity spacing for an asymmetric pair, trio, etc. The array elements cover the different depths and are driven at different levels (an A + B combination). The procedure works for either plane. The spacing goal is a custom minimum-variance line between ONAX A, XAB and ONAX B (Fig. 14.38).

Procedure 14.8: Find unity spacing (AB asymmetric)
Speaker elements: A and B (A is the longer throw system)
Processing: channels A and B for EQ, level and delay
Mic #1 (ONAX A): on axis to A at mid-point depth
Mic #2 (XAB): mid-point between A and B (exact location TBD)
Mic #3 (ONAX B): on axis to B at mid-point depth
Procedure: 1. Follow steps 1–3 of procedure 14.7 (compensated unity splay AB). 2. @XAB: Adjust spacing until B solo = A solo @XAB (-6 dB re. ONAX A). 3. @XAB: Delay the B speaker (if needed) to phase align B to A.
Outcomes and actions:
PASS: If XAB = ONAX A then spacing is optimized
FAIL: If XAB > ONAX A then increase spacing. If < ONAX A then decrease spacing

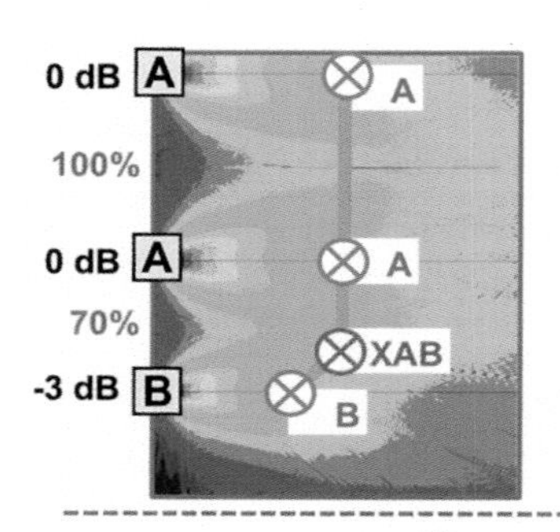

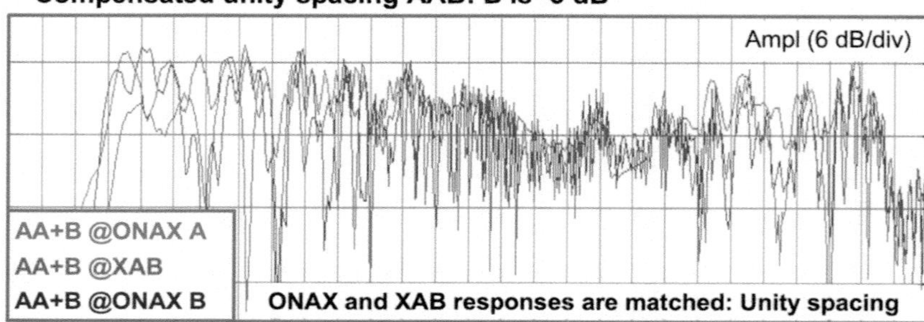

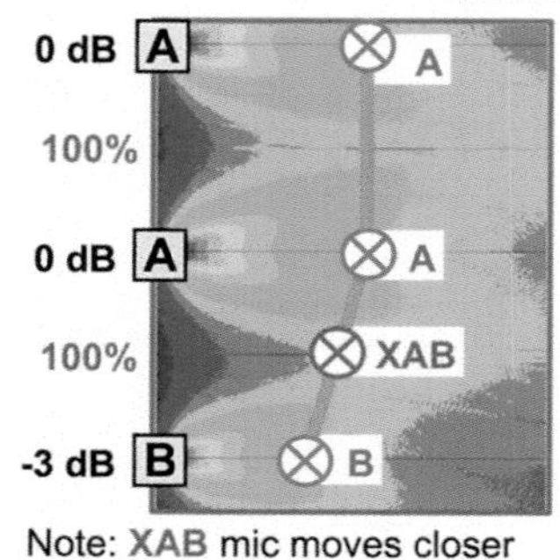

FIGURE 14.38
Compensated unity spacing for the asymmetric uncoupled line source (AB)

Procedure 14.9: Unity spectral crossover: main + sub (A + S)

Goal: Join a main element(s) and subwoofer(s). Assumed that elements would overlap in frequency if left unfiltered. Responses are filtered to create a unity gain combination in the crossover region. Mic location is usually ONAX to mains. Ground plane placement helps clarify the phase response without compromising the procedure (Fig. 14.39).

Procedure 14.9: Unity spectral crossover (A + S)
Speakers elements: Main (A, HF) + Subs (S, LF)
Signal processing: LF and HF channels for LPF, HPF, level and delay
Mic #1: ONAX, mid-point depth, on axis horizontally to mains. Can be ground plane
Procedure: 1. Prerequisite: Solo EQ is complete on both LF and main systems. If crossover frequency is already decided then go to step 5. 2. Drive main (A) solo. Determine approximate LF cutoff frequency (where the response starts to slope downward). 3. Drive subwoofer (S) solo. Determine approximate HF cutoff frequency. 4. Select crossover frequency (F_X), which must be inside the overlapped frequency range, preferably near the mid-point (e.g. if overlapping from 60–120 Hz then the mid-point frequency would be around 90 Hz. 5. Adjust levels to match individual systems in the crossover frequency range. 6. Drive main (A) solo and adjust processor HPF until F_X response is -6 dB. 7. Drive subs (S) solo and adjust the processor LPF until F_X response is -6 dB. 8. Phase align the crossover. Determine which arrives first (the one with less downward phase slope by comparing the solo phase responses). 9. Delay the leading system until phase matched in the crossover range. 10. Combine systems together and verify coupling.
Outcomes and actions:
PASS: Crossover is optimized when combination adds to unity
FAIL: If combination rises above unity then corner frequencies are too close together. Lower the LPF corner frequency and/or raise it on the HPF, then re-align phase. If below unity then vice versa.

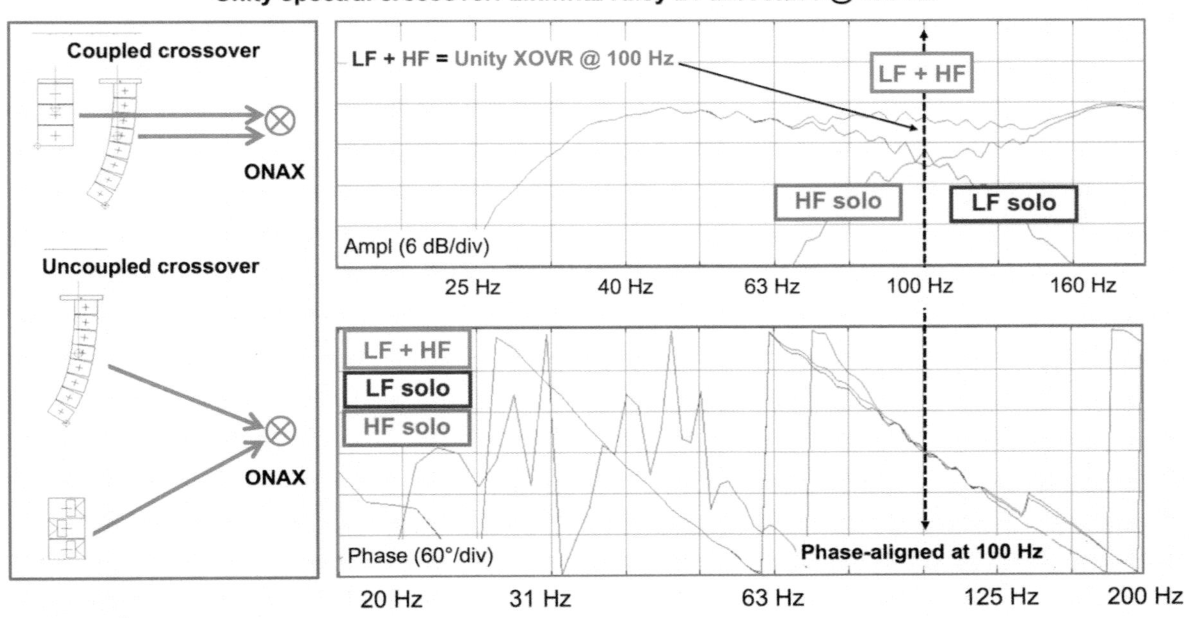

FIGURE 14.39
Unity spectral crossover (LF + Mains @100 Hz)

Procedure 14.10: Overlap spectral crossover: main + sub (A + S)

Goal: Join a main element(s) and subwoofer(s) with overlap in the crossover range. Most aspects of the previous main/sub combination apply (Fig. 14.40).

Procedure 14.10: Overlap spectral crossover (A + S)
Procedure: 1. Perform steps 1–5 of the unity crossover procedure (14.9). 2. Phase align the crossover. Determine which arrives first (the one with less downward phase slope by comparing the solo phase responses). 3. Combine systems and observe the approximately 6 dB peak at F_X. The bandwidth depends on the size of the overlap range. 4. Add combined EQ to both channels to flatten the peak (6 dB cut at F_X).
Outcomes and actions:
PASS: Crossover is optimized when equalized combination adds to unity
FAIL: If combination has peaks off center from F_X then EQ filter bandwidth is too narrow. If combination has dips, then BW is too wide

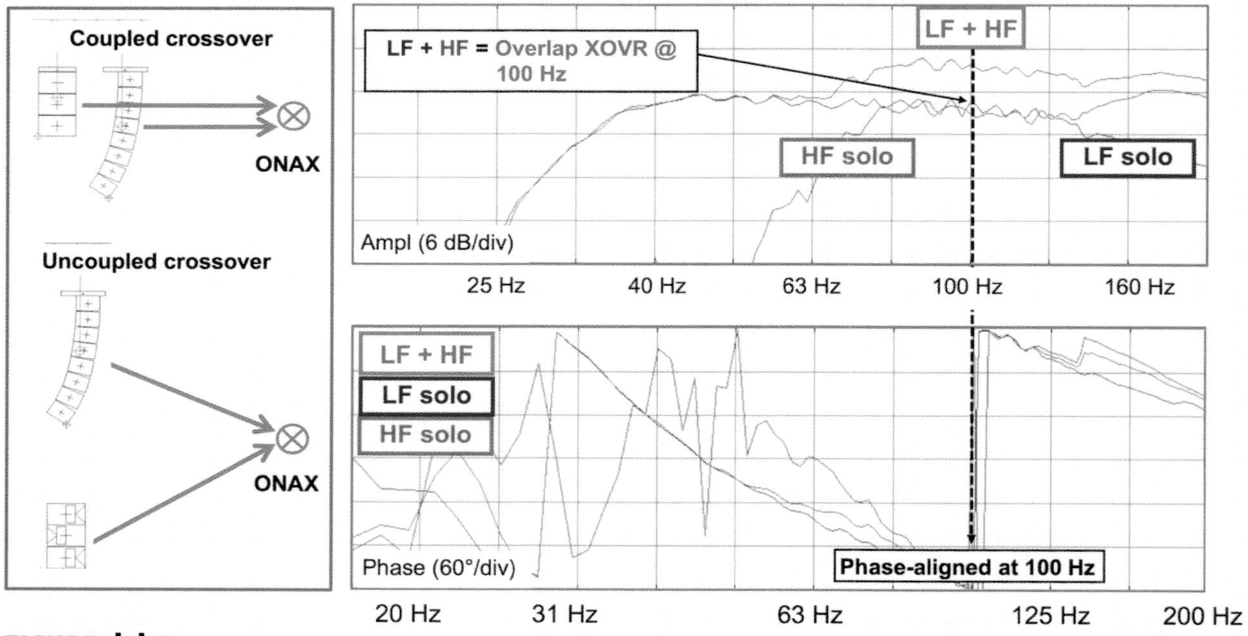

FIGURE 14.40
Overlap spectral crossover (LF + mains @100 Hz)

Procedure 14.11: Relative-level setting of isolated systems (A vs. B)

Goal: Set the unity level for the lesser system (B) of an asymmetric pair. The B element covers a different depth and/or may be a different model or quantity, and therefore may need drive-level adjustment to scale the level with the A system. The procedure works for either plane. The level-setting goal is a unity link between ONAX A and ONAX B (Fig. 14.41).

Procedure 14.11: Level setting of isolated systems
Speaker elements: A and B (B is the minority system)
Processing: channels A and B for level
Mic #1 (ONAX A): on axis to A at mid-point depth
Mic #2 (ONAX B): on axis to B at mid-point depth
Procedure: 1. @ONAX A: solo EQ speaker A. 2. @ONAX B: solo EQ speaker B. 3. Compare B solo response to A solo response. 4. @ONAX B: adjust level until B solo @ONAX B = A solo @ONAX A.
Outcomes and actions:
PASS: Level is optimized when ONAX A (A solo) = ONAX B (B solo)
FAIL: If processor levels appear wrongly scaled for distances/devices then check for wiring errors (unbalanced lines) or physical issues (blockage, poor aiming, etc.)

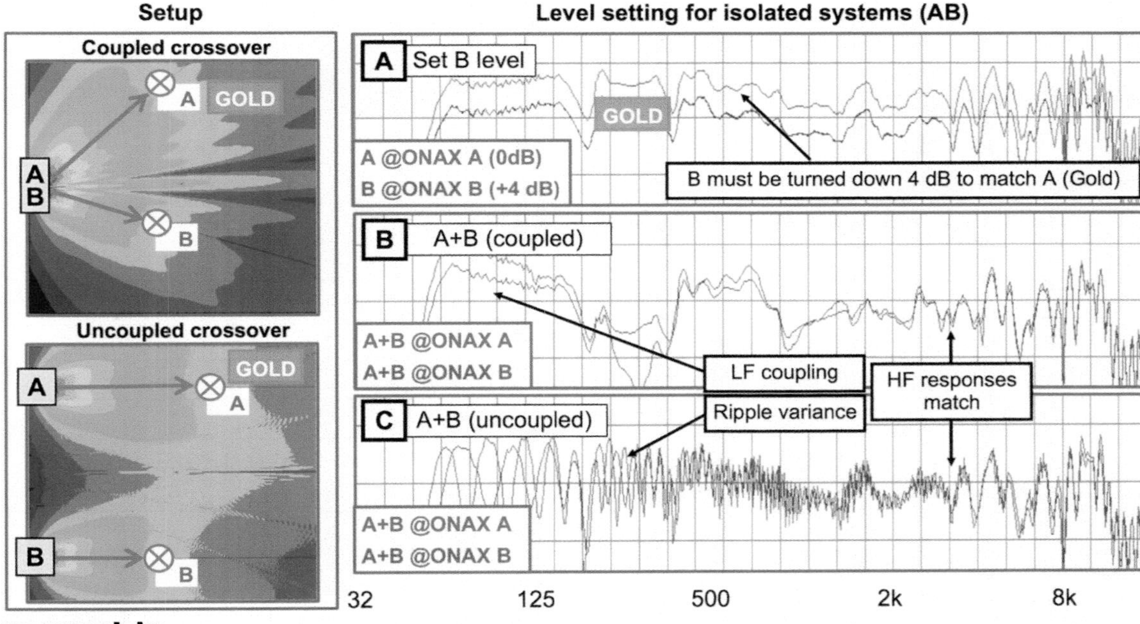

FIGURE 14.41
Level setting of isolated systems (AB)

Procedure 14.12: Relative-level setting of non-isolated systems (A vs. B)

Goal: Set the unity level for a fill system (B) that supplements coverage of a dominant system. The goal is that the combined level of A + B in the fill area (ONAX B) will equal the solo response of speaker A at ONAX A. ONAX B functions also as XAB because of the high overlap. This contrasts with the isolated approach (procedure 14.11 above) where A and B each reached unity as soloists at their ONAX locations. The combined response of A + B has very little effect in the ONAX A area and a very large effect in the ONAX B/XAB area (e.g. we expect to hear the mains under the balcony but not the underbalcony speakers out in the center of the main system's coverage). This procedure works for either plane (Fig. 14.42).

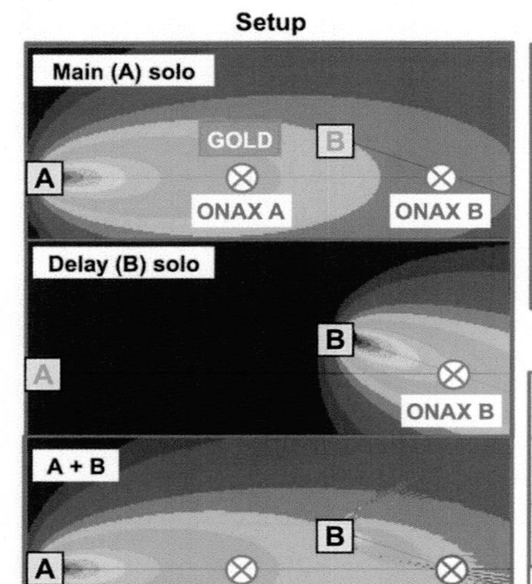

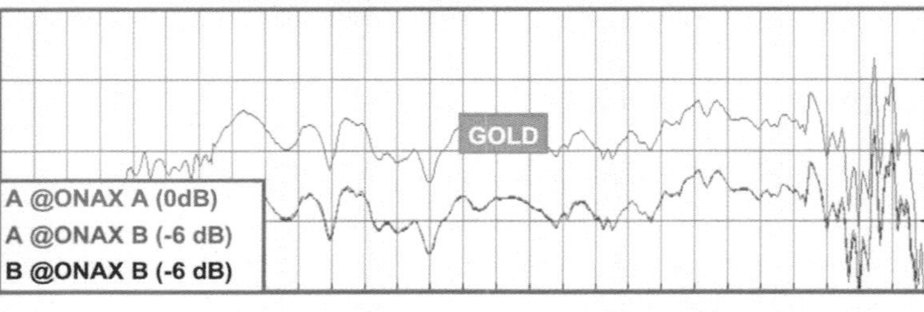

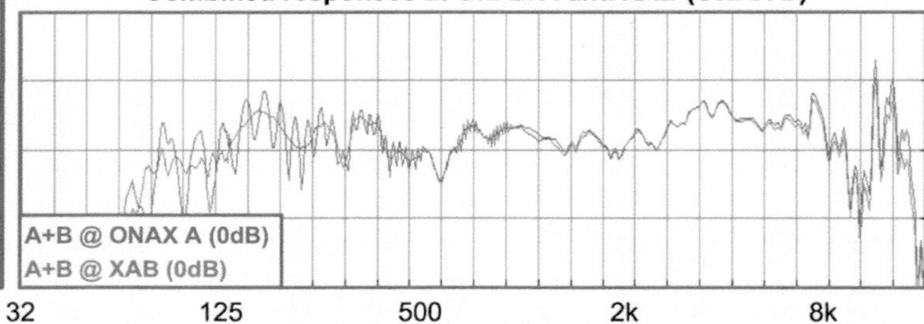

FIGURE 14.42
Level setting for non-isolated (overlapping) systems

Procedure 14.12: Level setting of overlapped systems
Speaker elements: A and B (B is the fill system)
Processing: channels B for level and delay (A level is already set)
Mic #1 (ONAX A): on axis to A at mid-point depth (for reference only)
Mic #2 (ONAX B): on axis to B at mid-point depth (also functions as XAB)
Procedure:
1. Prerequisite 1: Main system (A) has been previously level set and EQ'd.
2. Prerequisite 2: If the fill system is an uncoupled array, the element spacing/splay has already been set.
3. @ONAX B: Measure B solo, EQ. Set level to match A solo reference.
4. @ONAX B: Compare A solo to A solo @ONAX A. This tells us how much fill we need (e.g. -6 dB from A needs -6 dB from B to reach 0 dB, -3 dB from A needs -10 dB from B etc.)
5. @ONAX B: Set delay on speaker B to synchronize A and B (procedure 14.13).
6. @ONAX B: Combine A + B.
7. @ONAX B: Adjust B level until A + B @ONAX B = A solo @ONAX A.
Outcomes and actions:
PASS: Level is optimized when ONAX A (A solo) = ONAX B (A + B combined)
PASS/FAIL: Coherence response @ONAX B (A + B) should be a large improvement from A solo there. Should be comparable to coherence @ONAX A

Procedure 14.13: Delay setting for the spatial crossover (A + B)

Goal: Synchronize arrivals of two system elements covering approximately the same frequency range. Applicable on multiple levels, from individual speakers to large arrays. Intended to create the most efficient and least detectable transition between the main (A) and fill (B) systems at their crossover (Fig. 14.43).

Procedure 14.13: Delay setting of spatial crossovers
Speaker elements: A and B (A is the longer throw system)
Processing: channels A and B for delay
Mic #1 (XAB): mid-point between A and B (exact location TBD)
Procedure: 1. Prerequisite: A and B are level matched and EQ'd in their respective zones. 2. @XAB: Measure the A solo impulse response. The peak shows the arrival of the bulk of the HF range. This is the "0 ms" relative target response. 3. @XAB: Measure B solo. Compare the B and A arrival times here. 4. @XAB: Adjust the B delay to match B solo impulse to A solo. 5. @XAB: Measure the A and B solo frequency responses. 6. @XAB: Combine A + B and compare with the A and B solo responses. Combined response should increase +6 dB over the soloists. 7. @XAB: Fine adjust delay the B speaker (if needed) to phase align B to A.
Outcomes and actions:
PASS: Delay is optimized when the arrivals match and coupling extends full range
PASS/FAIL: If the phase responses do not match over the full range (e.g. different speaker models) then a single delay time cannot sync all frequencies

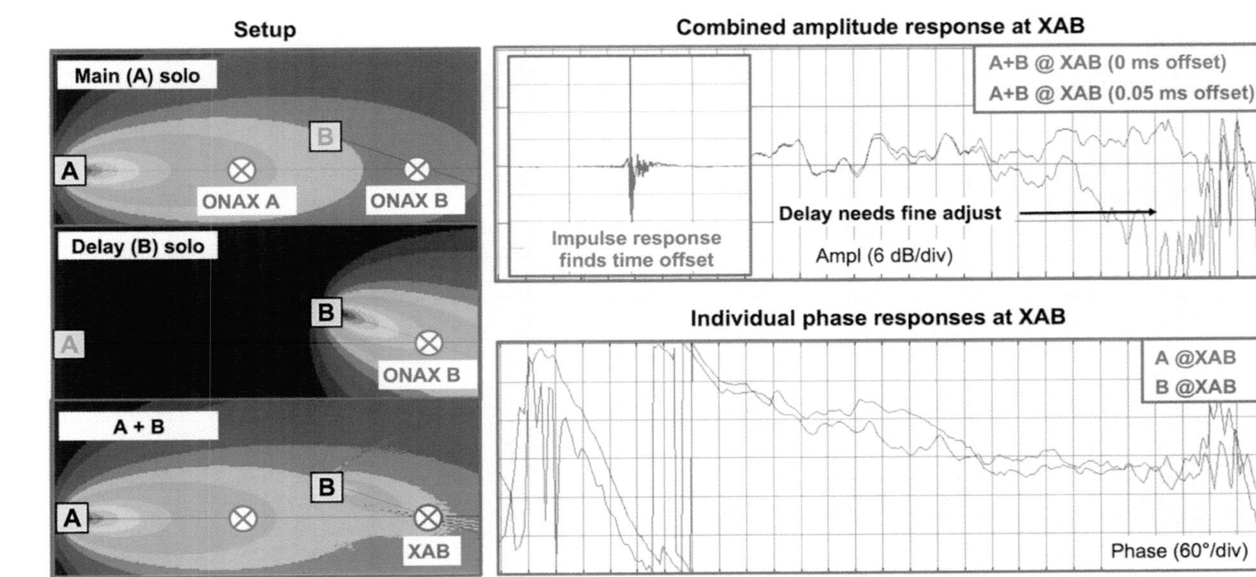

FIGURE 14.43
Delay setting for spatial crossover alignment

Procedure 14.14: Phase align a spectral crossover (HF + LF)

Goal: Synchronize a particular frequency range shared by two elements There are two basic versions that differ in element spacing and mic placement.

The "coupled" spectral crossover combines elements within the same enclosure, or close proximity. This can be repeated for three-way (or four-way) systems. Goal is the most efficient and least detectable transition between the (HF) and (LF) elements at their F_X. Mic position should be near field (1–2 meter depth typical). Mic is located at the mid-point between the drivers in the plane in which the drivers are offset and on axis to the shared plane. Minimize reflections as they complicate the phase alignment process. If not possible to get a near-field position, the closest ONAX position is the best alternative.

The uncoupled version synchronizes a frequency range shared by two elements in different enclosures. The primary application for modern systems is subwoofers separated from the mains. A common example from ancient times is the flying junkyard of horns separated from woofer(s). Our goal is to centralize the synchronization in the room and thereby distribute the errors most evenly over the space. Therefore the mic is placed in the far field (mid-point depth is typical). We seek to distribute the errors as evenly as possible because an uncoupled crossover cannot remain synchronized over the whole room. It helps to minimize reflections in the measurement as they complicate the phase alignment process. Ground plane placement can be a good option for subwoofer to mains alignment (Fig. 14.44).

Procedure 14.14: Phase align a coupled crossover
Speaker elements: LF and HF (or MF and HF etc.)
Processing: LF and HF channels for delay
Mic #1 (ONAX): coupled version—mid-point between HF and LF (near field)
Mic #1 (ONAX): uncoupled version—mid-point depth of room
Procedure: 1. Measure the HF solo phase response. Set analyzer internal delay for clear view of phase slope in the acoustic crossover range (F_X). This is the "0°" relative target response. Do not reset the internal delay during this procedure. 2. Measure the LF solo. Compare the LF solo phase to HF solo phase. 3. Determine which channel arrives earlier at F_X (the steeper phase slope). 4. Delay the earlier channel (on processor) to phase match LF and HF solo. 5. Combine LF + HF and compare with the solo responses. Combined response should couple in the crossover and add up to 6 dB over the soloists. 6. Coupled: Move mic in the offset plane to observe response over angle. 7. Uncoupled: Move mic around room to observe response over space.
Outcomes and actions:
PASS: Delay is optimized when the arrivals match and coupling through crossover
FAIL: Coupled: If the phase responses cannot stay within the coupling zone (<120°) within the coverage angle then rework the crossover frequency/slope
FAIL: Uncoupled: If too many transitions between coupling and cancelling over the space then consider reworking the speaker positions to reduce variance

Procedure 14.15: Equalization of a single system (A or AA)

Goal: Compensate frequency response effects of transmission and room/speaker summation to match the target curve shape. Where transmission losses have filtered the HF response we will boost the sound up; where

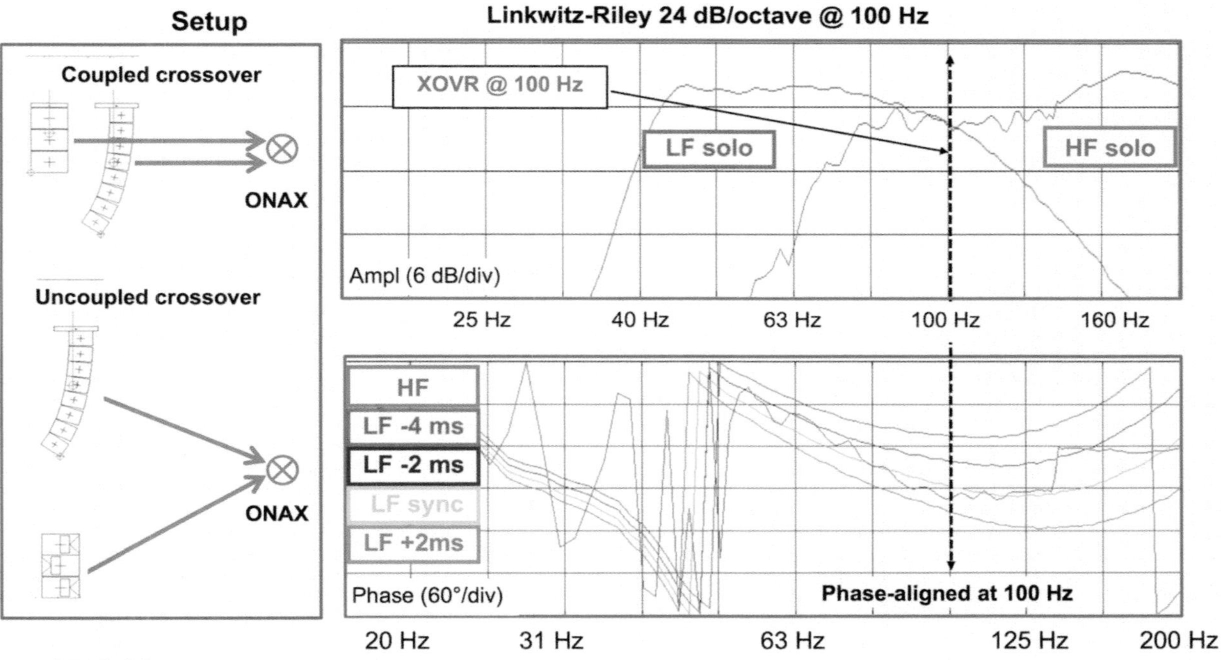

FIGURE 14.44
Phase alignment for the spectral/spatial crossover

summation has put peaks in the response we will flatten them down. The mic location(s) may be spatially averaged around the ONAX area (Fig. 14.45).

Procedure 14.15: Equalization of a single system (A or AA)
Speaker elements: A solo
Processing: channel A for EQ
Mic #1 (ONAX): coverage center, mid-point depth (ground plane may be applicable)
Mic #2 ($ONAX_2$): additional mic(s) or positions for spatial averaging
Procedure: 1. Prerequisite: known near-field response of speaker (e.g. flat response). 2. @ONAX A: Measure the unequalized frequency response. Option: XFR function "Room + Spkr" measurement (processor output vs. mic). 3. Vertically center the response in the plateau range (typically 2–8 kHz). 4. (Optional) Spatially average mics (or positions) to find raw response. 5. Use cut filters to EQ the most substantial spkr/room summation peaks. 6. Restore desired amount of HF air loss with a VHF boost filter.
Outcomes and actions:
PASS: EQ is optimized when the response matches the target curve
FAIL: If frequency response has excessively high ripple variance then seek other solutions before attempting EQ

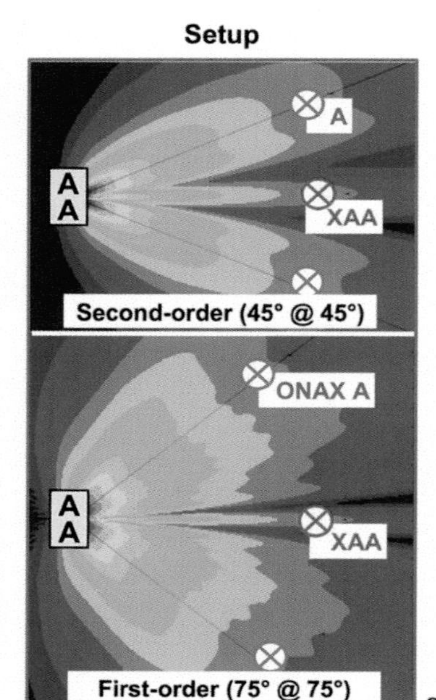

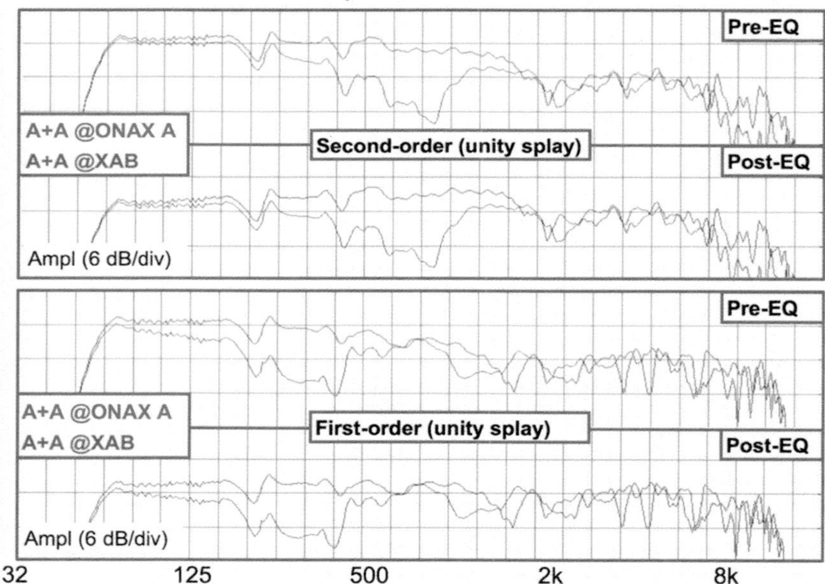

FIGURE 14.45
Single system equalization (A or AA)

Procedure 14.16: Equalization of a combined system (A + B)

Goal: Compensate frequency response effects of speaker/speaker summation to match the target curve shape. The coupling of previously EQ'd systems varies with frequency This second-stage EQ compensates for the coupling. Both the A and B systems are affected by the process, so mics are needed at both ONAX A and ONAX B, and optionally XAB.

The goal of combined system equalization is to restore the ONAX areas to as close as possible to their single-system response and to minimize the differences between the three locations. Combined equalization is, for the most part, performed similarly for both the A and B systems, but the EQ responses will diverge more substantially if the level differences are large between the systems (Fig. 14.46).

Procedure 14.16: Equalization of a combined system (A + B)
Speaker elements: A and B
Processing: A and B channels for EQ
Mic #1 (ONAX A): on axis to A at mid-point depth
Mic #2 (XAB): mid-point between A and B (exact location TBD)
Mic #3 (ONAX B): on axis to B at mid-point depth
Procedure: 1. Prerequisite: Position, level, solo EQ and delay are set on both systems. 2. @ONAX A, XAB and ONAX B: Measure and store response of A solo, B solo and A + B. 3. Compare each A + B response to the solo responses at the same locations. 4. Compare the A + B response to each other at the three locations. 5. Apply asymmetric EQ to restore maximum conformity to the target curve. (see Fig. 14.22)
Outcomes and actions:
PASS: EQ is optimized when combined response matches target curve at A, XAB and B
PASS/FAIL: If combined EQ filters differ more than 6 dB from solo system filters, then reset and restart. It may not be possible to achieve minimum variance
PASS/FAIL: If variance is unavoidable then favor A system over B

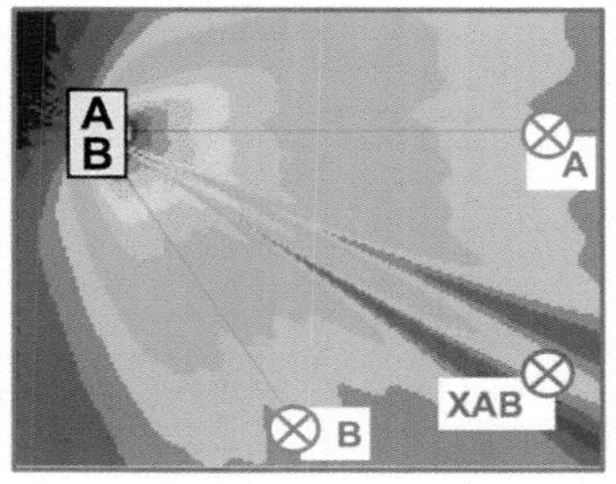

Second-order (80° @ 56°)
B speaker is - 3dB
Compensated unity splay
Asymmetric EQ:
B channel has -6 dB
shelving filter below 250 Hz.

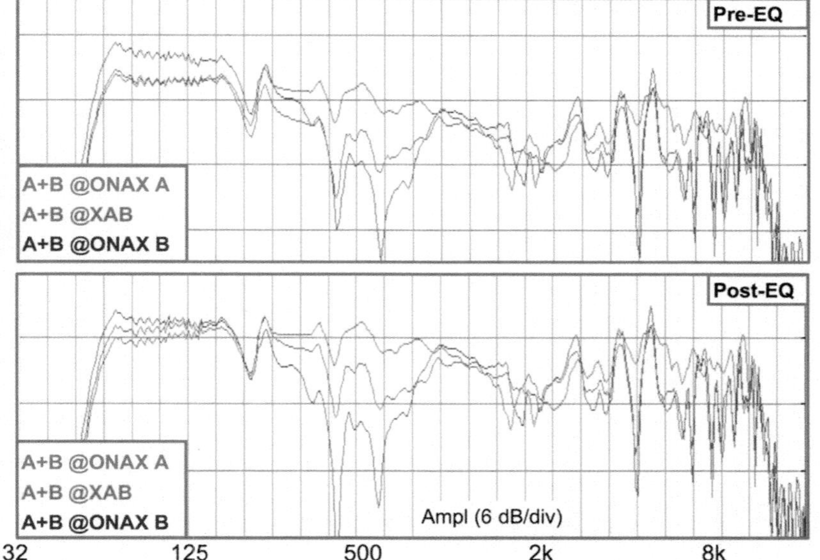

FIGURE 14.46
Combined system equalization (AB)

Procedure 14.17: Phase align the gradient in-line cardioid subwoofer array

Goal: Phase align the rear radiation of a subwoofer for maximum cancellation and creation of a cardioid coverage pattern (Fig. 14.47).

Procedure 14.17: Phase align the gradient in-line array
Speaker elements: front (+) and rear (-)
Processing: 2 channels (+ and -) for polarity and phase adjustment (+ and -)
Mic #1 (REAR): on axis to the front/back line of the array (1–2 m behind rear speaker typical but distance is not critical). Ground plane OK
Mic #2 (FRONT): This is optional for verification. Tuning is done at rear. One to two meters ahead of front speaker is typical, on the same axis as the speakers
Procedure: 1. @REAR: Store phase response of front (solo). 2. @REAR: Mute front and measure phase of rear (solo). 3. @REAR: Adjust delay on rear until its phase response matches front solo. 4. @REAR: Reverse polarity of rear channel (phase is 180° from front solo). 5. @REAR: Combine front + rear and verify cancellation. 6. @FRONT: Store both solo responses (with delay settings) - (optional). 7. @FRONT: Combine front + rear and observe - (optional).
Outcomes and actions:
PASS: @REAR: Alignment is optimized when rejection in rear is maximized
PASS/FAIL: Combined level @FRONT should be around 20 dB above REAR, but may show less if mics are close. If unsure then move mics farther away and compare
FAIL: @FRONT: If the phase responses @F_X reach 360° then spacing is too high. Consider spacing reduction and restart

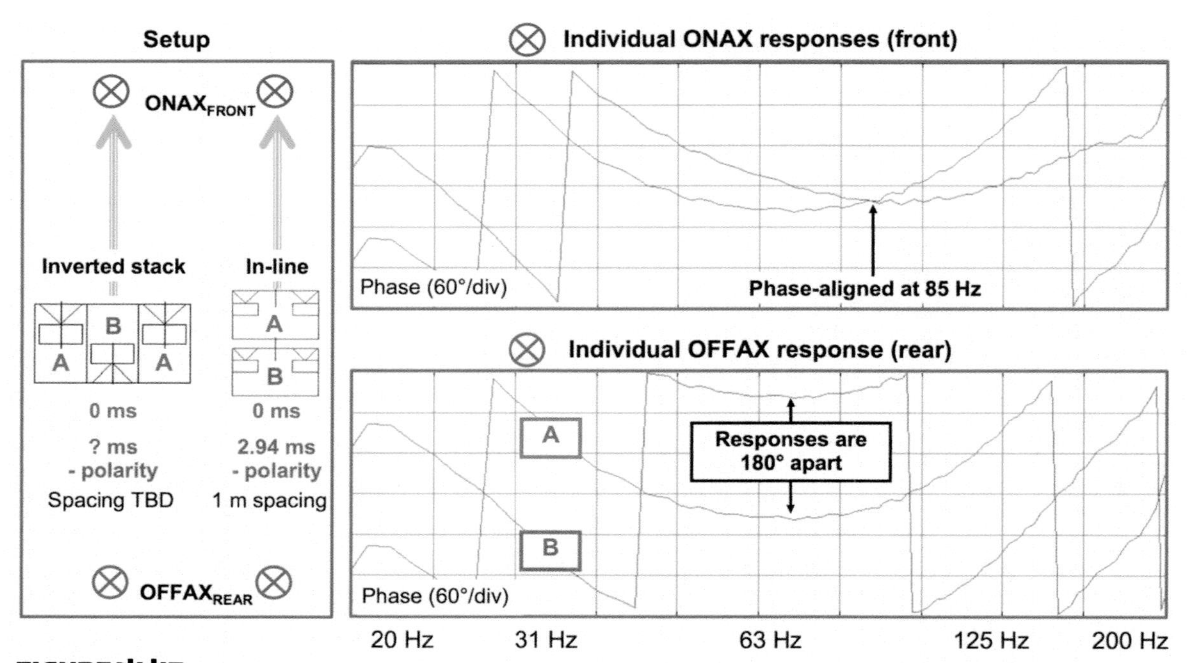

FIGURE 14.47
Phase alignment of the gradient subwoofer array (in-line and inverted stack)

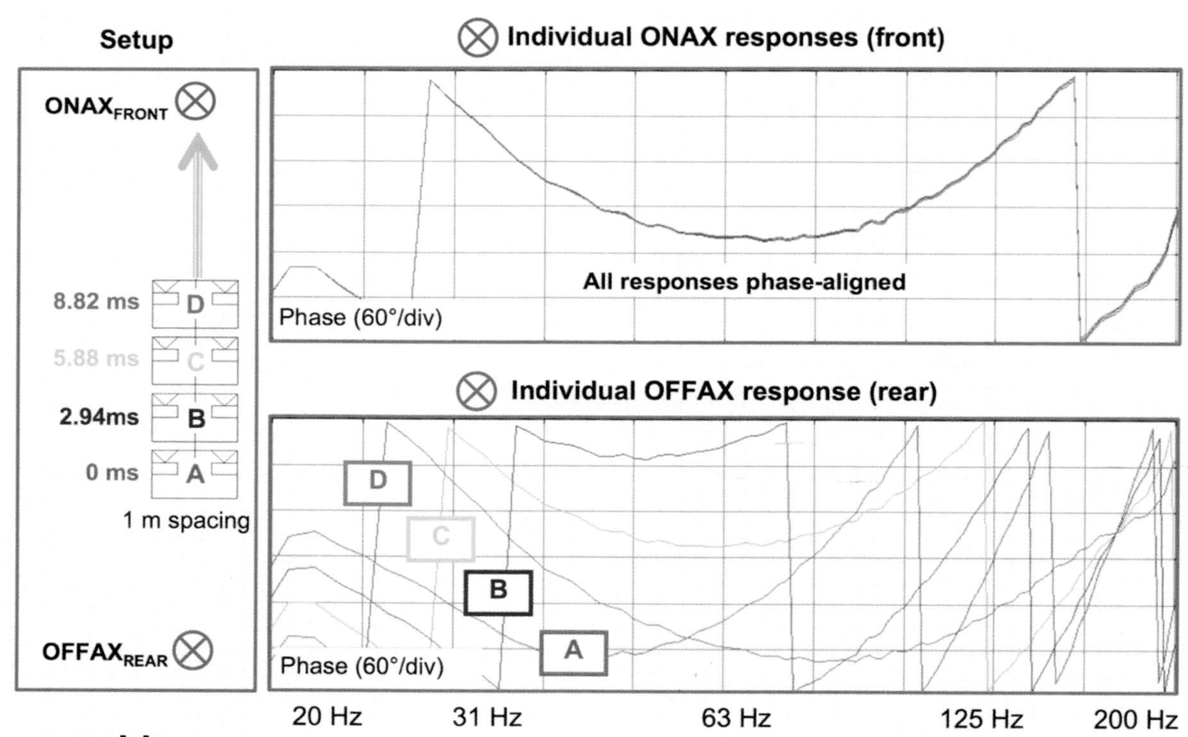

FIGURE 14.48
Phase alignment of the end-fire array

Procedure 14.18: Phase align the gradient inverted stack cardioid subwoofer array

The tuning of the inverted stack gradient array is almost step-by-step identical to the in-line array above so it won't be repeated. The mic placement is slightly affected because the gradient is usually stacked vertically. A ground plane placement will be closer to the bottom than the top element. This is a minor factor if the mic is placed in the far field but will alter the settings if used in the near field. In the near field it is advised to place the mic at the height of the rear-firing middle box (-).

There is one additional consideration for the inverted stack: the relative level at the rear (which should approximate unity from the forward and rearward elements). If not (a) try moving the mic back and see if the doubling distance takes care of it or (b) check to see if something is blocking the circular paths. For example, subs with one side against the proscenium wall will have half of their return path blocked. An opening of 0.3 m will provide enough air gap (Fig. 14.47 again).

Procedure 14.19: The standard four-element end-fire array

Goal: Phase align the front radiation of a subwoofer for maximum addition and creation of a cardioid coverage pattern by cancellation in the rear (Fig. 14.48).

Procedure 14.19: Phase align the standard four-element end-fire array
Speaker elements: A (at rear) B, C, D (front)
Processing: 4 channels (A,B,C,D) for phase adjustment
Mic #1 (FRONT): on axis to the front/back line of the array (1–2 m ahead of front speaker typical but distance is not critical). Ground plane OK
Mic #2 (REAR): This is optional for verification. Tuning is done in front. One to two meters behind rear speaker is typical, on the same axis as the speakers
Procedure: 1. @FRONT: Store phase response of A (rear speaker) (solo). 2. @FRONT: Mute A and measure phase of B (solo). 3. @FRONT: Adjust delay on B until its phase response matches A solo. 4. @FRONT: Repeat steps 2 and 3 for C and D solo and set delays. 5. @FRONT: Combine A + B + C + D. Verify unchanged phase and added level. 6. @REAR: Store A, B, C, D solo responses (with delay settings) (optional). 7. @REAR: Combine A, B, C, D and observe (optional).
Outcomes and actions:
PASS: @FRONT— Alignment is optimized when coupling is maximized
PASS/FAIL: Combined level @FRONT should be around 20 dB above REAR, but may show less if mics are close. If unsure then move mics farther away and compare
FAIL: If incremental delays between ABCD >3 ms then reduce spacing and restart
PASS: Delay is optimized when the arrivals match and coupling extends full range

Procedure 14.20: The staggered four-element end-fire array

The tuning of the staggered end-fire array is step-by-step identical to the four-element standard. The difference is mic placement, which moves off the central axis to a location approximately 45° off the centerline of the array. The mic needs to be in the far field (10 m from the array is sufficient).

Procedure 14.21: Ongoing equalization of a combined system (A + B)

Goal: Compensate frequency response effects of audience presence and room acoustic/weather during live performance to match the pre-show response. This is third-stage equalization: compensating for changes in a live performance. The changes occur as layers: empty stage vs. band on stage, empty house vs. full house, temperature and humidity variations. In practice these may be presented in sequence or all at the same time. The sound check allows us to isolate the first change layer. It is difficult to isolate the other layers because audience presence, temperature and humidity change happen together.

Mic location is critical to maintaining continuity between the original calibration data and the ongoing equalization. In the ideal world we keep mics in the same spots as room empty but this is extremely rare. The fallback positions require some key features to be of substantial help during the show. Each mic position must be clearly within the coverage of a subsystem. We can't establish cause or make clear decisions from ambiguous positions. If we see a change in the area between A and B, which system do we modify? Without an independent look at A and B we are guessing.

Procedure 14.21: Ongoing equalization of a combined system (A + B)
Speaker elements: all (all speakers on at all times)
Processing: all
Mic #1, 2, 3 . . . (SHOW): as many as possible as close to ONAX locations as possible
Procedure: 1. Prerequisite: fully calibrated system. 2. Store full system combined response with mics in original ONAX positions. 3. Move mics to best available SHOW positions and store full system response. 4. Compare data from the pre-show ONAX positions with closest comparable SHOW positions dedicated to the same subsystems. If the divergence is too large then consider repositioning or deleting the particular SHOW mic. 5. Store the frequency response of the final SHOW positions with all speakers on. These become the mono reference (REF) traces. 6. During the soundcheck compare the live responses to the REF traces. Stable changes may be the result of busing changes (such as stereo or matrix sends). Unstable changes may be the result of leakage from the stage or busing. 7. Store the most stable final REF traces during the soundcheck. 8. Showtime: Compare the live SHOW traces to the REF traces.
Outcomes and actions:
PASS: Equalization is optimized if it maintains minimum variance to the target curve
PASS/FAIL: Only take action if the readings show plausible, physically possible changes due to new conditions. If the changes are out of scale then consider all possibilities for error before taking action

14.8 FINISHING THE PROCESS

How can we end an iterative process that approaches perfection but can never reach it? If the fully combined system is not acceptable, we can disassemble the pieces, adjust them, and recombine them again. Alternative approaches might have equal merit. We can first try dividing the space and combining the systems in one way, and then try the other. Nothing ends a discussion better than "Let's test both theories."

Here is how it ends: We run out of time. Seriously. As long as there is the time there is more that can be learned. Anything we learn today is applicable in advance for tomorrow.

Listening

Our ears are trained professional audio tools but unlike the FFT analyzer they are neither stable nor objective. Each ear (and set of ears) has inherent physiological differences. They are not stable physically over our lifespan or over the short term. We need time to adjust to the local atmospheric pressure when we fly into town for a gig. Our dynamic thresholds are temporarily shifted when exposed to high stage levels. Long-term high-level exposure can cause permanent shifts in both dynamic and spectral responses. Those lucky enough to grow old lose dynamic and spectral range in the normal course of aging. These factors and others make the ear a subjective partner in the optimization process.

The ears connect to our brain, which contains our unique personal reference library of accrued aural experience. Our expectations about how a violin *should* sound come from the thousand references to violins in our aural library. We accrue a map of the expected response of a single violin, sets of violins, and details such as how they sound when plucked or bowed. We use this ear training to perform moment-to-moment comparisons, internal "transfer functions" as it were, against our memory maps and make conclusions. The question then becomes how closely it matches the expected response.

The ear/brain system also brings context into the equation. We evaluate a sound with respect to the surroundings. Is this normal for this distance? Is the reverberation level in scale with this space's size and surfaces? Our trained ear/brain system can make contextually derived conclusions such as "acceptable for a basketball arena's back rows."

Our personal "reference" program material is the ultimate sonic memory map. We have heard it so many times in so many different places and systems that it's permanently burned into our brains. The newly tuned system must satisfy our ears against this standard.

Ear/brain training is a lifelong endeavor. The complex audio analyzer can be a great aid to this process through ear/eye training. The linkage between what we see on the screen, to our perceptions of a known source such as our reference tracks close the loop on the learning process. If something looks really good on the analyzer (or really bad, strange or unexpected), go there, look and listen. Correlate what is seen on the analyzer, seen in the room and heard. Conversely, if we hear something "strange" somewhere, we can move a mic there and see what "strange" looks like. This helps learning to read traces in context and identify transitional trends in the room. At the end of the optimization process I have often made adjustments to the sound system based on the walkthrough while listening to reference material. In most cases the adjustments are minor, but in all cases I try to find what I missed in the data interpretation. The answer is there, somewhere in the data. This becomes part of the learning process we carry forward to the next job.

14.9 ONGOING OPTIMIZATION

14.9.1 Using program material as the source

A powerful feature of dual-channel FFT analysis is the capability to perform transfer function measurements using an unknown source material. We are "source independent" (the origin of the acronym SIM™), which means we can continue to analyze the response in an occupied hall during performance. The independence has its limits. The source must excite all frequencies, eventually. We'll have to wait a long time to get data if the source has low spectral density. If the data is dense we can move almost as quickly as pink noise.

The result is an ongoing optimization process that provides continual unobtrusive monitoring of the system response. Once the sound system is in operation, the control is in the hands of the artistic sector: the mix engineer. Ongoing optimization allows objectivity to remain in place to aid the artistic process in a number of ways. The first is the detection of changes in the response. Restorative action can be taken in some cases, which minimizes the remixing required for the show in progress. When remedial options are limited it can be helpful to keep the mixer informed of differences so this can be considered in the mix.

Much can change between the setup and showtime, and even over the course of a performance. Only a portion of those changes is detectible with our analyzer, and a portion of those, in turn, will be treatable with ongoing optimization.

CHANGES FROM SETUP TO SHOWTIME FOR A SINGLE PROGRAM CHANNEL

- **Dynamic room acoustics**: absorption/reflection changes due to the audience presence.
- **Dynamic transmission and summation**: changes due to temperature and humidity.
- **Leakage from stage emission sources**: the band etc.
- **Leakage from stage transmission sources**: stage monitors etc.
- **Re-entry summation**: the speaker system returning into the stage mics.
- **Duplicate entry summation**: source leakage into multiple stage mics.

14.9.2 Audience presence

We know that adding live bodies to a lively concert hall can deaden the acoustics, but adding dead bodies to a dead hall has not proven to liven up the room. The audience absorption acoustical effects are not evenly distributed. Floor reflections undergo the largest change, and ceilings the least. The changes largely depend upon the seating area prior to occupancy. The most extreme case is the transition from a flat unfurnished hard floor to a densely packed standing audience (e.g. disco, rave, arena floor). The least effect will be found with cushioned seats. In any case the strongest local effects are modifications of the early reflections. This changes the local ripple variance, which may require an equalization adjustment. Reduction of the late reflections also affects the local ripple variance structures and signal/noise ratio. The frequency range of practical equalization shrinks downward as time offset rises, but the upper frequency ranges are likewise affected. Increased absorption corresponds to reduced fine-grain ripple and improved coherence as frequency rises. Reflections that arrive too late for our FFT time window, those seen as noise, are reduced in the room, creating the rise in signal/noise ratio (coherence).

Audiences create noise of their own: Screaming fans, singing congregations and the mandatory coughing during classical music all degrade our data quality. Generally speaking the louder the audience gets, the less it matters what our analyzer says. If they are screaming for the band we are probably OK. If they are screaming at us, then looking at the analyzer might not be the wisest course.

14.9.3 Temperature and humidity

Temperature and humidity changes affect both the direct and reflected paths. Temperature rise changes the reflection structure as if the room has shrunk. Direct sound arrives faster to the listener, as do the reflections (see Fig. 4.65). A temperature change of 5.6°C rescales each path by 1%, which results in a different amount of time for each. The deck is thereby reshuffled for all path length-based ripple variance, which is driven by time offset as an absolute number, NOT a percentage. The timing relationship between mains and delay speakers changes with transmission speed (the larger the difference in paths, the greater the time offset with temperature). Speaker delay offset over temperature is more easily visualized than reflection changes, even though they are the same mechanism. A delayed system changes from synchronized to early (or late). A reflection changes from late to not as late (or later). Think of it this way: The main/underbalcony timing relationship will still change with temperature even if we forget to set the delay in the first place. Expect the analyzer to see the following response differences: a redistribution of ripple structure center frequencies, a change in the amplitude range of some portions of the ripple variance, modified coherence and delay system responses that need some fine-tuning.

Humidity change acts like a moving HF filter. The changes scale with transmission distance to each local area. Longer throws have proportionally stronger air absorption effects, and so the changes are more severe over distance. A short throw speaker sees only minimal differences even with large relative humidity variance. This precludes the option of a global master filter to compensate for the complete system. Corrective measures must take the relative distance into account. Expect to see the following response differences on the analyzer: a change in the HF response.

14.9.4 Stage leakage

Stage performers present a leakage path into the sound system coverage area (section 6.1.3) that changes on a moment-to-moment basis. Listeners may have difficulty telling whether they are hearing stage leakage or the sound system (a desirable outcome when sound image is of high importance). Alternatively, out-of-control band gear and stage monitors are the mix engineer's worst nightmare. We can mute the mic in front of the Marshall but it is still our fault that the guitar is too loud (somewhere) and too soft (somewhere else). The mix engineer can't mix what's not in the system.

Stage leakage seriously compromises the ability to get accurate measurement data for the same reasons. Our electronic reference is what's in the mix. If it came from Planet Claire (and not from our system) we fail the correlation test and the response on our analyzer is no longer stable and repeatable. Leakage contaminates our data and reduces our reliability and treatment options. If the show begins and a peak appears in some frequency range, we are put on high alert. But it's a false alarm if the peak is due to leakage from stage sources. An inverse filter won't remove it if it was sent by something that doesn't pass through the equalizer. To make matters worse, our polygraph detector (the coherence response) may be fooled by the leakage. The waveform contained in the leaked acoustical signal will be present in our electrical response from the console. How did it get there? It leaked into the microphones. Since the waveform is recognized, the coherence can still remain high, and therefore the peak has the appearance of a treatable response modification, except that it will not go away, no matter how much equalization we throw at it.

There is no single means of detecting stage leakage. One is to try to equalize it. If it responds, we are fine. If not, it might be leakage. Fishing expeditions like this, however, are not appreciated by the mixer, and should be a measure of last resort. The first consideration is the obvious: What do our ears and eyes tell us? We are in a small club and the guitarist has four Marshall stacks. This is going to be a leakage issue. The next item to consider is plausibility. What mechanism could cause a 10 dB peak to arise between setup and showtime? If this was the result of the audience presence, then why do the peaks keep changing shape? Before we grab an equalizer knob or touch a delay we must consider how the change we see could be attributed to the changes in transmission and summation due to the audience presence and environmental conditions. Large-scale changes and song-to-song instability point toward the band. Less dramatic and more stable changes point to the sound system in the room. Those are the ones worth venturing after.

14.9.5 Stage microphone summation

A well-tuned system can sound great in every seat when we listen to our reference track, and take on a completely different character when the band comes on stage. We just discussed stage source leakage into the house, but there is an even more insidious source of trouble: re-entry and duplicate entry mic summation (section 6.1.3). Leakage from stage sources, monitor speakers or main speakers back into the stage mics becomes part of the sound system's source signal. Massive comb filtering can be introduced into the mixed signal by the stage mics. We can all hear it but our analyzer cannot see it. Why? Because our measurement reference point begins at the console output and the damage happens in (or even before) the mix console. This is best detected by logical deduction: The analyzer doesn't see it but we can hear it. Therefore it must be upstream. Recall the ripple variance progression in the room. If we stay put, the ripple stays put. If we move, it moves. The opposite can be true when we have mic/mic (duplicate) or speaker/mic (re-entry) summation. The ripple progression upstream is governed by stage relationships. If stage sources move, the ripple moves. If we're sitting still and yet the ripple is moving, the source of the combing is upstream of the sound system (unless you feel the wind blowing).

The most effective means of isolating this effect is to keep good records. Store the data at a given location before the performance (or even the sound check). If the transfer function response remains stable while the sound is changing, there is strong evidence that the solution lays on the upstream side of the art/science line. Letting the mix engineer know that we cannot solve this should initiate a search on the front end for the cause and possible solutions. Leakage in a primary mic channel (like the lead singer) can make the entire PA sound like it's underwater but can't be solved by a bilge pump on the sound system. It must be plugged at the source or the ship will continue to sink. We have to use our ears to do this, because the analyzer can't see it. Using our ears is not a problem, but tuning the whole PA around a combing stage mic is. Don't do it. Every other source into the sound system will be detuned. We have joined the artistic staff when we're using our ears to make changes not indicated by the analyzer. We'll need to inform the mix engineer that we are also mixing the show (which may or may not be welcome news). We should always use our ears, but must be able to discern which side of the art/science line we are hearing. Our side of the line is primarily concerned with audible changes over the spatial geometry of the room. If it sounds bad, somewhere, fix it. If it sounds the same bad everywhere, don't fix it. Tell the mixer to fix it.

EAR–EYE TRAINING TIPS FOR FINDING UPSTREAM/DOWNSTREAM PROBLEMS

- **Live audio is radically different from playback**: problem and solution are upstream.
- **Sound is changing but measurement is not**: problem and solution are upstream.
- **Measurement is changing but sound is not**: cross-channel leakage (such as stereo).
- **Sound and measurement change**: problem and (possible) solution are downstream.
- **Modulating combing**: If you are moving, the measurement mic is moving, the PA is moving, the wind is blowing or there's a strong HVAC vent in front of the speakers then it's probably downstream. Hint: It shows up on the analyzer. If there are moving stage mics, actors/musicians or stage monitors leaking into mics then it is probably upstream. This will not show up on the analyzer.

14.9.6 Feedback

The worst-case scenario of re-entry summation into the microphones is feedback. Feedback is very difficult to detect in a transfer function measurement (it's present in both the electrical reference and acoustic signals). The only hint is that the frequency in question may suddenly have perfect coherence. Feedback detection can be conducted in the FFT analyzer in single-channel mode and time spectrograph, where we are looking at the spectrum in absolute terms. The feedback frequency rises above the crowd and can be identified after it is too late. The time spectrograph display is the most effective analysis display for feedback detection because it can see ringing over time that has not fully developed into a full-blown howl.

14.9.7 Multichannel program material

Everything just discussed pertains to a single channel of program. Stereo and other multichannel formats mix the sound in the acoustic space. As if it were not already difficult enough to get clear and stable data, we now add the complication of leakage between sound system channels. Our electrical reference is one particular isolated channel whereas our acoustical data includes related (our channel) and unrelated material (other channels).

14.9.7.1 STEREO

Stereo is a changing mix of related and unrelated signals. It's a wonderfully desirable listening experience. Unfortunately it makes for a very challenging optimization environment.

The stereo dilemma. Which channel do we use as the reference?

- **Left channel**: Measure as far to the left as possible to maximize isolation for right. Data will never stabilize if the mic is placed near or at center. Coherence can be fooled by opposite-side panned signals (recognizes the waveform but mostly it is coming from a system it can't control). EQ changes will not respond linearly if based on panned signals.
- **Right channel**: analogous and opposite to left.
- **Mono sum of L/R**: stable at the exact centerpoint between the systems. Cannot tell left from right. Systems must be perfectly matched, mic must be perfectly centered and all actions must be taken symmetrically on L and R. Unstable at all positions off center.

The relevant question is: What can we do with unstable data? Should we be turning equalizer and level controls to stabilize it? Of course not. It's stereo. It's supposed to be changing. A mic position has very limited use (and can lead to wild goose chases) if it cannot provide a stable frequency response. Maximum stability is found with a centered mic (XOVR LR) and mono sum reference. The secondary route is ONAX L (as far off center as reasonable) with L as the reference (or vice versa).

14.9.7.2 MULTICHANNEL

Systems with highly overlapped multichannel interaction are the utmost challenge for in-show measurement. Musical theater often uses separate music and vocal systems. The signals don't share an electrical reference but are mixed acoustically and cannot be separated without stopping the show. The overture of a musical is likely to be the last shot at measuring the music system. Dialog portions without music would be the best chance for the vocals.

14.9.8 Show mic positions

Most of the optimal mic positions will be unavailable when the hall is occupied (I suspect it has something to do with people sitting there). This greatly reduces the quantity and quality of data available during performance. Fortunately we need fewer mic locations. We're finished with OFFAX, XOVR, VTOP and VBOT because we won't be adjusting speaker positions or adding acoustic treatment during the show.

The principal task of ongoing optimization is monitoring the effects of audience presence, and changing temperature/humidity on EQ and delay settings. The best-case scenario for mic placement is an ONAX (A, B, C, etc.) mic position that's highly isolated from other subsystems and other channels of sound. Local response changes can be detected and acted on by adjusting the affected subsystem without risking the whole system tuning.

SHOW MIC POSITION PRIORITY

- **ONAX A**: the most important position for EQ decisions of the majority system. The most global adjustments can be made based on this location.
- **ONAX B**: important position with semi-isolated EQ capability. Depends on how much isolation (over frequency) this system has from the A system. Often the mix location.
- **ONAX D (delays)**: useful for delay time adjustment. Insufficient isolation for EQ.
- **ONAX C or fills**: Nice to have but minimal isolation leaves few options for EQ.
- **WHATVR**: A mic placed somewhere in the room with no particular system in control of its coverage. Useful for feedback detection and spectral viewing. Otherwise unusable.

The empty room response at each usable ONAX position is stored for reference during soundcheck. Performance data is compared with this and changes made if necessary to restore the response. We use fallback positions when we can't get an isolated ONAX. The first level is a less isolated "ONAX-ish" position. This might have substantial overlap from a related subsystem (e.g. between the ideal ONAX A and XAB). Adjustments made here will have only a partial effect on the combined response due to leakage from the nearby system. XOVR positions are the least useable because they are far too volatile to serve an EQ function during the show, just as they were not used for EQ during setup.

14.9.8.1 LESSER SUBSYSTEMS

This is a show. There will be no speaker muting to observe a particular subsystem. It's a major challenge to measure the main (A) system with the other subsystems adding in. The challenge level goes up tremendously for all of the lesser subsystems (B, C, delays and fills) because of their lack of isolation from the main (A) and other subsystems up the chain. Even the best ONAX position for a lesser subsystem has a very limited range of use in show conditions. Only the frequency range where that system enjoys dominance will respond to independent adjustment (i.e. a 3 dB cut on a filter reduces the acoustic response by 3 dB). The farther down the hierarchy that our subsystems fall, the less we can do for them in a show context. The LF is the least likely range for the subsystem to maintain local control. We cannot expect an LF filter change in the subsystem equalizer to exert much control over the combined response. The VHF is the most likely range with enough isolation to use a local filter.

Delayed systems can be viewed during performances in the hope of maintaining synchronicity under changing environmental conditions. This requires a mic positioned at the spatial XOVR for the delay and the mains. Such mics are more useful for delay monitoring than equalization. Time offsets between the main and delay can be detected and remedied, but EQ is difficult to conclusively adjust because the response in the fill area is a combination of the two subsystems. It is inadvisable to adjust the main system equalization to compensate for changes in the fill area. If any adjustments are to be made here they should be done sparingly and exclusively on the fill system.

Detecting the time offset errors between the speakers without muting either system can be a difficult practice in the field. It is easiest when the main and delay have closely matched level and HF content. The impulse response is much more sensitive to HF content, so if the HF is strongly rolled off in the mains then it can be difficult to differentiate the two impulses during the show. There is an alternative method for situations where the time offset cannot be discerned or where a show mic is not practical in the delay area: Estimate the time offset based on a known sound speed change. Make note of the propagation delays before the show at a representative location. Compare this with the current propagation delay. Compute the percentage of time change. Modify the delay lines by the percentage change.

14.9.8.2 MEASUREMENT MIC AT THE MIX POSITION

The location with the highest probability of showtime access, the mix position, is also one of the most challenged. In stereo systems a central mix position lacks L/R channel isolation and will therefore be hard-pressed to establish a stable response. Positions on the outermost horizontal edges of the mix position can find a more isolated response, and provide a clearer picture of "before and after" responses. These side positions will also include the opposite side arrival but, if the system has limited stereo overlap, some degree of control can be established. None of these is perfect, and often the best we can do is monitor a response solely in relative terms (house empty, house full).

Key considerations for show measurement microphone positions

- **Actionable intelligence**: Is the mic in a location where we can get meaningful data?
- **Scope of work**: probably just ongoing EQ and delay adjustments.
- **Security**: Will the mics be stolen or tampered with?
- **Orientation**: Is the mic aimed at the speakers or hanging down (affects the HF)?
- **Plausibility**: Does the data confirm the laws of physics or is it showing magic?
- **Local/global**: How much of the room is changing in the same way as we see here?

14.9.9 Subwoofers as a program channel

There is no point in arguing with somebody who wants to run subs on an aux. However, if we are tasked with ongoing optimization of the system during the concert we have a duty to explain that the frequency range below 150 Hz will not be analyzable. The range covered by the aux subs cannot be reliably measured with a transfer function analyzer because there is no electrical reference that tracks the full range of acoustical response. Like the music/voice systems discussed earlier, the subwoofer range cannot be untangled from the other signals. A transfer function using the mains as a source sees the subs as an unstable, uncontrolled contamination. If the subwoofer send is used as the source the opposite occurs. In either case the electrical reference extends to the full range, so there is no way to separate the acoustic responses.

IF USING THE SUB AUX FEED AS THE TRANSFER FUNCTION REFERENCE

- **Usable data**: only range with LF ≥10 dB louder than mains (isolation zone).
- **Unstable, unusable data**: all frequencies above or equal to the spatial XOVR.
- **Coherence is invalid**: Can be fooled by correlated data arriving from other speakers.
- **Non-linear EQ tracking**: EQ changes in the LF band only track in the isolated range.

IF USING THE MAIN CHANNEL FEED AS THE TRANSFER FUNCTION REFERENCE

- **Usable data**: only frequency range ≥10 dB louder than subs (isolation zone).
- **Unstable, unusable data**: all frequencies below or equal to the spatial XOVR.
- **Coherence is invalid**: Can be fooled by correlated data arriving from other speakers.
- **Non-linear EQ tracking**: EQ changes only track in the isolated range.

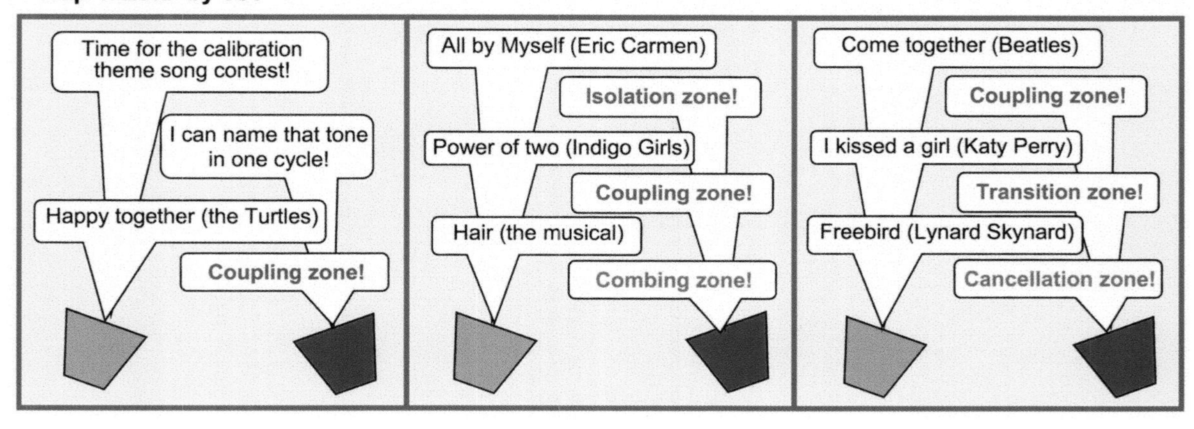

CHAPTER 15

Application

application n. 1. putting of one thing to another. 2. employment of means. apply v. make use of; put to practical use.
Concise Oxford Dictionary

Congratulations and thanks for making it to this point. This chapter is the summation of all our efforts so far. Our goal is to couple all of the various information sources together for maximum power and coherence. We will explore the design and tuning of eighteen sound systems for sixteen halls using the principles and procedures outlined previously. Although this is not a complete library of every application you might encounter, a wide variety is represented here. The general flow is toward increasing size and complexity. Each application has unique aspects similar to those you may encounter in the field.

Eight of these application examples include field data from the SIM3 audio analyzer captured during optimization. There are over 340 frequency and phase response measurements compiled into 96 screen shots. The measurements are by no means a complete record of the data taken at any of the venues, but provide some insight into the process. This chapter differs from the previous ones in that it purposefully details my personal perspective and decision-making process.

The figures for this chapter are packed with a tremendous amount of technical information. The text expands upon these figures and provides context and background. So much of the text in each application is linked to its related figures that virtually every sentence could include "as shown in panel x of Fig. 15.x." Read the text and keep an eye on the figures and it should hold together. For the most part, when I do refer to a figure it will be one from a previous chapter.

Be warned that there is repetition here and much of it is done on purpose. That's in the nature of the work. Every system has frontfills. We don't need all the details for eighteen of them, but we will do enough variations on the themes to make these processes clear enough to assist you in the field. There is a huge body of work in this chapter, much of it provided by others. I am grateful for all the support provided by sound designers, mix engineers, system techs, integrators, acousticians, riggers, digital network experts and more who made these designs and optimizations possible.

I have attempted to use consistent formatting, conventions, symbol set and acronyms for this chapter in order to pack the maximum amount of information into the figures. The figures are a series of panels designated with letters and numbers. The letters show a common link, whereas the numbers designate a series of steps. For example panels A1 and A2 might be plan and section views for mains speaker aiming whereas B1 and B2 provide the same function for frontfills. The MAPP Online™ plots are 3 dB/color @4 kHz, 1 octave unless otherwise stated. Speaker locations are labeled by function with a simple rectangle. Frontfills (FF) are the same size as subwoofers (LF) and so on. A muted speaker is shown with its name grayed out. The speaker symbols are not scaled but the drawings are. Plan and section views are almost always identically scaled and placed so that they line up. A speaker or mic placed in plan lines up directly with itself in section. Most of the section views are complete from floor to ceiling. By contrast, most of the plan views are cropped to include just over half of the hall. This saves an enormous amount of layout space and reduces redundancy. Obviously all operations mirror to the opposite side. The mic locations shown correspond to our placement strategy for calibration.

All of the SIM3 traces have the same vertical scale for amplitude (±30 dB) and ±180° for phase. Leaving these off the graphs saves valuable space for traces. I realize that the vertical scale may appear compressed to you. It is. I show it this way here because that is how I view it to make decisions in the field. In my experience it is far too easy to get lost in small details. In practice I look for the trends I can see at this resolution. If you see it here, you can hear it there and it is fair game for action. Note that the EQ trace is sometimes normal, and other times shown inverted. The coherence scale is 0 to 1 (as always). The top of the graphs is always 1 but the zero is not consistent in the SIM3 trace set (sorry). Sometimes coherence uses half the vertical height and other times ⅓.

You may have been wondering whether or not anyone actually places mics at all these locations and follows the methodical procedures outlined in this book. Can it *really* be done this way or do we just wing it once the first shot of pink noise has been fired? The answer is in the SIM3 traces.

One final note, before we begin. This chapter reads like a cypher code if you skipped the rest of the book to get to the really good stuff here. The decryption key for this chapter starts on page one of the book.

15.1 VENUE #1: SMALL HALL

We start simple with a small hall and small budget. The system is L/R mains, an LF system and frontfills.

15.1.1 Mains

The horizontal shape is nearly square, just barely wider than deep. The FAR value is <1 so we would need a very wide speaker (>180°) to cover the whole room from center. That is a definite no-go, leaving the option for a coupled point source of two or more elements at center, if desired. Instead an L/R system is used, so we bisect the room and find the FAR value for each side (1.8), which means we need at least an 70° speaker/side. The available positions are offstage in the corners so the speakers are aimed inward to form a symmetric uncoupled point destination. We follow the horizontal aim method (middle/middle), which yields 25° inward.

The horizontal aim for the left main can be verified by measuring its custody edges at mid-depth (OFFAX L and XLR). The L main solo response should be equal at these locations. The horizontal coverage can be verified by comparing ONAX L to the previous measurements. The speaker is not wide enough if the HF range at OFFAX L and XLR are more than 6 dB down from ONAX L. In this case we needed a minimum of 70° horizontal coverage and specified an 80° model, so we have a little bit of extra width.

The vertical shape is a simple rake with a 2:1 range ratio, the limit of coverage for a single speaker. Mains vertical coverage begins at the third row and continues to the last. We need 25° of vertical coverage so the compensated coverage angle is 50° (25° × 2:1 range ratio). The speaker is aimed at VTOP (-11°), which we do any time the range ≥2:1.

The vertical aim for the left main can be verified by measuring its upper and lower custody edges (VTOP and VBOT). The L main solo response should be equal at these locations. Vertical coverage can be verified by comparing ONAX L with the previous measurements. The speaker is not wide enough if the HF range at VTOP and VBOT are more than 6 dB down from ONAX L.

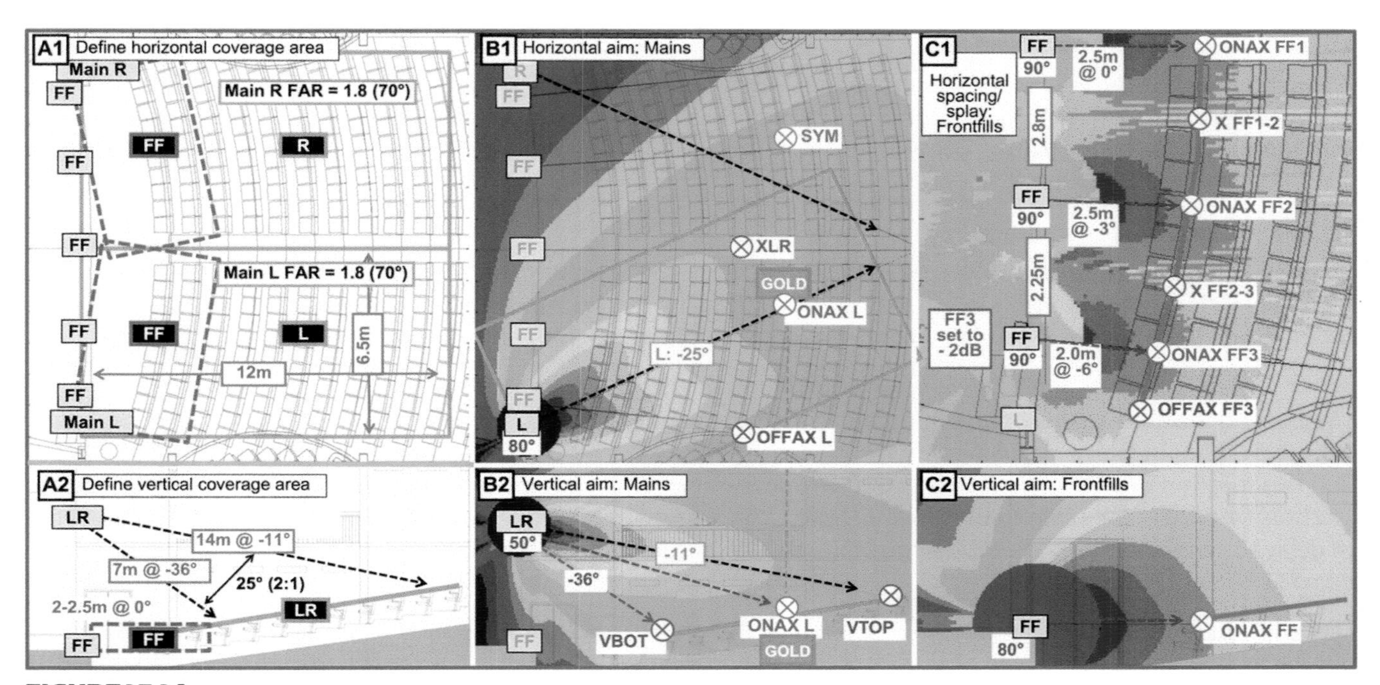

FIGURE 15.1A
Design process for Venue #1: Coverage partition strategy, solo mains and frontfills

Solo equalization is performed around the ONAX L location, which is on the horizontal axis between mid-point and the VTOP. This then becomes the GOLD standard.

15.1.2 Frontfills (FF)

The horizontal shape is a slightly rounded stage but the seating is on a different arc, making the outermost seats closer. The difference is 2 dB (2.5 m/2.0 m) so we will compensate this with level and spacing. The outers are turned down 2 dB and their spacing is reduced to 80% of normal. The spacing math works out such that four frontfills can just barely cover the space. The outer and innermost seats are both at the coverage edge (-6 dB). This means that we will have to get exactly the positions we need when construction occurs, which is as probable as winning the lottery. The four-box scenario leaves a gap at the center aisle (which is OK) but risky because we are a long way from the L/R mains. I chose to add a fifth box on the centerline, which assures we have coverage where we most need it (near the center). All the spacings are reduced proportionally, which fills the entire row and leaves some room for real-world positions.

Field verification of the spacing is performed along the first row (the unity line) with mics placed at the ONAX FF and XFF positions. Level setting and EQ are performed at the ONAX locations. The spacing is optimized when the combined responses at ONAX and XFF are equal.

The vertical aspect *should* be a non-event. Fight for the highest possible location (footfill prevention) and minimize the risk of HF missing the ears by using first- or second-order speakers.

Equalization for frontfill systems is usually carried out as a combined block (not as soloists). This is simply a practical matter, and could be done in two stages (solo then combined) if time permits. EQ is done at the ONAX locations, not the volatile XOVR areas.

You may wish to high-pass the frontfills to reduce leakage and maximize headroom. As a general rule I roll them off gently (second order @160 Hz), which simulates a unity-class spectral crossover to the subwoofers. Excessive range reduction can increase the risk of spectral sonic image separation (section 4.3.5.11).

15.1.3 Low-frequency system (LF)

We have enough room at overhead center to implement an in-line gradient cardioid subwoofer array. This takes only two speakers so it is the same money and L/R but yields superior uniformity and reduced leakage. Field calibration is done by measuring behind the array (procedure 14.17).

15.1.4 Combined systems

The frontfills are combined with the mains along the third row. Timing is set to sync the two systems there and then the level is set to sum together to the unity GOLD standard. We can separately delay each frontfill to the main on their side if we have independent channels (center would be the longest). Use the middle frontfills (FF2) if there is only one processing channel (delay dilemmas, section 14.4.6.2).

The L/F will meet the mains on the floor. We will reuse the ONAX L location and sync the systems and set level there. Use the phase delay method for the spectral crossover calibration (procedures 14.9 and 14.10). Note that this location is the L main center (not the room center). This prevents three sources from sync'ing at center (section 14.4.6.2). The mains will probably have to be delayed because they would likely arrive first. If the amount of delay to sync is too high then we may decline and let the subs fall behind rather than having the whole system do so (TANSTAAFL/triage).

Let's consider three options for main + LF spectral crossover: overlap, low-order unity or high-order unity. Overlap has the highest amount of ripple variance between the uncoupled sources, but has the highest immunity to spectral separation (i.e. distinctly localizing the LF and HF as discussed in section 5.5.4.1). High-order unity has the best immunity from ripple variance, but is the most vulnerable to spectral splitting. A low-order unity is the middle ground in both categories. The risk of spectral separation is very high here because the sources are uncoupled in *both* the vertical and horizontal planes, which would lean me toward overlap (procedure 14.10). Room reflections in such a small space will likely have a stronger effect in the overlapping frequency range than the speaker/speaker interaction.

The combined equalization will be very minor. There is nothing to do with L + R. The main + LF crossover may require some EQ to reduce the bump in the overlap region. Frontfills are too small a contributor to create a combined response worth treating.

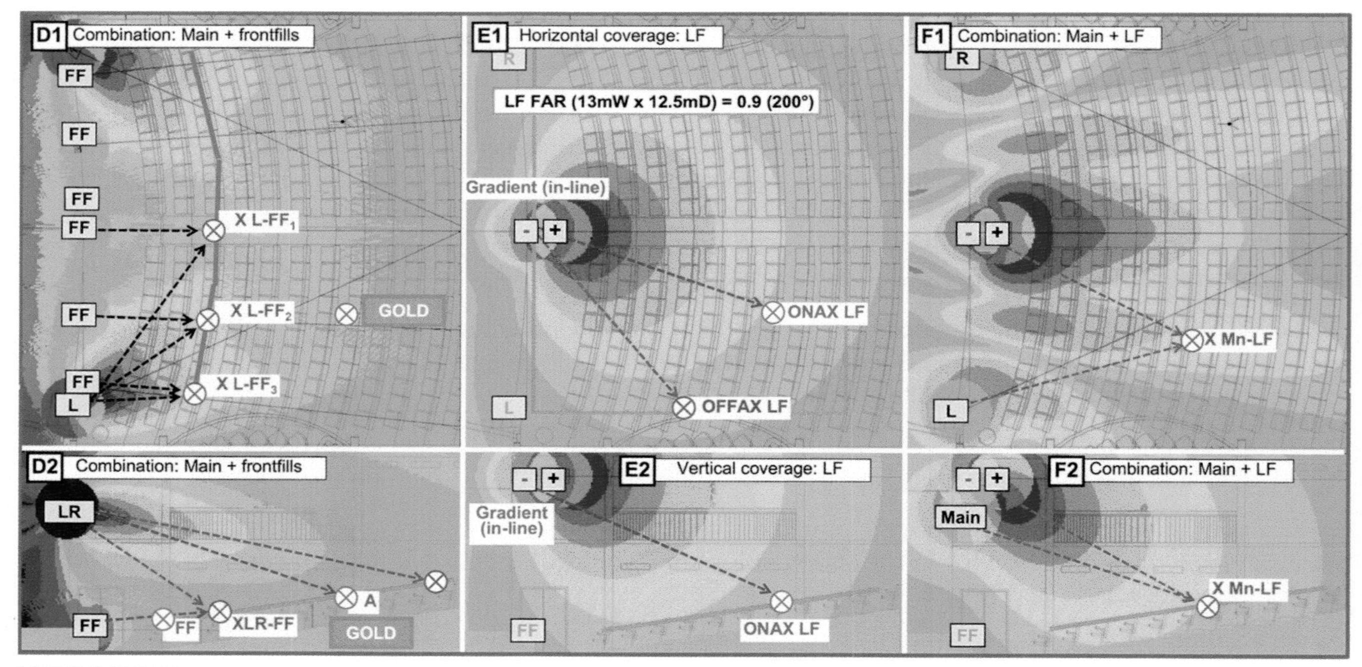

FIGURE 15.1B
Design process for Venue #1: main + frontfills, cardioid subs and main + subs

15.2 VENUE #2: SMALL RECTANGLE, AB MAIN ARRAY

This venue requires vertical coverage beyond the capabilities of a single element. The 3:1 shape would do well with a line array of third-order proportional elements but this is not in the budget. Two elements/side is the lower limit for quantity and the upper limit for budget, so that's what it's going to be. The system design consists of L/R mains (A), L/R downfills (B), frontfills and subwoofers. Our study will be limited to the first three.

15.2.1 L/R mains (AB)

The room is narrow enough to cover with a 90° center cluster (with frontfills and some small sidefills). The system design will be L/R mains because (a) that is what engineers want and (b) L/R is not impossible. Just your daily reminder that center clusters are great when no other option is possible, but otherwise we default to L/R. Each main will be a single horizontal element and an AB-type asymmetric coupled point source in the vertical. The mains horizontal aim is calculated the same way as we would without a downfill. Its propagation begins at the speaker and it has to make it to the end of the room. The mid-depth calculation yields an inward aim of -16°. We only need 45° of coverage/side. We have the option of cutting it close with 50° speakers or allowing more overlap with an 80° model. In this case the walls are covered with heavy drapes so we are not afraid of them and will go with 80°. This is a TANSTAAFL/triage decision that gains more spatiality at a cost of ripple variance. The non-hostile walls tipped the scales in favor of added spatiality.

The vertical plan is a steady incline that yields a 3:1 range ratio from the minimum allowable height. The coverage strategy is to make the mains carry the coverage from the rear to the point where their underside gets too much spectral tilt. The upper main (A) is aimed at the rear seating ($ONAX_{FAR}$) and then we start down from there. A wide vertical speaker could almost reach the front in the HF range because the coverage target keeps getting closer underneath. The stopper is spectral variance because the LF range will be much louder in front than back. Therefore a 50° model was chosen that leaves us with 25° underneath before calling for backup. The total range to be covered is 45° so we will be leaving the bottom 20° for the downfill to connect to the frontfill. The downfill (B) will be aimed at the off-axis edge of the main (A), which means that ONAX B and XAB are the same location. This is

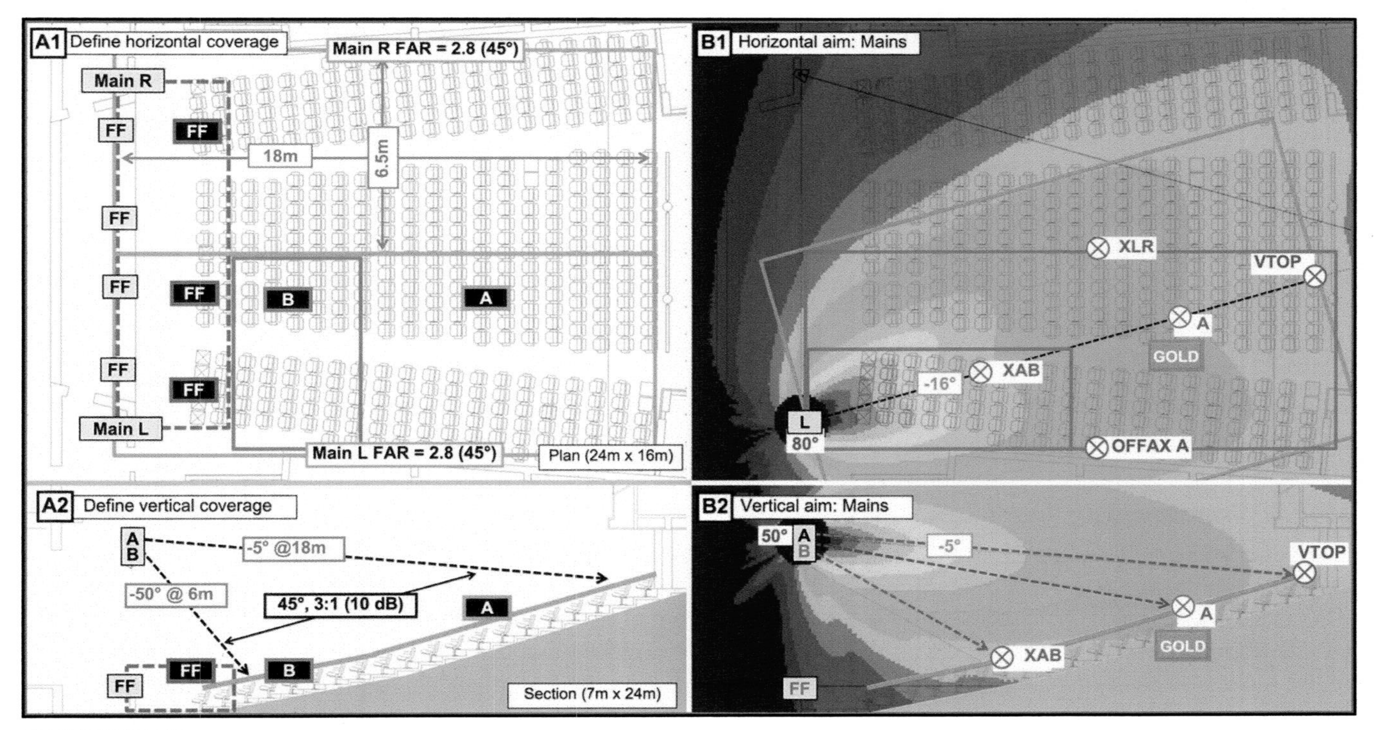

FIGURE 15.2A
Design process for Venue #2: coverage partition strategy, solo mains (A)

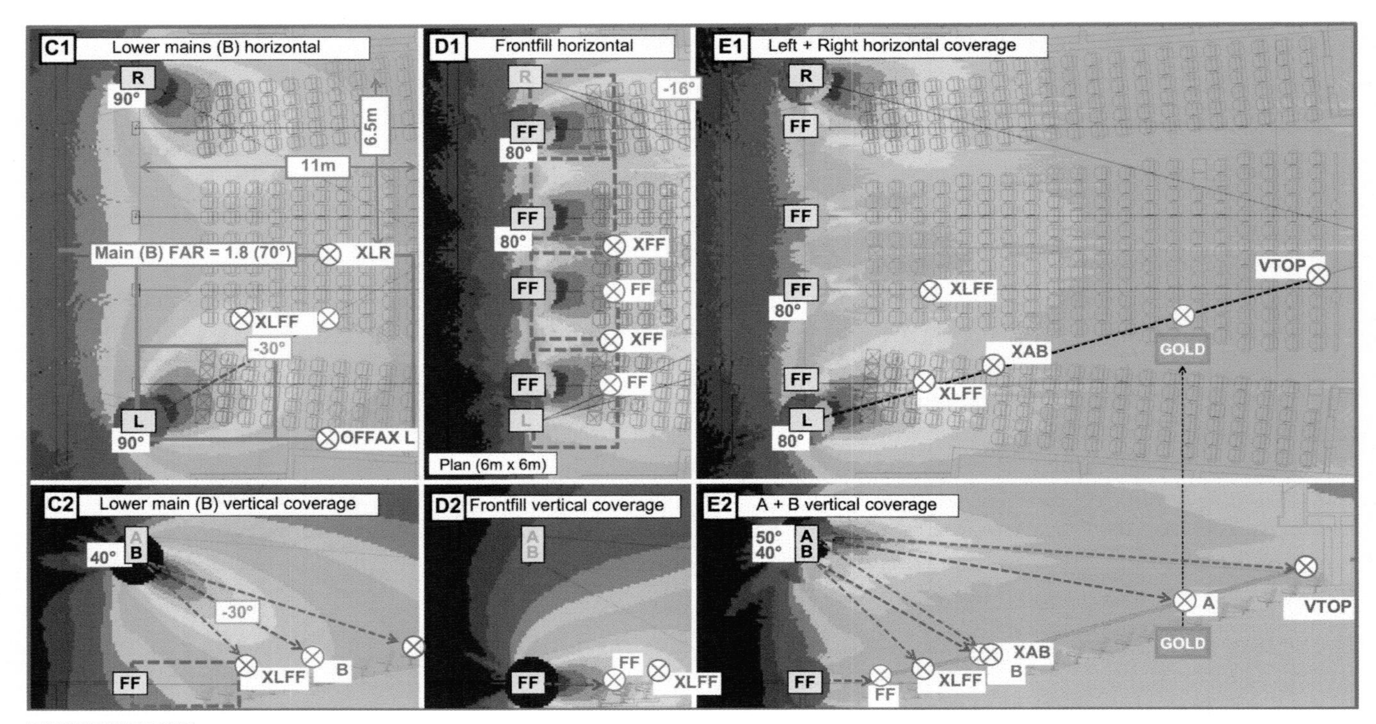

FIGURE 15.2B
Design process for Venue #2: solo mains (B), solo frontfill, solo and combined mains AB + FF

typical of highly asymmetric combinations (which is the case here). Element A is aimed at -5°, which puts XAB at -30°. Element B is a 40° device that can make it to -50° (that's 20° below its aim at -30°). We can conclude from this that we have enough vertical coverage for the shape.

The B element does not need to copy the horizontal aim of the unit above. Its range is limited to the front half of the room so we should truncate the shape and recalculate the horizontal aim based on the reduced depth. This leads to a 30° inward tilt. The coverage angle is also reconsidered here because the FAR shape is different (same width but smaller depth). Therefore a wider speaker is warranted here (a 50° model is definitely not an option here) and a 90° speaker was selected.

The last item is power scaling. Element B is covering around half the range of A (6 dB) and is aiming its on-axis response into the off-axis edge of element A (another 6 dB advantage). It would be a waste of money to use the same element for A and B because the B element can be 6–10 dB less powerful without worries.

15.2.2 Frontfill (FF)

The frontfill spacing is set by the distance to the front row. The stage is a straight line so the calculation is easy. The elements are 80° so the spacing is 1.25 × the distance to the first row.

15.2.3 Combined systems

The upper and lower systems cross at XAB. Element B will be delayed to sync with A here and its level adjusted to combine to approximate the GOLD reference (the element A response in its prime area). We are combining a highly tilted response (A solo) with a flat one (B solo), so the combined response will likely be somewhat louder than the GOLD response. The combination effects will be felt more strongly below XAB than above due to the level dominance of the A element. Asymmetric EQ may improve things by reducing the B element's LF contribution with shelving filters.

The frontfills join the AB mains at XB–FF, which is at the second or third row (depends on blockage). The frontfills are delayed to the arrival of mains element B and level set to combine to match (or slightly exceed) the GOLD reference.

15.3 VENUE #3: SMALL SHOEBOX, L/R, DOWNFILL AND DELAYS

This is such an extreme version of the shoebox hall that it can only fit a single shoe. The horizontal shape has an FAR value of 3:1, which means a 40° center cluster could cover the room. The vertical plane is a steady rake with a flat ceiling. The ceiling height is very low by the time we reach the back of the hall. This type of vertical shape is beautifully solved with the modern line array (asymmetric composite coupled point source). The client wanted an L/R system so a standard line array seems like the standard solution, except for one small thing: These are first-order systems in the horizontal plane, typically around 90° of coverage. That's 180° of speaker in a 40° room. The video also has a strong impact as this wiped out the standard sound reinforcement locations. The mains were given the uppermost corners with not even enough room to hang a downfill (those were uncoupled from the mains and hung several meters below).

15.3.1 Mains (L/R, AA)

Because our main speakers were coupled to the sidewalls we wanted to ensure they didn't spill a large amount of their pattern outward. An array comprised of two second-order elements with 20° of horizontal coverage provided a sharp edge in the horizontal plane. The L and R mains each had 40° of horizontal coverage so the overage was perfectly reasonable. Because the hall is so skinny there is actually a sizable percentage of listeners within the "stereo possible" window, so our overlap should work well.

The L/R main was designed as an AB pair but the calibration settings turned out to be identical so it's now an AA symmetric pair. Flexibility is better to have and not need, than need and not have.

The mains have a wide vertical coverage (60°) and cover the majority of the depth. Delays finish coverage at the top. L/R downfills and stage lip frontfills finish coverage at the bottom.

15.3.2 Downfill (DF)

The horizontal aim for the downfills was evaluated differently than the mains as was done previously in Venue #2. The coverage depth was much shallower than the mains so the elements were turned inward 25°. The vertical aim was directed at XAB where the flat response of the downfills meets the spectrally tilted response of the main's underside. Delay and level are set to achieve a combined response that matches GOLD as much as possible.

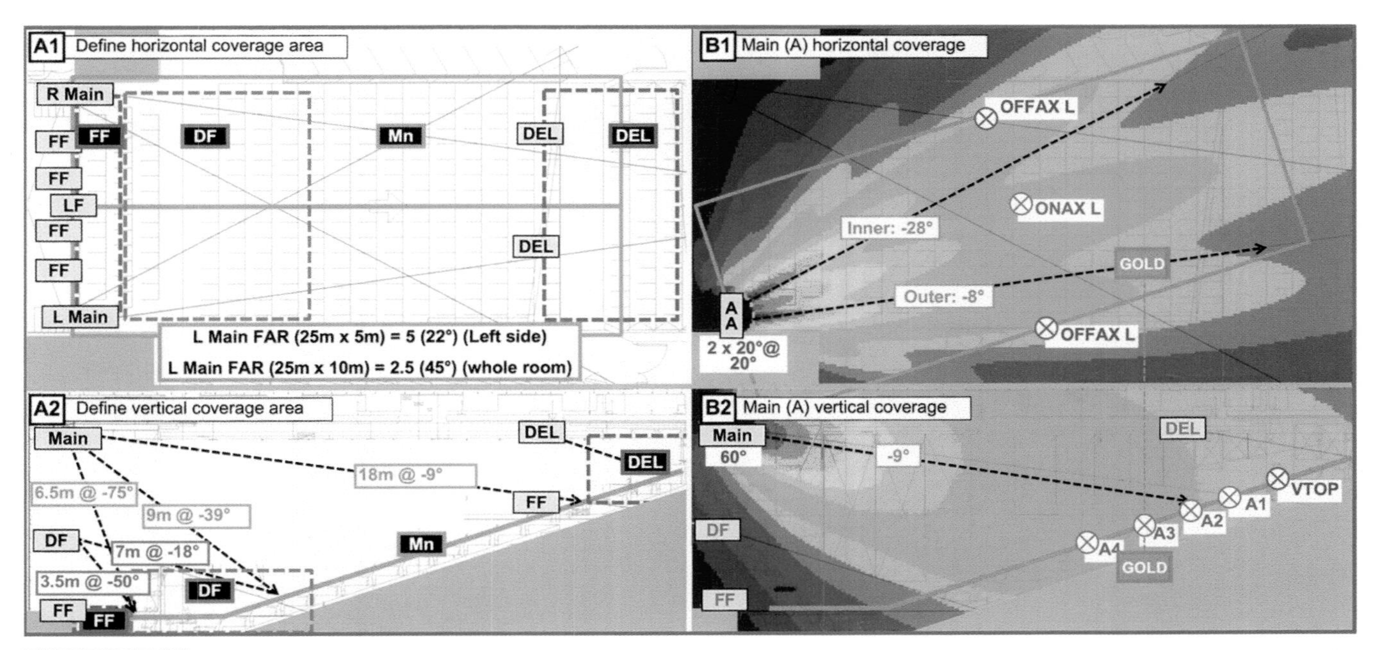

FIGURE 15.3A
Design process for Venue #3: coverage partition strategy, solo mains (A)

15.3.3 Frontfill (FF)

The horizontal spacing was determined through the uncoupled line source design reference (Fig. 11.20). The unity line was the first row. Unfortunately this also turned out to be the limit line because the speakers were mounted too low to reach the second row. Fortunately the downfill system was able to begin its coverage there. The frontfills were timed to sync to the downfills. The outer frontfills were closer to the downfill so they received a shorter time than the inners. This is another case of delay dilemma, which weighs in favor of the larger number people affected (only one seat/side is in the FF1–FF2 crossover range). You might ask why we didn't time things to a fictitious stage source (another player in the delay dilemma game). The digital age has largely handled this concern for us. Console and signal processor latency give the stage a healthy head start. By the time the downfills come down and in to the meet the frontfill they are definitely behind a nominal stage source (which is a guesstimate at best). Therefore this dilemma solves itself and leaves us to set to delay to the fixed and stable relationship between DF and FF.

15.3.4 Delays (DEL)

Once again the horizontal spacing was determined through the uncoupled line source design reference (Fig. 11.20). The unity line was designed to fall at an advantageous location: the first row following a cross-aisle. This put the most volatile part of the combination (the underside of the delays) at a location where no listeners will sit. The vertical aim was the last row (VTOP) because the range ratio is >2:1. Timing and level were set at ONAX DEL, which also served as XL–DEL

15.3.5 Combined systems

The most challenging aspect of the combination was fitting the downfills into the box between the mains and the frontfills. The mains overreach their coverage, so the downfills muscle their way in to stop the spectral tilt. The result was minimum spectral variance with a small cost in level variance (a rise of about 2 dB in the DF seating area). This is a trade I will take, especially in a small hall where the stage sound is a strong presence (it will be louder in front as well). On the other end the frontfills did not reach their original meeting point. We needed all of the downfill's 90° × 40° coverage to meet with the inner frontfills.

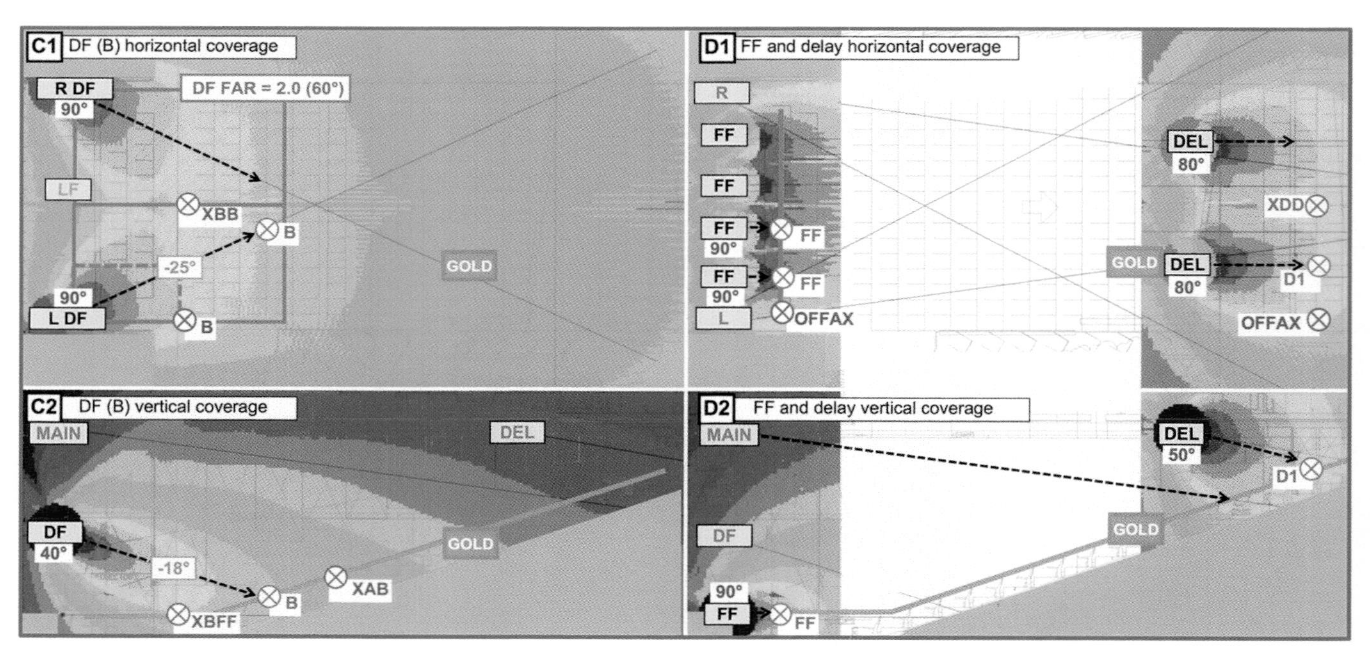

FIGURE 15.3B
Design process for Venue #3: solo mains (B), solo frontfill and solo delay

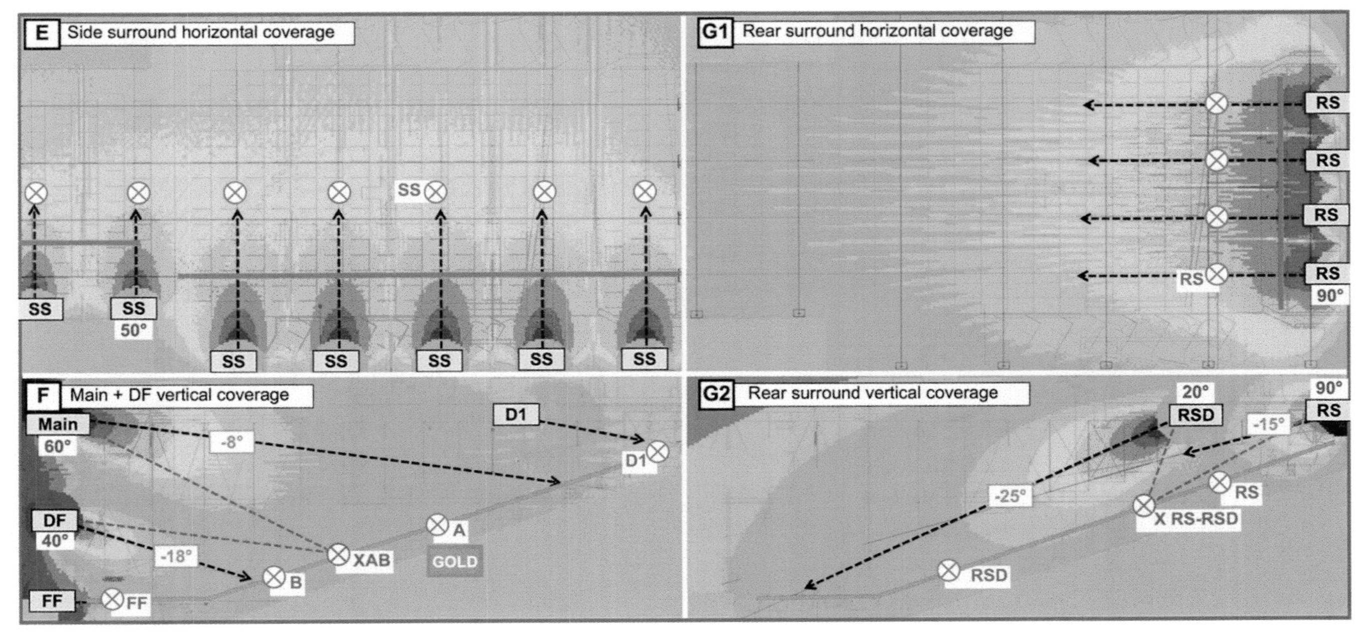

FIGURE 15.3C
Design process for Venue #3: combined mains AB + FF + delay, side and rear surrounds

15.3.6 SIM3 optimization data

Here are some details on the SIM3 traces acquired during the tuning. Refer to Fig. 15.3D. Panel A1 shows the response of ONAX A1 and A2 in the approximate positions shown in Fig. 15.3A(B2). The responses are spectrally matched but A_1 is 2 dB below A_2 (the traces are shown offset by 2 dB). Recall that the upper main system (A) covers most of the room. Therefore, multiple mics were used to ensure the aim was correct. We expanded our perspective downward by including mics A_3 and A_4 (Fig. 15.3D, panel A2). The responses were still spectrally matched, which gave confidence to the EQ settings. The minimal ripple variance (and very high coherence) occurs because the room was acoustically optimized for amplified sound reinforcement (highly absorptive) and was designed from the start to utilize Meyer Sound's variable acoustic Constellation™ system. This is the best possible environment for system optimization. The total level variance was 3 dB from A1–A3. ONAX A3 became the GOLD reference for the rest of the tuning.

Panels B1–B4 show the process of adding the downfill. Variations of this sequence are repeated for many of the remaining examples so I will fully detail it here, being more brief as we go on. Do we need the downfill? This is assessed in panel B1 by seeing what we have without it and comparing that to GOLD. We are comparing the underside of the mains with its ONAX response. Do we need it? For level variance, no (we are even). For spectral variance, yes. The reality here is that it's going to get louder in front because we have a large-range ratio (10 dB) and limited vertical steering. We can, and will, make it spectrally consistent. Panel B2 shows what will be combined: the spectrally tilted underside of A with the flat-on axis signal from B. This reverses the spectral variance down there. Panel B3 shows the AB combination, which reveals a rise in the HF with negligible effects elsewhere, i.e. spectral un-tilting. The final step is comparison to the GOLD standard, which reveals that we have minimized the spectral variance. The level variance is 3 dB (the XAB trace is shown offset -3 dB).

We tackle the frontfills in Fig. 15.3E, panels C1–C3. We know we need them so we move on to preparing for combination. Panel C1 includes the phase data of the two speakers to be combined (after the delay has been set). Are they phase compatible? Yes, extremely. Don't take this for granted. Not all speakers are created equal (or compatible). Panel C2 shows the three parts of the combination: B solo, FF solo and B + FF. Panel C3 verifies the combination by comparing it with GOLD. We maintained a matched spectrum to GOLD and added no more level variance. The combined level here is +3 dB over GOLD (same as XAB). Our final stop is up top where we will add the delays. The process is the same (surprise!): assess, prepare, combine and verify. These are laid out in panels

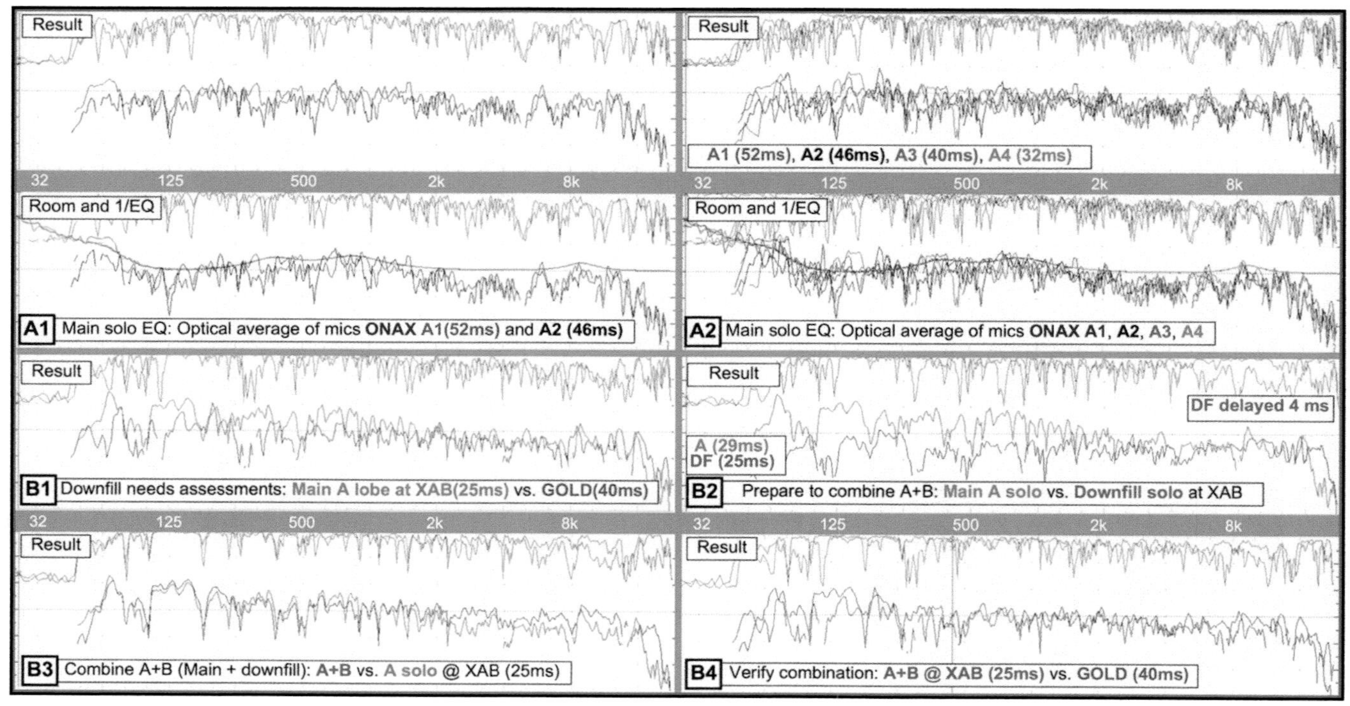

FIGURE 15.3D
SIM3 optimization data for Venue #3

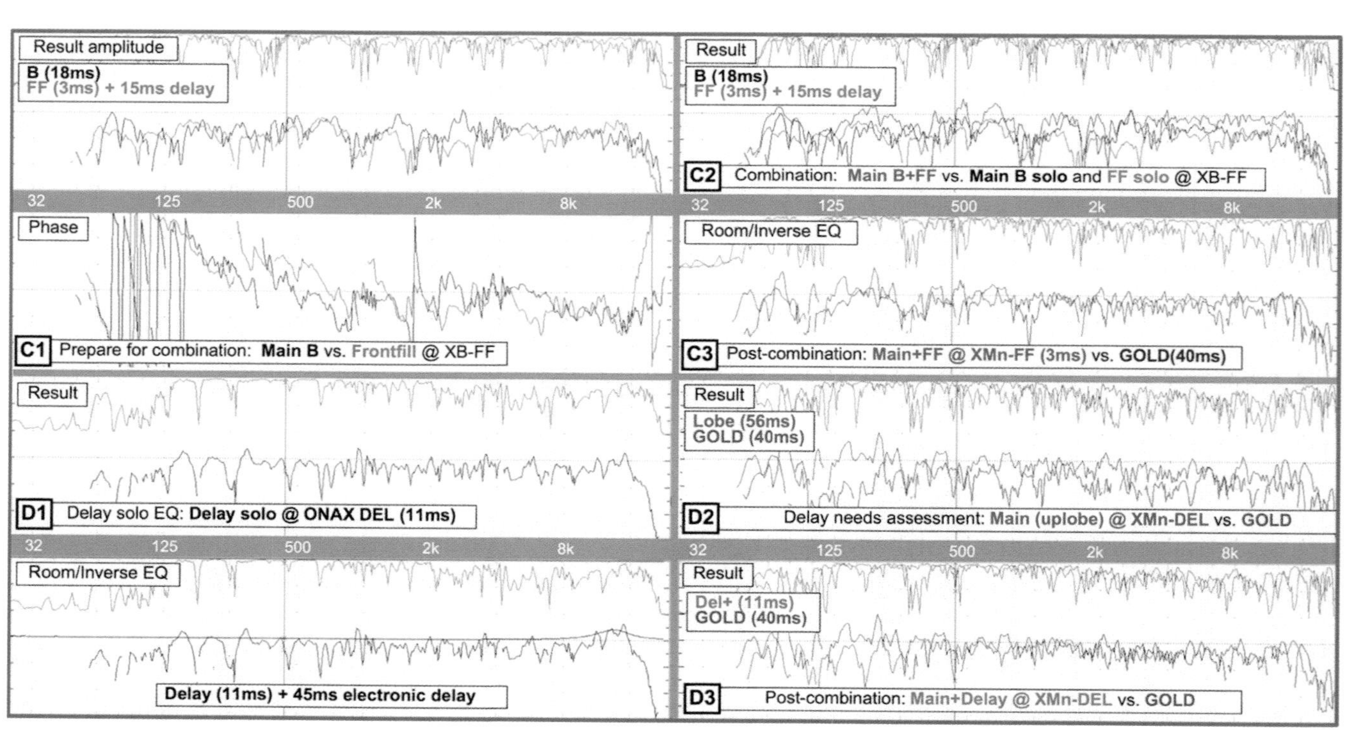

FIGURE 15.3E
SIM3 optimization data for Venue #3

D1–D3. The equalization that was applied is shown in panel D1 (a single filter in the VHF range). This is a case where we can't improve it much so just don't &*%# it up. The needs assessment shows that the mains are -6 dB down in the rear rows compared with GOLD. The coherence is also low there, so we have double confirmation that we should use the delays. The final panel verifies that we have achieved minimum spectral, level and ripple variance between the last rows and GOLD. This family of SIM3 traces shows an unbroken line of minimum variance through four subsystems (DEL–A–B–FF) from the last to the first row.

15.4 VENUE #4: MEDIUM HALL, TWO LEVELS, L/C/R WITH FILLS

This is a theater with two sound system designs in the same room: L/R and C. The L/R system covers the complete room on its own, as does the center system. The application is musical theater, an environment that can take full advantage of an L/C/R system. The L/R system is placed in proscenium towers. It's an asymmetric point source in both planes: upper and lowers, each of which are inners and outers. Subwoofers are also designed and tuned as part of the L/R system. The center main system is a composite point source over the proscenium along with sidefills down low in the towers and underbalcony delays.

15.4.1 Left/right upper/lower mains (L/R, Up, Lo)

We've got inners and outers, upper and lowers. Where do we start? The order of operations favors coupled pairings first over uncoupled. Inners and outers are directly coupled. Uppers and lowers are separated by a few meters within the towers. Therefore we marry inner and outer on each floor before connecting them together at the balcony front.

The inner/outer connection is in the horizontal plane. The outer speaker sees the two levels quite similarly because they are nearly the same depth. The inners see totally different things: people on the floor and air upstairs. The lower inners were level tapered to prevent overheating the center. If allowed to remain at full level, the center seating area could easily rise 10 dB above GOLD. It's a very square room so the center is much closer to the speakers than

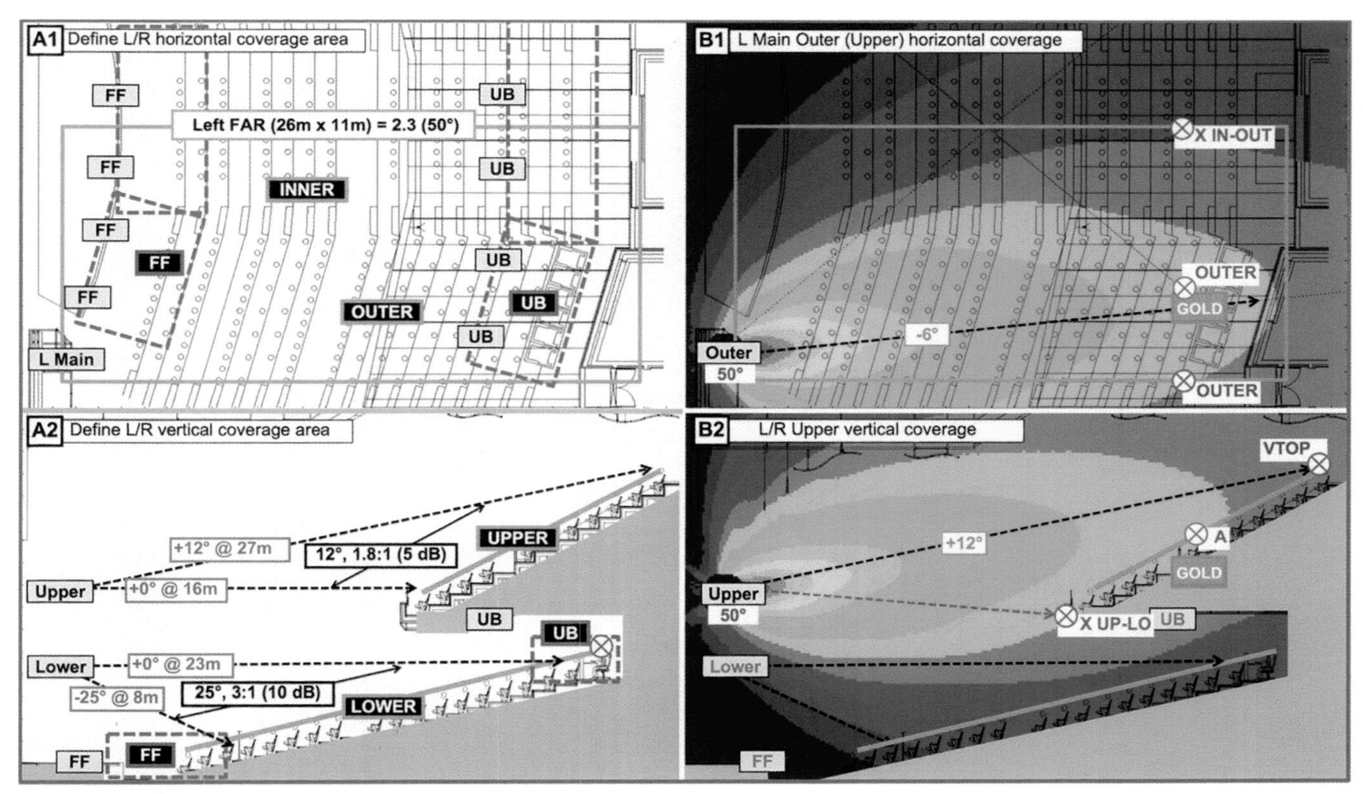

FIGURE 15.4A
Design process for Venue #4: coverage partition strategy (L/R), solo mains (upper)

the rear. Both L and R's inner speaker cover the center seating, whereas the outer pretty much goes it alone in its coverage area. Therefore a center-panned signal will overload the center area if we don't take steps (i.e. taper down the inner level 3 dB). The splay angle was adjusted by the compensated unity splay method to 35° (50° elements × 70%). The combined shape squares off the coverage to minimize the level variance in the center.

Upstairs we can see that the inner speaker covers the opposite side as much as its own. Its role is largely spatial enhancement, which is acceptable for this application because the L/R system is for music reinforcement, not voice. We could have concerns if the walls were highly reflective, but we knew they were acoustically absorptive. An alternative option would have been to reduce the splay, which would increase the overlap, add ripple variance and couple for more power. TANSTAAFL would have leaned this way if the room was reflective and the client was Megadeth.

The vertical roles of the upper and lower systems are clear-cut with a balcony to divide them. The upper/lower range ratio was 2 dB, which was level compensated (lowers turned down 2 dB). The vertical aim requires minimal calculation for either system. The lower's range ratio exceeds 2:1 and the upper is facing almost straight up the rake. The conclusion is the same: Aim them at their last rows.

15.4.2 Center main (ABC)

The center main is a single asymmetric composite coupled point source. The overall vertical coverage spans 52° with a 2.5:1 range ratio (8 dB). We would love to avoid the balcony front but we can't not get there from here (a twist on an old phrase). Balcony avoidance is unworkable here due to the smiles and frowns effect (Fig. 11.24). We are high above a balcony that's almost straight, which makes our speaker very sad. Instead the strategy is to approach the vertical plane as a single slope. It's not perfect, but it's not a foolish overreach for something made of unobtanium. We have ten boxes so our average splay is 5° (52°/10) and the splay ratio is 2.5 (9°/4°), a match for the range ratio. Our vertical coverage extends from the top to the third row where we cross over to the frontfills. The underbalcony area is shadowed from the mains and therefore has mandatory mini-mains to cover there.

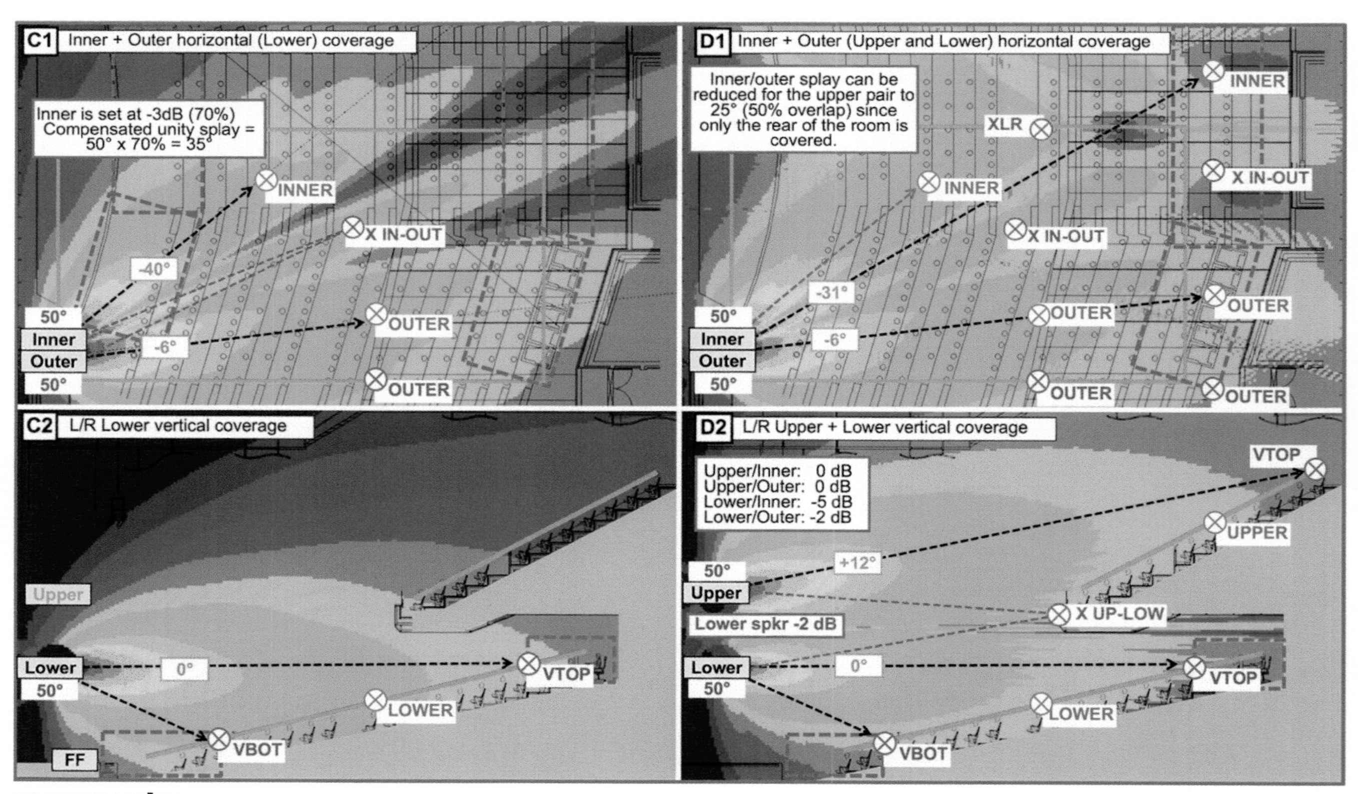

FIGURE 15.4B
Design process for Venue #4: solo mains (lower), combined mains inner/outer + upper/lower

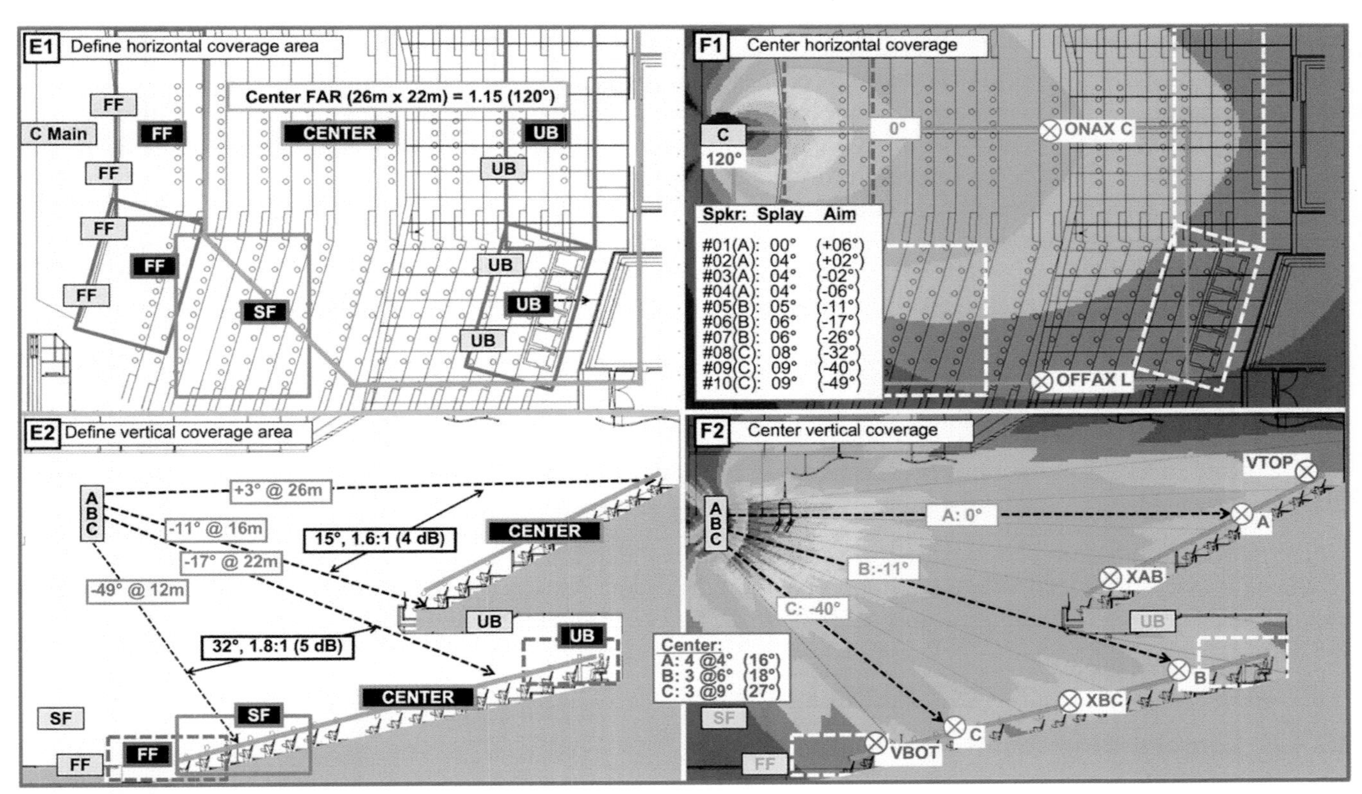

FIGURE 15.4C
Design process for Venue #4: coverage partition strategy for center (C), solo mains (ABC)

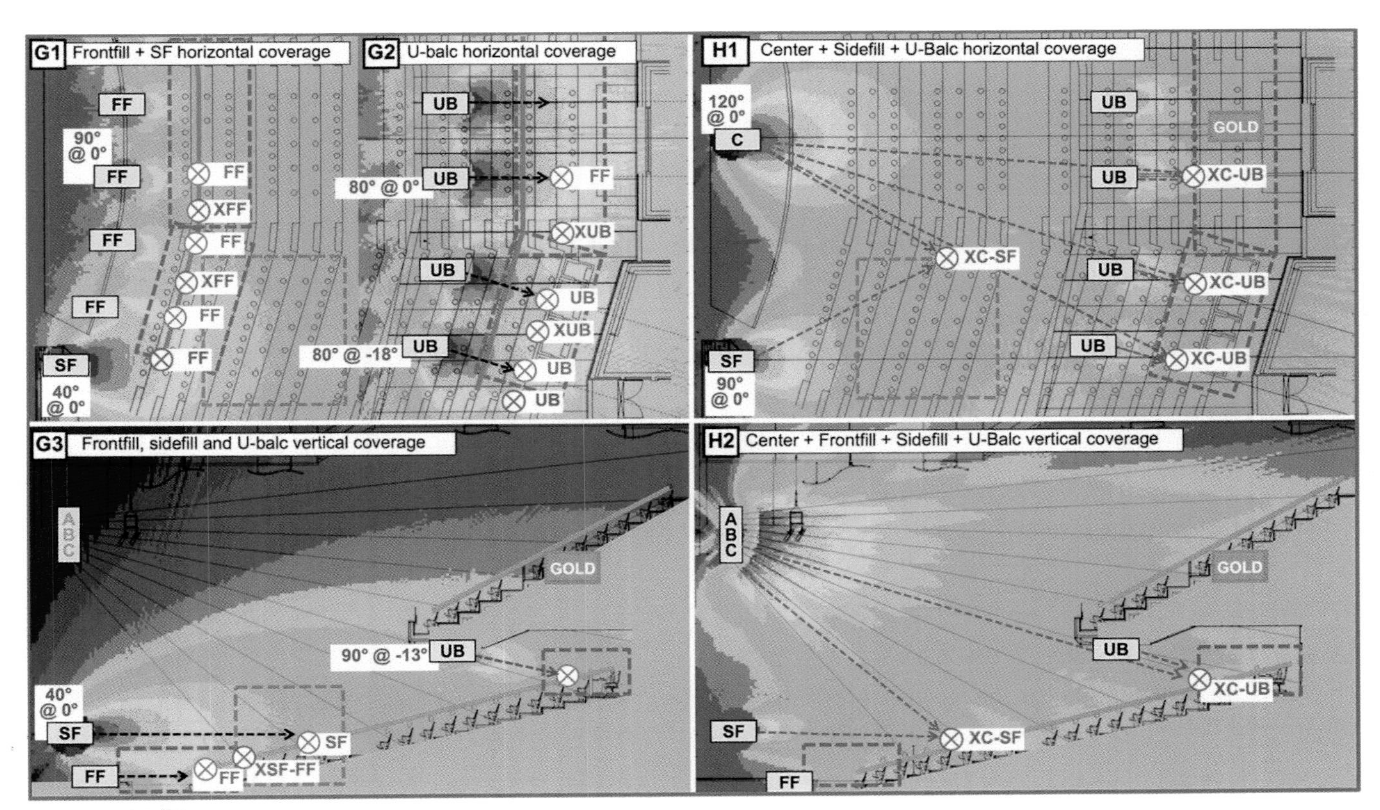

FIGURE 15.4D
Design process for Venue #4: sidefill + frontfill, solo u-balc, mains + sidefill + frontfill + u-balc

15.4.3 Center channel sidefills (SF)

The outer corners near the front need help to counter level, spectral and sonic image variance. The sidefills cover this area and improve all three scores. Sidefill level depends on the depth of the coverage gap, which is determined by the lateral width of the center mains on the floor. This hall was used as an example application for the technique back in Fig. 11.10. In this case we can finish the sidefill coverage at the cross-aisle, which provides a convenient exit point. The vertical aim is at the desired end point, which ensures that the level and spectrum blend gracefully and prevent it from getting too loud in front. The safe approach to these types of fill applications is to use low-order speakers rather than attempt precise surgical maneuvers with sharp edges. Proper level, aim and delay can soften the blends and prevent shocks to the sonic image, spectrum or level.

15.4.4 Underbalcony delays (UB)

The mix position is in the occulted area under the balcony. The most critical audio signal (the vocal center channel) will be mixed in a bunker on a tiny fill speaker used as a main. Nice. Spacing and aim are set as usual, but we can make allowances here to ensure the mix position is as close as possible to GOLD.

Underbalcony occultation begs the question of where to set the delay. There really is no crossing in the crossover, a whole new category of delay dilemma. I don't have a right answer (except a jackhammer) but will tell you what I do in the field: move the mic out to the last seat before the blockage and set the time there. Level and EQ are set back at ONAX UB.

15.4.5 Combined systems

The sidefills and frontfills are all that remain for us to combine for the center channel system. The SF level is set to match GOLD in its isolated area of coverage (ONAX SF). Timing and splay angle are adjusted at XC–SF to achieve a combined response that matches GOLD. Frontfill level and delay are set at the crossover XC–FF to achieve a combined level matched to GOLD.

The L/R system has already combined inner and outer, upper and lower. All that is left for it is the subwoofers. They are just a pair on each side, located very close to the lower system. Essentially we have a coupled crossover to the lowers and uncoupled to the uppers, so the timing choice highly favors going with the lowers.

15.5 VENUE #5: MEDIUM FAN SHAPE, SINGLE LEVEL, L/C/R

The venue is a simple fan shape with proscenium stage. The program material is theatrical/showroom productions that will run for an extended period of time. This makes an L/C/R system a viable alternative because the shows will have ample time to properly matrix the channels. In this case the principle channels would be center for vocals and L/R for music.

The main center cluster required two elements in the horizontal plane because a single unit could not fill the entire fan. Therefore the main center array is a type AA symmetric coupled point source (horizontal) and a type ABCD asymmetric composite coupled point source (vertical). The center channel also includes uncoupled sidefills near the stage level, which provided coverage and reduced vertical image distortion in the near outer seats.

15.5.1 Left/right mains (ABCD)

The horizontal aim strategy is the same as previous examples. We need at least 70° of coverage so the 90° model chosen will have only small amounts of overlap onto the sides and across the center.

The vertical target is a steady rake from a low position that yields a 5:1 range ratio over 36°. That's a lot of range but fortunately not a lot of angle. We have eleven boxes broken into successive splay doublings of 1.5–3° and 6°, and then finish at 9°. This gives us a 6:1 splay ratio to counter the 5:1 range ratio so we are in good shape.

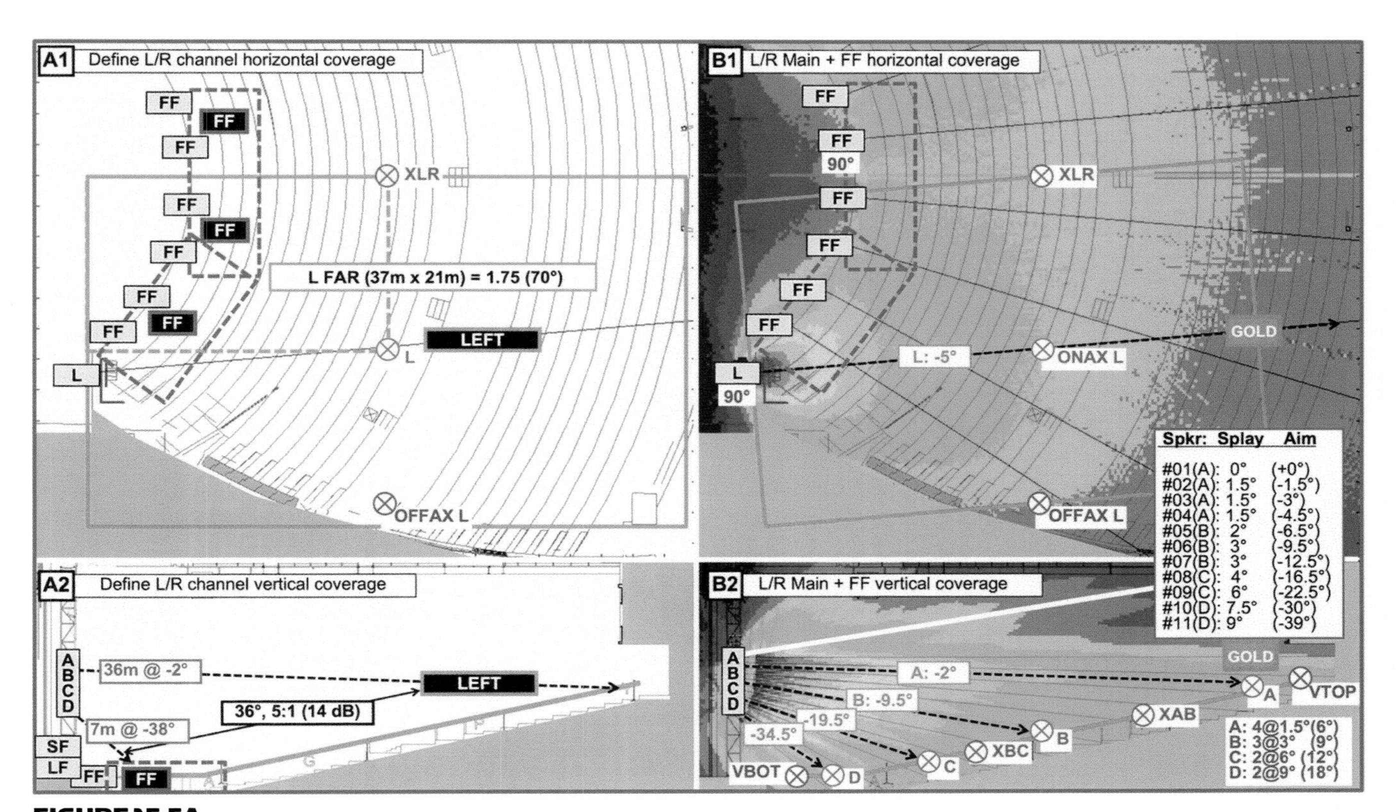

FIGURE 15.5A
Design process for Venue #5: L/R coverage partition strategy, combined L/R main (ABCD) + frontfill

15.5.2 Center main (ABC)

The horizontal splay between the two 90° elements of the center cluster is a compromise value. The unity splay of 90° gives more coverage than needed and splashes too much on the side walls. We also knew we had the center channel sidefills to cover the outer extremes so we did not need to go all the way out with the center cluster. Reducing the splay to 60° balanced the overage between the center and the outside edges.

The center's vertical coverage target had the same VTOP and VBOT as the L/R mains but from a very different perspective. The high location cut the range ratio in half but increased the coverage spread from 36° to 60°. Again the solution was a steady progression of increasing splay from A to D, but with wider elements at top and bottom and a reduced splay ratio.

15.5.3 Center channel sidefills (SF)

This is a wide hall and we did not want to have to splay the center mains out so far that they faced into the side walls. A powerful sidefill deck system with strong horizontal control would ensure that we could penetrate vocal content along the outer edges of the fan with minimal risk. Arrays with 3 × second-order 20° elements were used that had level taper flexibility to custom shape as needed on site. The area they need to cover could be found by the lateral width method (shown in Fig. 15.5B (D1)).

15.5.4 Combined systems

The L/R and center mains are combined as ABCD coupled arrays. Each layer adds some low-frequency buildup to the combination, which can be incrementally compensated. The center differs from the L/R arrays in that the AA combination is fully correlated (i.e. mono). The combination of the center cluster's two sections is a coupled fully correlated (i.e. mono) AA combination. Therefore the combined spectral effects are compensated. This contrasts to the L/R clusters, which are both uncoupled and semi-correlated (i.e. stereo), and therefore are left with their combined effects uncompensated.

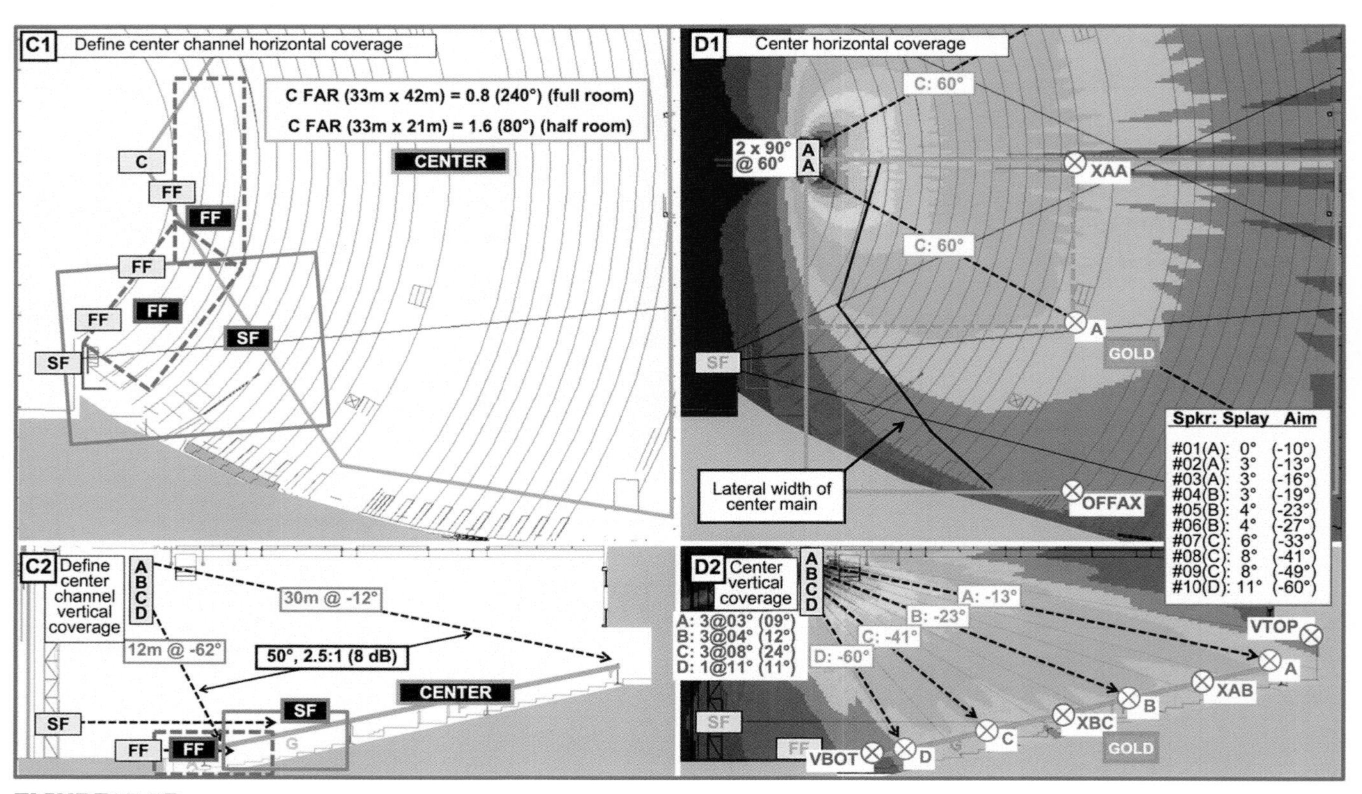

FIGURE 15.5B
Design process for Venue #5: center coverage partition strategy, solo center (C) mains (AA, ABCD)

The L/R system connects to frontfills and subwoofers while the center cluster connects to frontfills and sidefills. Obviously the optimal delay time for the frontfills would be different for the L/R system and the center, so which do we choose? The center system gets priority because it is carrying the vocals. I am a big fan of intelligible tuba but vocals take priority when I have to choose.

The grayest decision in the combination set is the merger between the center cluster and its sidefills. The sidefill level setting has a strong effect on the crossover location, which determines the delay setting. This is best determined on site during calibration by searching around to find the locations where the center mains drop in level or become excessively tilted. The crossover is expected to lie between the house center and the on-axis line of the sidefills (typically closer to the sidefills than house center). This is an uncoupled combination so there is probably very little (if any) combined equalization.

This is a musical theater showroom with in-house productions. These types of applications almost always drive their subwoofers from an auxiliary feed. In such cases I just time the subs for mid-depth and set a level that would meet the target curve if they sent a unity level signal through the aux send. In such cases I leave the mains and subs with an overlapping crossover.

15.5.5 SIM3 optimization data

Panels A1–A3 of Fig. 15.5D show the family of EQ curves used for the L/R mains (ABCD). SIM3 has three transfer functions: room (EQ out vs. mic), EQ (EQ in vs. out) or result (EQ in vs. mic). These responses are shown for the each composite element (solo) at its ONAX location. The electrical responses are shown in panel A2. The VHF range is different for each, with A getting a boost and D getting a cut. This is because the longest throw system (A) has the most air loss. The EQ in the LF range also varies over the set of four composites but the result (panel A3) shows very well-matched solo spectrums.

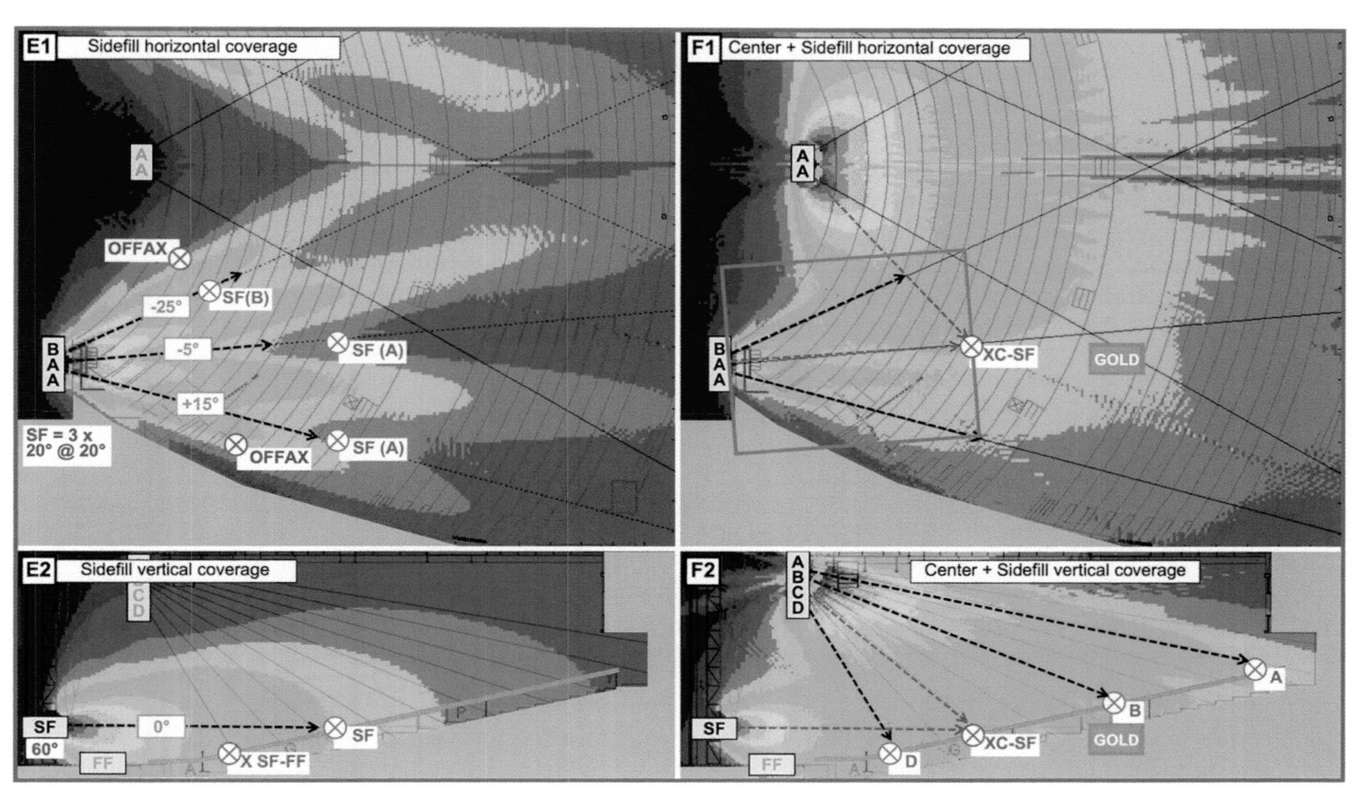

FIGURE 15.5C
Design process for Venue #5: solo sidefill, combined center (C) mains + sidefill

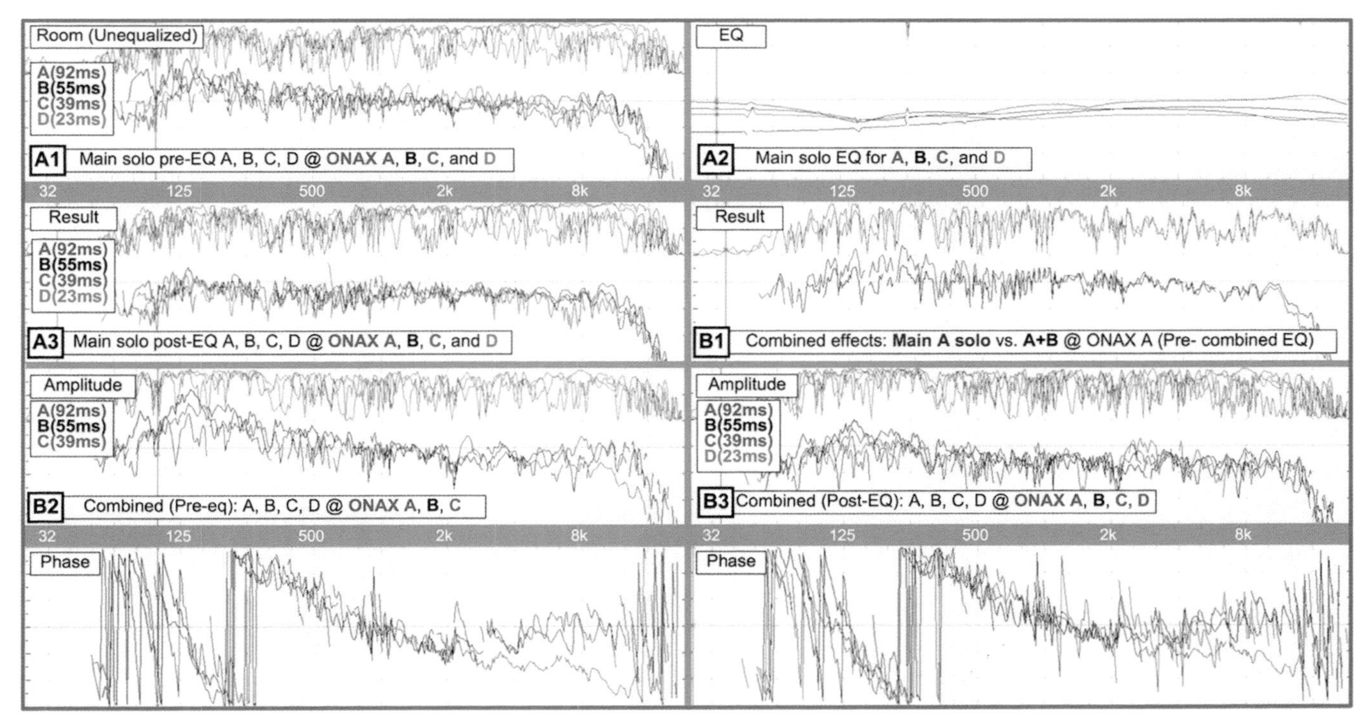

FIGURE 15.5D
SIM3 optimization data for Venue #5

The next step is combination, which begins with A + B and continues from there. Panel B1 shows the beginning of the process by showing the spectral tilt that happens at ONAX A when we add the B section. The combined EQ will seek to control the tilt. The next panel (B2) shows the tilt of the full ABCD combination at various positions. The final panel in the series (B3) shows the final result after combined EQ. Notice that the response has more tilt overall than the equalized solo responses (which were fairly flat). My EQ strategy here was to keep the soloist flat and then let the combined response tilt us into the "target curve" area for this application. The combination tilted more than desired (shown in B2) and then was brought back down to the target (+6 to 10 dB in the LF range) with combined EQ. The tilt is greatest in the middle of the vertical coverage (ONAX B and C). This results from the beam concentration behavior described back in Section 9.4.2.2. The level variance is 1 dB (it's louder at the bottom), which is notable in light of the fact that the range ratio from ONAX A (92 ms) to ONAX D (23 ms) is 4:1 (12 dB).

I have included the phase traces here for those who might fear that unmatched EQ would cause them some problems. Notice that the combined phase traces are well matched.

The second set of traces (Fig. 15.5E) shows the calibration process for the center mains. This array has less range ratio and more angular spread than the L/R mains. Panels C1–C4 detail the combination effects at a single position (ONAX A) as elements are combined (without combined EQ applied). It begins with A solo, adds B, then C and D, and then finally the entire other half of the AA point source (horizontal) array (ABCD). The strongest effects are the closest neighbors (A + B, vertically) and A + A horizontally. The folks near ONAX are only minimally affected by the C and D sections of the array.

The final set (D1–2) shows the combined EQ from top to bottom and the end result. A total of seven mic positions span the two drawings (GOLD appears on both for reference). The amount of equalization is substantial because the boxes start with individually flat responses. Large quantity at wide angles set the spectral tilt mechanism to maximum. The combined spectral variance is very small after EQ. Our target curve is flatter in this array because this will be the vocal system. The combined level variance is just 1 dB over the 9 dB range ratio between ONAX A (92 ms) and VBOT (37 ms).

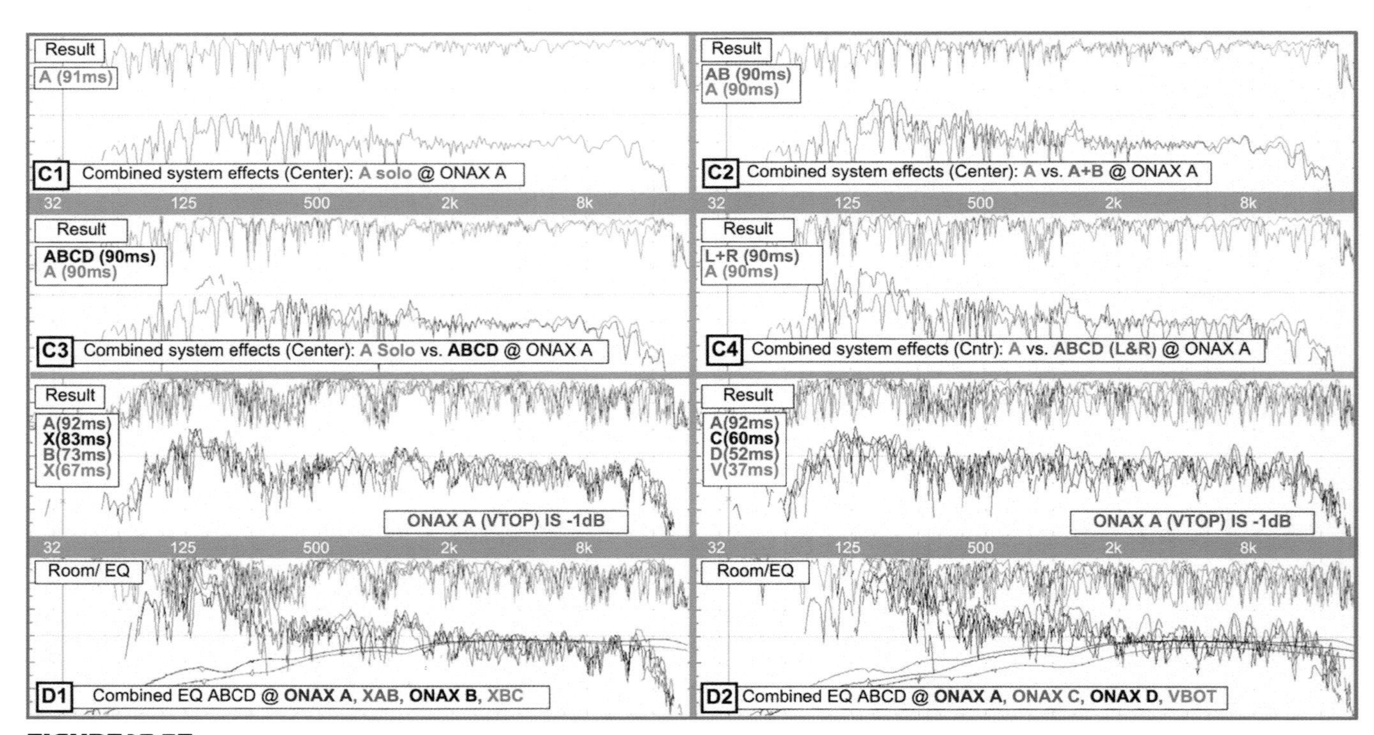

FIGURE 15.5E
SIM3 optimization data for Venue #5

15.6 VENUE #6: MEDIUM HALL, L/R WITH CENTERFILL

This is another fan-shaped venue but the application is a rental with a simple high-power L/R system. In this case the design begins with "we have twelve boxes/side" and we go from there. The range ratio is fairly high (7:1, around 15 dB) but the rake is a simple diagonal so we can approach this with a series of sequential doublings. The horizontal coverage expands with distance and our 70° speakers are wide enough to fill the width (we need at least 60°). There is a gap in the middle that will require centerfill, but none on the sides because we are at the outer edge of the fan. The remaining parts of the system are frontfills and subwoofers.

15.6.1 Left/right mains (ABCD)

The horizontal aim follows the middle/middle guidelines shown in Fig. 11.25. Coverage and aim can be verified by measuring across the width of the room at mid-depth. The vertical approach is to create a diagonal line of minimum variance, which is accomplished by progressive splay doubling. We are seeking to counter a 7:1 range ratio in twelve boxes so we have to start from a small angle (1°) and work up to 7° at the bottom. The upper composite elements have more overlap and more quantity, which maximizes our ability to cut the diagonal contour.

15.6.2 Centerfill (CF)

The gap area that needs centerfill is a triangle whose bottom touches the frontfills and top meets L and R. The centerfill is needed most at its bottom (which is closest) and least at its top, where we will see it fade away under the mains. The centerfill cluster is also a line array (an AB composite) whose more important member is the smaller one (B). We don't need or want this array to lay a surgical unity line from front to back. The sharp edges that we can make with tight splay angles are a disadvantage here. Help is on the way as we move back, so we can allow this array to gracefully retreat at the rear. This is why the splay ratio is less than 2:1 between the elements.

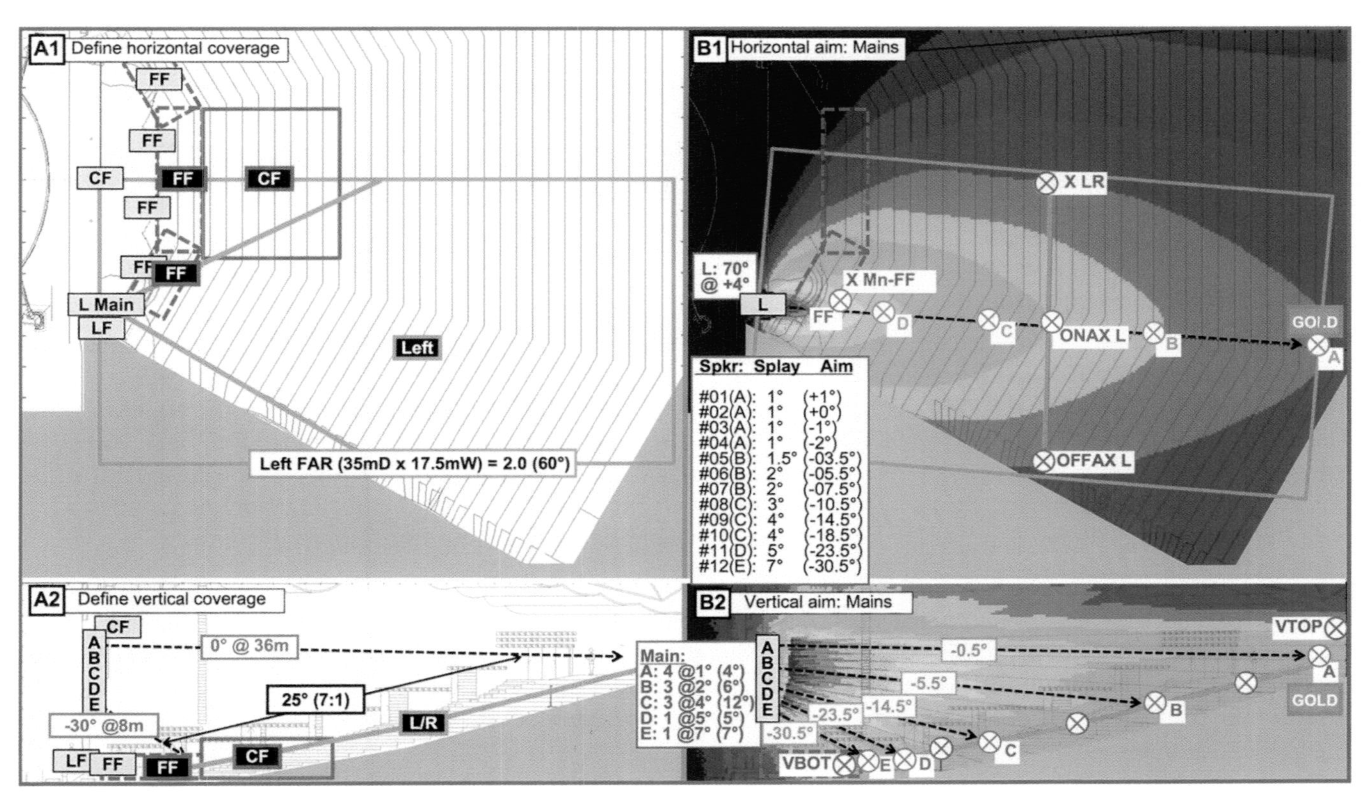

FIGURE 15.6A
Design process for Venue #6: coverage partition strategy, solo L/R main (ABCD)

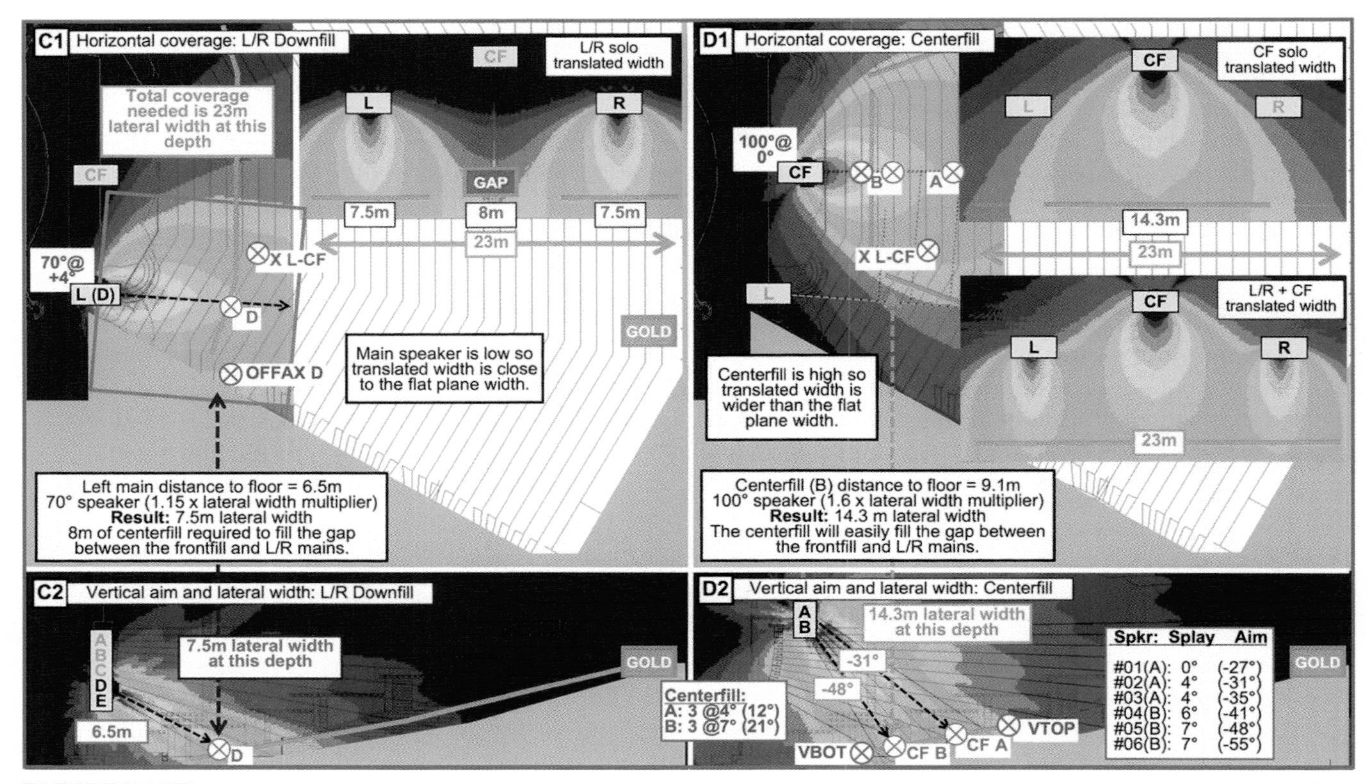

FIGURE 15.6B
Design process for Venue #6: solo L/R downfill (D), solo centerfill, main + centerfill

15.6.3 Combined systems

The low-frequency range rises as the composite elements of the L/R mains are added together. In this case I EQ'd the individual composites fairly flat and then let the combined response provide the desired spectral tilt of the "target curve." The LF array for this application was a simple L/R pair of gradient-inverted stack arrays. They were in close proximity to the mains and timed to sync with them at the middle depth. The crossover was a second-order unity type because there were no worries about image separation.

15.6.4 SIM3 optimization data

We are going to move right to the combined EQ because we have just analyzed a similar L/R main array in the previous venue. This one has an even greater range ratio challenge (101 ms vs. 18 ms, i.e. around 15 dB). The biggest difference in EQ between the five subsystems is found in the VHF region again to counter the air loss. There are also variations in the mid-high EQ to maintain a constant level in that range from front to back. The unequalized response (room) on Panel A1 shows how far apart the mid-high and VHF responses were to start with. The equalized response (result) shows how close together we got them after asymmetric EQ. Panel A2 compares top with bottom. The 15 dB range ratio is reduced to a 2 dB level variance with minimum spectral variance.

Now onto the really glamorous part: frontfills (panels B1–3). The notable feature in panel B1 is the filter at 16 kHz that cuts down a perfectly innocent VHF response in the solo frontfill. This is done because our mission is to match (not exceed) GOLD. The frontfill has a tiny tweeter and only 2 meters of air to travel through. GOLD (and everywhere else) can't talk to the dogs so we have to penalize the frontfill and limit its range. If you don't think to do this here (and underbalconies) you will get to experience "bacon fills," the sizzling sound of speakers playing in the VHF range that the mix engineer doesn't hear in the mains.

The centerfill is calibrated in Fig. 15.6D. Panel C1 assesses the situation at crossover XL–CF. This should be the edge of the mains and hopefully a healthy area for the centerfill. The mains appear edgy, but not so much because of

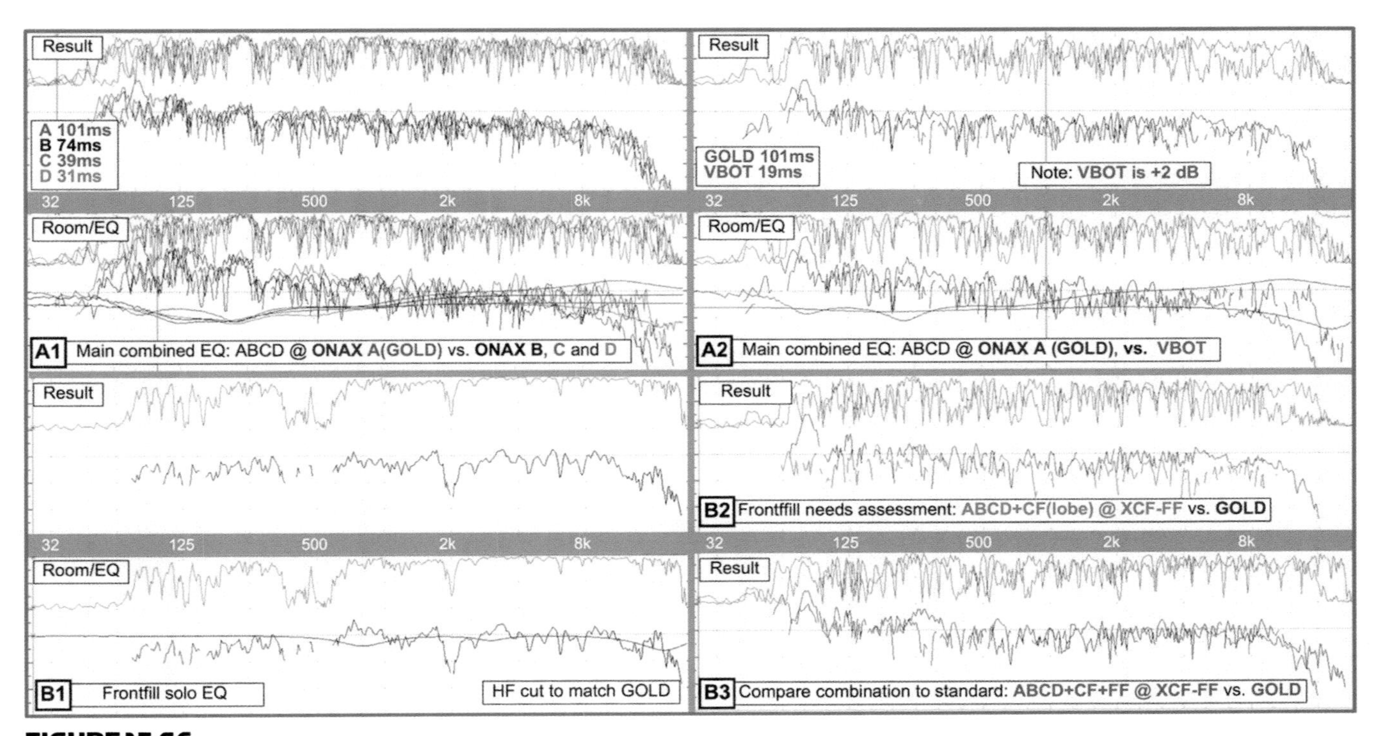

FIGURE 15.6C
SIM3 optimization data for Venue #6

level, or spectrum. There is energy there; it's just lumpy and low in coherence. Ripple variance is what tips the scale here. Next we prepare for combination by setting delay and examining the amplitude and phase. The amplitude gives us the flat CF vs. the tilted mains that we were looking for. Are they phase compatible? Above 500 Hz, yes. Below, not so much. This will not be a big issue because we need only a minimal-level contribution from the

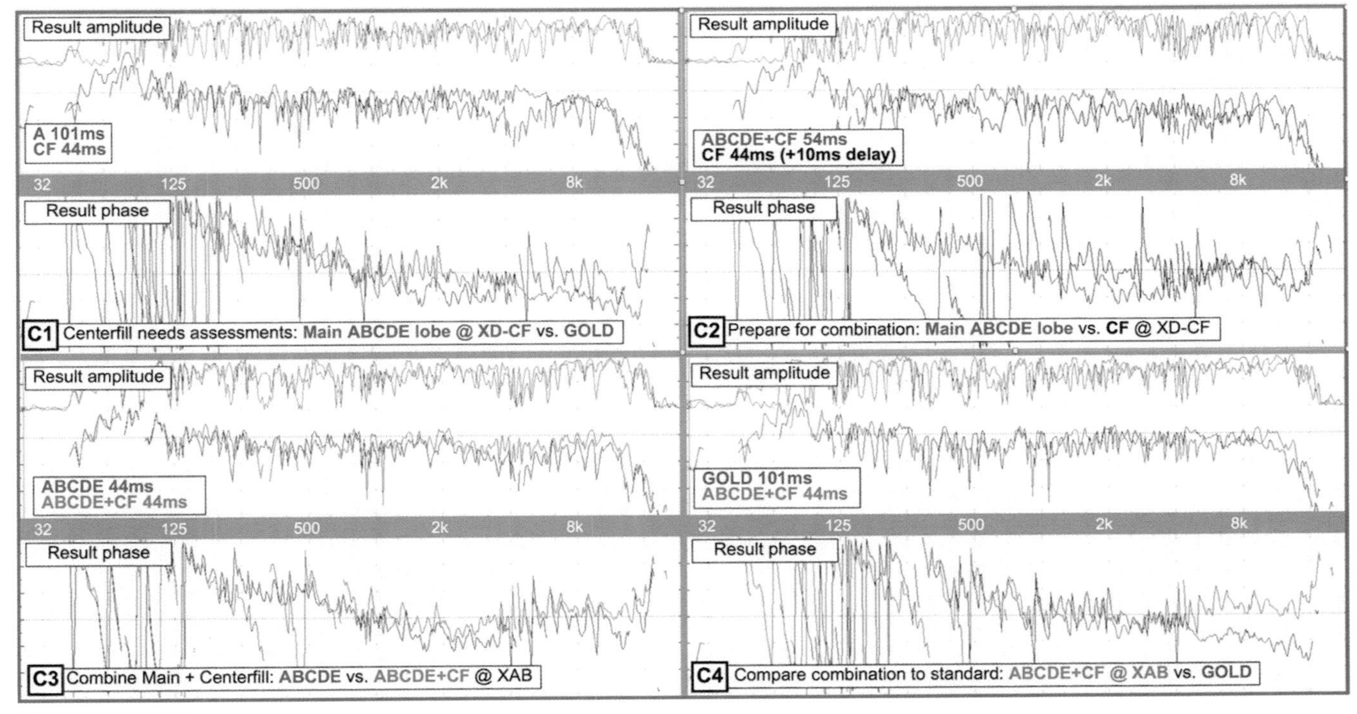

FIGURE 15.6D
SIM3 optimization data for Venue #6

centerfill. Remember that we are looking for some fresh, direct sound, not a power boost. The combined response (main + CF) is shown in C3. There is modest addition above 1 kHz (desired) and we've smoothed out the ripple (desired) and only lost a dB or 2 at 500 Hz, which we can live with. The final panel verifies that the centerfill + mains @XL–CF are matched in level, spectrum and ripple variance to GOLD.

15.7 VENUE #7: MEDIUM HALL, L/R WITH MANY FILLS

This hall looks deceptively simple but required a large number of coverage partitions. The architectural team originally insisted on hiding the speakers in soffit panels far offstage. In fact, the proposed positions were so far offstage and so deeply recessed that they would not have line of sight to many of the seats on their own side. Listeners at the back of the house would hear the opposite side louder than their own. The fight for a usable position was hugely important for this venue. A compromise position was granted: far offstage but in plain view with unbroken sight lines. The mains cover most of the room but needed help in the center, nearby sides and under the balcony. A centerfill and gradient inverted-stack cardioid array is located above the proscenium. These are hidden behind scrim, which is no problem for their performance.

The coverage partitions are shown in Fig. 15.7A (A1 and A2). The L/R mains cover the largest shares overall, but only a small portion of the front area (which is divided among the frontfill, centerfill and sidefill). We can visualize the mains vertical coverage as an ABXCDE array configuration because it's split in the middle to minimize the balcony front reflection.

15.7.1 Left/right mains (ABCDE)

The L/R mains need to provide at least 60° of horizontal coverage (FAR 2.0), once we subtract the areas to be covered by the sidefills. The preferred speaker model is 90° due to power budget reasons, so we have more than the minimum (which means it won't be 6 dB down at the outer edges). The extra coverage is distributed to the outer

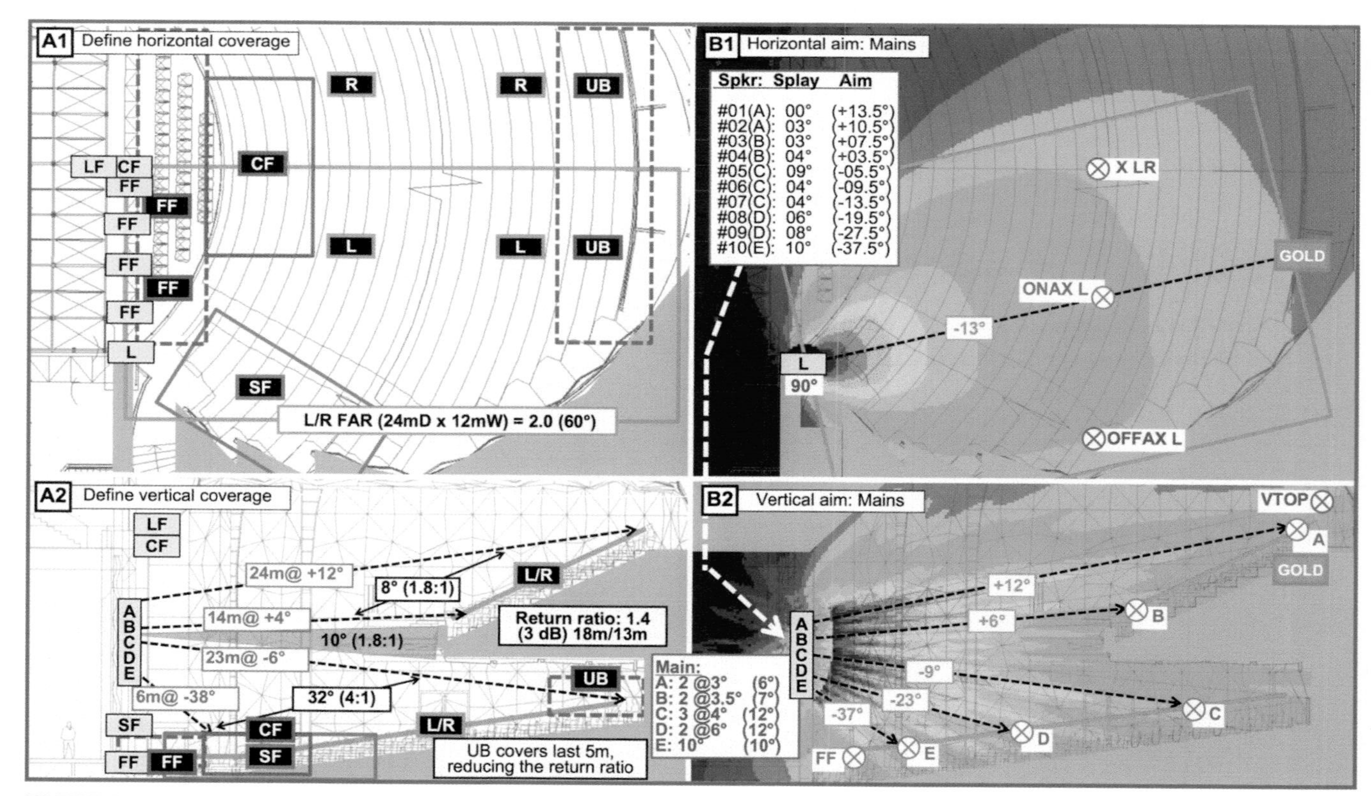

FIGURE 15.7A
Design process for Venue #7: coverage partition strategy, solo L/R mains (ABCDE)

walls and across the centerline. The overage effects are largely favorable (reduced level variance vs. increased ripple variance) because the walls are fairly HF absorptive.

We were given no restrictions on the vertical height so we looked to balance three factors: level variance (move it up), sonic image variance (move it down) and balcony front ripple variance (keep it in the middle). Recall that gapping the balcony requires a fairly straight shot to minimize the smiles and frowns effect (Fig. 11.24). The compromise middle placement provided reasonable solutions for all three.

One of the initial design decisions was between coupled vs. uncoupled (i.e. upper/lower) mains. The return ratio without underbalcony speakers was high, almost 2:1 (23 m/13 m). This falls to 3 dB once we factor in the underbalcony speakers (18 m/13 m), which lessens the case for splitting. The decision was made to keep the mains coupled because we were able to place the array at the most favorable location for coupling (i.e. the middle of the array at the balcony level).

The vertical shape is now defined in three parts: upper (8° @1.8 range ratio), gap (10°) and lower (32° @4:1 range ratio). The upper section is covered by a two-element composite (AB), whereas the lower section is covered in three elements (CDE). We had a budget of ten elements/side for the mains, which were apportioned as four upper and six lower. The lower section was given more elements even though it was a shorter throw because it had both a wider angular spread and greater range ratio.

15.7.2 Centerfill (CF)

The offstage locations for the mains made it a sure thing that we would need a centerfill, the question being how many (one or two). The gap area is a triangle that connects from the frontfill limit line to the off-axis edges of the L/R systems. This hall was used as the example for centerfill design so refer to Fig. 11.40 for more details. The only available position was above the proscenium and the chosen speaker was 80 H × 50 V aimed very steeply down. We didn't need quite all of the 80° of horizontal coverage but my preference is towards soft edges in these types of transitions. The reasons are (a) the coverage shapes come together like a twisted-up pretzel so it's safer to overlap a bit than leave potential for gaps, and (b) the level of the centerfill can be set low enough to fill the needed areas and then serve as an image aid as it fades out. The power scale for the centerfill was nearly comparable to the mains because it had to be hung quite high, making it a comparable range to the mains.

15.7.3 Sidefill (SF)

The need for sidefills was assessed during the design process using the lateral width evaluation. This hall was used as the example application (Figs 11.38 and 11.39) so we won't repeat it here. The advantage in this instance is that the sidefill can cover the outer area to a limited depth and be finished before the wall turns inward. The mains could not cover this area without over-coverage as side walls closed in. The advantage also translates in the vertical plane because an outward turn of the mains would bring the entire horizontal coverage outward just to help this small outcropping of seats. This can be viewed through the triage perspective as a ripple variance tradeoff: Aiming the mains outward adds ripple variance for the whole room versus the uncoupled sidefill adding ripple variance in only its crossover range with the mains. That's an easy choice.

15.7.4 Underbalcony delays (UB)

The needs assessment for underbalcony speakers was evaluated as described in section 11.7.2. The clearance ratio is 25% (2.25 m/9 m) and the return ratio is 1.65 (23 m/14 m), yielding a composite value of 40%. Therefore underbalcony speakers were specified. The selected speaker was 80 H × 50 V. The vertical aim point was the last row (since the range ratio ≥2:1), which is -15°. This sets the coverage start at -40°, the off-axis edge of the vertical pattern (-25° from the vertical aim). The coverage start (the unity line) sets the horizontal spacing by the uncoupled line/point source spacing method (Figs 11.20 and 11.22). The 80° speaker (1.25 × lateral multiplier) has a vector range to the unity line of 3.5 m. This would yield a 4.5 m spacing on a straight line, but is slightly reduced by the 6° splay that follows the arc of the balcony front (net spacing is 4.2 m).

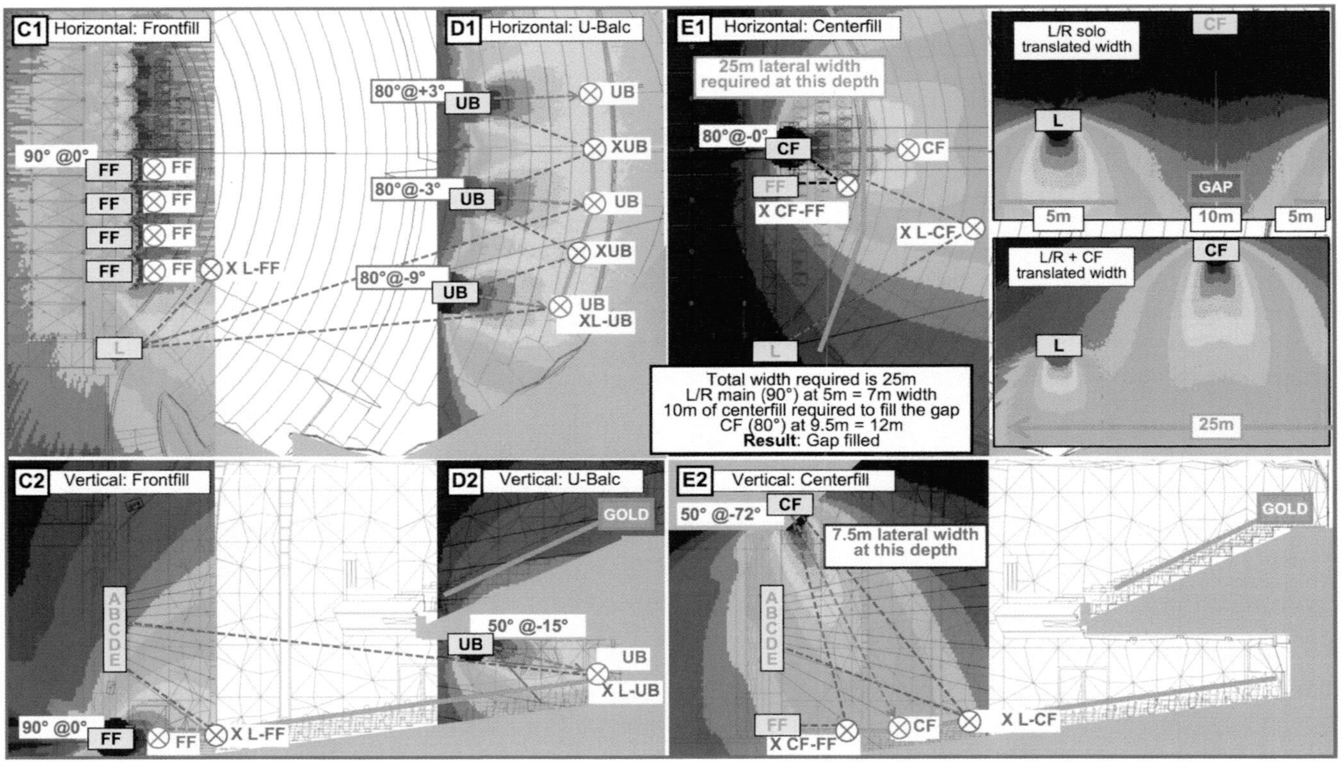

FIGURE 15.7B
Design process for Venue #7: solo frontfill, solo underbalcony, solo centerfill

15.7.5 Low-frequency system (LF)

What happens if we place subwoofers at left and right with the mains? The upside is coupling with the mains. We get a low image variance (+) and horrendous level variance (-) if we place them on the deck. Fly them at the top of the mains and the level variance improves at a cost of image. The down side is ripple variance from the uncoupled L/R relationship, also known as "power alley." A center location provides the opportunity for minimum-level variance front to back and side to side. We used a five-element cardioid configuration (gradient inverted stack) centered over the proscenium. The array is aimed downward so it's centered on the shape in both planes. The result is minimum-level variance and ripple in exchange for a high sonic image (a trade I will usually take). The worst-case sonic image and level variance would be found at the near seats on the sides (the exact same situation we find with the center channel of L/C/R systems). This could be remedied by "sidefill subs" placed on the deck if the budget and aesthetic concerns allow. This subwoofer subsystem (ba-da-boom) must be operated with proper level scaling so it doesn't get blown up trying to keep up with the center LF system. Adult supervision is required.

15.7.6 Combined systems

There are a lot of pieces to put together in this system. Each requires proper aim and level to correctly partition the coverage and delay setting in order to minimize ripple variance around the crossovers. Combined EQ for the mains follows the pattern we have seen in previous applications. The underbalcony delays are timed and level set to combine with the mains to unity (GOLD). There is very little combined EQ here, other than possibly reducing the LF response of the delays. The sidefills have isolated dominance within their coverage (unlike the UB system) and therefore level-setting and delay-setting operations are separated. Level setting is based on the ONAX SF response whereas splay and timing refer to the crossover (XL–SF). The intent is to match GOLD in both the isolated ONAX SF and XL–SF locations.

The centerfill (CF) bridges the gap between frontfills, L and R. It's a triangle with crossovers on all three sides and ONAX in the middle (XFF–CF, XL–CF and XR–CF respectively). Its level is set at ONAX CF (to match GOLD) and aim can be adjusted until the most uniform level distribution is found between these locations. Timing is the final step. Centerfill gets delayed to the mains at XL–CF (if it's not already too late) and then frontfills (at least the inner ones) are delayed to CF at XFF–CF.

15.7.7 SIM3 optimization data

This is the "room of many fills" and we will get a look at all of them here. The mains come first in Fig. 15.7E (A1–4). This is an ABCDE mains, so we have done several variations on this theme already. This time I will show the combined traces in different locations as a one-on-one series of comparisons with GOLD. This is how the work is done in the field. The multi-trace pileups are only done *after* we have sorted through them one-on-one like this. These plots are pretty self-explanatory but I will add the following notes. Notice that the room curve (the unequalized response) strongly differs for each but the result (post-EQ) hangs closely with GOLD in each position. The VHF extension is stronger at the bottom (E) but that is one of the hazards of battling a 4:1 range ratio on the floor with only six boxes (the other four are upstairs). ONAX C shows the most spectral tilt. Is this normal? Yes, for two reasons: (a) It's under the balcony, (b) it's the middle of the array. I would be concerned if it *didn't* have the most tilt.

Before we get to the fills we will verify the horizontal aim for the mains (Fig. 15.7E). Panel B compares the outermost seat (OFFAX L) and center of the hall (XLR) at mid-depth. This process has been a part of all the optimizations but here at last we can see one. The responses are matched in the highs and mids, indicating a proper aim. The low mids are stronger (and have more ripple) at the OFFAX edge near the physical wall. The XLR location will get it share of that when the other side is turned on.

The underbalcony fill is seen in the next three panels (C1–3). We see the parties to be joined in (C1). The mains solo response is a typical underbalcony response: strong spectral tilt and compromised coherence (ripple variance).

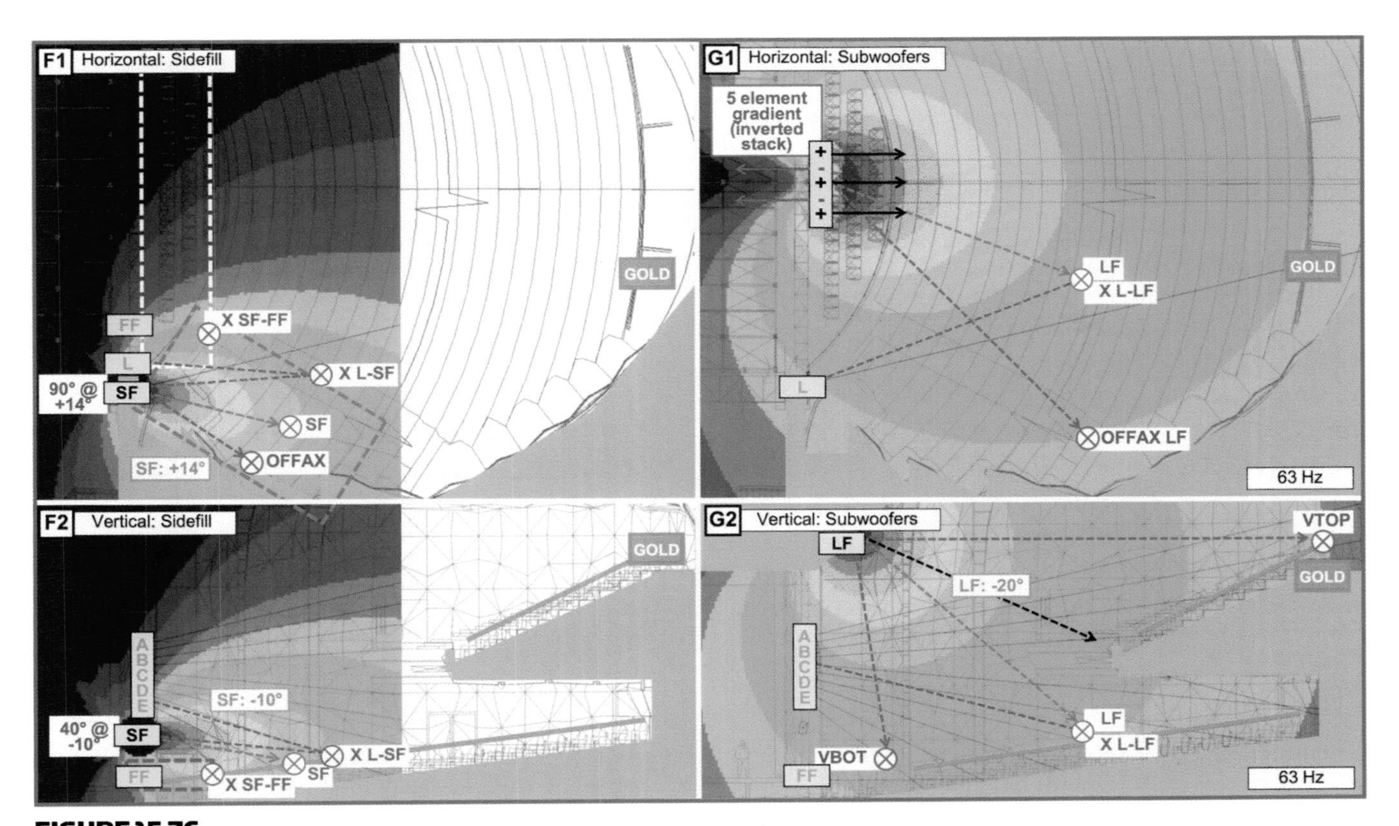

FIGURE 15.7C
Design process for Venue #7: solo sidefill, solo cardioid subwoofers

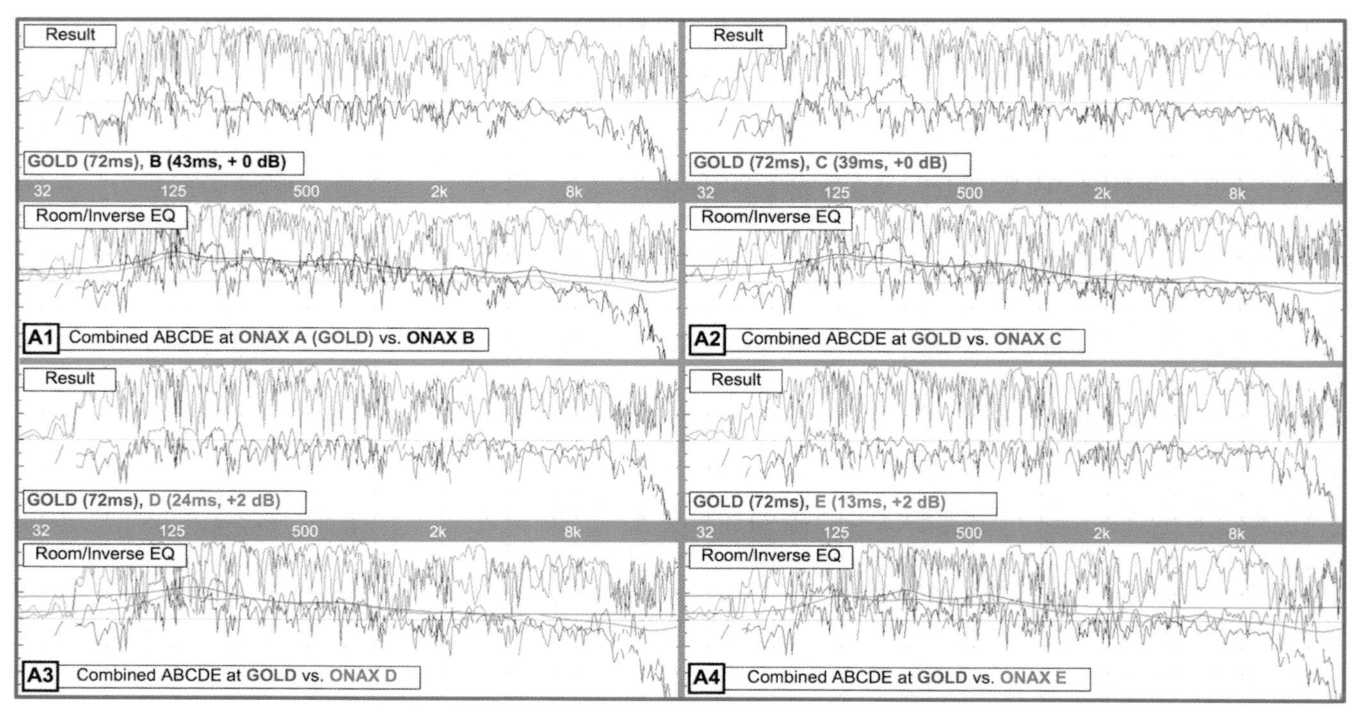

FIGURE 15.7D
SIM3 optimization data for Venue #7

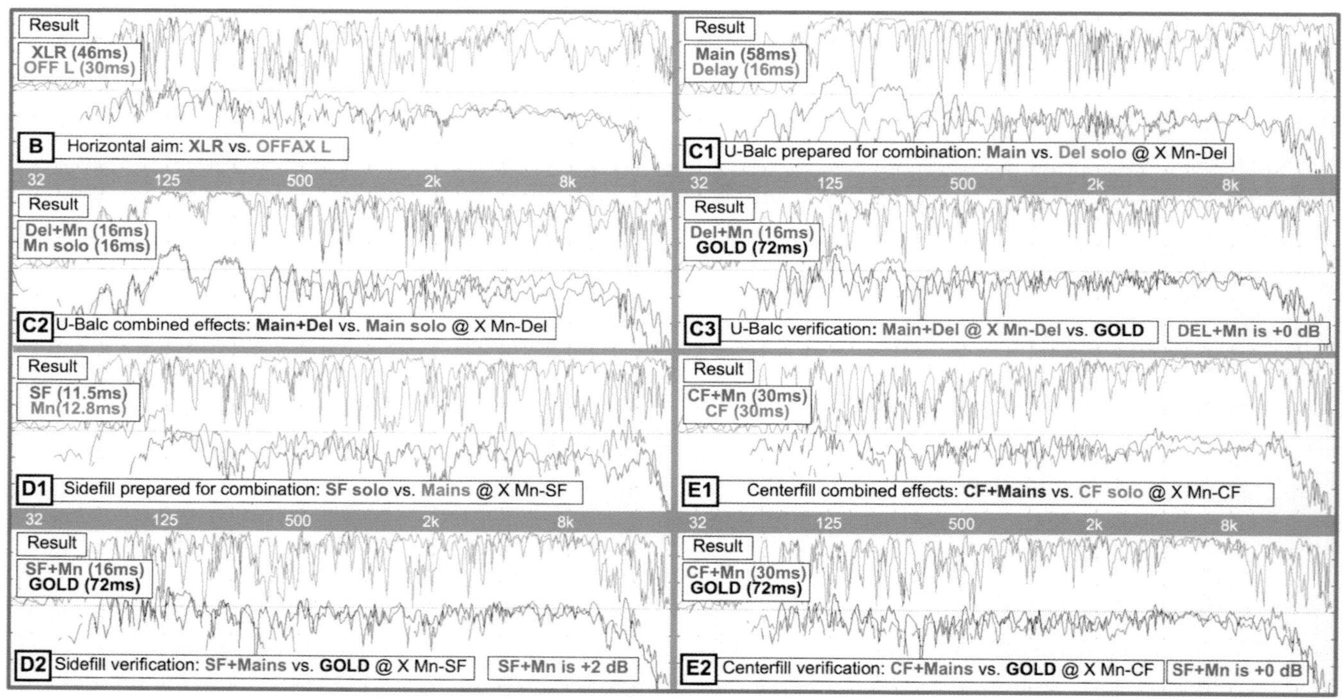

FIGURE 15.7E
SIM3 optimization data for Venue #7

The underbalcony fill can bring flat, coherent sound to counter both effects. Panel C2 shows how we balance the spectrum above 400 Hz with negligible addition below. Note that this is done without strangling the delay, i.e. cutting its lows and low-mids a million dB. It's all about level, aim, delay and placing the delay to start at the right place. The final is (you guessed it) verification that the combined response is a match for GOLD (panel C3).

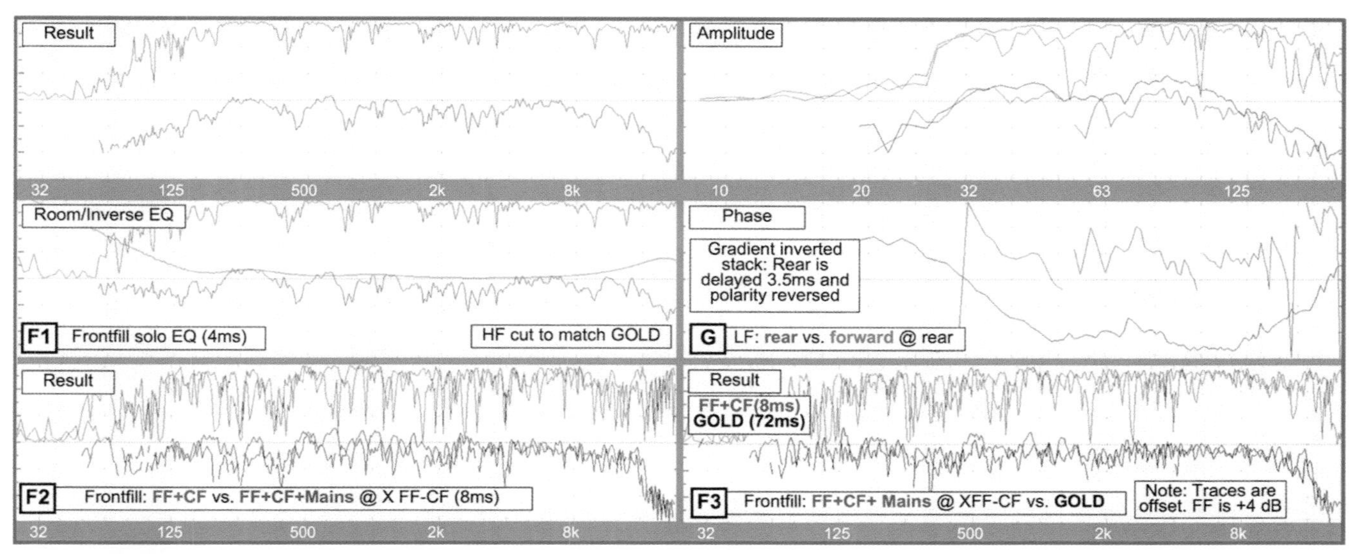

FIGURE 15.7F
SIM3 optimization data for Venue #7

Let's move on to the sidefills (Panels D1–2). It's the same process of course, and we've done it enough times now to start to abbreviate. Panel D1 shows the parties to be joined at crossover and D2 verifies that the combination matches GOLD. The centerfill is next (E1–2). Panel E1 compares the combined response at crossover with the solo centerfill. In E2 we again verify that the combination equals GOLD.

The frontfills are handled in Fig. 15.7F. Panels F1–3 tell the story. The solo EQ (including a second-order HPF) is shown in F1. Also notable is the bacon prevention filter at 18 kHz. Panel F2 shows a three-way crossover combination: mains, centerfill and frontfill. We finish with GOLD verification in F3.

The final set in the series is the tuning of the gradient cardioid subwoofer array. This was done up in the fly space above the proscenium, which is just as much fun as it sounds. The delay is set to match the front and back response, the rear driver polarity is reversed then we get the heck out of there.

15.8 VENUE #8: MEDIUM HALL, L/R, VARIABLE ACOUSTICS

This is a multi-purpose hall designed for everything from symphonic to rock and roll. Multi-purpose is a term that has gained a justifiable reputation for making everybody miserable, but this hall proves that we need to update the file. The physical acoustics are extremely dead: RT60 of 700 ms in a hall that holds 1760 people. Nonetheless the reverberation can be extended to achieve the response needed for unamplified symphonic and beyond (RT60 > 3 seconds). The physical acoustics were designed from the start to include Meyer Sound's Constellation system to provide the reverberation enhancement (variable acoustic systems were described in section 6.3.3). The result is an optimized environment for purely amplified sound (rock, EDM), reinforced sound (theater, jazz), as well as Bach. Engineers can add in the appropriate amount of reverberation character for the application. The huge advantage here is that we can get the upside of room reverberation (spatial envelopment) without the tonal coloration and coherence loss that comes from hard walls near our speakers. The SIM3 plots of the optimization show very high coherence compared with other halls at comparable depths. The tonal coloration from the room was very minor, which left us with minimal EQ requirements.

15.8.1 Left/right mains (ABCD)

The optimal horizontal aim follows the middle–middle strategy (again). The room is a trapezoid: a fan in front and rectangle in the rear. The horizontal width is calculated from the middle depth. We need at least 60° of coverage. The chosen main system is 90° so we will have some overage in the rear. The trapezoidal shape minimizes the need for sidefills, because the hall is widening with depth along with the speaker coverage. We can also breathe easy knowing that the walls are highly absorptive, thereby reducing our concerns about over coverage.

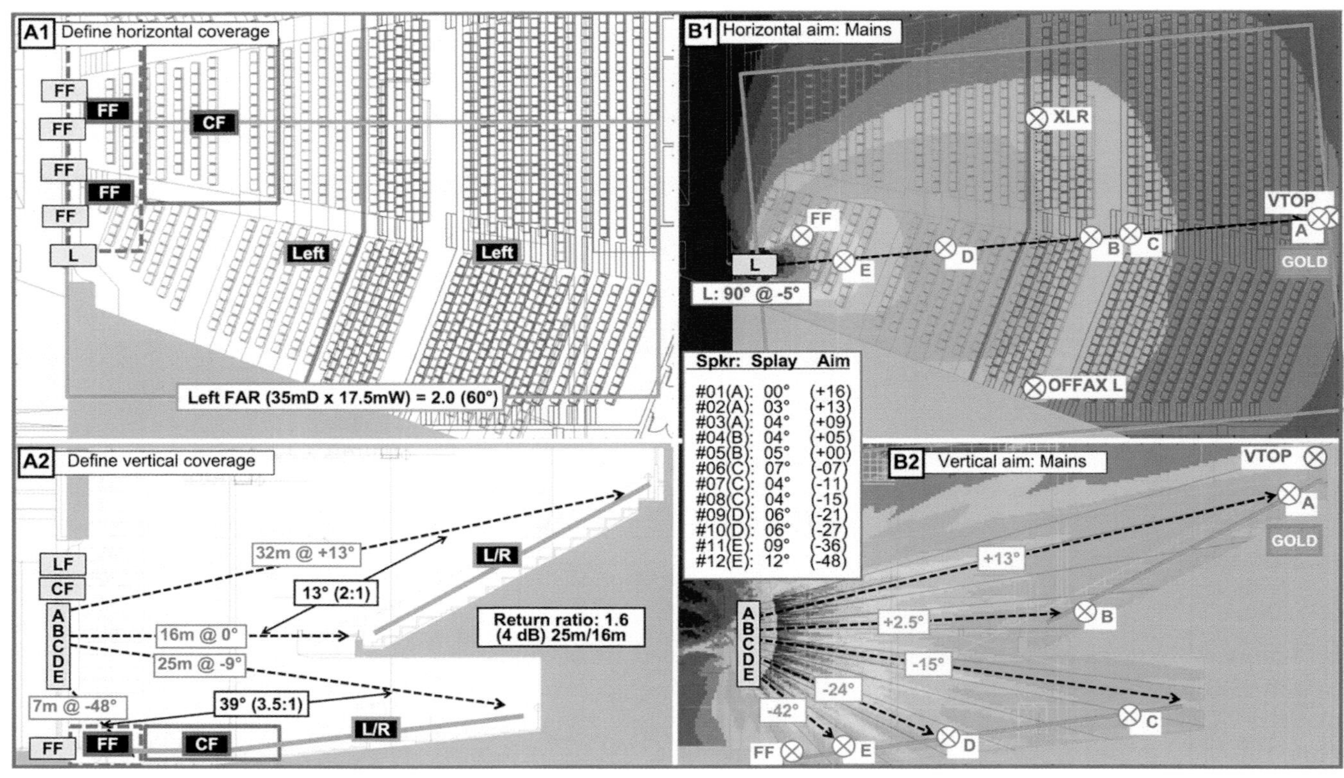

FIGURE 15.8A
Design process for Venue #8: coverage partition strategy, solo L/R mains (ABCDE)

15.8.2 Centerfill (CF)

We have done several centerfills already and this one follows a similar pattern. The challenge here was that we had very little margin for error in the mains–CF connection. The arrays ended up lower than originally planned, which shrinks the lateral width on the sides, and increases the central gap area to be covered by the centerfill. The centerfill is hidden in the proscenium at a fixed height, so we can't raise it up to expand its width (we can only tilt it up or down). The connection was made successfully here but I am passing this on as a cautionary tale. It's safer to plan for some overage in these areas and use speakers with relatively soft edges for centerfill and sidefill. We want to leave the mains with the flexibility to move up or down to get their best shot without worrying about opening up a gap on the floor.

15.8.3 Low frequency (LF)

Once again we meet a center subwoofer array: a gradient inverted stack in the proscenium. The vertical aim is downward because (a) that's where the room is and (b) we can.

15.8.4 Combined systems

The main array combined EQ sequence followed the process of previous examples. The only fill systems here were small and uncoupled from the mains and therefore required minimal combined EQ effort.

15.8.5 SIM3 optimization data

In many ways this room is similar to the last venue (#7). Its macro shape is fairly similar but its details make us able to keep the fills down to a minimum. Frontfills, centerfills and done. The mains are the same speaker model, nearly the same quantity and we will again gap the array around the balcony. There is one very notable difference about the rooms: This one has far less (or far more) reverberation than the other. This room was built with the

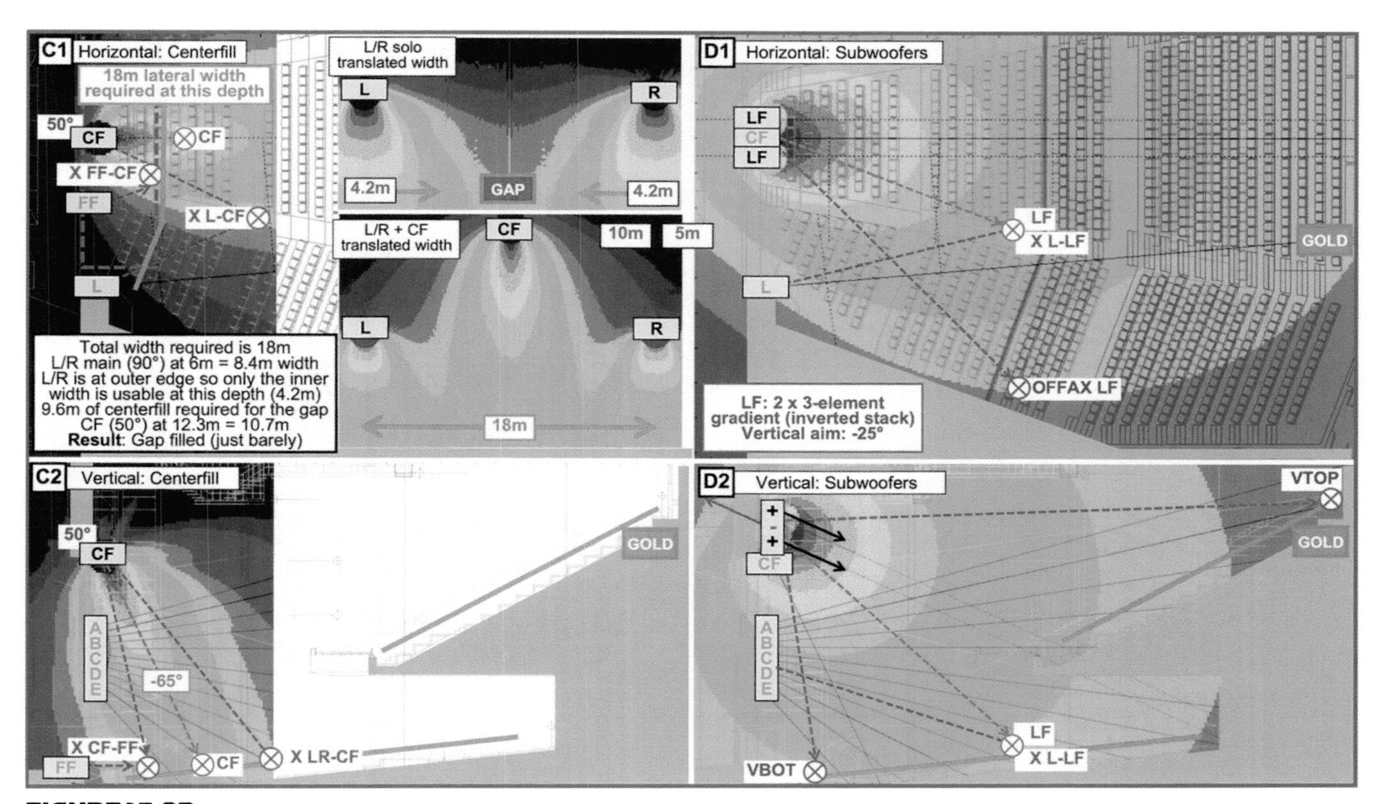

FIGURE 15.8B
Design process for Venue #8: solo centerfill, solo cardioid subwoofers

optimal physical acoustical properties to utilize Meyer Sound's Constellation variable acoustic system, i.e. *very* dead. The contrast between traces taken at comparable distances with matched speakers in the two halls is striking. Figure 15.8C shows the fully combined main system (ABCDE) from top to bottom in six steps. Notice that the coherence between 125 Hz and 2 kHz hugs the top of the screen for the entire run from top to bottom. Compare this with Fig. 15.7E, which is a very typical room in acoustical terms.

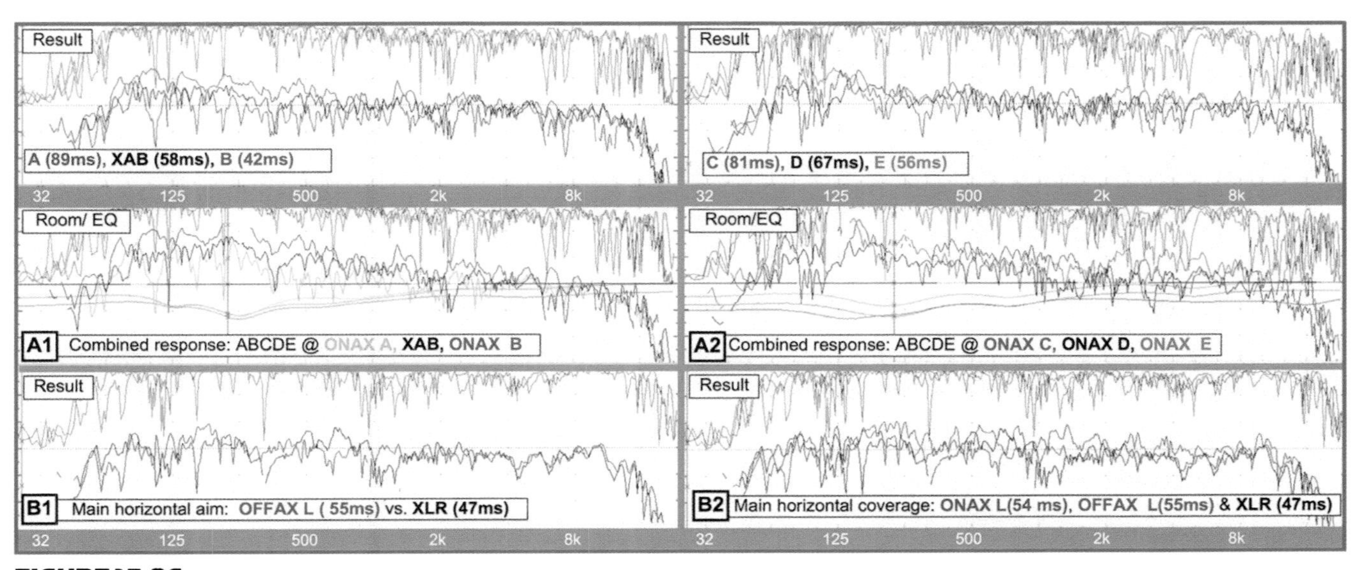

FIGURE 15.8C
SIM3 optimization data for Venue #8

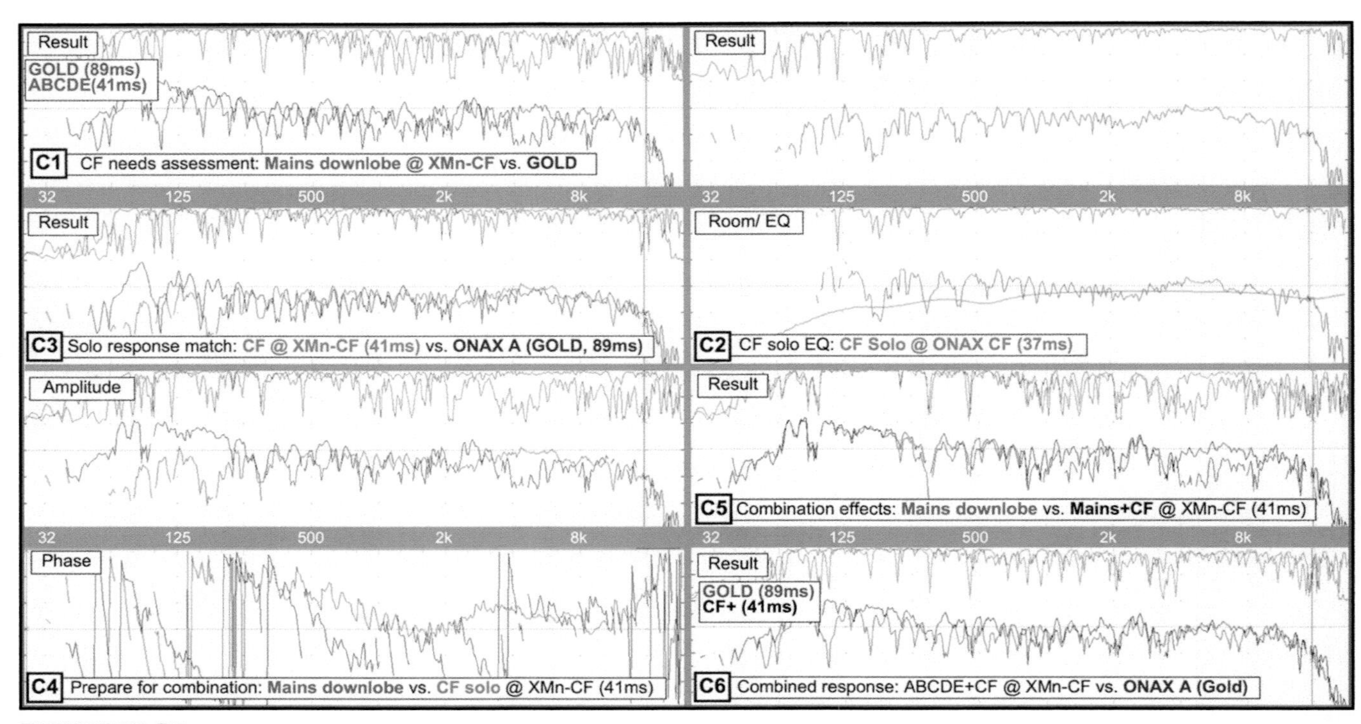

FIGURE 15.8D
SIM3 optimization data for Venue #8

Needless to say, this made for a very easy tuning and very consistent response, as can be seen in the traces. Constellation can then add the appropriate reverberation as needed for the program material (even unamplified symphonic).

Let's get back to the tuning. The B1 panel of Fig. 15.8C shows the horizontal aiming process again (OFFAX L vs. XLR). Panel B2 shows the process of coverage verification, i.e. the answer to the question: Is the speaker wide enough? This is done by comparing ONAX L with OFFAX L and XLR. We know our speaker is too narrow if OFFAX and XLR fall behind ONAX by more than 6 dB. It's too late, of course, because the speakers are already hung, but isn't it good to know for the next time?

The remaining traces for this venue are for the centerfill (Fig. 15.8D). The needs assessment (C1) shows only minimal help is necessary at crossover XMn–CF. Panel C2 shows the solo EQ, which includes a mild HPF and some VHF range matching. Panel C3 verifies that the equalized response at the crossover matches GOLD. Next we prepare for combination by setting delay, level and checking for phase compatibility. The phase compatibility answer is yes, no, maybe. This is not what we want but we have to look at the situation and be realistic. The downlobe from the mains is the product of a line of speakers at different distances from here. The centerfill is a single element facing right here. Expecting perfect phase compatibility is not realistic (and wouldn't hold for more than one seat anyway if we *did* have it). We combine then in panel C5 where they are compared to the downlobe alone. The final verification is found in panel C6 where the combined response matches GOLD.

15.9 VENUE #9: LARGE HALL, UPPER/LOWER MAINS

This rental system was loaded into a historic hall. As with most historic halls we are facing highly reflective surfaces with limited options for speaker placement, particularly in the underbalcony area. The hall shape is more like two halls stacked on top of each other: wide, deep, tall and steep (above) and wide, deep, low and flat (below). A tall room on top of a ballroom. This hall is the poster child for upper/lower mains and is used for the example in Fig. 11.30 D.

15.9.1 Upper mains (ABCD)

The horizontal coverage was determined by the FAR method and aimed by the standard middle depth method. The vertical shape is a perfect diagonal line, which can be covered with sequential doublings of the composite elements

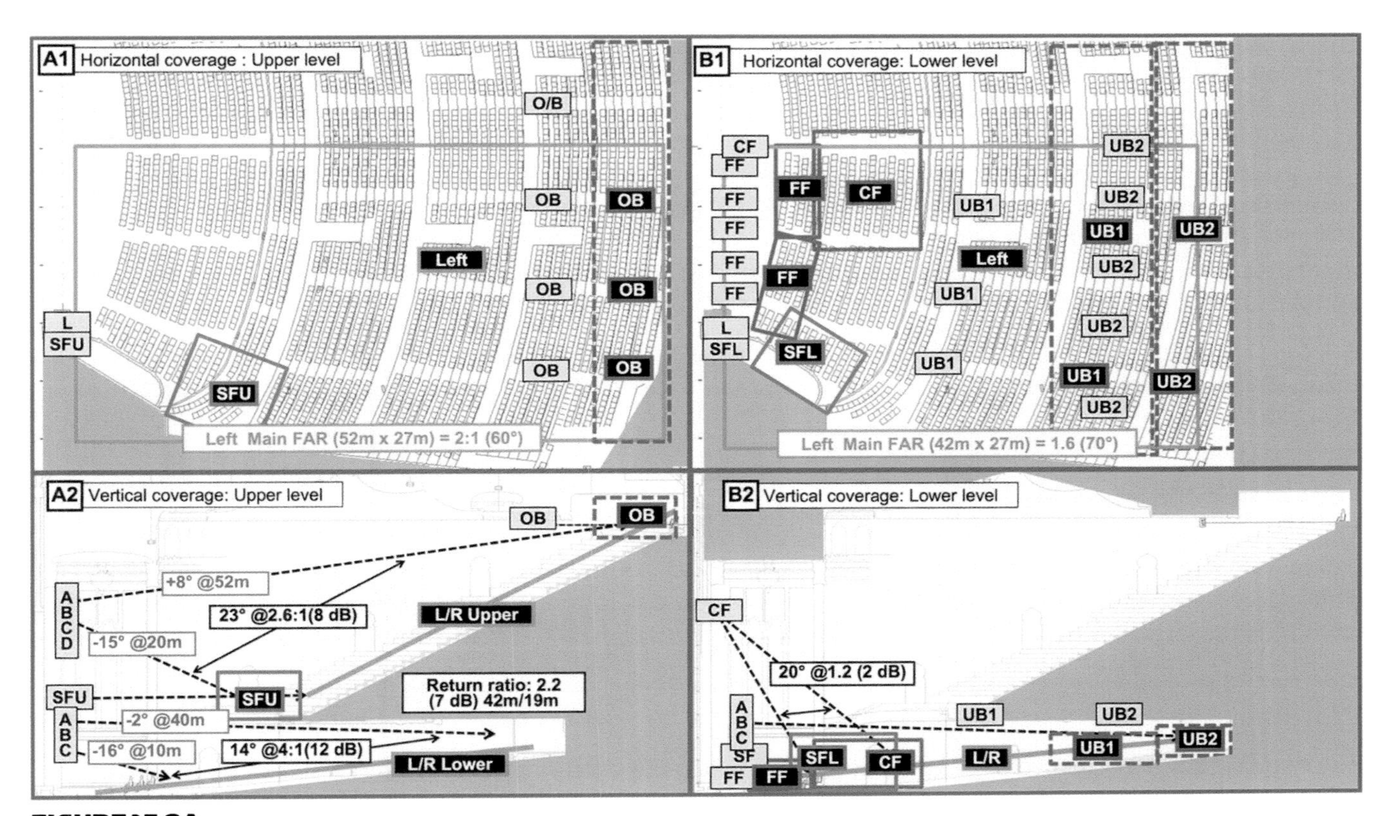

FIGURE 15.9A
Design process for Venue #9: coverage partition strategy for upper and lower levels

(4 × 1°, 4 × 2°, etc.). In this case we were given a quantity of sixteen elements/side for the upper system. With this quantity we could have started the doublings at 0.5° splays but I chose not to attempt such precise laser targeting due to the smiles and frowns effect (Fig. 11.24) that will curl the pattern up into the plaster at the top. Our upper mains were firing upward into the rake, which means the longest paths (rear center and rear opposite side) would have a beam climbing up the very lively plaster back wall and ceiling. We can't make the smile go away without raising the cluster. Sometimes it's better to resist the urge to use a super-precision laser beam. This is a perfect case since our upper speakers are throwing curveballs up there. It's better to open up our splay and, with it, our margin for error. Here we started with two sets of 4 × 1° composites, which created a gentler nose along the rear. The key consideration here is that smiles and frowns must be seriously considered before setting extremely narrow splay angles.

The frown side of the equation also comes into play in this case. The array aims down at the front of the balcony, which is curved. If the mains were a center cluster then the curve would follow the frown then the effects would be minimal. The mains, however, are off to the side (closest to the balcony there), and farthest from the balcony at center. A laser line of coverage at the bottom of the array will fall lower at center than the side so we can see that a single angular solution cannot work for all. There will be some leakage to the floor when the coverage falls short but the center balcony seats still get covered (by the next box above). By contrast the seats on the near side can have the coverage fly over them without anybody to step in. We added an upper sidefill speaker to cover this area, which reduced the vertical (and horizontal) load for the upper mains.

15.9.2 Lower mains (ABC)

The room is slightly shallower on the lower level, which means we could optimize the horizontal aim by turning it inward a few degrees. I did the calculation and determined it was not enough to make it worthwhile, especially since we had a substantial centerfill cluster. This can be a worthwhile consideration in some venues, especially when the depths of upper and lower systems vary greatly (such as Venue #3 in this chapter).

The vertical coverage has an extreme challenge in terms of an unbeatable range ratio dilemma. A low and flattop location is the most advantageous for underbalcony penetration and separation from the uppers. A flat shot at our VTOP allows us to narrow the splay without fear of smiles or frowns. The cost is level variance (too loud in front). Raising the mains upward causes underbalcony seats to lose sight of the mains and rely solely on fill speakers as their mains. The historic building aspect now comes into the picture because there are only limited locations for underbalcony speakers, leaving spotty coverage at best. Did I mention that the mix position was deep under the balcony? This is a case where we have to go to TANSTAAFL and triage to aid the decision process. There are a lot more people affected by the underbalcony limitations than the hot spot area near the front. The low position wins and the underbalcony speakers will only be used as needed.

A secondary aspect is that the lower system was comprised of elements whose individual responses were narrower (vertically) and higher power than the upper system. This allowed us to penetrate better underneath with fewer boxes and still remain properly power scaled.

15.9.3 Combined systems

There is surprisingly little to say about the biggest combination (upper + lower mains) because (a) the hall so thoroughly separates them and (b) both systems had enough control to render the combination effects barely noticeable. The crossover between the systems was at one of my favorite positions (the balcony front), which allows us to hide the largest mold mark in the system design.

The frontfill system was set to cover the first three rows. Its timing was set to sync to the strongest local system at its limit depth. This was the lower L/R mains on the sides and the centerfill system in the middle.

The centerfill system intended to provide unity level (match the GOLD response) in its most isolated area (the center of the triangular gap) as well as the crossover areas to the L/R mains. The timing is set to sync to the mains where the inner edge of L and R meet the outer edge of CF (the same as previously shown in Venue #7) and CF level is set to raise the combined CF + L (or R) level to match GOLD.

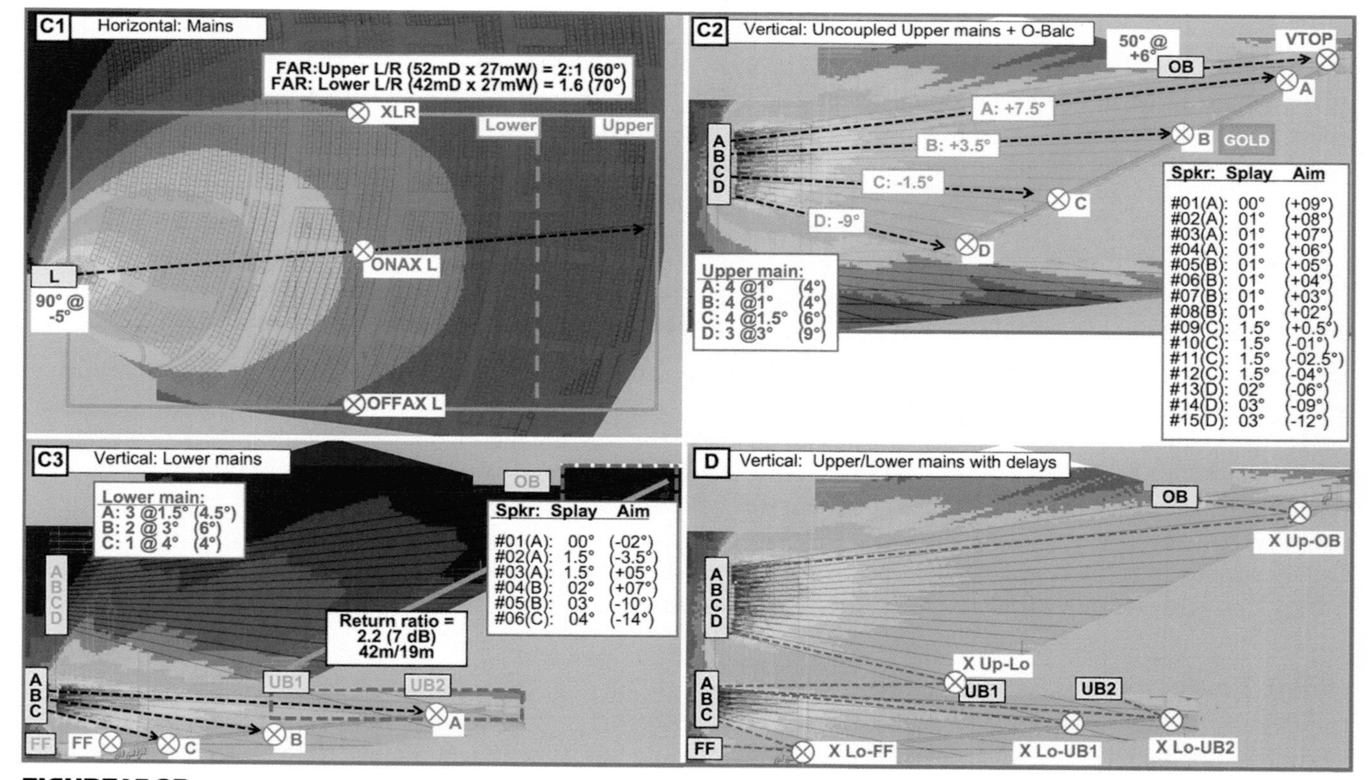

FIGURE 15.9B
Design process for Venue #9: solo upper(ABCD) and lower (ABC) L/R mains, upper + lower

15.9.4 SIM3 optimization data

The mains are a variation on the familiar themes. Actually two variations (upper and lower). We focus here on the solo EQ process for the A, B, C and D sections of the upper array. These are found in Fig. 15.9C where the room, inverse EQ and result are shown for each. I show the EQ inverted because that is the way I have done solo EQ since 1984. Viewing the EQ inverted makes it easier to complement the unequalized response (i.e. matching the shape rather than mirroring it). Panel B1 of Fig. 15.9D shows the same four solo EQ responses overlaid. The matched results are clear in B1 as are the unmatched EQ in B2. I included the phase response in B2 to allay the fears of the phase police. No phase demons were released by this equalization. The most probable cause of phase problems during equalization is the wrong amplitude settings (which are most likely narrow, deep filters). Seen any here?

The last panels show the combined response. Panel C1 shows what was done when A was added to B. The low mids came up and we brought them down with filters in both A and B. The complete combined EQ family is shown in C2, which shows a level variance of 2 dB over a range ratio of 9 dB (144 ms to 56 ms).

We move on to the lower mains in Fig. 15.9E. We must maintain a link with the GOLD standard set earlier for the upper system (upper ONAX B). We will also use ONAX B on the lower system for consistency and because ONAX A is so deep under the balcony. Panel D1 shows a comparison the upper and lower systems, and verifies our link. This is very important in this case because the mix position is deep under the balcony and they will have no idea what is happening upstairs. This trace shows they need not worry. We will stick with this same reference point (lower ONAX B) and check out this array from back to front. Panel D2 compares this to the response near the back at ONAX A. We had lost 1 dB back there but our coherence is still holding up extremely well. Our aiming strategy seems to be providing good penetration under the balcony. We move forward to ONAX C (Panel D3), where we have gained 2 dB of level. Our level variance is only 3 dB so far even though we are spanning 10 dB of range ratio. VBOT (panel D4) is found in the fourth row (17 ms). It has lost the 2 dB we had gained earlier and returns to unity level with ONAX B and GOLD. We have overcome over 15 dB of range ratio but there is a price to be paid for steering over the heads of these near seats. Look at the HF coherence for TANSTAAFL in action. We are at the fourth row though, so help is nearby in the form of the frontfills.

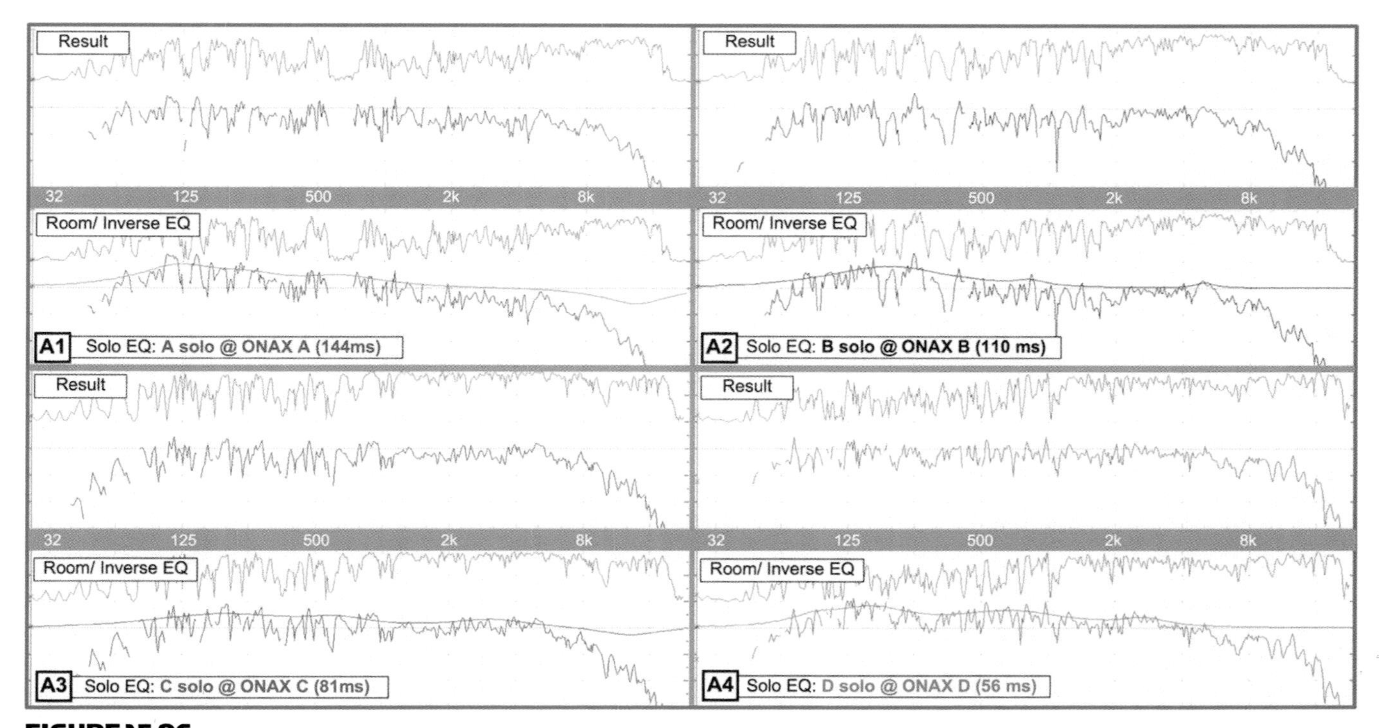

FIGURE 15.9C
SIM3 optimization data for Venue #9

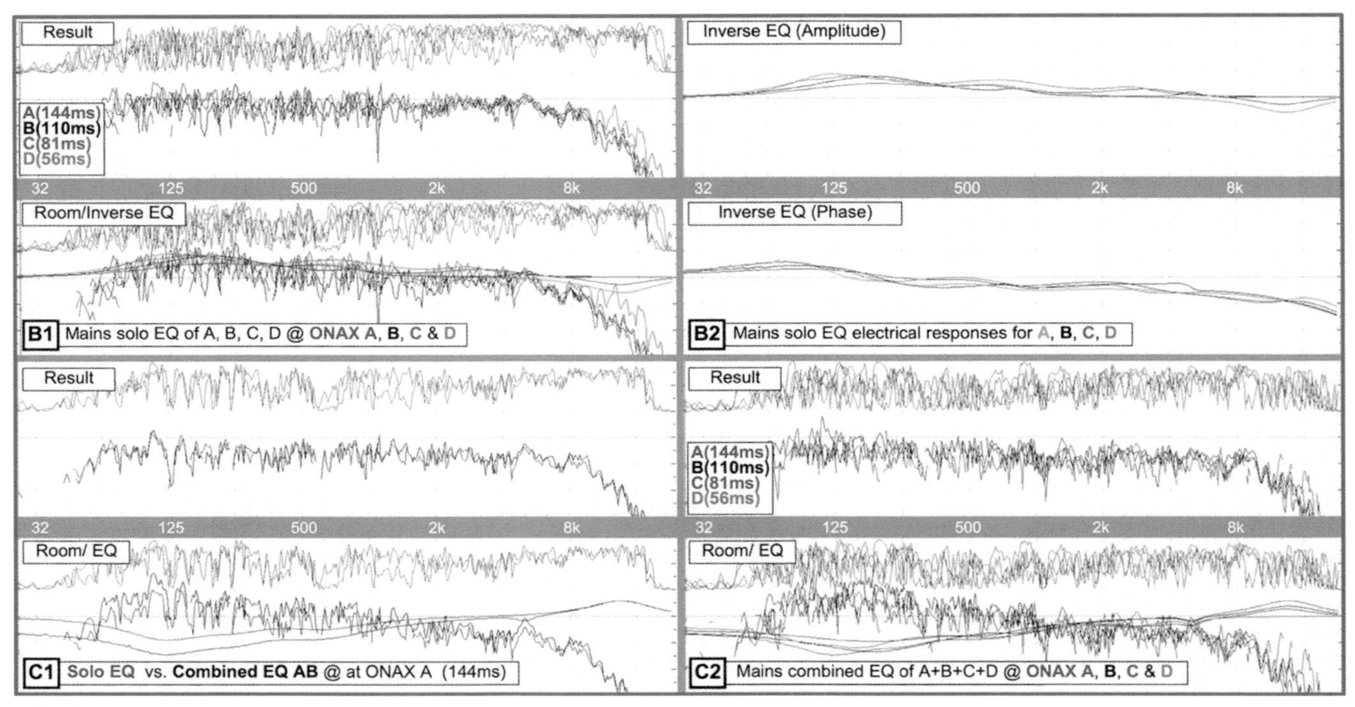

FIGURE 15.9D
SIM3 optimization data for Venue #9

The LF array for this application was a simple L/R pair of gradient inverted-stack arrays. They were in close proximity to the lower mains, so they were timed to be in sync with them. The crossover was a second-order unity type because there were no worries about image separation.

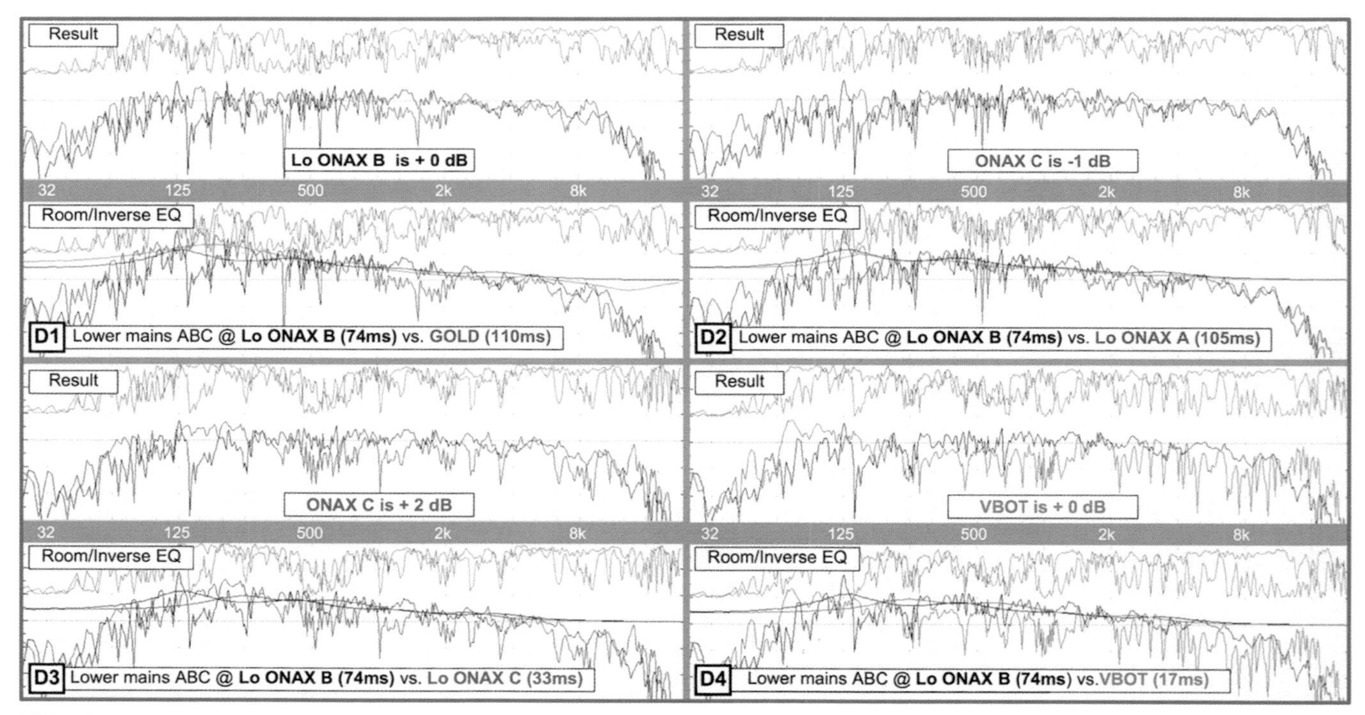

FIGURE 15.9E
SIM3 optimization data for Venue #9

15.10 VENUE #10: WIDE FAN SHAPE WITH MULTIPLE MAINS

Fan-shaped rooms are very popular venues for houses of worship. The most pressing sound design questions are often about the video: the locations left for us to fly speakers and amount of money left in budget. This leaves us to figure out how many mains are needed to fill the shape. Lower clusters, wider fans and extended stages lead us to more mains, while higher trims, narrow fans and shallow stages bring the number down. We get bonus points if we can bring the number down to two because the client feels like they have stereo, which is highly valued. Once we go beyond two clusters we need to be prepared to discuss why they can't have stereo (the real answer is because they wanted the big fan and the huge stage and giant video but I will leave the diplomacy on that up to you). This is followed by discussions of why don't we try cross-matrix delays and DSP magic and every kind of trick to try to make stereo out of three or four or more clusters. The correct answer is TANSTAAFL: because it should be more important for people to understand the spoken word than be entertained with panning effects of musical instruments. OK. Time for me to stop preaching.

The main cluster quantity can be found by using the lateral aspect ratio to find the coverage width required at the start of coverage, e.g. at the end of the frontfills. This application example shows two ways to solve the same shape: simple and complex. The simple solution is three clusters of mono. The complex version is L/R mains with centerfill and sidefills. Either will work and depends entirely on the extent of the customer's attachment to the L/R configuration. Both system options include the same frontfills and underbalcony fills.

15.10.1 Left/center/right mains option

The height of the mains is set by the video projection and sightlines from the last row. This puts the bottom of our clusters at nearly the same height as the uppermost seat in the balcony. Needless to say this would not be our first choice. The height is 11 m, and the speaker is 90° (horizontal). Therefore the lateral width on the ground is 15 m (11 m × 1.4 = 15 m).

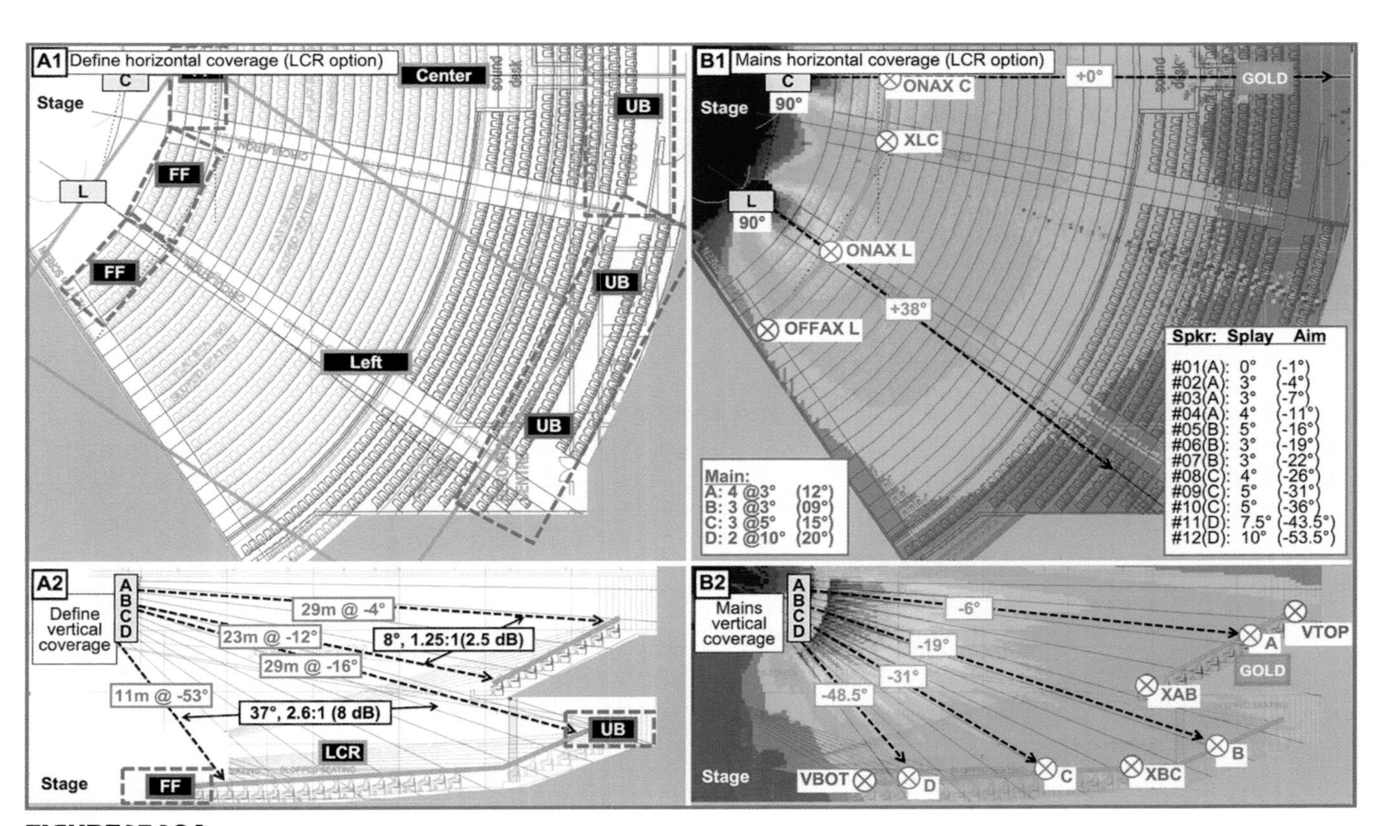

FIGURE 15.10A
Design process for Venue #10: Coverage partition strategy (L/C/R), solo mains (ABCD)

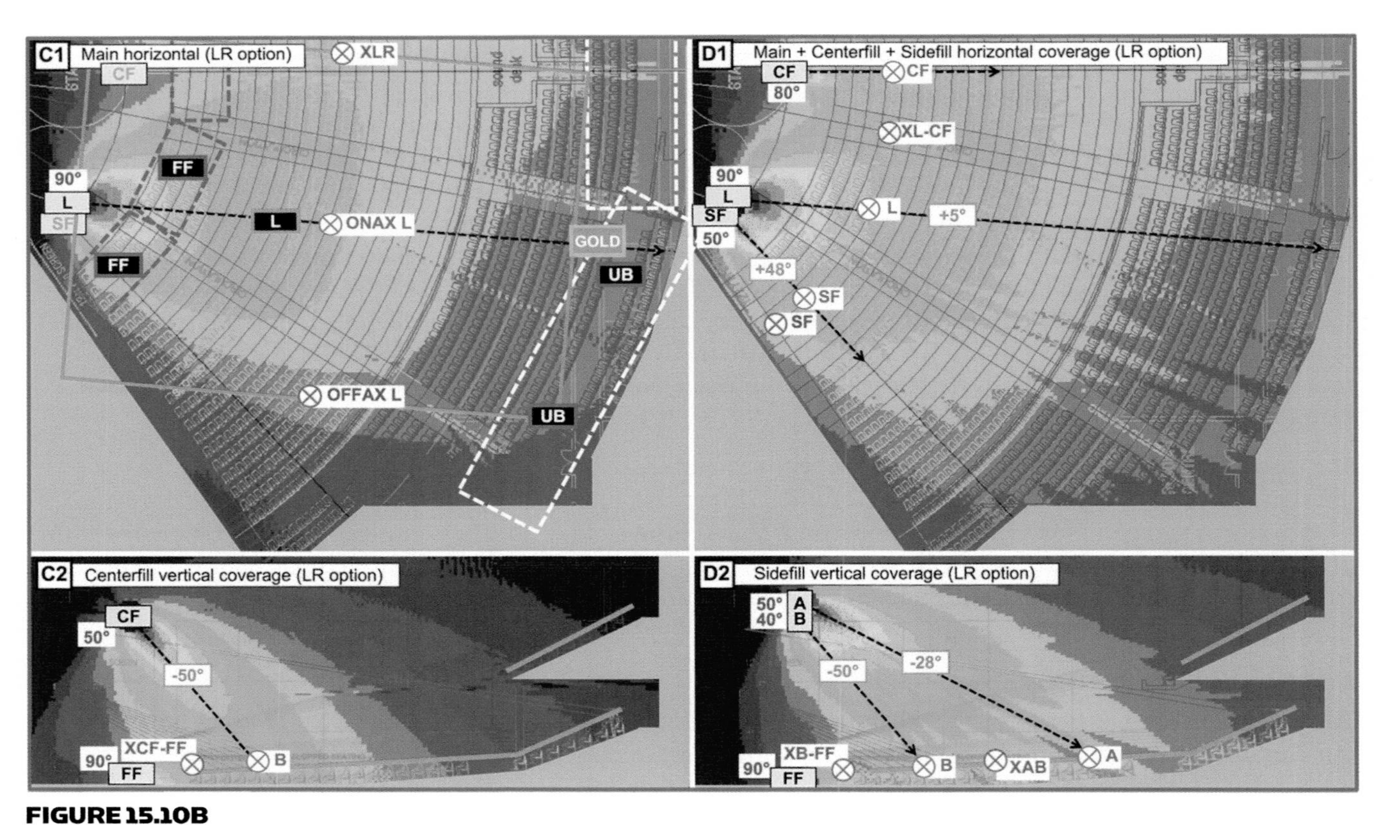

FIGURE 15.10B
Design process for Venue #10: solo L/R mains (ABCD), main + centerfill + sidefill

We would have to walk 45 meters to go from end to end at the third row, our start of coverage. This is the coverage target's lateral width, the minimum needed to make the connection. We need three clusters (45 m/15 m = 3) to cover the width. This venue was used as an example for multiple mains in Fig. 11.27 (A1–3) so you can review the process there.

The vertical coverage has one notable feature. Notice that very little effort is made here to gap the balcony front. It's a sad situation (i.e. it's about the frown). The cluster is aimed steeply down, so there is no opportunity to precisely place the gap in the balcony front. The gap will slip below the balcony front at all but the centerline of the speaker's horizontal aim.

15.10.2 Left/right mains option with sidefill and centerfill

The L/R option is more complicated so we will break it down. First we have to look at the hall from an L/R perspective, which starts with re-aiming the mains. The wide spacing makes it a certainty that we can't close the gap at center without a fill.

The fan is too wide to cover inside and out so we can shave off the outer edges and rectangulate the macro shape. Notice that the hall depth gets shorter at the outer edges, which allows us to slightly reduce the throw for the sidefill system. It still takes a two-element coupled point source to reach our required depth.

15.11 VENUE #11: CONCERT HALL, THREE LEVELS

This application features two approaches to the same hall (shape). There is a twist. Two highly reverberant halls were built from the same plans and renovated their systems with new mains at the same time. The leftovers from the existing systems differed, which opened the option to approach the design and optimization differently for each. Two plans were developed: one that gave the mains custody of all three floors (option 3F) and one that delegated the third floor to the overbalcony speakers alone (option 2F). The vertical splay angles within the array and the composite element subdivision were the primary differences between the two designs. The optimization proved that both approaches could provide minimum variance with only minor differences in power capability, ripple variance and sonic image. Option 2F had slightly more power capability (tighter splay angles), less ripple (less reach into the reflective upper level) and more image distortion (the sonic image was in the ceiling for the third-floor listeners).

The coverage partition strategy for option 2F is shown in Fig. 15.11A (A1–A2). Notice how the two overbalcony delay systems (four and two elements respectively) provide exclusive coverage for the third floor. Option 3F is shown in 15.11C (A1–A2), where the A section of the mains covers the first half of the balcony by itself and then joins the delays in the rear. All other subsystems are functionally equivalent.

The room is complicated and merits a brief note. The side seating rises up gradually in a "stairway to heaven" fashion while the central area has a low rake and jumps up vertically at the balcony. This gives us an impossible target shape for elements that are vertical lasers spread over 90° of horizontal. Two lines of vertical coverage are shown in different colors on the figures to denote the central and outer seating levels.

The complete systems include cardioid subwoofers, sidefills, a centerfill array and frontfills, all of which change depending upon the stage apron settings. We have already done many of these subsystems so we will focus on the unique aspect here: the same hall with two different main system strategies.

15.11.1 Option 2F mains (ABCDE)

The top element in the L/R main was aimed at 0°, the best position for precise horizontal coverage lines with no smiles or frowns. This allowed us to avoid the balcony front just above this line (one of the main reasons to implement option 2F). The range ratio is 2.6 (8 dB) and we were able to slightly exceed the minimum splay ratio with angles ranging from 2–7° (splay ratio of 3.5). The choice of eleven elements was budget driven but was totally capable of spanning the coverage angle we needed over the given range ratio.

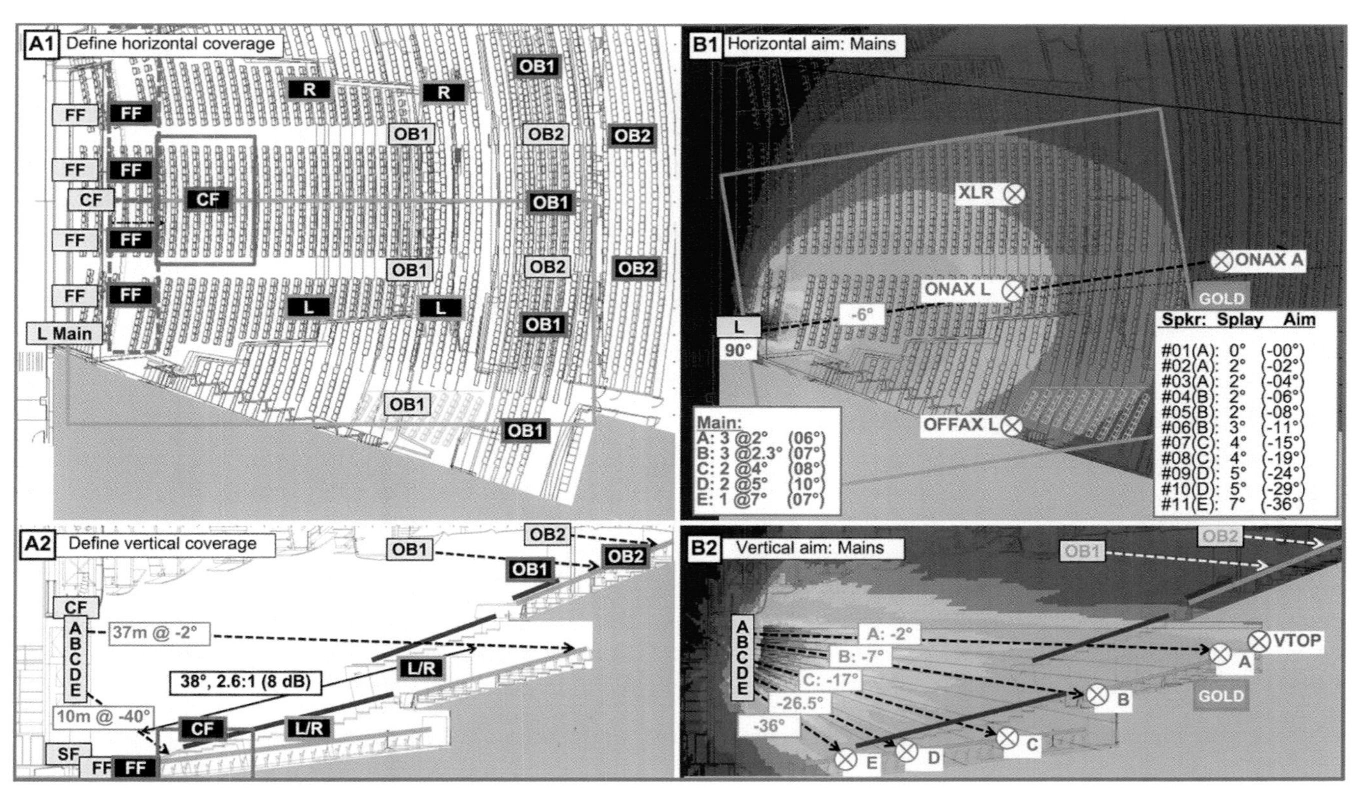

FIGURE 15.11A
Design process for Venue #11: coverage partition strategy (option 2F), solo main (ABCDE)

Horizontal aim was handled by the same method as in previous examples and the coverage needs were more than met with the 90° elements.

The vertical progression is quite similar to many of the previous examples with the notable exception of the B section of the composite elements. This is a three-element composite with internal splay angles of 2° and 3°. Yes, I have broken my own guidelines. My preferred splay at this location was 2.5°, an option that was not physically available, leaving 2° or 3° as the options. I bent the rule (and the composite) because I wanted to center the element where I could place a mic. The 2° option compressed the coverage upward too much and the 3° option put ONAX B in the air below the balcony. The resulting compromise averages out to the 2.5° I wanted and got the mic where I could measure it without a ladder.

15.11.2 Option 2F delay 1 (ABBA)

Yes, it's an ABBA array but this is not Mamma Mia. It is fairly standard to break such an array into outers and inners for delay setting but in this case we had different vertical aims, depths and levels. The balcony is much deeper in the center section than the sides (eleven vs. five rows). The inners pass the fourth row and need to keep moving up to connect to the second wave at the tenth or eleventh row. The outers pass the fourth row and run smack into a solid wood reflector panel, thank you. ABBA: different aim, different level, EQ and delay.

The spacing for these speakers was already set and not adjustable. We could, and did, adjust the splay angles to find the optimal horizontal coverage and crossovers. I don't have an exact number for you but it was found soon after I had said "too close," "too far" and "split the difference."

The vertical aim required on-site measurement as well because we needed to ensure that the balcony front was fully covered (because we were not aiming the mains here). The first-wave delays needed to make it from the front row to the tenth or eleventh row where we wanted to hand over custody to the second-wave delays. This one's not a delay dilemma. It's a splay dilemma and we have to look at TANSTAAFL for guidance. The front row takes priority because there is virtually nothing from the mains for us there. At the upper end we can aim the second-wave delays down a bit and hopefully still make it close at the top. The top rows are less likely to be populated so they lose priority. As it turns out we made the connection without triage (but just barely).

15.11.3 Option 2F delay 2 (AA)

The last rows are aided by an uncoupled pair of third-order line source elements (vertical). These are extremely narrow in the HF range and therefore only able to cover the last three rows. In the ideal world the second-wave delays would be riding on top of distant on-axis energy from the previous systems. In this case the mains are not even on the same floor and the first wave had to aim down enough to cover the front of the balcony. The role of these delay speakers was much more than direct/reverberant ratio helpers. They function here more like a distributed mini-main system, more relay than delay.

15.11.4 Option 2F combined systems

The composite elements of the main arrays are added together and their combined EQ brings them to match the GOLD standard. The mains to delays crossover is on the balcony front (XA–D1). The delay timing is set in this gap area, above which the delay speakers move into level dominance. The level for the D1 area is set to match the GOLD standard as a soloist, because we are not counting on help from below. Therefore the combined response here was not much different than the solo response. There is certain to be low-frequency information coming from below but extensive LF reduction of the relay-delays increases their perceived disconnection from the rest of the system (not to mention that it does little to level the spectral tilt anyway). It is extremely rare for me personally to strangle speakers, which I define as high-passing above 200 Hz, but to each their own.

We have now merged the main and first-wave delays so it's on to the second wave. Many second-wave delay applications require an investigation as to whether to delay to the mains or first wave. This one was easy: a delay-to-delay connection at XD1–D2. Level setting was based on the combined response matching the GOLD standard.

We learned something interesting during the calibration of the gradient inverted-stack subwoofer array. The subs were rolled out on carts and put in place at the proscenium edge. We measured behind them to get the timing and saw strange readings. The rearward level of the front pair could not keep up with rearward energy from the single backwards box. The reason was that the proscenium was closing off too much of the rearward energy. Once we opened a 0.3 m gap between the speakers and the wall, we were fine. The lighting designer, not so much.

15.11.5 Option 3F mains (ABCDE)

The top two elements in the L/R mains were aimed upward into the front of the third floor. It is unusual (for me) to create the composite module A with fewer elements than B or C. This case calls for it because we need to cover the front of the third floor and then gap the balcony before resuming the coverage area we met previously with option 2F. A coverage gap should always be at a composite transition (in this case A to B). From here we need to cover the identical shape as previously, but with nine elements instead of eleven. This is done by proportionally expanding all the angles by 20% (as close as possible using the actual available splays). This process was shown in Fig. 11.11 where a consistent 80° shape was created with various element quantities.

15.11.6 Option 3F delay 1 (OBBO)

This time it's an OBBO array, which seems even stranger than ABBA. This is the official array designation for a four-element uncoupled line source where the outer elements got turned off and moved into "special event" storage. Option 3F made the outers redundant and just a source of ripple variance. The mains coverage there was a perfect match for GOLD, so thanks and goodbye.

This venue did not have the second-wave delays so we must make it all the way to the back with the two inner units of delay one (and only). This was doable because we did not need to cover the early balcony seating, instead being able to merge with the mains halfway up like a traditional delay. The delays could be aimed at the deepest spot and

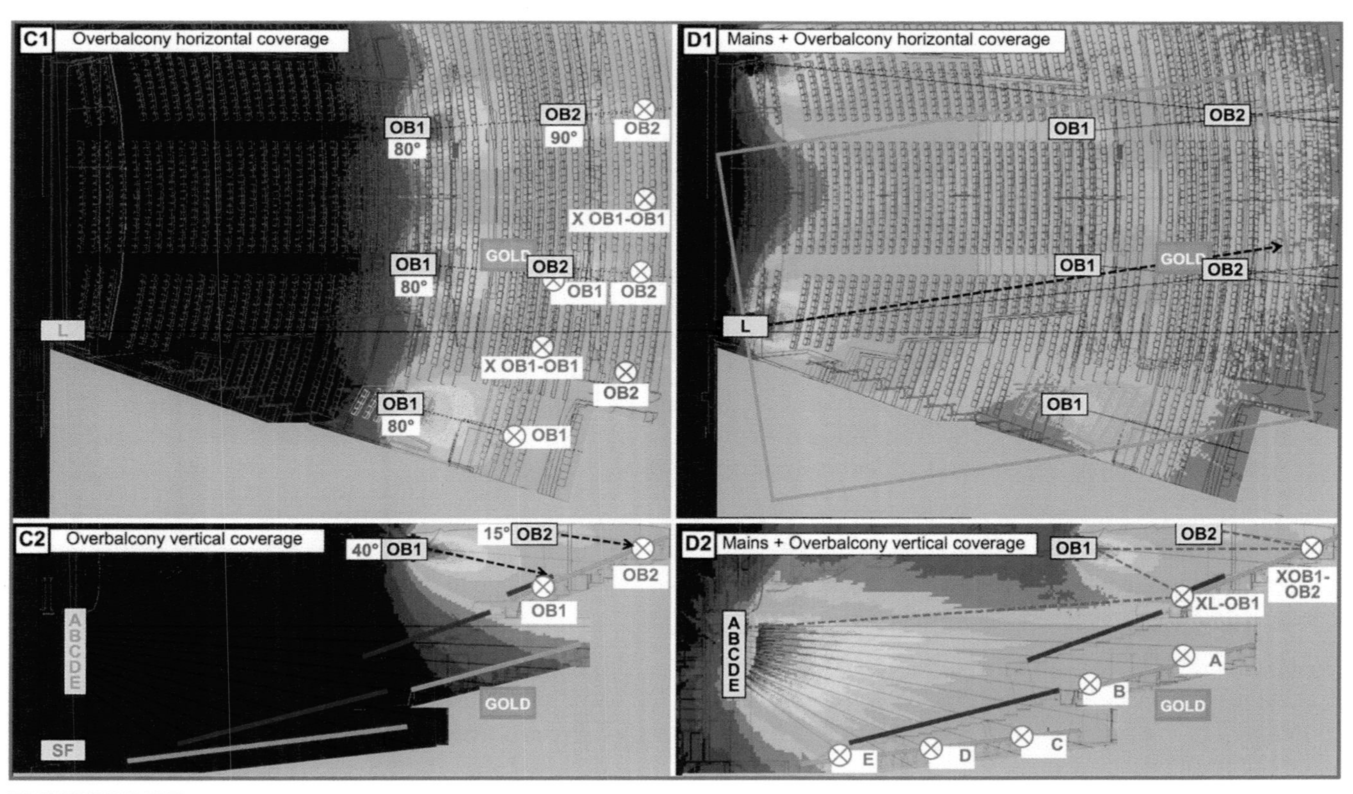

FIGURE 15.11B
Design process for Venue #11: solo o-balc delays, main + o-balc (option 2F)

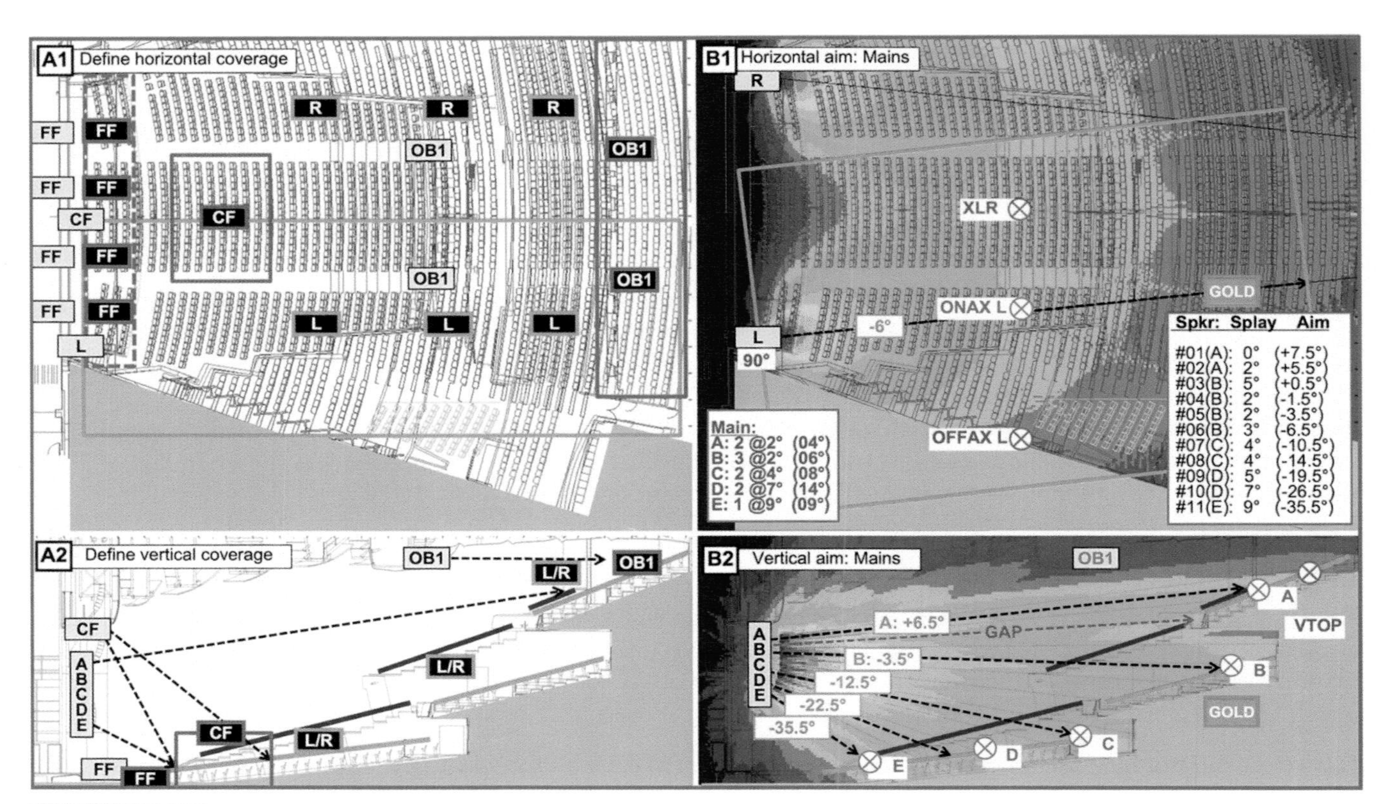

FIGURE 15.11C
Design process for Venue #11: coverage partition strategy (option 3F), solo main (ABCDE)

allow their underside to meet the mains topside and gently extend the coverage and minimize the image distortion. The crossover (XA–D1) is located higher (and lower) than those we used in option 2F. In this case the level setting provided restoration to GOLD as a combination rather than as a soloist.

15.11.7 SIM3 optimization data

Yes it's time to tune yet another ABCDE main array. Is there another way to show the process without being insanely redundant? I'll try. Panels A1 and A2 of Fig. 15.11E contrast the solo and combined EQs over the span of ONAX A, B, C and D. This is a reverberant hall so I knew I could count on the room for plenty of spectral tilt. Notice that all of the solo responses are brought to nearly flat (panel A1) and yet the combined responses create a "target curve" shape with around 10 dB of extra low end (panel A2). This was done by only partially compensating the combined system coupling. This is evident by contrasting the solo EQs (A1) with the combined EQs, which differ only in the very gentle reduction in the low end added during combination.

Panels C1 and C2 break some new ground. These are traces from the game called "hotter or colder." This is how we find ONAX in a composite element. Place the mic, take a trace of the solo element and then move to the next row. Hotter or colder? The center of the composite (the real ONAX) should have the strongest response in the VHF range (because this is where the phase offsets are smallest).

We haven't done a frontfill in a while (Fig. 15.11F). Panel D1 shows that we need this for more than imaging. The coherence above 4 kHz is terrible compared with GOLD. Actually it is just plain terrible. We are deep in the underbelly of the array in a reverberant hall so this is no surprise. We can make two conclusions from panel D2: (a) we definitely can help (FF has flat response and high coherence) and (b) phase compatibility is a very relative thing (multiple arrivals from the array and reflection make the phase response very difficult to read (and hear). We finish the job in panel D3 where we can see a closer matching to GOLD and improved coherence.

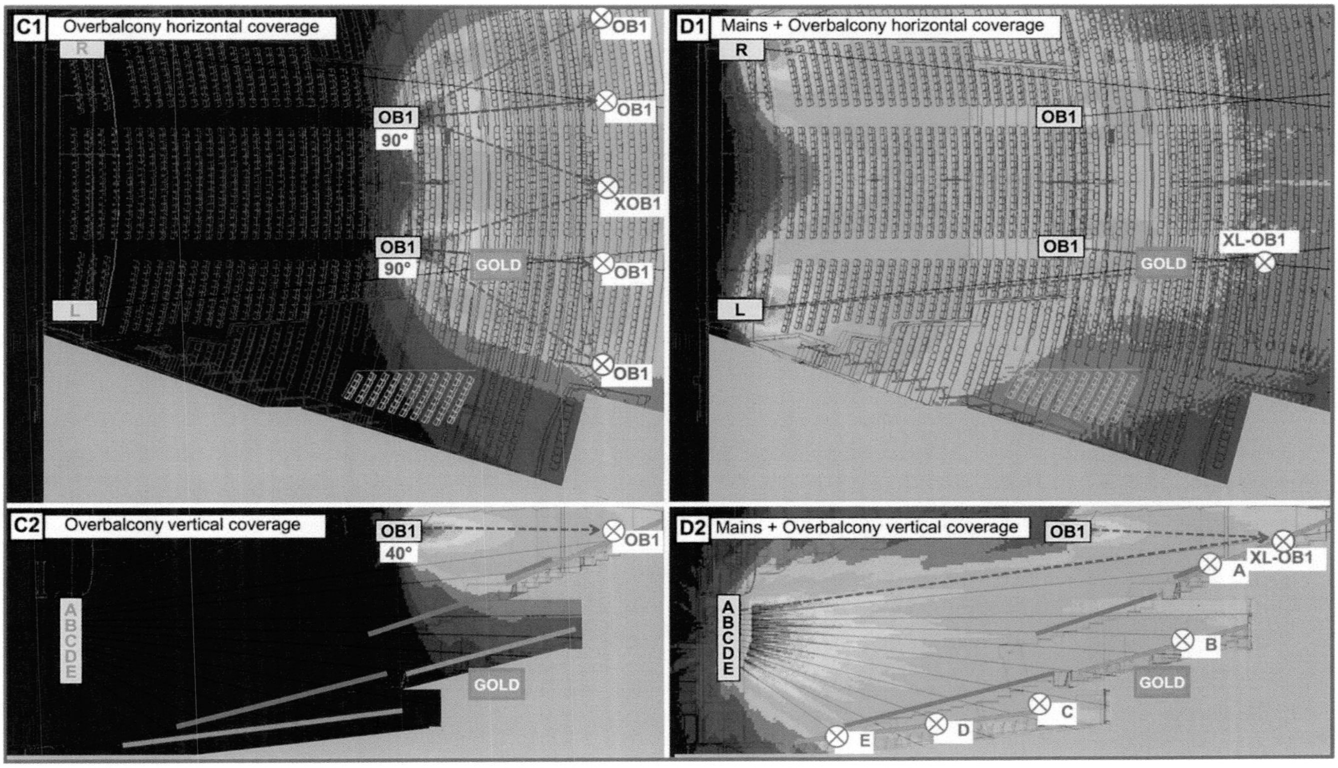

FIGURE 15.11D
Design process for Venue #11: solo o-balc delays, main + o-balc (option 3F)

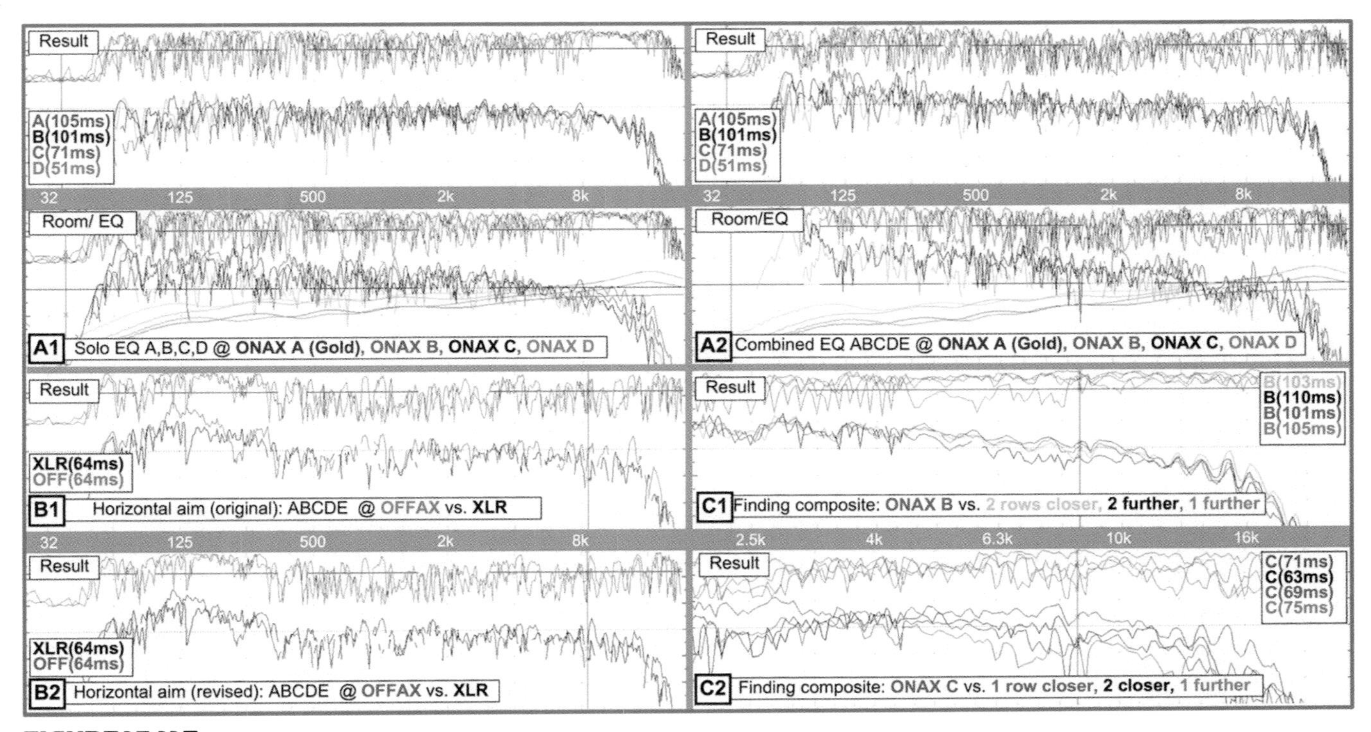

FIGURE 15.11E
SIM3 optimization data for Venue #11

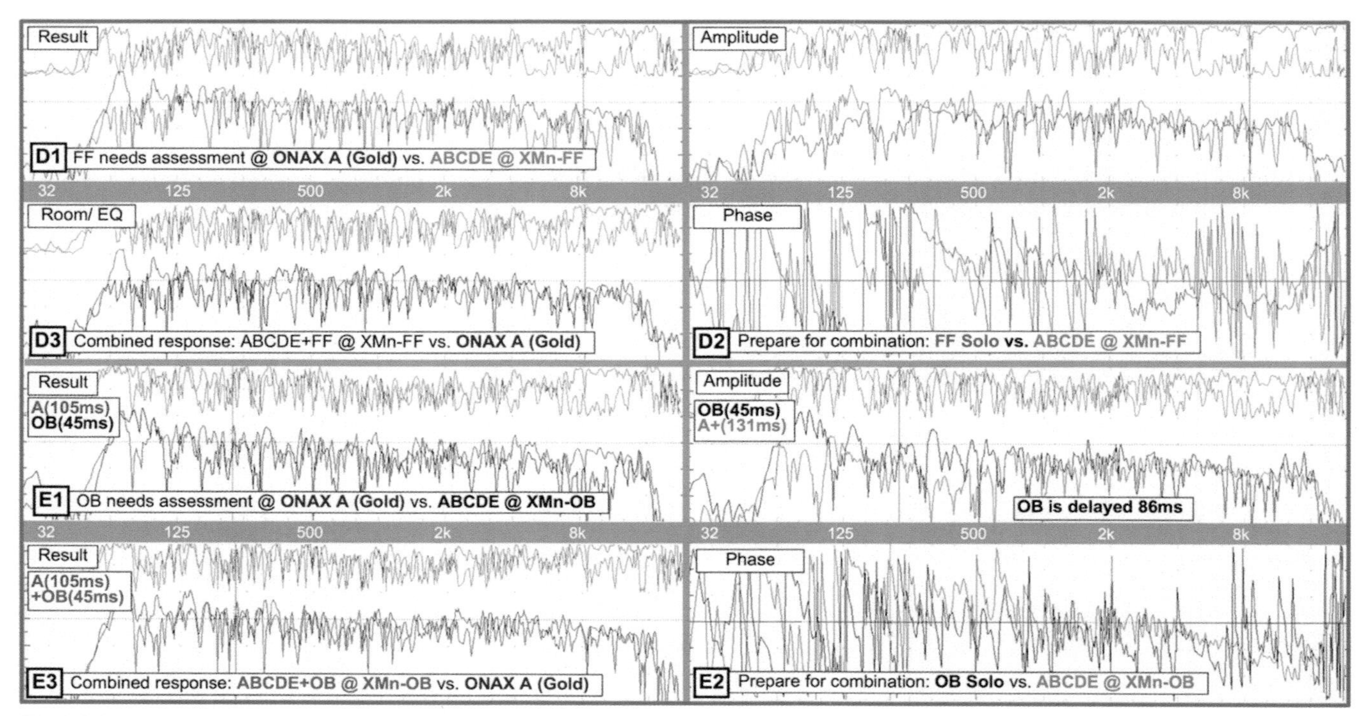

FIGURE 15.11F
SIM3 optimization data for Venue #11

The final addition is the overbalcony system (E1–E3). The needs assessment (E1) shows the spectrum tilting, level falling and coherence dropping off. We prepare for combination in E2. Our overbalcony speaker can deliver a flat response to counter the spectral tilt and has an advantage in coherence due to its short(er) throw. It's not exactly a short throw at 45 ms, but that's much closer to the mains at 120 ms. The phase is easy to read and they are clearly compatible (probability is always better when the speakers are both aimed in your direction). The final step is verification (E3), which shows our combined response matches GOLD.

15.12 VENUE #12: ARENA SCOREBOARD

This application calls for 360° of coverage from a central scoreboard. The room's macro horizontal shape has a slightly rectangular 5:4 aspect ratio so the system will need only minor asymmetry to adapt to that shape. The intent is to have all sound (playback and announce only) come from this central point with no delays or fills. These applications often have limitations of various sorts that impact our available options. This one had a very compressed vertical space requirement, which eliminated the obvious choice of a modern line array. We need power, so we'll have to spread horizontally. The solution is an asymmetric coupled point source in both the horizontal and vertical planes. The basic plan is three levels in the vertical plane (ABC) with decreasing power scale from top to bottom.

The coverage partition plan is shown in Fig. 15.12A (A1, A2). The upper system covers the huge majority of the seating. The middle system extends the range to the floor and the lower system is only used when the central area of the floor is seated.

15.12.1 Horizontal plane (N/E/S/W, ABC)

The upper array (A) is a circle in a square (or nearly a square). The array is comprised of a ring of 18 × 20° elements. There is 2 dB of level taper to rectangulate the response but otherwise it is simply a 360° continuous ring of sound. It helps that the room's corners are rounded as these are otherwise the most distant and hardest seats to reach.

The middle-level array (B) is 6 × 80° with slightly asymmetric splays to create the rectangulation. The bottom level covers the court and is a simple pair of speakers aimed to make a circle on the flat floor.

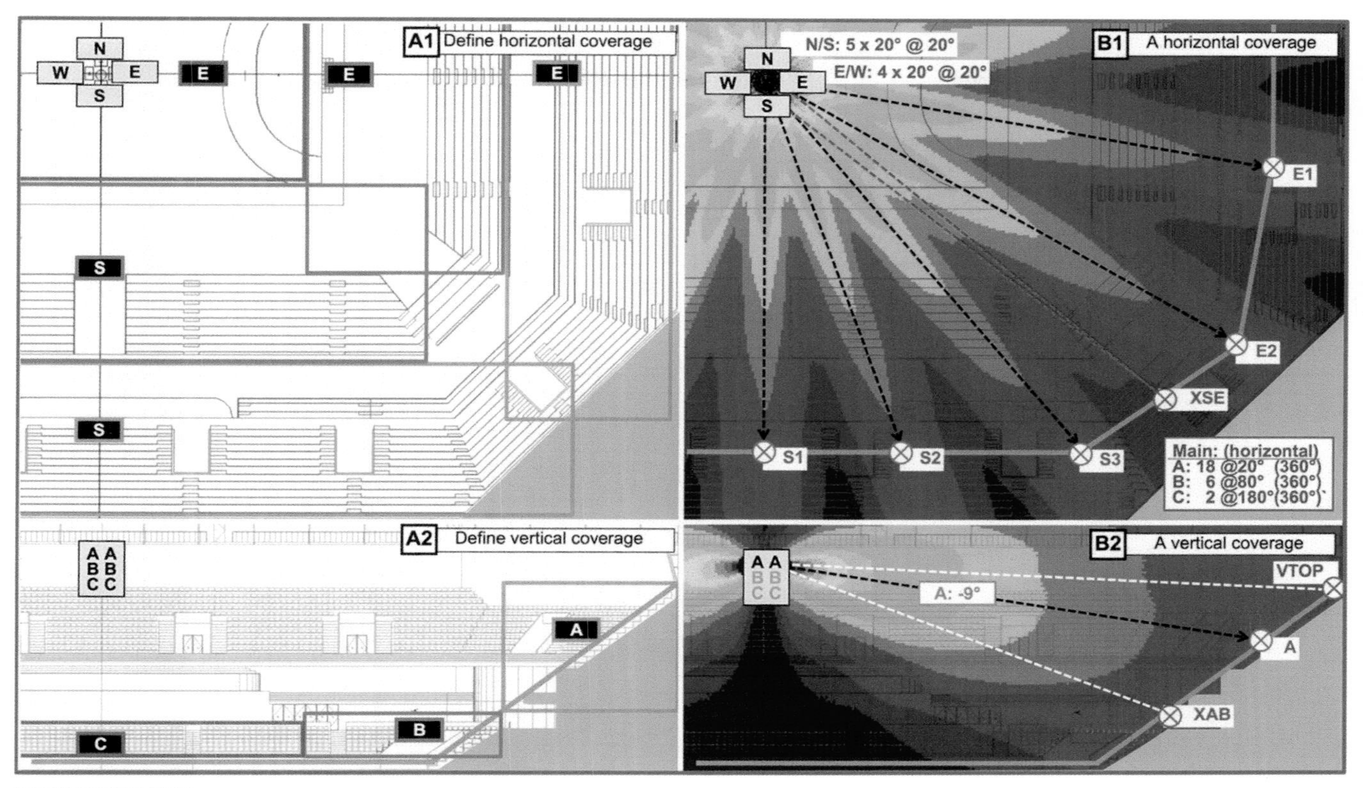

FIGURE 15.12A
Design process for Venue #12: coverage partition strategy, E/W main A solo

15.12.2 East/west mains (E/W, ABC)

East and west are the longer sides. The overall coverage target is 87° with a 3:1 (10 dB) range ratio. This is oversimplified because this shape is clearly double-sloped (section 11.6.5.1). More than 90% of the listeners are in the upper slope (the raised seating area) while the remainder is found on the flat floor. The raised area is a simple target: 31° with a 1.6 range ratio (4 dB). This could be handled with a single element if we only needed to cover the raised seating, but the application calls for full floor coverage as well. We know that the upper element can cover this because it has 60° of vertical range (we need 48° (1.6 × 31°)). This 60° vertical coverage is more than we need but recall how the height limitations forced us to go horizontal. This speaker was chosen primarily for the horizontal power it could pack into the circle. The ±6° of extra coverage are a minor concern (and better than not enough). The vertical aim is found by the compensated unity aim method (Fig. 11.9). We have less than a 2:1 range ratio so we won't be aiming at VTOP (-3°). The vertical center of the shape is 15° below this at -18°. The aim moves 6° down from VTOP to -9°. The best way to turn the corner on the double slope is to have the B element aimed at the transition point (-34°). This allows it to fade quickly above (moving off axis and away) and hold out for a long time on the floor (getting closer as it moves off axis).

How can we make a circular pattern on the floor? We could face a 90° × 90° speaker straight down. An alternative is a pair of 45° by 90° speakers splayed at 45° and facing down. We almost have that (2 × 40° × 90°) to make an 80° × 90° spotlight on the floor.

15.12.3 North/south mains (N/S, ABC)

The short sides are only marginally shorter and require only minor adjustments to bring us to the same place. Both the A and B sections are aimed lower to achieve the same basic shape. The revised angles are seen in the figures.

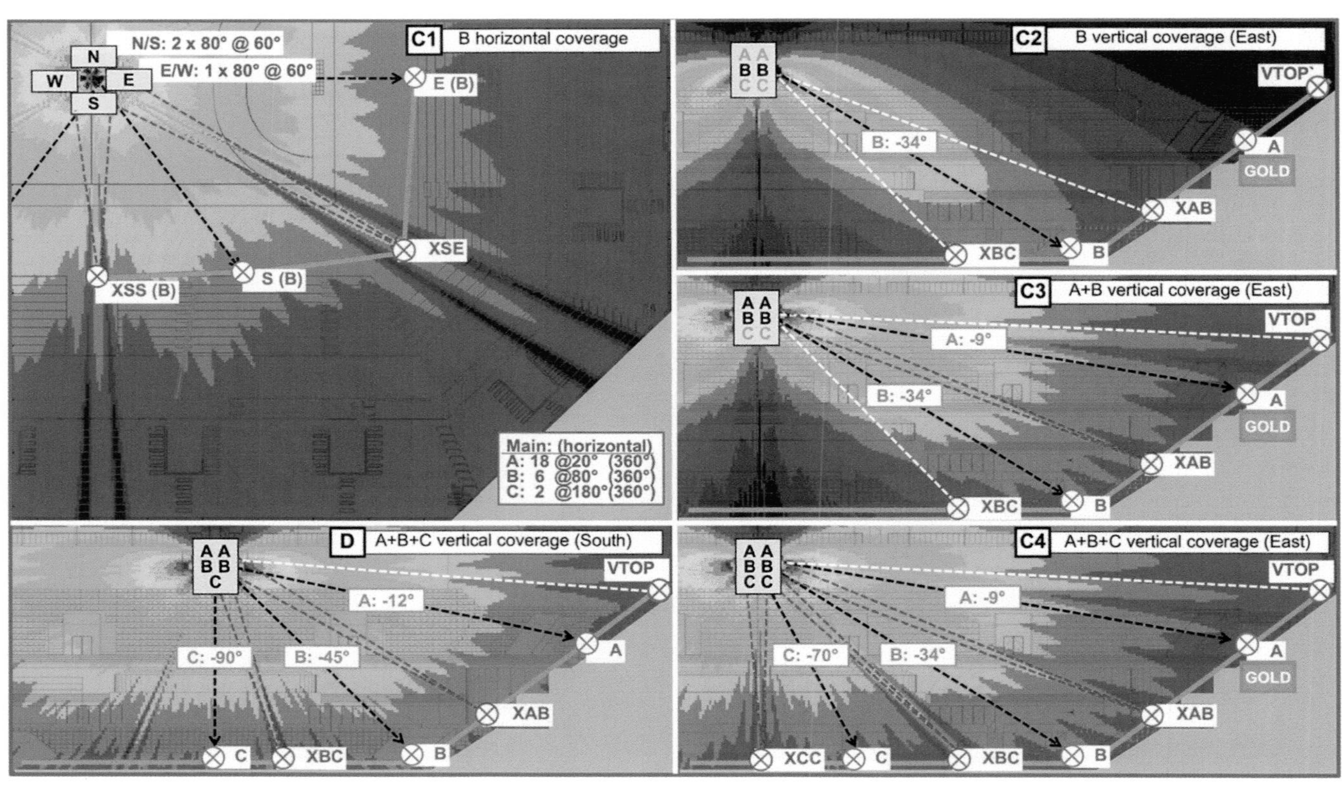

FIGURE 15.12B
Design process for Venue #12: E/W main B solo, A + B, A + B + C, N/S main A + B + C

15.12.4 Combined systems

The most dramatic combination effects will be the summation of a circle of eighteen low-frequency devices. The effect will be largely symmetrical (because there is only 2 dB of level taper), so the changes will be uniform over the space and therefore highly treatable. The vertical combined response makes a clearly defined double slope that follows the raised and floor seating. Delay for the B section can be set at crossover XAB and the same follows for the C section at XBC.

15.13 VENUE #13: ARENA, L/R MAINS

We return to the same arena to do a concert. The system will be L/R mains, coupled sidefills, LF and frontfills. Our focus here is on the mains and coupled sidefill. The stage is placed at one end of the long side. The mains cover the central long shot and the sidefills handle the leftovers.

15.13.1 L/R mains (ABCD)

The horizontal aim was calculated by the "we always do it this way" method, i.e. straight ahead. Yes it's true that we would have higher uniformity if we splayed them outward, but I never said that every design decision in this chapter was mine to make. We move on to the vertical where we see the classic double slope of an arena floor. People need to see the game and athletes need a level playing field. The raised seating area is a coverage target of 15° at 1.4:1 range ratio (3 dB). It's far, but it's easy. The floor is 30° at 3.5:1 range ratio (11 dB). It's far and close and definitely not easy. Overall we are facing around a 5:1 ratio that is reflected in our splay ratio spanning 9° to 2° (4.5×). Half of our sixteen boxes go to the upper shape and half to the lower. The upper set runs from 2° to 3°, a splay ratio of 1.5 to counter the range ratio of 1.4. The lower set expands the splays and finishes at the maximum allowable angle for the system.

15.13.2 Coupled sidefills (ABCD)

We will use the compensated unity splay calculation to find the horizontal aim for the coupled sidefill. The on-axis throw of the 90° horizontal mains is 70 m. The off-axis edge of the mains (45°) reaches its last seat at 50 m.

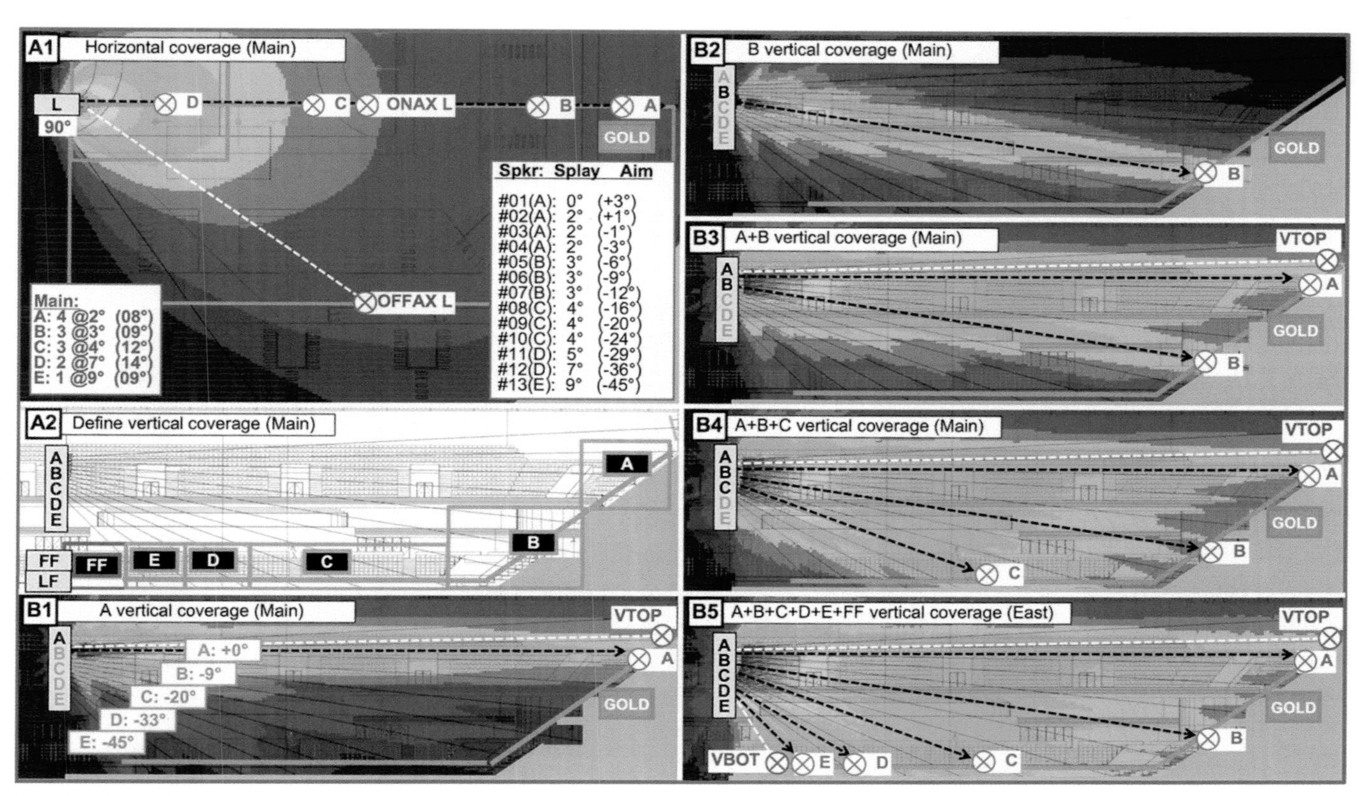

FIGURE 15.13A
Design process for Venue #13: L/R mains coverage partition strategy, main A solo, B solo, A + B, A + B + C + D + E + FF

This OFFAX position loses -6 dB at this point (by angle) and gains +3 dB (by distance) for a net -3 dB below ONAX A. This is 3 dB of level variance, but still 6 dB of spectral variance. We will use the sidefill to decrease them both. We will use the 3 dB range ratio (70%) between ONAX and OFFAX to set the compensated splay angle. Both speakers are 90° so the result is 90 × .7 = 63°. Now we know where the sidefill is aimed. To set its level we look to

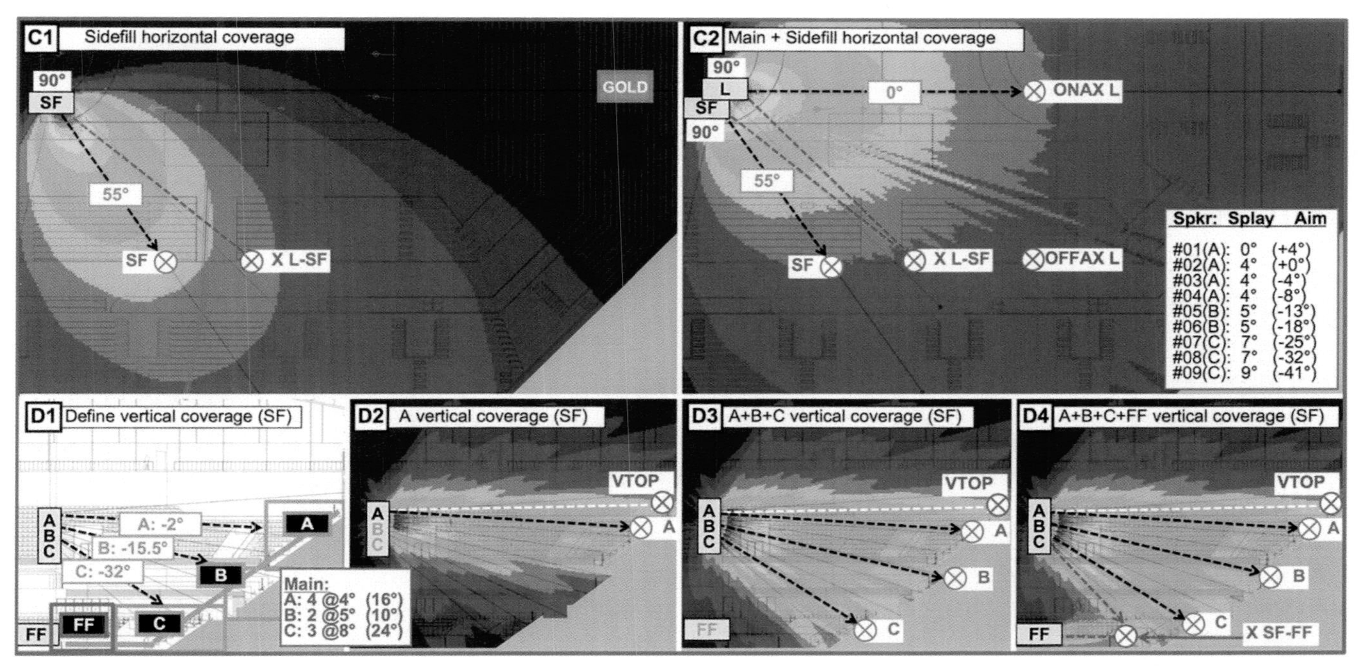

FIGURE 15.13B
Design process for Venue #13: sidefill coverage partition strategy, sidefill A solo, A + B + C + FF, main + SF

see how far it has to go, which is 37 m along its on axis line. The ONAX A to ONAX B range ratio is 1.9:1 (5.5 dB), which gives us the sidefill's level setting.

The vertical target for the coupled sidefills is basically the upper portion of the mains double-slope coverage without the flat floor. This can be approached as a single slope of 52° at 2.2:1 range ratio (7 dB). We have nine boxes with an average splay of 6°, and a splay ratio of 2.5 (9° to 4°).

15.13.3 Combined systems

The mains and sidefills are coupled in the low-mid range so we can expect to see some addition in both directions. The interaction is asymmetric because the sidefills are set at a lower level than the mains (3 dB shorter throw). Therefore the low-mid buildup is likely to gravitate toward the sidefills, which can reduce their low-mid response by more than the mains (Fig. 14.22 provides guidance regarding asymmetric EQ).

The mains–sidefill crossover should be phase aligned at XMn–SF. Typically the sidefills are delayed.

15.14 VENUE #14: LARGE TENT, MULTIPLE DELAYS

The venue is an 80 meter-deep tent with weight restrictions and limited placement options. The design featured L/R mains with centerfill, sidefill and frontfill. The mains were hung from a ground-based frame while the delays were hung from the ribs of the tent. The tent's low clearance and the client's requirement for high vocal intelligibility led us to use two levels of delays. We have already done lots of centerfills, sidefills and frontfills so the focus here will be on the two levels of delays.

Each of the delay rings is comprised of three delay clusters arrayed as an uncoupled point source. Another unique aspect of the tent shape is that it's much taller in the center than the sides. Sightline considerations forced the center cluster up high so we couldn't hang L/C and R at the same height. We could aim the centers down more steeply, but this is poor for imaging and shortens the usable range (see Fig. 9.48). The better option was to move the centers back to the previous truss, which kept their aim angle consistent with the outers. This made the coverage start and stop points close to the same, i.e. a straight unity line depth.

The timing sequence is clear for the first delay ring, i.e. they are sync'd at their start of coverage to the L/R mains. The second set faces the delay dilemma as to whether to delay them to the mains or the first delay ring. I don't remember which we did in this case, but I can tell you how I determined it on site: measure at the coverage start location for delay 2 and sync to whoever is stronger (mains or delay 1).

The delay ring levels are set in two stages. First make the solo responses of the three systems in each ring (L, C and R) land at the same level in their respective zones. Each of these is an AB asymmetric composite coupled point source so the A and B levels can be adjusted for best uniformity over their range. Then set their overall level to sum with the mains to match the GOLD reference level. The process is repeated on the second ring but now the level is set so that the summation of main, delay 1 and delay 2 matches GOLD.

15.14.1 Mains (ABCD)

We have done enough ABCD main systems to limit our discussion to the unique aspects of this one. It's too low, for starters. This is hardly unique for our field of work, but we have yet to face this in this application series. This is totally standard for outdoor applications such as festivals where trim height is limited by outdoor staging realities, crane capacity, weather safety concerns and more. We have similar limitations inside this tent as well as traffic from video and staging. Too low means we have to give in to the limits of our throw capacity, and tolerate more level variance than we would from a higher origin. In short: It's going to get loud in front. The GOLD reference for most venues in this chapter is the ONAX A position. In this case we used ONAX B because the ONAX A location sees a combined response of the mains and delays. We still have a fully dominant main system at ONAX B, which makes it the better choice for GOLD. Closer areas (e.g. ONAX C) are calibrated to match GOLD as part of the ABC coupled array. Areas beyond ONAX B are calibrated to match GOLD as a combination of the ABC mains and delays. Recall the driving force behind this: Mains are too low to reach anywhere near the back without delays.

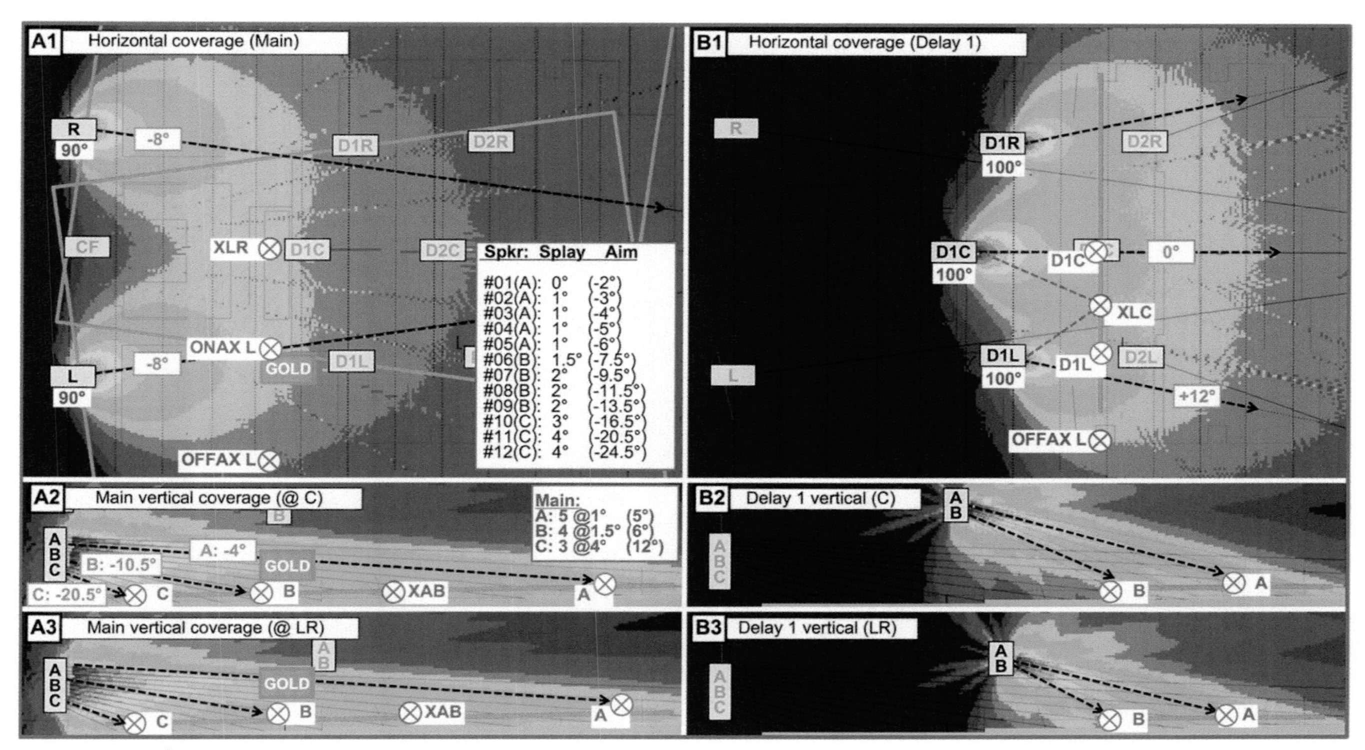

FIGURE 15.14A
Design process for Venue #14: coverage partition strategy, L/R main solo, delay 1

The next unique aspect is that there is very little (if any) listening area where our long throw system (A) will enjoy strong level dominance. We are basically throwing the top of the system out there as a coverage starter for the delays to ride along with. The mains and both delay systems all have something in common: They are aiming their top boxes to the very back of the room. The combined result is improved level uniformity and reduced sonic image variance over the alternative. The possible downside is increased ripple variance, but that is why we are using an extremely well-controlled system at small splay angles.

So what's the alternative mentioned above? It's a dive and relay strategy. Angle the main systems down and let them die into the floor. Just make sure that the coverage is picked up by the relay system before it falls too low. The relay strategy can reduce the ripple off the sides and ceiling but the level variance suffers as each system goes through the startup of its propagation-doubling distances.

15.14.2 Delay 1 (D1, LCR, AB)

This is the largest-scale delay system we have seen here, and the first composite delay array(s). The design factors were described above so let's concentrate on the tuning. The A and B sections are coupled composite elements. We can apply asymmetric EQ and level as needed to minimize variance between A and B but only a single delay time and overall level will join the system with the mains. Internal level and EQ operations use the ONAX A and B positions but the global level and delay setting is done at crossover (XL–D1). The choice for crossover depth is more complicated for a delay array than a solo delay speaker (in the vertical plane). There are five milestone locations in this type-AB delay array that we could choose for global delay and level setting (front to back): VBOT, ONAX B, XAB, ONAX A and VTOP. We can eliminate VBOT and VTOP because the combination will end up too loud in the range between ONAX A and B. The delay system (solo) should have constant level from ONAX B through to ONAX A. We can maintain a nearly equal combined level between ONAX B and ONAX A because the loss rate for the mains is very low (long doubling distance) and the loss rate for the delays is zero. Therefore any location between ONAX B and A can get an acceptable result. Using ONAX B as the crossover to the mains (sync'd with level set to combine to match GOLD) keeps the combined level equal to or less than GOLD. The delay level would be set

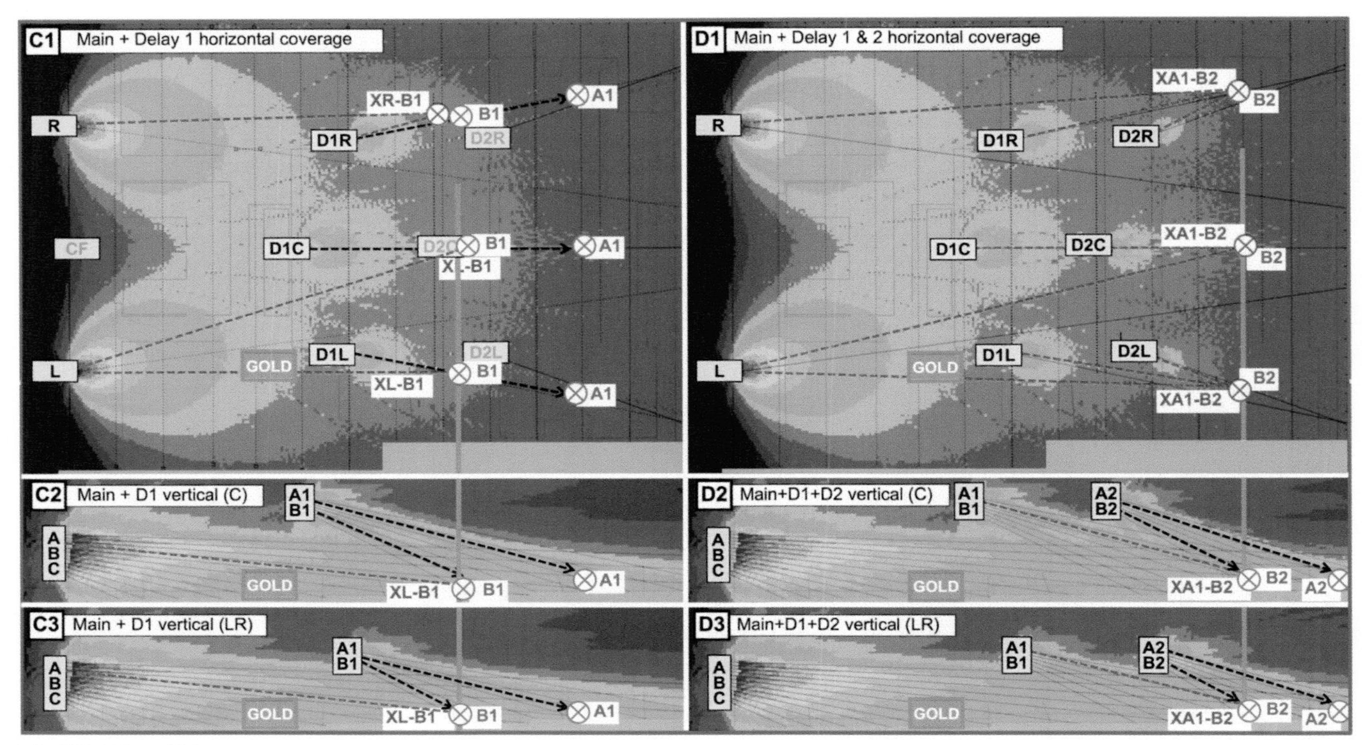

FIGURE 15.14B
Design process for Venue #14: main + delay 1, main + delay 1 + delay 2

higher if ONAX A is used, which would make the combined level at ONAX B greater than GOLD. My strategy here is delay (not relay) so I look to ensure that distant areas don't exceed the GOLD level. The choices are quieter, earlier delays (ONAX B) or louder, later (ONAX A). The net effect on imaging should be a wash. Or you can compromise at XAB. What did I do? ONAX B.

15.14.3 Delay 2 (D1, LCR, AB)

We previously discussed a second delay ring in Venue #11. In this case we are in the coverage of all three systems (mains, delay 1 and delay 2). The crossover is set at delay 2's ONAX B (XL–AI–B2). Do we delay to the mains or the delay 1? We will have to measure and see who is strongest.

15.15 VENUE #15: STADIUM SCOREBOARD, SINGLE LEVEL

We now move outdoors to a large football stadium. The main system will live in the scoreboard and cover all the seating except a few pockets behind glass or blocked from sight. The shape is quite irregular, especially from the speaker system's point of view. House left is a single high-raked level. House right is two levels and more than double the height. Coverage in the center area doesn't start till we have crossed over 150 meter of open field and is extremely short in height. Scoreboards look huge from a distance but the space left for us is amazingly small after video and various sponsors have taken their places. It's not necessarily shaped with speaker array-friendly openings, and full of solid steel beams. Not little ones. The "what the &%$# happened to my horn?" type beams.

Coverage division is shown in Fig. 15.15A (A) which clarifies the minority of seats covered by the sidefills and the long gap before beginning centerfill coverage.

The horizontal aim for the L/R mains highlights the asymmetric aiming strategy shown back in Figs 11.7 and 11.25. The approach here is to aim at the far corner of the shape ($ONAX_{FAR}$). On the outside edge we find the audience getting closer as we move on axis. This keeps the line of minimum variance moving in the right direction as we move off axis. We need the SF system at the near outer edges because we are too far off axis (excess spectral variance). The opposite occurs as we move from ONAX toward center. We are getting farther and more off axis so the dropoff is rapid. This is why we add the centerfill there, but only a small vertical slice because the stands are very low there. Aiming the L/R mains toward the corners allowed us to get the longest throw from the mains while wasting the least amount of them over the heads of folks in the low center area.

15.15.1 Right mains (R, ABCDEF)

This system had to cover upper and lower levels, while (hopefully) avoiding the line of glass boxes in between. The goal was to position the array so that we could gap the coverage along the glass box line. This is only viable if the array is placed at the right height to avoid smiles and frowns. There was not enough vertical room in the scoreboard for the full array so we had to break it in half and hang them side by side. This turned out to be a favorable event because it allowed us to position the lower main at the best height to skim across the lower level and gap the glass (Fig. 15.15A (C2)). Coverage moves gradually down into the bowl from the sharp edge at the top of the array.

The upper mains were essentially a five-element symmetric composite configuration (5 × 1.5°). The fifth box (B) was on a separate channel just in case level taper was warranted (it wasn't). The angles were kept constant and symmetric because of the expected smile effects from our upwardly aimed array as the coverage spins outward (and the stands get closer).

15.15.2 Centerfill (CF, AB)

The centerfill system is a parabolic dish loudspeaker with an extremely tight pattern (6°) and flat beamwidth over most of its range. The speakers don't go as low in the spectrum as the mains (250 Hz and up) but we knew we would get plenty of low-mid help from the mains (whether we wanted it or not). The dish speakers form a two-dimensional symmetric coupled point source: two rows of 6 × 6° (Fig. 15.15B).

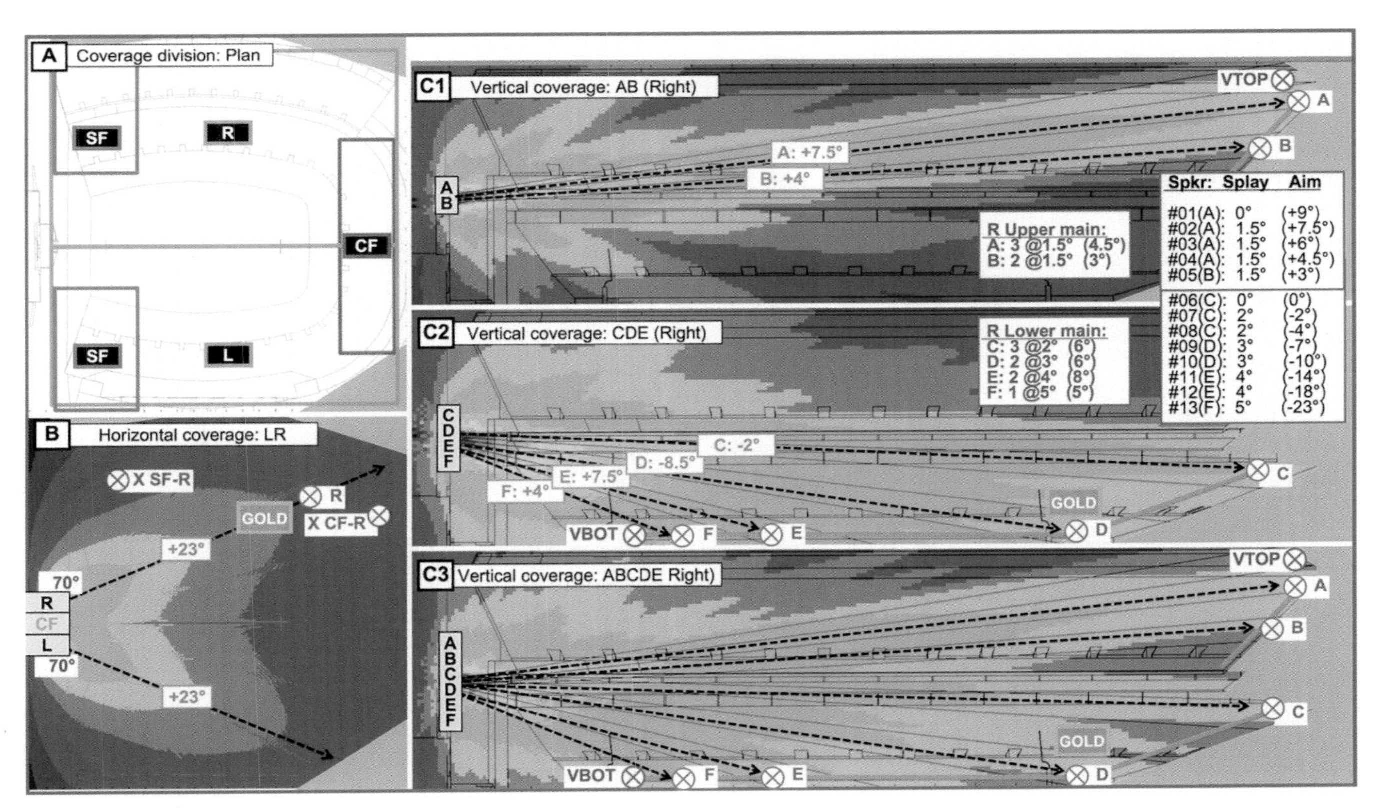

FIGURE 15.15A
Design process for Venue #15: coverage partition strategy, R main AB solo, CD solo, ABCDE

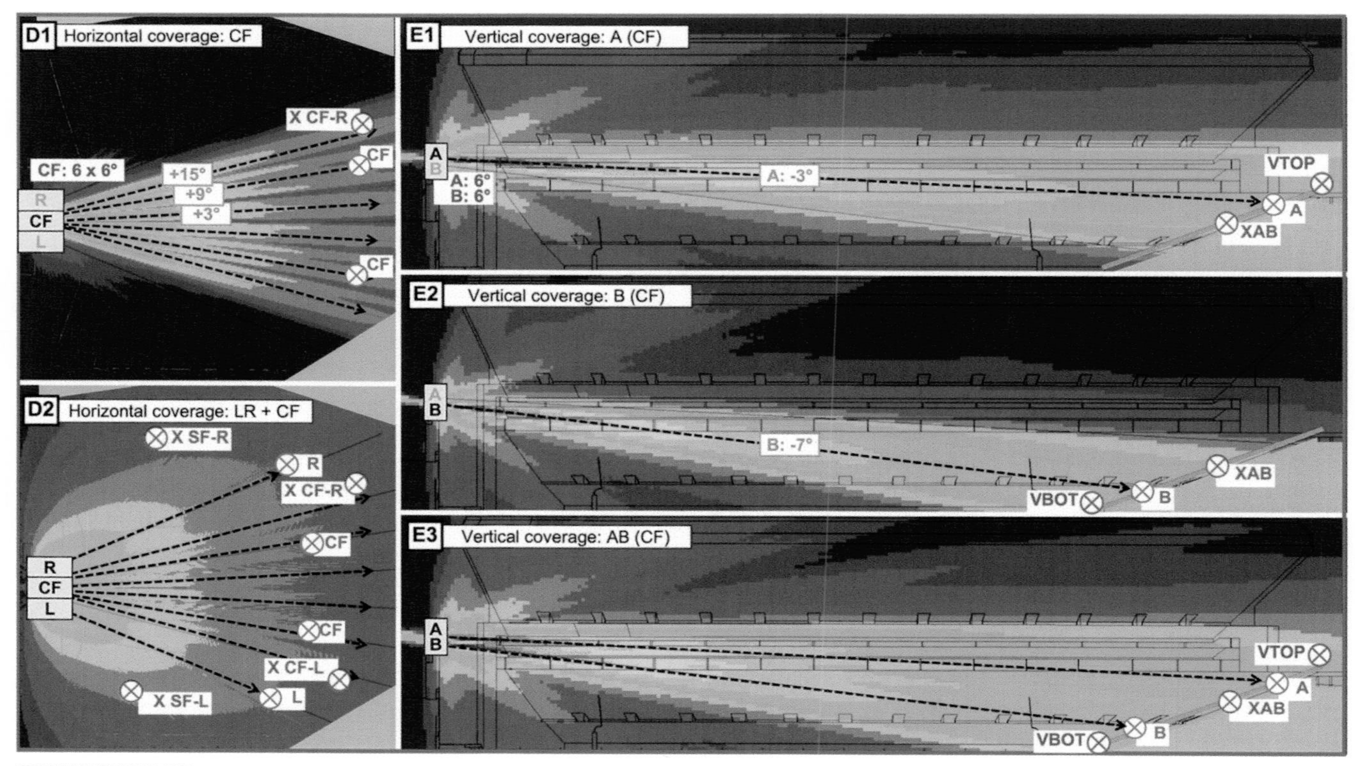

FIGURE 15.15B
Design process for Venue #15: centerfill A solo, B solo, A + B, L + R + center

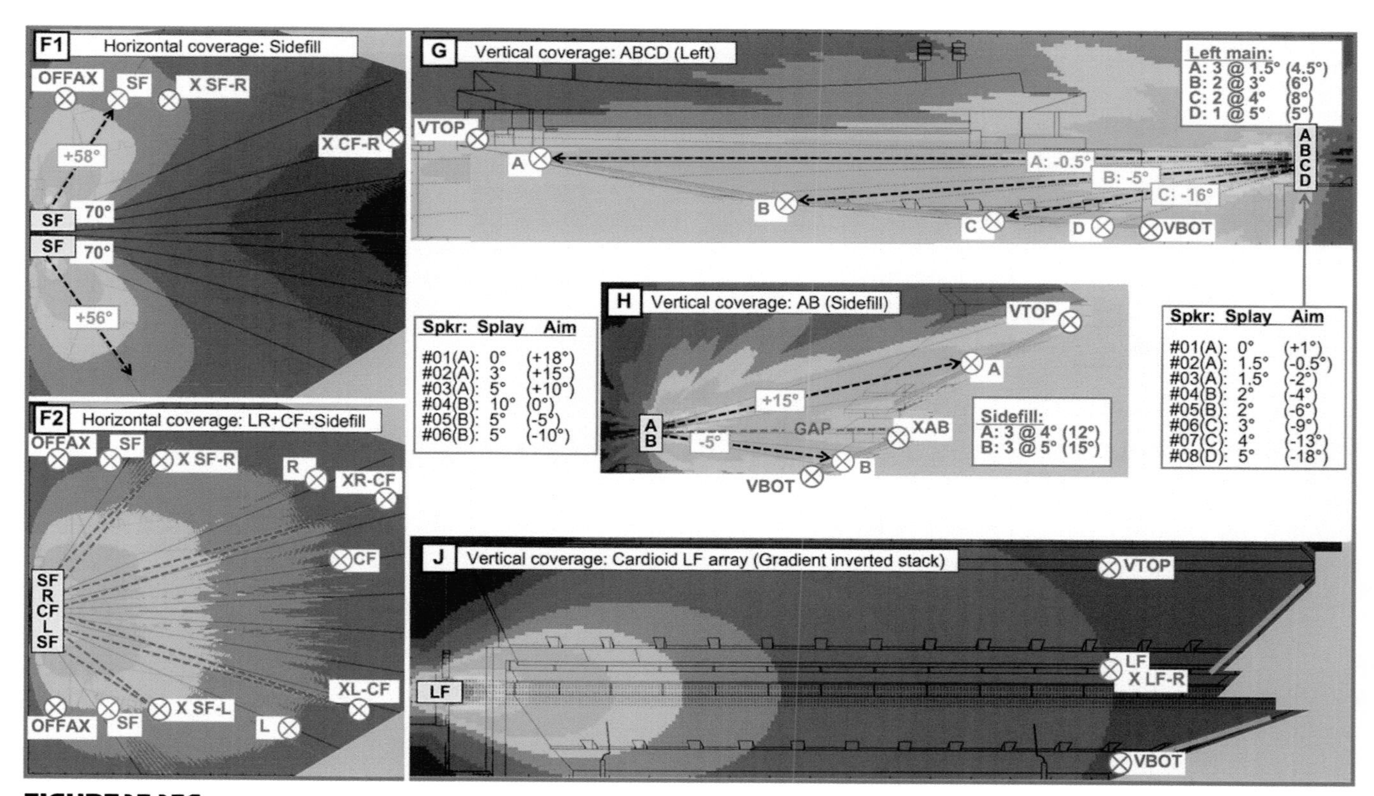

FIGURE 15.15C
Design process for Venue #15: solo sidefill, solo cardioid subs, solo left main, L + C + R + sidefills

15.15.3 Sidefills (SF, AB)

The sidefills were also different on left and right. The right side had to cover upward as well. The throw distance to the upper area was still quite long so the same speaker model was used here as in the mains. The area covered by the lower system was close enough to merit a power scale reduction.

15.15.4 Subwoofers (LF, +-)

A cardioid subwoofer array was employed here, and the neighbors behind the scoreboard are grateful. Cardioid steering also reduces the LF level near the scoreboard, which helps the overall LF level variance.

15.15.5 Left mains (L, ABC)

The left mains were essentially a copy of the lower section of the right mains. They also face press boxes and employed the same coverage avoidance strategy as the other side.

15.16 VENUE #16: CONCERT HALL, CENTER, 360°, MULTIPLE LEVELS

Congratulations for making it this far. The ultimate test of your endurance is at hand. If M.C. Escher had ever designed a concert hall, this would be it. This is the most complicated mind-bending design and tuning I have yet to face in my thirty-one years of design and optimization. It begins with a highly reverberant room with 360° of multiple seating levels that stagger vertically and horizontally, guarded by highly reflective surfaces over, under and between them. The main system will be a center cluster with extensive limitations on height, width and weight.

The front of house is covered on three vertical levels that wrap around with 180° of horizontal coverage. The sides are covered with two vertical levels and the rear with a single one. The remaining subsystems are sidefills and subwoofers on the deck and underbalcony speakers on the first floor.

The twist here is that the hall is twisted, which means we will need the speakers to be as well. Simple solutions go out the window once we see that the height changes so much that seats on the third floor sides are the same elevation as those at the center of the second floor. The balcony fronts are large and reflective as are the surfaces behind them. Speaker coverage tends to run in straight horizontal lines. Our approach must find a way to bend the coverage down as we move around the front of the house.

The coverage partition plan is shown in Fig. 15.16A (A1–A5). This hall takes five panels whereas all the others have needed only two. Take your time with that and come back when you're ready to start.

15.16.1 Upper mains (A)

The upper mains array cannot be a modern line array. It won't fit (too long) and it creates straight, thin, vertical lines. And we'd need two of them splayed apart horizontally to make the 180° radial shape. Instead we will go with fine slices in the horizontal plane (20°) that can be staggered vertically as we spin around the room. We got lucky because most of the seating sections were 20° slices, which allowed us to drop the vertical aim in sync with the seating sections. There is 15° of vertical stagger along the nine elements, which makes a huge difference in regards to getting the sound onto the seats and off the reflective surfaces. The elements have sharp horizontal edges, and soft vertical (60°). The combined EQ had to factor in the low-mid addition as cabinets were added, but isolation was dominant in the upper ranges. Minor level tapering was done in the horizontal plane because the side speakers had a shorter throw than the center (around 2 dB). Delays were set to compensate for the stagger between elements at different vertical aims (the lower aimed boxes leaned forward). These may seem like small details but they all add up in the combined system.

The vertical aim(s) follow the aiming guidelines (Figs 11.8 and 11.9). The shape is fairly symmetric because we are covering two levels so the aim is not far above the middle.

15.16.2 Middle mains (B)

The upper mains covered the top two floors. The middle system (B) lands the response on the ground and moves it most of the way to the stage. These are wider elements (80° H) with a slightly reduced power scale that can cover the

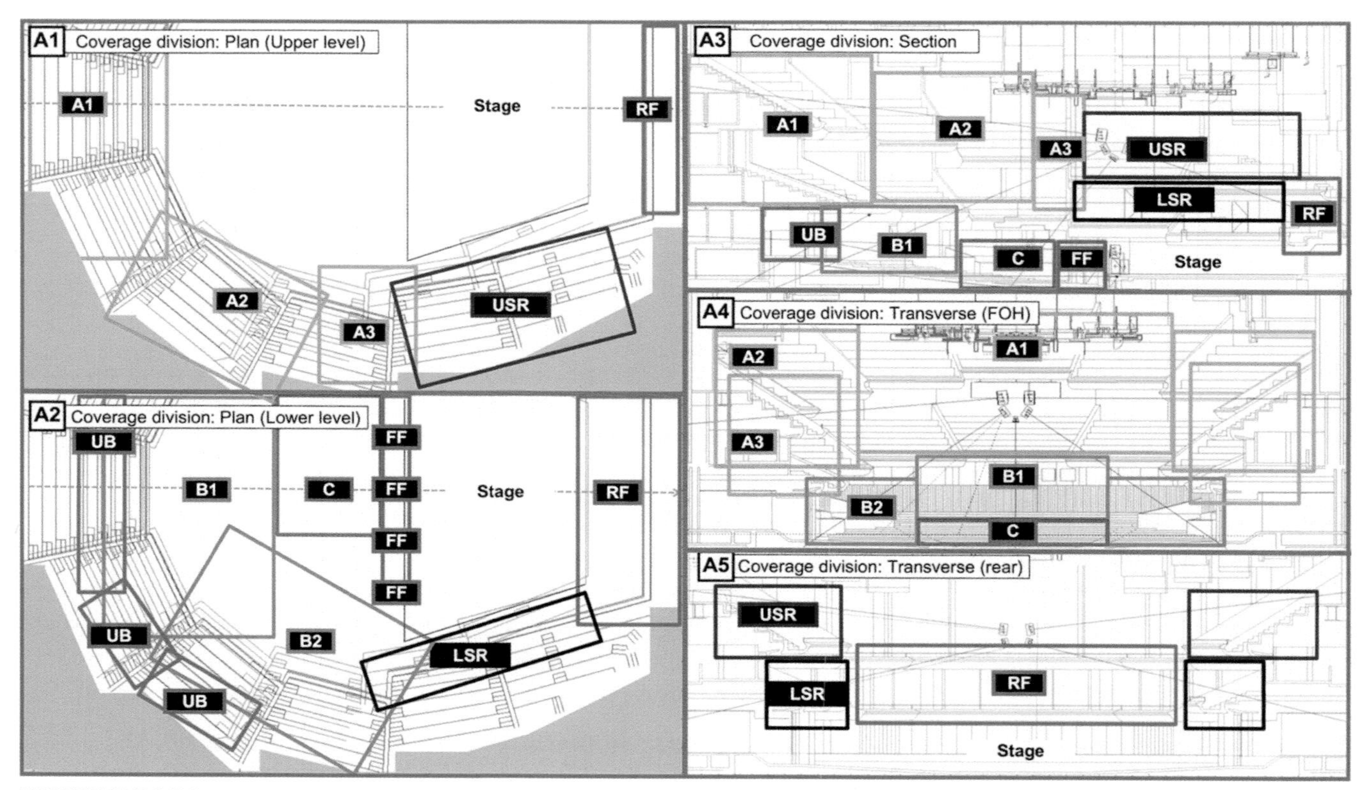

FIGURE 15.16A
Design process for Venue #16: coverage partition strategy (upper, lower, section, transverse)

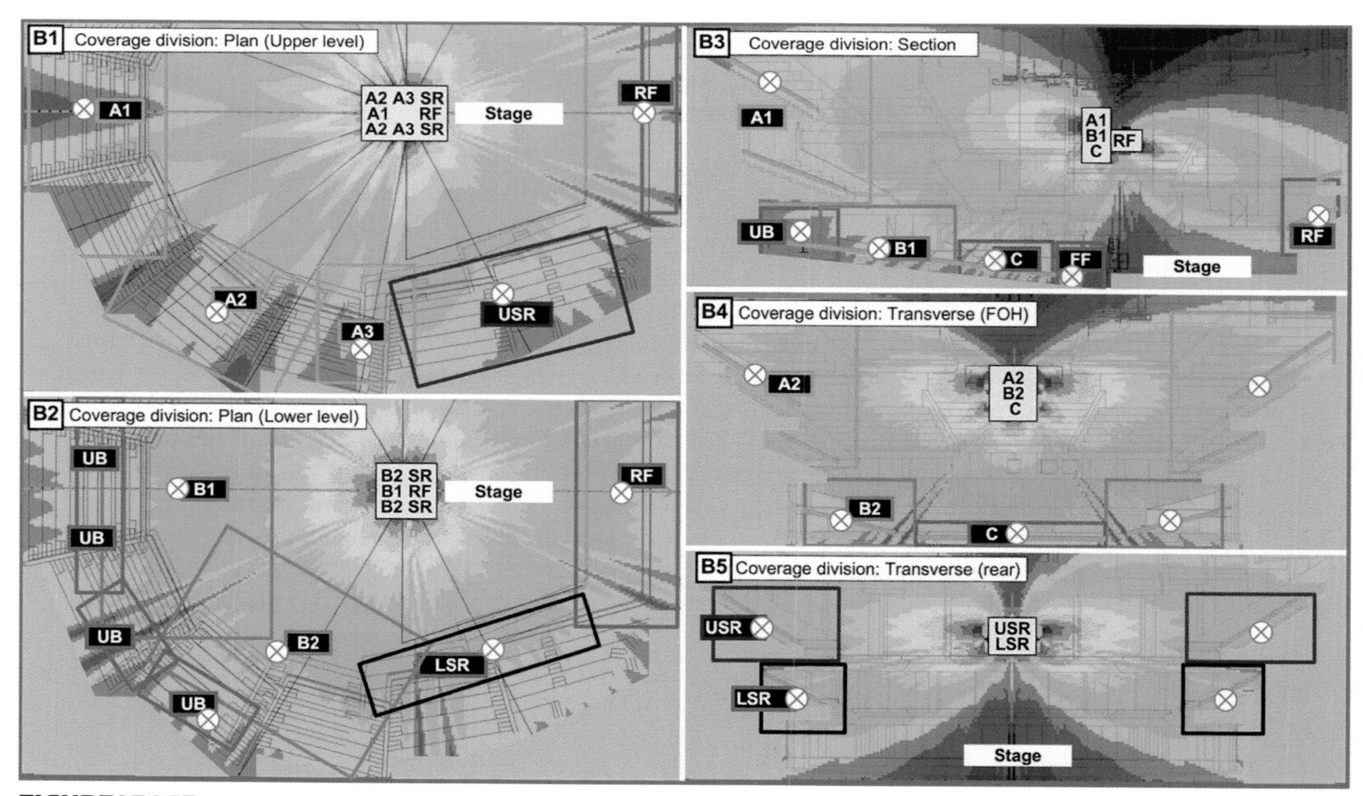

FIGURE 15.16B
Design process for Venue #16: coverage overview (upper, lower, section, transverse)

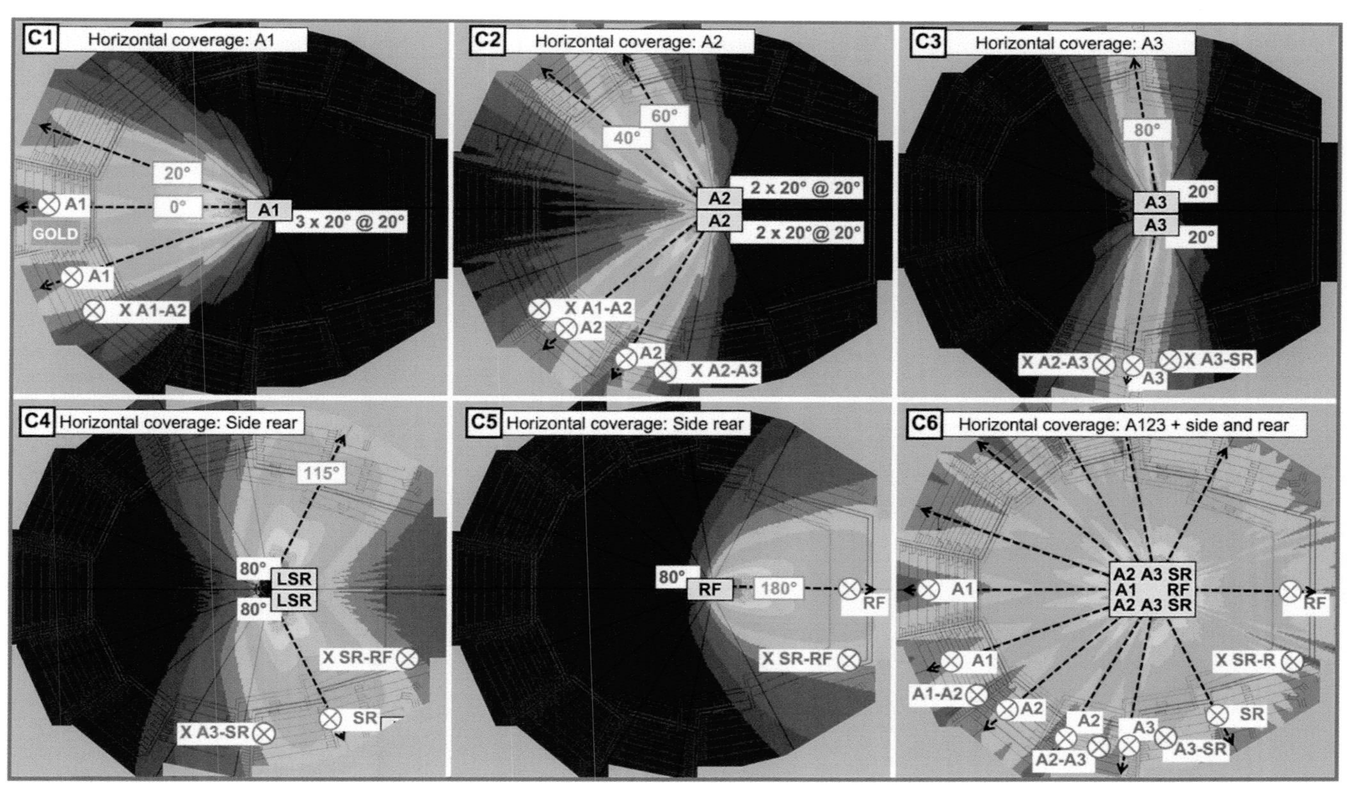

FIGURE 15.16C
Design process for Venue #16: (horizontal) upper mains (A), sides and rears (solo and combined)

180° target in three slices (splayed at 60°). The vertical shape is another double slope, in this case a fairly straight upper slope and angled lower slope. The vertical aim of the B element links its coverage to the VBOT edge of mains A. The underbalcony speakers are located at the joint between the slopes, which helps to soften the transition. The horizontal assembly for both the A and B sections was used for the symmetric point source example in Fig. 11.12.

15.16.3 Lower mains (C)

The lower section of the first floor is covered by a single speaker (90° × 40°) at the bottom, which connects to the frontfill and sidefill speakers below. We now have connected a coverage line from VTOP A on the third floor to the frontfill.

15.16.4 Deck sidefills (SF)

Let's finish the floor while we are here. The deck fills function as a sonic image anchor and coverage extension at the outer near edges. These speakers could be power scaled to fill only their local area, which would be less than or equal to the lowest section (C) of the mains. In this case, however, the L/R sidefills were actually scaled at equal level to the upper mains (A), far more than the minimum. The reason is acute "monophobia," i.e. the addiction to mixing on L/R mains even if it means it will be 12 dB louder in the front rows. A mono-center permanent install that services a wide range of visiting acts must build in the capability to provide boom-boom on the floor or face an endless demand for rental equipment.

15.16.5 Upper and lower side rearfills (USR and LSR)

Our next destination is the side area along the depth of the stage. There are two levels here and they are quite close and only a few rows deep. These are the kind of shapes we normally like to cover with uncoupled elements, but we are all hanging out together at the mother of all clusters. It takes a surprising amount of vertical and horizontal

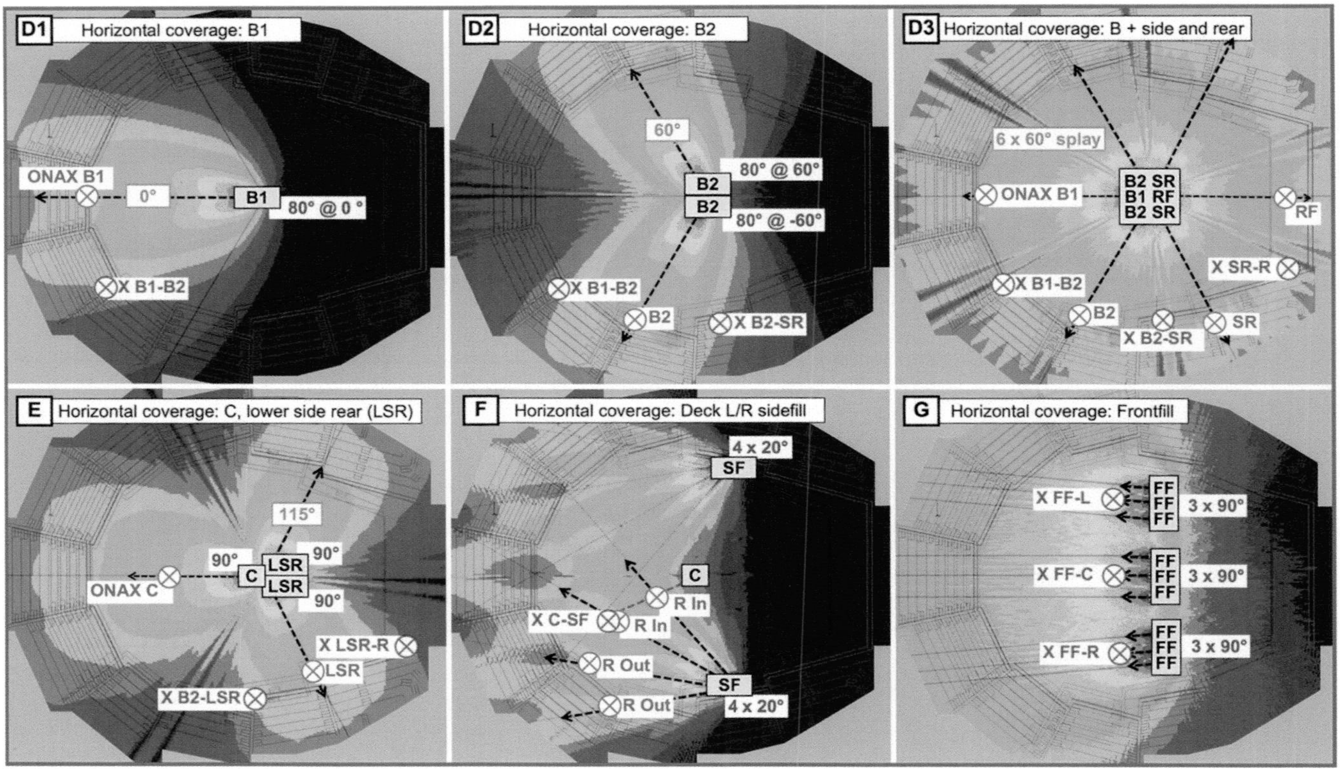

FIGURE 15.16D
Design process for Venue #16: (horizontal) middle mains (B), sides and rear, deck sidefill, frontfill

coverage to hit these targets, which calls for precise aiming in both planes. The upper and lower side rearfills connect horizontally to the outermost elements of main A and B respectively. Splay is verified and delays are set at XA–USR and XB–LSR respectively. The vertical crossover is found at XUSR–LSR, which is on a balcony front. A delay dilemma between arises here. Do we sync horizontally or vertically if there is a conflict? The vertical crossover is on a balcony front where nobody sits (for long). That's the one we can let slide.

15.16.6 Rearfills (RF)

The last stop is directly behind the stage. There is only one speaker here, so no delay dilemmas or question about horizontal aim. Vertical aim follows the usual guidelines. Level is set at ONAX RF to match GOLD. The delay is set at the crossover to the lower-side rearfill (XRF–LSR). It might be tempting to delay the rearfills to a fictitious stage source. This would be viable if the rearfill was uncoupled from the sides and front system (it's not). The coupled array that delays together, stays together.

15.16.7 Combined systems

We have three levels of various speaker models covering 360° at different heights and depths hung together as close as Geppetto the puppeteer can rig them. Combined system EQ should be a breeze! It can be if we are careful at each step of the way to set the proper aim, solo EQ, level and delay. The predictable low-mid addition can be monitored and compensated globally. The key to the process is a proper order of operations. First we linked all the horizontal elements of the A system (A1–A2–A3). Horizontal for B was next (B1–B2) and only then did A and B come together in the vertical plane (XAB). The C section follows and we have completed the front system. Our focus moves to the side where the upper side rears (USR) join the mains horizontally at XA3–USR. The lower side rears (LSR) have to fit like a puzzle piece: horizontally joined to the mains B2 at XB2–USR and vertically joined to USR at XUSR–LSR. Still with me? The circle is completed when LSR meets the rearfills at XLSR–RF.

Of course we still have the frontfills, deck sidefills and UB delays to add in but those seem like the easiest things in the world right now, eh?

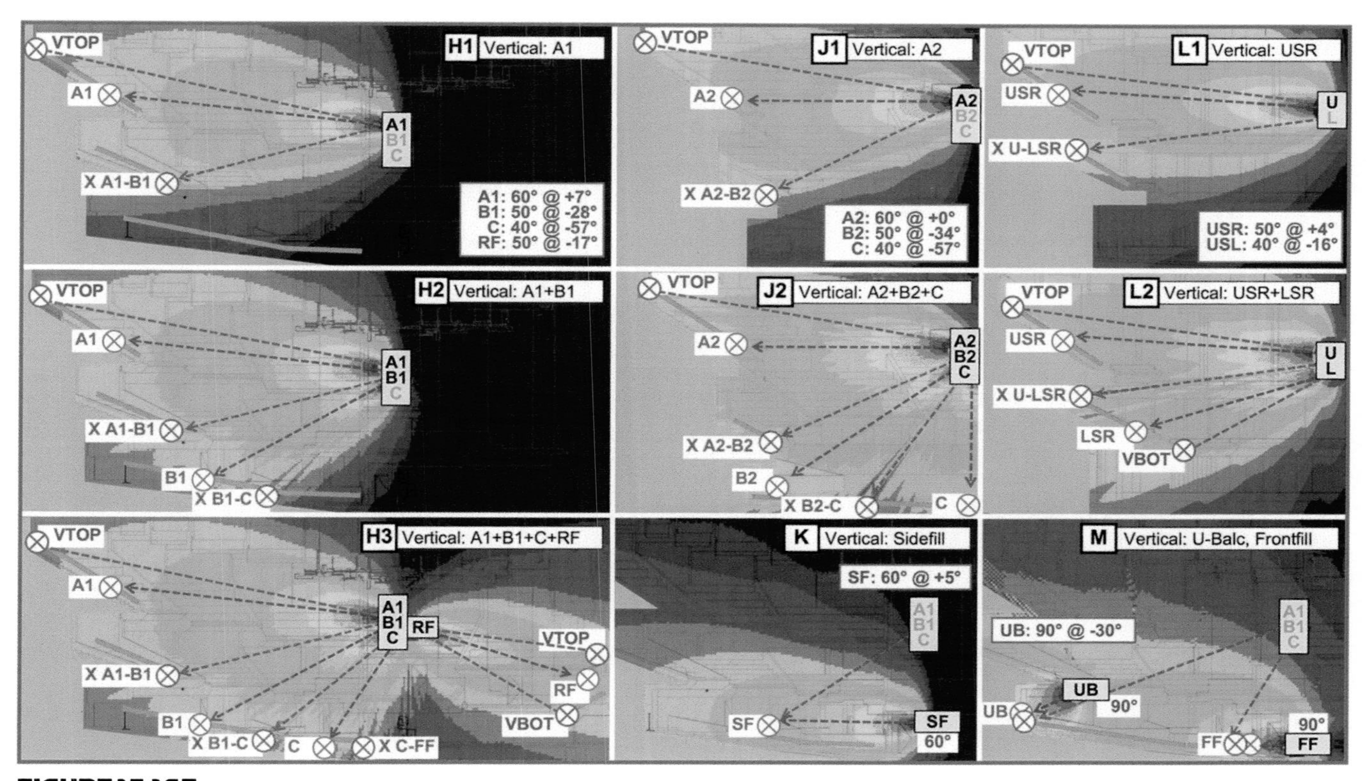

FIGURE 15.16E
Design process for Venue #16: Vertical, main (ABC) sides, rears, frontfill, deck and u-balc

15.17 VENUE #17: MEDIUM HALL, SINGLE LEVEL

We just completed the most complicated design. And now for something completely different: total simplicity (and not in a good way). Old folks like me remember when line arrays hit full stride in the marketplace and they were the cure for everything audio. Anyone caught using those old-fashioned speakers must have been sleeping under a rock for the last twenty months! This application (or better named mis-application) includes a five-element/side line array in a room with a 2.1 m (7′2″) ceiling at its *highest* point. One of the selling points was that they won't need delays in the 14 m deep room because line arrays can throw so well. The clients were using the side and rear surrounds as delays by the time I was called in to attempt to tune it. Obviously the solution was 2.7 ms of delay and -4 dB at 2 kHz.

The primary program material was spoken-word reinforcement with tons of headroom so they can get it as loud as possible. Live music and playback held the next level of importance there.

15.17.1 Original system

The main systems were hung from the ceiling. We can't just duct tape the top speaker to the ceiling. We need to tie in to the beam and then shackle down to the hanging grid, and only then does the sound-making part begin. In this case the uppermost speaker was at standing head level and the array finished at the knees. Did I forget to mention that the listeners (or should I say *attempted* listeners) like to stand when the system is in use? At best case we have two usable elements/side when people sit down. When they stand up the sound stops at the third row. There is no vertical plane to discuss. Closer listeners are louder by the inverse square law. In this case the range ratio exceeded 20 dB, as did the level variance. The system was not optimizable as such, and we worked out a plan to go back to the dinosaur era and use the old-school speakers and try again a year later.

15.17.2 Revised mains (L/R, A/B)

Priority was given to ensuring that a speaker could be mounted horn up as close to the ceiling as possible. This left some clearance all the way to the back of house even with standing patrons. It was also determined to partition the

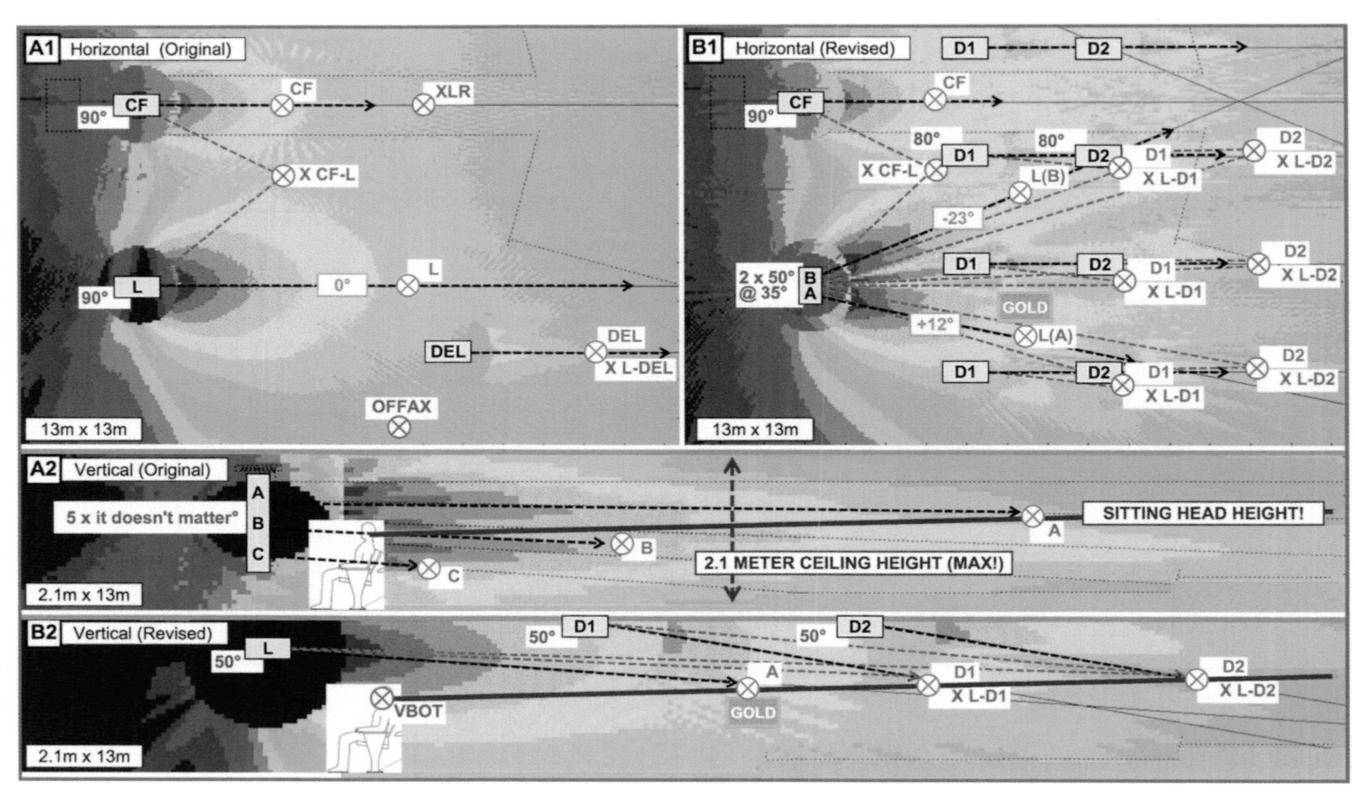

FIGURE 15.17A
Design process for Venue #17: coverage overview (original vs. revised)

room with two levels of delay, which meant the mains would hand over the baton after 6.5 meters. The intent is for the front mains to remain a partial presence all the way back to minimize image distortion and hold together as a system, rather than a scatter of ceiling speakers.

The frontline consisted of L/R and a centerfill, which was delayed back to the podium. The mains were set up as inner/outer coupled point sources to allow for maximum flexibility and level tapering to maximize gain before feedback. We could have used a single 80° speaker but the pair of 50° units offered sharper edges and the option of asymmetric shaping. The inners were turned down 3 dB (due to their merger with the centerfill) and therefore the compensated unity splay angle of 35° was used (50° × 70% = 35°). The coupled crossover is asymmetric, so we need to delay the inners (because they were turned down). The vertical aim for the mains was toward the rear of the room (>2:1 range ratio). The equalized response at ONAX A becomes the GOLD reference for the rest of the calibration.

15.17.3 Revised delays (D1, D2)

The first ring of delays was placed 4 meters ahead of the mains and scheduled to land a unity line at 6 m (2.2 m vector distance from delay 1). The fixed height and known unity line set the spacing as follows: 2.2 m × 80° (1.25) = 2.75m. This gave us a quantity of six elements to extend the lateral line across the room. The vertical aim was also set to hit the rear of the room (>2:1 range ratio).

The second delay ring followed a similar process to the previous one in terms of spacing, quantity and aim. The floor rises slightly as we go back in the room so the clearances get a bit smaller. The difference in calculated spacing was small enough that is was not worth trying to get that implemented in the install. Instead both delay rings follow the same model, spacing and quantity. Vertical aim is at the last seat.

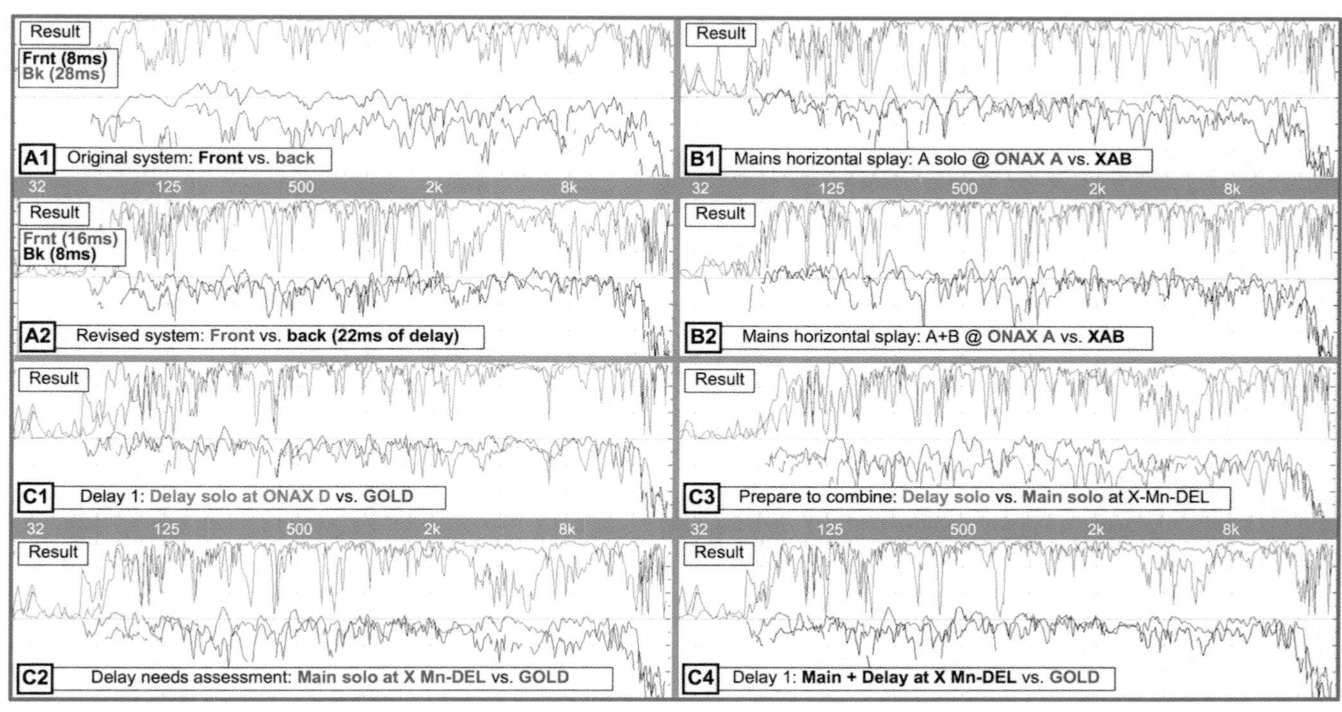

FIGURE 15.17B
SIM3 optimization data for Venue #17

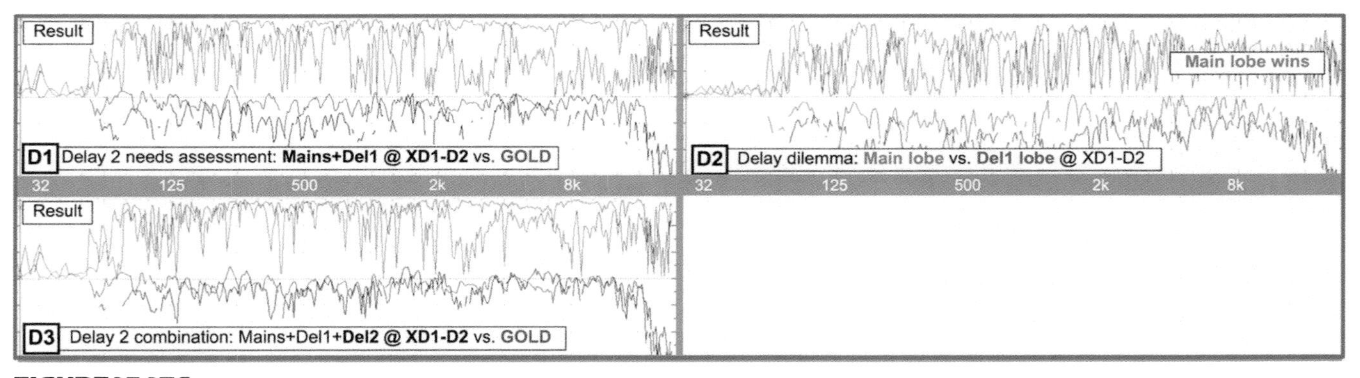

FIGURE 15.17C
SIM3 optimization data for Venue #17

15.17.4 Revised combined systems

We have already combined the inner/outer mains (AB). The next step is to join them to the centerfill by phase aligning the systems at XL–CF (delay whoever gets there first). This completes the first wave. The second wave (delay ring one) is timed to the first wave and the level is set to make the combined response of the first and second waves match GOLD at XL–D1.

The third wave (delay ring two) will join the combined energy of the first two. The delay is set to the stronger of the two at XL–D1–D2. Is it the left main or delay 1? It was the left main in this case.

15.17.5 SIM3 data

The calibration data from the original installation was extremely minimal. The trace shown in panel A1 of Fig. 15.17B gives an optimistic view of the situation. The hall was unoccupied, which meant the speaker transmission blockers, also known as "audience," were not in place. The "front" and "back" mic positions in my measurements were not all the way at either. Nonetheless the traces show more than 10 dB of level variance. The revised system at comparable locations is shown in panel A2. This is recognizable as a member of our minimum-variance family.

Let's tune the revised system. The L/R mains are AB coupled point sources. The process of setting their splay is shown in panels B1–2. Here we see GOLD (ONAX A) compared to crossover XAB (without the B speaker on). The beamwidth plateau is visible 6 dB down from ONAX. The response restores back to 0 dB (panel B2) when the B speaker is splayed properly to create unity gain.

The first set of delays is described in Panels C1–4. The first data shows the equalized solo delay. It matches GOLD in spectrum and coherence. Do we need delays? I think that question was answered a year earlier, but we will prove it again (panel C2). The mains are not getting it done out here. They are 6 dB down and losing coherence. We prepare for combination in panel C3, which shows the familiar pairing of tilted main vs. flat fill. Delay is set and the level is set to create combined unity. The last panel verifies that the crossover connection has matched the response to GOLD.

Onward and outward to delay ring 2 (Fig. 15.17C). The needs assessment is crystal clear (because the sound coming from the mains and delay 1 isn't). This brings us to the delay dilemma. Do we sync to the mains or delay 1. We will ride with the strongest. Panel D2 shows the results: mains by a landslide. The final data (D3) verifies that we have arrived at the end of the room and still match GOLD. You have arrived at the end of this book, and I sincerely hope you still match GOLD.

AFTERWORD

This concludes my transmission, for the moment. The cycle of design and optimization has been and will continue to be a learning experience. This is still a young field and the potential for growth is huge. For me personally, I feel that each day has the potential for some new discovery. Rare is the day when that potential is not reached. Over thirty years later "still learning" is the order of the day. I understand why doctors refer to themselves as "practicing" medicine.

The most common question I receive goes something like this: "Unlike you, I don't have perfect clients with unlimited budgets, unlimited tools, and unlimited time and support for system tuning. If we can't do it all, then what should we do?"

First of all, I have never had this client, but would love to. Jobs happen in the real world, and require prioritization and "triage." I have never had the opportunity, on any single job, to perform all of the design and optimization steps shown in this book. I have, however, used all of them over the course of time. They are all in my playbook, ready to use, as the situation requires. Coaches are not required to use all players on their team. They must, however, be prepared to read the situation on the field and be ready to deal with whatever contingencies arise. This book strives to bring into focus the nature of the forces at work in the practical world of our sound reinforcement environment. This knowledge alone is a powerful ally, even without an analyzer.

The end product we provide to our clients is a complete combined system. If we must streamline the process to fit in the time allowed, we should do so consciously. A skipped step is a leap of faith and a calculated gamble. We must maintain clear knowledge of where the leaps are, lest they come back to haunt us.

What is the most important? This is a matter of opinion. For me personally, it is like food. We need a variety of the highest-quality ingredients. Good speakers and signal processing. The next level is like real estate: location, location, location. Good placement, good angles, good spacing and good architecture. Level, delay and EQ setting are the finishing processes.

Even more important, however, is maintaining perspective of our role in the big picture. We are members of a multifaceted team. We are there to provide a service to the clients on many levels. The importance of personal relations cannot be overstated. In some cases we are our own clients, stepping out of our lab coats, putting on the artist's beret and mixing the show.

The meeting point between the scientific and artistic sides of our world is the optimized design.

GLOSSARY

Absorption coefficient: A specification to indicate the sound energy lost during the transition at a surface. The range runs from a maximum of 1.00 (an open window) to 0.00 (100% reflection).

Active balanced interconnection: A balanced-line connection to or from a powered (active) input or output device.

Active electronic device: An audio device that receives power from an external source (or battery) in order to carry out its signal-processing functions. An active device is capable of providing amplification of the signal.

AES/EBU: The standard protocol for packetized digital audio transmission.

Air absorption loss: High-frequency attenuation that accrues over transmission distance in air. The humidity, ambient temperature and atmospheric pressure all play a part in the parameters of this filter function.

Amplifier (power): An active electronic transmission device with line-level input and speaker-level output. The power amplifier has sufficient voltage and current gain to drive a loudspeaker.

Amplitude: The level component of the audio waveform, also referred to as **magnitude**. Amplitude can be expressed in absolute or relative terms.

Amplitude threshold: An optional feature of transfer analyzers that allows for the analysis to be suspended when insufficient data are presented at the analyzer inputs.

Array: A configuration of sound sources defined by their element quantity, displacement and angular orientation.

Aspect ratio: A common term in architecture to describe a space as a ratio of length vs. width (or height). This term is also applied for the coverage shape of speakers (interchangeably termed the **forward aspect ratio** here).

Asymmetric: Having dissimilar response characteristics in either direction from a defined center line.

Averaging (optical): The finding of a representative response over an area by viewing the individual responses from various locations.

Averaging (signal): A mathematical process of complex audio analyzers that takes multiple data samples and performs complex division to acquire a statistically more accurate calculation of the response.

Averaging (spatial): The finding of a representative response over an area by averaging the individual responses from various locations into a single response.

Balanced: The standard two-conductor audio signal transmission configuration chosen for its noise immunity. This is suitable for long distances.

Bandwidth: Describes the frequency span of a filter function (in Hz).

Beam concentration: The behavior of speaker array elements when they have a high proportion of overlap. Beam concentration is characterized by a narrowing of the coverage area with maximum power addition.

Beam spreading: The behavior of speaker array elements when they have a high proportion of isolation. Beam spreading is characterized by a widening of the coverage area with minimal power addition.

Beam steering: A technique of asymmetric delay tapering used in subwoofer arrays to steer the coverage pattern.

Beamwidth: A characterization of speaker directional response over frequency. The beamwidth plot shows coverage angle (-6 dB) over frequency.

Binaural localization: Horizontal localization mechanism driven by the arrival difference between the two ears.

Bit depth: The resolution of the digital quantization, which sets the dynamic range of the system, approximately equal to 6 dB × number of bits.

Cancellation zone: The inverse of the coupling zone. The combination is subtractive only. Phase offset must be between 120° and 180° to prevent addition.

Cardioid (microphones): Unidirectional microphones commonly used on stage. The cardioid action is the result of cancellation zone summation behind the microphone derived from the combination of forward and rear entry of sound at the diaphragm.

Cardioid (subwoofers and arrays): A configuration of standard low-frequency elements (separated or within an enclosure) configured and aligned to create a cardioid pattern.

Channel: A distinct audio waveform source, such as left and right, surrounds or a special source effect. Each channel must be optimized separately.

Clipping: An audio waveform distortion that occurs when the signal is driven beyond its linear operating range.

Coherence: A measure of the ratio of signal to noise in an FFT transfer function measurement.

Combing zone: The summation zone having less than 4 dB of isolation and an unspecified amount of phase offset. Combing zone interaction has the highest ripple variance.

Combining zone: See **Transition zone**.

Compensated unity splay angle: The unity splay angle between array elements with asymmetric relative levels.

Complex audio analyzer: An analyzer that provides both amplitude and phase data.

Composite point source: The combination of multiple array elements into a virtual single symmetric array element. A symmetric composite point source element (matched levels and splays) will resemble the response of a single speaker.

Compression: A slow-acting reduction of audio signal dynamic range, typically to prevent clipping or protect drivers.

Constant bandwidth: A linear rendering of bandwidth, with each filter (or frequency spacing) having the same bandwidth expressed in Hz. The FFT calculates filters with constant bandwidth.

Constant percentage bandwidth: A logarithmic rendering of bandwidth, with each filter (or frequency spacing) having the same percentage bandwidth expressed in octaves, e.g. ⅓ octave. The RTA filters are constant percentage bandwidth.

Coupled (arrays): Arrays with elements within close proximity, i.e. within a single wavelength over a majority of its operational frequency range.

Coupling zone: The summation zone where the combination of signals is additive only. Phase offset must be <120° to prevent subtraction.

Coverage angle: The angular spread from on-axis (0 dB) to the -6 dB points on either side.

Coverage pattern: The shape of equal relative level around a sound source for a given propagation plane.

Crest factor: The term used to describe the peak to RMS ratio of a waveform.

Critical bandwidth: The frequency resolution to which tonal character is audible. One-sixth octave is a typical published value.

Crossover (acoustic): The frequency and/or location where two separate sound sources combine together at equal level.

Crossover (asymmetric): An acoustic crossover where one of the combined elements has different properties than the other. For spectral crossovers this includes level, filter type and speaker parameters; for spatial crossovers this includes level, angle and speaker type.

Crossover (class): A classification based on the combined level at the crossover relative to the isolated areas. This includes unity (0 dB), gap (less than 0 dB) and overlap (more than 0 dB).

Crossover (order): The slope rates of the individual elements which combine at a crossover. As the slope rates rise, the crossover order increases. Crossovers may be asymmetric, containing elements with different slope orders.

Crossover (phase-aligned): An acoustic crossover that is matched in both level and phase.

Crossover (spatial): An acoustic crossover in a spatial domain, i.e. the location where elements combine at equal level.

Crossover (spectral): An acoustic crossover in the frequency domain, i.e. the frequency where HF and LF drivers operate at equal level.

Cycles per second: The frequency of an audio signal measured in hertz (Hz).

dB SPL (sound pressure level): The quantity of sound level relative to the threshold of human hearing.

dBV: A measure of voltage relative to a standard value of 1 volt RMS.

Decibel (dB): The decibel is a logarithmic scaling system used to describe ratios (e.g. the output/input of an audio device).

Dedicated speaker controller: An active signal processing device with optimal settings for a particular loudspeaker model.

Delay line: An active electronic transmission device (usually digital) that delays the signal for a selected time period.

Diffraction: The ability of sound to pass around objects or through openings.

Diffusion: The reflection of sound in a manner in which it is scattered in different directions over frequency.

Digital signal processor (DSP): A signal-processing device with a wide variety of capabilities, including level, equalization, delay, limiting, compression and frequency division.

Displacement (source): The physical distance between two sound sources. This fixed displacement will affect the interactions in a given orientation by a fixed amount of time offset.

Displacement (wavelength): The distance between two sound sources expressed as a function of proportional displacement over frequency. The fixed-source displacement will affect all interactions in a given orientation by different amounts of wavelength displacement over frequency.

Driver: An individual loudspeaker component that covers only a limited frequency range. A speaker enclosure may contain various drivers with dedicated ranges that combine to become a speaker system.

Dynamic range: The range between the maximum linear operational level and the noise floor.

Echo perception: The listener's subjective experience of the direct sound and late arrivals as being distinct and separate entities.

Element: A single sound source within an array.

Emission: The origination of sound from a natural source.

End-fire array: An array of multiple of subwoofers, placed in a line, one behind the other, with a specific spacing and delay strategy in a timed sequence that creates forward addition and rearward subtraction.

Energy–time curve (ETC): A logarithmic expression (vertical scale) of the impulse response.

Envelope: The audible shape of the spectrum, the tonal character. The envelope follows the widest and highest spectral characteristics of the response and does not incorporate narrow dips and nulls.

Equal-level contours (isobaric contours): A radial rendering of a speaker coverage pattern. The on-axis response is normalized to 0 dB and the equal pressure point over angle is plotted radially.

Equal-loudness contours: (Fletcher–Munson curves) The non-linear nature of the ear's frequency response over level is expressed by this family of curves.

Equalization: The process of tonal compensation with an active (or in ancient times, passive) filter set. Equalization is used here primarily to control spectral tilt and thereby minimize spectral variance.

Equalizer: An active electronic transmission device with a set of user-settable filters.

False perspective: Any of the various ways in which the listener is made aware that they are listening to loudspeakers rather than a natural sound source.

FFT: The acronym for Fast Fourier Transform, describing the process of converting the time record data into frequency response data. Also known as the Discrete Fourier Transform (DFT).

Filter (frequency): The action of a system that causes some frequencies to rise (or fall) above (or below) others. Filters in electronic circuits have a wide variety of types, such as shelving, high pass, band pass and band reject. Examples of filters in acoustic systems include axial attenuation (directional control) over a space, and air absorption loss.

Filter order: The gross classification of filter behavior over frequency, designated as first order, second order, third order, etc. As filter order rises the roll-off slope becomes steeper. Each filter order connotes an additional 6 dB of roll-off per octave.

Fixed points per octave FFT (Constant Q Transform): A quasi-log expression of frequency resolution derived from multiple time records of different lengths.

Forward aspect ratio (FAR): A rectangular rendering of the speaker's forward coverage shape expressed as a ratio of length vs. width (or height). The aspect ratio is the fundamental representation of the building block of speaker design: the single speaker element.

Fourier theorem: Any complex waveform can be characterized as a combination of individual sine waves, with defined amplitude and phase components.

Frequency: The number of cycles per second given in hertz (Hz).

Frequency divider: An active (or passive) electronic device that separates the spectrum into frequency bands which are delivered to different speaker drivers for transmission. The waveform will be reconstituted in the acoustical space at the acoustic crossover.

Frequency response: The response of a system in various categories over frequency. Here, these include amplitude, relative amplitude, relative phase and coherence.

Full-scale digital: The maximum peak voltage before overload for digital audio systems. The actual voltage is model dependent and is often user adjustable.

Graphic equalizer: An active (or passive) electronic transmission device with parallel filters having fixed center frequency and bandwidth, and variable level.

Gradient array: A cardioid configuration commonly used for subwoofer arrays with front and rear elements. The rear element is delayed and polarity reversed to effectively cancel behind the speakers.

Hygrometer: An atmospheric instrument that measures humidity.

Impedance: The combination of DC resistance and capacitive (or inductive) reactance for a given source or receiver.

Impulse response: A rendering of the calculated response of a system as if it were excited by a perfect impulse. This relative amplitude vs. time display is derived from the FFT transfer function measurement.

Inclinometer: A device that measures the vertical angle of a device or surface.

Inter-aural level difference (ILD): The difference in level at our ears of a horizontally displaced sound source. This is one of the prime factors in horizontal localization.

Inter-aural time difference (ITD): The difference in time arrival at our ears of a horizontally displaced sound source. This is one of the prime factors in horizontal localization.

Inverse square law: Sound propagation in free field loses 6 dB of SPL for each doubling of distance from the source.

Isobaric contours: See **Equal-level contours**.

Isolation zone: The summation zone where there is greater than 10 dB of isolation and an unspecified amount of phase offset. The ripple variance is less than 63 dB.

Latency: The transit time through any device, independent of the user-selected settings.

Limit line: The maximum recommended range for an uncoupled system (the distance where combing becomes dominant due to three-way speaker interaction.

Line level: Standard operating audio transmission signal level. The nominal value is 1 volt (0 dBV) with maximum level around 124 dBV.

Line source: A speaker array configuration in which the axial (angular) orientation is identical.

Linear frequency axis: An equal space per frequency (Hz) rendering of the frequency axis.

Log frequency axis: An equal space per octave rendering of the frequency axis.

Maximum acceptable variance: A value of 6 dB in level or spectral variance.

Measurement microphone: The mic type used for system optimization. They should be free-field omnidirectional type, very flat, stable and consistent with low distortion and high dynamic range.

Mic level: Low-level audio signal transmission. The nominal values are typically at least 30 dB below line level.

Minimum variance: The spatial response of a sound system with <6 dB differences in level, spectral tilt and ripple.

Mix position: The location of the mix console and engineer. One of the 15,000 most important seats in an arena.

Noise (causal): A late-arriving copy of the source signal (e.g. late reflections or late speaker arrivals) that is too late to meet the stable summation criteria. Possible solutions are within the sound system and acoustical scope.

Noise (non-causal): A secondary signal that is not a related waveform and therefore will meet none of the stable summation criteria. The solutions to non-causal noise will not be found in the sound system.

Noise floor: The level of ambient non-causal noise in an electronic device or a complete system.

Nyquist frequency: The highest frequency that can be captured at a given sample rate (<0.5 × the sampling rate).

OFFAX: A mic position used for calibration located at or near the horizontal coverage edge of a given speaker or array.

Offset (level): The level difference (dB) between two sources at a given point.

Offset (phase): The difference in arrival time (degrees of phase shift) between two sources at a given frequency and location. Phase offset (for a given amount of delay) is frequency dependent.

Offset (time): The difference in arrival time (ms) between two sources at a given point. Time offset is frequency independent, but its effects on phase offset are not.

Offset (wavelength): The difference in arrival time (complete cycles) between two sources at a given frequency and location. Wavelength offset (for a given amount of delay) is frequency dependent.

ONAX: A mic position used for calibration located at or near the horizontal and vertical center of a given speaker or array.

Oscilloscope: A waveform analysis device with electrical signals displayed as voltage over time.

Panoramic field: The horizontal width of the stereo field. The maximum width of the panoramic field at any given location is the angular spread between the stereo sources.

Panoramic perception: The experience of apparent sound sources along the horizontal plane between two speakers, also known as **stereo perception**.

Parallel pyramid: The behavior of coupled loudspeakers with matched angular orientation. The spatial crossovers are stacked sequentially in the form of a pyramid.

Parametric equalizer: An active electronic transmission device with parallel filters having variable center frequency, bandwidth and level recommended for system optimization.

Pascal: A unit of pressure equal to one newton per square meter. In acoustical terms, one pascal equals 94 B SPL. Microphone sensitivity is commonly defined in mV/Pa.

Peak limiting: A fast-acting reduction of the audio signal dynamic range. This is done to prevent clipping or for driver protection.

Peak voltage (V_{PK}), peak-to-peak voltage (V_{PK-PK}): A waveform's highest voltage level. V_{PK-PK} spans the positive and negative peaks.

Percentage bandwidth: The frequency span of a filter function (in octaves).

Perception: The subjective experience of the human hearing system.

Period: The time length of a complete cycle is the reciprocal of frequency, commonly expressed in milliseconds (ms).

Phase: The radial component of the audio waveform expressed in degrees. For a given frequency, the phase value can be converted to time.

Phase delay: A delay value (in ms) describing the frequency-dependent delay over a limited frequency span.

Pink shift: See **Spectral tilt**.

Pinna (outer ear): The primary mechanism for vertical localization of sound.

Point destination: A speaker array configuration in which an inward splay creates a virtual source at a destination in front of the elements.

Point source: A speaker array configuration in which an outward splay creates a virtual source at a point behind the elements.

Polar plot: A rendering of a speaker coverage pattern in radial form. The on-axis response is normalized to 0 dB and the dB loss over angle is plotted radially.

Polarity: The measure of waveform orientation above or below the median line. A device that is "normal" polarity has the same orientation from input to output. A device with reverse polarity has opposite orientations from input to output.

Precedence effect: A description of the offsetting relationships in our binaural localization mechanisms (relative level and relative time). A sound source's perceived location can be manipulated by time and level offsets within a limited range.

Pressurization: The "positive" portion of an acoustic transmission, where the pressure is higher than ambient.

Propagation delay: The transit time from a source through a medium to a destination. Our primary focus is on acoustic propagation and refers to the transit time between speaker and a listening position.

Push–pull: An output stage in an active balanced output with two identical signals with opposite polarities.

Q (filter): The quality factor of a filter circuit. It is a linear-scale representation of the filter bandwidth. As the filter narrows, the Q rises.

Range ratio: The relative lengths between a source (or sources) to different destinations. The range ratio (in dB) is used to evaluate coverage requirements and match the relative level and power scaling between speakers.

Rarefaction: In acoustic transmission, the half of the cycle that is lower than the ambient pressure.

Ray-tracing model: The rendering of predicted speaker response as rays of light emitting from a source.

Real-time analyzer (RTA): An acoustic measurement device that uses a bank of log-spaced parallel filters to characterize the spectral response.

Refraction: The bending of a sound transmission as it passes through layers of media.

Resolution: The detail level presented in the measured data that is the basis of the predictive or measured model.

Resolution (angular): The spacing (in degrees) between the measured data points that comprise the speaker data for an acoustic prediction.

Resolution (frequency): The width of each evenly spaced FFT frequency *"line"* or *"bin"*, which is calculated by dividing the *sampling rate* by the number of samples in the *time window*.

Return ratio: A dedicated range ratio calculation used to evaluate balcony and underbalcony coverage options (coupled vs. upper/lower mains and needs assessment for delays). Return ratio is the range to the farthest seat vs. the balcony front. The difference (in dB) between the distance from a sound source to the front of the balcony and the farthest area angularly adjacent to that balcony front.

Ripple variance: The range (dB) between the peaks and dips.

Root mean square (RMS): The voltage in an AC circuit that would provide an equivalent to that found in a DC circuit.

Sampling rate: The clock frequency of the analog-to-digital conversion.

Sensitivity (microphone): The acoustical to electrical conversion ratio for microphones; the voltage output for a given SPL at the diaphragm, typically rated in mV/Pa.

Sensitivity (speaker): The electrical to acoustical power conversion ratio for speakers; the acoustic output for a given power input from the amplifier, typically rated as dB SPL @1 watt/1 meter.

Signal processor: Any active electronic transmission device charged with the jobs of equalization, level setting or delay.

Sonic image: The perceived location of a sound source, regardless of whether a natural source or loudspeaker are present at that location.

Sonic image distortion: The extent to which the perceived sound image differs from the intended sound source.

Spatial perception: A subjective characterization of sound source location for summed signals such as multiple speaker arrivals and/or reflections in which the sound source is perceived as having an indistinct point of origin.

Speaker level: Analog electronic power transmission between amplifiers and speakers, typically expressed in watts, rather than voltage.

Speaker order: The gross classification of speakers by coverage angle (first order, second order, etc). This is analogous to filter order (spatial vs. spectral) with higher-order speakers exhibiting steeper spatial roll-offs.

Spectral tilt (pink shift): A spectral distribution in favor of more low-frequency energy than high-frequency energy, e.g. an off-axis response.

Spectral variance: A substantial difference in spectral response between two locations. Locations with matched spectral tilt would be considered minimum spectral variance.

Splay angle: The angular orientation between two array elements.

Stable summation criteria: The conditions required for summation behavior to be perceived as an ongoing effect on the system response. These include common ancestry of the waveform and some duration of overlap at the summing junction.

Stereo perception: See **Panoramic perception**.

Subsystems: One of a family of systems that together comprise a single channel of the sound system transmission. Subsystems transmit a related waveform to local areas.

Summation duration: The length of time that two frequencies share the same location. In related waveforms the summation duration will depend upon the time offset and the transient nature of the waveform.

Summation zones: The five categories of summation interaction behavior (coupling, transition, isolation, cancellation and combing) based upon the relative level and phase.

Symmetric: Having similar response characteristics in either direction from a defined center line.

Systems (fills): Secondary sound sources for a given signal channel designed to cover the listening areas not handled by the mains. These less powerful subsystems operate under lower-priority status to the mains.

Systems (mains): The primary sound source for a given signal channel. These powerful systems cover the largest percentage of the listening area. Main systems operate with higher priority over the fill subsystems.

TANSTAAFL "There ain't no such thing as a free lunch." The concept, attributed to Robert Heinlein, teaches that no action or solution can occur in isolation without affecting some other action or solution.

Time bandwidth product (FFT): The relationship between the length of the time record and the bandwidth. The relationship is reciprocal, therefore the combined value is always 1. A short time record creates a wide bandwidth, whereas a long time record creates a narrow bandwidth.

Time record (FFT): (also called the time window) The period of time over which a waveform is sampled, expressed in ms. FFT lines (bins) are the number of samples (division operations) of the time record.

Tonal perception: The subjective characterization of the spectral response of the signal. For summed signals, the listener's subjective experience of the direct sound and late arrivals is as a single spectrally modified response.

Total harmonic distortion (THD): A measure of the presence of harmonics added to the original signal (the fundamental) by a particular device.

Transfer function: The quantification of a system's response to a signal passing through it. Audio transfer functions examples include gain/loss, frequency-dependent responses (peaks, dips), etc.

Transfer function measurement: A dual-channel system of audio measurement that compares one channel (the reference) to a second channel (measurement). Transfer function measurement illustrates the difference between the two signals.

Transformer balanced: A balanced-line connection to or from a passive (transformer) input or output device.

Transition zone: (formerly termed combining zone) The summation zone having between 4 and 10 dB of isolation and an unspecified amount of phase offset. The ripple variance is less than 66 dB.

Triage: A resource allocation strategy of prioritizing choices in cases where a solution provides unequal benefits.

Unbalanced line: Single conductor analog electronic audio transmission configuration suitable only for short distances due to its lack of noise immunity.

Uncoupled (arrays): Sound sources that are not within close proximity, i.e. displaced beyond a single wavelength over the majority of its operational frequency range.

Uniformity: The extent to which we can create a similar experience for all listeners in the hall.

Unity gain: The condition of an electronic device whose output to input ratio is 1 (0 dB).

Unity line The recommended starting range for an uncoupled system (the distance where the displacement gap is filled due to two-way speaker interaction).

Unity spacing: The recommended spacing for an uncoupled system (the spacing that places the unity line at the desired depth).

Unity splay angle: The splay angle between speaker elements that produces minimum-level variance between the individual on-axis points and the spatial crossover.

Variable acoustics: An architectural design having adjustable acoustical features, which can change to suit different uses and sonic requirements. This can be accomplished mechanically and/or electronically.

VBOT: A mic position used for calibration located at or near the bottom of the vertical coverage of a given speaker or array

Voltage gain: The ratio of output to input voltage through a device or stage. This can be described linearly (2×, 4×) or logarithmically (16 dB, 112 dB, etc.).

Volt/ohm meters (VOMs): An electronic test instrument that provides AC and DC voltage testing, continuity and short-circuit detection.

VTOP: A mic position used for calibration located at or near the top of the vertical coverage of a given speaker or array.

Wavelength: The physical distance required to complete a cycle while propagating through a particular medium. Typically expressed in meters or feet.

Weighting (averages): The extent to which an individual data sample is given preference in the averaged total. An unweighted averaging scheme treats all samples equally, whereas weighted schemes give certain samples a larger proportion.

Weighting (frequency response): The extent to which individual spectral regions are given preference in the averaged total level. An unweighted frequency response scheme gives all frequencies equal treatment in the averaged total.

Window function (FFT): A form of waveform shaping of the time record that prevents odd multiples of the time record length distorting the frequency response calculation.

Wraparound (phase): An artifact of the display properties of the FFT analyzer. The cyclical nature of phase requires the display to recycle the phase values such that 0° and 360° occupy the same vertical-scale location. The wraparound occurs at the edges of the rectangular display of the circular function.

BIBLIOGRAPHY

Beranek, L. L. (1962), *Music, Acoustics & Architecture*, Wiley.

Beranek, L. L. (2004), *Concert Halls and Opera Houses: Music, Acoustics, and Architecture*, Springer.

Cantu, L. (1999), Monaural Hearing and Sound Localization, Austin State University, http://hubel.sfasu.edu/courseinfo/SL99/monaural.html.

Cavanaugh, W. J. and J. A. Wilkes (1999), *Architectural Acoustics: Principles and Practice*, Wiley.

Duda, R. O. (1998), Sound Localization Research, San Jose State University, http://www-engr.sjsu.edu/~duda/ Duda.html.

Everest, F. A. (1994), *The Master Handbook of Acoustics*, TAB Books (a division of McGraw-Hill).

Giddings, P. (1990), *Audio System: Design and Installation*, Sams.

Herlufsen, H. (1984), *Dual Channel FFT Analysis (Part 1)*, Technical Review, advanced techniques in acoustical, electrical and mechanical measurement, No. 1, Bruel & Kjaer.

Huntington, J. (2012), *Show Networks and Control Systems*, Zircon Designs Press.

McCarthy, B. (1998), *Meyer Sound Design Reference for Sound Reinforcement*, Meyer Sound Laboratories.

Martin, K. D. (1994), A Computational Model of Spatial Hearing, Massachusetts Institute of Technology, http://xenia.media.mit.edu/~kdm/proposal/ chapter2_1.html.

Tremaine, H. (1979), *Audio Cyclopedia*, Sams.

INDEX

D

E

F

G

H

S

T

U

V

W